Handbook of Analytical Techniques

edited by Helmut Günzler and
Alex Williams

Volume II

For more information about analytical techniques please visit our website (www.wiley-vch.de) or register for our free e-mail alerting service (www.wiley-vch.de/home/pas)

Handbook of Analytical Techniques

edited by Helmut Günzler and
Alex Williams

Volume II

Chapter 19
to
Chapter 30
Subject Index

Weinheim · New York · Chichester · Brisbane · Singapore · Toronto

Prof. Dr. Helmut Günzler
Bismarckstr. 4
D-69469 Weinheim
Germany

Alex Williams
19 Hamesmoor Way, Mytchett
Camberley, Surrey GU16 6JG
United Kingdom

> This book was carefully produced. Nevertheless, authors, editors and publisher do not warrant the information contained therein to be free of errors. Readers are advised to keep in mind that statements, data, illustrations, procedural details or other items may inadvertently be inaccurate.

Library of Congress Card No. applied for.

British Library Cataloguing-in-Publication Data:
A catalogue record for this book is available from the British Library.

Deutsche Bibliothek – CIP Cataloguing-in-Publication-Data:
A catalogue record for this publication is available from Die Deutsche Bibliothek.

ISBN 3-527-30165-8

© WILEY-VCH Verlag GmbH, D-69469 Weinheim (Federal Republic of Germany), 2001

Printed on acid-free paper.

All rights reserved (including those of translation in other languages). No part of this book may be reproduced in any form – by photoprinting, microfilm, or any other means – nor transmitted or translated into machine language without written permission from the publishers. Registered names, trademarks, etc. used in this book, even when not specifically marked as such, are not to be considered unprotected by law.

Composition: Rombach GmbH, D-79115 Freiburg
Printing: Strauss Offsetdruck GmbH, D-69509 Mörlenbach
Bookbinding: Wilhelm Osswald & Co., D-67433 Neustadt (Weinstraße)
Cover Design: Gunter Schulz, D-67136 Fußgönheim
Printed in the Federal Republic of Germany.

Contents

Volume I

1. Analytical Chemistry: Purpose and Procedures 1

1.1.	The Evolution of Analytical Chemistry.................	1	1.5.	Analytical Tasks and Structures ...	8
1.2.	The Functional Organization of Analytical Chemistry...........	4	1.6.	Definitions and Important Concepts.	13
			1.7.	"Legally Binding Analytical Results"....................	20
1.3.	Analysis Today...............	5	1.8.	References..................	20
1.4.	Computers	7			

2. Quality Assurance in Instrumentation 23

2.1.	Introduction.................	23	2.5.	Routine Maintenance and Ongoing Performance Control	30
2.2.	Selecting a Vendor	24			
2.3.	Installation and Operation of Equipment	25	2.6.	Handling of Defective Instruments .	34
			2.7.	References..................	35
2.4.	Qualification of Software and Computer Systems.............	29			

3. Chemometrics 37

3.1.	Introduction.................	37	3.7.	Signal Processing	49
3.2.	Measurements and Statistical Distributions.................	38	3.8.	Basic Concepts of Multivariate Methods.....................	51
3.3.	Statistical Tests...............	40	3.9.	Factorial Methods.............	53
3.4.	Comparison of Several Measurement Series	44	3.10.	Classification Methods..........	56
			3.11.	Multivariate Regression	58
3.5.	Regression and Calibration.......	45	3.12.	Multidimensional Arrays	59
3.6.	Characterization of Analytical Procedures	47	3.13.	References..................	61

4. Weighing 63

4.1.	Introduction.................	63	4.7.	Gravity and Air Buoyancy	67
4.2.	The Principle of Magnetic Force Compensation.................	63	4.8.	The Distinction Between Mass and Weight.....................	68
			4.9.	Qualitative Factors in Weighing ...	68
4.3.	Automatic and Semiautomatic Calibration	65	4.10.	Governmental Regulations and Standardization.................	69
4.4.	Processing and Computing Functions	66			
4.5.	Balance Performance...........	66	4.11.	References..................	69
4.6.	Fitness of a Balance for Its Application...................	67			

5. Sampling ... 71

5.1.	Introduction and Terminology	71	5.4.	Acceptance Sampling	74
5.2.	Probability Sampling	72	5.5.	Conclusions	76
5.3.	Basic Sampling Statistics	73	5.6	References	76

6. Sample Preparation for Trace Analysis ... 77

6.1.	Introduction	78	6.3.	Sample Preparation in Organic Analysis	96
6.2.	Sample Preparation and Digestion in Inorganic Analysis	80	6.4	References	104

7. Trace Analysis ... 109

7.1.	Subject and Scope	110	7.4.	Calibration and Validation	113
7.2.	Fields of Work	110	7.5.	Environmental Analysis	117
7.3.	Methods of Modern Trace Analysis	111	7.6	References	125

8. Radionuclides in Analytical Chemistry ... 127

8.1.	Introduction	127	8.4.	Isotope Dilution Analysis	136
8.2.	Requirements for Analytical Use of Radionuclides	131	8.5.	Radioreagent Methods	140
8.3.	Radiotracers in Methodological Studies	134	8.6	References	145

9. Enzyme and Immunoassays ... 147

9.1.	Enzymatic Analysis Methods	147	9.3	References	171
9.2.	Immunoassays in Analytical Chemistry	158			

10. Basic Principles of Chromatography ... 173

10.1.	Introduction	174	10.7.	Band Broadening	186
10.2.	Historical Development	175	10.8.	Qualitative Analysis	189
10.3.	Chromatographic Systems	176	10.9.	Quantitative Analysis	192
10.4.	Theory of Linear Chromatography	177	10.10.	Theory of Nonlinear Chromatography	194
10.5.	Flow Rate of the Mobile Phase	182	10.11.	Reference Material	196
10.6.	The Thermodynamics of Phase Equilibria and Retention	183	10.12	References	197

11. Gas Chromatography ... 199

11.1.	Introduction ...	200	11.7.	Practical Considerations in Qualitative and Quantitative Analysis ... 242
11.2.	Instrumental Modules ...	201	11.8.	Coupled Systems ... 244
11.3.	The Separation System ...	201	11.9.	Applicability ... 250
11.4.	Choice of Conditions of Analysis ...	212	11.10.	Recent and Future Developments ... 254
11.5.	Sample Inlet Systems ...	215	11.11.	References ... 258
11.6.	Detectors ...	231		

12. Liquid Chromatography ... 261

12.1.	General ...	262	12.8.	Sample Preparation and Derivatization ... 301
12.2.	Equipment ...	266	12.9.	Coupling Techniques ... 305
12.3.	Solvents (Mobile Phase) ...	283	12.10.	Supercritical Fluid Chromatography ... 308
12.4.	Column Packing (Stationary Phase) ...	285	12.11.	Affinity Chromatography ... 316
12.5.	Separation Processes ...	288	12.12.	References ... 323
12.6.	Gradient Elution Technique ...	297		
12.7.	Quantitative Analysis ...	298		

13. Thin Layer Chromatography ... 327

13.1.	Introduction ...	327	13.6.	Development ... 337
13.2.	Choice of the Sorbent Layer ...	327	13.7.	Visualization ... 339
13.3.	Sample Cleanup ...	330	13.8.	Quantitation ... 341
13.4.	Sample Application ...	332	13.9.	References ... 344
13.5.	The Mobile Phase ...	334		

14. Electrophoresis ... 345

14.1.	Introduction ...	345	14.8.	Two-Dimensional Maps (Proteome Analysis) ... 356
14.2.	Basic Principles ...	346	14.9.	Isotachophoresis ... 358
14.3.	Electrophoretic Matrices ...	346	14.10.	Immunoelectrophoresis ... 360
14.4.	Discontinuous Electrophoresis ...	350	14.11.	Staining Techniques and Blotting ... 362
14.5.	Isoelectric Focusing ...	351	14.12.	Immobilized pH Gradients ... 362
14.6.	Sodium Dodecyl Sulfate Electrophoresis ...	355	14.13.	Capillary Zone Electrophoresis ... 363
14.7.	Porosity Gradient Gels ...	355	14.14.	Preparative Electrophoresis ... 364
			14.15.	References ... 369

15. Structure Analysis by Diffraction ... 373

15.1.	General Principles ...	373	15.5.	Electron Diffraction ... 413
15.2.	Structure Analysis of Solids ...	374	15.6.	Future Developments ... 413
15.3.	Synchrotron Radiation ...	412	15.7.	References ... 414
15.4.	Neutron Diffraction ...	412		

16. Ultraviolet and Visible Spectroscopy ... 419

- 16.1. Introduction ... 420
- 16.2. Theoretical Principles ... 421
- 16.3. Optical Components and Spectrometers ... 430
- 16.4. Uses of UV–VIS Spectroscopy in Absorption, Fluorescence, and Reflection ... 443
- 16.5. Special Methods ... 452
- 16.6. References ... 459

17. Infrared and Raman Spectroscopy ... 465

- 17.1. Introduction ... 466
- 17.2. Techniques ... 466
- 17.3. Basic Principles of Vibrational Spectroscopy ... 470
- 17.4. Interpretation of Infrared and Raman Spectra of Organic Compounds ... 474
- 17.5. Applications of Vibrational Spectroscopy ... 489
- 17.6. Near-Infrared Spectroscopy ... 502
- 17.7. References ... 504

18. Nuclear Magnetic Resonance and Electron Spin Resonance Spectroscopy ... 509

- 18.1. Introduction ... 510
- 18.2. Principles of Magnetic Resonance ... 511
- 18.3. High-Resolution Solution NMR Spectroscopy ... 514
- 18.4. NMR of Solids and Heterogeneous Systems ... 546
- 18.5. NMR Imaging ... 547
- 18.6. ESR Spectroscopy ... 548
- 18.7. References ... 557

Volume II

19. Mössbauer Spectroscopy ... 561

- 19.1. Introduction ... 561
- 19.2. Principle and Experimental Conditions of Recoil-free Nuclear Resonance Fluorescence ... 561
- 19.3. Mössbauer Experiment ... 564
- 19.4. Preparation of Mössbauer Source and Absorber ... 567
- 19.5. Hyperfine Interactions ... 568
- 19.6. Evaluation of Mössbauer Spectra ... 573
- 19.7. Selected Applications ... 574
- 19.8. References ... 577

20. Mass Spectrometry ... 579

- 20.1. Introduction ... 580
- 20.2. General Techniques and Definitions ... 580
- 20.3. Sample Inlets and Interfaces ... 585
- 20.4. Ion Generation ... 590
- 20.6. Analyzers ... 597
- 20.7. Metastable Ions and Linked Scans ... 603
- 20.8. MS/MS Instrumentation ... 604
- 20.9. Detectors and Signals ... 607
- 20.10. Computer and Data Systems ... 610
- 20.11. Applications ... 613
- 20.12. References ... 622

21. Atomic Spectroscopy ... 627

- 21.1. Introduction ... 628
- 21.2. Basic Principles ... 629
- 21.3. Spectrometric Instrumentation ... 642
- 21.4. Sample Introduction Devices ... 660
- 21.5. Atomic Absorption Spectrometry ... 673
- 21.6. Atomic Emission Spectrometry ... 688
- 21.7. Plasma Mass Spectrometry ... 704
- 21.8. Atomic Fluorescence Spectrometry ... 713
- 21.9. Laser-Enhanced Ionization Spectrometry ... 716
- 21.10. Comparison With Other Methods ... 718
- 21.11. References ... 721

22. Laser Analytical Spectroscopy ... 727

- 22.1. Introduction ... 727
- 22.2. Tunable Lasers ... 730
- 22.3. Laser Techniques for Elemental Analysis ... 732
- 22.4. Laser Techniques for Molecular Analysis ... 744
- 22.5. Laser Ablation ... 750
- 22.6. References ... 751

23. X-Ray Fluorescence Spectrometry ... 753

- 23.1. Introduction ... 753
- 23.2. Historical Development of X-ray Spectrometry ... 755
- 23.3. Relationship Between Wavelength and Atomic Number ... 755
- 23.4. Instrumentation ... 757
- 23.5. Accuracy ... 760
- 23.6. Quantitative Analysis ... 761
- 23.7. Trace Analysis ... 762
- 23.8. New developments in Instrumentation and Techniques ... 763
- 23.9. References ... 765

24. Activation Analysis ... 767

- 24.1. Introduction ... 767
- 24.2. Neutron Activation Analysis ... 768
- 24.3. Photon Activation Analysis ... 779
- 24.4. Charged-Particle Activation Analysis ... 780
- 24.5. Applications ... 781
- 24.6. Evaluation of Activation Analysis ... 783
- 24.7. References ... 783

25. Analytical Voltammetry and Polarography ... 785

- 25.1. Introduction ... 785
- 25.2. Techniques ... 788
- 25.3. Instrumentation ... 803
- 25.4. Evaluation and Calculation ... 808
- 25.5. Sample Preparation ... 810
- 25.6. Supporting Electrolyte Solution ... 812
- 25.7. Application to Inorganic and Organic Trace Analysis ... 814
- 25.8. References ... 823

26. Thermal Analysis and Calorimetry ... 827

- 26.1. Thermal Analysis ... 827
- 26.2. Calorimetry ... 836
- 26.3. References ... 849

27. Surface Analysis ... 851

- 27.1. Introduction ... 852
- 27.2. X-Ray Photoelectron Spectroscopy (XPS) ... 854
- 27.3. Auger Electron Spectroscopy (AES) ... 874
- 27.4. Static Secondary Ion Mass Spectrometry (SSIMS) ... 889
- 27.5. Ion Scattering Spectroscopies (ISS and RBS) ... 898
- 27.6. Scanning Tunneling Methods (STM, STS, AFM) ... 910
- 27.7. Other Surface Analytical Methods ... 917
- 27.8. Summary and Comparison of Techniques ... 940
- 27.9. Surface Analytical Equipment Suppliers ... 940
- 27.10. References ... 944

28. Chemical and Biochemical Sensors ... 951

- 28.1. Introduction to the Field of Sensors and Actuators ... 952
- 28.2. Chemical Sensors ... 953
- 28.3. Biochemical Sensors (Biosensors) ... 1032
- 28.4. Actuators and Instrumentation ... 1051
- 28.5. Future Trends and Outlook ... 1052
- 28.6. References ... 1053

29. Microscopy ... 1058

- 29.1. Modern Optical Microscopy ... 1061
- 29.2. Electron Microscopy ... 1077
- 29.3 References ... 1125

30. Techniques for DNA Analysis ... 1131

- 30.1. Introduction ... 1131
- 30.2. Primary Molecular Tools for DNA Analysis ... 1133
- 30.3. Methods of DNA Detection ... 1135
- 30.4. Applications of DNA Analysis ... 1144
- 30.5. References ... 1150

Sucject Index ... 1151

Symbols and Units

Symbols and units agree with SI standards. The following list gives the most important symbols used in the handbook. Articles with many specific units and symbols have a similar list as front matter.

Symbol	Unit	Physical Quantity
a_B		activity of substance B
A_r		relative atomic mass (atomic weight)
A	m^2	area
c_B	mol/m^3, mol/L (M)	concentration of substance B
C	C/V	electric capacity
c_p, c_v	$J\ kg^{-1} K^{-1}$	specific heat capacity
d	cm, m	diameter
d		relative density (ϱ/ϱ_{water})
D	m^2/s	diffusion coefficient
D	Gy (= J/kg)	absorbed dose
e	C	elementary charge
E	J	energy
E	V/m	electric field strength
E	V	electromotive force
E_A	J	activation energy
f		activity coefficient
F	C/mol	Faraday constant
F	N	force
g	m/s^2	acceleration due to gravity
G	J	Gibbs free energy
h	m	height
h	$W \cdot s^2$	Planck constant
H	J	enthalpy
I	A	electric current
I	cd	luminous intensity
k	(variable)	rate constant of a chemical reaction
k	J/K	Boltzmann constant
K	(variable)	equilibrium constant
l	m	length
m	g, kg, t	mass
M_r		relative molecular mass (molecular weight)
n_D^{20}		refractive index (sodium D-line, 20 °C)
n	mol	amount of substance
N_A	mol^{-1}	Avogadro constant ($6.023 \times 10^{23}\ mol^{-1}$)
p	Pa, bar *	pressure
Q	J	quantity of heat
r	m	radius
R	$J\ K^{-1} mol^{-1}$	gas constant
R	Ω	electric resistance
S	J/K	entropy
t	s, min, h, d, month, a	time
t	°C	temperature
T	K	absolute temperature
u	m/s	velocity
U	V	electric potential
U	J	internal energy
V	m^3, L, mL	volume
w		mass fraction
W	J	work
x_B		mole fraction of substance B
Z		proton number, atomic number
α		cubic expansion coefficient

Symbols and Units

Symbol	Unit	Physical Quantity
α	W m^{-2}K^{-1}	heat-transfer coefficient (heat-transfer number)
α		degree of dissociation of electrolyte
$[\alpha]$	10^{-2} deg cm^{2}g^{-1}	specific rotation
η	Pa · s	dynamic viscosity
θ	°C	temperature
κ		c_p/c_v
λ	W m^{-1}K^{-1}	thermal conductivity
λ	nm, m	wavelength
μ		chemical potential
ν	Hz, s^{-1}	frequency
ν	m^2/s	kinematic viscosity (η/ϱ)
π	Pa	osmotic pressure
ϱ	g/cm^3	density
σ	N/m	surface tension
τ	Pa (N/m^2)	shear stress
φ		volume fraction
χ	Pa^{-1} (m^2/N)	compressibility

* The official unit of pressure is the pascal (Pa).

19. Mössbauer Spectroscopy

P. GÜTLICH, Institut für Anorganische Chemie und Analytische Chemie, Johannes Gutenberg-Universität, Mainz, Germany

J. ENSLING, Institut für Anorganische Chemie und Analytische Chemie, Johannes Gutenberg-Universität, Mainz, Germany

19.	Mössbauer Spectroscopy	561
19.1.	Introduction	561
19.2.	Principle and Experimental Conditions of Recoil-free Nuclear Resonance Fluorescence	561
19.2.1.	Nuclear Resonance	561
19.2.2.	Recoil Effect	562
19.2.3.	Mössbauer Effect	563
19.3.	Mössbauer Experiment	564
19.3.1.	Doppler Effect	564
19.3.2.	Mössbauer Spectrometer	565
19.4.	Preparation of Mössbauer Source and Absorber	567
19.5.	Hyperfine Interactions	568
19.5.1.	Isomer Shift	568
19.5.2.	Magnetic Hyperfine Interaction	570
19.5.3.	Quadrupole Splitting	571
19.5.4.	Combined Quadrupole and Magnetic Hyperfine Interactions	573
19.5.5.	Relative Intensities of Resonance Lines	573
19.6.	Evaluation of Mössbauer Spectra	573
19.7.	Selected Applications	574
19.7.1.	Magnetic Materials	574
19.7.2.	Corrosion Products	575
19.7.3.	Catalysis	576
19.7.4.	Atmospheric Aerosol Samples	576
19.7.5.	US-$ Bill	577
19.8.	References	577

19.1. Introduction

Since the discovery of the Mössbauer effect by RUDOLF MÖSSBAUER [1]–[3] in 1958 this nuclear spectroscopic method has found a wide variety of applications in materials science, solid state physics, chemistry, metallurgy, and earth sciences.

Up to the present time, the Mössbauer effect has been observed for almost 100 nuclear transitions in more than 40 different elements. Not all of these transitions are suitable for actual studies, for reasons that will be discussed below. But ca. 15–20 elements remain that are appropriate for applications, of which the best known Mössbauer nuclide is iron, because on the one hand the Mössbauer experiment is easy to perform with the 14.4 keV transition in ^{57}Fe, and on the other hand the element iron is very abundant in nature.

It is the purpose of this article to introduce the reader unfamiliar with the field to the principles of Mössbauer spectroscopy, and to the various types of chemical information that can be extracted from the electric and magnetic hyperfine interactions reflected in the Mössbauer spectrum. For pedagogical reasons we shall focus on the principles of ^{57}Fe Mössbauer spectroscopy in the introductory part, and shall then present examples of typical applications. The more seriously interested reader is advised to consult the various text books, reviews, and special compilations of original communications [4]–[13].

19.2. Principle and Experimental Conditions of Recoil-free Nuclear Resonance Fluorescence

19.2.1. Nuclear Resonance

The resonance absoption of electro-magnetic radiation is a phenomenon well known in many branches of physics. The scattering of sodium light by sodium vapor, the excitation of a tuning fork by

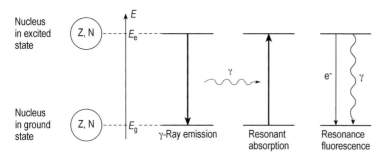

Figure 1. Schematic representation of nuclear resonance absorption of γ-rays (Mössbauer effect) and nuclear resonance fluorescence

sound, and of a dipole by radiofrequency radiation are some familiar examples. The possible observation of the phenomenon in nuclear transitions involving γ-radiation was suggested in 1929 by KUHN [14] but for many years unsuccessful attempts were reported. It was the great merit of RUDOLF MÖSSBAUER who established the experimental conditions as well as the theoretical description of the occurrence of the nuclear resonant absorption (fluorescence) of γ-rays [1]–[3]. The following conditions have to be fulfilled for the observation of the Mössbauer effect.

Consider a nucleus in an excited state of energy E_e with the mean life time τ. Transition to the ground state of energy E_g occurs by emitting a γ-quantum with the energy

$$E_0 = \hbar\omega = E_e - E_g \tag{1}$$

This γ-quantum may be absorbed by a nucleus of the same kind in its ground state, thereby taking up the energy difference E_0. This phenomenon is called nuclear resonant absorption of γ-rays, and is schematically sketched in Figure 1.

The fact that the lifetime τ of an excited nuclear state is finite means that its energy has a certain distribution of width Γ, which, due to the Heisenberg uncertainty relationship, is connected to τ by

$$\tau\Gamma = \hbar \tag{2}$$

Consequently, the probability for the emission of a quantum is a function of the energy, or, in other words, the energy spectrum exhibits an intensity distribution $I(E)$ about a maximum at E_0 with a width at half maximum, Γ, which corresponds to the uncertainty in energy of the excited nuclear state. The explicit shape of this distribution is Lorentzian according to

$$I(E) \frac{\Gamma/2\pi}{(E - E_0)^2 + (\Gamma/2\pi)^2} \tag{3}$$

The mean lifetime τ of the excited states of the Mössbauer isotopes are in the range of 10^{-6} to 10^{-10} s; the corresponding uncertainties of the transition energies are between 10^{-9} and 10^{-5} eV. Nuclear resonance fluorescence is known to possess an amazingly high energy resolution given by the ratio Γ/E_0, which is of the order of 10^{-12}. As an example, the first excited state of ^{57}Fe has a mean lifetime τ = 141 ns, which leads to a natural linewidth $\Gamma = \hbar/\tau = 4.7 \cdot 10^{-9}$ eV and hence to a relative energy resolution of $\Gamma/E_0 = 3.3 \cdot 10^{-13}$.

The absorption line is also described by a Lorentzian curve. Nuclear resonant absorption can only be observed when the difference of the transition energies for emission and absorption are not much greater than the sum of the linewidths, i.e., when the emission and absorption lines overlap sufficiently.

After resonance absorption, the excited nucleus will decay by either emitting isotropically a γ-quantum (as in the primary γ-ray emission of Figure 1) or a conversion electron e⁻, preferentially from the K-shell. This phenomenon is termed *nuclear resonance fluorescence* and may be used in Mössbauer scattering experiments (surface investigations).

19.2.2. Recoil Effect

Consider the emission of a γ-quantum with energy E_γ from an excited nucleus of mass M in a free atom or molecule (in the gas or liquid phase). The energy of the excited state is E_0 above

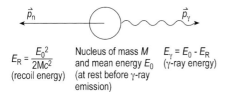

$E_R = \dfrac{E_0^2}{2Mc^2}$ (recoil energy)

Nucleus of mass M and mean energy E_0 (at rest before γ-ray emission)

$E_\gamma = E_0 - E_R$ (γ-ray energy)

Figure 2. Recoil of momemtum $\to p_n$ and energy E_R imparted to an isolated nucleus upon γ-ray emission

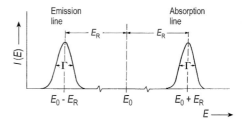

Figure 3. Consequences of the recoil effect caused by γ-ray emission and absorption in isolated nuclei. The transition lines for emission are separated by $2E_R \approx 10^6\,\Gamma$. There is no overlap between emission and absorption line and hence no resonant absorption is possible

that of the ground state. If the excited nucleus is supposed to be at rest before the decay, a recoil effect will be imparted to the nucleus upon γ-emission, causing the nucleus to move with velocity v in opposite direction to that of the emitted γ-ray (see Fig. 2).

The resulting recoil energy is

$$E_R = \tfrac{1}{2}Mv^2 = \dfrac{p_n^2}{2M} \qquad (4)$$

Momentum conservation requires that

$$p_\gamma = E_\gamma/c = -p_n \qquad (5)$$

where p_γ and p_n are the linear momenta of the γ-quantum and the recoiled nucleus, respectively, and c is the velocity of light. Thus the energy of the emitted γ-quantum is

$$E_\gamma = E_0 - E_R \approx E_0 \qquad (6)$$

Since E_R is very small compared to E_0, we may write for the recoil energy (in a nonrelativistic approximation)

$$E_R = \dfrac{p_n^2}{2M} = \dfrac{E_\gamma^2}{2Mc^2} \approx \dfrac{E_0^2}{2Mc^2} \qquad (7)$$

For many nuclear transitions $E_R \approx 10^{-2} \ldots 10^{-3}$ eV. The recoil effect causes the energy of the emitted photon to be decreased by E_R and the emission line appears at an energy $E_0 - E_R$ rather than at E_0. In the absorption process, the γ-ray to be absorbed by a nucleus requires the total energy $E_0 + E_R$ to make up for the transition from the ground to the excited state and the recoil effect, which causes the absorbing nucleus to move in the same direction as the incoming γ-ray photon. As shown schematically in Figure 3 the transition lines for emission and absorption are separated by $2E_R$, which is about 10^6 times larger than the natural line width Γ.

The condition for overlap of emission and absorption line, viz. $2E_R \leq \Gamma$ is not fulfilled at all. Thus, the cross section of nuclear resonant absorption in isolated atoms and molecules in the gaseous and liquid phases is extremely small.

In optical spectroscopy, however, the recoil energy is negligible compared with the linewidth and the resonance condition is certainly fulfilled for freely emitting and absorbing particles.

Before the discovery of the Mössbauer effect some successful experiments on nuclear resonance absorption in the gas phase had been carried out utilizing the Doppler effect. High velocities were required ($\approx 10^2 - 10^3$ ms^{-1}). This could be achieved either by mounting a source of emitting nuclei on a high speed rotor or by heating the source and the absorber, so that some overlap of the lines could take place.

Rudolf Mössbauer, while working on his Ph.D. thesis, carried out experiments of this kind with ^{191}Ir. But surprisingly, on lowering the temperature he found an increase in the absorption effect rather than a decrease. Mössbauer was able to explain this unexpected phenomenon, and was awarded the Nobel Prize in 1961 for the observation and correct interpretation of the *recoilless nuclear resonance absorption (fluorescence)* which is now known as the *Mössbauer Effect* and which provided the basis for a powerful technique in solid state research.

19.2.3. Mössbauer Effect

As explained in Section 19.2.2, nuclear resonance absorption and fluorescence of γ-rays is not possible for nuclei in freely emitting or absorbing particles. This is different in the solid state. The nuclei are more or less rigidly bound to the lattice. If a γ-ray photon is emitted from the excited nucleus, the concomitant recoil energy may now be assumed to consist of two parts,

$$E_R = E_{tr} + E_{vib} \quad (8)$$

E_{tr} refers to the translational energy transferred through linear momentum to the crystallite of mass M. As M is $\approx 10^{20}$ times larger than the mass of a single atom, the corresponding recoil energy E_{tr} imparted to the crystallite becomes negligibly in comparison with γ [Equation (7)].

Most of the recoil energy E_R will be dissipated into processes other than linear momentum, viz. to the lattice vibrational system. E_{vib}, the vibrational energy part, still being of the order of 10^{-3} eV and thus several orders of magnitude smaller than the atom displacement energy (ca. 25 eV) and also much smaller than the characteristic lattice phonon energies (ca. $10^{-1} - 10^{-2}$ eV), is dissipated by heating the nearby lattice surroundings. As a consequence of this phonon creation with frequencies up to w_i, the resulting energy of the emitted γ-ray will be

$$E_\gamma = E_0 - \sum_i n \hbar w_i \quad (9)$$

This again would destroy the resonance phenomenon by shifting the emission and absorption lines too far apart from each other on the energy scale. Fortunately, however, the lattice vibrations are quantized and E_{vib} changes the vibrational energy of the lattice oscillators by integral multiples of the phonon energy, $n\hbar w_i$ ($n = 0, 1, 2, ...$). This means that there is a certain probability f that no lattice excitation (energy transfer of $0 \cdot \hbar w_i$, called the zero-phonon-process) occurs during the γ-ray emission or absorption process. Mössbauer has shown that only to the extent of this probability f of unchanged lattice vibrational states during γ-ray emission and absorption does nuclear resonance absorption become possible. This has been termed the *Mössbauer effect*. The probability f is called the recoil-free fraction and denotes the fraction of nuclear transitions which occur without recoil.

The recoil-free fraction in Mössbauer spectroscopy is equivalent to the fraction of X-ray scattering processes without lattice excitation; this fraction of elastic processes in X-ray and neutron diffraction is described by the Debye–Waller factor:

$$f = e^{-k^2 \langle x^2 \rangle} \quad (10)$$

where $\langle x^2 \rangle$ is the mean square amplitude of vibration of the resonant nucleus in the photon propagation direction and $k = 2\pi/\lambda = E_\gamma/(\hbar c)$ is the wavenumber of the γ-quantum. The value of f depends on the lattice properties, described by $\langle x^2 \rangle$, and on the photon properties, represented by k^2. The ratio of the mean square displacement and wavelength λ of the photon must be small ($\ll 1$), in order to have a sufficiently high Debye–Waller factor f. It is also obvious that the Mössbauer effect cannot take place in liquids, where the molecular motion is characterized by unbound $\langle x^2 \rangle$, causing the recoil-free fraction to vanish. The following consequences emerge.

1) f decreases exponentially with k^2, i.e., E_γ^2. The Mössbauer effect can hardly be detected with γ-ray energies above 150 keV.
2) The mean square displacement $\langle x^2 \rangle$ should be small, in other words, the Mössbauer atom should be bound tightly to the crystal lattice. However, crystallinity of the material is not a necessary condition for observation of the Mössbauer effect. The effect is also detectable in amorphous materials, glasses, and in frozen solutions.
3) The recoil-free fraction f is temperature dependent, because $\langle x^2 \rangle$ decreases with decreasing temperature and hence forces f to increase. However, f does not reach unity even at $T = 0$ due to the fact that $\langle x^2 \rangle \neq 0$ at $T = 0$ because of the quantum-mechanical zero-point motion of the atoms (nuclei). In order to express f in terms of the usual experimental variables, the mean square amplitude $\langle x^2 \rangle$ is calculated using lattice dynamic models (e.g., Einstein, Debye) [6], [15].
4) $\langle x^2 \rangle$ and hence f can be anisotropic. As a consequence, a dependence of the intensity of absorption lines on the observation direction in single crystal experiment may be observed.

19.3. Mössbauer Experiment

19.3.1. Doppler Effect

The main components of a Mössbauer spectrometer are the γ-ray source containing the *Mössbauer active* nuclides, the absorber (or scatterer), and a detector for low-energy γ-radiation (see Fig. 4) plus electronics for automatic recording of the spectrum. The source and the absorber are

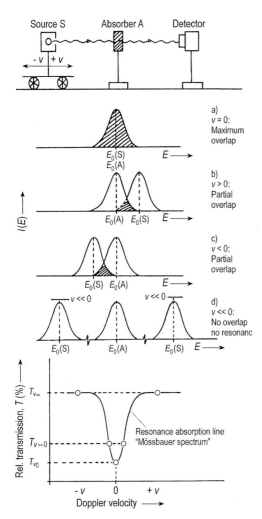

Figure 4. Schematic illustration of the experimental arrangement for recoilless nuclear resonance absorption, and relative transmission of recoilless γ-quanta as a function of Doppler velocity

where E_0 is the energy of the γ-quantum emitted from the same source at rest and is the velocity of light. This relationship is valid for a moving source, the velocity being positive for the source moving towards the absorber, and negative for moving away from the absorber. Owing to the very narrow linewidth Γ of the resonance line, it is generally sufficient to produce a small Doppler energy change E_D of the order of the line width Γ by Doppler shifting the source in order to sweep over the resonance. The Doppler velocities needed in the case of ^{57}Fe spectroscopy are in the range 0 to ± 10 mm s^{-1}, and for most of the other Mössbauer nuclides velocities of less than ± 100 mm s^{-1} are sufficient.

The Mössbauer spectrum, a plot of the relative transmission as a function of Doppler velocity, shows maximum resonance and therefore minimum relative transmission at relative velocities where emission and absorption lines overlap ideally (cf. Fig. 4). At high positive or negative velocities the overlap of emission and absorption lines is negligible, the resonance effect being practically zero, i.e., the relative transmission yields the *base line*.

19.3.2. Mössbauer Spectrometer

The Mössbauer spectrometers which are in use nowadays generate the spectrum by the velocity-sweep method. The drive system moves the source (or absorber) repeatedly over a range of velocities, while simultaneously counting the γ-quanta behind the absorber into synchronized channels.

The essential components of a modern Mössbauer spectrometer as illustrated in the block-diagram of Figure 5 are: the velocity transducer, the wave form generator and synchronizer, the multichannel analyzer, γ-ray detection system, a cryostat or oven for low and temperature dependent measurements, a velocity calibration device, the source and the absorber, and a read-out unit.

The source (and in a few cases the absorber) is mounted on the vibrating axis of an electromagnetic transducer (loudspeaker system) [16] which is moved according to a voltage waveform applied to the driving coil of the system. The usual velocity functions are of the triangular, sawtooth, or sinusoidal form. In special cases the source may be moved with a constant velocity.

A function generator produces the desired waveform of the motion. A feedback system (feedback amplifier) operates in such a manner that the

moved relative to each other (either by moving the source and keeping the absorber fixed or vice versa). The transmitted (scattered) γ-quanta are registered as a function of the relative velocity. In this way, it is possible to trace, stepwise, the absorption line by the emission line utilizing the Doppler effect. A γ-quantum which is emitted from the source moving at a velocity v receives a Doppler energy E_D modulation such that

$$E_\gamma = E_0 + E_D = E_0\left(1 + \frac{v}{c}\right) \quad (11)$$

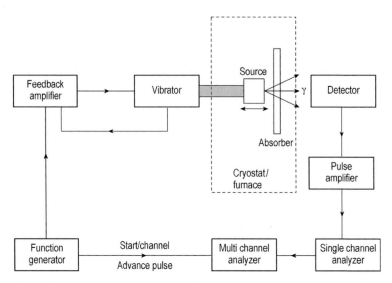

Figure 5. Block diagram illustrating the principle of a Mössbauer spectrometer

actual velocity of the source follows the reference waveform of the function generator with an accuracy better than 0.1 %. Using the triangular waveform one obtains an undistorted spectrum, because the velocity is a linear function of time. In extreme cases, where high velocities are required or the mass to be moved is large, it is preferable to use a sinusoidal waveform. For convenience, this type of spectrum is converted after the measurement into the linear form.

The counts detected behind the absorber are stored in a memory (mostly in the memory of a mutlichannel analyzer containing typically 512 or 1024 channels). Synchronization of the channel number in the memory and the instantaneous velocity of the source (absorber) is achieved by advancing the address of the memory one by one through an external clock which subdivides the period of the waveform into the number of available channels. A start pulse, which coincides with the beginning of the waveform, triggers the multichannel analyzer to start with channel No. 1, which is subsequently advanced by the clock pulses. In this way synchronization is achieved, i.e., during each period of motion a certain channel of the analyzer always corresponds to the same instantaneous velocity of the source. The whole velocity range of interest is scanned many (typically 10–50) times per second. As each instantaneous velocity occurs twice during one scan period, the spectrum is registered twice in the memory, where both halves of the memory appear as mirror images, symmetrically about the center of the channel series.

It is essential to have a standard reference line position which other signals can be referred to, particularly in Mössbauer spectroscopy, where different host metals for a particular Mössbauer isotope are often utilized, e.g., ^{57}Co/Cu, ^{57}Co/Pt, ^{57}Co/Rh in case of ^{57}Fe measurements with emitting γ-rays of slightly different energy. The most widely used reference material for velocity calibration in Mössbauer spectroscopy is a natural iron foil of which the six line positions of the magnetic hyperfine split spectrum are accurately known. One uses the distance between the two outermost lines of the iron spectrum as a calibration basis for the measurements of the hyperfine interactions with strengths corresponding to a Doppler velocity region of $0-10$ mm s^{-1}. This region is suitable for Mössbauer effect measurements in, e.g., ^{57}Fe, ^{119}Sn, ^{61}Ni, ^{197}Au, and others.

The usual detectors for low-energy γ-radiation are thin NaI(Tl) scintillation crystals coupled to a photomultiplier, and gas-filled proportional counters. Considerable improvements in energy resolution can be achieved with solid-state detectors such as the Li drifted Ge counter.

It is only with low-energy γ-emitters, such as the excited state of ^{57}Fe, that the Mössbauer effect can be observed at room temperature and even above. With most Mössbauer isotopes it is necessary to cool the absorber and sometimes also the source to liquid-nitrogen or liquid-helium temper-

Table 1. Some important isotopes in Mössbauer spectroscopy

Mössbauer isotope	Abundance of stable element, %	γ-Ray energy, keV	Half-life of Mössbauer transition, 10^{-9} s	Parent isotope	Half-life of parent isotope
^{57}Fe	2.14	14.4	97.8	^{57}Co	270 d
^{119}Sn	8.58	23.83	17.8	^{119m}Sn	245 d
^{121}Sb	57.25	37.15	3.5	^{121m}Sn	76 y
^{129}I	0 (β^-)	27.77	16.8	^{129}Te	33 d
^{151}Eu	47.82	21.54	9.5	^{151}Sm	90 y
^{197}Au	100	77.35	1.89	^{197}Pt	18 h

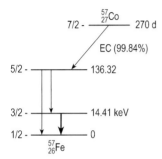

Figure 6. Decay scheme of ^{57}Co

state lifetime. By radioactive disintegration, the isotope populates a nuclear excited state which decays to the ground state by emitting low-energy γ-radiation. Table 1 lists a selection of isotopes which are suitable for practical Mössbauer effect measurements.

As a typical example, Figure 6 shows the decay scheme of ^{57}Co which populates the 14.4 keV Mössbauer level of ^{57}Fe with a lifetime of τ = 140 ns. The isotope ^{57}Co can be produced in a cyclotron by the nuclear reaction ^{56}Fe(d,n)^{57}Co. The decay of ^{57}Co occurs essentially by electron capture (99.8 %) from the K-shell leaving a hole in this shell which is filled from higher shells under emission of a 6.4 keV X ray. Sources of ^{57}Co are usually prepared by electrochemically depositing the carrier-free isotope on metallic supports and then diffusing it into the metal at high temperatures.

A Mössbauer source should meet the following requirements.

The emission line should be as narrow and intense as possible, unsplit and unbroadened

The recoil-free fraction should be as high as possible

The source material should be chemically inert during the lifetime of the source and resistant against autoradiolysis

The host material should not give rise to interfering X rays, and Compton scattering and photoelectric processes should be insignificant

A uniform absorber with randomly oriented crystallites can be prepared easily by sandwiching the finely ground material between the thin windows of a Perspex holder. In order to obtain undistorted line shapes it is desirable that the absorber is "thin". The natural line width Γ_{nat} of the Mössbauer line having Lorentzian shape [see Section 19.2.1, Eq. (3)] is determined, as already discussed, by the half-life of the excited nuclear state and the Heisenberg uncertainty principle. The ex-

atures in order to obtain a measurable effect. Apart from these practical requirements it is often desirable to vary the temperature of the absorber in order, e.g., to obtain information about phase transitions or the electronic state and the molecular symmetry of the compound of interest. Many types of cryostats are nowadays commercially available. The temperature range varies from 1.2 to 300 K in He-bath cryostats and from 6 to 300 K in He-flow cryostats.

A He-bath cryostat equipped with a superconducting magnet can be utilized for Mössbauer investigations of materials in an applied magnetic field. Magnetic fields up to 10 T are very common today.

Above room temperature electrical furnaces are used. In order to keep the risk of oxidation of the material low, a moderate vacuum of about $10^{-2} - 10^{-3}$ is appropriate.

19.4. Preparation of Mössbauer Source and Absorber

The source for a Mössbauer experiment is a radioactive isotope of reasonable half-life and appropriate nuclear transition energy and excited perimental line (Γ_{exp}), however, is broadened for several reasons. The hyperfine fields may have a distribution which is typically observed in amorphous materials and small particles (nano particles). The narrow hyperfine field distribution caused by imperfections and strains in molecular crystals is typically not observed by a line broadening. Dynamic behavior of hyperfine fields lead to line broadenings if the eneries $\hbar\tau_f$ corresponding to the fluctuation time τ_f of the fields at the nucleus are of the same order of magnitude as the difference of the hyperfine energies belonging to the field values. The sample thickness gives rise to the so-called thickness broadening which, however, can be taken into account evaluating the spectra by the fitting routines.

Heavy elements in the compound under investigation will scatter and absorb the Mössbauer γ-rays, and/or emit photons of similar energy which are detected along with the Mössbauer photons leading to a decrease in the signal to noise ratio. Mössbauer lines of appropriate intensity can therefore be obtained by decreasing the absorber thickness and increasing the recording time. It is therefore preferable to estimate an optimum absorber thickness for which the measuring time is as short as possible in order to achieve the maximum information from the experiment [17], [18].

19.5. Hyperfine Interactions

Owing to the narrow linewidth of the Mössbauer nuclear transition (of the order of 10^{-8} eV), the resonance spectrum is extremely sensitive to energy variations of the γ-radiation. Interactions between the nucleus and the surrounding electrons with energies comparable to the width of the transition lines manifest themselves in the Mössbauer spectrum. It is the influence of the electronic environment on the emission and absorption lines which determines the hyperfine structure of the spectrum. This interaction of the positively charged nucleus with the electric and magnetic fields caused by the orbital electrons in the region of the nucleus is called *hyperfine interaction*. This perturbation shifts or splits degenerate nuclear levels.

The Hamiltonian for the nucleus in a solid usually is expressed in a multipole expansion up to the second order

$$H = H_{E_0} + H_{M_1} + H_{E_2} + \ldots \quad (12)$$

Higher order terms are negligibly small and cannot be detected by Mössbauer effect measurements. The first term, the so-called monopole interaction, E_0, gives rise to the *isomer shift* (δ) and shifts the energy levels of both the ground state and the excited state. It originates from the Coulombic interaction between the nucleus and electrons at the nuclear site. The magnetic dipole (M_1) interaction removes the degeneracy of the nuclear levels if the nucleas has a magnetic moment. It is commonly known as the magnetic hyperfine interaction and describes the influence of a magnetic field on the nuclear spin (*magnetic splitting*). The third term in Equation (12) represents the electric quadrupole interaction, i.e., the interaction between the nuclear quadrupole moment and an electric field gradient at the nuclear site. The resulting Mössbauer parameter is called *quadrupole splitting*.

The different interactions and the kind of information one can obtain from the respective Mössbauer parameters are summarized in Table 2. In the following we shall discuss the three hyperfine interactions separately.

19.5.1. Isomer Shift

The isomer shift (other names occasionally used are chemical isomer shift or center shift) arises from the fact that the nucleus of an atom possesses a finite volume, and s-electrons have the ability of penetrating the nucleus and spending a fraction of time inside the nuclear region. p-, d-, and f-electrons do not have this ability.

The electric monopole energy of a uniform spherical nucleus with radius and charge in a constant electron charge density $-e|\psi(0)|^2$ is given by

$$H_{E_0} = \frac{2}{5}\pi Ze^2 |\psi(0)|^2 R^2 \quad (13)$$

The nuclear radii of the ground and excited states, in general, differ by a small amount δR. This results in a gain or loss of Coulombic energy δE of the nucleus, when it emits a γ-quantum. This energy is transferred to the γ-ray and therefore produces a shift. The same argument holds for the absorption process, so that the total shift, δ, is expressed by

$$\delta = \frac{4}{5}\pi Ze^2 R^2 (\delta R/R) \left\{ |\psi_a(0)|^2 - |\psi_s(0)|^2 \right\} \quad (14)$$

Table 2. The interactions between the nucleus and surrounding electrons, the respective Mössbauer parameters, and the type of information

Type of interaction	Mössbauer parameter	Information
Electric monopole interaction between nucleus and electrons at the nuclear site	Isomer shift δ	a. Oxidation state (nominal valency) of the Mössbauer atom b. Bonding properties in coordination compounds (covalency effects between central atom/ion and ligands. Delocalization of d-electrons due to back-bonding effects, shielding of s-electrons by p- and d-electrons) c. Electronegativity of ligands
Electric quadrupole interaction between electric quadrupole moment of the nucleus and electric field gradient at the nuclear site	Quadrupole splitting ΔE_Q	a. Molecular symmetry b. Oxidation state (nominal valency) c. Spin state d. Bonding properties
Magnetic dipole interaction between magnetic dipole moment of the nucleus and a magnetic field at the nucleus	Magnetic splitting ΔE_M	Magnetic properties (e.g., ferro-, antiferro-, para-, diamagnetism, absolute value and direction of local magnetic fields)

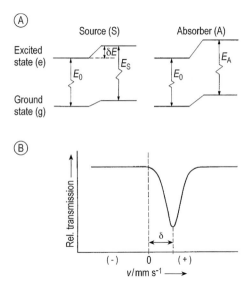

Figure 7. Origin of the isomer shift.
A) Electric monopole interaction in source and absorber shifts the nuclear energy levels without affecting the degeneracy; B) resultant Mössbauer spectrum (schematic)

where $\delta R = R_e - R_g$ is the change in nuclear radius on going from the excited (e) to the ground state (g). The subscripts a and s refer to absorber and source, respectively. Figure 7 shows schematically the origin of the isomer shift. From Equation (14) it is obvious that the isomer shift is always a relative shift, between absorber and source.

Therefore, the measured isomer shift is always given with respect to a standard material; this can be the Mössbauer source used in the particular experiment or any conventional absorber material.

Thus, the isomer shift is expressed by a nuclear term $\delta R/R$ and an electronic term $\{|\psi_a(0)|^2 - |\psi_s(0)|^2\}$, which measures the electron density at the nucleus of an absorber *relative to a given source*. For a given Mössbauer atom the nuclear factor is constant; the isomer shift is thus exclusively dependent on the difference of the electron densities between absorber and source. For ^{57}Fe $\delta R/R$ is negative, i.e., the nuclear radius in the excited state is smaller than in the ground state; in the case of ^{119}Sn the reverse is true.

The isomer shift can reflect minute changes in the electron density at the nucleus. In a non-relativistic approximation, the electron density at the nucleus is large only for electrons with zero angular momentum (s-electrons) and can be approximated by $|\psi(0)|^2$. Thus, the isomer shift of the Mössbauer spectra is a relative measure of the total s-electron density at the nuclear probe in a compound.

The total s-electron density of an atom in a chemical compound is the sum of all contributions from the inner (filled) and outer (valence electron) shells. The contribution from the outer electron shells is sensitive to changes in the chemical environment of the element (e.g., change of charge state, change of bonding properties by electron delocalization) and will exert a *direct* influence on the isomer shift. The valence p-, d-, and f-electrons have only an *indirect* influence on isomer shift by their shielding effect on the outer s-electron shells against the positive nuclear charge.

The measurement of the isomer shift gives information on

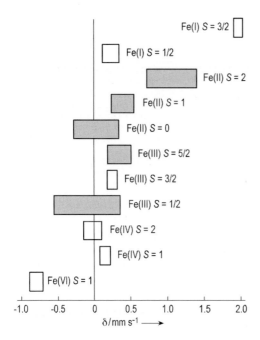

Figure 8. Approximate ranges of isomer shifts observed in iron compounds (relative to metallic iron). S refers to the electronic spin quantum number. Note that $(\delta R/R) < 0$ for ^{57}Fe

- the oxidation state (e.g., Fe(II), Fe(III), Fe(IV)),
- the spin state (high spin, low spin),
- the bond properties (covalency, ionicity),
- the electronegativity of the ligands

of a Mössbauer atom in solid material (coordination compounds, alloys, intermetallic phases, amorphous materials, etc.).

Figure 8 shows approximate ranges of the isomer shifts determined in iron compounds with different oxidation and spin states of the central metal ion.

It was shown [19] that the electronic density difference on going from the configuration $3d^6$ (Fe^{2+}) to $3d^5$ (Fe^{3+}) essentially originates from changes in the 3s shell. The filled 1s and 2s orbitals remain practically constant for both configurations. The removal of the 3d electron leads to an increase in the electron density at the nucleus due to the less effective shielding of the 3s electrons from the nuclear charge by the 3d electrons in Fe^{3+} as compared with Fe^{2+}. As $\delta R/R$ is negative for ^{57}Fe the isomer shift becomes more negative for Fe^{3+} than for Fe^{2+} (with respect to the same standard). The more or less wide-spread ranges of the δ values for each oxidation state is a direct consequence of the nature of the chemical bond. The electron distribution in the molecular orbitals is the result of the variable abilities of the ligands to donate electrons to the metal ion via σ-bonding and to accept electrons from the metal ion via π-bonding.

Similar correlation diagrams as shown in Figure 8 are available for numerous Mössbauer atoms [20].

19.5.2. Magnetic Hyperfine Interaction

An atomic nucleus, either in the ground or excited state, with nuclear spin quantum number $I > 0$ possesses a magnetic dipole moment $\vec{\mu}$ and interacts with a magnetic field $\vec{B}$ at the nuclear site. The magnetic field $\vec{B}$ may arise from the electronic environment or from an external magnet. This interaction is called *magnetic dipole interaction* and is described by the Hamiltonian

$$H_{M_1} = -\vec{\mu}\cdot\vec{B} = -g_N\mu_N\hat{I}\cdot\vec{B} \qquad (15)$$

where g_N is the nuclear Landé factor and $\mu_N = e\hbar/2Mc$ (M is the mass of the nucleus) is the nuclear magneton. The magnetic dipole interaction splits a nuclear state $|I\rangle$ into $2I+1$ equally spaced substates $|I, m_I\rangle$, each of these being characterized by the nuclear magnetic spin quantum number $m_I = -I, -I+1, ..., I$ (nuclear Zeeman effect). The Zeeman energies of the substates are $|I, m_I\rangle$

$$E_M(m_I) = -\mu B m_I / I = -g_N\mu_n B m_I \qquad (16)$$

Since the Mössbauer transition in ^{57}Fe is of the magnetic dipole type (M_1), there are only transitions between nuclear sublevels with $\Delta m_I = 0, \pm 1$ and $\Delta I = \pm 1$. This selection rule yields only six transitions between the two ground state sublevels ($I = 1/2$) and the four excited state sublevels ($I = 3/2$). Figure 9 illustrates the splitting of the nuclear levels, the allowed transitions, and the resulting ^{57}Fe Mössbauer spectrum, a sextet due to the magnetic hyperfine interaction.

The magnetic hyperfine field $\vec{B}$ at the nucleus can originate from the own electron shell ($\vec{B}_{\text{ion}}$) or can be generated by paramagnetic ions in the nearby lattice ($\vec{B}_{\text{lat}}$), or can be due to an applied magnetic field ($\vec{B}_{\text{app}}$):

$$\vec{B} = \vec{B}_{\text{ion}} + \vec{B}_{\text{lat}} + \vec{B}_{\text{app}} \qquad (17)$$

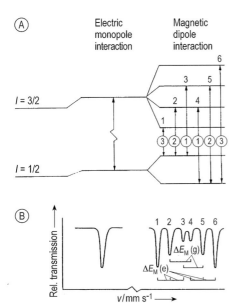

Figure 9. Origin of the magnetic hyperfine splitting
A) Magnetic dipole interaction in ^{57}Fe in absorber;
B) resultant Mössbauer spectrum (schematic)

The dipolar field of the moments of the surrounding paramagnetic ions $\rightarrow B_{lat}$ is a very small contribution.

$\rightarrow B_{FC}$ is the so-called Fermi contact field [21] and represents the most important contribution to the field $\rightarrow B_{ion}$. This field is the direct coupling between the nucleus and the unpaired s-electron density at the nucleus [22], [23],

$$\rightarrow B_{FC} = -\frac{16\pi}{3} \mu_B \left(\sum (\uparrow \psi_s^2(0) - \downarrow \psi_s^2(0)) \right) \quad (18)$$

where $\uparrow \psi_s^2(0)$ is the spin density at the nucleus of s-electrons with spin parallel to the total spins of the valence d-electrons spin and $\downarrow \psi_s^2(0)$ that with spin antiparallel to the d-electrons spins.

Owing to the fact that the exchange interaction between the spin-up polarized d-shell and the spin-up s-electron is attractive, while that between the spin-up d-shell and the spin-down s-electron is repulsive, the radial parts of the two s-electrons distort, one being closer to the nucleus and the other being more distant. As a consequence, the spin density with spin-down is enhanced at the nucleus. The direction of the Fermi contact field is antiparallel to this spin, and consequently the Fermi contact field is negative by defintion, which means anti-parallel to the magnetic moments of the d-electrons.

For high spin ferric compounds with electron configuration 3d^5, ^{6}S, the dipolar field due to its own electrons and the contribution of the orbital momentum to the total field are zero. The Fermi contact term, estimated to be -11 T per unpaired 3d electron, dominates over the contributions of the neigboring dipoles and the applied field. In high spin ferrous compounds, however, the negative Fermi contact term is opposed by a positive orbital contribution of almost the same order of magnitude. The positive dipolar contribution is in general quite small. In this case it is impossible to predict the sign of the internal hyperfine field, B_{ion}. This problem can be solved by applying an external magnetic field that aligns the 3d moments. Sign and absolute value of the internal field can then be obtained from the vector sum of the applied and the internal field.

19.5.3. Quadrupole Splitting

The third term in the multipole expansion [cf. Equation (12)] describes the *electric quadrupole interaction*. This type of nuclear–electronic interaction is also electrostatic in nature and occurs if the following conditions are fulfilled.

1) The Mössbauer nucleus must possess a measurable electric quadrupole moment eQ, which arises from non-spherical nuclear charge distribution leading to spin quantum numbers $I > 1/2$. A measure of the deviation from spherical symmetry is given by the electric quadrupole moment, eQ, which can be calculated by the spatial integral

$$eQ = \int p^2 (3\cos^2\theta - 1) \, dV \quad (19)$$

where e is the proton charge and p is the charge density at the spherical coordinates r and θ. The sign of the nuclear quadrupole moment refers to the shape of the distorted nucleus: Q is negative for a flattened and positive for an elongated nucleus.

2) The electric field at the Mössbauer nucleus must be inhomogeneous, measured by the electric field gradient (EFG) $V_{ij} = \partial^2 V/\partial i \partial j$ ($i, j = x, y, z$). This arises from a non-cubic charge distribution of the electrons and/or of the neigboring ions around the nucleus.

With respect to the principal axes, the EFG tensor is described by only two independent pa-

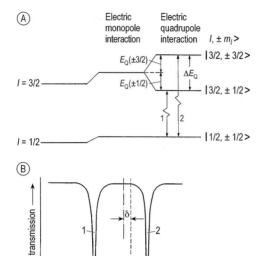

Figure 10. Origin of the quadrupole splitting in ^{57}Fe
A) electric quadrupole interaction in case of $eQV_{zz} > 0$;
B) resultant Mössbauer spectrum (schematic) with equal intensities as for a powder sample; the inevitable shift of the nuclear levels due to electric monopole interaction giving rise to the isomer shift is also shown

rameters, usuallly chosen as $V_{zz} = eq$ and $0 \leq \eta = (V_{xx} - V_{yy})/V_{zz} \leq 1$.

The spin Hamiltonian of the nucleus in the principal axes system of the EFG is given by:

$$H_{E_2} = \frac{eQV_{zz}}{4I(2I-1)} \left\{ 3\hat{I}_z^2 - I(I+1) + \frac{\eta}{2}\left(\hat{I}_+^2 + \hat{I}_-^2\right) \right\} \quad (20)$$

where $\hat{I}_{z,+,-}$ are the usual spin operators.

As an example, the effect of the electric quadrupole interaction in a Mössbauer nucleus with $I = 3/2$ in the excited state and $I = 1/2$ in the ground state, as is the case in ^{57}Fe and ^{119}Sn, is given in Figure 10. The quadrupole interaction gives rise to a doublet with the splitting

$$\Delta E q = \frac{1}{2} e \left| QV_{zz} \right| \left(1 + \frac{\eta^2}{3}\right)^{\frac{1}{2}} \quad (21)$$

The quadrupole splitting in a Mössbauer spectrum is a highly sensitive probe for the chemical environment that determines the EFG at the nucleus of the Mössbauer atom. In general, there are two fundamental sources which can contribute to the total EFG.

1) Charges on ions surrounding the Mössbauer probe atom in noncubic symmetry give rise to the so-called *lattice contribution*, $(EFG)_{lat}$ or V_{zz}^{lat} (for axial symmetry).
2) Noncubic distribution of the electrons in the partially filled valence orbitals of the Mössbauer atom generate the *valence contribution*, $(EFG)_{val}$ or V_{zz}^{val} (for axial symmetry).

According to the different origins (valence/lattice contributions) it is expected that observed quadrupole splittings may reflect information about the electronic structure (oxidation state, spin state), bond properties, and molecular symmetry. As an example the spectra of three iron coordination compounds are shown in Figure 11.

a) The observed large quadrupole splitting of $\Delta E_Q = 3.4$ mm s^{-1} for FeSO$_4 \cdot$ 7H$_2$O essentially arises from the valence contribution V_{zz}^{val}. This compound is a typical high spin compound where the Fe(II) ion is octahedrally surrounded by six water molecules. This type of ions with a $^5T_{2g}$ ground state are subject to Jahn–Teller distortion, which lifts the degeneracy of the t_{2g} orbitals partly or completely and leads then to a noncubic distribution of the six electrons in the 3d shell of the [Fe(H$_2$O)$_6$]$^{2+}$ ions. The doubly occupied lowest d_{xy} orbital gives rise to the relatively large quadrupole splitting in the tetragonally compressed [Fe(H$_2$O)$_6$]$^{2+}$ octahedron. Further details can be found in reference [12].

b) K$_4$[Fe(CN)$_6$] $\cdot$ 3H$_2$O is a typical low spin Fe(II) compound. The single line Mössbauer spectrum results from the absence of a quadrupole interaction due to the cubic charge distribution of the electron configuration t_{2g}^6 which consequently leads to zero valence contribution V_{zz}^{val} [12]. The lattice contribution V_{zz}^{lat} also vanishes since the [Fe(CN)$_6$]$^{4-}$ complex ion forms a regular octahedron.

c) The relatively large quadrupole splitting in the case of Na$_2$[Fe(CN)$_5$NO] $\cdot$ 2H$_2$O is solely due to the fact that the presence of different ligand molecules can no longer form a regular octahedron as in the case of [Fe(CN)$_6$]$^{4-}$. The magnitude of the quadrupole splitting is mainly determined by the lattice contribution.

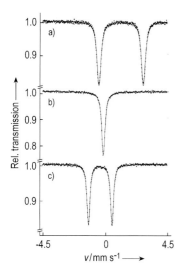

Figure 11. Mössbauer spectra of: a) FeSO$_4$·7H$_2$O; b) K$_4$[Fe(CN)$_6$]·3H$_2$O; and c) Na$_2$[Fe(CN)$_5$NO]·2H$_2$O

19.5.4. Combined Quadrupole and Magnetic Hyperfine Interactions

Quite frequently electric quadrupole interactions and magnetic dipole interactions may be present in addition to the electric monopole interactions which are always active. The total Hamiltonian H_{HFS} for the combined electric quadrupole and magnetic dipole interaction cannot be solved analytically in the general case of comparable interaction strengths and arbitrary direction of the magnetic hyperfine field in the PAS of the EFG. In case of a small EFG and large magnetic field, a situation often met, a perturbation treatment may be sufficient. For an axially symmetric EFG with the polar angle ϑ of the direction of the magnetic field in the PAS of the EFG the eigenvalues of the nuclear sublevels of the $I = 3/2$ state are given by

$$E_{M,Q}(I, M_I) = -g_N \mu_N B m_I + (-1)^{|m_I| + \frac{1}{2}} (eQV_{zz}/8)(3\cos^2\vartheta - 1) \quad (22)$$

We find in the case of ^{57}Fe, which is depicted in Figure 12, that the sublevels $|3/2, \pm 3/2\rangle$ are shifted by an amount $E_Q(\pm m_I)$ to a higher energy and the sublevels $|3/2, \pm 1/2\rangle$ are shifted by E_Q to a lower energy, if V_{zz} is positive. These energy shifts by E_Q are reversed if V_{zz} is negative. As a result, the sublevels of the excited state $I = 3/2$ are no longer equally spaced, unless $\cos\vartheta = \sqrt{1/3}$. Combined electric quadrupole and magnetic dipole interactions generally manifest themselves in the Mössbauer spectrum as an asymmetrically split sextet (for $I = 3/2 \longleftrightarrow I = 1/2$ transitions as in ^{57}Fe) as described in Figure 12. As the sublevel spacings and thus the asymmetry of the spectrum is directly correlated with the sign of V_{zz}, one can determine the sign of the EFG of a polycrystalline material from the magnetically split hyperfine spectrum. The effect of a magnetic dipole interaction (applied magnetic field) as a perturbation of the electric quadrupole splitting is described in detail in reference [24].

These three types of hyperfine interactions with the relevant Mössbauer parameters are most important in solid state research. In addition, one often extracts further helpful information from the temperature and pressure dependence of the Mössbauer parameters, the shape and width of the resonance lines (relaxation phenomena), and the second-order Doppler shift (lattice dynamics).

19.5.5. Relative Intensities of Resonance Lines

The relative intensities of the lines in a Mössbauer spectrum provide additional information about the orientation of the hyperfine fields with respect to the crystal system in the case of single crystals or partially oriented (textured) samples. For pure magnetic or electric hyperfine fields the intensities are analytical expressions dependent on the direction of the fields with respect to the direction of γ-radiation. For powder samples the intensity ratios of pure transitions do not depend on the strength of the fields.

For the case of a magnetically split spectrum of a $3/2 \rightarrow 1/2$ transition the intensity ratios are $3:2:1:1:2:3$ for the hyperfine components of the Zeeman pattern. The corresponding quadrupole spectrum has equal intensities as already mentioned above. For more details the reader is referred to the references [25], [26].

19.6. Evaluation of Mössbauer Spectra

The evaluation of a Mössbauer spectrum implies the determination of the physical parameters, the isomer shift, Debye–Waller factor, and electric and magnetic hyperfine fields. When fluc-

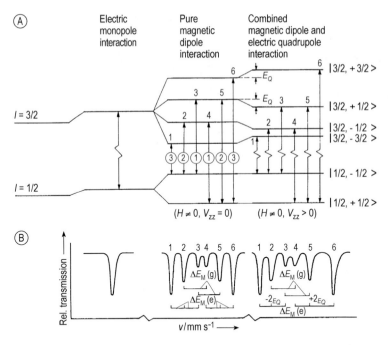

Figure 12. Combined quadrupole and magnetic hyperfine interaction: (a) splitting and shifts, (b) resultant Mössbauer spectrum (schematic)

tuation times of the hyperfine fields of the order influencing the shape of the spectra are present the fluctuation times themselves may be a parameter of interest.

Fitting of Mössbauer spectra requires the existence of a model for the hyperfine interaction at the Mössbauer nuclei in the material under study. Computer programs are available for a variety of models. The interested reader is referred to the homepage of the Mössbauer community: http://www.kfki.hu/mixhp/.

19.7. Selected Applications

Since the discovery more than 40 years ago, Mössbauer spectroscopy has become an extremely powerful analytical tool for the investigation of various types of materials. In most cases, only two parameters are needed, viz. the isomer shift and the quadrupole splitting, to identify a specific sample. In case of magnetically ordered materials, the magnetic dipole interaction is a further helpful parameter for characterization. In the following we shall discuss some applications selected from different research areas.

19.7.1. Magnetic Materials

As an example we will consider the magnetic ordering in magnetite, Fe_3O_4.

Magnetite is an inverse spinel, with iron atoms occupying interstitial sites in the close-packed arrangement of oxygen atoms. The tetrahedral sites (A) are occupied by ferric cations, and the octahedral sites (B) half by ferric and half by ferrous cations. In the unit cell there are twice as many B-sites as A-sites. The structural formula can therefore be written: $(Fe(III))_{tet}[Fe(II)Fe(III)]_{oct}O_4$. There is little distortion in the cubic close packing of the oxygens, so that the structure is close to ideal.

Magnetite has a Curie temperature of 840 K. In the magnetically ordered phase the iron atoms on A-sites are antiferromagnetically coupled to those on B-sites resulting in a complete cancelling of the moments of the Fe(III) ions. Thus, a ferromagnetic moment, originating in the Fe(II) ions on A-sites, remains. On the whole, magnetite shows ferrimagnetic behavior.

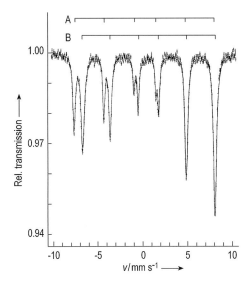

Figure 13. Room temperature Mössbauer spectrum of magnetite; subspectra of lattice sites A and B are indicated by the stick diagram

The statistical distribution of the di- and trivalent ions of iron on equivalent lattice sites (B) explains many unusual properties of magnetite, in particular its extremely high electrical conductivity comparable to that of metals. Investigations have shown that the conductivity is caused by electron and not by ion transport. A fast electron-transfer process (electron hopping) between ferrous and ferric ions on the octahedral B-sites takes place [27]. The Mössbauer spectrum of magnetite at room temperature (cf. Figure 13) confirms this interpretation.

The spectrum consists of two more or less overlapping six-line spectra with internal magnetic fields of 49 and 46 T and a relative area ratio of approximately 1:2, which originate from the iron ions on sites A (Fe(III)) and B (average field of Fe(II) + Fe(III)). The Mössbauer parameters (δ, ΔE_Q and H_{int}) for the sextet stemming from the site B ions are intermediate between those of oxidic Fe(II) and Fe(III) compounds; i.e., concerning the iron ions on lattice site B, we are dealing with a mixed-valent system of Class III in the Robin–Day classification. At ca. 120 K, magnetite shows the so-called Verwey transition. The rate of the electronic hopping between the Fe(II) and Fe(III) ions on the lattice B sites slows down such that it becomes slower than the reciprocal of the Mössbauer time window. This causes a significant change in the magnetically split subspectra [28], [29].

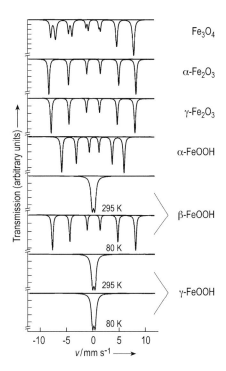

Figure 14. Mössbauer spectra of the most important corrosion products

19.7.2. Corrosion Products

Mössbauer spectroscopy of ^{57}Fe has proven to be an extremely powerful tool for the analysis of iron containing corrosion products on steel and technical alloys. Depending on the corrosion conditions a variety of iron oxides and oxyhydroxides may be formed. Mössbauer spectroscopy enables one to differentiate between the various phases on the basis of their different magnetic behavior. Figure 14 illustrates this with six different corrosion products. The Fe$_3$O$_4$ (magnetite), shows the superposition of two six-line spectra as described in Section 19.7.1.

The α-Fe$_2$O$_3$ (haematite), γ-Fe$_2$O$_3$ (maghemite), and α-FeOOH (goethite) differ in the size of the internal magnetic field yielding different spacings between corresponding resonance lines. A problem seems to arise with β- and γ-FeOOH, because they have nearly identical Mössbauer parameters at room temperature and cannot be distinguished by their Mössbauer spectra. However, lowering the temperature solves the problem: β-FeOOH appears to be magnetically ordered at 80 K, whereas γ-FeOOH still remains in the paramagnetic state.

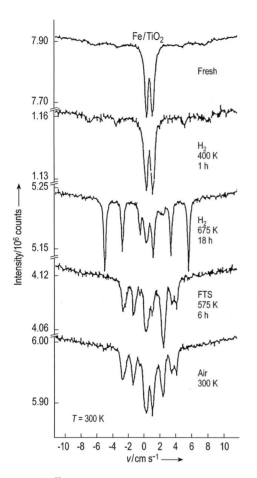

Figure 15. ^{57}Fe Mössbauer spectra at room temperature (from top to bottom) of the catalyst system Fe/TiO$_2$ after different treatments (cf. text). Reproduced from Applied Catalysis [29]

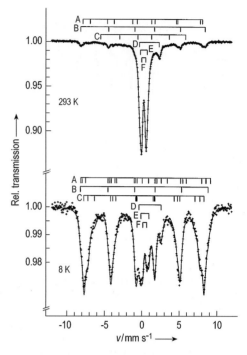

Figure 16. 293 K (top) and 8 K (bottom) Mössbauer spectra of the aerosol sample; the decomposition into the spectral components is indicated by the stick bars, where A stands for magnetite (Fe$_3$O$_4$), B for hematite (α-Fe$_2$O$_3$), C for goethite (α-FeOOH), D for Fe^{2+}-, E and F for Fe^{3+}-compounds

Such Mössbauer studies may be carried out in a nondestructive manner and are even possible with highly dispersed particles which are no longer amenable to X-ray diffraction measurements.

19.7.3. Catalysis

The successful use of Mössbauer spectroscopy for studies of catalysts is demonstrated in an important technical example (catalyst system). In Figure 15 a series of spectra of the Fe/TiO$_2$ catalyst system [30] is shown. The top spectrum represents the freshly prepared catalyst; it is obvious that iron is present in the 3+ oxidation state, possibly in the form of finely dispersed oxide or oxyhydroxide on TiO$_2$. Under a reducing atmosphere (H$_2$ at 675 K) this iron phase is essentially converted into metallic iron (six line pattern). Some residues of Fe^{2+} and Fe^{3+} contribute to the spectrum as doublets. Under the conditions of the Fischer–Tropsch synthesis (FTS) at 575 K in CO and H$_2$ the metallic iron is transformed into the Hägg carbide χ-Fe$_5$C$_2$. The unreduced iron is now present as Fe^{2+}, which is oxidized to ferric iron when the catalyst is exposed to air at ambient temperature. The carbide phase is left unchanged.

19.7.4. Atmospheric Aerosol Samples

Iron is one of the most abundant elements in solid and aqueous atmospheric samples and is usually introduced into the atmosphere as soil dust, fly ash from power plants, exhaust from combustion engines, and from industrial operations. The role of iron in important atmospheric chemical reactions is subject to numerous investigations. Mössbauer spectroscopy proves to be a useful analytical method to determine the iron containing species in the atmosphere. In Figure 16

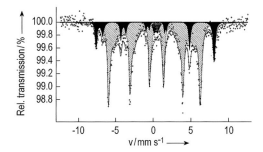

Figure 17. ^{57}Fe Mössbauer spectrum of a one US-$ bill. The different spectral components are marked as: black = magnetite (Fe$_3$O$_4$), dashed lines = goethite (α-FeOOH), white = spm Fe^{3+} component

the result of the Mössbauer analysis [31] of an aerosol sample taken from an air-conditioning device of a building is shown.

The typically very small crystals of the different components in the aerosol sample give rise to the so-called superparamagnetic (spm) behaviour (cf. BRONGER et al. [32] and references therein) resulting in the pronounced resonance doublet for Fe^{3+} compounds in the room temperature spectrum. The low temperature spectrum (8 K) revealed that this doublet component mainly originates in superparamagnetic microcrytals of goethite and hematite.

19.7.5. US-$ Bill

Figure 17 shows the Mössbauer spectrum of a one dollar bill. For the measurement the note was folded in such a manner that approximately 10 layers of the bill were analyzed. The result of the analysis: the banknote contains about 20 wt-% of magnetite and 80 wt-% of goethite as iron containing color pigments.

19.8. References

[1] R. Mössbauer, *Z. Phys.* **151** (1958) 124.
[2] R. Mössbauer, *Naturwissenschaften* **45** (1959) 538.
[3] R. Mössbauer, *Z. Naturforsch.* **14a** (1959) 211.
[4] H. Frauenfelder: *The Mössbauer Effect,* Benjamin, New York 1962.
[5] G. K. Wertheim: *Mössbauer Effect: Principles and Applications,* Academic Press, New York 1964.
[6] H. Wegener: *Der Mössbauer-Effekt und seine Anwendung in Physik und Chemie,* Bibliographisches Institut, Mannheim 1965.
[7] V. I. Goldanskii, R. Herber (eds.): *Chemical Applications of Mössbauer Spectroscopy,* Academic Press, New York 1968.
[8] N. N. Greenwood, T. C. Gibb: *Mössbauer Spectroscopy,* Chapman and Hall, London 1971.
[9] G. M. Bancroft: *Mössbauer Spectroscopy, An Introduction for Inorganic Chemists and Geochemists,* McGraw-Hill, New York 1973.
[10] U. Gonser (ed.): "Mössbauer Spectroscopy," in *Topics in Applied Physics,* vol. 5, Springer, Berlin, Heidelberg, New York 1975.
[11] T. C. Gibb: *Principles of Mössbauer Spectroscopy,* Chapman and Hall, London 1976.
[12] P. Gütlich, R. Link, A. X. Trautwein: "Mössbauer Spectroscopy and Transition Metal Chemistry," in *Inorganic Chemistry Concepts,* vol. 3, Springer, Berlin, Heidelberg, New York 1978.
[13] J. G. Stevens: Mössbauer Effect Reference and Data Journal, Mössbauer Data Center, University of North Carolina, Asheville, USA.
[14] W. Kuhn, *Philos. Mag.* **8** (1929) 625.
[15] D. Barb: *Grundlagen und Anwendungen der Mössbauerspektroskopie,* Akademie-Verlag, Berlin 1980.
[16] E. Kankeleit: *Rev. Sci. Instrum.* **35** (1964) 194.
[17] U. Shimony, *Nucl. Instrum. Methods* **37** (1965) 348.
[18] S. Nagy, B. Levay, A. Vertes, *Acta. Chim. Acad. Sci. Hung.* **85** (1975) 273.
[19] R. E. Watson, *Phys. Rev.* **118** (1960) 1036.
[20] G. K. Shenoy, F. E. Wagner: *Mössbauer Isomer Shifts,* North Holland Publishing Company, Amsterdam, New York, Oxford 1978.
[21] A. Abragam: *The Principles of Nuclear Magnetism,* Oxford University Press, Oxford 1961.
[22] W. Marshall, *Phys. Rev.* **110** (1958) 1280.
[23] W. Marshall, *Phys. Rev.* **105** (1957) 158.
[24] R. L. Collins, *J. Chem. Phys.* **42** (1965) 1072.
[25] T. Viegers: *Thesis,* University of Nijmegen, Netherlands 1976.
[26] A. Gedikli, H. Winkler, E. Gerdau, *Z. Phys.* **267** (1974) 61.
[27] E. J. W. Vervey, *Nature (London)* **144** (1939) 327.
[28] R. S. Hargrove, W. Kundig, *Solid State Commun.* **8** (1970) 330.
[29] M. Rubinstein, D. M. Forester, *Solid State Commun.* **9** (1971) 1675.
[30] A. M. van der Kaan, R. C. H. Nonnekens, F. Stoop, J. W. Niemantsverdriet, *Appl. Catal.* **27** (1986) 285.
[31] P. Hoffmann, A. N. Dedik, J. Ensling, S. Weinbruch, S. Weber, T. Sinner, P. Gütlich, H. M. Ortner, *J. Aerosol. Sci.* **27** (1996) 325.
[32] A. Bronger, J. Ensling, P. Gütlich, H. Spiering, *Clays Clay Miner.* **31** (1983) 269.

20. Mass Spectrometry

MICHAEL LINSCHEID, Department of Chemistry, Humboldt University, Berlin, Germany

20.	Mass Spectrometry 579	20.6.	Analyzers. 597
20.1.	Introduction. 580	20.6.1.	Electromagnetic Sector Fields . . . 597
20.2.	General Techniques and Definitions 580	20.6.2.	Quadrupoles. 598
20.2.1.	Resolution 580	20.6.3.	Time-of-Flight (TOF) Mass Spectrometer 599
20.2.2.	Tools for Structure Elucidation. . . 582	20.6.4.	Fourier Transform Mass Spectrometry 599
20.2.2.1.	Full Spectra: Low Resolution. . . . 582		
20.2.2.2.	Elemental Compositions of Ions . . 583	20.6.5.	Ion Traps. 601
20.2.3.	Fragmentation in Organic Mass Spectrometry 583	20.6.6.	Isotope Mass Spectrometer. 601
		20.6.7.	Accelerator Mass Spectrometry (AMS). 602
20.2.4.	Quantitative Analysis 584		
20.3.	Sample Inlets and Interfaces 585	20.6.8.	Other Analyzers 603
20.3.1.	Direct Probe. 585	20.7.	Metastable Ions and Linked Scans 603
20.3.2.	Batch Inlets 586	20.7.1.	Detection of Metastable Ions 603
20.3.3.	Pyrolysis 586	20.7.2.	Mass Selected Ion Kinetic Spectrometry 603
20.3.4.	GC/MS Interfaces. 586		
20.3.5.	LC/MS Interfaces 587	20.7.3.	Linked Scans 603
20.3.5.1.	Moving Belt. 587	20.8.	MS/MS Instrumentation 604
20.3.5.2.	Continuous Flow FAB. 588	20.8.1.	Triple Quadrupoles 605
20.3.5.3.	Direct Liquid Introduction 588	20.8.2.	Multiple-Sector Instruments 605
20.3.5.4.	Supercritical Fluid Interface. 588	20.8.3.	Hybrids (Sector – Quadrupoles). . . 606
20.3.5.5.	Particle Beam Interface 588	20.8.4.	Ion-Storage Devices (FTMS, Ion Traps) 606
20.3.5.6.	Chemical Ionization at Atmospheric Pressure. 589		
		20.8.5.	Quadrupole Time-of-Flight Tandem Mass Spectrometers 606
20.3.5.7.	Thermospray 589		
20.3.5.8.	ESI Interface 589	20.8.6.	Others 607
20.3.6.	TLC/MS 590	20.9.	Detectors and Signals 607
20.4.	Ion Generation. 590	20.9.1.	Faraday Cage 607
20.4.1.	Electron Impact 590	20.9.2.	Daly Detector. 607
20.4.2.	Chemical Ionization 591	20.9.3.	Secondary Electron Multiplier (SEM). 608
20.4.3.	Negative Chemical Ionization (Electron Capture) 591		
		20.9.4.	Microchannel Plates 609
20.4.4.	Field Ionization (FI) 592	20.9.5.	Signal Types 609
20.4.5.	Plasma Ionization 592	20.10.	Computer and Data Systems 610
20.4.6.	Thermal Ionization 594	20.10.1.	Instrument Control, Automation . . 610
20.4.7.	Optical Methods 594	20.10.2.	Signal Processing 610
20.4.8.	Desorption Methods 594	20.10.3.	Data Handling 611
20.4.8.1.	Secondary Ion Mass Spectrometer (SIMS) 594	20.10.4.	Library Searches. 611
		20.10.5.	Integration into Laboratory Networks. 612
20.4.8.2.	Field Desorption (FD) 595		
20.4.8.3.	Fast Atom Bombardment (FAB) . . 595	20.10.6.	Integration in the World Wide Web 613
20.4.8.4.	^{252}Cf Plasma Desorption 595	20.11.	Applications. 613
20.4.8.5.	Laser Desorption/Ionization 596	20.11.1.	Environmental Chemistry. 613
20.4.9.	Electrospray 596		

20.11.2. Analysis of Biomedical Samples . . 615
20.11.3. Determination of High Molecular Masses 617
20.11.4. Species Analysis 618
20.11.5. Determination of Elements 618
20.11.6. Surface Analysis and Depth Profiling 619
20.12. References 622

Abbreviations

AMS   accelerator mass spectrometry
API   atmospheric pressure ionization
APCI   atmospheric pressure chemical ionization
CAD   collision activated dissociation
CIMS   chemical ionization mass spectrometry
CZE   capillary zone electrophoresis
DADI   direct analysis of daughter ions
EIMS   electron impact (ionization) mass spectrometry
ESI   electrospray ionization
FABMS   fast atom bombardment mass spectrometry
FDMS   field desorption mass spectrometry
FIMS   field ionization mass spectrometry
FTMS   Fourier transform mass spectrometry
GC/MS   gas chromatography/mass spectrometry
GDMS   glow discharge mass spectrometry
HPLC   high performance liquid chromatography
ICPMS   inductively coupled plasma mass spectrometry
ICR   ion cyclotron resonance
IDMS   isotope dilution mass spectrometry
IMMA   ion microprobe mass analyzer
IRMS   isotope ration mass spectrometry
LC/MS   liquid chromatography/mass spectrometry
LD   laser desorption
LAMMA   laser microprobe mass analyzer
LAMS   laser ablation mass spectrometry
LSIMS   liquid secondary ion mass spectrometry (see FAB)
MALDI   matrix assisted laser desorption
MID   multiple ion detection
MIKES   mass analyzed ion kinetic energy spectrometry
MS/MS   mass spectrometry/mass spectrometry
MUPI   multiphoton ionization
NCI   negative chemical ionization
PD   plasma desorption
PFK   perfluorokerosene
REMPI   resonance multi photon ionization
RIMS   resonance ionization mass spectrometry
SEM   secondary electron multiplier
SIMS   secondary ion mass spectrometry
SIR   selected ion recording
SNMS   sputtered neutral mass spectrometry
TCDD   tetrachlorodibenzodioxin
TIMS   thermal ionization mass spectrometry
TOFMS   time-of-flight mass spectrometry

20.1. Introduction

A mass spectrometer can be defined as any instrument capable of producing ions from neutral species and which provides a means of determining the mass of those ions, based on the mass-to-charge ratio (m/z, where z is the number of elemental charges) and/or the number of ions. Therefore, a mass spectrometer has an ion source, an analyzer, and a detector. All further details are largely dependent on the purpose of the mass spectrometric experiment.

Historically, mass spectroscopy was developed to separate atoms and to determine the masses of isotopes accurately [1]. The predecessor of magnetic sector field mass spectrometers was built in 1918 by DEMPSTER [2]. Later, some of the first instruments for the analysis of molecules [3] were used by analytical chemists in refineries to determine hydrocarbons (the first article in the first issue of the journal *Analytical Chemistry* dealt with mass spectrometry of a mixture of hydrocarbons [4]). Combinations of infrared and mass spectroscopy were reported shortly afterwards [5].

Today, mass spectrometers may be used to determine the isotopic distribution of an element, the elemental or molecular composition of a sample, or the structure of a compound or its molecular mass. They also make it possible to study the kinetics and thermodynamics of gasphase processes or the interactions at phase boundaries. Mass spectrometers can also serve to accurately determine physical laws and natural constants.

In recent years, mass spectrometry has become one of the key technologies in proteome and genome research because of the rapid development of new components—from the ion sources to the detector—and intergration into the World Wide Web has also helped to make this possible.

20.2. General Techniques and Definitions

20.2.1. Resolution

Generally, resolution is defined by $R = m/\Delta m$, where m is the mass under consideration. This can be separated from a mass $m + \Delta m$, assuming a peak overlap of 10 %. The equation can be used either to calculate the resolution of two signals in a spec-

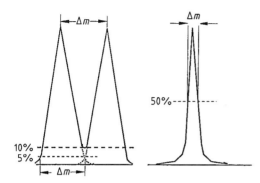

Figure 1. Resolution of two peaks, defined for the 10% height of the valley, for one signal at the 5% (sector field MS), or the 50% (all others) height

trum or to calculate the resolution necessary to separate two species of different elemental composition. For example, consider C_7H_{16} and $C_6H_{12}O$. The mass difference at $m/z = 100$ is 0.0364 amu. Thus, a resolution of ca. 3000 is required to resolve the two signals. The same result is obtained when the peak width at 5% overlap is used for the calculation.

A different definition uses the peak width at half height as Δm (50% definition) for the calculation (see Fig. 1). This approach is used often for Fourier transform mass spectroscopy (FTMS) and time-of-flight (TOF) instruments. In modern mass spectrometers, the resolution is calculated directly by the computer from the peak width.

More information can be extracted if the elemental composition of fragments or molecular ions can be determined. For this purpose two different requirements have to be fulfilled. First, the signal must be homogeneous, i.e., even isobaric ionic species have to be separated (resolved) and, secondly, the accuracy of the mass measurement has to be sufficient to allow a calculation with small errors.

High Resolution. The separation of isobaric ionic species (i.e., high resolution, with $R > 10\,000$), is a relative requirement. It depends on various factors, such as the type of heteroatoms, but with $R = 10\,000$ most common species can be resolved. Exceptions can be ions containing, e.g., sulfur. Increasing the resolution requires reducing the width of the ion beam in the mass spectrometer. Thus, in theory, the height of the signal is inversely proportional to the resolution. In reality, this relation is not linear over the full range of possible resolutions, and generally the loss of peak height becomes much greater when the resolution is increased above 10 000. Merely instruments like the ion trap (with very large molecules) or especially the FT mass spectrometers are capable of achieving resolution exceeding 10^5, even up to 10^6 under optimal circumstances without deterioration in detection power. The scan speed and the scan width may be affected. Thus, for applications in analytical problem solving, the resolution should be set as high as necessary yet as low as possible.

Accurate Mass Measurements. For accurate mass measurements a signal containing sufficient ions must be recorded on a precisely calibrated mass scale. In the past this was a cumbersome undertaking because each peak had to be determined separately by comparison to a known reference signal. The accelerating voltage U_0 was changed by the amount necessary to shift the signal of the mass m_x to the position of the reference m_0 on the monitor. This can be performed very accurately by switching between the two voltages U_0 and U_x (for the unknown). Since $U_0 m_0 = U_x m_x$, m_x can be calculated from $m_x = m_0 \cdot (U_0/U_x)$. This technique, known as peak matching, is still the most precise method, and today computerized versions are available.

Techniques were then developed that allow accurate mass measurement of many signals in a spectrum simultaneously. A reference compound such as perfluorokerosene (PFK) is introduced into the ion source together with the analyte in such a concentration that signals of both can be acquired simultaneously. The reference compound is chosen so that its signals are easily resolved from those of the analyte. The computer then searches for the reference signals in the spectrum, interpolates between them, and calculates the correct position of the unknowns on the mass scale. Sufficient precision can be achieved to allow the calculation of possible elemental compositions for the signal, provided the number of ions is sufficient. It has been shown [6] that the number of ions N required to define the peak position is a function of resolution R, where σ is the standard deviation of the mass measurement in parts per million

$$N = \frac{1}{24}\left(\frac{10^6}{R\sigma}\right)^2$$

Thus, to obtain an accuracy of 2 ppm, the number of ions required for $R = 10\,000$ is $N = 105$, but for

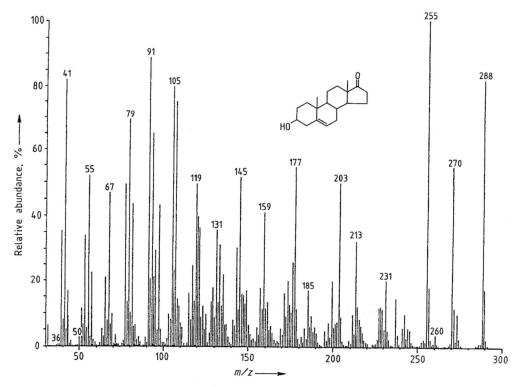

Figure 2. Low resolution ($R = 1000$) spectrum of dehydro-epi-androsteron, direct inlet using a VG 7070E, electron impact mass spectrometer (EIMS)

$R = 5000$, $N = 420$. As long as the reduction of signal intensity with increasing resolution is linear, higher resolution can be beneficial, because fewer ions are required for a given accuracy. However, if the intensity of the signal is also of relevance (e.g., for quantitative determinations) more ions must be detected. Thus, in this case the optimal resolution may be lower to allow better detection power.

20.2.2. Tools for Structure Elucidation

20.2.2.1. Full Spectra: Low Resolution

In the structure determination of natural products, electron impact ionization not only produces ions of the compound itself ("molecular ions") but also results in formation of fragments, often with typical patterns (Fig. 2).

The spectra can be used to identify a compound or at least the class of compound. The spectrum of, for example, a steroid can reflect its chemical nature to such an extent that the probability of correct identification of the type of compound or even the compound itself may be acceptable. To achieve this it is sufficient to register the spectrum of fragments with integer precision. However, this means that isobaric molecules containing different heteroatoms are not separated. The success of this method for the elucidation of structures of, say, terpenes, steroids, complex lipids, and saccharides is documented in many publications and textbooks [7], [8]. The typical features of such spectra are described, and explanations are given for the characteristic fragmentation patterns. This is still the most widely used application of mass spectrometry, for which there are several reasons. First, most mass spectrometers are low resolution instruments such as quadrupoles. Second, the most common inlet systems or interfacing techniques, namely the direct probe and GC, are compatible with electron impact, giving highly reproducible spectra, largely independent of the type of mass spectrometer. Even the more modern interface for supercritical fluid chromatography or the particle-beam LC/MS interface can be used. Finally, modern library searches or intelligent data handling systems make efficient use of such data and can retrieve much of the information.

20.2.2.2. Elemental Compositions of Ions

The determination of the elements contained in an ion is a major aid in structure analysis. This information could be derived in principle from the isotope pattern of the molecular ion, because the isotope distribution is typical for each element. The ratios of $^{12}C/^{13}C$ and sometimes $^{16}O/^{18}O$ are the most significant parameters as well as the distinctive patterns of sulfur, silicon, the halogens, and metals in the mass range below ca. 1000 amu. Generally, however, the accuracy of the intensity measurement is not sufficient to distinguish between the various possibilities because the differences may be very small. As a rough indication, the height of the $M+1$ signal can be used to estimated the number of carbon atoms, since ca. 1.1 % of carbon is ^{13}C. Only elements such as sulfur, silicon, bromine, and chlorine with typical isotope patterns can be identified easily. For the calculation of the isotope pattern from a known elemental composition, many programs are available and can be very helpful.

Another way of obtaining information about the elemental composition is to accurately determine the mass of an ion. It is then possible to propose possible elemental compositions. These proposals can be used to support suggestions from the low-resolution data and from other means or even to elucidate the structure of a compound. However, with increasing mass (above about $m/z = 500$) the total number of possible elemental compositions for the measurements even under the assumption of very small errors becomes so great that without further information interpretation becomes impossible.

In such cases, information about the unknown compound from other sources is required. Other spectroscopic techniques or chemical evidence may be used, because the knowledge about the absence of certain elements can reduce the number of possible combinations to a manageable level. Generally, ions above $m/z = 1000$ cannot be analyzed in this way, even when the errors are very small, because the number of possible elemental compositions is far too large. There are many successful examples in the literature, mostly involving the structure elucidation of natural compounds. Accurate mass measurement has also been used in GC/MS analyses with a resolution of 10 000; in one example the structures of photoadducts of psoralens to nucleosides were determined. Since a portion of the molecules was known (thymine was the nucleoside) it was possible to ascertain the nature of the previously unknown product [9].

20.2.3. Fragmentation in Organic Mass Spectrometry

The formation of fragments is one of the most useful characteristics of organic mass spectrometry. This becomes particularly apparent when no fragments are available, as with soft-ionization techniques, which produce almost exclusively molecular ions with negligible fragmentation. With electron impact, still the most important ionization technique, the majority of molecules form a large number of fragments.

Organic mass spectrometry can be regarded as gas-phase chemistry. Knowledge about the chemical reactions leading to specific fragments is largely based on isotopic labeling studies, often employing metastable ions, linked scans, or MS/MS experiments. These reactions are described by the so-called quasi-equilibrium theory. This statistical theory was formulated originally in 1952 [10] to characterize qualitatively the fragmentation behavior of gas-phase ions by using physical parameters. In Figure 3 a brief summary is given in terms of the so-called Wahrhaftig diagram. A treatment based on thermal reactions gave a similar description, today often referred to as the Rice–Ramsberger–Kassel–Marcus (RRKM) theory [11]. The most important unimolecular reactions can be explained in terms of reaction kinetics and thermochemistry.

More simplified concepts for explaining the fragmentation patterns of complex molecules such as steroids and terpenes are based on charge- or radical-induced reactions ("charge localization"). By considering the stability of intermediate products it is possible to construct reasonable fragmentation pathways, even though the true energetic, kinetic, and thermochemical background is unclear. The greatest problem here is that many rearrangements take place, some of which are well known and discernable [e.g., the McLafferty rearrangement (Fig. 4), retro-Diels–Alder rearrangement, hydrogen shifts], but others are not obvious in the mass spectra (e.g., hydrogen scram-

584   Mass Spectrometry

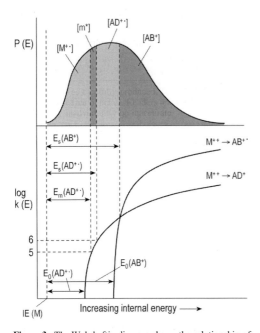

Figure 3. The Wahrhaftig diagram shows the relationship of probability function $P(E)$ (upper half) and two hypothetical functions of the rate constants $k(E)$ (lower half) for unimolecular descompositions of a molecule $ABCD^+$
$IE(M)$ = Ionization energy; $E_0(AD^{+\cdot})$ = Critical reaction value for the rearrangement $M^{+\cdot}$ to $AD^{+\cdot}$; and $E_0(AB^+)$ = Critical reaction value for the decomposition $M^{+\cdot}$ to AB^+
For energies between $E_m(AD^{+\cdot})$ and $E_s(AD^{+\cdot})$ corresponding to values between $\log k = 5$ and 6, metastable ions are formed (lifetimes longer than 10^{-6} s; from then on the rearrangement dominates until the function for the decomposition crosses at $E_s(AB^+)$; the formation of the fragment AB^+ is now favored. Adapted from [13] with permission

Figure 4. The McLafferty rearrangement with examples for R = methyl or phenyl and the corresponding ionization energies IE for charge retention (α cleavage) and charge migration (inductive fission) [13]

bling, skeleton rearrangements, silyl group shifts), since the mass-to-charge ratio remains the same. Some of them can be studied by using techniques such as isotope labeling.

The large number of rearrangement reactions, controlled by small differences in activation energy, make the prediction of mass spectra and the understanding of the bidirectional relation between spectra and structures impossible. For a thorough discussion of these problems and related topics see [12] and [13].

20.2.4. Quantitative Analysis

Generally stated, mass spectrometers are unstable detectors. The rather high noise level of the baseline and the signal is a problem. The noise can be of chemical origin (i.e., undefined ions throughout the spectrum) or may have its origin in electronic components such as the voltages of the detection system. In the long term, drift of the calibration, contamination of the source or the analyzer, and decreasing efficiency of the detector can slowly degrade the performance of the instrument, resulting in poor reproducibility over long periods. In addition, especially for small signals, the ion statistics may be poor. If N is the number of ions in a peak, than the percentage variation coefficient for the measurement of the area is

$$v = 100\sqrt{\frac{1}{N}}$$

Even with 1000 ions in the signal, the area can only be measured with a variation coefficient of 1 %. Clearly the only way to improve the ion statistics for a given ion current is to increase the time spent at the top of the peak. This is possible if the associated chromatography allows slow scanning or the use of selected ion recording. For quantitative determinations, often the peak areas are used. If the separation of signals is not complete, peak heights rather than peak areas may give better results, because the overlap should be smaller.

Generally speaking, quantitative measurements require most meticulous attention to obtain high precision and good reproducibility. This requires carefully chosen acquisition conditions and, in most cases, use of internal standards.

Selected Ion Recording. Most mass spectrometers in practical work are scanning type instruments, which screen the mass range for ions by varying the voltage, frequency, or magnetic field. Therefore, a major fraction of the signal remains unrecorded, giving away the sensitivity necessary to obtain signals with a satisfactory number of ions. Techniques are available which circumvent this problem by observing only a few ions or even only one type of ion. This increases the sensitivity by two orders of magnitude but also results in a considerable loss in specificity, since a signal with a single m/z value makes distinction of unknown compounds impossible. Therefore, chromatographic information from the GC or LC is necessary. The technique, termed selected ion recording (SIR) or multiple (single) ion detection (MID), is in general applicable for known compounds with known chromatographic properties. In some cases, the observation of isotopic patterns can improve the certainty of identification and may help to rule out cross contamination. When the mass of one of the ions in the mass spectrum is accurately known, high resolution can be used. This brings back some specificity, although the power of detection of the mass spectrometer is reduced. Often, the improved signal-to-noise ratio due to the reduction of chemical noise more than makes up for the reduction in detection power. In cases where the compound shows strong fragmentation it may be possible to record the metastable transition signal for this reaction. The technique, termed "reaction monitoring," is specific for a given compound.

Internal and External Standards. Quantitative determinations require standards, either external or internal. External standardization using samples with known amounts of analyte ("standard addition") is possible, if a reproducibility and accuracy of ca. 10% is acceptable. In pharmaceutical studies or routine environmental work such an accuracy is sufficient, since individual variations are much larger. When the required precision must be higher, internal standards must be used. As internal standard a compound with similar chromatographic and, if possible, similar mass spectrometric properties as the analyte can be employed. This may be a homologue, an analogue with, e.g., a different heteroatom, or a positional isomer. The best choice is an isotopically labeled compound containing stable isotopes.

Isotope Dilution. This technique is based on the fact that most elements have several stable isotopes; if only one stable isotope exists, as for phosphorus or fluorine, the technique cannot be used. Let the ratio of two isotopes of an element in the sample be R_s, and the ratio for the same isotopes in the internal standard R_I. From the ratios, the percentage peak heights can be calculated for the sample (s_1 and s_2) and the standard (i_1 and i_2). Then, a known amount I of the standard is added to the sample S and the two signals are measured (Fig. 5). The ratio of the two peaks is then

$$R = \frac{Ss_1 + Ii_1}{Ss_2 + Ii_2}$$

Rearranging for the unknown amount S of the sample gives

$$S = I \frac{i_1 - Ri_2}{Rs_2 - s_1}$$

Isotope dilution for the analysis of organic compounds is often accomplished by deuterium labeling at positions not prone to rapid exchange in the ion source. The synthesis of such compounds is normally straightforward (e.g., by exchange processes). Care must be taken, however, where the mass difference between the labeled and the unlabeled compound must be large because the isotope cluster is broad (several mass units). If many hydrogens are replaced with deuterium, the labeled compound may have a different chromatographic behavior and the retention times can differ. If deuterium labeling is difficult or not possible, ^{13}C, ^{15}N, or ^{18}O enrichment or combinations thereof can be used although the synthesis may be laborious.

20.3. Sample Inlets and Interfaces

20.3.1. Direct Probe

For many years the direct probe for introducing samples into the spectrometer has dominated all the others. The sample is transferred into a small cup with a volume of a few microliters, made of gold, silica, or aluminum. Some of them have a cap with a small hole to exploit the Knudsen effect for controlled evaporation. The cup is placed in the tip of the removable probe. The probe is introduced into the ion source housing with the sam-

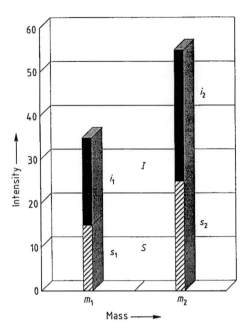

Figure 5. Isotope dilution

ple cup inside the ion volume without interfering with the electron beam. It can be heated and cooled directly under computer control, when necessary as a function of the total ion current or of a single ion current.

20.3.2. Batch Inlets

Batch inlets used to be important for the analysis of gases, volatile liquids, and mixtures such as hydrocarbons in oil samples. The inlet consists of a capillary, which connects a heated reservoir with a volume of several hundred milliliters, normally glass-lined to prevent cracking and pyrolysis, to the ion source. The capillary acts as a throttle, controlling the flow into the ion source; alternatively, a diaphragm can be used. The sample is introduced in the reservoir through a septum by means of a syringe or via a small sample inlet connected to the reservoir. This type of inlet has been largely replaced by GC/MS interfaces.

20.3.3. Pyrolysis

Pyrolysis is a powerful tool for the analysis of very high molecular mass samples, where there is no possibility for separating mixtures and determining the components or where the compounds are too complex to give mass spectra [14]. Examples are the attempts to characterize different fractions of humic acids, strains of bacteria, and unusual components in intact DNA of a variety of animals. In all such cases, mixtures of smaller molecules are prepared by carefully controlled pyrolysis (e.g., Curie point pyrolysis). The distribution of the resulting mixture of thermally stable compounds is characteristic of the starting material. Since the identity of the pyrolytic decomposition products is not known in most instances and the chemistry involved is complex, only the appearance of the particular distribution is significant, and techniques for cluster analysis based on pattern recognition must be used. Field desorption, field ionization, direct chemical ionization, and low-energy EI are mainly used with pyrolysis, because molecular ions are mostly formed and the pattern is not further complicated by fragmentation in the ion source. The desorption process in FDMS (see Section 20.4.8) can be considered a pyrolysis technique. Often, the pyrolysis products are injected into a GC, separated, and soft ionization techniques such as FIMS, low energy EI, or photoionization [15] are used to detect the components of the mixture.

20.3.4. GC/MS Interfaces

The most successful hyphenated method continues to be GC/MS, generally with EI ionization, but particularly the inductively coupled plasma MS (ICPMS) has become a successful detector as well. In the form of benchtop instruments, this combination is found in the chromatography laboratory rather than in mass spectrometry laboratories. Instruments such as mass selective detectors and ion traps are dedicated devices for gas chromatography and have developed into robust, dependable instruments with high detection power and selectivity. With modern computerized instruments and the extensive storage capabilities of data systems, series of spectra can be acquired rapidly, and elaborate data extraction procedures have been devised for the identification of important spectra with relevant data.

The reasons for this success are numerous: Gas chromatography is a well-developed technique with high separation power and excellent stability, and its specificity can easily be adjusted to analytical problems (see → Gas Chromatography). Further, the gas flow in capillary GC is a reasonable match to the pumping requirements in the mass spectrometer. The GC/MS interface has now been developed to such an extent that compromises are no longer necessary, allowing both to be optimized separately. Today, the mass spectrometer can be regarded as a GC detector with high detection power and specificity. Compared with flame ionization or electron capture detectors, for example, it has many advantages, and the disadvantages such as higher expenses and lack of long-term stability have become less significant.

Separators. As long as packed columns with high gas-flow rates were used, it was necessary to separate the carrier gas from the analytes. This was accomplished by a variety of devices called separators; examples include the Biemann–Watson separator and the jet separator. The latter was subsequently used in LC/MS interfaces. Due to the replacement of packed GC columns by capillary columns, separators are no longer used; they are described in detail in [16].

Direct Coupling. In direct coupling, the GC capillary is introduced directly into the ion source. This technique has the advantage of simplicity in construction and it avoids losses of analyte. The disadvantage is that the chromatographic separation is influenced by the vacuum of the ion source. Furthermore, the flow of gas into the ion source is a function of the GC oven temperature. The ion source pressure and the sensitivity may vary with the temperature program. Recently, this problem has been addressed using flow and temperature programming simultaneously. It is therefore not straightforward to transfer a separation developed for GC alone to the GC/MS system, because the parameters may be vastly different. But the simplicity of construction makes this version very popular.

Open Coupling. In open coupling, the end of the capillary from the chromatograph is inserted into a small glass tee piece and the interface capillary, with a slightly wider inner diameter, is located opposite to it, separated by a small gap. Through the third entrance of the tee piece, an additional flow of inert gas (the make-up gas) protects the gap from air and adds to the flow into the spectrometer. If the dimensions are right, the complete effluate from the separation column, together with some make-up gas, enters the ion source and no loss of analyte occurs. Since the end of the GC capillary remains at atmospheric pressure, the separation integrity is maintained and analytical procedures can be transferred.

Reaction Interfaces. For special purposes it can be advantageous to add a reaction zone between the GC and the MS to convert the analytes into species that give characteristic mass spectra or more precise quantitative information. Examples are the hydrogenation of double bonds in unsaturated hydrocarbons and the combustion of organic materials to CO_2 for the accurate determination of the $^{13}C/^{12}C$ ratio. Generally, the analyte is allowed to contact a reactive surface or catalyst, or a gaseous reagent is mixed with the gas stream from the GC. The analyzer can be any conventional sector or quadrupole spectrometer. For accurate isotope analysis, specially designed versions can be used. Since this method is important in analytical applications such as the determination of the origin of sugars in wine, dedicated instruments are commercially available. For a recent review with well chosen instructive examples, see reference [17].

20.3.5. LC/MS Interfaces

LC/MS interfaces can be divided into three groups: the first comprises sputtering and desorption processes, the second group has an aerosol generating step and separate ionization by EI or CI, and the third group—thermospray and electrospray—has an integrated ionization/aerosol generation step. The areas of applications for the three groups are: natural products for the first; smaller molecules, metabolites, and industrial compounds for the second; and biological oligomers and polymers for the third. Recent years have seen dramatic developments towards very stable and sensitive interfaces based on electrospray and atmospheric pressure ionization, whereas all others, except for the particle beam interface, have disappeared almost completely. The latter is still in use in industrial laboratories and for environmental studies where a detector for less polar compounds, is required.

20.3.5.1. Moving Belt

The moving belt system, invented by W. H. McFadden, was one of the first LC/MS interfaces. The stream of eluent from the LC is directed onto the surface of a belt made of pure nickel or polyimide, which runs through a series of vacuum stages into the ion source of the mass spectrometer. To minimize memory effects (signals from not completely removed compounds) and to clean the belt, infrared heaters and washing steps are included. The analyte was transferred into the source by thermal desorption using direct or indirect heating, by SIMS, FAB, or laser desorption. However, the technical requirements and the rather high costs (additional vacuum systems, maintenance of the belt), and the high background and memory effects limited the number of users from the beginning. The interface has the advantage of a complete separation of chromatography and mass spectrometry, allowing a wide range of chromatographic techniques (HPLC, micro HPLC, SFC) to be used.

20.3.5.2. Continuous Flow FAB

The continuous-flow FAB interface (online FAB, frit FAB) was developed by R. CAPRIOLI et al. [18] and Y. ITO et al. [19]. The technique, based on the conventional FAB source (see Section 20.4.8.3), is straightforward and is compatible with many mass spectrometers. The exit of a capillary, which is coupled to a micro HPLC system, is connected to an FAB target and the effluent is sputtered from the target as usual. Since the gas load to the ion source is high due to the evaporating effluent, glycerol, and the FAB gas, the sensitivity is enhanced by adding high pumping capability to the source housing using an additional cryopump. Glycerol is used to enhance the desorption, as with FAB; it can be added directly to the eluent (1–5%) or after separation, by means of concentric capillaries [20]. Flow rates of 15 µL/min are acceptable, but the performance is better with lower rates. Recently, capillary zone electrophoresis (CZE) has been interfaced [21], [22], although serious problems arise due to the very small amounts of sample and narrow peaks from the electrophoretic separation. Mass spectrometers with simultaneous detectors compare very favorably [23] (see Section 20.8.4).

20.3.5.3. Direct Liquid Introduction

Direct introduction of liquids was probably the first LC/MS interface to be used in modern organic mass spectrometry [24], [25]. A flow of typically 10–20 µL/min is allowed to enter the ion source without separation of eluent and analyte. The construction is relatively simple, and conventional electron impact ion sources can be used, even without modifying the pumping systems [26]. The main disadvantage is the inflexibility with respect to the ionizing method (for CI only the eluent is available as reactant gas), while the compounds, that can be analyzed are similar to those amenable to GC/MS.

20.3.5.4. Supercritical Fluid Interface

The SFC/MS interface is a direct-introduction system with careful temperature control up to the very tip of the capillary [27]. The end of the capillary contains a frit or a diaphragm that keeps the phase supercritical until it leaves the capillary and enters the ion source. Otherwise, the dissolved analyte would clog the end of the capillary. Since the gas load is rather high, chemical ionization is normally used with capillary SFC, although EI is possible. When packed columns are used, momentum separators can reduce the pressure in the ion source, making electron impact possible. Limitations of the method with respect to classes of compounds are mainly due to SFC itself, not to the interface or the MS.

20.3.5.5. Particle Beam Interface

The particle beam interface (Fig. 6) was created under the acronym MAGIC (monodisperse aerosol generator interface for chromatography) [28]. Now, the aerosol is produced by a variety of means (with auxiliary gas, thermospray, or ultrasonic nebulizers) at atmospheric pressure and a uniform distribution of the droplets results in particles of a narrow size distribution, which can be handled more efficiently by the separator. The droplets are dried to particles in a heated expansion chamber, and a momentum separator isolates the particles from the gas. In the source, the particles are destroyed by impact and the sample is released and ionized by using EI, CI, or even FAB. The appearance of the EI spectra is almost identical to conventional EI spectra obtained by direct probe or GC/MS. Therefore, library searches are possible, which is the major advantage of this interface.

The main problem of the technique is its unsatisfactory basic power of detection. Also, sensitivity can vary widely even for similar compounds, and the difficulties in obtaining a linear response over a wider range of concentrations make quantitative determinations difficult. In addition, components eluting at the same time may interfere

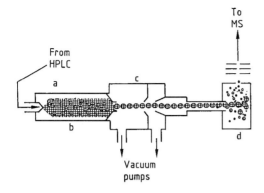

Figure 6. Particle beam interface
a) Nebulizer; b) Desolvation chamber; c) Momentum separator; d) Ion source

with one another, leading to unpredictable effects. For applications where the sensitivity is not of importance (e.g., monitoring of synthesis), this technique can be employed with success, because typical, interpretable mass spectra can be obtained.

20.3.5.6. Chemical Ionization at Atmospheric Pressure

Atmospheric-pressure chemical ionization (APCI) was described by HORNING et al. in 1960 [29], and they recognized the potential of this approach for LC/MS applications. An aerosol is generated by spraying an LC eluate within a heated sheath gas flow to give a spray stable at atmospheric pressure. In the spray, ions are generated via a chemical ionization plasma, created by a corona discharge from a needle. The analyte is ionized by the reactant ions in the plasma and subsequently transferred into the mass spectrometer. This is achieved by a sampler–skimmer system with an intermediate pumping stage to separate the high vacuum from the atmosphere. A curtain of drying gas is used in front of the sampler to reduce the background ions.

In general, basic compounds with high proton affinity can be detected with the highest limit of detection [30], and in some cases, a strong temperature dependence has been observed [31]. In principle, APCI is compatible with every separation technique which allows the effluent to be sprayed in an aerosol. Thus, not only HPLC but also SFC can be interfaced to this ion source.

20.3.5.7. Thermospray

Thermospray, developed in 1980 by M. VESTAL [32] rapidly found its way into routine applications, but now the atmospheric pressure interfaces have replaced it completely. The eluent of an LC is rapidly heated in a steel capillary either by direct resistive heating or by indirect heating using a cartridge system. A fine aerosol of charged droplets is generated in the ion source at reduced pressure, although the details of the formation of ions in the source is still subject to discussion [33]–[36]. The ions are transferred into the mass spectrometer by a sampling cone reaching into the center of the inner chamber of the ion source, but most of the solvent is rapidly pumped away. The source has to be very tight to stabilize the pressure inside and the vacuum outside. The method is shown schematically in Figure 7 [37].

It is possible to include a Townsend discharge or a filament and a repeller electrode to allow plasma chemical ionization to be performed. For some compounds, the sensitivity, selectivity, and fragmentation behavior can be modified. The advantage of the technique is the simplicity of transferring previously developed LC procedures since the normal flow rates and some of the common buffers (e.g., ammonium acetate, formate) can be used.

20.3.5.8. ESI Interface

The combination of the electrospray ion source with HPLC has without a doubt become "the" LC/MS interface in recent years. It is a particularly powerful combination, since this ionization technique covers a wide range of samples [38] that are commonly separated by HPLC [39] or electrophoresis [40]–[44]. The ESI source exhibits concentration-dependent behavior and thus gives optimal signals at most flow rates. The principle of the ionization process is discussed in Section 20.4.9. The most important feature of this interface is a spray needle which can be connected directly to the separation column, if the flow rates are compatible. Initially the major limitation was that only low flow rates (a few μL/min) could be used, but now flow rates of 1 mL/min or more are possible by using heated sprayers or ultrasonic devices. Splitting of the flow is possible as well, allowing two detectors to be used simultaneously. Since buffers can be used as long as they are volatile and not too concentrated, a sheath flow

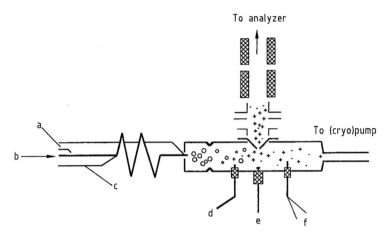

Figure 7. Thermospray interface/ion source
a) Connections for direct heating; b) HPLC flow; c) Thermocouple; d) Townsend discharge; e) Repeller; f) Thermocouple (aerosol)

of methanol can be added to enhance the ion yield and reduce background due to buffer salt clusters. Recently, interfacing with capillary zone electrophoresis (CZE) has been demonstrated [45] and this combination appears to be a powerful tool for the analysis of trace organic compounds [46], although the amount of analyte which can be separated in a CZE capillary and transferred to the mass spectrometer is very small and may even lie below the detection capabilities of the mass spectrometer [44], [47]–[50]. The near future will see devices used in many fields based on a chip with separation by CE or HPLC and an integrated sprayer, especially in biochemical and clinical analysis [51], [52]. In Figure 8 such a device is shown (a), and a photograph of the Taylor cone developed from the chip (b).

20.3.6. TLC/MS

Two-dimensional separations such as thin layer chromatography (TLC) or gel electrophoresis in combination with SIMS [53], FAB [54], and laser desorption [55] have recently given impressive results and are promising for the future [56]. In principle the technique is identical to FAB or laser desorption, with the crucial difference being that the target is the device used for the separation of the sample mixture. There are problems associated with this difference: the sample may be rather dilute, not in an appropriate matrix and, because the layers are rather thick, much of the sample is protected from the sputtering beam. It is therefore necessary to optimize the TLC procedure, e.g., by using micro rough surfaces for the separation or blotting techniques to transfer the analyte to a different matrix. To allow the scanning of larger TLC plates, the incorporation of sample stages with x,y-adjustments would be necessary. The major advantage of the technique is that method development of TLC methods is rapid, efficient, and cost effective, and the plates can be investigated with several spectroscopic techniques sequentially after separation, making this setup very attractive for routine analysis (e.g., in environmental chemistry) [57], [58]. For biochemistry, matrix assisted laser desorption of peptides, proteins, and oligonucleotides separated in thin-layer electrophoresis and often transferred to a more suitable matrix by blotting techniques appears promising, e.g., for screening purposes and the identification of modifications and mutants [59]–[62], even though excision of the spots and off line analysis is often favored. This strategy allows direct enzymatic digest ion in the gel avoiding sample loss in the course of transfers.

20.4. Ion Generation

20.4.1. Electron Impact

Electron impact is still the most widely used technique for generating ions. It uses electrons, emitted from a rhenium or tungsten wire or ribbon and accelerated with typically 70 V, to ionize the

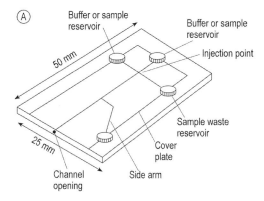

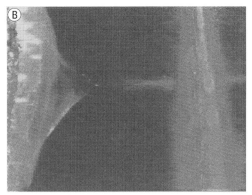

Figure 8. (A) Schematic diagram of microchip used to pump fluids electroosmotically and to generate the electrospray; (B) Photomicrograph of the Taylor cone and electrospray generated from a 60 % H_2O – 40 % methanol solution that was electroosmotically pumped using a microchip (reproduced from RAMSEY [51])

vapor of a compound. In this process, molecular ions are formed together with a pair of electrons after impact. The molecular ion may be stable or it may dissociate immediately into a neutral and an ionic species (dissociative ionization) or into two ions of opposite charge (ion-pair formation). The positive ions may have sufficient internal energy to undergo further fragmentation. Generally, the fragments are formed in the ion source within a few microseconds and are responsible for the normal mass spectrum. If ions are stable enough to dissociate later in one of the field-free regions between analyzers, they appear as "metastable ions" or may be detected by using MS/MS experiments. These metastable ions have been studied in great detail, since they facilitate the understanding of many gas-phase processes. The EI ion source is a small chamber with a filament (cathode), an electron trap to regulate the emission, and a pusher or repeller, which controls the space charge distribution in the chamber. Small permanent magnets perpendicular to the electron beam broaden the electron cloud, which enhances the sensitivity. The ions are extracted by means of a set of extraction plates and focused with several lenses and steering plates onto the entrance of the analyzer.

20.4.2. Chemical Ionization

Chemical ionization can be carried out either at atmospheric pressure or under reduced pressure in an ion source chamber; in both cases ion/molecule reactions are used to produce the ions of the analyte [63]. At atmospheric pressure, the plasma is generated with a discharge needle and the ions come mainly from the surrounding air, water, and from components of the analyte spray, which can contain organic solvents and buffers. Mainly protonation occurs. Under reduced pressure, a reactant gas must be added to the mixture. For this purpose, a gas such as methane, isobutane, ammonia, or water is introduced into a gas-tight ion source. This gas is then ionized by electron impact (or, sometimes, by a Townsend discharge) to give a reactive plasma. The nature of the plasma ions determines which type of spectrum is formed; therefore, the different gases can be used to control the specificity of the ionization process. The reactions most often observed are proton transfer from the reagent ion to the analyte, charge exchange, electrophilic addition of a plasma ion to the analyte, and abstraction of a negative leaving group from the analyte. When ammonia is used, only substances with high proton affinity are protonated, and hydrocarbons, for example, do not show up in the mass spectra. With CO_2 as reagent gas, the spectrum resembles a normal EI spectrum due to the high energy associated with the electron transfer in the CI plasma. The important feature of chemical ionization with respect to analytical applications is that the ratio of molecular ions to fragments is usually much higher than with EI, thus yielding information about the molecular mass.

20.4.3. Negative Chemical Ionization (Electron Capture)

Negative chemical ionization is similar to CI only in so far as the same gases are used in the same ion source, although the pressure is usually

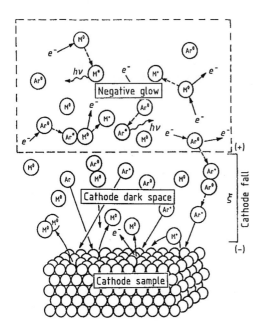

Figure 9. Principles of GDMS illustrating the main processes of sputtering, excitation, and ionization
M^0 = Sample atom; M^*/Ar^* = Excited states; M^+ = Sample atom
Only neutrals can cross into the negative glow, where ions are formed via electron impact or in collisions with excited argon molecules (Reproduced with permission from [68])

lower than in CI. The purpose of the gas is to moderate the energy of the electrons emitted from the filament to thermal energies. With such slow electrons, electron capture becomes possible, resulting in the formation of negative ions. If the analyte can stabilize the negative charge either in electronegative moieties, by charge delocalization, or by dissociation to give a stable fragment, the cross section of the ionizing process can be high, and the technique very sensitive. For the analysis of substances unable to accommodate negative charge, special derivatization techniques such as electrophoric labeling are used. An example is the derivatization of adducts of carcinogenic metabolites to nucleosides with perfluorobenzyl bromide yielding GC/MS detection limits in the attomole range [64].

20.4.4. Field Ionization (FI)

Field ionization, developed by H. D. BECKEY [65], was of importance only for a short time. In the ion source, a strong electric field gradient (10^7 V/cm) is formed by means of a potential applied to a sharp edge, a tip, or an emitter with whiskers; the potential difference between the acceleration voltage and field potential is ca. 11 kV. In the gradient, the vaporized molecules of the analytes travel to the regions of increasing field strength. When the probability of an electron leaving the molecule by tunneling through the internal energy barrier becomes sufficiently high, ions can be observed. The ionization process is very soft. Since the technique is technically difficult and needs skilled operators, only a few laboratories have used FI for practical work. Its strength is that only molecular ions are formed, allowing, for example, the distribution analysis of a series of homologous hydrocarbons in oil samples. A special version of the technique, called field ion kinetics, has been used for some time to study the very fast fragmentation processes of several classes of compounds [65].

20.4.5. Plasma Ionization

Plasma ionization can, in general, only be applied to the analysis of elements, since the energy of the plasma is too high for most molecules to survive. However, molecules formed from the plasma gas and the analytes in the plasma are one of the main obstacles of this ion source type. Interferences in the lower mass regions — others come from doubly charged ions and oxides — complicate the spectra and may even obscure some elements completely if only trace amounts are present. High-resolution mass spectrometers can be used to solve this problem. Today, there are three major types of plasma source in use: the glow discharge source, the inductively coupled plasma source, and sputtered neutral mass spectrometry [66].

Glow Discharge (GDMS). Glow discharge ion sources have been built for quadrupole and sector field instruments. The solid sample is part of the source chamber and operates as cathode. The source chamber serves as anode and is filled with noble gas (generally argon) at a reduced pressure (1 kPa) that allows a stable plasma to be formed. A high voltage is applied between cathode and anode resulting in a self-sustaining discharge. Two main regions exist in the plasma: the cathode fall next to the sample and the negative glow (Fig. 9).

In the cathode fall, argon ions are accelerated toward the sample surface and atoms from the sample are sputtered into the plasma. Upon elec-

Table 1. Some ICPMS interferences and the resolution required to resolve the signals

Analyte species	Exact mass	Interfering species	Exact mass	Resolution required for separation
^{55}Mn	54.938	$^{40}Ar^{14}N^1H$	55.090895	359
^{28}Si	27.976929	$^{14}N_2$	28.006146	960
^{31}P	30.9737634	$^{14}N^{16}O^1H$	31.005813	966
^{48}Ti	47.9479467	$^{32}S^{16}O$	47.986986	1 228
^{44}Ca	43.95549	$^{12}C^{16}O^{16}O$	43.98982	1 280
^{32}S	31.9720727	$^{16}O_2$	31.989828	1 802
^{56}Fe	55.934938	$^{40}Ar^{16}O$	55.957299	2 504
^{51}V	50.944	$^{35}Cl^{16}O$	50.96376	2 580
^{62}Ni	61.9283	$^{46}Ti^{16}O$	61.94754	3 220
^{75}As	74.921596	$^{40}Ar^{35}Cl$	74.9312358	7 771
^{80}Se	79.91647	$^{40}Ar_2$	79.92477	9 638
^{48}Ti	47.94795	^{48}Ca	47.95253	10 466
^{58}Ni	57.9353	^{58}Fe	57.9333	29 000
^{115}In	114.903871	^{115}Sn	114.903346	218 900
^{87}Rb	86.90916	^{87}Sr	86.908892	325 000

tron impact, the sputtered atoms are ionized and extracted into the mass spectrometer by application of a potential [67]. The technique is considered to suffer only minor matrix effects thus giving reliable quantitative results. Since the sample is one of the discharge electrodes, the thermal and electrical conductivity of the sample may influence the plasma and the sputtering process. For the analysis of nonconducting materials it may be necessary to mix the sample with a good conductor.

Sputtered Neutral MS. In sputtered neutral MS (SNMS) an electron cyclotron resonance in a static magnetic field is used. Under these conditions, an inductively coupled RF plasma is formed in high vacuum with argon as plasma gas; since no electrodes are involved the electrical properties of the sample do not influence the plasma, allowing the analysis of metals, particulate materials, and ceramics. The sample is connected to an adjustable high voltage, and argon ions from the plasma are accelerated onto the surface of the sample. The intensity of the sputtering process can be varied by means of the high voltage and thus the depth resolution obtainable can be optimized for the analytical problem. Since only atoms can penetrate into the potential—sputtered ions being reflected—the electron impact ionization step is virtually independent of the sputtering process and the technique is therefore free of matrix effects. In addition, ions formed at plasma potential are suppressed by an energy filter. Therefore, the technique gives reliable quantitative results when calibrated with standard reference materials. The best resolution that has been achieved is 4 nm. Since the sputtering process is slow, the usable range for maximum depth is in the micrometer range. Consequently, the technique has some attraction for the analysis of semi- and superconductors [69].

Inductively coupled plasma mass spectrometry (ICPMS) is very sensitive and allows the analysis of solutions of metal ions at pg/mL levels [66]. In principle, the ion source consists of an aerosol generator, a desolvation unit, the inductively coupled plasma, and the ion transfer and focusing optics. Several types of aerosol generator, such as the Babington and Meinhard types, ultrasonic nebulizers, and thermospray are used [70]. For interfacing ICPMS to LC/MS, efficient drying of the aerosol is necessary to stabilize the plasma. The most effective method appears to be heating and subsequent cooling of the aerosol, sometimes supported by a counterflow of gas. The ions are generated in an argon plasma, which burns in the high-frequency field of the induction coil just in front of the transfer optics. Since this all takes place at atmospheric pressure, the sampling into the mass spectrometer by means of a skimmer sampler arrangement, similar to that of other API sources, requires very effective pumping to ensure sufficient sensitivity, and the outer skimmer must withstand the temperature of the plasma. ICPMS instruments are available with quadrupole and sector field analyzers [71]; the latter is of interest, when higher resolution (typically ca. 7000–10 000) is necessary to separate molecular interferences due to oxides and argides [66], [72]–[74] (see Table 1 and Fig. 10).

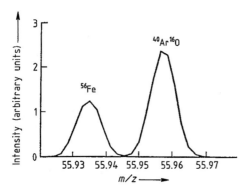

Figure 10. The spectrum shows the ^{56}Fe and ^{40}Ar^{16}O, resolved with a resolution of $R = 3000$, using a laboratory prototype of the ELEMENT, Finnigan MAT (high resolution ICPMS) [75]

20.4.6. Thermal Ionization

Thermal ionization is a technique for elemental analysis that uses thermal energy to ionize elements. The sample is deposited on a ribbon which can be heated to very high temperature, and a second ribbon a few millimeters from the first is used to ionize the atoms in the gas phase. This ribbon is made from rhenium or tungsten, sometimes treated with carbon or thorium, depending on the nature of the element under investigation. Heavy metals such as lead, uranium, thorium, and others have been determined as positive ions in very different matrices such as geological samples, microelectronic chips, and ice cores from the antarctic, in conjunction with isotope dilution (IDMS) techniques. Iodine, bromine, chlorine, selenium, and also some heavy metals have been determined by using negative thermal ionization and IDMS [76]. Thermal ionization with isotope dilution is considered one of the reference techniques allowing the most precise quantitative determination of elements.

20.4.7. Optical Methods

Photo-Ionization. Photo-ionization has been used mainly to study basic principles of the ionization process. The light source is usually a helium lamp emitting 21.22 eV photons; other noble gases can be used to produce other photon energies [77]. When the energy deposited in the molecules is sufficient to exceed the first ionization potential, ions of the analyte are formed. By analyzing the energy distribution of the photoelectrons, the energy distribution of the molecular ions can be deduced. It has been shown that the distribution formed by electron impact may be similar.

Multiphoton Ionization. The specificity of multiphoton ionization (MUPI, sometimes termed resonance multiphoton ionization, REMPI) is based on the existence of well-defined electronic states at low temperature [78], [79]. If a molecule is cooled to a few kelvin, a first photon can be absorbed upon resonance and excites the electrons to a state below the ionization potential. This absorption process is specific for a compound or class of compounds; for example, isomeric methylphenols can be distinguished with such a technique. A second photon is necessary to excite the molecules above the ionization threshold. By varying the laser power, the fragmentation can be controlled (see the example given in [80]). The more energy is available the greater the fragmentation.

In the ion source, a supersonic beam of noble gas is generated and the sample is seeded into it (e.g., by laser desorption). Together with the beam, the sample is carried through a skimmer system that allows only the isentropic core to enter the vacuum chamber where the ions are generated by two tunable lasers. The technique is still in the development stage.

20.4.8. Desorption Methods

Desorption methods have been developed since around 1970. They have not only expanded the scope of mass spectrometry, but have also changed the strategies of sample introduction and brought attention to solid and liquid phase chemistry as well as to particle and gas phase chemistry. They have also stimulated the rapid development of new ion sources, mass analyzers, and detectors with higher mass range capabilities and improved sensitivity.

20.4.8.1. Secondary Ion Mass Spectrometer (SIMS)

Secondary ion mass spectrometry, pioneered by A. BENNINGHOVEN et al. [81], can be regarded as one of the first desorption techniques. A primary ion beam (typically ions as Cs$^+$ and O$^-$) is focused onto a target; the sample can be deposited on the surface of the target or the surface is the

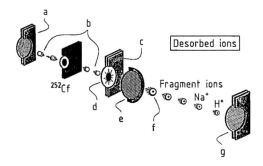

Figure 11. Plasma desorption MS
a) Fission fragment detector; b) Fission fragments; c) Excitation area (100 nm^2); d) Sample foil; e) Acceleration grid; f) Molecular ion; g) Desorbed ions detector

sample itself. Upon impact, the sample is sputtered and ionized simultaneously, and the secondary ions are extracted into a time-of-flight instrument. Since the primary beam can be focused to a small spot, the technique can be used in principle to achieve spatial resolution as well (ion micro probe) [82]. SIMS is used in material science to analyze the surface of metals, but due to the combination of the sputtering and ionization processes, matrix effects can be strong and atomic mixing occurs. Although this approach is used primarily for the determination of elements, the investigation of organic surfaces is also possible [83], and organic materials deposited on a appropriate surface such as silver have been studied [84].

20.4.8.2. Field Desorption (FD)

Field desorption is technically closely related to field ionization, although the desorption/ionization process is different [85]. The sample is deposited directly on the emitter to produce the high field gradient and, therefore, materials with virtually no vapor pressure can be investigated, in contrast to field ionization. The emitter can be heated either by direct current or by using external sources (e.g., lasers). The sample is generally desorbed as so-called "preformed ions" (protonated or with alkali cations attached, both or either type are always present in samples). The ion currents are generally very weak and often erratic; therefore, integrating devices (photoplates, multichannel analyzers) have been used to overcome these problems. Since FDMS and FIMS are similar methods, they have the same technical problems. FDMS was the first desorption technique to allow mass spectrometric analysis of peptides, oligonucleotides, and polysaccharides. Today the technique is used for the analysis of industrial polymers.

20.4.8.3. Fast Atom Bombardment (FAB)

Fast atom bombardment, developed by M. BARBER in 1980 [86], was the first technique which could be used with virtually every mass spectrometer, since the ion source can be retrofitted to a given mass spectrometer without problems. Furthermore, sample preparation is straightforward. The technique is based on SIMS, with the major difference being that the sample on the target is dissolved in a liquid matrix, usually glycerol. In the meantime, many other matrices have been examined for the analysis of different classes of compounds. The most widely used are glycerol, thioglycerol, "magic bullet" (a 5:1 mixture of dithiothreitol/dithioerythritol), 3-nitrobenzyl alcohol, 2-nitrophenyl octyl ether and triethanolamine. It can be beneficial to add acid (e.g., phosphoric acid, sulfonic acids) to glycerol to increase the abundance of analyte ions. The second difference to SIMS, the use of accelerated atoms instead of ions, is of importance only in so far as an atom gun can be added to the source housing of even sector field instruments. Because atoms can penetrate the high electric field but cannot be focused on a small area, the atom gun has often been replaced by a high-energy ion gun in a technique known as liquid SIMS (LSIMS). The advantage is that the primary ions can be focused and deliver a higher ion density in comparison to the saddle field atom guns commonly used with FAB MS.

20.4.8.4. ^{252}Cf Plasma Desorption

Plasma desorption (PD), first described as a tool for protein analysis in 1976 [87], exploits the fact that ^{252}Cf "explodes" spontaneously into smaller nuclides, the lighter ones centered around ^{108}Tc, and the heavier ones around ^{144}Cs. The products from each unsymmetric fission contain most of the high energy as kinetic energy and travel in opposite directions. The californium source is located between a thin foil loaded with the sample and a target, which serves to start the clock of a detector in a time-of-flight mass spectrometer [88] (see Fig. 11). When the heavy atom hits the foil, a cloud of atoms and neutrals, including intact large molecules, are desorbed in a very short pulse. The ions observed are preformed ions (protonated or with alkali metal ions), which are either desorbed directly or formed in the dense cloud of desorbed material.

The technique gives poor resolution, but good mass accuracy and high detection capability, although strong background problems must be solved. The time required to record a spectrum depends on several factors (sample preparation, amount of ^{252}Cf available) and can vary from a few minutes to several hours. The technique has already experienced the same fate as FDMS. When it was developed, it was highly praised for its unprecedented high molecular mass determination capabilities, its sensitivity, and simplicity. The results for peptides [89] and other high mass polymers were equally impressive. This type of work is now largely done by matrix-assisted laser desorption and electrospray.

20.4.8.5. Laser Desorption/Ionization

Lasers have been used in mass spectrometry for many years. Trace elements in biological samples [90] can be determined by using laser microprobes (LAMMA, laser microprobe mass analyzer) or a combination of laser ablation with ICPMS. For the analysis of bulk materials, techniques such as resonance ionization mass spectrometry (RIMS) and laser ablation MS (LAMS) are employed; for a review see [91].

The desorption of organic substances from surfaces without destruction is difficult and the duration of ion production is short. Therefore, the mass limitation of this approach without a matrix is around 1000 amu [92]. Recently, molecular ions of complete proteins with $M_r > 200\,000$ [93] have been generated, which is mainly due to the use of appropriate matrices for matrix-assisted laser desorption/ionization [94]. For sample preparation, the analytes are dissolved and crystallized in a matrix on a target. The matrix for the first experiment using a UV laser for protein analysis was nicotinic acid [95], but the search for other matrices is continuing, since the nature of the matrix seems to be crucial. A similar approach with a metal powder (10 nm diameter) mixed with a liquid matrix (glycerol) to form a colloid gave comparable results [96]. The crucial point is the absorption of the matrix, which protects the large molecules from the laser energy. Apparently, a variety of ions and radicals are formed initially from the matrix by photoionization and, in the dense expanding plume of desorbed material, protons become attached to the analyte molecules.

These ions are then extracted in the TOF analyzer. The first lasers used were UV lasers tuned to the absorption of nicotinic acid, but later it was shown that IR lasers give similar spectra [97]. Due to the pulsed ion generation, TOF instruments are the analyzers of choice, although recent attempts have used sector field instruments with simultaneous detectors. The signals obtained are rather broad, probably due to matrix–molecule adducts of the analytes, but the detection potential is very good. Since the accuracy of the mass measurement is ca. $1\times10^{-3} - 5\times10^{-3}$, yielding an error of ca. 100–500 amu at $M_r = 100\,000$, the precision of the technique is superior to other methods such as gel chromatography or centrifugation. Today, benchtop instruments are available with easy sample handling and short analysis times, making the technique suitable for biochemical laboratories. The development of delayed extraction has improved the resolution considerably [98]–[100] and with post-source decay [101]–[103] the detection of fragments has also become possible.

20.4.9. Electrospray

Electrospray was described in 1968 by DOLE et al. [104] and the ions were detected by ion mobility. FENN and coworkers [105], [106], and, independently, ALEXANDROV et al. [107] re-adopted the idea as a means of creating ions for mass spectrometry and since then rapid development of this technique has occurred. The fundamental difference to thermospray is that the spray is formed by charging droplets to such an extent that they explode by coulomb repulsion into smaller and smaller droplets. The droplets are further decreased in size by collisions with gas and, finally, highly charged ions are liberated. The process in which ions are created from the droplets is still not clear and several partial models exist [108].

The electrospray source is shown schematically in Figure 12. A capillary is held under high voltage (3–5 kV) at atmospheric pressure to generate a spray of charged droplets. To obtain a dry aerosol, either a heated steel capillary [109] or a hot gas curtain [110] is used in conjunction with collisions in the first vacuum stage. These collisions are necessary to decluster the ionic species, but may also be used to make fragments in a process equivalent to collision-induced dissociations for MS/MS experiments.

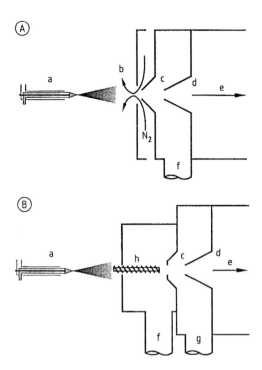

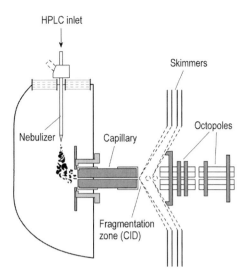

Figure 13. Schematic of the orthogonal sprayer assembly ESI Ion Source and transfer optics (Hewlett Packard)

Figure 12. Electrospray ion source types
A) Desolvation region with gas curtain; B) Desolvation region with heated capillary
a) Sprayer with three concentric capillaries: the inner flow is the eluent from the separation (CZE, HPLC), the next flow is sheath liquid such as methanol or isopropanol to enhance desolvation, and the outer flow is sheath gas, normally nitrogen; b) Gas curtain; c) Nozzle; d) Skimmer; e) To ion optics and analyzer; f) To roughing pump; g) To turbomolecular pump; h) Heated capillary

For quadrupole mass spectrometers, the ions from the first stage are extracted and pass directly into the mass spectrometer through a skimmer. For sector field mass spectrometers, a second vacuum stage is required to minimize high-energy collisions in the acceleration region, which could destroy the ions. The main advantage of electrospray is that even molecules with very high molecular mass can be detected at low m/z values, because the ions are highly charged (charges of up to 100 have been observed). Thus, the use of inexpensive quadrupole instruments is possible, but the more powerful analyzers such as FTMS or the new hybrid spectrometers have demonstrated superior performance (vide infra). The range of commercially available ion sources with new design features such as the orthogonal sprayer, developed by Hewlett Packard, (Fig. 13) or similar devices allow connection with HPLC even with high salt loads. The new low flow sprayer has increased the sensitivity for microHPLC and capillary electrophoresis and the nanospray sources for biomedical work [111] have become pivotal in proteomic research.

20.5. Analyzers

20.5.1. Electromagnetic Sector Fields

The first mass spectrometer [112] was a single focusing instrument with a magnet deflecting the ions through 180°. Since then, magnetic fields alone and in combination with electrostatic fields have been used most successfully to separate ions. When an ion of mass m has the charge ze where z is the number of charges and e is the elemental charge, its kinetic energy after acceleration through a potential V is

$$\frac{1}{2}mv^2 = zeV$$

where v is the velocity after acceleration. In a magnetic field of strength B, the force is $Bzev$. For the stable radius r

$$Bzev = \frac{mv^2}{r}$$

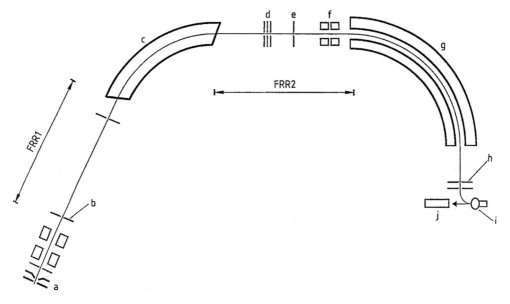

Figure 14. Double focusing MS (reversed Nier–Johnson geometry)
a) Ion source; b) Entrance slit; c) Magnetic sector; d) Focusing elements; e) Energy slit; f) Beam Shaping static quadrupoles; g) Electrostatic analyzer; h) Exit slit; i) Conversion dynode; j) Multiplier; FFR1, FFR2, Field-free regions

Combining the equations yields

$$\frac{m}{z} = \frac{B^2 r^2 e}{2V}$$

Thus, for any given m/z value a certain r is valid, if B and V are constant. Scanning the magnetic field or the voltage, on the other hand, allows different ions to travel the same path through the magnet and reach the detector consecutively. The resolution attainable with a magnet is limited to ca. 5000 by the initial energy spread of ions due to the Boltzmann distribution and as a result of inhomogeneous and fringe fields. It can be increased by adding an electrostatic field so that the ions are focussed in both direction and energy ("double focusing"). An example is given in Figure 14.

Various designs exist, with the electrostatic field preceding or following the magnet [Nier–Johnson (BE), reversed geometry (EB), Herzog–Mattauch], and all of them differ in performance. The advantage in all cases is a far superior resolution (10 000 – 100 000) compared to magnets alone. But another degree of complexity is added to the mass spectrometer and there are some other handicaps as well. In comparison, the instruments are expensive, and more difficult to operate and maintain. In addition the long-term stability of calibration is rather poor and some of the interfacing techniques are quite cumbersome to carry out. Nevertheless, sector field instruments are, generally speaking, still the most versatile mass spectrometers with strengths in detection power, mass accuracy, resolution, and flexibility [113].

20.5.2. Quadrupoles

Quadrupoles are the most successful concept from a broad range of so-called dynamic mass spectrometers [114]. They differ from static analyzers in that they use a high-frequency voltage to disperse the ions. In a quadrupole four poles (rods) form a hyperbolic field through which the ions travel with low kinetic energies. The superposition of high frequency and a variable d.c. voltage enables the quadrupole to separate the ions according to their mass. The regions of stability and instability are shown in Figure 15. The coefficients a and q are derived from the Matthieu equations. They are functions of the d.c. voltage U and the a.c. voltage $V_0 \cos wt$. For a fixed frequency, m is proportional to both U and V_0 and when the ratio $a/q = 2U/V_0$ is held constant, the masses follow stable paths and a spectrum may be recorded.

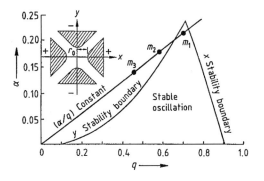

Figure 15. Quadrupole stability diagram indicating the conditions for stable and unstable oscillations

Although mathematical treatment of the separation process is complicated, the operation of such a spectrometer is simple. They have quickly become established in GC/MS systems, since they can be very compact, easy to handle, and less expensive than other instruments. In addition, computer control is simple. This advantage became even more significant with the development of triple quadrupole instruments (see Section 20.7.1).

20.5.3. Time-of-Flight (TOF) Mass Spectrometer

A few years ago, time-of-flight instruments were used almost exclusively for surface analysis. Then, the development of plasma desorption and laser desorption/ionization changed the situation dramatically. It is now conceivable that in the future TOF analyzers may become routine analyzers. The mass analysis is based on the fact that after uniform acceleration in the ion source, small ions arrive earlier at the detector than heavy ions. This is because all particles acquire the same energy resulting in different velocities. If the acceleration voltage is V, the energy is eV. For an analyzer of length L, the time to travel to the detector is

$$t = \sqrt{\left(\frac{m}{z}\right)\left(\frac{1}{2eV}\right)} L$$

and, solved for m/z

$$\frac{m}{z} = \frac{2eVt^2}{L^2}$$

The detectors, normally multichannel arrays, are capable of resolving the resulting small differences in time. Almost every ion leaving the source can be recorded, giving TOF instruments excellent sensitivity. To enhance resolution, a reflector device can be added to the linear flight tube. This reflectron focuses the ion beam on the detector, compensating for the initial energy spread of the ions, because the faster ions penetrate more deeply into the reflector than the slower ions and thus have a somewhat longer path to the detector. Resolutions in the range of 5000 – 10 000 have been obtained and ions with $m/z > 100 000$ have been recorded. Alternatively, the reflector can be used to determine the sequence of biopolymers by means of laser desorption MS. Many metastable ions dissociate after acceleration, and fragments can be generated by collisions (psd, post-source decay). These normally arrive simultaneously with the precursor ions, since no field is present to induce separation. If the voltage of the reflectron is then changed stepwise, such ions can be separated prior to arrival at the detector [115], since their masses are different. Recently orthogonal injection of ions into the TOF instrument was used to interface electrospray and electron impact sources, also giving excellent performance (for a review see [116]). The acquisition speed of the TOF instruments renders them excellent detectors for rapid separations, such as microCE. An impressive example has been published recently [117] and in Figure 16 spectra of such a combination are shown.

20.5.4. Fourier Transform Mass Spectrometry

Fourier Transform MS was developed on the bases of ion cyclotron (ICR) mass spectrometry [118]. The principle is shown in Figure 17. The cubic cell comprises two trapping plates (T), two excitation plates (E), and two detector plates (D) and is located in the center of a high-field magnet, generally a superconducting magnet with up to 7 Tesla. Trapping voltages applied to two opposite plates keep the ion in the center. After application of a radiofrequency (RF) voltage at two plates perpendicular to the others, the ions are excited and travel on circular trajectories with the same radius, but with frequencies that depend on their mass. The third pair of plates detects the RF image current induced by the rotating ions. The different frequencies and amplitudes in the image current

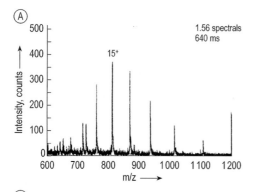

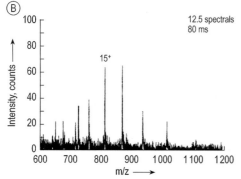

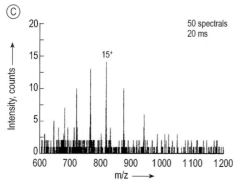

Figure 16. TOFMS of cytochrome c, concentration 0.1 µM in the solvent; Data aquisition: 5000 Hz; Spectrum A 3200, B 400, and C 100 summed scans

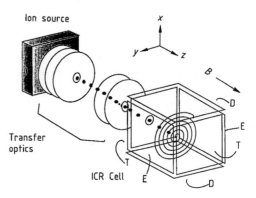

Figure 17. Schematic of a Fourier transform ion cyclotron resonance mass spectrometer with external ion source, transfer optics, and cubic FTMS cell
Ions can be generated using EI (even in combination with GC), FAB (LC/MS), or laser desorption/ionization; the transfer optics allow the required vacuum to be maintained.

signal can be extracted by Fourier transformation and converted into mass spectra. Theoretically, FT ICR MS has almost unmatched high-mass and high-resolution capabilities, the latter at the expense of very long acquisition times. CsI clusters up to $m/z = 31\,830$ have been observed [119], with rather low resolution, using an external ion source and quadrupole transfer ion optics; with a different instrument, a resolution of 60 000 at $m/z = 5922$ has been demonstrated using laser desorption [120]. For small molecules, resolutions in excess of 10^8 have been established. Further, MS/MS is rather easy to perform and operation is straightforward [121]. FTMS is the only competitor for sector field instruments with respect to resolution and accuracy.

However, its usefulness in analytical mass spectrometry still seems uncertain, partly because the very high cost prevents widespread use. Furthermore, optimal vacuum conditions ($< 10^{-6}$ Pa) are necessary in the ICR cell, which is difficult to maintain during GC/MS analyses. Attempts to solve the vacuum problem have been made with dual cell systems, where the ions are transported from an ionizer cell to an analyzer cell [122]. Two quadrupoles for mass selection and an RF-only quadrupole for focused injection [123] have been used, as has a time-of-flight arrangement [124]. They all serve to isolate the "dirty" ion source or the sample matrix from the analyzer to maintain optimal pressure conditions. With the last mentioned arrangement, the GC inlet replaced by a FAB target, and an atom gun, a spectrum of an enkephalin (Tyr-GlyGly-Phe-Leu) was obtained [125] with normal FAB features, but with almost no matrix background. Newly designed instruments seem to have solved the technical problems to a great extent and both, MALDI and electrospray spectra have been published with exorbitantly high resolutions (in excess of 500 000) and with impressive sensitivity.

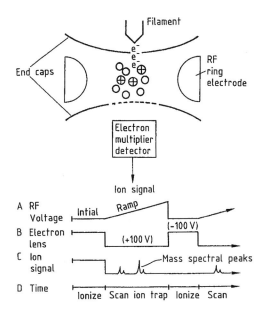

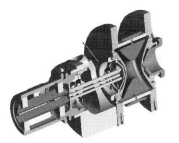

Figure 18. Schematic representation of the ion trap components, the scan function, the electron beam, and the resulting ion signals on the time axis (Adapted from [113] and [127])

Figure 19. Cut view of the LCQ$_{Deca}$ (Thermoquest)

20.5.5. Ion Traps

Another ion storage device in rapid development is the ion trap. Although it has long been used by physicists [126] it was only much later that it was converted into a powerful GC detector [127]. In this case, only high frequencies and no magnetic field are necessary to create a cage for the ions in a closed quadrupole field. The ions are ejected several times per second from the trap, hitting the detector sequentially. The principles of operation are depicted in Figure 18. The first commercial version had a limited mass range of 600 amu and low resolution, but quite convincing detection capabilities.

The mass range has since been dramatically increased (up to 40 000 amu), although the resolution and particularly the mass accuracy is still fairly poor. In recent versions, an additional scan ramp is used to evaluate the ion population of the trap for optimization of the residence time. However, high-resolution experiments using electrospray have now been described [128] and the detection capabilities have been steadily improved by optimizing the ion transfer into the trap. A modern version that is used as a detector for HPLC is shown in Figure 19. A weak point is still quantitative reproducibility. The storage capability is limited because when many ions are stored, coulomb repulsion destroys the performance of the trap. Therefore, software has been developed to monitor the number of ions in the trap and adjust the storage time accordingly to avoid unwanted ion–molecule reactions. Ion traps are the analyzer of choice in all cases where structure elucidation is the major application and several commercial versions from different verndors are on the market today.

20.5.6. Isotope Mass Spectrometer

The reliable determination of isotope ratios of elements normally needs specialized instrumentation. The ion sources depend on the type of element or molecule; for many elements the thermal ionization source has been employed, but for others and for small molecules, EI can be used as well. Although conventional analyzers can be used in such instruments, for precise determination, simultaneous detection of more than one ion beam is necessary and the instruments must be optimized for this purpose. Isotope ratio mass spectrometry (IRMS) is a specialized technique designed for precise determination of isotope ratios such as $^{12}C/^{13}C$, $^{1}H/^{2}H$ in organic compounds. The analysis requires the separation of compounds by GC, combustion of the compound to small molecules such as CO_2 and H_2O, ionization and accurate measurement of the signals due to the isotopically pure fragments. The analyzers allow simultaneous detection of the signals at low levels with high precision by using several Faraday cups [129] (Fig. 20). With such instruments, an abundance sensitivity greater than 10^{-8} was obtained for $^{230}Th/^{232}Th$ ratios [129].

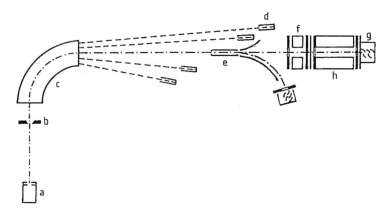

Figure 20. Isotope mass spectrometer with variable faraday cups and a quadrupole mass filter for MS/MS experiments where a deceleration lens is necessary between the analyzers
a) Ion source; b) Aperture slit; c) Magnet; d) Variable Faraday cups; e) Center slit; f) Deceleration lenses; g) Counter-SEM; h) Quadrupole mass filter

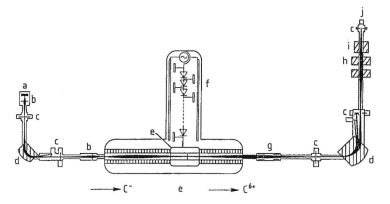

Figure 21. Accelerator mass spectrometry
a) Ion source; b) Lens; c) Slits; d) Beam bending magnet; e) Stripper foil charged to 2.5×10^6 V; f) High voltage source; g) Electrostatic lens; h) Magnetic lens; i) Velocity filter; j) Ion detector
(Reproduced with permission from [131])

To obtain the highest precision, further precautions must be taken with sample handling, instrumental parameters, and electronics to minimize the noise levels. This makes the procedure highly complicated, and only a few laboratories in the world are able to perform such precise measurements. An example is the correction of the Avogadro number, based on isotope mass spectrometry with silicon isotopes [130].

20.5.7. Accelerator Mass Spectrometry (AMS)

Accelerator mass spectrometry (Fig. 21) is designed for the most precise atom counting of cosmogenic radionuclides such as ^{10}Be, ^{14}C, ^{26}Al, ^{32}Si, and ^{36}Cl. It can measure these elements with unparalleled detection power in, for example, polar ice cores, deep sea sediments, or wood. The nuclide ^{14}C is used in the geosciences for dating of organic matter. The ions from the source (different sources may be used) are selected with a 90° magnet, accelerated toward the stripper foil, and converted into highly charged positive ions. These ions are focused again into a magnetic sector and detected within a gas-filled chamber by means of secondary ions produced there. The differentiation of, for example, ^{14}C^{6+} and N^{7+} is possible, since both species penetrate into the chamber with a different depth.

The detector is designed to identify species of nearly the same mass and velocity but different charges with very high selectivity.

20.5.8. Other Analyzers

The number of additional analyzer systems similar to those discussed above, but with only minor modifications, and others making use of different principles, is large, but most of them have no practical application. An exception may be the Wien filter, an analyzer with crossed electric and magnetic fields [132]. It has high-mass capabilities and high sensitivity but very poor resolution, which has inhibited its widespread application as an analyzer, although it is used to control ion energies in some instruments. The use of Wien filters in hybrid MS/MS instrumentation has been proposed [133].

Aside from quadrupoles and ion traps, other dynamic analyzers [114] such as the monopoles have been merely of theoretical interest.

20.6. Metastable Ions and Linked Scans

The determination of metastable ions has found much interest, because of their structural and theoretical significance. Several techniques have been developed and some of them are discussed here. Some of the techniques can be made more powerful by means of forced fragmentation in a field-free region (collision-induced dissociation). This process is essential for MS/MS experiments (see Chapter 20.7). Metastable ions are gaining new relevance in time-of-flight instruments with reflectrons (see Section 20.5.3).

20.6.1. Detection of Metastable Ions

In a magnetic sector instrument the signals of metastable ions appear as broad, mostly gaussian peaks in normal mass spectra. The location m^* may be calculated from the precursor ion mass m_p and the product ion mass m_d using the relation:

$$m^* = \frac{m_d^2}{m_p}$$

These signals are normally lost in double focusing instruments, although in Nier–Johnson type mass spectrometers, reactions in the field-free region between electrostatic analyzer and magnet deliver signals at the same positions as well. In one experiment the acceleration voltage is scanned. To detect a metastable transition, the spectrometer is focused for the product ion m_d with reduced acceleration voltage V_d. The acceleration voltage is then increased to V_p without changing the electrostatic analyzer and the magnet until the signal of the unknown precursor mass m_p appears. The ratio V_d/V_p is equal to the ratio of the masses m_d/m_p and m_p can be calculated. The resolution obtained with this so-called Barber–Elliott technique [134] or accelerating voltage scan is low and the instrument is continuously detuned, but for processes with rather small losses the detection capabilities can be great and kinetic energy release information is retained. This means that part of the excess energy of the ion is released as kinetic energy, leading to a range of translational energies yielding broad signals.

20.6.2. Mass Selected Ion Kinetic Spectrometry

Double-focusing instruments can be regarded as a combination of two mass spectrometers. For instruments with a reversed Nier–Johnson geometry, the magnet can serve as the first mass spectrometer (MS-1) that isolates one ion species from the mixture. This ion may undergo collisions in the field-free region following the magnet to form fragments, which can be separated by means of the electrostatic sector (MS-2). This technique, called mass selected ion kinetic spectrometry (MIKES) or direct analysis of daughter ions (DADI), has only the very limited resolution of the energy analyzer for the fragments and requires reversed geometry. Therefore, the number of applications in analytical problems is small; its importance lies in determining the energy released during fragmentation.

20.6.3. Linked Scans

A simple and inexpensive way of obtaining MS/MS information is the use of linked scans in sector field instruments. The advantage over the MIKES technique is that the resolution is generally better and that all double-focusing instruments can be used. The two sectors and the acceleration voltage are linked together by functions that allow

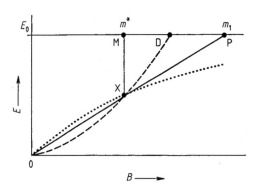

Figure 22. The B/E plane
—— for $B/E = $ const.; – – – for $B^2/E = $ const.; ····· for neutral loss scan $B(1-E)^{1/2}/E = $ const.

only specific fragment types to pass both analyzers. These may be the product ions of a precursor, all precursor ions of a particular product, or all ions with a preselected mass difference of precursor and product ions (neutral loss scan). In Figure 22, the relation between the most common linked scans is shown.

Product (Daughter Ion) Scans. When the ratio of magnetic field B and acceleration voltage E is held constant ($B/E = 1/v$, with v as velocity) during a mass scan downward from a precursor ion, only those fragments of this preselected ion which are formed in the first field-free region are transmitted through the electrostatic sector. This is because the velocity is unique for each precursor and thus for all the fragments formed after acceleration. The resolution of the precursor ion is low, but the resolution of the fragment ions can be quite high and the kinetic energy release information is lost. If a collision cell is available, high-energy collisions may be employed to induce fragmentation.

This type of scan is often used to analyze the fragmentation pattern of a substance. The spectra obtained may serve as a "finger print" of an ion and may allow a fragmentation scheme to be established. For a known reaction of a known compound the instrument can be set to detect only this one signal on this particular $B/E = $ const. surface, enhancing the specificity of detection (see "reaction monitoring," Chap. 20.2).

Precursor (Parent) Ion Scans. When a product ion appears with high frequency and the precursor is to be identified, a different scan is necessary. This precursor or parent ion scan requires a scan law in which $B^2/E = m$ with constant m. The velocity spread of the ions is not filtered, thus having poor resolution for the precursors, but good detection capabilities. This approach has been employed successfully for the screening analysis of drugs and their metabolites (metabolic profiling).

Neutral Loss Scans. Among several other possibilities for linking the fields and voltages in a double-focusing mass spectrometer, the neutral loss scan is the only one which deserves a brief discussion here. The scan function is

$$\frac{B^2(1-E)}{E^2} = \text{const.}$$

This type of scan has a distinct analytical potential because the loss of a common neutral fragment can be taken as an indication of a particular class of compounds. Therefore, it has been applied in the search for specific compounds derived from a certain precursor compound (drug/metabolite monitoring).

20.7. MS/MS Instrumentation

The combination of two or more analyzers, commonly known as MS/MS or tandem mass spectrometry, is a highly specific means of separating mixtures, studying fragmentation processes, and analyzing gas-phase reactions. The field is reviewed thoroughly in [135]. With the first analyzer, one ion is isolated from all the others. In the next, reactions of that ion are studied further. Fragments may be formed in unimolecular reactions from metastable ions, but generally fragmentation is induced by a variety of means. Collisions with gases in a collision chamber (CAD: collision-activated decomposition), laser photodissociation [136], or surface collisions [137] can be employed to induce fission. Since MS/MS can be combined with chromatographic separation techniques, it has found a broad range of applications in analytical chemistry. Indeed, LC/MS interfaces such as thermospray or the API techniques depend largely on collision-induced dissociations and MS/MS. Since they produce only molecular ions, little structural information is provided. With CAD and MS/MS these details are obtainable. A recent review gives a fairly detailed discussion of the thermochemical aspects of the various fragmentation induction methods used in mass spectrometry [138].

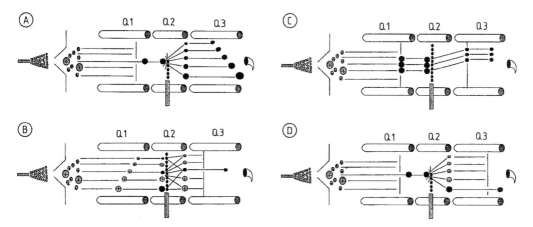

Figure 23. MS/MS experiments with the triple quadrupole MS
A) Daughter scan; B) Parent scan; C) Neutral loss scan; D) Reaction monitoring
Q1, Q3: mass analyzers; Q2: collision cell

20.7.1. Triple Quadrupoles

Triple quadrupoles for analytical MS/MS experiments have been built since the pioneering work of YOST and ENKE in 1978 [139] and they have found applications in all areas of mass spectrometry. The first and the third quadrupole are MS-1 and MS-2, whereas the second in the middle is an RF-only quadrupole acting as collision cell. In the collision cell, the transmitted ions undergo low energy collisions with a gas. For the product ion scan (daughter scan), only one precursor (parent) is selected and the second analyzer scans for the product ions. For the precursor scan, the first quadrupole is scanned and only the selected product ion is recorded. In the neutral loss scan, both analyzers are scanned with the selected mass difference, and for reaction monitoring only one precursor and one product ion species is permitted to travel through MS-1 and MS-2, respectively. In the meantime, instruments with different collision regions have been built. It was surprising that fragments are formed under the low energy regime in the collision cell. The fragment ion reactions are similar to those from high-energy collisions; exceptions include side-chain reactions of amino acids in peptides. The collision efficiency is quite high, particularly when the collision cell is an RF-only quadrupole [140] or octapole with nearly 100 % transmission [141]. The resolution of both quadrupoles as MS-1 and MS-2 can be set independently allowing, for example, unit resolution for precursor and product ions. Many different experiments with scan functions similar to the linked scans have been made and the set up and computer control is quite straightforward. Figure 23 shows schematically the widely used MS/MS experiments. The triple quadrupole is the most common MS/MS instrument today.

20.7.2. Multiple-Sector Instruments

In cases where high-resolution MS/MS experiments are required, multiple-sector field instruments are the only choice. They have the advantages of potentially high resolution in both stages, a wide mass range, and high sensitivity. However, there is a considerable trade off due to complexity, space requirements, and high cost. This is true particularly for the four-sector field mass spectrometers. As an example, the schematic of a four-sector field instrument is given in Figure 24.

Such mass spectrometers are highly flexible since the possible combinations of sectors and the number of field-free regions allow many different experiments to be performed [142]. The limitation is that the fragmentation process becomes less probable with increasing mass and the upper limit seems to be around 2500 amu. To increase the dissociation efficiency not only collision cells, but also photo-dissociation and surface-collision-induced dissociation have been investigated as well as simultaneous recording with a zoom lens integrated into the second MS [143].

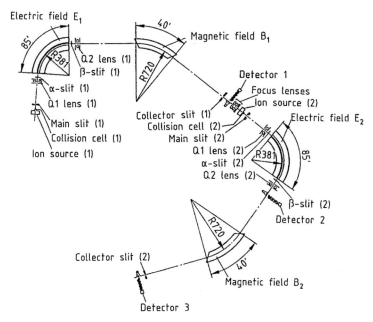

Figure 24. The schematic diagram and the ion optical components of the four-sector field mass spectrometer JEOL HX110/HX110 (Reprinted from [142])

20.7.3. Hybrids (Sector – Quadrupoles)

The apparently logical combination of the high resolution capabilities of sector instruments with the simplicity of quadrupoles in hybrid instruments is complicated in practice. Since the quadrupole only accepts ions of low kinetic energy, it must be floated at the potential of the sector instrument or, if high-energy collisions are to be possible, a zoom lens allowing the deceleration of the ions must be incorporated between the mass spectrometers. The latter has been realized in only one commercial version, whereas the simpler first combination has been made by several manufacturers. Mostly the sequence is sector – quadrupole, but it is possible to arrange the spectrometers the other way around. Combinations of double focusing spectrometers with time-of-flight instruments [156] or with ion traps have been built and both are hybrids with great analytical potential. However, in general the sector instruments made these combinations too complicated and too expensive.

20.7.4. Ion-Storage Devices (FTMS, Ion Traps)

Two types of ion-storage device (ion traps) are in use today as mass spectrometers, as discussed already, with built in MS/MS capabilities: the magnetic trap as ion cyclotron resonance spectrometer, which is, in its modern version, the fourier transform mass spectrometer [144] – [147] and the electric field version, called ion trap [148] – [152]. Both types of instrument allow all but the one selected ion to be ejected, which then undergoes ion – molecule reactions, collisions, or photodissociation to yield fragments. This process may be repeated, allowing not only MS/MS but also MS/MS/MS and even MS^n, which allows the detailed study of fragmentation reactions. However, currently only daughter-ion scans are possible.

20.7.5. Quadrupole Time-of-Flight Tandem Mass Spectrometers

The development of combination sector field/TOF led almost instantaneously to the combination of quadrupoles with TOF spectrometers, since at the same time the TOF spectrometers were con-

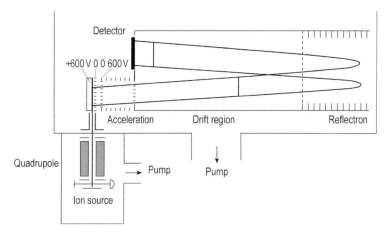

Figure 25. Q-TOF tandem mass spectrometer, schematic.

verted into high-resolution high-precision instruments (Fig. 25). A new aspect of this is that the ion injection into the TOF spectrometer is orthogonal, yielding an MS/MS combination with extremely high sensitivity for the fragments, without compromising scanning speed, resolution, mass range or the mass accuracy [153].

Thus this spectrometer has become one of the most successful new developments in the history of mass spectrometry (produced by two vendors) and today the instrument is one of the key tools in protein, peptide, and DNA research all over the world [154].

20.7.6. Others

Several TOF/TOF instruments have recently been described with impressive sensitivities, although the resolution in the second spectrometer is low due to the energy spread (similar to the situation in MIKE) [155]. The same is true of BE/TOF combinations, which seem to be highly suitable for electrospray and MALDI techniques [156], but it is too early to predict the future importance of these instruments.

20.8. Detectors and Signals

The detectors in mass spectrometry can be divided into two general types. One group is designed to detect one ion beam at a single location only (e.g., Faraday cage, secondary ion multiplier), and the second group can detect several ion beams simultaneously (e.g., photoplates and multiplier arrays). It is clear that the second group can be adapted only to specific analyzers, but the advantage in terms of detection power is so great that such instruments have been developed. The most common types of detector are shown in Figure 26.

20.8.1. Faraday Cage

One of the first detectors in mass spectrometry was the faraday cage. In principle it is simply a cup, which converts the current induced by impacting ions into a voltage pulse, the height of which is a measure of the number of ions reaching the detector per unit time. Since at that stage no amplification is achieved, its detection power is limited, whereas its linearity over the full dynamic range is still unsurpassed. The latter characteristic makes it even today a valuable tool, when high precision must be maintained (e.g., isotope mass spectrometry). In early instruments it was possible to switch between a faraday cup and a secondary electron multiplier. Conversion dynodes normally included into modern spectrometers, may be used as faraday cups as well, if they are positioned in line with the ion beam [157].

20.8.2. Daly Detector

The Daly detector [158] is still in use; the first version was able to distinguish between metastable and normal ions. The ions are converted into photons by means of a scintillator and the photons

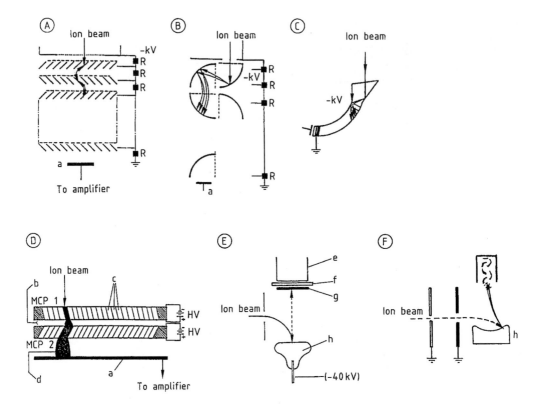

Figure 26. Types of detectors
A) Venetian blind multiplier; B) Box and grid multiplier; C) Channeltron multiplier; D) Microchannelplates; E) Daly detector; F) Postacceleration/conversion dynode
a) Anode; b) Chevron; c) Channels; d) Electrons; e) Photomultiplier; f) Window; g) Scintillator (+ 10 kV); h) Dynode

are measured by a photomultiplier which is located outside the vacuum (see Fig. 26 E). It is easy to integrate a post-acceleration/conversion dynode opposite to the scintillator plate to enhance the sensitivity. The originalversion was arranged in beam and had an additional plate in front of the scintillator, serving as conversion dynode. With a potential at the ion energy, the normal ion would only gently touch the scintillator, giving no signal. Metastable ions with less energy will be reflected to the intermediate plate, giving secondary ions which are in turn accelerated toward the scintillator; thus, it is possible to distinguish metastable ions from "normal" ions.

20.8.3. Secondary Electron Multiplier (SEM)

The most common detector today is the secondary electron multiplier, which amplifies the first impulse from the impacting ion by orders of magnitude. Several different types have been developed, with specific properties. Multipliers with open dynodes, the surfaces of which are made from beryllium oxide, have high amplification, can be baked out and, depending on the number of stages, can have short blind times allowing ion counting. This is the period of time during which the electron cascade runs through the dynodes and when no new impact is seen. Similar are the venetian blind multipliers, which, however, cannot be used for ion counting. Each dynode consists of slats arranged like a venetian blind with a light grid in front to ensure a homogeneous field in the dynode. Due to the circular, wide open active area, the positioning is not critical, although the ion beam should not be too narrow (see Fig. 26 A). Multipliers based on semiconducting surfaces (channeltrons) are widely used in high-current systems and are incorporated in quadrupole and ion-trap instruments. Instead of discrete dynodes with

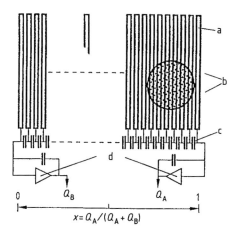

Figure 27. The Patric detector
a) Conductive strips; b) Electron cloud; c) Condenser chain; d) Charge sensitive amplifiers (position and time information for every ion)
(Reproduced with permission)

spectrometers use fiber optics to transmit the signals to photodiode arrays [160]. Detectors with zoom capabilities [143], static or scanning modes, and switching between analog and counting modes have been designed and successfully used [161], [162]; the gain in sensitivity can be two orders of magnitude depending on the system employed, although the resolution and sometimes the dynamic range is limited. The Patric detector (see Fig. 27) is a simultaneous detector with scanning option. The detector is capable of locating the impinging ion on the surface as long as the events are not exactly simultaneous; therefore, the dynamic range of the device is somewhat limited. The resolution obtainable is ca. 5000; since the mass spectrometer can switch between analog recording with a conventional SEM and the Patric detector even in a scan, the enhancement in sensitivity (ca. 100) can be fully utilized.

20.8.5. Signal Types

The shape of the signals depends largely on the type of instrument. Signals in sector field instruments are ideally triangular when the magnetic field is scanned and resolution as well as intensity are under optimal conditions. A rectangular beam profile (from the entrance slit) is moved across the exit slit, the width of which should be adjusted to match the width of the ion beam. This means in practice that the height of the signal is adjusted by the entrance slit. Then, the width of the final triangle is optimized by means of the exit slit without reducing the intensity further.

In quadrupole instruments the signals are in principle rectangular, since no slit can be used to define the beam shape precisely. In Fourier transform mass spectrometers, the peak profile should be gaussian when the signals are detected by resonance phenomena. The signals in time-of-flight instruments depend on the shape of the primary pulse (ions, neutrals, or laser photons), but even more on the desorption characteristics of the sample. Since the time for desorption and ionization of the sample is usually much longer than the primary pulse, the resulting signals have rather wide, poorly defined peak shapes with poor resolution. It is possible to shape the secondary pulse by means of ion optics to enhance the performance of the TOF instruments with respect to resolution.

resistor chains, the channeltron uses a semiconducting material for the same effect and many different types and shapes are in use, giving similar performance as the dynode types (see Fig. 26 C).

When heavy molecules had to be detected, the need for post-acceleration to increase the probability for secondary electrons arose. Since then, conversion dynodes have been included in the detection systems to accelerate slow heavy ions (typical potentials between 15 and 30 kV), and thus enhance the sensitivity for both positive and negative ions.

20.8.4. Microchannel Plates

Microchannel plates are based on the same principle as the channeltron (Fig. 26 C) and can be regarded as an array of small channeltrons of rather short length (a few mm instead of cm) arranged side by side in one plate similar to the photoplates of Mattauch–Herzog instruments. They are used in TOF instruments and for the simultaneous detection of ions in sector field mass spectrometers. The major problem to be solved in the latter case was the fast readout of the signals, and various techniques have been developed for this purpose. A first, very elaborate version of such a device was described as early as 1974 [159]; more modern ones suitable for tandem mass

20.9. Computer and Data Systems

The first applications of computers in mass spectrometry were merely the recording of signals from the multipliers, conversion to spectra, and some data handling. Today, powerful computers are integrated in the spectrometer. They control the scan, define the scan function, set the required voltages, tune the source, and supervise the experiments. The data are stored and interpreted with advanced tools, although at that stage the fully automated procedures are at an end.

20.9.1. Instrument Control, Automation

Modern mass spectrometers are completely dependent on computers and data systems. All the instrumental parameters of the inlet systems such as direct probe, GC, or LC (temperatures, gradients, times) are supervised by the computer. The ion source is tuned by automated procedures, the instrumental parameters (voltages, frequencies, resolution, etc.) are set by software procedures, and even the design of experiments (EI, CI, positive/negative ions, linked scans, MS/MS, etc.) can be started automatically, if specified conditions are met (e.g., signal height, retention times). Under such circumstances, the spectrometers can run a large number of samples unattended, since the analytical methods can be specified first, stored and activated upon request of the software. One advantage of this development is that a single man–machine interface is sufficient to work with such an MS. Further, most of the information necessary to describe an experiment from an instrumental point of view is stored automatically and is therefore available for reports and documentation.

20.9.2. Signal Processing

Most mass spectrometers produce analog signals at the detector. Thus it is necessary to convert the analog current or voltage signal into digital signals by means of analog-to-digital converters. This step has been a problem in the past, because the dynamic range and the speed of such devices was not always sufficient to maintain the integrity of resolution and dynamic range delivered by SEMs. In cases where ion counting is possible, different problems must be solved: the maximum counting speed is the limiting factor for the dynamic range. Since the design of the multiplier defines the pulse width and blind period, the dynamic range is a function of the performance of the multiplier and the pulse counting device. When pulses overlap, errors may occur, although correction is possible if the frequency is not to high.

Analog-to-Digital Conversion and Mass Calibration. As mentioned above, the analog signal is digitized by fast and accurate A/D converters (typically above 200 kHz) with sufficient dynamic range (≥ 16 bit). The limiting factor for the dynamic range is often the noise of the baseline. The digitized information about the ion current is fed into the computer, where the signals are analyzed and converted from "raw data" into peaks or noise, the intensities are calculated and placed on a mass scale. The conversion can be performed on the basis of time, magnetic field, or frequency, depending on the type of spectrometer. The mass scale must be calibrated to achieve the required precision. The most reliable way to calibrate the mass scale for all mass spectrometers is by means of reference compounds, whose mass spectrum in terms of accurate masses and approximate intensities should be known to the software. In addition, they should have signals evenly spaced over a large mass range.

Some reference compounds for mass calibration are listed below:

EI, CI, positive and negative
Perfluorokerosene (PFK, up to 1000 amu)
Perfluorotributylamine (FC-43, heptacosa, M_r 671)
Tris(perfluoroheptyl)-*s*-triazine (up to 1200 amu)
Ultramark 3200F (Fomblin oil, perfluoro polyethers, up to 3200 amu)
Ultramark 1621 (perfluorinated phosphazine, up to 1600 amu)

LC/MS (Thermospray)
Polyethylene glycols (up to 2000 amu)
Polypropylene glycols (up to 2500 amu)
Acetic acid cluster (up to 1500 amu)

FDMS
Polystyrenes (up to 10 000 amu)

FABMS
Glycerol (useful up to 1000 amu)
CsI (wide mass range, rather wide spacings)
CsI/RbI and NaI/CsI/RbI (wide mass range)

Ultramark 3200F (up to 3200 amu)
Phosphoric Acid (5%) in glycerol (up to 1500 amu, negative ions)

Electrospray
Polypropylene glycols (up to 2500 amu)

After acquisition of the spectrum, the computer tries to identify some starting pattern and then searches for all the other peaks. When the number of signals identified is sufficient, a calibration table is created which allows raw data to be converted into mass/intensity pairs. If high-resolution data have been acquired and accurate masses are to be calculated, additional internal standards are necessary for precise calculation of the signal location relative to known signals of the internal standard.

Peak Detection. One of the problems to be solved is the detection of true signals. It is necessary to determine the base line and to discriminate between noise signals of low intensity or insufficient width and real peaks. For this purpose, sophisticated software has been developed. Several parameters must be defined under a given set of instrumental conditions to ensure that no signal is overlooked and that as few noise signals as possible are recognized as peaks, obscuring the information contained in the mass spectrum. To do this the peak width, the minimum height and minimal number of counts, and, for unresolved signals, the depth of the valley between two adjacent peaks are calculated for a given scan speed, scan function, and number of digital points ("counts"). Only signals meeting all such requirements are accepted as peaks and if a problem is encountered during assignment, the peak can be flagged for classification.

Dynamic Range. The dynamic range of a multiplier in a mass spectrometer covers several orders of magnitude (up to 1×10^6) and is defined as the ratio between the smallest and the largest signal which can be produced. The digitalization system in the computer interface has to comply with this adequately, meaning the dynamic range of the A/D converter and its speed must be sufficient. With modern converters and sample and hold devices coupled with autoranging systems, this problem is solved. The simultaneous recording devices have rather low dynamic ranges due to saturation effects and readout problems. Since they are specifically designed to acquire signals of very small ion currents, this limitation is acceptable.

20.9.3. Data Handling

When the mass spectral data have been stored in the computer, the interpretation of the data is the next step. In most cases the data are spectra with peak intensities on a *m/z* scale. However, chromatograms from selected ion recordings and data from multichannel analyzers must also be handled. Sophisticated software has been written to develop increasingly automated procedures for quantitative analyses, mathematical treatment of data, display of spectra and structures, and to transport the data into laboratory data systems for integration into reports. While this aspect of data handling can be considered rather satisfactory today, the extraction of information contained in the spectra is under development. The deduction of possible structures of unknown compounds from mass spectra is demanding and often fails, and the calculation of spectra based on the structure is in its infancy. It is still uncertain whether this is at all possible, because the competitive reactions of the gas-phase ions are controlled by minute differences in energy between intermediate structures, and the kinetic behavior is difficult to model for complex molecules. Initial attempts have been published for both the modeling of reactions in the mass spectrometer [163] and the prediction of mass spectra from structures [164]. In both cases the results are far from application for analytical purposes.

20.9.4. Library Searches

So far, the most successful approach for extracting structural information from spectra is the library search. Libraries have been collected from different sources with large numbers of spectra; the most widely used are the Wiley data base with ca. 160 000 and the NIST data base with ca. 60 000 spectra; others, specializing in topics such as toxicology, agricultural chemicals, or foodstuffs, are also available [165]. The quality of the data contained in some of the large databases is still a substantial problem and must be improved, since spectra often strongly reduced in the number of peaks, incomplete, or incorrect have been taken from old collections and included in the new compilations. It is not trivial to identify and to replace them with better spectra, because wrong spectra

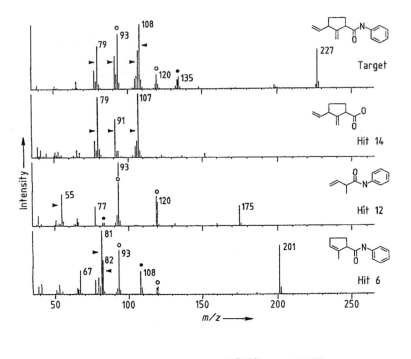

Figure 28. Spectra of similar compounds found by a similarity search for structures, not for spectra
▶ = Hydrocarbon fragment; ○ = Anilide moiety; ● = Loss of aniline radical or aniline molecule
The fragmentation pattern (bottom) explains why the analog reaction in Hit 6 yields m/z 82 instead of m/z 108; m/z 108 in spectrum Hit 6 must have a different origin. (Adapted from [171], with permission).

may be never found (similar spectra may no longer show up) and the definition of better spectra is also difficult. Quality is a relative attribute of a spectrum and sometimes two completely different spectra must be maintained, if they were recorded under diverse conditions. Library searches, with a variety of algorithms, have been developed, and the best known are SISCOM [166], [167], PBM [168], [169], and the INCOS type search [170]. The SISCOM approach also allows the identification of unknowns and the deduction of structural features, thus giving information about similar compounds. This is illustrated by the example given in Figure 28.

Recently the chemical structures of the compounds have been integrated as a means to enhance the power of the information retrieval. It has become possible to identify not only similar spectra, but also the spectra of similar structures, revealing much of the complex relation between spectra and structures [171].

20.9.5. Integration into Laboratory Networks

Since in many laboratories, particularly in industrial environments, mass spectrometers are merely one tool among others such as NMR and IR spectrometers, it is becoming increasingly necessary to integrate the spectrometers into laboratory networks [172]. Advantages of this approach include:

1) The customer as a producer of a sample can use his terminal in the network to inspect all the spectroscopic data of a compound submitted for analysis or structure elucidation.

2) "Intelligent" interpretation, aided by computer software, can be initiated prior to examination.
3) It is possible to follow the fate of sample not only on the spectroscopic environment, but in other analytical, biological, or toxicological examinations as well. In the following step, all the data can be combined and the information may be integrated into a single report.
4) The thorough documentation and the assurance of quality becomes possible. The integrity of the data from the sample to the final report becomes increasingly important and is only possible with adequate use of the data systems.

20.9.6. Integration in the World Wide Web

As with many others examples, mass spectrometric data have been published through the WWW from many publishers and it is fairly easy today to access new bibliographic data via the internet. The homepages of most publishers provide either the bibliographic data or even the abstracts free and often the costs for the online publications are more or less the same as for the printed versions. Even more relevant may be the fact that in protein and nucleic acid research the most important databases are accessible over the net. MS datasystems have built-in software to search the databases directly or even via an automated strategy for known sequences in peptides or oligonucleotides, thus allowing the rapid identification of known proteins based on MALDI data of peptide mixtures and electrospray MS/MS data ("sequence tags"). It can be stated that the stunning speed in proteomic research is largely due to the immediate access to such databases and the retrieval software.

20.10. Applications

Typical areas for the application of mass spectrometry are:

1) Analysis of trace elements and compounds in environmental samples
2) Problem solving in biomedical research (protein identification/DNA), particularly the structure elucidation of natural compounds and determination of molecular mass
3) Control of synthetic procedures by rapid analysis of intermediates and products
4) Analysis of bulk materials and surfaces

One of the major driving forces for the rapid development of mass spectrometry during the last years is trace analysis in environmental chemistry and biochemistry. The search for sensitive, reliable methods that allow the determination of compounds at very low levels together with the option for mass spectra of highly polar and labile compounds motivated the development of new interfaces, ion sources, analyzers, and detectors. The second impetus is the search for techniques for the determination of very high molecular masses with better precision than is attainable by other known methods.

20.10.1. Environmental Chemistry

Mass spectrometry is a powerful tool in environmental chemistry because of its detection power, specificity, and structure analysis capabilities. Techniques have been developed for the determination of catabolites and metabolites of chemicals in practically all relevant matrices since no other comparable tool is available, even though technical problems can occur. Generally, sample preparation uses at least one type of chromatography, connected with MS either off-line or on-line. Of the on-line combinations ("hyphenated techniques"), GC/MS is the most successful, even though LC/MS is rapidly catching up in this area. The development of benchtop GC/MS and LC/MS instruments has made it possible to use mass spectrometry in routine laboratories and even in the field [173].

Air. The investigation of reactions in the atmosphere is one of the major fields of modern environmental chemistry. The processes responsible for the formation of smog in cities [174] as well as the interactions of man-made or natural hydrocarbons with ozone [175] are examples where mass spectrometric techniques have been applied. Usually, the chemicals under investigation must be trapped on a stationary phase such as charcoal or tenax. The samples are then transferred by thermodesorption to a GC/MS system [176]. For the investigation of the organic contents of single particles from aerosols, laser desorption has been employed successfully [177]. The potential of the new ion sources such as the atmospheric pressure ion (API) source for monitoring reactions directly or for detecting air pollutants at very low levels has yet to be fully explored; however, initial experiments are encouraging [178]. The process of gas to particle conversion has been investigated using an adapted API source [179]. Volatile organic compounds in air and in water have been successfully analyzed by using membranes in various configurations to give enrichment factors in the range 40–780 [180]; with nitrogen as matrix, several halogenated or aromatic compounds have been detected in the ppb range. Recently, the specificity and sensitivity of REMPI mass spectrometry has been used to detect organic compounds in gases directly without trapping [181]–[183]; one additional benefit of this technique is the superior time resolution allowing rapid changes in concentration, to be followed.

Water. For water the situation is similar in that usually in the first step pollutants must be extracted from the matrix. Derivatization is often necessary to make MS analysis possible. Extraction and preconcentration is the critical step in the procedure, because the nature of the chemicals to be monitored can vary widely. LC/MS techniques that avoid derivatization have been used successfully. Pesticides such as triazines, organophosphorus compounds, and phenylureas have been determined in water samples by thermospray LC/MS [184]. Even though the well-developed GC/MS procedures are mostly used, they fail to detect some classes of compounds, such as plasticizers in drinking water, where they were first found by LC/MS [185]. A broad range of pollutants have been identified in wastewaters by particle beam LC/MS [186]. For the direct analysis of organic compounds in water an instrument has been designed with a membrane inlet, which allows certain chemicals to pass directly into the detection system, in this case, an ion trap. The detection limits for some of the compounds are low enough (e.g., benzene 1 ppb, toluene 10 ppb, ether 10 ppb; acetone and tetrahydrofuran 100 ppb) for routine monitoring of water [187].

For the determination of elements in water, ICPMS has demonstrated superior performance due to its sensitivity and multielement capabilities (see Section 20.11.6).

Soil. The analysis of herbicides, insecticides, and fungicides in soil is commonly based on extraction with liquids or supercritical fluids. The samples are then concentrated, derivatized if necessary, and transferred into the detection system. Often, GC/MS methods are the first step, although MS can be replaced with a less complicated detector in the final routine version of the procedure. For developing new procedures and in cases where the identification of a compound is uncertain due to contamination, GC/MS may be the only choice. One of the most important applications is the analysis of tetrachlorodioxins (TCDDs), especially the toxic isomer 2,3,7,8-tetrachlorodibenzo-p-dioxin, and related compounds. These are formed as byproducts of pesticide production, and occur in fly ashes from municipal incinerators and other sources [188]. All standard procedures for the detection of such compounds have as the key step GC/MS, normally with high resolution (10 000) [189], [190] to increase the specificity. Mass spectrometers have been designed solely for this purpose by several manufacturers and, since the detection limits are very low and the reliability of the procedures must be high, the use of reference compounds with ^{13}C labels is mandatory [191]. Another method of enhancing the specificity is the use of API sources [192]. For the investigation of metabolism or catabolism of herbicides (e.g., by photooxidation), LC/MS has become increasingly important [193]. The fate of such compounds (e.g., phenoxyalkanoic acids and their derivatives [194]) can be monitored in soil samples by analysis of their stability and by the emergence of catabolic and metabolic products. A further example is the analysis of phenols and their biological conjugates (e.g., sulfates, glucuronides) with LC/MS [195], which was previously very difficult and time consuming. The fate of elements in soil is difficult to investigate, since the interactions with fulvic and humic acids are complex and not well understood.

Waste Materials and Dump Sites. The major difficulty in MS analysis of wastes is often the highly complex matrix, which must be excluded from the analysis. One possibility is the use of multiple chromatographic steps prior to MS analysis. Alternatively the compounds of interest may have very different physical properties from the matrix. This is the case with polycyclic aromatic compounds, which have been determined directly by using thermal desorption with GC/MS [196] detection. Recently, small portable quadrupole mass spectrometers have been built as a means to identify contaminated sites such as abandoned industrial plant locations and old waste dumps.

N-terminal ions

a_n: $\text{H}-(\text{NH}-\text{CHR}-\text{CO})_{n-1}-\overset{+}{\text{N}}\text{H}=\text{CHR}_n$

or

$\overbrace{\text{H}-(\text{NH}-\text{CHR}-\text{CO})_{n-1}-\text{NH}-\overset{\overset{\displaystyle CR_n^a R_n^b}{\|}}{\text{CH}}}^{\text{H}^+}$

$a_n + 1$: $\overbrace{\text{H}-(\text{NH}-\text{CHR}-\text{CO})_{n-1}-\text{NH}-\overset{\displaystyle R_n}{\text{CH}}\cdot}^{\text{H}^+}$

b_n: $\text{H}-(\text{NH}-\text{CHR}-\text{CO})_{n-1}-\text{NH}-\text{CHR}_n-\text{C}\equiv\text{O}^+$

c_n: $\overbrace{\text{H}-(\text{NH}-\text{CHR}-\text{CO})_n-\text{NH}_2}^{\text{H}^+}$

d_n: $\overbrace{\text{H}-(\text{NH}-\text{CHR}-\text{CO})_{n-1}-\text{NH}-\overset{\overset{\displaystyle CR_n^b}{\|}}{\text{CH}}}^{\text{H}^+}$

C-terminal ions

v_n: $\overbrace{\text{HN}=\text{CH}-\text{CO}-(\text{NH}-\text{CHR}-\text{CO})_{n-1}-\text{OH}}^{\text{H}^+}$

w_n: $\overset{\overset{\displaystyle CR_n^b}{\|}}{\text{CH}}-\text{CO}-\overbrace{(\text{NH}-\text{CHR}-\text{CO})_{n-1}-\text{OH}}^{\text{H}^+}$

x_n: $^+\text{O}\equiv\text{C}-\text{NH}-\text{CHR}_n-\text{CO}-(\text{NH}-\text{CHR}-\text{CO})_{n-1}-\text{OH}$

or

$\overbrace{\text{O}=\text{C}=\text{N}-\text{CHR}_n-\text{CO}-(\text{NH}-\text{CHR}-\text{CO})_{n-1}-\text{OH}}^{\text{H}^+}$

y_n: $\overbrace{\text{H}-(\text{NH}-\text{CHR}-\text{CO})_n-\text{OH}}^{\text{H}^+}$

$y_n - 2$: $\overbrace{\text{HN}=\text{CR}_n-\text{CO}-(\text{NH}-\text{CHR}-\text{CO})_{n-1}}^{\text{H}^+}$

z_n: $\overset{\overset{\displaystyle CR_n^a R_n^b}{\|}}{\text{CH}}-\text{CO}-\overbrace{(\text{NH}-\text{CHR}-\text{CO})_{n-1}-\text{OH}}^{\text{H}^+}$

$z_n + 1$: $\cdot\,\overbrace{\text{CHR}_n-\text{CO}-(\text{NH}-\text{CHR}-\text{CO})_{n-1}-\text{OH}}^{\text{H}^+}$

Figure 29. Most common peptide fragments from positive ions
R = Side chain of an amino acid; R^a and R^b = Beta substituents of the nth amino acid (From [207], with permission)

20.10.2. Analysis of Biomedical Samples

In the last few years the biomedical field has experienced a revolution of mass spectrometric possibilities [197].

MS has long been important for the detection of trace amounts of, for example, steroids [198], [199], drugs [200], and prostaglandins [201] in biological fluids such as urine, plasma, or saliva. Another area with many applications is the structure analysis of a wide variety of compounds, sometimes after derivatization or chemical modification. Micro-scale methylation [202], acetylation, and silylation have been used for the MS analysis of large molecules. For oligosaccharides, often not only the sequence and the type of monomer must be analyzed, but also the position of the glycosidic linkage and derivatization/cleavage schemes have been developed for this purpose [203], [204]; for details of the Lindberg method [205], see [197]. The sequence of peptides can be determined by cleavage to short oligomers and GC/MS, often in combination with FABMS for the determination of large fragments made enzymatically [206]. The reactions of the protonated ions produced with FAB (or liquid SIMS, LSIMS) have been studied extensively. Chemically labile bonds have the tendency to break more often, leading to analytically significant ions. Some of the most important are shown in Figure 29. The same type of ions are observed with electrospray CID (MS/MS) or MALDI PSD spectra which have revolutionized the analysis of proteins in the last few years, and proteomics without mass spectrometry would be unthinkable. The general strategy applied is shown in Figure 30.

Since such ionizing techniques are compatible with many MS/MS experiments, sequence information can be deduced from the spectra in great detail. When high-energy collisions are used, even leucine and isoleucine can be distinguished due to the different fragmentation of the side chain.

The analysis of complex lipids such as glycosphingolipids [208] and lipopolysaccharides [209] became even more complete with the advent of desorption techniques. But the localization of the position of double bonds in unsaturated fatty acids from complex lipids is still difficult. Several techniques were conceived using pyrolidides [210] or dimethylsulfides [211] and EIMS, CIMS [212], or MS/MS techniques [213]. As another example, electrophoretic labeling (perfluoroacylation) allows the determination of modifications in nucleobases at the attomole level [64] or the characterization of prostaglandins in biological fluids by means of negative chemical ionization (NCI) [214].

Field desorption mass spectrometry (FDMS) indicated for the first time [215] that the analysis

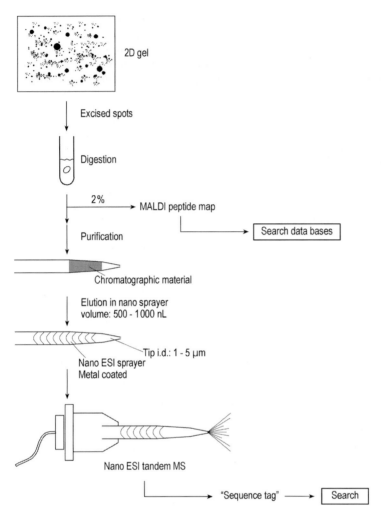

Figure 30. Strategy for the identification of proteins (after Protana, Odense)

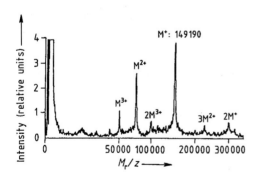

Figure 31. Matrix-assisted laser desorption (MALDI) spectrum of a carefully purified monoclonal antibody (IgG of mouse against a specific human lymphokine) (Reproduced with permission)

of large, labile, and highly polar compounds such as nucleotides [216] and oligosaccharides [217] might become possible. The next step was the development of fast atom bombardment, although it seems in retrospect that plasma desorption was the first method to allow molecular ions of very large molecules to be observed. Thus the whole range of biological polymers became accessible to mass spectrometry. This expectation was indeed largely fulfilled with laser desorption and electrospray, which appeared shortly afterwards. As an instructive early example, the spectrum of an intact protein is shown in Figure 31.

With these methods, large proteins, enzymes, or polynucleotides yield molecular ions, allowing the precise determination of molecular mass. In combination with chemical modification of amino acid side chains, the secondary and even the tertiary structure of proteins could be studied [218]. The synthesis of peptides, for example, became much more efficient, because the time-limiting step was formerly the analysis of the products, which can now be performed in minutes.

In addition, electrospray is also well suited to the analysis of small, highly polar substances (e.g., drugs and metabolites [219]). Recently, even noncovalent, highly specific interactions [220] have been studied. Apparently, the conformation of some biological polymers can be retained in the gas phase, allowing receptor–ligand complexes to survive the ionizing step. Examples include the interaction of the synthetic macrolide FK506 with its cytoplasmic receptor [221] and the detection of selfcomplementary DNA duplexes [222].

Since some of the techniques can be interfaced to liquid chromatography (LC/MS) or capillary zone electrophoresis (CZE/MS), the analysis of complex mixtures is also possible. Thus, these methods are suitable for the efficient design of pharmaceuticals and agrochemicals where careful analysis of the metabolic processes is necessary, but was formerly difficult and time consuming.

The detection capability of this type of instrumentation allows the reactions of xenobiotic compounds and drugs with macromolecules such as proteins or nucleotides [223] to be studied and very low concentrations of platinum-drug-modified biopolymers [224] in living organisms to be investigated. The metabolic or catabolic fate of drugs can be studied (e.g., the abuse of steroid hormones by athletes).

20.10.3. Determination of High Molecular Masses

One of the most important applications of mass spectrometry has always been the determination of molecular mass. Since this can be difficult with EIMS, the search for soft ionization techniques is almost as old as organic mass spectrometry. The first step was chemical ionization and particularly direct chemical ionization (DCI), which allowed the samples to be desorbed directly from a thin wire into the CI plasma. It has been successfully used for natural compounds such as glycerides. Later, field ionization (FI) and especially field desorption (FD) opened up new possibilities for the analysis of labile natural compounds such as saccharides, peptides, and nucleotides. When fast atom bombardment (FAB), plasma desorption (PD), electrospray ionization (ESI), and laser desorption (LD) mass spectrometry came into use, molecular ions of labile and very large molecules in excess of 200 000 amu could be determined with unparalleled accuracy. The electrospray spectrum of a protein allows its molecular mass to be calculated with high precision from a series of differently charged molecular ions (Fig. 32).

In this context, the discussion about resolution became a new relevance [225]. The use of nominal integral masses (e.g., 12 for C, 1 for H, 16 for O) is meaningless here as is true for the monoisotopic masses (e.g., 1.0078 for ^{1}H, 14.0030 for ^{14}N, 15.9949 for ^{16}O). Even with resolutions exceeding 10 000 it is impossible to resolve all the possible isobaric species in the molecular ion pattern of molecules such as peptides or oligonucleotides with hundreds of carbons and hydrogens, and tens of oxygen, nitrogen, and sulfur atoms. This is even more difficult with multiply charged ions as occurs regularly with electrospray. In such a case, the averaged chemical mass of the elements can be used for the calculation such as 12.011 for C, 1.008 for H, 14.007 for N, 15.999 for O, 32.060 for S, and 30.974 for P. The resolution power of the instrument should always be set to give maximum sensitivity and, if possible, only sufficient separation power to resolve the major peaks due to carbon isotopes for singly charged molecules (see Fig. 33). For multiply charged species, even this is impossible but, nevertheless, information about the chemical molecular mass can be deduced from such spectra (all the isotopes are essentially merged into one peak). It is only necessary to distinguish between different biological modifications (e.g., phosphorylation, glycosylation).

If instruments can, however, achieve much higher resolution (100 000 or more) without sacrificing sensitivity, the gain in information can be striking: since from m/z, the distance between isotope signals is $1/z$, the charge state of a multiply charged ion signal can be read directly simply by counting the peaks within one mass unit. For example, 10 peaks at m/z 800–801 means that this ion carries 10 charges. The distance between the isotope signals is, therefore, $1/10$ of the mass. This can be very important for MS/MS work, since the charge state of fragments from multiply charged ions is generally unknown and the information about the "true" molecular mass is obscured. When the charge is known, sequencing of peptides on the basis of MS/MS analysis of electrospray ions becomes feasible. This is the reason why in

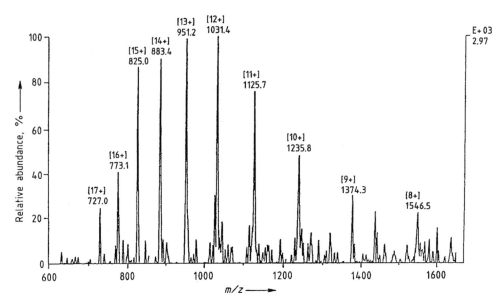

Figure 32. Electrospray spectrum of ca. 180 pmol cytochrome c using an MAT 90 sector field MS; homemade ESI source

the last few years FTMS with MALDI and electrospray have begun to be surprisingly employed in biomedical laboratories. The results reported on large peptides [145]–[147], [227]–[229] as well as on large oligonucleotides [147], [230]–[233] are so promising that even the high costs of such instruments cannot stop the development and not only research institutes but also industry have bought the FTMS instruments. As an illustration, data from the second isotope signal of a peptide show a resolution of 420 000, separating the isotope signals in this peak (Fig. 34).

In Figure 35, the product ions from a doubly charged tryptic digestion product illustrate the use. The product ion spectrum of the doubly protonated oligo-peptide Phe-Ser-Trp-Gly-Ala-Glu-Gly-Glu-Arg (m/z = 520) contains several significant singly charged sequence ions; this type of data is used for "mass tagging" in protein identification.

20.10.4. Species Analysis

The storage, transport, and action of metals in the environment is largely dependent on the chemical form such as oxidation state and association with organic ligands. Heavy metals in the environment are stored in complexes with humic acids, can be converted by microbes into different complexes, and may be transported into living animals and human beings. This is true for many elements such as lead, mercury [234], arsenic [235], [236], astatine, tin, and platinum [237]. Several mass spectrometric techniques have been employed in the study of the fate of metals in the human body and other organic environments by the determination of the species formed in the biological or environmental matrices.

For example, tin and lead alkylates have been determined in soil, water, or muscle tissue with GC/MS after exhaustive alkylation or with thermospray [239], API [240], and ICP–LC/MS [244] methods. The ESI spectra of some tin compounds is given in Figure 36. The APIMS techniques have proven to be very successful, even with metals bound to proteins and enzymes [241] and, interfaced to microseparation techniques, even elements such as iodine [242] or phosphorus [243] can be determined quantitatively.

20.10.5. Determination of Elements

Mass spectrometry as a multielement technique has the advantage that many metal ions can be detected and quantitatively determined at once. In bulk materials such as steel [245] or refractory metals [246], elements can be determined by means of low-resolution glow-discharge

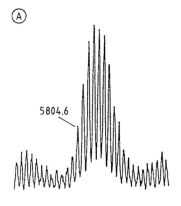

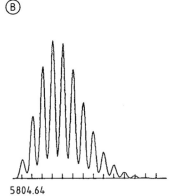

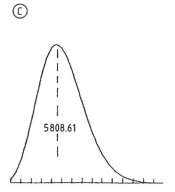

Figure 33. Effect of resolution on peak shapes
A) The protonated molecular cluster of human insulin ($C_{257}H_{384}N_{65}O_{77}S_6$) recorded with FABMS; B) Calculated at a resolution of 6000 (10% valley definition) with the monoisotopic mass at $m/z = 5804.6455$; C) The unresolved cluster with an average mass at $m/z = 5808.61$
(Reproduced from [226], with permission)

mass spectrometry. High-resolution GDMS has been used to study semiconducting materials [247]. GDMS is regarded as being virtually free of matrix effects [248]. The state of the art in glow discharge mass spectrometry has been reviewed recently and from the data presented there it is evident that the technique is a mature tool for the material sciences [249]. If part of the material can be dissolved prior to mass spectral analysis, ICPMS or, for high precision isotope determination, thermal ionization mass spectrometry (TIMS) can be applied. Detection limits in ICPMS are listed in Table 2.

ICPMS and TIMS are also suitable for the analysis of solutions such as biological fluids or water samples. In combination with laser ablation, bulk materials or small biological compartments can be analyzed with ICPMS (see Section 20.4.5). The desorbed neutrals, ions, and clusters are entrained in the gas, which carries the aerosol directly into the plasma. One important area of application is geochronology where today ICPMS and AMS are in use; for a recent comprehensive description of this field see [250].

20.10.6. Surface Analysis and Depth Profiling

The characterization of material surfaces and of the underlying layers is one of the most important problems in modern material science, since the properties of a material are not only dependent on the bulk concentration of elements, but also on the distribution on the surface, in the upper layers, and at grain boundaries. Therefore, techniques have been developed which can analyze the surface and the next layers with very high resolution. The first two monolayers, ca. 0.5 nm in depth, can be investigated by secondary ion mass spectrometry (SIMS) [81]. Positive and negative ions from the elements in the surface can be detected with a sensitivity in the ppm range. It has found many applications in the analysis of metals, alloys, semiconductors, ceramics, and glasses. The problem with SIMS is that matrix effects can cause difficulties in quantitation. It is possible to employ SIMS for the analysis of organic compounds on a metal surface. This technique, called static SIMS, makes use of the sputtered organic ions from material deposited on the surface (typically Ag), in monolayers. Since the ion current is weak and lasts for only a few seconds, time-of-flight instruments are generally used.

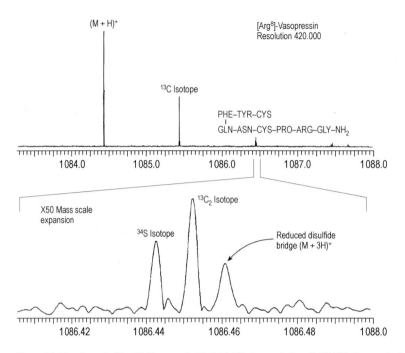

Figure 34. Top: 1 pmol of [arg8]-Vasopressin MALDI MS at a resolution of 420 000. Bottom: isobaric 34S (m/z 1086.4415) and 13C2 (m/z 1086.4524) and a third peak due to the reduced disulfide bridge are clearly distinguished

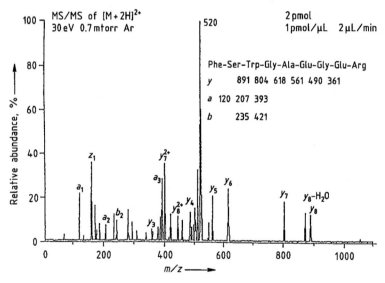

Figure 35. Doubly charged ions from tryptic digests of proteins
The fragments are labeled according to the scheme given in Figure 29 (Reproduced with permission from [238])

For the depth profiling of a metal, of particulate matter, or a ceramic with a resolution of a few nanometers, sputtered neutral mass spectrometry (SNMS) [251] is the best choice since the sputtering process is completely decoupled from the ionization, and quantitation is possible by calibrating the instrument with standard reference materials. The technique has found many applications for the

Table 2. Detection limits in ICPMS [65]

Element	Detection limit, pg/mL	Element	Detection limit, pg/mL
Li	100	Sn	60
B	400	Sb	60
F	1.1×10^5	Te	80
Mg	130	I	10
Al	160	Cs	20
P	2×10^4	Ba	150
S	1×10^5	La	10
Cl	1×10^3	Ce	10
Ti	320	Pr	10
V	80	Nd	45
Cr	10	Sm	30
Mn	30	Eu	15
Co	10	Gd	50
Ni	40	Tb	10
Cu	20	Dy	35
Zn	10	Ho	10
Ge	20	Er	35
As	40	Tm	10
Se	800	Yb	30
Br	1×10^3	Lu	10
Y	50	Hf	10
Zr	30	Ta	50
Mo	40	W	50
Ru	60	Au	60
Ag	30	Hg	20
Cd	30	Pb	10
In	60	Th	20
		U	10

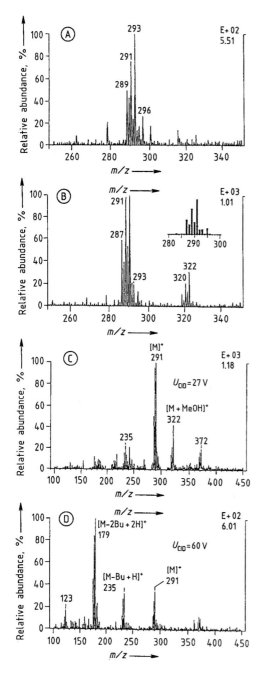

Figure 36. ESI spectra of some tin compounds
A) Dibutyl tin acetate (DBT) $(C_4H_9)_2Sn^{2+}(OCOCH_3)^-$;
B) Tributyl tin (TBT) $(C_4H_9)_3Sn^+$ cations with theoretical distribution given in the inset; C) TBT with small CAD voltage; D) TBT with large CAD voltage showing fragments due to loss of the side chains
Instrument: MAT 90, electrospray MS, flow injection
Upper spectrum: CAD voltage 27 V; Lower spectrum: CAD voltage 60 V

characterization of semiconductors and implanted atoms. Figure 37 gives an example compared to a result using GDMS.

The determination of the spatial distributions of elements is becoming increasingly important since the grain boundaries in ceramics or the metal distribution in microelectronic devices have a major influence on the stability of the material and the performance of the device. With a highly focused primary ion beam it is possible to scan the surface in two dimensions and the distribution of the elements on the surface. Combinations of sputtering processes and microprobing allow this distribution to be investigated in depth. Instruments used for this type of analysis are called ion microprobe mass analyzers (IMMA). The spot of the primary beam has a diameter of less than 10 μm, thus defining the resolution obtainable.

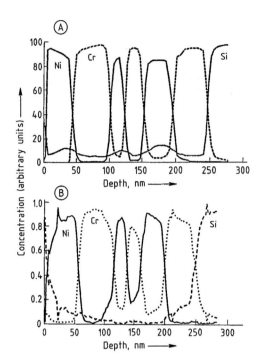

Figure 37. Depth profiles of a multilayer sample (Ni, Cr 50nm, Ni, Cr 25nm, Ni, Cr 50nm)
A) SNMS; B) GDMS
(Reproduced with permission from [252])

20.11. References

[1] F. W. Aston: *Mass Spectra and Isotopes*, Edward Arnold & Co, London 1933.
[2] A. J. Dempster, *Phys. Rev.* **11** (1918) 316.
[3] H. W. Washburn, H. F. Wiley, S. M. Rock, *Ind. Eng. Chem. Anal. Ed.* **15** (1943) 541–547.
[4] R. A. Brown, R. C. Taylor, F. W. Melpolder, W. S. Young, *Anal. Chem.* **20** (1948) 5–9.
[5] M. J. O'Neal, Jr., *Anal. Chem.* **22** (1950) 991–995.
[6] A. J. Campbell, J. S. Halliday in R. M. Elliott (ed.): *13th Annual Symposium on Mass Spectrometry*, vol. 2, Pergamon Press, Oxford 1962.
[7] H. Budzikiewicz, C. Djerassi, D. H. Williams: *Structure Elucidation of Natural Products by Mass Spectrometry*, Holden Day, San Francisco 1964.
[8] G. R. Waller (ed.): *Biochemical Applications of Mass Spectrometry*, Wiley Interscience, New York 1972.
[9] D. Kanne, K. S. Straub, J. E. Hearst, H. Rapoport, *J. Am. Chem. Soc.* **103** (1982) 6754–6764.
[10] H. M. Rosenstock, M. B. Wallenstein, A. L. Wahrhaftig, H. Eyring, *Proc. Natl. Acad. Sci. USA* **38** (1952) 667.
[11] R. A. Marcus, O. K. Rice, *J. Phys. Colloid. Chem.* **55** (1951) 894.
[12] K. Levsen: *Fundamental Aspects of Organic Mass Spectrometry*, VCH Verlagsgesellschaft, Weinheim 1978.
[13] F. W. McLafferty, F. Turecek: *Interpretation of Mass Spectra*, University Science Books, Mill Valley, California 1993.
[14] J. J. Boon, *Int. J. Mass Spectrom. Ion Processes* **118/119** (1992) 755–787.
[15] W. Genuit, J. J. Boon, O. Faix, *Anal. Chem.* **59** (1987) 508–513.
[16] W. H. McFadden: *Techniques of Combined Gas Chromatography/Mass Spectrometry*, Wiley & Sons, New York 1973.
[17] W. A. Brand, *J. Mass Spectrom.* **31** (1996) 225.
[18] R. M. Caprioli, T. Fan, *Biochem. Biophys. Res. Commun.* **141** (1986) 1058.
[19] Y. Ito, T. Takeuchi, D. Ishii, M. Goto, *J. Chromatogr.* **346** (1985) 161.
[20] M. A. Moseley et al., *Anal. Chem.* **61** (1989) 1577.
[21] R. M. Caprioli et al., *J. Chromatogr.* **480** (1989) 247.
[22] M. A. Moseley, L. J. Deterding, K. B. Tomer, J. W. Jorgenson, *Rapid Commun. Mass Spectrom.* **3** (1989) 87.
[23] N. J. Reinhoud et al., *J. Chromatogr.* **516** (1990) 147.
[24] M. A. Baldwin, F. W. McLafferty, *Org. Mass Spectrom.* **7** (1973) 1111.
[25] V. L. Tal'roze, G. V. Karpov, I. G. Gordetskii, V. E. Skurat, *Russ. J. Phys. Chem. (Engl. Transl.)* **42** (1968) 1658.
[26] J. D. Henion, *J. Chromatogr. Sci.* **19** (1981) 57.
[27] R. D. Smith, W. D. Felix, J. C. Fjedsted, M. L. Lee. *Anal. Chem.* **54** (1982) 1885.
[28] R. C. Willoughby, R. F. Browner, *Anal. Chem.* **56** (1984) 2625.
[29] E. C. Horning, M. G. Horning, D. I. Carroll, R. N. Stillwell, I. Dzidic, *Life Sci.* **13** (1973) 1331.
[30] J. Sunner, G. Nicol, P. Kebarle, *Anal. Chem.* **60** (1988) 1300.
[31] J. Sunner, M. G. Ikonomou, P. Kebarle, *Anal. Chem.* **60** (1988) 1308.
[32] C. R. Blakley, J. C. Carmody, M. L. Vestal, *Clin. Chem. (Winston Salem N.C.)* **26** (1980) 1467.
[33] J. V. Iribarne, B. A. Thomson, *J. Chem. Phys.* **64** (1976) 2287.
[34] B. A. Thomson, J. V. Iribarne, *J. Chem. Phys.* **71** (1979) 4451.
[35] G. Schmelzeisen-Redeker, L. Bütfering, F. W. Röllgen, *Int. J. Mass Spectrom. Ion Processes* **90** (1989) 139.
[36] M. L. Vestal, *Int. J. Mass Spectrom. Ion Phys.* **46** (1983) 193.
[37] J. Hau, W. Nigge, M. Linscheid, *Org. Mass Spectrom.* **28** (1993) 223–229.
[38] M. Linscheid, D. Westmoreland, *Pure Appl. Chem.*, (1994) in press.
[39] M. A. Niessen, J. van der Greef: *Liquid Chromatography-Mass Spectroscopy*, Marcel Dekker, New York 1992.
[40] R. D. Smith et al., *Anal. Chem.* **65** (1993) 574A–584A.
[41] W. M. Niessen, *J. Chromatogr. A* **856** (1999) 179.
[42] W. M. A. Niessen, A. P. Tinke, *J. Chromatogr. A* **703** (1995) 37.
[43] M. R. Emmett, F. M. White, C. L. Hendrickson, S.-H. Shi, A. G. Marshall, *J. Am. Soc. Mass Spectrom.* **9** (1997) 333.
[44] I. M. Lazar, A. L. Rockwood, E. D. Lee, J. C. H. Sin, M. L. Lee, *Anal. Chem.* **71** (1999) 2578.

[45] R. D. Smith, J. A. Olivares, N. T. Ngyen, H. R. Udseth, *Anal. Chem.* **60** (1988) 436–441.
[46] C. G. Edmonds et al., *Biochem. Soc. Trans.* **19** (1991) 943.
[47] S. A. Hofstadler, F. D. Swanek, D. C. Gale, A. G. Ewing, R. D. Smith, *Anal. Chem.* **67** (1995) 1477.
[48] J. C. Severs, S. A. Hofstadler, Z. Zhao, R. T. Senh, R. D. Smith, *Electrophoresis* **17** (1996) 1808.
[49] I. M. Johannson, E. C. Huang, J. D. Henion, J. Zweigenbaum, *J. Chromatogr.* **554** (1991) 311.
[50] A. L. Burlingame, J. A. McCloskey, *Biol. Mass Spectrom., Proc. Int. Symp. Mass Spectrom. Health Life Sci.*, Elsevier, Amsterdam 1990.
[51] R. S. Ramsey, J. M. Ramsey, *Anal. Chem.* **69** (1997) 1174.
[52] R. D. Oleschuk, D. J. Harrison, *Trends Anal. Chem.* **19** (2000) 379.
[53] K. L. Busch, *J. Planar Chromatogr.-Mod. TLC* **2** (1989) 355.
[54] A. Hayashi, T. Matsubara, Y. Nishizawa, T. Hattori, *Iyo Masu Kenkyukai Koenshu* **11** (1986) 147.
[55] L. Li, D. M. Lubman, *Anal. Chem.* **61** (1989) 1911.
[56] K. L. Bush, *TrAc Trends Anal. Chem. (Pers. Ed.)* **11** (1992) 314.
[57] I. D. Wilson, *J. Chromatogr. A* **856** (1999) 429.
[58] J. T. Mehl, D. M. Hercules, *Anal. Chem.* **72** (2000) 68.
[59] Y. Liu, J. Bai, X. Liang, D. M. Lubman, *Anal. Chem.* **67** (1995) 3482.
[60] M. Schreiner, K. Strupat, F. Lottspeich, C. Eckerskorn, *Electrophoresis* **17** (1996) 954.
[61] C. Eckerskorn et al., *Anal. Chem.* **69** (1997) 2888.
[62] D. Schleuder, F. Hillenkamp, K. Strupat, *Anal. Chem.* **71** (1999) 3238.
[63] A. G. Harrison: *Chemical Ionization Mass Spectrometry*, CRC Press, Boca Raton 1982.
[64] G. B. Mohamed et al., *J. Chromatogr.* **314** (1984) 211–217.
[65] H. D. Beckey: *Principles of Field Ionization and Field Desorption Mass Spectrometry*, Pergamon Press, Frankfurt 1977.
[66] G. M. Hieftje, L. A. Norman, *Int. J. Mass Spectrom. Ion Processes* **118/119** (1992) 519–573.
[67] W. W. Harrison, *J. Anal. At. Spectrom.* **7** (1992) 75–79.
[68] W. W. Harrison et al., "Glow Discharge Techniques in Analytical Chemistry," *Anal. Chem.* **62** (1990) 943A–949A.
[69] H. Oechsner, *Pure Appl. Chem.* **64** (1992) 615.
[70] A. Montaser, H. Tan, I. Ishii, S. H. Nam, M. Cai, *Anal. Chem.* **63** (1991) 2660.
[71] N. Jakubowski, L. Moens, F. Vanhaecke, *Spectrochim. Acta B: At. Spectrosc.* **53** (1998) 1739.
[72] M. Morita, H. Ito, M. Linscheid, K. Otsuka, *Anal. Chem.* **66** (1994) 1588.
[73] K. E. Milgram et al., *Anal. Chem.* **69** (1997) 3714.
[74] J. R. Eyler, R. S. Houk, C. H. Watson, K. E. Milgram: "Development and Application of Inductively Coupled Plasma (ICP)-FTICR Mass Spectrometry," *46th ASMS Conference*, Orlando, Fl 1998.
[75] N. Jakubowski, D. Stuewer, unpublished results.
[76] K. G. Heumann, *Int. J. Mass Spectrom. Ion Processes* **118/119** (1992) 575–592.
[77] W. Genuit, Chen He-Neng, A. J. H. Boerboom, J. Los, *Int. J. Mass Spectrom. Ion Phys.* **51** (1983) 207–213.
[78] U. Boesl, J. Grotemeyer, K. Walter, E. W. Schlag, *Anal. Instrum. (N. Y.)* **16** (1987) 151.
[79] J. Grotemeyer, U. Boesl, K. Walter, E. W. Schlag, *Org. Mass Spectrom.* **21** (1986) 645.

[80] J. Grotemeyer, E. W. Schlag, *Acc. Chem. Res.* **22** (1989) 399.
[81] A. Benninghoven, *Z. Phys.* **230** (1970) 403.
[82] L. Schwieters et al.: "Secondary Ion Mass Spectrometry," in A. Benninghoven, K. T. F. Janssen, J. Tümpner, H. W. Werner (eds.): *Proc. 8th Int. Conf. (SIMS VIII)* **1992**, 497.
[83] P. M. Thompson, *Anal. Chem.* **63** (1991) 2447–2456.
[84] B. Hagenhoff et al., *J. Phys. D, Appl. Phys.* **25** (1992) 818.
[85] F. W. Roellgen in A. Benninghoven (ed.): *Ion Formation from Organic Solids*, Springer-Verlag, Berlin 1983, p. 2.
[86] M. Barber, R. S. Bordoli, R. D. Sedgwick, A. N. Tyler, *J. Chem. Soc. Chem. Commun.* 1981, 325–327.
[87] R. D. McFarlane, D. F. Torgerson, *Science (Washington, D.C. 1883)* **191** (1976) 920.
[88] R. D. McFarlane: "Methods in Enzymology," vol. 193 in J. A. McCloskey (ed.): *Mass Spectrometry*, Academic Press, San Diego 1990, p. 263–280.
[89] G. Jonsson et al., *Rapid Commun. Mass Spectrom.* **3** (1989) 190.
[90] A. H. Verbueken, F. J. Bryunseels, R. E. van Grieken, *Biomed. Mass Spectrom.* **12** (1985) 438.
[91] J. S. Becker, H. J. Dietze, *Fresenius J. Anal. Chem.* **344** (1992) 69–86.
[92] F. Hillenkamp: "Ion Formation from Organic Solids II," in A. Benninghoven (ed.): *Springer Series in Chem. Phys.* **1983**, no. 25, p. 190.
[93] M. Karas, U. Bahr, A. Ingendoh, F. Hillenkamp, *Angew. Chem.* **101** (1989) 805–806.
[94] M. Karas, D. Bachmann, F. Hillenkamp, *Anal. Chem.* **57** (1985) 2935.
[95] M. Karas, F. Hillenkamp, *Anal. Chem.* **60** (1988) 2299–2301.
[96] K. Tanaka et al., *Rapid Commun. Mass Spectrom.* **8** (1988) 151.
[97] F. Kirpekar, S. Berkenkamp, F. Hillenkamp, *Anal. Chem.* **71** (1999) 2334.
[98] R. Kaufmann, T. Wingerath, D. Kirsch, W. Stahl, H. Sies, *Anal. Biochem.* **238** (1996) 117.
[99] P. Juhasz et al., *Anal. Chem.* **68** (1996) 941.
[100] U. Bahr, J. Stahl-Zeng, E. Gleitsmann, M. Karas, *J. Mass Spectrom.* **32** (1997) 1111.
[101] R. Kaufmann, *J. Biotechnol.* **41** (1995) 155.
[102] B. Sprengler, *J. Mass Spectrom.* **32** (1997) 1019.
[103] P. Chaurand, F. Luetzenkirchen, B. Spengler, *J. Am. Soc. Mass Spectrom.* **10** (1999) 91.
[104] M. Dole et al., *J. Chem. Phys.* **49** (1968) 2240.
[105] M. Yamashita, J. B. Fenn, *J. Phys. Chem.* **88** (1984) 4451–4459.
[106] C. M. Whitehouse, R. N. Dreyer, M. Yamashita, J. B. Fenn, *Anal. Chem.* **57** (1985) 675–679.
[107] M. L. Alexandrov, L. N. Gall', N. V. Krasnov, V. I. Nikolaev, V. A. Shkurov, *Zh. Anal. Khim.* **40** (1985) 1227–1236.
[108] L. Tang, P. Kebarle, *Anal. Chem.* **65** (1993) 3654–3668.
[109] S. K. Chowdhury, V. Katta, B. T. Chait, *Rapid Commun. Mass Spectrom.* **4** (1990) 81–87.
[110] A. P. Bruins, T. R. Covey, J. D. Henion, *Anal. Chem.* **59** (1987) 2642–2646.
[111] M. Wilm, M. Mann, *Anal. Chem.* **68** (1996) 1.
[112] A. J. Dempster, *Phys. Rev.* **11** (1918) 316.
[113] C. Brunnée, *Int. J. Mass Spectrom. Ion Processes* **76** (1987) 121–237.

[114] E. W. Blauth: *Dynamische Massenspektrometer,* Vieweg & Sohn, Braunschweig 1965.
[115] R. Kaufmann, B. Spengler, F. Lützenkirchen, *Rapid Commun. Mass Spectrom.* **7** (1993) 902.
[116] I. V. Chernushevich, W. Ens, K. G. Standing, *Anal. Chem.* **71** (1999) 452A.
[117] I. M. Lazar, R. S. Ramsey, S. Sundberg, J. M. Ramsey, *Anal. Chem.* **71** (1999) 3627.
[118] M. B. Comisarow, A. G. Marshall, *Chem. Phys. Lett.* **26** (1974) 489;M. B. Comisarow, A. G. Marshall, *J. Chem. Phys.* **64** (1976) 110.
[119] C. B. Lebrilla, D. T. S. Wang, R. L. Hunter, R. T. McIver, Jr., *Anal. Chem.* **62** (1990) 878–880.
[120] C. F. Ijames, C. F. Wilkins, *J. Am. Chem. Soc.* **110** (1988) 2687–2688.
[121] A. G. Marshall, *Acc. Chem. Res.* **18** (1985) 316–322.
[122] C. Giancaspro, F. R. Verdun, *Anal. Chem.* **58** (1986) 2097.
[123] R. T. McIver, R. L. Hunter, W. D. Bowers, *Int. J. Mass Spectrom. Ion Processes* **64** (1985) 67.
[124] P. Kofel, M. Allemann, H. Kellerhals, K. P. Wanczek, *Int. J. Mass Spectrom. Ion Processes* **72** (1986) 53.
[125] P. Grossmann, P. Caravatti, M. Allemann, H. P. Kellerhals, *Proc. 36th ASMS Conf. Mass Spectrom. All. Top.* 1988, 617.
[126] W. Paul, H. Steinwedel, *Z. Naturforsch. A*: *Phys. Phys. Chem. Kosmophys.* 8 (1953) 448.
[127] G. C. Stafford et al., *Int. J. Mass Spectrom. Ion Processes* **60** (1984) 85.
[128] B. D. Nourse, R. G. Cooks, *Anal. Chim. Acta* **228** (1990) 1.
[129] H.-J. Laue, K. Habfast, *Proc. 37th ASMS Conf. Mass Spectrom. All. Top.* 1989, p. 1033.
[130] P. De Bievre, *Fresenius J. Anal. Chem.* **337** (1990) 767.
[131] R. E. M. Hodges in J. F. J. Todd (ed.): *Proc. 10th Int. Mass. Spectrom. Conf. Swansea 1985*, Wiley, London 1986, p. 185.
[132] W. Aberth, *Biomed. Mass Spectrom.* **7** (1980) 367.
[133] H. Matsuda, H. Wollnik, *Int. J. Mass Spectrom. Ion Processes* **86** (1988) 53.
[134] M. Barber, R. M. Elliott, paper presented at ASTM E-14 Conference on Mass Spectrometry, Montreal 1964.
[135] K. L. Busch, G. L. Glish, S. A. McLuckey: *Mass Spectrometry/Mass Spectrometry: Techniques and Applications of Tandem Mass Spectrometry,* VCH Verlagsgesellschaft, Weinheim 1988.
[136] M. A. Seeterlin, P. R. Vlasak, D. J. Beussmann, R. D. McLane, C. G. Enke, *J. Am. Soc. Mass Spectrom.* **4** (1993) 751–754.
[137] S. R. Horning, J. M. Wood, J. R. Gord, B. S. Freiser, R. G. Cooks, *Int. J. Mass Spectrom. Ion Processes* **101** (1990) 219.
[138] S. A. McLuckey, D. E. Goeringer, *J. Mass Spectrom.* **32** (1997) 461.
[139] R. A. Yost, C. G. Enke, *J. Am. Chem. Soc.* **100** (1978) 2274.
[140] R. E. March, R. J. Hughes, *Quadrupole Storage Mass Spectrometry,* Wiley Interscience, New York 1989.
[141] C. Hagg, I. Szabo, *Int. J. Mass Spectrom. Ion Processes* **73** (1986) 295.
[142] K. Sato et al., *Anal. Chem.* **59** (1987) 1652.
[143] J. A. Hill et al., *Int. J. Mass Spectrom. Ion Processes* **92** (1989) 211.
[144] I. J. Amster, *J. Mass Spectrom.* **31** (1996) 1325.
[145] V. C. M. Dale et al., *Biochem. Mass Spectrom. Biochem. Soc. Trans.* **24** (1996) 943.
[146] M. W. Senko, J. P. Speir, F. W. McLafferty, *Anal. Chem.* **66** (1994) 2801.
[147] E. R. Williams, *Anal. Chem.* **10** (1998) 179A.
[148] A. Ingendoh et al., *J. Mass Spectrom. Soc. Jpn.* **45** (1997) 247.
[149] R. E. March, *J. Mass Spectrom.* **32** (1997) 351.
[150] R. J. Strife, M. Bier, J. Zhou, *Proc. 43rd ASMS Conference on Mass Spectrometry and Allied Topics, (Atlanta)*, (1995) 1113.
[151] J. C. Schwartz, M. Bier, D. M. Taylor, J. Zhou, *Proc. 43rd ASMS Conference on Mass Spectrometry and Allied Topics, (Atlanta)*, (1995) 1114.
[152] E. R. Badman, J. M. Wells, H. A. Bui, R. G. Cooks, *Anal. Chem.* **70** (1998) 3545.
[153] H. R. Morris et al., *Rapid Commun. Mass Spectrom.* **10** (1996) 889.
[154] C. Borchers, C. E. Parker, L. J. Deterding, K. B. Tomer, *J. Chromatogr. A* **854** (1999) 119.
[155] T. J. Cornish, R. J. Cotter, *Anal. Chem.* **65** (1993) 1043–1047.
[156] F. H. Strobel, L. M. Preston, K. S. Washburn, D. H. Russell, *Anal. Chem.* **64** (1992) 754–762.
[157] M. Linscheid, *Fresenius Z. Anal. Chem.* **337** (1990) 648–661.
[158] N. R. Daly, *Rev. Sci. Instrum.* **34** (1963) 1116.
[159] C. E. Griffin, H. G. Boettger, D. D. Norris, *Int. J. Mass Spectrom. Ion Phys.* **15** (1974) 437.
[160] J. S. Cottrell, S. Evans, *Anal. Chem.* **59** (1987) 1990–1995.
[161] C. E. D. Ouwerkerk, A. J. H. Boerboom, T. Matsuo, T. Sakurai, *Int. J. Mass Spectrom. Ion Phys.* **70** (1986) 79–96.
[162] R. Pesch, G. Jung, K. Rostand, K.-H. Tietje, *Proc. 37th ASMS Conf. Mass Spectrom. All. Top.* 1989, 1079.
[163] S. Bauerschmidt, W. Hanebeck, K.-P. Schulz, J. Gasteiger, *Anal. Chim. Acta* **265** (1992) 169–182.
[164] J. Gasteiger, W. Hanebeck, K.-P. Schulz, *I. Chem. Inf. Comp. Sci.* **32** (1992) 264–271.
[165] K. Pfleger, H. H. Maurer, A. Weber: *Mass Spectral and GC Data of Drugs, Poisons, Pesticides, Pollutants and their Metabolites,* 2nd ed., VCH Verlagsgesellschaft, Weinheim 1992.
[166] H. Damen, D. Henneberg, B. Weimann, *Anal. Chim. Acta* **103** (1978) 289.
[167] D. Henneberg, *Anal. Chim. Acta* **150** (1983) 37.
[168] G. M. Pesyna, R. Venkataraghavan, H. G. Dayringer, F. W. McLafferty, *Anal. Chem.* **48** (1976) 1362.
[169] B. L. Atwater, D. B. Stauffer, F. W. McLafferty, D. W. Peterson, *Anal. Chem.* **57** (1985) 899.
[170] S. Sokolov, J. Karnovsky, P. Gustafson, *Finnigan Appl. Rep.* no. 2, 1978.
[171] D. Henneberg. B. Weimann, U. Zalfen, *Org. Mass Spectrom.* **28** (1993) 198–206.
[172] M. J. Hayward, P. V. Robandt, J. T. Meek, M. L. Thomson, *J. Am. Soc. Mass Spectrom.* **4** (1993) 742–750.
[173] T. Kotiaho, *J. Mass Spectrom.* **31** (1996) 1.
[174] P. Ciccioli, A. Cecinato, E. Brancaleoni, M. Frattoni, A. Liberti, *HRC CC J. High Resolut. Chromatogr.* **15** (1992) 75–84.
[175] T. Hoffmann, P. Jacob, M. Linscheid, D. Klockow, *Int. J. Environ. Anal. Chem.* **52** (1993) 29–37.
[176] A. Robbat, C. J. Liu, T. Y. Liu, *J. Chromatogr.* **625** (1992) 277–288.

[177] P. J. McKeown, M. V. Johnston, D. M. Murphy, *Anal. Chem.* **63** (1991) 2073–2075.

[178] K. G. Asano, G. L. Glish, S. A. McLuckey, *Proc. 36th ASMS Conf. Mass Spectrom. All. Top.* 1988, 637.

[179] T. Hoffmann, R. Bandur, U. Marggraf, M. Linscheid, *J. Geophys. Res. D: Armospheres* **103** (1998) 25.

[180] M. A. LaPack, J. C. Tou, C. G. Enke, *Anal. Chem.* **62** (1990) 1265–1271.

[181] R. Zimmermann, H. J. Heger, A. Kettrup, U. Boesl, *Rapid Commun. Mass Spectrom.* **11** (1997) 1095.

[182] R. Zimmermann et al., *Rapid Commun. Mass Spectrom.* **13** (1999) 307.

[183] U. Boesl, *J. Mass Spectrom.* **35** (2000) 289.

[184] G. Durand, N. de Bertrand, D. Barcelo, *J. Chromatogr.* **554** (1991) 233–250.

[185] H. Fr. Schröder, *J. Chromatogr.* **554** (1991) 251–266.

[186] M. A. Brown, I. S. Kim, F. I. Sasinos, R. D. Stephens: "Liquid Chromatography/Mass Spectrometry, Applications in Agricultural, Pharmaceutical and Environmental Chemistry," in M. A. Brown (ed.): *ACS Symp. Ser.* **420** (1990) 199–214.

[187] A. K. Lister, R. G. Cooks, *Proc. 36th ASMS Conf. Mass Spectrom. All. Top.* 1988, 646.

[188] C. Rappe, S. Marklund, R.-A. Berqvist, M. Hanson in G. Choudhary, L. H. Keith, C. Rappe (eds.): *Chlorinated Dioxins and Dibenzofurans in the Total Environment*, Butterworth, Stoneham 1983, chap. 7.

[189] R. M. Smith et al., *Environ. Sci. Res.* **26** (1983) 73–94.

[190] L. Q. Huang, A. Paiva, H. Tond, S. J. Monson, M. L. Gross, *J. Am. Soc. Mass Spectrom.* **3** (1992) 248–259.

[191] E. R. Barnhart et al., *Anal. Chem.* **59** (1987) 2248–2252.

[192] W. A. Korfmacher, G. F. Moler, R. R. Delongchamp, R. K. Mitchum, R. L. Harless, *Chemosphere* **131** (1984) 669–685.

[193] M. Linscheid, *Int. J. Environm. Anal. Chem.* **49** (1992) 1–14.

[194] T. L. Jones, L. D. Betowski, J. Yinon: "Liquid Chromatography/Mass Spectrometry, Applications in Agricultural, Pharmaceutical and Environmental Chemistry," in M. A. Brown (ed.): *ACS Symp. Ser.* **420** (1990) 62.

[195] W. M. Draper, F. R. Brown, R. Bethem, M. J. Miille: "Liquid Chromatography/Mass Spectrometry, Applications in Agricultural, Pharmaceutical and Environmental Chemistry," in M. A. Brown (ed.): *ACS Symp. Ser.* **420** (1990) 253.

[196] A. Robbat, T. Y. Liu, B. M. Abraham, *Anal. Chem.* **64** (1992) 1477–1483.

[197] J. A. McCloskey (ed.): "Mass Spectrometry," *Methods in Enzymology*, **vol. 193**, Academic Press, San Diego 1990.

[198] H. M. Leith, P. L. Truran, S. J. Gaskell, *Biomed. Environm. Mass Spectrom.* **13** (1986) 257–261.

[199] C. H. L. Shackleton in H. Jaeger (ed.): *Glass Capillary Chromatography in Clinical Medicine and Pharmacology*, Marcel Dekker, New York 1985.

[200] A. Frigerio, C. Marchioro, A. M. Pastorino in D. D. Breimer, P. Speiser (eds.): *Top. Pharm. Sci. Proc. Int. Congr. Pharm. Sci. F.I.P.* **45th** (1985) 297–310.

[201] H. Schweer, C. O. Meese, O. Fuerst, P. K. Gonne, H. J. W. Seyberth, *Anal. Biochem.* **164** (1987) 156–163.

[202] H. R. Morris, D. H. Williams, R. P. Ambler, *Biochem. J.* **125** (1971) 189.

[203] V. N. Reinhold in S. J. Gaskell (ed.): *Mass Spectrometry in Biomedical Research*, J. Wiley & Sons, Chichester 1986, p. 181.

[204] D. Rolf, G. R. Gray, *Carbohydr. Res.* **137** (1985) 183–216.

[205] H. Björndal, C. G. Hellerqvist, B. Lindberg, S. Svensson, *Angew. Chem.* **82** (1979) 643.

[206] K. Biemann, *Pract. Spectrosc.* **8** (1990) 3–24.

[207] K. Biemann in J. A. McCloskey (ed.): "Mass Spectrometry," *Methods in Enzymology*, **vol. 193**, Academic Press, San Diego 1990, p. 886–887.

[208] S. A. Carr, V. N. Reinhold, *Biomed. Mass Spectrom.* **11** (1984) 633–642.

[209] A. Dell, C. E. Ballou, *Carbohydr. Res.* **120** (1983) 95–111.

[210] B. A. Andersson, W. W. Christie, R. T. Holman, *Lipids* **10** (1975) 215–219.

[211] P. Sperling, M. Linscheid, S. Stöcker, H.-P. Mühlbach, E. Heinz, *J. Biol. Chem.* **268** (1993) 26935–26940.

[212] M. Suzuki, T. Ariga, M. Sekine, E. Araki, T. Miyatake, *Anal. Chem.* **53** (1981) 985–988.

[213] N. J. Jensen, K. B. Tomer, M. L. Gross, *J. Am. Chem. Soc.* **107** (1985) 1863.

[214] H. Schweer, H. W. Seyberth, C. O. Meese, O. Fürst, *Biomed. Environm. Mass Spectrom.* **15** (1988) 143–151.

[215] H.-R. Schulten in H. R. Morris (ed.): *Soft Ionization of Biological Substrates*, Heyden & Sons, London 1981.

[216] M. Linscheid, G. Feistner, H. Budzikiewicz, *Isr. J. Chem.* **17** (1978) 163.

[217] M. Linscheid, J. D'Angona, A. L. Burlingame, A. Dell, C. Ballou, *Proc. Natl. Acad. Sci. USA* **78** (1981) 1471–1475.

[218] D. Suckau, M. Mak, M. Przybylski, *Proc. Natl. Acad. Sci. USA* **89** (1992) 5630–5634.

[219] L. Weidolf, T. R. Covey, *Rapid Commun. Mass Spectrom.* **6** (1992) 192–196.

[220] R. D. Smith, K. J. Light-Wahl, *Biol. Mass Spectrom.* **22** (1993) 493–501.

[221] B. Ganem, Y.-T. Li, J. D. Henion, *J. Am. Chem. Soc.* **113** (1991) 6294–6296.

[222] B. Ganem, Y.-T. Li, J. D. Henion, *Tetrahedron Lett.* **34** (1993) 1445–1448.

[223] M. G. Ikonomou, A. Naghipur, J. W. Lown, P. Kebarle, *Biomed. Environm. Mass Spectrom.* **19** (1990) 434–446.

[224] C. E. Costello, K. M. Comess, A. S. Plaziak, D. P. Bancroft, S. J. Lippard, *Int. J. Mass Spectrom. Ion Processes* **122** (1992) 255–279.

[225] J. Yergey, D. Heller, G. Hansen, R. J. Cotter, C. Fenselau, *Anal. Chem.* **55** (1983) 353–356.

[226] K. Biemann in J. A. McCloskey (ed.): "Mass Spectrometry," *Methods in Enzymology*, vol. 193, Academic Press, San Diego 1990, p. 295–305.

[227] J. Yao, M. Dey, S. J. Pastor, C. L. Wilkins, *Anal. Chem.* **67** (1995) 3638.

[228] Y. Li, R. T. McIver, R. L. Hunter, *Anal. Chem.* **66** (1994) 2077.

[229] C. C. Stacey et al., *Rapid Commun. Mass Spectrom.* **8** (1994) 513.

[230] Y. Li, K. Tang, D. P. Little, H. Köster, R. L. Hunter, R. T. M. Jr, *Anal. Chem.* **68** (1996) 2090.

[231] E. A. Stemmler, M. V. Buchanan, G. B. Hurst, R. L. Hettich, *Anal. Chem.* **67** (1995) 2924.

[232] E. A. Stemmler, M. V. Buchanan, G. B. Hurst, R. L. Hettich, *Anal. Chem.* **66** (1994) 1274.

[233] H. Yoshida, R. L. Buchanan, *Radiat. Res.* **139** (1994) 271.
[234] R.-D. Wilken, R. Falter, *Appl. Organomet. Chem.* **12** (1998) 551.
[235] L. Ebdon, A. Fischer, N. B. Roberts, M. Yaqoob, *Appl. Organomet. Chem.* **13** (1999) 183.
[236] Y. Inoue et al., *Appl. Organomet. Chem.* **13** (1999) 81.
[237] C. Siethoff, B. Pongraz, M. Linscheid, *Proc. 44th ASMS Conference on Mass Spectrometry and Allied Topics* (Portland, USA), (1996) 987.
[238] I. Jardine in J. Villafranca (ed.): *Current Research in Protein Chemistry,* Academic Press, San Diego 1991.
[239] N. Blaszkiewicz, G. Baumhoer, B. Neidhart, R. Ohlendorf, M. Linscheid, *J. Chrom.* **439** (1988) 109–119.
[240] K. W. M. Siu, G. J. Gardner, S. S. Berman, *Rapid Commun. Mass Spectrom.* **2** (1988) 201–204.
[241] Y. Morita, K. Takatera, T. Watanabe, *Seisan Kenkyu* **43** (1991) 367.
[242] B. Michalke, P. Schramel, *Electrophoresis* **20** (1999) 2547.
[243] C. Siethoff, I. Feldmann, N. Jakubowski, M. Linscheid, *J. Mass Spectrom.* **34** (1999) 421.
[244] I. Tolosa, J. M. Bayona, J. Albaigés, L. F. Alencastro, J. Taradellas, *Fresenius J. Anal. Chem.* **339** (1991) 646–653.
[245] N. Jakubowski, D. Stuewer, W. Vieth, *Anal. Chem.* **59** (1987) 1825.
[246] N. Jakubowski, D. Stuewer, W. Vieth, *Mikrochim. Acta* **1** (1987) 302.
[247] A. P. Mykytiuk, P. Semeniuk, S. Berman, *Spectrochim. Acta Rev.* **13** (1990) 1.
[248] D. Stüwer in G. Holland, A. Eaton (eds.): *Application of Plasma Source Mass Spectrometry,* Royal Society of Chemistry, Cambridge 1990, p. 1 ff.
[249] F. L. King, J. Teng, R. E. Steiner, *J. Mass Spectrom.* **30** (1995) 1061.
[250] J. R. d. Laeter, *Mass Spectrom. Rev.* **17** (1998) 97.
[251] H. Oechsner, W. Gerhard, *Phys. Lett.* **40A** (1972) 211.
[252] N. Jakubowski, D. Stuewer in D. Littlejohn, D. T. Burns (eds.): *Rev. Anal. Chem., Euroanalysis VIII,* Royal Society of Chemistry, (1994) 121–144.

21. Atomic Spectroscopy

José A. C. Broekaert, Universität Leipzig, Institut für Analytische Chemie, Leipzig, Germany

E. Hywel Evans, University of Plymouth, Plymouth, United Kingdom

21.	Atomic Spectroscopy	627
21.1.	Introduction	628
21.2.	Basic Principles	629
21.2.1.	Atomic Structure	629
21.2.2.	Plasmas	631
21.2.3.	Emission and Absorption of Radiation	631
21.2.4.	Ionization	635
21.2.5.	Dissociation	637
21.2.6.	Sources and Atom Cells	638
21.2.7.	Analytical Atomic Spectrometry	641
21.3.	Spectrometric Instrumentation	642
21.3.1.	Figures of Merit of an Analytical Method	642
21.3.2.	Optical Spectrometers	644
21.3.2.1.	Optical Systems	645
21.3.2.2.	Detectors	649
21.3.2.3.	Nondispersive Spectrometers	652
21.3.3.	Mass Spectrometers	654
21.3.3.1.	Types of Mass Spectrometer	654
21.3.3.2.	Ion Detection	656
21.3.3.3.	Ion Extraction	657
21.3.4.	Data Acquisition and Processing	658
21.4.	Sample Introduction Devices	660
21.4.1.	Pneumatic Nebulization	660
21.4.2.	Ultrasonic Nebulization	663
21.4.3.	Hydride Generation	663
21.4.4.	Electrothermal Evaporation	664
21.4.4.1.	The Volatilization Process	664
21.4.4.2.	Types of Electrothermal Device	665
21.4.4.3.	Temperature Programming	666
21.4.4.4.	Analytical Performance	667
21.4.5.	Direct Solid Sampling	667
21.4.5.1.	Thermal Methods	667
21.4.5.2.	Slurry Atomization	668
21.4.5.3.	Arc and Spark Ablation	668
21.4.5.4.	Laser Ablation	669
21.4.6.	Cathodic Sputtering	670
21.5.	Atomic Absorption Spectrometry	673
21.5.1.	Principles	673
21.5.2.	Spectrometers	674
21.5.2.1.	Spectrometer Details	674
21.5.2.2.	Primary Sources	675
21.5.3.	Flame Atomic Absorption	676
21.5.3.1.	Flames and Burners	676
21.5.3.2.	Nebulizers	677
21.5.3.3.	Figures of Merit	677
21.5.4.	Electrothermal Atomic Absorption	678
21.5.4.1.	Atomizers	679
21.5.4.2.	Thermochemistry	679
21.5.4.3.	Figures of Merit	680
21.5.5.	Special Techniques	681
21.5.5.1.	Hydride and Cold Vapor Techniques	681
21.5.5.2.	Direct Solid Sampling	682
21.5.5.3.	Indirect Determinations	682
21.5.6.	Background Correction	682
21.5.6.1.	Deuterium Lamp Technique	683
21.5.6.2.	Zeeman Effect Technique	684
21.5.6.3.	Smith–Hieftje Technique	685
21.5.6.4.	Coherent Forward Scattering	686
21.5.7.	Fields of Application	686
21.6.	Atomic Emission Spectrometry	688
21.6.1.	Principles	688
21.6.2.	Spectrometers	690
21.6.3.	Flame Emission	691
21.6.4.	Arcs and Sparks	691
21.6.5.	Plasma Sources	694
21.6.5.1.	Direct Current Plasmas	694
21.6.5.2.	Inductively Coupled Plasmas	695
21.6.5.3.	Microwave Plasmas	699
21.6.6.	Glow Discharges	700
21.6.6.1.	Hollow Cathodes	701
21.6.6.2.	Furnace Emission Spectrometry	701
21.6.6.3.	Glow Discharges with Flat Cathodes	701
21.6.6.4.	New Developments	702
21.6.7.	Laser Sources	703
21.7.	Plasma Mass Spectrometry	704
21.7.1.	Principles	704
21.7.1.1.	Instrumentation	704
21.7.1.2.	Analytical Features	704
21.7.1.3.	Applications	710
21.7.2.	Glow Discharge Mass Spectrometry	710

21.7.2.1.	Instrumentation	711	21.9. Laser-Enhanced Ionization Spectrometry	716
21.7.2.2.	Analytical Performance	712		
21.7.2.3.	Applications	712	21.10. Comparison With Other Methods	718
21.8.	**Atomic Fluorescence Spectrometry**	713	21.10.1. Power of Detection	719
21.8.1.	Principles	714	21.10.2. Analytical Accuracy	720
21.8.2.	Instrumentation	714	21.10.3. Cost	720
21.8.3.	Analytical Performance	715	**21.11.** **References**	721

Symbols :

A	absorbance, $A = \log I_0/I$ with I_0 the intensity of the incident beam and I the intensity of the sorting beam
A_{qp}	Einstein transition probability for spontaneous emission of radiation corresponding with a transition from the higher level q to the lower level p (s^{-1})
B_{pq}	Einstein transition probability for absorption corresponding with a transition from the lower level p to the higher level q (s^{-1})
B_{qp}	Einstein transition probability for stimulated emission corresponding with a transition from the higher level q to the lower level p (s^{-1})
c	velocity of light in vacuum (3×10^{10} m/s)
c_L	detection limit in concentration units or in terms of absolute amounts of analyte
c_U	background equivalent concentration, calibration constant in optical atomic spectrometry for trace analysis
C	capacitance (Farad)
d_0	Sauter diameter; droplet diameter for an aerosol at which the volume-to-surface ratio of the particles equals the volume-to-surface ratio for the whole aerosol
D_{hkl}	lattice constant of crystal
f_k	focal length of collimator (m)
g_q	statistical weight of an atomic (molecular) energy level q
G	optical conductance
h	isotopic abundance in %
j	total internal quantum number
k	Boltzmann's constant (1.38×10^{-23} J/K)
k_{12}	velocity constant for collisional excitation (s^{-1})
k_{21}	velocity constant for collisional de-excitation (s^{-1})
l	orbital quantum number
L	inductance (Henry)
L	Lorentz unit ($e/4\pi m \cdot c^2$)
m_l	magnetic quantum number
M	magnetic quantum number
n	principle quantum number
$P(v)$	profile function of a line
q	sputtering rate (µg/s)
QE	quantum efficiency of a photocathode
R	Rydberg constant (109.677 cm^{-1})
R_0	theoretical resolving power of optical spectrometer
RSF	relative sensitivity factor in elemental mass spectrometry
s	spin quantum number
s_a	exit slit width (µm)
s_e	entrance slit width (µm)
s_{eff}	effective slit width of an optical spectrometer in µm; $s_{eff}^2 = s_e^2 + s_L^2 + s_0^2 + s_z^2$ with s_L and s_z the contributions of the physical width of the line measured and the optical aberrations, respectively (all in µm)
s_0	diffraction slit width (µm)
$s_r(c_x)$	relative standard deviation of the estimate
S	blackening of a photographic emulsion
S	sputtering yield in sputtered ions/incident ion
$S_{nj}(T)$	Saha function for neutrals n of an element j
S/N	signal-to-noise ratio
t_r	response time of a decay process for an excited level (s)
t_p	pulse duration (s)
U	potential (V)
v'	wavenumber (cm^{-1})
Δv	frequency broadening contribution (cm^{-1})
w	penetration rate (µm/s)
Z	charge of nucleus or ion (in elementary charge units)
Z_{ij}	partition function of an element j in the ith ionization state
α_D	divergency of laser beam with $\alpha_D = \lambda/D$ with λ the wavelength and D the beam diameter
γ	contrast of photographic emulsion
$\Delta\lambda_L$	physical halfwidth of a spectral line (nm or pm)
$\theta(V)$	amplification of photomultiplier
$\lambda/\Delta\lambda$	practical resolving power
ϱ_v	radiant density of a spectral feature with frequency v
$\sigma(v)$	collision cross section for species with velocity v
φ	radiant flux $\varphi = \tau B G$ with τ the transmittance and B the radiant density of the source in W m^{-2} sterad^{-1}

21.1. Introduction

Atomic spectroscopy is the oldest principle of instrumental elemental analysis, going back to the work of BUNSEN and KIRCHHOFF in the mid-nineteenth century [1]. The optical radiation emitted from flames is characteristic of the elements present in the flame itself or introduced by various means. This discovery depended on the availability of dispersing media, such as prisms, which spectrally resolved the radiation to produce the line spectra of the elements. It was also found that radiation with the same wavelength as the emitted lines is absorbed by a cold vapor of the same element. This observation is the basis for atomic emission and absorption spectrometry.

Flames proved to be suitable sources for the analysis of liquids, but the need for direct chemical analysis of solids soon arose, including the need for rapid analysis of more than one element in a solid sample. The development of electrical discharges fostered the growth of atomic spectrometry; arc and spark sources proved useful for analyte ablation and excitation. Low-pressure discharges proved to be powerful radiation sources (a similar development has been observed more recently, with the availability of laser sources).

The limitations stemming from the restricted temperatures of flames led to the development of high-temperature plasma sources for atomic emission spectrometry. The development of high-frequency inductively coupled plasmas and microwave plasmas has led to the widespread use of these methods for routine analytical work.

Parallel with these developments, there have been advances in spectrometer design, from the spectroscope, through photographic spectrographs, sequential spectrometers, and simultaneous multichannel spectrometers with solid state detectors.

The use of flames as atomizers for atomic absorption spectrometry (AAS) has also made its impact on analytical methodology, largely as the result of work by WALSH [2]. L'VOV and MASSMANN [3], [4] introduced the graphite furnace atomizer, now widely available in commercially available AAS with electrothermal atomization, especially for ultratrace analysis.

The sources used for optical atomic spectrometry are also powerful sources for elemental mass spectrometry, from the classical spark source mass spectrometer to present-day plasma mass spectrometric methods such as glow discharge and inductively coupled plasma mass spectrometry.

21.2. Basic Principles

21.2.1. Atomic Structure

The basic processes in atomic spectrometry involve the outer electrons of atoms. Accordingly, the possibilities and limitations of atomic spectroscopy depend on the theory of atomic structure.

On the other hand, the availability of optical spectra was decisive in the development of the theory of atomic structure and even for the discovery of a number of elements.

In 1885, BALMER reported that for a series of atomic emission lines of hydrogen, a simple relationship existed between the wavelengths:

$$\lambda = k \frac{n^2}{n^2 - 4} \tag{1}$$

where $n = 2, 3, 4, \ldots$ for the lines $H_\alpha, H_\beta, H_\gamma, \ldots$

Equation (1) can be written in wavenumbers:

$$v' = \frac{1}{\lambda} = R\left(\frac{1}{2^2} - \frac{1}{n^2}\right) \tag{2}$$

where v' is the wavenumber (cm^{-1}) and R the Rydberg constant (109.677 cm^{-1}). The wavenumbers of all series in the hydrogen spectrum are given by:

$$v' = R\left(\frac{1}{n_1^2} - \frac{1}{n_2^2}\right) \tag{3}$$

where n_2 is a series of numbers $> n_1$, with $n_1 = 1, 2, 3, 4, \ldots$ for the Lyman, Balmer, Paschen, and Brackett series, respectively.

Rydberg extended the Balmer formula to:

$$v' = RZ^2\left(\frac{1}{n_1^2} - \frac{1}{n_2^2}\right) \tag{4}$$

where Z is the effective charge of the atomic nucleus, allowing calculation of wavelengths for elements other than hydrogen. The wavenumber of each line in an atomic spectrum can thus be calculated from a formula involving two positive numbers, called spectral terms.

The significance of the spectral terms emerges from Bohr's theory of atomic structure, in which the atom has a number of discrete energy levels related to the orbits of the electrons around the nucleus. As long as an electron is in a defined orbit no energy is emitted, but when a change in orbit occurs, another energy level is reached and the excess energy is emitted in the form of electromagnetic radiation. The wavelength is given by Planck's law:

$$E = hv = hc/\lambda \tag{5}$$

Here $h = 6.626 \times 10^{-34}$ J·s, v is the frequency (s^{-1}), $c = 3 \times 10^{10}$ cm/s is the velocity of light, and λ is the wavelength (cm). Accordingly:

$$\bar{\nu} = \frac{1}{\lambda} = \frac{E}{hc} = \frac{E_1}{hc} - \frac{E_2}{hc} = T_1 - T_2 \quad (6)$$

where T_1 and T_2 are the Bohr energy levels, the complexity of emission spectra reflects the complex structure of atomic energy levels.

For an atom with nuclear charge Z and one valence electron, the energy of this electron is given by:

$$E = -\frac{2\pi Z^2 e^4 \mu}{n^2 h^2} \quad (7)$$

where $\mu = mM/(m+M)$, with m the mass of the electron and M the mass of the nucleus; n is the principal quantum number ($n = 1, 2, 3, \ldots$) on which the order of the energy levels depends. For the orbital angular momentum L:

$$\|L\| = \frac{h}{2\pi}\sqrt{l(l+1)} \quad (8)$$

l is the orbital quantum number and has values of: $0, 1, 2, \ldots, (n-1)$; $l = 0$ for a circular orbit, $l = 1, 2, \ldots$ for elliptical orbits.

The possible orientations of the elliptical orbits with respect to an external electric or magnetic field:

$$L_z = \frac{h}{2\pi} m_l \quad (9)$$

where L_z is the component of the orbital angular momentum along the field axis and $m_l = \pm 1, \pm(l-1), \ldots, 0$ is the magnetic quantum number; for each value of l it has $(2l+1)$ values.

When a spectral line source is subject to a magnetic field, the spectral lines display hyperfine structure (Zeeman effect). In order to explain hyperfine structure it is postulated that the electron rotates on its axis with spin angular momentum S:

$$\|S\| = \frac{h}{2\pi}\sqrt{S(S+1)} \quad (10)$$

The spin quantum number m_s determines the angles between the axis of rotation and the external field as:

$$s_z = \frac{h}{2\pi} m_s \quad (11)$$

where $m_2 = \pm 1/2$.

The orbital and spin angular momenta determine the total angular momentum of the electron J:

$$\vec{J} = \vec{L} + \vec{S} \text{ with } \|J\| = \frac{h}{2\pi}\sqrt{j(j+1)} \quad (12)$$

where $j = l \pm s$ is the total internal quantum number.

Atomic spectral terms differ in their electron energies and can be characterized by the quantum numbers through the term symbols:

$$n^m l_j \quad (13)$$

Here $l = 0, 1, 2, \ldots$ and the corresponding terms are given the symbols s (sharp), p (principal), d (diffuse), f (fundamental), etc., relating originally to the nature of different types of spectral lines; n is the principal quantum number, m is the multiplicity ($m = 2s \pm 1$), and j is the total internal quantum number. The energy levels of each element can be given in a term scheme, in which is also indicated which transitions between energy levels are allowed and which forbidden. This is reflected by the so-called selection rules; only those transitions are allowed for which Δn has integral values and at the same time $\Delta l = \pm 1$, $\Delta j = 0$ or ± 1, and $\Delta s = 0$. The terms for an atom with one outer (valence) electron can easily be found, e.g., for Na ($1s^2 2s^2 2p^6 3s^1$), in the ground state: $3^2 S_{1/2}$ [$l = 0$ (s), $m = 2(1/2) + 1 = 2$ ($s = 1/2$), and $j = 1/2, (j = |l \pm s|)$]. When the 3s electron is promoted to the 3p level, the term symbol for the excited state is: $3^2 P_{1/2, 3/2}$ ($l = 1$ (p), $m = 2(1/2) + 1 = 2$, and $j = 1/2, 3/2$).

The term schemes of the elements are well documented in the work of GROTRIAN [5]. The term scheme for Na is shown in Figure 1.

For atoms with more than one valence electron, Russell–Saunders (L–S) coupling applies. The orbital momenta of all electrons have to be coupled to the total orbital momentum, like the spin momentum. The fact that each electron has a unique set of quantum numbers is known as the Pauli exclusion principle. The total quantum number L is obtained as $L = l$, $S = s$, and $J = L - S, \ldots, L + S$. The term symbol accordingly becomes:

$$^M L_J \quad (14)$$

For the case Mg ($1s^2 2s^2 2p^6 3s^2$), the ground state is $3^1 S_0$ ($L = 0$ as $l_1 = 0$ and $l_2 = 0$, $S = 0$ as $s_1 = 1/2$ and $s_2 = -1/2$, and $J = 0$ as both L and $S = 0$). The first excited state ($1s^2 2s^2 2p^6 3s 3p$) is characterized by the terms: $3^1 P_1$ ($L = 1$ as $l_1 = 0$ and $l_2 = 1$, $S = 0$ as $s_1 = 1/2$ and $s_2 = -1/2$, and $J = |L \pm S| = 1$), but also

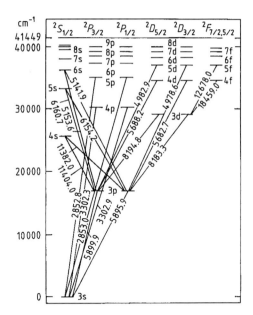

Figure 1. Atomic energy level diagram for the sodium atom [5]

3^3P_2, 3^3P_1, and 3^3P_0 (as for the parallel spins $s_1 = 1/2$ and $s_2 = 1/2$, $S = 1$, and further $J = 0, 1, 2$).

As the number of electrons increases, the coupling becomes more complex, increasing the number of spectral terms and thus the number of lines in the spectrum.

21.2.2. Plasmas [6]

Partially ionized gases are usually referred to as plasmas. They contain molecules, radicals, atoms, ions, and free electrons, and result from the interaction of energy with matter in the gaseous state. Just as for atoms, the radicals, molecules, and ions in a plasma can be in their ground or excited states and radiation can be emitted or absorbed when transitions occur from one state to another. The wavelengths of spectral lines can be obtained from Planck's law, but their intensities depend on the number densities of the species and states involved.

Energy transfer in a plasma results from nonradiative processes (involving collisions) as well as from radiative processes (emission, absorption, and fluorescence). The efficiency of collision processes is described by the cross section $\sigma(v)$. It reflects the change in momentum of a particle with mass m and velocity v when it collides with a particle with mass M. Apart from the cross section, the velocity distribution for a given species affects energy transfer in a plasma.

For a Maxwell distribution, the velocity distribution is given by:

$$\frac{dn}{n} = \frac{2}{\sqrt{\pi}} \sqrt{u'} \exp(-u') du' \qquad (15)$$

For the so-called Druyvenstein distribution which pertains at low pressure:

$$\frac{dn}{n} = 1.039 \sqrt{u'} \exp(-0.548 u'^2) du' \qquad (16)$$

where $u' = E/kT$, E is the mean energy of the particles, and T the absolute temperature.

For a plasma contained in a closed system and in thermal equilibrium, the population of the excited levels for all particles is given by Boltzmann's law:

$$\frac{n_q}{n_0} = \frac{g_q}{g_0} \exp(-E_q/kT) \qquad (17)$$

where n_q is the number density of particles in the excited state, n_0 is the number density of particles in the ground state, g_q and g_0 are the statistical weights of the corresponding levels, E_q is the excitation energy of the state q, k is Boltzmann's constant (1.38×10^{-23} J/K), and T is the absolute temperature. Equation (17) can be transformed when n_q is expressed as a function of the total number density of the species. Then:

$$\frac{n_q}{n} = \frac{g_q \exp(-E_q/kT)}{\sum_m g_m \exp(-E_m/kT)} \qquad (18)$$

as $n = \sum_m n_m$. The sum $Z_m = \sum_m g_m \exp(-E_m/kT)$ is the partition function. It is a function of temperature and its coefficients for a large number of neutral and ionized species are listed in the literature, e.g., [7]. When E_q is expressed in electron volts, Equation (18) can be written:

$$\log n_{aq} = \log n_a + \log n_q - (5040)/TV_q - \log Z \qquad (19)$$

21.2.3. Emission and Absorption of Radiation [8]

In a steady-state plasma, the number of particles leaving an energy level per unit time equals the number returning to this level. The most important energy exchange processes in a plasma are:

1) Collisions where atoms are excited to a higher level by collision with energetic neutrals (collisions of the first kind)
2) Collisions where excited neutrals lose energy through collisions without emission of radiation (collisions of the second kind)
3) Excitation by collision with electrons
4) De-excitation where energy is transferred to electrons
5) Excitation of atoms by the absorption of radiation
6) De-excitation of atoms by spontaneous or stimulated emission

When n is the number density of one type of particle and N is that of a second species, present in excess ($n << N$), the following equilibria are to be considered:

$$\alpha N n_0 = \beta N n_q \qquad (20)$$

$$\alpha_e n_e n_0 = \beta_e n_e n_q \qquad (21)$$

$$B' \varrho_v n_0 = (A + B\varrho_v) n_q \qquad (22)$$

where n_e is the electron number density; A, B, and B' are the Einstein transition probabilities for spontaneous emission, stimulated emission, and absorption, respectively; and α_e, α, β_e, and β are functions of the cross sections for the respective processes, as well as of the velocity distribution of the particles involved. The radiation density at frequency v is ϱ_v. When the system is in thermodynamic equilibrium, the neutrals and electrons have the same Maxwell velocity distribution and at temperature T:

$$n_q/n_0 = \alpha/\beta = \alpha_e/\beta_e = B'/(A/\varrho_v + B)$$
$$= \frac{g_q}{g_0} \exp(-E_q/kT) \qquad (23)$$

Thus, each process is in equilibrium with its inverse and the Boltzmann distribution of each state is maintained by collisions of the first and second kind, including collisions with electrons.

In a real radiation source, emission and absorption of radiation have to be considered as well, but contribute little to the energy balance. The system is in so-called local thermal equilibrium (LTE) and:

$$\alpha N n_0 + \alpha_e n_e n_0 + B' \varrho_v n_0$$

$$= \beta N n_q + \beta_e n_e n_q + (A + B\varrho_v) n_q \qquad (24)$$

from which n_q/n_0 can be calculated. The population of excited states is determined by the excitation processes in the radiation source, as reflected by the coefficients in Equation (24). For a d.c. arc, $\alpha N >> \alpha_e n_e + B' \varrho_v$ and $\beta N >> \beta_e n_e + (A + B\varrho_v)$. This leads to:

$$\frac{n_q}{n_0} = \frac{\alpha}{\beta} = \frac{g_q}{g_0} \exp(-E_q/kT) \qquad (25)$$

As the radiation density is low, the d.c. arc plasma can be considered to be in thermal equilibrium. The simplification which leads to Equation (24) does not apply for low-pressure discharges, where collisions with electrons are very important as are radiative processes; moreover, the velocity distributions are described by the Druyvensteyn equation. These sources are not in thermal equilibrium.

Excited states are prone to decay because of their high energy. This can take place by collisions with molecules, atoms, electrons, or ions, or by emission of electromagnetic radiation. In the latter case, the wavelength is given by Planck's law. For levels q and p, the number of spontaneous transitions per unit time is given by:

$$-dN_q/dt = A_{qp} N_q \qquad (26)$$

where A_{qp} is the Einstein coefficient for spontaneous emission (s^{-1}). When several transitions can start from level q, Equation (26) becomes:

$$-dN_q/dt = N_q \sum_p A_{qp} = N_q v_q \qquad (27)$$

where v_q is the lifetime of excited state q. If decay can take place by an allowed radiative transition, the lifetime is of the order of 10^{-8} s. When no radiative transitions are allowed levels are metastable (e.g., Ar 11.5 and 11.7 eV), which can decay only by collisions. In low-pressure discharges such levels may have very long lifetimes (up to 10^{-1} s).

For absorption of electromagnetic radiation with frequency v_{qp} and radiation density ϱ_v, the number density of N_q increases as:

$$dN_q/dt = B_{qp} N_q v_\varrho \qquad (28)$$

For stimulated radiation, atoms in excited state q decay only when they interact with radiation of wavelength λ_{qp} and:

$$-dN_q/dt = B_{qp}N_q\varrho_v \tag{29}$$

For thermal equilibrium:

$$g_q B_{qp} = g_p B_{pq} \tag{30}$$

where g_p and g_q are the degeneracies of the respective levels.

The intensity I_{qp} of an emitted spectral line (a) is proportional to the number density of atoms in state q:

$$I_{qp} = A_{qp} n_{aq} h v_{qp} \tag{31}$$

or after substitution of n_{aq} (n_q for the atomic species according to Eq. 18):

$$I_{qp} = A_{qp} h v_{qp} n_a (g_q/Z_a) \exp(-E_q/kT) \tag{32}$$

where T is the excitation temperature which can be determined from the intensity ratio for two lines (a and b) of the same ionization state of an element as:

$$T = [5040(V_a - V_b)]/\{\log[(gA)_a/(gA)_b] - \log(\lambda_a/\lambda_b) - \log(I_a/I_b)\} \tag{33}$$

Often the line pair Zn 307.206/Zn 307.59 nm is used. This pair is very suitable because ionization of zinc is low as a result of its relatively high ionization energy, the wavelengths are close to each other, which minimizes errors introduced by changes in the spectral response of the detector, and the ratio of the gA values is well known.

The excitation temperatures can also be determined from the slope of the plot $\ln[I_{qp}/(g_q A_{qp} v_{qp})]$ or $\ln(I_{qp} \lambda/gA_{qp})$ against E_q, which is $-1/kT$. The λ/gA values for a large number of elements and lines are available [8]. Spectroscopic measurement of temperatures from line intensities may be hindered by deviations from ideal thermodynamic behavior in real radiation sources, and by inaccuracies of transition probability estimates. Determination of excitation temperatures in spatially inhomogeneous plasmas is treated extensively by BOUMANS [9].

According to classical dispersion theory, the relation between absorption and the number density of the absorbing atoms is given by:

$$\int K_v \, dv = \frac{\pi e^2}{mc} N_v f \tag{34}$$

where K_v is the absorption coefficient at frequency v, m is the mass and e the charge of the electron, c is the velocity of light, N_v is the density of atoms with frequency between v and $v+dv$ (practically the same as N), and f is the oscillator strength. This relation applies strictly to monochromatic radiation, so use of a primary source which emits very narrow lines is advantageous. The relation between absorption A and the concentration of the absorbing atoms in an atomizer is given by the Lambert–Beer law. If I_0 is the intensity of the incident radiation, l the absorption path length, and I the intensity of the transmitted radiation, the change in intensity dI resulting from absorption within the absorption path length dl is:

$$-dI = k I_0 c \, dl \tag{35}$$

or

$$\int_{t_0}^{t} -dt/t = kc \int_0^l dl \tag{36}$$

$A = \log(I_0/I) = \log(1/T)$, in which A is the absorbance and T the transmittance. Accordingly, Equation (36) becomes:

$$A = kcl \tag{37}$$

The absorbances are additive. The Lambert–Beer law, however, is valid only within a restricted concentration range; in atomic absorption spectrometry deviations from linearity are common.

Line Broadening. Atomic spectral lines have a physical width resulting from several broadening mechanisms [10]. The *natural width* of a spectral line results from the finite lifetime of an excited state, τ. The corresponding half-width in terms of frequency is:

$$\Delta v_N = 1/(2\pi\tau) \tag{38}$$

This results for most spectral lines in a halfwidth of the order of 10^{-2} pm.

The *Doppler spectral width* results from the velocity component of the emitting atoms in the observation direction. The corresponding half-width is:

$$\Delta v_D = \left[2\sqrt{(\ln 2)}/c\right] v_0 \sqrt{[2(RT)/M]} \tag{39}$$

where c is the velocity of light, v_0 is the frequency of the line maximum, R is the gas constant, and M

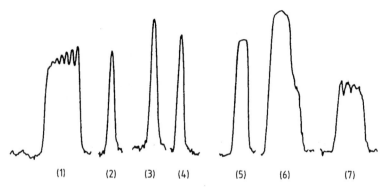

Figure 2. Profiles of some rare earth atomic emission lines in inductively coupled plasma atomic emission spectrometry (photographic measurements obtained with a high-resolution grating spectrograph; theoretical resolving power 460 000 [12]

the atomic mass. Doppler broadening thus depends strongly on temperature. Accordingly, it is often referred to as temperature broadening and reflects the kinetic energy of the radiating atoms, ions, or molecules. For the Ca 422.6 nm line in the case of a hollow cathode discharge at pressures of the order of 0.1 kPa, the Doppler broadening at 300 K is 0.8 pm whereas at 2000 K it is 2 pm [11].

Lorentz or *pressure broadening* results from the interaction between the emitting atoms and atoms of other elements. The half-width is:

$$\Delta v_L = (2/\pi)\sigma_L^2 N \sqrt{\left[2\pi RT\left(\frac{1}{M_1} + \frac{1}{M_2}\right)\right]} \quad (40)$$

where M_1 and M_2 are the atomic masses, N is the concentration of the foreign atoms, and σ_L is their cross section. Pressure broadening is low in low-pressure discharges. For the Ca 422.6 nm line, this type of broadening at 300 K and 0.9 kPa is only 0.02 pm. At atmospheric pressure, however, this type of line broadening is predominant.

Isotopic and *hyperfine structure*, *resonance broadening* (resulting from interaction between radiating and nonradiating atoms of the same species), and *Stark broadening* (resulting from interaction with electric fields), contribute to the widths of spectral lines.

Natural and Lorentz broadening have a Lorentzian distribution:

$$I(v) = I_0 / \left\{1 + [2(v - v_0)/\Delta v]^2\right\} \quad (41)$$

Doppler broadening has a Gaussian distribution:

$$I(v) = I_0 \exp -\left\{[2(v - v_0)/2\Delta v_D]\sqrt{(\ln 2)}\right\} \quad (42)$$

The combination of both types of profile (normally both pressure and temperature broadening are important) results in a so-called Voigt profile:

$$V(\alpha,\omega) = \alpha/\pi \int_{-\infty}^{+\infty} \left\{[\exp(-y^2 \, dy)]/[\alpha^2 + (\omega - y)^2]\right\}$$
(43)

where $\omega = 2(v - v_0)/\Delta v_D \sqrt{(\ln 2)}$ and $\alpha = (\Delta v_L/\Delta v_D)\sqrt{(\ln 2)}$, when the contribution of the natural width is neglected. From the widths of spectral lines, in most cases between 1 and 20 pm, and knowledge of the distribution functions, the contributions of the different broadening processes can be calculated by deconvolution methods. This allows calculation of the so-called gas temperature from the Doppler broadening component, giving an estimate of the kinetic energy of neutrals and ions.

The profiles of spectral lines from plasma sources at atmospheric pressure are illustrated by the high-resolution spectra of a number of rare-earth elements in an inductively coupled plasma (ICP) (Fig. 2) [12].

Self-Absorption. The radiation emitted by the source is absorbed by ground state atoms of the same species. This phenomenon is known as self-absorption [13]. As the chance that an absorbed photon is re-emitted is less than unity, the observed radiation is reduced. When I_0 is the intensity emitted at the line maximum and $P_E(v)$ is the profile function, the intensity distribution for an emission line is $I_0 P_E(v)$ and the intensity observed after the radiation passes through a layer with number density of absorbing atoms of n_A is:

$$I(v) = I_0 P_E(v) \exp[-\varrho P_A(v)/P_A(v_0)] \tag{44}$$

where v_0 is the frequency at the line center, $P_A(v)$ is the absorption profile function, $P_A(v_0)$ is its value at the line center, and ϱ is an absorption parameter:

$$\varrho \sim B P_A(v) n_A \tag{45}$$

ϱ increases with the transition probability for absorption and thus is larger for resonance lines which involve transitions to the ground state. It is also larger in sources with high analyte number density n_A. As the absorption maximum is in the center of the line, self-absorption always leads to flatter line profiles. When a minimum occurs in the absorption profile, the line undergoes so-called self-reversal and $\varrho > 1$. Self-reversal only occurs when there is a strong temperature gradient in the radiation source and when the analyte number densities in both the hot and the cooler zones of the plasma are high.

21.2.4. Ionization

If sufficient energy is transferred to a plasma, atoms may be ionized. This depends on the temperature of the plasma, but also on the ionization energy of the elements. As ions have discrete energy levels between which transitions are possible, ionic spectra are also important when considering the emission of radiation by a plasma. The ionization of atoms a of the element j into ions i is an equilibrium:

$$n_{aj} \rightleftarrows n_{ij} + n_e \tag{46}$$

and the equilibrium constant $S_{nj}(T)$, known as the Saha constant, is:

$$S_{nj}(T) = (n_{ij} n_e)/n_{aj} \tag{47}$$

The degree of ionization α_j for element j is:

$$\alpha_j = n_{ij}/n_j = n_{ij}/(n_{aj} + n_{ij}) \tag{48}$$

where n_{aj} and n_{ij} are the concentrations of atoms and ions and can be expressed as a function of the total number of atoms n_j by:

$$n_{aj} = (1 - \alpha_j) n_j$$

and

$$n_{ij} = \alpha_j n_j \tag{49}$$

In the notation used by BOUMANS [9], the intensity of an atom line can be written:

$$I_{qp} = A_{qp} h v_{qp} (g_q/Z_{aj})(1 - \alpha_j) n_j \exp(-E_q/kT) \tag{50}$$

and the intensity of an ion line is:

$$I_{qp}^+ = A_{qp}^+ h v_{qp}^+ (g_q^+/Z_{ij}) \alpha_j n_j \exp(-E_q^+/kT) \tag{51}$$

These expressions for the intensities contain three factors which depend on temperature; i.e., the degree of ionization, the Boltzmann factor, and the partition functions. The degree of ionization can be written as a function of the electron number density and the Saha function:

$$[\alpha_j/(1 - \alpha_j)] n_e = S_{nj}(T) \tag{52}$$

However, the latter is also given by the Saha equation. When the Saha function is expressed in partial pressures, the Saha equation is:

$$S_{pj}(T) = [p_{ij} p_e]/[p_{aj}] = \frac{\left[(2\pi m)^{3/2} (kT)^{5/2}\right]}{h^3 (2Z_{ij}/Z_{aj}) [\exp(-E_{ij}/kT)]} \tag{53}$$

The factor 2 is the statistical weight of the free electron (two spin orientations), $k = 1.38 \times 10^{-23}$ J/K, the mass of the electron $m = 9.11 \times 10^{-28}$ g, $h = 6.62 \times 10^{-34}$ J·s, and 1 eV = 1.6×10^{-19} J. This leads to the following expression for the Saha equation:

$$\log S_{pj} = \frac{5}{2} \log \frac{T - 5040}{T} V_{ij} + \log \frac{Z_{ij}}{Z_{aj}} - 6.18 \tag{54}$$

where V_{ij} is the ionization energy (in eV).

The Saha equation is valid only for a plasma in local thermal equilibrium; then the temperature in the equation is the ionization temperature. When this condition is not fulfilled, the equilibrium between the different states of ionization is given by the so-called Corona equation [14].

Accordingly, the degree of ionization in a plasma can be determined from the intensity relation of atom and ion lines of the same element:

$$\log[(\alpha_j)/(1 - \alpha_j)] = \log\left(I_{qp}^+/I_{qp}\right) - \log\left[\left(g_q^+ A_{qp}^+ v_{qp}^+\right)/(g_q A_{qp} v_{qp})\right] + (5040/T)(V^+ - V_q) + \log(Z_{ij}/Z_{aj}) \tag{55}$$

This method can only be applied for a plasma in local thermal equilibrium at known temperature. The partition functions Z_{aj} and Z_{ij} for the atom and ion species are functions of the temperature, and their coefficients have been calculated for many elements [7]. The accuracy of the gA values and of the temperatures affects the accuracy of the degree of ionization. The line pairs Mg II 279.6 nm/Mg I 278.0 nm, and Mg II 279.6 nm/Mg I 285.2 nm are often used for determinations of the degree of ionization of an element in a plasma.

Once α_j is known, the electron pressure can be determined. From

$$\log\left[\alpha_j/(1-\alpha_j)\right] = \log\left[S_{pj}(T)/p_e\right] \qquad (56)$$

one can calculate:

$$\log p_e = -\log\left[\alpha_j/(1-\alpha_j)\right] + \log S_{pj}(T) \qquad (57)$$

Taking into account Equations (54) and (55), this results in:

$$\log p_e = -\log\left(I_{qp}^+/I_{qp}\right) + \log\left(g_q^+ A_{qp}^+ v_{qp}^+/g_q A_{qp} v_{qp}\right)$$
$$- (5040/T)\left(V_{ij} + V_q^+ - V_q\right) + 5/2\,(\log T) - 6.18 \qquad (58)$$

This reflects the fact that the intensity ratio of the atom and ion lines of an element changes considerably with the electron pressure in the plasma, particularly for elements with low ionization energy, such as Na. This is analytically very important as it is the cause of so-called ionization interferences in classical d.c. arc emission spectrometry, atomic absorption, and plasma optical emission, as well as mass spectrometry.

When the plasma is not in LTE, the electron number densities cannot be determined from the Saha equation, but have to be derived directly from the Stark broadening of the H_β line or of a suitable argon line. This contribution to broadening can be written [15]:

$$\delta\lambda = 2\left[1 + 1.75\alpha(1 - 0.75\varrho)\right]\omega \qquad (59)$$

where ϱ is the ratio of the distance between the ions and the Debye path length, ω is the broadening due to the interaction of the electrons ($\sim n_e$), and α is the contribution of the interaction with the quasi-static ions ($\sim n_e^{1/4}$). The value of $\delta\lambda$ can be calculated as a function of n_e. The electron number density can be determined directly from the widths of the Ar I 549.59 or Ar II 565.07 nm lines, which are mainly due to Stark broadening. The temperature can also be determined by combining measurements of the intensities of atom and ion lines of the same element:

$$\log n_i/n_a = -\log n_e + 3/2\,(\log T) - (5040/T)V_{ij} + \log\left(Z_{ij}/Z_{aj}\right) + 15.684 \qquad (60)$$

which can be combined with Equation (58).

From Equations (50) and (51), which give the intensity of a line, and from the Saha equation (54), it can be seen that, for each spectral line emitted by a plasma source, there is a temperature where its emission intensity is maximum. This is the so-called standard temperature. To a first approximation [16], it can be written:

$$T = (0.95V_{ij} \times 10^3)/\left[1 - 0.33\,(V_{ij}/10) + 0.37\log\,(V_{ij}/10) - 0.14\log P_e^*\right] \qquad (61)$$

where V_{ij} is the ionization energy and V_e the excitation energy; P^*_e is the electron pressure (in atm; 1 atm ≈ 101 kPa) and is a function of T and n_e. In the case under discussion, the standard temperature for a line of an element which is present as an impurity in a plasma (e.g., in a noble gas), dilution in the plasma must also be considered. The standard temperatures indicate which types of line will be optimally excited in a plasma of given temperature, electron pressure, and gas composition, and are thus important indications for line selection. Atom lines often have standard temperatures below 4000 K, especially when analyte dilution in the plasma is high, whereas ion lines often have values of 10 000 K. Atom and ion lines are called "soft" and "hard" lines, respectively.

Because of the interaction of free and bound electrons in a plasma continuous radiation is also emitted over a wide wavelength range. The intensity distribution for this background continuum can be expressed:

$$I_\nu \, d\nu = K n_e n_r r^2 / \left(T_e^{1/2}\right) \exp(-h\nu/kT_e) \cdot d\nu \, (\text{free} - \text{free})$$
$$+ K'(1/j^3) n_e n_Z Z^4 / \left(T_e^{3/2}\right) \cdot \exp(U_j - h\nu)/(kT_e)(\text{free} - \text{bound}) \quad (62)$$

where n_r is the concentration of atoms with charge r, Z is the nuclear charge, and U_j is the ionization energy of the level with quantum number j. This background continuum consists of so-called bremsstrahlung, caused by the interaction of free electrons with each other, and of recombination radiation, which is due to the interaction of free and bound electrons. The latter is especially important in the UV region. It is continuous radiation with spectral lines superimposed. As T_e is the electron temperature, absolute measurements of the background continuum emission in a plasma, e.g., for hydrogen, allow the determination of the electron temperature in a plasma, whether in local thermal equilibrium or not.

21.2.5. Dissociation

The dissociation of molecular plasma gases or analyte molecules within the radiation source is an equilibrium reaction. Accordingly, highly stable radicals or molecules are always present in a radiation source. They emit molecular bands which are superimposed on the atomic and ionic line spectra in the emission spectrum. Radicals and molecules may also give rise to cluster ions which may be detected in mass spectra. Common species in plasma gases are: CN, NH, NO, OH, and N_2 (or N_2^+). From the analytes, highly stable oxides may persist (e.g., AlO^+, TiO^+, YO^+). A thorough treatment of molecular spectra is available [17], [18].

Vibrational and Rotational Hyperfine Structures. Molecules or radicals have various electronic energy levels ($^1\Sigma$, $^2\Sigma$, $^2\Pi$, ...), which have vibrational fine structure ($v = 0, 1, 2, 3, ...$), which, in turn, have rotational hyperfine structure ($J = 0, 1, 2, 3, ...$). The total energy of a state may be written:

$$E_i = E_{el} + E_{vib} + E_{rot} \quad (63)$$

E_{el} is of the order of 1–10 eV, the energy difference between two vibrational levels of the same electronic state is of the order of 0.25 eV, and the separation of rotational levels is of the order of 0.005 eV. When the rotational levels considered belong to the same electronic level, the emitted radiation is in the infrared region. When they belong to different electronic levels, they occur in the UV or visible region. Transitions are characterized by the three quantum numbers of the states involved: n', v', j' and n'', v'', j''. All lines which originate from transitions between rotational levels belonging to different vibrational levels of two electronic states form the band: $n', v' \longrightarrow n'', v''$. For these band spectra the selection rule is $\Delta j = j' - j'' = \pm 1, 0$. Transitions for which $j'' = j' + 1$ give rise to the P-branch, $j'' = j' - 1$ to the R-branch, and $j' = j''$ to the Q-branch of the band. The line corresponding to $j' = j'' = 0$ is the zero line of the band. When $v' = v'' = 0$ it is also the zero line of the system. The difference between the wavenumber of a rotation line and the wavenumber of the zero line in the case of the P- and the R-branch is a function of the rotational quantum number j and the rotational constant B_v for which:

$$E_j/(hc) = B_v j(j+1) \quad (64)$$

The functional relation is quadratic known as the Fortrat parabola.

As in the case of atomic spectral lines, the intensity of a rotational line can be written:

$$I_{nm} = (N_m A_{nm} h \nu_{nm})/2\pi \quad (65)$$

where N_m is the population of the excited level and ν_{nm} the frequency of the emitted radiation. The transition probability for dipole radiation is:

$$A_{nm} = (64\pi^4 \nu_{nm}^3)/(3k)[1/(g_m)] \sum \|Rn_i m_k\|^2 \quad (66)$$

where i and k are degenerate levels of the upper (m) and the lower state (n), $Rn_i m_k$ is a matrix element of the electrical dipole moment and g_m is the statistical weight of the upper state; N_m is given by the Boltzmann equation:

$$N_m = N(g_m)/Z(T) \exp(-E_r/kT) \quad (67)$$

where E_r is the rotational energy of the excited electronic and vibrational level, given by:

$$E_r = hcB_{v'}J'(J'+1) \quad (68)$$

where $B_{v'}$ is the rotational constant and J' is the rotational quantum number of the upper state m.

For a $^2\Sigma_g - {}^2\Sigma_u$ transition, between a so-called "gerade" (g) and "ungerade" (u) level, the term $\Sigma |Rn_j m_k|^2 = J' + J'' + 1$, where J' and J'' are the rotational quantum numbers of the upper and lower state. Accordingly:

$$I_{nm} = (16\pi^3 c N v_{nm}^4)/3Z(T)(J' + J'' + 1) \exp(-hc B_v J'(J' + 1)/kT) \tag{69}$$

or

$$\ln[I_{nm}/(J' + J'' + 1)] = \ln[16\pi^3 c N v_{nm}^4]/[3Z(T)] - [hc B_v J'(J' + 1)]/kT \tag{70}$$

By plotting $\ln[I_{nm}/(J'+J''+1)]$ against $J'(J'+1)$ for a series of rotational lines a so-called rotational temperature can be determined, reflecting the kinetic energy of neutrals and ions in the plasma.

Molecular Bands. Spectral lines of molecular bands emitted by molecules and radicals in a plasma often interfere with atomic emission lines. In atomic absorption spectrometry, absorption by molecular bands arising from undissociated molecules in the atomizer may also interfere. Therefore, it is important to study the dissociation of molecular species in high-temperature radiation sources or atomizers. In plasma sources, band-emitting species may stem from the working gas. In this respect, N_2, N_2^+, CN, OH, and NH band emission are important. Undissociated sample and analyte species are also present in the plasma. Highly stable molecules such as Al_2O_3, La_2O_3, BaO, AlF_3, CaF_2, MgO, may be present in atomic spectrometric sources. It is important to know their dissociation as a function of plasma temperature and composition. This dependence can be described by a dissociation equation, similar to the Saha equation:

$$K_n = \left[(2\pi/h^2)(m_X m_Y/m_{XY})(kT)^{3/2}\right] \cdot (Z_X Z_Y/Z_{XY})[\exp(-E_d/kT)] \tag{71}$$

where:

$$K_n = (n_X n_Y)/n_{XY} \tag{72}$$

Thus, for a metal oxide the degree of dissociation can be calculated when the plasma temperature, the partial pressure of oxygen in the plasma, and the dissociation constant are known.

21.2.6. Sources and Atom Cells

For the purposes of the following discussion it is useful to distinguish between the terms "source" and "atom cell". Sources are the generators of the radiation (narrow line or continuum) which is eventually to be measured by the detector. Atom cells are the devices for producing the population of free atoms (or ions) from a sample. In atomic absorption and fluorescence spectrometry these two devices are separate. In atomic emission spectrometry the source is also the atom cell. In atomic mass spectrometry the terms source and atom cell are often used interchangeably to denote the "ion source".

In atomic spectrometry, the sample is introduced, by means of a sampling device, into a high-temperature source or atom cell (plasma, flame, etc.). Here, the sample is vaporized, e.g., by thermal evaporation or sputtering. It is important to supply as much energy as possible, so that the volatilization processes, which involve a physical or chemical equilibrium, result in complete atomization, irrespective of the state of aggregation, solid state structure, or chemical composition of the sample. This is very important, both to ensure maximum sensitivity and to minimize matrix interference in the analysis. The effectiveness of the volatilization processes involved, the plasma temperature, and the number densities of the various plasma components will all influence sample atomization.

Rotational temperatures are relevant for all processes in which molecules, radicals, and their dissociation products are involved. They can be obtained from the intensity distribution for the rotational lines in rotation–vibration spectra. Molecules such as OH or CN have often been used as "thermometers" (Eqs. 63–70) [19]–[21].

Table 1. Temperatures of sources and atom cells in atomic spectrometry

Source	Temperature, K				State
	Rotational, T_{rot}	Excitation, T_{ex}	Electron, T_e	Ion T_i	
Arc (d.c.)	5000	5000	5500	5000	LTE
Spark		20000	20000	20000	LTE
Inductively coupled plasma	4800	5000	6000	6000	≈ LTE
Microwave plasma	2000	4000	6000	6000	non-LTE
Low-pressure discharge	600	20000	30000	30000	non-LTE

The *gas temperature* is determined by the kinetic energy of neutral atoms and ions. It can be estimated from the Doppler broadening (Eq. 39). However, the contributions of Doppler and temperature broadening have to be separated, which involves the use of complicated deconvolution procedures [22].

The *electron temperature* is a measure of the kinetic energy of the electrons. It is relevant in the study of excitation and ionization by collisions with electrons, which is an important process for analyte signal generation. The electron temperature can be determined from the intensity of the recombination continuum or of the bremsstrahlung (Eq. 62).

The *excitation temperature* describing the population of the excited levels of atoms and ions is important in studies of the dependence of analyte line intensities on various plasma parameters in analytical emission spectrometry. It can be determined from the intensity ratio of two atomic emission lines of the same element and state of ionization (Eq. 33) or from plots of the appropriate function for many atomic emission lines against their excitation energies.

The *ionization temperature* is relevant for all phenomena involving equilibria between analyte atoms, ions, and free electrons in plasmas. In the case of thermal equilibrium, it occurs in the Saha equation (Eqs. 53, 54) and can be determined from the intensity ratio of an ion and an atom line of the same element. In all other cases, ionization temperatures can be determined from the n_e value obtained from Stark broadening (Eqs. 59, 60). Temperatures for the most important sources in atomic spectrometry are listed in Table 1.

In a plasma which is at least in local thermal equilibrium, all these temperatures are equal. This implies that the velocity distribution of all types of particle in the plasma (molecules, atoms, ions, and electrons) at any energy level can be described by the Maxwell equation (15). For all species, the population of the different levels is then given by the Boltzmann equation (Eq. 17). Further, the ionization of atoms, molecules, and radicals can be described by the Saha equation (Eqs. 53, 54) and the related equations of chemical equilibrium. Finally, the radiation density in the source obeys Planck's law, and the exchange of both kinetic energy between particles and electromagnetic radiation are in equilibrium with each other. Real plasma sources used in atomic spectrometry are, at best, approximations to local thermal equilibrium. Although processes between the particles may be in thermal equilibrium, those involving electromagnetic radiation are not, as the plasma cannot be considered as a closed system. Moreover, real plasma sources are extremely inhomogeneous with respect to temperature and number density distributions. Accordingly, the equilibria occur only within small volume elements of the sources, in which gradients can be neglected. Many plasmas, however, have cylindrical symmetry and can be observed laterally, integrating the information from many volume elements along the observation direction. From this integral value, the values at a defined radial distance from the plasma center can be calculated. This necessitates side-on observation along several axes, which are equidistant with respect to each other, and a so-called Abel inversion. If $I'(x)$ is the first derivative of function $I(x)$, describing the variation of a measured value as a function of the lateral position x during side-on observation (Fig. 3), the radial values at distance r are:

$$I(r) = -(1/\pi) \int_r^\infty [I'(x)\,dx]/\sqrt{(r^2 - x^2)} \qquad (73)$$

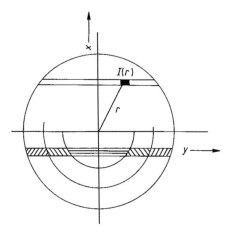

Figure 3. Abel inversion procedure for the determination of radial distributions in plasma sources
$I(r)$ = Intensity at radial distance r in volume element $dx \times dy$

This integral, known as the Abel integral, can only be solved numerically, writing each laterally measured intensity as a sum of the different volume elements with coefficients denoting their contribution. For certain numbers of lateral observation positions (3, 10, ...) these Abel coefficients are listed in the literature [23]. By inversion of the corresponding matrix the radial values are obtained. This allows determination of radial distributions of emissivities, absorbances, temperatures, or number density distributions in plasmas. Repeating this procedure at different heights in a plasma is equivalent to plasma tomography. Similar results, which do not require the assumption of cylindrical symmetry in the source can be obtained by imaging spectrometry with two-dimensional detectors [24].

In real plasmas, departures from thermal equilibrium often occur. In the extreme case, as encountered in plasmas under reduced pressure, emission or absorption of radiation becomes so important that there is no longer a clear relation between the mean kinetic energies of the species and the excitation temperatures. The latter then lose the physical meaning of temperature. The absence of local thermal equilibrium in these non-LTE plasmas relates to the existence of high field gradients or a.c. fields; only the electrons, not the heavy atoms and ions, can follow the field changes and fully take up the dissipated energies. Accordingly, the mean kinetic energy of the electrons and thus the electron temperature is much higher than the gas temperature.

Sources and atom cells for atomic spectrometry include flames, arcs, sparks, low-pressure discharges, lasers as well as d.c., high-frequency, and microwave plasma discharges at reduced and atmospheric pressure (Fig. 4). They can be characterized as in Table 2. *Flames* are in thermal equilibrium. Their temperatures, however, are at most 2800 K. As this is far below the standard temperature of most elemental lines, flames have only limited importance as sources for atomic emission spectrometry, but are excellent atom cells for absorption, fluorescence, and laser-enhanced ionization work. *Arcs and sparks* are well-known sources for atomic emission spectrometry. The high temperature obtained in spark sources leads to the excitation of ion lines, for which the standard temperatures are often beyond 10 000 K, whereas in arc sources atom lines predominate. In *plasma sources* under reduced pressure the gas kinetic temperatures are low and their atomization capacity is limited. When these sources are used as atomic emission sources or as primary sources in atomic absorption, line widths are very narrow. Moreover, especially with gases having high ionization energy such as helium, high-energy lines such as those of the halogens can be excited. Low-pressure discharges are valuable ion sources for mass spectrometry as analyte ionization takes place. Because of the low pressure, coupling with a mass spectrometer is possible. So-called plasmajet and plasma sources at atmospheric pressure are especially useful for the emission spectrometric analysis of solutions. Their gas kinetic temperatures are high enough to achieve complete dissociation of thermally stable oxides and both atom and ion lines occur in the spectra. As reflected by the fairly high ionization temperature, they are powerful ion sources for mass spectrometry; plasma mass spectrometry is now one of the most sensitive methods in atomic spectrometry. Lasers are very suitable for ablation of solids. Owing to their high analyte number densities, plasmas are subject to high self-absorption so it is better to use them solely for volatilization, and to introduce the ablated material in a second source for signal generation.

Table 2. Applications of sources and atom cells for atomic spectrometry

Source	Spectroscopy	Sample	Concentration range[a]
Flames	AES[b]	liquid	t (alkali)
	AAS[c]	liquid (solid)	t
	AFS[d]	liquid	ut
	LEI[e]	liquid	ut
Arc (d.c.)	AES[b]	solid, liquid	t
	MS[f]	solid	ut
Spark	AES[b]	solid	t
	MS[f]	solid	ut
Glow discharge	AES[b]	solid, liquid, gas	t
	AAS[c]	solid, liquid, gas	t
	AFS[d]	solid, liquid, gas	ut
	MS[f]	solid, gas	ut
	LEI[e]	solid, gas	ut
Laser	AES[b]	solid	m
	AAS[c]	solid	t
	AFS[d]	solid	ut
	MS[f]	solid	ut
	LEI[e]	solid	ut
Inductively coupled plasma	AES[b]	liquid (solid, gas)	t
	AFS[d]	liquid (solid, gas)	t
	MS[f]	liquid (solid, gas)	ut
Microwave plasma	AES[b]	gas (liquid)	t
	MS[f]	gas (liquid)	ut
Furnace	AES[b]	liquid, solid	t
	AAS[c]	liquid, solid	ut

[a] m = Minor; t = Trace; ut = Ultra-trace.
[b] AES = Atomic emission spectroscopy.
[c] AAS = Atomic absorption spectroscopy.
[d] AFS = Atomic fluorescence spectroscopy.
[e] LEI = Laser-enhanced ionization.
[f] MS = Mass spectrometry.

21.2.7. Analytical Atomic Spectrometry

Analytical atomic spectrometry nowadays includes the use of flames and plasma discharges for optical and for mass spectrometry. These are used directly as emission sources, as atomizers for atomic absorption or atomic fluorescence, or for ion production. In a source for atomic spectrometry atomic vaporization and signal generation take place together. The first processes require high energy for complete atomization, whereas the signal generation processes often require discrete, selective excitation. Therefore, the use of so-called tandem sources is preferred, where analyte vapor generation and signal generation take place separately [26].

For atomic emission spectrometry, selectivity is achieved by isolation of the spectral line at the exit slit of the spectrometer. This puts high demands on the optical quality of the spectral apparatus. In atomic absorption and fluorescence, selectivity is partly controlled by the radiation source delivering the primary radiation, which in most cases is a line source (hollow cathode lamp, laser, etc.). Therefore, the spectral bandpass of the monochromator is not as critical as it is in atomic emission work. This is especially true for laser-based methods, where in some cases of atomic fluorescence a filter is sufficient, and in laser-induced ionization spectrometry where no spectral isolation is required at all. In the case of glow discharges and inductively coupled high-frequency plasmas, ion generation takes place in the plasmas. In the first case, one can perform direct mass spectrometry on solids, and in the second case on liquids or solids subsequent to sample dissolution. In the various methods of atomic spectrometry, samples have to be delivered in the proper form to the plasma source. Therefore, in the treatment of the respective methods attention will be given to techniques for sample introduction.

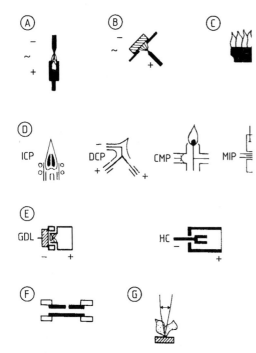

Figure 4. Sources for atomic spectrometry [25]
A) Arc; B) Spark; C) Flame; D) Plasma sources; E) Low-pressure discharges; F) Graphite furnace; G) Laser plume
ICP = Inductively coupled plasma; DCP = Direct current plasmajet; CMP = Capacitively-coupled microwave plasma; MIP = Microwave-induced plasma; GDL = Glow discharge lamp; HC = Hollow cathode

21.3. Spectrometric Instrumentation

Atomic spectrometric methods of analysis essentially make use of equipment for spectral dispersion to achieve their selectivity. In optical atomic spectrometry, this involves the use of dispersive as well as nondispersive spectrometers, whereas in the case of atomic spectrometry with plasma ion sources mass spectrometric equipment is used. In both cases, suitable data acquisition and processing systems are built into the instruments.

In the design of instruments for atomic spectrometry the central aim is to achieve the full potential of each method. This involves the power of detection and its relation to precision, freedom from spectral interference, and the price–performance ratio.

21.3.1. Figures of Merit of an Analytical Method

In practice, atomic spectrometric methods are relative methods and need to be calibrated. The calibration function describes the relationship between the analytical signals (absorbances, absolute or relative radiation intensities, or ion currents). In its simplest form, the calibration function can be written (→Chemometrics):

$$Y = ac + b \tag{74}$$

The inverse function is often used:

$$c = a'Y + b' \tag{75}$$

This is known as the analytical evaluation function. Calibration curves are often nonlinear, so a polynomial function is needed:

$$Y = a_0 + a_1 c + a_2 c^2 + \ldots \tag{76}$$

where $a_0, a_1, a_2, \ldots$ are determined by multivariate regression. Normally, a second-degree polynomial is sufficient to describe the calibration function over a large concentration range. Alternatively, segmented calibration curves can be used if the calibration function is nonlinear.

For a series of analytical signal measurements for a well-defined analyte concentration, a statistical uncertainty exists which stems from fluctuations in the analytical system. The precision achievable is an important figure of merit of an analytical procedure, for the statistical evaluation of analytical data see, →Chemometrics, and classical textbooks [27]–[29]. The precision of a measurement procedure is expressed in terms of the standard deviation:

$$\sigma = \sqrt{\left[\sum (Y_m - Y_i)^2 / (n-1)\right]} \tag{77}$$

where $Y_m = \sum (Y_i / n)$ is the mean value, Y_i an individual measurement, and n the number of measurements. The precision of a certain measurement system depends on the noise in the system. Different types of noise can be distinguished [30].

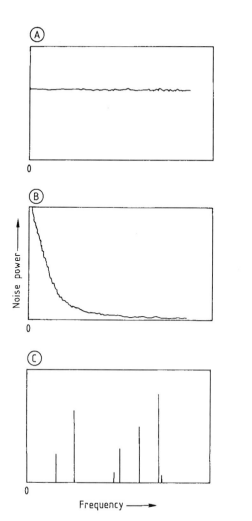

Figure 5. Noise power spectra commonly found in chemical instrumentation
A) White noise; B) Flicker ($1/f$) noise; C) Interference noise [31]

Fundamental or Random Noise. This is statistically distributed and its amplitude as a function of frequency can be written as a sum of sinusoidal functions. This type of noise is related to the discrete nature of matter and the quantization of radiation, and cannot be completely eliminated.

Nonfundamental, Flicker, or Excess Noise. Here, the sign or the magnitude can correlate with well-defined phenomena.

Interference Noise. This is observed at well-defined frequencies and mostly stems from components in the system.

Both nonfundamental and interference noise can often be eliminated by appropriate filtering.

In a noise spectrum (Fig. 5) the noise amplitude is plotted as a function of frequency [31]. White noise (Fig. 5 A) occurs over all frequencies and is almost always fundamental in origin, whereas for $1/f$ noise (Fig. 5 B) the amplitude decreases with the frequency and it is nonfundamental in origin. Discrete noise bands with well-defined causes may also occur (Fig. 5 C). These may stem from the source, caused by gas flow dynamics or contributions from the vacuum line, etc. Noise spectra are a powerful diagnostic tool to trace the sources of noise, and to study instrumental limitations to the power of detection. In atomic spectrometry, it is important to determine if noise from the detector is predominant; if so, it can be described by a Poisson distribution:

$$\sigma^2 = n \qquad (78)$$

where n is the number of events per unit time and σ the standard deviation. Alternatively, the background noise of the source may be much more important, or Flicker noise or frequency-dependent noise are predominant. In the latter case, overtones often occur.

For the precision of an analytical method, not only the reproducibility of single measurements, but also calibration errors have to be considered. This is a complex problem as, depending on the nature of the noise, precision may vary considerably with concentration.

When a linear regression is performed, the standard deviation of the regression can be defined:

$$s(Y) = \sqrt{\left[\sum (Y_i - Y)^2/(n - 2)\right]} \qquad (79)$$

where Y_i is the signal obtained for a standard sample with concentration c from the regression equation. The latter is calculated by a least-squares procedure from the pairs (c, Y), where Y are the measured signals, and n is the total number of measurements. The standard deviation for the concentration of the analytical sample c_x can be calculated through propagation of errors:

$$s_r(c_x) = \ln_{10} s(c_x) = \ln_{10} a' s(Y) \qquad (80)$$

where a' is the slope of the calibration curve. The magnitude of the concentrations of the analytical sample with respect to those of the standard samples has to be considered, and can be included in Eq. (80):

$$s_r(c_x) = \ln_{10} a' s(Y) \cdot \sqrt{\left\{(1/n) + (1/m) + (c_x - c_m)^2 / \left[\sum (c - c_m)^2\right]\right\}} \qquad (81)$$

where m is the number of replicates for the analytical sample and c_m is the mean of all the standard concentrations measured.

Equations (74) and (75) are valid only for a limited concentration range, known as the linear dynamic range. This range is limited at the upper end by physical phenomena, such as detector saturation, and at the lower end by the limit of determination. This limit is typical of a given analytical procedure and is the lowest concentration at which a determination can be performed with a certain precision [32] (→Chemometrics; →Trace Analysis, →Trace Analysis). For 99.86% certainty, and provided the fluctuations of the limiting noise source can be described by a normal distribution, the lowest detectable net signal Y_L is three times the relevant standard deviation:

$$Y_L = 3\sigma^* \qquad (82)$$

The net signal is determined from the difference between a background signal and a gross signal including analyte and background contributions, so a factor of $\sqrt{2}$ has to be introduced.

In many cases the limiting standard deviation is often the standard deviation obtained for a series of blank measurements [33]. From the calibration curve the detection limit is:

$$c_L = a' \left(3\sqrt{2}\right)\sigma \qquad (83)$$

The detection limit is thus closely related to the signal-to-background and the signal-to-noise ratios. It is the concentration for which the signal-to-background ratio is $3\sqrt{2}$ times the relative standard deviation of the background, or at which the signal-to-noise ratio is $3\sqrt{2}$. The signal-to-noise ratio itself is related to the types of noise occurring in the analytical system. From a knowledge of the limiting noise sources, well-established signal acquisition measures can improve the signal-to-noise ratio and hence the power of detection.

The analytical signals measured often include contributions from constituents other than the analyte (e.g., matrix constituents). This is known as spectral interference and can be corrected by subtracting contributions calculated from the magnitude of interference and the concentration of the interfering species. A special type of interference influences the background signal on which the analyte signals are superimposed; a number of correction methods exist. Freedom from interference is an important figure of merit for an analytical method.

21.3.2. Optical Spectrometers

In optical atomic spectrometry, the radiation emitted by the radiation source, or the radiation which comes from the primary source and has passed through a separate atomizer, must be fed into a spectrometer; this radiation transfer should be as complete as possible. The amount of radiation passing through an optical system is expressed by its optical conductance G_0:

$$G_0 = \int_A \int_B (\cos\alpha_1 \cos\alpha_2 \, dA \, dB)/a_{12}^2 \qquad (84)$$

$$\sim (AB)/a_{12}^2 \qquad (85)$$

where dA and dB are surface elements of the entrance and exit apertures, a_{12} is the distance between them, and α_1, α_2 are the angles between the normals of the aperture planes and the radiation. If n is the refractive index of the medium, the optical conductance is:

$$G = G_0 n^2 \qquad (86)$$

The radiant flux through an optical system is:

$$\varphi = \tau B G \qquad (87)$$

where τ is the transmittance, determined by reflection or absorption losses at the different optical elements, and B is the radiant density of the source (W m^{-2} sterad^{-1}). For optimal optical illumination of a spectrometer, the dispersive element should be completely illuminated to obtain full resolution. However, no radiation should bypass the dispersive element, as stray radiation may produce anomalous signals. Further, the optical conductance at every point in the optical system should be maximum.

21.3.2.1. Optical Systems

The type of illumination system needed to fulfill these conditions depends on the source and detector dimensions, the homogeneity of the source, the need to fill the detector homogeneously with radiation, the distance between the source and the entrance aperture of the spectrometer, and the focal length of the spectrometer.

In conventional systems, lenses and imaging mirrors are used. Glass lenses can be used only at wavelengths above 330 nm. For quartz, radiation with wavelengths down to 165 nm is transmitted, but evacuation or purging the illumination system and spectrometer with nitrogen or argon is required below 190 nm, as a result of the absorption of short-wavelength radiation by oxygen. At wavelengths below 160 nm, the system must be evacuated and MgF$_2$ or LiF optics must be used. With lenses, three main illumination systems are used.

Imaging on the Entrance Collimator. In this system a lens is placed immediately in front of the entrance slit to image the relevant part of the radiation source on the entrance collimator (Fig. 6 A). This has the advantage that the entrance slit is homogeneously illuminated; however, stray radiation may occur inside the spectrometer. The distance between the source and the entrance slit a is given by the magnification required:

$$x/W = a/f_k \qquad (88)$$

where x is the width of the source, W is the width of the entrance collimator, and f_k its focal length. The f-number of the lens is:

$$1/f = 1/a + 1/f_k \qquad (89)$$

The focal length of a lens depends on the wavelength:

$$f(\lambda_1)/F(\lambda_2) = n_1/n_2 \qquad (90)$$

where n_1, n_2 are the refractive indices of the lens material at wavelengths λ_1, λ_2. For quartz, a factor of 0.833 has to be applied when passing from the Na 583 nm D-line to 200 nm.

Illumination with Intermediate Image. In the second illumination system, a field lens is used to produce an intermediate image on a diaphragm. A suitable zone can be selected to fully illuminate the collimator mirror with the aid of a lens placed immediately in front of the exit slit. A third lens is used to illuminate the entrance slit homogeneously (Fig. 6 B). The magnification is divided over all three lenses, so that chromatic aberrations are minimized, but the set-up is inflexible. The distances between the three lenses must be selected to achieve the required magnifications:

$$x/I = a_1/a_2 \qquad (91)$$

$$D/s_h = a_2/a_3 \qquad (92)$$

and

$$I/W = a_3/f_k \qquad (93)$$

where I is the diameter of the intermediate image, D the diameter of the field lens, s_h the entrance slit height, and a_1, a_2, a_3 are the distances between the lenses. Further

$$a_1 + a_2 + a_3 = A \qquad (94)$$

is mostly fixed by constructional factors, and:

$$1/f_1 = 1/a_1 + 1/a_2 \qquad (95)$$

$$1/f_2 = 1/a_2 + 1/a_3 \qquad (96)$$

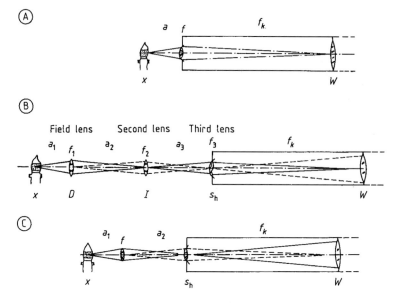

Figure 6. Illumination of the optical spectrometer with lenses
A) Imaging on the entrance collimator; B) Illumination with intermediate image; C) Imaging on the entrance slit
For explanation of symbols see text

$1/f_3 = 1/a_3 + 1/f_k$ \hfill (97)

give the *f*-numbers of the respective lenses. Only one parameter (e.g., the width of the intermediate image) can be freely selected because x, W, s, A, and f_k are fixed.

Image on the Entrance Slit. The third illumination system produces an image on the entrance slit with the aid of one lens; the structure of the source is imaged on the entrance collimator and the detector. This allows spatially resolved line intensity measurements when a detector with two-dimensional resolution, such as a photographic emulsion or an array detector (see Section 21.3.2.2), is employed. This type of imaging is often used for diagnostic studies.

Fiber optics are very useful for radiation transmission. Small lenses are necessary to accommodate the entrance angle of the fiber, which depends on the refractive index of the material (usually quartz), and is typically 30–40°. A typical illumination of a spectrometer with an optical fiber (Fig. 7) uses a lens (diameter d) for imaging the source on the fiber. Bundles of fiber beams with diameter $D = 600$ μm are often used. The magnification x/D as well as the entrance angle $\alpha = $ arc-tan $(d/2)/a_2$ and the lens formula determine the *f*-number of the lens and d. At the fiber exit, a lens allows the radiation to enter the spectrometer, without causing stray radiation.

With quartz fibers, radiation is efficiently fed into spectrometer, but the transmittance decreases seriously below 220 nm. This may lead to detector noise limitations for analytical lines at lower wavelengths.

Spectrometers are used to produce the spectrum or to isolate narrow spectral ranges, without further deconvolution. In dispersive apparatus, the spectrum is produced with a prism or diffraction grating. In nondispersive apparatus, spectral regions are isolated from the radiation beam by reflection, interference, or absorption in interferometers or filter monochromators. The latter are of use only in flame emission spectrometry.

A *dispersive spectrometer* contains an entrance collimator, a dispersive element, and an exit collimator [34], [35].

The entrance collimator produces a quasi-parallel beam from the radiation coming through the entrance aperture, of width s_e and height h_e. The entrance collimator has focal length f_k and width W. The diffraction slit width s_0 and height h_0 are the half-widths:

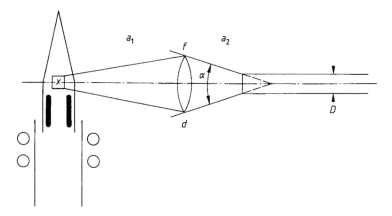

Figure 7. Use of optical fibers for illumination of an optical spectrometer
α = Entrance angle of the fiber; d = Lens diameter; D = Fiber diameter; x = Diameter of the zone in the radiation source to be selected; a_1 = Distance between lens and radiation source; a_2 = Distance between fiber entrance and lens

$$s_0 = \lambda f / W$$

and

$$h_0 = \lambda f / h \tag{98}$$

The entrance aperture dimensions should not be smaller than s_0 and h_0, so that diffraction does not limit resolution. The value of f/W is a measure of the amount of radiant energy entering the spectrometer. The exit collimator images the monochromatized radiation leaving the dispersive element on the exit slit, giving a series of monochromatic images of the entrance slit. In a monochromator, a single exit slit allows isolation of one line after another by turning the dispersive element. In a polychromator the dispersive element is fixed and there are many exit slits placed where the images of the lines of interest occur. They are often on a curved surface with a radius of curvature R (Rowland circle). Simultaneous measurement of several lines and thus simultaneous multielement determinations are possible. With a spectrographic camera, lines are focused in a plane or on a slightly curved surface, where a two-dimensional detector can be placed. With such a detector (photographic plate, diode array detector, etc.), the signal intensities of part of the spectrum over a certain wavelength range or intensities of a spectral feature at several locations in the source can be simultaneously recorded. The energy per unit area at the detector is given by the irradiance:

$$E = \varphi \cos\alpha / A \tag{99}$$

Diffraction Gratings. Diffraction gratings are used almost exclusively as the dispersive element in modern apparatus; prisms are still used as predispersors and crossdispersors. Both plane and concave gratings are used; the latter have imaging qualities. Because of the profile of the grooves, holographically ruled gratings have a rather uniform radiant output over a large spectral area compared with mechanically ruled gratings which always have a so-called blaze angle, and hence a blaze wavelength where radiant energy is maximum. Modern spectrometers tend to use reflection gratings. As a result of interference, a parallel beam of white radiation at incident angle φ_1 with the grating normal is dispersed through angle φ_2; radiation with wavelength λ is sorted (Fig. 8) according to Bragg's equation:

$$\sin\varphi_1 + \sin\varphi_2 = m\lambda/a \tag{100}$$

where m is the order of interference and $a = 1/n_G$ is the grating constant, where n_G is the number of grooves per unit length. When B is the grating width, the total number of grooves $N = B n_G$ determines the theoretical resolving power R_0:

$$R_0 = B n_G m \tag{101}$$

The angular dispersion can be obtained by differentiation of Equation (101) with respect to λ:

$$d\varphi_2/d\lambda = m/(a\cos\varphi_2) \tag{102}$$

The reciprocal linear dispersion is:

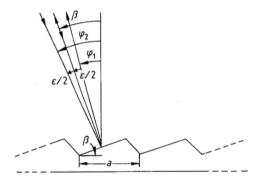

Figure 8. Spectral dispersion at a diffraction grating
α = Grating constant; β = Blaze angle; φ_1 = Angle of incident radiation; φ_2 = Angle of diffracted radiation with wavelength λ; $\varepsilon = \varphi_1 - \varphi_2$

$$dx/d\lambda = (dx/d\varphi_2)(d\varphi_2/d\lambda)$$
$$= (f_k/\cos\theta')(m/a\cos\varphi_2) \quad (103)$$

where θ' is the slope of the surface of the radiation detector.

The form and depth of the grooves determine the blaze angle β defined earlier. The blaze wavelength for order m is:

$$\sin\beta = \lambda_B m / [2a\cos(\varepsilon/2)] \quad (104)$$

where $(\varphi_1 - \varphi_2) = \varepsilon$.

In stigmatic spectrometers the height and width of the slit are imaged in the same plane. Several mountings (Fig. 9) are used with a plane gratings.

In the Czerny–Turner mounting (Fig. 9 B), two spherical mirrors with slightly different focal lengths are positioned at slightly different angles to correct for spherical aberration. This mounting is used for high-luminosity monochromators with short focal length and highly dispersive gratings (more than 1800 grooves per millimeter). In the Ebert mounting (Fig. 9 A), there is only one mirror serving for both entrance and exit collimator; accordingly, aberrations occur and require the use of curved slits, to achieve maximum resolution. The Fastie–Ebert mounting is used for large spectrographs. The entrance collimator lies under the exit collimator plane. The spectrum is focused in a plane and the f/W value can be as low as 1/30, which may require the use of a cylindrical lens to increase the detector irradiance. Photographic plates or films are used as detectors.

Normally, gratings are used at low orders; different orders can be separated by special photomultipliers. With a solar-blind photomultiplier

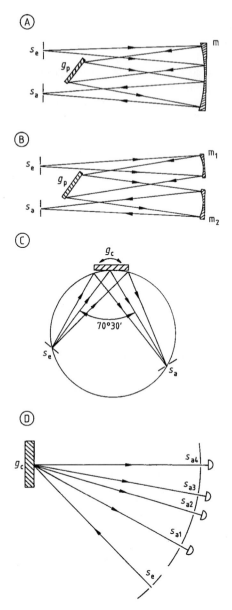

Figure 9. Optical mountings for optical spectrometers with a plane grating
A) Ebert; B) Czerny–Turner; C) Seya–Namioka; D) Paschen–Runge
s_e = Entrance slit; s_a = Exit slit; g_P = Plane grating; g_c = Concave grating; m = Mirror

only radiation with wavelength below 330 nm can be detected. This allows separation of the first order radiation at 400 nm from the second order at 200 nm. This can be used in polychromators to double the effective resolution. So-called echelle gratings have low groove density (up to 1/100 mm) and therefore enable the use of orders up to 70 [36]. Here, the orders overlap and must be separated by a second dispersive element (e.g., a prism) either with its axis parallel to that of the echelle grating or in the so-called crossed-dispersion mode. In the latter case, the spectrum occurs as a number of dots of height equal to that of the entrance slit. The radiant flux is:

$$\varphi = (B_\lambda \tau s h W H \cos\beta)/f^2 \qquad (105)$$

where B_λ is the spectral radiance, τ is the transmittance of the optics, s is the slit width and h is height, W is the width, H the height of the grating, and β is the angle between the collimator plane and the radiation beam. Echelle spectrometers often use an Ebert mounting for high resolution at low focal length. They are thermally very stable and are used as both monochromators and polychromators.

For concave gratings, the radius of the grating determines the Rowland circle. In the direction of dispersion spectral lines are focused on the Rowland circle as monochromatic images of the entrance slit. In the Paschen–Runge mounting (Fig. 9D), grating, entrance slit, and all the exit slits are fixed on the Rowland circle. This mounting is most often used in large simultaneous polychromators with photoelectric detection. In the Eagle mounting, the grating can be simultaneously turned around its axis and moved along the radiation direction. This mounting can be easily housed in a narrow tank; it is mechanically rather simple and is much used in vacuum monochromators. In the Seya–Namioka mounting (Fig. 9C), the directions of incident and dispersed radiation beams are kept constant at an angle of 70°30′. Wavelength selection at the exit slit is performed by rotating the grating. At large angles serious defocusing occurs and the aberrations are larger than in the Eagle mounting. The Seya–Namioka system is often used in vacuum monochromators.

In most sequential spectrometers, switching from one line to another is achieved by turning the grating. From Equation (100) it can be seen that, as the spectral lines have widths of 1–3 pm, angle selection must be very accurate. For a grating with $a = 1/2400$ mm and a line width of 1 pm, the grating must be positioned with an accuracy of 0.0001°.

It is possible to step through the line profile or to achieve random access by using angle encoders. Computer-controlled stepper motors are used to turn the grating. Their angular resolution is often above 1000 steps per turn. For a so-called sinebar drive, the number of steps performed by the motor is directly proportional to the wavelength displacement. A pen recorder is running off on the spindle driven by the motor. A further approach is to use a Paschen–Runge spectrometer with equidistant exit slits and a detector moving in the focal plane. Fine adjustment of the lines is by computer-controlled displacement of the entrance slit.

21.3.2.2. Detectors

Photographic emulsions or photoelectric devices can be used as detectors for electromagnetic radiation between 150 and 800 nm. Among the latter, photomultipliers are the most important, but solid state devices are now becoming very common.

Photographic emulsions allow recording of the whole spectrum simultaneously and, accordingly, they are useful for multielement survey analysis. However, their processing is slow, precision is low, and they are not capable of on-line data processing. Quantitative treatment of photographically recorded spectra is of limited importance for analytical use and current interest in photographic detection is limited to research in which documentation of the whole spectrum is required. Therefore, it is treated only briefly.

When a photographic emulsion is exposed to radiation blackening S occurs, which is a function of the radiant energy accumulated during the exposure time. It is given by:

$$S = \log(1/\tau) = -\log\tau = \log\varphi_0/\varphi \qquad (106)$$

where τ is the transmission, φ_0 is the flux through a nonirradiated part of the emulsion, and φ is the flux for an exposed part of the emulsion, when, in a densitometer, a beam of white light passes through the emulsion. The emulsion characteristic for a selected illumination time gives the relation between the logarithm of the intensity ($Y = \log I$) and the blackening, and is S-shaped (Fig. 10). Originally, only the linear part of the characteristic was used for quantitative work. However, techniques have been developed, which linearize the characteristic at low exposures. The P-transformation has been described by KAISER [37]:

$$P = \gamma(Y - Y_0) = S - \varkappa D \qquad (107)$$

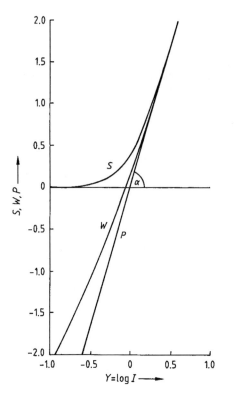

Figure 10. Emulsion characteristic (S) and transformation functions (P, W).

where $D = \log[1/(1-\tau)]$.

The Seidel function is:

$$W = \log(10^S - 1) \tag{108}$$

and thus:

$$P = (1 - \varkappa)S + \varkappa W \tag{109}$$

where Y_0 is the inertia (speed) of the emulsion and constants $\varkappa$ and γ describe the properties of the emulsion, but also depend on the densitometer. In most cases, Y_0 is unknown so ΔP or ΔY values are used. $\gamma = \tan \alpha$ is the contrast of the emulsion. The emulsion can be calibrated with the aid of a step filter placed in front of the spectrometer [38]. When the ratio of the intensities passing through two sectors of the filter is:

$$\Delta Y_m = \log\left(I_{m(1)}/I_{m(2)}\right) \tag{110}$$

the transformation equation becomes:

$$\Delta S = \gamma \Delta Y_m + \varkappa \Delta D \tag{111}$$

where ΔD is the independent variable and ΔS the function. A linear regression for a number of line pairs gives κ as slope and $\gamma \Delta Y_m$ as intercept on the x-axis. Values of $Y = \log I$ for a given blackening can be calculated:

$$Y = (\varkappa/\gamma) - \log(1 - \tau) - (1/\gamma)\log\tau + Y_0 \tag{112}$$

and

$$\Delta Y = -(\varkappa/\gamma)\Delta D + (1/\gamma)\Delta S \tag{113}$$

Errors in the intensity measurement stem from the graininess of the emulsion. According to SIEDENTOPF [39], the standard deviation at low exposure is:

$$\sigma(S) = \sqrt{[0.5\,(a/F)\,S]} \tag{114}$$

where a is the surface area of a grain and F the area of the densitomer slit (e.g., 10 µm×1 mm). From $\sigma(S)$ it is possible to calculate the relative standard deviation of the intensities $\sigma_r(I)$:

$$\sigma_r(I) = \ln 10 \sigma(S)(1/\gamma\{1 + [\varkappa/(10^S - 1)]\}) \tag{115}$$

Typical values for γ are 1–2.5 and for κ, 0.5–2. Nonsensitized emulsions can be used in the wavelength range 220–450 nm. At longer wavelengths green- or red-sensitized emulsions can be used. They have a large γ and accordingly a small dynamic range. At UV or vacuum UV wavelengths the gelatin of the emulsion absorbs significantly, and low-gelatin emulsions (which are rather insensitive) are used; alternatively, radiation can be transformed to longer wavelengths with the aid of a scintillator, such as sodium salicylate.

Photomultipliers [40] are most commonly used for precise measurements of radiant intensities. The photomultiplier has a photocathode and a series of dynodes. The radiation releases photoelectrons from the photocathode as a result of the Compton effect. These photoelectrons impact after acceleration on the dynodes so that a large number of secondary electrons is produced, which after successive action on a number of dynodes lead to a measurable photocurrent I_a at the anode. For a photon flux N_φ through the exit slit of the spectrometer and impacting on the photocathode, the flux of photoelectrons produced N_e is:

$$N_e = N_\varphi \, \text{QE}(\lambda) \tag{116}$$

where QE is the quantum efficiency (up to 20%). The cathode current I_c is:

$$I_c = N_e \, e \tag{117}$$

where $e = 1.9 \times 10^{-19}$ A · s and:

$$I_a = I_c \theta(V) \tag{118}$$

where $\theta(V)$ is the amplification factor (up to 10^6). The dark current I_d is the value of I_a measured when no radiation enters the photomultiplier (ca. 1 nA). The quantities QE, $\theta(V)$, and I_d are characteristics for a certain type of photomultiplier. The photocurrents are fed into a measurement system. First, I_a is fed to a preamplifier containing a high-ohmic resistor. The voltage produced is fed to an integrator. By selection of the resistors and capacitors, a high dynamic measurement range can be realized. When I_a is accumulated in a capacitor with capacitance C, during an integration time t the potential obtained is:

$$U = I_a t / C \tag{119}$$

This voltage is digitized. The dynamic range is limited at the lower end by the dark current of the photomultiplier and at the high end by saturation at the last dynode.

For a photomultiplier, the relative standard deviation of the measured intensities $\sigma_r(I)$ is determined by the noise of the photoelectrons and the dark current electrons. The quantum efficiency and the amplification at the dynodes are also important:

$$\sigma_r(I) \approx 1.3 \left(N_{e,d}\right)^{-1/2} \tag{120}$$

where $N_{e,d}$ is the total number of photoelectrons and dark current electrons in measurement time t.

Photomultipliers for the spectral range 160–500 nm usually have an S 20 photocathode; QE$(\lambda) \approx 5-25\%$, the amplification factors are 3–5 at dynode voltages 50–100 V, and there are 9–15 dynodes mounted in head-on or side-on geometry. For selected types, the dark current may be below 100 photoelectrons per second ($I_a < 10^{-10}$ A). Red-sensitive photomultipliers have a so-called bialkali photocathode. Their dark current is higher, but can be decreased by cooling. For wavelengths below 160 nm, photomultipliers with MgF$_2$ or LiF windows are used. Solar-blind photomultipliers are only sensitive for short-wavelength radiation (e.g., below 330 nm).

Solid-state detectors are increasingly important for optical atomic spectrometry. They are multichannel detectors and include vidicon, silicon-intensified target (SIT) vidicon, photodiode array detectors, and image dissector tubes [41]. Charge-coupled devices (CCD) and charge-injection devices (CID) [42] have been introduced. Photodiode arrays (PDA) may consist of matrices of up to 512 and even 1024 individual diodes (e.g., Reticon) with individual widths of 10 μm and heights of up to 20 mm. The charge induced by photoelectric effects gives rise to photocurrents which are sequentially fed into a preamplifier and a multichannel analyzer. They are rapid sequential devices, with the advantage that memory effects due to the incidence of high radiant densities on individual diodes (lag) as well as cross-talk between individual diodes (blooming) are low. Their sensitivity in the UV and vacuum UV is low compared with photomultipliers. Nevertheless, they are of interest for atomic emission spectrometry, especially as they can be coupled with microchannel plates, giving additional amplification. Signal-to-noise ratios can be considerably improved by cooling the array with the aid of a Peltier element or liquid nitrogen. Photodiode arrays have been used successfully in a segmented echelle spectrometer. Here, spectral segments around the analytical lines are sorted out by primary masks subsequent to dispersion with an echelle grating, and are detected after a second dispersion on a diode array. More than ten analytical lines and their spectral background contributions can be measured simultaneously. This is shown for a glow discharge lamp in Figure 11 [43], but restrictions arise due to detector noise limitations at low wavelengths.

Image dissector tubes (Fig. 12) make use of an entrance aperture (d) behind the photocathode (a), by which photoelectrons from different locations on the photocathode can be scanned and measured after amplification in the dynode train, as in a conventional photomultiplier. Though used in combination with an echelle spectrometer with crossed dispersion for flexible rapid sequential analyses [45], the system has not been successful. This may be due to its limited cathode dimensions, but also to stability problems.

Charge transfer devices (CTDs) are solid state, silicon-based detectors which store charge as a result of photons impinging on the individual

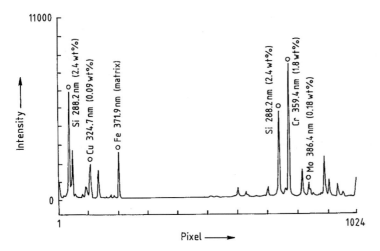

Figure 11. Segmented spectrum of glow discharge [43]
Grimm-type glow discharge lamp with floating restrictor 8 mm diameter, 50 mA, 3.5 torr, 2 kV [44], plasma-array spectrometer (LECO), steel sample 217 A

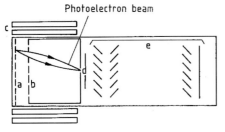

Figure 12. Image dissector tube
a) Photocathode; b) Accelerating electrode (mesh); c) Deflection coils; d) Aperture; e) Electron multiplier

CTD elements, called pixels, and can be used to monitor large portions of the spectrum in multiple orders by taking an "electronic photograph". There are two main types, namely charge injection device (CID) and charge-coupled device detectors (CCD). They differ in that the CID can be read randomly and non-destructively, whereas the CCD can only be read sequentially and destructively. A typical CCD detector [46] is the segmented array CCD used in the PerkinElmer Optima ICP-AES instrument. This consists of 224 linear photodetector arrays on a silicon chip with a surface area of 13 mm by 18 mm (Fig. 13). The array segments detect three to four analytical lines of high analytical sensitivity, large dynamic range and which are free from spectral interferences. Each subarray is comprised of pixels. The pixels are positioned on the detector at $x-y$ locations that correspond to the locations of the desired emission lines, usually generated by an echelle spectrometer. The emission lines are detected by means of their location on the chip and more than one line may be measured simultaneously. The detector can then be electronically wiped clean and the next sample analyzed. A number of newer CCD detectors are capable of a pixel resolution of 0.009 nm and can record the continuous first order spectrum from 175 to 800 nm, giving simultaneous access to 10 000 emission lines. Compared with many PMTs, a CCD offers improvement in quantum efficiency and lower dark current.

21.3.2.3. Nondispersive Spectrometers

Filter monochromators are of use only for flame photometry. They make use of interference filters, which often have a spectral bandpass of a few nanometers or less. Multiplex spectrometers include Hadamard transform spectrometers and Fourier transform spectrometers, and are especially useful where very stable sources are needed. Hadamard transform instruments make use of a coding of the spectrum produced by recombining the information with the aid of a slit mask which scans the spectrum [48].

Fourier transform spectrometry [49] often uses a Michelson interferometer (Fig. 14) to produce the interferogram. With the aid of a beam splitter, the radiation is split into two parts which are each directed to a mirror. Shifting the mirror in one of

Table 3. Evaluation of detectors for atomic emission spectrometry

Detector	Dimensions, mm	Spectral region, nm	Sensitivity vs. photomultiplier
Vidicon [41]	12.5×12.5[a]	with scintillator down to 200	poorer, especially at < 350 nm; dynamics < 10^2; lag/blooming
Diode array [47]	up to 25 (up to 1024 diodes)	especially for > 350	poorer in UV; dynamics 10^3
Diode array MCP[b] [41]	12.5	200–800	similar; linear; dynamic range > 10^3
Dissector tube [45]	up to 60 mm[a]	200–800	similar; linear; dynamic range > 10^3
CCD [42]	up to 13×18	200–800	similar; linear; dynamic range > 10^5

[a] Coupling with crossed-dispersion echelle spectrometer possible.
[b] MCP = Microchannel plate.

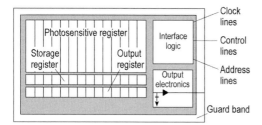

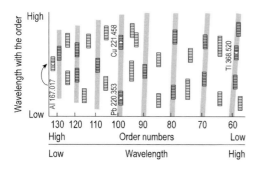

Figure 13. Schematic of a CCD detector (top) and a segmented two-dimensional detector array with respect to wavelength and order (bottom)

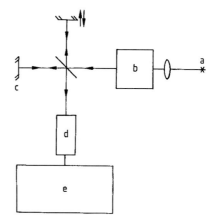

Figure 14. Fourier transform emission spectrometry
a) Source; b) Predispersion; c) Interferometer; d) Detector; e) FT in computer

the side arms, gives the interference for each wavelength:

$$I(x) = B(\sigma)[1 + \cos(2\pi\sigma x)] \quad (121)$$

where x is the optical path difference, σ is the wavenumber of the radiation, I is the intensity measured with the detector, and B is the radiation density of the source. A polychromatic radiation source gives an interferogram where the intensity of each point is the sum of all values resulting from Equation (121). The central part contains the low-resolution information and the ends contain high-resolution information. The resolution depends on the recording time, the spectral bandwidth, and the number of scans. By applying a Fourier transform to the signal for each point of the interferogram:

$$I(x) = \int_{-\infty}^{+\infty} B(\sigma)\,d\sigma + \int_{-\infty}^{+\infty} B(\sigma)\cos(2\pi\sigma x)\,d\sigma \quad (122)$$

$$= C + \int_{-\infty}^{+\infty} B(\sigma)\cos(2\pi\sigma x)\,d\sigma \quad (123)$$

where C is a constant which must be subtracted before the transformation. The end result is:

$$B(\sigma) = \int_{-\infty}^{+\infty} I(x)\cos(2\pi\sigma x)\,dx \quad (124)$$

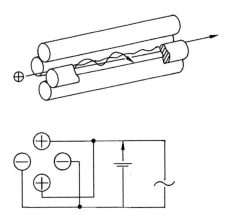

Figure 15. Quadrupole mass spectrometry

By digitizing the interferogram, rapid Fourier transform is possible with a powerful computer. For complex spectra, this is possible by small, but highly accurate stepwise displacements of the side arm. Repetitive scanning intensifies the image, and a reference laser is used to make the mirror positioning reproducible. The technique is well known in infrared spectroscopy and can now be used in the visible and UV regions. Fourier transform atomic emission spectrometry is suitable for sources with a low radiance with a high stability detector to avoid detector noise. Wavelength calibration must be very accurate to achieve maximum resolution. Reasonable signal-to-noise ratios are only achieved with low-noise sources.

21.3.3. Mass Spectrometers

In the sources used in optical atomic spectrometry, considerable ionization takes place, so they are useful ion sources for mass spectrometry [50]. Although, an overall treatment of instrumentation for mass spectrometry is given elsewhere (→ Mass Spectrometry), the most widely used types of mass spectrometer are briefly outlined here. In particular, the new types of elemental mass spectrometry have to be considered, i.e., glow discharges, and inductively coupled and microwave plasmas. In contrast to classical high voltage spark mass spectrometry [51] or thermionic mass spectrometry [52], which are beyond the scope of this article, the plasma sources are operated at considerably higher pressure between some 10 Pa and atmospheric pressure than the pressure in the mass spectrometer itself (10^{-3} Pa). Consequently, there has to be an interface with the appropriate apertures and high capacity vacuum pumps to bridge the pressure difference between source and spectrometer.

In commercial instrumentation both low-cost quadrupole and time-of-flight (TOF) mass spectrometers, and expensive double-focusing sector field mass spectrometers are used [53].

21.3.3.1. Types of Mass Spectrometer

In all mass spectrometers a vacuum of 10^{-3} Pa or better is maintained to avoid further ion formation from residual gas species or collisions of analyte ions with these species. Nowadays, turbomolecular pumps are preferred over diffusion pumps as their maintenance is easier and oil backflow does not occur.

Quadrupole Mass Spectrometers. In a quadrupole instrument (Fig. 15), the spectrometer consists of four equidistant, parallel rods (diameter 10–12 mm) between which a d.c. voltage and a high-frequency field (up to 1 MHz) are maintained. The d.c. voltage at the quadrupole should be slightly below the energy of the entering ions (usually below 30 eV). The ions enter through the so-called skimmer and have to pass the ion lenses and then a mask which prevents UV from entering the spectrometer (beam-stop). By changing the voltages at the ion lenses, at the beam-stop, at the quadrupole rods, and on the spectrometer walls, both the transmission of the spectrometer and the resolution of ions of a given mass and energy, can be optimized. As these parameters are all interdependent, optimization is complex. By changing the quadrupole field, the transmission for a certain ion changes. Accordingly, ions of a certain mass can be filtered out by manual setting of the field, or the spectrum can be scanned. The scanning velocity is usually limited to 30 000 mass numbers per second, because of the high-frequency and d.c. components in the spectrometer. Accordingly, the mass interval of 0–300 can be scanned in about 30 ms. Mass resolution is mainly determined by the quality of the field, and in quadrupoles for plasma mass spectrometry it is optimally about 1 atomic mass unit. Line profiles are approximately triangular, but sidewings occur as well; together with the isotopic abundances they determine the magnitude of spectral interference. Quadrupole mass spectrometers are rapid sequential measurement systems. In the case of isotopic dilution techniques for calibration, both precision and figures of merit can be limited by noise in the source. The price of quadrupole instruments is reasonable and their transmission is high. This is important because it allows their use in plasma mass spectrometry, where the source itself can be kept at earth potential. This is an advantage in inductively coupled plasmas.

Time-of-Flight (TOF) Mass Spectrometers. In quadrupole and sector field mass analysers the ion signal is a continuous beam. In TOF mass spectrometry the ion beam is pulsed so that the ions are either formed or introduced into the analyser in "packets". These ion packets are introduced into the field free region of a flight tube 30–100 cm long. The principle behind TOF analysis is that, if all ions are accelerated to the same kinetic energy, each ion will acquire a characteristic velocity dependent on its m/z ratio [54]. The ion beam reaches its drift energy (2700 eV) in less then 2 cm. The ions are then accelerated down the TOF tube with whatever velocity they have acquired. Because all ions have essentially the same energy at this point, their velocities are inversely proportional to the square roots of their masses. As a result, ions of different mass travel down the flight tube at different speeds thereby separating spatially along the flight tube with lighter, faster, ions reaching the detector before the heavier ions. Hence, the m/z ratio of an ion, and its transit time (T, in microseconds) through a flight distance (L, in cm) under an acceleration voltage (V) are given by

$$t = L\sqrt{\left[\left(\frac{m}{z}\right)\left(\frac{1}{2V}\right)\right]} \quad or \quad \frac{m}{z} = \frac{2Vt^2}{L^2} \quad (124a)$$

TOF mass analyzers are calibrated using two ions of known mass, so exact values of L and V need not be known. The mass calibration is based on the conditions of the analyzer during the entire period of measurement.

While simple in theory, the TOF analyzer caused numerous problems when coupled with a plasma source such as an ICP. The spread in ion kinetic energies caused by the ion sampling process resulted in the ions entering the field free region of the flight tube at different angles [55]–[57]. Another difficulty with ICP-TOF-MS is that ions have to be introduced in "packets". This can be achieved, for example, by using an orthogonal interface with a pulsed repeller plate. Background noise can be reduced by using a combination of quadrupole ion optics and an energy discriminator before the detector.

Sector Field Mass Spectrometers. In a magnetic sector analyzer, ions are subjected to a magnetic field which causes the ions to be deflected along curved paths. The ions are introduced into the analyzer via a series of electrostatic slits which accelerate and focus the ions into a dense ion beam. The velocity of the ions is controlled by the potential (V) applied to the slits. As the ions enter the magnetic sector analyzer they are subjected to a magnetic field parallel to the slits but perpendicular to the ion beam. This causes the ions to deviate from their initial path and curve in a circular fashion. A stable, controllable magnetic field (H) separates the components of the beam according to momentum. This causes the ion beam to separate spatially and each ion has a unique radius of curvature, or trajectory (R), according to its m/z. Only ions of a single m/z value will posses the correct trajectory that focuses the ion on the exit slit to the detector. By changing the magnetic field strength, ions with differing m/z values are brought to focus at the detector slit.

The ion velocity (v) in the magnetic field is given by:

$$\frac{1}{2}mv^2 = zV \quad or \quad v = \sqrt{\frac{2zV}{m}} \quad (124b)$$

As the ions enter the magnetic field, they are subjected to a magnetic force at right angles to both the magnetic lines of force and their line of flight. This leads to a centrifugal force leading to curvature of the ion beam.

$$\frac{mv^2}{R} = Hzv \quad (124c)$$

The radius of curvature of the flight path is proportional to its momentum and inversely proportional to the magnetic field strength:

$$R = \frac{mv}{zH} \quad (124d)$$

By eliminating the velocity term (v) between Equations (124b) and (124d) we get

$$R = \frac{1}{H}\sqrt{2V\left(\frac{m}{z}\right)} \quad (124e)$$

Hence, ions accelerated through an electrostatic field, uniform in nature, and then subjected

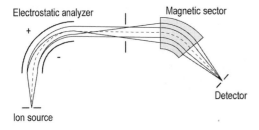

Figure 16. Sector field mass spectrometer

to a uniform magnetic field have different radii of curvature. This leads to ions of a specific m/z value being focused on to the detector slit while all other ions hit the side of the analyzer. Thus, the magnetic field classifies and separates ions into a spectrum of beams with each part of the spectrum having a different m/z ratio, where:

$$\frac{m}{z} = \frac{H^2 R^2}{2V} \quad (124f)$$

To obtain a complete mass spectrum from a magnetic sector analyzer, either the accelerating voltage (V) or the magnetic field strength (H) is varied. Each m/z ion from light to heavy is sequentially focused on the detector, and hence a mass spectrum is obtained.

Single focusing magnetic sectors, as described above, have the disadvantage that ion energies vary depending on their point of formation in the ion source. The difference in ion energy is accentuated by the accelerating voltage, which leads to peak broadening and low resolution in the single focusing mass analyser.

Double focusing magnetic/electrostatic sector instruments use magnetic and electrical fields to disperse ions according to their momentum and translational energy (Fig. 16). An ion entering the electrostatic field travels in a circular path of radius (R) such that the electrostatic force acting on it balances the centrifugal force. The equation of motion or transmission is:

$$\frac{mv^2}{R} = Ez \quad (124g)$$

where (E) is the electrostatic field strength. Hence the radius of curvature of the ion path in the electrostatic sector is dependent on its energy and not its mass [58]. A narrow slit placed in the image plane of the electrostatic sector can be used to transmit a narrow band of ion energies. If this sort of analyzer was placed in front of a magnetic sector analyzer an increase in resolution would result but a decrease in detection would be inevitable due to the decrease in ions exiting the electrostatic sector in comparison with ions exiting the ion source. This loss in sensitivity can be compensated for by the choice of a suitable combination of electrostatic and magnetic sectors, such that the velocity dispersion is equal and opposite in the two analyzers. The narrow ion energy range transmitted the full length of a sector instrument would suggest a loss in sensitivity, compared with quadrupole mass analyzers when both are operated at the same resolution, this has not been observed in practice. This has been mainly attributed to the ion focusing requirements for the different analyzers. Double focusing mass analyzers require a high energy (3–8 kV) ion beam for effective ion transmission and resolution. Whereas, if such a high energy beam was focused down a quadrupole, the ions would have obtained a velocity profile too great to be affected by the hyperbolic energy fields of the quadrupole analyzer, which generally requires ion energies of less than 10 eV for effective mass separation. The higher ion energy beam is less affected by space charge interference, which causes scattering of the ion beam, which has lead to increased sensitivity of double focusing instruments.

Other Mass Spectrometers. Spectrometers such as ion traps and ion cyclotron resonance have also been used in atomic mass spectrometry, but have not yet been incorporated into a commercial instrument. The former instrument has the advantage of ion storage and ion–molecule reaction capability, while the latter instrument is capable of extremely high resolution.

21.3.3.2. Ion Detection

For ion detection, several approaches are possible. Classical spark source mass spectrometry has developed the use of photographic plates. Very hard emulsions with low gelatin content are required. Emulsion calibration is similar to that described in optical emission spectrography, but usually varying exposure time are applied instead of using optical step filters. Provided an automated microdensitometer is used, mass spectrography is still a useful tool for survey analysis of solids down to the sub-µg/g level.

Faraday cup detectors consist of a collector electrode which is surrounded by a cage. The elec-

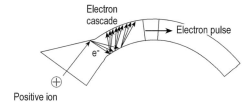

Figure 17. Continuous dynode electron multiplier

trode is positioned at an angle with respect to the ion beam, so that ions exiting the analyzer strike the electrode but secondary emissions are reflected away from the detector entrance. The function of the cage is to prevent detected ions and secondary electrons escaping from the detector. The electrode is connected to ground via a resistor. The ion current striking the electrode is neutralized by electron flow from ground through the resistor. This causes a potential drop which is amplified to create a signal. Currents as low as 10^{-15} A have been successfully measured in this way.

Electron Multipliers. For currents of less then 10^{-15} A an electron multiplier is necessary for detection. When the ion beam exits the analyzer it strikes a conversion plate which converts ions into electrons. The ions are drawn towards the plate by a strong voltage applied to the conversion plate. On striking the conversion plate the ions stimulate the ejection of electrons which are accelerated by the voltage applied to the plate. The electrons are multiplied using either continuous or discrete dynodes. A continuous dynode channel electron multiplier shown in Figure 17. This consists of a curved glass tube of approximately 1 mm in internal diameter with an inner resistive coating and a flared end. The multiplier can be operated in one of two modes. In pulse counting mode — the most sensitive mode of operation — a high voltage of between -2600 V and -3500 V is applied to the multiplier, which attracts ions into the funnel opening. When a positive ion strikes the inner coating the collision results in the ejection of one or more secondary electrons from the surface, which are accelerated down the tube by the potential gradient, and collide with the wall resulting in further electron ejection. Hence, an exponential cascade of electrons rapidly builds up along the length of the tube, eventually reaching saturation towards the end of the tube, resulting in a large electron pulse and a consequent gain of $10^7 - 10^8$ over the original ion collision. The electron pulses are read at the base of the multiplier and are approximately 50–100 mV and 10 ns in duration. Continuous dynode multipliers consist of a leaded glass tube which contains a series of metal oxides. The tube is curved and electrons are drawn down the tube by the potential gradient established by the resistivity of the glass. The tube is curved to stop electron feedback. A $10^5 - 10^7$ increase in signal is expected; however, the major constricting factor is the background noise of the system. Alternatively, the multiplier can be operated in analogue mode with a gain of only $10^3 - 10^4$ so that the multiplier does not become saturated and the pulses vary greatly in size. In this mode the applied voltage is between -500 V and -1500 V and the electron pulses are read at the collector electrode where they are amplified and averaged over a short time interval to allow rapid data acquisition. The greatest sensitivity is achieved with the detector in pulse counting mode, but the detector will become saturated at counting rates above 10^6 Hz, which are encountered when the analyte is at a high concentration in the sample. If the detector is switched into analogue mode it is less sensitive, but can be used for analyte concentrations which are much higher, typically up to three orders of magnitude higher than for pulse counting. Such dual mode operation results in an extremely large linear dynamic range of up to nine orders of magnitude. A discrete dynode detector consists of an array of discrete dynode multipliers, usually containing 15–18 dynodes, coated with a metal oxide that has high secondary electron emission properties. The dynodes are placed in one of two configurations, either venetian blind or box and grid fashion. Secondary electrons emitted by the metal oxide are forced to follow a circular path, by a magnetic field, so they strike successive dynodes thereby multiplying the signal.

21.3.3.3. Ion Extraction

An important aspect of plasma mass spectrometry is the sampling interface. Ions are physically extracted from the plasma into a mass spectrometer which is required to be at extremely low pressure, so the sampling interface must be in direct contact with the plasma (usually an ICP). The problem of extracting ions from an extremely hot plasma at atmospheric pressure into a mass spectrometer at $\sim 10^{-4}$ Pa (10^{-9} atm) is overcome by making use of a series of differentially pumped vacuum chambers held at consecutively lower pressures. A schematic diagram of the ICP–MS

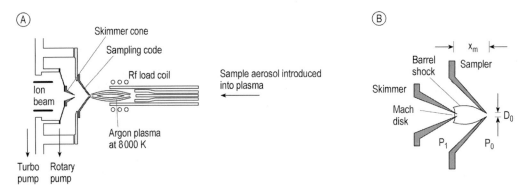

Figure 18. Ion extraction interface for ICP–MS

sampling interface is shown in Figure 18. The ICP is aligned so that the central channel is axial with the tip of a water-cooled, sampling cone, made of nickel or copper, which has an orifice of approximately 1 mm in diameter. The pressure behind the sampling cone is reduced, by means of a vacuum pump, to ~ 200 Pa $(2 \times 10^{-3}$ bar) so the plasma gases, together with the analyte ions, expand through the sampling orifice to form a shock-wave structure. This expansion is isentropic (i.e., no change in the total entropy) and adiabatic (i.e., there is no transfer of energy as heat), resulting in a supersonic expansion accompanied by a fall in temperature. This supersonic expansion takes the form of a cone with a shock-wave stucture at its base called a Mach disk. The region within the expansion cone is called the "zone of silence" which is representative of the ion species to be found in the ICP, i.e., the ionization conditions have been "frozen". The skimmer cone is another metal cone, the tip of which has an orifice of approximately 0.7 mm in diameter, that protrudes into the "zone of silence", and is axially in line with the sampling orifice. The ions from the zone of silence pass through the orifice in the skimmer cone, into a second intermediate vacuum chamber held at $< 10^{-2}$ Pa $(10^{-7}$ bar), as an ion beam. The ion beam can then be focused by means of a series of ion lenses, which deflect the ions along a narrow path and focus them onto the entrance to the mass analyzer.

21.3.4. Data Acquisition and Processing

Modern optical and mass spectrometers are controlled by computers, responsible for wavelength or ion mass presetting, parameter selection, as well as control of the radiation or ion source and the safety systems. The software for data acquisition and processing may comprise routines for the calculation of calibration functions with the aid of linear or higher order regression procedures, for calculation of sample concentrations by standard addition techniques or by calibration with synthetic samples, for calculation of statistical errors, and drift correction. Computers also offer graphical display of parts of the spectrum, or superimposition of spectra of several samples for recognition of spectral interference or background correction.

In most cases, a linear calibration function can be used (see Section 21.3.1). If c_i is the concentration of a standard sample and x_i is the radiation intensity, absorption, or ion current for the element to be determined, the linear regression has the form:

$$x_i = a c_i \qquad (125)$$

where a is the sensitivity, i.e., the slope of the calibration curve. To overcome source fluctuations and for maximum precision, it may be useful to plot signal ratios as the analytical signal:

$$x_i / R = a c_i \qquad (126)$$

where R is the signal obtained for a reference element, which may be the matrix element, a major component, or a foreign element added in known concentration to all samples to be analyzed. This approach is known as internal standardization. If the calibration function is nonlinear, it may be described by a second- or third-degree polynomial:

$$x_i / R = a_0 + a_1 c_i + a_2 c_i^2 \qquad (127)$$

or

$$x_i/R = a_0 + a_1 c_i + a_2 c_i^2 + a_3 c_i^3 \quad (128)$$

The coefficients a are calculated by regression procedures. Alternatively, different linear calibration functions may be used for different concentration ranges.

Interference or matrix effects occur when the analytical signals depend not only on the analyte element concentrations, but also on the concentrations of other sample components. Additive interference may be avoided by estimation of the magnitude of the interfering signal from a scan of the spectral background in the vicinity of the analytical line. Mathematical interference correction expresses the net analytical signal as:

$$x_i/R = (x_i/R)_{\text{measured}} - [d(x_i/R)/dc_{\text{interferent}}] \times c_{\text{interferent}} \quad (129)$$

where $c_{\text{interferent}}$ must be determined separately and $d(x_i/R)/dc_{\text{interferent}}$ is the interference coefficient for a certain sample constituent, determined from a separate series of measurements. One can also express the calibration function as:

$$x_i/R = a_1 c_1 + a_2 c_2 \quad (130)$$

where a_1 and a_2 are determined by regression, which is straightforward for a linear relationship.

Certain matrix constituents may produce signal enhancement or reduction. These are multiplicative in nature and can be due to influences on sampling efficiency, analyte transport into the source, or generation of the species delivering the analytical signals. They can be written as:

$$Y = a(c_{\text{interferent}})c \quad (131)$$

where the sensitivity a is a function of the concentration of the interferent $c_{\text{interferent}}$. This relation can be linear or of higher order. Calibration by standard addition allows correction of errors arising from signal reduction or enhancement insofar as the spectral background is fully compensated.

When calibrating by standard addition, the concentration of the unknown sample can be determined graphically. This is done by extrapolating to zero the curve of the signals from samples to which known amounts of analyte have been added plotted against the added amounts. Standard addition can be very useful in atomic absorption and plasma mass spectrometry, which are zero-background methods. However, for atomic emission, where the lines are superimposed on a spectral background from the atomic emission source itself, and which is highly dependent on the sample matrix, it is more difficult. Here, calibration by standard addition can be used only when a highly accurate background correction is applied. It is important that the curve remains within the linear dynamic range of the method. This can be problematic atomic absorption spectrometry.

The precision obtainable in standard addition methods is optimum when the added amounts of analyte are of the same order of magnitude as the analyte concentrations present in the sample. An estimate of the analytical error can be made from the propagation of the errors of the measured intensities [64]. If X is a function of n measured values, each resulting from N replicate measurements, the standard deviation of the estimate $s(X)$ can be calculated from their estimates x_k ($k = 1, 2, 3, \ldots, n$) and standard deviations s_k:

$$s^2(X) = \sum_{k=1}^{n} [(\delta X/\delta x_k) s_k]^2 \quad (132)$$

For one standard addition [65], the intensities I_x and $I_{x+\Delta}$ for the unknown sample c_x and the standard addition sample $c_{x+\Delta}$ are:

$$c_x = a I_x \quad (133)$$

$$c_x + \Delta c = a I_{x+\Delta} \quad (134)$$

and

$$c_x = \Delta c I_x / (I_{x+\Delta} - I_x) \quad (135)$$

The standard deviation is obtained by applying Equation (132) to c_x. If I_x and $I_{x+\Delta}$ scatter independently and the added concentration Δc is free of error [66]:

$$s^2(c_x) = \{c_x[I_{x+\Delta}/I_x][s(I_x)]/[I_{x+\Delta} - I_x]\}^2 + \{c_x[s(I_{x+\Delta})]/[I_{x+\Delta} - I_x]\}^2 \quad (136)$$

or

$$s_r^2(c_x) = [I_{x+\Delta}^2/(I_{x+\Delta} - I_x)][s_r^2(I_x) + s_r^2(I_{x+\Delta})] \quad (137)$$

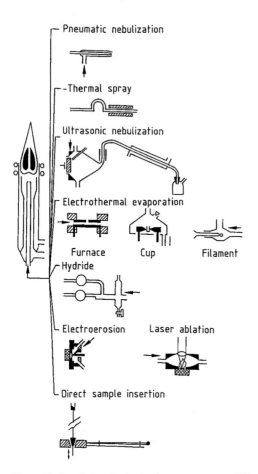

Figure 19. Sample introduction in plasma spectrometry [72]

21.4. Sample Introduction Devices

In atomic spectrometry, the sample must be converted to aerosol form and atomized before excitation. Accordingly, sample introduction is a very important step in all atomic spectrometric methods. The approach varies for different types of sample (size, state of aggregation stability, etc.) [67]–[71].

If sample volatilization and signal generation take place in a single source, both thermal evaporation and cathodic sputtering may occur. The latter is especially important in low-pressure discharges, as the ions can reach high energies when passing through the high-energy zones of the discharge and energy losses are small, because of the low number of collisions. In plasma sources at atmospheric pressure or flames, aerosol generation is often performed in a separate device. Figure 19 shows a number of sampling techniques for the analysis of solutions and solids. The techniques used for plasma spectrometry [73] and for sampling by cathodic sputtering, as performed in low-pressure discharges, are treated in detail in what follows.

21.4.1. Pneumatic Nebulization

For analysis of solutions, pneumatic nebulization is well known from early work on flame emission spectrometry. Pneumatic nebulizers must produce a fine aerosol with a gas flow, low enough to transport the aerosol droplets through the source with low velocity; this is essential for complete atomization. Nebulization of liquids is based on the viscosity drag forces of a gas flow passing over a liquid surface, producing small, isolated droplets. This may occur when the liquid is passed through a capillary tube, with the exit gas flows concentric around the tube, or perpendicular to the liquid stream. It may also occur at a frit, which is continuously wetted and through which the gas passes. Nebulizers are mounted in a nebulization chamber, which separates off the larger particles. Impact surfaces are often provided, fractionate the aerosol droplets further, and increase the aerosol production efficiency. The liquid flow is typically a few milliliters per minute and the maximum efficiency (a few percent) can be achieved at liquid flows of 2 mL/min; even below this flow rate, droplet diameters of the order of 10 µm and injection velocities of the order of 10 m/s are obtained. A number of nebulizer types are used. For flame emission, atomic absorption, and for plasma spectrometry, they include concentric nebulizers, cross-flow nebulizers, Babington nebulizers, and fritted-disk nebulizers (Fig. 20) [68].

Concentric nebulizers date back more than a century, to GOUY [75], and in principle are self-aspirating. They can be made of metal, plastic, or glass. In atomic absorption spectrometry, Pt–Ir capillaries are often used to allow the aspiration of highly acid solutions. In plasma atomic spectrometry, concentric glass nebulizers (Meinhard) [74] are well known. They use a rather lower gas flow (1–2 L/min) than atomic absorption types (up to 5 L/min). In both cases, aspiration rates are 1–2 mL/min and aerosol efficiencies 2–5%. Direct injection nebulizers are a useful alternative for the introduction of small-volume samples [76].

Cross-flow nebulizers are less sensitive to salt deposits at the nozzle. They are used in plasma

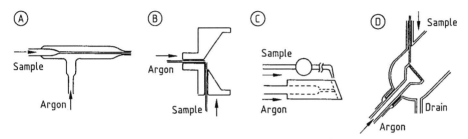

Figure 20. Pneumatic nebulizers for plasma spectrometry [68]
A) Concentric glass nebulizer [74]; B) Cross-flow nebulizer; C) Babington nebulizer; D) Fritted-disk nebulizer

spectrometry and flame work, and require a peristaltic pump for sample feeding. Cross-flow nebulizer capillaries can also be made of metal, glass, or plastic and have similar characteristics to those of concentric nebulizers. Materials such as poly(*p*-phenylene sulfide) (Ryton) enable aspiration of solutions containing HF.

In a *Babington nebulizer*, the sample solution is pumped through a wide tube and over a small gas orifice, so slurries can be analyzed without the risk of clogging.

Fritted-disk nebulizers have high nebulization efficiencies and low sample consumption (down to 0.1 – 0.5 mL/min). They are especially useful for plasma spectrometry of organic solutions, or for coupling with liquid chromatography [77]. However, in comparison with other types of nebulizer they suffer from memory effects; rinsing times are much higher than the 10 – 20 s required for the other pneumatic nebulizers.

With pneumatic nebulizers, samples can be taken up continuously. After a few seconds of aspiration, a continuous signal is obtained, and can be integrated over a certain time. Alternatively, small sample aliquots can be injected directly into the nebulizer [78] – [81] or into a continuous flow of liquid. The latter is known as *flow injection analysis* (*FIA*) [82]. The signals in both cases are transient; they have a sharp rise and some tailing, resulting from displacement of the aerosol cloud at the signal onset and dilution of the aerosol at its end. Signals last for a few seconds and with sample volumes 10 – 20 µL are capable of giving sufficient signal intensities. These techniques enable analysis of small samples (down to some 10 µL), reduce salt deposits, and maximize the absolute power of detection. FIA are also amenable to automatation and on-line sample preparation steps, such as preconcentration and matrix removal by coupling to column chromatography [83] – [85].

Optimization of pneumatic nebulizers is concentrated on selection of the working conditions for optimum droplet size and efficiency. The so-called Sauter diameter d_0, i.e., the diameter for which the volume-to-surface ratio equals that of the complete aerosol, is given by the Nukuyama – Tanasawa equation [86]:

$$d_0 = (C/v_G)(\sigma/\varrho)^{0.5} + C'\{\eta/[(\sigma\varrho)^{0.5}]\}^{C'} [1000\,(Q_L/Q_G)]^{C'''} \quad (138)$$

where v_G is the gas velocity, Q_G the gas flow, Q_L the liquid flow, η the viscosity, ϱ the density, σ the surface tension of the liquid, and C, C', and C'' and C''' are constants.

When the nebulizer gas flow increases, d_0 becomes smaller, the sample introduction efficiency increases, and so do the signals. However, as more gas is blown through the source, it is cooled, and the residence time of the droplets decreases, so that atomization, excitation, and ionization also decrease. These facts counteract increase of the signals as a result of the improved sampling; maximum signal intensity and power of detection are achieved at a compromise gas flow.

The physical properties of the sample liquid (σ, η, ϱ) also influence d_0, the efficiency, and the analytical signals. Differences in the physical properties of the sample liquids thus lead to so-called nebulization effects. Viscosity has a large influence through the second term in Equation (138). However, in free sample aspiration, it also influences d_0 through Q_L, which is given by the Poiseuille law:

$$Q_L = [(\pi R_L^4)/(8\eta l)]\,\Delta P \quad (139)$$

where R_L is the diameter and l the length of the capillary. ΔP is the pressure difference:

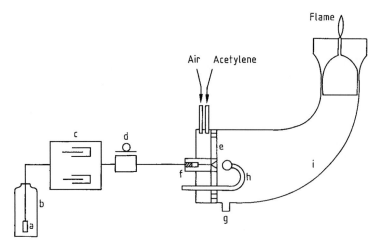

Figure 21. Hydraulic high-pressure nebulization system [94]
a) Solvent filter (20 µm); b) Solution reservoir (water, organic solvents;c) High-pressure pump; d) Sample feed valve/sample loop; e) Plate with concentric holes; f) High-pressure nebulization nozzle (10–30 µm) with integrated protection filter; g) Drain; h) Impact bead; i) Gas mixing chamber

$$\Delta P = \Delta P_U - g h \varrho - f(\sigma) \quad (140)$$

where ΔP_U is the difference between the pressure inside and outside the nebulizer, h is the height and $f(\sigma)$ a correction factor. In order to minimize nebulization effects, forced feeding with a peristaltic pump can be applied, so that Q_L is no longer a function of η. The nebulizer can also be operated at high gas flow, so that Equation (138) reduces to:

$$d_0 = C/v_G(\sigma/\varrho)^{0.5} \quad (141)$$

when $Q_G/Q_L > 5000$. However, the gas flow may cool the source, so that incomplete atomization, or signal reduction/enhancement may lead to even stronger interference.

Aerosol droplet size distributions can be determined by stray-light measurements [87], [88]. Data obtained from sampling with cascade impactors may suffer from evaporation of the droplets on their way through the device, especially as sampling may require the application of underpressure [89]. Data for different nebulizers are available [90]. Particle size distributions depend on the nebulizer and its working conditions, on the liquid nebulized, and on the nebulization chamber. These relations have been studied in depth for organic solutions [91], [92], for which, at low gas flow, pneumatic nebulizers often produce an aerosol with a lower mean particle size compared with aqueous solutions, and thus a higher efficiency and power of detection.

Apart from pneumatic nebulization by gas flow over a liquid surface, other methods have been developed. It is possible to use the impact of a high pressure jet of liquid on a surface (*jet impact nebulization*) [93], or of the expansion of a liquid jet under high pressure at a nozzle (*high-pressure nebulization*), as described by BERNDT [94] (Fig. 21). Much higher nebulization efficiencies can be obtained, even for solutions with high salt content or oils. This is due to the small droplet size, ca. 1–10 µm [95]. The system is ideally suited for speciation work as it is the appropriate interface for HPLC and atomic spectrometry [96]. A further advantage lies in the fact that the aerosol properties are independent of the physical properties of the liquid. In plasma spectrometry, however, it is necessary to desolvate the aerosol, as high water loadings cool the plasma too much and decrease the analytical signals; they may also lead to increased matrix effects. Desolvation can be done by heating the aerosol followed by solvent condensation, as applied in atomic absorption by SYTI [97]. New techniques involve the use of Peltier elements for cooling in aerosol desolvation. High-pressure nebulization is very promising for on-line preenrichment or separation by high-performance liquid chromatography.

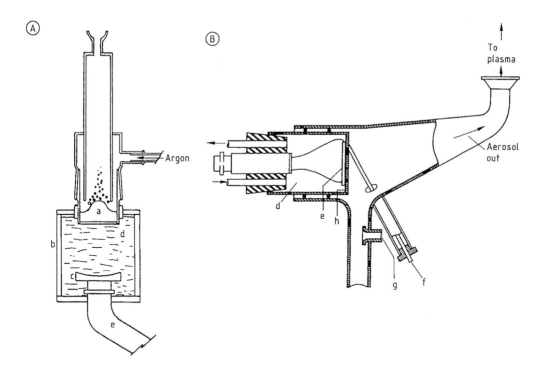

Figure 22. Ultrasonic nebulization
A) Discontinuous system [231]; B) System with liquid flowing over the transducer [99] a) Sample solution; b) Focusing chamber; c) Focusing lens; d) Cooling water; e) Ultrasonic transducer; f) Sample introduction; g) Gas inlet; h) Transducer plate

21.4.2. Ultrasonic Nebulization

By the interaction of sufficiently energetic acoustic waves of suitable frequency with the surface of a liquid, vortices are formed which produce an aerosol. The diameter of the aerosol droplets depends on the frequency and the physical properties of the liquid. For water, aerosol droplets formed at a frequency of 1 MHz have a diameter of ca. 4 µm. The energy can be focused with the aid of a liquid lens on the surface of the sample solution, or the sample liquid can be pumped continuously over the transducer, which must be cooled efficiently (Fig. 22).

Ultrasonic nebulization has two advantages over pneumatic nebulization. The aerosol particles have a lower diameter and a narrower particle size distribution compared with pneumatic nebulization (< 5 compared with 10 – 25 µm). Therefore, aerosol production efficiency may be up to 30 %, and analyte introduction efficiency is high. No gas flow is required for aerosol production, the transport gas flow can be selected completely freely. However, when applying ultrasonic nebulization to plasma spectrometry, it is necessary to desolvate the aerosol, to prevent too intensive cooling of the plasma. After this measure, ultrasonic nebulization leads to an increase in the power of detection.

Memory effects in ultrasonic nebulization are generally higher than in the case of pneumatic nebulization. The nebulization of solutions with high salt concentrations may lead to salt deposition; special attention has to be paid to rinsing, and precision is generally lower than in pneumatic nebulization. Its state of the art is summarized in [98].

21.4.3. Hydride Generation

For elements with volatile hydrides, sampling efficiency can be increased by volatilizing the hydrides. This applies to elements As, Se, Sb, Te, Bi,

Sn, and some others. By in-situ generation of the hydrides of these elements from sample solutions, sampling efficiency can be increased from a few percent in the case of pneumatic nebulization to virtually 100%.

Hydride generation can be efficiently performed by reduction with nascent hydrogen, produced by the reaction of zinc or $NaBH_4$ with dilute acids. In the latter case, the sample can be brought into contact with a solid pellet of $NaBH_4$, which may prove useful as a microtechnique [100]. Alternatively, a stream of $NaBH_4$ solution stabilized with NaOH can be added to an acidified sample. This can be done in a reaction vessel, as in atomic absorption work [101], or in a flow cell, as in plasma atomic emission spectrometry [102], [103]. The hydrides produced are separated from the liquid and subsequently transported by a carrier gas into the source. The hydrides are accompanied by excess hydrogen. In weak sources, such as microwave discharges, this may disturb discharge stability. To avoid this the hydrides can be separated off, e.g., by freezing them out and sweeping them into the source by subsequent heating [104]. Permselective membranes may also be useful [105].

Hydride generation increases the power of detection of atomic spectrometric methods for the determination of certain elements, and allows their matrix-free determination. However, the technique is prone to a number of systematic errors. First, the hydride-forming elements must be present as inorganic compounds in a well-defined valence state. This may require sample decomposition prior to analysis. In water analysis, treatment with H_2SO_4/H_2O_2 may be effective [106]. Traces of heavy metals such as Cu^{2+} may have a catalytic influence on the formation and dissociation of the hydrides, as investigated by WELZ et al. [107] in atomic absorption with quartz cuvettes. These interferents can be masked by complexation with tartaric acid or coprecipitated with $La(OH)_3$. Calibration by standard addition is advisable.

Mercury compounds can also be reduced to metallic mercury with nascent hydrogen or Sn(II). The released Hg can be transferred to an absorption cell or plasma source. It can also be trapped on a gold substrate, from which it can be swept into the source after release by heating. This approach allows effective preconcentration and, in the case of optical and mass spectrometry, leads to very sensitive determinations of Hg.

21.4.4. Electrothermal Evaporation

Thermal evaporation of analyte elements from the sample has long been used in atomic spectrometry. In 1940, PREUSS [108] evaporated volatile elements from a geological sample in a tube furnace and transported the released vapors into an arc source. Thermal evaporation has also been used in double arc systems, where selective volatilization gives many advantages for direct solid analysis. Electrothermal evaporation became especially important with the work of L'VOV [3] and MASSMANN [4], who introduced electrothermally heated systems for the determination of trace elements in dry solution residues by atomic absorption spectrometry of the vapor cloud.

21.4.4.1. The Volatilization Process

Electrothermal evaporation can be performed with dry solution residues as well as with solids. In both cases, the analyte evaporates and the vapor is held for a long time inside the atomizer, from which it diffuses. The high analyte concentration in the atomizer results from a formation and a decay function. The formation function is related to the production of the vapor cloud. After matrix decomposition, elements are present in the furnace as salts (nitrates, sulfates, etc.). They dissociate into oxides as a result of the temperature increase. In a device made of carbon (graphite), the oxides are reduced in the furnace:

$$MO_{(s/l)} + C_{(s)} \rightleftharpoons M_{(g)} + CO_{(g)}$$

However, a number of metals tend to form carbides, which can be very stable (thermodynamic control) or refractory (kinetic control). In this case, no analyte is released into the vapor phase. Decay of the vapor cloud is influenced by several processes:

1) Diffusion of the sample liquid into the graphite, which can often be prevented by the use of tubes, pyrolytically coated with carbon
2) Diffusion of the sample vapor
3) Expansion of the hot gases during the temperature increase (often at a rate of more than 1000 K/s)
4) Recombination processes, minimized by transferring the sample into the electrothermal device on a carrier platform with low heat capacity [109]
5) Action of purging gases

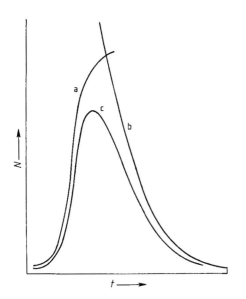

Figure 23. Signal form in graphite furnace atomic absorption spectrometry
a) Supply function; b) Removal function; c) Overall response function

Therefore, in electrothermal devices, transient signals are obtained. They have a sharp increase and a more or less exponential decay, lasting 1–2 s. Their form has been studied for volatilization processes [110]. The signal form (Fig. 23) is also influenced by adsorption and resorption of analyte inside the electrothermal device.

21.4.4.2. Types of Electrothermal Device
(Fig. 24)

Graphite Furnaces. In most cases, the furnace is made of graphite, which has good thermal and corrosion resistance [3], [111]. As a result of its porosity, it can take up the sample without formation of appreciable salt deposits at the surface. Atomizers made of refractory metals such as tungsten have been used [112], [113].

Graphite tubes with internal diameters of some 4 mm, wall thickness 1 mm, and a length up to 30–40 mm are mostly used. Filaments enabling the analysis of very small sample volumes and mini-cups of graphite held between two graphite blocks have also been described [111]. When graphite tube furnaces are used for optical spectrometry, the optical path coincides with the axis of the tubes, whereas in filaments and cups, the radiation passes above the atomizer. Furnace currents up to 100 A and potentials of a few volts are

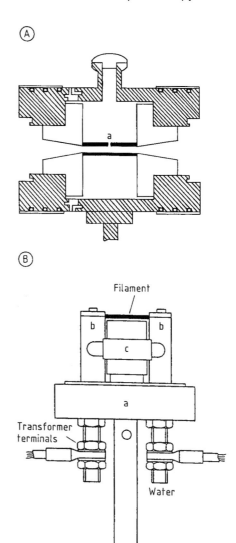

Figure 24. Graphite atomizers used in atomic spectrometry
A) Original graphite tube furnace according to MASSMANN [4]; B) Carbon-rod atomizer system according to WEST [109] a) Sampling hole; b) Heating blocks; c) Graphite rod or cup

used, giving temperatures of 3000 K or more. The tube is shielded by an argon flow, assuring stable working conditions for up to several hundreds of firings. With filaments, a lower power is used and the system can be placed in a quartz enclosure to keep out the air. Cups and tubes are nowadays often made of pyrographite, which prevents the analyte solution entering the graphite. Accordingly, chromatographic effects leading to selective volatilization are avoided, and the formation of refractory carbides, which hampers the volatilization of elements such as Ti or V, decreases.

Refractory Metal Furnaces. Electrothermal furnaces made of refractory metals (tungsten especially) have been described by SYCHRA et al. [112] for use in atomic absorption work. Their heat capacity is generally smaller that of graphite tubes, which results in steeper heating and cooling. This may be extreme in the case of tungsten probes and cups [113], which are mechanically more stable than graphite probes. The signals are extremely short and the analyte is released in very short times, which leads to high signal-to-background ratios and extremely low detection limits, as in wire loop atomization in atomic absorption spectrometry [114] or wire loops used for microsample volatilization in plasma spectrometry [115]. Therefore, metal devices used for electrothermal evaporation find it hard to cope with high analyte loading. Small amounts of sample are easily lost during heating, as the sample is completely at the surface and thermal effects cause stress in the salt crystals, formed during the drying phase. In graphite, the sample partly diffuses into the graphite, suppressing this effect. Graphite is further advantageous, as for many elements, reduction occurs, which leads to free element formation as required for atomic absorption spectrometry.

21.4.4.3. Temperature Programming

When electrothermal evaporation is used for sample introduction, the development of a suitable temperature program is of prime importance. In liquid samples, a small sample aliquot (10–50 µL) is transferred to the electrothermal device with a syringe or with the aid of an automated sampler and several steps are performed.

Drying. The solvent is evaporated and the vapor allowed to escape, e.g., through the sampling hole in the case of a graphite furnace. This step may last from tens of seconds to several minutes, and may take place at a temperature near the boiling point, e.g., 105 °C for aqueous solutions. This procedure often comprises several steps, advisable in the case of serum samples to avoid splashing.

Matrix Destruction. During this step, the matrix is decomposed and removed by volatilization. Chemical reactions are often used to enable volatilization of matrix constituents or their compounds. The temperature must be selected so that the matrix, but not the analyte, is removed. This is often achieved by applying several temperatures or even by gradually increasing the temperature at one or more ramping rates. Temperatures during this step are between 100 and 1000 °C, depending on the matrix to be removed and the analyte elements. Thermochemical reagents are often used to assist matrix destruction chemically and to achieve complete matrix removal at a lower temperature. Quaternary ammonium salts are often employed for the destruction of biological matrices at relatively low temperatures. Matrix decomposition is very important to avoid the presence of analyte in different chemical compounds, which can lead to transient peaks with several maxima.

Evaporation. The temperature selected for analyte evaporation depends strongly on the analyte elements and may range from 1000 K for volatile elements (e.g., Cd, Zn) to 3000 K for refractory elements (Fe). This step normally lasts no longer than 10 s, to maximize the number of firings which can be performed with a single tube.

Heating. The temperature is increased to its maximum (e.g., 3300 K over ca. 5 s with graphite tubes) to remove any sample residue from the evaporation device and to minimize memory and cross-over effects.

Direct solid sampling in electrothermal evaporation can be performed by dispensing an aliquot of a slurry prepared from the sample into the furnace. The analytical procedure is completely analogous to that with solutions. Powders can also be sampled with special dispensers, e.g., as described by GROBENSKI et al. [116]. They allow sampling of a few milligrams of powder reproducibly. Analyses with such small amounts make high demands on sample homogeneity. In direct powder sampling, the temperature program often starts with matrix decomposition and proceeds as for dry solution residues. In modern atomic spectrometers, sampling temperature programming, selection of gas flow, and visualization of the temperature program and signals are mostly controlled by computer.

The development of the temperature program is most important in the methodology of a spectrometric method using electrothermal evaporation. It should be documented completely in an analytical prescription for the determination of a given series of elements in well-defined samples.

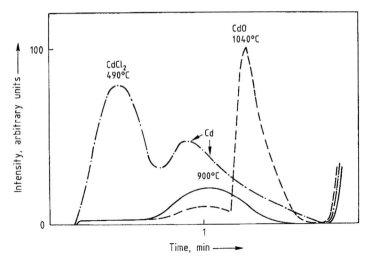

Figure 25. Thermochemical behavior of Cd and its compounds in graphite furnace evaporation [66]
ICP3kW argon–nitrogen; samples 10 µg/mL Cd(NO$_3$)$_2$ in aqueous solution (solid line), added to 0.1 g/mL of SRM 1571 (dashed line) and SRM 1577 (NaCl) (dash-dotted line); Cd II 226.5 nm line; sampling volume 50 µL; temperature program: drying 95 °C, gradual temperature increase from 350–2400 °C (600 °C/min)

21.4.4.4. Analytical Performance

Electrothermal atomization, because of its high analyte vapor generation efficiency (in theory 100%), can yield extremely high absolute and relative detection power with any type of atomic spectrometry. In two-stage procedures, where the analyte vapor has to be transported into the signal generation source, diffusional losses of analyte vapor may occur. This has been described in detail for Cd [117], but is a general problem.

Owing to the transient nature of the analytical signals, precision is generally lower than with nebulization of liquids. Relative standard deviations are 3–5% for manual injection of microaliquots, and 1–3% for automated sample dispensation, whereas in pneumatic nebulization they are below 1%.

Interference in electrothermal evaporation may stem from differences in the physical properties of the sample liquids from one sample to another. This may influence the wetting of the graphite or the metal of the electrothermal device. When the latter is temperature dependent, it leads to differences in volatilization. Differences in the anions present may severely influence evaporation (chemical matrix effect). The boiling points of the compounds formed dictate the volatilization. Accordingly, the occurrence of double peaks is easily understood. In the case of Cd, this is documented for a sample rich in chloride (CdCl$_2$ evaporates at 480 °C) and for a sample rich in nitrate (Cd volatilizes first at 900 °C and then at 1200 °C as CdO; Fig. 25) [66].

21.4.5. Direct Solid Sampling

These techniques are especially useful in two-stage procedures for solid analysis. A separate source is used for the generation of a sufficiently concentrated sample vapor cloud, the composition of which truly reflects the sample. Several approaches may be used according to sample properties.

21.4.5.1. Thermal Methods

For powders, electrothermal evaporation may be used. It is often hampered by sample inhomogeneity, as it is generally a micro method, as well as by anion effects, causing chemical interference [118]. Transport losses may also occur. These vanish when the sample is inserted directly in the signal generation source and evaporated thermally. This approach is known from work with graphite or metal probes in atomic absorption, where tungsten wire cups and loops are used. In spectrometry with inductively coupled plasmas, the technique is used both in atomic emission [119]–[121] and in mass spectrometry [122]. Its absolute power of detection is extremely high and the technique can be used both for the analysis of dry solution residues as well as for the volatilization of micro-amounts of solids.

In-situ thermal destruction of the matrix or selective volatilization can be applied. The latter has proved useful in geological samples [123]. Selective volatilization of volatile elements from refractory matrices is useful; spectral interference from matrices with complex spectra can be avoided. The approach can also be used for the volatilization of refractory oxide- or carbide-forming elements, which form volatile halogenides (e.g., Ti). Here, substances such as AgCl or polytetrafluoroethylene (PTFE) powder can be used as thermochemical reagents [121]. In the case of Ti in the presence of C and PTFE, the following reactions have to be considered:

$TiO_2 + 2 C \rightleftharpoons Ti + 2 CO$
$Ti + C \rightleftharpoons TiC$
$Ti + 4 F \rightleftharpoons TiF_4$

At high temperature, the latter reaction is favored thermodynamically as well as kinetically (TiF_4 is volatile).

21.4.5.2. Slurry Atomization

Powders often form stable slurries in suitable dispersing fluids. The latter should be able to wet the powder and to form stable suspensions, both of which depend on their physical properties (viscosity, density, surface tension). Surface tension, which is generally a function of pH, is known to influence the surface charge of suspended particles and hence the stability of the slurry (for ZrO_2 powders see [124]). The stability of suspensions depends greatly on the particle size of the powder and its tendency to form agglomerates. The latter depends on the preparation method applied, as known from synthetic ceramic powders [125], [126]. The particle size should be in the low micrometer range at most. The particle size distribution can be determined by laser diffractometry or by automated electron probe microanalysis [127], and is particularly critical when the aerosol is produced by pneumatic or ultrasonic nebulization. Measurements with powders of different particle size have shown that particles larger than 12–15 µm cannot be held in aerosol form with gas flows ≤ 2 L/min, as in the nebulizers used in plasma spectrometry [198]. In the other techniques of sample introduction, particle size is not so critical. This applies when slurries are analyzed by graphite furnace atomic absorption spectrometry. Volatile elements can be evaporated selectively from some types of sample, as recently shown for ceramic powders [128]. In any case, it is often necessary to apply ultrasonic stirring to destroy agglomerates and to disperse powders optimally, prior to slurry analyses. The addition of surface active substances such as glycols [129] has been proposed, but may introduce contamination when trace determinations are required.

In pneumatic nebulization, it is advisable to stir the sample continuously during slurry aspiration. A Babington nebulizer should be used, as other types are prone to clogging. The aerosol loaded with the solid particles, with or without a solvent layer, is transported to the source. Complete drying and evaporation on its way through the source is necessary for calibration by standard addition with aqueous solutions of the analyte. From knowledge of the temperature distribution in the source, the trajectory of the particles through the source, and the thermochemical properties of the sample material, it is possible to calculate the degree of volatilization for particles of a given size and composition at a certain point in the source, or the maximum size for particles to be completely volatilized [126], [130], [131].

Slurry analysis can also be applied to compact samples which are difficult to dissolve after pulverization. Attention must be paid to possible contamination resulting from abrasion in the mill. For mills of very hard materials such as tungsten carbide abrasion in the case of ceramics may amount to serveral percent of the sample mass. Slurry sampling has been reviewed comprehensively in [132].

21.4.5.3. Arc and Spark Ablation

Arc ablation has long been used for producing an aerosol at the surface of electrically conducting samples, in combination with various sources. In the version described by JOHNES et al. [133], the metal sample acts as the cathode of a d.c. arc discharge. With an open circuit voltage of 600 V and currents of 2–8 A, a broad pulse spectrum (mean frequency up to 1 MHz) is observed and rapid movement of the discharge across the cathode produces uniform sampling over a well-defined area. A gas stream can then transport the aerosol particles to the signal generation source. Remote sampling at up to 20 m makes the technique attractive for the analysis of large samples. On the other hand, it is also useful for precision analyses, both by atomic absorption and plasma spectrometric methods. The analysis of electrically insulating powders is possible. A mixture of the powder to be analysed and copper powder in a ratio of about 1:5 can be made and formed into a pellet at pressures up to 80 t/cm² (ca. 80 MPa) [134]. The mass ratio of analyte to copper may differ considerably from one sample to another.

Spark Ablation. When sparks are used, the ablation of electrically conducting solids is less dependent on matrix composition. This applies both to high-voltage sparks (up to 10 kV) and to medium-voltage sparks, as used for emission spectrometric analysis of metals. Voltages of 500–1000 V with repetition rates up to 1 kHz are used. Increase in voltage and spark repetition rate leads to a considerable increase of the ablation rate. For aluminum samples, trapping the ablated material in concentrated HNO_3 increases the ablation rate from 1–5 µg/min for a 25-Hz spark. However, the particle diameter also increases significantly hampering volatilization of the material in a high-temperature source such as ICP, and leading to flicker noise [135]. It is better to use a medium-voltage spark at a high sparking frequency [136]. After aerosol sampling on Nucleopore filters, X-ray fluorescence spectrometry shows that the composition of the aerosol produced in aluminum samples agrees with the composition of compacted samples [137]. The particle size in the case of 400-Hz spark is 1–2 µm and the particles are mostly spherical, which argues for their formation by condensation outside the spark channel. In the case of supereutectic Al–Si alloys (c_{Si} > 11 wt %) some elements (especially Si) are enriched in the smaller particles.

Spark ablation is very abrasive when performed under liquid. BARNES and MALMSTADT [138] used this effect to reduce errors stemming from sample inhomogeneity in classical spark emission spectrometry. The approach is also useful for the dissolution of refractory alloys, which are highly resistant to acids [139]. When sparking under liquid, ablation rates up to ca. 3 mg/min can easily be achieved. In high-alloy Ni–Cr steels, electron probe micrographs show that selective volatilization can become problematic. Very stable colloids can be formed when a ligand such as ethylenediaminetetraacetic acid is present in the liquid.

21.4.5.4. Laser Ablation (→ Laser Analytical Spectroscopy)

The interaction of the radiation from high-power laser sources with solids can lead to evaporation. As this is independent of the electrical conductivity of the sample, laser ablation is increasingly important for solid analysis [140]–[142].

Sources. Laser sources make use of population inversion. When radiation enters a medium, both absorption and stimulated emission of radiation may occur and the change in flux at the exit is:

$$dF = \sigma F(N_2 - N_1)dz \qquad (142)$$

where dz is the length of the volume element in the z direction, σ is the cross section for stimulated emission or absorption and F is the flux. The sign of $(N_2 - N_1)$ is normally negative, as the population of the excited state N_2 is:

$$N_2 = N_1 \exp[-(E_2 - E_1)/kT] \qquad (143)$$

and is smaller than the population of the ground state N_1. In the case of population inversion $N_2 > N_1$, the medium acts as an amplifier. When it is positioned between two mirrors, one of which is semitransparent, energy can leave the system. If R_1 and R_2 are the reflectances of the mirrors, the minimum theoretical population inversion is:

$$(N_2 - N_1)_{th} = 1/(2\sigma l) \ln[1/(R_1 R_2)] \qquad (144)$$

as the losses by reflection in a double pathway $2l$ are compensated. Population inversion requires the existence of a three- or four-level energy diagram. Laser radiation is monochromatic and coherent. The beam has a radiance–divergence relation α_D ($\approx \lambda/D$, where D is the beam diameter). The medium is in a resonator of which the length d determines the resonance frequency $v = nc/(2d)$, where n is the mode.

Solid state lasers are especially of interest for laser ablation. The laser medium is a crystal or a glass, doped with a transition metal. The medium is pumped optically by flash-lamps (discharges of 100–1000 J over a few ms) or continuously with a tungsten–halogen lamp. The resonator may be the space between two flat mirrors, or an ellipsoid, with the laser rod at one focus and the flash-lamp at the other. The ruby laser emits in the visible region (694.3 nm). It is thermally very robust, but requires a high pumping energy and its energy conversion efficiency is low. The neodymium–yttrium aluminum garnet (Nd–YAG) laser is very widely used. Its wavelength is 1.06 µm and it has a much higher energy conversion. Gas lasers with CO_2 or Ar can be pumped electrically have a high power output, whereas dye and semiconductor lasers are mostly used for selective excitation.

Lasers can be operated in the Q-switch mode, so as to deliver their energy spikes very reproducibly. Optoacoustic or electroacoustic switches are often used. The first make use of a pressure-dependent change of refractive index of gases such as SF_6, which may be produced periodically at a suitable frequency (Boissel switch), whereas the latter are based on periodic changes of the transmittance of crystal in an a.c. electric field. As the interaction of laser radiation with solids highly depends on wavelength, frequency doubling, resulting from nonlinear effects, e.g., in $LiNbO_3$, is often used.

Interaction with Solids [142], [143]. When a laser beam with small divergence impinges on a solid surface, part of the energy (10–90%) is absorbed and material evaporates locally. The energy required varies between ca. 10^4 W/cm² (for biological samples) and 10^9 W/cm² (for glasses). A crater is formed, of minimum diameter determined by diffraction of the laser radiation:

$$d \approx 1.2 f \alpha_D v$$

where f is the focal length of the lens used for focusing the laser radiation on the sample (5–50 mm, minimum dictated by the risk of material deposits) and α_D is the divergency (2–4 mrad). Typical crater diameters are of the order of 10 µm. They also depend on the energy and the Q-switch used. The depth of the crater relates to the laser characteristics (wavelength, radiant density, etc.) as well as to the properties of the sample (heat conductance, evaporation energy, reflectance, etc.). The ablated material is ejected from the surface at high velocity (up to 10^4 cm/s); it condenses and is volatilized again by absorption of radiation. In this way, a laser vapor cloud is formed of which the temperature and the optical density are high, and the expansion velocity, composition, and temperature depend on the laser parameters and the gas atmosphere and pressure. Favorable working conditions in laser ablation work involve reduced pressures (1–10 kPa) [144]–[146]. Optical and mass spectrometry can be performed directly on the laser vapor cloud, or the ablated material can be transferred to another source.

Analytical Performance. Laser ablation is a microsampling technique and thus enables local distribution analyses. Laterally resolved measurements can be made with a resolution of ca. 10 µm [144]. Sampling depth can be varied from 1 to ca. 10 µm by adjusting the laser. The amounts of material sampled can be of the order of 0.1–10 µg. As a result of the high signal generation efficiencies possible with mass spectrometry or laser spectroscopy, its absolute power of detection is very high. The laser sources now available allow high precision (relative standard deviations ca. 1%), limited only by variations in reflectivity along the sample surface, and sample homogeneity.

21.4.6. Cathodic Sputtering

In low-pressure discharges, the atoms and ions undergo few collisions in the gas phase. Therefore, an electric field can induce in them the high energies required to remove material from solid lattices by impact and momentum transfer. Positive ions in particular can acquire the necessary energies; the phenomenon takes place when the sample is used as cathode and is known as cathodic sputtering. Its nature can be understood from the properties of low-pressure discharges. Models developed in physical studies reveal the analytical features of cathodic sputtering [147].

Low-Pressure Discharges. When a voltage is applied across two electrodes in a gas under reduced pressure, the ionization in the vicinity of the electrodes produces electrons, which may gain energy in the field and cause secondary ionization of the gas by collision. Secondary electron emission occurs when they impact on the electrodes. Field emission may take place near the electrodes (at field strengths above 10^7 V/cm) and, when the cathode is hot, glow emission can take place as well. In such a discharge, energy exchange takes place as a result of elastic and inelastic collisions, of which the latter may lead to ionization and recombination. With increasing voltage across the electrodes, a current builds up.

The characteristic (Fig. 26) includes the region of the corona discharge and the normal region, where the discharge starts to spread over the elec-

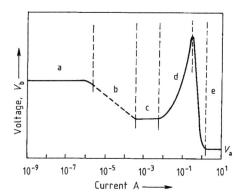

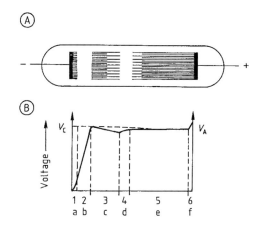

Figure 26. Current–voltage characteristic of a self-sustaining d.c. discharge
a) Townsend discharge; b) Corona range; c) Glow discharge with normal cathode fall; d) Glow discharge with abnormal cathode fall; e) Arc discharge
V_a = Arc voltage [148]; V_b = Breakdown voltage

Figure 27. Geometry (A) and potential distribution (B) of a d.c. electrical discharge under reduced pressure
a) Aston dark space; b) Hittorf dark space; c) Negative glow; d) Faraday dark space; e) Positive column; f) Anode region
V_A = Anode fall; V_C = Cathode fall

trodes. Here, the current can increase at practically constant voltage, whereas once the discharge covers the entire electrodes (restricted discharge), the current can only increase when the voltage is increased drastically (abnormal part of the characteristic). Here, the burning voltage, the positive space charge, and the field gradient in front of the cathode are very high.

Once the cathode has been heated to a sufficiently high temperature, thermal emission may start and the characteristic enters the arc discharge region, where burning voltage and space charge decrease rapidly. Such a discharge displays a cathode dark space where the energies are too high for efficient collisions, the cathode layer where intensive emission takes place, a further dark space, and the negative glow where the negative space charge is high (Fig. 27). The plasma is not in local thermal equilibrium, as the number of collisions is too low to thermalize the plasma. Electron temperatures are high (5000 K for the electrons involved in recombination and >10 000 K for the high-energy electrons responsible for excitation by electron impact), but the gas temperatures are low (<1000 K). In radio frequency (r.f.) discharges, which are increasingly important for atomic spectrochemical analysis, a barrier-layer and a bias voltage are built up in the vicinity of a metal electrode as a result of the different mobilities of electrons and ions in the r.f. field. Accordingly, the energy situation becomes very similar to that of d.c. discharges and similar sputtering phenomenona can occur.

Ablation. The models developed for cathodic sputtering start from ideal solids, i.e., single, perfect crystals; whereas real samples in atomic spectrochemical analyses are polycrystalline or even chemically heterogeneous. The available models are valid only for monoenergetic beams of neutrals impacting on the sample, whereas in fact both ions and atoms of widely different energies impact at different angles.

The ablation is characterized by a sputtering rate q (µg/s) and a penetration rate w (µm/s). The latter is the thickness of the layer removed per unit time and relates to the sputtering rate as:

$$w = (10^{-2} q)/(\varrho s) \tag{145}$$

where s is the target area (cm^2) and ϱ its density (g/cm^3). The sputtering yield S indicates the number of sputtered atoms per incident particle:

$$S = (10^{-6} q N_A e)/(M i^+) \tag{146}$$

where N_A is the Avogadro constant, e the charge on the electron (C), M the atomic mass, and i^+ the ion current (A).

Classical sputtering experiments with a monoenergetic ion beam in high vacuum show that the sputtering yield first increases with the mass of the incident ions and the pressure, but then decreases. For polycrystals, it is maximum at incident angle 30°. For single crystals, it is maximum in the direction perpendicular to a densely packed plane.

The results of these experiments can only be explained by the impulse theory. According to this theory a particle can be removed from a lattice site when the displacement energy E_d (the sum of the covalent or electrostatic binding energies) is delivered by momentum transfer from the incident particles. The maximum fraction of the energy transferrable from an incident particle is:

$$E_{max}/E = 4mM/\left[(m+M)^2\right] \qquad (147)$$

The ablation rate is thus proportional to the number of particles which deliver an energy equal to the displacement energy. It should, however, be taken into account that some incident particles are reflected (f_r) or adsorbed at the surface. Particles of low mass can penetrate into the lattice and be captured (f_p). Other particles enter the lattice and cause a number of collisions until their energy is below the displacement energy. The overall sputtering yield, accounting for all processes mentioned, is:

$$S = [(\alpha E)/E_d]^{1/2} (f_p + Af_r)\varphi$$

$$\alpha = 2mM/(m+M)^2$$

and

$$\varphi = f(m, M) \qquad (148)$$

Accordingly, cathodic sputtering increases with the energy of the incident particles and is inversely proportional to the displacement energy E_d. It is maximum when $m = M$. This explains why sputtering by analyte particles which diffuse back to the target is very efficient.

The dependence of sputtering yield on the orientation of the target with respect to the beam can be easily explained. In a single crystal, there is a focusing of momentum along an atom row in a direction of dense packing. If D_{hkl} is the lattice constant and d the smallest distance between atoms (or ions) during a collision, the angle θ_0, at which particles are displaced from their lattice sites relates to the angle θ_1, between the direction of the atom row and the line between the displaced atom and its nearest neighbor in the next row, according to the equation:

$$\theta_1 = \theta_0(D_{hkl}/d - 1) \qquad (149)$$

This focusing of momentum, referred to as Silsbee focusing, takes place when:

$$f = \theta_{n+1}/\theta_n$$

or

$$D_{hkl}/d < 2$$

Analytical Performance. The model described explains the features of cathodic sputtering as a technique for sample volatilization and is very helpful for optimizing the performance of cathodic sputtering sources [149]. When using sputtering for sample volatilization, it should be noted that some features can be realized only when working at sputtering equilibrium. On initiating a discharge, the burning voltage is normally high in order to break through the layer of oxides and of gases adsorbed at the electrode surface. When these species are sputtered off and the breakdown products are pumped away, the burning spot (crater) can start to penetrate with a constant velocity in the sample, so that the composition of the ablated material may become constant. The time required to reach sputtering equilibrium (burn-in time) depends on the nature and pretreatment of the sample as well as on the filler gas used and its pressure. All measures which increase ablation rate will shorten burn-in times.

The topography of the burning crater depends on the solid-state structure of the sample. It reflects the graininess, chemical homogeneity, and the degree of crystallinity. Inclusions and defects may locally disturb the sputtering. These effects can be observed on micrographs, comparing the craters obtained with a glow discharge and a spark (Fig. 28) [150]. The roughness of the burning crater can be measured with sensing probes; it imposes the ultimate limitation of the depth resolution obtainable when applying sputtering to study variations of sample composition with depth from the sample surface.

Achievable ablation rates depend on the sample composition, the discharge gas, and its pressure. As filler gas, a noble gas is normally used. With nitrogen or oxygen, chemical reactions at the sample surface occur and disturb the sputtering, as electrically insulating oxide or nitride layers are formed. Reactions with the ablated material produce molecular species which emit molecular band spectra in optical atomic spectrometry, or produce cluster ion signals in mass spectrometry. In both cases, severe spectral interference can hamper the measurement of the analytical signals. The relation between ablation rates and sample composition can be understood from impulse theory. In most cases argon ($m = 40$) is used as sput-

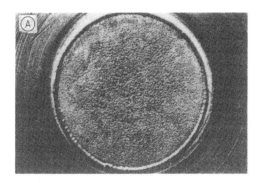

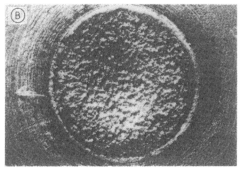

Figure 28. Burning spots obtained with a glow discharge (A) and a medium-voltage spark (B) Sample: aluminum [150]

tering gas, and the sequence of the ablation rates follows the mass sequence: C < Al < Fe < steel < Cu < brass < Zn

Helium is not suitable, as its small mass renders its sputtering efficiency negligible. The sputtering rates further increase in the sequence Ne < Ar < Kr < Xe

The last two gases are rarely used because of their price. Neon is attractive because of its high ionization potential; argon may cause spectral interference. Gas pressure has a very important influence on electrical characteristics. At low gas pressure the burning voltage is high, as is the energy of the incident particles. At high pressure, the number density of potential charge carriers is higher and the voltage decreases. The number of collisions also increases, so the energy of the incident particles decreases. The resulting decrease of sputtering rate with gas pressure for a glow discharge with planar cathode and abnormal characteristic [151], can be described:

$$q = c/\sqrt{p} \qquad (150)$$

where c is a constant and p is the pressure.

Many studies have been performed on volatilization by cathodic sputtering in analytical glow discharges used as sources for atomic emission, atomic absorption, and mass spectrometry [152]. Also, studies on the trajectories of the ablated material have been performed, as described, e.g., for a pin glow discharge [153].

21.5. Atomic Absorption Spectrometry

21.5.1. Principles

As a method for elemental determinations, atomic absorption spectrometry (AAS) goes back to the work of WALSH in the mid-1950s [2]. In AAS, the absorption of resonant radiation by ground state atoms of the analyte is used as the analytical signal. This process is highly selective as well as very sensitive, and AAS is a powerful method of analysis, used in most analytical laboratories. Its methodological aspects and applications are treated in several textbooks [154]–[156].

A primary source is used which emits the element-specific radiation. In the beginning, continuous sources were used with a high-resolution spectrometer to isolate the primary radiation. However, due to the low radiant densities of these sources, detector noise limitations occurred, or the spectral band width was too large for sufficiently high sensitivity.

Generally, a source which emits discrete lines was found more suitable, with low-cost monochromators to isolate the radiation. It was also necessary in early work to use sources in which the physical widths of the emitted analyte lines are low. This is necessary to get high absorbances (Fig. 29). When the band width of the primary radiation is low with respect to the absorption profile of the line, a higher absorption results from a given amount of analyte as compared with a broad primary signal. Narrow lines in the primary radiation are obtained with low-pressure discharges, as in hollow cathode lamps or low-pressure r.f. discharges. Recently, the availability of narrow-band, tunable laser sources, such as the diode lasers, has opened new perspectives [157], [158]. Only the analytical line is present and no monochromator is necessary. If tuning is applied, it is possible to scan across the absorption profile, to increase the dynamic range or to perform back-

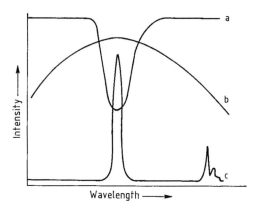

Figure 29. Importance of physical line widths in atomic absorption spectrometry
a) Absorption signal for elemental line; b) Spectral bandpass of monochromator; c) Emission of hollow cathode lamp

ground correction. Switching from one analyte element to another becomes feasible, certainly when several diode lasers are provided. However, laser sources are restricted by the limited spectral range which can be covered — at the time of writing, only down to the green.

The use of line sources generally leads to high analytical selectivity, sensitivity, and power of detection as well as a high dynamic range within which the Lambert – Beer law (Eq. 38) applies. A linear relationship between absorption and concentration can only apply when all radiation passing to the detector is absorbed to the same extent by the analyte atoms (ideal case). In the normal case, the curve of concentration and absorption falls off toward the concentration axis, as a result of the presence of nonabsorbed radiation. For a primary source emitting narrow lines (widths below one-fifth of those of the absorption lines) this consists mainly of contributions from nonabsorbed lines of the cathode material or of the filler gas, which fall within the spectral bandwidth of the monochromator. At high concentrations, a decrease in dissociation gives rise to lower absorbances, and suppression of ionization leads to higher absorbances.

For AAS, the analyte must be present as an atomic vapor, i.e., an atomizer is required. Both flames and furnaces are used, and the corresponding methodologies are known as flame AAS and graphite furnace AAS. Special methods of atomization are based on volatile compound formation, as with the hydride technique (Sections 21.4.3 and 21.5.5). AAS is generally used for the analysis of liquids, so solids must first be dissolved. Wet chemical decomposition methods are of use, but involve all the care required in trace analysis to prevent losses or contamination, and the resultant systematic errors. For direct solid analysis, a few approaches exist and are discussed in Chapter 21.4.

21.5.2. Spectrometers

Atomic absorption spectrometers contain a primary source, an atomizer with its sample introduction system, and a monochromator with a suitable detection and data acquisition system (Fig. 30).

21.5.2.1. Spectrometer Details (Fig. 30)

Radiation from the primary source (a) is led through the absorption volume (b) and subsequently into the monochromator (c). As a rule, radiation densities are measured with a photomultiplier and processed electronically. Czerny – Turner or Ebert monochromators with low focal length (0.3 – 0.4 m) and moderate spectral bandpass (normally not below 0.1 nm) are frequently used.

In most instruments, the radiant flux is modulated periodically. This can be achieved by modulating the current of the primary source, or with the aid of a rotating mirror (d) in the radiation beam. Accordingly, it is easy to differentiate between the radiant density emitted by the primary source and that emitted by the flame by lock-in amplification. Both single- and dual-beam instruments are used. In the latter, part of the primary radiation is fed directly into the monochromator, while the rest passes first through the flame. In this way, fluctuations and drift can be compensated insofar as they originate from the primary radiation source or the measurement electronics.

The spectrometer can be equipped for quasi-simultaneous measurement of the line and background absorption [159]. Radiation from a second, continuous source, such as a deuterium arc lamp, is rapidly switched with the radiation of the primary source and fed through the absorption volume, by a semitransparent mirror. The radiant flux of the continuous source is not decreased significantly by atomic absorption, owing to the low spectral resolution of the monochromator; however, it will be weakened by broad-band absorption of molecules, or stray radiation. By this means, nonelement-specific background absorption can

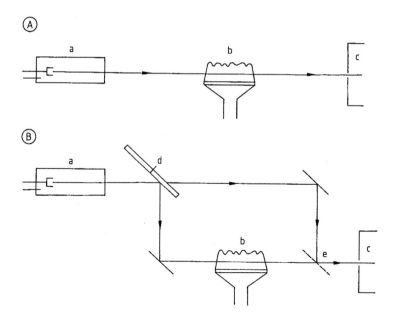

Figure 30. Flame atomic absorption spectrometer
A) Single beam; B) Dual-beam system
a) Source; b) Burner; c) Monochromator; d) Rotating mirror; e) Half-silvered mirror

be compensated. This and other approaches to background correction are discussed in detail in Section 21.5.6.

Dual-Channel Instruments. Apart from single-channel instruments, with which only measurements at a single wavelength in one channel can be performed, dual-channel instruments are also used. They contain two independent monochromators which enable simultaneous measurements at two wavelengths. They are of use for the simultaneous determination of two elements of which one element may be the analyte and the second a reference element. It is also possible to use two lines with widely different intensity to determine one element over a wide concentration range.

Multichannel spectrometers which allow the simultaneous determination of a large number of elements, as in atomic emission spectrometry have not encountered a breakthrough in AAS yet. However, over a number of years, work with high-intensity continuous sources and high-resolution echelle spectrometers for multielement AAS determinations has aroused some interest [160]. Fourier transform spectrometry and multichannel detection with photodiode arrays has opened new perspectives for the simultaneous detection of larger numbers of spectral lines, and high-intensity sources have improved considerably.

21.5.2.2. Primary Sources

The primary radiation sources used in AAS have to fulfill several conditions:

1) They should emit the line spectrum of one or more analyte elements with line widths smaller than those of the absorption lines in the atomizer
2) They should have high spectral radiance density (radiant power per unit area, solid angle, and unit wavelength) in the center of the spectral line
3) Their optical conductance (radiant area multiplied by usable solid angle) must be high
4) The radiances of the analytical lines must be constant over a long time

These conditions are fulfilled by low-pressure discharge such as hollow cathode discharges and, for some elements, high-frequency discharges.

In a commercially available hollow cathode lamp (Fig. 31), the cathode (a) has the form of a hollow cylinder closed at one side. The lamp is sealed and contains a noble gas at a pressure of the order of 100 Pa. At discharge currents up to

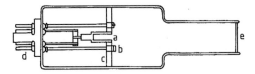

Figure 31. Hollow cathode source for atomic absorption spectrometry
a) Hollow cathode; b) Anode; c) Mica insulation; d) Current supply; e) Window (e.g., quartz)

10 mA (at ca. 500 V), a glow discharge is sustained between this cathode and a remote anode (b). The atomic vapor is produced by cathodic sputtering and excited in the negative glow contained within the cathode cavity. Lines of the discharge gas are also emitted, which may lead to interference in AAS. In most cases, high-purity argon or neon are used. For mechanical reasons, it may be necessary to manufacture the cathode mantle from a material other than that of the internal part of the hollow cathode, which is usually made of the analyte, so a further atomic spectrum may be emitted.

For a number of elements, lamps in which the hollow cathode consists of several elements may also be used. The number of elements contained in one lamp is limited because of the risk of spectral interference.

For a few elements, electrodeless discharge lamps are to be preferred over hollow cathode lamps. This applies for volatile elements such as As, Se, and Te. In the hollow cathode lamps of these relatively volatile elements self-absorption at low discharge currents may be considerable, and self-reversal may even occur. This is not the case with electrodeless discharge lamps. The latter consist of a quartz balloon in which the halide of the element is present. The analyte spectra are excited with the aid of a high-frequency (MHz) or microwave field (GHz), which may be supplied through an external antenna.

21.5.3. Flame Atomic Absorption

In flame AAS, the sample solutions are usually nebulized pneumatically in a spray chamber and the aerosol produced is transferred, together with a mixture of a combustion gas and an oxidant, into a suitable burner.

21.5.3.1. Flames and Burners

Flames may use propane, acetylene, or hydrogen as combustion gases, with air or nitrous oxide as oxidant. With these mixtures, temperatures of 2000 – 3000 K can be obtained. Temperatures of flames (in kelvin) used in AAS are given below [157]:

Air – propane	1930
Air – acetylene	2300
Air – hydrogen	2045
Nitrous oxide – acetylene	2750
Oxygen – acetylene	3100

The *air – acetylene flame* is most often used. Its temperature is high enough to obtain sufficient atomization for most elements which can be determined by AAS, but not so high that ionization interference becomes significant. The latter occur only with the alkali and alkaline earth metals. The analytical conditions can be optimized by changing the composition of the gas mixture. The sensitivity for the noble metals (Pt, Ir, Au, etc.) can be improved by increasing the oxidative properties of the flame (excess air), which also lowers interference. The use of a reducing flame is more advantageous for determination of the alkali metals.

For more than 30 elements, however, the temperature of this flame is too low. This applies for elements which have very stable oxides, such as V, Ti, Zr, Ta, and the lanthanides. For these elements, *nitrous oxide – acetylene flame*, as introduced in flame AAS by AMOS and WILLIS in 1966 [161], is more suitable. However, this flame has a higher background emission, especially due to the radiation of the CN bands.

The *hydrogen – air flame* is of litte use. For the determination of Sn, however, its sensitivity is higher than that of the acetylene – air flame. Because of its higher transmission at short UV wavelengths the hydrogen – air flame is also used for the determination of As (193.7 nm) and Se (196.0 nm).

The *propane – air flame* has the lowest temperature and is used for the determination of easily atomized elements such as Cd, Cu, Pb, Ag, Zn, and especially the alkali metals.

In flame work, the burners have one or more parallel slits of length 5 – 10 cm, providing an absorption path for the primary radiation. In special laminar burners, the density of the analyte atoms at relatively high observation positions (up to more than 5 cm) can be kept high, which is ad-

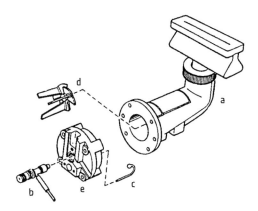

Figure 32. Burner–nebulizer assembly for flame atomic absorption spectrometry
a) Burner head with mixing chamber; b) Nebulizer; c) Impactor bead; d) Impact surfaces; e) Nebulizer socket

vantageous as measurements are then possible in these zones where more complete atomization occurs. Multislit burners may be useful to prevent deposits in solutions with high salt content. High-temperature flames require the use of special burners. Detailed information on flames is available in classical textbooks [162]–[164].

21.5.3.2. Nebulizers

The nebulizer is mounted in the spray-chamber burner assembly, as shown for the nitrous oxide or air–acetylene burner system (Fig. 32). Air or nitrous oxide are fed to the mixing chamber through the nebulizer. The combustion gas is fed directly to the burner. A self-aspirating concentric nebulizer is often used, but other types, including the Babington and cross-flow types (Section 21.4.1) are also applied. Nebulizers may be made of glass, corrosion-resistant metals such as Pt–Ir, or plastics such as Ryton. Sample solution consumption usually amounts to ca. 5 mL/min in the case of gas flows of 1–5 L/min.

In flame AAS, the nebulization system must be able to generate small droplets (< 10 μm), as only such droplets can be transported and completely vaporized in the flame. This is a prime condition for the realization of high sensitivity and low volatilization interference. Nebulizers for AAS generate not only small droplets, but also larger ones (up to 25 μm or more). For fragmentation of the latter, the nebulizer should be positioned in a mixing chamber provided with impact surfaces. In pneumatic nebulization, the sampling efficiency remains restricted (a few percent) as fragmentation is limited, and all particles larger than 10 μm are led to waste. In flame AAS with pneumatic nebulization, the sample solution is usually aspirated continuously. After 5–7 s, a stable signal is obtained and signal fluctuations depending on concentration are of the order of 1%.

Apart from continuous sample aspiration, flow injection and discrete sampling can be applied (see Section 21.4.1), which both deliver transient signals. In the latter, 10–50 μL aliquots can be injected manually or with a sample dispenser into the nebulization system [78], [79]. This approach is especially useful for preventing clogging in the case of sample solutions with high salt content, for the analysis of microsamples as required in serum analysis, or when aiming for maximum absolute power of detection in combined analytical procedures for trace analysis.

21.5.3.3. Figures of Merit

Power of Detection. The lowest elemental concentrations which can be determined by flame AAS are often given in terms of the so-called characteristic concentrations (μg/mL). For aqueous solutions of such concentration an absorption of 1% corresponds to an extinction of 0.0044. The noise comprises contributions from the nebulizer as well as from the flame. Detector noise limitations may also occur. The latter can be minimized by operating the hollow cathode lamp at sufficiently high current.

In order to obtain maximum power of detection, the atomization efficiency should be as high as possible. Optimization of the geometry of the spray chamber and the nebulizer gas flow is required. The primary radiation should be well isolated by the monochromator and the amount of nonabsorbed radiation reaching the detector should be minimized by selection of the appropriate observation zone with the aid of a suitable illumination system.

The detection limits of flame AAS are especially low for rather volatile elements, which do not form thermally stable oxides or carbides, and have high excitation energies, such as Cd and Zn. Apart from these and some other elements such as Na and Li, the detection limits in flame AAS are higher than in ICP–AES (see Table 9).

Analytical Precision. By integrating the extinction signals over 1–3 s, relative standard deviations down to 0.5% may be achieved. When injecting discrete samples into the nebulizer or applying flow injection analysis, slightly higher relative standard deviations are obtained. They increase rapidly on leaving the linear part of the calibration curve and applying linearization by software.

Interference. Interference in flame AAS comprises spectral, chemical, and physical interference.

Spectral Interference. Spectral interference of analyte lines with other atomic spectral lines is of minor importance compared with atomic emission work. It is unlikely that resonance lines emitted by the hollow cathode lamp coincide with an absorption line of another element present in the atomizer. However, it may be that several emission lines of the hollow cathode are within the spectral band width or that flame emission from bands or continuum occurs. Both contribute to the non-absorbed radiation, and decrease the linear dynamic range. Nonelement-specific absorption (Section 21.5.6) is a further source of spectral interference.

Chemical Interference. Incomplete atomization of the analyte causes chemical interference, due to the fact that atomic absorption can only occur with free atoms. Reactions in the flame which lead to the formation of thermally stable species decrease the signal. This is responsible for the depression of calcium signals in serum analysis by proteins, as well as for the low sensitivities of metals which form thermally stable oxides or carbides (Al, B, V, etc.) in flame AAS. A further example of chemical interference is the suppression of the extinction of alkaline earth metals as a result of the presence of oxyanions (OX) such as aluminates or phosphates. This well-known "calcium phosphate" interference is caused by the reaction:

$$MOX \rightleftharpoons X + MO \xrightarrow{high\ T} M + O \xrightarrow{+C} M + CO$$

In hot flames, such as the carbon-rich flame, the equilibrium lies to the right, but with excess oxyanions it is shifted to the left, and no free M atoms are formed. This can be corrected by adding a metal R, the so-called "releasing agent", which forms yet more stable oxysalts and releases metal M again. La and Sr compounds can be used:

$$MOX + RY \xrightleftharpoons{excess} ROX + MY$$

When $LaCl_3$ is added to the sample solutions, phosphate can be bound as $LaPO_4$.

For alkali metals, the free atom concentrations in the flame can decrease as a result of ionization, especially in hot flames. This leads to a decrease of extinction for alkali metals. It may also lead to false analysis results, as the ionization equilibrium for the analyte element is changed by changes in concentration of easily ionized elements. In order to suppress these effects, ionization buffers can be added. The addition of excess (low ionization potential) is most effective to suppress ionization changes of other elements, as it provides a high electron number density in the flame.

Physical Interference. Physical interference may arise from incomplete volatilization, and occurs especially in strongly reducing flames. In steel analysis, the depression of Cr and Mo signals as a result of excess Fe is well known. It can be reduced by adding NH_4Cl. Further interference is related to nebulization effects, and arises from the influence of the concentration of acids and salts on the viscosity, density, and surface tension of the analyte solutions. Changes in physical properties from one sample solution to another influence aerosol formation efficiencies and the aerosol droplet size distribution, as discussed in Section 21.4.1. However, related changes of nebulizer gas flow also influence the residence time of the particles in the flame.

21.5.4. Electrothermal Atomic Absorption

The use of furnaces as atomizers for quantitative AAS goes back to the work of L'vov, and led to the extension of AAS to very low absolute detection limits. In electrothermal AAS, graphite or metallic tube or cup furnaces are used and, with resistive heating, temperatures are achieved at which samples can be completely atomized. For volatile elements, this can be accomplished at 1000 K, whereas for more refractory elements the temperature should be up to 3000 K.

The high absolute power of detection of electrothermal AAS is due to the fact that the sample is completely atomized and vaporized, and the free atoms spend a long time in the atomizer. The signals obtained are transient, as discussed in Section 21.4.4.1.

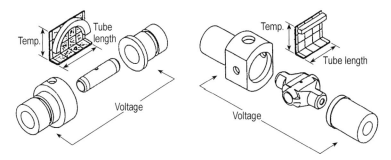

Figure 33. Temperature distribution in a longitudinally (left) and transversely (right) heated graphite furnace

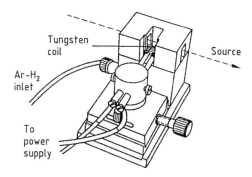

Figure 34. Tungsten filament atomizer [166]

21.5.4.1. Atomizers

Graphite furnaces, cups, and filaments are all used as atomizers in electrothermal AAS [165]. The models originally proposed by L'VOV et al. [3] and by MASSMANN [4] are described in Section 21.4.4.2. In the most widely used type the optical beam is led centrally through a graphite tube, closed on both sides, with a quartz viewing port mounted in the cooled tube holders. Sample aliquots are introduced with the aid of a micropipette or a computer-controlled dispenser, through a sampling hole in the middle of the tube. The normal temperature profile of a graphite furnace leads to differences in spreading of the analyte over the graphite surface and may produce changes in volatilization from one sample to another. This effect can be avoided by using a transversely heated furnace, where the temperature is constant over the whole tube length (Fig. 33). Especially in electrothermal AAS, problems may arise from recombination of the atomized analyte with oxygen and other nonmetals, as the free atom concentration measured in AAS decreases. This may occur when the volatilized analyte enters a fairly cool plasma, as is normally the case in a tube furnace. This can be prevented [109] by dispensing the sample on a low-mass graphite platform located in the tube furnace. The low heat content then allows very rapid heating and volatilization of the analyte. The analyte is surrounded by a furnace plasma of about the same temperature as the sample carrier, which lowers the risk of recombination. Tungsten furnaces (see Section 21.4.4.2) [112] are useful for the determination of refractory carbide-forming elements, but they are more difficult to use with real samples, and reduction of analyte oxides does not occur. BERNDT et al. [166] have shown that tungsten coils are suitable atomizers for dry solution aliquots, especially for matrix-free solutions obtained in combined analytical procedures involving separation of the analyte from the matrix elements. Owing to the relatively large coil surface area (Fig. 34), salt problems in real samples may be lower than with a metal tube furnace.

21.5.4.2. Thermochemistry

The dissociation equilibria between the analyte elements and their compounds are very important, as they determine the fraction of analyte available as free atoms for AAS. They affect both the analytical sensitivity, and the possibility of systematic errors. The thermochemical behavior of the sample is of prime importance for the height of the extinction signal as well as its form.

The acids present in the sample solution are normally removed during drying and matrix decomposition. Residues present during the absorption measurement lead to nonelement-specific absorption, especially in the case of acids with high boiling points, such as $HClO_4$, H_2SO_4, or H_3PO_4. The absorption stems from ClO, SO, SO_2, or PO molecular bands. These species may also be produced by the dissociation of salts. Further problems may be caused by the oxides of the analytes, which result from salt dissociation. This nondissociated oxide is lost for AAS, and may give rise to nonelement-specific absorption. When salt dissociation and reduction of analyte oxides is not complete before the absorption measurement, several peaks may occur in the absorption signal. This can often be avoided by the platform technique, which produces a sharp rise in heating of the furnace and lowers the risks of analyte deposition at the cool parts of the furnace.

Thermochemistry is especially important reliable matrix destruction and removal of matrix elements without analyte loss. Thermochemical reagents such as quarternary ammonium salts ($R_4N^+Cl^-$) mineralize organic samples at temperatures below 400 °C, and prevent losses of elements which are volatile or form volatile, perhaps organic, compounds. This may be helpful for Cd and Zn, which form volatile compounds in a number of organic matrices. Removal of NaCl, present in most biological samples, may be helpful to prevent matrix interference; this must be done at low temperature to prevent analyte loss, and can be achieved by the addition of NH_4NO_3:

$NH_4NO_3 + NaCl \quad NH_4Cl + NaNO_3$

Excess NH_4NO_3 dissociates at 350 °C, NH_4Cl sublimates and $NaNO_3$ decomposes below 400 °C, so NaCl is completely removed at temperatures below 400 °C. Without the addition of NH_4NO_3, this would be possible only at the volatilization temperature of NaCl (1400 °C), with inevitable analyte loss. In the graphite furnace, several elements (Ti, V, etc.) form thermally stable compounds such as carbides, producing negative errors some of the analyte is bound and does not contribute to the AAS signal. The use of pyrolytically coated graphite tubes is helpful, as diffusion of the analyte solution into the graphite decreases the risk of carbide formation. Flushing the furnace with nitrogen can be helpful. In the case of Ti, a nitride is formed which, in contrast to the carbide, can be easily dissociated. Other thermochemical means to decrease interference are known as matrix modification. Addition of Pd compounds, $Mg(NO_3)_2$, etc., has been successful for matrix-free vapor cloud formation [167]. The mechanisms involved also relate to surface effects in the furnace [168] and are an active field of research.

Development of the temperature program is vital in furnace AAS. Selection of the different temperature steps, and the use of all kinds of thermochemical effect are most important to minimize matrix influence without analyte loss.

Radiotracers are helpful for understanding and optimizing analyte volatilization in furnace AAS; element losses and their causes at all stages of the atomization process can be followed quantitatively. This has been studied in detail for elements such as As, Pb, Sb, and Sn by KRIVAN et al. [169].

21.5.4.3. Figures of Merit

The *analytical sensitivity* and *power of detection* of electrothermal AAS are orders of magnitude higher than those of flame AAS. This is due to the fact that, in the furnace, a higher concentration of atomic vapor can be maintained compared with flames. Dilution of the analyte by the solvent is avoided, as it has already evaporated before atomization and large volumes of combustion gases are not involved. For most elements, the characteristic mass, i.e., the absolute amount for which an extinction of 0.0044 (1 % absorption) is obtained, are orders of magnitude lower than in flame AAS [165].

Detection limits in flame AAS: 0.1 – 1 ng/mL (Mg, Cd, Li, Ag, Zn); 1 – 10 ng/mL (Ca, Cu, Mn, Cr, Fe, Ni, Co, Au, Ba, Tl); 10 – 100 ng/mL (Pb, Te, Bi, Al, Sb, Mo, Pt, V, Ti, Si, Se); > 100 ng/mL (Sn, As)

Detection limits in furnace AAS, (20 µL): 0.005 – 0.05 ng/mL (Zn, Cd, Mg, Ag); 0.05 – 0.5 ng/mL (Al, Cr, Mn, Co, Cu, Fe, Mo, Ba, Ca, Pb); 0.5 – 5 ng/mL (Bi, Au, Ni, Si, Te, Tl, Sb, As, Pt, Sn, V, Li, Se, Ti, Hg)

When sample aliquots of 20 – 50 µL are used, the concentration detection limits are often in the sub-ng/mL range. They are especially low for elements such as Cd, Zn, and Pb which have high volatility and high excitation energies, but also for Mg, Cu, Ag, Na, etc.

Interference in furnace AAS is much higher than in flame AAS. This applies both for physical and chemical interference.

Physical Interference. Physical interference may stem from differences in viscosity and surface tension, resulting in different wetting of the graphite tube surface with the sample solution, and changes in diffusion of the sample solution into the graphite. The first may be suppressed by the addition of surfactants to the sample solutions, and the second can be avoided by using pyrolytically coated graphite tubes, where diffusion of the analyte into the tubes is lower.

The influences of concomitants on volatilization can be minimized by the use of platform techniques and of an isothermal furnace. Both measures are particularly helpful in direct solid sampling.

Undissociated molecules, which may be oxides formed by dissociation of ternary salts (MgO, ZnO, etc.) or radicals and molecular species from solvent residues (OH, PO, SO_2, etc.) may cause nonelement-specific absorption. Rayleigh scattering of primary radiation by nonevaporated solid particles may occur, which also leads to radiation losses. Both phenomena necessitate the application of suitable techniques for background correction for most analyses of real samples by furnace AAS (Section 21.5.6). Emission by the furnace itself may give continuous radiation, which may lead to systematic errors.

Chemical Interference. Chemical interference may be caused by losses of analyte during ashing, especially in the case of volatile elements such as As, Sb, Bi, Cd, but also for elements forming volatile compounds (e.g., halides). In this respect, removal of NaCl, which is present in most biological samples, is critical. Further element losses may occur from the formation of thermally stable compounds such as carbides or oxides which cannot be dissociated and hamper atomization of the analyte element; elements such as Ti or V are especially difficult to determine by furnace AAS. Some of these sources of systematic error can be reduced in furnace AAS by the use of thermochemical reactions.

Differences in the volatility of elements and their compounds may also be used for speciation work. It is possible to determine organolead compounds such as $Pb(CH_3)_4$ and $Pb(C_2H_5)_4$ directly by furnace AAS [170], [171] or after on-line coupling with chromatographic separation. This approach can be used for other species, but needs to be carefully worked out in each case.

Calibration in furnace AAS is usually done by standard addition, as many of the sources of interference change the slope of the calibration curve.

Recommendations to be followed to enable determinations to be as free from interferences as possible include: (i) isothermal operation; (ii) the use of a matrix (chemical) modifier; (iii) an integrated absorbance signal rather than peak height measurements; (iv) a rapid heating rate during atomization; (v) fast electronic circuits to follow the transient signal; (vi) the use of a powerful background correction system such as the Zeeman effect. Most or all of these recommendations are incorporated into virtually all analytical protocols nowadays and this, in conjunction with the transversely heated tubes, has decreased the interference effects observed considerably.

21.5.5. Special Techniques

21.5.5.1. Hydride and Cold Vapor Techniques

Determination of traces and ultra-traces of Hg, As, Se, Te, As, and Bi is often by formation of either volatile mercury vapor, or of the volatile hydrides. This gives a high sampling efficiency and power of detection. The absorption measurement is often performed in a quartz cuvette. Hg, for instance, can be transported after reduction by means of a carrier gas. For the absorption measurement no heating is required. The other elements are reduced to the volatile hydride, which is then transported with a carrier gas flow (argon or nitrogen) into a cuvette. In order to dissociate the hydrides thermally, the quartz cuvette must be heated to a temperature of 600–900 °C.

In both cases, flow cells enable continuous formation of the volatile compound and subsequent separation of the reaction liquid from the gaseous compound [172]. This is very useful (Section 21.4.3) for routine work and has practically replaced earlier systems, where $NaBH_4$ solution was reacted with a given volume of acidified sample [173]. In any case, trapping of the analyte vapor released may increase the absolute power of detection. In the case of Hg, this may be achieved by passing the Hg vapor over a gold sponge and binding it as an amalgam [174]. After collection, Hg can be liberated completely by heating. In the case of the hydrides, cold trapping can be applied. Here, the hydrides can be condensed by freezing in a liquid nitrogen trap, and freed from the excess hydrogen formed during the reaction [175]. The isolated hydrides can be released by warming. This removes hydrogen and gives trapping enrichment factors well above 100.

Hot trapping can also be used [176] as the hydrides have decomposition temperatures lower than the boiling points of the elements. Trapping can thus be performed in the graphite furnace itself. It is only necessary to heat the furnace to ca. 600–700 °C to decompose the volatile hydrides, and the elements condense in the furnace. Here they can be preenriched and volatilized at temperatures above 1200 °C during absorption measurement.

The cold vapor technique for Hg gives detectin limits < 1 ng with 50 mL of sample, and this can be further improved by trapping. The hydride technique gives detection limits below the ng/mL level for As, Se, Sb, Bi, Ge, Sn, etc., levels appropriate for analyses of drinking water. Hydride techniques, however, are prone to interference (Section 21.4.3). In AAS, interference may occur not only as a result of influences on the hydride formation reaction, but also as a result of influences of concomitants on the thermal dissociation of the hydride. Interference by other volatile hydride-forming elements may also occur [177].

21.5.5.2. Direct Solid Sampling

In both flame and furnace AAS, direct solid sampling is possible. In flames, sampling may involve a boat or cup, as introduced by DELVES [113]. New approaches such as the combustion of organic samples and the introduction of the vapors released into a flame are possible [178]. For volatile elements such as Pb, detection limits are in the µg/g range. Direct powder sampling in furnaces was introduced in atomic spectrometry by PREUSS in the 1940s, and in furnace AAS it can be easily done with a powder sampling syringe [116]. For direct powder analyses with flame and furnace AAS, however, it is preferable to work with slurries. Slurry atomization was introduced in flame AAS work by EBDON and CAVE [179] in 1982. It has been shown to be useful for flame work, and to improve powder sampling in furnace AAS. The latter has been successfully used for direct analysis of SiC powders [180]. Direct solid sampling, however, should be used very carefully. The amount of sample used is often of the order of a few milligrams, and sampling errors may occur as a result of sample inhomogeneity. Therefore, the use of furnaces in which larger powder amounts can be handled has been proposed. In direct powder analysis, calibration may be troublesome because the particle size affects analyte volatilization and transport.

In compacted samples, direct analysis by AAS is feasible. The use of cathodic sputtering combined with AAS of the atomic vapor cloud produced has been recently reviewed [181], with jet-enhanced sputtering to give high analyte number densities. The approach has been made commercially available [182]. Its feasibility for direct analyses of steels has been shown, especially for samples which are difficult to dissolve, in which refractory oxide-forming elements such as Zr need to be determined [183]. For both electrically conducting and insulating samples, laser ablation combined with AAS may be useful. AAS measurements can be performed directly in the laser plume. Measurement of nonelement-specific absorption is very important, because of the presence of particles, molecules, and radicals, and the emission of continuous radiation. The absorption measurement should be performed in the appropriate zone. When using laser ablation for direct solid sampling, the atomic vapor produced can be fed into a flame for AAS work, as described by KANTOR et al. [184].

21.5.5.3. Indirect Determinations

For the determination of elements of which the most sensitive lines are in the vacuum UV conventional AAS cannot be applied. For this, and for the determination of chemical compounds, indirect methods can be used. For instance, sulfates can be determined by precipitation as $BaSO_4$ and a determination of excess barium with AAS. A similar approach can be used for the determination of halides after precipitation with silver.

Further indirect determinations can be performed by making use of chemical interference. Calcium phosphate interference in flame AAS can be used to determine phosphate by measuring the decrease of the absorption signal for calcium. However, for real samples, indirect techniques should be used very carefully in view of possible systematic errors.

21.5.6. Background Correction

In AAS, systematic errors are often due to nonelement-specific absorption, which necessitates the use of background correction procedures. They can only be omitted in the analysis of sample solutions with low matrix concentrations by flame AAS, and are especially necessary in furnace AAS. Determination of nonelement-specific absorption can be performed in several ways.

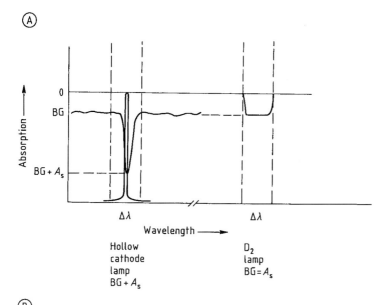

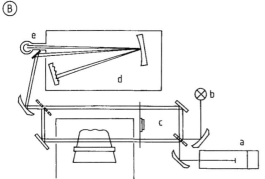

Figure 35. Background correction with the D_2 lamp technique
A) $\Delta\lambda$ = Spectral bandwidth of monochromator; B) Optical diagram of the spectrometer (Model 2380, Bodenseewerk Perkin-Elmer)
a) Hollow cathode lamp; b) D_2 lamp; c) Chopper; d) Monochromator; e) Photomultiplier
BG = Nonelement-specific background absorption; A_S = Element-specific absorption signal

21.5.6.1. Deuterium Lamp Technique

This approach was introduced by KOIRTHYOHANN and PICKETT in 1965 [159] and is now provided in almost every AAS system. The total absorption resulting from the presence of the element and the background absorption are measured with hollow cathode lamp radiation, but, in addition, a continuum source is used which measures only the background absorption. This is possible as the monochromators used in AAS have a large spectral band width compared with the physical width of the resonance line emitted by the hollow cathode source, and the width of the absorption line. Accordingly, all radiation emitted by the hollow cathode lies within the absorption profile of the line, and elemental absorption does not contribute significantly to the absorption of the continuous source (Fig. 35), which extends homogeneously over the whole spectral band with. As the continuous source, an electrical discharge in deuterium can be used, as it emits a rather smooth continuum in the UV range up to 400 nm.

Alternative measurements of the absorption of the hollow cathode lamp radiation and of the radiation of the continuous source can be performed in a rapidly alternating mode to give a quasi-simultaneous measurement of line and background absorption, as required with transient signal. The latter can be realized by a rotating mirror and is rapid enough for high time resolution [185].

This type of background correction assumes that the background absorption is continuous within the spectral band width of the monochromator. This is not the case when the background absorption stems from molecular bands, which have a rotation–vibration hyperfine structure. They can arise from radicals produced by dissociation of the solvent (OH, SO_2, SO_3, N_2^+, CN, etc.) or from molecular oxides (MO). As such contributions occur especially with the complex matrices of real samples, background correction with a D_2 lamp often leads to systematic errors, which cannot be corrected by standard addition methods. Furthermore, the radiant density of the D_2 lamp over a large part of the spectrum is rather low. Accordingly, there are limits to the number of analytical lines which can be used and the number of elements which can be determined. As the spectral radiance of the D_2 lamp is lower than that of the hollow cathode, the latter must be operated at low current, which leads to detector noise limitations and poor detection limits. Finally, the two primary radiation sources are difficult to align, so they have to pass through the same zone of the atomizer, and this may lead to further systematic errors.

21.5.6.2. Zeeman Effect Technique

Zeeman AAS makes use of the splitting of atomic spectral lines by a magnetic field. When a magnetic field B (≤ 1 T) is applied, the shift in wavenumber (ΔT_m) of the so-called σ-components, where the π-components remain constant, is:

$$\Delta T_m = M g L B \qquad (151)$$

where M is the magnetic quantum number, L is the Lorentz unit (e/(4 πmc^2)), and g the Landé factor, a function of the total quantum number J, the orbital momentum quantum number L, and the spin quantum number S. The intensities of the σ-components (for which $\Delta M = \pm 1$) and the π-components (for which $\Delta M = 0$) are a function of ΔJ (0, 1) and ΔM (0, ±1) for the transitions. In the normal Zeeman effect, which occurs for singlet transitions (e.g., with alkaline earth metals and metals of group 12 such as Cd and Zn) $g = 1$, whereas in all other cases multiplets occur (anomalous Zeeman effect). For a transverse magnetic field (perpendicular to the observation direction) there is a π-component at the original wavelength. For this component, $\Delta M = 0$ and it is polarized parallel to the field. There are two σ-components (σ^+ and σ^-) for which $\Delta M = \pm 1$. They are polarized perpendicular to the field. For a longitudinal field (parallel to the observation direction) there is no π-component ($\Delta M = 0$) and the σ-components ($\Delta M = \pm 1$) are circularly polarized. In order to use the Zeeman effect for background correction, several approaches are possible (Fig. 36) [186]. A constant transverse field may be used, measuring the π- and the σ-components alternately with the aid of a polarizer and a rotating analyzer. Alternatively, an a.c. longitudinal field can be used, however, measuring σ-components only with the aid of a static polarizer, at zero and at maximum field strength.

In both cases, substracting one signal from the other eliminates the background absorbance. This assumes that both signals have constant intensities through the whole analytical system and that both have the same absorption coefficients for the background. The latter technique has the additional advantage that the magnetic field can be controlled since it is produced by an electromagnet. Accordingly, the splitting can be optimized with respect to the element determined and to the background structure. The magnetic field can be applied in the atomizer, which is possible both with permanent and electromagnets. In principle, the magnetic field can also be placed at the primary source, which is optimal in the case of a permanent magnet; both flames and furnaces can be used as atomizers and the exchange is easier. Moreover, a larger furnace can be used which is very useful for direct solid sampling.

Zeeman AAS has several analytical advantages. The accuracy of correction of a structured background is better than with the D_2 lamp technique. However, when the background structure stems from molecular bands, it must be remembered that molecular bands may also display a Zeeman effect. Systematic errors resulting from this fact may be larger when one line component is measured in an alternating field.

The *detection limits* in Zeeman AAS should be lower than in the case of background correction with a D_2 lamp. The system uses only one source; it can be operated at high intensity, so detector noise limitations are avoided. This advantage is most pronounced when one component is measured in an alternating field.

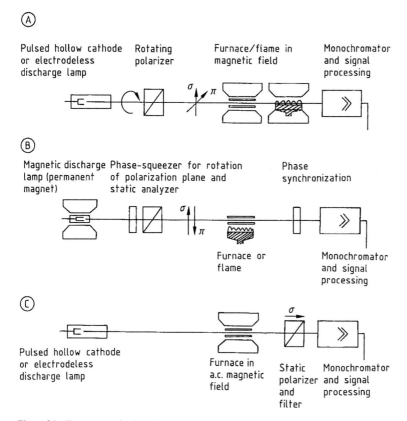

Figure 36. Zeeman atomic absorption
A) Rotating polarizer and permanent magnet applied to the atomizer; B) Permanent magnet around the primary source; C) Longitudinal field of a.c. magnet applied to the atomizer [186]

Another consequence of the use of one primary source is improved stability. The analytical sensitivity in Zeeman AAS, however, is inferior to that of conventional AAS, is the linear dynamic range. The latter relates to the fact that a difference between two absorbances is involved, which may even lead to bending of the calibration curve. These effects are less pronounced when measuring one component in an alternating field. Zeeman AAS is widely used, for instance, for trace determinations in biological samples.

21.5.6.3. Smith–Hieftje Technique [187]

This technique of background correction makes use of the fact that resonant atomic spectral lines emitted by a hollow cathode lamp may display self-reversal when the lamp is operated at high discharge current. A hollow cathode lamp operates during the first part of the measurement cycle, at low current. Self-reversal does not occur and the resonant radiation is absorbed both by the analyte atoms and by background-producing species. In the second part of the measurement cycle, the current is briefly pulsed to >500 mA, and a pronounced self-reversal occurs. The intensity at the analytical wavelength decreases, but the intensities in the side-wings remain high (mostly background absorption). By subtraction of the absorbances, the net atomic absorption signal is obtained.

Similar to Zeeman AAS, the Smith–Hieftje technique can be used for lines over the entire spectral range, and uses only one primary radiation source, so that alignment and stability are optimum. The technique is simple and much cheaper than Zeeman AAS. It does not suffer from limitations due to Zeeman splitting of molecular bands. However, as self-reversal is not complete, the technique can be used only for rather low background absorbance, sensitivity is decreased as the self-absorption is 40% at most, and special provisions have to be made for pulsing the lamps at high currents.

21.5.6.4. Coherent Forward Scattering

The intensity of scattered radiation is particularly high when monochromatic radiation is used as primary radiation, the wavelength of which is equal to that of a resonance line of the scattering atoms. When the latter are subjected to a magnetic field, the scattered radiation becomes coherent in the direction of the primary beam and the scattering atomic vapor becomes optically active (magneto-optic effect). Depending on whether a transverse or longitudinal magnetic field is used, a Voigt or Faraday effect is observed. In coherent forward scattering, radiation from the primary source is led through the atomizer across which a magnetic field is applied. When the atomizer is placed between crossed polarizers, scattered signals for the atomic species occur against zero background. When a line source, such as a hollow cathode lamp or a laser, is used, determinations of the respective element can be performed. For a continuous source, such as a xenon lamp, and a multichannel spectrometer, simultaneous multielement determinations can be performed. The method is known as coherent forward scattering (CFS) atomic spectrometry [188]–[190].

The *intensities of the scattering signals* in both Voigt and Faraday effects can be calculated [188]. In the *Faraday effect*, two waves are polarized parallel to the magnetic field. If n^+ and n^- are the refractive indices, $n_m = (n^+ + n^-)/2$ and $\Delta n = (n^+ - n^-)/2$, the intensity $I_F(k)$ at wavenumber k is:

$$I_F(k) = I_0(k) F[\sin(k\, l\, \Delta n)] \exp(k\, l\, n_m) \qquad (152)$$

where $I_0(k)$ is the intensity of the primary radiation at wavenumber k, and l is the length of the atomizer. The sinusoidal term relates to the rotation of the polarization plane and the exponential term to the atomic absorption. As both n_m and Δn are functions of the density of the scattering atoms, $I_F(k)$ is proportional to the square of the density of scattering atoms N:

$$I_F(k) = I_0 N^2 l^2 \qquad (153)$$

In the *Voigt effect*, the scattered radiation has two components. One is polarized parallel to the magnetic field (normal component) and the other perpendicular (abnormal component). If n_0 and n_e are the respective refractive indices, their intensities can be calculated as for the Faraday effect.

Coherent forward scattering atomic spectrometry is a multielement method. The instrumentation required is simple and consists of the same components as a Zeeman AAS spectrometer. As the spectra contain only a few resonance lines, a spectrometer with low spectral resolution is sufficient. The detection limits depend considerably on the primary source and on the atomizer. With a xenon lamp as primary source, multielement determinations can be performed, but the power of detection is low as the spectral radiances are low compared with those of a hollow cathode lamp. By using high-intensity laser sources, the intensities of the signals and the power of detection can be considerably improved. When furnaces are used as atomizers, typical detection limits in the case of a xenon arc are: Cd 4 ng; pb 0.9; Tl 1.5; Fe 2.5; and Zn 50 ng [188]. These are considerably higher than in furnace AAS.

The sensitivity of CFS atomic spectrometry is high as the signals are proportional to the square of the atom number densities. The dynamic range is similar to atomic emission spectrometry, of the order of three decades. Information on matrix effects in real samples is still scarce. As scattering by molecules and undissociated species is expected to be low, background contributions may be low compared with AAS.

21.5.7. Fields of Application

The different methods of AAS are very powerful for analysis of solutions. The instruments are simple and easy to operate; they are in use in almost any analytical laboratory. Especially when one or only a few elements have to be determined in a large number of samples, as is the case in clinical or food analysis, AAS methods are of great importance compared with other methods of elemental analysis.

When solids need to be analyzed, samples have to be dissolved. Sample decomposition methods range from simple dissolution in aqueous solutions to treatment with strong and oxidizing acids. In all sample dissolution and pretreatment work for ASS, attention must be paid to all the problems which be set trace elemental analytical chemistry. This includes precautions for avoiding contamination from the reagents, the vessels used, and from the laboratory atmosphere (→Trace Analysis). All factors which may cause analyte loss as a result of adsorption or volatilization must be studied in detail [191].

Clinical chemistry, the metals industry, the analysis of high-purity chemicals and pure substances, environmental analysis, and the life sciences are important fields for the application of AAS. Applications in all these areas are regularly reviewed [192].

Clinical Chemistry. In clinical analysis, flame AAS is very useful for serum analysis. Here, Ca and Mg can be directly determined in serum samples after 1:50 dilution, even in micro-aliquots of 20–50 µL [193]. In the case of Ca, La^{3+}, or Sr^{2+} is added to avoid phosphate interference. Na and K are mostly determined in the flame emission mode, which can be carried out with any flame AAS instrument. Here, the burner head is often turned to shorten the optical path, avoiding self-reversal. For the direct determination of Fe, Zn, and Cu, flame AAS can be used, but at lower sample dilution. For trace elements such as Al, Cr, Co, Mo, and V, determination with flame AAS often requires preenrichment, but in serum and other body fluids as well as in biological matrices, some elements can be directly determined with furnace AAS. This applies to toxic elements such as Ni, Cd, Pb, etc., in screening for workplace exposure.

For *metal samples*, flame and furnace AAS are important analysis methods. They find applications in the characterization of incoming material, and for product analysis. In combination with matrix removal, they are indispensable for the characterization of laboratory standards. The latter are used to calibrate direct methods such as X-ray spectrometry and spark emission spectrometry for production control. However, relevant elements such as B, the rare earths, Hf, Zr, etc., are much better determined by plasma emission spectrometry. For the determination of environmentally relevant elements in slags and ores, dissolution by fusion and subsequent AAS may be applied.

In the analysis of *high-purity substances*, matrix removal is often very important for preconcentration. For this, separation techniques such as ion exchange, liquid–liquid extraction of metal complexes with organic solvents, fractional crystallization, precipitation, coprecipitation, and electrochemical methods may be used [194].

For analytical problems in the *life sciences* and *environmental analyses*, AAS is widely used. For the analysis of wastewater, flame and furnace AAS are complementary to plasma emission spectrometry. The latter enables determination of a large number of elements per sample, but is more expensive than flame AAS in terms of instrumentation, and less sensitive than furnace AAS. For analysis of drinking water, furnace AAS is very useful, but because of its single-element character, it is increasingly being replaced by plasma mass spectrometry, especially in large laboratories.

Atomic absorption spectrometry is a useful tool for *speciation* work. This applies for combinations of species-specific extraction procedures and subsequent determination of the respective species by AAS. Such a method has been worked out for Cr(III) and CR(VI) compounds [195]. On-line coupling of gas or liquid chromatography with AAS in increasingly used [196]. In furnace AAS, coupling with gas chromatography has been shown to be very useful for determination of organolead [170] and organotin compounds [197]. By combining the cold vapor technique with HPLC, a very sensitive determination of Hg species at the sub-nanogram level becomes possible [198]. For speciation, coupling of flow injection analysis and column chromatography with flame AAS, and direct coupling of HPLC with flame AAS (possible with high-pressure nebulization) are very powerful, as shown by the speciation of Cr(III) and Cr(VI) for wastewater analysis [96].

Manufacturers. The most important manufacturers of AAS equipment are:

Baird Atomic, USA: arc/spark AES, ICP-AES, AAS
Analytic Jena AG, Germany: AAS
Fisher Scientific, USA: AAS
GBC, Australia: TOF-ICP-MS, AAS
Hitachi, Japan: ICP-AES, AAS, ICP-MS
PerkinElmer Instruments (incorporating Princeton Applied Research, ORTEC, Signal Recovery and Berthold products), USA: AAS, ICP-OES, ICP-MS, GC, GC/MS
Shimadzu, Japan: arc/spark AES, ICP-AES, AAS
TJA Solutions (incorporating Thermo Jarrell Ash, VG Elemental and Unicam AA), USA: AAS, ICP-AES, ICP-MS, SF-ICP-MS, GD-AES, GD-MS
Thermoquest (incorporating Finnigan), USA: SF-ICP-MS, AAS
Varian Associates, Australia: ICP-AES, ICP-MS, AAS

21.6. Atomic Emission Spectrometry

21.6.1. Principles

Atomic emission spectrometry (AES) goes back to BUNSEN and KIRCHHOFF, who reported in 1860 investigations of the alkali and alkalineearth metals with the aid of their spectroscope [1], which is still the heart of any dispersive spectrometer. The new elements cesium, rubidium, thorium, and indium were discovered from their atomic emission spectra. By the beginning of the twentieth century, HARTLEY, JANSSEN, and DE GRAMONT further developed spectrochemical analysis, and flames, arcs, and sparks became powerful analytical tools. To overcome the difficulties arising from unstable excitation conditions, GERLACH in 1925 introduced the principle of the internal standard [199], where the ratio of intensities of the analytes to that of the matrix or of an added element is used. This was the breakthrough in atomic spectroscopy for elemental analysis. In AES, a reproducible and representative amount of the sample is introduced into a radiation source. Its role in AES is similar to the dissolution procedures of wet chemical analysis; continuous endeavors are made to improve existing radiation sources, and to develop new sources, to perform analyses with enhanced figures of merit. The role of the radiation source in AES does not end at the atomization stage (as in AAS), but includes excitation of states from which emission or even ionization occur.

Qualitative analyses can be performed with AES; the unambiguous detection and identification of a single, interference-free atomic spectral line of an element is sufficient to confirm its presence. The most intense line under a given set of working conditions is known as the most sensitive line. Elemental lines for different elements are situated in widely different spectral regions and may differ from one radiation source to another, as a result of the excitation and ionization processes. Despite the fact that spectral lines are very narrow (a few picometers), line coincidence can occur. High-resolution spectrometers are necessary for the qualitative evaluation of spectra, and more than one line should be used to decide on the presence of an element. Qualitative emission spectral analysis is easy when the spectra are recorded photographically or with a scanning monochromator. For the first case, atlases are available, in which the spectra are reproduced with the most sensitive lines indicated. Spectral line tables are also very useful. They are available for arc and spark sources [200]–[202], and in much less complete form for newer radiation sources such as glow discharges [203] and inductively coupled plasmas [204]–[206].

Quantitative AES. In quantitative AES, the intensity of an elemental atomic or ionic line is used as analytical signal. As AES is a relative method, calibration has to be performed.

Calibration Function. The determination of the calibration function is important; it relates the intensity of a spectral line I to the concentration c of an element in the sample. SCHEIBE [207] and LOMAKIN [208], propose the following relation between absolute intensities and elemental concentrations:

$$I = a c^b \quad (154)$$

where a and b are constants. Absolute intensities are practical only in flame work or in plasma spectrometry, where very stable radiation sources are used. Normally, the intensity ratio of an analyte line to a reference signal is used, as proposed by GERLACH. This leads to calibration functions of the form:

$$I/I_R = a' c^{b'} \quad (155)$$

The inverse calibration function is the analytical evaluation function:

$$c = c_R (I/I_R)^\eta \quad (156)$$

or:

$$\log c = \log c_R + \eta \log (I/I_R) \quad (157)$$

In trace analyses, the slope of the analytical evaluation curve (in logarithmic form) is usually 1; at higher concentrations η may be > 1 as a result of self-reversal. In trace analysis by AES, the intensity of the spectral background I_U can be used as reference signal and c_R should be replaced by c_U, the concentration at which the line and background intensities are equal (background equivalent concentration or BEC value):

$$c_U = [(I_X/I_U)/c]^{-1} \quad (158)$$

The ratio (I_X/I_U) and c_U for a given element and line depend on the radiation source and its working conditions, and on the spectral apparatus [209], [210]:

$$c_U = \frac{(dB_U/d\lambda)\,\Delta\lambda_L}{B_0}\,\frac{s_{eff}R_L}{s_0 R_0} \cdot [(\pi/(\ln 16))]^{1/2}\left\{\left[(s_a/s_{eff})^4 + 11\right]\right\}^{1/4}$$

$$c_U = c_{U,\infty} A_1 A_2 \tag{159}$$

where $c_{U,\infty}$ describes the influence of the radiation source, $(dB_U/d\lambda)$ reflects the spectral distribution of the background intensity, B_0 is the radiant density for an analytical line at $c=1$, and $\Delta\lambda_L$ is the physical width of the analysis line. The term A_1 describes the influence of the spectral apparatus, and:

$$s_{eff} = \left(s_e^2 + s_0^2 + s_L^2 + s_z^2\right)^{1/2} \tag{160}$$

where s_L is the slit width corresponding to the physical width of the spectral line $\Delta\lambda_L$, and $R_L = \lambda/\Delta\lambda_L$, the resolution required to resolve the line; s_z is the slit width corresponding to the optical aberrations in the spectral apparatus:

$$s_z = \Delta\lambda_z\, dx/d\lambda \tag{161}$$

where $\Delta\lambda_z$ can be determined by measuring the practical resolution of the monochromator from deconvolution of the two components of the Hg 313.1 nm doublet and subtraction of the contributions of the diffraction slit width $s_0 = \lambda f/W$ and the entrance slit width s_e. The contribution of the natural width of the Hg lines can be neglected, as it is very low in the case of a hollow cathode lamp. In monochromators, s_z predomates, but at very high resolution it becomes less significant and $A_1 \longrightarrow 1$.

The factor A_2 describes the influences of the profile of the analysis line and the effective measurement slit. For photoelectric measurements, the latter is the exit slit width s_a of the spectrometer. This contribution is relevant only when $s_a/s_{eff} < 2$, which is practically the case only for photographic measurements where the line profile is scanned with a very narrow densitometer slit. For photoelectric measurements, $A_2 = s_a/s_{eff} \gg 1$, as the exit slit width must be larger than the effective line width because of the thermal and mechanical stability of the system. The detection limit is a relevant figure of merit for an analytical method (Section 21.3.1). It can be written for AES:

$$c_L = c_U(I_X/I_U) \tag{162}$$

where I_X/I_U is the smallest line-to-background ratio which can be measured. When the definition is based on a 99.86% probability, it is three times the relative standard deviation of the background intensities.

For photographic or diode array registration, the intensities of the line and the spectral background are recorded simultaneously, and:

$$\sigma^2 = \left[\sigma_r^2(I_X + I_U) + \sigma_r^2(I_U)\right]^{1/2} \tag{163}$$

As the intensities of the line and the spectral background are almost equal near the detection limit, their standard deviations are equal and the relevant relative standard deviation reduces to:

$$\sigma^2 = \sigma_r(I_U)\,3\sqrt{2} \tag{164}$$

Its value can be measured directly, but in photographic measurements it can be calculated from the standard deviation of the blackening with the aid of Equation (115). The latter shows that for trace analyses, emulsions with a high γ must be used. In photoelectric measurements, σ is the relative standard deviation of the background intensity, measured at a wavelength adjacent to the spectral line or in a blank sample at the wavelength of the analytical line. It may comprise several contributions:

$$\sigma^2(I_U) = \sigma_P^2 + \sigma_D^2 + \sigma_A^2 + \sigma_V^2 \tag{165}$$

where σ_P is the photon noise of the source or the noise of the photoelectrons and $\sigma_P \approx \sqrt{n}$, where n is the number of photoelectrons; σ_D is the dark current noise of the photomultiplier and $\sigma_D \approx I_D$, where I_D is the dark current; σ_A is the flicker noise of the radiation source and $\sigma_A \approx \varphi$, where φ is the radiant density; and σ_V is the amplifier noise.

In most cases, σ_V can be neglected. For very constant sources, σ_A is small. Owing to the proportionality between σ_D and I_D it is advisable to select a photomultiplier with a low dark current, to avoid detector noise limitations. When these points are taken care of, the photon noise limits the power of detection.

When signals include a blank contribution, the latter also has fluctuations, which limit the power of detection. If I_X is the lowest measurable signal without blank contribution, I'_X the lowest analytical signal including a considerable blank value, I_U the background signal, and I_{Bl} the blank signal:

$$(I'_X + I_{Bl} + I_U) - (I_{Bl} + I_U) = 3\sqrt{[\sigma^2(I'_X + I_{Bl} + I_U) + \sigma^2(I_{Bl} + I_U)]}$$

and

$$(I_X + I_U) - I_U = 3\sqrt{[\sigma^2(I_X + I_U) + \sigma^2(I_U)]} \qquad (166)$$

This leads to:

$$(I'_X/I_U) = (I_X/I_U)(1 + I_{Bl}/I_U)[\sigma_r(I_{Bl} + I_U)/\sigma_r(I_U)] \qquad (167)$$

and the detection limit becomes [211]:

$$c'_L = c_U\{(I_X/I_U)(1 + I_{Bl}/I_U)[\sigma_r(I_{Bl} + I_U)/\sigma_r(I_U)]\}$$

or:

$$c_L[\sigma_r(I_{Bl} + I_U)/\sigma_r(I_U)](1 + I_{Bl}/I_U) \qquad (168)$$

To obtain maximum power of detection in AES, the standard deviations of the background, the blank contributions, and the c_U values must be minimized. The latter are a function of the radiation source, the elements, and the lines (reflected in $c_{U,\infty}$), and of the spectrometer (through A_1 and A_2). In order to keep A_1 and A_2 as close to unity as possible, it is necessary to use a spectrometer with high resolving power (high R_0), a narrow entrance slit ($s_e = 1.5 s_0$), and to take $s_a = s_{\text{eff}}$. On the other hand, the spectral background intensities must still be measurable. Indeed, the background intensities obtained are proportional to the entrance slit width, and detector noise limitations can arise when slit widths become too narrow. Also thermal and mechanical stabilities become limiting at narrow slit widths.

Apart from high power of detection, maximum analytical accuracy is very important. This relates to the freedom from interference. Whereas interference from influences of the sample constituents on sample introduction or volatilization and excitation in the radiation source differ widely from one source to another, most sources emit complex spectra and the risks of spectral interference in AES are much more severe than in absorption or fluorescence. Therefore, it is advisable to use high-resolution spectrometers. This is especially the case when trace determinations are performed in matrices emitting complex spectra. Knowledge of the atomic spectra is also very important so as to be able to select interference-free analysis lines for a given element in a well-defined matrix at a certain concentration level. This requirement is also much more stringent than in absorption or fluorescence.

21.6.2. Spectrometers

As all elements present in the radiation source emit their spectra at the same time, AES is by definition a multielement method. As well as simultaneous determinations, so-called sequential analyses are possible, provided the analytical signals are constant. Sequential and simultaneous multielement spectrometers have their own advantages and limitations.

Sequential spectrometers usually include Czerny–Turner or Ebert monochromators with which lines can be rapidly selected and measured one after another. Owing to the complex emission spectra, focal lengths are often up to 1 m and a grating with width up to 100 mm and spacing of 1/2400 mm is used. The instrument requires very accurate wavelength presetting, which is difficult for random access. The half-width of a spectral line corresponds to a grating angle of 0.0001°. Therefore, it is possible to work in a scanning mode and integrate stepwise at different locations across the line profile; alternatively, optical encoders and direct grating drives may be used. Systems employing fixed optics with a multislit exit mask, a movable photomultiplier, and fine wavelength adjustment by computer-controlled displacement of the entrance slit are also used. Sequential spectrometers are flexible because they can measure any elemental line and any background wavelength within the spectral range (usually 200–600 nm); however, they are slow when many elements need to be determined. They are especially of use in the case of stable radiation sources such as plasmas at atmospheric pressure or glow discharges operated at steady-state sputtering conditions.

Simultaneous Spectrometers. Modern simultaneous spectrometers use photoelectric radiation detection rather than the photographic plates formerly used for qualitative and semi-quantitative analyses. Simultaneous photoelectric spectrometers mostly have a Paschen–Runge optical mounting. They have many exit slits and are especially suitable for rapid determinations of many elements in a constant, well-characterized matrix; the achievable precision is high. However, the analysis program is fixed and simultaneous spectrometers are not suitable for the analysis of random samples. Owing to stability requirements, larger exit slits are often used to overcome thermal drift, but spectral resolution is reduced. Especially for trace analysis, background correction may be required. This can be achieved by computer-controlled displacement of the entrance slit or by rotation of a quartz refractor plate behind the entrance slit.

Echelle spectrometers [36] are also often used. By a combination of an order-sorter and an echelle grating, in either parallel or crossed-dispersion mode, high practical resolution (up to 300 000) can be realized with an instrument of rather low focal length (down to 0.5 m). Therefore, stability and luminosity are high. By using an exit slit mask with a high number of preadjusted slits or a solid-state, two-dimensional array detector highly flexible and rapid multielement determinations are possible.

As the most sensitive lines of a number of elements such as S, P, C, As, are in the vacuum, it is desirable to work in this spectral range, which necessitates the elimination of oxygen from the entire optical path. This can be achieved by continuous evacuation of the spectrometer, purging with nitrogen or argon, or by the use of sealed gas-filled spectrometers. It is often helpful to measure low wavelengths in the second order, to increase spectral resolution and wavelength coverage (into the region where the alkali metals have their most sensitive lines). Solar blind photomultipliers can be used to filter out long-wavelength radiation of the first order.

21.6.3. Flame Emission

The oldest spectroscopic radiation sources operate at low temperature (Section 21.5.3.1), but have good spatial and temporal stability. They readily take up wet aerosols produced by pneumatic nebulization. Flame atomic emission spectrometry [162] is still a most sensitive technique for determination of the alkali metals, e.g., for serum analysis. With the aid of hot flames such as the nitrous oxide–acetylene flame, a number of elements can be effectively excited, but cannot be determined at low concentration. Interference arising from the formation of stable compounds is high.

21.6.4. Arcs and Sparks

Arc and spark AES have been widely used for analysis of solids and remain important methods for routine analysis.

Electrical arc sources for AES use currents between 5 and 30 A and burning voltages between the electrodes, for an arc gap of a few millimeters, of 20–60 V. Direct current arcs can be ignited by bringing the electrodes together, which causes ohmic heating. Under these conditions, d.c. arcs have a normal (decreasing) current–voltage characteristic owing to the high material ablation by thermal evaporation. A d.c. arc displays the cathode fall near the cathode (cathode layer) the positive column, and the anode fall area. The cathode glow delivers a high number of electrons carrying the greater part of the current. This electron current heats the anode to its sublimation or boiling temperature (up to 4200 K in the case of graphite) and the cathode reaches temperatures of ca. 3500 K as a result of atom and ion bombardment. An arc impinges on a limited number of small burning areas. The plasma temperature in the carbon arc is of the order of 6000 K and it can be assumed to be in local thermal equilibrium. Arcs are mostly used for survey trace analysis requiring maximum power of detection, but they may be hampered by poor precision (relative standard deviation $\geq 30\%$).

Direct Current Arc Spectrometry. The sample is presented to an electrode with suitable geometry and volatilizes as a result of the high electrode temperature. Cup-shaped electrodes made of graphite are often used. In the case of refractory matrices, anodic evaporation may be applied and the positive column or, for very sensitive methods, the cathode layer, where the volatilized sample ions are enriched (cathode layer effect) is selected as analytical zone.

Solid samples must be powdered and sample amounts up to 10 mg (powder or chips) are brought to the electrode after mixing with graphite powder and pelleting. Thermochemical reagents such as NH_4Cl may be added, to create similar volatilization conditions for all samples. Spectrochemical buffers such as CsCl may be added for controlling the arc temperature or the electron pressure. In most cases internal standards are used. The plasma is operated in noble gas atmospheres or in O_2 or CO_2 to create favorable dissociation and excitation conditions or to avoid intense bands such as those of CN. Procedures for the analysis of oxide powders by d.c. arc analysis are well known [212]; d.c. arcs are also of use for the analysis of metals, as is still done for copper with the so-called globule arc. Here distillation of volatile elements from the less volatile matrix is applied.

Stabilized d.c. arcs may be obtained by sheathing gas flows or by wall stabilization [213]. Magnetic fields can be used to produce rotating arcs, which have higher precision.

Dry solution analysis can be performed with metal electrodes (e.g., copper) or with pretreated graphite electrodes.

Direct current arc spectrography is still a most powerful method for trace determinations in solids, even with difficult matrices such as U_3O_8. Detection limits for many elements are in the sub-µg/g range [214]. It is still in use for survey analysis.

Alternating Current Arc Spectrometry. Alternating current arcs have alternating polarity at the electrodes. Arc reignition is often by a high-frequency discharge applied at the a.c. discharge gap. Thermal effects are lower and the burning spot changes more frequently, so reproducibility is better than in d.c. arcs. Relative standard deviations of the order of 5–10% may be obtained.

Spark Emission Spectrometry. Spark emission spectrometry is a standard method for direct analysis of metallic samples and is of great use for production as well as for product control in the metals industry [215], [216].

Sparks are short capacitor discharges between at least two electrodes. The spark current is delivered by a capacitor charged to voltage U_C by an external circuit (Fig. 37 A). When the voltage is delivered directly by an a.c. current, an a.c. spark is produced. The voltage can also be delivered after transforming to a high voltage and rectifying, to produce d.c. spark discharges. In both cases, the voltage decreases within a very short time from U_C to U_B, the burning voltage. During the formation of a spark, a spark channel is first built up and a vapor cloud is produced at the cathode. This plasma may have a temperature of 10 000 K or more, whereas in the spark channel even 40 000 K may be reached. This plasma is in local thermal equilibrium, but its lifetime is short and the afterglow is soon reached. Here the temperature and the background emission are low, but owing to the longer lifetime of atomic and ionic excited states, line emission may still be considerable.

In a discharge cirucit with capacitance C, inductance L, and resistance R, the maximum discharge current is:

$$I_0 = (U_0 - U_B)\sqrt{(C/L)} \approx U_0\sqrt{(C/L)} \qquad (169)$$

The spark frequency within the spark train is:

$$f = 1/(2\pi)\sqrt{[1/LC - R^2/(4L^2)]} \qquad (170)$$

or, when R is small:

$$f = 1/(2\pi)\sqrt{(1/LC)} \qquad (171)$$

The capacitor may store energy:

$$Q = 1/2 U_C^2 C \qquad (172)$$

of which 20–60% is dissipated in the spark and the rest in the so-called auxiliary spark gaps in series with the analytical spark, or as heat. Depending on R, L, and C the spark has different properties; e.g., when L increases the spark becomes harder. As shown in Figure 37B, there may be an oscillating (a), critically damped (b), or an overcritically damped spark (c).

High-voltage sparks with $10 < U_C < 20$ kV were frequently used in the past. Their time cycle was regulated mechanically or electrically. Medium-voltage sparks ($0.5 < U_C < 1.5$ kV) are now used for metal analysis. They need a high-frequency discharge for ignition and are mostly unipolar discharges in argon. Accordingly, band emission is low, the most sensitive lines of elements such as C, P, S, and B can be excited, and the vacuum UV radiation is not absorbed. Often, a high energy is used to prespark, by which the sample melts locally, is freed from structural effects, and becomes homogeneous. Working frequencies are often up to 500 Hz, so high energies can be dissipated and high material ablation occurs.

Analytical Features. Spark emission spectrometry enables rapid and simultaneous determination of many elements in metals, including C, S, B, P, and even gases such as N and O. Therefore, spark emission AES is complementary to X-ray spectrometry for metallurgical analyses [217]. Analysis times including presparking (5 s) may be down to 10–20 s and the method is of great use for production control. Special sampling devices are used for liquid metals, e.g., from blast

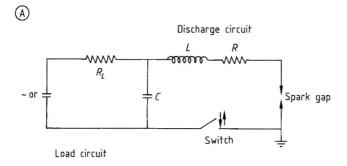

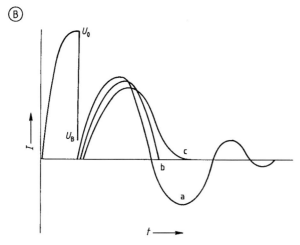

Figure 37. Circuitry in spark generator for optical emission spectrometry (A) and types of spark discharge (B)
a) Oscillating spark ($R < 2\sqrt{L/C}$); b) Critically damped spark ($R = 2\sqrt{L/C}$); c) Overcritically damped spark ($R > 2\sqrt{L/C}$)

furnaces or convertors, and special care must be taken that samples are homogeneous and do not include cavities. The detection limits obtained in metal analysis differ from one matrix to another, but for many elements they are in the µg/g range (Table 4). The power of detection of spark analysis is lower than that of d.c. arc analysis. Analytical precision is high. By using a matrix line as internal standard, relative standard deviations at the 1% level and even lower are obtained.

Matrix interference is high and relates to the matrix composition and the metallurgical structure of the sample. Therefore, a wide range of standard samples and matrix correction procedures are used (Section 21.5.2.1). Better selection of the analytical zone of the spark is now possible with the aid of optical fibers and may decrease matrix interference.

Acquisition of the spectral information from each single spark is now possible with advanced measurement systems, using integration in a small capacitor and rapid computer-controlled readout combined with storage in a multichannel analyzer. With the aid of statistical analysis, sparks with outlier signals can be rejected and the precision improves accordingly. Pulse differential height analysis enables discrimination between dissolved metals and excreted oxides for a number of elements such as Al and Ti. The first give low signals and the latter high intensities as they stem from locally high concentrations [218]. By measurements in the afterglow of single sparks, or by the application of cross-excitation in spark AES [219], there is still room for improvement of the power of detection.

Spark AES has also been used for oil analysis. The oil is sampled by a rotating carbon electrode. Electrically insulating powders can be analyzed after mixing with graphite powder and pelleting.

Table 4. Detection limits (µg/g) for steels in spark and glow discharge AES and glow discharge MS

	Spark AES[a]	GD–AES[b]	GD–MS[c]
Al	0.5	0.1	
B	1	0.3	
Cr	3	0.05	
Cu	0.5	0.3	
Mg	2	0.9	
Mn	3	0.2	
Mo	1	0.8	
Nb	2	0.6	
Ni	3	0.1	
Pb	2		0.004
Si	3	0.4	
Ti	1	0.6	
V	1	1	
Zr	2	1.5	

[a] Spark discharge in argon with polychromator [215].
[b] Grimm-type glow discharge with 1 m Czerny–Turner monochromator [290].
[c] Grimm-type glow discharge and quadrupole mass spectrometer [334].

21.6.5. Plasma Sources

In order to overcome the disadvantages related to the low temperature of chemical flames, but to have sources with similarly good temporal stability and versatility with respect to analyte introduction, efforts were directed toward electrically generated plasmas. At atmospheric pressure, these discharges have a temperature of at least 5000 K and, provided their geometry is properly optimized, they allow efficient introduction of analyte aerosols. In most cases, they are operated in a noble gas to avoid chemical reactions between analyte and working gas.

Among these plasma sources, d.c. plasmajets, inductively coupled plasmas (ICP), and microwave discharges have become important sources for AES, mainly for the analysis of solutions [220], [221].

21.6.5.1. Direct Current Plasmas

Direct current plasmajets (DCPs) were first described as useful devices for solution analysis by KOROLEW and VAINSHTEIN [222] and by MARGOSHES and SCRIBNER [223]. Sample liquids were converted to aerosols by pneumatic nebulization, and arc plasmas operated in argon at temperatures of ca. 5000 K.

Types of Plasmajets. It is customary to distinguish between so-called current-carrying and current-free plasma jets. With *current-carrying plasmajets*, cooling as a result of changes in aerosol introduction has little influence. The current decrease resulting from a lower number density of current carriers can be compensated by an increase of the burning voltage, and this stabilizes the energy supply. The discharge, however, is very sensitive to changes in the electron number density, caused by varying concentrations of easily ionized elements. When the latter increase, the burning voltage energy supply and break down.

In *current-free* or *transferred plasmas*, the observation zone is situated outside the current-carrying zone. This can be achieved by the use of a supplementary gas flow directed perpendicular to the direction of the arc current, and observation of the tail flame. In this observation zone, there is no current. This type of plasma reacts strongly to cooling as no power can be delivered to compensate for the temperature drop. Therefore, it is rather insensitive to the addition of easily ionized elements. They do not cause a temperature drop, but shift the ionization equilibrium and give rise to so-called ambipolar diffusion. When the ionization in the observation zone increases, the pressure (proportional to the total particle number density — ions and electrons) increases, and ions as well as electrons move toward the outer zones of the plasma. Owing to the higher mobility of the electrons, a field is built up which amplifies the force on the positive ions directed to the outer plasma zones and thus increases the plasma volume. The d.c. plasmajet described by MARGOSHES and SCRIBNER [223] is a current-carrying plasma, whereas the plasma described by KRANZ [224] is a transferred plasma.

Three-Electrode Plasmajet. The three-electrode d.c. plasmajet developed by Spectrometrics (Fig. 38) [225] uses two graphite anodes protected by sheathing gas flows, each having a separate supply (2×10 A, 60 V). There is a common shielded cathode made of tungsten. The sample solution is usually nebulized by a cross-flow nebulizer, and the aerosol is aspirated by the gas flows around the anodes. The observation zone is small (0.5×0.5 mm^2), but very intense and located just below the point where the two positive columns meet. Optimally, this plasma is combined with a high-resolution echelle monochromator with crossed dispersion, as the latter has an entrance slit height ca. 0.11 mm. It has the advantages of a current-free plasma, as the observation zone is just outside the current-carrying part of the

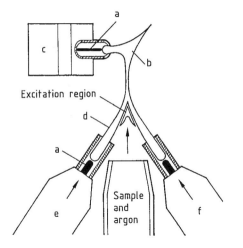

Figure 38. Three-electrode plasmajet [225]
a) Electrode; b) Plume; c) Cathode block; d) Plasma column (background) continuum; e) Argon anode block front; f) Argon anode block back

discharge, but still close enough to be compensated for strong cooling. The plasma temperature is of the order of 6000 K, $n_e = 10^{14}$ cm^3 [226], and the plasma is near local thermal equilibrium.

Detection limits for most elements are of the order of 5–100 ng/mL. For elements with very sensitive atomic lines such as As, B, and P, the detection limits are slightly lower than in ICP–AES [227]. The high power of detection also relates to the high resolution of the echelle spectrometer. Different concentrations of alkali metals, however, cause higher matrix effects than in ICP–AES and even may necessitate the use of spectrochemical buffers. The analytical precision achievable is high and relative standard deviations below 1 % can be reached. The system can cope with high salt content (>100 g/L), and has found considerable use in water analysis [228]. Its multielement capacity is limited, as the optimal excitation zones for most elements differ slightly; as a result of the small observation zones, considerable losses in power of detection occur. Owing to the high resolution of the echelle spectrometer, DCP–AES is very useful for the analysis of materials with complex emission spectra, such as high-alloy steels, tungsten, molybdenum, and other refractory metals.

21.6.5.2. Inductively Coupled Plasmas

Inductively coupled plasmas in their present form date back to the work of REED [229], who used these sources for crystal growth in the 1960s. They were introduced in spectrochemical analysis by GREENFIELD et al. in 1964 [230] and WENDT and FASSEL [231], who used them in their present form as sources for AES.

In inductively coupled plasmas, the energy of a high-frequency generator (6–100 MHz) is transferred to a gas flow at atmospheric pressure (usually argon) in a quartz tube system, with the aid of a working coil (Fig. 39). The electrons take up energy and collide with atoms to form a plasma with a temperature of up to 6000 K. At a suitable gas flow, torch geometry, and frequency, a ring-shaped toroidal plasma occurs, where the sample aerosol passes centrally through the hot plasma. The burner consists of three concentric quartz tubes. The aerosol is fed, with its carrier-gas, through the central tube. Between the outer and intermediate tubes, a gas flow is introduced tangentially. It takes up the high-frequency energy, and prevents the torch from melting. The Fassel torch has a diameter of ca. 18 mm and can be operated at 0.6–2 kW with 10–20 L/min of argon. In so-called minitorches, gas consumption is ca. 6 L/min, by using a special gas inlet or by further reducing the torch dimensions [232]. GREENFIELD used a larger torch (outer diameter up to 25 mm), in which an ICP can be operated at higher power and where argon, oxygen, or air can be used as outer gas flows (up to 40 L/min) with an intermediate argon flow (ca. 8 L/min). With the Fassel torch, an intermediate gas flow (ca.1L/min) is useful when analyzing solutions of high salt content or organic solutions, to prevent salt or carbon deposition.

The ICP is near local thermal equilibrium. Excitation temperatures (from atomic line intensity ratios) are about 6000 K [233] and rotation temperatures (from the rotation lines in the OH bands) are 4000–6000 K [234], [235]. From the broadening of the H$_\beta$-line, an electron number density of 10^{16} cm^{-3} is obtained, whereas from the intensity ratio of ionic and atomic lines of the same element the electron number density is 10^{14} cm^{-3}. Measured line intensity ratios of ionic to atomic lines were a factor of 100 higher than those calculated for a temperature of 6000 K and an electron number density of 10^{16} cm^{-3}. This indicates the existence of overionization. The excitation processes taking place include:

1) Excitation through electrons or processes involving radiation:
   Ar + e $\longrightarrow$ Ar*, Ar^m (metastable levels $E \approx 11.7$ eV)

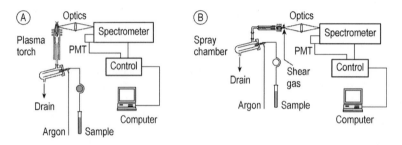

Figure 39. ICP-AES with (a) side and (b) axial viewing of the plasma

and
Ar$^+$ ($E_i = 15.7$ eV)
Ar + $h\nu$ ⟶ Ar$^+$ (radiation trapping)
M + e ⟶ M*, M^{+*} + e (electron impact)
M$^+$ + e ⟶ M + $h\nu$ (radiative recombination)

2) Excitation through argon species:
Ar + M ⟶ Ar + M*
Ar^m + M ⟶ Ar + M$^+$ (Penning effect)
Ar$^+$ + M ⟶ Ar + M^{+*} (charge transfer)

Accordingly, an overpopulation of the argon metastable levels would explain both the overionization as well as the high electron number density in the ICP. Indeed, it may be that argon metastables act as ionizers, but at the same time are easily ionized [236]. This could explain the rather low interference caused by easily ionized elements, and the fact that ionic lines are excited very efficiently, despite the fact that their standard temperatures are much higher than the plasma temperatures. However, discrepancies are not encountered to such a great extent when the temperatures from the Saha equation are used in the calculations [237]. Nevertheless, a number of processes may be considered to predominate in well-defined zones of the plasma, as indicated by spatially resolved measurements of various plasma parameters [238].

Instrumentation (Fig. 39). Inductively coupled plasma generators are $R-L-C$ circuits producing a high-frequency current at high power. As end-stage a triode or transistor is used. The high-frequency energy can be led to the remote plasma stand by coaxial cable. This requires the use of a matching unit to tune the plasma with the generator before and after ignition. In other systems, the working coil may be an integral part of the oscillator. Generators with a fixed frequency and generators where the frequency may change with load are used. In the first case, a change of load changes the ratio of the forward power and the power reflected to the generator, and requires retuning; in the second case, a small frequency shift occurs. Both types of generator may perform equally, provided the power supply to the plasma, usually ca. 50% of the energy dissipated in the generator, remains constant. Both the power stability and constant gas flows are essential to obtain maximum analytical precision.

In plasma spectrometry, the sample must be introduced into the plasma as a fine aerosol, usually by pneumatic nebulization. The aerosol should enter the hot plasma zones with low injection velocity for residence times in the millisecond range, required for efficient atomization and excitation. Therefore, the carrier gas flow should not exceed 1–2 L/min. With advanced pneumatic nebulizers for ICP spectrometry, aerosols with a mean droplet diameter ca. 5 µm can be obtained with an efficiency of a few percent [90], [91] for a gas flow of 1–2 L/min and a sample uptake of 1–2 mL/min. The concentric glass Meinhard nebulizer is self-aspirating, the cross-flow nebulizer may have capillaries made of different materials and is pump-fed, but less sensitive to clogging in the case of high-salt solutions. The Babington nebulizer can be used for slurry nebulization [179]. It has been used in ICP–AES for many types of powder analyses, including ceramic powders [125]. The fritted disk nebulizer has not achieved widespread use. New approaches for ICP–AES lie in the use of the jet-impact nebulizer [93], direct injection nebulization [94] (Section 21.4.1). In pneumatic nebulization for ICP–AES, continuous sample feeding requires a sample aspiration time of ca. 30 s, to give a stable signal, a measurement time of some 5–10 s, and a minimum rinsing time of 30 s. Discrete sampling is also possible with injection systems developed for flame AAS [78], [79], and by flow injection. Work with sample aliquots down to 10 µL becomes possible — especially useful with microsamples [91], or for the analysis of high-salt solutions [239].

For ICP–AES, sequential, simultaneous, and combined instruments are used. In sequential spectrometers, special attention is given to the speed of wavelength access and, in simultaneous spectrometers, to the provision of background correction. In combined instruments, a number of frequently used channels are fixed and a moving detector or an integrated monochromator gives access to further lines. Echelle instruments allow maximum resolution at low focal length and can be combined with advanced detectors such as CCDs.

Analytical Performance. *Power of Detection.* The commonest configuration for an ICP-AES instrument is for the monochromator and detector to view the plasma side-on as shown in Figure 39 (a). This means that there is an optimum viewing height in the plasma which yields the maximum signal intensity, lowest background, and least interferences, and it is common practise to optimize this parameter to obtain the best performance from the instrument. Criteria of merit commonly used include the signal-to-noise ratio (SNR), signal-to-background ratio (SBR) and net signal intensity. Different elements, and different analytical lines of the same element, will have different optimum viewing heights so a set of compromise conditions is usually determined which gives satisfactory performance for a selected suite of analytical lines. An alternative configuration which is gaining popularity is end-on viewing of the plasma [Fig. 39 (b)]. This has the advantage that an optimum viewing height does not have to be selected because the analytical signal from the whole length of the central channel is integrated. However, any interferences present are also summed, so that interference effects are expected to be more severe with this configuration.

The nebulizer gas flow and power, and the observation height are also important parameters to optimize. Such multivariate optimizations can be done by single-factor studies or Simplex optimization [240], [241]. The nebulizer gas influences the aerosol droplet size, nebulizer efficiency, and plasma temperature. For each element and line there is a rather sharp optimum for the aerosol gas flow where the SBR is maximum. The power determines mainly the plasma volume and is optimum for soft lines (atomic and ionic lines of elements with low ionization potential) at rather low values (1–1.2 kW), whereas for hard lines (lines of elements with high excitation potential and high-energy ionic lines) and for organic solutions, the optimum value is higher (1.5–1.7 kW). The optimum observation height is a compromise between the analyte number densities (highest in the lower zones) and completeness of atomization and excitation (a few millimeters above the working coil). Only elements such as the alkali metals are measured in the tail flame. As the analytical lines in ICP–AES have a physical width of 1–20 pm and the spectra are complex, a high-resolution spectrometer is required.

The detection limits for most elements are at the 0.05 ng/mL (Mg, Ca, etc.) to 10 ng/mL level (As, Th, etc.). Especially for elements which have refractory oxides (Cr, Nb, Ta, Be, rare earths, etc.), fairly low ionization potential, and sensitive ionic lines, the detection limits are much lower than in AAS. For P, S, N, O, and F, the most sensitive lines are at vacuum wavelengths.

Interference. Most interference in ICP–AES is additive and relates to coincidences of spectral background structures with the elemental lines used. Interference is minimized by the use of high-resolution spectrometers and background correction procedures. In the wavelength range 200–400 nm more than 200 000 atomic emission lines of the elements are listed [200] and molecular bands also occur. Accordingly, coincidences often occur within the spectral line widths themselves. In order to facilitate the selection of interference-free spectral lines in ICP–AES, BOUMANS worked out a system which gives the interference for about 900 prominent lines [205] for spectrometers with a given spectral slit width, in terms of critical concentration ratios. Further, line tables for ICP–AES [242] and atlases [204], [206] are available, although incomplete. In ICP–AES multiplicative interference also occurs. This may have several causes:

1) Nebulization effects: as discussed earlier, differences in physical properties of the sample solutions lead to variations in the aerosol droplet size and thus in the efficiency of nebulization and sample introduction. This effect is strongest in the case of free sample aspiration and relatively low nebulizer gas flow, and can be minimized (Section 21.4.1).
2) Easily ionized elements have a complex influence. First they may cause a decrease in excitation temperature, as energy is consumed for ionization. Further, they may shift the ion-

ization equilibrium for partially ionized elements as their easy ionization can influence the electron number densities. They may also cause changes in plasma volume as a result of ambipolar diffusion or they may even quench excited species such as metastables [243]. After correcting for the influence of easily ionized elements on the spectral background, signal enhancements or suppressions are negligible up to concentrations of 500 µg/mL for Na, K, Ca, or Mg. At larger differences in concentration matrix matching of the standards should be applied.

Analytical Precision. As sample introduction in the plasma is very stable, relative standard deviations in ICP–AES are ca. 1% and the limiting noise is proportional, including flicker and nebulizer noise. Below 250 nm detector noise limitations may occur, especially with high-resolution spectrometers. At integration times beyond 1 s, the full precision is normally achieved.

With the aid of an internal standard, fluctuations in sample delivery and, to a certain extent, changes in excitation conditions can be compensated. To gain the full benefit of internal standardization, the intensities for the analytical and internal standard lines (and in trace determinations that of the spectral background near the analytical line) should be measured simultaneously. The selection of line pairs for ICP–AES has been investigated by FASSEL et al. [248] and the background correction by MEYERS and TRACY [249]. With internal standardization relative standard deviations below 0.1% can be obtained.

Multielement Capacity. The multielement capacity of both sequential and ICP–AES is high, as is the linear dynamic range (up to five orders of magnitude). It is limited at the lower end by the limits of determination and at higher concentrations by the onset of self-reversal. This starts at analyte concentrations of ca. 20 000 µg/mL, as analyte atoms hardly occur at all in the cooler plasma zones.

Special Techniques. The analytical performance of ICP–AES can be considerably improved by using alternative techniques for sample introduction.

Organic solutions (methyl isobutyl ketone, xylene, kerosene) can be nebulized pneumatically and determinations in these solvents can be performed directly with ICP–AES [250]. The addition of oxygen or the use of higher power is helpful in avoiding carbon deposits. Oils can be analyzed after dilution 1:10 with xylene ($c_L = 0.5 - 1$ µg/g) and heavy metals can be extracted from wastewater as complexes with ammonium pyrrolidinedithiocarbonate (APDTC) and separated from more easily ionized elements [251].

Ultrasonic nebulization has been used right from the beginning of ICP–AES [99]. Nebulizer types where the sample liquid flows over the nebulizer transducer crystal and types where ultrasound at 1 MHz is focused through a membrane on the solution have both been used. With aerosol desolvation the power of detection of ICP–AES can be improved by a factor of ten by using ultrasonic nebulization. This applies especially to elements such as As, Cd, and Pb, which are of environmental interest. However, because of the limitations discussed in Section 21.4.2, the approach is of practical use only for dilute analytes, as in water analysis [98]. New approaches such as thermal spray nebulization [252] and high-pressure nebulization [94] increase analyte introduction efficiency and thus the power of detection, but again only when aerosol desolvation is applied. They are especially interesting for speciation by coupling of ICP–AES to HPLC.

Electrothermal evaporation (ETV) and direct sample insertion increase sampling efficiency so that work with microsamples becomes possible, and the power of detection can be further improved. ETC–ICP–AES has been applied with graphite cups [117], graphite furnaces [66], and tungsten filaments [253]. Direct sample insertion enables the direct analysis of waste oils [124], and of microamounts of biological samples [122], as well as the determination of volatile elements in refractory matrices, as shown for Al_2O_3 [121]. Difficulties lie in calibration and in signal acquisition. In trace analysis, the latter necessitates simultaneous, time-resolved measurement of the transient signals for the line and background intensities.

Hydride generation with flow cells [102] decreases the detection limits for As, Se, etc., by 1.5 orders of magnitude down to the ng/mL level — useful in water analysis. However, possible systematic errors from heavy metals and analytes in nonmineralized form remain.

Gas and liquid chromatography coupled to ICP–AES are useful for speciation work [196], [254], e.g., for Cr(III)/Cr(VI) speciation.

Direct solid sampling is of special interest, even though ICP–AES is of use mainly for the analysis of liquids. In a number of cases the same precision and accuracy can be obtained as in solutions, but without the need for time-consuming sample dissolution, involving analyte dilution. For compacted solids, arc/spark ablation is a viable approach in the case of metals [255], [135]. Aerosols with particle sizes at the µm level [137] and detection limits at the µg/g level are obtained [256]. Owing to the separate ablation and excitation, matrix influences are especially low, as shown for aluminum [135] and steels [256]. Laser ablation can be used with similar figures of merit [257], even for electrically insulating samples with lateral resolution (up to ca. 10 µm).

High-frequency discharges at low power and with capacitive power coupling, e.g., stabilized capacitively coupled plasmas (SCP) [255] for element-specific detection in chromatography may be a useful alternative, especially when advanced sample introduction techniques are used.

Applications. Nowadays, ICP–AES is a routine analytical method, of special use when a large number of elements has to be determined in many samples (solutions or dissolved solids). In many cases it is complementary to AAS. ICP–AES is of special interest for the analysis of geological samples, environmental analysis, clinical analysis, food analysis, metals analysis, chemical analysis, and in the certification of reference materials [220], [221].

In the analysis of *geological samples* [243], sample dissolution is important with respect to both nebulization effects and ionization interference.

The analysis of *environmentally relevant samples* is a major field of application [243] and many standard methods have been developed, e.g., by bodies such as the US Environmental Protection Agency [244], [245]. This covers procedures for sample preparation and determination of 25 metals in waters, wastes, and biological samples. Details of quality assurance procedures, and interference correction are included.

In *clinical analysis*, Ca, Fe, Cu, Mg, Na, and K can be directly determined in serum samples even of microsize [211]. Electrothermal evaporation and direct sample insertion are particularly useful from this point of view. Generally, however, the power of detection of ICP–AES is too low for most analytical problems encountered in the life sciences.

In the field of industrial products, especially *metals analysis* deserves mention [246]. Apart from solution analysis, direct metal analysis by spark ablation is very useful [256]. Analysis of refractory powders by slurry nebulization, even for ZrO_2 matrices [125] or analysis subsequent to sample decomposition, is very powerful. Laser ablation of solid samples is an important technique which has benefited from the advent of solid-state detectors [247].

21.6.5.3. Microwave Plasmas

Microwave plasmas are operated at 1–5 GHz. They are mostly produced in a magnetron. Electrons emitted by a glowing cathode are led through a chamber with a series of radially arranged resonance chambers to which an ultrahigh-frequency field is applied. The currents produced, can be coupled to the anode and transported by coaxial cable. The microwave current is led to a waveguide or a cavity. Only when the latter have dimensions below certain critical values can microwave radiation be transported. The transport efficiency can be regulated with the aid of movable walls and screws. In a cavity, a standing wave is produced and the microwave energy is coupled to the resonator by a loop or antenna; $\lambda/2$, $\lambda/4$, etc., cavities have been proposed [267] and allow tuning with respect to a favorable ratio of forward to reflected power.

The role of microwave plasmas in optical spectrometry has recently been reviewed [268].

Capacitively-Coupled Microwave Plasmas. This type of single-electrode discharge goes back to the work of MAVRODINEANU and HUGHES [269]. The microwave radiation is led to a pin-shaped electrode, and the surrounding burner housing acts as counter-electrode. A bush-shaped discharge burns at the top of the electrode which enters the sample aerosol concentrically. The plasma, which can be operated with 2–4 L/min argon, helium or nitrogen (at 400–800 W), is not in local thermal equilibrium ($T_{ex} = 5000-8000$ K, $T_g = 3000-5700$ K, $n_e = 10^{14}$) and it is a current-carrying plasma. As the sample enters, with fairly low efficiency, into a hot zone where the background emission is also high, the detection limits are poor, especially for elements with thermally stable oxides [270], e.g., Na 589 nm, 0.05 µg/mL; Be 234.9 nm, 0.006 µg/mL; Mg 285.2 nm, 0.04 µg/mL; Al 396.1 nm, 0.16 µg/mL. Alkali and alkaline-earth metals cause high matrix effects; e.g., for 500 µg/mL they amount to 100% in the case of Mg. Therefore, spectrochemical

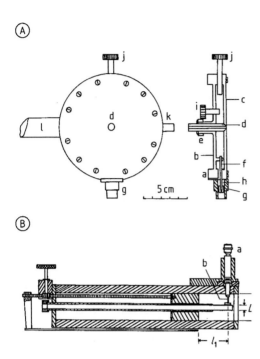

Figure 40. Microwave-induced plasma discharges
A) MIP in a TM_{010} resonator according to BEENAKKER [272]
a) Cylindrical wall; b) Fixed bottom; c) Removable lid;
d) Discharge tube; e) Holder; f) Coupling loop; g) PTFE insulator; h) Microwave connector; i, j) Tuning screw; k) Fine-tuning stub; l) Holder
B) Surfatron MIP, original version 20 mm $< l_1 <$ 250 mm, D_1 11 mm, D_2 40 mm, cut-off frequency 2 GHz [278]
a) Standard coaxial connector; b) Coupler

buffers such as CsCl are often used. Because it can work with He, the source is also useful for the speciation of widely diverse elements [271].

Microwave-Induced Plasmas. These electrodeless discharges are operated in a noble gas at rather low power (usually < 200 W). They became important as sources for atomic spectrometry when BEENAKKER et al. [272] succeeded in operating a stable plasma at atmospheric pressure with less than 1 L/min argon or helium and power below 100 W (Fig. 40). The so-called TM_{010} resonator is used. It has a cylindrical resonant cavity (diameter ca. 10 cm) where the power is introduced with a loop or antenna and the filament plasma is contained in a centrally located capillary. At higher power toroidal argon [273] and diffuse helium [274] discharges can be obtained. The plasma is not in local thermal equilibrium and metastable argon or helium species are thought to contribute to the excitation. MOUSSANDA et al. [275] studied the spectroscopic properties of various microwave-induced plasma (MIP) discharges and reported rotational temperatures of ca. 2000 K. These plasmas are very useful for coupling with electrothermal evaporation, where detection limits in the upper picogram range can be obtained. This can be achieved with both graphite furnace [81] and tungsten filament atomizers [115]. Excitation of desolvated aerosols is also possible. Wet aerosols can only be taken up with the toroidal argon MIP. This allows element-specific detection in liquid chromatography [276]. Helium MIPs are excellent for element-specific detection in gas chromatography, and are commercially available [277]. Not only the halogens and other elements relevant in pesticide residue analysis, but also organolead and organotin compounds can be determined down to low concentrations. This makes MIP–AES very useful for speciation work [197]. The MIP is also useful for excitation of volatile hydride-forming elements after removal of excess hydrogen. With different trapping techniques detection limits down to the sub-nanogram level are easily obtained [104]. A further useful type of MIP is the so-called surfatron described by HUBERT et al. [278] (Fig. 40 B). Here, the power is introduced by a perpendicular antenna directed near the plasma capillary; a plasma is formed by microwave coupling through a slit and power propagates along the surface of the discharge. Compared with the MIP obtained in a TM_{010} resonator this plasma is more stable and allows more variability in sampling, as shown by direct comparative studies with the same tungsten filament atomizer [279]. The MIP torch described by QUEHAN JIN et al. [280] can operate both argon and helium discharges and make use of hydride generation, electrothermal evaporation, etc.

A strip-line microwave plasma has been described by Barnes and Reszke [281], [282]. A particular advantage of this design is that it is possible to sustain more than one plasma in the cavity at the same time. Such a plasma has compared favorably with other designs [282], though it has not yet been widely used.

21.6.6. Glow Discharges

Glow discharges have long been recognized as unique sources for AES [283]. Their special features relate to the possibility of analyte introduction by sputtering as well as to the advantages of

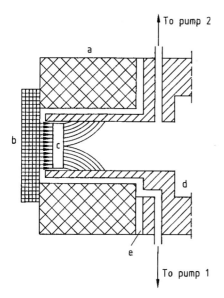

Figure 41. Classical GRIMM glow discharge lamp [149]
a) Cathode body; b) Sample; c) Negative glow; d) Anode body; e) PFTE sheath (0.2 mm thick)

plasmas not in local thermal equilibrium, enabling the discrete excitation of terms, even those of high energy. These advantages are realized in the hollow cathode and related sources, in glow discharges with flat cathodes, with d.c. or r.f. powering, and in special sources such as gas-sampling glow discharges.

21.6.6.1. Hollow Cathodes

Here, the energetically important parts of the discharge (cathode layer, dark space, and negative glow) as well as the sample, are inside the cathode cavity. The volatilization results from cathodic sputtering and/or thermal evaporation and the analyte is excited in the negative glow. The characteristic depends on the discharge gas (at 100–500 Pa Ar, $i = 0.2-2$ A, $V = 1-2$ kV; at 1–2 kPa He, $i = 0.2-2$ A, $V < 1$ kV), and also on the cathode mounting. In a cooled cathode, the characteristic is abnormal and the analyte volatilizes by cathodic sputtering only, whereas in a hot hollow cathode, thermal evaporation also takes place, and the characteristic becomes normal especially at high currents.

Owing to the long residence times in the excitation zone and the high line-to-background ratios, resulting from the nonequilibrium character of the plasma (gas temperatures of 600–1000 °C

[19] and much higher electron temperatures), the hollow cathode is the radiation source with the lowest absolute detection limits. For this reason, it is widely used for dry solution residue analysis, which is useful subsequent to matrix separation in the analysis of high-purity substances [284]. The Doppler widths of the lines are low and this source can even be used for isotopic analysis (determination of $^{235}U/^{238}U$). It is still employed for the determination of volatile elements in a nonvolatile matrix, such as As, or Sb in high-temperature alloys [285], where detection limits below 1 µg/g can be obtained. A treatment of the hollow cathode lamp as an AES source is given in [286].

21.6.6.2. Furnace Emission Spectrometry

Similar to hollow cathodes, electrically heated graphite furnaces in a low-pressure environment can be used for sample volatilization. The released analyte can then be excited in a discharge between the furnace and a remote anode. This source is known as furnace atomic nonresonance emission spectrometry (FANES), and was introduced by FALK et al. [287]. Owing to the separation of volatilization and excitation, its absolute detection limits are in the picogram range. For real samples, however, volatilization and excitation interference may be considerable.

21.6.6.3. Glow Discharges with Flat Cathodes

With this source, which is normally a d.c. discharge, compact electrically conducting samples can be analyzed. They must be flat, are used as the cathode, and ablated by cathodic sputtering. The sputtered material is excited in the negative glow of the discharge, which is usually operated at 100–500 Pa of argon. As the sample is ablated layer-by-layer, both bulk and in-depth profiling analysis are possible.

Glow Discharge Sources. A useful device was first described by GRIMM in 1968 [288]. The anode is a tube (diameter 8–10 mm) with the distance from its outer side to the cathode block and the sample less than the mean free path of the electrons (ca. 0.1 mm). The anode–cathode interspace is kept at a lower pressure than the discharge lamp. According to the Paschen curve, no discharge can take place at these locations, and the glow discharge is restricted to the cathode surface (Fig. 41). In later versions, a lamp with floating restrictor and remote anode is used, so that no second vacuum is necessary. The Grimm lamp has an abnormal characteristic, the current is set at 40–200 mA and the burning voltage is usually below 1 kV. In both krypton and neon, the burning voltage is higher than in argon, but it also depends on the cathode material. The ablated material is deposited both at the edge of the burning crater, limiting the discharge time to a few minutes, and inside the anode tube, making regular rinsing necessary. In AES, the glow discharge is viewed end on.

Characteristics. After ignition of the discharge, a burn-in time is required to achieve sputtering equilibrium. First, the surface layers have to be removed, but once equilibrium is reached the discharge penetrates into the sample at constant velocity. Burn-in times are usually up to 30 s for metals (at 90 W in argon, zinc 6 s; brass 3–5 s; steel 20 s; Al 40 s), but depend on the sputtering conditions (shorter at high voltage, low pressure, etc.). The burning crater has a structure depending on the material structure or inclusions, and the electrical field may induce a small curvature, especially in the classical Grimm lamp [289]. Material volatilization is of the order of mg/min and increases in the sequence: C < Al < Fe < steel < Cu < brass < Zn. It also depends on the gas used, normally a noble gas. Helium is not used; because of its low mass sputtering is too limited and the sputtering rate increase in the sequence: neon < argon < krypton.

Sputtering is proportional to $1/\sqrt{p}$ [151], where p is the gas pressure. The sputtering rates of alloys can be predicted to a certain extent from the values for the pure metals [149].

For excitation, collision with high-energy electrons, but also other processes such as charge transfer or excitation by metastables are important. Electron temperatures are 5000 K (slow electrons for recombination) and >10 000 K (high-energy electrons for excitation), and gas temperatures are below 1000 K, so line widths are low. Owing to the low spectral background intensities, the limits of detection arising from the source are low and, for metals, range down to µg/g in a classical Grimm lamp [290]. Elements such as B, P, S, C and As can also be determined. The plasma is optically thin, but for resonance lines of matrix elements, self-reversal may occur. However, the linear dynamic range is high compared with arc or spark sources.

Owing to the stable nature of the discharge, noise is low. There is hardly any flicker noise and only a few frequency-dependent components from the vacuum line are superimposed on a white noise background [44]. Relative standard deviations with no internal standard can be below 1%. However, the low radiant densities may lead to shot noise limitations. Because of its stable nature, the glow discharge source can be coupled with Fourier transform spectrometry. As sample volatilization is due to cathodic sputtering only, matrix interference as a result of the thermochemical properties of the elements do not occur.

Applications. Glow discharge AES is widely used for *bulk analysis of metals*. This applies less to production control (the burn-in time required is long compared with sparks) than to product control. Here, the easy calibration due to the absence of volatilization interference and low spectral interference, as well as the high linear dynamic range are advantageous. For elements with high sputtering rates, such as copper, the method is especially useful. Samples have to be available as flat and vacuum-tight disks (thickness 10–30 mm), but with special holders threads and metal sheets can be analyzed. The samples must be polished and freed from oxide layers prior to analysis.

Depth-profile analysis is also possible, as the sample is ablated layer-by-layer with a penetration rate of 1–3 µm/min. Here, the intensities of the analyte lines are measured as a function of time. However, the sputtering rates of alloys with varying composition must be known to convert the time scale into a depth scale. The intensities must be related to concentrations, which can be done by using theoretical models and sputtering constants [291]. The power of detection may be quite good and depth resolution is of the order of 5 nm, when elemental concentrations >0.1% are monitored. Depth profile analysis is now a main field of application of glow discharge emission spectrometry in the metals industry (Fig. 42) [291].

Electrically insulating powders may be analyzed after mixing with Cu powder and pelleting [134]. Performance depends greatly on the particle size of the powders, but detection limits in the µg/g range are possible for the light elements (Al, Be, B, etc.).

21.6.6.4. New Developments

Since it was found that a great part of the analyte released by sputtering consists of ground-state atoms, cross-excitation by d.c. [292],

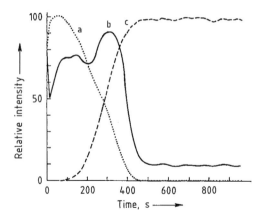

Figure 42. Depth profile through a galvanized steel sheet surface. The Zn coating (a) is ca. 18 µm thick, and the Al (b) concentration in the Fe bulk (c) is 0.049%, the voltage increases from 510 V in the Zn to 740 V in the bulk steel [291]

high-frequency [293], and microwave discharges [290] was investigated. The latter was found to improve the power of detection by up to a factor of ten, e.g., for steels. Further, it was found that simple types of d.c. discharges were capable of exciting vapors and even molecular species such as volatile hydrides [294].

Radio-frequency glow discharges are very useful sources for atomic spectrometry. By means of a bias potential in the vicinity of the sample, insulating samples such as ceramics can be directly ablated and analyzed by AES [295].

21.6.7. Laser Sources (→ Laser Analytical Spectroscopy)

High-power lasers have proved useful sources for the direct ablation of solids. In AES, ruby and Nd–YAG lasers have been used since the 1970s for solid ablation. When laser radiation interacts with a solid, a laser plume is formed. This is a dense plasma, containing both atomized material and small solid particles, evaporated from the sample. The processes occurring and the figures of merit in terms of ablation rate, crater diameter (ca. 10 µm), depth at various energies, and the expansion velocities of the plume are well known. The plume is optically thick and seems of little interest as an emission source. By applying cross-excitation with a spark or microwave energy, however, the power of detection is considerably improved [141], [142]. Absolute detection limits go down to 10^{-15} g, but relative standard deviations are no better than 30%. The method, however, can be applied for identification of mineral inclusions and microdomains in metallurgical analysis, forensic samples, archeology, and remote analysis in the nuclear industry.

The method has evoked renewed interest, mainly due to the availability of improved Nd–YAG laser systems. Different types of detector, such as microchannel plates coupled to photodiodes and CCDs in combination with multichannel analyzers, make it possible to record an analytical line and an internal standard line simultaneously, so that analytical precision is considerably improved. By optimizing the ablation conditions and the spectral observations, detection limits obtained with the laser plume as source for AES are in the range 50–100 µg/g with standard deviations ca. 1% [142].

Manufacturers. Equipment for atomic emission spectrometry now is available from many different manufacturers. Several types of sources—arc, spark, ICP, MIP, glow discharge (GD), etc.—are available:

Agilent Technologies (formerly Hewlett Packard), USA: ICP-MS, MIP-AES
Baird Atomic, USA: arc/spark AES, ICP-AES, AAS
Analytic Jena AG, Germany: AAS
Hitachi, Japan: ICP-AES, AAS, ICP-MS
Instruments SA/Jobin Yvon Emission, France: ICP-AES, GD-AES
LECO Corporation, USA: TOF-ICP-MS, GD-AES
Leeman Labs Inc., USA: ICP-AES
PerkinElmer Instruments (incorporating Princeton Applied Research, ORTEC, Signal Recovery and Berthold products), USA: AAS, ICP-AES, ICP-MS, GC, GC/MS
Shimadzu, Japan: arc/spark AES, ICP-AES, AAS Spectro Analytical Instruments, Germany: ICP-AES, arc/spark AES, ICP-MS
TJA Solutions (incorporating Thermo Jarrell Ash, VG Elemental and Unicam AA), USA: AAS, ICP-AES, ICP-MS, SF-ICP-MS, GD-AES, GD-MS
Varian Associates, Australia: ICP-AES, ICP-MS, AAS

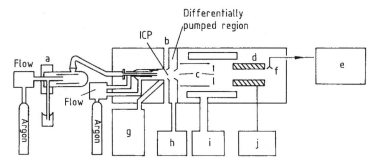

Figure 43. Instrumentation for low-resolution ICP–MS
a) Nebulizer; b) Sampler, skimmer; c) Ion lenses; d) Quadrupole mass filter; e) Computer; f) Electron multiplier; g) Radio-frequency supply; h) Mechanical pump; i) Cryosystem; f) Quadrupole supply

21.7. Plasma Mass Spectrometry

The plasma sources developed for AES have also proved very suitable ion sources for mass spectrometry. This is especially true for electrical discharges at pressures in the 100–500 Pa range and sources at atmospheric pressure with the advent of powerful vacuum systems with which to bridge the pressure difference between the mass spectrometer (of the order of 10^{-4} Pa) and the source. Elemental mass spectrometry, however, goes back to the use of high-vacuum arcs and sparks, for ultratrace and survey analyses of metal samples [50]. Spark source mass spectrography, with high-resolution sector field mass spectrometers, though still useful for characterization of electrically conducting solids down to the ng/g level, needs expensive instruments, and the analytical precision is low. In this chapter, only the newer mass spectrometric methods, which make use of plasma sources from optical atomic spectrometry are treated, namely, ICP mass spectrometry (ICP–MS) and glow discharge mass spectrometry (GD–MS). For a detailed treatment of mass spectrometric methods see →Mass Spectrometry.

21.7.1. Principles

Plasma mass spectrometry developed on the basis of the experience gathered from the extraction of ions from flames, as performed for diagnostic purposes in the mid-1960s [296]. In 1980 GRAY, HOUK, and coworkers [297] first showed the analytical possibilities of ICP–MS. The method has the same ease of sample introduction as ICP–AES, but has much lower detection limits, covers more elements, and allows the determination of isotope concentrations; ICP–MS rapidly became an established analytical method.

The principles and practice of ICP–MS have been treated in a number of excellent texts [220], [298], [299].

21.7.1.1. Instrumentation (Fig.43)

In ICP–MS, the ions formed in the ICP are extracted with the aid of a sampler into the first vacuum stage, where a pressure of the order of 100 Pa is maintained. Here, a supersonic beam is formed and a number of collision processes take place as well as an adiabatic expansion. These processes, together with their influence on the ion trajectories in the interface and behind the second aperture (skimmer), are very important for the transmission of ions and for related matrix interference [300]. Today's commercially available instrumentation falls into two groups: quadrupole and high-resolution instruments.

21.7.1.2. Analytical Features

The analytical features of ICP–MS are related to the highly sensitive ion detection and the nature of the mass spectra themselves.

ICP Mass Spectra. In low-resolution ICP–MS, resolution is at best 1 atomic mass units and, especially in the lower mass range, signals from cluster ions occur, which may cause spectral interference with analyte ions [301]. Cluster ions may be formed from different types of compounds:

1) Solvents and acids: H^+, OH^+, H_2O^+, NO^+, NO^{2+}, Cl^+, SO^+, SO_2^+, SO_3H^+ (when residual H_2SO_4 is present)
2) Radicals from gases in the atmosphere: O_2^+, CO^+, CO_2^+, N_2^+, NH^+, NO^+
3) Reaction products of the above-mentioned species with argon: ArO^+, $ArOH^+$, $ArCl^+$, Ar_2^+

Cluster ions give strong signals in the mass range up to 80 atomic mass units and may hamper the determination of light elements as a result of spectral interference. At higher masses, singly charged ions of the heavier elements occur, as well as doubly charged ions of the light elements, especially those with low ionization energies.

Signals from compounds of the analyte with various other species occur, such as MO^+, MCl^+, MOH^+, MOH_2^+. These ions may be formed by the dissociation of nitrates, sulfates, or phosphates in the plasma, or they may result from reactions between analyte ions and solvent residuals or oxygen in the plasma, or at the interface between the plasma and the mass spectrometer. Selection of the sampling zone in the plasma and the aerosol carrier gas flow may strongly influence the signal intensities of cluster ions and doubly charged analyte ions [302], [303].

For each element, signals from its different isotopes are obtained. Their intensity ratios correspond to the isotopic abundances in the sample. This fact can be used in calibration by isotopic dilution with stable isotopes, and in tracer experiments. Isotopic abundance is also useful for the recognition of spectral interference. The spectra in ICP–MS are simpler than in ICP–AES.

In ICP–MS, the background intensities are mostly low and are mainly due to the dark current of the detector, and to signals produced by ions scattered inside the mass spectrometer as a result of collisions with residual gas species or reflection. In order to shield the direct UV radiation, a beam stop is supplied often in the mass spectrometer. The plasma contributes little to the background, in contrast to ICP–AES where the background continuum stems from interactions between electrons and ions in the plasma, from molecular bands, from the wings of broad matrix lines, and from stray radiation.

A particular problem with quadrupole instruments is that they have insufficient resolution to resolve molecular ion interferences. Resolution is defined as:

$$R = \frac{M}{\Delta M} \qquad (172a)$$

where: R is resolution; M is mass (strictly speaking m/z); and ΔM is peak width at 5% peak height. The resolution obtainable with quadrupoles is limited by the stabiliy and uniformity of the r.f./d.c. field, and by the spread in ion energies of the ions. Quadrupoles used in ICP–MS are typically operated at resolutions between 12 and 350, depending on m/z, which corresponds to peak widths of between 0.7 and 0.8. In comparison, sector field instruments (Fig. 44) are capable of resolution up to 10 000, resulting in peak widths of 0.008 at 80 m/z. For most applications the resolution provided by a quadrupole is sufficient, however, for applications when spectroscopic interferences cause a major problem, the resolution afforded by a magnetic sector may be desirable. Table 5 gives examples of some common spectroscopic interferences that may be encountered, and the resolution required to separate the element of interest from the interference. For example, a particular problem is the determination of arsenic in a matrix which contains chloride (a common component of most biological or environmental samples). Arsenic is monoisotopic (i.e., it only has one isotope) at m/z 75 and the chloride matrix gives rise to an interference at m/z 75 due to $^{40}Ar^{35}Cl^+$, so an alternative isotope is not available for analysis. A quadrupole has insufficient resolution to separate the two species but a magnetic sector could do so easily (Table 5). The ion transmission of sector field instruments is much greater, compared with a quadrupole, for comparable resolution, so sensitivity is much higher and detection limits are lower by three orders of magnitude or so (i.e., detection limits of 1–10 fg ml^{-1} providing sufficiently clean reagents can be procured). Also, simultaneous isotopic detection is possible using an array of detectors, so extremely precise isotope ratio measurements can be made, allowing application of the technique for geochronology.

Recently, ICP time-of-flight (ICP–TOF) mass spectrometers have been commercialized (Figure 45). The TOF analyser has a number of advantages over other analyzers used for ICP–MS. The fact that all masses in the spectrum can be monitored simultaneously is the main attraction for many ICP–MS users. Even for simple multielement analysis, scanning analyzers suffer from a decrease in sensitivity as the number of ions to be monitored is increased. The fact that over 30 000 simultaneous mass spectra per second can

Table 5. Examples of spectroscopic interferences caused by molecular ions, and the resolution that would be necessary to separate the analyte and interference peaks in the mass spectrum

Analyte ion		Interfering ion		
Nominal isotope	Accurate m/z	Nominal isotope	Accurate m/z	Resolution required
^{51}V	50.9405	^{16}O^{35}Cl	50.9637	2580
^{56}Fe	55.9349	^{40}Ar^{16}O	55.9572	2510
^{63}Cu	62.9295	^{40}Ar^{23}Na	62.9521	2778
^{75}As	74.9216	^{40}Ar^{35}Cl	74.9312	7771

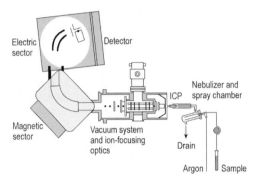

Figure 44. A sector-field ICP–MS instrument

be obtained by the TOF analyzer has many advantages for monitoring transient signals, such as those produced by laser ablation, chromatographic, capillary electrophoresis, and electrothermal vaporization sample introduction techniques. Also, improved isotope ratio measurements have been reported owing to the elimination of noise due to temporal fluctuations in the plasma and sample introduction systems.

Optimization. For the optimization of ICP–MS with respect to maximum power of detection, minimal spectral interference, signal enhancement or depression, and maximum precision, the most important parameters are the power of the ICP, its gas flows (especially the nebulizer gas), the burner geometry, the position of the sampler, and the ion optical parameters. These parameters determine the ion yield and the transmission, and thus the intensities of analyte and interference signals. At increasing nebulizer gas flow, the droplet size decreases (Section 21.4.1) and thus the analyte introduction efficiency goes up, but at the expense of the residence time in the plasma, the plasma temperature, and the ionization [304]. However, changes of the nebulizer gas flow also influence the formation and breakdown of cluster ions as well as the ion energies (as shown for ^{63}Cu$^+$ and ArO$^+$ [305]), and the geometry of the aerosol channel. Normally, the aerosol gas flow is 0.5–1.5 L/min. It must be optimized together with the power, which influences the plasma volume, the kinetics of the different processes taking place, and the position of the sampler. By changing the voltages of the different ion lenses, the transmission for a given ion can be optimized, enabling optimization of its detection limit and minimization of interference. In multielement determinations, a compromise must always be selected.

Power of Detection. For optimum power of detection, the analyte density in the plasma, the ionization, and the ion transmission must be maximized. The necessary power is 0.6–2 kW with the sample ca. 10–15 mm above the tip of the injector. The detection limits, obtained at single element optimum conditions, differ considerably from those at compromise conditions, but are still considerably lower than in ICP–AES (Table 6). For most elements they are in the same range, but for some they are limited by spectral interference. This applies to As (^{75}As$^+$ with ^{40}Ar^{35}Cl$^+$), Se (^{80}Se$^+$ with ^{40}Ar^{40}Ar$^+$), and Fe (^{56}Fe$^+$ with ^{40}Ar^{16}O$^+$). The acids present in the measurement solution and the material of which the sampler is made (Ni, Cu, etc.) may have considerable influence on these sources of interference and the detection limits for a number of elements. The detection limits for elements with high ionization potential may be even lower when they are detected as negative ions (for Cl$^+$ $c_L = 5$ ng/mL and for Cl$^-$ $c_L = 1$ ng/mL).

Precision and Memory Effects. The constancy of the nebulizer gas flow is of prime importance for the precision achievable in ICP–MS. After stabilizing the nebulizer gas flows, relative standard deviations may be below 1%. They can be further improved by internal standardization

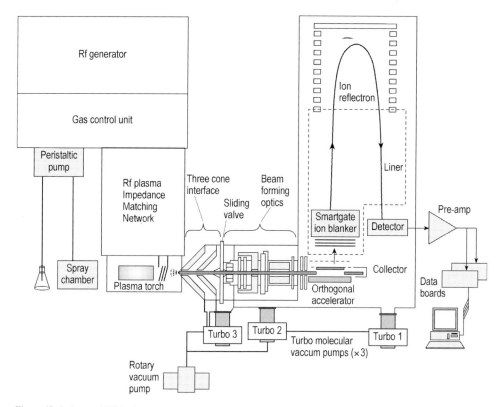

Figure 45. Orthogonal TOF-ICP-MS instrument

[306]. The tolerable salt concentrations (1–5g/L) are much lower than in ICP–AES. Memory effects are low and 1–2 min rinsing times are appropriate.

Interference. ICP–MS suffers from two principal types of interference [301]. Spectroscopic interferences arise when an interfering species has the same nominal m/z as the analyte of interest. The interfering species can be either an isotope of another element (which are well documented and hence easily accounted for), or a molecular ion formed between elements in the sample matrix, plasma gas, water, and entrained atmospheric gases. The molecular ions are less easy to correct for since they will vary depending on the nature of the sample matrix. Some common molecular ion interferences are shown in Table 7. Many of these interferences can be overcome by choosing an alternative isotope of the analyte which is free from interference, though a sacrifice in sensitivity may result. If a "clean" isotope is not available then one recourse is to separate the analyte from the matrix before the analysis using chemical extraction and chromatography, or use a magnetic sector instrument which is capable of resolving the interfering species from the analyte. Many of these interferences are thought to be formed in the interface due to a secondary discharge. This discharge can be eliminated by operating the plasma at low power, typically 600 W, and modifying the torch by inserting a grounded shield between it and the coil, or by using a center-tapped r.f. coil. Under these so-called "cool plasma" conditions interferences due to ArO^+, for example, can be eliminated, though the sensitivity for refractory elements and those with high ionization potentials may be reduced. An alternative method of reducing molecular ions is to use a collision cell prior to the quadrupole. A collision gas, such as helium, ammonia or water vapor, is introduced into the cell where it either reacts with, collisionally dissociates, or neutralizes the polyatomic species or precursors. It remains to be seen whether new interferences, formed from the matrix species and the gas, outweigh the benefits.

Non-spectroscopic interferences are caused by the sample matrix, and are manifested as an ap-

Table 6. Detection limits in ICP–AES and quadrupole ICP–MS

Element	Detection limit, ng/mL	
	ICP–AES [204]	ICP–MS [156]
Ag	7	0.005
Al	20	0.05
As	50	0.005
Au	20	0.005
B	5	0.05
Cd	3	0.005
Ce	50	0.001
Co	6	0.001
Cr	6	0.005
Fe	5	0.05
Ge	50	0.05
Hg	20	0.001
In	60	0.001
La	10	0.005
Li	80	0.001
Mg	0.1	0.05
Mn	1	0.001
Ni	10	0.005
Pb	40	0.001
Se	70	0.05
Te	40	0.05
Th	60	0.001
Ti	4	0.05
U	250	0.001
W	30	0.005
Zn	2	0.005

Table 7. Some commonly occurring spectroscopic interferences caused by molecular ions derived from plasma gases, air, and water, and the sample matrix

Molecular ion interference	Analyte ion interfered with	Nominal m/z
O_2^+	S^+	32
N_2^+	Si^+	28
NO^+	Si^+	30
NOH^+	P^+	31
Ar^+	Ca^+	40
ArH^+	K^+	41
ClO^+	V^+	51
CaO^+	Fe^+	56
ArO^+	Fe^+	56
ArN^+	Cr^+, Fe^+	54
$NaAr^+$	Cu^+	63
Ar_2^+	Se^+	80

parent enhancement or suppression in the analyte signal in the presence of a concentrated sample matrix. Such effects are thought to be caused primarily by space charge in the ion beam, whereby positive analyte ions are repelled from the ion beam by the high poisitive charge of the matrix ions, with low mass ions being relatively more affected than high mass ions. Such interferences are usually compensated for by using an internal standard, or by separating the matrix from the analyte before analysis. Because ICP–MS has such low detection limits it is also possible to dilute the sample to such an extent that the interference becomes negligible.

Isotope Ratio Analysis. Isotope ratio measurements are performed whenever the exact ratio, or abundance, of two or more isotopes of an element must be known. For example, the isotopic ratios of lead are known to vary around the world, so it is possible to determine the source of lead in paint, bullets, and petrol by knowing the isotopic abundances of the four lead isotopes 204, 206, 207, 208. Another example is the use of stable isotopes as metabolic tracers, where an animal is both fed and injected with an element having artificially enriched isotopes and the fractional absorption of the element can be accurately determined. An important field is that of geological dating, where extremely precise isotope ratio measurements must be made.

In order to perform the isotope ratio experiment correctly it is necessary to compensate for a number of biases in the instrumentation. Mass spectrometers and their associated ion optics do not transmit ions of different mass equally. In other words, if an elemental solution composed of two isotopes with an exactly 1:1 molar ratio is analyzed using ICP–MS, then a 1:1 isotopic ratio will not necessarily be observed. This so-called mass bias will differ depending on mass, and even very small mass-biases can have deleterious effects on the accuracy of isotope ratio determinations, so a correction must always be made using an isotopic standard of known composition, as shown in Equation (173):

$$C = \frac{R_t}{R_o} \tag{173}$$

where C is the mass bias correction factor, R_t is the true isotopic ratio for the isotope pair, and R_o the observed isotopic ratio for the isotope pair. For best results, such a correction should be applied to each individual isotopic pair that is to be ratioed, though this is not always possible in practice since a large number of isotopic standards would be needed to cover every eventuality. Also important is the effect of detector dead time. When ions are detected using a pulse counting (PC) detector the resultant electronic pulses are approximately 10 ns long. During and after each pulse there is a period of time during which the detector is effectively "dead" (i.e., it cannot detect any ions). The dead time is made up of the time for each pulse and recovery time for the detector and associated electronics. Typical dead times vary between 20 and 100 ns. If dead time is not taken into account there will be an apparent reduction in the number of pulses at high count rates, which would cause an inaccuracy in the measurement of isotope ratios when abundances differ markedly. However, a correction can be applied as shown in Equation (174):

$$R_t = \frac{R_o}{1 - R_o \cdot D} \quad (174)$$

where R_t is the true count rate and R_o is the observed count rate.

The relative abundances of the isotopes and operating conditions will affect the precision of any isotope ratio measurement. The rate at which the mass spectrometer scans or hops between masses the time spent monitoring each mass (the dwell time), and the total time spent acquiring data (total counting time) are important parameters which affect precision. The best precision achievable with a quadrupole instrument is ca. 0.1 % for a 1 : 1 isotope ratio at high count rate. This can be reduced by at least an order of magnitude if a magnetic sector instrument with an array of detectors is used to facilitate simultaneous detection. A long total counting time is desirable, because the precision is ultimately limited by counting statistics and the more ions that can be detected the better.

Isotope Dilution Analysis. A result of the ability to measure isotope ratios with ICP–MS is the technique known as isotope dilution analysis. This is done by spiking the sample to be analyzed with a known concentration of an enriched isotopic standard, and the isotope ratio is measured by mass spectrometry. The observed isotope ratio (R_m) of the two chosen isotopes can then be used in the isotope dilution equation (Eq. 174a) to calculate the concentration of the element in the sample:

$$C_x = \frac{C_s W_s M_x}{W_x M_s} \times \frac{A_s - R_m B_s}{R_m B_x - A_x} \quad (174a)$$

where R_m is the observed isotope ratio of A to B; A_x and B_x the atom fractions of isotopes A and B, respectively, in the sample; A_s and B_s the atom fractions of isotopes A and B, respectively, in the spike; W_x the weight of sample; W_s the weight of spike; C_x and C_s the concentration of element in sample and spike respectively; M_x and M_s the molar mass of element in sample and spike, respectively.

The best precision is obtained for isotope ratios near one, however, if the element to be determined is near the detection limit when the ratio of spike isotope to natural isotope should be between 3 and 10 so that noise contributes only to the uncertainty of natural isotope measurement. Errors also become large when the isotope ratio in the spiked sample approaches the ratio of the isotopes in the spike (overspiking), or the ratio of the isotopes in the sample (underspiking). The accuracy and precision of the isotope dilution analysis ultimately depends on the accuracy and precision of the isotope ratio measurement, so all the precautions that apply to isotope ratio analysis also apply in this case. Isotope dilution analysis is attractive because it can provide very accurate and precise results. The analyte acts as its own *de facto* internal standard. For instance, if the isotopic spike is added prior to any sample preparation then the spike will behave in exactly the same way as the analyte because it is chemically identical, providing it has been properly equilibrated by a series of oxidation/reductions. Hence, any losses from the sample can be accounted for because the analyte and spike will be equally affected.

Alternative Methods for Sample Introduction. Apart from continuous pneumatic nebulization, all sample introduction techniques known from ICP–AES have been used for ICP–MS. Similar to ICP–AES, the analysis of organic solutions is somewhat more difficult [308].

The use of *ultrasonic nebulization* as in ICP–AES increases sampling efficiency. The high water loading of the plasma has to be avoided by desolvation, not only to limit the cooling of the ICP, but also to minimize the formation of cluster ions and the related spectral interference and it should also be borne in mind that increased matrix loading in the plasma will occur. The use of high-pressure nebulization in ICP–MS has similar advantages [309], and is suitable for coupling ICP–MS to HPLC.

With the formation of *volatile hydrides*, the detection limits for elements such as As, Se, Sb, and Pb can be improved [310]. The improvement is due to enhanced analyte sampling efficiency and to the decrease of cluster ion formation.

Electrothermal evaporation (ETE) in addition to its advantages for microanalysis and improving sample transport, has the additional advantage in ICP–MS of introducing a dry analyte vapor into the plasma. Accordingly, it has been found useful for elements of which the detection limits are high as a result of spectral interference with cluster ions. In the case of ^{56}Fe, which is interfered by ^{40}ArO$^+$, PARK et al. [311] showed that the detection limit could be considerably improved by ETE. For similar reasons, the direct insertion of samples in ICP–MS maximizes the absolute power of detection [312], [313].

Although ICP–MS is mainly a method for the analysis of liquids and dissolved solids, techniques for *direct solid sampling* have been used [142]. They are especially needed when samples are difficult to dissolve or are electrically insulating and thus difficult to analyze with glow discharge or spark techniques or spatial analysis over the sample surface is required. In particular, the use of laser ablation for geological applications using UV or frequency doubled lasers is attracting considerable interest [314], [315]. For powders such as coal, slurry nebulization with a Babington nebulizer can be used in ICP–MS as well [316]. For the direct analysis of metals, spark ablation can be applied and the detection limits are in the ng/g range, as shown for steels [317]. For the analysis of metals [318] as well as of electrically insulating samples, laser ablation combined with ICP–MS is very useful. A Nd–YAG laser with a repetition rate of 1–10 pulses per second and an energy of ca. 0.1 J has been used. For ceramic materials such as SiC, the ablated sample amounts are of the order of 1 ng, and the detection limits are down to the 0.1 µg/g level.

21.7.1.3. Applications

Applications for ICP–MS are broadly similar to those for ICP-AES, though the better sensitivity of the former has resulted in a greater number of applications where ultra-low levels of detection are required. Sample preparation methods are similar to those used for FAAS and ICP-AES, however, nitric acid is favoured for sample digestion since the other mineral acids contain elements that cause spectroscopic interferences.

For geological applications ICP–MS is extremely well suited to the determination of the lanthanide elements because of its relatively simple spectrum. When coupled with UV laser ablation, it is an extremely powerful technique for spatial analysis and the the analysis of geological inclusions [319], especially with the extremely low limits of detection afforded by a sector field instrument. If simultaneous array detection is used then extremely precise isotope ratios can be measured, suitable for geochronology [320]. In environmental analysis the sensitivity of ICP–MS makes it ideal for applications such as the determination of heavy metals [243] and long-lived radionuclides [321] in the environment, and the determination of transition metals in seawater with on-line preconcentration [322]. Because of its capability for rapid multielement analysis ICP–MS is particularly suited to sample introduction methods which give rise to transient signals. For example, electrothermal vaporization, flow injection, and chromatographic methods for the speciation of organometallic and organohalide compounds [299]. A major attraction is the ability to perform isotope dilution analysis [320] which is capable of a high degree of accuracy and precision, and approaches the status of a primary method of analysis.

ICP–MS is the method of choice for routine analysis of samples containing ppb levels of analyte, and for speciation studies. Most quadrupole instruments are now benchtop-sized, and sector-field instruments are becoming easier to use and maintain, though high resolution operation still requires considerable skill and experience.

21.7.2. Glow Discharge Mass Spectrometry

Glow discharges [152] are known from their use as radiation sources for AES, and have been recognized as powerful ion sources for mass spectrometry. This development started from spark source mass spectrometry, where there was a continuous search for more stable sources to reduce the matrix dependence of the analyte signals [51].

In glow discharge mass spectrometry (GD–MS), the analyte is volatilized by sputtering and the ions are produced in the negative glow of the discharge, mainly as a result of collisions of the first kind with electrons and Penning ionization. Subsequently, an ion beam of composition representative of the sample is extracted. Between the glow discharge and the spectrometer a reduction in gas pressure is required. As the glow discharge is operated at a pressure around 100 Pa, a two-stage vacuum system is required. Cathodic extraction can be done at the cathode plume, by making

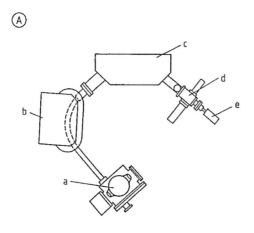

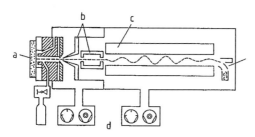

Figure 46. Glow discharge mass spectrometry
A) Sector field instrument VG 9000 [324]
 a) Source with sample introduction; b) Magnetic sector field; c) Electrostatic energy analyzer; d) Daly detector; e) Faraday cup
B) Quadrupole-based, low-resolution glow discharge mass spectrometer [325]
 a) Source; b) Ion optics; c) Quadrupole mass filter; d) Differential pumping system; e) Detector

3) Interference of analyte ions with cluster ions formed from analyte and gas species, e.g., $^{40}Ar^{16}O^+$ with $^{56}Fe^+$
4) Interference by signals from residual gas impurities, e.g., $^{14}N^{16}O^1H^+$ with $^{31}P^+$

Separation of the interfering signals depends on instrument resolution. In the case of quadrupoles, the mass resolution is at best unity, and interference can be corrected only by mathematical procedures. However, with high-resolution instruments with resolution of 5000 interference remains, especially for hydrides in the mass range 80 – 120.

21.7.2.1. Instrumentation

Glow discharges operated at pressures in the 10 – 100 Pa range can be coupled to mass spectrometers by using an aperture between both chambers with a size of ca. 1 mm. The glow discharge may use pin-shaped samples which can be introduced into the source without admitting air. BENTZ and HARRISON [323] coupled discharges with a pin cathode and hallow cathode plumes to a high-resolution sector feld mass spectrometer. Various electrode geometries were studied and diagnostic studies performed. This approach dates back to the VG 9000 spectrometer (Fig. 46 A), a double-focusing instrument in so-called reversed Nier – Johnson geometry, with a spectral resolution of ca. 10^4. The mass range to 280 can be covered with an acceleration voltage of 8 kV. As detectors, a Faraday cup and a so-called Daly multiplier are used, with the possibility of continuous switching. The vacuum of 10^{-4} Pa is maintained with the aid of oil diffusion or turbomolecular pumps. To reduce background, the source housing can be cooled to act as a cryopump.

Quadrupole mass spectrometers have also been used to detect the ions produced in a glow discharge. A high voltage is not required to achieve adequate ion extraction, and rapid scans can easily be performed, which enables quasi-simultaneous, multielement measurements. An electrostatic analyzer is often used in front of the spectrometer as an energy filter. JAKUBOWSKI et al. [325] described a combination of a Grimm-type glow discharge with a quadrupole mass spectrometer (Fig. 46 B) and studied the basic influence of the working and construction parameters. Glow discharge mass spectrometers using quadrupoles are commercially available.

the sample the cathode and drilling a hole in it. The aperture should reach the hot plasma center, and in most cases the extraction is done anodically.

Owing to impurities in the filler gas and the complexity of the processes taking place, not only analyte and filler gas ions, but many others may occur in the mass spectrum. Spectral interference may occur from several causes:

1) Isobaric interference from isotopes of different elements, e.g., $^{40}Ar^+$ with $^{40}Ca^+$, or $^{92}Zr^+$ with $^{92}Mo^+$
2) Interference of analyte ion signals with doubly charged ions, e.g., $^{56}Fe^{2+}$ with $^{28}Si^+$ This type of interferences is much rarer than in earlier work with high-vacuum sparks

21.7.2.2. Analytical Performance

The discharge is usually operated at about 100 Pa, a gas flow rate of a few mL/min, current ca. 2–5 mA, with a burning voltage of 500–1500 V. Normally the power is stabilized and pressure, current, voltage, or power are taken as optimization parameters. All electrically conducting samples, and even semiconductors, can be directly analyzed. After a burn-in phase, the discharge can be stable for hours. Detection limits down to the ng/g range are obtained for most elements.

Mass spectrometric methods are relative methods, and need to be calibrated with known reference samples; so-called relative sensitivity factors (RSF) are used:

$$RSF = (x_{el}/x_{ref})(c_{ref}/c_{el}) \qquad (175)$$

where x_{el} and x_{ref} are the isotope signals for the element to be determined and a reference element, and c are their concentrations. When matrix effects are absent RSF = 1. In Gd–MS, RSFs are much closer to unity as compared for instance with spark source mass spectrometry [326] and secondary ion mass spectrometry.

Especially in quadrupole systems, spectral interference may limit the power of detection as well as the analytical accuracy. This is shown by a comparison of high- and low-resolution scans of a mass spectrum (Fig. 47) [326], [327]. In Figure 47 A the signals of $^{56}Fe^{2+}$, $^{28}Si^+$, $^{12}C^{16}O^+$, and $^2N_2^+$ are clearly separated, but not in the quadrupole mass spectra. In the latter (Fig. 47 B), interference can be recognized from comparisons of the isotopic abundances. Physical means, such as the use of neon as discharge gas [328], can be used to overcome the spectral interference problem.

21.7.2.3. Applications

Glow discharge mass spectrometry is of use for the direct determination of major, minor, and trace bulk concentrations in electrically conducting and semiconductor solids down to concentrations of 1 ng/g [329]. Because of limited ablation (10–1000 nm/s), the analysis times are long, especially when samples are inhomogeneous. The technique has been applied especially for the characterization of materials such as Al, Cd, Ga, In, Si, Te, GaAs, and CdTe. Both for high-purity substances [330], and semiconductor-grade materials [331], up to 16 elements can be determined in 5 N and 6 N samples. At high resolution, determination of Ca and K is possible [332]. In high-purity Ga, the detection limits are below the blank values of chemical analysis [333]; SiC analysis results agree with those of neutron activation analysis [331]. In Al samples, Th and U concentrations of 20 ng/g can be determined [330]. The capabilities of low-resolution GD–MS have been shown in steel analysis [333], where detection limits are down to 1 ng/g and as many as 30 elements can be determined. Electrically insulating powder samples can be analyzed after mixing with a metal powder such as Cu, Ag, or Au and pelleting, as known from glow discharge–optical emission spectrometry (GD–OES). However, degassing and oxygen release may necessitate long burn-in times and induce instabilities. Blanks arising from the metal powder may be considerable, e.g., for Pb, in the case of copper powder.

Depth profiles can be recorded just as in GD–OES, but with such high resolution that much thinner multilayers can be characterized [334]. Conversion of the intensity scale to a concentration scale is easier and the sensitivity higher. By the use of different working gases [335], the scope of the method can be continuously adapted.

Dry solution residue analysis can be performed with GD–MS. This has been shown with the aid of hollow cathodes, in which Sr, Ba, and Pb can be determined [336] and where in samples of 10–20 μL, up to 70 elements can be determined with detection limits in the picogram range [337]. With a Grimm-type discharge, solution residues can be formed on the cathode by evaporating microvolumes of analyte solutions with low total salt content. With noble metals [338], cementation can be used to fix the analyte on a copper target, so that it is preconcentrated and can be sputtered reproducibly. This has been found useful in fixing picogram amounts of Ir on a copper plate prior to analysis with GD–MS.

Manufacturers. Inductively coupled plasma (ICP) and glow discharge (GD) mass spectrometers with high (h) and low (l) resolution are available from several manufacturers:

Agilent Technologies (formerly Hewlett Packard), USA: ICP-MS, MIP-AES
Extrel, USA: GD-MS
GBC, Australia: TOF-ICP-MS, AAS
Hitachi, Japan: ICP-AES, AAS, ICP-MS
LECO Corporation, USA: TOF-ICP-MS
Micromass, UK: S-ICP-MS, ICP-MS

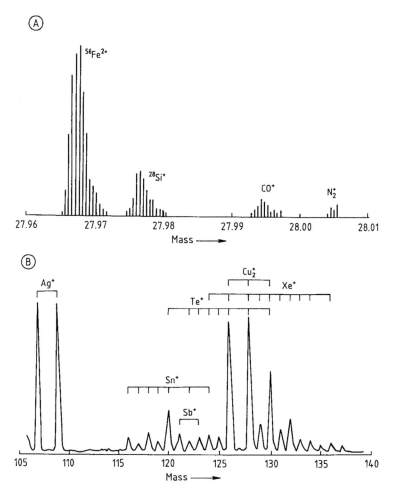

Figure 47. Glow discharge mass spectrum
A) High resolution [326]; B) Low resolution [327]

Nu Instruments Ltd., UK: SF-ICP-MS
PerkinElmer Instruments (incorporating Princeton Applied Research, ORTEC, Signal Recovery and Berthold products), USA: AAS, ICP-OES, ICP-MS, GC, GC/MS
Spectro Analytical Instruments, Germany: ICP-AES, arc/spark AES, ICP-MS
TJA Solutions (incorporating Thermo Jarrell Ash, VG Elemental and Unicam AA), USA: AAS, ICP-AES, ICP-MS, SF-ICP-MS, GD-AES, GD-MS
Thermoquest (incorporating Finnigan), USA: SF-ICP-MS, AAS
Varian Associates, Australia: ICP-AES, ICP-MS, AAS

21.8. Atomic Fluorescence Spectrometry

In atomic fluorescence spectrometry (AFS), the analyte is introduced into an atomizer (flame, plasma, glow discharge, furnace) and excited by monochromatic radiation emitted by a primary source. The latter can be a continuous source (xenon lamp) or a line source (hollow cathode lamp, electrodeless discharge lamp, or tuned laser). Subsequently, the fluorescence radiation, which may be of the same wavelength (resonance fluorescence) or of longer wavelength (nonresonance fluorescence), is measured.

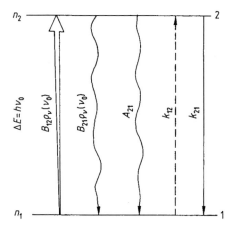

Figure 48. Transitions in a two-level system for atomic fluorescence spectrometry [339]

21.8.1. Principles [339]

For a two-level system (Fig. 48), with excitation only as a result of absorption of radiation with radiant density ϱ_v, the population of the excited level n_2 is:

$$dn_2/dt = n_1 B_{12} \varrho_v - n_2 [A_{21} + k_{21} + B_{21} \varrho_v] \quad (176)$$

where B_{12} and B_{21} are the transition probabilities for absorption and stimulated emission, v the frequency, and n_1 the population of the lower level; A_{21} is the transition probability for spontaneous emission, and k_{21} is the coefficient for decay by collisions. The exciting radiation is supposed to be constant over the absorption profile. When the statistical weights of n_1 and n_2 are equal, $B_{21} = B_{12}$, and when $n_T = n_2 + n_1$, Equation (176) can be solved:

$$n_2 = n_T B \varrho_v t_r [1 - \exp(-t/t_r)] \quad (177)$$

where:

$$t_r = [2B\varrho_v + A_{21} + k_{21}]^{-1} \quad (178)$$

Accordingly, when the absorption of radiation increases to a certain value, A_{21} and k_{21} become negligible. The $n_2 = n_T/2$, and is independent of the radiant density. This state of saturation can be obtained when lasers are used as primary sources. As the fluorescence intensities are low, especially in the case of nonresonant AFS, special precautions must be taken for signal acquisition. Signal-to-noise ratios can be considerable improved by using pulsed signals and phase-sensitive amplification. The noise is partly due to flickering (proportional to the product of measurement time t and the mean radiance B^s_{ave} of the source and to background contributions (proportional to background mean intensity B^b_{ave} and time t) and:

$$S/N = K B^s_{ave} / (B^s_{ave} + B^b_{ave})^{1/2} \quad (179)$$

By pulsing the source (with pulse duration t_p, frequency f, and pulse intensity B^s_p) as in the case of a laser, the signal-to-noise ratio is:

$$S/N = K B^s_p f t_p / (B^s_p f t_p + B^b_{ave})^{1/2} \quad (180)$$

but by phase-sensitive amplification, in the case of background noise limitations, it can be improved to:

$$S/N' = K B^s_p f t_p / \left[f t_p \left(B^s_p + B^b_{ave} \right) \right]^{1/2} = S/N (f t_p)^{1/2} \quad (181)$$

This improvement can be considerable when "boxcar" integrators with small time windows are used, to give $f t_p \approx 10^{-8}$.

21.8.2. Instrumentation

An AFS system comprises a primary source, an atomizer, and a detection system for measuring the fluorescence radiation, which in some cases can be done without spectral dispersion.

Primary Source. As primary sources, continuous sources, such as tungsten halogenide or deuterium lamps, can be used. They have the advantage that multielement determinations are possible, but owing to the low radiant densities saturation is not obtained and the power of detection not fully exploited. With line sources, such as hollow cathode sources and electrodeless discharge lamps, much higher radiances can be obtained. Even ICPs into which a concentrated solution is introduced can be used as primary sources, so that multielement determinations become possible. With laser sources, saturation can be obtained. In most cases, dye lasers are used. They have a dye (mostly a substituted coumarin) as active medium. This liquid is continuously pumped through a cuvette and has a broad wavelength band (ca. 100 nm), within which a wavelength can be selected by a grating or prism. By using several dyes, the whole wave-

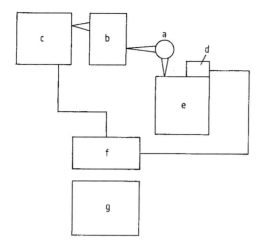

Figure 49. Experimental set-up for laser-induced atomic flourescence
a) Flame, ICP, etc.; b) Dye laser; c) Pumping laser; d) Photomultiplier; e) Monochromator; f) Boxcar integrator; g) Data treatment and display

length range down to 440 nm can be covered. The laser is often pumped by a second laser (e.g., an excimer laser) and operated in a pulsed mode. As the energy output is very high, it is possible to apply frequency doubling and Raman shifting, to access the wavelength range down to 200 nm. Accordingly, saturation can be flexibly achieved for almost any term. In the case of resonant transitions (with an oscillator strength: 0.1–1) the energy required is 10–100 mW, whereas for nonresonant transitions (oscillator strength 10^{-3}) it can reach 1 W. Tunable dye lasers can deliver frequencies up to ca. 10 Hz and the pulse length can be down to 2–10 ns.

Atomizers. As fluorescence volumes, chemical flames are very useful, because of their good sample uptake and temporal stability. There has been interest in the low radiative background, low quenching, argon–hydrogen diffusion flame. The temperature of this flame is too low to prevent severe chemical interferences and therefore the argon separated air–acetylene flame has been most widely used. The hot nitrous oxide–acetylene flame (argon separated) has been used where atomization requirements make it essential. In all cases, circular flames, sometimes with mirrors around them, offer the preferred geometry. These problems do not occur when an ICP is used as atomizer [340]. In the ICP, measurements should be made in the higher part of the tail flame, otherwise the ionized and excited analyte fraction is too high. To minimize the formation of oxides and sideways diffusion of analyte, a burner with an extended outer tube should be used for AFS. In a graphite furnace, the atomization efficiency is high and, owing to the limited expansion of the vapor cloud, the residence time of the analyte atoms is long; accordingly, graphite furnace atomization laser AFS has extremely low absolute detection limits. Low-pressure discharges can also be used as the fluorescence volume. They have the advantage that quenching of the fluorescent levels, which may limit the power of detection and cause matrix effects, is lower [341]. Laser-produced plasmas may be used as the fluorescence volume, [145], and are especially useful for direct trace determinations in electrically insulating solids.

Detection of fluorescence differs in resonant and nonresonant AFS. In *resonant AFS*, the radiation is measured perpendicular to the direction of the exciting radiation. The system suffers from stray radiation and emission of the flame. The latter can be eliminated by using pulsed primary sources and phase-sensitive detection. In *nonresonant fluorescence*, stray radiation problems are not encountered, but fluorescence intensities are lower, which necessitates the use of lasers as primary sources, and spectral apparatus to isolate the fluorescence radiation. A set-up for laser-excited AFS (Fig. 49) may make use of a pulsed dye laser pumped by an excimer laser. Here, pulses of 2–10 ns occur at a frequency up to 25 Hz. A mirror can be positioned behind the source to pass the exciting radiation twice through the fluorescence volume and to lead the largest possible part of the fluorescence radiation into the monochromator. The photomultiplier signal is then amplified and measured with a box-car integrator with a nanosecond window synchronized with the pumping laser.

21.8.3. Analytical Performance

Detection limits in AFS are low, especially for elements with high excitation energies, such as Cd, Zn, As, Pb, Se, and Tl.

Elements such as As, Se, and Te can be determined by AFS with hydride sample introduction into a flame or heated cell followed by atomization of the hydride. Mercury has been determined

Table 8. Detection limits in laser atomic fluorescence and laser-enhanced ionization spectrometry

	Detection limit, ng/mL		
	Laser AFS ICP [340]	Laser AFS furnace [341]	LII flame [348]
Ag		0.003	0.07
Al	0.4		0.2
B	4		
Ba	0.7		0.2
Ca			0.03
Co		0.002	0.08
Cu		0.002	0.7
Eu			10
Fe		0.003	0.08
Ga	1		0.04
In			
Ir		0.2	
K			0.1
Li			0.0003
Mn		0.006	0.02
Mo	5		10
Na		0.02	0.003
Pb	1	0.000025	0.09
Pt		4	
Rb			0.1
Si	1		40
Sn	3		0.3
Ti	1		1
Tl	7		0.008
V	3		0.9
Y	0.6		10
Zr	3		

by cold-vapor AFS. A non-dispersive system for the determination of Hg in liquid and gas samples using AFS has been developed commercially. Mercury ions in an aqueous solution are reduced to mercury using tin(II) chloride solution. The mercury vapour is continuously swept out of the solution by a carrier gas and fed to the fluorescence detector, where the fluorescence radiation is measured at 253.7 nm after excitation of the mercury vapor with a high intensity mercury lamp (detection limit 0.9 ng/L). Gaseous mercury in gas samples (e.g., air) can be measured directly or after preconcentration on an absorber comprising, for example, gold-coated sand. By heating the absorber, mercury is desorbed and transferred to the fluorescence detector. This preconcentration method can also be applied to improve the detection limits when aqueous solutions (or dissolved solid samples) are analyzed.

In *flame AFS*, elements which form thermally stable oxides, such as Al, Mg, Nb, Ta, Zr, and the rare earths, are hampered by insufficient atomization. This is not the case when an ICP is used as the fluorescence volume. Detection limits in the case of laser excitation and nonresonant fluorescence are lower than in ICP–AES (Table 8) [340]. For both atomic and ionic states, ICP–AFS can be performed [342].

In the case of a *graphite furnace* as fluorescence volume, and laser excitation, detection limits are very low; e.g., Ag 3×10^{-3} ng/mL; Co 2×10^{-3} ng/mL; Pb 2×10^{-3} ng/mL [339], [343].

Dry solution residue analysis with laser-excited AFS and a *hollow cathode discharge* atomizer offers very sensitive determinations of Mo [347]. BOLSHOV et al. [345] showed that very low levels of lead in Antarctic ice samples could be determined by laser-excited AFS of dry solution residues with graphite furnace atomization.

When performing laser-excited AFS at a *laser plume*, it seems useful to produce the laser plasma below atmospheric pressure, as then the ablation depends little on the matrix [146], [346].

In AFS, spectral interference is low as the fluorescence spectra are not very complex. The linear dynamic range of AFS is similar to that of OES, or even larger. There is no background limitation nor a serious limitation by self-reversal, both of which limit the linear dynamic range in OES, at the lower and at the higher end, respectively. It well may occur that the exciting radiation gets less intense when penetrating into the fluorescence volume, and that fluorescence radiation is absorbed by ground-state atoms before leaving the atomizer. These limitations, however, are less stringent than self-reversal in OES.

The analytical application of laser-excited AFS has so far, been seriously hampered by the complexity and cost of tunable lasers; this may change with the availability of less expensive but powerful and tunable diode lasers.

21.9. Laser-Enhanced Ionization Spectrometry (→Laser Analytical Spectroscopy)

Laser-enhanced ionization spectrometry (LEI) is based on the optogalvanic effect, known from early work on gas discharges. Here, an increase of the current occurs in a low-pressure discharge when irradiating with monochromatic radiation at a wavelength corresponding to a transition in the energy level diagram of the discharge gas. As an analytical method, LEI is based on measurement of the increase of analyte ionization in an

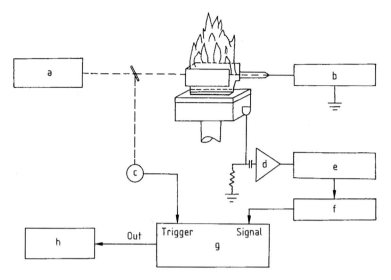

Figure 50. Flame laser-enhanced ionization spectrometry [0]
a) Flashlamp/dye laser; b) High voltage; c) Trigger photodiode; d) Preamplifier; e) Pulse amplifier; f) Active filter; g) Boxcar averager; h) Chart recorder

atomizer, resulting from selective excitation by laser radiation [347], [348].

Alternatively, two lasers can be used to both excite and photoionize the analyte. This is known as direct laser ionization (DLI) and the two methods (LEI and DLI) are collectively termed laser-induced ionization (LII).

Ionization of atoms in an atomizer may result from collisions with high-energy atoms of another species (Fig. 50) and the rate constant for this reaction can be written:

$$k_{1i} = n_X [8kT/(\pi m_r)]^{1/2} Q_{MX} \exp(-E_i/kT) \quad (182)$$

where X refers to analyte atoms, Q_{MX} is the cross section for collisions between X and M (another element), k is Boltzmann's constant, E_i is the ionization energy, and T is the temperature in the atomizer. When the atoms X are in an excited state, being 1 eV above the original level k_{1i} changes to:

$$k_{2i} = \exp(1/kT) k_{1i} \quad (183)$$

and the increase in ionization for a flame with $T = 2500$ K is $e^5 = 150$. In order to get sufficiently high population of the excited level, a laser must be used. For a closed system with two levels n_1 and n_2, the equilibrium is characterized by:

$$B_{12}\varrho_v n_1 = B_{21}\varrho_v n_2 + A_{21} n_2 + k_{21} n_2 \quad (184)$$

where:

$$B_{21} = (g_1/g_2) B_{12}$$

and:

$$n_T = n_1 + n_2 + n_i$$

B_{12} and B_{21} are the transition probabilities for absorption and stimulated emission, and n_1, n_2, n_i, and n_T the analyte atom densities for the lowest state, the excited state, the ionized state, and the total number densities; g_1 and g_2 are the statistical weights, A_{21} is the transition probability for spontaneous emission and k_{21} the rate coefficient for collisional decay. Accordingly:

$$n_2 = \frac{B_{12}\varrho_v (n_T - n_i)}{[(1 + g_1/g_2) B_{12}\varrho_v + A_{21} + k_{21}]} \quad (185)$$

and n_2 will increase until $A_{21} + k_{21}$ becomes negligible and $n_2/n_1 = g_2/g_1$. The required radiation densities are ca. 10–100 mW/cm³ for a resonant transition (with oscillator strength 0.1–1) and 2 W/cm³ for a nonresonant transition (with oscillator strength 0.001). The use of a continuous wave laser maximizes n_i, and the use of multistep excitation may be helpful, especially when high-energy terms (Rydberg states) have to be populated.

As an analytical technique, LEI requires a tunable laser, an atomizer, a galvanic detector and read-out electronics, including a boxcar integrator. It was first realized experimentally with a flame, with pneumatic nebulization for sample uptake and atomization [347].

A dye laser is used in most cases, as the selection of different dyes, frequency doubling, and Raman shifting allow coverage of the spectral range down to 200 nm. In the case of a flame (Fig. 50), the laser enters a few millimeters above the burner head, and a large collector electrode is situated a similar distance above the laser beam. This allows high-efficiency extraction of the ions formed. Apart from flames, furnaces and glow discharges are also very useful atomizers. The measurement unit includes phase-sensitive amplification to give high signal-to-noise ratios, but does not need any spectrometric equipment as only an ion current has to be measured.

The sensitivity in LEI depends on the properties of the laser system, the terms populated, the atomizer, and the detection system. It can be written:

$$S = c X_A \beta(A) [\exp(-E_i^*/kT) B_{12}(A) \varrho_v^p(A)] \quad (186)$$

where c is a constant, X_A is the number density of analyte atoms A in the lower level according to Boltzmann, β is the atomization efficiency in the flame, $\varrho_v^p(A)$ is the spectral radiation density per laser pulse, E_i^* is the ionization energy of the excited level, and S is the sensitivity in nA ng^{-1} mL^{-1}. The noise includes contributions from flicker noise, nebulizer noise, and detector noise, and with phase-sensitive amplification, detection limits are in the sub-ng/mL range (Table 8) [347], [349]. Because of the maximum atom number densities and residence times, detection limits are lowest for furnace atomizers. In pulsed lasers, the radiation densities favor sensitivity, whereas in continuous wave lasers the n_i values increase as a result of improved excitation efficiencies. In the thermionic diode, a heated collector electrode is placed in the vicinity of the laser beam and isolated from the vessel walls. No extraction voltage is applied and the ion current flowing as a result of decreasing negative space charge around the collector, caused by laser ionization, is taken as analytical signal. The whole system is usually operated at reduced pressure and can detect as few as 10^2 atoms [350].

In LEI, matrix effects are low. This applies especially for samples with complicated atomic spectra, as spectral interference is absent. Elements with low ionization energies may cause matrix effects, possibly due to a change of the ion current as a result of changes in the electron number density in the flame. This can be compensated by measuring alternately with and without laser irradiation. Ionization of the matrix by the laser may also falsify the ion currents measured. This can be eliminated by modulation of the frequency of the exciting laser radiation around the frequency v, which corresponds to $E_2 - E_1 = h v$ when E_2 is the intermediate level (Rydberg state) and E_1 the ground state. Ion production as a result of laser irradiation may influence the collection efficiency at the electrode as the electrical field around the electrode changes. This effect may be minimized by using a large electrode and by collecting all the ions produced. This sets limits to the pulsing frequency. For a distance of 5 mm between the collector and the burner, a field of 100 V/cm and a mobility of 20 cm^2 V^{-1} s^{-1} (typical for a flame in which $T = 2500$ K), the highest allowed pulsing rate is 4000 Hz. It has been found that for Na concentrations of 100 µg/mL, matrix effects are still negligible.

The linear dynamic range in flame LEI is of the order of three decades. However, the multi-element capacity is limited as monochromatic laser radiation is required.

When using two lasers and applying two-photon spectroscopy, only those atoms which do not have a velocity component in the observation direction will undergo LEI. The absorption signals become very narrow (Doppler-free spectroscopy), enhancing the selectivity and the power of detection, and making isotope detection feasible.

The LEI technique has been applied successfully for certification purposes [351], [352], and represents a powerful approach for combinations of laser evaporation with all other atomization techniques. Apart from the galvanic detection of the ion currents, direct mass spectrometric detection of the ions can be applied, as in resonance ionization mass spectrometry [353]. Here, ionization can also be performed by multiphoton absorption, which requires very intense primary sources.

21.10. Comparison With Other Methods

Atomic absorption, optical emission, atomic fluorescence, plasma atomic spectrometry, and

Table 9. Figures of merit of analytical methods

Method	Power of detection	Matrix effects	Cost
Atomic absorption			
Flame	++	+	+
Furnace	+++	+++	++
Atomic emission			
d.c. Arc	++	+++	+++
Spark	+	++	++
d.c. Plasma	++	++	+
ICP	++	+	++
MIP	++	+++	+
Glow discharge	++	+	++
Laser	+	+	++
Atomic fluorescence			
(laser)	+++	++	+++
LEI	+++	+	+++
X-ray spectrometry			
XRF	++(+[a])	+	++
Electron microprobe	++	++	+++
PIXE	+++	++	+++
Auger electron spectroscopy	++	++	+++
Mass spectrometry			
Spark	+++	+++	+++
ICP	+++	++	+++
Glow discharge	+++	+	+++
SIMS	+++	++	+++
Laser	+++	++	+++
Activation analysis	+++	+	+++
Electrochemistry	+++	+++	+
Spectrophotometry	++	+++	+
Spectrofluorimetry	+++	+++	+
Chromatography			
Gas	+++	+++	++
Liquid	++	+++	++

+ Low; ++ medium; +++ very high
[a] TYXRF = Total reflection X-ray fluorescence.

newer approaches such as laser-enhanced ionization now represent powerful tools for elemental analysis, including speciation, and are to be found in many analytical laboratories. Their power of detection, reliability in terms of systematic errors, and cost are compared with those of other methods of analysis in Table 9.

21.10.1. Power of Detection

Atomic absorption spectrometry using furnaces remains a most sensitive method for easily volatilized elements with high excitation energies, and this also applies to AFS using laser excitation. The development of diode lasers should make it possible to eliminate the use of spectrometers in AAS. Dynamic range problems can be decreased by measuring in the wings of the absorption lines, and the background measurement can be performed by tuning the laser [157]. This requires lasers with analytical wavelengths down to the vacuum UV. Laser-enhanced ionization is a further promising approach with respect to power of detection, especially when applying Doppler-free spectroscopy. However, these methods have limited multielement capacity. From this point of view, plasma atomic spectrometric methods offer the better possibilities; ICP–OES is likely to remain a powerful tool for multielement determinations in liquid samples and its power of detection is between that of flame and furnace AAS. It is the appropriate method for tasks such as wastewater analysis. Inductively coupled plasma–mass spectrometry offers further progress, provided the matrix loading of the sample solutions remains low. It may well have limitations in the low mass range (below 80), owing to interference by cluster ions. The use of sample introduction techniques for generating dry aerosols has great potential, e.g., for Fe, As, Se, etc. Further progress is possible by the use of high-resolution mass spectrometry with sector field instruments. It is likely to occur as now lower cost instruments make the technique available to a wider circle of users.

Whereas ICP atomic spectrometry is certainly a most useful method for multielement determinations in liquid samples at low concentrations, glow discharges are very useful for the analysis of solids. Glow discharge atomic emission has established itself, especially for depth profiling analysis, and is useful for bulk analysis, together with standard methods such as arc and spark emission spectrometry. Glow discharge mass spectrometry first found a role in the analysis of ultra-pure metals, as requird in microelectronics. However, its quantification is difficult and requires certified samples for calibration. In dry solution residue analysis, GD–MS has extremely low absolute detection limits. This approach is very useful for many applications in the life sciences, as it combines high power of detection with the possibilities of isotope dilution. In mass spectrometry, this approach offers new possibilities, as more classical thermionic techniques are only applicable to elements which form volatile compounds. Optical atomic and mass spectrometric methods can be used for the determination of light elements, which is an advantage over X-ray spectrometric methods. Apart from this restriction, total reflection X-ray fluorescence (TRXRF) offers high powers of detection, accuracy, and multielement capacity, with a minimum of sample preparation [354].

21.10.2. Analytical Accuracy

The analytical accuracy of methods can only be discussed in view of the complete analytical procedure applied. It is necessary to tune sample preparation and trace–matrix separations to the requirements of the analytical results in terms of accuracy, power of detection, precision, cost, number of elements, and, increasingly, the species to be determined. However, the intrinsic sensitivity of the different determination methods to matrix interference remains important. In optical emission and mass spectrometry, spectral interference remains an important limitation to the achievable analytical accuracy. In atomic emission, this applies especially to the heavier elements, as they have the more complex atomic spectra. Especially when they are present as the matrix (as is often the case in metals analysis) the necessitity of trace matrix separation is obvious when trace analyses are to be performed. In order to overcome limitations due to spectral interference, high-resolution echelle spectrometers are increasingly used. They are compact and thus combine high resolution with excellent stability, and further enable multielement determinations, by using advanced detector technology, e.g., CCDs.

In mass spectrometry, spectral interference especially limits the accuracy in the low mass range. Progress may be expected from the use of sector field and time-of-flight mass spectrometry. However, in the first case transmission and in the second case dynamic range problems must be given attention. Signal depression and enhancement are a further cause of interference in ICP–MS. They can be taken care of by standard addition, as the sprectral background is low. Internal standardization may compensate for the effects of easily ionized element. This is more difficult in ICP–AES, where the spectral background (especially in trace analysis) is considerable and may be strongly influenced by changes in the concentrations of easily ionized elements.

In AAS and AFS, limitations in the analytical accuracy are mostly related to physical and chemical interference and less to spectral interference. Especially in furnace AAS, thermochemical processes limit the achievable accuracy and necessitate carefully worked out temperature programs. In AFS and LEI, it is necessary to control matrix influences relating to quenching when analyzing real samples.

21.10.3. Cost

The power of detection and accuracy of analytical methods cannot be discussed without considering the costs of instruments and their operation, including the costs of laboratory personnel. Methods allowing multielement analyses and high throughput are advantageous for routine analyses. The ICP spectrometric methods offer particularly good possibilities, despite the high instrument costs and the high consumption of power and gases. In many cases, however, the costs arising from sample preparation are decisive, and favor X-ray spectrometric methods, provided the limitations mentioned earlier are not encountered. Further progress is likely to depend on the availability of on-line sample treatment, e.g., with flow injection and on-line sample dissolution, which now becomes possible with microwave-assisted heating. In any case, the method selected has to be discussed in detail for any new analytical task.

21.11. References

[1] G. R. Kirchhoff, R. Bunsen, *Philos. Mag.* **20** (1860) 89–98.
[2] A. Walsh, *Spectrochim. Acta* **7** (1955) 108–117.
[3] B. V. L'vov, *Spectrochim. Acta* **17** (1961) 761–770.
[4] H. Maßmann, *Spectrochim. Acta Part B* **23 B** (1968) 215–226.
[5] W. Grotrian: *Graphische Darstellung der Spektren von Atomen mit ein, zwei und drei Valenzelektronen,* Springer Verlag, Berlin 1928.
[6] H. R. Griem: *Plasma Spectroscopy,* McGraw-Hill, New York 1964.
[7] L. de Galan, R. Smith, J. D. Winefordner, *Spectrochim. Acta Part B* **23 B** (1968) 521–525.
[8] W. J. Pearce, P. J. Dickerman (eds.): *Symposium on Optical Spectrometric Measurements of High Temperatures,* University of Chicago Press, Chicago 1961.
[9] P. W. J. M. Boumans: *Theory of Spectrochemical Excitation,* Hilger & Watts, London 1966.
[10] J. Junkes, E. W. Salpeter, *Ric. Spettrosc. Lab. Astrofis. Specola Vaticana* 2 (1961) 255.
[11] C. F. Bruce, P. Hannaford, *Spectrochim. Acta Part B* **26 B** (1971) 207–235.
[12] J. A. C. Broekaert, F. Leis, K. Laqua, *Spectrochim. Acta Part B* **34 B** (1979) 73–84.
[13] W. Lochte-Holtgreven (ed.): *Plasma Diagnostics,* North-Holland Publ., Amsterdam 1968.
[14] G. Elwert, *Z. Naturforsch. A: Astrophys. Phys. Phys. Chem.* **7 A** (1952) 703–711.
[15] J. M. Mermet in P. W. J. M. Boumans (ed.): *Inductively Coupled Plasma Emission Spectroscopy,* Part II, Wiley-Interscience, New York 1987, p. 353.
[16] R. Diermeier, H. Krempl, *Z. Phys.* **200** (1967) 239–248.
[17] G. Herzberg: *Molecular Spectra and Molecular Structure,* 2nd ed., Van Nostrand Reinhold, New York 1950.
[18] R. W. B. Pearce, A. G. Gaydon: *The Identification of Molecular Spectra,* Chapman and Hall, London 1953.
[19] J. A. C. Broekaert, *Bull. Soc. Chim. Belg.* **86** (1977) 895–906.
[20] B. Raeymaekers, J. A. C. Broekaert, F. Leis, *Spectrochim. Acta Part B* **43 B** (1988) 941–949.
[21] I. Ishii, A. Montaser, *Spectrochim. Acta Part B* **46 B** (1991) 1197–1206.
[22] N. P. Ferreira, H. G. C. Human, L. R. P. Butler, *Spectrochim. Acta Part B* **35 B** (1980) 287–295.
[23] K. Bockasten, *J. Opt. Soc. Am.* **51** (1961) 943–947.
[24] G. M. Hieftje, *Spectrochim. Acta Part B* **47 B** (1992) 3–25.
[25] J. A. C. Broekaert, *Anal. Chim. Acta* **196** (1987) 1–21.
[26] M. Borer, G. M. Hieftje, *Spectrochim. Acta Reviews* **14** (1991) 463–486.
[27] K. Doerffel: *Statistik in der analytischen Chemie,* Verlag Chemie, Weinheim 1984.
[28] L. Sachs: *Statistische Auswertungsmethoden,* Springer Verlag, Berlin 1969.
[29] V. V. Nalimov: *The Application of Mathematical Statistics to Chemical Analysis,* Pergamon Press, Oxford 1963.
[30] J. D. Ingle, Jr., S. R. Crouch: *Spectrochemical Analysis,* Prentice-Hall International, Englewood Cliffs 1988.
[31] G. M. Hieftje, *Anal. Chem.* **44** (1972) 81 A–88 A.
[32] H. Kaiser, *Spectrochim. Acta* **3** (1947) 40–67.
[33] H. Kaiser, H. Specker, *Fresenius Z. Anal. Chem.* **149** (1956) 46–66.
[34] E. L. Grove: *Analytical Emission Spectrometry,* vol. 1, Part I and II, Decker Inc., New York 1971.
[35] C. S. Williams, O. A. Becklund: *Optics,* Wiley-Interscience, New York 1972.
[36] P. N. Keliher, C. C. Wohlers, *Anal. Chem.* **48** (1976) 333 A–340 A.
[37] H. Kaiser, *Spectrochim. Acta* **3** (1948) 159–190.
[38] P. W. J. M. Boumans, *Colloq. Spectrosc. Int. Proc. 16th* **II** (1971) 247–253.
[39] H. Siedentopf, *Z. Phys.* **35** (1934) 454.
[40] V. K. Zworykin, E. G. Ramberg: *Photoelectricity and Its Application,* Wiley, New York 1949.
[41] Y. Talmi: "Multichannel Image Detectors," *ACS Symp. Ser.* **102** (1979).
[42] J. V. Sweedler, R. F. Jalkian, M. B. Denton, *Appl. Spectrosc.* **43** (1989) 954–962.
[43] J. A. C. Broekaert, K. R. Brushwyler, G. M. Hieftje, unpublished results.
[44] J. A. C. Broekaert, C. A. Monnig, K. R. Brushwyler, G. M. Hieftje, *Spectrochim. Acta Part B* **45 B** (1990) 769–778.
[45] A. Danielson, P. Lindblom, E. Södermann, *Chem. Scr.* **6** (1974) 5–9.
[46] R. B. Bilhorn, P. M. Epperson, J. V. Sweedler, M. B. Denton, *Appl. Spectrosc.* **41** (1987) 1114–1124.
[47] N. Furuta, K. R. Brushwyler, G. M. Hieftje, *Spectrochim. Acta Part B* **44 B** (1989) 349–358.
[48] P. J. Treado, M. D. Morris, *Anal. Chem.* **61** (1989) 723 A–734 A.
[49] L. M. Faires, *Anal. Chem.* **58** (1986) 1023 A–1043 A.
[50] F. Adams, R. Gijbels, R. Van Grieken (eds.): *Inorganic Mass Spectrometry,* Wiley, New York 1988.
[51] J. R. Bacon, A. Ure, *Analyst (London)* **109** (1984) 1229–1254.
[52] K. G. Heumann, F. Beer, H. Weiss, *Mikrochim. Acta I* 1983, 95–108.
[53] C. Brunnee, *Int. J. Mass Spectrom. Ion Processes* **76** (1987) 125–237.
[54] H. H. Willard, L. L. Merrit, A. J. Dean, A. F. Settle: *Instrumental Methods of Analysis,* 7th ed., Wadsworth, California 1998, Chap. 16.
[55] D. P. Myers, G. L. P. Yang, G. M. Hieftje, *J. Am. Soc. Mass Spectrom.* **5** (1994) 1008.
[56] D. P. Myers, G. Li, P. P. Mahoney, G. M. Hieftje, *J. Am. Soc. Mass Spectrom.* **6** (1995) 411.
[57] D. P. Myers, G. Li, P. P. Mahoney, G. M. Hieftje, *J. Am. Soc. Mass Spectrom.* **6** (1995) 400.
[58] J. R. Chapman: *Practical Organic Mass Spectrometry,* 2nd ed., John Wiley & Sons, Chichester 1993.
[59] A. Benninghoven, F. G. Rüdenauer, H. W. Werner: *Secondary Ion Mass Spectrometry,* Wiley, New York 1986.
[60] E. Denoyer, R. Van Grieken, F. Adams, D. F. S. Natusch, *Anal. Chem.* **54** (1982) 26 A–41 A.
[61] D. P. Meyers, G. M. Hieftje: "Use of a Time-of-Flight Mass Spectrometer for Plasma Mass Spectrometry," *1992 Pittsburgh Conference,* New Orleans, March 9–13, 1992, Paper 606.

[62] A. L. Gray, A. R. Date, *Analyst (London)* **108** (1983) 1033–1050.
[63] F. E. Lichte, A. L. Meier, J. G. Crock, *Anal. Chem.* **59** (1987) 1150–1157.
[64] F. Kohlrausch: *Praktische Physik,* 21st ed., vol. 1, Taubner Verlagsgesellschaft, Stuttgart 1961.
[65] F. Rosendahl, *Spectrochim. Acta* **10** (1957) 201–212.
[66] A. Aziz, J. A. C. Broekaert, F. Leis, *Spectrochim. Acta Part B* **37 B** (1982) 369–379.
[67] J. Sneddon (ed.): *Sample Introduction in Atomic Spectroscopy,* Elsevier, Amsterdam 1990.
[68] J. A. C. Broekaert: "Sample Introduction Techniques in ICP-AES," in P. W. J. M. Boumans (ed.): *Inductively Coupled Plasma Emission Spectroscopy,* vol. I, Wiley, New York 1987, pp. 296–357.
[69] A. Montaser, D. W. Golightly (eds.): *Inductively Coupled Plasmas in Analytical Atomic Spectrometry,* 2nd ed., VCH Publishers, Weinheim 1992.
[70] R. F. Browner, A. W. Boorn, *Anal. Chem.* **56** (1984) 786 A–798 A.
[71] R. F. Browner, A. W. Boorn, *Anal. Chem.* **56** (1984) 875 A–888 A.
[72] J. A. C. Broekaert, G. Tölg, *Fresenius' J. Anal. Chem.* **326** (1987) 495–509.
[73] J. A. C. Broekaert, *Anal. Chim. Acta* **196** (1987) 1–21.
[74] J. E. Meinhard in R. M. Barnes (ed.): *Applications of Plasma Emission Spectroscopy,* Heyden, London 1979, p. 1.
[75] G. L. Gouy, *Ann. Chim. Phys.* **18** (1887) 5.
[76] E. L. Kimberley, G. W. Rice, V. A. Fassel, *Anal. Chem.* **56** (1984) 289–292.
[77] L. R. Layman, F. E. Lichte, *Anal. Chem.* **54** (1982) 638–642.
[78] E. Sebastiani, K. Ohls, G. Riemer, *Fresenius' Z. Anal. Chem.* **264** (1973) 105–109.
[79] H. Berndt, W. Slavin, *At. Absorpt. Newsl.* **17** (1978) 109.
[80] S. Greenfield, P. B. Smith, *Anal. Chim. Acta* **59** (1972) 341.
[81] A. Aziz, J. A. C. Broekaert, F. Leis, *Spectrochim. Acta Part B* **37 B** (1982) 381–389.
[82] J. Ruzicka, E. H. Hansen: *Flow Injection Analysis,* Wiley, New York 1988.
[83] S. D. Hartenstein, J. Ruzicka, G. D. Christian, *Anal. Chem.* **57** (1985) 21–25.
[84] Z. Fang: *Flow Injection Separation and Preconcentration,* VCH Verlagsgesellschaft, Weinheim 1993.
[85] P. Schramel, L. Xu, G. Knapp, M. Michaelis, *Fresenius J. Anal. Chem.* **345** (1993) 600–606.
[86] S. Nukuyama, Y. Tanasawa, *Nippon Kikai Cakkai Ronbunshu* **4** (1938) 86, **5** (1939) 68.
[87] R. Nießner, *Angew. Chem.* **103** (1991) 542–552.
[88] T. Allen: *Particle Size Measurement,* Chapman & Hall, London 1981.
[89] W. Van Borm et al., *Spectrochim. Acta Part B* **46 B** (1991) 1033–1049.
[90] S. D. Olsen, A. Strasheim, *Spectrochim. Acta Part B* **38 B** (1983) 973–975.
[91] A. W. Boorn, R. Browner, *Anal. Chem.* **54** (1982) 1402–1410.
[92] G. Kreuning, F. J. M. J. Maessen, *Spectrochim. Acta Part B* **42 B** (1987) 677–688.
[93] M. P. Doherty, G. M. Hieftje, *Appl. Spectrosc.* **38** (1984) 405–411.
[94] H. Berndt, *Fresenius' J. Anal. Chem.* **331** (1988) 321–323.
[95] J. Posta, H. Berndt, *Spectrochim. Acta Part B* **47 B** (1992) 993–999.
[96] J. Posta, H. Berndt, S. K. Luro, G. Schaldach, *Anal. Chem.* **65** (1993) 2590–2595.
[97] A. Syti, *CRC Crit. Rev. Anal. Chem.* **4** (1974) 155.
[98] V. A. Fassel, B. R. Bear, *Spectrochim. Acta Part B* **41 B** (1986) 1089–1113.
[99] P. W. J. M. Boumans, F. J. De Boer, *Spectrochim. Acta Part B* **30 B** (1975) 309–334.
[100] N. W. Barnett, L. S. Chen, G. F. Kirkbright, *Spectrochim. Acta Part B* **39 B** (1984) 1141–1147.
[101] D. C. Manning, *At. Absorpt. Newsl.* **10** (1971) 123.
[102] M. Thompson, B. Pahlavanpour, S. J. Walton, G. F. Kirkbright, *Analyst (London)* **103** (1978) 568–579.
[103] J. A. C. Broekaert, F. Leis, *Fresenius' Z. Anal. Chem.* **300** (1980) 22–27.
[104] E. Bulska, P. Tschöpel, J. A. C. Broekaert, G. Tölg, *Anal. Chim. Acta* **271** (1993) 171–181.
[105] H. Tao, A. Miyazaki, *Anal. Sci.* **7** (1991) 55–59.
[106] C. S. Stringer, M. Attrep, Jr., *Anal. Chem.* **51** (1979) 731–734.
[107] B. Welz, M. Melcher, *Analyst (London)* **108** (1983) 213–224.
[108] E. Preuß, *Z. Angew. Mineral.* **3** (1940) 1.
[109] B. V. L'vov, L. A. Pelieva, A. I. Sharnopolsky, *Zh. Prikl. Spektrosk.* **27** (1977) 395.
[110] J. McNally, J. A. Holcombe, *Anal. Chem.* **59** (1987) 1105–1112.
[111] T. S. West, X. K. Williams, *Anal. Chim. Acta* **45** (1969) 27.
[112] V. Sychra et al., *Anal. Chim. Acta* **105** (1979) 105.
[113] H. T. Delves, *Analyst (London)* **95** (1970) 431–438.
[114] H. Berndt, J. Messerschmidt, *Spectrochim. Acta Part B* **34 B** (1979) 241–256.
[115] E. I. Brooks, K. J. Timmins, *Analyst (London)* **110** (1985) 557–558.
[116] Z. Grobenski, R. Lehmann, R. Tamm, B. Welz, *Mikrochim. Acta I* 1982, 115–125.
[117] D. L. Millard, H. C. Shan, G. F. Kirkbright, *Analyst (London)* **105** (1980) 502–508.
[118] J. A. C. Broekaert, F. Leis, *Mikrochim. Acta II* 1985, 261–272.
[119] D. Sommer, K. Ohls, *Fresenius' Z. Anal. Chem.* **304** (1980) 97–103.
[120] E. D. Salin, G. Horlick, *Anal. Chem.* **51** (1979) 2284–2286.
[121] G. Zaray, J. A. C. Broekaert, F. Leis, *Spectrochim. Acta Part B* **43 B** (1988) 241–253.
[122] V. Karnassios, G. Horlick, *Spectrochim. Acta Part B* **44 B** (1989) 1345–1360.
[123] V. Karnassios, M. Abdullah, G. Horlick, *Spectrochim. Acta Part B* **45 B** (1990) 119–130.
[124] L. Blain, E. D. Salin, *Spectrochim. Acta Part B* **47 B** (1992) 205–217.
[125] R. Lobinski et al., *Fresenius' J. Anal. Chem.* **342** (1992) 563–568.
[126] B. Raeymaekers et al., *Spectrochim. Acta Part B* **43 B** (1988) 923–940.
[127] B. Raeymaekers, P. Van Espen, F. Adams, *Mikrochim. Acta II* 1984, 437–454.
[128] B. Docekal, V. Krivan, *J. Anal. At. Spectrom.* **7** (1992) 521–528.
[129] H. Min, S. Xi-En, *Spectrochim. Acta Part B* **44 B** (1989) 957–964.
[130] R. M. Barnes, S. Nikdel, *J. Appl. Phys.* **47** (1976) 3929.

[131] G. M. Hieftje, R. M. Miller, Y. Pak, E. P. Wittig, *Anal. Chem.* **59** (1987) 2861–2872.

[132] L. Ebdon, M. Foulkes, K. Sutton, *J. Anal. At. Spectrom.* **12** (1997) 213–229.

[133] J. L. Johnes, R. L. Dahlquist, R. E. Hoyt, *Appl. Spectrosc.* **25** (1971) 629–635.

[134] S. ElAlfy, K. Laqau, H. Maßmann, *Fresenius' Z. Anal. Chem.* **263** (1973) 1–14.

[135] A. Aziz, J. A. C. Broekaert, K. Laqua, F. Leis, *Spectrochim. Acta Part B* **39 B** (1984) 1091–1103.

[136] J. A. C. Broekaert, F. Leis, K. Laqua: "A Contribution to the Direct Analysis of Solid Samples by Spark Erosion Combined to ICP-OES," in B. Sansoni (ed.): *Instrumentelle Multielementanalyse*, Verlag Chemie, Weinheim 1985.

[137] B. Raeymaekers, P. Van Espen, F. Adams, J. A. C. Broekaert, *Appl. Spectrosc.* **42** (1988) 142–150.

[138] R. M. Barnes, H. V. Malmstadt, *Anal. Chem.* **46** (1974) 66–72.

[139] N. Bings, H. Alexi, J. A. C. Broekaert, Universität Dortmund, unpublished results.

[140] N. Omenetto: *Analytical Laser Spectroscopy*, Wiley, New York 1979.

[141] L. Moenke-Blankenburg: *Laser Micro Analysis*, Wiley, New York 1989.

[142] J. Sneddon, T. L. Thiem, Y. Lee: *Lasers in Analytical Spectrometry*, VCH, New York 1997.

[143] K. Laqua: "Analytical Spectroscopy Using Laser Atomizers," in S. Martellucci, A. N. Chester (eds.): *Analytical Laser Spectroscopy*, Plenum Publishing Corp., New York 1985, pp. 159–182.

[144] F. Leis, W. Sdorra, J. B. Ko, K. Niemax, *Mikrochim. Acta II* 1989, 185–199.

[145] W. Sdorra, A. Quentmeier, K. Niemax, *Mikrochim. Acta II* 1989, 201–218.

[146] J. Uebbing et al., *Appl. Spectrosc.* **45** (1991) 1419–1423.

[147] M. Kaminsky: *Atomic and Ionic Impact Phenomena on Metal Surfaces*, Springer Verlag, Berlin 1965.

[148] F. M. Penning: *Electrical Discharges in Gases*, Philips Technical Library, Eindhoven 1957, p. 41.

[149] P. W. J. M. Boumans, *Anal. Chem.* **44** (1973) 1219–1228.

[150] J. A. C. Broekaert, *J. Anal. At. Spectrom.* **2** (1987) 537–542.

[151] M. Dogan, K. Laqua, H. Maßmann, *Spectrochim. Acta Part B* **27 B** (1972) 631–649.

[152] R. Marcus (ed.): *Glow Discharge Spectrometries*, Plenum Press, New York 1993.

[153] M. Van Straaten, A. Vertes, R. Gijbels, *Spectrochim. Acta Part B* **46 B** (1991) 283–280.

[154] B. Welz: *Atomic Absorption Spectrometry*, Verlag Chemie, Weinheim 1985.

[155] W. J. Price: *Spectrochemical Analysis by Atomic Absorption*, Heyden, London 1979.

[156] S. J. Hill, A. Fisher: in L. Ebdon, E. H. Evans (eds.): *An Introduction to Analytical Atomic Spectrometry*, Wiley, Chichester 1998.

[157] A. Zybin, C. Schneurer-Patschan, M. A. Bolshov, K. Niemax, *Trends. Anal. Chem.* **7** (1998) 513.

[158] N. Omenetto, *J. Anal. At. Spectrom.* **13** (1998) 385–399.

[159] R. S. Koirthyohann, E. E. Pickett, *Anal. Chem.* **37** (1965) 601–603.

[160] G. P. Moulton, T. C. O'Haver, J. M. Harnly, *J. Anal. At. Spectrom.* **5** (1990) 145–150.

[161] M. D. Amos, J. B. Willis, *Spectrochim. Acta* **22** (1966) 1325–1343.

[162] R. Hermann, C. T. J. Alkemade: *Chemical Analysis by Flame Photometry*, 2nd ed., Interscience Publishers, New York 1983.

[163] C. T. J. Alkemade, T. Hollander, W. Snelleman, P. J. Th. Zeegers: *Metal Vapours in Flames*, Pergamon Press, Oxford 1982.

[164] J. A. Dean, T. C. Rains (eds.): *Flame Emission and Atomic Absorption Spectrometry*, vol. 1, Marcel Dekker, New York 1969.

[165] W. Slavin, *Anal. Chem.* **58** (1986) 589 A–597 A.

[166] H. Berndt, G. Schaldach, *J. Anal. At. Spectrom.* **3** (1988) 709–713.

[167] G. Schlemmer, B. Welz, *Spectrochim. Acta Part B* **41 B** (1986) 1157–1166.

[168] B. Welz et al., *Spectrochim. Acta Part B* **41 B** (1986) 1175–1201.

[169] V. Krivan, *J. Anal. At. Spectrom.* **7** (1992) 155–164.

[170] W. R. A. De Jonghe, F. C. Adams, *Anal. Chim. Acta* **108** (1979) 21–30.

[171] D. Chakraborti et al., *Anal. Chem.* **56** (1984) 2692–2697.

[172] M. Yamamoto, M. Yasuda, Y. Yamamoto, *Anal. Chem.* **57** (1985) 1382–1385.

[173] E. Jackwerth, P. G. Willmer, R. Höhn, H. Berndt, *At. Absorpt. Newsl.* **18** (1978) 66.

[174] G. Kaiser, D. Götz, P. Schoch, G. Tölg, *Talanta* **22** (1975) 889–899.

[175] J. Piwonka, G. Kaiser, G. Tölg, *Fresenius' Z. Anal. Chem.* **321** (1985) 225–234.

[176] R. E. Sturgeon, S. N. Willie, S. S. Berman, *Anal. Chem.* **57** (1985) ff 2311.

[177] B. Welz, M. Melcher, *Anal. Chim. Acta* **131** (1981) ff 17.

[178] H. Berndt, *Spectrochim. Acta Part B* **39 B** (1984) 1121–1128.

[179] L. Ebdon, M. R. Cave, *Analyst (London)* **107** (1982) 172–178.

[180] B. Docekal, V. Krivan, *J. Anal. At. Spectrom.* **7** (1992) 521–528.

[181] D. S. Gough, *Spectrochim. Acta Part B* **54 B** (1999) 2067–2072.

[182] A. E. Bernhard, *Spectroscopy* **2** (1987) 118.

[183] K. Ohls, *Fresenius' Z. Anal. Chem.* **327** (1987) 111–118.

[184] T. Kantor, L. Polos, P. Fodor, E. Pungor, *Talanta* **23** (1976) 585–586.

[185] H. Berndt, D. Sopczak, *Fresenius' Z. Anal. Chem.* **329** (1987) 18–26.

[186] M. T. C. De Loos-Vollebregt, L. De Galan, *Prog. Anal. At. Spectrom.* **8** (1985) 47–81.

[187] S. B. Smith, Jr., G. M. Hieftje, *Appl. Spectrosc.* **37** (1983) 419–414.

[188] M. Yamamoto, S. Murayama, M. Ito, M. Yasuda, *Spectrochim. Acta Part B* **35 B** (1980) 43–50.

[189] P. Wirz, H. Debus, W. Hanle, A. Scharmann, *Spectrochim. Acta Part B* **37 B** (1982) 1013–1020.

[190] G. M. Hermann, *Anal. Chem.* **64** (1992) 571A–579A.

[191] P. Tschöpel, G. Tölg, *J. Trace Microprobe Tech.* **1** (1982) 1–77.

[192] Analytical Spectrometry Updates, *J. Anal. At. Spectrom.*

[193] H. Berndt, E. Jackwerth, *J. Clin. Chem. Clin. Biochem.* **19** (1979) 71.

[194] A. Mizuike: *Enrichment Techniques for Inorganic Trace Analysis*, Springer Verlag, Berlin 1983.

[195] K. S. Subramanian, *Anal. Chem.* **60** (1988) 11–15.
[196] J. Szpunar, *Analyst* **125** (2000) 963–988.
[197] W. M. R. Dirkx, R. Lobinski, F. C. Adams, *Anal. Sci.* **9** (1993) 273–278.
[198] C. E. Oda, J. D. Ingle, *Anal. Chem.* **53** (1981) 2305–2309.
[199] W. Gerlach, E. Schweitzer: *Die chemische Emissionsspektralanalyse,* vol. 1, L. Voss, Leipzig 1930.
[200] G. R. Harrison: *Wavelength Tables,* The M.I.T. Press, Cambridge, Mass., 1991.
[201] A. N. Saidel, V. K. Prokofiev, S. M. Raiski: *Spektraltabellen,* V.E.B. Verlag Technik, Berlin 1955.
[202] A. Gatterer, J. Junkes: *Atlas der Restlinien,* **vols. 3,** Specola Vaticana, Citta del Vaticano, 1945, 1947, 1959.
[203] E. W. Salpeter: *Spektren in der Glimmentladung von 150 bis 400 nm,* vols. 1–5, Specola Vaticana, Citta del Vaticano 1973.
[204] R. K. Winge, V. A. Fassel, V. J. Peterson, M. A. Floyd: *ICP Atomic Emission Spectroscopy,* Elsevier, Amsterdam 1985.
[205] P. W. J. M. Boumans: *Line Coincidence Tables for ICP-OES,* vols. 1 and 2, Pergamon Press, Oxford 1984.
[206] B. Huang et al.: *An Atlas of High Resolution Spectra for Rare Earth Elements for ICP–AES,* Royal Society of Chemistry, Cambridge 2000.
[207] G. Scheibe: *Chemische Spektralanalyse, physikalische Methoden der analytischen Chemie,* **vol. 1,** 1933.
[208] A. Lomakin, *Z. Anorg. Allg. Chem.* **75** (1930) 187.
[209] *Ullmann,* 4th ed., **5,** 441–500.
[210] K. Laqua, W.-D. Hagenah, H. Wächter, *Fresenius' Z. Anal. Chem.* **221** (1967) 142–174.
[211] A. Aziz, J. A. C. Broekaert, F. Leis, *Spectrochim. Acta Part B* **36 B** (1981) 251–260.
[212] N. W. H. Addink: *DC Arc Analysis,* MacMillan, London 1971.
[213] M. Z. Riemann, *Fresenius' Z. Anal. Chem.* **215** (1966) 407–424.
[214] R. Avni in E. L. Grove (ed.): *Applied Atomic Spectroscopy,* vol. 1, Plenum Press, New York 1978.
[215] K. Slickers: *Die automatische Atom-Emissions-Spektralanalyse,* Brühlsche Universitätsdruckerei, Gießen 1992.
[216] K. Slickers: *Automatic Atomic Emission Spectroscopy,* Brühlsche Universitätsdruckerei, Gießen 1993.
[217] K. H. Koch, *Spectrochim. Acta Part B* **39 B** (1984) 1067–1079.
[218] K. Tohyama, J. Ono, M. Onodera, M. Saeki: *Research and Development in Japan,* Okochi Memorial Foundation, 1978, pp. 31–35.
[219] D. M. Coleman, M. A. Sainz, H. T. Butler, *Anal. Chem.* **52** (1980) 746–753.
[220] A. Montaser: *Inductively Coupled Plasmas in Analytical Atomic Spectrometry,* 2nd ed., Wiley-VCH, New York 1998.
[221] S. J. Hill (ed.): *Inductively Coupled Plasma Spectrometry and its Applications,* Sheffield Academic Press, Sheffield 1999.
[222] V. V. Korolev, E. E. Vainshtein, *J. Anal. Chem. USSR (Engl. Transl.)* **14** (1959) 658–662.
[223] M. Margoshes, B. Scribner, *Spectrochim. Acta* **15** (1959) 138–145.
[224] E. Kranz in: *Emissionsspektroskopie,* Akademie-Verlag, Berlin 1964, p. 160.
[225] J. Reednick, *Am. Lab.* 1979, no. 5, 127–133.
[226] R. J. Decker, *Spectrochim. Acta Part B* **35 B** (1980) 19–35.
[227] F. Leis, J. A. C. Broekaert, H. Waechter, *Fresenius' Z. Anal. Chem.* **333** (1989) 2–5.
[228] G. W. Johnson, H. E. Taylor, R. K. Skogerboe, *Spectrochim. Acta Part B* **34 B** (1979) 197–212.
[229] T. B. Reed, *J. Appl. Phys.* **32** (1961) 821–824.
[230] S. Greenfield, I. L. Jones, C. T. Berry, *Analyst (London)* **89** (1964) 713–720.
[231] R. H. Wendt, V. A. Fassel, *Anal. Chem.* **37** (1965) 920–922.
[232] R. Rezaaiyaan et al., *Appl. Spectrosc.* **36** (1982) 626–631.
[233] D. J. Kalnicki, V. A. Fassel, R. N. Kalnicky, *Appl. Spectrosc.* **31** (1977) 137–150.
[234] B. Raeymaekers, J. A. C. Broekaert, F. Leis, *Spectrochim. Acta Part B* **43 B** (1988) 941–949.
[235] I. Ishii, A. Montaser, *Spectrochim. Acta Part B* **46 B** (1991) 1197–1206.
[236] P. W. J. M. Boumans, F. J. de Boer, *Spectrochim. Acta Part B* **32 B** (1977) 365–395.
[237] L. De Galan, *Spectrochim. Acta Part B* **39 B** (1984) 537–543.
[238] M. Huang, D. S. Hanselmann, P. Y. Yang, G. M. Hieftje, *Spectrochim. Acta Part B* **47 B** (1992) 765–786.
[239] J. A. C. Broekaert, F. Leis, *Anal. Chim. Acta* **109** (1979) 73–83.
[240] F. H. Walters, L. R. Parker, Jr., S. L. Morgan, S. N. Deming: *Sequential Simplex Optimization,* CRC Press, Inc., Boca Raton 1991.
[241] G. L. Moore, P. J. Humphries-Cuff, A. E. Watson, *Spectrochim. Acta Part B* **39 B** (1984) 915–929.
[242] M. L. Parsons, A. R. Forster, D. Anderson: *An Atlas of Spectral Interferences in ICP Spectroscopy,* Plenum Press, New York 1980.
[243] M. R. Cave, O. Butler, J. M. Cook, M. S. Cresser, L. M. Garden, D. L. Miles, *J. Anal. At. Spectrom.* **15** (2000) 181–235.
[244] U.S. Environmental Protection Agency: *Methods for the Determination of Metals in Environmental Samples,* EPA600491010, Washington DC 1991.
[245] U.S. Environmental Protection Agency: *Methods for the Determination of Metals in Environmental Samples,* Supplement 1, EPA600R94111, Washington DC 1994.
[246] B. Fairman, M. Hinds, S. M. Nelms, D. M. Penny, P. Goodall, *J. Anal. At. Spectrom.* **14** (1999) 1937–1969.
[247] R. E. Russo, X. L. Mao, O. V. Borisov, *Trends Anal. Chem.* **17** (1998) 461.
[248] W. B. Barnett, V. A. Fassel, R. N. Kniseley, *Spectrochim. Acta Part B* **23 B** (1968) 643–664.
[249] S. A. Meyers, D. H. Tracy, *Spectrochim. Acta Part B* **38 B** (1983) 1227–1253.
[250] P. W. J. M. Boumans, M. C. Lux-Steiner, *Spectrochim. Acta Part B* **37 B** (1982) 97–126.
[251] J. A. C. Broekaert, F. Leis, K. Laqua, *Talanta* **28** (1981) 745–752.
[252] K. Vermeiren, C. Vandecasteele, R. Dams, *Analyst (London)* **115** (1990) 17–22.
[253] K. Dittrich et al., *J. Anal. At. Spectrom.* **3** (1993) 1105–1110.
[254] A. G. Cox, L. G. Cook, C. W. McLeod, *Analyst (London)* **110** (1985) 331–333.
[255] H. G. C. Human, R. H. Scott, A. R. Oakes, C. D. West, *Analyst (London)* **101** (1976) 265–271.

[256] A. Lemarchand, G. Labarraque, P. Masson, J. A. C. Broekaert, *J. Anal. At. Spectrom.* **2** (1987) 481–484.
[257] T. Ishizuka, Y. Uwamino, *Spectrochim. Acta Part B* **38B** (1983) 519–527.
[258] I. B. Brenner et al., *Spectrochim. Acta Part B* **36B** (1981) 785–797.
[259] J. R. Garbarino, M. E. Taylor, *Appl. Spectrosc.* **33** (1979) 167–175.
[260] U.S. Environmental Protection Agency: Inductively Coupled Plasma-Atomic Emission Spectrometric Method for Trace Element Analysis of Water and Wastes, Washington, DC, 1979.
[261] DIN 38 406,*Bestimmung der 24 Elemente Ag, Al, B, Ba, Ca, Cd, Co, Cr, Cu, Fe, K, Mg, Mn, Mo, Na, Ni, P, Pb, Sb, Sr, Ti, V, Zn und Zr durch Atomemissionsspektrometrie mit induktiv gekoppeltem Plasma (ICP-AES) (E 22)*, Beuth Verlag, Berlin 1987.
[262] M. Thompson, B. Pahlavanpour, S. J. Walton, G. F. Kirkbright, *Analyst (London)* **103** (1978) 705–713.
[263] A. Miyazaki, A. Kimura, K. Bansho, Y. Umezaki, *Anal. Chim. Acta* **144** (1982) 213–221.
[264] H. Berndt, U. Harms, M. Sonneborn, *Fresenius' Z. Anal. Chem.* **322** (1985) 329–333.
[265] M. Hiraide et al., *Anal. Chem.* **52** (1980) 804–807.
[266] J. A. C. Broekaert, S. Gucer, F. Adams: *Metal Speciation in the Environment*, Springer Verlag, Berlin 1990.
[267] R. K. Skogerboe, G. N. Coleman, *Anal. Chem.* **48** (1976) 611 A–622 A.
[268] A. E. Croslyn, B. W. Smith, J. D. Winefordner, *Crit. Rev. Anal. Chem.* **27** (1997) 199–255.
[269] R. Mavrodineanu, R. C. Hughes, *Spectrochim. Acta* **19** (1963) 1309–1307.
[270] A. Disam, P. Tschöpel, G. Tölg, *Fresenius' Z. Anal. Chem.* **310** (1982) 131–143.
[271] S. Hanamura, B. W. Smith, J. D. Winefordner, *Anal. Chem.* **55** (1983) 2026–2032.
[272] C. I. M. Beenakker, *Spectrochim. Acta Part B* **32B** (1977) 173–178.
[273] D. Kollotzek, P. Tschöpel, G. Tölg, *Spectrochim. Acta Part B* **37B** (1982) 91–96.
[274] G. Heltai, J. A. C. Broekaert, F. Leis, G. Tölg, *Spectrochim. Acta Part B* **45B** (1990) 301–311.
[275] P. S. Moussanda, P. Ranson, J. M. Mermet, *Spectrochim. Acta Part B* **40B** (1985) 641–651.
[276] D. Kollotzek et al., *Fresenius' Z. Anal. Chem.* **318** (1984) 485–489.
[277] P. Uden: *Element-Specific Chromatographic Detection by Atomic Emission Spectroscopy*, American Chemical Society, Washington 1992.
[278] J. Hubert, M. Moisan, A. Ricard, *Spectrochim. Acta Part B* **33B** (1979) 1–10.
[279] U. Richts, J. A. C. Broekaert, P. Tschöpel, G. Tölg, *Talanta* **38** (1991) 863–869.
[280] Q. Jin, C. Zhu, M. W. Borer, G. M. Hieftje, *Spectrochim. Acta Part B* **46B** (1991) 417–430.
[281] R. M. Barnes, E. E. Reszke, *Anal. Chem.* **62** (1990) 2650.
[282] K. A. Forbes, E. E. Reszke, P. C. Uden, R. M. Barnes, *J. Anal. At. Spectrom.* **6** (1991) 57.
[283] R. K. Marcus (ed.): *Glow Discharge Spectroscopies*, Plenum Press, New York 1993.
[284] K. I. Zil'bershtein: *Spectrochemical Analysis of Pure Substances*, A. Hilger, Bristol 1977, pp. 173–227.
[285] B. Thelin, *Appl. Spectrosc.* **35** (1981) 302–307.
[286] S. Caroli: *Improved Hollow Cathode Lamps for Atomic Spectroscopy*, Ellis Horwood, Chichester 1985.
[287] H. Falk, E. Hoffmann, C. Lüdke, *Spectrochim. Acta Part B* **39B** (1984) 283–294.
[288] W. Grimm, *Spectrochim. Acta Part B* **23B** (1968) 443–454.
[289] J. B. Ko, *Spectrochim. Acta Part B* **39B** (1984) 1405–1423.
[290] F. Leis, J. A. C. Broekaert, K. Laqua, *Spectrochim. Acta Part B* **42B** (1987) 1169–1176.
[291] A. Bengtson, *Spectrochim. Acta Part B* **40B** (1985) 631–639.
[292] R. M. Lowe, *Spectrochim. Acta Part B* **31B** (1978) 257–261.
[293] N. P. Ferreira, J. A. Strauss, H. G. C. Human, *Spectrochim. Acta Part B* **38B** (1983) 899–911.
[294] J. A. C. Broekaert, R. Pereiro, T. K. Starn, G. M. Hieftje, *Spectrochim. Acta Part B* **48B** (1983) 1207–1220.
[295] M. R. Winchester, C. Lazik, R. K. Marcus, *Spectrochim. Acta Part B* **46B** (1991) 483–499.
[296] T. M. Sugden in R. L. Reed (ed.): *Mass Spectrometry*, Academic Press, New York 1965, p. 347.
[297] R. S. Houk et al., *Anal. Chem.* **52** (1980) 2283–2289.
[298] A. Montaser (ed.): *Inductively Coupled Plasma Mass Spectrometry*, Wiley-VCH, New York 1998.
[299] E. H. Evans, J. J. Giglio, T. Castillano, J. A. Caruso: *Inductively Coupled and Microwave Induced Plasmas for Mass Spectrometry*, Royal Society of Chemistry, Cambridge 1995.
[300] H. Niu, R. S. Houk, *Spectrochim. Acta Part B* **51B** (1996) 119–815.
[301] E. H. Evans, J. J. Giglio, *J. Anal. At. Spectrom.* **8** (1993) 1–8.
[302] A. L. Gray, R. S. Houk, J. G. Williams, *J. Anal. At. Spectrom.* **2** (1987) 13–20.
[303] M. A. Vaughan, G. Horlick, *Appl. Spectrosc.* **40** (1986) 434–445.
[304] G. Horlick, S. H. Tan, M. A. Vaughan, C. A. Rose, *Spectrochim. Acta Part B* **40B** (1985) 1555–1572.
[305] N. Jakubowski, B. J. Raeymaekers, J. A. C. Broekaert, D. Stüwer, *Spectrochim. Acta Part B* **44B** (1989) 219–228.
[306] C. Vandecasteele, M. Nagels, H. Vanhoe, R. Dams, *Anal. Chim. Acta* **211** (1988) 91–98.
[307] J. S. Crain, R. S. Houk, F. G. Smith, *Spectrochim. Acta Part B* **43B** (1988) 1355–1364.
[308] D. Hausler, *Spectrochim. Acta Part B* **42B** (1987) 63–73.
[309] N. Jakubowski, I. Feldman, D. Stüwer, H. Berndt, *Spectrochim. Acta Part B* **47B** (1992) 119–129.
[310] X. Wang, M. Viczian, A. Lasztity, R. M. Barnes, *J. Anal. At. Spectrom.* **3** (1988) 821–828.
[311] C. J. Park, J. C. Van Loon, P. Arrowsmith, J. B. French, *Anal. Chem.* **59** (1987) 2191–2196.
[312] D. Boomer, M. Powell, R. L. A. Sing, E. D. Salin, *Anal. Chem.* **58** (1986) 975–976.
[313] G. E. M. Hall, J. C. Pelchat, D. W. Boomer, M. Powell, *J. Anal. At. Spectrom.* **3** (1988) 791–797.
[314] T. E. Jeffries, S. E. Jackson, H. P. Longerich, *J. Anal. At. Spectrom.* **13** (1998) 935–940.
[315] E. K. Shibuya, J. E. S. Sarkis, J. Enzweiler, A. P. S. Jorge, A. M. G. Figueiredo, *J. Anal. At. Spectrom.* **13** (1998) 941–944.
[316] J. G. Williams, A. L. Gray, P. Norman, L. Ebdon, *J. Anal. At. Spectrom.* **2** (1987) 469–472.

[317] N. Jakubowski, I. Feldman, B. Sack, D. Stüwer, *J. Anal. At. Spectrom.* **7** (1992) 121–125.
[318] P. Arrowsmith, *Anal. Chem.* **59** (1987) 1437–1444.
[319] D. L. Miles: "Geological Applications of Plasma Spectrometry," in S. J. Hill (ed.): *Inductively Coupled Plasmas in Analytical Atomic Spectrometry,* 2nd ed., Wiley-VCH, New York 1999.
[320] F. Vanhaecke, L. Moens, P. Taylor: "Use of ICP–MS for Isotope Ratio Measurements," in S. J. Hill (ed.): *Inductively Coupled Plasmas in Analytical Atomic Spectrometry,* 2nd ed., Wiley-VCH, New York 1999.
[321] J. B. Truscott, L. Bromley, P. J. Jones, E. H. Evans, J. Turner, B. Fairman, *J. Anal. At. Spectrom.* **14** (1999) 627–631.
[322] LM. J. Bloxham, S. J. Hill, P. J. Worsfold, *J. Anal. At. Spectrom.* **9** (1994) 935–938.
[323] W. W. Harrison, B. L. Bentz, *Prog. Anal. At. Spectrosc.* **11** (1988) 53–110.
[324] D. J. Hall, K. Robinson, *Am. Lab.* 1987, no. 8, 14.
[325] N. Jakubowski, D. Stüwer, G. Tölg, *Int. J. Mass Spectrom. Ion. Processes* **71** (1986) 183–197.
[326] P. M. Charalambous, *Steel Res.* **5** (1987) 197.
[327] W. Vieth, J. C. Huneke, *Spectrochim. Acta Part B* **46 B** (1991) 137–154.
[328] N. Jakubowski, D. Stüwer, *Fresenius' Z. Anal. Chem.* **335** (1988) 680–686.
[329] A. Bogaerts, R. Gijbels, *Spectrochim. Acta Part B* **53** (1998) 1.
[330] VG Instruments, Technical Information GD 012, Manchester, U.K.
[331] VG Instruments, Application Note 02.681, Manchester, U.K.
[332] VG Instruments, Technical Information GD 701, Manchester, U.K.
[333] N. Jakubowski, D. Stüwer, W. Vieth, *Anal. Chem.* **59** (1987) 1825–1830.
[334] N. Jakubowski, D. Stüwer, W. Vieth, *Fresenius' Z. Anal. Chem.* **331** (1988) 145–149.
[335] M. Hecq, A. Hecq, M. Fontignies, *Anal. Chim. Acta* **155** (1983) 191–198.
[336] W. A. Mattson, B. L. Bentz, W. W. Harrison, *Anal. Chem.* **48** (1976) 489–491.
[337] G. O. Foss, H. J. Svec, R. J. Conzemius, *Anal. Chim. Acta* **147** (1983) 151–162.
[338] N. Jakubowski, D. Stüwer, G. Tölg, *Spectrochim. Acta Part B* **46 B** (1991) 155–163.
[339] N. Omenetto, J. D. Winfordner, *Prog. Anal. At. Spectrom.* **2** (1979) 1–183.
[340] N. Omenetto, H. G. C. Human, P. Cavalli, G. Rossi, *Spectrochim. Acta Part B* **39 B** (1984) 115–117.
[341] V. A. Bolshov, A. V. Zybin, II. Smirenkins, *Spectrochim. Acta Part B* **36 B** (1981) 1143–1152.
[342] N. Omenetto et al., *Spectrochim. Acta Part B* **40 B** (1985) 1411–1422.
[343] B. W. Smith, P. B. Farnsworth, P. Cavalli, N. Omenetto, *Spectrochim. Acta Part B* **45 B** (1990) 1369–1373.
[344] S. Grazhulene, V. Khvostikov, M. Sorokin, *Spectrochim. Acta Part B* **46 B** (1991) 459–465.
[345] M. A. Bolshov, C. F. Boutron, A. V. Zybin, *Anal. Chem.* **61** (1989) 1758–1762.
[346] W. Sdorra, K. Niemax, *Spectrochim. Acta Part B* **45 B** (1990) 917–926.
[347] G. C. Turk, J. C. Travis, J. R. DeVoe, *Anal. Chem.* **51** (1979) 1890–1896.
[348] R. B. Green, M. D. Seltzer: "Laser Induced Ionization Spectrometry," in J. Sneddon (ed.): *Advances in Atomic Spectroscopy,* **vol. I,** JAI Press, London 1992.
[349] G. J. Havrilla, S. J. Weeks, J. C. Travis, *Anal. Chem.* **54** (1982) 2566–2570.
[350] K. Niemax, *Appl. Phys. B* **B 38** (1985) 147–157.
[351] N. Omenetto, T. Berthoud, P. Cavalli, G. Rossi, *Anal. Chem.* **57** (1985) 1256–1261.
[352] G. C. Turk, M. DeMing, in W. F. Koch (ed.): *NBS Special Publication 260–106,* US Government Printing Office Washington DC 1986, pp. 30–33.
[353] J. P. Young, R. W. Shaw, D. H. Smith, *Anal. Chem.* **61** (1989) 1271 A–1279 A.
[354] R. Klockenkämper, J. Knoth, A. Prange, H. Schwenke, *Anal. Chem.* **64** (1992) 1115 A–1123 A.

22. Laser Analytical Spectroscopy

MICHAEL A. BOLSHOV, Institute of Spectroscopy, Russian Academy of Sciences, Troitzk, Russia; Institute of Spectrochemistry and Applied Spectroscopy, Dortmund, Germany

YURI A. KURITSYN, Institute of Spectroscopy, Russian Academy of Sciences, Troitzk, Russia

22.	Laser Analytical Spectroscopy . . .	727
22.1.	**Introduction**.	727
22.1.1.	Properties of Laser Light	728
22.1.1.1.	Tunability	728
22.1.1.2.	Intensity	728
22.1.1.3.	Time Resolution	728
22.1.1.4.	Temporal Coherence	728
22.1.1.5.	Spatial Coherence	728
22.1.2.	Types of Laser Analytical Techniques.	728
22.1.3.	Peculiarities of Resonance Laser Radiation Absorption.	729
22.2.	**Tunable Lasers**.	730
22.2.1.	Dye Lasers	730
22.2.2.	Semiconductor Diode Lasers	730
22.3.	**Laser Techniques for Elemental Analysis**.	732
22.3.1.	Laser-Excited Atomic Fluorescence Spectrometry	732
22.3.1.1.	Physical Principles	732
22.3.1.2.	Atomizers	733
22.3.1.3.	Limits of Detection.	734
22.3.1.4.	Real Sample Analysis	734
22.3.2.	Laser-Enhanced Ionization	735
22.3.2.1.	Physical Principles	735
22.3.2.2.	Atomizers	737
22.3.2.3.	Limits of Detection.	737
22.3.2.4.	Real Sample Analysis	737
22.3.3.	Resonance Ionization Spectroscopy	738
22.3.3.1.	Physical Principles	738
22.3.3.2.	Atomizers	740
22.3.3.3.	Limits of Detection: Real Samples	740
22.3.4.	Diode Lasers in Elemental Analysis	741
22.3.4.1.	Physical Principles of Absorption Spectrometry	741
22.3.4.2.	Atomic Absorption Spectrometry with Diode Lasers	741
22.3.4.3.	Element Specific Detection by DLAAS.	743
22.3.4.4.	Isotope Selective Detection with Diode Lasers	743
22.3.4.5.	Commercial System	743
22.4.	**Laser Techniques for Molecular Analysis**	744
22.4.1.	Molecular Absorption Spectroscopy with Diode Lasers.	744
22.4.1.1.	Detection of Molecular Gas Impurities with Lead-Salt Diode Lasers	744
22.4.1.2.	Near IR Gas Sensors	744
22.4.1.3.	Advanced Developments	745
22.4.2.	Laser Optoacoustic Spectroscopy .	745
22.4.3.	Thermal Lens Spectroscopy	747
22.4.4.	Fluorescence Analysis of Organic Molecules in Solid Solutions	748
22.5.	**Laser Ablation**	750
22.6.	**References**	751

22.1. Introduction

Modern technologies in electronics, in high-purity materials and chemical reagents, in toxicology, biology, and environmental control and protection need extremely sensitive analytical techniques. Impurity concentrations at and below pg/mL (pg/g) levels are state-of-the-art for many analyses. In many cases the sample volume is very small; if an impurity concentration of the order of ng/g has to be detected in a sample of mass 1 µg, the analytical technique should have an absolute detection limit of ca. 1 fg (10^{-15} g). Even techniques such as atomic absorption spectrometry (AAS), optical emission spectrometry with inductively coupled plasma (ICP-OES), neutron activation spectroscopy (NAS), and X-ray fluorescence spectroscopy (XRFS) can not detect directly trace

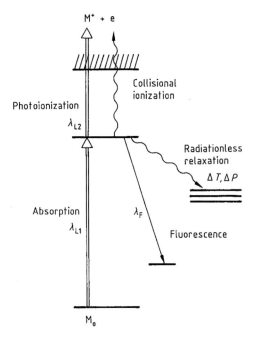

Figure 1. The basic laser analytical techniques

elements at such low levels of concentration and/or absolute mass. In most cases, these techniques need elaborate chemical pretreatment of samples. Contamination or loss of analyte are very probable during any chemical procedure at such low concentration.

During the last 30 years the development of laser technology has opened up new possibilities in analytical spectroscopy, both by improvements in traditional analytical techniques, such as absorption and fluorescence or optoacoustic spectroscopy, and by the introduction of new techniques such as multistep photoionization spectroscopy, thermal lens spectroscopy, and site-selection spectroscopy.

22.1.1. Properties of Laser Light

22.1.1.1. Tunability

Practically all laser analytical techniques are based on systems with tunable wavelength. By using different types of lasers and the methods of nonlinear optics, one can obtain intense tunable laser radiation over the spectral range of 0.2–40 μm and in some regions of the vacuum ultraviolet and far infrared. Almost any strong atomic or molecular transition is thus accessible to laser excitation.

22.1.1.2. Intensity

The intensities of lasers exceed by many orders of magnitude those of conventional light sources. At high intensity levels, nonlinear interaction of light with atomic and molecular systems becomes pronounced. New analytical techniques involving multistep and multiphoton excitation and ionization, optical saturation, and excitation of forbidden transitions become possible. High radiation intensity also increases the sensitivity of laser analytical techniques.

22.1.1.3. Time Resolution

Pulsed lasers with extremely short pulse durations (down to 10^{-15} s) are now available, enabling investigations of chemical reaction kinetics, energy and electron transfer in complex excited molecules, etc. The efficiency of excitation of higher states increases because the duration of the laser pulse is often shorter than the decay time of the intermediate state of atoms or molecules.

22.1.1.4. Temporal Coherence

The spectral linewidth of modern lasers is often less than the homogeneous linewidth of a given atom or molecule. This results in very high spectral intensity and thus very efficient excitation. It also allows highly selective excitation of a particular atom or molecule in a complex mixture. High selectivity is an important advantage of laser analytical techniques.

22.1.1.5. Spatial Coherence

The ability to focus a laser beam to a very small spot (a few μm diameter) provides high spatial resolution in sample analysis. High spatial coherence also enables collimated laser beams to be used for remote sensing.

22.1.2. Types of Laser Analytical Techniques

Laser analytical techniques can be classified on the basis of the physical process caused by the interaction of laser radiation with matter (Fig. 1).

The primary process is absorption of photons by atoms or molecules. If the absorption lines are strong and the number density of the absorbing species is sufficient, the difference between incident and transmitted energies may be measurable for a single pass of the laser beam through the sample. In this case the signal can be measured by a simple light detector, such as a photodiode. For weaker absorption lines and lower number densities of the absorbing species multipass schemes are used.

Another set of analytical techniques (optocalorimetric methods) is based on indirect measurement of the energy absorbed by the sample. Absorbed energy is transformed into kinetic energy of the sample particles followed by an increase in the local temperature of the sample. The resultant changes in pressure and refractive index can be measured by optoacoustic methods or by laser scattering techniques.

The absorption of laser photons leads to excitation of atoms or molecules. In many cases, radiative decay of the excited state is very efficient, i.e., quantum yields of luminescence or fluorescence are very high. Measurement of the intensity of emitted photons as a function of analyte concentration is a widespread analytical method with conventional light sources. The use of intense laser sources greatly increased the sensitivity and selectivity of traditional luminescence methods.

Excitation of atoms or molecules may be followed by collisional or multistep (multiphoton) ionization of the excited species. The charged particles so formed may be recorded by conventional methods. Several techniques involve a combination of optical excitation of a neutral particle and its subsequent ionization.

22.1.3. Peculiarities of Resonance Laser Radiation Absorption

The interaction of intense laser radiation with matter differs from that of conventional radiation sources. The high incident irradiances available with lasers can lead to a significant depletion of the initial level population. As a result, the optical absorption coefficient decreases as a function of the excitation energy and the absorbed energy tends towards a constant value. This so called optical saturation effect leads to a nonlinear dependence of the absorption signal on the light intensity. Because the amount of energy released by radiative and nonradiative relaxation processes depends on the amount of energy absorbed, nonlinear intensity-dependent signals are also obtained in other spectrometric techniques, such as fluorescence, optoacoustic, thermal lens, etc.

In the simplest case for the interaction of a two-level system with cw (continuous wave) laser radiation (steady-state mode), the number of transitions per unit time from level 1 to level 2 (absorption) is equal to the number of transitions from level 2 to level 1 (stimulated and spontaneous emission):

$$\sigma(N_1 - N_2)I/h\nu = N_2/\tau$$

where σ_0 is the cross-section of the radiation (stimulated) transition, N_1 and N_2 are the number densities of particles at the levels 1 and 2, respectively, I is the intensity of laser radiation, $h\nu$ is the energy of the resonance photon, τ is the radiative decay time of upper state 2 (assuming that nonradiative transitions due to collisions are negligible). Both N_1 and N_2 are dependent on laser intensity, so that

$$N_2 = \frac{1}{2} N \frac{\frac{I}{I_{sat}}}{1 + \frac{I}{I_{sat}}}, \quad N_1 = N - N_2$$

where N is the total number density of particles, $I_{sat} = h\nu/2\sigma_0\tau$ is the saturation intensity, defined as the intensity for which the number density of the excited particles N_2 reaches one half of its maximum value $1/2\,N$.

If the intensity of laser radiation is much lower than I_{sat}, the absorption energy and the number density of the excited particles N_2 are directly proportional to the light intensity, and hence the analytical signal for different techniques (absorption, optocalorimetric, fluorescence, etc.) has a linear dependence on the intensity of the radiation. This linear regime is typical for conventional radiation sources used in elemental analysis: hollow cathode lamps (HCLs), electrodeless discharge lamps (EDLs) and for low intensity lasers.

In the opposite case, when the intensity of the resonance laser radiation is high enough ($I \gg I_{sat}$), the absorbed laser energy reaches its maximal value and the population of the upper level 2 approaches $1/2\,N$. The analytical signal for the absorption technique

$$\Delta I \approx \sigma_0(N_1 - N_2)lI = \frac{\sigma_0 N l I}{1 + I/I_{sat}}$$

(where l is the absorption pathlength) tends towards a constant value. Because the shot noise of radiation is directly proportional to $\sqrt{I}$, then optimum detection of the absorbing particles is achieved at $I = I_{sat}$.

If deep saturation is achieved ($I \gg I_{sat}$), two very important advantages of different laser analytical techniques (fluorescence, optocalorimetric, ionization) are gained: Maximum analytical signal and minimal influence of intensity fluctuations of the excitation source on the precision of the analysis. The ultimate sensitivity down to single atom/molecule detection has been demonstrated under saturation conditions in many analytical applications with lasers.

In real analytes the scheme of the energy levels is rather complicated. Thus, there may be many channels of radiative and collisional transfer of the absorbed laser energy. The complex structure of the energy levels scheme offers different combinations of excitation and detection of a particular species, but nevertheless the simplest scheme described above gives correct qualitative presentation of the analytical signal behavior.

Details on the optimization of laser parameters for different analytical techniques can be found in [1], [2], and [3].

22.2. Tunable Lasers

In most analytical applications, tunable lasers are used. Tunability allows selective excitation of a particular analyte in a complex mixture and relatively simple "switching" from one analyte to another. For conventional sources with fixed spectral lines it is often necessary to use different lamps for different analytes.

For a few laser analytical techniques (molecular fluorimetry of liquids, spontaneous Raman scattering), fixed wavelength lasers are used. The following discussion is limited to the two most widely used types of tunable laser.

22.2.1. Dye Lasers [4]

For a long time the dye lasers were the most popular type for atomic and molecular analysis. The active medium of these lasers is an organic dye in an appropriate solvent. About 20 different dyes cover the spectral range from ca. 320 to ca. 900 nm.

Typically, dye lasers are pumped by an intense fixed-frequency laser, the most often used being nitrogen (337 nm), argon ion (488.0, 514.5 nm), XeCl excimer (308 nm), and frequency-doubled and frequency-tripled Nd-YAG (530, 353.3 nm).

Dye molecules have broad emission bands. The optical resonator of a dye laser includes a spectrally selective element (grating, prism, interferometer, Lyot filter, etc.) which narrows the dye laser spectral line. By appropriate adjustment (e.g., rotation of the grating) one can smoothly tune the laser wavelength within the luminescence band of a particular dye. The tuning range may be 30–40 nm. At the extremes of the tuning range the energy of a dye laser falls off.

The organic dyes used for lasers can be classified on the basis of molecular structure: terphenyls (spectral range: 312–400 nm); coumarines (420–580 nm); rhodamines (570–720 nm); oxazines (680–860 nm). Each class consists of several dyes with slightly modified chemical structure and, hence, slightly shifted absorption and luminescence bands. There are special dyes for UV and near IR.

The most utilized solvents are ethanol, methanol, cyclohexane, ethylene glycol, and dioxane. For different solvents, the tuning curves of the same dye may be shifted by 10–15 nm.

The dyes for the UV spectral range (below 380–400 nm) are not photochemically stable. In most analytical applications, intense UV radiation below 360 nm is obtained by nonlinear optics, i.e., the generation of harmonics and sum frequencies of the visible dye laser radiation in nonlinear crystals. The most widely employed crystals are KDP, ADP, LiF, $LiIO_3$, and BBO. For example, the second harmonic of a rhodamine dye laser (fundamental band near 640 nm) is a more practical way to obtain tunable UV radiation near 320 nm rather than direct lasing of an unstable UV dye laser.

The efficiency of energy conversion of the pumping laser radiation in commercially available lasers varies from a few percent to ca. 20 %, depending on the type of dye and the spectral range.

22.2.2. Semiconductor Diode Lasers [5]

In the mid-1960s, tunable semiconductor diode lasers (DLs) were introduced into spectroscopic and analytical research. In conventional DLs the laser action can occur when a forward bias current is applied to a $p-n$ junction. Photons are generated in the process of electron–hole recombination. The wavelength of laser radiation is determined by the energy gap between conduction and valence bands, which, in turn, is determined by the chemical composition of the semiconductor. Wavelength tuning is possible by varying the injection current and by the temperature of the laser.

At first, DLs based on the IV–VI (Pb-salt) materials were used for analytical applications. These lasers operate in the 3–30 μm region, where the fundamental absorption bands of most molecules lie. Linewidths as small as 2×10^{-6} cm^{-1} were observed for the DLs, but in the commercial systems used for analytical applications the DL linewidth is determined by the quality of temperature stabilization. Usually the temperature is stabilized at a level of 10^{-2}–10^{-3} K, which corresponds to a linewidth of $\sim 10^{-3}$ cm^{-1}. This width is much smaller than the widths of molecular absorption lines at atmospheric pressure and even smaller than Doppler broadening for light molecules. The IV–VI DLs operate at cryogenic temperatures. In laboratory high-resolution spectrometers operating temperatures of 10–120 K are usually provided by closed-cycle Stirling microrefrigerators. Now lasers operating above 77 K are available at most wavelengths, and in analytical instruments temperature-controlled liquid-N$_2$ Dewars are the cooling method of choice. Typical emitting powers are 100–300 μW, and cw lasers with powers of 1 mW can also be fabricated.

Recently, the characteristics of DLs operating in the visible/near-IR (NIR) region have been greatly improved. High quality lasers, made from the III-V group of semiconductor materials, are now commercially available in a number of discrete spectral ranges between 630 nm and 2.0 μm. In principle, lasers can be made at any wavelength, but because of the market demand the DL are commercialized mainly for specific spectral ranges. These lasers can work in cw mode at temperatures accessible with thermoelectric coolers, generally −40 to 60 °C. The typical linewidth of a free-running DL is 20–100 MHz. Commercially available edge-emitting Fabry–Perot (FP) DLs yield singe-mode radiation up to 150 mW in the 670–870 nm spectral range and up to 30 mW in the 630–670 nm range. DLs, working in the 630–780 nm region, have been shown to be very attractive for atomic absorption analyzers.

Compared with other lasers, the output radiation from DLs is highly divergent. The emission angles are typically 10° and 30° parallel and perpendicular to the laser junction, respectively, and may vary from sample to sample. These beam features can be corrected with appropriate collimating and anamorphic beam-conditioning optics.

The drawbacks of the FP-DLs are the frequency drift and mode jumping, therefore other types of lasers such as distributed feedback (DFB), distributed Bragg reflector (DBR), external cavity (ECDL), and vertical cavity surface emitting lasers (VCSELs) have been developed. In a typical DFB laser, an index grating is fabricated directly on a chip alongside the active layer, providing distributed feedback only for a selected wavelength specified by the grating. A DBR laser uses the same principle, but the grating is placed beyond the active region. DFB/DBR lasers provide single-mode operation, can be continuously tunable by current and temperature, and exhibit no mode hops. The drawbacks are limited tuning range and higher cost compared with simple FP lasers.

Wavelengths longer than 1 μm are attractive for molecular analysis. Commercially available DLs for this region include: InGaAsP/InP (1.3, 1.55 μm FP structures, 1.2–1.8 μm DFB and EC lasers). For wavelengths longer than 1.8 μm, DLs are made from compounds containing antimonides such as AlGaAsSb, InGaAsSb, and InAsSb. Lasing in the 2–2.4 μm region at room temperature is available, and the spectral range up to 3.7 μm at liquid N$_2$ is achieved.

Vertical cavity surface emitting lasers (VCSELs), in contrast to conventional FP lasers, emit from their top surface. A conventional VCSEL consists of a multi-quantum-well (MQW) active region placed in a cavity between two DBR mirrors. These mirrors control the wavelength and beam divergence. For typical VCSELs, the far field beam characteristic will be a near Gaussian shape, with a beam divergence less than 12°. The narrow beam divergence, circular symmetry, and lack of astigmatism greatly simplify the optical design with these type of DLs. Currently, VCSELs operating in cw mode at room temperature in the visible/NIR range up to 2.3 μm can be produced.

An alternative approach for tunable single-mode operation is to build an external cavity. In this case, a FP laser is placed into an optical cavity, and the wavelength is tuned by using a dispersive element, such as a grating. The wavelength can be tuned continuously over many wavenumbers (20–50 cm^{-1}) by rotating the grating. The linewidth depends on the construction of the external cavity and its stability and is usually less than 1 MHz.

To obtain generations in the regions of mode-hops for conventional FP lasers a simple design with a small external cavity was proposed. A glass plate mounted in front of the diode chip acts as the low-reflectivity mirror of an extended resonator and helps to select particular longitudinal modes within gain profile, even where laser action is not possible without the feedback.

Recently developed multi-electrode lasers are very promising devices for multielement sensors. These lasers separate the wavelength selection, gain, and phase control regions of the laser cavity and are available in a number of different configurations. A tuning range over 100 nm around 1.5 µm has been demonstrated. These tuning ranges are equivalent to EC devices, thereby simplifying present multi-laser sensor concepts by allowing comparable wavelength coverage with a single laser.

Frequently conversion is used to extend the available spectral range of DLs. Second harmonic generation (SHG) in bulk non-linear crystals, such as $LiIO_3$ and $KNbO_3$, can provide up to 0.1 µW of radiation in the 335–410 nm range and up to 1–3 µW in the 410–430 nm range. The generation of sum frequencies (SFG) in non-linear crystals can open the way to tunable radiation in the UV/visible part of the spectrum. Over the past several years, progress in the development of quasi-phase matching structures (QPMS) and waveguides, based on non-linear crystals, has been achieved. Conversion efficiencies in both types of such non-linear converters are much higher than in bulk crystals, thus single-mode powers up to some mW in the UV/blue range can be produced. Mid-IR radiation can be also produced by difference-frequency mixing of two NIR lasers. Difference-frequency generation in periodically poled $LiNbO_3$ and bulk $AgGaSe_2$ or $AgGaS_2$ has been demonstrated for generating tunable radiation between 3.2 and 8.7 µm.

In the mid-IR, very promising substitutes to lead-salt lasers are quantum-cascade lasers (QCLs), made using III-V compounds. In QCL the active region consists of several gain quantum well (QW) layers separated by the digitally graded electron-injecting regions. Electrons are injected into the high energy level of the first gain layer, fall down to the lower level with photon emission, then through the digitally graded region are injected into the high level of the next gain layer and the process repeats. Quantum efficiencies higher than 100 % are possible in such structures. It is important that such cascades can be organized between quantized states in the conduction band only. The close spacing of these quantized states allows light to be emitted in the mid-IR range. Currently, room temperature lasing up to 8.2 µm in structures fabricated from AlInAs/InGaAs can be obtained. QCLs with wavelength as long as 11 µm at cryogenic temperatures have been demonstrated.

In 1999 blue DLs became commercially available. InGaN MQW DLs were demonstrated to have a lifetime of more than 10 000 h under conditions of room temperature cw operation. Single-mode emissions were produced in the region of 390–450 nm. Maximum output power is about 30 mW. With an external cavity the narrowband 1 MHz laser linewidth can be coarsely adjusted over a tuning range of 2.5 nm and then fine-tuned by a piezo stack over a more than 20 GHz range without mode hops.

22.3. Laser Techniques for Elemental Analysis

In this chapter we shall discuss laser analytical techniques for elemental analysis, i.e., for detection of impurity atoms of a particular element in a sample.

22.3.1. Laser-Excited Atomic Fluorescence Spectrometry [2], [3], [6]

22.3.1.1. Physical Principles

The principles of the laser-excited atomic fluorescence (LEAF) technique are very simple. A liquid or solid sample is atomized in an appropriate device. The atomic vapor is illuminated by laser radiation tuned to a strong resonance transition of an analyte atom. The excited analyte atoms spontaneously radiate fluorescence photons and a recording system registers the intensity of fluorescence (or total number of fluorescent photons). The extremely high spectral brightness of lasers makes it possible to saturate a resonance transition of an analyte atom. Therefore, the maximum fluorescence intensity of the free analyte atoms can be achieved while the effect of intensity fluctuations of the excitation source are minimized. Both factors provide the main advantage of LEAF—extremely high sensitivity. The best absolute detection limits achieved in direct analysis by LEAF

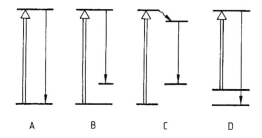

Figure 2. Classification of LEAF schemes
A) Resonance fluorescence; B) Direct-line fluorescence; C) Collisionally assisted direct-line fluorescence; D) Anti-Stokes direct-line fluorescence

lie in the femtogram-range, which corresponds to concentrations down to pg/mL.

Several different excitation–recording schemes used in LEAF are shown in Figure 2. The choice of a particular scheme is dictated by the structure of the atomic energy levels, the sources of background and noise, and spectral interference with the major sample components.

The resonance scheme (Fig. 2 A) has the great advantage of a possible cyclic interaction of analyte atoms with the laser radiation and, hence, a high yield of fluorescence photons. The highest sensitivity of LEAF, the detection of a single free atom, has been demonstrated with Na atomic beams and cw laser excitation. This scheme has a serious disadvantage — a high level of scattered laser light and, hence, high background. For a few elements (Na, Cd, etc.) the resonance scheme is the only possible one. For the most elements this scheme is not optimal in real analytical conditions when pulsed dye lasers are used for excitation.

The Stokes direct-line or collisionally assisted Stokes direct-line schemes (Fig. 2 B, C) are often employed (Tl, Pb, Au, Ag, Co, etc.). Their main advantage is that different wavelengths are used for excitation and fluorescence. Because of this, the laser scattered light can be easily separated from the weak fluorescence radiation by cut-off optical filters or monochromators. The analytical potential of collisionally assisted detection schemes depends strongly on the efficiency of collisional mixing between the neighboring levels. Typically this efficiency is high enough for energy gaps between the levels of ca. 1000 cm^{-1}. For some elements with low-lying excited states (Ga-type) the anti-Stokes direct-line scheme (Fig. 2 D) is very useful. An advantage of this scheme is a blue shift of the fluorescence and an appropriate reduction in background, caused by scattered laser light. In some cases, double-resonance excitation by two lasers with different wavelengths is used.

The possibility to select the optimal excitation–detection scheme often simplifies the problem of spectral interference with the major components, providing the second advantage of LEAF—high selectivity.

22.3.1.2. Atomizers

Graphite electrothermal atomizers (GETA), flames, glow discharges, and laser plumes are the most widely used types of atomizers in modern LEAF analytical practice.

GETA. Graphite electrothermal atomizers are of two types: open and closed. The main unit is a graphite crucible (cup, profiled rod, boat, furnace) held between massive water-cooled electrodes. Typical sample volumes are 2–100 µL. The maximum temperature of a crucible depends on its mass, the electrical power supply, and the type of graphite used. In most applications, it is 2700–2900 °C. In open and closed atomizers an inert gas atmosphere is used to minimize quenching of the excited analyte atoms and sublimation of the hot graphite crucible. Graphite furnaces have proved preferable because of the high temperature of the analytical zone and, hence, less chemical interference. Low-pressure noble gas in a closed GETA has been used to minimize matrix interference.

Flames. The flames used in LEAF are very similar to those used in AAS. In most applications, cylindrically symmetric burners are used since the dimensions of the analytical zone in LEAF do not exceed 10 mm. Quenching of the excited atoms by components of the flame is a serious problem. As oxygen is one of the most effective quenching agents, flames with minimum oxygen concentration should be used (e.g., an oxygen–argon–hydrogen flame). The high temperature of the analytical zone increases not only the fluorescence signal, but also the background, caused by flame emission. The optimal position of the analytical zone on the flame axis maximizes the signal-to-noise ratio. The background radiaton is usually much higher in flames as compared with GETA because of fluctuations in the optical density of the analytical zone (Rayleigh scattering of the excited laser radiation), luminescence of organic compounds in the flame, Mie scattering by dust particles, etc. As a result detection limits are typically much worse for flame-LEAF than for GETA-LEAF. On the other hand, the relative simplicity of design and the stationary analytical regime are definite advantages of flames.

Glow Discharges. Glow discharges (GD) have proved very useful atomizers for direct analysis of solid samples by LEAF. In GD the surface of a solid sample is bombarded by ions of a noble gas, which is continuously pumped through the atomizer at low pressure. Depending on the type of GD, the gas pressure ranges from 10^{-1} Pa to 10 kPa. At such pressures collisions between analyte or matrix atoms and atoms in the gas phase are relatively scarce, which substantially reduces the probability of gas phase reactions and fluorescence quenching. The sample to be analyzed is used as the cathode. The anode–cathode voltage is typically several hundred volts, the energy range of the bombarding ions is $10^{-17} - 10^{-15}$ J (0.1–10 keV). Both direct current and pulsed discharges are used. In the latter case the discharge is synchronized with the laser pulse, with an appropriate time delay, depending on the distance between the sample surface and the analytical zone and the mean energy of the sputtered atoms.

The process has substantial advantages over thermal atomization. The sputtering yields of different elements differ by no more than one order of magnitude, whereas their vapor pressures may differ by several orders of magnitude. Sputtering yields are much less sensitive to the chemical composition of the matrix as compared to thermal atomization efficiencies. Ion sputtering avoids the drawbacks of GETA (e.g., chemical interference, recombination of free atoms into molecules and clusters, carbide formation). The problem of surface contamination can be easily solved by preliminary cleaning in a discharge. After the surface layer is removed, sample sputtering stabilizes and the bulk composition of a homogeneous sample or the layer-by-layer composition of an inhomogeneous one can be analyzed.

22.3.1.3. Limits of Detection

Like most spectroscopic analytical techniques, LEAF needs a calibration procedure. In common with other modern analytical techniques, determination of limits of detection (LOD) poses a new problem. Commonly, LOD are defined as the concentration (or absolute amount) of analyte that gives an analytical signal three times the standard deviation of the blank. The analyte concentration in the purest solvent used for blank measurements can often be detected by modern analytical techniques. One should differentiate the LOD determined by the limited purity of the solvent from the limit determined by the intrinsic characteristics of the analytical technique (nature of background, electronic noise, recording sensitivity, etc.). The first LOD is measured with the laser tuned to the resonance line of the analyte (on-line LOD). The second is measured with the laser detuned (off-line LOD). The detection power of most laser analytical techniques is very high and most published LOD are off-line values.

The best LEAF LODs currently available are listed in Table 1 [7].

22.3.1.4. Real Sample Analysis

Trace amounts of Pb, Cd, and Bi at femtogram levels have been directly determined in ancient ice and recent snows in the Antarctic by LEAF–GETA. Toxic metal concentrations in deep Antarctic ice cores give unique information about the preindustrial tropospheric cycles of these metals and about recent anthropogenic sources of heavy metals. Concentrations of Pb and Cd in the samples analyzed were 0.3–30 pg/mL and 0.08–2 pg/mL, respectively. The values measured by LEAF–GETA were in good agreement with those measured by isotope dilution mass spectrometry (Pb) and graphite furnace AAS (Cd). Both traditional techniques required elaborate chemical pretreatment and large sample volumes (50–200 mL). The sample volume in LEAF experiments was 20–50 µL.

The most successful analysis of real samples by LEAF was done with sample atomization in a graphite furnace. The high detection power (absolute LODs at and below femtogram levels) and wide linear range (up to 5–7 orders of magnitude) of LEAF were combined with modern furnace technology:Fast heating,platform atomization, matrix modification, and direct solid sampling. As a result low LODs, reasonable accuracy and precision, less matrix interference, and simpler sample pretreatment (if necessary) were achieved (Table 2).

The potential of LEAF with glow discharge atomizers (LEAF-GD) have been demonstrated. The LODs for Pb and Ir as low as 0.1 and 6 ng/mL, respectively, were achieved with hollow cathode GD and aqueous standards. With a similar atomizer 40 ng/g LOD was obtained for direct analysis of Pb traces in Cu-based alloys. With

Table 1. Best detection limits for LEAF-GETA [7]

Element	λ_{11}, nm	λ_{12}*, nm	λ_{fl}, nm	Limit of detection fg	Limit of detection pg/mL**
Ag	328.1		338.3	8.0	0.4
Al	308.2		394	100	5.0
Au	242.8		312.3	10	0.5
Bi	306.8		472.2	800	40
Cd	228.8	643.8	361	18	0.9
	228.8		228.8	3.5	0.175
Co	308.4		345.4	4.0	0.2
Cu	324.7		510.5	60	3.0
Eu	287.9		536.1	3.6×10^5	1.8×10^4
Fe	296.7		373.5	70	3.5
Ga	403.3	641.4	250.0	1.0	0.05
In	410		451	20	1.0
Ir	295.1		322.1	6×10^3	300
Li	670.8		670.8	10	0.5
Mn	279.5		279.5	90	4.5
Mo	313.3		317.0	10^5	5×10^3
Na	589.6		589.6	30	1.5
Ni	322.2		361.9	10^3	50
Pb	283.3		405.7	0.2	0.01
Pt	265.9		270.2	10^3	50
Rb	780		780	20	1.0
Sb	212.7		259.8	10	0.5
Sn	286.3		317.5	30	1.5
Te	214.3		238	20	1.0
Tl	276.8		353	0.1	0.005
V	264.8		354.4	2.2×10^6	1.1×10^5
Yb	398.8	666.7	246.4	220	11

* λ_{12} is the wavelength of the second laser in a two-step excitation scheme.
** The pg/mL values are calculated assuming a sample volume of 20 μL.

the more efficient magnetron GD even lower LODs were achieved: 3 ng/g for Pb in Cu alloys, 0.8 ng/g and 2 ng/g for Si in In and Ga, respectively.

Analysis of pure gases by LEAF is much less developed. Nevertheless, some very impressive results have been demonstrated. Traces of Pb and Tl in the air of a normal laboratory and a class-100 clean room were detected by LEAF. Air from the room was pumped through a tantalum jet, fixed in a hole inside a graphite tube. The heavy particles impacted onto the tube wall opposite the jet. After pumping a certain volume of ambient air in a particular time interval the impacted metal traces were detected by graphite furnace LEAF. LODs of 0.1 pg/m^3 and 0.01 pg/m^3 for Pb and Tl, respectively, were achieved. The Pb LOD is two orders of magnitude better than the LOD for graphite furnace AAS. The measured concentrations of Pb and Tl in laboratory air were 1.25 ng/m^3 and 4.3 pg/m^3, respectively. The same values in clean-room air were about 0.3–0.6 ng/m^3 and 0.9–1.0 pg/m^3. The measured concentrations of the same metals inside a clean bench, in the clean room were 6.5 pg/m^3 and 0.035 pg/m^3.

Detection of atoms and molecules which have transitions in the vacuum UV region is a serious challenge for any spectroscopic technique. One way to solve the problem is excitation of a species, not from the ground state, but from a higher metastable state which can be efficiently populated by electron impact in a gas discharge. The population of a metastable level by electron impact followed by LEAF detection was successfully applied for the detection of Ne and N$_2$ traces in He, and for detection of NO and NO$_2$ traces in He, Ar, N$_2$, and air. LODs of ca. 10^{-7}–10^{-8} vol % were obtained. These LODs are 100–1000 times better as compared with any traditional analytical technique.

22.3.2. Laser-Enhanced Ionization [3]

22.3.2.1. Physical Principles

Laser-enhanced ionization (LEI) spectroscopy is based on the difference in collisional ionization rates of atoms in their ground and excited states. If

Table 2. Real sample analysis by LEAF

Analyte	Matrix[a]	Certified value	Measured value	LEAF LOD (aqueous standard)	Comments
Co	ASRMs	ng/g	ng/g	fg	digested, diluted samples
	wheat grains (SBMP)	60 ± 20	45 ± 2	60	
	dried potato (SBMK)	100 ± 30	90 ± 5		
	dried grass (SBMT)	60 ± 20	48 ± 2		
	Sn alloys			pg	vacuum atomizer
	OSCh		4.4 ± 1.5	4	
	OVCh		2.5 ± 0.8		
Mn	ASRMs	µg/g	µg/g	fg	slurried samples
	citrus leaves (SRM 1572)	23.0 ± 2	23.4 ± 0.7	100	
	bovine liver (SRM 1577a)	9.9 ± 0.8	9.4 ± 0.5		
	milk powder (SRM 1549)	0.26 ± 0.06	0.29 ± 0.01		
Pb	Ni alloys	µg/g	µg/g	fg	direct solid sampling
	SRM 897	11.7 ± 0.8	12.6 ± 1.2	1	
	SRM 898	2.5 ± 0.6	2.2 ± 0.4		
	SRM 899	3.9 ± 0.1	3.7 ± 0.7		
	ASRMs				slurried samples
	citrus leaves (SRM 1572)	13.3 ± 2.4	14.1 ± 0.5		
Tl	ASRMs	ng/g[b]	ng/g	fg	dissolved samples
	pine needles (SRM 1575)	50	48 ± 2	3	
	tomato leaves (SRM 1573)	50	51 ± 4		
	bovine liver (SRM 1577a)	3 ± 0.3	3.2 ± 0.2		
	Ni alloys	µg/g	µg/g	fg	direct solid sampling
	SRM 897	0.51 ± 0.03	0.56 ± 0.07	10	
	SRM 898	2.75 ± 0.02	2.5 ± 0.5		
	SRM 899	0.252 ± 0.003	0.24 ± 0.03		
Te	Ni alloys	µg/g	µg/g	fg	direct solid sampling
	SRM 897	1.05 ± 0.07	0.97 ± 0.15	20	
	SRM 898	0.54 ± 0.02	0.52 ± 0.05		
	SRM 899	5.9 ± 0.6	5.4 ± 0.7		
Sb	Ni alloys	µg/g[c]	µg/g	fg	dissolved samples
	PW 1A	5	4.8 ± 0.5	10	
	PW 2A	9	8.3 ± 0.5		
	PW 3A	17	16.5 ± 0.8		
P	Ni alloys	µg/g	µg/g	pg	digested, diluted samples
	SRM 349	30 ± 10	23 ± 3	8	
	SRM 865	120 ± 30	100 ± 20		
	SRM 126c	40[c]	37 ± 5		
	ASRMs	mg/g	mg/g		
	SRM 1577a	11.1 ± 0.4	11.6 ± 0.9		
	SRM 1549	10.6 ± 0.2	10.9 ± 0.9		
	SRM 1575	1.2 ± 0.2	1.1 ± 0.2		

[a] ASRM = agricultural standard reference materials (Russian State Reference Materials). [b] Noncertified value. [c] Reference value.

a d.c. voltage is applied across the atom reservoir and analyte atoms are optically excited by laser radiation, the enhanced probability of collisional ionization can be detected as a change of current (or a change of voltage across the load resistor). The LEI signal is linearly proportional to the ionization rate of the excited analyte atom, which in turn depends on its ionization potential, the nature of its collision partners, and the temperature of the analytical zone. The maximum sensitivity of LEI occurs when both the probabilities of ionization and ion collection are maximal (close to unity). The ionization efficiency depends exponentially on the energy difference between the upper excited state and ionization potential of the analyte atom. The main advantage of LEI as compared with LEAF is the very high probability of ion collection (close to unity in the optimal scheme); the collection efficiency of spontaneously emitted fluorescence photons is much lower.

Different excitation schemes for LEI are shown in Figure 3. The simplest and most widely used is the one-step excitation scheme (Fig. 3A). Elements with low ionization potential (mainly alkali metals) can be efficiently ionized by the one-step scheme with dye lasers in the visible range. For a number of elements (Cd, Mg, Pb, etc.) the one-step scheme can be efficient with

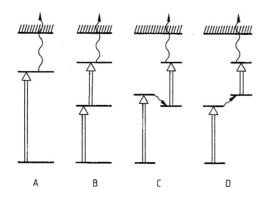

Figure 3. Classification of LEI schemes
A) One-step excitation; B) Two-step excitation; C, D) Collisionally assisted two-step excitation

excitation in the UV (harmonics of visible dye lasers). Excitation by UV photons below 300 nm may increase the background caused by nonselective one- or two-step photoionization of the thermally excited species in the analytical zone. The two-step excitation scheme (Fig. 3 B) is more complicated—thus more expensive—as it utilizes two tunable lasers, but it has significant advantages. Ionization efficiency increases up to three orders of magnitude compared with one-step excitation. The background does not change significantly if both lasers operate in the visible range. Selectivity increases dramatically as the selectivities of the single steps are multiplied in the two-step process. LEI is an efficient technique only if collision rates are high. Thus, the excitation–ionization scheme (Fig. 3 C), with noncoinciding levels of the first and second steps, might be efficient owing to effective collisional coupling of both levels.

22.3.2.2. Atomizers

The most commonly used atomizers for LEI are atmospheric flames. Flames have relatively high temperatures in the analytical zone and, hence, high efficiency of analyte atomization. Thermally excited molecular compounds in flames provide efficient ionization of the optically excited analyte atoms. All types of flames used in atomic absorption and emission spectrometry have been investigated in LEI. Relatively hot hydrocarbon flames such as C_2H_2/air or C_2H_2/N_2O are preferable for LEI because of the high degree of atomization of the elements. A large number of elements can be detected by LEI with C_2H_2/air flames; the hotter C_2H_2/N_2O flame is more suitable for the detection of refractory elements.

Few numerical values can be given for LEI in flames. In a flame at 2500 K, the rate of ionization of an element is enhanced ca. 100-fold for each electron volt (1.6×10^{-19} J) of laser excitation energy. If the energy gap between the excited level and the ionization continuum is less than 1 eV, the probability of ionization within 10 ns is close to unity. If cw laser excitation is used 100 % ionization can be achieved for an energy gap of ca. 3 eV (and less).

Other atomizers such as graphite furnaces, ICP, and GD have also been investigated for LEI. Argon buffer gas, typically used in graphite furnaces, is a much poorer collisional partner than the molecular constituents of flames, so collisional ionization is less efficient. On the other hand, good analytical results have been obtained with a combined flame–rod atomizer. A small volume of sample (positioned in a hole in a graphite rod) combined with the high temperature of the analytical zone (located above the rod in the surrounding flame) provided a low absolute detection limit in the picogram–femtogram range. ICP and GD proved to be electrically noisy for LEI.

22.3.2.3. Limits of Detection

The best LODs for one- and two-step LEI in flames are listed in Table 3.

22.3.2.4. Real Sample Analysis

The potential of LEI has been investigated by analysis of Standard Reference Materials (SRMs) at the National Institute of Standards and Technology (NIST). Traces of Co, Cu, Cd, Mn, Ni, and Pb were measured in different SRMs of water; Mn and Ni traces in bovine serum; Mn, Ni, and Fe traces in apple and peach leaves; Tl and Ni traces in river sediments; and Na, K, Ca, and Mg traces in rainwater.

The concentrations of the alkali metals Cs, Li, and Rb were detected by LEI in different types of reference rock samples. Rock samples were dissolved by standard methods and analyses performed without preconcentration, with pure aqueous standards for calibration. Concentrations were determined by both calibration curve and standard additions methods. The results obtained by the two techniques coincided within experimental errors. Concentrations varied in the range

Table 3. Best detection limits for one- and two-step flame LEI of aqueous solutions

Element	λ_1, nm	λ_2, nm	Flame*	Detection limit, ng/mL
Ag	328.1		AA	1
	328.1	421.1	AA	0.05
Al	309.3		AN	0.2
As	278.0		AA	3000
Au	267.7		AA	1.2
	242.8	479.3	AA	1
Ba	307.2		AA	0.2
Bi	227.7		AA	0.2
Ca	227.6		AA	0.006
	422.6	585.7	HA	0.03
Cd	228.8		AA	0.2
	228.8	466.2	AA	0.1
Co	240.8		AA	0.06
	252.1	591.7	AA	0.08
Cr	240.9		AA	0.2
	298.6	483.6	AA	0.3
Cs	455.5		AA	0.002
Cu	324.8		AA	3
	324.8	453.1	AA	0.07
Fe	302.1		AA	0.08
	271.9	468.1	AA	0.1
Ga	294.4		AA	0.01
In	271.0		AA	0.001
	451.1	571.0	AA	0.0009
K	766.5		AA	0.1
Li	670.8		AA	0.001
	670.8	610.4	AA	0.0002
Mg	285.2		AA	0.003
	285.2	470.3	AA	0.001
Mn	279.5		AA	0.04
	279.5	521.5	AA	0.02
Mo	319.4		AN	10
Na	285.3		AA	0.001
	589.0	568.8	AA	0.0006
Ni	229.0		AA	0.02
	282.1	501.4	AA	0.04
Pb	283.3		AA	0.2
	283.3	600.2	AA	0.09
Rb	780.0		HA	0.09
Sb	287.8		AA	50
Si	288.2		AN	40
Sn	284.0		AN	0.4
	284.0	597.0	HA	0.3
Sr	230.7		AA	0.003
	460.7	554.3	AA	0.2
Ti	320.0		AN	1
Tl	276.8		AA	0.006
	291.8	377.6	AA	0.008
V	318.5		AN	0.9
W	283.1		AA	300
Yb	267.2		AA	1.7
	555.6	581.2	AA	0.1
Zn	213.9	396.6	AA	1

* AA = acetylene – air; AN = acetylene – nitrous oxide; HA = hydrogen – air.

0.4 – 17.0 ppm and were in a good agreement with the certified values for the reference samples.

The LEI technique has been applied to the analysis of pure materials for the semiconductor industry. The concentrations of Cr, Fe, and Ni in Cr-doped GaAs disks were measured with the simplest pretreatment (dissolution in high-purity nitric acid and dilution with pure water). After dilution, a concentration of 0.1 ng/g in solution corresponded to 1 µg/g in the solid. Impurity concentrations were 4, 8, and 2.3 µg/g, respectively.

The rod–flame atomizer avoids the general disadvantage of flame-based analytical techniques—the necessity to dissolve the solid sample. The potential of two-step LEI with the rod–flame atomizer was investigated by analysis of Cu traces in Ge and In traces in CdHgTe. The LODs as low as 1 pg/g for In and 0.5 ng/g for Cu were achieved in direct analysis of 10-mg solid samples. Measured values in real samples varied in the range 30 – 70 ng/g for Cu and 2.6 – 3.4 ng/g for In.

Laser-enhanced ionization is a powerful, versatile and relatively simple technique which provides LODs in the ng/g – pg/g range with large linear dynamic range. A serious drawback of the technique is its susceptibility to interference: Thermal ionization of easily ionized matrix components (mainly alkali metals); photoionization of the thermally excited major components (especially if UV laser light is used for excitation); thermal ionization from levels excited by the spectral wing of an intense laser line; and ionization of the molecular components of the flame (especially NO). Interference may be greatly reduced or avoided by the appropriate choice of excitation wavelength, by the two-step excitation scheme, by specially constructed collecting electrodes; by optimal location of the analytical zone within the flame. To reduce interference one should investigate precisely the origins of noise and background for a particular analyte and matrix.

22.3.3. Resonance Ionization Spectroscopy [2]

22.3.3.1. Physical Principles

Resonance ionization spectroscopy (RIS) is based on resonance multistep excitation of high-lying levels of free analyte atoms and their subsequent ionization. Ionization may be caused by the photons of the final laser step (photoioniza-

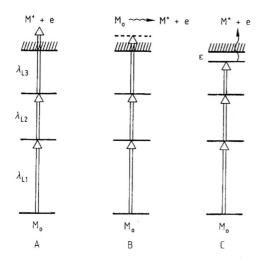

Figure 4. Classification of RIS schemes
A) Nonresonance photoionization to a continuum; B) Photoionization via an autoionization state; C) Electric field ionization of a Rydberg state

tion) or by an external electric field (field ionization). Sometimes the latter technique is referred to as field ionization laser spectroscopy (FILS). Multistep excitation of atoms may be effective if the rates of stimulated transitions to each intermediate level are much higher than the total decay rates (radiative and collisional) of any intermediate level. In contrast to LEI, RIS is efficient only if negligible collisions occur in the analytical zone. As in the case of LEI, the high sensitivity of RIS is based on the high efficiency of ion collection (close to unity). Thus, the overall efficiency of neutral atom detection by RIS depends on the efficiency of stepwise excitation and final ionization. Maximum sensitivity is achieved if all the intermediate transitions are saturated and the ionization efficiency is close to unity. In optimal conditions, a single free atom may be detected with a probability close to unity.

The three most widely used schemes for RIS are shown in Figure 4. They differ from each other in the final ionization step. Nonresonant photoionization to a continuum (Fig. 4 A) can be efficient if the energy flux of the ionizing laser is high enough (0.01 – 1 J/cm^2), as the nonresonant ionization cross sections are relatively low ($10^{-17} - 10^{-19}$ cm^2). For comparison, typical cross-sections of resonance optical transitions are $10^5 - 10^6$ times higher.

The photoionization scheme via the so-called autoionization state (AIS) is shown in Figure 4 B. The AISs are discrete states of the inner excited electrons with energies above the ionization potential of the atom. On the one hand, the cross section of a transition to AIS may be several orders of magnitude higher than direct photoionization to the continuum (Fig. 4 A); on the other hand, the AIS may decay to the continuum (i.e., the atom is ionized) with a typical lifetime of a few nanoseconds. Thus, in the field of an intense laser pulse with a duration of ca. 10^{-8} s, the probability of excited AIS depletion and ionization may be close to unity. Ionization via AIS is highly efficient for some elements (Gd, Lu).

Ionization through high-lying Rydberg states (Fig. 4 C) has been found to be very effective. Rydberg states close to the ionization limit can be efficiently ionized by an electric field pulse of moderate amplitude. The dependence of ionization yield of a particular Rydberg state on the electric field amplitude has a characteristic Z-profile with a pronounced threshold E_{crit}. For electric fields above E_{crit} the ionization yield is nearly constant and close to unity. Both the excitation cross section and E_{crit} depend strongly on the principal quantum number of the Rydberg state (n^*). For higher n^*, E_{crit} is lower (an advantage), but the excitation cross section is much lower (a disadvantage). Therefore, compromise values of n^* are used, which for most elements lie between 15 and 30. For such n^*, excitation cross sections are of the order $10^{-14} - 10^{-15}$ cm^2 and E_{crit} values are 1 – 10 kV/cm. For analytical applications, ionization via Rydberg states has proved to be an ideal scheme for RIS. Because of the relatively high excitation cross sections, maximum sensitivity can be realized with tunable lasers of moderate pulse duration and average power, and with reasonable electric fields. The effectiveness of AIS is limited by the absence of AIS near the ionization potential of many elements and the lack of appropriate spectroscopic information.

The fact that RIS measures charged particles is an important advantage. Highly efficient signal ion collection is possible by appropriate electromagnetic focusing. Second, discrimination of signal ions from background ions can be achieved by mass separation methods. These two factors improve both the sensitivity of RIS and its accuracy.

In most analytical RIS applications two- or three-step excitation is used. This provides extremely high selectivity (up to $10^{10} - 10^{12}$) as the overall selectivity of a multistep process is a product of one-step selectivities.

22.3.3.2. Atomizers

Vacuum or low-pressure atomizers are used in RIS to obtain collisionless interaction of the free analyte atoms with resonance laser radiation. Different types of GD and vacuum electrothermal atomizers (VETA) are most commonly used in RIS. In some specific analytical applications (detection of traces of rare radionuclides) RIS is used on-line with fast atomic beams.

In many applications of RIS–VETA the generated ions are detected by a secondary-emission multiplier incorporated in the atomizer. The vacuum should be better than 10^{-3} Pa throughout the analytical volume and the path of free ions from the interaction zone to the multiplier. This limits the speed of sample evaporation and dictates the efficiency of the pumping system. Both unified vacuum systems and systems with differential pumping of the analytical volume and free flight zones are used. In these respects the strategy of electrothermal atomization of a sample in RIS differs significantly from the traditional one used in graphite furnace AAS. In the latter technique fast atomization (1–10 s) gives the best analytical results. In RIS, very slow atomization (up to several minutes) is typical.

In some applications, a proportional counter operating near atmospheric pressure is used. In this case, the electrons created in the process of atomic resonance ionization are detected. By appropriate choice of gas mixture and accelerating voltage, an electron cascade with a magnification factor of about 10^4 can be achieved. This gain is high enough for single photoelectron detection.

For maximum sensitivity and isotope selectivity RIS is often combined with time-of-flight (RIS-TOF) or quadrupole mass spectrometry (RIMS).

22.3.3.3. Limits of Detection: Real Samples

The high sensitivity of RIS has been demonstrated in the analysis of pure semiconductor materials, biological samples, seawater, and environmental samples. In some cases, direct analysis was possible; for some matrices, sample pretreatment was used.

Traces of Na, Al, and B in high purity Ge were detected by RIS with direct electrothermal atomization of solid Ge chips in vacuum. For all these elements, two-step excitation of Rydberg states and electric field ionization were used. The measured concentrations of Na, Al, and B in real samples were 0.2, 1.0, and 0.2 ng/g while the LODs were 0.05, 0.01, and 0.05 ng/g, respectively. In the case of B, the strong ion background seriously limited the sensitivity and selectivity of traditional RIS. This background was caused by efficient two-step nonresonance photoionization of Ge atoms (the energy difference between the excited $2s^23s$ level of B and the $4s^24p5s$ level of Ge is only 20 cm^{-1}). Both analytical characteristics were greatly improved by a special technique, with two sequential electrical pulses of different duration, polarity, and voltage.

The distribution of ruthenium and rhodium traces in the ocean has been investigated by RIS–VETA. Both elements are rare and their distribution is of great geological and geochemical importance. The abundance of Ru and Rh in the Earth's crust is estimated to be 1–100 pg/g and can not be directly measured by current analytical techniques. The fire assay method and concentration by polymer sorbents were used prior to RIS analysis. Concentrations of Ru varied from 1.3 pg/mL in ocean water to 7.6 ng/g in Red Sea sediments. Concentrations of Rh in ocean water were 3–10 pg/mL.

The detection of ^{81}Kr, produced by cosmic rays, is of great importance for dating samples from the Earth's crust, groundwater, and polar ice cores. Measurement of ^{81}Kr in groundwater is a unique analytical problem as its concentration is ca. 1400 atoms per liter. The problem is further complicated by isotopic interference of stable krypton isotopes (^{82}Kr/^{81}Kr is about 10^{12}). Prior to final analysis by RIS–TOF [8] several consecutive steps of field sampling and isotope concentration were necessary. All steps required extreme care to avoid sample contamination. Water samples of 50–100 L were necessary to detect ^{81}Kr above the background. This value is more than 100 times less than for the routine technique of ^{39}Ar dating.

The environmentally important radionuclides ^{90}Sr and ^{210}Pb have been detected by RIMS–VETA. Two-step excitation of Rydberg states by two cw dye lasers, followed by nonresonant ionization by a cw infrared CO$_2$ laser, was used. The solid samples were dissolved by standard methods. The LODs were 60 ag (6×10^{-17} g) for ^{210}Pb and 30 ag for ^{90}Sr. Absolute ^{210}Pb concentrations of 0.12–0.14 fg were measured in brain tissue. Isotope selectivity of 10^9 for ^{90}Sr/^{89}Sr was achieved.

Recently the diode lasers were successfully used in a multiphoton RIMS scheme for isotopically selective detection of strontium. The two-step resonance excitation of ^{90}Sr was performed with two single-mode cw DLs ($\lambda_1 = 689.45$ nm, $\lambda_2 = 688.02$ nm). The excited ^{90}Sr atoms were photoionized by a powerful Ar$^+$ laser ($\lambda = 488.0$ nm). The ions were detected in a commercial quadrupole MS. Using graphite crucible atomization the absolute LOD of 0.8 fg and overall (optical + mass spectrometer) isotopic selectivity of $> 10^{10}$ in the presence of stable ^{88}Sr was achieved for ^{90}Sr. No other non-radiochemical technique can realize such sensitivity and selectivity of ^{90}Sr detection.

A similar scheme was used for the spectroscopic studies on hyperfine components and isotope shifts of rare stable isotopes of calcium and the radionuclide ^{41}Ca. Calcium atoms in an atomic beam were excited with single-frequency cw dye and titanium sapphire lasers and then photoionized with the 363.8 nm or 514.5 nm line of an Ar$^+$ laser.

The use of RIMS in chemical vapor deposition diagnostics has been investigated.

22.3.4. Diode Lasers in Elemental Analysis [6], [8]

22.3.4.1. Physical Principles of Absorption Spectrometry

Quantitative absorption measurements are based on the Lambert–Beer law:

$$P_t(\lambda) = P_0 \exp[-k(\lambda)l]$$

where P_t is the laser power at wavelength λ, transmitted through a layer with absorption coefficient $k(\lambda)$ and absorption pathlength l. The absorbed power ΔP is $P_0 - P_t = P_0[1 - \exp(-k(\lambda)l)]$. For small values of the absorption $k(\lambda)l \ll 1$, this is approximated with $\Delta P(\lambda) \approx P_0 k(\lambda) l$.

Theoretically, the minimum detectable absorption (MDA) is determined by the shot noise of detected radiation. In single pass measurements ($l = L$, where L is a sample absorption pathlength)

$$(kL)_{min} = \Delta P_{min}/P_0 = (2Bh\nu/\eta P_0)^{1/2}$$

where η is the quantum efficiency of the photodetector, and B is a detection bandwidth. To obtain noise equivalent relative power variations, $\Delta P_{min}/P_0$, smaller than 2×10^{-8} with a measurement time of 1 s, the laser photon flux should be at least 5×10^{15} photons/s. This corresponds to laser powers of 1 mW at 1 µm and 100 µW at 10 µm. In practice there are additional noise factors affecting the detection sensitivity, such as intensity fluctuations of the laser (laser excess noise) and technical fluctuations. Thus, for typical DLs, the value of $\Delta P_{min}/P_0 = 1 \times 10^{-8}$ can be considered as the lowest, fundamentally limiting one.

Usually, absorption measurements are not considered to be very sensitive since the signal $P_0 k(\lambda) l$ is a subtle variation of a large baseline P_0. In order to have zero baseline, sample or wavelength/frequency modulation (WM/FM) techniques can be used. In the case when λ is modulated with frequency f, a signal at the n-th harmonics of f is simply proportional to $P_0 k_{nf}(\lambda) l$, where k_{nf} is the amplitude of the corresponding harmonics of k. As $k(\lambda) = \sigma(\lambda) N$, where $\sigma(\lambda)$ is the absorption cross section, $\sigma(\lambda) = \sigma_0(\lambda)/(1 + I/I_{sat})$ (see Section 22.1.3), then the signal after the phase-sensitive detector is directly proportional to the concentration of absorbing particles N. Another important advantage of WM/FM is the detection of a signal at high frequency (> 10 kHz), where the low-frequency flicker noise is greatly reduced.

Besides reducing noise, improvements in the MDA can be achieved by enhancing the signal. This can be done by increasing the absorption pathlength using multipass cells. By an appropriate mirror arrangement a DL beam can be directed many times through the absorbing layer, increasing the overall absorption with an appropriate decrease in MDA. The number of passes is limited by the quality of the mirrors, which determines the cell transmission, or by interference effects.

22.3.4.2. Atomic Absorption Spectrometry with Diode Lasers [6], [8]

UV/visible/NIR semiconductor diode lasers (DLs) as sources of tunable narrowband resonance radiation have been discussed in Section 22.2.2. The practical and technical advantages of visible/NIR DLs are ease of operation, small size, room-temperature operation, and low price. In comparison with hollow cathode lamps (HCLs), they have such advantages as high spectral power, spatial coherence, and tunability. Owing to these features, detection limits, dynamic range, and selectivity of analysis are all significantly improved.

The presently available powers of DLs are several orders of magnitude higher than the powers of the best commercial HCLs. In contrast to an HCL, a DL emits a prominent single narrow line. This feature dramatically simplifies the construction of an analytical device because one does not need a monochromator, which is used in commercial AAS instruments for isolation of the analytical line from the spectral lines of the HCL buffer gas and unwanted lines of the cathode material. In the case when optimal experimental conditions are realized, more than 1 mW of laser radiation power can be obtained at the photodetector. It has been shown that a near fundamental shot-noise limit can be achieved and relative power variations $\Delta P_{min}/P_0$, which can be measured with DLs, can be as low as 10^{-8}.

The LODs of diode laser atomic absorption spectrometry (DLAAS) have been greatly improved using a WM technique. In conventional atomic absorption spectrometry with HCLs, low-frequency (10–60 Hz) amplitude modulation (mainly by mechanical choppers) is used. In contrast, the wavelength of the DL can be easily modulated with much higher frequencies, in DLAAS usually up to 10–100 kHz. By recording the analytical signal at the harmonics of the modulation frequency, usually the second one (2f-mode), one can dramatically suppress the 1/f noise, thus decreasing the LOD. Even in the first simple measurements with a DL power of ca. 100 µW, a modulation frequency of 5 kHz, and with 2f-mode of recording, absolute LODs in the femtogram range have been achieved experimentally. These LODs correspond to measured values of $\Delta P_{min}/P_0 \approx 10^{-4}-10^{-5}$ (compared with $10^{-2}-10^{-3}$ for conventional AAS).

Further improvement of WM-DLAAS has been demonstrated for the case when both the radiation source and the absorption are modulated. Such a situation was realized with a microwave induced plasma (MIP) or a direct current plasma (DCP) as the atom reservoir. The double-beam double-mudulation approach with logarithmic signal processing (LSP) and 2f-mode recording permitted a complete elimination of the background, caused by non-selective absorption, low-frequency fluctuations, and drift of the baseline and laser intensity. A shot-noise limited MDA as low as 3×10^{-8} for a 1 s time constant was experimentally realized. The linear dynamic range can be also substantially extended with this technique, in the best cases up to six–seven orders of magnitude as compared with less than two orders of magnitude for conventional AAS.

Applications of DLs in analytical atomic spectrometry are collected in the following list (numbers in parentheses are analytical wavelengths in nanometers):

Diode laser atomic absorption spectrometry (DLAAS):
  Graphite furnace
    Al (396.15), Ba (791.13), Cs (894.35), K (769.90), La (657.85), Li (670.78), Rb (780.03), Sr (689.26)
  Flame
    Cs (425.44, 852.11), Cr (425.44), K (766.49), Li (670.78), Rb (780.03), Ti (399.97, 842.65)
  Discharge (direct current plasma (DCP) or microwave induced plasma (MIP))
    Cl (837.60), F (739.87), O (777.18)
Laser induced fluorescence (LIF):
  Li (670.78), Rb (780.03)
Isotope selective detection:
  Li, Pb, wavelength modulation diode laser atomic absorption spectrometry (WMDLAAS)
  $^{235}U/^{238}U$ in glow discharge, optogalvanic detection
  $^{235}U/^{238}U$ in solid samples, laser ablation + LIF with DL excitation
Physical vapor deposition process control:
  Y (668), external cavity diode laser (ECDL), high frequency modulation spectroscopy
  Al (394), second harmonic generation (SHG) of ECDL
Control of semiconductor films growth:
  Al (394), Ga (403), In (410), SHG in bulk crystals
Specific detection of haloform components:
  Cl (837.60), Br (827.24), MIP or DCP, GC-WMDLAAS, double-modulation
Environmental analysis:
  Cr^{III}/Cr^{VI} speciation in deionized and drinking water
    Cr (427.48), SHG in LiIo$_3$, flame, HPLC-WMDLAAS
  Cr^{VI} in tap water
    Cr (427.48), SHG in KNbO$_3$, flame, HPLC-WMDLAAS, double-beam, LSP
  Chlorinated hydrocarbons in oil from plastic material recycling
    Cl (837.60), MIP, GC-WMDLAAS
  Chlorophenols in plant extracts
    Cl (837.60), MIP, GC-WMDLAAS, double-beam, logarithmic signal processing (LSP)
  Speciation of methylcyclopentadienyl manganese carbonyl (MMT) in gasoline, human urine and tap water
    Mn (403.1), flame, SHG in LiIO$_3$, HPLC-WMDLAAS, 4f, LSP

Analysis of chlorine in polymers
  Cl (837.60), MIP, laser ablation + WMDLAAS, double-modulation, LSP

Resonance ionization mass spectrometry (RIMS):
  La (657.85), three-step RIMS, DL – first step
  Sr (689.45, 688.02), double-resonance excitation by two DLs

The detection capabilities for various species differ considerably depending on the type of atomizer, oscillator strength of a transition, detection technique used. Multielement analysis with DLs has also been investigated. Several laser beams may be collimated and directed simultaneously through an absorbing layer in a conventional atomizer, either by a system of mirrors, or by optical fibers. The monochromator–photomultiplier system can be replaced by a simple semiconductor photodiode. Appropriate modulation of a wavelength of a particular DL, followed by Fourier transform or lock-in amplification analysis of the photodiode signal, makes it possible to process analytical signals of individual analytes. A six-element analysis with DLs and a commercial graphite tube atomizer has been demonstrated.

22.3.4.3. Element Specific Detection by DLAAS

Diode laser based analytical devices can be used as element selective detectors for gas/liquid chromatography (GC/LC). The atomic constituents of different molecular species in a sample can be measured in an atomizer, such as a flame or low pressure plasmas, e.g., an MIP or DCP. The GC-MIP/DCP in combination with DLAAS enables the detection of non-metals, such as H, O, S, noble gases, and halogens. Most of these elements have metastable levels, which are efficiently populated in a plasma, and which have strong absorption transitions in the red–NIR spectral region.

Recently it has been possible to observe a definite trend in elemental analysis, from detection of the total content of a heavy metal in a sample to the detection of its different chemical compounds or oxidation states (speciation). This is because the essentiality or toxicity of the metals should not be attributed to an element itself but to element compounds with distinct biological, physical, and chemical properties. A compact and relatively cheap DL based instrument could be a reasonable alternative to the powerful but expensive ICP-MS, presently used for speciation. The potential of DLAAS as an element selective detector for speciation has been demonstrated by its coupling with high performance liquid chromatography (HPLC). This system has been used for speciation of chromium (III)/(VI) and organo-manganese compounds. An LOD of 30 pg/mL for Cr^{III} and Cr^{VI} was obtained by use of 3 µW power frequency-doubled laser light, a double-beam configuration, and logarithmic processing of the signals. In the case of chromium, the LOD of HPLC-DLAAS is several times lower, than the LOD for HPLC-ICP-MS.

22.3.4.4. Isotope Selective Detection with Diode Lasers

The narrow spectral line of a DL enables isotope selective analysis. For light and heavy elements (such as Li and U) the isotope shifts in spectral lines are often larger than the Doppler widths of the lines, in this case isotopically selective measurements are possible using simple Doppler-limited spectroscopy–DLAAS or laser induced fluorescence (LIF). For example, ^{235}U and ^{238}U ratios have been measured by Doppler-limited optogalvanic spectroscopy in a hollow cathode discharge. DLAAS and LIF techniques have been combined with laser ablation for the selective detection of uranium isotopes in solid samples. This approach can be fruitful for development of a compact analytical instrument for rapid monitoring of nuclear wastes.

22.3.4.5. Commercial System

The DLAAS devices are now commercially available from Atomica Instruments (Munich, Germany). The first system was prepared for the measurement of ultra-trace levels of Al in the semiconductor industry. The commercial laser diode module includes the DL with a heat sink for temperature tuning, the microoptics, and the non-linear crystal for SHG. The module has the size of a HCL. Since delivery, the module has successfully worked routinely under the conditions of an industrial analytical laboratory without any repair or maintenance. The semiconductor industry in particular is interested in the measurements of light elements, such as Al, K, Ca, Na, in ultra pure water or chemicals. For example, the module designed for the detection of K provides an LOD of 0.5 pg/mL (10 µL aliquot). Similar compact and sensitive devices with electrothermal atomizers, flames or micro-plasma are currently available for a number of elements, such as Mn, Ni, Pt, Pd, Ca, Cu, Fe, Ga, the alkali elements, the halogens, C, O, and H. In forthcoming years, one can also expect the commercialization of specific element selective DLAAS detectors for GC or LC.

22.4. Laser Techniques for Molecular Analysis [6], [8]

22.4.1. Molecular Absorption Spectroscopy with Diode Lasers

22.4.1.1. Detection of Molecular Gas Impurities with Lead-Salt Diode Lasers [2], [9]

The Pb-salt-based DLs (Section 22.2.2) efficiently generate tunable laser radiation in the region 3–30 µm, where the great majority of molecules have intense vibrational rotational absorption lines. As the Doppler-broadened absorption linewidths of molecules are of the order of $(1-3) \times 10^{-3}$ cm^{-1}, DLs provide the best selectivity for linear absorption techniques.

Both cw and pulsed (sweep) modes of DL generation have been widely tested in analytical applications. Many well-developed modulation techniques have been used with consequent improvement of analytical sensitivity. The best results were achieved with a high FM technique ($f \sim 100-1000$ MHz). The minimal shot-noise-limited relative power variations $\Delta P_{min}/P_0$ which can be experimentally detected with FM DLs are about 10^{-7}. Typical values of $\Delta P_{min}/P_0 \approx 10^{-5}$ (1 Hz detection bandwidth) are mainly determined by instrumental factors (in most cases by interference effects). For this minimal detectable value of $\Delta P_{min}/P_0$ and for molecular species at atmospheric pressure the corresponding concentration LODs are ca. 0.1–10 ppb m (0.1–10 parts per billion by volume oner an optical path of 1 m). Evidently, if 10 m absorption length could be realized, the minimal detectable concentration of absorbing molecules would be 10 times lower.

The most important analytical applications of mulecular absorption spectroscopy with mid-IR DLs are environmental monitoring and atmospheric studies [9]. Two different approaches are used: (1) measurements in sample cells at reduced pressure, when the absorption linewidths are of ca. $(1-3) \times 10^{-3}$ cm^{-1}, and (2) open-path measurements at atmospheric pressure, when the pressure broadening leads to ca. 10^{-1} cm^{-1} linewidths.

In a cell, optimal gas pressure and appropriate dilution of primer gas by a buffer can be used. A number of mulecules have been measured with lead-salt DLs: CO, NO, NO_2, N_2O, HNO_3, NH_3, SO_2, O_3, HCl, H_2CO, CH_4, C_2H_4, C_2H_6, CF_2Cl_2, etc. The effective absorption pathlength of 20–300 m was realized with multipass cells with an appropriate decrease of the LOD down to a record value of 10 ppt (5 min integration time). Several Pb-salt DL-based gas monitors have been used to perform *in situ* trace gas measurements in the troposphere and lower stratosphere from airborne or balloon-borne platforms. The systems were designed to measure CH_4, CO, N_2O, NO_2, N_2O, HCl and H_2O with LODs at sub-ppb concentration levels (3–30 s integration time). A typical system includes a liquid-N_2 Dewar for diode lasers and detectors, a multipass cell, and can simultaneously measure up to four gases.

Owing to pressure broadening, ground-based remote detection of trace gases with DLs was not as effective as measurements in sample cells. Long-path atmospheric gas analyzers have been exploited mainly to monitor carbon monoxide concentrations. Mirror retroreflectors have been used to reflect the DL probing beam back to the detector providing a round-trip absorption pathlength of up to 1 km. Sensitivities achieved are in the range from 0.06 to 40 ppm · m.

Besides atmospheric trace measurements, Pb-salt DLs have successfully been used for detection of impurities in the process control of semiconductor materials, analysis of high purity gases, sensing of toxic gases, explosive detection, isotope analysis of gases, and medical diagnosis of various diseases by analysis of expired air. The use of a DL provides high time resolution (0.1–1 µs), facilitating sensitive control of transient processes (e.g., intermediates in chemical reactions, discharges, molecular beams). So far the high price, complexity, and limited reliability of the Pb-salt DL spectrometers limit their applicability to the research laboratories when utmost sensitivity and selectivity have to be realized.

22.4.1.2. Near IR Gas Sensors

Recent advances in room-temperature NIR lasers has led to a new generation of gas sensors. A number of DL-based instruments have been developed and are now commercially available.

These sensors have rather simple constructions, may be very compact, have low cost, are reliable, and allow completely autonomous operation for a long time.

Peculiarities of developing NIR-DL sensors are defined by characteristics of molecular absorption in this spectral region. There exists a large quantity of absorption lines that arise from overtones and combination bands. Typical linestrengths of these transitions are two to three orders weaker than those for the fundamental bands in the mid-IR. The strongest absorptions in the 1.3–1.6 µm region are provided by the H_2O, HF, HCl, and C_2H_2 molecules. The LODs which are achieved for these molecules in pure gas can be at the ppt level. However, if the absorption of the matrix gas is larger than that of the detected impurity, the LOD will be determined by the interfering absorption of the matrix gas molecules.

There are a number of quite successful applications of NIR-DL sensors for monitoring of gas molecules, especially H_2O. A few types of hygrometers based on a 1.4 µm InGaAsP-laser are commercially available. A very small LOD of 65 ppt for water in air was achieved in laboratory test measurements with a hygrometer, designed to measure moisture in semiconductor feed gas. Another DL sensor, capable of measuring low levels of water vapor in the presence of methane, is designed to be attached to a sampling station where pipeline gas is extracted at near-ambient pressures (1–2 atm).

Flight versions of NIR-DL instruments for trace water vapor measurements have been demonstrated on research aircraft with a sensitivity of the order of 10 ppm. Examples of such systems flown on NASA aircraft include both open-path and closed-path configurations designed for fast-response measurements of H_2O from the Earth's surface to an altitude of 20 km.

Similar instruments developed for monitoring O_2, NH_3, HCl, HF, H_2S, CO, CO_2 are suited to industrial stack monitoring, process and environmental control, where high temperatures, continuous vibration, thermal cycling, or high-speed air flow are present. The automated sensors provide fast and accurate ppm-level measurements and are used for optimizing incineration, sewage and waste treatment, power generation, and so on. Fiber optics are often preferred for their simplicity and ability to place the instrument at distances hundreds of meters from the hazardous zone.

22.4.1.3. Advanced Developments

Rapid improvements of the DL characteristics make these devices very attractive for absorption spectrometry not only of gas species (atomic or molecular) but also of liquid and condensed-phase samples. Much attention is now focused on using NIR-DLs for molecular fluorescence spectrometry, Raman spectroscopy, thermal lens spectrometry, medical diagnostics, development of fiber sensors, and so on.

It is unlikely that the NIR-DLs will permit measurement of all molecular trace constituents in the atmosphere with sensitivities sufficient for environmental control. New, room temperature operated lasers have been tested for trace measurements in mid-IR region. Potentially the most significant technology is quantum-cascade lasers (QCLs, see Section 22.2.2). Spectroscopic characteristics and noise properties of these structures enable measurements of trace impurities with LODs achieved using cryogenic Pb-salt lasers.

New absorption methods, like intracavity spectroscopy, cavity-ring-down and cavity-enhanced spectroscopy, have demonstrated very high sensitivities in laboratory measurements with DLs. An ultrasensitive technique that combines external cavity enhancement and FM spectroscopy has been developed recently. This method, which has been called NICE-OHMS, or noise-immune cavity-enhanced optical heterodyne molecular spectroscopy, is based on frequency modulation of the laser at the cavity free-spectral-range frequency or its multiple. The MDA of 5×10^{-13} ($k_{min} = 1 \times 10^{-14}$ cm^{-1}) in the detection of narrow saturated absorption spectra of C_2H_2 and C_2HD has been reported, with a cavity finesse of 10^5. This result is one of the most sensitive absorption measurement ever reported.

22.4.2. Laser Optoacoustic Spectroscopy
[1], [2], [10]

The optoacoustic effect, based on partial nonradiative dissipation of the absorbed light energy, followed by generation of an acoustic wave, was discovered about a hundred years ago. The effect is very weak and, before the advent of lasers, it was applied mainly for spectroscopic investigations of gases. Lasers and improvements in acoustic vibration detection have led to the development of laser optoacoustic spectroscopy (LOAS). Several physical processes are involved in LOAS: (1)

absorption of laser radiation by one or more compounds in a medium; (2) radiationless conversion of a part of the absorbed energy into local heating of the sample; (3) formation of an acoustic wave due to a temperature gradient; and (4) detection of the acoustic vibrations.

The main advantages of LOAS are high sensitivity, applicability to samples of any state of aggregation, and relatively simple instrumentation. Its drawback is poor selectivity, due to spectral interference of the molecular compounds in a complex mixture. To improve selectivity, LOAS is often combined with a preselecting technique such as gas or liquid chromatography. This discussion is limited to the applications of LOAS for sensitive analysis of simple mixtures.

The LOAS instrument consists of the following main units: A tunable laser, a sample cell with an acoustic transducer (sometimes called a spectrophone or SP), an amplifier, and a recording system. The acoustic response of the medium can be stimulated by amplitude or frequency modulation of the laser light or by Stark or Zeeman modulation of the absorption line of the analyte. Both pulsed and cw lasers with mechanical or electro-optical amplitude modulation are often employed in LOAS analytical applications.

The key component of an LOAS instrument is the spectrophone. Its construction depends critically on the state of aggregation of the sample. The SP for gas samples consists of a gas cell with input and output windows for laser radiation and a microphone, fixed on an inside or outside wall of the cell. As in the case of MAS with SDLs, the sensitivity can be improved by the use of multipass SPs. Plane capacitor or electronic microphones are the most widely used. The sensitive element of these microphones is a thin (1–10 μm) elastic membrane made of Mylar, Teflon, or metallic foil. The membrane serves as one electrode of a dielectric capacitor (membranes made of dielectric materials have metallic coatings). Thus, the acoustic vibrations of a gas mixture in the cell can be directly converted to an electrical signal. Commercial microphones with sensitivities of 5–50 mV/Pa are utilized in routine applications of LOAS; specially constructed and optimized ones are used for ultrasensitive analyses.

Resonant and nonresonant SPs can be used in LOAS. In the former type, the acoustic resonances are formed inside the cell. Resonant SPs provide higher sensitivity, but they are less flexible, as the acoustic resonances depend on the physical dimensions of the cell, the type of gas mixture, its temperature, and pressure. The modulation frequency of cw laser radiation should be matched to the acoustic resonance frequency (typically of the order of several kilohertz). The dimensions of nonresonant cells are commonly 5–10 mm diameter and 5–20 cm length, while some resonant cells have a diameter up to 10 cm.

For the analysis of liquid and solid samples, more complicated SPs are used. For liquid samples, the simplest and most sensitive detection of acoustic waves is by direct contact of the microphone and a liquid. The acoustic vibrations of irradiated solid samples can be transduced to a microphone via a buffer gas or a "pure" liquid. Piezoelectric microphones are the most widely used type for solids and liquids.

Some important molecular impurities in pure gases have been successfully detected by LOAS. Typical gas pressures range from 1–100 kPa; sample volumes 0.5–3 cm^3; routine accuracy 5–15%; linear dynamic range up to 10^5. The most impressive LODs are: 0.1 ppb for SO_2 (multipass, resonant SP); 10 ppb for NO_2 (resonant SP); 1 ppm for HF (nonresonant SP, pulse laser, time resolution); 100 ppb for CH_4 (nonresonant SP); about 100 ppb for CO (resonant SP); 0.1 ppb for NO (nonresonant cell, six microphones); 10 ppb for SF_6 (resonant SP).

The best sensitivity of LOAS for liquid samples (up to 10^{-6} cm^{-1}) was achieved with pulsed lasers of ca. 1 mJ pulse energy. With modulated cw lasers of ca. 1 W average power, a minimum absorption of 10^{-5}–10^{-6} cm^{-1} was detected. A few examples of real analyses include the LODs for β-carotene, Se, and Cd in chloroform of 0.08, 15, and 0.02 ng/mL, respectively; the LODs for bacteriochlorophyll, chlorophyll b, and hematoporphyrin in ethanol of 1, 0.3, and 0.3 ng/mL, respectively; the LODs for cytochrome c, vitamin B_{12}, uranium(IV), and uranium(VI) in water of 30 ng/mL, 4 ng/mL, 8×10^{-7} mol/L and 10^{-6} mol/L, respectively. Typically, the LODs for direct analysis of liquids are worse as compared with "pure" gases because of higher background absorption of the solvent. To reduce this background, frequency modulation of lasers and differential two-beam detection schemes are used.

LOAS has been successfully applied for the detection of thin films and layers deposited on a solid substrate. Both direct measurement by a piezoelectric microphone connected to the substrate and detection of acoustic waves via an intermediate gas or liquid were used. For example, the absorption of thin Al_2O_3 layers (up to monolayer)

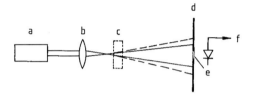

Figure 5. Schematic diagram of a single-beam thermal lens spectrometer
a) Laser; b) Lens; c) Sample; d) Screen; e) Pinhole; f) Detector
The dashed lines show the beam divergence due to the thermal lens effect

on an aluminum substrate was measured with a cw CO_2 laser, modulated by a mechanical chopper. Measurements of thin layers of liquids and powders as well as aerosols, suspensions, colloids, and small dispersed particles are important applications of LOAS.

22.4.3. Thermal Lens Spectroscopy [1], [3]

Thermal lens spectroscopy (TLS) is another laser-based technique, associated with dissipation of absorbed laser energy followed by the formation of temperature gradients in a medium. It was observed for the first time in 1964 when a 1-cm Brewster cell containing benzene was placed inside the resonator of a cw He–Ne laser. Contrary to the expected behavior, the laser intensity began to fluctuate and in a few seconds, when the steady state was reached, the beam diameter at the resonator mirrors had increased. The benzene cell acted as a spherical negative lens with a focal length of ca. 1 m.

The physical principles of TLS are rather simple. If a laser beam with a Gaussian-like intensity distribution propagates in a weakly absorbing medium the local temperature will increase and the transverse temperature profile will match the intensity profile of the laser beam. Since most liquids have a negative temperature coefficient of their refractive index ($dn/dT < 0$) the induced transverse profile of the refractive index acts as a negative (diverging) lens. The focal length of the lens depends on the parameters of the laser beam (power, diameter) and the characteristics of the medium (absorption coefficient, thermal conductivity, thermal diffusivity, dn/dT, cell length). Calculations show that in benzene a 0.01-W laser beam of 0.1 cm radius induces a negative thermal lens with focal length -1.14×10^4 cm. If a beam of the same power is focused to radius 0.01 cm, the focal length is -2.27×10^2 cm [2].

The absorption coefficient of a medium and the concentration of the absorbing species can be evaluated from the experimentally measured focal length value of an induced thermal lens. The simplest way to determine focal length is to measure the "far field" diameter of a laser beam with and without the sample. Since the optical strength of a thermal lens is not very high, the distance between the sample and the plane of measurement should be several meters. The maximum TLS signal can be detected optimally when the laser beam is focused just before the sample cell. This distance (the confocal length) is defined by the parameters of the laser beam and the focal length of the lens.

An optimal TLS scheme detects the fraction of laser beam power P passed through a pinhole in a screen, located in the farfield, being detected by a photomultiplier (or vacuum photodiode) and lock-in amplifier (Fig. 5). The TLS signal can be expressed as a normalized fractional change in the detected laser power:

$$[P(t=0) - P(t=\infty)]/P(t=\infty)$$

where $P(t=0)$ is the detected part of a laser beam not affected by the thermal lens effect, and $P(t=\infty)$ is the equilibrium value of P in the presence of the thermal lens. The absorption coefficient of a sample can be directly evaluated from this ratio and the parameters of the sample and laser beam. Both single- and two-beam versions of TLS are widely used in analytical applications. In the latter version, one beam of appropriate power is used as the heating beam while the second, low intensity, stable probe beam is used for signal detection.

From the introduction of TLS, cw lasers with discrete wavelengths (e.g., He–Ne, He–Cd, Ar^+, Kr^+, CO_2) are the most widely used. These lasers have the best power stability and beam profile. In most cases, the absorption bands of the trace species of interest do not coincide with a discrete wavelength of the heating laser. Chemical pretreatment is necessary for formation of an analyte derivative with an appropriate absorption band. As in the case of LOAS, TLS has poor selectivity because of spectral interference of sample components. Selectivity depends mainly on the appropriate choice of reagent for analyte derivative formation.

The signal also depends on the solvent. Carbon tetrachloride, benzene, and acetone are preferred.

Table 4. Best detection limits for pure solutions and some examples of real sample analysis by TLS

Analyte	Solvent	Sample*	Laser	Detection limit, wt %
Co^{2+}	$CHCl_3$	PS	He–Ne	1.2×10^{-6}
Cu^{2+}	water	PS	He–Ne	3.8×10^{-6}
Fe^{2+}	$CHCl_3$	PS	Ar^+	1.7×10^{-10}
Nd^{3+}	D_2O	PS	dye	3.0×10^{-5}
Pr^{3+}	water	PS	dye	7.2×10^{-3}
Tb^{3+}	butanol	PS	Ar^+	7.0×10^{-6}
UO_2^{2+}	water	NW	dye	4.7×10^{-7}
U^{6+}	$HClO_4$	NW	dye	7.0×10^{-6}
Am^{3+}	$HClO_4$	PS	dye	7.0×10^{-7}
Pu^{4+}	HCO_3^-	PS	dye	2.5×10^{-6}
PO_4^{3-}	water–acetone	PS	dye	5.0×10^{-10}
AsO_3^{3-}	2-butanol	NW	He–Ne	5.0×10^{-8}
SO_3^{2-}	water	PS	Ar^+	1.3×10^{-7}
S^{2-}	2-butanol	water, gas	He–Ne	5.0×10^{-8}
I_2	CCl_4	PS	Ar^+	1.0×10^{-16}**
Formaldehyde	water	PS	dye	4.5×10^{-8}
Nitroaniline	methanol–water	PS	He–Cd	1.2×10^{-9}**
Trimethylpentane	CCl_4	PS	He–Ne	8.0×10^{-9}
Azulene	liquid CO_2	PS	Kr^+	3.0×10^{-10}
NO_2	air	air	Ar^+	5.0×10^{-7}
Methanol	N_2	PS	CO_2	1.2×10^{-6}
CCl_2F_2	Ar	PS	CO_2	1.0×10^{-7}
CS_2	Ar	PS	CO_2	1.0×10^{-7}

* PS = pure solution; NW = natural water.
** Absolute detection limit, g.

For example, the TLS signal of a given analyte in water is 45 times smaller than in carbon tetrachloride. Suspended dust and air bubbles in a sample may perturb the laser beam, affecting the signal. To minimize these perturbations, liquids should be filtered and allowed to equilibrate in the cell before beginning the experiment. Absorption by the cell windows followed by a heating of the sample should also be minimized, as well as the formation of interference fringes due to the reflections of the laser beam by the optical components of the apparatus.

Originally, TLS was expected to be an absolute (standardless) technique because of the direct relationship between absorption coefficient and the experimental parameters. It was later recognized that the accuracy of the experimental and molecular parameters is not sufficient for absolute quantitative evaluation of the trace concentration of an analyte; calibration by standard reference solutions is necessary for quantitative analysis.

The minimum absorption coefficients measured by TLS are of the order of $10^{-6} - 10^{-7}$ cm^{-1}, which corresponds, for strong absorbers ($\varepsilon \geq 10^5$ M^{-1} cm^{-1}), to minimal concentrations of ca. $10^{-11} - 10^{-12}$ M. Some recent data from real sample analysis by TLS are listed in Table 4.

22.4.4. Fluorescence Analysis of Organic Molecules in Solid Solutions [2]

Fluorescence is one of the most sensitive techniques for the detection of complex organic molecules such as aromatic hydrocarbons, heterocyclic compounds, porphyrins (including chlorophyll and its derivatives), organic dyes, etc. Its high sensitivity depends on the relatively high fluorescence quantum yields for these classes of molecules and well-developed methods of photon detection. If a pure, single-component solution is analyzed, extremely low LODs can be achieved, e.g., highly luminescent dyes in aqueous solution can be easily detected with LODs down to 10^{-13} M. For many organic compounds in liquid or solid solution form, single molecules can be detected by modern experimental techniques. In many cases, the main analytical problem is not sensitivity, but selectivity. At room temperature, the absorption and luminescence spectra of organic molecules consist typically of one or more bands whose spectral width is hundreds or thousands of wavenumbers; severe spectral interference is a key problem in the analysis of multicomponent mixtures.

The Shpol'skii method is a powerful improvement. The spectra of many organic molecules embedded in a crystalline matrix of short-chain nparaffins consist, at helium temperature, of well-defined quasi-lines with linewidth 2–20 cm^{-1}. In spite of many successful analytical applications the Shpol'skii technique is not universally applicable because many organic molecules have low solubility in *n*-paraffins and, moreover, many compounds in *n*-paraffin matrices have broad-band spectra even at helium temperature.

The discovery that in many cases the spectra of organic compounds at low temperature are inhomogeneously broadened was the next and very important step in the understanding of the spectroscopy of complex molecules. In most cases, the broad spectra have an intrinsic line structure, which can be revealed by selective laser excitation. This new technique, laser fine-structure selective spectroscopy of complex molecules in frozen solutions is often called site-selection spectroscopy (SSS). A physically more realistic term is energy selection spectroscopy. By now the SSS technique is so far developed that inhomogeneous broadening of organic molecules at low temperature can be eliminated by selective laser excitation [2].

The fundamentals of SSS are based on the theory of impurity centers in a crystal. The optical spectrum of an organic molecule embedded in a matrix is defined by electron–vibrational interaction with intramolecular vibrations (vibronic coupling) and interaction with vibrations of the solvent (electron–phonon coupling). Each vibronic band consists of a narrow zero-phonon line (ZPL) and a relatively broad phonon wing (PW). ZPL corresponds to a molecular transition with no change in the number of phonons in the matrix (an optical analogy of the resonance γ-line in the Mössbauer effect). PW is determined by a transition which is accompanied by creation or annihilation of matrix phonons. The relative distribution of the integrated intensity of a band between ZPL and PW is characterized by the Debye–Waller factor:

$$\alpha = I_{ZPL}/(I_{ZPL} + I_{PW})$$

and depends critically on the temperature and the strength of electron–phonon coupling. At liquid helium temperature, if this coupling is not too strong, the vibronic bands of an isolated impurity molecule consist of a strong, narrow ZPL and a broad, weak PW. The observed broad band of an assembly of organic molecules in solution at low temperature is an envelope of a great number of narrow ZPLs (accompanied by PWs) of individual molecules which are surrounded by different local fields (mainly because of small variations of the relative orientation of an analyte molecule and the solvent molecules). Statistical variations cause shifts in ZPL positions and, hence, inhomogeneous broadening of a spectral band. A narrow-band laser excites only a small fraction of the molecules with an appropriate ZPL position. Thus, well-separated, narrow ZPLs, accompanied by PWs, are observed in the fluorescence spectra of a multicomponent cryogenic solid mixture. This effect is often called fluorescence line narrowing (FLN). Informative spectra may be obtained for many types of molecule (polar and nonpolar), in different media (glasses, polymers, and crystals). The efficiency of FLN for a particular compound is well defined by the factor α. The higher the value of α, the better is excitation selectivity. For most systems, α varies within the interval 0.2–0.8. For compounds with $\alpha \ll 1$, selective excitation is inefficient. Examples of such molecules are ionic dyes (proflavines and rhodamines in ethanol).

Fluorescence line narrowing depends strongly on temperature. For most organic molecules, ZPL disappears completely at temperatures above 40–50 K. This is the main disadvantage of FLN for analytical applications; the sample temperature should be below 15–30 K.

The main components of FLN instruments for analysis are a cryogenic system, a tunable laser, a medium resolution spectrometer, and a recording system. The cryogenic system consists of a liquid He optical cryostat with a He circulation system. In some cases, commercially available closed-cycle cryostats can be used. The optical system consists of a sample holder, optics for the exciting beam input, and a high aperture collector of the fluorescent radiation.

The most intense 0–0 ZPLs of organic molecules of interest occupy the spectral range 650–250 nm. Thus, lasers for FLN should be tunable within this range. The laser linewidth should match the FLN vibronic linewidth of the analytes, typically of the order of 1–3 cm^{-1}. To avoid sample heating, saturation effects, and/or stimulated two-photon photoreactions, the average laser power should not exceed 1–10 mW. The cw dye lasers, pumped by Ar$^+$ or Kr$^+$ ion lasers are relatively cheap and reliable sources for the visible spectrum. For UV, frequency-doubled dye lasers pumped by copper or quasi-continuous Nd-YAG lasers (typical repetition rates 1–10 kHz) may be preferable.

As the distance between the wavelengths of excitation and fluorescence in FLN is ca. 100–200 cm^{-1}, background scattering of the excitinglaser radiation may be a serious problem. Therefore, high-quality double- or even triple-grating spectrometers are used in FLN analysis. New multichannel recording systems (photodiode arrays, image intensifiers with linear CCDs) are potentially very useful detectors for FLN.

Polycyclic aromatic hydrocarbons (PAH) and heterocyclic aromatic compounds are the most widely analyzed organic compounds by FLN techniques. Both types of compound are widespread in nature. Many have strong carcinogenic and mutagenic properties and these properties depend on the isomeric structure of the compound. Analysis of PAH in a glassy matrix by FLN has many advantages compared with other techniques; simple sample preparation, high optical quality of the frozen matrices, etc., provide higher sensitivity and selectivity. LODs down to $10^{-12}-10^{-13}$ g/cm^3 have been achieved for single-component solutions. For example, a concentration of perylene in ethanol of 10^{-12} g/mL (ca. 10^5 molecules in the sample) was easily detected in a frozen solution by FLN. The calibration curve was linear over the concentration range $10^{-6}-10^{-9}$ g/cm^3. Well-characterized fine-structure FLN spectra were detected and identified in an artificial mixture of 14 PAHs in a water–glycerol solution. In another experiment, a mixture of 7 PAHs with concentrations of individual components ranging from 10^{-7} to 10^{-9} g/cm^3 was quantitatively analyzed and all compounds were correctly identified.

In some cases, direct analysis of raw products (without sample pretreatment or dissolution) is possible. To avoid self-absorption and energy transfer effects, low analyte concentrations and low optical density of the sample are necessary. Several PAH were successfully detected in gasoline when a frozen sample ($T = 4.2$ K) was selectively excited by laser radiation of appropriate wavelength. The same sample excited by a conventional Hg-lamp exhibited a smooth, unresolved spectrum with no possibility of identifying different compounds. As an example, the active carcinogen 3,4-benzpyrene was easily detected in gasoline by FLN. The concentration of 10^{-7} g/mL was measured by the standard additions method. The estimated LOD depended on the background luminescence of a gasoline sample which is mainly determined by its composition. The LODs varied within $10^{-10}-10^{-11}$ g/mL for different gasoline samples. By the same technique, perylene at concentration 5×10^{-8} g/mL was detected in solid paraffin.

22.5. Laser Ablation [3], [6], [11]

Laser ablation (LA) is a technique in which focused laser radiation is used to release material from a sample. The advantages of LA are now well known: reduced or no sample preparation is needed; conducting or non-conducting samples may be analyzed; spatial resolution up to a few micrometers can be obtained.

Different analytical techniques are used for detection of the elemental composition of the ablated material. The simplest one is direct detection of the emission of a plasma plume formed above a sample surface. This technique is generally referred to as LIBS (laser induced breakdown spectroscopy). Strong continuous background radiation of the hot plasma plume does not allow detection of atomic and ionic lines of specific elements during the first few hundred nanoseconds of plasma evolution. One can achieve reasonable signal-to-noise ratio for the measurement of atomic and ionic spectral line intensities by optimization of experimental parameters: type and pressure of a buffer gas, time delay, and distance from the sample surface for detection of line intensities. The LODs in (mg–µg)/g range have been realized for direct analysis of metals, glasses, and ceramics by LIBS techniques for a variety of trace elements. Better LODs were demonstrated when LA was followed by detection of the ablated atoms or ions by laser atomic absorption or fluorescence spectroscopy (LAAS, LIF). Even though matrix-effects are significant and detection limits are not as good as for LA-ICP-MS, LIBS has one major advantage. It allows fast and on-line analysis in industrial environment, including the nuclear industry. When fibers are used both for excitation and emission collection, a very robust system can be built that allows on-line quality control.

The intrinsic drawback of LIBS is a short duration (less than a few hundred microseconds) and significant non-stationary conditions of a laser plume. Much higher sensitivity was realized by transport of the ablated material into a secondary atomic reservoir such as an MIP. Owing to the much longer residence time of ablated atoms and ions in a stationary MIP (typically several ms as compared with at most a hundred microseconds in a laser plume) and due to additional excitation of the radiating upper levels in the low pressure plasma, the line intensities of atoms and ions are greatly enhanced. Because of both factors the LODs of the LA-MIP technique have been improved by 1–2 orders of magnitude as compared with LIBS.

Recently, analytical capabilities of LIBS and LA-MIP-OES were noticeably improved by use of an advanced detection scheme based on an Echelle spectrometer combined with a high sensitivity ICCDs (intensified charge-coupled devices) detector. This detection scheme enables simultaneous detection of a large spectral range from the UV to NIR in a single laser shot. It allows estimation of temperature of a laser plume by constructing Boltzman plots and correction for plasma temperature variations. The advantages of this technique are: complete sample analysis in a single laser shot, improved accuracy and precision, the possibility to detect sample inhomogeneities.

The sensitivity, accuracy, and precision of solid sample analysis were greatly improved by coupling of LA with ICP-OES/MS. The ablated species are transported with a carrier gas (usually argon) into the plasma torch. Additional atomization, excitation and ionization of the ablated species in a stationary hot plasma provide a dramatic increase in the sensitivity of emission detection (LA-ICP-OES) or detection of ions (LA-ICP-MS). The efficiency of the transport of ablated species into an ICP strongly depends on the size of the particles. The optimal conditions for ablation in the case of LA-ICP differ significantly from the optimal conditions for LIBS because the efficient transport of the ablated matter to an ICP requires a fine aerosol (with solid particle diameters less than a few micrometers), whereas direct optical emission spectroscopy of the laser plume needs excited atoms and ions.

The LODs in the sub-µg/g–ng/g range were realized for different materials (metals, glasses, polymers) by LA-ICP-OES. Even higher sensitivity was achieved for LA-ICP-MS due to the high efficiency of ion collection and detection. Under optimal conditions of ablation (choice of gas, diameter of a crater, flow rate of a carrier gas) the LODs in the ng/g–pg/g range are now routinely realized for most elements of the periodic table. Today the LA-ICP-MS technique is accepted as a most powerful technique for direct analysis of solid samples.

As a result of its attractive characteristics, LA-ICP-MS is currently being used for a large variety of applications, such as the *in situ* trace element analysis of glasses, geological samples, metals, ceramic materials, polymers, and atmospheric particulates. Fingerprinting (characterization of the trace element composition) of diamonds and gold and glass and steel using LA-ICP-MS has been reported to be very useful for provenance determination and for forensic purposes, respectively.

So far the pulsed lasers with a pulse duration of some nanoseconds have been most widely used for LA – Nd: YAG (1064, 532, 354.7, 266 nm), excimer lasers XeCl (308 nm), KrF (248 nm). For these lasers the mechanisms responsible for the material removal are thermal melting and evaporation or some kind of explosive evaporation. As a result, so called fractional evaporation could occur — the elements with different melting and boiling temperatures evaporate from a melt at different rates. Because of this composition of a laser plume does not match the bulk composition. This was a serious problem as inadequate probing of a sample could not be corrected for by any sensitive detection scheme. The optimal conditions for laser ablation and advanced methods of data processing were found to avoid the problem of fractionation for different classes of samples in the case of nanosecond pulses.

The use of the lasers with femtosecond duration opens the way to overcome the problem of fractionation in the most radical way. The first generation of all-solid-state femtosecond lasers is now commercially available, although expensive. For such a fast energy deposition the non-thermal mechanisms of material removal (e.g., ion repulsions due to Coulomb forces) can be significant. The non-thermal character of material removal with fs pulses enables the depth profiling of a solid sample with a few nm resolution, which was impossible with ns pulses because of melting and mixing of different layers.

22.6. References

[1] D. S. Kliger (ed.): *Ultrasensitive Laser Spectroscopy*, Academic Press, New York 1983.
[2] V. S. Letokhov (ed.): *Laser Analytical Spectrochemistry*, Adam Hilger, Bristol 1986.
[3] J. Sneddon, T. L. Thiem, Y. Lee (eds.): *Lasers in Analytical Atomic Spectrometry*, John Wiley & Sons, New York 1997.
[4] F. P. Schäfer (ed.): *Dye Lasers*, Springer Verlag, Heidelberg 1973.
[5] M. Ch. Amann, J. Buus: *Tunable Laser Diodes*, Artech House, Boston 1998.

[6] B. Smith: "Lasers in Analytical Chemistry," *Trends Anal. Chem.* **17** (1998) nos. 8+9.
[7] S. Sjöström, *Spectrochimica Acta Rev.* **13** (1990) 407–465.
[8] K. Niemax: "Diode Laser Spectroscopy," *Spectrochim. Acta Rev.* **15** (1993) no. 5.
[9] H. I. Schiff, G. I. Mackay, J. Bechara, in M. W. Sigrist (ed.): *Air Monitoring by Spectroscopic Techniques*, John Wiley & Sons, New York 1994.
[10] S. Martellucci, A. N. Chester (eds.): *Optoelectronics for Environmental Science*, Plenum Press, London 1991.
[11] R. Russo: "Laser Ablation," *Appl. Spectr.* **49** (1995) no. 9, 14 A–28 A.

23. X-Ray Fluorescence Spectrometry

RON JENKINS, International Centre for Diffraction Data, Newtown Square, Pennsylvania, United States

23.	X-Ray Fluorescence Spectrometry . 753	23.5.	Accuracy	760
23.1.	Introduction 753	23.5.1.	Counting Statistical Errors	760
23.1.1.	Properties of X Rays 753	23.5.2.	Matrix Effects	760
23.1.2.	Scattering and Diffraction 754	23.6.	Quantitative Analysis	761
23.1.3.	Absorption of X Rays 754	23.6.1.	Internal Standards	761
23.1.4.	X-ray Fluorescence 755	23.6.2.	Type Standardization	761
23.2.	Historical Development of X-ray Spectrometry 755	23.6.3.	Influence Correction Methods	762
		23.6.4.	Fundamental Methods	762
23.3.	Relationship Between Wavelength and Atomic Number 755	23.7.	Trace Analysis	762
		23.7.1.	Analysis of Low Concentrations . . .	763
23.3.1.	Characteristic Radiation 755	23.7.2.	Analysis of Small Amounts of Sample	763
23.3.2.	Selection Rules 756			
23.3.3.	Nomenclature for X-ray Wavelengths 757	23.8.	New developments in Instrumentation and Techniques . .	763
23.4.	Instrumentation 757			
23.4.1.	Sources 757	23.8.1.	Total Reflection Spectrometry	763
23.4.2.	Detectors 758	23.8.2.	Spectrometers with Capillary Optics	764
23.4.3.	Types of Spectrometer 758	23.8.3.	Applications of the Synchrotron . . .	764
23.4.4.	Wavelength Dispersive Systems . . . 759			
23.4.5.	Energy Dispersive Systems 760	23.9.	References	765

23.1. Introduction

X-ray spectrometric techniques provided important information for the theoretical physicist in the first half of the 19th century, and since the early 1950s they have found increasing use in the field of materials characterization. While most of the early work in X-ray spectrometry was carried out with electron excitation [1], today most stand-alone X-ray spectrometers use X-ray excitation sources rather than electron excitation. X-ray fluorescence spectrometry typically uses a polychromatic beam of short wavelength X-radiation to excite longer wavelength characteristic lines from the sample to be analyzed. Modern X-ray spectrometers use either the diffracting power of a single crystal to isolate narrow wavelength bands, or a proportional detector to isolate narrow energy bands, from the polychromatic radiation (including characteristic radiation) excited in the sample. The first of these methods is called wavelength dispersive spectrometry and the second, energy dispersive spectrometry. Because the relationship between emission wavelength and atomic number is known, isolation of individual characteristic lines allows the unique identification of an element, and elemental concentrations can be estimated from characteristic line intensities.

23.1.1. Properties of X Rays

X rays are a short wavelength form of electromagnetic radiation discovered by WILHELM ROENTGEN in the late 19th century [2]. When a high-energy electron beam is incident upon a specimen, one of the products of the interaction is the emission of a broad-wavelength band called the continuum, also referred to as white radiation or bremsstrahlung. This white radiation is produced as the atomic electrons of the elements making up the specimen decelerate the impinging high-energy electrons. Characteristic X-ray photons are produced following the ejection of an inner orbital electron from an excited atom and subsequent transition of atomic orbital electrons from states of high to low energy.

A beam of characteristic X rays passing through matter is subject to three processes, absorption, scatter, and fluorescence. The absorption of X rays involves mainly inner orbital atomic electrons and varies as the third power of the atomic number of the absorber. X rays are scattered mainly by the loosely bound outer electrons of an atom. Scattering of X rays may be coherent (same wavelength) or incoherent (longer wavelength). Coherently scattered photons may undergo subsequent mutual interference, leading in turn to the generation of diffraction maxima. Fluorescence occurs when the primary X-ray photons are energetic enough to create electron vacancies in the specimen, leading in turn to the generation of secondary (fluorescence) radiation produced from the specimen. This secondary radiation is characteristic of the elements making up the specimen. The technique used to isolate and measure individual characteristic wavelengths following excitation by primary X-radiation is called X-ray fluorescence spectrometry.

23.1.2. Scattering and Diffraction

Scattering occurs when an X-ray photon interacts with the electrons of the target element. Where this interaction is elastic, i.e., no energy is lost in the collision process, the scattering is referred to as coherent (Rayleigh) scattering. Since no energy change is involved, the coherently scattered radiation retains exactly the same wavelength as that of the incident beam. It can also happen that the scattered photon gives up a small part of its energy during the collision. In this instance, the scatter is referred to as incoherent (Compton scattering).

X-ray diffraction is a combination of two phenomena — coherent scatter and interference. At any point where two or more waves cross one another, they are said to interfere. Under certain geometric conditions, wavelengths that are exactly in phase may add to one another, and those that are exactly out of phase may cancel each other. Under such conditions, coherently scattered photons may constructively interfere with each other, giving diffraction maxima. Since a crystal lattice consists of a regular arrangement of atoms, with layers of high atomic density throughout the crystal structure, a crystal can be used to diffract X-ray photons. Since scattering occurs between impinging X-ray photons and the loosely bound outer orbital atomic electrons, when a monochromatic beam of radiation falls onto the high atomic density layers, scattering will occur. In order to satisfy the requirement for constructive interference, it is necessary that the scattered waves originating from the individual atoms, that is, the scattering points, be in phase with one another. The geometric conditions for this are that:

$$n\lambda = 2d\sin\theta \quad (1)$$

where λ is the wavelength, d the interplanar spacing of the crystal, and n is an integer. Equation (1) is a statement of Bragg's law. The diffraction phenomenon is the basis of wavelength dispersive spectrometry, since by using a crystal of fixed $2d$, each wavelength is diffracted at a unique diffraction (Bragg) angle θ. Hence, by measuring the diffraction angle, knowledge of the d-spacing of the analyzing crystal allows the determination of the wavelength.

23.1.3. Absorption of X Rays

When a beam of X-ray photons of intensity $I_0(\lambda)$ falls on a specimen, a fraction of the beam passes through, this fraction being:

$$I(\lambda) = I_0(\lambda)\exp(\mu \varrho x) \quad (2)$$

where μ is the mass attenuation coefficient of absorber for wavelength λ, ϱ is the density of the specimen, and x the distance traveled by the photons through the specimen. It can be seen from Equation (2) that a number (I_0-I) of photons has been lost in the absorption process. Although a significant fraction of this loss may be due to scatter, by far the greater loss is due to the photoelectric effect. Photoelectric absorption occurs at each of the energy levels of the atom; thus the total photoelectric absorption is determined by the sum of each individual absorption effect within a specific shell. Where the absorber is made up of a number of different elements, as is usually the case, the total absorption is the sum of the products of the individual elemental mass attenuation coefficients and the weight fractions of the respective elements. This product is referred to as the total matrix absorption. The value of the mass attenuation in Equation (2) is a function of both the photoelectric absorption and the scatter. However, the photoelectric absorption influence is usually large in comparison with the scatter, and to all intents and purposes the mass absorption coefficient is equivalent to the photoelectric absorption. A plot of the mass attenuation coefficient as a function of wavelength contains a number of discontinuities, called absorption edges, at wavelengths corresponding to the binding energies of the electrons in the various subshells. Between absorption edges, as the wavelength of the incident X-ray photons increases, the absorption increases. This effect is very important in quantitative X-ray spectrometry, because the intensity of a beam of characteristic photons leaving a specimen is dependent on the relative absorption effects of the different atoms making up the specimen. This is called a matrix effect and is one of the reasons why a curve of characteristic line intensity as a function of element concentration may not be a straight line.

23.1.4. X-ray Fluorescence

X-ray fluorescence spectrometry provides the means for the identification of an element by measurement of its characteristic X-ray emission wavelength or energy. The method allows the quantitative estimation of a given element by first measuring the emitted characteristic line intensity and then relating this intensity to elemental concentration. The inherent simplicity of characteristic X-ray spectra make the process of allocating atomic numbers to the emission lines relatively easy, and the chance of making a gross error is rather small. Within the range of the conventional spectrometer, each element gives, on average, only half a dozen lines. A further benefit of using the X-ray emission spectrum for qualitative analysis is that, because transitions arise from inner orbitals, the effect of chemical combination, or valence state is almost negligible.

23.2. Historical Development of X-ray Spectrometry

While the roots of the X-ray spectrometric method go back to the early part of this century, it is only during the last thirty years or so that the technique has gained major significance as a routine means of elemental analysis. The first use of the X-ray spectrometric method dates back to the classic work of HENRY MOSELEY [3]. One of the first papers on the use of X-ray spectroscopy for real chemical analysis appeared in 1922, when HADDING [4] described its use for the analysis of minerals. The use of X rays, rather than electrons, to excite characteristic X-radiation [5] avoids the problem of heating of the specimen. Use of X-rays rather than electrons represented the beginnings of X-ray fluorescence as we know it today. The fluorescence method was first employed on a practical basis in 1928 by GLOCKER and SCHREIBER [6]. However, widespread use of the technique had to wait until the mid-1940s when X-ray fluorescence was rediscovered by FRIEDMAN and BIRKS [7]. The basis of their spectrometer was a diffractometer, originally designed for the orientation of quartz oscillator plates. A Geiger counter was used to measure the intensities of the diffracted characteristic lines, and quite reasonable sensitivity was obtained for a very large part of the atomic number range. The first commercial X-ray spectrometers became available in the early 1950s and, although these operated only with an air path, they were able to provide qualitative and quantitative information on all elements above atomic number 22 (titanium). Later versions allowed use of helium or vacuum paths that extended the lower atomic number cut-off.

23.3. Relationship Between Wavelength and Atomic Number

23.3.1. Characteristic Radiation

If a high-energy particle, such as an electron, strikes a bound atomic electron, and the energy of the particle is greater than the binding energy of the atomic electron, it is possible that the atomic electron will be ejected from its atomic position. The electron departs from the atom with a kinetic energy $E - \varphi$, equivalent to the difference between the energy E of the initial particle and the binding energy φ of the atomic electron. When the exciting particles are X-ray photons, the ejected electron is called a photoelectron, and the interaction between primary X-ray photons and atomic elec-

Table 1. Atomic structures of the first three principal shells

Shell (electrons)	n	m	l	s	Orbitals	J
K (2)	1	0	0	$\pm\frac{1}{2}$	1s	$\frac{1}{2}$
L (8)	2	0	0	$\pm\frac{1}{2}$	2s	$\frac{1}{2}$
	2	0	0	$\pm\frac{1}{2}$	2p	$\frac{1}{2}$
	2	−1	1	$\pm\frac{1}{2}$	2p	$\frac{1}{2}, \frac{3}{2}$
	2	+1	1	$\pm\frac{1}{2}$	2p	$\frac{1}{2}, \frac{3}{2}$
M (18)	3	0	0	$\pm\frac{1}{2}$	3s	$\frac{1}{2}$
	3	1	0	$\pm\frac{1}{2}$	3p	$\frac{1}{2}$
	3	−1	1	$\pm\frac{1}{2}$	3p	$\frac{1}{2}, \frac{3}{2}$
	3	+1	1	$\pm\frac{1}{2}$	3p	$\frac{1}{2}, \frac{3}{2}$
	3	0	0	$\pm\frac{1}{2}$	3d	$\frac{1}{2}$
	3	−1	1	$\pm\frac{1}{2}$	3d	$\frac{1}{2}, \frac{3}{2}$
	3	+1	1	$\pm\frac{1}{2}$	3d	$\frac{1}{2}, \frac{3}{2}$
	3	−2	2	$\pm\frac{1}{2}$	3d	$\frac{3}{2}, \frac{5}{2}$
	3	+2	2	$\pm\frac{1}{2}$	3d	$\frac{3}{2}, \frac{5}{2}$

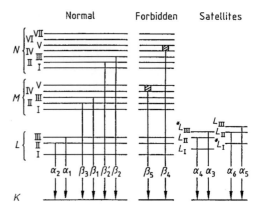

Figure 1. Observed lines in the K series (asterisks indicate ionized states)

trons is called the photoelectric effect. As long as the vacancy in the shell exists, the atom is in an unstable state, and there are two processes by which it can revert to its original state. The first of these involves a rearrangement that does not result in the emission of X-ray photons, but in the emission of other photoelectrons from the atom. This is known as the Auger effect [8], and the emitted photoelectrons are called Auger electrons.

The second process by which the excited atom can regain stability is by transfer of an electron from one of the outer orbitals to fill the vacancy. The energy difference between the initial and final states of the transferred electron may be given off in the form of an X-ray photon. Since all emitted X-ray photons have energies proportional to the differences in the energy states of atomic electrons, the lines from a given element are characteristic of that element. The relationship between the wavelength of a characteristic X-ray photon and the atomic number Z of the element was first established by MOSELEY. Moseley's law is written:

$$1/\lambda = K(Z - \sigma)^2 \quad (3)$$

in which K is a constant that takes different values for each spectral series; σ is the shielding constant with a value just less than unity. The wavelength of the X-ray photon λ in Å is inversely related to the energy E of the photon in keV according to the relationship:

$$\lambda = 12.4/E \quad (4)$$

Since there are two competing routes by which an atom can return to its initial state, and since only one of these processes produces a characteristic X-ray photon, the intensity of an emitted characteristic X-ray beam depends on the relative effectiveness of the two processes within a given atom. As an example, the number of quanta of K series radiation emitted per ionized atom is a fixed ratio for a given atomic number, this ratio being called the fluorescent yield. The fluorescent yield varies as the fourth power of atomic number and approaches unity for higher atomic numbers. Fluorescent yield values are several orders of magnitude less for very low atomic numbers.

23.3.2. Selection Rules

An excited atom can revert to its ground state by transferring an electron from an outer atomic level to fill the vacancy in the inner shell. An X-ray photon is emitted from the atom as part of this de-excitation step, the emitted photon having an energy equal to the energy difference between the initial and final states of the transferred electron. The selection rules for the production of normal (diagram) lines require that the principal quantum number n must change by at least one, the angular quantum number l must change by only one, and the J quantum number [the total momentum J of an electron is given by the vector sum of $(l + s)$ where s is the spin quantum number] must change by zero or one (Table 1). Transition groups can now be constructed, based on the appropriate number of transition levels. Figure 1 shows the lines that are observed in the K series. Three groups of lines are indicated with the normal lines being shown on the left-hand side of the

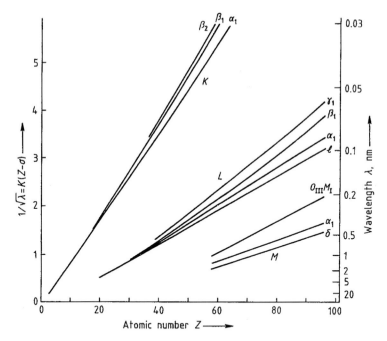

Figure 2. Moseley diagrams for K, L, and M series

figure. While most of the observed fluorescent lines are normal, certain lines may also occur in X-ray spectra that at first sight do not fit the basic selection rules. These lines are called forbidden lines and are shown in the center portion of the figure. Forbidden lines typically arise from outer orbital levels where there is no sharp energy distinction between orbitals. A third type of line may also occur, namely, satellite lines, which arise from dual ionizations. Neither forbidden transitions nor satellite lines have great analytical significance; they may cause confusion in qualitative interpretation of spectra, and may even be misinterpreted as coming from trace elements.

23.3.3. Nomenclature for X-ray Wavelengths

The classical nomenclature system for the observed lines is that proposed by SIEGBAHN in the 1920s. Figure 2 shows plots of the reciprocal of the square root of the wavelength as a function of atomic number for the K, L, and M series. As indicated by Moseley's law (Eq. 3), such plots should be linear. A scale directly in wavelength is also shown, to indicate the range of wavelengths over which a given series occurs. In practice, the number of lines observed from a given element depends on the atomic number of the element, the excitation conditions, and the wavelength range of the spectrometer. Generally, commercial spectrometers cover the K series, the L series, corresponding to transitions to K, L, and M levels respectively. There are a much larger number of lines in the higher series and for a detailed list of all of the reported wavelengths the reader is referred to the work of BEARDEN [9]. In X-ray spectrometry most analytical work is carried out using either the K or the L series wavelengths. The M series may also be useful, especially in the measurement of higher atomic numbers.

23.4. Instrumentation

23.4.1. Sources

Several different types of source have been employed for the excitation of characteristic X-radiation, including those based on electrons, X rays, γ rays, protons, and synchrotron radiation. Sometimes a bremsstrahlung X-ray source is used to generate specific X-radiation from an intermediate pure element sample, called a secondary fluorescer. By far the most common source today is the X-ray photon source. This source is used in primary mode in the wavelength and primary energy dispersive systems, and in secondary fluorescer mode in secondary target energy dispersive spectrometers. A γ-source is typically a radioisotope that is used either directly, or in a mode equivalent to the secondary fluorescer mode in energy dispersive spectrometry. Most conventional wavelength dispersive X-ray spectrometers use a high-power (2–4 kW) X-ray bremsstrahlung source. Energy dispersive spectrometers use either a high- or low-power (0.5–1.0 kW) primary source, depending on whether the spectrometer is used in the secondary or primary mode. In all cases, the primary source unit consists of a very stable high-voltage generator, capable of providing a potential of typically 40–100 kV. The current from the generator is fed to the filament of the X-ray tube, which is typically a coil of tungsten wire. The applied current causes the filament to glow and emit electrons. A portion of this electron cloud is accelerated to the anode of the X-ray tube, which is typically a water-cooled block of copper with the required anode material plated or cemented to its surface. The impinging electrons produce X-radiation, a significant portion of which passes through a thin beryllium window to the specimen.

23.4.2. Detectors

An X-ray detector is a transducer for converting X-ray photon energy into voltage pulses. Detectors work by a process of photoionization, in which interaction between the incoming X-ray photon and the active detector material produces a number of electrons. The current produced by these electrons is converted to a voltage pulse by a capacitor and resistor, such that one digital voltage pulse is produced for each entering X-ray photon. In addition to being sensitive to the appropriate photon energies, i.e., being applicable to a given range of wavelengths or energies, there are two other important properties that an ideal detector should possess: proportionality and linearity. Each X-ray photon entering the detector produces a voltage pulse, and if the size of the voltage pulse is proportional to the photon energy, the detector is said to be proportional. Proportionality is needed where the technique of pulse height selection is to be used. Pulse height selection is a means of electronically rejecting pulses of voltage levels other than those corresponding to the characteristic line being measured. X-ray photons enter the detector at a certain rate and, if the output pulses are produced at the same rate, the detector is said to be linear. Linearity is important where the various count rates produced by the detector are to be used as measures of the photon intensity for each measured line.

In the case of wavelength dispersive spectrometers, the gas flow proportional counter is generally employed for the measurement of longer wavelengths, and the scintillation counter for shorter wavelengths. Neither of these detectors has sufficient resolution to separate multiple wavelengths on its own, and so they have to be employed along with an analyzing crystal. However, in the case of energy dispersive spectrometry, where no dispersing crystal is employed, a detector of higher resolution must be used, generally the Si(Li) detector.

The Si(Li) detector consists of a small cylinder (ca. 1 cm diameter and 3 mm thick) of p-type silicon, compensated by lithium to increase its electrical resistivity. A Schottky barrier contact on the front of the silicon disk produces a p-i-n-type diode. To inhibit the mobility of the lithium ions and to reduce electronic noise, the diode and its preamplifier are cooled to liquid-nitrogen temperature. Incident X-ray photons interact to produce a specific number of electron–hole pairs. The charge produced is swept from the diode by the bias voltage to a charge-sensitive preamplifier. A charge loop integrates the charge on a capacitor to produce an output pulse. Special energy dispersive detectors have been used in miniaturized instrumentation for in vivo applications [10] and for in situ analysis [11].

23.4.3. Types of Spectrometer

The basic function of the spectrometer is to separate the polychromatic beam of radiation coming from the specimen in order that the intensity of each individual characteristic line can be measured. A spectrometer should provide sufficient resolution of lines to allow such data to be taken, at the same time providing a sufficiently large response above background to make the measurements statistically significant, especially at low analyte concentrations. It is also necessary that the spectrometer allow measurements over the wavelength range to be covered. Thus, in the selection of a set of spectrometer operating variables, four factors are important: resolution, response, background level, and range. Owing to many factors, optimum selection of some of these characteristics may be mutually exclusive; as an example, attempts to improve resolution invariably cause lowering of absolute peak intensities. There is a wide variety of instrumentation available today for the application of X-ray fluorescence techniques and it is useful to break the instrument types down into three main categories: wavelength dispersive spectrometers (sequential and simultaneous); energy dispersive spectrometers (primary or secondary); and special spectrometers (including total reflection, synchrotron source, and proton-induced). The wavelength dispersive system was introduced commercially in the early 1950s and probably around 25 000 such instruments have been supplied commercially. Energy dispersive spectrometers became commercially available in the early 1970s and today there are around 15 000 of these units in use. There are far fewer of the specialized spectrometers in use and these generally incorporate an energy dispersive rather than a wavelength dispersive spectrometer.

X-ray spectrometers differ in the number of elements measurable at one time, and the speed at which they collect data. All of the instruments are, in principle at least, capable of measuring all elements from $Z=9$ (fluorine) upwards, and most modern wavelength dispersive spectrometers can do useful measurements down to $Z=6$ (carbon). Most systems can be equipped with multisample handling facilities and can be automated by use of minicomputers. Typical spectrometer systems are capable of precision on the order of a few tenths of a percent, with sensitivities down to the low ppm level. Single-channel wavelength dispersive spectrometers are typically employed for routine and nonroutine analysis of a wide range of products, including ferrous and nonferrous alloys, oils, slags and sinters, ores and minerals, thin films, etc. These systems are very flexible, but somewhat slower than multichannel spectrometers. Multichannel wavelength dispersive instruments are used almost exclusively for routine, high-throughput analyses where the great need is for fast, accurate analysis, but where flexibility is of no importance. Energy dispersive spectrometers have the great advantage of being able to display information on all elements at the same time. They lack somewhat in resolution compared with wavelength dispersive spectrometers, but the ability to reveal elements absent, as well as elements present, makes the energy dispersive spectrometer ideal for general trouble-shooting problems. They have been particularly effective in the fields of scrap alloy sorting, in forensic science, and in the provision of elemental data to supplement X-ray powder diffraction data.

23.4.4. Wavelength Dispersive Systems

A wavelength dispersive spectrometer may be a single-channel instrument in which a single crystal and a single detector are used for the sequential measurement of a series of wavelengths; or a multichannel spectrometer in which many crystal–detector sets are used to measure elements simultaneously. Of these two basic types, the sequential systems are the most common. A typical sequential spectrometer system consists of the X-ray tube, a specimen holder, a primary collimator, an analyzing crystal, and a detector. A portion of the characteristic fluorescence radiation from the specimen is passed via a collimator or slit onto the surface of an analyzing crystal, where individual wavelengths are diffracted to the detector in accordance with the Bragg law. A goniometer is used to maintain the required $\theta-2\theta$ relationship between crystal and detector. Typically, about six different analyzing crystals and two different collimators are provided in this type of spectrometer, giving the operator a wide choice of dispersion conditions. The separating power of a crystal spectrometer depends on the divergence allowed by the collimators, which to a first approximation determines the width of the diffracted lines, and the angular dispersion of the crystal [12]. While the maximum wavelength covered by traditional spectrometer designs is about 2 nm, recent developments allow extension of the wavelength range significantly beyond this value.

The output from a wavelength dispersive spectrometer may be either analog (a) or digital (d). For qualitative work, an analog output is traditionally used; the digital output from the detector amplifier is fed through a d/a converter, called a rate meter, to an x/t recorder synchronously coupled with the goniometer scan speed. The recorder thus records an intensity–time diagram, in terms of an intensity–2θ diagram. It is generally more convenient to employ digital counting for quantitative work, and a timer–scaler combination is provided which allows pulses to be integrated over a period of several tens of seconds and then displayed as count or count rate. In more modern spectrometers, a scaler–timer may take the place of the rate meter, using step-scanning. Some automated wavelength dispersive spectrometers provide the user with software programs for the interpretation and labeling of peaks [13].

23.4.5. Energy Dispersive Systems

The energy dispersive spectrometer consists of the excitation source and the spectrometer/detection system. The spectrometer/detector is typically a Si(Li) detector. A multichannel analyzer is used to collect, integrate, and display the resolved pulses. While similar properties are sought from the energy dispersive system as with the wavelength dispersive system, the means of selecting these optimum conditions are very different. Since the resolution of the energy dispersive system equates directly to the resolution of the detector, this feature is of paramount importance. The output from an energy dispersive spectrometer is generally displayed on a cathode ray tube, and the operator is able to dynamically display the contents of the various channels as an energy spectrum. Provision is generally made to allow zooming in on portions of the spectrum of special interest, to overlay spectra, to subtract background, etc. Generally, some form of minicomputer is available for spectral stripping, peak identification, quantitative analysis, and other useful functions.

All of the earlier energy dispersive spectrometers were operated in what is called the primary mode and consisted simply of the excitation source, typically a closely coupled low-power, end-window X-ray tube, and the detection system. In practice there is a limit to the maximum count rate that the spectrometer can handle and this led, in the mid-1970s, to the development of the secondary mode of operation. In the secondary mode, a carefully selected pure element standard is interposed between primary source and specimen, along with absorption filters, such that a selectable energy range of secondary photons is incident on the sample. This allows selective excitation of certain portions of the energy range, thus increasing the ratio of useful to unwanted photons entering the detector. While this configuration does not completely eliminate the count rate and resolution limitations of the primary system, it certainly does reduce them.

23.5. Accuracy

23.5.1. Counting Statistical Errors

The production of X rays is a random process that can be described by a Gaussian distribution. Since the number of photons counted is nearly always large, typically thousands or hundreds of thousands, rather than a few hundred, the properties of the Gaussian distribution can be used to predict the probable error for a given count measurement. The random error $\sigma(N)$ associated with a measured value of N, is equal to $\sqrt{N}$. For example, if 10^6 counts are taken, the standard deviation σ is $\sqrt{10^6} = 10^3$, or 0.1 %. The measured parameter in wavelength dispersive X-ray spectrometry is generally the counting rate R, and the magnitude of the random counting error associated with a given datum can be expressed as:

$$\sigma(\%) = 100/\sqrt{N} = 100/\sqrt{(Rt)} \qquad (5)$$

Care must be exercised in relating the counting error (or indeed any intensity-related error) with an estimate of the error in terms of concentration. Provided that the sensitivity of the spectrometer in count rate (c/s) per percent analyte is linear, the count error can be directly related to a concentration error. However, where the sensitivity of the spectrometer changes over the range of measured response, a given fractional count error may be much greater when expressed in terms of concentration. In many analytical situations, the peak lies above a significant background and this adds a further complication to the counting statistical error.

23.5.2. Matrix Effects

In the conversion of net line intensity to analyte concentration, it may be necessary to correct for absorption and/or enhancement effects. Absorption effects include both primary and secondary absorption. Primary absorption occurs because all atoms of the specimen matrix absorb photons from the primary source. Since there is competition for these primary photons by the atoms making up the specimen, the intensity — wavelength distribution of the photons available for excitation of a given analyte element may be modified by other matrix elements. Secondary absorption refers to the absorption of characteristic analyte radiation by the specimen matrix. As characteristic radiation passes out from the specimen in which it was generated, it is absorbed by all matrix elements by amounts relative to their mass attenuation coefficients. The total absorption of a specimen depends on both primary and secondary absorption. Enhancement effects occur when a nonanalyte matrix element emits a characteristic line, of energy just in excess of the absorption edge of the analyte element. This means that the nonanalyte element is able to excite the analyte, giving characteristic photons in addition to those produced by the primary continuum. This gives an enhanced signal from the analyte.

23.6. Quantitative Analysis

The simplest quantitative analysis situation to handle is the determination of a single element in a known matrix. In this instance, a simple calibration curve of analyte concentration versus line intensity is sufficient for quantitative determination. A slightly more difficult case might be the determination of a single element where the matrix is unknown. Three basic methods are commonly employed in this situation: use of internal standards, addition of standards, and use of a scattered line from the X-ray source. The most complex case is the analysis of all, or most, of the elements in a sample, about which little or nothing is known. In this case a full qualitative analysis would be required before any attempt is made to quantitate the matrix elements. Once the qualitative composition of the sample is known, one of three general techniques is typically applied: type standardization, influence coefficient methods, or fundamental parameter techniques. Both the influence coefficient and fundamental parameter techniques require a computer for their application.

In principle, an empirical correction procedure can be described as the correction of an analyte element intensity for the influence of interfering elements, using the product of the intensity from the interfering element line and a constant factor, as the correction term [14]. This constant factor is today generally referred to as an influence coefficient, since it is assumed to represent the influence of the interfering element on the analyte. Commonly employed influence coefficient methods may use either the intensity or the concentration of the interfering element as the correction term. These methods are referred to as intensity correction and concentration correction methods, respectively. Intensity correction models give a series of linear equations which do not require much computation, but they are generally not applicable to wide ranges of analyte concentration. Various versions of the intensity correction models found initial application in the analysis of nonferrous metals where correction constants were applied as look-up tables. Later versions [15] were supplied on commercially available computer-controlled spectrometers, and were used for a wider range of application. A concentration model (e.g., [16]) requires the solving of a series of simultaneous equations by regression analysis or matrix inversion techniques. This approach is more rigorous than the intensity models, and so it became popular in the early 1970s as suitable low-cost minicomputers became available.

23.6.1. Internal Standards

One of the most useful techniques for the determination of a single analyte element in a known or unknown matrix is to use an internal standard. The technique is one of the oldest methods of quantitative analysis and is based on the addition of a known concentration of an element that gives a wavelength close to that of the analyte. The assumption is made that the effect of the matrix on the internal standard is essentially the same as the effect of the matrix on the analyte element. Internal standards are best suited to the measurement of analyte concentrations below about 10%. Care must also be taken to ensure that the particle sizes of specimen and internal standard are about the same, and that the two components are adequately mixed. Where an appropriate internal standard cannot be found, it may be possible to use the analyte itself as an internal standard. This method is a special case of standard addition, and it is generally referred to as spiking.

23.6.2. Type Standardization

Provided that the total specimen absorption does not vary significantly over a range of analyte concentrations, and provided that enhancement effects are absent, and that the specimen is homogeneous, a linear relationship will be obtained between analyte concentration and measured characteristic line intensity. Where these provisos are met, type standardization techniques can be employed. Type standardization is probably the oldest of the quantitative analytical methods employed, and the method is usually evaluated by taking data from a well-characterized set of standards and, by inspection, establishing whether a linear relationship is indeed observed. Where this is not the case, the analyte concentration range may be further restricted. The analyst of today is fortunate in that many hundreds of good reference standards are commercially available. While the type standardization method is not without its pitfalls, it is nevertheless extremely useful, especially for quality control applications where a finished product is compared with a desired product.

23.6.3. Influence Correction Methods

Influence coefficient correction procedures can be divided into three basic types: fundamental, derived, and regression. Fundamental models are those which require starting with concentrations, then calculating the intensities. Derived models are based on some simplification of a fundamental method but still allow concentrations to be calculated from intensities. Regression models are semiempirical in nature and allow the determination of influence coefficients by regression analysis of data sets obtained from standards. All regression models have essentially the same form and consist of: a weight fraction term W (or concentration C); an intensity (or intensity ratio) term I; an instrument-dependent term which essentially defines the sensitivity of the spectrometer for the analyte in question; and a correction term, which corrects the instrument sensitivity for the effect of the matrix. The general form is:

$$W = I(\text{instrument})[1 + \{\text{model}\}] \qquad (6)$$

Equation (6) shows that the weight fraction W of analyte is proportional to the product of the measured intensity I of the analyte, corrected for instrumental effects, and a matrix correction term. Different models vary only in the form of this correction term.

The major advantage of influence coefficient methods is that a wide range of concentration ranges can be covered by using a relatively inexpensive computer for the calculations. A major disadvantage is that a large number of well-analyzed standards may be required for the initial determination of the coefficients. However, where adequate precautions have been taken to ensure correct separation of instrument- and matrix-dependent terms, the correction constants are transportable from one spectrometer to another and, in principle, need only be determined once.

23.6.4. Fundamental Methods

Since the early work of SHERMAN, there has been a growing interest in the provision of an intensity–concentration algorithm which would allow the calculation of the concentration values without recourse to the use of standards. Sherman's work was improved upon, first by the Japanese team of SHIRAIWA and FUJINO [17] and later, by the Americans, CRISS and BIRKS [18], [19] with their program NRLXRF. The same group also solved the problem of describing the intensity distribution from the X-ray tube [20]. The problem for the average analyst in the late 1960s and early 1970s, however, remained that of finding sufficient computational power to apply these methods. In the early 1970s, DE JONG suggested an elegant solution [21], in which he proposed the use of a large main-frame computer for the calculation of the influence coefficients, then use of a minicomputer for their actual application using a concentration correction influence model.

23.7. Trace Analysis

One of the problems with any X-ray spectrometer system is that the absolute sensitivity (i.e., the measured c/s per % of analyte element) decreases significantly as the lower atomic number region is approached. The three main reasons for this are: first, the fluorescence yield decreases with atomic number; second, the absolute number of useful long-wavelength X-ray photons from a bremsstrahlung source decreases with increasing wavelength; and third, absorption effects generally become more severe with increasing wavelength of the analyte line. The first two of these problems are inherent to the X-ray excitation process and to constraints in the basic design of conventional X-ray tubes. The third, however, is a factor that depends very much on the instrument design and, in particular, upon the characteristics of the detector.

23.7.1. Analysis of Low Concentrations

The X-ray fluorescence method is particularly applicable to the qualitative and quantitative analysis of low concentrations of elements in a wide range of samples, as well as allowing the analysis of elements at higher concentrations in limited quantities of materials. A measurement of a line at peak position gives a counting rate which, in those cases where the background is insignificant, can be used as a measure of the analyte concentration. However, where the background is significant, the measured value of the analyte line at the peak position now includes a count-rate contribution from the background. The analyte concentration in this case is related to the net counting rate. Since both peak and background count rates are subject to statistical counting errors, the question now arises as to the point at which the net peak signal is statistically significant. The generally accepted definition for the lower limit of detection is that concentration is equivalent to two standard deviations of the background counting rate. A formula for the lower limit of detection LLD can now be derived [22]:

$$\text{LLD} = 3/m\sqrt{R_b/t_b} \tag{7}$$

where R_b is the count rate at the background angle, t_b the time spent counting the background, and m the sensitivity of the spectrometer.

It is important to note that not only does the sensitivity of the spectrometer vary significantly over the wavelength range of the spectrometer, but so does the background counting rate. In general, the background varies by about two orders of magnitude over the range of the spectrometer. From Equation (7), the detection limit is optimal when the sensitivity is high and the background is low. Both the spectrometer sensitivity and the measured background vary with the average atomic number of the sample. While detection limits over most of the atomic number range lie in the low ppm range, the sensitivity of the X-ray spectrometer falls off quite dramatically towards the long-wavelength limit of the spectrometer due mainly to low fluorescence yields and the increased influence of absorption. As a result, poorer detection limits are found at the long-wavelength extreme of the spectrometer, which corresponds to lower atomic numbers. Thus, the detection limits for elements such as fluorine and sodium are at the levels of hundredths of one percent rather than parts per million. The detection limits for very low atomic number elements such as carbon ($Z=6$) and oxygen ($Z=8$) are very poor, typically on the order of $3-5\%$.

23.7.2. Analysis of Small Amounts of Sample

Conventional X-ray fluorescence spectrometers are generally designed to handle rather large specimens with surface areas on the order of several square centimeters, and problems occur if the sample to be analyzed is limited in size. The modern wavelength dispersive system is especially inflexible in the area of sample handling, mainly because of geometric constraints arising from the need for close coupling of the sample to the X-ray tube and the need to use an airlock to bring the sample into the working vacuum. The sample to be analyzed is typically placed inside a cup of fixed external dimensions that is, in turn, placed in the carousel. This presentation system places constraints not only on the maximum dimensions of the sample cup, but also on the size and shape of samples that can be placed into the cup itself. Primary source energy dispersive systems do not require the same high degree of focusing, and to this extent are more easily applicable to any sample shape or size, provided that the specimen fits into the radiation-protected chamber. In some instances, the spectrometer can even be brought to the object to be analyzed. Because of this flexibility, analyses of odd-shaped specimens have been almost exclusively the purview of the energy dispersive systems.

In the case of secondary target energy dispersive systems, while the geometric constraints are still less severe than those of the wavelength system, they are much more critical than in the case of primary systems. Where practicable, the best solution for the handling of limited amounts of material is invariably found in one of the specialized spectrometer systems.

23.8. New developments in Instrumentation and Techniques

23.8.1. Total Reflection Spectrometry

One of the major problems that inhibits the obtaining of good detection limits in small samples is the high background due to scatter from the sample substrate support material. The suggestion to overcome this problem by using total reflection of the primary beam was made as long ago as 1971 [23], but the absence of suitable instrumentation prevented real progress being made until the late 1970s [24], [25]. The TRXRF method is essentially an energy dispersive technique in which the Si(Li) detector is placed close to (ca. 5 mm) and directly above the sample. Primary radiation enters the sample at a glancing angle of a few seconds of arc. The sample itself is typically presented as a thin film on the optically flat surface of a quartz plate. A beam of radiation from a sealed X-ray tube passes through a fixed aperture onto a pair of reflectors placed very close to each other. Scattered radiation passes through the first aperture to impinge on the sample at a very low glancing angle. Because the primary radiation enters the sample at an angle barely less than the critical angle for total reflection, this radiation barely penetrates the substrate medium; thus scatter and fluorescence from the substrate are minimal. Because the background is so low, picogram amounts can be measured or concentrations in the range of a few tenths of a ppb. can be obtained without recourse to pre-concentration [26]. Ideally, the radiation falling onto the specimen surface should be monochromatic. Unfortunately, in earlier executions of TXRF spectrometers the use of monochromatic source radiation was not possible because of the high intensity loss. This problem was solved with he advent of multilayer monochromators, since these exhibit high reflectivity and produce an intense almost parallel beam [27]. A TXRF spectrometer has been designed to operate with a standard X-ray tube and a multilayer monochromator, and this was found to give a two- to threefold increase in intensity over the normal geometry [28].

23.8.2. Spectrometers with Capillary Optics

Curved glass capillaries with channel diameters on the order of a few microns can efficiently scatter X rays through large angles by multiple grazing incidence reflection. This technology generates a very high specific intensity, yielding an increase in intensity of two to three orders of magnitude [29]. The use of capillary optics has opened up tremendous possibilities in energy dispersive X-ray fluorescence. A new type of X-ray spectrometer has now evolved, broadly categorized as microscopic X-ray fluorescence (MXRF). Earlier versions of MXRF systems simply used an aperture X-ray guide to control the size of the irradiated spot. Use of a monolithic polycapillary optic in place of the original aperture was found to provide more than two orders of magnitude increase in intensity [30]. There are two basic types of MXRF based on capillary optics: the first is based on classical sealed X-ray tube sources, and the second on the use of the synchrotron. As an example, one low-power, portable XRF analyzer, based on this technology [31], allowed the analysis of specimens as small as 30 mm in diameter with detection limits down to 5×10^{-5} %. GORMLEY [32] has described a series of experiments, based on the use of a CCD camera, to better quantify the focusing and transmission characteristics of these capillaries. Another instrument was designed with an 80 mm^2 detector [33] and allowed the analysis of small (5 mm) areas of bulky objects. Each of these spectrometers was designed to work with low-power X-ray tubes. At the other end of the source power scale, VEKEMANS et al. described an instrument based on a rotating anode X-ray tube. This high-efficiency, microscopic X-ray fluorescence system has been utilized for the analysis of 16-20th century brass statues originating from Northern India and Nepal and for the imaging analysis of enamel paints on 17th century Chinese porcelain [34].

23.8.3. Applications of the Synchrotron

The availability of intense, linearly polarized synchrotron radiation beams [35] has prompted workers in the fields of X-ray fluorescence [36] to explore what the source has to offer over more conventional excitation media. In the synchrotron, electrons with kinetic energies on the order of several billion electron volts (typically 3 GeV at this time) orbit in a high-vacuum tube between the poles of a strong (ca. 1 T) magnet. A vertical field accelerates the electrons horizontally, causing the emission of synchrotron radiation. Thus synchrotron source radiation can be considered as magnetic bremsstrahlung as opposed to the normal electronic bremsstrahlung that is produced when primary electrons are decelerated by the electrons of an atom. It has been found that because the primary source of radiation is so intense, it is possible to use a high degree of monochromatization between source and specimen, giving a source that is wavelength- and, therefore, energy-tunable, as well as being highly monochromatic. There are several different excitation modes that can be used for synchrotron-source X-ray fluorescence spectroscopy (SSXRF) including direct excitation with continuum, excitation with absorber-modified continuum, excitation with source crystal monochromatized continuum, excitation with source radiation scattered by a mirror, and reflection and transmission modes.

Most SSXRF is carried out in the energy dispersive mode. Since all solid-state detectors are count-rate limited, the tremendous radiation flux available from the synchrotron is, in general, not directly useful. The real power of the synchrotron comes from the use of highly conditioned beams, which provide a highly focused, monochromatic beam at the sample. A newer development in the area of beam conditioning for the synchrotron is the use of laterally graded multilayers [37]. The use of X-ray optical multilayers in X-ray instrumentation became widespread in the 1990s, and newer technology based on pulsed laser deposition is resulting in the production of multilayers of uniform thickness with a minimum of columnar thin-film growth. The intensity of the synchrotron beam is probably four to five orders of magnitude greater than the conventional bremsstrahlung source sealed X-ray tubes. This, in combination with its energy tunability and polarization in the plane of the synchrotron ring, allows very rapid bulk analyses to be obtained on small areas. Because the synchrotron beam has such a high intensity and small divergence, it is possible to use it as a microprobe of high spacial resolution (about 10 μm). Absolute limits of detection around 10^{-14} μg have been reported for such an arrangement [38]. Use of monolithic polycapillary X-ray lenses has added an even wider dimension to synchrotron microbeam analysis, and beam sizes down to 4–10 μm have been reported [34]. Such systems have been successfully applied in trace element determination [39].

Synchrotron source X-ray fluorescence has also been used in combination with TRXRF. The small divergence of the continuous beam can be caused to strike the sample at less than the critical angle. Very high signal-to-background ratios have been obtained on employing this arrangement for the analysis of small quantities of aqueous solutions dried on the reflector, with detection limits of < 1 ppb, i.e., 1 pg. [40]. Additional advantages accrue because synchrotron radiation is highly polarized, and background due to scatter can be greatly reduced by placing the detector at 90° to the path of the incident beam and in the plane of polarization. A disadvantage of the SSXRF technique is that the source intensity decreases with time, but this can be overcome by bracketing analytical measurements between source standards and/or by continuously monitoring the primary beam.

23.9. References

[1] G. v. Hevesey: *Chemical Analysis by X-Rays and its Application,* McGraw-Hill, New York 1932.
[2] W. C. Roentgen, *Ann. Phys. Chem.* **64** (1898) 1.
[3] H. G. I. Moseley, *Philos. Mag.* **26** (1912) 1024;**27** (1913) 703.
[4] A. Hadding, *Z. Anorg. Allg. Chem.* **122** (1922) 195.
[5] D. Coster, J. Nishina, *Chem. News J. Ind. Sci.* **130** (1925) 149.
[6] R. Glocker, H. Schreiber, *Ann. Phys. (Leipzig)* **85** (1928) 1085.
[7] L. S. Birks: *History of X-Ray Spectrochemical Analysis,* American Chemical Society Centennial Volume, ACS, Washington, D.C., 1976.
[8] P. Auger, *Compt. R.* **180** (1925) 65;*Journal de Physique* **6,** 205.
[9] J. A. Bearden: *X-Ray Wavelengths,* U.S. Atomic Energy Commission Report NYO-10586, 1964, 533 pp.
[10] J. Iwanczyk, B. E. Patt, *Adv. X-ray Anal.,* **41** (1997) 951–957.
[11] W. T. Elam, et.al., *Adv. X-ray Anal.,* **42** (1999) 137–145.
[12] R. Jenkins: *An Introduction to X-Ray Spectrometry,* Chap. 4, Wiley/Heyden, London 1974.
[13] M. F. Garbauskas, R. P. Goehner, *Adv. X-Ray Anal.* **26** (1983) 345.
[14] M. J. Beattie, R. M. Brissey, *Anal. Chem.* **26** (1954) 980.
[15] H. J. Lucas-Tooth, C. Pyne, *Adv. X-Ray Anal.* **7** (1964) 523.
[16] G. R. Lachance, R. J. Traill, *Can. Spectrosc.* **11** (1966) 43.
[17] T. Shiraiwa, N. Fujino, *Bull. Chem. Soc. Jpn.* **40** (1967) 1080.
[18] J. W. Criss, L. S. Birks, *Anal. Chem.* **40** (1968) 1080.
[19] J. W. Criss, *Adv. X-Ray Anal.* **23** (1980) 93.
[20] J. V. Gilfich, L. S. Birks, *Anal. Chem.* **40** (1968) 1077.
[21] W. K. de Jongh, *X-Ray Spectrom.* **2** (1973) 151.
[22] R. Jenkins, J. L. de Vries: *Practical X-Ray Spectrometry,* 2nd ed., Springer Verlag, New York 1970.
[23] Y. Yoneda, T. Horiuchi, *Rev. Sci. Instrum.* **42** (1971) 1069.
[24] J. Knoth, H. Schwenke, *Fresenius Z. Anal. Chem.* **301** (1978) 200.
[25] H. Schwenke, J. Knoth, *Nucl. Instrum. Methods Phys. Res.* **193** (1982) 239.
[26] H. Aiginger, P. Wobrauschek, *Adv. X-Ray Anal.* **28** (1985) 1.
[27] E. Zeigler, O. Hignette, M. Lingham, A. Souvorov, *SPIE,* vol. 2856/61 (1996).
[28] R. Schwaiger, P. Wobrauschek, C. Streli, *Adv. X-ray Anal.,* **41** (1997) 804–811.

[29] M. A. Kumakhov, *Nucl. Instr. Methods,* **B48** (1990) 283.
[30] G. Worley, G. Havrilla, N. Gao, Q. F. Xiao, *Adv. X-ray Anal.* **42** (1998) 26–35.
[31] M. A. Kumakhov, *Adv. X-ray Anal.* **41** (1997) 214–226.
[32] J. Gormley, *Adv. X-ray Anal,* **41** (1997) 239–242.
[33] J. A. Nicolosi, E. Scrugge, *Adv. X-ray Anal.* **41** (1997) 227–233.
[34] *Adv. X-ray Anal.* **41** (1997) 278–290.
[35] C. J. Sparks Jr. in H. Winnick, S. Doniach (eds.): *Synchrotron Radiation Research,* Plenum Press, New York 1980, p. 459.
[36] J. V. Gilfrich et al., *Anal. Chem.* **55** (1983) 187.
[37] T. Holz et al., *Adv. X-ray Anal.* **41** (1997) 346–355.
[38] W. Petersen, P. Ketelsen, A. Knoechel, R. Pausch, *Nucl. Instrum. Methods Phys. Res. Sect. A* **246** (1986) no. 1–3, 731.
[39] K. W. Jones, B. M. Gordon, *Anal. Chem.* **61** (1989) 341A.
[40] A. Iida, Y. Gohshi, *Adv. X-ray Anal.* **28** (1985) 61–68.

24. Activation Analysis

RICHARD DAMS, Gent University, Laboratory Analytical Chemistry, Gent, Belgium
KAREL STRIJCKMANS, Gent University, Laboratory Analytical Chemistry, Gent, Belgium

24.	Activation Analysis	767	24.3.	Photon Activation Analysis	779
24.1.	Introduction	767	24.4.	Charged-Particle Activation Analysis	780
24.2.	Neutron Activation Analysis	768	24.5.	Applications	781
24.2.1.	Basic Principles	768	24.5.1.	High-Purity Materials	781
24.2.2.	Methods of Neutron Activation	769	24.5.2.	Environmental Materials	781
24.2.3.	Standardization	771	24.5.3.	Biological Materials	782
24.2.4.	Sources of Errors	772	24.5.4.	Geo- and Cosmochemistry	782
24.2.5.	Measuring Equipment	774	24.5.5.	Art and Archaeology	782
24.2.6.	Procedures for Activation Analysis	776	24.6.	Evaluation of Activation Analysis	783
24.2.7.	Sensitivity and Detection Limits	778	24.7.	References	783

In addition to the standard symbols defined in the front matter of this volume the following symbols and abbreviations are used.

A	activity (count rate, decay rate)
AA	activation analysis
B	background
C, D	correction terms
CPAA	charged-particle activation analysis
E	element
E	energy
ENAA	epithermal neutron activation analysis
FNAA	fast-neutron activation analysis
FWHM	full width at half maximum
I	beam intensity
I_0	resonance integral
INAA	instrumental neutron activation analysis
IPAA	instrumental photon activation analysis
l	length
MCA	multichannel analyzer
n	particle density
N	number of nuclides, counts
NAA	neutron activation analysis
PAA	photon activation analysis
R	reaction rate; range
RBS	Rutherford backscattering
S	saturation function; signal; stopping power
SCA	single-channel analyzer
TNAA	thermal neutron activation analysis
γ	relative intensity of gamma radiation
ε	detection efficiency
Θ	abundance of isotope
κ	gamma quantum; particle
λ	disintegration constant
σ	cross section
τ	dead time of detector
φ	flux density of radiation (fluency rate)
Φ	flux (fluency)

24.1. Introduction

Activation analysis has, in a few decades, become one of the most important methods for determination of minor (0.1 – 10 mg/g), trace (1 – 100 µg/g), and ultratrace (1 – 1000 ng/g) elements in solid samples. The main advantages of activation analysis are its accuracy and sensitivity (detection limit). Moreover, it is an independent method, i.e., not subject to the same systematic errors as other, more commonly used analytical methods. When a radiochemical separation is applied (after irradiation and prior to measurement) it is not subject to systematic errors due to reagent blanks, sample contamination (after irradiation), and non-quantitative (and even non-reproducible) yield. Precision goes down to a few percent in most favorable cases. The method is applied in the semiconductor industry, medicine, biology, criminology, archaeology, geochemistry, and environmental studies or control. Activation analysis is based on the exposure of a sample to a flux of activating particles or radiation. Irradiation is performed mainly with thermal neutrons (90%), but fast neutrons (3%), charged particles (4%), or photons (3%) can also be used (figures based on publications in Chemical Abstracts over the last two decades). This induces a nuclear reaction, whereby an excited intermediate is formed, which deexcites in 10^{-14} s by emission of prompt gamma rays, which can be measured, or the prompt emission of another particle [neutron(s), proton, alpha] i.e., nuclear reaction analysis (NRA). However, in activation analysis (AA) the resulting radioactive nuclide is monitored by means of the emitted alpha, beta, gamma or X-ray quanta, or delayed neutrons. The radiation emitted during the decay is measured for the identification and quantification of the target elements.

24.2. Neutron Activation Analysis

For a number of reasons, neutrons have found the widest applications as bombarding particles in activation analysis. First, the cross sections for reactions of nuclides with neutrons, yielding primarily (n, γ) reactions, but also (n, α); (n, p), (n, 2 n), and (n, n') reactions, are often large. The type of reaction depends on the energy of the neutron and is chosen individually for each case, on the basis of the following considerations:

1) A suitable reaction product with not too long a half-life (preferably below one year) and with decay parameters compatible with the available counting equipment
2) A sufficiently high cross section
3) The absence of important interfering reactions yielding the same radionuclide
4) Easy treatment of the sample after irradiation (limited matrix activity)

Second, intense neutron sources (nuclear reactors, neutron generators, isotopic sources) were generally available. Third, the high penetration depth of neutrons allows a nearly homogeneous flux density of the entire sample to be analyzed.

24.2.1. Basic Principles

If a target with n^A nuclides of A per cubic centimeter is placed in a neutron beam φ (number of neutrons $cm^{-2} s^{-1}$) then the number of interactions to form nuclei B is given by the reaction rate R $(cm^{-3} s^{-1})$

$$R = \varphi \sigma n^A \quad (1)$$

where σ is the cross section.

The general nuclear reaction can be written as follows:

$$A(n, \kappa) \lambda \rightarrow C$$

where n is a neutron; κ the promptly emitted gamma quantum or particle; λ is the disintegration constant of the radioactive nuclide B produced (equal to $\ln 2/t_{1/2}$; $t_{1/2}$ denotes the half-life of the radionuclide B); and C is the nuclide produced after radioactive decay of B.

From Equation (1) the definition of the cross section σ for this reaction is obtained:

$$\sigma = \frac{\text{Number of interactions}}{\varphi n^A} \quad (2)$$

The cross section is expressed as an area in square centimeters or, more practically, in barn (1 b = 10^{-24} cm^2).

The number N^A of target nuclei is assumed to remain constant, as is the flux density φ throughout the target. Since the nuclides B decay at a rate equal to $-\lambda N^B$,

$$\frac{dN^B}{dt} = \varphi \sigma N^A - \lambda N^B \quad (3)$$

If no nuclides B were present before irradiation ($N^B = 0$ for $t = 0$), integration gives, after an irradiation time t,

$$N^B = \frac{\varphi \sigma N^A}{\lambda} (1 - e^{-\lambda t}) \quad (4)$$

The absolute decay rate A_0, in terms of the number of decays per second, at the end of the irradiation is given by

$$A_0 = N^B \lambda = \varphi \sigma N^A (1 - e^{-\lambda t}) \quad (5)$$

The expression

$$S = 1 - e^{-\lambda t} = 1 - e^{\ln(2) t/t_{1/2}} \quad (6)$$

is termed the saturation function. After an irradiation time equal to one half-life of B, a saturation of 50% is obtained. For irradiation times that are

Table 1. Typical neutron fluxes at irradiation facilities in research reactors

Irradiation site	Neutron flux, cm^{-2} s^{-1}		
	Thermal	Epithermal	Fast
Large reactor			
Reactor core	2×10^{14}	3×10^{13}	4×10^{13}
Surrounding moderator	1–2×10^{14}	0.5–2×10^{13}	0.2–2×10^{13}
Small reactor			
Reactor core	5×10^{12}	4×10^{11}	3×10^{12}
Reflector	0.3–2×10^{12}	0.2–8×10^{10}	0.3–30×10^{10}

much lower than the half-life, the saturation function can be considered approximately linear:

$$S \approx \ln(2)\, t/t_{1/2} \quad (\text{for } t < 0.1 t_{1/2}) \tag{7}$$

A correction term ($D = e^{-\lambda t_d}$) must be applied for the decay during the time t_d between the end of irradiation and the measuring time.

When the counting (measuring) time is not very short ($t_m > 0.3\, t_{1/2}$) compared with the half-life, a correction C for decay during the counting period t_m can also be applied. It equals

$$C = \frac{1 - e^{-\lambda t_m}}{\lambda t_m} \tag{8}$$

Finally the counting equipment has a detection efficiency ε, which is usually less than unity. Thus the number of counts N measured after an irradiation time t, a waiting time t_d, and during a counting time t_m becomes

$$N = \varphi \sigma N^A \left(1 - e^{-\lambda t}\right) e^{-\lambda t_d} \frac{1 - e^{-\lambda t_m}}{\lambda} \varepsilon \tag{9}$$

$$= \frac{\varphi \sigma m \theta N_A\, S D C \varepsilon}{A_r} t_m \tag{10}$$

where m is the mass of the element A; A_r is its relative atomic mass; θ is the abundance of the isotope yielding the radioactive isotope; and N_A is the Avogadro constant.

24.2.2. Methods of Neutron Activation

Reactor Neutrons. Nuclear reactors are the largest and most often used neutron sources. Neutrons formed in the reactor have a continuous energy spectrum extending from nearly 0 to 15 MeV. Conventionally the neutron spectrum is divided in three components:

1) *Fast or fission neutrons*, produced by the fission of ^{235}U with energies ranging from 0.1 to 15 MeV. The mean energy is ca. 2 MeV. These neutrons give rise to so-called threshold reactions, e.g., (n, p), (n, α), and (n, 2n).

2) *Epithermal or resonance neutrons* have energies varying from 0.1 MeV down to about 0.5 eV. In the ideal case the epithermal flux Φ_{epi} is inversely proportional to the neutron energy E

$$\Phi_{epi}(E) = \frac{\Phi_{epi}}{E} \tag{11}$$

The highest resonance neutron flux is in the reactor core.

3) *Thermal or moderated neutrons* are in thermal equilibrium with the atoms of the moderator, and their velocities of 2200 m/s at 20 °C correspond to an energy of 0.0253 eV.

In some research reactors, pulses of very high neutron flux may be generated by removing the control rods. The neutron flux increases immediately to several times 10^{16} cm^{-2} s^{-1} and falls after ca. 10 ms to a very low value.

Table 1 shows that at most irradiation facilities the highest fluxes belong to thermal neutrons. Fortunately for activation analysis, they are the most useful ones. As shown in Figure 1 for the lower-energy neutrons (up to several hundred electronvolts) the decrease in cross section with increasing energy is roughly proportional to $1/v$ (where v is the velocity of the neutrons). Sometimes superimposed on this smoothly dropping curve are sharp resonance peaks in the epithermal region just above 1 eV. The total cross section for capture of epithermal neutrons is, therefore, the sum of the $1/v$ probability and that associated with these superimposed resonance peaks, and is called the resonance integral I_0. In neutron activation analysis, thermal and epithermal flux are separated by the so-called cadmium cutoff energy E_{Cd} of 0.55 eV. Cadmium in fact strongly absorbs neutrons with energies of less than 0.55 eV.

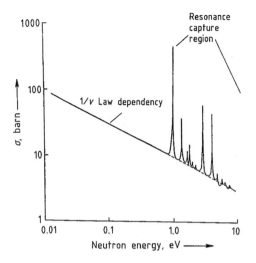

Figure 1. Typical excitation function for neutron-induced reaction, illustrating $1/v$ dependency and resonance peaks in epithermal region

In Equation (1) and following, the reaction rate can thus be split up according to the Høgdahl convention [29]

$$R = N(\Phi_{th}\sigma_0 + \Phi_{epi} I_0) \quad (12)$$

where σ_0 is the tabulated thermal neutron cross section for the corresponding energy (0.0253 eV). The conventional thermal flux is, with a thermal neutron density n

$$\varphi_{th} = n v_0$$

The second term in Equation (12) is the somewhat more complicated solution for the integration of the reaction probability in the epithermal region. The resonance integrals I_0 have been tabulated:

$$I_0 = \int_{E_{Cd}}^{\infty} \sigma_{epi}(E) \frac{dE}{E} \quad (13)$$

Values of φ_{th} and φ_{epi} can be found by irradiating a nuclide (monitor) with known σ_0 and I_0 in a cylindrical cadmium box with 1-mm wall thickness. Often, gold is used as a monitor.

The cross sections for slow neutron reactions are often on the order of several barns and can in some cases be as large as 10^4 barn. The predominant reaction for slow neutrons is the (n, γ) reaction.

Equation (10) allows calculation, to a first approximation, of the detection limit attainable by neutron activation analysis. If a count rate of $A = 1\,\text{s}^{-1}$ can be detected and $A_r = 100$, $S = 0.5$, $\sigma = 1$ barn, $\varphi = 3 \times 10^{12}\,\text{cm}^{-2}\text{s}^{-1}$, $\varepsilon = 0.33$, with θ, D, and C close to unity, a mass of 10^{-9} g of a typical element can be detected; for many elements with larger cross sections or when irradiating at high neutron fluxes, much lower masses can be detected.

When determining a specific element, the *half-life* of the radionuclide to be measured is a very important parameter. For the production of a radioisotope with a half-life of only a few hours or less, the irradiation time should be limited to a saturation of ca. 0.5–0.75, to avoid production of much longer-lived radionuclides. The short-lived radionuclides must be measured soon after the end of irradiation. Also, the counting period should not greatly exceed the half-life of the radionuclide. This however, results in poor counting statistics. For really short-lived radionuclides, this drawback can to some extent be overcome by *cyclic activation analysis* which is based on the use of repetitive short irradiations and counting periods, and summing of the γ-ray spectra. Also *pulsed neutron irradiation* enhances the detection efficiency of radionuclides with $t_{1/2} < 20$ s. In contrast, to determine elements producing longer-lived radionuclides (e.g., $t_{1/2} > 5$ d), much longer irradiation times are required although activation to a high saturation factor is not meaningful. Treatment of the sample and counting should be postponed several days to allow for decay of the shorter-lived radionuclides.

An additional selectivity factor is the neutron energy. By covering the sample with cadmium or boron carbide, the thermal neutrons are filtered out and the sample is activated largely with epithermal neutrons. (Boron has a nearly pure $1/v$ excitation function and is therefore also chosen to filter out thermal low-energy resonance neutrons.) A distinction is made between *thermal (TNAA)* and *epithermal neutron activation analysis (ENAA)*. In the latter, selective activation of a number of elements relative to other interfering elements can be enhanced. ENAA is often advantageous because most of the matrix elements in geological and environmental samples (e.g., Na, Al, P, K, Ca, Sc, Cr, Mn, and Fe) are activated primarily with thermal neutrons (σ_0), whereas a number of trace or ultratrace elements (e.g., As, Sb, Se, Sn, Co, Ni, Ga, Mo, Ag, Pd, In, Cd, Br, I, Rb, Sr, Cs, Ta, W, Th, U, Pt, and Au) have large resonance integrals

(I_0). Not only is the relative activation of many trace elements enhanced, but the overall activity is also generally reduced so that the sample can be measured much closer to the detector, which improves the counting geometry and thus the precision. Since fast neutrons induce threshold reactions, *fast-neutron activation analysis (FNAA)* is used primarily for the determination of light elements (O, N, F, P, S, Cl, Si, Mg, Al, Fe), several of which cannot (or can barely) be determined by thermal neutron activation. Although the reactor neutron spectrum covers a wide range of fast-neutron energies, neutron generators are much more appropriate for FNAA.

Neutron Generators. Neutrons can also be obtained by bombarding an appropriate target with ions accelerated to high kinetic energies. In cyclotrons, a typical neutron flux of 2×10^{11} s^{-1} is obtained with a 10-µA beam of 30-MeV deuterons on a thick beryllium target. Much more important, however, are the Cockroft–Walton accelerators, called 14-MeV neutron generators, in which neutrons are produced by the ^{3}H(d, n)^{4}He reaction. The neutrons obtained have an energy of 14.7 MeV and are emitted nearly isotropically from the target. The total neutron yield varies from 10^8 to 10^{12} s^{-1}, which gives a usable flux density for irradiation ca. 20–100 times lower. Since neutron generators are rather compact, are not too expensive, and do not need very large shielding, they are used in many research and industrial laboratories. Cross sections, especially for light elements, often range from 10 mb to a few hundred millibarn for (n, p) and (n, α) reactions on light elements, and even up to 2–3 b for some (n, 2n) reactions on heavy elements. With a neutron flux of 10^9 cm^{-2} s^{-1} typical detection limits of 10^{-4} to 10^{-6} g, and for a few elements (F, Si, P, Cu, Br, Ag, Sb) ca. 10^{-7}, can be reached. Disadvantages of FNAA with neutron generators are the many possibilities for producing the same indicator radionuclide from different elements and the large flux gradients.

The determination of oxygen by the reaction ^{16}O(n, p)^{16}N is most important. The half-life of the indicator radionuclide ^{16}N is only 7.1 s, and it emits high-energy gamma rays (>6 MeV), which can be measured without the interference of other radionuclides. A detection limit of about 0.005 % can be obtained by using large samples (10–100 g).

Isotopic Neutron Sources [30]. Neutrons can also be produced from isotopic sources. They have the unique advantage of being small, portable, and completely reliable, and of having a constant neutron flux and low cost. There is no need for complicated controlling mechanisms, and the shielding can be very limited. Isotopic neutron sources offer interesting analytical perspectives for accurate determination of minor and major constituents.

Their operation is based on releasing neutrons from a target nucleus either by the action of an α particle or a γ quantum, or during spontaneous fission of a transuranic element. Beryllium is mostly used as target material. As an α source, historically ^{226}Ra was used; it has now been replaced by ^{227}Ac, ^{241}Am, or ^{238,239}Pu, and in the future possibly ^{224}Cm. As a γ source, ^{124}Sb is employed ($t_{1/2}$ = 60.2 d). The neutrons emitted have a broad energy spectrum that peaks at ca. 3–6 MeV. Paraffin wax is often used as a moderator. The neutron outputs range from 10^6 to 1.5×10^7 s^{-1} Ci^{-1}. In practice, sources of 10^8 neutrons/s are available commercially. The isotope ^{252}Cf disintegrates not only by alpha decay, but also by spontaneous fission, releasing several neutrons in the process. It has a number of advantages over (α, n) sources: the small size of the source, and the close similarity of the neutron spectrum to that obtained from the fission of uranium. The main disadvantage is obviously its relatively short effective half-life of 2.65 a.

Since the obtainable neutron flux density is generally less than 2×10^6 cm^{-2} s^{-1}, elements with cross sections > 10 barn and half-lifes of less than 5 d have typically been determined. Examples are In, Dy, Rh, V, Ag, La, Eu, Mn, I, Br, Co, Ga, Cu, Ir, and Au. They are determined preferentially in matrices that do not activate very much so that nondestructive analyses are possible. Attainable precisions are better than 1 %, sometimes as good as 0.1 %.

24.2.3. Standardization

Standardization (i.e., calibration) is performed by Equation (10), which shows that m, the mass of the element to be determined, can be calculated from the measured activity. In practice, this is rarely done because although the values for θ, N_A, A_r, S, D, $C(\lambda)$, and t_m are known with good accuracy for all radionuclides considered, this is not the case for φ (φ_{th} and φ_{epi}), σ (σ_0 and I_0), and ε. Therefore a *relative or comparator method* is normally applied, although more recently absolute methods have also been used. The relative method is essentially very simple. Each sample is coirradiated with a standard containing known amounts of the elements to be determined. When samples and standards are counted under identical conditions, all these factors may be assumed to cancel out, leaving only the following very simple equation:

$$\frac{m_x}{m_{st}} = \frac{A_x}{A_{st}} \cdot \frac{D_{st}}{D_x} = \frac{A_x}{A_{st}} \frac{(e^{-\lambda t_d})_{st}}{(e^{-\lambda t_d})_x} \quad (14)$$

where the subscripts x and st denote the unknown and the standard, respectively. This simple relationship is valid only when all other factors are the same or are held constant. For short-lived radionuclides, the sample and the standard are irradiated sequentially. Since the neutron flux cannot be relied on to remain entirely constant with time, the activities measured must be corrected for any flux variation by coirradiating flux monitors. Second, the sample and the standard, even when irradiated simultaneously, are not always exposed to the same neutron flux because vertical or lateral flux gradients may occur. This can be corrected by means of flux-gradient monitors. The standards containing known amounts of the elements to be determined can be mixtures of pure elements or compounds of known stoichiometry. Since in NAA the samples to be irradiated are mostly solid, the mixed standard solutions are spotted on clean paper filters or ultrapure graphite, dried, and finally pressed into pellets with similar geometry to the sample. To avoid problems associated with the preparation of these standards, reference materials or other "in-house standard materials" are often used in NAA.

k_0-Method. The k_0-method has been developed to overcome the labor-intensive and time-consuming work of preparing such multielement standards when routine multielement or panoramic analyses are required. Single comparators have been used for a long time, but they are applicable only to constant experimental parameters of activation and counting. Therefore, the k_0-method is being implemented increasingly in NAA laboratories. It is intended to be an absolute technique in which uncertain nuclear data are replaced by a compound nuclear constant, the k_0-factor, which has been determined experimentally for each radionuclide. This k_0 is given by

$$k_0 = \frac{A_r^* \theta \sigma_0 \gamma}{A_r \theta^* \sigma_0^* \gamma^*} \quad (15)$$

where γ is the relative intensity of the gamma radiation in the decay scheme of the radionuclide and * denotes the comparator.

For determination of the k_0-factor, gold was used as comparator, which was coirradiated as dilute Au–Al wire. This Au–Al wire should be coirradiated with each sample. It can, however, be converted to any comparator that is found suitable for coirradiation and that has been coirradiated with gold before. To determine the epithermal-to-thermal flux ratio, coirradiation of a zirconium monitor is suitable. The k_0-factors for 68 elements and their relevant gamma lines of 135 analytically interesting radionuclides have been determined and published by DE CORTE and coworkers. For nearly all of them, the uncertainly is < 2 %. An overview is given in Table 2 [31]–[33].

24.2.4. Sources of Errors

Besides typical analytical errors in sample preparation (contamination, inhomogeneity, inaccurate standards) and errors during counting (pulse pileup, instability, spectrometric interference, electronic failure, etc.), some systematic bias may also occur during the irradiation step, such as neutron shielding and nuclear interference. By taking the proper measures, these can be avoided or corrected for. *Flux gradients* can be determined experimentally and corrected for. For accurate work, vertical and horizontal gradients should be measured for the different types of flux (e.g., thermal, epithermal, and fast). When relatively large samples or standards with large absorption cross sections or resonance integrals are irradiated, nonnegligible *neutron shadowing* or *self-shielding* may also occur.

In addition, *flux hardening* (preferential absorption of low-energy neutrons) and *self-moderation* (unmoderated neutrons are further moderated inside the sample) may occur. Since the correction is difficult to apply for samples with mixed composition and irregular shape, the effect is often avoided by using as small a sample and a standard as possible or by diluting the sample with a material having a low absorption cross section (graphite, cellulose). An internal standard can also be applied.

In thermal neutron activation, the determination of an element $^M_Z E$ is often based on an (n, γ)

Table 2. Elements and their analyte radionuclides that can be determined by k_0 INAA

Element	Analyte radionuclides
F	^{20}F
Na	^{24}Na
Mg	^{27}Mg
Al	^{28}Al
Si	^{31}Si
S	^{37}S
Cl	^{38}Cl
K	^{42}K
Ca	^{47}Ca
	^{47}Ca → ^{47}Sc
	^{49}Ca
Sc	^{46}Sc
Ti	^{51}Ti
V	^{52}V
Cr	^{51}Cr
Mn	^{56}Mn
Fe	^{59}Fe
Co	^{60}Co
Ni	^{65}Ni
Cu	^{64}Cu
	^{66}Cu
Zn	^{65}Zn
	^{69m}Zn
	^{71}Zn
Ga	^{72}Ga
Ge	^{75m}Ge
	^{75m}Ge → ^{75}Ge
	^{77m}Ge
	^{77m}Ge → ^{77}Ge
As	^{76}As
Se	^{75}Se
	^{77m}Se
Br	^{80m}Br
	^{80m}Br → ^{80}Br
	^{82m}Br → ^{82}Br
Rb	^{86m}Rb → ^{86}Rb
	^{88}Rb
Sr	^{85m}Sr
	^{85m}Sr → ^{85}Sr
	^{87m}Sr
Y	^{90m}Y
Zr	^{95}Zr
	^{95}Zr → ^{95m}Nb
	^{95}Zr → ^{95m}Nb → ^{95}Nb
	^{97}Zr
	^{97}Zr → ^{97m}Nb
	^{97}Zr → ^{97m}Nb → ^{97}Nb
Nb	^{94m}Nb
Mo	^{99}Mo
	^{99}Mo → ^{99m}Tc
Mo	^{101}Mo
	^{101}Mo → ^{101}Tc
Ru	^{97}Ru
	^{103}Ru
	^{105}Ru
	^{105}Ru → ^{105m}Ru
	^{105}Ru → ^{105m}Ru → ^{105}Ru
Rh	^{104m}Rh
	^{104m}Rh → ^{104}Rh
Pd	^{109m}Pd
	^{109m}Pd → ^{109}Pd
	^{109m}Pd → ^{109}Pd → ^{109m}Ag
	^{111m}Pd
Ag	^{108}Ag
Cd	^{110m}Ag
	^{110}Ag
	^{115}Cd
	^{115}Cd → ^{115m}In
In	^{114m}In
	^{116m}In
Sn	^{113m}Sn → ^{113}Sn → ^{113m}In
	^{117m}Sn
	^{123m}Sn
	^{125m}Sn
	^{125m}Sn → ^{125}Sn
	^{125m}Sn → ^{125}Sn → ^{125}Sb
Sb	^{122m}Sb → ^{122}Sb
	^{124m}Sb → ^{124m1}Sb → ^{124}Sb
	^{124m2}Sb → ^{124m1}Sb
Te	^{131m}Te → ^{131}Te → ^{131}I
I	^{128}I
Cs	^{134m}Cs
	^{134m}Cs → ^{134}Cs
Ba	^{131m}Ba → ^{131}Ba
	^{133m}Ba
	^{139}Ba
La	^{140}La
Ce	^{141}Ce
	^{143}Ce
Pr	^{142m}Pr → ^{142}Pr
Nd	^{147}Nd
	^{149}Nd
	^{149}Nd → ^{149}Pm
	^{151}Nd
	^{151}Nd → ^{151}Pm
Sm	^{153}Sm
	^{155}Sm
Eu	^{152m}Eu
	^{152m}Eu → ^{152}Eu
	^{154m}Eu → ^{154}Eu
Gd	^{153}Gd
	^{159}Gd
	^{161}Gd
Tb	^{160}Tb
Dy	^{165m}Dy
	^{165m}Dy → ^{165}Dy
Ho	^{166}Ho
Er	^{171}Er
Tm	^{170}Tm
Yb	^{169}Yb
	^{175}Yb
	^{177}Yb
Lu	^{177}Lu
	^{176m}Lu
Hf	^{175}Hf
	^{179m}Hf
	^{180m}Hf
	^{181}Hf
Ta	^{182m}Ta → ^{182}Ta
W	^{187}W
Re	^{186}Re
	^{188m}Re
	^{188m}Re → ^{188}Re
Os	^{185}Os
	^{191m}Os → ^{191}Os
	^{193}Os
Ir	^{194}Ir
Pt	^{199m}Pt → ^{199}Pt → ^{199}Au

Table 2. continued

Au	^{198}Au
Hg	^{197m}Hg
	^{203}Hg
	^{205}Hg
Th	^{233}Th $\rightarrow$ ^{233}Pa
U	^{239}U
	^{239}U $\rightarrow$ ^{239}Np

reaction producing the radionuclide $^{M+1}_{Z}$E. An (n, p) reaction on nuclide $^{M+1}_{Z+1}$E' or an (n, α) reaction on nuclide $^{M+3}_{Z+2}$E'' will produce the same indicator radionuclide, resulting in positive errors. This *nuclear interference* depends on the concentration ratios of the elements in the sample, the fast-to-thermal flux ratio, and the cross sections involved, and can be estimated by a simple calculation. The cross sections for (n, p) and (n, α) reactions are usually much smaller than the (n, γ) cross sections. Well-known examples are ^{27}Al (n, γ) ^{28}Al interference by ^{28}Si (n, p) ^{28}Al, and ^{55}Mn (n, γ) ^{56}Mn interference by ^{56}Fe (n, p) ^{56}Mn.

If the interference is not too important it can be corrected for. If uranium or other fissionable material is present in the sample, the fission products with high fission yields (Sr, Mo, Zr, Ce, Ba) can induce important positive errors. The interference can easily be estimated when the uranium content is known. In FNAA, *secondary interference reactions* may occur when fast neutrons interact with other elements and produce particles that induce a nuclear reaction that forms the same indicator nuclide. These particles are usually protons ejected by fast neutrons from a matrix with a high hydrogen content. Examples are:

^{19}F (n, 2 n) ^{18}F interference by: ^{17}O (p, γ) ^{18}F and ^{18}O (p, n) ^{18}F

^{14}N (n, 2 n) ^{13}N interference by: ^{13}C (p, n) ^{13}N

24.2.5. Measuring Equipment

For identification and quantification of the radionuclides produced during the activation step a radiation detector is needed. The major way to select a specific radionuclide is of course by *radiochemical separation*. At the end of the irradiation a nonradioactive carrier of the element to be measured can be added prior to separation. A second parameter used for selection is the *half-life* of the radionuclide. Selection on the basis of the *energy of the emitted radiation* is, however, the most powerful technique. Obviously, detection of the γ and X rays is much more suitable for identification and selective for quantification purposes than β-ray detection. Alpha emitters are almost never produced in activation analysis.

Gas-filled detectors, such as ionization chambers, proportional counters, and Geiger–Müller counters are mainly sensitive to β-radiation. They rarely allow any selection on the basis of energy. Gamma counting with a selection for energy is called gamma spectrometry and is performed by means of *scintillation* or *semiconductor detectors*. Currently the gamma spectrometric measurements in neutron activation analysis are only performed with semiconductor detectors. Scintillation counters will therefore be discussed only briefly.

Interaction of γ or X rays with matter leads to three effects: (1) A photoelectric effect in which the entire energy is absorbed by the detector and the result is a photopeak in the spectrum, with a pulse height proportional to the full energy of the gamma radiation. (2) Compton scattering results in a partial absorption of the γ-ray energy. In the spectrum a continuous background is seen, which can interfere with the measurement of other lower-energy γ or X rays. (3) Pair production may result in the escape of one or two annihilation photons, each with an energy of 0.511 MeV. The *efficiency* ε of a detector refers to the ratio of the number of radiations actually detected to the number emitted by the source. In addition a photopeak-efficiency ε_p may be defined as the number of counts recorded under the photopeak. The *energy resolution R* is the ability of the detector to discriminate between two γ or X rays of different energy. This resolution is defined as the full width ΔE of the peak at half-maximum (FWHM). For scintillation detectors, it is expressed mostly as a percentage of the energy E corresponding to the centroid of the peak. For semiconductor detectors, it is expressed in energy units (kiloelectronvolts or electronvolts). Finally, the *dead time* τ of the detector is the amount of time required before the detector can recover from an incoming radiation and respond to the next event. In gamma spectrometry the main cause of dead time is not the detector itself but the electronic equipment and, more specifically, the amplifier (i.e. pulse pile up) and the multichannel analyzer (MCA).

Scintillation Detectors. In scintillation detecors the incoming radiation interacts with the material by ionization or excitation. The excited atoms or

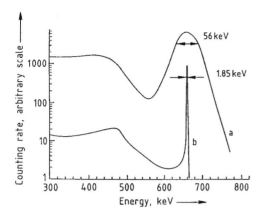

Figure 2. Comparison of gamma spectrum of ^{137}Cs (gamma energy 661.6 keV) obtained with 7.5×7.5-cm NaI(Tl) scintillation detector (a) and with a 50-cm^3 Ge(Li) semiconductor detector (b)

molecules undergo deexcitation by the emission of a photon of light. This light passes through the optically transparent scintillator material and is guided through a light pipe to a photomultiplier, which converts the light to an electrical signal. The signal is then amplified and fed to a counting or storage system. This can be a single-channel analyzer (SCA) or a multichannel analyzer (MCA). The detection efficiency is excellent. The major drawback of NaI(Tl) scintillators is their poor energy resolution, which is typically 7 % for the 661.6-keV line of ^{137}Cs (Fig. 2). Dead times are important at high count rates. More selective counting can be performed with NaI(Tl) detectors when coincidence or anticoincidence spectrometry is applied by using two or even more detectors.

In charged-particle activation analysis (Chapter 24.4) pure positron emitters have to be measured by their annihilation photons, emitted simultaneously in a 180° geometry. The sample is then introduced between two NaI detectors, coupled to an SCA to select the photopeak of the 511-keV annihilation photons. The fast SCA output is fed into a coincidence circuit, which selects only coincident signals (typically within 40 ns). The poor energy resolution of the NaI detector is no drawback, while the high detection efficiency (as compared with semiconductor detecors) is an advantage. The NaI detectors are much cheaper than semiconductor detectors (for the same detection efficiency) and do not require liquid nitrogen cooling. The signal-to-noise ratio is much higher for the NaI coincidence set-up, than for a Ge spectrometer, mainly because the noise (background) is much lower.

Semiconductor Detectors. These have a $p-i-n$ diode structure in which the intrinsic region is formed by depletion of charge carriers by applying a reverse bias across the diode. The energy of an electron in a semiconductor material is confined to the conduction and the valence bands, both states separated by forbidden energies. The passage of radiation through the semiconductor may inject sufficient energy to raise an electron from the valence band to the conduction band, thus creating an electron–hole pair. In order for the material to be used as a radiation detector, this electrical charge must be collected. Therefore, several thousand volts are applied to the detector. Owing to the small amount of energy needed to create an electron–hole pair, a large number of pairs (>300 000 pairs/MeV) is created, with small relative statistical fluctuations. Since the energy resolution of the detector depends on this fluctuation, which can be predicted by Poisson statistics, much better energy resolution is obtained than with scintillation counters. For a highly efficient detector, a large, nearly intrinsic or depletion region must be present. Therefore, the $p-n$ junction is created within a single p or n crystal. Ways to achieve this are the surface barrier detector (used only for detection of charged particles because the depletion region is too small), the lithium-drifted detector [Ge(Li) and Si(Li)], and at present the high-purity or intrinsic germanium detector (HP-Ge). *Ge(Li) detectors* must be cooled constantly at liquid nitrogen temperature, they are still in use but are no longer available. They are more efficient for γ-ray detection than Si(Li) detectors, owing to their higher atomic number Z. *Si(Li) detectors* are sufficiently efficient for detection of X rays and low-energy γ rays and must be cooled only during operation to ensure optimum energy resolution. Typical resolutions of Si(Li) detectors are 140–160 eV for the 5.9-keV Mn K line from a ^{55}Fe source.

HP-Ge detectors have the advantage that they can be stored at room temperature and are cooled only during operation to reduce the problem of thermal excitation of electrons. Coaxial *p*-type detectors and *n*-type detectors are available. In the *n-type detector* the inner contact is made by diffused lithium, while the outer contact is achieved by ion implantation, which ensures a very thin entrance window (0.3 µm). Thin *planar germanium detectors* (5–20 mm) can also be used to measure low-energy radiation and have the advantage of not stopping the more energetic γ rays, thereby reducing interference effects such as Compton radiation. *Well-type detectors* are also available with well sizes up to 15 mm in diameter and a depth of 40 mm. For cooling, a cold finger and a large cryostat are needed, which makes semiconductor detectors bulky and difficult to maneuver.

Germanium detectors are characterized by three parameters: resolution, peak-to-Compton ratio, and efficiency. The *resolution* is typically given for the 1332-keV ^{60}Co line and varies from 1.8 keV for the very best to 2.3 keV for the very large detectors. The *peak-to-Compton ratio* is measured as the ratio of the number of counts in the 1332-keV peak to the number of counts in a region of the Compton continuum. Values vary from 30 to 90 for the most expensive model. The efficiency is expressed as a relative efficiency compared with the 7.5×7.5-cm NaI(Tl) scintillation detector. Relative efficiencies of HP-Ge detectors vary from 10 % up to 150 %. The dead time of semiconductor detectors is low, so the count rate is limited largely by the electronic circuit.

Gamma spectrometry is performed mainly by using MCAs, and the relation between the obtained pulse height or corresponding channel number and gamma energy must be established. When an absolute method of standardization is applied, the peak detection efficiency ε_p of the detector with surrounding material must also be measured.

For *peak area calculation*, two procedures can be applied in principle: integration or fitting. In *integration*, the number of counts recorded in the channels in which about 90 % of the peak is located is corrected for Compton contribution from other peaks by subtracting the content of an equal number of channels before and after the peak. The method fails, of course, when other peaks are located in the immediate neighborhood or in the case of superimposed peaks. *Fitting procedures* assume a defined shape, either Gaussian or with non-Gaussian distortions (determined from the shape of a well-defined peak in the spectrum).

The overall procedure for the evaluation of γ-ray spectra as part of activation analysis consists of several steps as defined by ERDTMANN und PETRI [34]:

1) Search for the peaks
2) Calculation of γ-ray energy
3) Calculation of peak areas
4) Calculation of γ-ray emission rates
5) Identification of radionuclides
6) Calculation of decay rates
7) Calculation of amounts of elements

The amount of data to be handled during this evaluation procedure requires the use of computers. At present, the multichannel analyzer not only carries out pulse height analysis but also does most of the data processing. The manufacturers of computer-based MCAs usually provide built-in or software programs for the different aforementioned steps.

Obviously the use of a personal computer is an important step toward automation of the analysis. The computer can control the automatic sample changer, the timing of the analysis, counting, storing of the spectra, and finally, data processing.

24.2.6. Procedures for Activation Analysis

Sample Preparation. As in any other analytical technique, the quality of the final results depends strongly on the care taken during sampling ($\rightarrow$ Sampling). Since sample sizes can often be small (2 mg up to 20 g), owing to the high sensitivity of NAA, a well-homogenized material is required. In a number of cases, some preirradiation treatment will be necessary, such as: (1) the removal of water by drying, freeze drying, or ashing for biological materials, or evaporation for liquid samples, to avoid pressure buildup due to evaporation or radiolysis during irradiation; (2) the removal of surface contamination; or (3) perhaps even preconcentration. All of these treatments should be performed in clean environments.

For the irradiation itself, the sample is packed in polyethylene vials for total neutron doses of less than 10^{17} cm^{-2} and in high-purity quartz ampoules for longer irradiations at higher fluxes. The packed samples are transferred to the reactor in containers (rabbits) of aluminum, polyamide, graphite, or high-pressure polyehtylene for low-dose irradiation.

The major advantage of activation analysis, however, is that after irradiation, contamination with nonradioactive material does not introduce more errors. On the contrary, when radiochemical separations are required, an amount of nonradioactive carrier of the element to be separated is added, after irradiation.

Instrumental Neutron Activation Analysis (INAA). A purely instrumental approach is applicable for the determination of many elements (i.e., after irradiation, the sample can be unpacked and the induced radioactivity measured without further treatment). Such a technique has a number of advantages. After decay of the induced activities, the sample is essentially unchanged (nondestructive activation analysis), and is therefore available, in principle, for further investigation. In addition, short-lived radionuclides can be used for the analysis. Tedious chemical separations are not required, which implies reduction of analysis time and avoidance of errors associated with separation. INAA becomes feasible when the activity induced in the sample matrix is not prohibitively high and no single major activity is produced that overshadows that of the other radionuclides. Of course, the differences in half-lifes of various radionuclides can also be exploited by limiting the saturation factor S of interfering long-lived activities or by allowing the short-lived matrix activity to decay before counting. Virtually all instrumental multielement activation analysis is based on high-resolution gamma spectrometry using semiconductor detectors after thermal or epithermal activation of the sample in a reactor or a neutron generator.

INAA is favored especially for many organic, biological, and environmental samples, because the elements O, H, C, N, S, and P are not activated or do not produce gamma emitters. The most abundant minor or trace elements such as Si, Fe, Al, Ca, Mg, and Cl have relatively small cross sections or short half-lifes, the nuclides with the highest activity often being ^{24}Na ($t_{1/2}$ = 15.0 h) or ^{82}Br ($t_{1/2}$ = 36 h). Also, a number of metallic matrices can be analyzed purely instrumentally with thermal neutrons. Examples of these metals are Be, Mg, Al, Ca, Ti, V, Ni, Y, Zr, Nb, Rh, Pb, and Bi, some of them after decay of the short-lived matrix activity.

In multielement analysis, choosing the optimal irradiation, cooling, and counting conditions for each element is obviously not feasible, so a compromise must be found. The balance between working up a reasonable number of samples and the desired quality of the analytical results will dictate the mode of the irradiation–counting scheme. A typical scheme has been described for atmospheric aerosols (Fig. 3) [35]. For the *short-lived radionuclides*, irradiation of a 2- to 25-mg sample for a few minutes, together with a flux monitor, is employed. Typically, 10–15 elements are determined from this irradiation. *Longer irradiation* of a number of samples, together with a multielement standard or a flux monitor, is followed by two to three counts after decay times varying from 1 to 30 d. In total, the scheme allows detection of up to 45 elements in favorable cases although, more realistically, about 30 elements can be determined with precisions varying from 2 to 10 %. Shorter and less complicated schemes can be applied at the expense of precision and the number of elements determined. The selectivity for the determination of some elements can be enhanced by applying epithermal, pulsed, or cyclic activation and by counting with low-energy photon detectors or anticoincidence spectrometers.

Radiochemical Neutron Activation Analysis. The ultimate sensitivity and the lowest detection limits are generally obtained when radiochemical separations are performed after irradiation. Chemical treatment of the samples and the standards is aimed at separating the radionuclides formed. The separated fractions are then used in a chemical and physical form suitable for counting. Because the actual amount of radionuclides is very small, a chemical carrier is added before the decomposition step to ensure chemical equilibration with the radionuclide. The next stage is chemical separation using classical methods such as precipitation, distillation, solvent extraction, chromatography, ion exchange, or electrodeposition. Since the amount of carrier added is known and greatly exceeds the amount of the element originally present, the yield of the chemical separation can be determined. Correction of the activity measured is possible.

If clean separation of one radionuclide is achieved, scintillation counting or even β-counting can be performed in principle. Nevertheless, high resolution HP-Ge spectrometry is applied to check the absence of any interfering radionuclide. For most types of samples, such as geological, environmental, and biological materials, typical group separation schemes have been developed and can be found in the literature. For biological samples, the emphasis is on the removal of major activities from ^{32}P, ^{24}Na, and ^{82}Br. In geological samples, rare earth elements or noble metals are often determined after separation. Sometimes carriers or so-called scavengers such as hydroxides of Fe^{3+}, Mn^{4+}, and Sn^{4+} are used to remove interfering activities or to separate the analyte radionuclides.

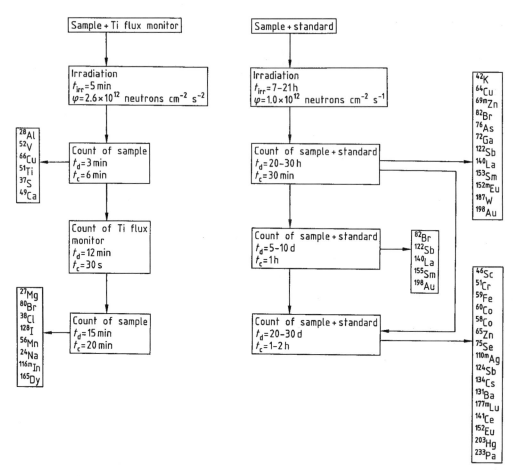

Figure 3. Irradiation–counting scheme for INAA of air particulates collected on filter paper [35]
φ = Thermal neutron flux; t_{irr} = Irradiation time; t_d = Decay time; t_c = Counting time

24.2.7. Sensitivity and Detection Limits

The smallest amount of an element that can be determined depends on the specific activity produced and the minimum activity measurable with sufficient precision. From Equation (10) the activity produced and measured per gram of an element can be calculated. However, the minimum activity that can be measured depends not only on the decay properties of the radionuclide and on the counting equipment, but also on the background of the detector or the Compton continuum on which the photopeak must be detected in gamma spectrometry. Thus the detection limit of an element for specific irradiation–counting conditions is not immutable since it depends on the presence of other radionuclides.

For measurement results giving a Poisson distribution, as in radioactive decay, the equation used to calculate the standard deviation σ is simply the square root of the number of counts recorded when the number is large. The net signal S is obtained as the difference between the measured signal $(S+B)$ minus the background B (e.g., Compton continuum in gamma spectrometry). Thus the statistically derived standard deviation is given as

$$\sigma_s = (S + B + B)^{1/2} \qquad (16)$$

The limit of detection depends on whether the background is well known or is not well known as in gamma spectrometry. CURRIE [36] defines-

Table 3. Detection limits (units: 10^{-12} g) for INAA under ideal conditions of gamma spectrometry [135 cm^3 Ge(Li) detector, lead shielded] [a] [35]

Na	0.1	Ge	30	Sb	0.02	Tm	2
Si	900 000	As	0.03	Te	3	Y	0.06
Cl*	10 000	Se	3	I*	1500	Lu	0.03
Ar*	30	Br	0.05	Cs	0.5	Hf	0.4
K	5	Rb	8	Ba	40	Ta	0.4
Ca	1 500	Sr*	50	La	0.02	W	0.04
Sc	0.04	Y	1 500	Ce	1.5	Re	0.015
Ti	600	Zr	15	Pr	0.5	Os	0.8
Cr	3	Nb	15 000	Nd	5	Ir	0.006
Mn*	0.4	Mo	0.5	Sm	0.002	Pt	1
Fe	300	Ru	2	Eu	0.003	Au	0.001
Co	0.8	Pd	40	Gd	0.9	Hg	1
Ni	200	Ag	2	Tb	0.15	Tl	3 000
Cu	2	Cd	0.6	Dy*	2	Pb	30 000
Zn	6	In*	4	Ho	0.3	Th	0.2
Ga	0.06	Sn	100	Er	0.5	U	0.06

[a] For detection of radionuclides with $t_{1/2} > 1$ h, irradiation: 48 h at 2×10^{14} cm^{-2} s^{-1} (cadmium ratio =11, i.e., ratio obtained by irradiation without and with cadmium cover); decay time 10 h; counting time 64 h. For detection of radionuclides with $t_{1/2} < 1$ h (marked with *), irradiation: 2 h at 3×10^{11} cm^{-2} s^{-1}; decay time 1 h; counting time 3 h.

three specific quality levels for a measurement (→ Chemometrics):

1) *Decision limit* L_c is the net signal above which an "observed" signal can be accepted (95% probability) as detected; L_c equals 1.64 σ_B or 2.33 σ_B; for well-known or not well-known background, respectively
2) *Detection limit* L_D is the "true" net signal that a priori may be expected to be detected; L_D equals 3.29 σ_B or 4.65 σ_B
3) *Determination limit* L_Q is the net signal on which quantitive determination with a precision of 10 % can be performed; L_Q equals 10 σ_B or 14.1 σ_B

Because the sources of errors in NAA are generally known, they can be kept under control. When good laboratory practice is applied, the reproducibility and precision can be nearly as good as the counting statistics, in favorable cases as low as 1–2%.

Detection limits obtainable by multielement INAA were determined by VERHEYKE and RATH [37] for compromise irradiation–counting conditions but for ideal gamma spectrometry conditions, namely, calculated from a background (Compton continuum) spectrum obtained from very pure irradiated graphite (Table 3). For several elements (Ar, Mn, Sr, Y, Dy) a better sensitivity can be calculated for a shorter decay time. For some elements (F, Mg, Al, V, and Rh) giving rise to radionuclides with half-lifes of less than 10 min no limits are given, although they can be determined with low detection limits after a short irradiation and a fast count. Elements such as S, P, and Bi produce only β-emitters and are therefore absent from the table.

24.3. Photon Activation Analysis [23], [24]

Photons—often obtained as bremsstrahlung produced in an electron accelerator—with energies varying from 10 to 45 MeV can overcome the threshold of (γ, n), (γ, p), $(\gamma, 2n)$, and (γ, α) reactions. Inelastic scattering with low-energy photons leads to short-lived metastable isotopes by (γ, γ') reaction. The light elements C, N, O, and F are determined by (γ, n) reactions, but the reaction leaves the nuclides proton rich, so that a positron emitter is formed. Radiochemical separation or decay curve analysis is needed. Detection limits of < 1 µg are obtainable. *Instrumental photon activation analysis (IPAA)* is possible for heavier elements in the presence of an excess of low atomic number material, because when a bremsstrahlung spectrum is used, the yields of the reactions increase monotonically with increasing atomic number. Typical detection limits range from 0.01 to 1 µg. IPAA has been applied to the analysis of many environmental samples. Typically, about 15 elements can be determined precisely. Photon activation analysis was considered a valuable complement to NAA for the determination of light elements and some heavy elements before charged-particle activation analysis

(CPAA) was developed. The major drawback is the limited availability of high-energy electron beams with sufficient intensity.

24.4. Charged-Particle Activation Analysis

Although NAA and PAA have a few characteristics in common with charged-particle activation analysis (CPAA), there are fundamental differences, owing to the electric charge of the bombarding particle.

In NAA there are endo-ergic reactions [e.g., (n, 2n) or (n, α) in FNAA] as well as exo-ergic reactions [e.g., (n, γ) in TNAA] and this is also the case for CPAA. However, owing to the Coulombic repulsion between a charged particle (CP) and the target nucleus, CPs always require a minimum energy, even for endo-ergic reactions. For CPAA, mono-energetic particles with energies between 5 and 50 MeV are utilized, leading to more nuclear reactions, and more possible nuclear interferences. For a proton energy below 10 MeV, e.g, in general (p, n) and (p, α) reactions are possible. Below 20 MeV also (p, 2n) (p, d) (p, t), (p, ^{3}He), (p, αn) ... reactions have to be considered. For TNAA only (n, γ) reactions have to be considered, nuclear interference from a (n, p) reaction is fairly infrequent.

CPs are slowed down in matter and are all stopped at depth, called the range, being 0.1 to 1 mm. As a consequence the beam intensity is hardly reduced, in contrast to neutron shadowing. Moreover, the induced activity is not related to the mass of the analyte, but to its mass fraction. So standardization (calibration) is very different from NAA.

The *stopping power (S)* is defined as the energy loss ($-\mathrm{d}E$) per unit pathlength ($\mathrm{d}l$) for a particular CP in a particular target:

$$S = \frac{-\mathrm{d}E}{\mathrm{d}l} \qquad (17)$$

It depends on the CP energy. As a result the CPs are stopped at a depth, called *range R*. The range is characteristic for a CP, a target, and the CP energy, being mono-energetic.

The equivalent of $\varphi \sigma N_A$ in Equations (3)–(10), i.e., the reaction rate in s^{-1}, can be written for an infinitively thin target

$$I \sigma n \, \mathrm{d}l \qquad (18)$$

with I = beam intensity (number of CPs per s), n = number of target nuclides per cm^3, and $\mathrm{d}l$ = thickness (cm) of the infinitively thin target. For a target thicker than the range R, equation (18) is integrated

$$In \int_0^R \sigma_l \, \mathrm{d}l = In \int_{E_i}^0 \frac{\sigma_E \mathrm{d}E}{\mathrm{d}E/\mathrm{d}l} = In \int_{E_i}^{E_i} \frac{\sigma_E \mathrm{d}E}{S_E} \qquad (19)$$

with E_i the incident CP energy.

The sample is supposed to be homogenous. The CP beam intensity is nearly unchanged when penetrating the target, while the CP energy decreases, and consequently the reaction cross section. As the reaction cross section σ is only known as a function of the energy, the stopping power S is used. For a target for which the thickness D is less than the range R, Equation (19) can be rewritten

$$In \int_0^D \sigma_l \, \mathrm{d}l = In \int_{E_i}^{E_0} \frac{\sigma_E \mathrm{d}E}{\mathrm{d}E/\mathrm{d}l} = In \int_{E_0}^{E_i} \frac{\sigma_E \mathrm{d}E}{S_E} \qquad (20)$$

with E_0 = outgoing energy.

It is clear that the induced activity is proportional to the mass fraction ($\sim n$ in cm^{-3}) of the analyte, in contrast to NAA where it is proportional to the mass of the analyte ($\sim N^A$ dimensionless) in Equations (3)–(10). This difference in dimensions may be suprising for two different modes of activation analysis. However, it is related to (1) the analysed depth, being limited to the range (cm) and (2) the use of beam intensity (in s^{-1}) versus fluency rate (in cm^{-2} s^{-1}) in NAA.

As relative standardization is applied in CPAA, there is no need for absolute activity and beam intensity measurements, nor for absolute data of the reaction cross section σ and stopping power S. If the (relative) cross section is not accurately known, approximations have been developed. The stopping power of any elemental matter can be calculated accurately; for mixtures or compounds the major matrix should be known. If not, an internal standardization method has been developed.

CPAA has been applied successfully for the determination of (trace) elements in the bulk of solid samples, such as pure metals and alloys, semiconductors, geological and solid environmental materials. The method has proven its unique capabilities and outstanding performance in the determination of light elements such as B, C, N, and O. Activation analysis has the unique possibility of removing surface contamination after irradiation and prior to measurement. CPAA has been a milestone, e.g., in the assessment of the bulk oxygen concentration in aluminum.

Recently CPAA has been extended to surface characterization. If it is possible to determine a trace element A in a matrix B at the µg/g level in the bulk of a sample, i.e., in a 1 mm thick layer, then it is also possible to determine the thickness of a mono-elemental layer A on a substrate B (that does not contain A) down to 1 nm. Alternatively, one can determine the composition of such a layer if the thickness is known, and consequently the stoichiometry without any foreknowledge of the thickness.

The principles of CPAA for bulk analysis are summarized in [25] and [26]; for surface characterization in [27] and in [28]. Applications of CPAA for bulk analyses are reviewed in [38]; the feasibility study on surface characterization in [39].

24.5. Applications

24.5.1. High-Purity Materials

Since the 1960s, activation analysis has been used intensively for determining trace elements in high-purity materials. Analysis of semiconductor materials, as well as of high-purity metals, obtained by zone refining, and of materials used in nuclear technology, has been one of the major fields of interest in activation analysis. Impurities are known to affect the properties of these materials, even when present in concentrations so low that they can barely be detected by other methods.

INAA is especially suited for the multielement analysis of matrices that, on neutron irradiation, do not produce intense gamma radiation or produce only short-lived ones. Examples of these matrices are graphite, silicon, and aluminum, or their oxides. A typical irradiation counting scheme allows the determination of 55 elements at submicrogram-per-gram concentrations in wafers of silicon [37]. Also, other semiconductor or alloying materials such as Se, Ge, Sb, Te, and ultrapure metals such as Cu, Fe, Ti, Mo, Nb, Ni, Ga Zn, Sn, Zr, Bi, Pb, and noble metals have been analyzed. Often a postirradiation separation of the matrix activity is required, or a radiochemical separation scheme must be applied based on ion exchange, extraction, or precipitation. Matrix activation can sometimes be avoided by irradiation in a well-thermalized neutron flux or under the cover of cadmium (ENAA).

Photon activation (PAA) is used for elements not activated by (n, γ) reaction, with the emphasis on carbon, nitrogen, and oxygen. Detection limits of the order of 20 µg can be obtained.

Even lower limits of detection can be obtained for determining the light elements in semiconductor materials or ultrapure metals by *CPAA*.

The major advantage of activation analysis over nearly all other techniques of analysis — namely, the absence of contamination problems after irradiation — is extremely valuable in this field of application. Preirradiation treatments should be carried out in clean environments, and to further avoid surface contamination, postirradiation washing or etching is highly recommended. Assessing the impurity levels of these typical industrial products by activation analysis is often used in the quality control of manufacturing companies and has therefore been entirely automated and computerized by using a single comparator for calibration.

24.5.2. Environmental Materials

Neutron activation analysis is one of the major techniques for the determination of many minor and trace elements in a large variety of solid environmental and pollution samples, such as atmospheric aerosols, particulate emissions, fly ash, coal, incineration ash, and sewage sludge. *Instrumental neutron activation analysis* of total, inhalable, or respirable airborne particulate matter collected on a cellulose or membrane filter, or in a cascade impactor on some organic substrate, allows the determination of up to 45 elements by an irradiation – counting scheme similar to the one given in Figure 3. *Radiochemical NAA* is applied only when extremely low limits of determination are required. *Instrumental photon activation analysis* is also complementary to INAA.

The problem in charged-particle activation analysis related to a matrix of unknown composition and degradation (votalization) of the organic matrix during irradiation, has been solved by the development of an internal standardization method.

When applied to water samples, NAA suffers from some severe drawbacks compared with other multielement techniques. Irradiation of water results in radiolysis. Evaporation, freeze drying, adsorption on charcoal, and ion exchange can be used to remove water. A final problem is the interference of intense radiation induced in matrix elements such as Na, K, Cl, Br, and P. Saline water in particular, yields difficult samples for INAA. The preconcentration step may include separation from these elements. A wealth of literature exists describing these environmental applications [40], [41].

24.5.3. Biological Materials

Activation analysis, particularly with neutrons, is a very effective method for elemental trace analysis of biological materials and plays an important role in studies of trace elements related to health. Because of the limited sensitivity of classical analytical methods, the trace and ultratrace elements that play an important role in biological systems can rarely be determined by the usual methods. Radiochemical NAA has contributed significantly to establishing the normal levels of these elements in blood (total blood, serum, packed cells, and erythrocytes), liver, kidney, lung, muscles, bones, teeth, nerves, and hair and in studies involving disease [42]. It is the method of choice for the determination of Sb, As, Cs, Co, Mn, Mo, Rb, Se, Ag, and V, and is very reliable for Br, Cr, Cu, Hg, I, and Zn, but it lacks sensitivity for Cd, Ni, and Sn [43]. Since radiochemical separations are generally required, a number of group separations using ion exchange or distillation procedures have been developed to remove at least the interfering activities from P, Na, and Br. For a limited number of elements (Co, Cs, Fe, Rb, Se, and Zn), INAA is possible only after a long decay period [44]. Activation analysis offers many advantages for the analysis of foods and plants. INAA of vegetables, meat, fish, and poultry allows the detection of about 20 trace elements [43].

In the last decade speciation (the determination of the species of an element, i.e., its oxidation state or its compounds) has become more important than trace element determination. Speciation analysis is obtained by hyphenated techniques, i.e., coupling of, for example, a chromatographic separation technique to a very sensitive elemental detector. As NAA cannot measure on-line, the role of NAA becomes less important. However, for the development of separation and preconcentration techniques, the use of radiotracers with very high specific activity is an outstanding tool, as these techniques can be developed at realistic concentration levels. Radiotracers produced by charged-particle activation, provide just such a specific activity, as the radionuclide formed is not an (radio)isotope of the element irradiated.

24.5.4. Geo- and Cosmochemistry

Activation analysis has probably most often been applied in the analysis of geological samples. The NAA results on rare earth elements in meteorites, rocks, and sediments were a significant contribution to the development of modern geo- and cosmochemistry. Also in geological surveys, hundreds of analyses are performed on soil and stream sediments by this technique. Silicate rock, nonsilicate rock, meteorites, minerals, and marble can often be analyzed purely instrumentally. *Epithermal activation* enhances the detectability of a number of interesting elements. When, however, up to 40 elements are to be determined — including the rare earths and the platinum-group elements — a preconcentration or some group separations based on ion exchange and liquid–liquid extraction are applied. A *14-MeV activation* can be applied for the determination of some major elements (O, F, Mg, Al, Si, P, Fe, Cu, Zn). Radioanalysis in geochemistry has been described in [45].

24.5.5. Art and Archaeology

Archaeologists and museum directors hope that small samples of art objects can be analyzed with high sensitivity, preferentially in a nondestructive way. INAA can usually achieve this. The wealth of information on the concentration of trace elements, obtained by the technique, can often serve as so-called fingerprints for identification or classification purposes. The concentrations may give clues to the provenance of the artifact, the methods of treatment, or the cultural or technological context in which it was made. The artifacts of archaeological interest analyzed range from metals (coins, medals, statuettes, utensils) to nonmetallic artifacts (clays, pottery, marble, obsidian, and paintings). An overview is given in [46].

24.6. Evaluation of Activation Analysis

When activation analysis is compared with other analytical techniques, it is most suited for the multielemental analysis of minor, trace, and ultratrace elements in solid samples. Often a purely instrumental approach is possible, making the technique nondestructive and leaving the sample intact for other investigations. About 65 % of the elements of the Periodic Table can be determined by *INAA* at concentrations lower than micrograms per gram and, in favorable cases, down to nanograms per gram. The detection of another 20 % is less sensitive, and about 10 – 15 % (mostly light elements) can rarely be detected at all. These can, however, be detected by *FNAA* (F) or *CPAA* (B, C, N, O). The lowest detection limits are always obtained after radiochemical separation. These low detection limits are possible because the technique can be made highly selective. Adjustable parameters are irradiation and waiting time before counting, the nature and energy of the bombarding particle or radiation, and above all the high resolving power of the measurement. A *major benefit* of activation analysis is that it can provide very accurate results for trace concentrations. Contamination from reagents or the laboratory environment is excluded after irradiation. Radiochemical separations can be performed after the addition of inactive carrier, avoiding the need for working with trace amounts. The absorption effects of neutrons during irradiation and of gamma rays during counting are small and generally negligible or can be corrected for. The information obtained is independent of matrix or chemical form. Other sources of error, such as interfering reactions, can generally be calculated and thus accounted for or avoided. Therefore the reproducibility and the precison can be as good as the counting statistics when good laboratory practice is applied. In favorable cases and for homogeneous samples, it can be as good as 1 – 2 %. For the detection of major contents, a precision of > 0.5 % is rarely achievable. The accuracy depends to a large extent on the calibration procedure. When appropriate standards are used, very accurate results can be obtained, making this a technique preeminently suited for certification purposes. The *major drawbacks* are probably that it is expensive, requires access to nuclear facilities such as a reactor, and is generally slow. Rapid analysis based on short-lived isotopes is possible only in exceptional cases. The cost of the counting equipment is of the order of or much less than most other instruments for trace analysis [atomic absorption spectrometry (AAS), inductively coupled plasma optical emission spectrometry (ICP-OES) and mass spectrometry (ICP-MS), X-ray fluoroescence (XRF), etc.]. To work with radioisotopes, a laboratory must meet legal requirements for radiological safety. An additional disadvantage is the lack of information on the chemical form of the element, which makes speciation studies difficult. Finally, analysis of strongly activated matrices renders the instrumental approach impossible and necessitates the application of important shielding material for the radiochemical separations.

24.7. References

General References

[1] W. D. Ehmann, D. E. Vance, J. D. Winefordner, I. M. Kolthoff (eds.): *Radiochemistry and Nuclear Methods of Analysis,* Wiley-Interscience, New York 1991.
[2] G. Erdtmann, H. Petri in I. M. Kolthoff, P. J. Elving, V. Krivan (eds.) *Treatise on Analytical Chemistry,* 2nd ed., part I, vol. 14, Wiley-Interscience, New York 1986, pp. 419 – 643.
[3] J. Hoste et al. in I. M. Kolthoff, P. J. Elving, V. Krivan (eds.): *Treatise on Analytical Chemistry,* 2nd ed., part I, vol. 14, Wiley-Interscience, New York 1986, pp. 645 – 775.
[4] S. J. Parry in J. D. Winefordner, I. M. Kolthoff (eds.): *Activation Spectrometry in Chemical Analysis,* Wiley-Interscience, New York 1991.
[5] Z. B. Alfassi (ed.): *Activation Analysis,* vols. I and II, CRC-Press, Boca Raton 1989.
[6] J. Tölgyessy, M. Kyrš: *Radioanalytical Chemistry,* vols. 1 and 2, Ellis Horwood, Chichester 1989.
[7] D. De Soete et al. (eds.): *Neutron Activation Analysis,* Wiley-Interscience, London 1972.
[8] Z. B. Alfassi (ed.): *Determination of Trace Elements,* VCH Verlagsgesellschaft, Weinheim 1994.
[9] Z. B. Alfassi (ed.): *Chemical Analysis by Nuclear Methods,* John Wiley, Chichester 1994.
[10] M. D. Glascock: Activation Analysis, in: Z. B. Alfassi (ed.): *Instrumental Multi-element Chemical Analysis,* Kluwer, Dordrecht 1999.
[11] O. Navrátil et al. (eds.): *Nuclear Chemistry,* Ellis Horwood, New York 1992.
[12] W. Seelmann-Eggebert, G. Pfennig, H. Münzel, H. Klewe-Nebenius: *Karlsruher Nuklidkarte – Charts of the Nuclides,* 5th ed., Kernforschungszentrum, Karlsruhe 1981.
[13] J. R. Parrington, H. D. Knox, S. L. Breneman, E. M. Baum, F. Feiner: *Chart of the Nuclides,* 15th ed., KAPL, Knolls Atomic Power Lab., New York 1996.
[14] R. B. Firestone, V. S. Shirley, C. M. Baglin, S. Y. F. Chu, J. Zipkin: *Table of Isotopes,* 8th ed. (book and CD-ROM + yearly update), John Wiley, Chichester 1996.

[15] M. D. Glascock: *Tables for Neutron Activation Analysis,* 4th ed., University of Missouri, Columbia 1996.
[16] http://nucleardata.nuclear.lu.se/database/se
[17] http://www.nndc.bnl.gov/nndc/nudat
[18] http://www.nndc.bnl.gov/nndcscr/pc_prog/ *Only* NDTxx_16.EXE *and* RADTIONS.ZIP *are required.*
[19] http://www.nndc.bnl.gov/nndc/exfor/
[20] http://www-nds.iaea.or.at/exfor/
[21] http://www.nea.fr/html/dbdata/x4/welcome.html *Only for users registered* at http://www.nea.fr/html/signon.html
[22] http://nucleardata.nuclear.lu.se./database/toi/
[23] C. Segebade, H. P. Weise, G. L. Lutz: *Photon Activation Analysis,* De Gruyter, Berlin 1987.
[24] A. P. Kushelevsky: "Photon Activation Analysis", in Z. B. Alfassi (ed.): *Activation Analysis,* **vol. II,** CRC Press, Boca Raton 1990, pp. 219–2237.
[25] K. Strijckmans: "Charged Particle Activation Analysis", in Z. B. Alfassi (ed.): *Chemical Analysis by Nuclear Methods,* Chap. 10, John Wiley, Chichester 1994, pp. 215–252.
[26] K. Strijckmans: "Charged Particle Activation Analysis", in A. Townshend, R. Macrae, S. J. Haswell, M. Lederer, I. D. Wilson, P. Worsfold (eds.): *Encyclopedia of Analytical Science,* Academic Press, London 1995, pp. 16–25.
[27] K. Strijckmans: "Charged Particle Activation Analysis", in D. Brune, R. Hellborg, H. J. Whitlow, O. Hunderi (eds.): *Surface Characterisation: a User's Sourcebook,* Wiley-VCH, Weinheim 1997, pp. 169–175.
[28] K. Strijckmans: "Charged Particle Activation Analysis", in R. A. Meyers (ed.): *Nuclear Methods – Theory and Instrumentation of the Encyclopedia of Analytical Chemistry (EAC), Part 2: Instrumentation and Applications,* Section: Nuclear Methods, John Wiley, New York 2000.

Specific References

[29] O. T. Høgdahl in IAEA (ed.): *Radiochem. Methods Anal. Proc. Symp. 1964* **I** (1965) 23.
[30] J. Hoste: "Isotopic Neutron Sources for Neutron Activation Analysis," *IAEA-TECDOC* 1988, 465.
[31] F. De Corte, F. Simonits, A. De Wispelaere, J. Hoste, *J. Radioanal. Nucl. Chem.* **113** (1987) 145–161.
[32] F. De Corte et al., *J. Radioanal. Nucl. Chem.* **169** (1993) 125–158.
[33] S. Van Lierde, F. De Corte, D. Bossus, R. Van Sluijs, S. S. Pommé, *Nucl. Instr. Meth.* **A422** (1999)874–879.
[34] G. Erdtmann, H. Petri in I. M. Kolthoff, P. J. Elving, V. Krivan (eds.): *Treatise on Analytical Chemistry,* 2nd ed., **part I, vol. 14.** J. Wiley & Sons, New York 1986, p. 446.
[35] R. Dams, J. A. Robbins, K. A. Rahn, J. W. Winchester, *Anal. Chem.* **42** (1970) 861–866.
[36] L. A. Curie, *Anal. Chem.* **40** (1968) 586–593.
[37] M. L. Verheyke, H. J. Rath in M. Grasserbauer, H. W. Werner (eds.): *Analysis of Microelectronic Materials and Devices,* J. Wiley and Sons, New York 1991, pp. 1–39.
[38] G. Blondiaux, J. L. Debrun, C. J. Maggiore: "Charged Particle Activation Analysis", in J. R. Temer, M. Nastasi, J. C. Barbour, C. J. Maggiore, J. M. Mayer (eds.): *Handbook of Modern Ion Beam Materials Analysis,* Material Research Society, Pittsburgh, PA 1995.
[39] K. De Neve, K. Strijckmans, R. Dams: "Feasibility Study on the Characterization of Thin Layers by Charged-Particle Activation Analysis," *Anal. Chem.* **72** (2000) 2814–2820.
[40] J. Tolgyessy, E. H. Klehr: *Nuclear Environmental Chemical Analysis,* Ellis Horwood, Chichester 1987.
[41] R. Dams in I. M. Kolthoff, P. J. Elving, V. Krivan (eds.): *Treatise on Analytical Chemistry,* 2nd ed., part I, vol. 14, J. Wiley & Sons, New York 1986, p. 685.
[42] J. Versieck, R. Cornelis: *Trace Elements in Human Plasma or Serum,* CRC Press, Boca Raton 1989.
[43] W. C. Cunningham, W. B. Stroube, *Sci. Total Environ.* **63** (1987) 29–43.
[44] R. Cesareo (ed.): *Nuclear Analytical Techniques in Medicine,* Elsevier, Amsterdam 1988.
[45] H. A. Das, A. Faanhof, H. A. van der Sloot: *Radioanalysis in Geochemistry,* Elsevier, Amsterdam 1989.
[46] J. Op de Beeck in I. M. Kolthoff, P. J. Elving, V. Krivan (eds.): *Treatise on Analytical Chemistry,* 2nd ed., part I, vol. 14, J. Wiley & Sons, New York 1986, p. 729.

25. Analytical Voltammetry and Polarography

GÜNTER HENZE, Institut für Anorganische und Analytische Chemie der Technischen Universität Clausthal, Clausthal-Zellerfeld, Germany

25.	Analytical Voltammetry and Polarography 785	25.2.6.	Stripping Techniques.	799
25.1.	Introduction. 785	25.3.	Instrumentation	803
25.2.	Techniques. 788	25.4.	Evaluation and Calculation	808
25.2.1.	Direct Current Polarography. 788	25.5.	Sample Preparation	810
25.2.2.	Pulse Techniques 791	25.6.	Supporting Electrolyte Solution . . .	812
25.2.3.	Alternating Current Polarography . . 794	25.7.	Application to Inorganic and Organic Trace Analysis.	814
25.2.4.	Linear-Sweep and Cyclic Voltammetry 795			
25.2.5.	Chronopotentiometry. 798	25.8.	References	823

25.1. Introduction

The term *voltammetry* is used to classify that group of electoanalytical techniques in which the current (ampere) that flows through an electrochemical cell is measured as the potential (volt) applied to the electrodes in the cell is varied. The term is derived from the units of the electrical parameters measured—*volt-am* (pere)-*metry*.

The essential difference between voltammetric and other potentiodynamic techniques, such as constant current coulometry, is that in voltammetry an electrode with a small surface area (< 10 mm^2) is used to monitor the current produced by the species in solution reacting at this electrode in response to the potential applied. Because the electrode used in voltammetry is so small, the amount of material reacting at the electrode can be ignored. This is in contrast to the case in coulometry where large area electrodes are used so that all of a species in the cell may be oxidized or reduced.

When mercury is used as the electrode in the form of small drops falling slowly from a fine capillary tube in the test solution, the technique has the special name *polarography*. This name is derived from the fact that the electrode can be polarized. An electrode is said to be polarized when no direct current flows across its interface with the solution even though there is a potential difference across this interface. In his work published in 1922 on "Electrolysis with a dropping mercury electrode", Jaroslav Heyrovsky referred to this phenomenon and called the recorded current–potential polarization curves *polarograms* [1].

The recommendation of IUPAC is that voltammetry is the general term to be used when current–potential relationships are being investigated and that only when a flowing conducting liquid electrode (such as a dropping mercury electrode) is used as the working electrode should the term polarography be used. Polarography, the original technique, is thus a special case of voltammetry.

The *voltammetric cell* is a multi-phase system, in the simplest form designed with two electronic conductors called *electrodes* immersed in an electrolytic conductor (ionic cell solution). The electrolytic conductor consists of the sample solution with the electrochemically active *analyte* and an excess of an inert *supporting electrolyte*. The analyte is an inorganic or organic species and can be present as a cation, an anion or a molecule.

The application of a voltage to the electrodes produces, as the result of an electrode process, an electrical response—a current—from the analyte in the cell solution. The nature and magnitude of this response may be used to both identify and quantify the analyte. Most polarographic and voltammetric procedures utilize electrode processes in which electrons are exchanged between the two phases. Such a process is referred to as a *charge transfer reaction*, because of the flow of charge, i.e., a current, through the electrode.

The small electrode used to monitor the response of the analyte is known as the *working electrode* (WE). Even though only a negligible amount of material is involved in the processes occurring at the working electrode, its small size ensures that a high current density develops at its surface. Consequently, it is the processes which occur at this small electrode that control the current flow through the cell. The WE may be constructed from a wide variety of conduction materials, preferably of mercury and graphite, or alternatively of gold and platinum. For polarographic experiments the dropping mercury electrode (DME) is important and for voltammetric investigations the thin mercury film electrode (TMFE), the hanging mercury drop electrode (HMDE), the glassy carbon electrode (GCE), the carbon paste electrode (CPE), the rotating platinum electrode (RPE), and chemically modified electrodes (CME's) may be used.

The second electrode in the simple voltammetric cell, called the *counter electrode* (CE), serves two purposes. It is used to control the potential applied to the working electrode and to complete the circuit for carrying the current generated at the WE. In the former role it must act as a reference electrode.

In modern measuring systems the current carrying role of the counter electrode is separated from its potential control role by introducing the *auxiliary electrode* (AE) as a third electrode of the cell. The addition of the auxiliary electrode means that the counter electrode is now used only to control the potential of the working electrode and so becomes a true *reference electrode* (RE). Two electrodes which are commonly used as reference electrodes for the precise control of the working electrode potential in aqueous media are the silver–silver chloride electrode in a solution of fixed chloride concentration and the saturated calomel electrode or SCE (a mercury–mercurous chloride electrode in a saturated KCl solution). These electrodes are robust, easily constructed and maintain a constant potential. The three electrode system will be discussed in more detail in Chapter 25.3.

When the voltage applied to the electrodes (WE and AE) is such that no charge transfer reactions are occuring, the working electrode is said to be "polarized". At higher voltages, the polarization of the working electrode disappears as soon as the conditions exist for a charge transfer reaction, $Ox + ne^- \rightleftarrows Red$, to occur at the interface between the electrolyte solution and the WE. The current flow that results from the oxidation or reduction of the analyte is known as a *Faradaic current*. Its magnitude depends on the concentration of the analyte in the sample solution and on the kinetics of all steps in the associated electrode process. Because of the small surface area of the working electrode ($1-10$ mm^2) the current flow is normally in the nA to µA range.

The temporal decrease of the analyte concentration in the interfacial region of the electrode, as a result of reduction or oxidation, is balanced chiefly by diffusion of the analyte from the bulk of the solution. The transport of charged analytes by migration is insignificant in the presence of an excess of supporting electrolyte in the test solution. Normally migration can be neglected if the supporting electrolyte concentration is at least 10^3 times higher than the analyte concentration. However, mass transport to the interface is assisted by convection, facilitated by movement of the solution relative to the electrode, e.g., by stirring or by the action of the mercury dropping from the capillary of the DME.

According to Nernst, diffusion occurs in a stable layer of thickness δ at the interface of the working electrode, i.e., from the electrode surface to some distance into the solution (Fig. 1). Within the Nernst diffusion layer, the decrease of analyte concentration c_a in the sample solution to c_s at the electrode surface is linear and voltage dependent. Convection within the layer is negligible.

For evaluating the diffusion-limited electron transfer current, the following equations are important:

1) The general equation for the cell current as a function of the number of moles converted:

$$i = \frac{dN}{dt} nFA \tag{1}$$

where N is the number of moles, t the time, n the number of electrons involved in the electrode reaction, F the Faraday constant, and A the surface area of the working electrode;

2) Fick's first law

$$\frac{dN}{dt} = D\left(\frac{dc}{dx}\right) \tag{2}$$

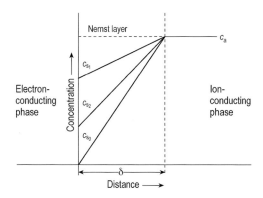

Figure 1. Variation of concentration in the Nernst diffusion layer

where D is the diffusion coefficient, and dc represents the change in concentration over the distance dx from the electrode surface. According to the ideas behind the Nernst diffusion layer, dc/dx can be replaced by $(c_a - c_s)/\delta$ (Fig. 1). From Equations (1) and (2):

$$i(t) = DnFA\frac{c_a - c_s}{\delta} \quad (3)$$

The concentration gradient increases with increasing voltage (see c_{s_1} and c_{s_2} in Fig. 1) and reaches its highest value when all active species reaching the electrode by diffusion immediately participate in the electron transfer reaction. Then $c_s = c_{s_0} = 0$ and Equation (3) becomes

$$i(t) = DnFA\frac{c_a}{\delta} \quad (4)$$

In this case the cell current is also called the *limiting diffusion current* i_d, whose value remains constant with further increase of voltage; $i_{\lim}$ is accordingly the maximum value of i and, being proportional to the concentration, is important for analytical voltammetry. However, these relations apply only if the thickness of the diffusion layer remains constant during the cell reaction (stationary state). This is true if the electrode is moved at constant velocity (rotating disk or ring electrode) or if the sample solution is stirred or flows past the electrode at constant velocity (flow cell). In such cases and under ideal experimental conditions, when i is plotted against the voltage E, a line parallel to the E axis is obtained on reaching the limiting diffusion current (Fig. 2 A). The relationships change if both electrode phases remain in the stationary state. Without additional convection the

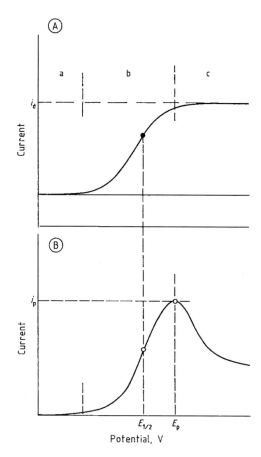

Figure 2. Current–voltage curve at electrodes
A) Stationary; B) Nonstationary diffusion layer thickness
Potential ranges for a) Electrode polarization; b) Electrode reaction; c) Limiting diffusion-controlled current i_d and peak current i_p
E_p = Peak potential; $E_{1/2}$ = Half-wave potential

thickness of the diffusion layer, and hence dc/dx, changes with time (Fick's second law). In the course of the electrode reaction, as a result of the electrochemical conversion, the diffusion layer thickness increases as $\sqrt{\pi D t}$. This means that the diffusion current, after reaching the limiting value i_p, decreases with time according to the Cottrell equation:

$$i(t) = DnFA\frac{c_a}{\sqrt{\pi D t}} \quad (5)$$

(Fig. 2 B).

In *polarography* with the DME, the relationships are even more complicated as both the concentration gradient and the diffusion layer width change with time. This is a consequence of the growth of the drop and hence the electrode surface area being in the opposite direction to the diffusion direction. With regard to the electrode reaction the two effects are superimposed, the time average being a constant limiting diffusion-controlled current corresponding to Figure 2 A. Further information on i_{lim} is provided by the Ilkovič equation (Section 25.2.1, Eq. 8).

The signals proportional to the analyte concentration, i_d and i_p, are analytically important.

Aside from the diffusion current, currents can occur in voltammetric/polarographic studies which are controlled by reaction kinetics, catalysis, and adsorption.

Kinetic Currents. In kinetic currents, the limiting current is determined by the rate of a chemical reaction in the vicinity of the electrode, provided this precedes the cell reaction. Electrochemically inactive compounds are converted into reducible or oxidizable forms (time-dependent protonation and deprotonation processes, formation and decomposition of complexes, etc.). Conversely, during a chemical reaction after the cell reaction, the product of the electrode reaction is converted to an electrochemically inactive form without influence on the current. However, owing to the changed equilibria between the concentrations of the oxidized and reduced forms at the electrode surface, the half-wave and peak potentials are shifted (Section 25.2.1). In evaluating kinetic effects, cyclic voltammetry can be helpful (Section 25.2.4).

Catalytic Currents. For analytical studies, catalytic currents, of which two types are known, are of great interest. In the first case the product of the cell reaction is returned, by a chemical reaction, to the initial state of the analyte. As a result, the analyte concentration at the electrode surface is always high, which results in a considerable increase of the limiting current.

The currents based on the catalytic evolution of hydrogen are the best known. The catalysts are electrochemically inactive heavy metals (primarily salts of the platinum group metals) or organic compounds. They lower the hydrogen overvoltage on the mercury working electrode and so shift hydrogen evolution to more positive potentials. The catalytic current associated with this depends, within limits, on the catalyst concentration. Because of the large number of substances that can lower the hydrogen overvoltage, catalytic currents can only be used to determine concentration under controlled conditions.

Adsorption currents arise when the analyte, in its original form or after chemical or electrochemical conversion, is adsorbed on the working electrode. Adsorption and desorption lead to capacitance changes in the electrochemical double layer and so to analytically usable current signals (Section 25.2.3).

The theoretical principles of voltammetry and polarography are covered in several monographs [1]–[7].

25.2. Techniques

In recent years, voltammetric measurement techniques have continued to improve. Some of the improvements have had to await the advance of electronics before being used in practical analysis. Methods with modulated direct current (d.c.) ramps, e.g., differential pulse polarography, as well as alternating current (a.c.) and square-wave polarography, etc., are worth mentioning. New possibilities for controlling experiments and evaluating current signals have followed from digital electronics. The basic principles of analytical voltammetry and polarography are still best explained by way of classical d.c. polarography.

25.2.1. Direct Current Polarography

Direct current polarography (DCP), introduced by JAROSLAV HEYROVSKY in 1923, is characterized by the recording of current–voltage curves in an unstirred sample solution with the DME as the working electrode. With a slow linear change of voltage, stepwise polarograms are obtained, with marked oscillations caused by the growth and fall of the mercury drop. For better evaluation, the curves can be damped with electronic filters (Fig. 3).

The mercury flows with a mass flow rate m (mg/s) and a periodic drop time t_d (s) from a glass capillary. The surface area A of the mercury drop is greatest at the end of the drop time and, at time t during the growth of the drop is:

$$A = k m^{2/3} t^{2/3} \qquad (6)$$

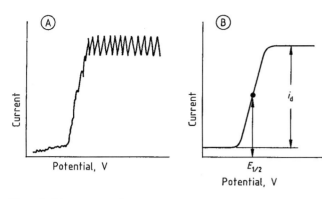

Figure 3. Direct current polarograms
A) Undamped; B) Damped

With this expression and the Cottrell equation (5), the following relation is obtained for the magnitude of the instantaneous current at time t_d for an individual drop:

$$i = k' m^{2/3} t_d^{1/6} n D^{1/2} c_a \qquad (7)$$

The constant k' includes all constant quantities, including the Faraday constant, and also takes account of the conditions for spherical diffusion. In 1935, it was calculated by ILCOVIČ to have the numerical value 0.708.

The mean of the limiting diffusion current in the DC polarogram is obtained from Equation (7) by integration over the entire drop time:

$$i_d = \frac{1}{t_d} \int_0^{t_d} i \cdot dt = 0.607 \cdot n \cdot m^{2/3} \cdot D^{1/2} \cdot t_d^{1/6} \cdot c_a \qquad (8)$$

This is the Ilcovic equation, and applies in the voltage range in which all analyte species reaching the electrode surface by diffusion are either reduced or oxidized. The product $m^{2/3} t_d^{1/6}$ is dependent on the particular glass capillary used and is called the capillary constant.

The characteristic shape of the DC polarogram can be described by the Nernst equation:

$$E = E'_0 + \frac{RT}{nF} \ln \frac{c_{a_{ox}}}{c_{a_{red}}} \qquad (9)$$

on condition that, for the particular electrode potential E, the mean current magnitude at an individual mercury drop is considered.

For the case of reduction $Ox + e^- \rightarrow Red$:

$$i = k\left(c_{a_{ox}} - c_{a_{ox}}^0\right) \qquad (10)$$

where $c_{a_{ox}}^0$ indicates the concentration of the analyte on the electrode surface (corresponding to c_s in Eq. 3). The same applies for the concentration $c_{a_{red}}^0$, which increases as $c_{a_{ox}}^0$ decreases; i.e., the sum $c_{a_{ox}}^0 + c_{a_{red}}^0$ corresponds to the total analyte concentration $c_{a_{ox}}$.

For the limiting diffusion current, at which $c_{a_{ox}}^0 = 0$ and $c_{a_{red}}^0 = c_{a_{ox}}$, it follows from Equation (10) that:

$$i_d = k c_{a_{ox}} \qquad (11)$$

Thus:

$$c_{a_{ox}}^0 = c_{a_{ox}} - \frac{i}{k} = \frac{i_d}{k} - \frac{i}{k} \qquad (12)$$

and:

$$c_{a_{red}}^0 = \frac{i}{k} \qquad (13)$$

When the expressions for $c_{a_{ox}}^0$ (Eq. 12) and $c_{a_{red}}^0$ (Eq. 13) are inserted in the Nernst equation (9):

$$E = E'_0 + \frac{RT}{nF} \ln \frac{i_d - 1}{i} \qquad (14a)$$

or:

$$E = E'_0 + \frac{0.059}{n} \log \frac{i_d - 1}{i} \qquad (14b)$$

For $i = i_d/2$, $E = E_{1/2}$ by definition. $E_{1/2}$, the half-wave potential in the DC polarogram is determined by the electrochemical properties of the redox system concerned.

According to Equation (14), the slope of the current curve depends on the electron transfer number n. Conversely, n can be calculated from

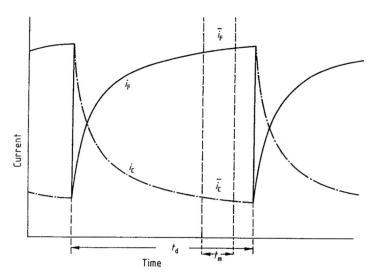

Figure 4. Change with time of i_F and i_C during the drop life t_d with the measurement or sampling interval t_m and the mean values $\bar{i}_F$ and $\bar{i}_C$ over the sampled range

the slope. However, the slope also depends on the reversibility of the electrode reaction.

If the activity coefficients f_{ox} and f_{red} for $c_{a_{ox}}$ and $c_{a_{red}}$ are inserted into the Nernst equation (9) as well as the diffusion coefficients $D_{ox}^{1/2}$ and $D_{red}^{1/2}$ (see Eq. 8), there follows for the half-wave potential:

$$E_{1/2} = E_0 - \frac{RT}{nF} \ln \frac{f_{red} D_{ox}^{1/2}}{f_{ox} D_{red}^{1/2}} \qquad (15)$$

As f and D are determined by the composition of the supporting electrolyte solution, the half-wave potential also depends on this composition.

Direct current polarography is a basic method, from which improved analytical techniques of polarography and voltammetry have been developed. With DCP, inorganic and organic analytes can be analyzed with a sensitivity of about 10^{-5} mol/L. For determining several analytes simultaneously, their half-wave potentials must be at least 100 mV apart.

On closer examination of DC polarograms, the current measured in the voltage range of the charge transfer reaction consists of two parts, the true measurement signal and an interference or noise signal. The measurement signal is the diffusion current, whose value depends on the electrode process and which, because of its origin, is called the Faradaic current i_F. The interference signal is the charging current or capacitive current i_C, which is a consequence of charging the electrochemical double layer. In the case of the DME, the charging current is given by the equation

$$I_C = (dE/dt)C_D A + (dA/dt)C_D E \qquad (16)$$

where C_D is the double layer capacitance per unit area, E the potential applied to the cell and A the area of the electrode.

The first term of this equation allows for the effect of the rate of change of the electrode potential on the charging current, while the second term gives the increase in i_C due to the growing mercury drop.

With deceasing analyte concentration i_F becomes smaller and approaches i_C. The $i_F : i_C$ signal-to-noise ratio is the determining factor for the sensitivity of DCP. If more sensitive polarographic determinations are to be performed, this ratio must be increased by using techniques in which either the Faradaic current is enhanced or the charging current is minimized, or where both can be achieved.

During drop growth (drop life t_d), i_F and i_C are subject to changes in opposite directions. As shown in Figure 4, i_F increases with increasing surface area of the mercury drop, while i_C becomes smaller.

If the current is not measured over the whole drop life (as in classical DCP), but over a short period of time at the end of the life of a mercury steps are synchronized with the drop formation and the current is sampled at the end of the drop life. Since the voltage is constant when the current is measured, there is no contribution to the charging current from this source (see Eq. 16). The second innovation which led to a reduction in the charging current is the *static mercury drop electrode* (SMDE). In this electrode the mercury drop is extruded rapidly and its growth then stopped after which the area of the mercury electrode remains constant. The area–time profile for the SMDE, the potential–time waveform with the sampling period, and the resulting DC polarogram is shown in Figure 5. The current sampled DC polarography (sometimes called Tast Polarography) is not only more sensitive than classical DCP (ca. 5×10^{-7} mol/L), but also leads to polarograms without oscillations.

25.2.2. Pulse Techniques

With voltammetric pulse methods, BARKER and GARDNER [31] found very effective means of reducing the unwanted capacitive (noise) current, markedly improving the sensitivity of polarographic and voltammetric determinations.

A periodically changing square-wave potential with constant or increasing amplitude ΔE_p is applied to the working electrode. During the pulse duration t_p, the currents i_F and i_C that flow in response to the potential change ΔE_p decay in different ways. The Faradaic current decreases according to $t^{-1/2}$, (Eq. 5) while the capacitive current decreases according to

$$i_C = \frac{\Delta E_p}{R} \exp\left(-\frac{t}{R \cdot C_D}\right) \qquad (17)$$

(Eq. 17 corresponds to the capacitor formula with discharging resistance R and double-layer capacitance C_D).

Figure 6 illustrates the variations of i_F and i_C during the duration t_p of the square pulse ΔE_p. If the current is measured only toward the end of the pulse, essentially only i_F is recorded since i_C has effectively decayed to zero.

The methods based on the application of square voltage pulses differ in the frequency and height of the pulses applied and in the principle of measurement.

Pulse methods may be used with the DME, the HMDE, the SMDE as well as with solid electrodes and the different types of modified electrodes.

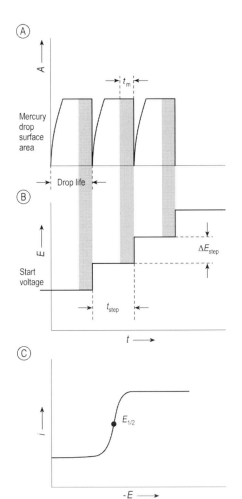

Figure 5. Measurement technique of current sampled DC polarography with the static mercury drop electrode as working electrode
(A) Variation of surface area with time for the SMDE; (B) Potential–time profile for the staircase voltage applied to the electrode; (C) Current–potential response (polarogram).
t_m Measuring interval; t_{step} Time interval of the voltage step; ΔE_{step} Voltage step of the applied staircase waveform

drop where the growth of surface area is least the current averaged over the measurement (sampling) interval t_m has an improved ratio of $i_F : i_C$. This is the basis of the *current sampled DC-polarography* using a linear scan.

Two recent developments in the measurement technique have resulted in a major reduction in the charging current. Firstly, with the introduction of digitally controlled instruments, the linear voltage ramp applied to the electrodes has been replaced with a stepped (staircase) waveform. The voltage

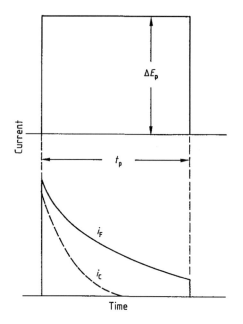

Figure 6. Principle of the pulse methods, according to BARKER

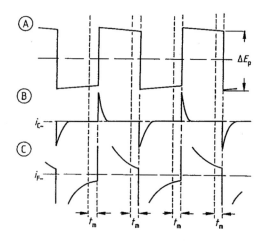

Figure 7. Square-wave alternating voltage and current signals in square-wave polarography
A) Waveform with pulse tilt; B) Capacitive current $i_{C\sim}$;
C) Faradaic current $i_{F\sim}$
t_m = Measuring interval

Except for square-wave polarography, normally only one pulse is applied to each drop in analytical studies with the DME or SMDE. This pulse is synchronized with the drop period and is applied in each case towards the end of the drop life.

Square-Wave Polarography. In the oldest technique of square-wave polarography (SWP), a periodic rectangular alternating voltage with a frequency of 125 Hz and an amplitude in the range of $\Delta E_p = 5-30$ mV is superimposed on the linear rising d.c. ramp. The current is sampled over a measurement interval t_m near the end of each square-wave half cycle. This current is then amplified and recorded as a function of the applied d.c. voltage. Peak-shaped polarograms are obtained in which the peak current corresponds to the DC polarographic half-wave potential.

Sampling the current near the end of each half cycle ensures that the charging current arising from the sudden step in potential at the beginning of the half cycle has decayed away and does not contribute to the measured current. However, when using the DME, there is still a charging current flowing during t_m as a result of the slight increase in area of the mercury drop (see Eq. 16). This is minimized by using a tilted square-wave as shown in Figure 7.

In the modern version of square-wave polarography the linear d.c. potential ramp is replaced by a d.c. staircase potential and a SMDE replaces the DME. The drop life is synchronized with each step in the staircase potential. The constant potential of each voltage step is modulated with a small amplitude (ΔE_p) alternating square-wave voltage of frequency f towards the end of the step as shown in Figure 8.

The current is measured over a number of cycles in the measurement period t_m, twice in each cycle, at point 1 at the positive end of the pulse and at point 2 at the negative end. The average of the differences between the current values $i_1 - i_2$ for each measurement cycle is plotted against the potential of the voltage ramp and gives a peak-shaped polarogram, as before.

For reversible electrode processes, the peak current is dependent on the superimposed rectangular voltage:

$$i_p = kn^2 D^{1/2} \Delta E_p c_a \qquad (18)$$

where k is constant at a given frequency.

The square-wave polarograms shown in Figure 9 highlight the influence of n and ΔE_p on the height and width of the peaks.

For reversible processes, sensitivities of about 10^{-8} mol/L can be achieved with SWP and the peak-to-peak resolution is 40–50 mV. The sensitivity decreases rapidly with increasing irreversibility of the electrode process.

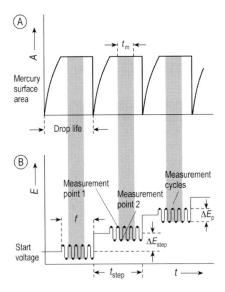

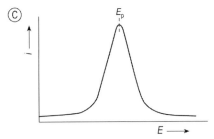

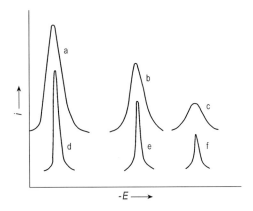

Figure 9. Dependence of the peak height on the square-wave voltage amplitude ΔE_p and on the number n of electrons transferred, n [4]
a)–c) $n=1$ (thallium); d)–f) $n=3$ (bismuth)
a), d) $\Delta E_p = 40$ mV; b), e) $\Delta E_p = 20$ mV; c), f) $\Delta E_p = 10$ mV

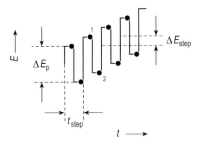

Figure 8. Digitally controlled SWP using the SMDE
A) Variation of the mercury drop surface area with the measurement period t_m; B) Potential–time waveform with the measuring points 1 and 2; C) Current potential response (polarogram). Typical values of parameters: $\Delta E_p \sim 25$ mV, $f \sim 125$ Hz (in 40 ms), $\Delta E_{step} \sim 10$ mV, $t_{step} \sim 10$ s

Figure 10. Measurement technique of square-wave voltammetry
ΔE_{step} = Voltage step of the staircase voltage ramp, (ca. 10 mV) ΔE_p = Amplitude of superimposed pulse, ca. 50 mV; t_{step} = wave form period, ca. 5–10 ms; 1 and 2 Measurement points

Normally the voltage scan rate of SWP measurements is about 20 mV/s, but much higher scan rates up to 1000 mV/s can readily be programmed making the method ideal for fast voltammetric measurements.

A high scan rate is the characteristic feature of square-wave voltammetry (SWV), first described by OSTERYOUNG and OSTERYOUNG [32]. In this technique the whole measurement process can be carried out with an extremely rapid change of voltage on a single mercury drop. SWV thus enables very short analysis times and is particularly important as a detection method in flow systems.

A relatively large square wave signal of amplitude $\Delta E_p = 50$ mV is superimposed on a stepped voltage ramp (staircase) with voltage steps of about 10 mV. The duration of the square wave cycle is equal to that of the staircase voltage steps t_{step} and is usually within the range 5–10 ms.

As can be seen from Figure 10, two values of current are determined for each square wave cycle. Measurements are made at point 1 at the positive end of the cycle and point 2 at the negative end. The difference between the two current values $i_1 - i_2$ is plotted against the potential of the voltage ramp and gives a peak-shaped current–voltage curve, as in SWP.

The pulse times, which are in the millisecond range, enable the potential to be scanned at extremely high rates of up to 1200 mV/s. Only one mercury drop is required for each measurement process. However, this rapid procedure is achieved

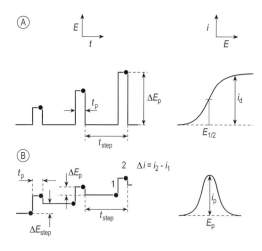

Figure 11. Potential–time waveforms and current–potential responses of A) normal pulse polarography and B) differential pulse polarography (typical values of parameter see text)

at the cost of sensitivity, as the ratio of Faradaic to capacitive currents is lowered by the short pulse times.

The state of development and capability of square-wave polarography/voltammetry has been reviewed [8], [33].

Normal Pulse Polarography. In normal pulse polarography (NPP), the voltage of the working electrode is changed, not by means of a direct voltage ramp (as in SWP), but by increasing rectangular pulses superimposed on an initial, constant voltage. The application of the pulses is synchronized with the mercury drop and occurs just before the end of the drop life. Only one voltage pulse with a duration of 30–60 ms is applied to each drop. The amplitude ΔE_p increases from one pulse to the next and reaches a maximum of 1000 mV. To eliminate i_C the current measurement is made ca. 10–15 ms before the end of the pulse (Fig. 11).

The measured current determined in pulse polarography is plotted against the pulse amplitude and, as in DCP, yields a curve in the form of a wave. The height of the wave is proportional to the analyte concentration, and the half-wave potential corresponds to the DCP value. Peak-shaped curves can be obtained when the current differences between successive pulses are plotted against the voltage. Sensitivities are ca. 10^{-7} mol/L; the peak resolution is reported to be ca. 100 mV.

Differential Pulse Polarography. The most important pulse method is differential pulse polarography (DPP). In this technique a stepped voltage rise (modern staircase technique) is used and a rectangular voltage pulse with constant amplitude ΔE_p of 10–100 mV is applied to each mercury drop at the end of its drop time; the pulse duration t_p is 40–60 ms.

As shown in Fig. 11 B two current measurements are made in DPP: the first i_1 at measurement point 1, immediately before the application of voltage pulse and the second i_2 at point 2 near the pulse end. Both measurements are made on the same mercury drop and also over the same surface area, when the SMDE is used as the working electrode. The difference $\Delta i_1 = i_2 - i_1$ is plotted against the applied d.c. voltage (staircase) and gives a peak-shaped polarogram, because $\Delta i_1/\Delta E$ reaches a maximum in the region of $E_{1/2}$.

Determinations by DPP are on average ten times more sensitive than determinations with NPP. The sensitivity of DPP is about 10^{-8} mol/L and for irreversible processes it falls off only slightly to 5×10^{-8} mol/L. The loss of sensitivity by irreversibility is thus smaller than in the case of other pulse methods. For reversible processes the peak height in DP (differential pulse) polarograms is proportional to the concentration to be analyzed, c_a:

$$i_p = \frac{n^2 F^2}{4RT} A c_a \Delta E_p \sqrt{\frac{D}{\pi t_p}} \qquad (19)$$

and depends not only on the quantities already mentioned, but on the amplitude of the voltage pulse ΔE_p and the pulse duration t_p. The DP polarograms shown in Figure 12 highlight the increase of peak heights with increasing ΔE_p.

The theory, measurement techniques, and analytical applications of pulse methods have been reported [9].

25.2.3. Alternating Current Polarography

In this technique, a linear voltage ramp is modulated with a sinusoidal alternating voltage of small amplitude ($\Delta E_\sim = 10–100$ mV) and low frequency ($f = 5–100$ Hz). The superimposed alternating voltage causes an alternating current $i_\sim$, whose size depends on the instantaneous value of

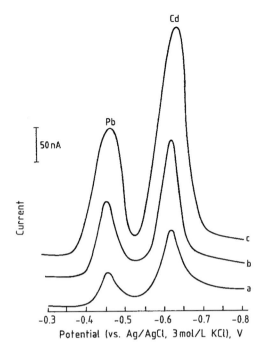

Figure 12. Influence of the pulse amplitude on DP polarograms; Pb and Cd (10 mg/L) in acetate buffer solution pH 2.8
a) $\Delta E_p = 10$ mV; b) $\Delta E_p = 25$ mV; c) $\Delta E_p = 75$ mV

the linear voltage ramp, and is greatest at the half-wave potential. On plotting the selectively measured alternating current against the applied voltage, the peak-shaped AC polarogram is obtained (Fig. 13).

In reversible processes, the peak potential of an AC polarogram is identical with the DC polarographic half-wave potential. The peak current is

$$i_p = \frac{n^2 \cdot F^2}{4RT} (2\pi \cdot D)^{1/2} \cdot A \cdot \Delta E_\sim \cdot f^{1/2} \cdot c_a \qquad (20)$$

and is thus dependent on the usual variables as well as being influenced by the frequency f, the amplitude $\Delta E_\sim$ of the superimposed alternating voltage and the area of the electrode A. The value of i_p also depends on the reversibility of the cell reaction and decreases with increasing irreversibility.

Owing to the high capacitive current component in AC polarography, caused by the periodic charging and discharging of the double layer, the sensitivity is limited to about 10^{-5} mol/L.

Compared with the alternating voltage applied to the electrode, the capacitive current shows a phase shift of 90° ($\pi/2$). Since the Faradaic alternating current $i_{F\sim}$ has a phase shift of only about 45°, the ratio $i_{F\sim} : i_{C\sim}$ and therefore the sensitivity of AC polarographic determinations can be improved by phase-selective rectification. Instruments designed for AC polarography generally have provision for phase selective current measurements to be made (*AC1 polarography*).

By using phase selective measurement of a higher-harmonic alternating current (e.g., 2nd harmonic with a frequency $2f$) increased sensitivity (10^{-7} mol/L) is obtained because of the marked reduction in the capacitive current component of the higher harmonic (*AC2 or second harmonic wave polarography*).

A phase-selective second harmonic wave polarogram is shown in Figure 14. In contrast to the AC1 polarograms, it has two current peaks. For an invariant residual current (Fig. 14 A), the distance between the two current peaks $i_p^+ + i_p^- = i_p$ can be used for analytical purposes. For a rising residual current (Fig. 14 B), the current components i_p^+ and i_p^- are measured over the extrapolated course of the residual current.

Tensammetry is a variant of AC polarography. In this method, the capacitive current component, whose value is determined by the double layer capacitance, is measured instead of the Faradaic alternating current. Capacitance changes are caused by the adsorption and desorption of surface active compounds at the electrode surface. Commonly, two almost symmetrical current peaks that are particularly well defined and known as tensammetric peaks are obtained. In the potential region between the peaks the compound is adsorbed. As a result, the capacitance of the electrochemical double layer is lowered and the capacitive current is smaller than the current of the supporting electrolyte solution (Fig. 15).

Tensammetric studies are important for the characterization of interfacial problems and for the analysis of surface-active compounds (e.g., surfactants) that are neither reducible nor oxidizable.

25.2.4. Linear-Sweep and Cyclic Voltammetry

Both linear-sweep voltammetry (LSV) and cyclic voltammetry (CV) are based on recording the current during a linear change of voltage at a stationary working electrode. The rate of change of voltage $v = dE/dt$ is relatively high, in the range

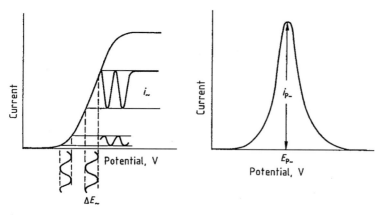

Figure 13. Waveform and response of AC polarography

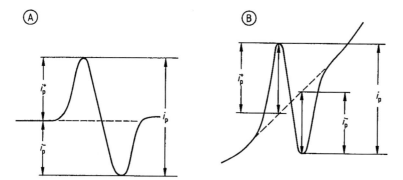

Figure 14. Current profile and evaluation of second harmonic wave polarograms
A) Horizontal residual current; B) Rising residual current

1–100 mV/s. The difference is that in the case of CV the voltage is returned to the starting potential; CV is therefore also known as triangle-sweep voltammetry.

Figure 16 compares the variation of voltage with time and the response for the two methods. The peak currents, for both LSV and CV, are proportional to the analyte concentration:

$$i_p = k c_a v^{1/2} \tag{21}$$

and are influenced by the rate of change of voltage v. The sensitivity of LSV is about 10^{-7} mol/L and the resolution is about 50 mV.

Cyclic voltammetry is mainly used for studying the reversibility of electrode processes and for kinetic observations, and only sometimes for analytical purposes. The voltage cycle illustrated in Figure 16 ensures that the reaction products formed at the potential $E_{P_{red}}$ on the cathodic path are reoxidized at $E_{P_{ox}}$ in the anodic sweep.

For a reversible redox process:

$$\Delta E_p = E_{P_{red}} - E_{P_{ox}} = \frac{-57}{n} \text{ mV} \tag{22}$$

the position of the current peak in this case being independent of the voltage scan rate. The two peaks have equal heights. With increasing irreversibility, ΔE_p becomes greater.

For quasi-reversible processes and for a slow change of voltage, the difference is about $(60/n)$ mV, but it becomes greater for a faster sweep. For totally irreversible processes, the reduction product is not reoxidized, so the anodic current peak is not seen.

If the reversible charge transfer is followed by a chemical reaction during which an electrochemically active product is formed, a cyclic voltam-

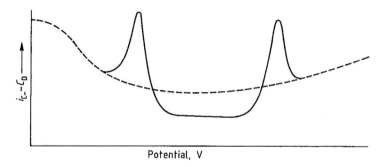

Figure 15. Tensammogram
Alternating current polarogram of the supporting electrolyte solution (---); Formation of current peaks after addition of a surfactant (——); $i_{c\sim}$ = Capacitive current; C_D = Double layer capacitance

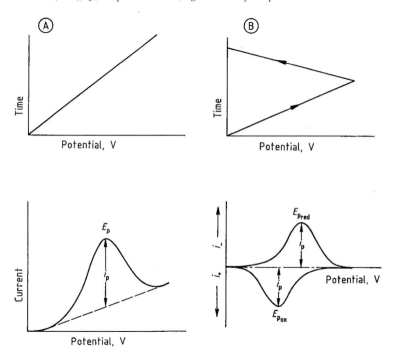

Figure 16. Principle of linear sweep voltammetry (A) and cyclic voltammetry (B)
Top: Variation of voltage with time; Bottom: Resulting current – potential curves

mogram of the type shown in Figure 17 is obtained.

In the first sweep the two peaks $E^1_{P_{red}}$ and $E^1_{P_{ox}}$ appear. The ratio of the peak heights is not, however, 1:1, as would be expected for a reversible process. Rather, $E^1_{P_{ox}}$ is smaller than $E^1_{P_{red}}$, since part of the electrolysis product is chemically converted and therefore no longer available for reoxidation. Since the formed product is electrochemically active, a second peak $E^2_{P_{ox}}$, which corresponds to the oxidation of this product, appears in the complete anodic sweep. In the second sweep an additional cathodic peak $E^2_{P_{red}}$, which indicates the reduction of the previously oxidized compound, can also be recorded.

Cyclic voltammetry gives information on the redox behavior of electrochemically active species and on the kinetics of electrode reactions as well as offering the possibility of identifying reactive intermediates or subsequent products [10].

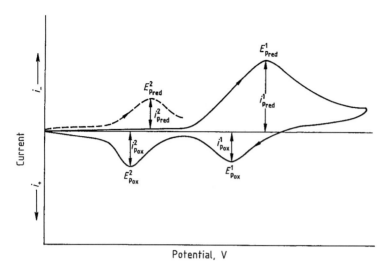

Figure 17. Cyclic voltammogram of a reaction with reversible charge transfer and subsequent chemical reaction
First sweep (——); Second sweep (– – –)

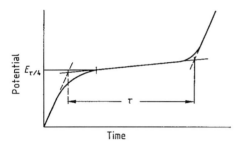

Figure 18. Chronopotentiometric potential–time curve with the transition time τ

25.2.5. Chronopotentiometry

In voltammetric studies, the current flowing through the working electrode is measured as a function of the applied voltage and the rate at which the voltage is scanned. In chronopotentiometry, the change of potential of the working electrode with time at constant current is followed. Chronopotentiometry has also been called "galvanostatic voltammetry" and can, therefore, be grouped with the methods described in this chapter.

Determinations are usually carried out under stationary conditions. The electrolytic processes occurring when current flows decide the ratio $c_{a_{ox}} : c_{a_{red}}$ of the analyte on the electrode surface, which leads to changes of potential according to the Nernst Equation (9). The potential–time plot is illustrated in Figure 18, and depends on the diffusion behavior and analyte concentration.

The time between the steep sections of the curve is called the transition time τ, which, in diffusion-controlled reversible electrode processes, is described by the Sand equation:

$$\tau^{1/2} = \frac{\pi^{1/2} n F D^{1/2}}{2 j_0} c_a \tag{23}$$

where j_0 is the current density (A/cm^2), and $\tau^{1/2}$ corresponds to the limiting diffusion current in DC polarography. At constant j_0, the potential–time function is:

$$E = E_{\tau/4} + \frac{RT}{nF} \ln \frac{\tau^{1/2} - t^{1/2}}{t^{1/2}} \tag{24}$$

with $E_\tau/4 = E_{1/2}$ for a reversible reaction.

Chronopotentiograms were traditionally evaluated graphically, which with small transition times is difficult and often inaccurate. It is more convenient to measure the time values—which are proportional to concentration—electronically via the first or second derivative. In this way empirical time values are obtained which always differ to some degree from the theoretical transition times [34].

Normal chronopotentiometry has little importance for analysis. Determinations are possible in the range $10^{-4} - 10^{-5}$ mol/L at best. However, the technique is of use in understanding the methods of stripping chronopotentiometry and potentiomet-

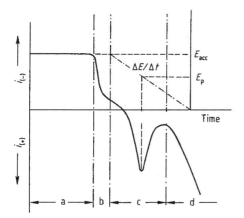

Figure 19. Stripping voltammetric determination
a) Accumulation time; b) Rest period; c) Determination or stripping step; d) Anodic dissolution of mercury

ric stripping analysis, which are important in trace analysis.

25.2.6. Stripping Techniques

The stripping techniques are so-called multi-stage combined procedures, with the characteristic that the voltammetric or chronopotentiometric determination is preceded by an electrochemical accumulation of analyte. The accumulation occurs on the surface of a stationary working electrode and leads to considerably improved performance. The methods enable analytical studies in the picotrace range [11]–[13].

Stripping Voltammetry. The analyte is accumulated, as the metal or in the form of a known compound, by controlled potential electrolysis or adsorption and with constant convection of the sample solution. The subsequent stripping of the preconcentrated species from the electrode occurs voltammetrically as a linear voltage ramp is applied to the electrode, generally using one of the DC, AC, or DP measurement modes. The current peaks in the voltammograms result from either the reduction or oxidation of the accumulated products. Depending on this electrode reaction, a distinction is made between *anodic stripping voltammetry* (ASV) and *cathodic stripping voltammetry* (CSV).

According to the Randles–Sevcik equation:

$$i_p = k n^{3/2} A D^{1/2} v^{1/2} c \qquad (25)$$

the peak height i_p in the stripping voltammogram is (analogous to i_p in the linear sweep and cyclic voltammogram) proportional to the concentration c and is also influenced by the working conditions, particularly by the surface area A of the working electrode and the rate of change of voltage v during the stripping process. The proportionality to concentration relates to the amount of accumulated material or to that part of the electrolysis product which takes part in the stripping process. Consequently, i_p depends on two further influences: the accumulation time and the degree of stripping of the electrolysis product. By careful selection of all parameters, limits of determination can be well below 10^{-8} mol/L.

Various means of trace accumulation and determination are used in stripping voltammetry. In the simplest case the analyte is reduced to the metal and accumulated as an amalgam at a stationary mercury electrode (HMDE or TMFE). The determination step proceeds in the reverse direction to the accumulation and is based on the anodic stripping of the metal (reoxidation). This process of ASV is also known as *inverse voltammetry* and can be illustrated as follows:

$$M^{n+} + n e^- + (Hg) \underset{\text{stripping}}{\overset{\text{accumulation}}{\rightleftarrows}} M^0 Hg$$

The current profile and the sequence of steps for inverse voltammetric determination of the analyte M^{n+} are shown in Figure 19.

The analyte is deposited from the stirred sample solution at a voltage E_{acc}, as metal on the mercury electrode. The accumulation time is followed by a rest period, during which the solution is unstirred and the cathodic current falls to a small residual value. During the determination step, the voltage is ramped to more positive values (linear sweep, potential scan rate $v = \Delta E/\Delta t$) and, at the appropriate potential, the accumulated metal is reoxidized. The resulting current–potential curve displays a peak, with peak potential E_p. The height of the peak is proportional to the analyte concentration (Eq. 25).

The results of stripping voltammetric determinations are reproducible if the operating conditions are maintained exactly. This includes reproducible renewal of the electrode surface, uniform stirring of the solution, reproducible electrolysis and rest times, and exact adjustment of the electrolysis voltage and its rate of change.

The electrolysis time, adjusted to the analyte concentration to be determined, is usually 1–5 min.

Useful guidance on the selection of a suitable electrolysis voltage E_{acc} is given by the DC or DP voltammogram of the analyte in the particular supporting electrolyte solution. For an electrode reaction with $n = 1$, the voltage should be about 0.15 V more negative than the half-wave or peak potential of the analyte. Consequently, E_{acc} should lie in the voltage range of the limiting diffusion current.

The potentiostatic principle of accumulation electrolysis enables analytes present simultaneously to be determined either individually or together by adjustment of the accumulation potential. Simultaneous determinations require a difference of peak potentials $\Delta E_p > 100$ mV.

Those metals which can be determined are soluble in mercury and form amalgams (Pb, Cu, Cd, Sb, Sn, Zn, Bi, In, Mn, and Tl). Nobler metals such as Hg, Ag, Au, and Pt are also determinable by ASV using inert solid electrodes (metal or carbon electrodes; Chap. 25.3).

Metal ions such as arsenic(III), selenium and tellurium (Me_a^{n+}) may be determined by stripping voltammetry after adding a second metal, such as copper (Me_b^{m+}), to the test solution and co-electrolysing the two metals onto the surface of the HMDE. The copper acts as a co-deposition agent and facilitates the deposition of the analyte, Me_a, on the electrode surface as an inter-metallic compound. The analyte may then be stripped from the electrode either by oxidation (ASV) or by further reduction (CSV) to an anionic species according to the following reaction scheme.

$$M_a^{n+} + M_b^{m+} + (m+n)e^- + (Hg) \longrightarrow M_a^0 M_b^0 (Hg) \quad \text{Accumulation}$$

$$M_a^0 M_b^0 (Hg) \begin{array}{c} \xrightarrow{\text{CSV}} M_a^{p-} + M_b^0 (Hg) - pe^- \\ \xrightarrow{\text{ASV}} M_a^{n+} + M_b^{m+} + (Hg_2^{2+}) + (m+n+2)e^- \end{array} \quad \text{Stripping}$$

Arsenic, selenium and tellurium are three such elements which may be determined by cathodic stripping voltammetry after having been reductively co-deposited with copper [38]. A characteristic of the cathodic stripping voltammograms of these three elements is that only a single current peak, which arises from the further reduction of the deposited analyte to As^{3-}, Se^{2-} or Te^{2-}, respectively, is observed. In this case the determination by CSV is more selective than by ASV, since in the anodic dissolution, additional current signals are obtained, which arise from the oxidation of the copper and possibly also of mercury [35]. Typical stripping voltammograms for the determination of arsenic by both ASV and CSV after deposition from a copper containing solution are shown in Figure 20. The determination limits for the CSV determination of As, Se or Te in the presence of copper were found to be 0.5 µg/L for As and 0.2 µg/L for Se and Te. Note that only As(III) is determined since under these experimental conditions As(V) is not electroactive. The small amount of arsenic present in water samples is mainly in the +5 oxidation state as a result of oxidation by oxygen. Recently it has been found that if the test solution contains D-mannitol, As(V) can be reduced electrolytically to As(0). This is the basis of a method for determining As(V) in the presence of As(III) by cathodic stripping voltammetry [36].

The technique of CSV can be applied to the determination of anions forming as accumulation products sparingly soluble Hg(I) salts on the electrode surface. These include the halides, pseudohalides, and oxometallates (vanadate, chromate, tungstate, molybdate).

During the subsequent cathode stripping step, the Hg_2^{2+} ion in the sparingly soluble compound is reduced with formation of a voltammetric current peak (Fig. 21). The peak height is proportional to the accumulated amount of the anion to be determined.

Organic substances can also be determined by this indirect procedure, provided they form insoluble Hg_2^{2+} or Hg^{2+} compounds. These are principally thiols, thioureas, thiobarbiturates, dithiocarbamates, and thioamides.

In adsorptive stripping voltammetry (AdSV) metal chelates and organic molecules are accumulated by adsorption at the surface of the working electrode. If these compounds are electrochemically active, i.e., if they are reducible or oxidizable, their subsequent voltammetric determination is possible. By this principle of so-called adsorptive stripping voltammetry, organic and organometallic compounds are determined in the ultra-trace range. This technique is particularly important for the trace analysis of metals that are not readily deposited as the element on mercury electrodes [13].

Two approaches have been used to effect the adsorption of metal ions onto the electrode surface

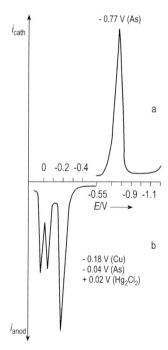

Figure 20. The differential pulse cathodic (a) and anodic (b) stripping voltammograms of arsenic. The cell solution was 0.1 mol/L HCl containing 2×10^3 mol/L Cu^{2+} and 10 µg/L As. Voltage scan rate = 100 mV/s, $E_{acc} = -0.55$ V; $t_{acc} = 1$ min

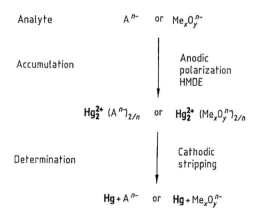

Figure 21. Determination of anions by CSV

as the metal chelate. The simplest is to add an excess of a suitable complexing agent to the test solution prior to the accumulation step. This is the most common approach and is used preferably with mercury or glassy carbon working electrodes. A selection of frequently used complexing agents is listed in Table 1.

An alternative approach is to modify the surface of a glassy carbon, carbon paste or graphite electrode with the complexing agent. The metal ion is then accumulated by reaction with this modified surface (see p. 807).

The potentiostatically accumulated species is determined voltammetrically by reduction or oxidation of the central atom or of the ligand of the metal chelate complex or via catalytic hydrogen evolution. Adsorbed organic molecules can be determined in an analogous way by the oxidation or reduction of their electroactive functional groups.

The quantity of the adsorbed species c_{ad} present after the accumulation time t_{acc} on the surface of the mercury drop area A and with the radius r determines the value of the voltammetric peak current i_p. Moreover, c_{ad} is proportional to the concentration c_a of the analyte in solution for low surface coverage. For diffusion-controlled processes, which are assumed here in order to simplify the relationships, the following dependence results for the peak in the stripping voltammogram:

$$i_p = kAc_{ad} = kAc_a\left(\frac{D}{r}t_{acc} + 2\sqrt{\frac{D}{\pi}}t_{acc}^{1/2}\right) \quad (26)$$

The peak current increases linearly with $t_{acc}^{1/2}$ until the electrode surface is saturated, and reaches its maximum value at:

$$i_{p(max)} = kAc_{ad(max)} \quad (27)$$

$c_{ad(max)}$ being proportional to the accumulation time $t_{acc(max)}^{1/2}$, and corresponds to the maximum surface coverage.

The linear proportionality

$$i_p \sim c_a t_{acc}^{1/2} \quad (28)$$

applies in AdSV only for the lower and middle µg/L range and for short accumulation times $< t_{acc(max)}^{1/2}$. In routine AdSV practice, departures from linearity at fairly high analyte concentrations are controlled, either by dispensing with the usual stirring of the sample solution during accumulation or by dilution of the sample solution.

The voltage dependence of the adsorption of neutral molecules is obviously connected with the interaction studied by A. N. FRUMKIN between the adsorption energy of a surface-active molecule and the value of the electrode potential determined by the surface charge. The zero charge potential in

Table 1. Complexing agents used for the determination of metal ions by adsorptive stripping voltammetry

Complexing agent	Element
Reduction of the central metal ion	
1,2-Dihydroxybenzene (catechol)	U, Cu, Fe, V, Ge, Sb, Sn, As
1,3-Butanedionedioxime (dimethylglyoxime)	Co, Ni, Pd
8-Hydroxyquinoline (oxine)	Mo, Cu, Cd, Pb, U
2-Hydroxy-2,4,6,-cycloheptatriene (tropolone)	Mo, Sn
2,5-Dichloro-3,6-dihydroxy-1,4-benzoquinone (chloranilic acid)	U, Mo, Sn, V, Sb
N-Nitroso-N-phenylhydroxylamine (cupferron)	U, Mo, Tl
Reduction of the ligand	
o-Cresolphthalexone (OCP)	Ce, La, Pr
5-Sulfo-2-hydrobenzeneazo-2-naphthol (Solochrome Violet RS - SVRS)	Al, Fe, Ga, Ti, Y, Zr, V, Tl, Mg, alkali and alkaline earth metals
1,2-Dihydroxyanthraquinone-3-sulfonic acid (Alizarin Red S)	Al
Chromazurol B (MB9, Mordant Blue 9)	Th, U
Catalytic hydrogen evolution	
Formazone	Pt

the electrocapillary curve of the mercury is critical for the adsorption of neutral molecules. Negatively charged surface-active ions are chiefly adsorbed at potentials corresponding to the positive branch of this curve, and the cations at potentials that lie in the negative range (see textbooks on electrochemistry).

Stripping Chronopotentiometry (SCP). With stripping chronopotentiometry, as with stripping voltammetry, the preliminary electrolytic accumulation of the analyte significantly increases the sensitivity of the technique. The difference between the two stripping techniques is that, in stripping chronopotentiometry the change of potential of the working electrode with time during the stripping of the electrolysis product (e.g., of a deposited metal) is recorded. Stripping is performed by application of an anodic current. The output is an "inverse" potential–time curve from which the transition time τ is obtained, as this is a function of concentration, analogous to the determination of the peak current i_p in the stripping voltammogram. From a comparison of the two curves in Figure 22 it can be seen that $E_{\tau/4}$ in the stripping chronopotentiogram corresponds to the peak potential E_p in the stripping voltammogram.

For the accumulation step, the HMDE and the TMFE are mainly used. The transition times for determinations by stripping chronopotentiometry at these electrodes are approximately:

$$\tau_{HMDE} = \frac{n \cdot F \cdot c_{amalg} \cdot r_d}{j_0} \tag{29}$$

and

$$\tau_{TMFE} = \frac{nF c_{amalg} \cdot l}{j_0} \tag{30}$$

Here, c_{amalg} is the amalgam concentration after accumulation, r_d the radius of the mercury drop, l the thickness of the mercury film on the electrode surface, and j_0 the current density during the stripping process.

In addition to the time for which the electrolysis potential is applied during the accumulation step, the current density during stripping is also of crucial importance for the limits of detection: as j_0 decreases, the determinations become more sensitive. Limits of detection of ca. 10^{-8} mol/L can be achieved [14].

Figure 23 illustrates the form of the curve for the determination of Cd, Tl, and Pb by stripping chronopotentiometry. The normal potential–time curve (A) is compared with the second derivative (B), from which the transition times can be more simply and accurately obtained. A TMFE coated in situ is used for the accumulation.

For the highest possible sensitivities in stripping chronopotentiometry, and to avoid systematic errors, the sample solutions must not contain oxidizing impurities. Care must therefore be taken to remove the oxygen present in the solution.

This precondition does not apply if the anodic stripping of the accumulated elements is replaced by chemical oxidative stripping. With this variant the technique is known as *Potentiometric Stripping Analysis* (PSA). The working electrode

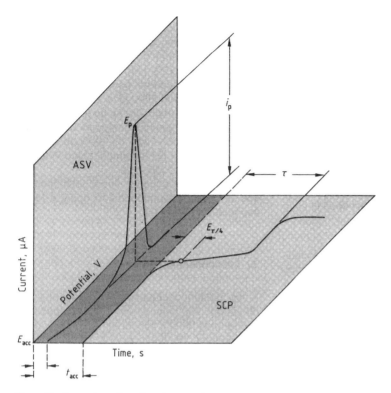

Figure 22. Comparison of anodic stripping voltammetry and stripping chronopotentiometry

chiefly used for this is the TMFE; the oxidants are Hg^{2+} salts or oxygen dissolved in the sample.

Potentiometric stripping analysis is carried out in several stages. After electrochemical generation of the mercury film on a graphite substrate, the elements to be determined are accumulated by electrolysis at constant potential. The next stage is the oxidation of the deposited elements by the oxidant present in the solution. For this, the current circuit is disconnected. The deposited analytes are stripped in the order of their electrochemical potentials. Anodically deposited precipitates can similarly be stripped by chemical reduction. In all cases, potential–time curves with transition times proportional to concentration result [39]–[41].

Recent developments have led to computer-assisted differential PSA $[E = f(dt/dE)]$ which has become important for the detection of trace elements [42]. Figure 24 compares the normal potential–time curve and the differential curve calculated from it for the simultaneous determination of the four elements.

The theory of stripping chronopotentiometry and differential potentiometric stripping analysis has been reviewed [43].

25.3. Instrumentation

In modern equipment, the measuring arrangement for polarographic or voltammetric studies consists of a measuring stand and a digital computer. The latter controls the more or less automated techniques and also analyzes the current signals.

The measuring stand includes the potentiostat locked into analog electronics, the current measurement amplifier, and the digital–analog and analog–digital converters. It also provides a holder for the electrochemical cell, which consists of measuring vessel, electrodes, stirrer, etc.

The *potentiostat* is the voltage source for the DC supply (ca. 4 V) as well as for the superimposition of voltage pulses (NPP and DPP modes) and

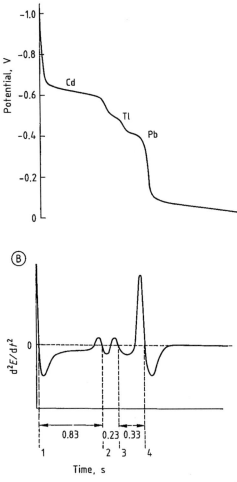

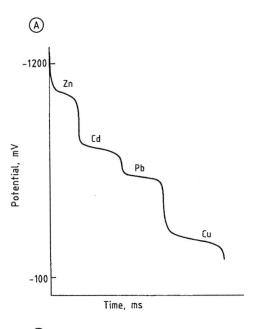

Figure 23. Simultaneous determination of Cd (5×10^{-7} mol/L), Tl (2×10^{-7} mol/L), and Pb (3×10^{-7} mol/L) in 0.05 mol/L HCl + 0.5 mol/L NaCl + Hg^{2+} 100 mg/L by stripping chronopotentiometry at a TMFE (electrolysis time 10 min; stripping current strength 10 μA)
A) Potential–time curve; B) Second derivative [14]

a sine-wave or square wave alternating voltage (ACP and SWP mode). The basic circuit diagram is shown in Figure 25.

When recording the voltammograms or polarograms using three electrodes, the current is flowing between the working electrode and auxiliary electrode. Compared with the working electrode, the auxiliary electrode has a much greater surface area and thus a smaller current density. The reference electrode is an electrode of the second kind, and the potential of the working electrode is measured against this reference electrode. Potential

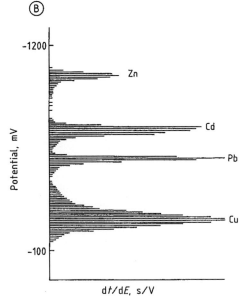

Figure 24. Potential–time curve (A) and differential plot (B) for the determination of zinc, cadmium, lead, and copper by PSA [42]

data for voltammetric and polarographic measurements are reported relative to calomel or silver–silver chloride reference electrodes.

The potentiostatic measuring arrangement with three electrodes (Fig. 25) has superseded the classical measuring arrangement with two electrodes,

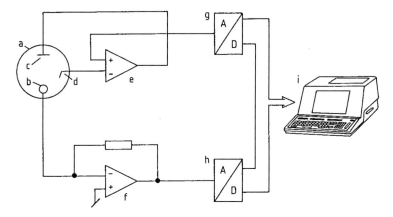

Figure 25. Circuit diagram of a polarograph for various polarographic/voltammetric measuring techniques (Courtesy of Metrohm)
a) Measuring cell; b) Working electrode; c) Auxiliary electrode; d) Reference electrode; e) Potentiostat; f) Current measurement amplifier; g) Digital–analog converter; h) Analog–digital converter; i) Computer

as with the latter the ohmic potential drop in the sample solution cannot be compensated adequately.

In addition to carbon and the noble metals, the principle material used for the working electrode is mercury, especially for capillary and film electrodes. Because of the high overvoltage of hydrogen on mercury, the latter electrodes ensure a relatively wide potential range for voltammetric/polarographic determinations in various supporting electrolyte solutions. If capillary electrodes are used, the electrode surface can be renewed easily and reproducibly as with the DME, the SMDE and the HMDE.

Dropping Mercury Electrode. The DME, originally proposed by HEYROVSKY and used for polarographic studies, consists of a glass capillary, 10–20 cm long, with internal diameter 0.04–0.1 mm, and is connected by a tube to the mercury storage vessel. The distance between the capillary opening and the mercury level determines the drop time and influences to different degrees the types of current useful for polarographic determinations [1]. The drop time also depends on the electrode potential and the surface tension of the mercury. The surface tension is mainly influenced by the medium, and particularly the solvent, into which the mercury flows. The drop time must be held constant during the measurement and should be 2–8 s. In modern instruments the mercury drop is knocked off mechanically by tapping on the capillary at a predetermined constant time.

Static Mercury Drop Electrode. Recent developments have led to the static mercury drop electrode (SMDE), which today is the preferred electrode for polarographic analysis. Compared with the DME, the drops from this type of electrode can be renewed more frequently, about once or twice per second. A microvalve with opening times of 20–200 ms controls the mercury inflow to the capillary. During each open period a mercury drop forms, whose surface area remains constant after the valve closes.

In Figure 26 the increase of the electrode surface area with time during the drop life is compared for the DME and the SMDE. The maximum electrode surface area is reached much more rapidly with the SMDE than with the DME so that the measurement can be carried out at constant electrode surface area. This results in a reduction of the capacitive current and increased analytical sensitivity.

The greatest advantage of this new type of electrode is its ability to function as a stationary and a nonstationary mercury electrode, as well as the possibility of synchronizing the electrode (drop) formation with the programm-control arrangement. The reproducibility of the mercury metering of an SMDE is important for its dependability in operation. In a commercial model (Multi-Mode electrode, Metrohm AG, Switzerland), the mercury is pushed through the capillary by a ni-

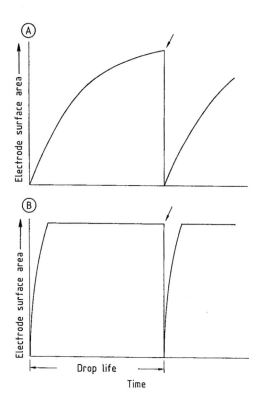

Figure 26. Drop growth as a function of time; comparison of the DME (A) with the SMDE (B)
Drop fall indicated by arrows

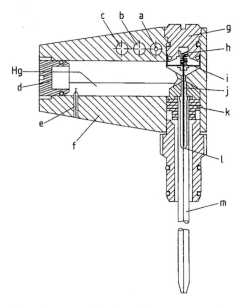

Figure 27. Construction of a static mercury electrode (Courtesy of Metrohm)
a) N_2 supply for membrane control; b) N_2 supply for overpressure; c) Blind hole; d) Stopper for topping up; e) Electrode connection; f) Hg reservoir; g) Adjusting screw; h) Steel spring; i) PTFE membrane; j) Steel needle; k) Rubber packing (damper); l) Seal; m) Capillary

trogen overpressure and a needle valve controls the mercury flow.

Figure 27 shows the construction of the pneumatically controlled Multi-Mode electrode. The electrode is suitable for all voltammetric/polarographic work for which a DME, SMDE, or HMDE is required.

Thin mercury film electrodes (TMFE) are used chiefly for ultra-trace analysis by stripping voltammetry. Glassy carbon has proved itself to be the best substrate onto which a mercury film can be formed. The mercury is deposited by electrolysis, either in a separate operation before the voltammetric determination or, in the case of stripping voltammetry, together with the amalgam-forming metal during the accumulation by electrolysis (in situ coated TMFE, Section 25.2.6; the mercury film generated should be 1 – 100 μm thick). Voltammetric determinations with the TMFE can be more sensitive than with the HMDE, with improved resolution of neighboring peaks.

Carbon and graphite electrodes are employed principally in the anodic potential range. Glassy carbon electrodes (GCE) and carbon paste electrodes (CPE) are mostly used and are available commercially. They can be used either as stationary or rotating electrodes. *Glassy carbon electrodes* consist of pins or rods 2 – 8 mm diameter, cemented or pressed into a glass or plastic holder. *Carbon paste electrodes* consist of mixtures of spectroscopically pure carbon or graphite powder and water-insoluble, highly viscous organic liquids, e.g., nujol, silicone oil, or naphthalene derivatives. The mixture is ground to a paste, transferred to a glass tube, and expressed from the tube with a piston. The surplus carbon paste must be rubbed off and the surface polished. In voltammetric studies the composition of CPE influences the peak potential and the peak current. Glassy carbon electrodes and carbon paste electrodes are also used as working electrodes in voltammetric detectors for liquid chromatography (→ Liquid Chromatography).

Noble Metal Electrodes. Working electrodes of noble metals are often used for studies in the anodic potential range, or for determining elements with which mercury electrodes interfere (e.g., voltammetric Hg determination at an Au electrode). The electrochemical properties of noble metal electrodes and the reproducibility of the analytical values obtained with them depends on their physical and chemical pretreatment. In comparison with mercury and carbon electrodes, noble metal electrodes are of secondary importance for voltammetric determinations.

Modified Electrodes. When the surface of a working electrode has been chemically changed or treated in order to change its electrochemical properties or to improve its sensitivity and selectivity in an analytical determination, the electrode is referred to as a *modified electrode*. The modification of a voltammetric working electrode, usually a carbon electrode, is done by fixing (immobilizing) reagents on the electrode surface. The reagent that is used to modify the surface is referred as the *modifier*. One of the reasons for the development of modified electrodes is to use them to replace mercury electrodes (for environmental reasons) and so facilitate the further development of voltammetry, and in particular stripping voltammetry, as a "leading edge" technique. A major role of the modifier is to improve the selectivity of the electrode and aid the binding and enrichment of the analyte on the electrode surface.

There are several methods used to modify electrode surfaces: adsorption of reagents on the surface, covalent bonding by reaction between the modifier and functional groups on the electrode surface, incorporation of the modifier within a gel or an electrically conducting polymer film, physically coating the electrode surface with the modifier, e.g., an enzyme, or mixing the modifier with carbon paste.

Carbon paste electrodes have the advantage that the surface can be quickly renewed by simply removing the used surface layer with a tissue or knife. Organic complexing agents, ion exchangers, solid phase extractants and biological materials may be used as modifiers. The modifiers must have a low solubility in the analyte solution and not be electroactive in the working potential range [44].

Modified screen printed electrodes may be prepared either by incorporating the modifier in the carbon containing "ink" or by fixing it to the screen printed electrode surface. A mercury film, electrolytically deposited on a screen electrode has been used to determine heavy metals by ASV or AdSV (e.g., uranium [45]).

Thick-film graphite electrodes can be modified in a similar way to screen printed electrodes to give *thick film modified graphite electrodes* (TFMGEs). After modifying the surface in different ways the electrodes can be used for the stripping voltammetric determination of a number of heavy metals [46].

Microelectrodes. In general, the carbon and mercury working electrodes commonly used in voltammetry have surface areas in the mm^2 range. This means that the linear dimensions of these electrodes are an order of magnitude greater than the thickness of the diffusion layer and the transport of the analyte to the electrode takes place mainly by linear diffusion. However, if the lateral dimensions of the electrode are similar to or less than those of the diffusion layer (< 0.1 mm), the diffusion of the analyte to the electrode is then spherical rather than linear. Electrodes with such small dimensions in the lower μm range are referred to as *microelectrodes*. Microelectrodes are becoming more important as working electrodes in voltammetric trace analysis.

The miniaturization of the working electrode results in a higher current density at the electrode surface, but also in small currents flowing in the electrochemical processes. The small area means that the charging current i_c, which is proportional to the area of the electrode (Eq. 16), is also small. Thus as the electrode size decreases, the signal to noise ratio (or sensitivity) increases. In addition, note the very fast response to potential changes (response time) at a microelectrode.

A method of overcoming the problem of measuring the small currents is to join many microelectrodes in parallel and so obtain a much larger current output. An arrangement with many (up to 10 000) single microelectrodes is called a *microelectrode array*. By wiring n single electrodes in any array together in parallel, the output current will be n times that of a single microelectrode, i.e., it can be in the μA range instead of in the pA–nA range. Such an array, while having the advantages of a single microelectrode, produces a current which is large enough to be measured with conventional polarographic equipment.

With microelectrode arrays (mainly made from platinum or carbon by various techniques) it is possible to extend the application of analytical voltammetry to the direct determination of analytes in non-aqueous samples without adding a supporting electrolyte. In stripping voltammetry, the accumulation of the analyte on the electrode can be achieved in a very short time without stirring the solution because of the higher mass transport [47].

Reference and Auxiliary Electrodes. The function of the reference electrode is to monitor the potential of the working electrode and the function of the auxiliary electrode is to complete the electrical circuit for the current generated at the working electrode. The voltage of the auxiliary electrode (Pt wire or graphite rod) with respect to the working electrode is electronically adjusted via the potentiostat to the desired voltage plus the iR drop in the solution so that the potential difference between the working and reference electrode is maintained at the required value. Usually the area of the auxiliary electrode is at least 50 times greater than the working electrode.

The potential of the reference electrode must be constant and independent of the composition of the solution. Therefore the reference electrodes used in polarographic and voltammetric practice are electrodes of the second kind, of which the silver–silver chloride electrodes (with potassium chloride of a defined concentration, for example Ag/AgCl, 3 M KCl) are the most important. Previously calomel electrode (for example the saturated calomel electrodes, SCE) were used.

Measuring Cells. The commonest cells have a maximum volume of 25 mL. The vessels are generally made of glass; for demanding trace analyses they consist of quartz or plastic (principally polytetrafluoroethylene, PTFE). Micro-cells with a capacity of 1 mL or less are used for special microanalytical purposes and are constructed individually. Flow cells for analyses in connection with liquid chromatography or for flow injection analysis (FIA) are becoming increasingly important.

25.4. Evaluation and Calculation

Voltammograms and polarograms are evaluated either graphically or by calculation. For accurate determination of the current value—which is proportional to concentration—from the current wave or from the current peak, the residual current must be subtracted from the signal. For routine analysis, it is often sufficient to interpolate the residual current in the potential range of the voltammetric/polarographic current signal.

Figure 28 compares various practical curves and the methods for their manual interpretation. Such current–voltage curves, and others, can be evaluated more rapidly and with improved accuracy with computer-controlled instruments. The best values are obtained when a polynomial describing the approximate shape of the baseline is used. The coefficients of the polynomial are determined from the experimental points which lie on both sides of the current peak [4].

This evaluation becomes straightforward with equipment in which the current–potential curves are recorded digitally at a high data rate. The first step in the data processing consists of the automatic search for stray experimental data points. These are then eliminated and replaced by the mean value of both neighboring data points. The line of best fit of the current–potential curve is then smoothed within a sliding window along the voltage axis. The smoothing factor that can be applied essentially depends on the number of data points available. The principle of the curve smoothing is illustrated in Figure 29.

After smoothing, the curve is automatically differentiated (first derivative) and in this form used for peak recognition. This is achieved by seeking out successive maxima and minima in the differentiated curve (Fig. 30). In the usual voltammetric presentation (increasing negative potentials to the right), a maximum followed by a minimum indicates a reduction peak, whereas the reverse order indicates an oxidation peak. The potentials of the maximum and minimum in the di/dE versus E plot are noted and the peak width is the difference between these values while the peak potential is the mid-point between them. The qualitative identification of the analyte is achieved by comparing the peak potential (± a pre-set tolerance) with data stored in the computer memory. For a peak to be identified it must be greater than a pre-determined minimum peak height. After the peak is recognized a baseline is constructed and its value at the peak potential is subtracted from the maximum value of current to give the peak height. The result is presented as a peak or wave height, or as the peak area, all of which are proportional to the concentration of the analyte.

When the baseline is curved the polynomial which approximately describes this baseline is used to establish it mathematically over the potential region across the base of the peak. This potential region is determined from the potentials at

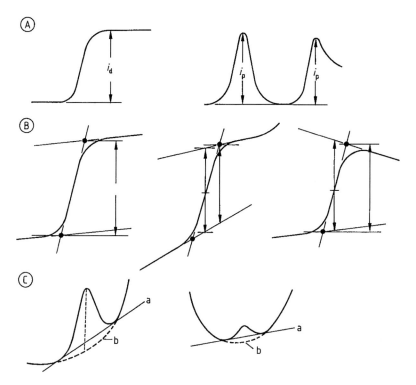

Figure 28. Interpretation of polarograms and voltammograms
A) Ideal conditions: linear trend of residual current; B) Evaluation of wave heights under nonideal conditions; C) Evaluation of peak currents with strongly curved residual current.
a) Tangent method; b) Evaluation with French curve [5]

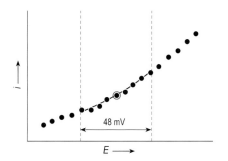

Figure 29. Curve smoothing using a voltage window with 9 data points

which the change in the gradient (becomes more positive) of the current–potential curve occur. The same procedure is used for the evaluation of closely adjacent peaks.

The peak or wave height (h) or peak area (A) calculated by the above procedure can be converted into the desired concentration units either by a *calibration curve* (c versus h or c versus A) or by the method of *standard additions*. When using a calibration curve its slope and linearity must be checked at least daily, also when experimental parameters are changed.

When using the method of standard additions, an aliquot, v, of a standard solution of known analyte concentration c is added (manually with a micropipette or automatically with a motorized burette) to a Volume V of the sample solution in which the analyte concentration is c_a. If h is the peak or wave height of the sample before the standard addition, and H the height after that, then the analyte concentration is given by

$$C_a = hcv / H(V + v) - Vh \qquad (31)$$

The highest accuracy can be obtained by using several (at least three) standard additions, provided the total concentration always lies within the linear range of the standard curve, when plotting h versus c, and extrapolating the straight line

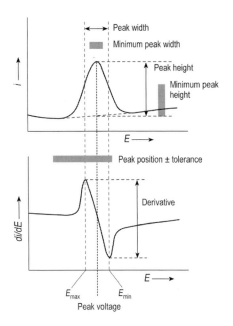

Figure 30. Peak recognition and characterization (Courtesy of Metrohm).
Peak potential = $(E_{max} + E_{min})/2$
Peak width = $E_{min} - E_{max}$

plot back to the c axis. The distance from the origin to the intercept corresponds to c_a.

The reproducibility of voltammetric/polarographic determinations depends not only on the evaluation, but also on the quality of the electrode and the measurement set-up. With modern equipment, reproducibility is 0.5–2.0 %.

A characteristic source of error is the temperature dependence of voltammetric currents. All equations for the proportionality of polarographic currents to concentration contain the diffusion coefficient, which increases with temperature. The rate constants of preceding and succeeding chemical reactions are also affected by temperature and this must be taken into account when evaluating kinetic and catalytic currents. Electrode reactions that are associated with analyte adsorption processes likewise show a specific temperature dependence. The temperature influence differs and leads to different temperature coefficients for the individual voltammetric methods [48].

In simple diffusion-controlled electrode reactions, deviations between +1.5 and +2.2 %/ °C occur. For kinetic and catalytic currents, the temperature coefficients can be up to +70 %/ °C. Currents that are based on adsorption processes can have negative temperature coefficients, because of decreasing adsorption with increasing temperature.

25.5. Sample Preparation

Before performing the voltammetric/polarographic determination, in some cases the analyte must be separated from the matrix and from other interfering components, and converted into a measurable form. For total element determinations, sample digestion is normally required. To separate organic analytes from the matrix, extraction methods and chromatographic separation are mainly used.

Heavy metals in association with organic materials are usually in a bound form, from which they must be liberated prior to analysis. This is carried out by dry or wet ashing, whereby the organic matrix must be completely destroyed. To avoid systematic errors, care must be taken in all forms of sample treatment to ensure that the elements to be determined remain completely in the sample and that sample contamination from digestion reagents and vessels is avoided.

Wet digestions are carried out only with high-purity acids. If there is a risk that an element will volatilize, the treatment is carried out under reflux, or better in pressure digesters with quartz or plastic inserts; heating is carried out electrically or by microwave radiation. Sample digestion with nitric acid at high temperatures is favored owing to the increased oxidation power. Figure 31 shows a high-pressure ashing apparatus for the digestion of organic and inorganic samples with acid (HNO_3, HCl, $HClO_4$, H_2SO_4, HF, or mixed acids) in sealed quartz vessels under high pressure (up to 13 MPa) and high temperature (up to 320 °C). The digester is contained in a heated autoclave and filled with nitrogen (ca. 10 MPa) in order to compensate the internal pressure resulting from the digestion reaction.

For the digestion of biological samples, combustion in oxygen (*dry ashing*) can also be useful under suitable conditions. The use of oxygen of highest purity ensures low contamination digestion. Some forms of equipment operate with reflux condensation to avoid losses of elements by volatilization; cooling with liquid nitrogen is particularly suitable. Organic samples can be digested at relatively low temperature in high-frequency oxygen plasmas.

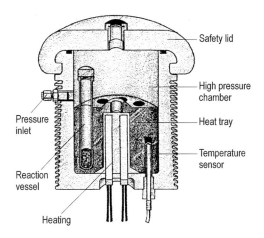

Figure 31. High-pressure ashing apparatus HPA-S (Courtesy of Perkin Elmer)

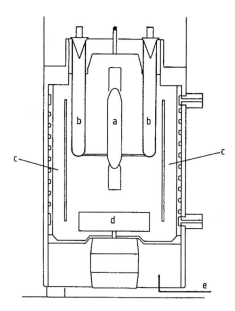

Figure 32. Ultraviolet digester (Courtesy of Metrohm)
a) Lamp; b) Digester vessels with stopcocks; c) Cooling chamber with water inlet and outlet; d) Fan; e) Cables

Voltammetry and polarography are particularly important for the analysis of water samples for heavy metals. In surface waters with organic contamination, they can only be analyzed after digestion. Sampling and storage is of prime importance for correct analytical results.

Water samples are stored in flasks of polyvinyl chloride (PVC) or PTFE, cleaned with 50 % nitric acid before use. The flask is subsequently pre-rinsed with the sample water. Contaminated samples are filtered through membrane filters directly into the sample flask and acidified, e.g., with high purity nitric acid to pH 1 – 2; the filter residue is analyzed separately. Water samples with little or no organic contamination are mixed with the chosen supporting electrolyte and immediately placed in the voltammetric/polarographic set-up.

Contaminated water samples are prepared for electrochemical heavy metal trace analysis either by UV photolysis or by oxidative microwave digestion.

Ultraviolet photolysis is based on the generation of OH radicals, which react with organic compounds and degrade them. Hydrogen peroxide is used as an initiator. The acidified water sample plus hydrogen peroxide is irradiated in quartz vessels. An arrangement for the digestion of one to six samples is shown in Figure 32.

When nitrate-containing samples are UV-irradiated, nitrite is formed by reduction. The voltammetrically active nitrite causes a broad current – voltage peak that can cause interference (e.g., with the determination of zinc). This must be taken into account when examining water samples acidified with HNO_3. Another acid should be used or the nitrite formed should be reduced to nitrogen with sulfamic acid (NH_2SO_3H) immediately after the end of the photolysis.

The complete destruction of organic substances in water samples by UV photolysis depends on the nature and extent of the contamination, and on the intensity of the irradiation. The irradiation time required for waste water samples is usually 1 h (see Fig. 33).

Microwave Digestion. Oxidative microwave digestion is suitable for the rapid degradation of interfering organic substances in surface waters. For this purpose the samples are heated with the oxidizing agent Oxisolv (Merck, Germany) in sealed PTFE vessels in a microwave oven. During the thermal treatment, the pH value of the sample passes through the wide range of pH values (from 11 to 2). Wastewater samples are generally completely digested after 50 – 60 s, so that the voltammetric/polarographic metal trace analysis can be carried out completely free of interference.

If the metals are determined by the highly sensitive method of adsorptive stripping voltammetry, the digestion time must be increased by 1 – 2 min.

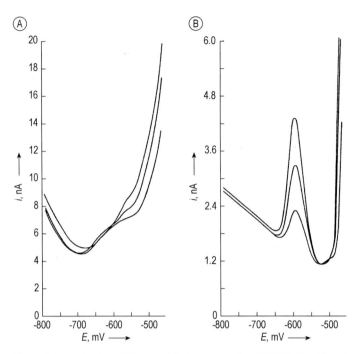

Figure 33. Determination of Cd in municipal wastewater by DPASV before (A) and after (B) 1 h UV irradiation [49]

Figure 34 shows the voltammograms for the determination of molybdenum in a river water sample; the digestion time was 2×50 s at 260 W microwave power [50].

25.6. Supporting Electrolyte Solution

The solution for the voltammetric/polarographic determination of elements and organic compounds always contains a supporting electrolyte in order to minimise the migration current and reduce the ohmic voltage drop. The supporting electrolytes used are mainly chlorides, nitrates, and sulfates of Li, Na, K, perchlorates of Li, Na, and salts of tetraalkylammonium bases (NR^+X^-; R = methyl, ethyl, butyl; X = Cl, ClO_4, Br, I). Moreover, acids (HCl, H_2SO_4), bases (LiOH, NaOH, $NR_4^+OH^-$), and frequently buffer solutions are used as they simultaneously provide for regulation of pH during electrode reactions. Supporting electrolyte concentrations are commonly between 0.1 and 1 mol/L.

If complexation occurs between the analyte and the supporting electrolyte or an added ligand, crucial changes in electrochemical behavior can occur. As a result of complexation, the analyte can become electrochemically inactive, or its behavior may change in such a way that the value of the polarographic half-wave or peak potential becomes more negative. For the simple complexing reaction of a hydrated cation $M(H_2O)_x$ with ligand L and coordination number i:

$$M(H_2O)_x + i L \rightleftharpoons ML_i + xH_2O$$

and with the stability constant K of the complex:

$$K = \frac{[ML_i]}{[M][L]^i}$$

the shift of the half-wave or peak potential is given by:

$$\Delta E_{1/2} = \Delta E_p = -\frac{0.059}{n} \log K - \frac{0.059}{n} i \log [L] \quad (32)$$

Stability constants and coordination numbers of complexes can be determined from the shift of the polarographic half-wave or peak potentials.

The complexing behavior of supporting electrolytes is analytically important for the resolution of overlapping current signals and for masking interfering sample components. Specific additions

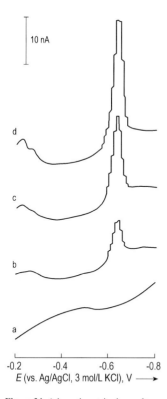

Figure 34. Adsorptive stripping voltammograms for Mo in river water after microwave digestion [50]
a) Untreated sample; b) After microwave digestion + Oxisolv; c) + 2 ng/mL Mo; d) + 4 ng/mL Mo Conditions: 10^{-3} mol/L chloranilic acid, pH 2.7, stirred 1 min, accumulation at −0.2 V
Calculated sample concentration of Mo 1.82 ng/mL

of suitable complexing agents are therefore used to improve the selectivity of polarographic determinations, as demonstrated in Figure 35 by the separation of the DPP peaks of Tl/Pb and Cd/In. Similar phenomena are known in multielement determinations by stripping voltammetry. By changing the composition of the supporting electrolyte solution after the simultaneous accumulation by electrolysis and use of suitable complexing agents, the metals are stripped at different potentials and can be determined from their separated peaks [5].

Simple ligands are used as complexing agents, e.g., CN^-, OH^-, NH_3, oxalate, halides, as well as stronger complexing agents, such as ethylenediaminetetraacetic acid (EDTA), citrate, tartrate, etc.

A high salt content, which can arise after sample digestion or as a result of preceding separation operations, changes the viscosity η of the supporting electrolyte solution and hence the diffusion behavior of the analyte. With increasing salt content the polarographic currents become smaller and displaced to more negative potentials. The peaks in the stripping voltammograms are subject to the same influence, except that the potentials become more positive. Small amounts of inert salts, as can occur in natural samples, generally have no influence on polarographic or voltammetric determinations. Sometimes, as in the case of heavy metal determination in seawater by stripping voltammetry, electrolyte addition becomes superfluous.

The useful voltage range for voltammetric/polarographic determinations depends significantly on the supporting electrolye used. The reduction of the cation of the supporting electrolyte restricts the cathodic working range, as hydrogen evolution from acidic supporting electrolyte solutions depends on the working electrode material. The high overvoltage of H^+ discharge on mercury is therefore particularly advantageous.

The anodic range is determined by oxidation of water or of the supporting electrolyte anion, as well as by oxidation of the electrode material. Anodic dissolution of mercury can even begin at ca. −0.4 V if anions that form poorly soluble Hg(I) salts are present.

Oxygen must be removed (by deaeration) from the supporting electrolyte solution before each voltammetric/polarographic determination. Oxygen is present in aqueous solutions at ca. 10^{-3} mol/L, depending on the solubility conditions, and is reduced at the working electrode in two stages. The course of the reduction is pH dependent and can be described by the following equations.

1) Reduction in acid solution:
$$O_2 + 2H^+ + 2e^- \longrightarrow H_2O_2$$
$$H_2O_2 + 2H^+ + 2e^- \longrightarrow 2H_2O$$
2) Reduction in alkaline solution:
$$O_2 + 2H_2O + 2e^- \longrightarrow H_2O_2 + 2OH^-$$
$$H_2O_2 + 2e^- \longrightarrow 2OH^-$$

In a DC polarogram the first stage lies between −0.1 and −0.3 V, depending on the pH value, and the second stage occurs between −0.7 and −1.3 V.

The currents caused by the reduction of oxygen interfere with the current–voltage signals of many analytes. In addition, there is interference due to possible subsequent reactions of the H_2O_2 and of the OH^- ions formed.

The dissolved oxygen is removed by introducing pure nitrogen or argon for 5–10 min. Oxygen

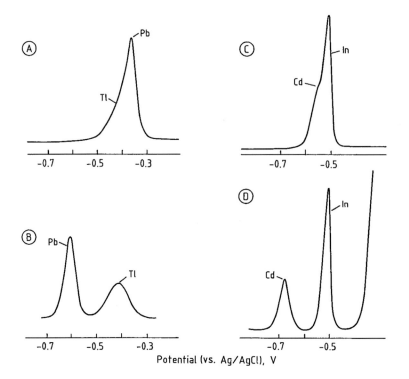

Figure 35. Influence of complexation on the peak separation in differential pulse polarographic curves [5]
10^{-4} mol/L Pb^{2+} and Tl^+ in 0.1 mol/L KCl (A) and 0.1 mol/L KCl + 0.1 mol/L NaOH (B) 10^{-4} mol/L Cd^{2+} and In^{3+} in 0.1 mol/L HCl (C) and 0.1 mol/L HCl + 2 mol/L KI (D)

can also be removed from alkaline solutions by addition of Na_2SO_3.

Inorganic ions are determined almost exclusively in aqueous supporting electrolyte solutions. For organic and organometallic compounds, either aqueous supporting electrolyte solutions mixed with organic solvents or pure organic solvents are used (e.g., alcohols, acetonitrile, propylene carbonate, dimethylformamide, and dimethylsulfoxide). Figure 36 illustrates the potential ranges that are available when using these solvents for anodic and cathodic studies at various working electrodes.

Organic solvents are also used for the extraction of metal chelates, which can then be analyzed in the extract, directly or after adding a polar solvent (in order to raise the solubility of the supporting electrolyte). The procedure known as *extractive voltammetry/polarography* combines the selectivity of partition methods with the possibility of carrying out the determination in organic solvents for many analytes. Using this procedure, the analyte can be simultaneously accumulated and separated from interfering components [15].

25.7. Application to Inorganic and Organic Trace Analysis

Voltammetry and polarography are used primarily for the trace analysis of inorganic and organic substances. Because of the introduction of the powerful pulse methods and stripping techniques, and of advances in instrumentation, these measurement techniques are also of interest for routine analysis. With modern polarographs the most important voltammetric or polarographic methods can be selected by simple switching. Investigations are therefore possible with the same measuring instrument in a concentration range extending over at least six decades. Ions and molecules can be determined and the resulting signals give qualitative and quantitative information as well as providing insight into the bonding state of the analyte.

Elements and Inorganic Ions. The determination of elements and inorganic ions is based chiefly on reduction and oxidation processes that

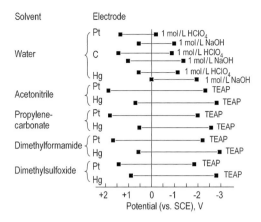

Figure 36. Polarographically usable voltage ranges with Hg, C, and Pt electrodes in various solvents and supporting electrolytes [16]
TEAP = Tetraethylammonium perchlorate

can take place in combination with kinetic, catalytic, or adsorptive processes. The reduction of simple or complexed ions of metals that are soluble in mercury is usually reversible and leads to readily interpretable current – potential curves. Examples of this are the polarographic determinations of cadmium, copper, thallium, zinc, and lead. Less important for routine analysis is the possibility of determining mercury, silver, and gold voltammetrically at noble metal or carbon electrodes.

Multivalent ions are often reduced in several steps, the mechanism and degree of reversibility of the electrode reaction in most cases being dependent on the composition of the supporting electrolyte solution. Two-step current – potential curves are obtained with chromium(VI), copper(II), iron(III), molybdenum(VI), and vanadium(V), among others, when suitable supporting electrolyte solutions are used. In a multi-step process, the shape and height of the polarographic wave or peak associated with each step is determined by the number of electrons involved and the reversibility of that step. For analytical purposes, the best sensitivity is obtained when a supporting electrolyte can be found in which the multi-electron exchange occurs in one step at a single potential, resulting in a large single wave or peak in the polarogram. This is illustrated in the DP-polarographic reduction of Cr(VI) shown in Figure 37. In 0.1 mol/L KNO$_3$ supporting electrolyte (curve A), Cr(VI) is reduced in two steps — firstly to Cr(III) at -0.3 V and then to Cr(II) at -1.2 V. In 0.1 mol/L Na$_2$EDTA solution (curve B), only one peak is observed corresponding to the four electron reduction of Cr(VI) to Cr(II). Because the three-electron reduction to Cr(III) is irreversible the first peak in curve A is smaller than second peak for the reversible one-electron reduction of Cr(III) to Cr(II). The largest peak, obtained for the four-electron reduction of Cr(VI) to Cr(II) in EDTA solution (curve B), gives a determination limit of 10^{-6} mol/L for the determination of chromium.

For those systems which give well behaved catalytic currents, fairly sensitive polarographic and voltammetric analyses are possible. Examples are the determination of Cr(VI) or Mo(VI) in the presence of nitrate ions or of Fe(III) in the presence of bromate ions. For example, the determination limit for Cr(VI) using DPP can be lowered to ca. 10^{-8} mol/L when Na$_2$EDTA containing nitrate ions is used as the supporting electrolyte [62]. The Cr(II) formed by the reduction of Cr(VI) at the working electrode is immediately oxidized back to the trivalent state by the nitrate ions and is available to be reduced again in the electrode reaction. Titanium, tungsten, uranium and vanadium can also be determined via catalytic currents in the presence of nitrate or chlorate ions with limits of detection in the µg/L (ca. 10^{-8} mol/L) range [4], [5], [63].

Voltammetric/polarographic methods are suitable both for the simultaneous determination of several analytes and for the determination of single analytes in a sample solution. The possibilities are a consequence of the different potentials at which the peak currents occur and the selection of a suitable voltage range for the analysis. Superimposition or overlap of responses can be eliminated by using another supporting electrolyte. Complexing compounds that are expected to change drastically the redox behavior of the analyte or of the interfering component are suitable for the purpose.

The half-wave potentials determined for selected elements in various supporting electrolyte solutions are compared in Figure 38.

Voltammetric/polarographic methods are not very important for *anion analysis*. Only bromate, iodate, and periodate can be determined by means of polarographic reduction currents. Indirect determinations are possible for those anions that form compounds of low solubility or stable complexes with the Hg$_2^{2+}$ ions formed in the anodic oxidation of the electrode mercury. The current caused by the dissolution of the mercury is proportional to the concentration of these anions in the sample

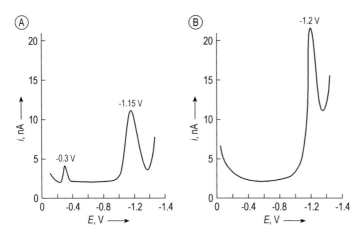

Figure 37. DP-polarogram of chromium(VI) in different supporting electrolytes
A) 0.5 mg/L Cr(VI) in 0.1 M KNO$_3$; B) 0.5 mg/L Cr(VI) in 0.1 M Na$_2$EDTA Electrode – SMDE; Potential scan from – 0.1 V to – 1.4 V (vs. Ag/AgCl, 3 M KCl)

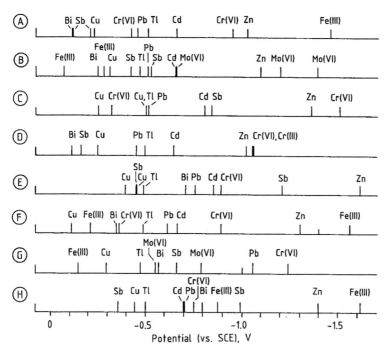

Figure 38. Polarographic half-wave potentials of selected elements measured in the following supporting electrolyte solutions
A) 1 mol/L KCl; B) 2 mol/L CH$_3$COOH + 2 mol/L CH$_3$COONH$_4$; C) 1 mol/L NH$_3$ + 1 mol/L NH$_4$Cl; D) 1 mol/L HCl;
E) 1 mol/L NaOH; F) 0.5 mol/L Na tartrate (pH 9); G) 0.1 mol/L Na$_2$EDTA; H) 0.1 mol/L Na citrate + 0.1 mol/L NaOH

solution. Halides, sulfide, and cyanide can be determined in this way [2], [5].

The most important and most successful application of voltammetry/polarography is the determination of traces of heavy metals in aquatic environmental samples. In only lightly polluted water samples, e.g., in rainwater, drinking water, and seawater, the determinations can often be carried out without further sample preparation, i.e., directly after sampling and adjustment to the required pH value. More heavily polluted samples must be digested, which is usually performed by

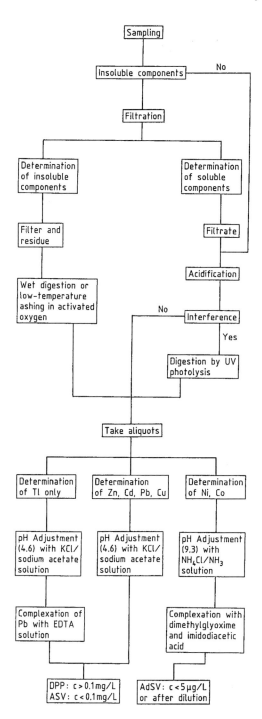

Figure 39. Flow diagram of the procedure for the polarographic/voltammetric determination of heavy metals in drinking water, groundwater, and surface water according to DIN 38 406 Part 16

UV irradiation or in a microwave oven (Chap. 25.5). There is particular interest in determining those metals whose concentration limits in natural waters or wastewaters are regulated by law and accordingly must be controlled (regulations governing purity of drinking water and wastewater discharge). In DIN 38 406, Part 16, for example (German Standard Method), the Zn, Cd, Pb, Cu, Tl, Ni, and Co concentrations are monitored with the aid of polarographic and voltammetric methods of determination: the procedure is shown in Figure 39.

The method illustrated in Figure 39 is suitable for determinations between the lower µg/L and middle mg/L ranges. It is used for monitoring the heavy metal content in polluted surface waters and identifies the metal content of solutions as well as the constituents that occur in association with suspended matter or colloidal particles. The generally very small amounts of Co and Ni are determined as dimethylglyoxime complexes by AdSV, while for the other metals DPP or ASV is used, depending on the concentration. Sample preparation depends on the state of the sample.

The methods of voltammetry/polarography are important for determining the concentrations of metals in different oxidation states and for characterizing physicochemical forms of bonding (*speciation analysis*) [17], [18].

The metal content determined by ASV in untreated water samples at natural pH values or in buffered solutions is usually lower than the total content and is described as the biologically available fraction. This is the fraction that is present in the form of labile inorganic complexes or hydrated ions. The metal content combined with organic ligands can only be determined voltammetrically after digestion by UV photolysis. In this way labile and nonlabile or combined metal species can be distinguished.

For the ultra-trace analysis of heavy metals in water the AdSV method, in addition to ASV, is especially powerful [19]–[21]. The limits of determination are generally in the middle to upper ng/L range; Table 2 gives a survey of the best-known methods.

Most AdSV determinations are carried out, after sample digestion by UV irradiation, by recording cathodic stripping voltammograms. Oxidative microwave digestion with Oxisolv (Chapter 25.5) is also suitable for sample preparation. Only rarely can trace concentrations of metals in natural samples such as seawater be determined without sample digestion. An example is the determination of

Table 2. Ultra-trace analysis of metals in natural waters by AdSV

Analyte	Ligand	Electrolyte	Sample pretreatment	Detection limit, ng/L	Reference
Aluminum	DASA[c]	BES[d] pH 7.1	UV irradiation (not seawater)	30	[52]
Antimony	chloranilic acid	pH 1.0 adjusted with HCl	UV irradiation	550	[53]
Gallium	solochrome violet RS	acetate buffer pH 4.8	UV irradiation (not seawater)	80	[54]
Molybdenum	chloranilic acid	pH 2.7 adjusted with HCl	UV irradiation (not seawater)	20	[50]
Platinum	formazone	formaldehyde + hydrazine + 0.5 mol/L H_2SO_4	UV irradiation	0.01	[55]
Thorium	mordant blue S	acetate buffer pH 6.5	UV irradiation	100	[56]
Tin	pyrocatechol	acetate buffer pH 4.2	UV irradiation (not seawater)	5	[57]
Titanium	mandelic acid	$KClO_3$ + NH_3 pH 3.2	UV irradiation	0.05	[58]
Uranium	pyrocatechol	acetate buffer pH 4.7	separation by ion exchange	240	[59]
		HEPES[a] + 0.5 mol/L NaOH pH 6.8	UV irradiation (not seawater)	70	[60]
	oxine	0.01 mol/L PIPES[b] + 0.5 mol/L NaOH pH 6.7–6.9	UV irradiation (not seawater)	500	[61]
	chloranilic acid	pH 2.5 adjusted with HCl	unnecessary	25	[64]
Vanadium	pyrocatechol	PIPES[b] pH 6.9	UV irradiation	5	[65]

[a] HEPES = N-2-hydroxyethylpiperazine-N-2-ethanesulfonic acid.
[b] PIPES = piperazine-N,N'-bis(2-ethanesulfonic acid) monosodium salt.
[c] DASA = 1,2-dihydroxyanthraquinone-3-sulfonic acid.
[d] BES = N,N'-bis-(2-hydroxyethyl)-2-aminoethanesulfonic acid

uranium using chloranilic acid (2,5-dichloro-3,6-dihydroxy-1,4-benzoquinone) (Fig. 40) as complexing agent, which has recently been developed. This method enables interference-free determinations in river water samples to be carried out [64].

In contrast, the determination of *platinum* by AdSV with formazone (a condensation product of formaldehyde and hydrazine) suffers interference from surface-active substances which can be present even in the purest drinking water at trace level. The unusual feature of the extremely sensitive determination of platinum by AdSV is that the complex of Pt(II) with formazone, after adsorptive accumulation at the HMDE, reduces the hydrogen overvoltage on the mercury surface. At the same time, catalytic currents are developed that can be used for the trace analysis of platinum in the ng/L range [55]. The Pt determination is carried out in water samples after prior UV photolysis, and in biological materials after digestion with HNO_3 in a high-pressure asher.

A further field of application of AdSV is the determination of heavy metal traces in high-purity chemicals. The methods originally developed for water analysis (Table 2) can be used without modification for the purity control of salts of the alkali and alkaline earth metals without having to take account of interference by the salt matrix [66].

Arsenic, selenium, and tellurium can be determined at the trace level by cathodic stripping voltammetry (CSV) after having been reductively co-deposited with copper. A characteristic of the stripping technique of these three elements is that only a single current peak which arises from the further reduction of the deposited analyte to As^{3-}, Se^{2-} or Te^{2-} is observed. This method is suitable for the determination and speciation of As(V) and As(III) in water samples [36], selenium in minerals [38] and biological samples [68], and for the simultaneous determination of selenium and tellurium (Fig. 41) [37].

Potentiometric stripping analysis (PSA) and chronopotentiometric stripping analysis (CPS) are an alternative techniques that may be used for the determination of metals such as Bi, Cd, Cu, Pb, Sn, Tl, and Zn in water samples [67]. With microprocessor controlled equipment, limits of detection are in the sub-µg/L range. The main advantage of PSA and CPS is that atmospheric oxygen dissolved in an aqueous sample does not have to be removed and that in many cases, in-

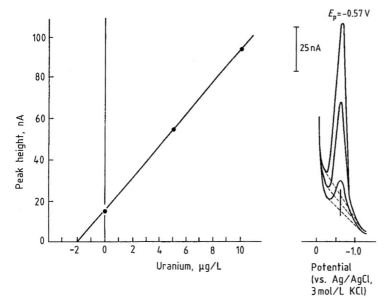

Figure 40. Direct determination of low concentrations of uranium in river water by AdSV [64]
Sample contains 2.4×10^{-4} mol/L chloranilic acid, adjusted with HCl to pH 2.5; $E_{acc} = +0.15$ V; $t_{acc} = 90$ s; standard additions method

cluding those of wine, urine, and blood, direct determination of cadmium and lead is possible [69]–[71].

The determination of trace concentrations of heavy metals by stripping voltammetry in body fluids and organ samples is used to assess their physiological function. Methods have been developed for determining toxic metals by ASV and PSA in blood, serum, and urine. Stripping methods are also important for monitoring the heavy metal content in food and drink. The samples are mineralized by dry ashing or by wet digestion. The choice of the appropriate decomposition procedure depends on the type of test and the form of the elements to be determined [4], [5].

Organic compounds. The polarographic and voltammetric determination of organic compounds usually depends on the reduction or oxidation of functional groups. In addition there are some applications where the determination is based on the reaction of organic molecules with the electrode material (mercury) or on the change in the double layer capacitance associated with the organic species being adsorbed on the electrode surface.

Organic functional groups which are electroactive and the oxidations or reductions of which are the bases of the analytical determinations are listed in Table 3 together with the most important classes of compounds containing these groups. The ranges for the half-wave or peak potentials of selected groups of organic compounds are presented in Figure 42.

Compounds with reducible functional groups predominate. Polyaromatic hydrocarbons, aromatic hydroxy compounds and amines, as well as amides and various nitrogen heterocyclic compounds can be determined anodically. These processes are in many cases pH dependent. Determinations based on redox processes are therefore carried out in buffered solutions. The composition and concentration of the buffer systems generally has no influence on the position of the half-wave or peak potentials. If its concentration is sufficiently high, the buffer simultaneously performs the function of the supporting electrolyte.

The polarographic analysis of aldehydes, organic acids, monosaccharides, and substituted pyridine derivatives is based on kinetic currents, if a rate-determining chemical reaction precedes or follows the actual electrode reaction. Alkaloids, proteins, and many organic sulfur compounds reduce the overvoltage of the hydrogen reaction on the mercury drop electrode and are determined via catalytic hydrogen waves. In native protein molecules, the sulfhydryl and disulfide groups are

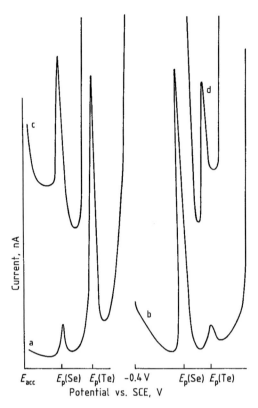

Figure 41. Cathodic stripping voltammograms of selenium and tellurium [37]
Supporting electrolyte solution: 0.5 mol/L $(NH_4)_2SO_4$ + 4×10^{-3} mol/L EDTA + H_2SO_4(pH 4.5) + 1 mg/L Cu^{2+}; t_{acc} = 1 min, E_{acc} = –0.4 V, sensitivity 200 nA fc
a) 1 µg/L Se, 10 µg/L Te; b) 1 µg/L Te, 10 µg/L Se; c) 1 µg/L Se standard addition to 1 mg/L Te; d) 1 µg/L Te standard addition to 1 mg/L Se

masked and therefore do not catalyze this electrode process. Not until denaturation by the action of acids, bases, UV radiation, etc., do the catalytic waves appear as a result of exposure of these groups. This phenomenon can be used to determine denatured DNA in natural DNA samples [22].

Indirect polarographic methods are used to determine nitrilotriacetic acid (NTA) and other *polyamino acids*, e.g., EDTA (auxiliary agents for detergents and cleansers). The methods are based on the formation of stable heavy metal complexes with polarographic properties that are different from those of the acids. Bismuth and cadmium salts are used, added in small excess to the sample. For determination of the acids, use is made of the current signals that are caused either by the excess of the metal ions or by a newly formed complex. Both NTA and EDTA are determined according to DIN 38 413 Part 5 after addition of bismuth(III) nitrate to the sample acidified to pH 2 with nitric acid. The polarographic curve is shown in Figure 43.

The determination of *surface-active substances* by tensammetry is particularly sensitive owing to adsorption on the electrode surface and the associated change of the double layer capacitance [23].

Voltammetric trace analysis of *alkaloids and other pharmaceuticals* is carried out after adsorptive accumulation at the electrode surface at constant potential [72]. The compounds that can be determined by AdSV in the submicromolar and nanomolar concentration range also include various pesticides [73]–[75]. The importance of AdSV for the trace analysis of adsorbable molecules has been summarized [24].

Organic substances that form slightly soluble or coordination compounds with mercury ions (Table 3) are determined polarographically via anodic waves or voltammetrically after anodic accumulation. Agrochemicals containing thiourea are also determined by this CSV technique; the limits of determination are reported to be 10^{-7} – 10^{-8} mol/L [76].

Electrochemically inactive organic compounds can be converted into an electrochemically active state by nitration, nitrosation, oxidation, hydrolysis, or some other preliminary chemical reaction, and can then be determined polarographically or voltammetrically. Examples of this are polarographic determinations of alkylbenzenesulfonates in natural water samples after extraction and nitration of the aromatic ring [77] and determinations of beta-receptor blocking agents in tablets after conversion to the corresponding *N*-nitroso derivatives [78].

Aliphatic and aromatic *nitro compounds* are reduced at relatively positive potentials via the hydroxylamines to the corresponding amines. The carcinogenic 4-nitroquinoline-*N*-oxide is determined by DPP in the presence of 4-hydroxyaminoquinoline-*N*-oxide and 4-aminoquinoline-*N*-oxide via the reduction of the nitro group [79]. Nitrazepam, parathion, nitrofurantoin, and the nitroimidazoles in blood plasma or urine are also determined via the reduction of the respective nitro groups.

The *nitroso group* is generally more easily reducible than the nitro group. *N*-Nitrosamines, e.g., *N*-nitroso-*N*-methylaniline in blood, serum, or albumin, are determined in the µg/kg range

Table 3. Functional groups for the polarographic and voltammetric determination of organic compounds

Reducible groups	Compounds
$\diagdown$C=C$\diagup$	unsaturated aliphatic and (poly)aromatic hydrocarbons with conjugated double or triple bonds and with multiple double bonds
$-C\equiv C-$	
$-\underset{\vert}{\overset{\vert}{C}}-X$	halogen-substituted aliphatic and aromatic hydrocarbons with the exception of fluoro compounds
$\diagdown$C=O$\diagup$	aliphatic and aromatic aldehydes, ketones, and quinones
$-O-O-$	aliphatic and aromatic peroxides and hydroperoxides
$-NO_2$	aliphatic and aromatic nitro and nitroso compounds
$-NO$	
$-N=N-$	azo compounds
$\diagdown$C=N$-$	benzodiazepines, pyridines, quinolines, acridines, pyrimidines, triazines, oximes, hydrazones, semicarbazones
$-S-S-$	disulfides
$\diagdown$C=S$\diagup$	thiobenzophenones
$-SO-$	diaryl and alkylaryl sulfoxides
$-SO_2-$	sulfones
$-SO_2NH-$	sulfonamides
$-\underset{\vert}{\overset{\vert}{C}}-Me$	organometallic compounds

Oxidizable groups	Compounds
$A-\underset{\vert}{\overset{\vert}{C}}-H$	poly(aromatic) hydrocarbons (A = aryl)
$-OH$	phenols
$-NH_2$	aromatic amines
$-CO-N\diagdown$	amides

Groups reacting with mercury ions	Compounds
$-SH$	thiols
$-N-C\diagup^S_{\diagdown S^-}$	dithiocarbamates
$-NH\diagdown_{C=S}\diagup-NH$	thioureas, thiobarbiturates
$-C\diagup^S_{\diagdown NH_2}$	thioamides
$-NH\diagdown_{C=O}\diagup-NH$	derivatives of barbituric acid and of uracil

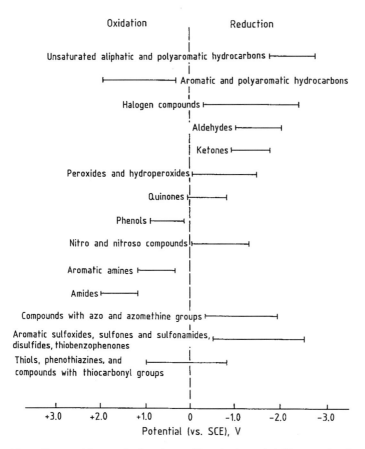

Figure 42. Potential ranges for the polarographic and voltametric half-wave and peak potentials of organic compounds

by DPP [80]. The polarographic and voltammetric behavior of the nitrosamines, quinones, steroid hormones, and imidazoles is of interest, particularly in connection with cancer research [25].

Knowledge about the bio-transformation of *drugs* and their excretion is important pharmacologically. Numerous methods for the polarographic and voltammetric determination of pharmaceuticals and their metabolites have been developed for the assessment of such relationships. The determination is generally preceded by separation from the matrix by liquid–liquid extraction with diethyl ether or ethyl acetate.

1,4-Diazepines are determined polarographically via their azomethine groups. In addition to flurazepam, further representatives of this class of pharmaceutically active agents can be determined polarographically. These include diazepam, chlordiazepoxide, bromazepam, lorazepam, and chlorazepam; the limits of determination with DPP are in the µg/L range [4], [5].

Organic compounds with oxidizable functional groups (see Table 3) are mostly determined, after separation by liquid (column) chromatography or by high-performance liquid chromatography, with the aid of a voltammetric (amperometric) detector (→ Liquid Chromatography). These compounds, which can be determined in liquid and solid biological samples, in foods, or in aquatic environmental samples, include catecholamines, sulfonamides, phenothiazines, dopamines, estrogens and other hormones, tocopherols, and various groups of agricultural pesticides. Liquid–liquid extraction or solid phase extraction are also used for sample preparation; the determinations are carried out in the ng/L range (→ Liquid Chromatography).

Polarography and voltammetry are of importance in the *analysis of organometallic* compounds

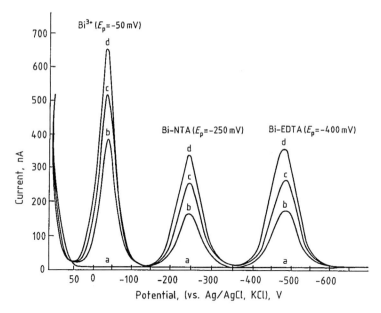

Figure 43. Polarographic determination of NTA and EDTA according to DIN 38 413 Part 5, standard additions method
a) Residual current; b) After Bi^{3+} addition; c) First addition of Bi-NTA/EDTA solution; d) Second addition of Bi-NTA/EDTA solution

in environmental samples. In contrast to spectroscopic methods which only give information on the total metal contents in a sample, the voltammetric techniques are capable of characterizing and determining the various forms of those compounds as they exist in the sample (speciation).

Representative of the organometallic compounds are the R–Me–compounds where Me = As, Hg, P, Pb, Sb, Se, Sn, Te, and Ti; the peak potentials are listed in [30].

Organomercury compounds of the type RHgX generally are reduced in two steps according to the mechanism.

$$RHg^+ + e^- \rightleftarrows RHg^\bullet + e^- \begin{cases} \xrightarrow{+H^+, pH<7} RH + Hg \\ \xrightarrow{pH>7} RHg^- \xrightarrow{+H^+} RH + Hg \end{cases}$$

$$\downarrow$$
$$(RHg)_2$$
$$\downarrow$$
$$R_2Hg + Hg$$

Likewise organo-arsenic, antimony, and tin compounds either singly or in admixture can be determined in the µg range by polarography and particularly mono-, di-, and tributyltin in the ng range, based on solid phase extraction and adsorptive cathodic stripping voltammetry [81].

The polarographic or voltammetric determination of organic compounds of biological, pharmaceutical, or environmental significance have been summarized [27] – [29].

25.8. References

General References

[1] J. Heyrovsky, J. Kuta: *Grundlagen der Polarographie*, Akademie Verlag, Berlin 1965.
[2] A. M. Bond: *Modern Polarographic Methods in Analytical Chemistry*, Marcel Dekker, New York 1980.
[3] F. G. Thomas, G. Henze: *An Introduction to Analytical Voltammetry – Theory and Practice*, CSIRO, Melbourne (2001).
[4] G. Henze: *Polarographie und Voltammetrie; Grundlagen und analytische Praxis*, Springer Verlag, (2001).
[5] G. Henze, R. Neeb: *Elektrochemische Analytik*, Springer Verlag, Berlin 1986.
[6] J. Koryta, J. Dvorak: *Principles of Electrochemistry*, J. Wiley & Sons, New York 1987.
[7] M. R. Smyth, J. G. Vos: *Analytical Voltammetry*, Elsevier, New York 1992.
[8] G. C. Baker, A. W. Gardner: "Forty Years of Square-Wave Polarography," *Analyst (London)* **117** (1982) 1811.

[9] J. G. Osteryoung, M. M. Schreiner: "Recent Advances in Pulse Voltammetry," *CRC Crit. Rev. Anal. Chem.* **19 (Suppl. 1)** (1988) 1.
[10] J. Heinze: "Cyclovoltammetrie – die Spektroskopie des Elektrochemikers," *Angew. Chem.* **96** (1984) 823.
[11] R. Neeb: *Inverse Polarographie und Voltammetrie,* Verlag Chemie, Weinheim 1969.
[12] J. Wang: *Stripping Analysis* VCH Verlagsgesellschaft, Weinheim 1985.
[13] R. Kalvoda, M. Kopanica: "Adsorptive Stripping Voltammetry in Trace Analysis," *Pure Appl. Chem.* **61** (1987) 97.
[14] G. Henze: "Stripping-Chronopotentiometrie," *Fresenius Z. Anal. Chem.* **315** (1983) 438.
[15] G. K. Budnikow, N. A. Ulakhovich: "Extraction Polarography and its Analytical Application," *Russ. Chem. Rev. (Engl. Transl.)* **49** (1980) 74.
[16] P. Rach, H. Seiler: *Polarographie und Voltammetrie in der Spurenanalyse,* Hüthig Verlag, Heidelberg 1985.
[17] G. Henze: "Application of Polarographic and Voltammetric Techniques in Environmental Analysis," *NATO ASI Ser. G* **23** (1990) 391.
[18] T. M. Florence: "Analytical Application to Trace Element Speciation in Water," *Analyst* **111** (1986) 489.
[19] C. M. G. van den Berg: "Adsorptive Cathodic Stripping Voltammetry of Trace Elements in Water," *Analyst* **114** (1989) 1527.
[20] C. M. G. van den Berg: "Electroanalytical Chemistry of Sea-Water," *Chem. Oceanogr.* **9** (1988) 195.
[21] C. M. G. van den Berg: "Potentials and Potentialities of Cathodic Stripping Voltammetry of Trace Elements in Natural Water," *Anal. Chim. Acta* **250** (1991) 265.
[22] E. Palecek: "Polarographic Techniques in Nucleic Acid Research," in W. F. Smyth (ed.): "Electroanalysis in Hygiene, Environmental, Clinical and Pharmaceutical Chemistry," *Anal. Chem. Symp.* **2** (1980) 79.
[23] P. M. Bersier, J. Bersier: "Polarographic Adsorption Analysis and Tensammetry: Toys or Tools," *Analyst* **113** (1988) 3.
[24] R. Kalvoda: "Polarographic Determination of Adsorbable Molecules," *Pure Appl. Chem.* **59** (1987) 715.
[25] "Electrochemistry in Cancer Research," *Anal. Proc. (London)* **17** (1980) 278 (Conference Proceedings)
[26] J. P. Hart: "Electroanalysis of Biologically Important Compounds," in *Analytical Chemistry,* Ellis Harwood Series.
[27] J. Wang: *Electroanalytical Techniques in Clinical Chemistry and Laboratory Chemistry,* VCH Verlagsgesellschaft, Weinheim 1988.
[28] W. F. Smyth: "Voltammetric Determination of Molecules of Biological, Environmental, and Pharmaceutical Importance," *CRC Crit. Rev. Anal. Chem.* **18** (1987) 155.
[29] W. F. Smyth, M. Smyth: "Electrochemical Analysis of Organic Pollutants," *Pure Appl. Chem.* **59** (1987) 245.

Specific References

[30] W. F. Smyth: *Voltammetric Determination of Molecules of Biological Significance,* John Wiley & Sons, New York 1992.
[31] G. C. Barker, A. W. Gardner, *Fresenius' Z. Anal. Chem.* **113** (1960) 73.
[32] J. G. Osteryoung, R. A. Osteryoung, *Anal. Chem.* **57** (1985) 101 A.
[33] E. J. Zachowski, M. Wojciechowski, J. Osteryoung, *Anal. Chim. Acta* **183** (1986) 47.
[34] G. Henze, R. Neeb, *Fresenius' J. Anal. Chem.* **310** (1982) 111.
[35] G. Henze, A. P. Joshi, R. Neeb, *Fresenius J. Anal. Chem.* **300** (1980) 267.
[36] G. Henze, W. Wagner, S. Sander, *Fresenius J. Anal. Chem.* **358** (1997) 741.
[37] G. Henze et al., *Fresenius' Z. Anal. Chem.* **295** (1979) 1.
[38] G. Henze, *Mikrochim. Acta* 1981, no. II, 343.
[39] D. Jagner, K. Aren, *Anal. Chim. Acta* **100** (1978) 375.
[40] D. Jagner, *Anal. Chem.* **50** (1978) 1924.
[41] L. Kryger, *Anal. Chim. Acta* **120** (1980) 19.
[42] A. Graneli, D. Jagner, M. Josefson, *Anal. Chem.* **52** (1980) 2220.
[43] T. Garai et al., *Electroanalysis* **4** (1992) 899.
[44] K. Kalcher, "Chemically Modified Carbon Paste Electrodes in Voltammetric Analysis," *Electroanalysis* **2** (1990) 419.
[45] J. Wang, T. Boamin, R. Setiadji, *Electroanalysis* **6** (1994) 317.
[46] Kh. Brainina, G. Henze, N. Stojko, N. Malakhova, Chr. Faller, "Thick-film Graphite Electrodes in Stripping Voltammetry," *Fresenius J. Anal. Chem.* **364** (1999) 285.
[47] J. Schiewe, A. M. Bond, G. Henze: "Voltammetrische Spurenanalyse mit Mikroelektroden" in G. Henze, M. Köhler, J. P. Lay (eds.): *Umweltdiagnostik mit Mikrosystemen,* Wiley-VCH, Weinheim 1999, p. 121.
[48] H. Y. Chen, R. Neeb, *Fresenius' Z. Anal. Chem.* **319** (1984) 240.
[49] M. Kolb, P. Rach, J. Schäfer, A. Wild, *Fresenius' J. Anal. Chem.* **342** (1992) 341.
[50] M. Karakaplan, S. Gücer, G. Henze, *Fresenius' J. Anal. Chem.* **342** (1992) 186.
[51] D. J. Myers, J. Osteryoung, *Anal. Chem.* **45** (1973) 267.
[52] C. M. G. van den Berg, K. Murphy, J. P. Riley, *Anal. Chim. Acta* **188** (1986) 177.
[53] W. Wagner, S. Sander, G. Henze, *Fresenius J. Anal. Chem.*(in press).
[54] J. Wang, J. M. Zadeii, *Anal. Chim. Acta* **185** (1986) 229.
[55] C. M. G. van den Berg, G. S. Jacinto, *Anal. Chim. Acta* **211** (1988) 129.
[56] J. Wang, J. M. Zadeii, *Anal. Chim. Acta* **188** (1986) 187.
[57] S. B. Adeloju, *Anal. Sci.* **7** (1991) 1099.
[58] K. Yokoi, C. M. G. van den Berg, *Anal. Chim. Acta* **245** (1991) 167.
[59] N. K. Lam, R. Kalvoda, M. Kopanica, *Anal. Chim. Acta* **154** (1983) 79.
[60] C. M. G. van den Berg, Z. Qiang Huang, *Anal. Chim. Acta* **164** (1984) 209.
[61] C. M. G. van den Berg, M. Nimmo, *Anal. Chem.* **59** (1987) 924.
[62] J. Zarebski, *Chemia Analityczna* **22** (1977) 1037.
[63] L. Meites: *Polarographic Techniques,* 2nd ed., Interscience, New York 1961.
[64] S. Sander, G. Henze, *Fresenius J. Anal. Chem.* **349** (1994) 654.
[65] C. M. G. van den Berg, Z. Q. Huang, *Anal. Chem.* **56** (1984) 2383.
[66] R. Naumann, W. Schmidt, G. Höhl, *Fresenius J. Anal. Chem.* **343** (1992) 746.
[67] J. M. Estela, C, Tomas, A. Cladera, V. Cerda, *Crit. Rev. Anal. Chem.* **25** (1995) 91.
[68] S. B. Adeloju, A. M. Bond, *Anal. Chem.* **56** (1984) 2397.

[69] D. Jagner, M. Josefson, S. Westerlund, K. Aren, *Anal. Chem.* **53** (1981) 1406.
[70] H. Huiliang, D. Jagner, L. Renman, *Talanta* **34** (1987) 539.
[71] C. Hua, D. Jagner, L. Renman, *Anal. Chim. Acta* **197** (1987) 25.
[72] R. Kalvoda, *Anal. Chim. Acta* **138** (1982) 11.
[73] M. Pedrero, F. J. Manuel de Villena, J. M. Pingarron, L. M. Polo, *Electroanalysis* **3** (1991) 419.
[74] Ch. Li, B. D. James, R. J. Magee, *Electroanalysis* **2** (1990) 63.
[75] M. A. Goicolea, J. F. Arrenz, R. J. Barrio, Z. G. de Balugera, *Fresenius' J. Anal. Chem.* **339** (1991) 166.
[76] M. R. Smyth, J. G. Osteryoung, *Anal. Chem.* **49** (1977) 2310.
[77] J. P. Hart, W. F. Smyth, B. J. Birch, *Analyst* **104** (1979) 853.
[78] M. A. Korany, H. Riedel, *Fresenius' Z. Anal. Chem.* **314** (1983) 678.
[79] J. S. Burmich, W. F. Smyth, *Anal. Proc. (London)* **17** (1980) 284.
[80] H. M. Pylypiw, G. W. Harrington, *Anal. Chem.* **53** (1981) 2365.
[81] J. Schwarz, G. Henze, F. G. Thomas, *Fresenius J. Anal. Chem.* **352** (1995) 474, 479.

26. Thermal Analysis and Calorimetry

STEPHEN B. WARRINGTON, Anasys, IPTME, Loughborough University, Loughborough, United Kingdom (Chap. 26)

GÜNTHER W. H. HÖHNE, Dutch Polymer Institute, Eindhoven University of Technology, Eindhoven, The Netherlands (Chap. 26.2)

26.	Thermal Analysis and Calorimetry	827
26.1.	Thermal Analysis	827
26.1.1.	General Introduction	827
26.1.1.1.	Definitions	827
26.1.1.2.	Sources of Information	828
26.1.2.	Thermogravimetry	828
26.1.2.1.	Introduction	828
26.1.2.2.	Instrumentation	828
26.1.2.3.	Factors Affecting a TG Curve	829
26.1.2.4.	Applications	829
26.1.3.	Differential Thermal Analysis and Differential Scanning Calorimetry	830
26.1.3.1.	Introduction	830
26.1.3.2.	Instrumentation	830
26.1.3.3.	Applications	831
26.1.3.4.	Modulated-Temperature DSC (MT-DSC)	833
26.1.4.	Simultaneous Techniques	833
26.1.4.1.	Introduction	833
26.1.4.2.	Applications	833
26.1.5.	Evolved Gas Analysis	833
26.1.6.	Mechanical Methods	834
26.1.7.	Less Common Techniques	835
26.2.	Calorimetry	836
26.2.1.	Introduction	836
26.2.2.	Methods of Calorimetry	836
26.2.2.1.	Compensation for Thermal Effects	837
26.2.2.2.	Measurement of a Temperature Difference	837
26.2.2.3.	Temperature Modulation	838
26.2.3.	Calorimeters	839
26.2.3.1.	Static Calorimeters	839
26.2.3.2.	Scanning Calorimeters	841
26.2.4.	Applications of Calorimetry	844
26.2.4.1.	Determination of Thermodynamic Functions	845
26.2.4.2.	Determination of Heats of Mixing	846
26.2.4.3.	Combustion Calorimetry	846
26.2.4.4.	Reaction Calorimetry	847
26.2.4.5.	Safety Studies	849
26.3.	References	849

26.1. Thermal Analysis

26.1.1. General Introduction

26.1.1.1. Definitions

Thermal analysis (TA) has been defined as "a group of techniques in which a physical property of a substance and/or its reaction products is measured as a function of temperature while the substance is subjected to a controlled temperature programme" [1]. The formal definition is usually extended to include isothermal studies, in which the property of interest is measured as a function of time.

The definition is a broad one, and covers many methods that are not considered to fall within the field of thermal analysis as it is usually understood. The present chapter will be restricted to the major techniques as currently practiced. Since all materials respond to heat in some way, TA has been applied to almost every field of science, with a strong emphasis on solving problems in materials science and engineering, as well as fundamental chemical investigations. TA is applicable whenever the primary interest is in determining the effect of heat upon a material, but the techniques can also be used as a means of probing a system to obtain other types of information, such as composition.

The aim of this chapter is to give an overview of the main TA methods and their applications,

Table 1. Processes that can be studied by thermogravimetry

Process	Weight gain	Weight loss
Ad- or absorption	*	
Desorption		*
Dehydration/desolvation		*
Sublimation		*
Vaporization		*
Decomposition		*
Solid–solid reactions		*
Solid–gas reactions	*	*

with sufficient references to provide reader access to more detailed information. The following list summarizes properties subject to investigation and the corresponding thermal analysis techniques:

Mass	Thermogravimetry (TG)
Temperature	Differential thermal analysis (DTA)
Enthalpy	Differential scanning calorimetry (DSC)
Evolved gas	Evolved gas analysis (EGA)
Physical dimensions	Thermodilatometry (TD)
Mechanical properties	Thermomechanical analysis (TMA)
	Dynamic mechanical analysis (DMA)
Optical properties	Thermoptometry
Electrical properties	Thermoelectrometry
Magnetic properties	Thermomagnetometry

26.1.1.2. Sources of Information

Two journals (the *Journal of Thermal Analysis and Calorimetry* and *Thermochimica Acta*) devote their contents entirely to TA; the *Proceedings* of the (now) four-yearly Conferences of the International Confederation for Thermal Analysis and Calorimetry (ICTAC) constitute an excellent additional source of research papers. Specific information regarding *Proceedings* volumes for the nine ICTAC Conferences between 1965 and 1991 is available in [1]. The most complete listing of worldwide TA literature is also found in the ICTAC handbook [1], which in addition gives addresses for national TA societies and important equipment suppliers. The most useful textbooks include [2]–[5].

There is a wealth of information available on the Internet; see, for example, http://www.users.globalnet.co.uk/~dmprice/ta/links/index.htm

26.1.2. Thermogravimetry

26.1.2.1. Introduction

Thermogravimetry (TG) is used to measure variations in mass as a function of temperature (or time). Processes amenable to study in this way are listed in Table 1. TG is one of the most powerful TA techniques from the standpoint of quantitative data, and for this reason it is often employed in combination with other measurements.

26.1.2.2. Instrumentation

The instrument used is a *thermobalance*. The schematic diagram in Figure 1 presents the main components of a typical modern unit. Component details vary according to the design, and choice of a particular instrument is usually dictated by requirements of the problem under investigation (temperature range, sensitivity, etc.). The balance mechanism itself is usually of the null-deflection type to ensure that the sample's position in the furnace will not change. The balance transmits a continuous measure of the mass of the sample to an appropriate recording system, which is very often a computer. The resulting plot of mass vs. temperature or time is called a *TG curve*. Balance sensitivity is usually of the order of one microgram, with a total capacity of as much as a few hundred milligrams. Furnaces are available that operate from subambient (e.g., $-125\,°C$) or room temperature up to as high as $2400\,°C$. A furnace programmer normally supports a wide range of heating and cooling rates, often in combination, as well as precise isothermal control. The programming functions themselves are increasingly being assumed by computers. Most applications involve heating rates of $5-20\,°C\,min^{-1}$, but the ability to heat a sample as rapidly as $1000\,°C\,min^{-1}$ can be useful in the simulation of certain industrial processes, for example, or in flammability studies. Special control methods, grouped under the term *controlled-rate thermal analysis* (CRTA), are receiving increasing attention [6], and offer advantages in resolving overlapping processes and in kinetic studies.

The quality of the furnace atmosphere deserves careful attention. Most commercial thermobalances operate at atmospheric pressure. Vacuum and high-pressure studies normally require specialized equipment, either commercial or home-

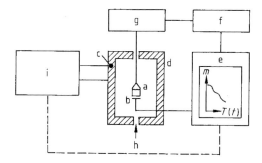

Figure 1. Schematic diagram of a thermobalance
a) Sample; b) Sample temperature sensor; c) Furnace temperature sensor; d) Furnace; e) Recorder or computer, logging sample mass, temperature, and time; f) Balance controller; g) Recording microbalance; h) Gas; i) Furnace temperature programmer
The computer may also control the furnace programmer

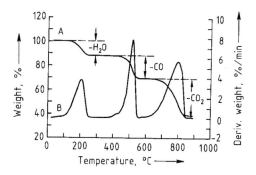

Figure 2. TG (A) and DTG (B) curves for calcium oxalate monohydrate
Sample mass 30 mg, heating rate 20 °C min^{-1}, argon atmosphere

made, as do experiments with corrosive gases [7]. The ability to establish an inert (oxygen-free) atmosphere is useful, as is the potential for rapidly changing the nature of the atmosphere.

The mode of assembly of the components varies; for example, the furnace might be above, below, or in line with the balance. Sample containers also vary widely in design; cylindrical pans are common, typically 5–8 mm in diameter and 2–10 mm high, though flat plates and semisealed containers may be used to investigate the effects of atmospheric access to a sample. Compatibility between the construction materials and the system under investigation must be carefully considered. Materials commonly available include aluminum, platinum, alumina, and silica. Temperature indication is normally provided by a thermocouple located near the sample container. Because of inevitable thermal gradients within the apparatus, an indicated temperature can never be taken as an accurate reflection of the temperature of the sample. Reproducible location of the thermocouple is vital; recommended calibration procedures have been described in [8]. Only in simultaneous TG–DTA instruments is direct measurement of the temperature possible.

26.1.2.3. Factors Affecting a TG Curve

WENDLANDT [2] has listed 13 factors that directly affect a TG curve, both sample- and instrument-related, some of which are interactive. The primary factors are heating rate and sample size, an increase in either of which tends to increase the temperature at which sample decomposition occurs, and to decrease the resolution between successive mass losses. The particle size of the sample, the way in which it is packed, the crucible shape, and the gas flow rate also affect the progress of a thermal reaction. Careful attention to consistency in experimental details normally results in good repeatability. On the other hand, studying the effect of deliberate alterations in such factors as heating rate can provide valuable insights into the nature of observed reactions. All these considerations are equally applicable to other techniques as well, including DTA (Section 26.1.3).

26.1.2.4. Applications

TG has been applied extensively to the study of analytical precipitates for gravimetric analysis [9]. One example is calcium oxalate, as illustrated in Figure 2. Information such as extent of hydration, appropriate drying conditions, stability ranges for intermediate products, and reaction mechanisms can all be deduced from appropriate TG curves. Figure 2 also includes the first derivative of the TG curve, termed the *DTG curve*, which is capable of revealing fine details more clearly.

A TG method has also been devised for the proximate analysis of coal, permitting four samples per hour to be analyzed with results of the same precision and accuracy as BS and ASTM methods [10]. A typical curve is shown in Figure 3. The approach taken here can be generalized to the compositional analysis of many materials [11].

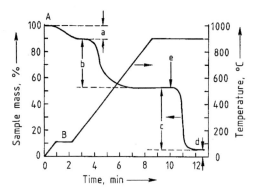

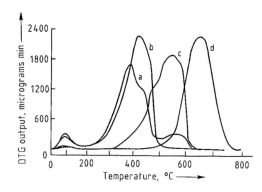

Figure 3. TG curve (A) illustrating a programmed-temperature method for the proximate analysis of coal, where the line B defines the temperature program
a) Moisture; b) Volatiles; c) Fixed carbon; d) Ash; e) Point at which the atmosphere was changed to air

Figure 5. DTG curves for four ASTM standard coal samples heated in air [12]
a) Lignite "A"; b) Sub-bituminous "A"; c) High-volatility bituminous "A"; d) Buckwheat anthracite

rapid assessment of new fuels (Fig. 5). The simulation of industrial processes, particularly heterogeneous catalysis, is another important area of application.

26.1.3. Differential Thermal Analysis and Differential Scanning Calorimetry

26.1.3.1. Introduction

Both differential thermal analysis (DTA) and differential scanning calorimetry (DSC) are concerned with the measurement of energy changes, and as such are applicable in principle to a wider range of processes than TG. From a practical standpoint DSC may be regarded as the method from which quantitative data are most easily obtained. The use of DSC to determine absolute thermodynamic quantities is discussed in Sections 26.2.3.2 and 26.2.4.1. Types of processes amenable to study by these methods are summarized in Table 2.

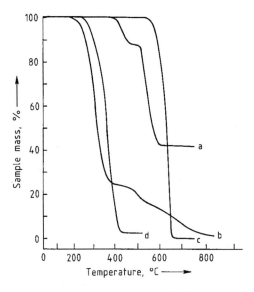

Figure 4. TG curves for four polymers in air, showing their relative stabilities
a) Silicone rubber; b) PVC; c) PTFE; d) Perspex

TG also facilitates comparisons of the relative stabilities of polymers (Fig. 4). An analysis of curves prepared at different heating rates makes it possible to extract kinetic data for estimating the service lifetimes of such materials. In another coal-related application, an empirical correlation has been established between characteristic temperatures obtained from DTG curves of coals examined under air and the performance of the same coals in steam-raising plants [12], permitting the

26.1.3.2. Instrumentation

Figure 6 shows the components of a typical DTA apparatus. The sample and an inert reference material (commonly alumina) are subjected to a common temperature program while the temperature difference between the two is monitored. In an ideally designed instrument, the temperature difference ΔT remains approximately constant if no reaction is taking place in the sample. When an endothermic process such as melting occurs, the

Table 2. Processes that can be studied by DTA/DSC

Process	Exothermic	Endothermic
Solid–solid transitions	*	*
Solid–liquid transitions	*	*
Vaporization		*
Sublimation		*
Adsorption	*	
Desorption		*
Desolvation		*
Decomposition	*	*
Solid–solid reactions	*	*
Solid–liquid reactions	*	*
Solid–gas reactions	*	*
Curing	*	
Polymerizations	*	
Catalytic reactions	*	*

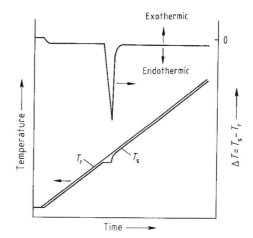

Figure 7. Typical DTA curve for the melting of a pure material

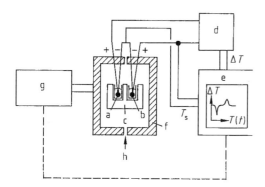

Figure 6. Schematic diagram of a classical DTA apparatus
a) Reference thermocouple; b) Sample thermocouple; c) Heating block; d) ΔT amplifier; e) Recorder or computer, logging T_s, ΔT, and time; f) Furnace; g) Temperature programmer, which may be linked to the computer; h) Gas inlet T_s, sample temperature; T_r, reference temperature; $\Delta T = T_s - T_r$

sample temperature T_s lags behind that of the reference T_r; $\Delta T = T_s - T_r$ is thus negative, so a (negative) peak is produced on the corresponding *DTA curve*, which is a record of ΔT against either temperature or time (Fig. 7). Exothermic processes lead to peaks in the opposite direction. The peak *area* is related to both the energy change and the sample size according to the relationship

$$\Delta H \cdot m = K \int \Delta T \cdot dt$$

where m is the sample mass, and ΔH is the enthalpy of reaction. The proportionality constant K is temperature-dependent, and with classical DTA it is also influenced by thermal properties of the sample and container. Certain apparatus designs minimize these influences, so that once the characteristics of $K = f(T)$ have been determined the ΔT signal can be conditioned (electronically or digitally) to give an output signal calibrated directly in power units; i.e., the sensitivity is nominally independent of temperature. This is the principle underlying heat-flux DSC. For a more complete description of this technique, as well as power-compensated DSC, see Section 26.2.3.2. A wide variety of commercial devices covers the temperature range from ca. −150–2500 °C, but high-temperature instruments are less sensitive than low-temperature designs due to constraints imposed by thermocouple technology. DSC instruments are usually limited to a maximum temperature of 700 °C, though heat-flux DSC has been extended to 1400 °C. DTA head designs vary enormously, but most accommodate samples in the range 5–200 mg in metal or ceramic containers. The thermocouples usually take the form of beads that fit into recesses in the containers, or plates upon which the containers rest. Nowadays the thermocouple is never actually placed in the sample, in contrast to the "classical" arrangement shown in Figure 6.

26.1.3.3. Applications

DSC is used extensively in polymer science [13]. A generalized DSC curve for a polymer is shown in Figure 8. Most polymers display a *glass transition*, in the course of which the material passes from a glassy to a rubbery state with a simultaneous increase in specific heat capacity.

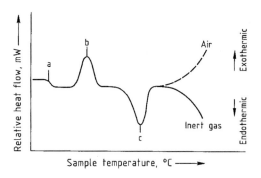

Figure 8. Generalized DSC curve for a polymer
a) Glass transition; b) Crystallization; c) Melting
The subsequent reactions may be endothermic or exothermic

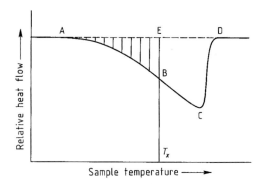

Figure 10. Generalized DSC curve for the melting of an impure material
Fraction melted at temperature T_x = (area ABE)/(area ADC)

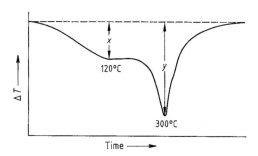

Figure 9. DTA curve for a partially converted high-alumina cement concrete from a test drilling
Degree of conversion = 100 $y/(x+y)$

Glass-transition temperature measurements are used for material characterization and comparison, often providing information about such factors as thermal history, crystallinity, extent of cure, and plasticizer content of a polymer. Amorphous polymers usually exhibit a crystallization exotherm, and thermosetting polymers a curing exotherm, both of which can be analyzed kinetically to establish the thermal treatment required to obtain desired properties in a particular product. The melting endotherm is a measure of the extent of crystallinity, an important parameter related to mechanical properties. The temperature at which exothermic degradation begins under an oxidizing atmosphere is used to compare oxidative stabilities of different polymers or different stabilizers. Discrimination can be improved by isothermal measurements, in which the material under investigation is heated initially in an inert gas, after which the atmosphere is changed to air or oxygen and the time lapse prior to exothermic deflection is measured. The same approach is applicable to studies of oils, greases, fats, etc., though high-pressure cells may be necessary to prevent volatilization.

The particular temperatures at which peaks are observed can be used for the identification of components in a mixture, and the size of a particular peak can be used for quantitative evaluation. Examples include the determination of quartz in clays and analysis of the constituents of polymer blends, both of which would be difficult to achieve by other means. DTA has been used to determine the presence and degree of conversion of high-alumina cement concrete (HAC), a material subject to severe loss of strength under certain atmospheric conditions due to solid-state reactions in the cement matrix. Figure 9 shows a DTA curve for partially converted HAC together with the way the data are used for a conversion measurement.

An analysis of the shape of the fusion peak from a relatively pure material (e.g., a pharmaceutical agent) can, with certain restrictions, lead to an estimate of purity [14]. Figure 10 shows an idealized melting curve for an impure material. By calculating the fraction (F) of material melted at a series of temperatures, and applying several simplifying assumptions, the mole fraction of an impurity can be calculated from the expression:

$$T_s = T_0 - \frac{RT_0^2 X_2}{\Delta H_f} \cdot \frac{1}{F}$$

where T_s = sample temperature, T_0 = melting point of the pure material, R = gas constant, X_2 = mole fraction of impurity, ΔH_f = enthalpy of fusion, and F = the fraction melted at temperature T_s. The cal-

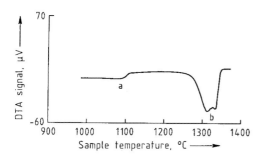

Figure 11. DTA curve for a nickel-based superalloy
a) Solution of the γ' phase; b) Melting 250 mg sample heated at 10 °C min^{-1} in argon

culations are readily performed by commercial software.

DTA curves can be used to construct phase diagrams for mixtures. Measurement of a single point on the phase diagram may be sufficient for quality-control purposes. Figure 11 shows the DTA curve for a nickel-based superalloy, where the solvus temperature (the temperature corresponding to complete solution of one solid phase in another) of the γ' phase characterizes the heat treatment to which the alloy has been subjected.

Solid–solid reactions have also been successfully studied by these techniques, as have decompositions of high explosives. The applicability of DTA/DSC to very small samples is of obvious value here. Similarly, preliminary screening of potentially hazardous reaction mixtures is conveniently carried out in this way as a guide to the likelihood of exothermic reaction, the corresponding temperature range and magnitude, and possible effects of pressure and atmospheric conditions (see also Section 26.2.4.5).

26.1.3.4. Modulated-Temperature DSC (MT-DSC)

MT-DSC is an important recent development (see also Section 26.2.2.3) which offers significant advantages in the quality of results and gives additional information, particularly for polymers [15]. A sinusoidal modulation is superimposed on the normally linear temperature ramp, and the resulting signals are analyzed by a Fourier Transform technique. This allows the separation of "reversing" and "nonreversing" thermal events and simplifies the analysis of overlapping transitions such as those often found in the glass transition region. Other benefits include the ability to measure heat capacity changes pseudo-isothermally, as in, e.g., curing systems, and improved signal-to-noise ratio and baseline linearity.

26.1.4. Simultaneous Techniques

26.1.4.1. Introduction

TG or DTA/DSC alone rarely gives sufficient information to permit a complete interpretation of the reactions in a particular system; results must usually be supplemented by other thermal methods and/or general analytical data. Alternative thermal methods are best applied simultaneously, leading, for example, to TG and DTA information from the same sample under identical experimental conditions. This avoids ambiguity caused by material inhomogeneity, but also problems attributable to different experimental conditions with different instruments, which sometimes markedly affects the correspondence between TG and DTA curves. Other advantages include: (1) indication of the thermal stability of materials examined by DTA, which in turn makes it possible to correct measured heats of reaction for partially decomposed samples; (2) accurate temperature measurement in TG work, which is vital in kinetic studies; and (3) the detection of unsuspected transitions in a condensed phase, which may help explain puzzling features of the corresponding TG curve.

The range of simultaneous techniques has been reviewed by PAULIK [16]. TG–DTA and TG–DSC are the commonest simultaneous methods, followed by evolved gas analysis (Section 26.1.5).

26.1.4.2. Applications

TG–DSC curves showing the curing and decomposition of a polyimide resin are presented in Figure 12. Following a glass transition (a) and subsequent fusion of the material at ca. 120 °C (b), which are detected by DSC, the TG curve reveals the initial stages of an exothermic curing reaction (c) that occurs with a 7% mass loss. Endothermic decomposition of this resin does not take place until ca. 430 °C, as shown again by TG (d), a finding that permitted assignment of the appropriate baseline for measuring the heat of curing.

26.1.5. Evolved Gas Analysis

Evolved gas analysis (EGA) measures the nature and/or quantity of volatile products released

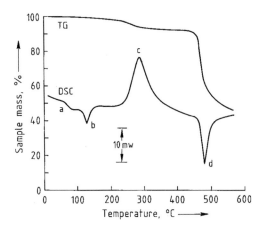

Figure 12. Simultaneous TG–DSC curves for an uncured polyimide resin
a) Glass transition; b) Melting; c) Curing; d) Degradation
10 mg sample heated at 10 °C min^{-1} in argon

on heating. In practice it is often performed simultaneously with TG or TG–DTA, and is particularly useful in furnishing direct chemical information to augment the physical measurements of TG or DTA. Additional advantages that have emerged include specificity — a single decomposition can be followed against a background of concurrent processes — and sensitivity, which can be far greater than with TG alone [17], [18].

Instrumentation. Almost every imaginable type of gas detector or analyzer has been utilized in EGA, including hygrometers, nondispersive infrared analyzers, and gas chromatographs. Absorption of the products into solution permits analysis by coulometry, colorimetry, ion selective electrode measurements, or titrimetry. The most important analyzers are Fourier transform infrared (FTIR) spectrometers and, preeminently, mass spectrometers [19]. The two latter methods can be used to record spectra repetitively, thereby producing a time-dependent record of the composition of the gas phase, from which EGA curves can be constructed for selected species.

Applications. Simultaneous TG–MS curves for a brick clay are shown in Figure 13, which simulates a set of firing conditions [20]. After an initial loss of moisture, combustion of organic matter at ca. 300 °C gives characteristic peaks for CO_2, H_2O, and SO_2. More SO_2 results from the oxidation of iron sulfides. Clay dehydroxyla-

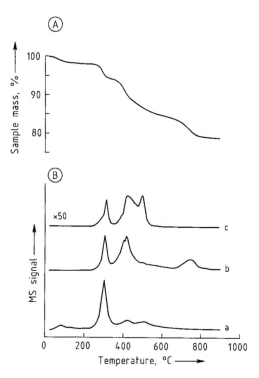

Figure 13. Simultaneous TG–MS curves for a brick clay
A) TG curve; B) MS data
Evolution of a) Water (18 amu); b) Carbon dioxide (44 amu); c) Sulfur dioxide (64 amu)
40 mg sample heated at 15 °C min^{-1} in air

tion occurs in the range 350–600 °C, and calcite dissociates to give the CO_2 peak observed at higher temperatures. EGA is of great value in the interpretation of complex TG and DTA curves.

In another example (Fig. 14), the evolution of benzene from two samples of PVC, one containing a smoke retardant, was compared quantitatively by injecting known amounts of benzene into the purge gas stream before and after the experiments, thereby providing a standard for estimating the amount of benzene produced by the polymer [20]. The smoke retardant was found in this case to reduce benzene evolution by a factor of ca. 20.

26.1.6. Mechanical Methods

This section covers a family of techniques leading to dimensional or mechanical data for a material as a function of temperature or time. For a succinct review of the application areas see [22].

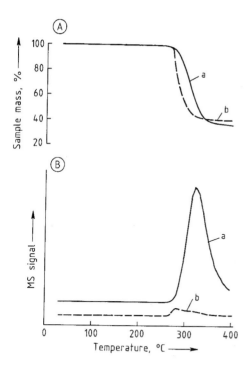

Figure 14. Simultaneous TG–MS curves showing the evolution of benzene (78 amu)
A) TG curve; B) MS curve
a) PVC without a smoke retardant; b) PVC with a smoke retardant (4 % MoO$_3$)
15 mg samples heated at 10 °C min^{-1} in argon

Thermodilatometry (TD) measures dimensional changes as a function of temperature in materials subject to negligible loads. A probe, which is held in light contact with the heated sample, is connected to a sensitive position sensor, usually a linear variable differential transformer (LVDT). In addition to providing expansion coefficients the technique can also indicate phase changes, sintering, and chemical reactions. Major application areas include metallurgy and ceramics.

Thermomechanical Analysis. The equipment used in thermomechanical analysis (TMA) is similar in principle to that for TD, but provision is made for applying various types of load to the specimen, so that penetration, extension, and flexure can be measured. This approach to analyzing such modes of deformation is illustrated schematically in Figure 15. The technique finds most use in polymer studies, as in the determination of glass-transition and softening temperatures for thin films and shrinkage characteristics of fibers.

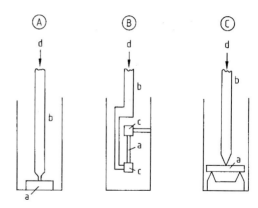

Figure 15. Modes of deformation subject to examination by TMA
A) Penetration; B) Extension; C) Flexure
a) Sample; b) Probe; c) Clamps; d) Load

Dynamic Mechanical Analysis. Polymers exhibit both elastic and viscous behavior under stress. In dynamic mechanical analysis (DMA) a sinusoidally modulated stress, often flexural (though measurement is also possible under tensile, compressive, shear, or torsional conditions) is applied to a specimen of material maintained under a specified temperature regime. Displacement transducers measure strain induced in-phase with the stress, as well as strain that lags behind. The former gives a measure of the sample's modulus, or stiffness, and the latter reflects damping characteristics. Like most other thermoanalytical techniques, DMA can be used to generate quantitative physical data, but it is also invoked in a comparative sense: to monitor the effects of additives, for example, or in quality control. It is particularly well suited to the examination of engineering composites, since gelation and curing behavior is sometimes more easily followed by DMA than by alternative techniques such as DSC.

26.1.7. Less Common Techniques

Thermoptometry. This group of techniques includes:

1) *Thermomicroscopy*, in which a material is observed under reflected, transmitted, or polarized transmitted light. The corresponding apparatus is referred to as a *hot stage*, which may also incorporate a DSC sensor for additional information.

2) *Thermophotometry*, in which the intensity of reflected or transmitted light is measured.
3) *Thermoluminescence*, which provides a measure of the intensity of light emitted by the material itself.

Electrical and Magnetic Techniques. Thermomagnetometry is used to study the magnetic properties of materials by revealing apparent weight changes that accompany phase transitions in the presence of an applied magnetic field, or variations in the amount of a magnetic substance.

Thermoelectrometry comprises methods for the measurement of resistance or capacitance during heating. Some equipment in this category provides remote probes for in situ monitoring of the curing of thermosets.

Micro-thermal Analysis (µ-TA) A family of techniques has recently been introduced that enables a range of thermal methods to be applied to regions of a material only a few micrometers wide. They are based on the use of an atomic force microscope (AFM) with a heatable tip. Many materials are heterogeneous, and the microstructure can influence the material properties on the macroscale. After acquiring the conventional AFM topographic image, selected features can be subjected to DTA and TMA with the probe tip. The tip can also be used to pyrolyze the selected area, and the vapors are then analyzed by GC-MS [23].

26.2. Calorimetry

26.2.1. Introduction [24]–[30], [31], [32]

Calorimetry in the broadest sense means the quantitative measurement of energy exchanged in the form of heat during a reaction of any type. By contrast, thermal analysis (Chap. 26.1) is concerned only with the measurement and recording of temperature-induced changes or temperature differences. Since all chemical reactions and many physical changes (e.g., deformation, phase transformations) are associated with the uptake or release of heat, the quantitative investigation of heat exchange is a relatively simple and universal method for characterizing particular processes both in an overall sense and with respect to time.

Nevertheless, only in recent decades has calorimetry emerged from the laboratories of a few thermodynamicists and specialists to become a widespread, convenient analytical method. The development of commercial calorimeters since the 1950s has led to rapid dissemination and application of the method even beyond the bounds of universities.

The explanation of this phenomenon, as with many other apparatus-based developments, lies in the refinement in measurement techniques made possible by modern electronics. As a rule, heat cannot be measured directly, but must be determined instead on the basis of a temperature change in the system under investigation. Very accurate classical thermometers tend to be very slow measuring devices. Resistance thermometers and thermocouples respond much more rapidly, but they require electronic amplification, and amplifiers with the required precision have become available only in recent decades.

A further explanation for the increasing importance of calorimetry is the development of personal computers, which relieve the user of the often laborious task of evaluating the experimental results, thereby opening the way for the method to become a routine laboratory technique.

In the sections that follow, calorimetry and the present state of instrument development will be described in such a way as to reveal both the possibilities and the limits of the method. The scope of the chapter rules out exhaustive treatment of the subject, so the goal is rather to delineate the method and guide the interested reader to other relevant technical literature.

26.2.2. Methods of Calorimetry

As noted above, heat that is released or consumed by a process cannot be measured directly. Unlike material quantities such as amount of substance, which can be determined with a balance, or the Volume of a liquid, which can be established with a liter measure, the quantity heat must be measured indirectly through its effect on a substance whose temperature it either raises or lowers. The fundamental equation for all of calorimetry describes the relationship between heat exchanged with a calorimeter substance and a corresponding temperature change:

$$\Delta Q = C(T) \cdot \Delta T \qquad (1)$$

where ΔQ = exchanged heat (J), ΔT = observed temperature change (K), and $C(T)$ = heat capacity (J/K).

For heats (and therefore temperature changes) that are not too large, the temperature change is directly proportional to the exchanged heat. The proportionality constant is the heat capacity of the calorimeter substance (previously referred to as the "water value"). However, if the temperature change exceeds a few Kelvin, the temperature dependence of heat capacity stands in the way of a linear relationship, and a knowledge of the temperature function of the particular heat capacity in question is required in order to determine heat on the basis of a measured temperature difference.

The relationship above leads directly to the standard calorimetric methods for determining the heat of a process: either temperature is held constant by appropriate compensation for the heat effect, and the required compensation power is measured, or a temperature change is determined and used to calculate a corresponding value for exchanged heat. A precondition for the latter approach is accurate knowledge of the heat capacity as well as the heat transport properties of the measurement system.

26.2.2.1. Compensation for Thermal Effects

In the first method of heat measurement, temperature changes in the calorimeter substance are avoided by supplying or dissipating heat exactly equal in magnitude (but opposite in sign) to that associated with the process under investigation. As a rule, electrical energy is used to provide this compensation, either by introduction in the form of Joulean heat or dissipation through the Peltier effect. A combination approach is also possible, with an appropriate constant level of cooling and simultaneous controlled electrical heating. The requisite amount of compensation power can today be determined with a high degree of precision. It was once customary to invoke for compensation purposes "latent" heat derived from a phase change; in other words, heat released in a process was measured by using it to melt a substance such as ice, with subsequent weighing of the resulting water and conversion of the data into energy units based on the known heat of fusion for ice.

The method of compensation is advantageous, since it permits measurements to be carried out under quasi-isothermal conditions, thereby avoiding heat loss from the calorimeter to the surroundings by heat transport processes. Furthermore, there is no need for a calibrated thermometer, only a sensor sufficiently sensitive to temperature changes to provide adequate control over the compensation power.

26.2.2.2. Measurement of a Temperature Difference

In this alternative method of heat measurement, which is also indirect, a measured temperature difference is used to calculate the amount of heat exchanged. A distinction can be made between temporal and spatial temperature-difference measurements. In *temporal* temperature-difference measurement the temperature of a calorimeter substance is measured before and after a process, and a corresponding heat is calculated on the basis of Equation (1) (which presupposes an accurate knowledge of the heat capacity). In the *spatial* method a temperature difference between two points within the calorimeter (or between the calorimeter substance and the surroundings) is the quantity of interest. The basis for interpretation in this case is from Fourier's law the equation for stationary heat conduction (Newton's law of cooling):

$$\Phi = \lambda(T) \cdot A \cdot \Delta T \cdot l^{-1} \qquad (2)$$

where Φ = heat flow rate (W), $\lambda(T)$ = thermal conductivity (W m^{-1} K^{-1}), A = cross-sectional area (m^2), ΔT = temperature difference (K), and l = length (m).

From this expression it follows that the heat flow rate through a heat-conducting material is proportional to the corresponding temperature difference. If this temperature difference is recorded as a function of time, then for a given thermal conductivity one acquires a measure of the corresponding heat flow rate, which can be integrated to give a total heat. The technique itself is very simple, but the result will be correct only if the measured temperature difference accurately reflects the total heat flow rate and no heat is lost through undetected "heat leaks". Careful calibration and critical tests of the calorimeter are absolutely necessary if one hopes to obtain reliable results.

A fundamental disadvantage of the temperature-difference method is the fact that nonisothermal conditions prevail, so the precondition of a stationary state is fulfilled only approximately. The potential therefore exists for nonlinear phenomena in the heat-transport process, which might result in conditions outside the range of validity for Equation (2). Furthermore, with nonisothermal studies it is impossible to rule out completely the presence of undetected leaks. It must also be pointed out that in the case of nonisothermal studies the thermodynamic quantities of interest cannot be established according to the methods of reversible thermodynamics, at least not without further consideration. Strictly speaking, *irreversible thermodynamics* should be applied to nonisothermal calorimetry. Although the distinction is generally unimportant in normal calorimetric practice, it becomes essential in cases where time-dependent processes are involved with timescales which are comparable to those of the experiment, e.g., in the *temperature-modulated* mode of operation.

26.2.2.3. Temperature Modulation

The temperature-modulated mode of operation has been well known for many decades in calorimetry [33], but became well established only during the 1990s, when commercial DSC was modified this way [34]. The idea is to examine the behavior of the sample for periodic rather than for isothermal or constant-heating-rate temperature changes. In this way it is possible to obtain information on time-dependent processes within the sample that result in a time-dependent generalized (excess) heat capacity function or, equivalently, in a complex frequency-dependent quantity. Similar complex quantities (electric susceptibility, Young's modulus) are known from other dynamic (dielectric or mechanical) measurement methods. They are widely used to investigate, say, relaxation processes of the material.

In calorimetry there are three established temperature modulation methods:

1) The sample is heated with a periodically changing heat flow from a electrical heater [35] or chopped light beam [36], and the temperature change of the sample (magnitude and phase shift) is measured as the response signal (*AC calorimetry*).
2) A thermal wave is sent into the sample from a metallic film that is evaporated onto its surface and which heats it periodically. The temperature change is measured by using the same metal film as a resistance thermometer. The third harmonic of the resulting temperature oscillations is proportional to the power input and depends on the product of the heat capacity and thermal conductivity of the sample in a characteristic way (*3ω method* [37]).
3) The sample (or furnace) temperature is controlled to follow a set course with superimposed periodical changes, and the heat flow rate is measured via the differential temperature between sample and reference (*temperature-modulated differential scanning calorimetry, TMDSC* [38]).

In all three cases, both the magnitude of the periodic part of the response signal and its phase shift with respect to the stimulating signal are measured, and this results in a complex quantity (temperature, heat flow rate, and heat capacity, respectively). As heat transport always requires time, the frequency of temperature modulation is normally limited to a maximum of 0.2 Hz to allow the heat to flow through the sample properly. For the 3ω method, however, frequencies of several kilohertz are possible because the heat source and the temperature probe are identical in this case. At the lower end, the frequency range is only limited by the noise threshold (sensitivity) of the sensors in question. As a rule only two decades are available. Consequently, only time-dependent processes with timescales within this window can be followed in the calorimeter.

For quasistatic conditions it follows from Equation (1) that:

$$\frac{dQ}{dt} = \Phi = C_p \cdot \frac{dT}{dt} \quad (3)$$

where the real quantities heat flow rate Φ and temperature change (heating rate) are strictly proportional, with the real heat capacity C_p as proportionality factor. This implies an infinite propagation velocity of thermal waves and an infinitely fast excitation time of the vibrational states that characterize the (static) heat capacity of a sample. Both conditions are fulfilled for normal solids within the above-mentioned frequency range. In cases where processes are involved which need time, the heat flow rate and the temperature change are no longer proportional, and the process in question causes the heat flow to lag behind the temperature. This results formally in a time-dependent apparent heat capacity, and Equation (3) no longer holds; instead:

$$\Phi(t) = C_p(t) * \frac{dT(t)}{dt} \quad (4)$$

This appears very similar, but the operator * stands for the convolution product, which is an integral operator rather than a product [39].

Equation (4) can be Fourier transformed and then reads:

$$\Phi(\omega) = C_p(\omega) \cdot \text{Fourier}\left(\frac{dT(t)}{dt}\right) \qquad (5)$$

The (complex) heat flow rate is given as a normal product of the complex heat capacity and the (complex) Fourier-transformed heating rate in the frequency domain. For a given $T(t)$ course (normally periodical with underlying constant heating rate), the complex heat capacity can be determined from the measured complex (magnitude and phase) heat flow rate function. The resulting C_p function is given either as real and imaginary part or as magnitude (absolute value) and phase. The two forms are mathematically equivalent. From the frequency dependence of these quantities the $C_p(t)$ behavior can be derived.

26.2.3. Calorimeters

There are many diverse types of calorimeters, differing with respect to measuring principle, mode of operation, and general construction. Measuring principles have been described in the preceding sections. In terms of mode of operation, two approaches can be distinguished: static calorimeters, in which the temperature remains constant or changes only in consequence of the heat of reaction, and dynamic calorimeters, in which the temperature of the calorimeter substance is varied deliberately according to a prescribed program (usually linear with time).

With regard to construction, a distinction can be made between single and differential calorimeters. Twin design is characteristic of the latter: that is, all components are present in duplicate, and the two sets are arranged as identical as possible. The reaction under investigation takes place in one side of the device, while the other side contains an inert reference substance. The output signal represents the difference between signals originating in the sample and reference side. With such a difference signal all symmetrical effects cancel (e.g., heat leaks to the surroundings).

26.2.3.1. Static Calorimeters

Considering first the static calorimeters, three types can be distinguished depending upon whether they are operated in an *isothermal, isoperibolic* (isoperibolic = uniform surroundings), or adiabatic manner.

26.2.3.1.1. Isothermal Calorimeters

In the *isothermal* mode of operation it is imperative that all thermal effects be somehow compensated. This is achieved either electrically or with the aid of a phase transition for some substance. Only phase-transition calorimeters can be regarded as strictly isothermal. In this case thermodynamics ensures that the temperature will remain precisely constant since it is controlled by a two-phase equilibrium of a pure substance. The most familiar example is the *ice calorimeter*, already in use by the end of the 18th century and developed further into a precision instrument about 100 years later by BUNSEN (Fig. 16). The liquid – gas phase transition has also been used for thermal compensation purposes; in this case a heat of reaction can be determined accurately by measuring the volume of a vaporized gas.

Phase-transformation calorimeters are easy to construct, and for this reason they are not available commercially. Such a device permits very precise determination of the heat released during a process. An important drawback is the fact that the isothermal method is limited to the few fixed temperatures corresponding to phase transitons of suitable pure substances.

This disadvantage does not apply to an isothermal device based on electrical compensation. Nevertheless, calorimeters of the latter type operate only in a quasi-isothermal mode, since electronic control systems depend for their response upon small deviations from an established set point, and a certain amount of time is required for changing the prevailing temperature. The use of modern circuits and components ensures that errors from this source will be negligible, however. Electrical compensation makes it possible to follow both endothermic and exothermic processes. In both cases the compensation power is readily measured and recorded or processed further with a computer. Isothermal calorimeters are used quite generally for determining heats of mixing and solution. Commercial devices are available that also support the precise work required for multiphase thermo-

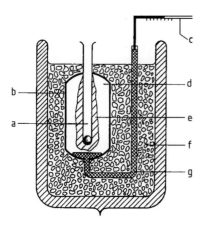

Figure 16. Ice calorimeter (based on [40])
a) Sample holder; b) Calorimeter vessel; c) Capillary with mercury thread; d) Water; e) Ice; f) Ice–water mixture; g) Mercury

dynamics. As a rule such calorimeters are of the single rather than the differential type.

26.2.3.1.2. Isoperibolic Calorimeters

An *isoperibolic calorimeter* is distinguished by the fact that it is situated within a thermostated environment. As a result, the temperature of the surroundings is kept constant, as is the effect of the surroundings on the calorimeter vessel. A well-defined thermal conduction path is maintained between the calorimeter vessel and its surroundings so that the temperature difference between the calorimeter vessel and the environment will be proportional to the heat flow rate. Temperature equilibration usually adheres to an exponential time function, and an integral taken over the entire course of this function is proportional to the heat of reaction. Such instruments often have exceptionally large time constants; that is, temperature equilibration occurs very slowly. Indeed, the greater the equilibration time, the more sensitive is the calorimeter. It is of course difficult with an exponential function to identify precisely the end point, so the integral over the measurement function and therefore the heat of interest are subject to corresponding degrees of uncertainty. The advantages lie in simplicity of construction and ease of producing a measurement signal (a straightforward temperature difference), which might for example be derived from a differential thermocouple and a simple set of amplifiers.

The best-known calorimeter of this type was developed by TIAN and CALVET (Fig. 17). Here the defined heat-conduction path to the thermostated surroundings (a large aluminum block) consists of a large number of differential thermocouples coupled in series (thermopile). This arrangement permits optimum determination of the heat flow rate to the surroundings, and such an instrument can be very sensitive (*microcalorimeter*).

If heat transport to the surroundings is restricted by suitable insulation, temperature equilibration becomes so slow that — with small corrections — temperature changes in the calorimeter substance alone can serve as the basis for determining the heat of a reaction. Calorimeters in this category include the classical *mixing calorimeter* (with heat exchange) in which the sample substance is brought into contact with a (usually) colder calorimeter substance, whereupon an intermediate "mixing temperature" is established (Fig. 18). Other examples of mixing calorimeters include the simpler types of *combustion calorimeters*, in which the calorimeter substance is water, and the *drop calorimeter* (see Fig. 24). The latter consists of a metal block whose temperature change subsequent to introduction of a sample of accurately known temperature provides information about the sample's heat capacity. The category of isoperibolic calorimeters also includes *flow calorimeters*, in which two fluid media (usually liquids containing the reactants) are first brought to the temperature of the thermostated surroundings and then combined. The ensuing reaction causes a measurable temperature change, the extent of which is proportional to the heat of the reaction (Fig. 19).

26.2.3.1.3. Adiabatic Calorimeters

An *adiabatic calorimeter* is equipped with an electronic control system designed to ensure to the greatest extent possible that the surroundings of the calorimeter cell remain at all times at precisely the same temperature as the substance under examination. The object is to prevent virtually all heat transport to the surroundings, thereby making it possible to establish the total heat effect from the change in temperature of the calorimeter substance. Good adiabatic calorimeters are difficult to construct, and they are not available commercially. Nevertheless, precision instruments of this type permit quantities such as specific heat capacities, as well as associated anomalies, to be meas-

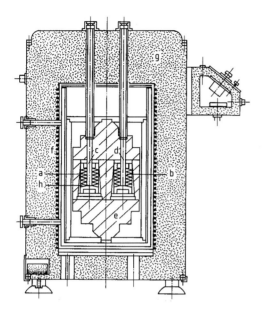

Figure 17. Tian–Calvet calorimeter
a, b) Heat conduction paths between measuring cells and block; c, d) Sample and reference containers; e) Isoperibolic block; f) Thermostatic jacket; g) Thermal insulation; h) Thermopiles

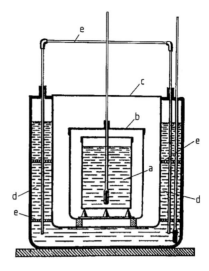

Figure 18. A classical mixing calorimeter (from Meyers Conversationslexicon, 1893)
a) Calorimeter vessel; b) Convection and radiation shield; c) Cover; d) Jacketing vessel (thermostated); e) Lifting mixer

ured very accurately (Fig. 20). The method entails an extraordinary level of effort, and it is practiced only by specialists working in a few laboratories in the world.

Simpler *adiabatic calorimeters* are distributed commercially in the form of adiabatic mixing and combustion calorimeters. These constitute very satisfactory routine instruments, especially since the adiabatic mode of operation (which entails measurement of only a single temperature difference) facilitates automated data acquisition and interpretation.

26.2.3.2. Scanning Calorimeters

Scanning calorimeters are devices in which the sample temperature is deliberately changed during the course of an experiment according to a prescribed program. As a rule, heating or cooling of the sample is caused to occur linearly with time, and the heat flow rate required to accomplish the desired change is determined and output as a function of time or temperature. Any chemical or physical change in the sample at a particular temperature will manifest itself as a change in the heat flow rate, leading to a "peak" in the measurement curve whose area is proportional to the heat of the process (see also Section 26.1.3.2).

Such instruments have been on the market since the 1950s. Their great advantage is that they permit heats of reaction and transformations of every type to be measured both rapidly and with reasonable precision. Heating rates usually fall in the range 5 – 20 K/min, so a measurement extending over 300 K can be completed within an hour, providing a reliable indication of all changes in the sample over this temperature range, including heats of reaction. The rapid development of calorimetry in recent years is due largely to the perfection of this type of calorimeter.

Scanning calorimeters are designed almost without exception as differential devices. With regard to the principle of measurement, two varieties can be distinguished: *differential-temperature* and *power-compensated calorimeters*. Although the two types rely upon very different approaches to determining heat flow rates with respect to the sample, and are therefore subject to different sources of error, they are marketed under the single heading *differential scanning calorimeter* (DSC). They can even be operated in *temperature-modulated* mode.

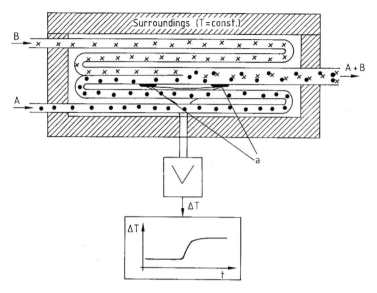

Figure 19. Flow calorimeter, with a chemical reaction occurring in the liquid A, B reactants, A+B reaction product; a) Differential thermocouple

26.2.3.2.1. Differential-Temperature Scanning Calorimeters

In a differential-temperature scanning calorimeter the heat necessary for raising the sample temperature reaches the sample position via one well-defined heat conduction path (cf. Fig. 21). The measured signal is derived from a temperature difference generated along this path, which is proportional to the heat flow rate [see Eq. (2)]. This measuring and operating principle leads to the alternative designation *heat-flux calorimeter* for such a device. In order to minimize errors from heat leaks and other influences of the surroundings, calorimeters of this type are always based on a differential design, with full duplex construction. Many commercial models are available.

Any temperature difference between the sample and the reference is directly proportional to the corresponding differential heat flow rate, and so this is the quantity to be measured. The appropriate functional relationship follows directly from Equation (2):

$$\Phi = K(T) \cdot \Delta T \qquad (6)$$

The heat flow rate of interest, Φ (in W), is proportional to the measured temperature difference ΔT (in K). Unfortunately, the proportionality constant K (the calibration factor for the calorimeter in K W^{-1}) is itself a function of the thermal conductivity of the sample [see Eq. (2)], and therefore temperature dependent. Moreover, temperature differences between the sample and the reference result in variable amounts of heat loss to the surroundings via inevitable heat leaks. Heat exchange with the environment also occurs via radiation and convection. Since the two transport processes are linked to temperature differences in a nonlinear way, the calibration factor in the presumably linear calorimeter equation necessarily depends on ΔT, which is the quantity to be measured. Consequently, the calibration factor depends upon such sample parameters as mass, thermal conductivity, and specific enthalpy difference [31], so heats measured with these instruments must realistically be assigned an uncertainty of about five percent unless considerable effort has been expended in calibration [42]. Repeatability of the results is generally many times better, but the user should not make the mistake of assuming an equally high level of certainty. Another source of systematic error may develop during the measurement if some reaction causes a change in the heat capacity and/or the extent of heat transfer to the sample holder. Both phenomena contribute to changes in the baseline, with the result that the course of the latter cannot be plotted accurately

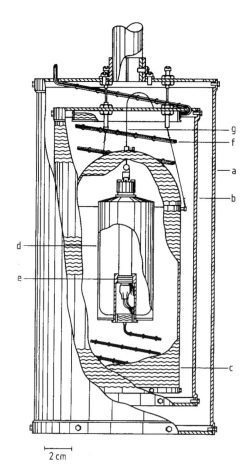

Figure 20. Construction of an adiabatic calorimeter (based on [41])
a) External radiation shield; b) Internal radiation shield; c) Adiabatic jacket; d) Calorimeter vessel; e) Sample suspension platform with resistance thermometer and heater; f) Electrical leads; g) Calorimeter suspension system

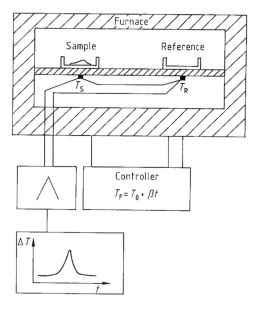

Figure 21. Operating principle of a temperature-difference scanning calorimeter (heat-flux calorimeter)
T_F furnace temperature, T_S sample temperature, T_R reference temperature, T_0 initial temperature, ΔT differential temperature $T_S - T_R$, β heating rate, t time

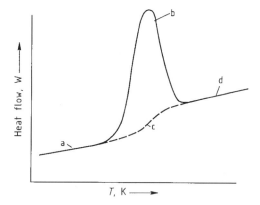

Figure 22. Output curve from a scanning calorimeter showing a shift in the baseline caused by variability in either c_p or the coefficient of heat transfer
a) Original baseline; b) Peak; c) Unknown baseline path; d) Final baseline

(Fig. 22). This in turn means that the area corresponding to the heat of reaction also cannot be determined exactly, leading to uncertainty in the associated heat [31]. Despite these restrictive comments, differential-temperature scanning calorimeters offer the advantages of a simple, transparent mode of operation and relatively low cost, and for this reason they account for a large share of the market. With appropriately critical supervision they are capable of fulfilling their assignment admirably.

26.2.3.2.2. Power-Compensated Scanning Calorimeters

Like differential-temperature devices, *power-compensated scanning calorimeters* always feature duplex construction, with sample and reference holders heated in such a way that at every

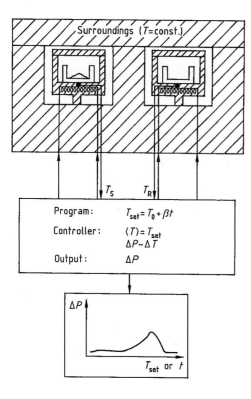

Figure 23. Operating principle of a power-compensated differential scanning calorimeter (Perkin–Elmer)
T_0 initial temperature; T_S sample temperature; T_R reference temperature; $\langle T \rangle$ mean temperature between sample and reference; ΔP differential power $P_S - P_R$; β heating rate; t time

instant the temperature of each corresponds almost exactly to a programmed set temperature. Any temperature differences that arise between sample and reference by reason of thermal events in the sample are compensated immediately by appropriate changes in the electrical heating (Fig. 23). The output signal is proportional to the instantaneous differential heat flow rate. Since temperature differences between sample and reference are essentially compensated at once, measurement errors due to undetected heat leaks can be largely avoided. Instruments of this type, the most widely distributed of which are manufactured by Perkin–Elmer, require a very complicated electronic control system, and are therefore rather expensive. The principal disadvantage lies in the fact that the electronic system is a black box for the average layman, thereby obscuring possible sources of error. These instruments also require that there be some deviation, albeit minimal, from the set temperature to guide the control circuit, so just as in the preceding case some systematic error must be anticipated (although certainly to a lesser degree), associated once again with temperature differences between sample and reference and nonlinear heat exchange. Accordingly, the level of systematic uncertainty for heats measured by power-compensated DSC can be estimated at 1–3 %.

26.2.3.2.3. Temperature-Modulated Scanning Calorimeters

Differential-temperature and *power-compensated scanning calorimeters* are now commercially available to run in the temperature modulated mode (TMDSC). In both cases the temperature is controlled to follow the normal (linear in time) course but with an additional periodical temperature change of particular amplitude and frequency. The measured heat flow rate function is the sum of the "underlying" part and the periodical part. The former can be extracted by calculating the "gliding average" integral over a period along the measured curve. It is identical to the curve that would be obtained in this DSC when the temperature modulation is switched off. The periodical part, on the other hand, is the difference of the measured and the underlying curves. From the periodic part the complex C_p can be calculated by using Fourier analysis or other mathematical techniques. This enables time dependent-processes to be investigated in cases where their timescale falls within the frequency window of the TMDSC method.

For precise measurements it is necessary to correct (calibrate) the measured heat flow rate magnitude and phase shift for influences from the apparatus, as well as from the heat transport, which also is a time-dependent process and therefore influences the measured heat flow rate function [43].

26.2.4. Applications of Calorimetry

Modern calorimeters permit relatively rapid and truly precise measurement of heat exchanges in a wide variety of reactions. Since heat evolution is proportional to the conversion (extent of reaction) in a chemical, physical, or biological reaction, calorimetric measurement constitutes one method for quantitative evaluation of the reaction itself. Measurement is possible not only of the total heat (and therefore the total conversion) of a reaction but also of the course of the reaction

26.2.4.1. Determination of Thermodynamic Functions

Various *thermodynamic potential functions* (enthalpy, entropy, free energy, etc.) can be determined by integration on the basis of *specific heat capacities*. The accurate establishment of specific heat capacities at constant pressure, c_p, as a function of temperature is thus of fundamental importance. Different types of calorimeters are used for determinations of c_p depending upon accuracy requirements and the temperature range in question.

The most accurate results are obtained with adiabatic calorimeters, which are useful over a temperature range from fractions of a Kelvin to about 1000 K. These instruments permit specific heat capacities to be determined with very small uncertainties (<1%). Nevertheless, such measurements are so complex that they are rarely applicable to normal laboratory practice in industry, especially given the lack of commercial instruments.

The standard approach to c_p measurement therefore relies on the less precise but more easily constructed and operated mixing calorimeters, in particular drop calorimeters (Fig. 24). A (mean) specific heat capacity is calculated in this case from the heat released when a sample of known temperature is introduced (dropped) into an isoperibolic calorimeter. Charging temperatures may exceed 2500 K. Determining the temperature dependence of c_p requires many individual measurements at various charging temperatures. The uncertainty in such measurements is usually 3–5%.

A truly elegant method for determining c_p at <1000 K takes advantage of a scanning calorimeter [31], preferably a power-compensated DSC. The procedure involves raising the temperature of a sample in the calorimeter at a defined rate (typically 10 K/min). The same measurement is subsequently repeated, this time with empty crucibles. The difference between the two measurement curves provides a record of the heat flow rate into the sample, from which the specific heat capacity can be calculated. This method makes it possible conveniently to measure the $c_p(T)$ of a material over a temperature range of several hundred Kelvins within the course of a single day. However, it does demand a very high level of repeatability (baseline stability) for the DSC in question. In the case of differential-temperature (heat-flux) DSC—even with careful calibration—uncertainties of 5–10% in the re-

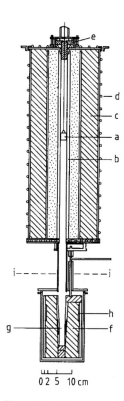

Figure 24. High-temperature drop calorimeter (based on [46])
a) Sample; b) Heated corundum tube; c) Magnesia insulation; d) Cooling coils; e) Sample mounting device; f) Copper block; g) Hole for receiving sample; h) Resistance thermometer; i) Level of the thermostated oil bath

with respect to time. This opens the way to studies of kinetics, problems related to stability and safety, and other areas in which temperature and time-dependencies in a reaction play a part.

The sections that follow describe several possible applications of calorimetry, although these should be regarded only as illustrative, since the scope of the present article rules out any treatment that is even approximately exhaustive. No such limits apply to the reader's imagination, however, and any process that generates (or consumes) heat can in principle be followed with a calorimeter. Most recently, enzymatic and bacterial manufacturing processes have assumed increasing importance in industry, and interest in biocalorimetry is growing accordingly [44], [45].

sults must be anticipated. With power-compensated DSC the systematic uncertainties tend to be in the range of 3–5%. Uncertainties in the calculated thermodynamic potential functions of interest are comparable.

The TMDSC enables another elegant possibility of c_p (magnitude) determination from the amplitude of the modulated part of the measured heat flow rate function both in the isothermal and scanning modes of operation. This method is especially advantageous in cases of noisy signals with low sample masses or low heating rates. Precise calibration of the heat flow rate amplitude is a prerequisite for obtaining reliable results [43].

However, a calorimeter operating dynamically always indicates a temperature that deviates to a greater or lesser extent from the true sample temperature, resulting in further uncertainty. It must in any case be borne in mind that potential functions determined on the basis of thermodynamic equilibrium, and thus defined statically, may differ from functions determined dynamically, since equilibrium conditions are not maintained in DSC, and in the case of a reaction even stationary conditions are lacking.

If time-dependent processes are involved, then the process is clearly outside the scope of classical thermodynamics. The time-dependent (apparent) heat capacity, measured with, say, TMDSC would lead to time-dependent potential functions which must be interpreted in terms of *irreversible thermodynamics*. In such cases, a nonzero imaginary part of the complex heat capacity exists which is linked to the entropy production of the process in question (for details see [32]). Thus, temperature-modulated calorimetry makes it possible to determine time-dependent (irreversible) thermodynamic quantities.

26.2.4.2. Determination of Heats of Mixing

For liquid systems consisting of several components, multiphase thermodynamics provides the theoretical background for computing phase diagrams from heats of mixing for the relevant components. Phase diagrams in turn are interesting not only to thermodynamicists; they are also of great importance in chemical process engineering practice.

The determination of heats of mixing is a classical field of investigation for calorimetry, one that at the required level of precision is today incomparably less laborious than previously. Enthalpies of mixing were formerly determined in some cases indirectly via vapor pressure measurements, but today they are measured directly in (quasi)isothermal or isoperibolic calorimeters.

The art of conducting such an experiment successfully lies in rigorously excluding vaporization during the measurement, since specific heats of vaporization are orders of magnitude larger than the effects of mixing that are to be measured. As excess quantities (which characterize deviations from ideal behavior), heats of mixing are invariably small, and they must be determined with high precision in order to permit the construction of sufficiently precise phase diagrams. Isothermal calorimeters are therefore definitely to be preferred, since here the problem of incomplete determination of a heat of mixing — owing perhaps to undetected heat leaks — is less severe than with an isoperibolic instrument. Moreover, the isothermal heat of mixing is defined more precisely, and it lends itself more readily to mathematical treatment than in the case of parallel temperature changes.

Phase transitions in solid mixtures can be followed directly in a scanning calorimeter, since vapor pressures associated with the components are here so low that vaporization plays virtually no role and therefore does not interfere. Corresponding phase diagrams (i.e., liquidus–solidus curves) can be inferred from heat flow rate curves obtained by controlled heating and cooling of appropriate mixtures (cf. Section 26.1.3.3). Thermodynamically relevant quantities can in turn be determined indirectly by comparing measured phase diagrams with calculated diagrams.

26.2.4.3. Combustion Calorimetry

The determination of heats of combustion for fuels of all types is prescribed in detail by government regulation, since the "calorific value" or "heat content" of a particular coal, oil, or natural or manufactured gas determines its economic value. Combustion calorimeters or "bomb calorimeters" (usually operated adiabatically) are used for such measurements, whereby the sample is combusted in a "calorimetric bomb" (likewise standardized; cf. Fig. 25) in pure oxygen at $3 \cdot 10^6$ Pa to a defined set of end products. Methods of measurement and interpretation are also standardized, and they have been simplified to such an extent that the entire procedure can be automated.

The same instruments are useful for determining heats of combustion for a wide variety of

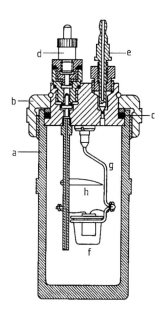

Figure 25. Example of a calorimetric bomb for use in a combustion calorimeter
a) Pressure vessel; b) Screw cap; c) Sealing ring; d) Inlet valve; e) Outlet valve; f) Sample crucible with sample; g) Crucible holder; h) Heat shield

(usually organic) substances, from which standard enthalpies of formation can be determined. For these calorimetric methods errors in measurement are very small ($< 1‰$), with a level of precision that is also appropriate for analytical purposes (e.g., purity or composition checks on batches of chemicals). Combustion calorimetry is useful in biology as well. For example, in determinations of the nutrient content of leaves and fruits as a function of season and place based on appropriate heats of combustion.

Many combustion calorimeters are available commercially. Instruments utilizing water as the calorimeter substance generally operate adiabatically, thereby avoiding the somewhat complicated Regnault–Pfaundler procedure (cf. [29]) for determining and interpreting prior and subsequent temperature patterns. "Dry" combustion calorimeters that avoid the use of water or other liquids have also recently come onto the market.

26.2.4.4. Reaction Calorimetry

Every chemical process is associated with some type of heat effect. Accurate knowledge of heat released or consumed is of great importance in process engineering. Moreover, heat evolution is strictly coupled with the particular course of a reaction. If one is able to determine unambiguously the time dependence of the heat flow rate associated with a particular sample, one also gains access to the corresponding reaction rate law, which in turn opens the way to the reaction kinetics [31]. Knowledge of the appropriate kinetic parameters permits one to predict the course of the reaction under other sets of conditions, which is of great practical value. Heat of reaction determinations as well as kinetic studies are best carried out under isothermal (and isobaric) conditions, and for this reason it is advisable to select a (quasi)isothermal calorimeter with compensation for effects of the measurement process. One practical difficulty lies in the fact that reaction begins immediately after introduction of the sample at the appropriate temperature, even though the calorimeter itself is not yet at thermal equilibrium (and therefore not ready for use). One way out of this dilemma is to devise a system for bringing the components of the reaction together only after isothermicity has been achieved. With liquids the challenge is not particularly great, since pumps and stirrers can ensure rapid (and thorough) mixing of the reactants at any desired time. The piercing of a membrane (or an ampoule) is also a common expedient. Another possibility is the use of a specially designed reaction calorimeter. With one device of this type the first reactant is present from the outset in a suitable glass or metallic vessel, to which a second reactant is added with mixing after stationary conditions have been established (Fig. 26). With the aid of electronic temperature control the reaction can then be carried out in any way desired: isothermally, adiabatically, or under the influence of some temperature program. Measurement of the relevant quantities, process control, and thermochemical data interpretation are all accomplished with the aid of a personal computer. A second type of special reaction calorimeter, suitable only for liquid components, is so constructed that each reactant is pumped in through its own system of tubing, with the streams uniting after temperature equilibration (Fig. 27). The ensuing reaction causes a temperature change, from which the corresponding heat of reaction can be calculated.

Transport and mixing of liquid or gaseous components is accomplished rather easily, but the situation is more complex for solid substances or mixtures. Here there is little hope of satisfactorily mixing the components inside a normal calorimeter, so one is forced to resort to a scanning

Figure 26. Operating principle of a reaction calorimeter (Contraves)
a) Reaction vessel; b) Heating and cooling bath; c) Stirrer; d) Balance; e) Pressure sensor for internal pressure; f) pH probe; g) Temperature sensor; h) Dosing valve; i) Control unit; j) Control and monitoring signals; k) Measured signal

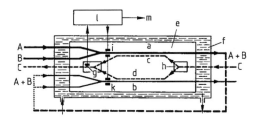

Figure 27. Schematic diagram of a flow calorimeter after P. PICKER (Setaram)
A, B) Reactants; A + B) Reaction mixture; C) Heat-exchange liquid
a) Reaction tube (sample tube); b) Reference tube; c, d) Heat-exchange tubes; e) Vacuum jacket; f) Thermostated liquid; g) Thermistor; h) Flow divider; i, k) Heating elements; l) Measurement and control electronics; m) Output signal

calorimeter. The calorimeter is charged at a temperature below that at which the substances react, after which the temperature is increased linearly with time until reaction is complete. From a heat flow rate–time function acquired at a well-defined heating rate it is possible to establish both the heat of reaction and, if need be, kinetic parameters for the process. One prerequisite, however, is that the recorded function reproduces in an unbiased way the true heat production of the sample. This presents serious problems, since heat must inevitably flow from the point at which it is generated to the site of the temperature sensors before any measurement signal can appear. In the case of a fairly rapid chemical process, where the rate constant is comparable to the time constant of the calorimeter, the measurement signal from the recording device must be treated mathematically ("desmeared") [31] before it can be utilized for a kinetic analysis. Owing to the smaller time constant and more precise linearity of the corresponding temperature–time program, a scanning calorimeter with power compensation is preferred over a temperature-difference DSC.

26.2.4.5. Safety Studies

Calorimetric methods applicable to safety studies are also concerned with clarifying the kinetic behavior of substances, but in this case from a safety engineering viewpoint. The important role assigned recently to such studies justifies their being treated as a separate topic.

The vast majority of chemical substances are endothermic materials; that is, they release heat when they decompose, and their decomposition is accelerated by heat. Chemical reaction kinetics teaches that any reaction is capable of proceeding at temperatures considerably lower than the temperature at which the maximum reaction rate is achieved. Usually, however, the reaction rate under these conditions is so low that the process is not apparent. The more unstable a substance, the lower is its decomposition temperature, and therefore the greater its decomposition rate at room temperature. In the event that heat evolved during storage — attributable to traces of reaction occurring under storage conditions — exceeds the level at which heat is emitted to the surroundings, then the decomposition rate may increase very rapidly, possibly to the point of an explosion.

Thorough knowledge of decomposition kinetics is therefore essential for ensuring safe storage and handling. The requisite information can, in principle, be gained through the methods described in Section 26.2.4.4, but it is often impossible to avoid an uncontrolled outcome and simultaneous destruction of the measuring system. This obviously rules out the use of expensive apparatus, which has led to the development of special safety calorimeters that are either inexpensive and disposable or designed to withstand the effects of violent reactions.

Simple arrangements of this type clearly preclude precision measurements, but they are entirely adequate from the specific viewpoint of assessing storage safety at particular temperatures or with particular types of packaging. The simplest example is an exceedingly primitive isoperibolic calorimeter consisting of a glass ampoule, either open or fitted with a pressure-tight closure, mounted together with a thermocouple inside a small furnace — a device that anyone can easily assemble and calibrate. More ambitious designs differ from standard calorimeters in that the sample holder (or measuring head) is incorporated into an autoclave of sufficient stability to permit studying the thermal behavior even of explosives. Included in this category are the so-called accelerating rate calorimeters (ARC), in which a sample is heated very slowly under quasiadiabatic conditions until the reaction intensifies by itself as a result of exothermic decomposition. The resulting temperature–time function can then be used to establish the corresponding heat effect and stability behavior.

26.3. References

[1] J. O. Hill (ed.): "For Better Thermal Analysis and Calorimetry," 3rd ed., International Confederation for Thermal Analysis and Calorimetry, 1991.
[2] W. Wendlandt: *Thermal Analysis*, 3rd ed., Wiley-Interscience, New York 1986.
[3] M. E. Brown: *Introduction to Thermal Analysis*, Chapman and Hall, London 1988.
[4] R. C. Mackenzie (ed.): *Differential Thermal Analysis*, Academic Press, London, vol. 1, 1970, vol. 2, 1972.
[5] M. E. Brown (ed.): *Handbook of Thermal Analysis and Calorimetry*, **vol. 1**, Elsevier, Amsterdam 1998.
[6] M. Reading in [5], pp. 423–443.
[7] S. P. Wolsky, A. W. Czanderna (eds.): *Microweighing in Controlled Environments*, Elsevier, Amsterdam 1980.
[8] ASTM Standard E 914-83.
[9] C. L. Duval: *Inorganic Thermogravimetric Analysis*, Elsevier, Amsterdam 1953.
[10] M. R. Ottaway, *Fuel* **61** (1982) 713.
[11] C. M. Earnest (ed.): *Compositional Analysis by Thermogravimetry*, ASTM STP 997, Philadelphia 1988.
[12] J. W. Cumming, J. McLaughlin, *Thermochim. Acta* **57** (1982) 253.
[13] E. Turi (ed.): *Thermal Characterization of Polymeric Materials*, Academic Press, San Diego 1997.
[14] R. L. Blaine, C. K. Schoff (eds.): *Purity Determination by Thermal Methods*, ASTM STP 838, Philadelphia 1984.
[15] M. Reading, *Trends Polym. Sci.* **1** (1993) 248.
[16] F. Paulik, J. Paulik, *Analyst (London)* **103** (1978) 417.
[17] W. Lodding (ed.): *Gas Effluent Analysis*, Marcel Dekker, New York 1967.
[18] S. B. Warrington in E. L. Charsley, S. B. Warrington (eds.): *Thermal Analysis – Techniques and Applications*, Royal Society of Chemistry, London 1992.
[19] M. R. Holdiness, *Thermochim. Acta* **75** (1984) 361.
[20] E. L. Charsley, C. Walker, S. B. Warrington, *J. Therm. Anal.* **40** (1993) 983 (Proc. 10th ICTAC Conf., Hatfield, 1992).
[21] V. Balek, J. Tolgyessy in C. L. Wilson, D. W. Wilson (eds.): *Comprehensive Analytical Chemistry*, vol. 12, part C, Elsevier, Amsterdam 1984.
[22] M. Reading in E. L. Charsley, S. B. Warrington (eds.): *Thermal Analysis – Techniques and Applications*, Royal Society of Chemistry, London 1992.
[23] T. Lever et al., *American Laboratory* **30** (1998) 15.

General References

[24] W. Hemminger, G. Höhne: *Calorimetry, Fundamentals and Practice,* Verlag Chemie, Weinheim 1984.
[25] E. Koch: *Non-Isothermal Reaction Analysis,* Academic Press, London 1977.
[26] J. P. McCullough, D. W. Scott (eds.): *Experimental Thermodynamics,* vol. 1: "Calorimetry of Non-reacting Systems," Butterworths, London 1968.
[27] B. Le Neindre, B. Vodar (eds.): *Experimental Thermodynamics,* vol. 2, Butterworths, London 1975.
[28] H. Spink, I. Wadsö: "Calorimetry as an Analytical Tool in Biochemistry and Biology," in D. Glinck (ed.): *Methods of Biochemical Analysis,* vol. 23, J. Wiley & Sons, New York 1976.
[29] S. Sunner, M. Mansson (eds.): "Combustion Calorimetry," *Experimental Chemical Thermodynamics,* vol. 1, Pergamon Press, Oxford 1979.
[30] H. Weber: *Isothermal Calorimetry,* Herbert Lang, Bern, Peter Lang, Frankfurt 1973.
[31] G. W. H. Höhne, W. Hemminger, H. J. Flammersheim, *Differential Scanning Calorimetry; An Introduction for Practitioners,* Springer-Verlag, Berlin, 1996.
[32] C. Schick, G. W. H. Höhne (eds.): "Temperature-Modulated Calorimetry", *Thermochim. Acta* **304/305** (1997) special issue.

Specific References

[33] E. Gmelin: "Classical Temperature-Modulated Calorimetry: A Review", *Thermochim. Acta* **304/305** (1997) 1–26.
[34] P. S. Gill, S. R. Sauerbrunn, M. Reading, *J.Thermal Anal.* **40** (1993) 931–939.
[35] P. F. Sullivan, G. Seidel *Phys. Rev.* **173** (1968) 679.
[36] P. Handler, D. E. Mapother, M. Rayl, *Phys. Rev. Lett.* **19** (1967) 356.
[37] O. M. Corbino, *Phys. Z.* **11** (1910) 413; **12** (1911) 292.
[38] H. Gobrecht, K. Hamann, G. Willers, *J. Physics E: Sci. Instr.* **4** (1971) 21–23.
[39] J. E. K. Schawe, G. W. H. Höhne, *Thermochim. Acta* **287** (1996) 213–223.
[40] R. W. Bunsen, *Ann. Phys.* **141** (1870) 1.
[41] E. D. West, E. F. Westrum Jr. in J. P. McCullough, D. W. Scott (eds.): *Experimental Thermodynamics,* **vol. 1,** Butterworths, London 1968, p. 354.
[42] S. M. Sarge et al.: "The Caloric Calibration of Scanning Calorimeters," *Thermochim. Acta* **247** (1994) 129–168.
[43] G. W. H. Höhne, *Thermochim. Acta* **330** (1999) 45–54.
[44] J. Lamprecht, W. Hemminger, G. W. H. Höhne (eds.): "Calorimetry in the Biological Sciences," *Thermochim. Acta* **193 (Special Issue)** (1991) 1–452.
[45] R. B. Kemp, B. Schaarschmidt (eds.): "Proceedings of Biological Calorimetry", *Thermochim. Acta* **251/252** (1995) special issues.
[46] T. B. Douglas, E. G. King in J. P. McCullough, D. W. Scott (eds.): *Experimental Thermodynamics,* **vol. 1,** Butterworths, London 1968, p. 297.

27. Surface Analysis

JOHN C. RIVIÈRE, Harwell Laboratory, AEA Technology, Didcot, United Kingdom

27.	Surface Analysis	851
27.1.	Introduction	852
27.2.	X-Ray Photoelectron Spectroscopy (XPS)	854
27.2.1.	Principles	854
27.2.2.	Instrumentation	856
27.2.2.1.	Vacuum Requirements	856
27.2.2.2.	X-Ray Sources	857
27.2.2.3.	Synchrotron Radiation	858
27.2.2.4.	Electron Energy Analyzers	859
27.2.2.5.	Spatial Resolution	861
27.2.3.	Spectral Information and Chemical Shifts	862
27.2.4.	Quantification and Depth Profiling	865
27.2.4.1.	Quantification	865
27.2.4.2.	Depth Profiling	866
27.2.5.	The Auger Parameter	867
27.2.6.	Applications	867
27.2.6.1.	Catalysis	867
27.2.6.2.	Corrosion and Passivation	868
27.2.6.3.	Adhesion	869
27.2.6.4.	Superconductors	871
27.2.6.5.	Interfaces	873
27.3.	Auger Electron Spectroscopy (AES)	874
27.3.1.	Principles	874
27.3.2.	Instrumentation	875
27.3.2.1.	Vacuum Requirements	875
27.3.2.2.	Electron Sources	875
27.3.2.3.	Electron Energy Analyzers	876
27.3.3.	Spectral Information	876
27.3.4.	Quantification and Depth Profiling	879
27.3.4.1.	Quantification	879
27.3.4.2.	Depth Profiling	881
27.3.5.	Scanning Auger Microscopy (SAM)	882
27.3.6.	Applications	883
27.3.6.1.	Grain Boundary Segregation	883
27.3.6.2.	Semiconductor Technology	883
27.3.6.3.	Thin Films and Interfaces	885
27.3.6.4.	Superconductors	887
27.3.6.5.	Surface Segregation	888
27.4.	Static Secondary Ion Mass Spectrometry (SSIMS)	889
27.4.1.	Principles	889
27.4.2.	Instrumentation	890
27.4.2.1.	Ion Sources	890
27.4.2.2.	Mass Analyzers	891
27.4.3.	Spectral Information	892
27.4.4.	Quantification	894
27.4.5.	Applications	895
27.4.5.1.	Oxide Films	895
27.4.5.2.	Interfaces	896
27.4.5.3.	Polymers	897
27.4.5.4.	Surface Reactions	898
27.5.	Ion Scattering Spectroscopies (ISS and RBS)	898
27.5.1.	Ion Scattering Spectroscopy (ISS)	899
27.5.1.1.	Principles	899
27.5.1.2.	Instrumentation	900
27.5.1.3.	Spectral and Structural Information	900
27.5.1.4.	Quantification and Depth Information	901
27.5.1.5.	Applications	904
27.5.2.	Rutherford Backscattering Spectroscopy (RBS)	906
27.5.2.1.	Principles	906
27.5.2.2.	Instrumentation	907
27.5.2.3.	Spectral and Structural Information	907
27.5.2.4.	Quantification	909
27.5.2.5.	Applications	909
27.6.	Scanning Tunneling Methods (STM, STS, AFM)	910
27.6.1.	Principles	910
27.6.2.	Instrumentation	912
27.6.3.	Spatial and Spectroscopic Information	913
27.6.4.	Applications	914
27.7.	Other Surface Analytical Methods	917
27.7.1.	Ultraviolet Photoelectron Spectroscopy (UPS)	917
27.7.2.	Light-Spectroscopic Methods (RAIRS, SERS, Ellipsometry, SHG, IBSCA, GDOS)	918
27.7.3.	Electron Energy Loss Methods (ELS, CEELS, HREELS, IETS)	921
27.7.4.	Appearance Potential Methods (SXAPS, AEAPS, DAPS)	926
27.7.5.	Inverse Photoemission (IPES, BIS)	928

27.7.6.	Ion Excitation Methods (IAES, INS, MQS, SNMS, GDMS)	928	27.7.10.	Diffraction Methods (LEED, RHEED) ... 938
27.7.7.	Neutral Excitation (FABMS)	932	27.8.	**Summary and Comparison of Techniques** ... 940
27.7.8.	Atom Probe Field-Ion Microscopy (APFIM, POSAP, PLAP)	932	27.9.	**Surface Analytical Equipment Suppliers** ... 940
27.7.9.	Desorption Methods (ESD, ESDIAD, TDS)	934	27.10.	**References** ... 944

27.1. Introduction

Wherever the properties of a solid surface are important, it is also important to have the means to measure those properties. The surfaces of solids play an overriding part in a remarkably large number of processes, phenomena, and materials of technological importance. These include catalysis; corrosion, passivation, and rust; adhesion; tribology, friction, and wear; brittle fracture of metals and ceramics; microelectronics; composites; surface treatments of polymers and plastics; protective coatings; superconductors; and solid surface reactions of all types with gases, liquids, or other solids. The surfaces in question are not always external; processes occurring at inner surfaces such as interfaces and grain boundaries are often just as critical to the behavior of the material. In all the above cases, the nature of a process or of material behavior can be understood completely only if information about both surface composition (i.e., types of atoms present and their concentrations) and surface chemistry (i.e., chemical states of the atoms) is available. Occasionally, knowledge of the arrangement of surface atoms (i.e., the surface structure) is also necessary.

First of all, what is meant by a *solid surface*? Ideally the surface should be defined as the point at which the solid terminates, that is, the last atom layer before the adjacent phase (vacuum, vapor, liquid, or another solid) begins. Unfortunately such a definition is impractical because the effect of termination extends into the solid beyond the outermost atom layer. Indeed, the current definition is based on that knowledge, and the surface is thus regarded as consisting of that number of atom layers over which the effect of termination of the solid decays until bulk properties are reached. In practice, this decay distance is of the order of 5–20 nm.

By a fortunate coincidence, the depth into the solid from which information is provided by the techniques described here matches the above definition of a surface almost exactly. These techniques are therefore *surface specific*; in other words, their information comes only from that very shallow depth of a few atom layers. There are other techniques that can be *surface sensitive*, in that they would normally be regarded as techniques for bulk analysis, but have sufficient sensitivity for certain elements that those elements can be analyzed with these techniques even if they are present only on the surface.

Why should surfaces be so important? The answer is twofold. Firstly, because the properties of surface atoms are usually different from those of the same atoms in the bulk, and secondly, because in any interaction of a solid with another phase the surface atoms are the first to be encountered. Even at the surface of a *perfect single crystal* the surface atoms behave differently from those in the bulk simply because they do not have the same number of nearest neighbors; their electronic distributions are altered and hence their reactivity. Their structural arrangement is often also different. When the surface of a *polycrystalline or glassy multielemental solid* is considered, such as that of an alloy or a chemical compound, the situation can be very complex. The processes of preparation or fabrication may produce a material whose surface composition is quite different from that of the bulk, in terms of both constituent and impurity elements. Subsequent treatments (e.g., thermal and chemical) will almost certainly change the surface composition to something different again. The surface is highly unlikely to be smooth, and roughness at both micro- and macro-levels may be present, leading to the likelihood that many surface atoms will be situated at corners and edges and on protuberances (i.e., in positions of increased reactivity). Surfaces exposed to the atmosphere, which include many of those of tech-

Table 1. Surface-specific analytical techniques using primary particle excitation*

Emission	Excitation**			
	Photons $h\nu$	Electrons e^-	Ions n^+	Neutrals n^0
$h\nu$	RAIRS SERS ellipsometry SHG	SXAPS DAPS CLS IPES BIS	IBSCA GDOS	
e^-	XPS XAES UPS SRPS	AES SAM ELS HREELS CEELS AEAPS LEED RHEED	IAES PAES INS MQS	
n^+, n^-	PSD	ESD ESDIAD	SSIMS ISS RBS GDMS	FABMS
n^0	PSD	ESD	SNMS	

* Some techniques in Table 1 have angle-resolved variants, with the prefix AR, thus giving ARUPS, ARAES, ARELS, etc.
** For meanings of acronyms, see Listing 1.

Table 2. Surface-specific analytical techniques using non-particle excitation

Emission	Excitation*		
	Heat kT	High electrical field F	Mechanical force
n^+	TDS	APFIM POSAP	
n^-	TDS		
e^-		IETS STM, STS	
(Displacement)			AFM

* For meanings of acronyms, see Listing 1.

nological interest, will acquire a contaminant layer 1–2 atom layers thick, containing principally carbon and oxygen but also other impurities present in the local environment. Atmospheric exposure may also cause oxidation. Because of all these possibilities the surface region must be considered as a separate entity, effectively a separate quasi-two-dimensional phase overlaying the normal bulk phase. Analysis of the properties of such a quasi phase necessitates the use of techniques in which the information provided originates only or largely within the phase (i.e., the surface-specific techniques described in this article).

Nearly all these techniques involve interrogation of the surface with a particle probe. The function of the probe is to excite surface atoms into states giving rise to emission of one or more of a variety of secondary particles such as electrons, photons, positive and secondary ions, and neutrals. Since the primary particles used in the probing beam may also be either electrons or photons, or ions or neutrals, many separate techniques are possible, each based on a different primary–secondary particle combination. Most of these possibilities have now been established, but in fact not all the resultant techniques are of general application, some because of the restricted or specialized nature of the information obtained and others because of difficult experimental requirements. In this article, therefore, most space is devoted to those surface analytical techniques that are widely applied and readily available commercially, whereas much briefer descriptions are given of the many others whose use is less common but that—in appropriate circumstances, particularly in basic research—can provide vital information.

Because the various types of particle can appear in both primary excitation and secondary emission, most authors and reviewers have found it convenient to group the techniques in a matrix, in which the columns refer to the nature of the exciting particle and the rows to the nature of the emitted particle [1]–[9]. Such a matrix of techniques is given in Table 1, in terms of the acronyms now accepted. The meanings of the acronyms, together with some of the alternatives that have appeared in the literature, are in Listing 1.

Listing 1. Meanings of surface analysis acronyms, and their alternatives, that appear in Tables 1 and 2.

1) Photon Excitation

RAIRS Reflection–absorption infrared spectroscopy (or IRAS IR reflection–absorption spectroscopy)
SERS Surface-enhanced Raman scattering
SHG Second harmonic generation
XPS X-ray photoelectron spectroscopy (or ESCA electron spectroscopy for chemical analysis)
XAES X-ray (excited) Auger electron spectroscopy

UPS   Ultraviolet photoelectron spectroscopy (or PES photoemission spectroscopy)
SRPS Synchrotron radiation photoelectron spectroscopy
PSD   Photon-stimulated desorption

2) Electron Excitation

SXAPS   Soft X-ray appearance potential spectroscopy (or APS appearance potential spectroscopy)
DAPS   Disappearance potential spectroscopy
CLS   Cathodoluminescence spectroscopy
IPES   Inverse photoemission spectroscopy
BIS   Bremsstrahlung isochromat spectroscopy
AES   Auger electron spectroscopy
SAM   Scanning Auger microscopy
ELS   (electron) Energy loss spectroscopy
HREELS   High-resolution electron energy loss spectroscopy (or LEELS low-energy electron loss spectroscopy)
CEELS   Core-electron energy loss spectroscopy (or ILS ionization loss spectroscopy)
AEAPS   Auger electron appearance potential spectroscopy
LEED   Low-energy electron diffraction
RHEED   Reflection high-energy electron diffraction
ESD   Electron-stimulated desorption (or EID electron-induced desorption)
ESDIAD   Electron-stimulated desorption ion angular distribution

3) Ion Excitation

IBSCA   Ion beam spectrochemical analysis (or SCANIIR surface composition by analysis of neutral and ion impact radiation) (or BLE bombardment-induced light emission)
GDOS   Glow discharge optical spectroscopy
IAES   Ion (excited) Auger electron spectroscopy
PAES   Proton (or particle) (excited) Auger electron spectroscopy
INS   Ion neutralization spectroscopy
MQS   Metastable quenching spectroscopy
SSIMS   Static secondary-ion mass spectrometry
ISS   Ion scattering spectroscopy (or LEIS low-energy ion scattering)
RBS   Rutherford backscattering spectroscopy (or HEIS high-energy ion scattering)
GDMS   Glow discharge mass spectrometry
SNMS   Secondary neutral mass spectrometry

4) Neutral Excitation

FABMS   Fast-atom bombardment mass spectrometry

5) Thermal Excitation

TDS   Thermal desorption spectroscopy

6) High Field Excitation

IETS   Inelastic electron tunneling spectroscopy
APFIM   Atom probe field-ion microscopy
POSAP   Position-sensitive atom probe
STM   Scanning tunneling microscopy
STS   Scanning tunneling spectroscopy

7) Mechanical Force

AFM   Atomic force microscopy

A few techniques, including one or two important ones, cannot be classified according to the nature of the exciting particle, since they do not employ primary particles but depend instead on the application either of heat or a high electric field. The latter techniques are listed in Table 2.

27.2. X-Ray Photoelectron Spectroscopy (XPS)

X-ray photoelectron spectroscopy is currently the most widely used surface analytical technique, and is therefore described here in more detail than any of the other techniques. At its inception by SIEGBAHN and coworkers [10] it was called ESCA (electron spectroscopy for chemical analysis), but the name ESCA is now considered too general, since many surface electron spectroscopies exist, and the name given to each one must be precise. Nevertheless, the name ESCA is still used in many places, particularly in industrial laboratories and their publications. Briefly, the reasons for the popularity of XPS are the exceptional combination of compositional and chemical information that it provides, its ease of operation, and the ready availability of commercial equipment.

27.2.1. Principles

The surface to be analyzed is irradiated with soft X-ray photons. When a photon of energy $h\nu$ interacts with an electron in a level with binding energy E_B, the entire photon energy is transferred to the electron, with the result that a photoelectron is ejected with kinetic energy

$$E_{Kin} = h\nu - E_B - e\varphi \qquad (1)$$

where $e\varphi$ is a small, almost constant, work function term.

Obviously $h\nu$ must be greater than E_B. The ejected electron may come from a core level or from the occupied portion of the valence band, but in XPS most attention is focused on electrons in core levels. Since no two elements share the same set of electronic binding energies, measurement of the photoelectron kinetic energies provides an *elemental analysis*. In addition, Equation (1) indicates that any changes in E_B are reflected in E_{Kin}, which means that changes in the chemical environment of an atom can be followed by monitoring

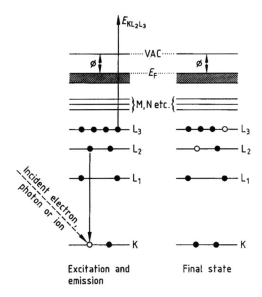

Figure 1. Auger process in an atom in a solid

changes in the photoelectron energies, leading to the provision of *chemical information*. XPS can analyze all elements in the periodic table except hydrogen and helium.

Although XPS is concerned principally with photoelectrons and their kinetic energies, ejection of electrons by other processes also occurs. An ejected photoelectron leaves behind a core hole in the atom. The sequence of events following the creation of the core hole is shown schematically in Figure 1. In the example, the hole has been created in the K shell, giving rise to a photoelectron whose kinetic energy would be $(h\nu - E_K)$, and is filled by an electronic transition from the L_2 shell. The energy $E_K - E_{L_2}$ associated with the transition can then either be dissipated as a characteristic X-ray photon or given up to an electron in a third shell, shown in this example as the L_3. The second of these possibilities is called the Auger process after its discoverer [11], and the resultant ejected electron is called an Auger electron and has an energy given by

$$E_{KL_2L_3} = E_K - E_{L_2} - E^*_{L_3} \qquad (2)$$

$E^*_{L_3}$ is starred because it is the binding energy of an electron in the L_3 shell in the presence of a hole in the L_2 shell, that is, not quite the same as E_{L_3}.

X-ray photon emission (i.e., X-ray fluorescence) and Auger electron emission are obviously competing processes, but for the shallow core levels involved in XPS and AES the Auger process is far more likely.

Thus in all X-ray photoelectron spectra, features appear due to both photoemission and Auger emission. In XPS, the Auger features can be useful but are not central to the technique, whereas in AES (see Chap. 27.3), Equation (2) forms the basis of the technique.

At this point the nomenclature used in XPS and AES should be explained. In XPS the *spectroscopic notation* is used, and in AES the *X-ray notation*. The two are equivalent, the different usages having arisen for historical reasons, but the differentiation is a convenient one. They are both based on the so-called $j-j$ coupling scheme describing the orbital motion of an electron around an atomic nucleus, in which the total angular momentum of an electron is found by summing vectorially the individual electron spin and angular momenta. Thus if l is the electronic angular momentum quantum number and s the electronic spin momentum quantum number, the total angular momentum for each electron is given by $j = l + s$. Since l can take the values 0, 1, 2, 3, 4, ... and $s = \pm \frac{1}{2}$, clearly $j = \frac{1}{2}, \frac{3}{2}, \frac{5}{2}$, etc. The principal quantum number n can take values 1, 2, 3, 4, ... In *spectroscopic notation*, states with $l = 0, 1, 2, 3, ...$ are designated $s, p, d, f, ...$, respectively, and the letter is preceded by the number n; the j values are then appended as suffixes. Therefore one obtains $1s, 2s, 2p_{1/2}, 2p_{3/2}, 3s, 3p_{1/2}, 3p_{3/2}$, etc.

In *X-ray notation*, states with $n = 1, 2, 3, 4, ...$ are designated K, L, M, N, ..., respectively, and states with various combinations of $l = 0, 1, 2, 3, ...$ and $j = \frac{1}{2}, \frac{3}{2}, \frac{5}{2}$, are appended as the suffixes 1, 2, 3, 4, ... In this way one arrives at K, $L_1, L_2, L_3, M_1, M_2, M_3$, etc. The equivalence of the two notations is set out in Table 3.

In X-ray notation the Auger transition shown in Figure 1 would therefore be labeled KL_2L_3. In this coupling scheme, six Auger transitions would be possible in the KLL series. Obviously, many other series are possible (e.g., KLM, LMM, MNN). These are discussed more fully in Chapter 27.3, dealing with AES.

The reasons why techniques such as XPS and AES, which involve measurement of the energies of ejected electrons, are so surface specific, should be examined. An electron with kinetic energy E moving through a solid matrix M has a probability of traveling a certain distance before losing all or part of its energy as a result of an inelastic collision. Based on that probability, the average dis-

Table 3. Spectroscopic and X-ray notation

Quantum numbers			Spectroscopic state	X-ray suffix	X-ray state
n	l	j			
1	0	1/2	$1s$	1	K
2	0	1/2	$2s$	1	L_1
2	1	1/2	$2p_{1/2}$	2	L_2
2	1	3/2	$2p_{3/2}$	3	L_3
3	0	1/2	$3s$	1	M_1
3	1	1/2	$3p_{1/2}$	2	M_2
3	1	3/2	$3p_{3/2}$	3	M_3
3	2	3/2	$3d_{3/2}$	4	M_4
3	2	5/2	$3d_{5/2}$	5	M_5
	etc.		etc.	etc.	etc.

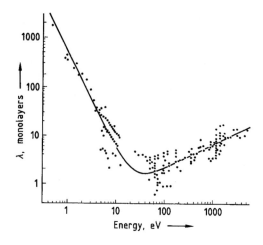

Figure 2. Compilation by SEAH and DENCH [12] of measurements of inelastic mean free path as a function of electron kinetic energy
The solid line is a least-squares fit

tance traveled before such a collision is known as the inelastic mean free path (imfp) $\lambda_M(E)$. The imfp is a function only of M and of E. Figure 2 shows a compilation of measurements of λ made by SEAH and DENCH [12], in terms of atomic monolayers as a function of kinetic energy. Note that both λ and energy scales are logarithmic. The important consequence of the dependence of λ on kinetic energy is that in the ranges of secondary electron kinetic energies used in XPS and AES, the values of λ are very small. In XPS, for example, typical ranges are 250–1500 eV, corresponding to a range of λ from about four to eight monolayers, while in AES, the energy range is typically 20 to 1000 eV, in which case λ would range from about two to six monolayers. What this means in practice is that if the photoelectron or the Auger electron is to escape into a vacuum and be detected, it must originate at or very near the surface of the solid. This is the reason why the electron spectroscopic techniques are surface specific.

27.2.2. Instrumentation

27.2.2.1. Vacuum Requirements

Electron spectroscopic techniques require vacua of the order of 10^{-8} Pa for their operation. This requirement arises from the extreme surface specificity of these techniques, mentioned above. With sampling depths of only a few atom layers, and elemental sensitivities down to 10^{-5} atom layers (i.e., one atom of a particular element in 10^5 other atoms in an atomic layer), the techniques are clearly very sensitive to surface contamination, most of which comes from the residual gases in the vacuum system. According to gas kinetic theory, to have enough time to make a surface analytical measurement on a surface that has just been prepared or exposed, before contamination from the gas phase interferes, the base pressure should be 10^{-8} Pa or lower, that is, in the region of ultrahigh vacuum (UHV).

The requirement for the achievement of UHV conditions imposes restrictions on the types of material that can be used for the construction of surface analytical systems, or inside the systems, because UHV can be achieved only by accelerating the rate of removal of gas molecules from internal surfaces by raising the temperature of the entire system (i.e., by baking). Typical baking conditions are 150–200 °C for several hours. Construction materials should not be distorted, lose their strength, produce excessive gas, or oxidize readily under such treatment, which should be repeated after each exposure of the inside to atmospheric pressure. Thus, the principal construction material is stainless steel, with mu-metal (76 % Ni, 5 % Cu, 2 % Cr) used occasionally where magnetic screening is needed (e.g., around electron energy analyzers). For the same reasons, metal seals, not elastomers, are used for the demountable joints between individual components — the sealing material usually being pure copper although gold is sometimes used. Other materials that may be used between ambient atmosphere and UHV are borosilicate glass or quartz for windows, and alumina for electrical insulation for current or voltage connections. In-

Table 4. Energies and linewidths of some characteristic low-energy X-ray lines

Line		Energy, eV	Width, eV
Y	M_ζ	132.3	0.47
Zr	M_ζ	151.4	0.77
Nb	M_ζ	171.4	1.21
Mg	K_α	1253.6	0.70
Al	K_α	1486.6	0.85
Si	K_α	1739.5	1.00
Y	L_α	1922.6	1.50
Zr	L_α	2042.4	1.70

side the system, any material is permissible that does not produce volatile components either during normal operation or during baking. Thus, for example, brass that contains the volatile metal zinc could not be used.

The production of vacua in the UHV region has been routine for many years now, but success still depends on attention to details such as the maintenance of strict cleanliness.

27.2.2.2. X-Ray Sources

The most important consideration in choosing an X-ray source for XPS is energy resolution. Equation (1) gives the relationship between the kinetic energy of the photoelectron, the energy of the X-ray photon, and the binding energy of the core electron. Since the energy spread, or linewidth, of an electron in a core level is very small, the linewidth of the photoelectron energy depends on the linewidth of the source, if no undue broadening is introduced instrumentally. Now in XPS the analyst devotes much effort to extracting chemical information by means of detailed study of individual elemental photoelectron spectra. Such a study needs an energy resolution better than 1.0 eV if subtle chemical effects are to be identified. Thus the linewidth of the X-ray source should be significantly smaller than 1.0 eV if the resolution required is not to be limited by the source itself.

Other considerations are that the source material, which forms a target for high-energy electron bombardment leading to the production of X rays, should be a good conductor—to allow rapid removal of heat—and should also be compatible with UHV.

Table 4 lists the energies and linewidths of the characteristic X-ray lines from a few possible candidate materials.

Not many sources with sufficiently narrow lines are available. Y $M\zeta$ and Zr $M\zeta$ would be suitable, but their line energies are too low to be useful for general application, although they have been used for special purposes. Silicon is a poor thermal conductor and is not easy to apply as a coating on an anode, so only the Mg K_α and Al K_α lines remain, which are in fact the two used universally in XPS.

For efficient production of X rays by electron bombardment, exciting electron energies that are at least an order of magnitude higher than the line energies must be used, so that in Mg and Al sources accelerating potentials of 15 kV are employed. Modern sources are designed with dual anodes, one anode face being coated with magnesium and the other with aluminum, and with two filaments, one for each face. Thus, a switch from one type of X irradiation to the other can be made very quickly.

To protect the sample from stray electrons from the anode, from heating effects, and from possible contamination by the source enclosure, a thin ($\approx$ 2-μm) window of aluminum foil is interposed between the anode and the sample. For optimum X-ray photon flux on the surface (i.e., optimum sensitivity), the anode must be brought as close to the sample as possible, which means in practice a distance of $\approx$ 2 cm. The entire X-ray source is therefore retractable via a bellows and a screw mechanism.

The X radiation from magnesium and aluminum sources is quite complex. The principal K_α lines are in fact unresolved doublets and should correctly be labeled $K_{\alpha_{1,2}}$. Many satellite lines also exist of which the more important in practice are the $K_{\alpha_{3,4}}$, separated from the principal line by $\approx$ 10 eV and roughly 8% of its intensity, and the K_β, lying about 70 eV higher in energy. In addition, the spectrum is superimposed on a continuous background called bremsstrahlung radiation, arising from inelastic processes.

Removal of satellites, elimination of the bremsstrahlung background, and separation of the $K_{\alpha_{1,2}}$ doublet can be achieved by monochromatization, shown schematically in Figure 3. The X-ray source is positioned at one point on a spherical surface, called a Rowland sphere after its originator, and a very accurately ground and polished quartz crystal is placed at another point. X rays from the source are diffracted from the quartz, and by placing the sample at the correct point on the Rowland sphere, the K_{α_1} component can be selected to be focused on it, according to

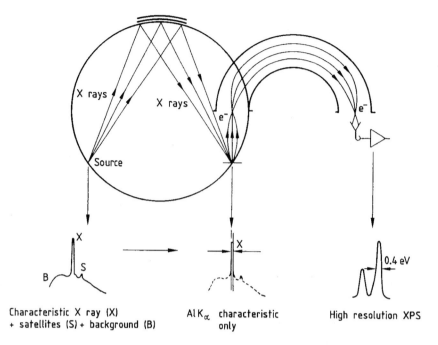

Figure 3. Schematic of X-ray monochromatization to remove satellites, eliminate bremsstrahlung background, and separate the Al $K_{\alpha_{1,2}}$ doublet
Courtesy of Kratos Analytical.

the well-known Bragg relation for diffraction. Quartz is a very convenient diffracting medium for Al K_α since the spacing between the $10\bar{1}0$ planes is exactly half the wavelength of the X radiation, and for first-order diffraction the Bragg angle is 78.5°. Since the width of the Al K_{α_1} line is < 0.4 eV, the energy dispersion needed around the surface of the sphere implies that the Rowland sphere should have a diameter of at least 0.5 m. Accurate positioning of source, crystal, and sample is essential. Although an XPS spectrum will be much "cleaner" when a monochromator is used, because satellites and background have been removed, the photon flux at the sample is much lower than that from an unmonochromatized source operating at the same power. Against this must be set the greatly improved signal-to-background level in a monochromatized spectrum.

27.2.2.3. Synchrotron Radiation

The discrete line sources described above for XPS are perfectly adequate for most applications, but some types of analysis require that the source be tunable (i.e., that the exciting energy be variable). The reason is to allow the photoionization cross section of the core levels of a particular element or group of elements to be varied, which is particularly useful when dealing with multielement semiconductors. Tunable radiation can be obtained from a synchrotron.

In a synchrotron, electrons are accelerated to near-relativistic velocities and constrained magnetically into circular paths. When a charged body is accelerated, it emits radiation, and when the near-relativistic electrons are forced into curved paths, they emit photons over a continuous spectrum. The general shape of the spectrum is shown in Figure 4. For a synchrotron with an energy of several gigaelectronvolts and a radius of some tens of meters, the energy of the emitted photons near the maximum is of the order of 1 keV (i.e., ideal for XPS). As can be seen from the universal curve, plenty of usable intensity exists down into the UV region. With suitable monochromators on the output to select a particular wavelength, the photon energy can be tuned continuously from about 20 to 500 eV. The available intensities are comparable to those from conventional line sources.

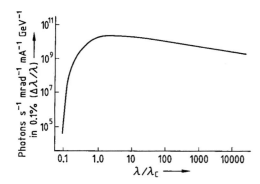

Figure 4. Normalized spectrum of photon energies emitted from a synchrotron
λ_c = wavelength characteristic of the individual synchrotron

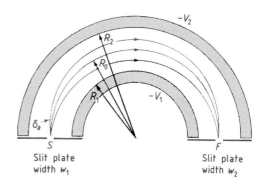

Figure 5. Schematic of a concentric hemispherical analyzer (CHA) [13]

27.2.2.4. Electron Energy Analyzers

In electron spectroscopic techniques — among which XPS is the most important — analysis of the energies of electrons ejected from a surface is central. Because of the low kinetic energies involved in the techniques, analyzers using magnetic fields are undesirable. Therefore the energy analyzers used are exclusively of the electrostatic deflection type. The two that are now universally employed are the concentric hemispherical analyzer (CHA) and the cylindrical mirror analyzer (CMA). Since both have been used in XPS, both are described here, although in practice the CHA is more suitable for XPS, and the CMA for AES.

Energy analyzers cannot be discussed without a discussion of energy resolution, which is defined in two ways. *Absolute resolution* is defined as ΔE, the full width at half-maximum (FWHM) of a chosen peak. *Relative resolution* is defined as the ratio R of ΔE to the kinetic energy E of the peak energy position (usually its centroid), that is, $R = \Delta E/E$. Thus absolute resolution is independent of peak position, but relative resolution can be specified only by reference to a particular kinetic energy.

In XPS, closely spaced peaks at any point in the energy range must be resolved, which requires the same absolute resolution at all energies. In AES, on the other hand, the inherent widths of peaks in the spectrum are greater than in XPS and the source area on the surface is far smaller, both of which lead to operation at constant relative resolution.

Concentric Hemispherical Analyzer (CHA). The CHA is shown in schematic cross section in Figure 5 [13]. Two hemispheres of radii R_1 (inner) and R_2 (outer) are positioned concentrically. Potentials $-V_1$ and $-V_2$ are applied to the inner and outer hemispheres, respectively, with V_2 greater than V_1. The source S and the focus F are in the same plane as the center of curvature, and R_0 is the radius of the equipotential surface between the hemispheres. If electrons of energy $E = eV_0$ are injected at S along the equipotential surface, they will be focused at F if

$$V_2 - V_1 = V_0 \left(\frac{R_2}{R_1} - \frac{R_1}{R_2} \right) \tag{3}$$

If electrons are injected not exactly along the equipotential surface, but with an angular spread $\delta\alpha$ about the correct direction, then the energy resolution is given by

$$\Delta E/E = (W_1 + W_2)/4R_0 + (\delta\alpha)^2 \tag{4}$$

where W_1 and W_2 are the respective widths of the entrance and exit slits. In most instruments, for convenience in construction purposes, $W_1 = W_2 = W$, whereupon the resolution becomes

$$\Delta E/E = W/2R_0 + (\delta\alpha)^2 \tag{5}$$

In XPS the photoelectrons are retarded to a constant energy, called the pass energy, as they approach the entrance slit. If this were not done, Equation (5) shows that to achieve an absolute resolution of 1 eV at the maximum kinetic energy of about 1500 eV (using Al K$_\alpha$ radiation), and with a slit width of 2 mm, would require an analyzer with an average radius of about 300 cm, which is impracticable. Pass energies are selected in the

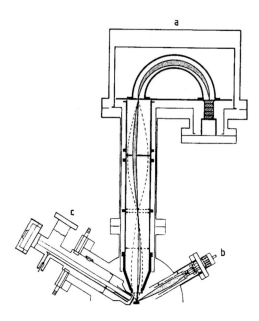

Figure 6. Typical configuration in a modern XPS spectrometer
Courtesy of Leybold
a) XPS–Auger analyzer; b) Auger electron gun; c) X-ray source

range 20–100 eV for XPS, which allows the analyzer to be built with a radius of 10–15 cm.

Modern XPS spectrometers employ a lens system on the input to the CHA, which has the effect of transferring an image of the analyzed area on the sample surface to the entrance slit of the analyzer. Such a spectrometer is illustrated schematically in Figure 6.

Cylindrical Mirror Analyzer (CMA). In the CMA shown schematically in Figure 7, two cylinders of radii r_1 (inner) and r_2 (outer) are accurately positioned coaxially. Annular entrance and exit apertures are cut in the inner cylinder, and a deflection potential $-V$ is applied between the cylinders. Electrons leaving the sample surface at the source S with a particular energy E on passing into the CMA via the entrance aperture are then deflected back through the exit aperture to the focus F. For the special case in which the acceptance angle $\alpha = 42.3°$, the first-order aberrations vanish, and the CMA becomes a second-order focusing device. The relationship between the electron energy E and the deflection potential is then

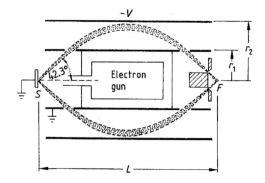

Figure 7. Schematic of a cylindrical mirror analyzer (CMA)

$$\frac{E}{eV} = \frac{1.31}{\ln(r_2/r_1)} \qquad (6)$$

As in the CHA, therefore, scanning the deflection potential $-V$ and recording the signal as a function of electron energy provide the distribution in energy of electrons leaving the sample surface.

If the angular spread of the acceptance angle is $\delta\alpha$, then the relative energy resolution of a CMA is

$$\Delta E/E = 0.18 W/r_1 + 1.39 (\delta\alpha)^3 \qquad (7)$$

where W is the effective slit width. For $\delta\alpha \approx 6°$, a typical value, W can be replaced by the source size (i.e., the area on the sample from which electrons are accepted).

In *AES* the inner cylinder is at earth potential, and primary electrons originate in an integral electron gun built into the middle of the analyzer. The source size in AES is very small, because of the focused primary beam, and the important property of the CMA for AES is therefore its very high transmission. In *XPS* the entire sample surface is normally flooded with X rays, in which case the source size is the entire acceptance area of photoelectrons, and Equation (7) shows that the relative energy resolution would be too poor. The important parameter in XPS is not the transmission alone, but the luminosity, which is the product of transmission, acceptance area, and acceptance solid angle.

The single-stage CMA depicted in Figure 7 is therefore not suitable for XPS, but a variation of the CMA that is suitable is the double-pass CMA shown in Figure 8 [14]. Two CMAs are placed in series, so that the exit aperture for the first is the entrance aperture for the second. Electrons from

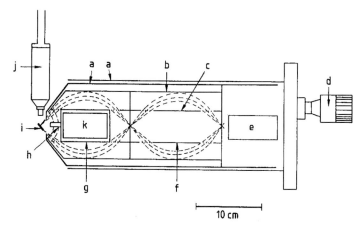

Figure 8. Schematic of double-pass CMA designed by PALMBERG [14] for either XPS or AES
a) Inner and outer magnetic shields; b) Outer cylinder; c) Inner cylinder; d) Rotary motion for aperture change; e) Electron multiplier; f) Second stage; g) First stage; h) Spherical retarding grids; i) Sample; j) X-ray source; k) Electron gun

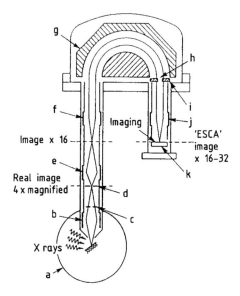

Figure 9. Schematic of the electron-optical arrangement designed by COXON et al. [15] for the production of energy-resolved two-dimensional images in XPS
a) Spherical mu-metal chamber; b) Lens 1; c) Objective aperture; d) Field aperture; e) Lens 2; f) Lens 3; g) 180° Hemispherical analyzer; h) Hole; i) Spectrum detector 1; j) Lens 5; k) Image detector 2

the sample surface are retarded to a constant pass energy, as in the CHA, by spherical retardation grids in front of the entrance aperture, centered on the source area on the sample. The first mesh is grounded, and the second mesh is at the potential of the inner cylinder, which in the XPS mode of operation is not grounded, as in AES, but floating at the scanning potential. Adequate energy resolution is achieved in the second stage because the effective source area, being defined not by the irradiated sample surface but by the size of the first stage exit aperture, is sufficiently small. Entrance and exit apertures can be changed in the second stage by an externally operated mechanism.

27.2.2.5. Spatial Resolution

The principal disadvantage of conventional XPS is the lack of spatial resolution; the spectral information comes from an analyzed area of several square millimeters and is, therefore, an average of the compositional and chemical analyses from that area. On the other hand, the majority of samples of technological origin are inhomogeneous on a scale much smaller than that of normal XPS analysis, and obtaining chemical information on the same scale as the inhomogeneities would be very desirable.

Major steps have been taken recently in the direction of improved XPS spatial resolution by making use of the focusing properties of the CHA described above. One of the more successful configurations that have evolved for this purpose is shown in Figure 9, from the work of COXON et al. [15]. The instrument incorporating it is called the ESCASCOPE. Compared to a conventional XPS system, it has additional lenses (3 and 5) on the input and output sides of the CHA. Lenses 1 and 2 collect electrons from the analyzed area and

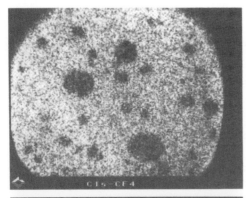

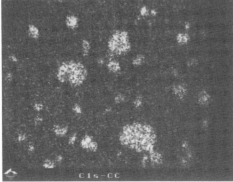

Figure 10. Chemical state images obtained with the ESCASCOPE in Figure 9, from a contaminated fluoropolymer [15]
A) Image in contribution to C 1s from C–F bonding;
B) Image in contribution to C 1s from C–C bonding

present a real image of the sample, magnified about 16 times, at the object plane of lens 3.

Lens 3 is so arranged that the distance to its object plane is equal to its focal length, so that electrons reach the object plane (also the entrance slit of the analyzer) traveling in straight lines. This means that electrons leaving a point on the sample surface within a narrow angular range, pass into the analyzer in a parallel beam inclined at a particular angle to the object plane, with no intensity variation in the plane; another way of putting this is to say that a Fourier transformation of the surface image exists at the entrance slit.

The CHA (lens 4) then forms an image of the Fourier-transformed surface image in its own image plane (the exit slit) and introduces energy dispersion at the same time, as discussed above. The angular information is conserved. Photoelectrons pass through a hole in the center of the image plane and into lens 5, which is arranged similarly to lens 3. The Fourier-transformed image is then

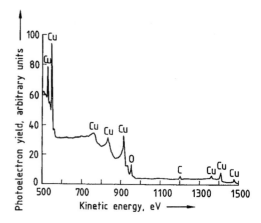

Figure 11. "Wide," or "survey," spectrum from almost clean copper, using Al K_α radiation

inverted so that a real image is formed in the image plane of lens 5, and this image will be an energy-filtered two-dimensional image of the sample surface. The image is detected via a position-sensitive detector consisting of two 40-mm-diameter channel plates, a phosphor screen, and a TV camera. Since the photoelectron kinetic energies can be selected at will in the CHA, images in electron energies characteristic of a particular element, or of a particular chemical state of an element, can be formed.

An example of chemical state imaging in the ESCASCOPE is given in Figure 10, also from [15]. The material is a contaminated fluoropolymer, the contamination being in the form of dark spots of 10–80-μm size. Both images in Figure 10 are in the C 1s photoelectron peak, but of different chemical states. The upper is in that contribution arising from carbon bound to fluorine, and the lower from carbon bound to carbon (i.e., graphitic in nature). The complementarity of the images indicates that the contamination is graphitic in nature.

27.2.3. Spectral Information and Chemical Shifts

Figure 11 shows a *wide* or *survey* XPS spectrum, that is, one recorded over a wide range of energies, in this case 1000 eV. The radiation used was unmonochromatized Al K_α, at 1486.6 eV, and the surface is that of almost clean copper. Such a spectrum reveals the major features to be found,

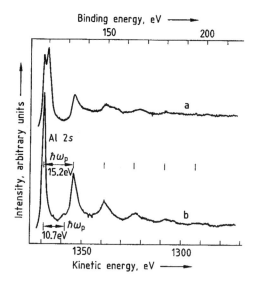

Figure 12. "Narrow," or "detail," spectra from aluminum [16]
a) Slightly oxidized aluminum; b) Clean aluminum
Successive bulk plasmon losses, and a surface plasmon loss are shown associated with the Al 2s peak

but to investigate minor or more detailed features, spectra are acquired over much more restricted energy ranges, at better energy resolution; the latter spectra are called *narrow* or *detail scans*.

The most prominent features in Figure 11, and in most photoelectron spectra, are peaks arising from excitation of core-level electrons according to Equation (1). Thus, at the left-hand end, two intense peaks are found at about 553 and 533 eV, corresponding to photoelectrons from the $2p_{3/2}$ and $2p_{1/2}$ levels of copper, respectively, the separation of 20 eV being the spin–orbit splitting. At the right-hand, high kinetic energy end, are three other copper photoelectron peaks at about 1363, 1409, and 1485 eV, which arise from the $3s$, $3p$, and $3d$ levels, respectively. The other prominent features associated with copper are the three Auger peaks $L_3M_{4,5}M_{4,5}$, $L_3M_{2,3}M_{4,5}$, and $L_3M_{2,3}M_{2,3}$ appearing at 919, 838, and 768 eV, respectively. As pointed out in Section 27.2.1, the creation of a core hole by any means of excitation can lead to ejection of an Auger electron, so that in XPS Auger features form a significant contribution to the spectrum. If Auger and photoelectron peaks happen to overlap in any spectrum, they can always be separated by changing the excitation (e.g., from Al K_α to Mg K_α, or vice versa) since the Auger peaks are invariant in energy whereas the photoelectron peaks must shift with the energy of the exciting photons according to Equation (1).

In addition to features due to copper in Figure 11 are small photoelectron peaks at 955- and 1204-eV kinetic energies, arising from the oxygen and carbon $1s$ levels, respectively, due to the presence of some contamination on the surface. Other minor features can be seen associated with the major peaks. Those situated about 10 eV toward higher kinetic energy occur because unmonochromatized X radiation was used; they are photoelectron peaks excited by the $K_{\alpha_{3,4}}$ satellite already mentioned (see Section 27.2.2.2.). They would disappear if a monochromator were used. Toward the lower kinetic energy of the major peaks (e.g., visible some 10–30 eV below the Cu $2p$ peaks) are the so-called *plasmon loss peaks*. Peaks of this type are shown more clearly in Figure 12, from BARRIE [16], in narrow scans from an aluminum surface. They arise when outgoing electrons of sufficient energy, originating in any way, lose energy by exciting collective oscillations in the conduction electrons in a solid. The oscillation frequency, and hence the energy involved, is characteristic of the nature of the solid and is called the plasmon frequency (or energy). The plasmon loss determined by the three-dimensional nature of the solid is called a "bulk" or "volume" loss, but when it is associated with the two-dimensional nature of the surface, it is called a "surface" loss. In Figure 12 (b), which is from clean aluminum, examples of both types of plasmon loss from the Al $2s$ peak can be seen—a bulk loss of 15.2 eV and a surface loss of 10.7 eV. Theoretically, the surface plasmon loss energy should be a factor of $2^{1/2}$ less than the bulk plasmon loss energy. Where the bulk loss is very pronounced, as in a free-electron-like metal such as aluminum, a succession of peaks can be found at multiples of the loss energy, as indicated in Figure 12, with progressively decreasing intensity. Spectrum (a) in Figure 12, from slightly oxidized aluminum, reveals the rapid quenching of the surface plasmon on oxidation and also shows a chemical shift in the Al $2s$ photoelectron peak.

Chemical shift is the name given to the observed shift in energy of a photoelectron peak from a particular element when the chemical state of that element changes. When an atom enters into combination with another atom or group of atoms, an alteration occurs in the valence electron density, which might be positive or negative according to whether charge is accepted or donated, causing a consequent alteration in the electrostatic potential affecting the core electrons. Therefore the

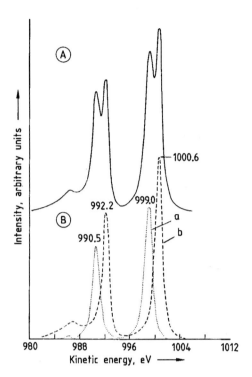

Figure 13. Example of a chemical shift in the Sn $3d$ peak, exhibited by a very thin layer of Sn oxide on Sn metal
A) Spectrum after linear background subtraction; B) Spectrum resolved into its respective components
a) Sn^{n+}; b) Sn^0

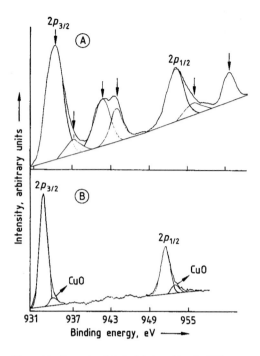

Figure 14. Example of Cu$2p$ "shake-up" peaks [17]
A) Present in a Cu^{2+} compound (CuO); B) Absent in a Cu^+ compound (Cu_2O)

binding energies of the core electrons change, giving rise according to Equation (1) to shifts in the corresponding photoelectron peaks. Tabulation of the chemical shifts experienced by any one element in a series of pure compounds of that element thus enables its chemical state to be identified during analysis of unknown samples. Many such tabulations have appeared, and a major collation of them is found in [2]. The identification of chemical state in this way is the principal advantage of XPS over other surface analytical techniques.

An example of a spectrum exhibiting a chemical shift is that of the tin $3d$ peaks in Figure 13. A thin layer of oxide on the metallic tin surface allows photoelectrons from both the underlying metal and the oxide to appear together. Resolution of the doublet $3d_{5/2}$, $3d_{3/2}$ into the components from the metal (Sn^0) and from the oxide Sn^{n+} is shown in Figure 13 B. The shift in this case is 1.6–1.7 eV. Curve resolution is an operation that can be performed routinely in data processing systems associated with photoelectron spectrometers.

Another type of satellite often seen in XPS spectra is the so-called *shake-up satellite*. Such satellites occur when formation of a compound leads to the presence of unpaired electrons in $3d$ or $4f$ levels; they can also occur in organic compounds according to the degree of conjugation in C=C double bonds, particularly in aromatic compounds. Figure 14, from SCROCCO [17], compares the copper $2p$ photoelectron spectra from CuO and Cu_2O. Since Cu^{2+} as in CuO has an open shell configuration, the $2p$ spectrum shows strong shake-up satellites, whereas that from Cu_2O does not because Cu^+ has a closed shell. In many cases, such satellite structures can be used to diagnose the oxidation state(s).

Other spectral features occasionally seen in XPS spectra arise from *multiplet splitting*. The latter occurs when an element has unpaired electrons either in the valence band or in localized core levels very close to the valence band (as in, for example, some of the rare-earth elements). In the $4f$ electron levels of the rare earths, correlation between spin and angular electron momenta can occur in many different ways, leading to multiplets at different energies, most of which are resolvable [18] in XPS as a multipeak spectrum.

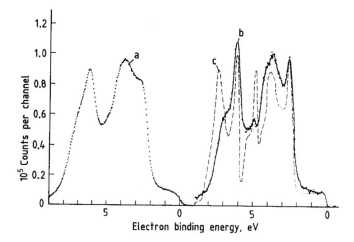

Figure 15. Valence band spectra of gold [19]
a) Using monochromatized Al K_α radiation, and high-energy resolution; b) Result of deconvoluting instrumental broadening from the raw data of a); c) Result of a theoretical calculation of the valence band structure

Photoelectrons of highest kinetic energy in a spectrum (i.e., those of lowest binding energy) originate in the shallowest levels of the solid, that is, in the valence band. Thus the photoelectron peak labeled 3d in the copper spectrum in Figure 11 corresponds to the valence band spectrum, although no detailed structure can be seen in such a "survey" spectrum because the energy resolution used there is not adequate. If much better resolution is used together with a monochromatized source, the valence bands of many metals reveal interesting detail, as shown [19] in Figure 15 for gold. However, for a systematic study of valence band structure, XPS is not ideal, because of both photoionization cross-section effects and the high kinetic energies involved, and the techniques of ultraviolet photoelectron spectroscopy (UPS) and synchrotron radiation photoelectron spectroscopy (SRPS) are preferred.

27.2.4. Quantification and Depth Profiling

27.2.4.1. Quantification

If an X-ray photon of energy $h\nu$ ionizes a core level X in an atom of element A in a solid, the photoelectron current I_A from X in A is

$$I_A(X) = K \cdot \sigma(h\nu:E_B) \cdot \beta(h\nu:X) \cdot \bar{N}_A \cdot \lambda_M(E_x) \cdot \cos\theta \qquad (8)$$

where $\sigma(h\nu:E_B)$ is the photoelectric cross section for ionization of X (binding energy E_B) by photons of energy $h\nu$; $\beta(h\nu:X)$ the asymmetry parameter for emission from X by excitation with photons of energy $h\nu$; $\bar{N}_A$ the atomic density of A averaged over depth of analysis; $\lambda_M(E_x)$ the inelastic mean free path in matrix M containing A, at kinetic energy E_x, where $E_x = h\nu - E_B$; and θ the angle of emission to surface normal. K is a constant of proportionality that contains fixed operational parameters such as incident X-ray flux, transmission of analyzer at kinetic energy E_x, and efficiency of detector at kinetic energy E_x, kept fixed during any one analysis.

Values of the total cross section σ for Al K_α radiation, relative to the carbon 1s level, have been calculated by SCOFIELD [20] (Fig. 16), and of the asymmetry parameter β by REILMAN et al. [21] (Fig. 17, plotted by SEAH [22]). SEAH and DENCH [12] have compiled many measurements of the inelastic mean free path, and for elements the best-fit relationship that they found was

$$\lambda_M(E) = 538E^{-2} + 0.41(aE)^{1/2} \text{ monolayers} \qquad (9)$$

where a is the monolayer thickness in nanometers.

In principle, therefore, the surface concentration of an element can be calculated from the intensity of a particular photoelectron emission, according to Equation (8). In practice, the *method of relative sensitivity factors* is in common use.

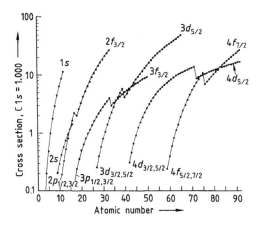

Figure 16. Photoionization cross-section values, for Al K_α radiation, relative to that of the C 1s taken as unity [20]

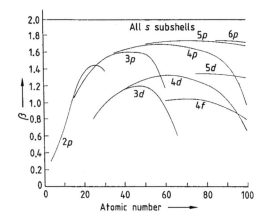

Figure 17. Values of the asymmetry parameter β (Eq. 8) for Al K_α radiation [21]

The intensity of a particular elemental photoelectron peak, normally the F 1s, is taken as standard and put equal to unity, and all intensity measurements are referred to it. Thus, from Equation (8), the ratio $I_A(X)/I_F(1s)$, with $I_F(1s)$ equal to unity, is

$$I_A(X) = \left[\frac{\sigma(h\nu:E_x) \cdot \beta(h\nu:X) \cdot \lambda_A(E_x)}{\sigma(h\nu:E_{1s}) \cdot \beta(h\nu:1s) \cdot \bar{N}_F \cdot \lambda_F(E_{1s})}\right]\bar{N}_A \quad (10)$$

When $I_A(X)$ is measured from pure A and a well-characterized fluorine standard is used, the expression in square brackets is the relative sensitivity factor of A. Tables of such empirically derived sensitivity factors have been published by several authors, the most comprehensive being that in [23].

27.2.4.2. Depth Profiling

Very often, in addition to analytical information about the original surface, information is required about the distribution of chemical composition to depths considerably greater than the inelastic mean free path. Nondestructive methods such as variation of emission angle or variation of exciting photon energy are possible, but they are limited in practice to evaluation of chemical profile within depths of only ca. 5 nm. To obtain information from greater depths requires destructive methods, of which the one used universally is removal of the surface by ion bombardment, also called sputtering. The combination of removal of the surface by sputtering and of analysis by either XPS, AES, or SSIMS is termed depth profiling. In AES and SSIMS the sputtering and analytical operations can occur simultaneously, so that a continuous composition profile is obtained if a single element is being monitored, or a quasi-continuous one if several elements are monitored. In XPS, which is inherently slower and where chemical information is required along with the depth, profiling is recorded in a stepwise fashion, with alternate cycles of sputtering and analysis. Examples of profiles through oxide films on pure iron and on Fe–12Cr–1Mo alloy are shown in Figure 18, in which the respective contributions from the metallic and oxide components of the iron and chromium spectra have been quantified [24]. In these examples the oxide films were only ≈5 nm thick on iron and ≈3 nm thick on the alloy.

Current practice in depth profiling is to use positively charged argon ions at energies between 0.5 and 10 keV, focused into a beam of 2–5-μm diameter, which is then rastered over the area to be profiled. In XPS the bombarded area should be significantly greater than the analyzed area so that only the flat bottom of the sputtered crater is analyzed. Ion current densities are variable between 5 and 50 μA/cm². Uniformity of sputtering, and therefore the depth resolution of the profile, can be improved by rotating the sample during bombardment.

During the sputtering process, many artifacts are introduced that can affect quantification; some of these derive from the nature of the sample itself (e.g., roughness, crystalline structure, phase distribution, electrical conductivity), and others are ra-

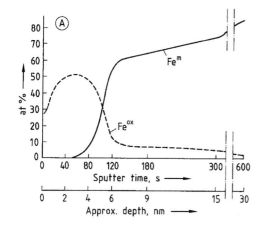

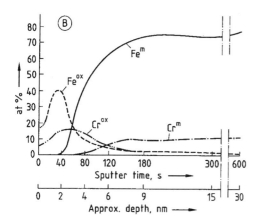

Figure 18. Depth profiles of thin (3–5-nm) oxide films [24] on
A) Pure iron; B) Fe–12Cr–1Mo alloy

diation induced (e.g., atomic mixing, preferential sputtering, compound reduction, diffusion and segregation, sputter-induced topography). This is an extensive subject in its own right; a discussion of all aspects is given in [25].

27.2.5. The Auger Parameter

Section 27.2.3 shows that in an XPS spectrum, X-ray-excited Auger peaks are often as prominent as the photoelectron peaks themselves. For many elements, the chemical shifts in Auger peaks are actually greater than the shifts in photoelectron peaks. The two shifts can be combined in a very useful quantity called the Auger Parameter α^*, first used by WAGNER [26] and defined in its modified form [27] as

$$\alpha^* = E_K(jkl) + E_B(i) \qquad (11)$$

where $E_K(jkl)$ is the kinetic energy of the Auger transition jkl, and $E_B(i)$ is the binding energy of an electron in level i. The Auger and photoelectron peak energies must be measured in the same spectrum.

The Parameter α^* is independent of photon energy and has the great advantage of being independent of any surface charging. It is therefore measurable for all types of material from metals to insulators. Extensive tabulations of α^* have appeared, based on a large number of standard materials, and the most reliable have recently been collected by WAGNER [28].

As defined in Equation (11), α^* is purely empirical; any prominent and conveniently situated Auger and photoelectron peaks can be used. If α^* is defined slightly more rigorously, by replacing $E_B(i)$ by $E_B(j)$ in Equation (11), it is related closely to a quantity called the extraatomic relaxation energy R^{ea}, since $\Delta\alpha^*$ can be shown to equal $2\Delta R^{ea}(j)$. In other words, the change in Auger Parameter between two chemical states is equal to twice the change in the extraatomic relaxation energies in the two states, associated with the hole in the core level j. Thus α^* has an important physical basis as well as being useful analytically.

27.2.6. Applications

XPS has been applied in virtually every area in which the properties of surfaces are important. Given below are a few selected examples from some typical areas; for more comprehensive discussions of the applications of XPS, see [1], [3]–[9], [23], [29]–[33].

27.2.6.1. Catalysis

An understanding of the basic mechanisms of catalysis has been one of the major aims of XPS from its inception. To this end the technique has been used in two ways: (1) in the study of "real" catalysts closely resembling those used industrially; (2) in the study of "model" catalysts, where the number of parameters has been reduced by the use of well-characterized crystal surfaces.

For industrial catalysts the chemical information obtainable from XPS is all important, since the course and efficiency of the catalyzed reaction will be governed by the states of oxidation of the elements at the surface. Thus XPS is used to an-

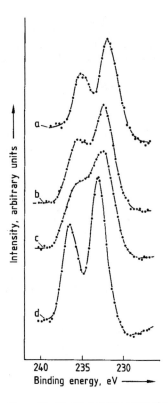

Figure 19. The Mo 3d XPS spectra of model catalysts in the Mo–Sm–O system [34]
Mo : Sm ratios are a) 0.25; b) 1.33; c) 4.00; d) 8.00

alyze both elemental and chemical composition as a function of bulk composition, of surface loading where the active material is deposited on a support, of oxidation and reduction treatments, and of time in reactor under operating conditions. Figure 19 demonstrates the effect on the chemical state of molybdenum in the surface of Mo–Sm–O catalysts of various compositions [34]. For the highest concentration of molybdenum (spectrum d) the Mo 3d doublet is very similar to that of MoO_3, but at the samarium-rich composition (spectrum a), it is more like that of a molybdate or a mixed oxide. The effect of successive reduction treatments on a cobalt–molybdenum–alumina catalyst is shown in the changes in the Mo 3d spectra of Figure 20 [35]. The air-fired catalyst in spectrum a contains entirely Mo(VI) at the surface, but with repeated hydrogen reduction the spectrum changes radically and can be curve resolved (Fig. 21) into Mo(VI), (V), and (IV) contributions. The chemical state of the surface in spectrum f contains approximately equal proportions of the three oxidation states.

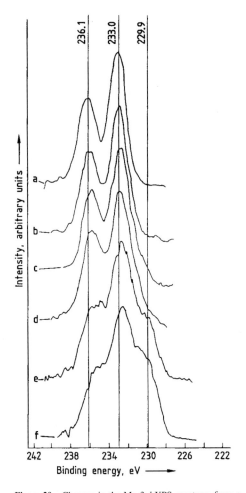

Figure 20. Changes in the Mo 3d XPS spectrum from a cobalt–molybdenum–alumina catalyst during successive reduction treatments in hydrogen at 500 °C [35]
a) Air-fired catalyst; b) Reduction time 15 min; c) 50 min; d) 60 min; e) 120 min; f) 200 min

Other recent applications of XPS to catalytic problems include catalyst–support interactions [36]–[38], variation of surface catalyst loading [39]–[42], effect of reduction [43]–[45], and characterization of catalyst active surface species [46]–[51].

27.2.6.2. Corrosion and Passivation

Corrosion products that are formed as thin layers on metal surfaces in either aqueous or gaseous environments, along with the nature and stability of passive and protective films on metals and alloys, have also been major areas of application of

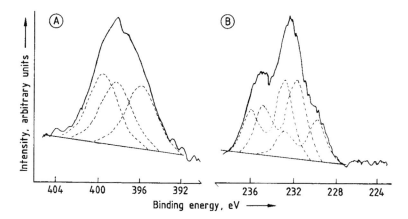

Figure 21. Examples of curve resolution of the Mo $3p$ (A) and $3d$ (B) spectral envelopes carried out by PATTERSON et al. [35] during the catalyst reduction treatment described in Figure 20

XPS. Just as in catalysis, XPS has been used in two ways, one in which materials corroded or passivated in the natural environment are analyzed, and another in which well-characterized, usually pure metal surfaces are studied after exposure to controlled conditions. Unlike catalysis, however, in which the two approaches tend to run parallel and not to overlap, in corrosion significant overlap and feedback from one to the other occur.

More effort has probably been devoted to study of the corrosion and passivation properties of Fe–Cr–Ni alloys such as stainless steels and other transition-metal alloys than to any other metallic system. The type of spectral information obtainable from an Fe–Cr alloy of technological origin, carrying an oxide and contaminant film after corrosion, is shown schematically in Figure 22 [52]. Actual XPS spectra recorded from a sample of the alloy Stellite-6 (a Co–Cr alloy containing some tungsten) treated in lithiated and borated water at 300 °C are shown as a function of treatment time in Figure 23. Stellite-6 is a hard alloy used in nuclear reactors. The high-temperature aqueous solution simulates light-water reactor conditions. The spectra reveal the removal of tungsten from the surface with increasing treatment time, a decrease in cobalt, and the appearance of nickel and iron deposited from stainless steel components elsewhere in the water circuit.

Spectra from an Fe–Cr alloy first polished, then passivated, and finally ion etched to remove the passive film are shown in Figure 24. Each spectrum has been resolved into the inelastic background and the various oxidation states of the element. The authors, MARCUS and OLEFJORD [53] were even able to separate the oxide and hydroxide contributions to the chromium spectrum from the passive film.

An example [54] of typical spectra recorded during comparison of the gaseous oxidation of pure chromium and of an Ni–Cr alloy is shown in Figure 25. There the Cr $2p$ spectra from the two materials have been recorded as a function of exposure to oxygen. Note that pure chromium does not oxidize as rapidly at room temperature as chromium in the alloy.

XPS has also been applied recently in this field to copper films for the study of microbial effects [55], the behavior of UO_2 in mineral water [56], corrosion-resistant amorphous alloys [57]–[59], and the hydrolysis of fluorozirconate glasses [60].

27.2.6.3. Adhesion

The adhesion of metal and ink to polymer, as well as the adhesion of paint and coating to metal, are of vital importance in several technologies. In the aircraft industry, aluminum-to-aluminum adhesion is being employed increasingly. The strength and durability of an adhesive bond depend absolutely on the way in which the adhesive compound interacts with the surfaces to which it is supposed to adhere, which in turn often involves pretreatment of the surfaces to render them more reactive. The nature and degree of the reactivity are functions of the chemical states of the adhering surfaces, states that can be monitored by XPS.

In electronic packaging technology, metal–polymer adhesion has been important for some years. Many studies have been devoted to the in-

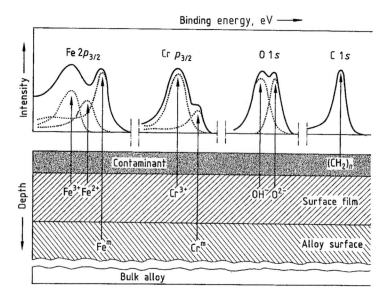

Figure 22. Schematic of type of information obtainable from XPS spectra from an Fe–Cr alloy with an oxide film underneath a contaminant film [52]

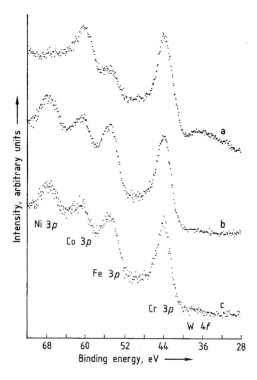

Figure 23. Transition-metal XPS spectra recorded from the surface of the alloy Stellite-6 after treatment for increasing lengths of time in lithiated and borated water at 300 °C [52]
a) 4 h; b) 168 h; c) 360 h

teractions between metals and polymers, and that of CHIN-AN CHANG et al. [61] is a good example. They deposited 200-nm-thick films of copper, chromium, titanium, gold, and aluminum on various fluorocarbon polymer films, and found that adhesion in terms of peel strength was highest for titanium. Figure 26 shows the reason why on the perfluoroalkoxy polymer tested, the C 1s spectra after deposition reveal that for titanium, and to a lesser extent chromium, a strong carbide bond is formed as evidenced by the peak at 282–283 eV. With the other metals, very low adhesion was found, and no carbide-like peaks occurred.

The path of failure of an adhesive joint can give information about the mechanism of failure if analysis of the elemental and chemical composition can be carried out along the path. Several authors have performed such analyses by loading the adhesive joint until it fractures and then using XPS to analyze each side of the fracture. Long-term weathering of the joint can be simulated by immersion in water at elevated temperature. The work of BOERIO and ONDRUS [62] demonstrates the nature of the results obtainable, an example of which is given in Figure 27. Silicon, carbon, and oxygen spectra were recorded from each side of the fracture face of an iron–epoxy resin joint before and after weathering in water for 7 d at 60 °C. Before weathering the failure path was as much

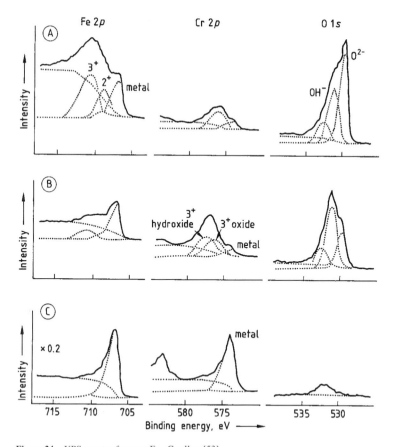

Figure 24. XPS spectra from an Fe–Cr alloy [53]
A) Mechanically polished, exposed in H$_2$O and air; B) Passivated in 0.5 M H$_2$SO$_4$ at 740-mV standard hydrogen electrode for 1 h; C) Ion etched ≈ 10 nm

through the adhesive itself as through the iron oxide–epoxy resin interface, but after weathering, failure occurred almost entirely near the interface.

Also studied by XPS recently have been the measurement of energies of adhesion of aluminum and silicon on molybdenum [63], the ion-enhanced adhesion of nickel to glassy carbon [64], the effects of pretreatments on paint–metal adhesion [65], and the adhesion of titanium thin films to various oxide surfaces [66].

27.2.6.4. Superconductors

In the years since the discovery of superconducting oxides in 1986, more papers have probably been published describing the application of XPS to superconductors than in any other field. One reason for this frenzied activity has been the search for the precise mechanism of superconductivity in new materials, and an essential piece of information concerns the chemical states of the constituent elements, particularly copper, which is common to all of them. Some theories predict that small but significant amounts of the Cu^{3+} state should occur in superconductors, but other models disagree. Comparison by STEINER et al. [67] of the Cu $2p_{3/2}$ spectra from copper compounds containing copper as Cu^0 (metal), Cu$^+$, Cu^{2+}, and Cu^{3+}, with the spectra from the superconducting oxide YBa$_2$Cu$_3$O$_7$ and from the base material La$_2$CuO$_4$, showed conclusively that the Cu^{3+} contribution to the superconducting mechanism must be negligible. Some of their comparative spectra are shown in Figure 28. The positions of the Cu $2p_{3/2}$ peak in YBa$_2$Cu$_3$O$_7$ and La$_2$CuO$_4$, along with the satellite structure, are very similar to that in CuO (i.e., Cu^{2+}, rather than in NaCuO$_2$, which contains Cu^{3+}).

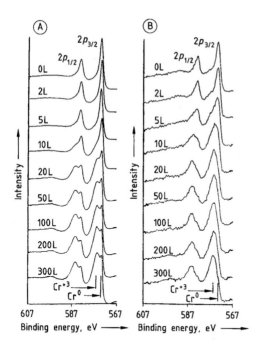

Figure 25. The Cr $2p$ XPS spectra during exposure to oxygen at room temperature [54] of
A) Pure Cr; B) Ni–Cr alloy
L = Langmuir (10^{-6} Torr · s)

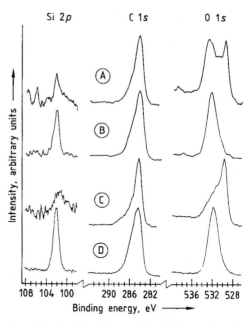

Figure 27. The Si $2p$, C $1s$, and O $1s$ XPS spectra from the adherend and adhesive sides of an iron–epoxy resin joint before and after weathering treatment [62]
A) and B) Adherend and adhesive before weathering; C) and D) Adherend and adhesive after immersion in water at 60 °C for 7 d

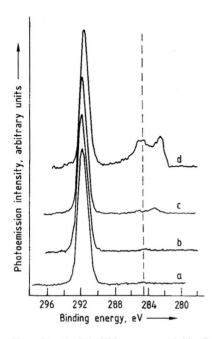

Figure 26. The C $1s$ XPS spectra recorded by CHIN-AN CHANG et al. [61] from perfluoroalkoxy polymer (PFA)
a) Before deposition; b) Deposition of copper; c) Deposition

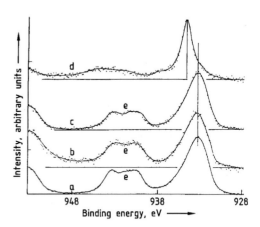

Figure 28. Comparison of the Cu $2p_{3/2}$ and satellite XPS spectra from several copper compounds with the spectrum from the superconducting oxide YBa$_2$Cu$_3$O$_7$ [67]
a) CuO, $\Gamma = 3.25$ eV; b) La$_2$CuO$_4$, $\Gamma = 3.30$ eV;
c) YBa$_2$Cu$_3$O$_7$, $\Gamma = 3.20$ eV; d) NaCuO$_2$, $\Gamma = 1.60$ eV;
e) Cu^{2+} satellites

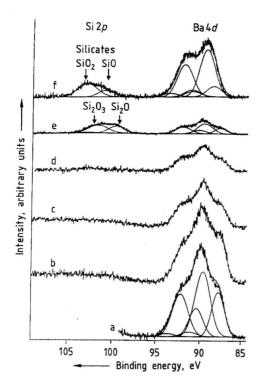

Figure 29. The Si $2p$ and Ba $4d$ XPS spectra [68] from the a) Clean surface of the superconducting oxide $YBa_2Cu_3O_{7-x}$; b)–e) Same surface after increasing deposition of b) 0.8 nm Si; c) 2.8 nm Si; d) 4.4 nm Si; e) 6.0 nm Si; f) Deposited surface after heating to 250 °C in oxygen (8×10^{-3} mbar)

Recent effort in superconductor research has been devoted mostly to attempts to fabricate thin films with the same superconducting properties as the bulk material, and to prepare suitable electrical contact to the surfaces of superconductors. XPS has been of great use in both areas through its ability to monitor both composition and chemical state during the processes either of thin superconducting film formation, or of the deposition of thin conducting and semiconducting films onto a superconducting surface. Typical of the latter studies is that of ZIEGLER et al. [68], who deposited increasing amounts of silicon onto epitaxial films of $YBa_2Cu_3O_{7-x}$. Figure 29 from their paper shows the Si $2p$ and Ba $4d$ spectra during the deposition of silicon and after heating the films in oxygen at 250 °C once 6 nm of silicon had accumulated. The Ba $4d$ spectrum at the bottom can be resolved into three $4d_{5/2,3/2}$ doublets, corresponding — with increasing energy — to barium in the bulk superconductor, in the surface of the superconductor, and as carbonate. With increasing silicon thickness the barium spectrum attenuates but does not change in character, but silicon does not appear in the spectrum until nearly 5 nm has been deposited, suggesting migration into the superconductor at room temperature. At a thickness of 6 nm, the Si $2p$ spectrum can be separated into suboxide contributions, indicating reaction of silicon with the $YBa_2Cu_3O_{7-x}$ surface. After heating in oxygen the Si $2p$ spectrum changes to that characteristic of SiO_2 and silicates, while the Ba $4d$ spectrum also changes by increasing in intensity and conforming mostly to that expected of a barium silicate. As a result of the latter changes the superconducting properties of the film were destroyed. The Y $3d$ and Cu $2p$ spectra establish that yttrium and copper oxides are formed as well.

Other surface reactions studied have been those of gold [69] and silver [70] with Bi_2Sr_2Ca-Cu_2O_{8+y}, of niobium with $YBa_2Cu_3O_{7-x}$ [71], and of lead with $YBa_2Cu_3O_{7-x}$ [72]. Characteristics of superconductors in the form of thin films prepared in several ways have been recorded by XPS for $YBa_2Cu_3O_{7-x}$ [73]–[77], $GdBa_2Cu_3O_{7-x}$ [78], $Nd_{2-x}Ce_xCuO_{4-y}$ [79], $Rb_xBa_{1-x}BiO_3$ [80], and $Tl_2Ca_2Ba_2Cu_3O_{10}$ [81]. Detailed electronic studies have been made of the bulk superconductors $Bi_2Sr_2Ca_{1-x}Nd_xCu_2O_y$ [82]–[83], $Nd_{2-x}Ce_x$-CuO_{4-y} [84]–[87], $ErBa_2Cu_4O_8$ [88], and $Tl_2Ba_2CaCu_2O_8$ [89].

27.2.6.5. Interfaces

The chemical and electronic properties of elements at the interfaces between very thin films and bulk substrates are important in several technological areas, particularly microelectronics, sensors, catalysis, metal protection, and solar cells. To study conditions at an interface, depth profiling by ion bombardment is inadvisable since both composition and chemical state can be altered by interaction with energetic positive ions. Normal procedure is therefore to start with a clean or other well-characterized substrate and deposit the thin film onto it slowly at a chosen temperature while XPS monitors the composition and chemical state by recording selected characteristic spectra. The procedure continues until no further spectral changes occur, as a function either of film thickness, of time elapsed since deposition, or of changes in substrate temperature.

A good example is the study of the reaction at the interface of the rare-earth element thulium with the silicon (111) surface, carried out by GOKHALE et al. [90]. Figure 30 shows the Si $2p$ spec-

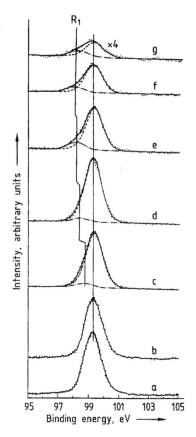

Figure 30. The Si $2p$ XPS spectra with Mg K_α excitation during deposition of thulium to a thickness of 12.5 nm at room temperature [90]
a) Clean silicon; b) 0.2 nm Tm; c) 0.5 nm Tm; d) 2 nm Tm; e) 4 nm Tm; f) 6.5 nm Tm; g) 12.5 nm Tm

trum from the substrate as thulium is deposited on it at room temperature. With the Mg K_α radiation used for the analysis, the kinetic energy of the Si $2p$ electrons is about 1155 eV, corresponding to an inelastic mean free path of about 2 nm. The Si $2p$ peak is still plainly detectable after 12.5 nm of thulium have been deposited, and in addition a new feature appears on the low binding energy side of the peak, increasing in magnitude and shifting to lower energy with thickness. Raising the temperature in stages of the thickest film and use of X-ray diffraction, established that the energetic final position labeled R_1 corresponded to Tm_5Si_3 and the initial energetic position found at a thickness of 0.5–1.0 nm, to $TmSi_2$. However, the magnitude of the residual Si $2p$ peak was too great at all stages to be accounted for by compound formation alone, indicating the formation

of thulium clusters followed by their coalescence into islands. Measurement of the Tm $4d_{5/2}$ binding energy during the interaction also showed that the valence of thulium was 3+ at all temperatures and coverages.

Recent studies of interfaces by XPS have also included those of SiO_2 [91], [92], samarium [93], and amorphous hydrogenated carbon [94] on silicon; of silver, indium, and aluminum on GaP [95]; of tin [96] and RbF–Ge interlayers [97] on GaAs; of platinum on TiO_2 [98]; of titanium, niobium, and nickel on Al_2O_3[99]; of chromium and iron on W(110) [100]; and of ZrO_2 on nickel [101].

Other important areas of application of XPS have included polymers [102]–[106], lubricating films [107], [108], glasses [109]–[113], ion implantation [114]–[118], and cleaning [119]–[121] and passivation [122]–[123] of semiconductor surfaces.

27.3. Auger Electron Spectroscopy (AES)

After XPS, AES is the next most widely used surface analytical technique. As an accepted surface technique, AES actually predates XPS by two to three years, since the potential of XPS as a surface-specific technique was not recognized immediately by the surface science community. Pioneering work was carried out by HARRIS [124] and by WEBER and PERIA [125], but the technique as it is known today is basically the same as that established by PALMBERG et al. [126].

27.3.1. Principles

The surface to be analyzed is irradiated with a beam of electrons of sufficient energy, typically in the range of 5–30 keV, to ionize one or more core levels in surface atoms. After ionization the atom can then relax by either of the two processes described in Section 27.2.1 for XPS, that is, ejection of a characteristic X-ray photon (fluorescence) or ejection of an Auger electron. Although these are competing processes, for shallow core levels ($E_B < 2$ keV) the probability of the Auger process is far higher. The Auger process is described schematically in Figure 1, which points out that the final state of the atom is doubly ionized and that the kinetic energy of the electron arising from the Auger transition ABC is given by

$$E_{ABC} = E_A - E_B - E_C^* \tag{12}$$

where E_A, E_B, and E_C are binding energies of the A, B, and C levels in the atom, E^*_C being starred because it is the binding energy of C modified by an existing hole in B.

Equation (12) shows that the Auger energy is a function only of atomic energy levels. Since no two elements have the same set of atomic binding energies, analysis of Auger energies provides elemental identification. Even if levels B and C are in the valence band of the solid, analysis is still possible because the dominant term in Equation (12) is always the binding energy of A, the initially ionized level.

Equation (12) also shows that the heavier the element (i.e., the greater the number of atomic energy levels), the more numerous are the possible Auger transitions. Fortunately, large differences exist in the probabilities between different Auger transitions, so that even for the heaviest elements, only a few intense transitions occur, and analysis is still possible.

In principle, chemical as well as elemental information should be available in AES, since the binding energies appearing in Equation (12) are subject to the same chemical shifts as measured in XPS. In practice, since the binding energies of three levels are involved, extracting chemical information from Auger spectra is, in most cases, still too difficult. However, significant progress is being made in this area, and improved theory and modern data processing methods are likely to enable valuable chemical information to be derived soon.

The reasons AES is one of the surface-specific techniques have been given in Section 27.2.1, with reference to Figure 2. The normal range of kinetic energies recorded in an AES spectrum would typically be from 20 to 1000 eV, corresponding to inelastic mean free path values of 2 to 6 monolayers.

The nomenclature used in AES has also been mentioned in Section 27.2.1. The Auger transition in which initial ionization occurs in level A, followed by the filling of A by an electron from B and ejection of an electron from C, would therefore be labeled ABC. In this rather restricted scheme, one would thus find in the KLL series the six possible transitions KL_1L_1, KL_1L_2, KL_1L_3, KL_2L_2, KL_2L_3, and KL_3L_3. Other combinations could be written for other series such as the LMM, MNN, etc.

27.3.2. Instrumentation

27.3.2.1. Vacuum Requirements

The same considerations discussed in Section 27.2.2.1 for XPS apply to AES.

27.3.2.2. Electron Sources

The energy of the exciting electrons does not enter into Equation (12) for the Auger energy, unlike that of the exciting X-ray photon in XPS; thus the energy spread in the electron beam is irrelevant. What is relevant is the actual energy of the primary electrons because of the dependence on that energy of the cross section for ionization by electron impact. For an electron in a core level of binding energy E_B, this dependence increases steeply with primary energy above E_B, passes through a maximum about 3–5 times E_B, and then decreases to a fairly constant plateau. It is illustrated in Figure 31, in which various theoretical and experimental cross sections are shown for ionization of the Ni K shell, from CASNATI et al. [127]. The implication of this dependence is that for efficient ionization of a particular core level, the primary energy should be about five times the binding energy of an electron in that level. Since in most samples for analysis, several elements are found in the surface region with core-level binding energies extending over a large range, choice of a primary energy that is sufficiently high to ionize all the core levels efficiently is advisable. Hence for *conventional AES*, primary energies are typically in the range 5–10 keV. In *scanning Auger microscopy (SAM)*, much higher primary energies are used, in the range of 25–50 keV, because in that version of AES as small an electron spot size must be achieved on the sample as possible, and the required focusing can be effected only at such energies.

For the primary energy range 5–10 keV used in *conventional AES*, the electron emitter is thermionic, usually a hot tungsten filament, and focusing of the electron beam is carried out electrostatically. Typically, such an electron source would be able to provide a spot size on the specimen of about 0.5 µm at 10 keV and a beam current of about 10^{-8} A. The beam can normally be rastered over the specimen surface, but such a source would not be regarded as adequate for SAM. Sources for *SAM* may be of either the thermionic or the field emission type, the latter being partic-

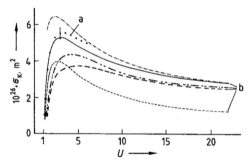

Figure 31. Electron impact ionization cross section for the Ni K shell, as a function of reduced electron energy U [127] $U = E_p/E_K$, E_p being the primary electron energy and E_K the binding energy of the K shell
a) Experimental points; b) Semiempirical or theoretical curves

ularly advantageous in achieving minimum spot sizes since it is a high-brightness source. Focusing is performed electromagnetically, and at optimum performance a modern electron gun for SAM would be capable of a spot size of 15 nm at 30 keV and a beam current of about 10^{-10} A. The beam is rastered over the surfaces at scan rates variable up to TV rates, with the dimensions of the scanned area variable as well.

27.3.2.3. Electron Energy Analyzers

The electron energy analyzer found to be most suitable for AES, the cylindrical mirror analyzer (CMA), has already been described in Section 27.2.2.4. The important property of this analyzer for AES is its very high transmission, arising from the large solid angle of acceptance; for a typical angular spread $\delta\alpha$ of 6°, the transmission is about 14 %. Since the source area (i.e., spot size) in AES is so small, as many Auger electrons as possible must be accepted into the analyzer, compatible with adequate resolution.

An example of the arrangement of a single-stage CMA in a modern system for AES analysis is shown in Figure 32, taken from [128]. The double-pass CMA described in Section 27.2.2.4 (see Fig. 8) can also be used for AES by grounding the spherical retardation meshes and the inner cylinders and by setting the variable internal apertures to minimum. Although the double-pass CMA is not ideal for either AES or XPS, use of the same analyzer, in its different modes of operation, for both techniques is very convenient. Similarly, the concentric hemispherical analyzer (CHA), also described in Section 27.2.2.4, is often used for both techniques for the sake of convenience, even though it is much more suitable for XPS than AES.

Until recently, all Auger spectra were presented in energy distributions differentiated with respect to energy, rather than in the direct energy distributions used in XPS [i.e., $dN(E)/dE$ rather than $N(E)$]. The effect of differentiating is to enhance the visibility of Auger features, as demonstrated in Figure 33, in which the differential distribution of secondary electron energies from boron is shown above the corresponding $N(E)$ distribution. The differentiation was performed by a synchronous detection technique, in which a small voltage modulation at high frequency was superimposed on the deflecting potential as the latter was scanned, and then the second harmonic of the detected signal was amplified preferentially. An indication of this technique is given in Figure 32. Nowadays the $N(E)$ distribution is usually recorded digitally, and the differentiation is carried out by computer. However, with the rapid improvement in the signal-to-noise characteristics of detection equipment, the trend in presenting Auger spectra is increasingly toward the undifferentiated $N(E)$ distribution.

Note from Figure 33 that although the true position of the boron KLL Auger peak in the $N(E)$ distribution is at 167 eV, the position in the $dN(E)/dE$ distribution is taken for purely conventional reasons to be that of the negative minimum (i.e., at 175 eV). The other spectral features in Figure 33 are described in Section 27.3.3.

27.3.3. Spectral Information

As stated above, the range of primary energies used typically in conventional AES is 5–10 keV, and the ionization cross section passes through a maximum at 3–5 times the binding energy of an electron in the ionized core level. Therefore the core levels that can be ionized efficiently are limited. Thus K-shell ionization efficiency is adequate up to $Z \approx 14$ (silicon), L_3 up to $Z \approx 38$ (strontium), M_5 up to $Z \approx 76$ (osmium), and so on (for the much higher primary energies used in SAM, the ranges are correspondingly extended). This means that in various regions of the periodic table, characteristic Auger transitions occur that are most prominent under typical operating conditions.

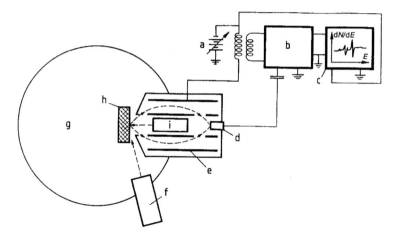

Figure 32. Typical experimental arrangement for AES using a single-pass CMA [128]
a) Voltage sweep; b) Lock-in amplifier; c) Display; d) Electron multiplier; e) Cylindrical mirror analyzer; f) Ion gun; g) UHV chamber; h) Sample; i) Electron gun

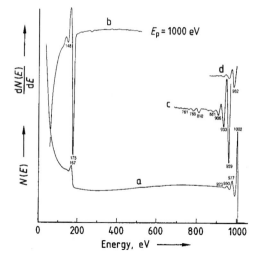

Figure 33. Secondary electron distribution from boron
a) In the $N(E)$ mode; b)–d) In the $dN(E)/dE$ mode, where the differentiation reveals the plasmon satellites

Figures 34–36 show some of these prominent Auger features. They have all been taken from [129], and recorded using a CMA, in the differential distribution. Examples of the KLL Auger series are shown in Figure 34, of the LMM series in Figure 35, and of the MNN series in Figure 36. In the KLL series the most prominent feature is a single peak arising from the $KL_{2,3}L_{2,3}$ transition, with minor features from other Auger transitions to lower energy. The LMM series is characterized by a triplet arising from $L_3M_{2,3}M_{2,3}$, $L_3M_{2,3}M_{4,5}$, and $L_3M_{4,5}M_{4,5}$ transitions in ascending order of kinetic energy, with another intense peak, the $M_{2,3}M_{4,5}M_{4,5}$, appearing at very low energy. The characteristic doublet seen in the MNN series arises from the $M_{4,5}N_{4,5}N_{4,5}$ transitions, in which the doublet separation is that of the core levels M_4 and M_5.

Note that AES transitions involving electrons in valence band levels are often written with a V rather than the full symbol of the level. Thus $L_3M_{2,3}M_{4,5}$ and $L_3M_{4,5}M_{4,5}$ would often appear as $L_3M_{2,3}V$ and L_3VV, respectively, and similarly $M_{2,3}M_{4,5}M_{4,5}$ as $M_{2,3}VV$. In Figure 35 the increase in intensity of the L_3VV peak relative to the other two, upon going from chromium to iron, is due to the progressive increase in the electron density in the valence band.

Chemical effects in Auger spectra are quite commonly observed, but difficult to interpret compared to those in XPS since additional core levels are involved in the Auger process. Some examples of the changes to be seen in the KLL spectrum of carbon in different chemical environments are given in Figure 37 [130].

Associated with prominent features in an Auger spectrum are the same types of energy loss feature, the plasmon losses, that are found associated with photoelectron peaks in an XPS spectrum. As described in Section 27.2.3, plasmon energy losses arise from excitation of modes of collective oscillation of the conduction electrons by outgoing secondary electrons of sufficient energy. Successive plasmon losses suffered by the backscattered

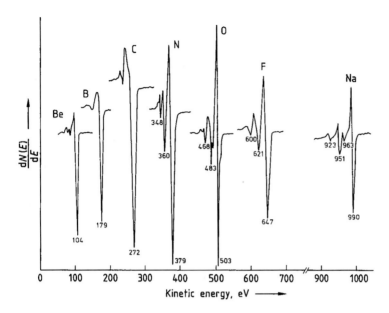

Figure 34. KLL Auger series characteristic of the light elements [129]

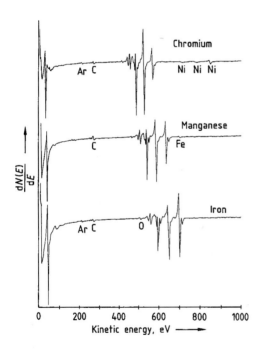

Figure 35. LMM and MMM Auger series in the middle of the first series of transition elements [129]

primary electrons can be seen in both the $N(E)$ and the $dN(E)/dE$ spectra of Figure 33, from boron. The same magnitude of plasmon loss, ≈ 27 eV, is also associated with the KLL Auger peak of boron. Plasmon losses are also present in Figures 34 35 36 37, but they are difficult to disentangle from minor Auger features.

Another type of loss feature not often recorded in Auger spectra can also be seen in Figure 33, in the $dN(E)/dE$ spectrum at 810 eV. This type arises from core level ionization and forms the basis for the technique of core electron energy loss spectroscopy (CEELS). A primary electron interacting with an electron in a core level can cause excitation either to an unoccupied continuum state, leading to complete ionization, or to localized or partly localized final states, if available. Since the primary electron need not give up all its energy in the interaction, the loss feature appears as a step in the secondary electron spectrum, with a tail decreasing progressively to lower energies. Differentiation, as in Figure 33, accentuates the visibility. In that figure, the loss at 810 eV corresponds to the K-shell ionization edge of boron at about 190 eV.

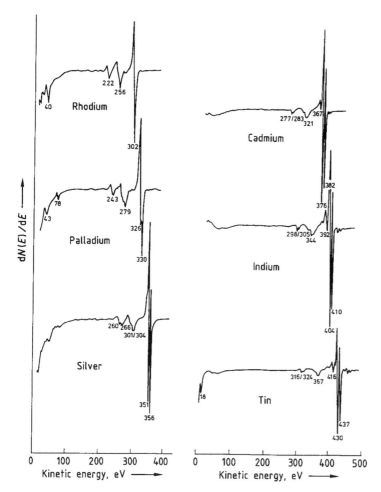

Figure 36. MNN Auger series in the second series of transition elements [129]

27.3.4. Quantification and Depth Profiling

27.3.4.1. Quantification

If ionization of a core level X in an atom A in a solid matrix M by a primary electron of energy E_p gives rise to the current I_{XYZ} of electrons produced by the Auger transition XYZ, then the Auger current from A is

$$I_{XYZ} = K \cdot \sigma(E_p{:}E_B)$$
$$\cdot \{1 + r_M(E_B{:}\alpha)\} \bar{N}_A \cdot \lambda_M(E_{XYZ}) \cdot \cos\theta \quad (13)$$

where $\sigma(E_p{:}E_B)$ is the cross section for ionization of level X with binding energy E_B by electrons of energy E_p; r_M the backscattering factor that takes account of the additional ionization of X with binding energy E_B by inelastic electrons with energies between E_p and E_B; α the angle of incident electron beam to surface normal; $\bar{N}_A$ the atomic density of A averaged over depth of analysis; $\lambda_M(E_{XYZ})$ the inelastic mean free path in matrix M containing A, at the Auger kinetic energy E_{XYZ}; and θ the angle of Auger electron emission to surface normal.

Similarly to Equation (8), K is a constant of proportionality containing fixed operational parameters such as incident electron current density, transmission of the analyzer at the kinetic energy E_{XYZ}, efficiency of the detector at the kinetic energy E_{XYZ}, and the probability of the Auger transition XYZ.

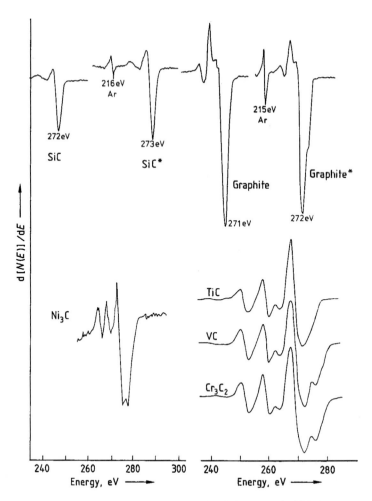

Figure 37. Examples of the effect of different chemical states on the KLL Auger spectrum of carbon [130] (SiC* and graphite* denote Ar$^+$-bombarded surfaces of SiC and graphite, respectively)

The cross section for electron impact ionization has already been mentioned in Section 27.3.2.2 in connection with electron sources, and various experimental and theoretical cross sections have been shown in Figure 31 for the particular case of the K shell of nickel. The expression for the cross section derived by CASNATI et al. [127] gives reasonably good agreement with experiment, with the earlier expression of GRYZINSKI [131] also being useful.

For the inelastic mean free path, Equation (9) given for XPS also applies.

The new factor in Equation (8) compared to that for XPS is the backscattering factor r_M. ICHIMURA and SHIMIZU [132] have carried out extensive Monte Carlo calculations of the backscattering factor as functions of primary beam energy and angle of incidence, of atomic number and of binding energy, and a selection of their results for $E_p = 10$ keV at normal incidence is shown in Figure 38. The best fit of their results to experiment gives the following relationship:

$$r = (2.34 - 2.1Z^{0.14})U^{-0.35}$$
$$+ (2.58Z^{0.14} - 2.98) \quad \text{for} \quad \alpha = 0°$$

$$r = (0.462 - 0.777Z^{0.20})U^{-0.32}$$
$$+ (1.5Z^{0.20} - 1.05) \quad \text{for} \quad \alpha = 30°$$

$$r = (1.21 - 1.39Z^{0.13})U^{-0.33}$$
$$+ (1.94Z^{0.13} - 1.88) \quad \text{for} \quad \alpha = 45° \quad (14)$$

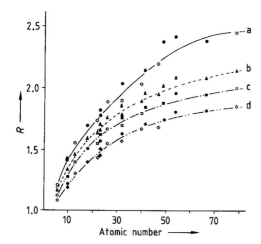

Figure 38. Values of the backscattering factor R (i.e., $1+r$) calculated as a function of atomic number for an electron beam of 10 keV at normal incidence by ICHIMURA et al. [132]
a) $E_B = 0.1$ keV; b) $E_B = 0.5$ keV; c) $E_B = 1.0$ keV; d) $E_B = 2.0$ keV

where Z is the atomic number and $U = E_p/E_B$.

Thus, as for XPS, in principle the average surface concentration $\bar{N}_A$ can be calculated from the measurement of the Auger current, according to Equation (13). Again, as in XPS, relative sensitivity factors are generally used. The Auger current $\bar{N}_A^s$ for the same transition XYZ in a standard of pure A is measured under the same experimental conditions as in the analysis of A in M, whereupon the ratio of Auger currents is

$$\bar{N}_A/\bar{N}_A^s = \frac{I_{XYZ}(1+r_S)\lambda_S}{I_{XYZ}^s(1+r_M)\lambda_M} \quad (15)$$

since the ionization cross section σ is the same for both standard and sample. Then, as before, Equation (15) can be reduced to

$$\bar{N}_A = S_A I_{XYZ} \quad (16)$$

where

$$S_A = \frac{\bar{N}_A^s(1+r_S)\lambda_S}{I_{XYZ}^s(1+r_M)\lambda_M} \quad (17)$$

S_A is then the relative sensitivity factor. Normally, values of S_A are derived empirically or semiempirically. Tables of such sensitivity factors have been published by PAYLING [133] and by MROCZKOWSKI and LICHTMAN [134] for differential Auger spectra, and for the direct, undifferentiated, spectra by SATO et al. [135]. A comprehensive discussion of the assumptions and simplifications involved and the corrections that should be applied is given in [136], [22]. Because many workers still measure Auger current as the peak-to-peak height in the differential distribution, which takes no account of peak width or area, quantification in AES is still not as accurate as that in XPS.

27.3.4.2. Depth Profiling

As in XPS, elemental distributions near the surface, but to depths greater than the analytical depth resolutions, are frequently required when using AES. Again, the universally employed technique is that of progressive removal of the surface by ion sputtering, with simultaneous or quasi-simultaneous analysis of the exposed surface. Since the peak-to-peak height of an Auger peak in the differential distribution can be recorded very quickly by a CMA, it is normal in AES depth profiling, when only 10–12 individual peaks need to be recorded, to perform simultaneous sputtering and analysis. Such an operation is called multiplexing; the CMA is programmed to switch rapidly from one energetic peak position to the next, spectral recording being within a chosen energy window at each position. Should many more peaks be needed, or an entire energy spectrum, the ion erosion must be stopped during acquisition, and the profiling is quasi simultaneous as in XPS.

In modern ion guns using differential pumping, the pressure of argon (the gas normally used) in the analysis chamber can be maintained below 5×10^{-5} Pa during depth profiling, thus reducing the possibility of damage to the electron detector and the electron gun filament in the CMA. Such a gun is shown schematically in Figure 39. The source region around the filament (g) and anode (f) is differentially pumped, and positive ions are extracted through a cone-shaped extractor (b), then accelerated through a small aperture (e) to an energy selected between 0.5 and 5 keV. Focusing occurs in two tubular lenses (c) down to a spot size of 10 µm at 5 keV, and the beam can be rastered by two pairs of deflector plates (d) at the end of the gun. The area of raster on the specimen can be chosen between 1×1 mm^2 and 10×10 mm^2. According to the emission from the coated nonburnout iridium filament, the beam current can be selected from 1 to 10 µA.

Since the ion intensity is not uniform across the beam from an ion gun, but approximately Gauss-

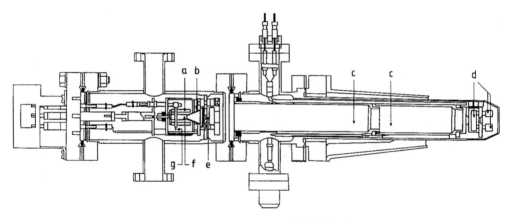

Figure 39. Ion source using a thermionic emitter in a collision chamber for ionization
Courtesy of Leybold
a) Electron repeller; b) Extractor; c) Tubular lenses; d) Deflector plates; e) Aperture; f) Anode; g) Filament

ian in distribution, in AES depth profiling either the area of the focused electron beam on the sample must be coincident with the region of maximum ion intensity in the focused ion beam or, preferably, the ion beam must be rastered. The analyzed area should then be centered within the rastered area to ensure that the flat bottom of the sputtered crater is analyzed and not the sloping sides.

The depth resolution achievable during profiling depends on many variables, and the reader is referred, as before, to the comprehensive discussion in [25].

A variation on depth profiling that can be performed by modern scanning Auger instruments (see Section 27.3.5) is to program the incident electron beam to jump from one preselected position on a surface to each of many others in turn, with multiplexing at each position. This is called *multiple point analysis*, which provides an elemental map after each sputtering step or each period of continuous sputtering. The set of maps can then be related to each other in a computer frame-store system to derive a three-dimensional analysis of a selected micro volume.

27.3.5. Scanning Auger Microscopy (SAM)

If an incident electron beam of sufficient energy for AES is rastered over a surface in a manner similar to that in a scanning electron microscope (SEM), and if the analyzer, usually a CMA, is set to accept electrons of Auger energies characteristic of a particular element, then a visual display whose intensity is modulated by the Auger peak intensity will correspond to the distribution of that element over the surface. The result is called an *Auger map or image*. Care must be taken in interpreting the intensity distribution, since the Auger intensity depends not only on the local concentration of the element but on the topography as well, because surface roughness can affect the inelastic background underneath the Auger peak. Therefore Auger maps are customarily presented as variations of the ratio of peak intensity divided by the magnitude of the background on the high-energy side, which can be carried out easily by computer.

In SAM the thrust of development is toward ever-better spatial resolution, which means the smallest possible focused spot size on the sample compatible with adequate Auger signal-to-noise ratio in the image, for acquisition within a reasonable length of time. Electrostatic focusing of the incident electron beam is inadequate on its own, and the electron guns used for SAM all employ electromagnetic focusing. Similarly, focusing electrons of low energy into a sufficiently small spot is difficult, so the energies that must be used in SAM are in the range of 25–50 keV, with beam currents of the order of 1 nA or less. An electron gun operating at such energies and currents with electromagnetic focusing would be capable of producing a minimum spot size of about 15 nm. An illustration of the application of SAM at moderate resolution is shown in Figure 40 [137]. The surface was that of a fractured compact of SiC to which boron and carbon had been added to aid the sintering process. The aim of the analysis was to establish the uniformity of distribution of the additives and the presence or absence of impurities. The *Auger maps* show not only very nonuniform distribution of boron (Fig. 40 A) but also a strong correlation of boron with sodium (Fig. 40 C), and a weaker correlation of boron with potassium (Fig. 40 B). Point analyses shown to the right at points A and B marked on the images reveal the presence of sulfur and calcium in some areas as well. In this case the sintering process had not been optimized.

27.3.6. Applications

Like XPS, the application of AES has been very widespread, particularly in the earlier years of its existence; more recently, the technique has been applied increasingly to those problem areas that need the high spatial resolution that AES can provide and XPS as yet cannot. Since data acquisition in AES is faster than in XPS, it is also employed widely in routine quality control by surface analysis of random samples from production lines of, for example, integrated circuits.

27.3.6.1. Grain Boundary Segregation

One of the original applications of AES, and still one of the most important, is the analysis of grain boundaries in metals and ceramics. Very small amounts of impurity or dopant elements in the bulk material can migrate under appropriate temperature conditions to the boundaries of the grain structure and accumulate there. In that way the concentration of minor elements at the grain boundaries can become much higher than in the bulk, and the cohesive energy of the boundaries can be so altered that the material becomes brittle. Knowledge of the nature of the segregating elements and their grain boundary concentrations as a function of temperature can be used to modify fabrication conditions and thereby improve the strength performance of the material in service.

GORETZKI [138] has discussed the importance and the analysis of internal surfaces such as grain boundaries and phase boundaries, and given several examples. He emphasizes that to avoid ambiguity in interpretation of the analysis, the internal surfaces must be exposed by fracturing the material under UHV conditions inside the electron spectrometer. If this is not done, atmospheric contamination would immediately change the surface condition irrevocably. One of the examples from GORETZKI's paper is given in Figure 41, in which Auger spectra in the differential mode from fracture surfaces in the embrittled (Fig. 41 A) and unembrittled (Fig. 41 B) states of a 12 % Cr steel containing phosphorus as an impurity, are compared. Embrittlement occurs when the steel is held at 400–600 °C for a long time. Note in the spectrum from the embrittled material not only the greatly enhanced phosphorus concentration at the grain boundary, but also increased amounts of chromium and nickel with respect to iron. By setting the analyzer energy first to that of the phosphorus peak, and then to that of the principal chromium peak, Auger maps could be recorded showing the distribution of phosphorus and chromium over the fracture surface. The two maps showed high correlation of the two elements, indicating association of phosphorus and chromium at the grain boundary, possibly as a compound.

An example of the application of SAM to fracture in ceramic materials has already been given in Section 27.3.5 (Fig. 40). Other recent applications in the study of grain boundary segregation in metals have included the effects of boron, zirconium, and aluminum on Ni_3Al intermetallics [139], the segregation of phosphorus and molybdenum to grain boundaries in the superalloy Nimonic PE 16 as a function of aging treatment [140], and the effect of addition of aluminum on the concentration of oxygen at grain boundaries in doped and sintered molybdenum rods [141].

27.3.6.2. Semiconductor Technology

With ever-increasing miniaturization of integrated circuits goes a need for the ability to analyze ever smaller areas so that the integrity of fabrication at any point on a circuit can be checked. This checking process includes not only microdetails such as continuity of components in the form of thin films and establishment of the correct elemental proportions in contacts, diffusion barriers, and Schottky barriers, but also the effectiveness of more macrotreatments such as surface cleaning. In all these types of analyses, depth profiling (see Section 27.3.4.2) is used extensively because, in general, elemental compositional information is required, not information about chemical state. As mentioned previously, chemical state information tends to become blurred or destroyed by ion bombardment.

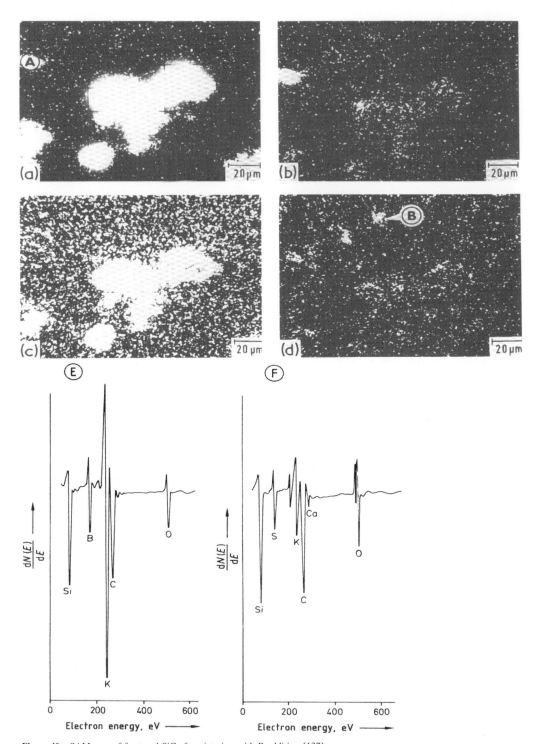

Figure 40. SAM map of fractured SiC after sintering with B addition [137]
a)–d) Elemental maps in boron, potassium, sodium, and oxygen, respectively; E), F) Point analyses at points ○A and ○B, respectively

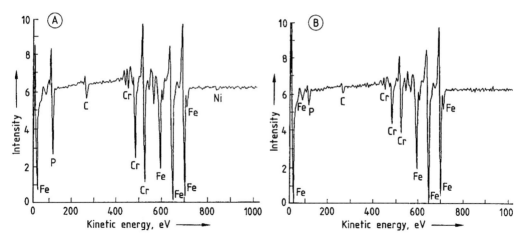

Figure 41. Auger spectra from fracture surfaces of a 12% Cr steel [138]
A) Embrittled state; B) Unembrittled state

Various materials have been prescribed as *diffusion barriers* to prevent one circuit material from diffusing into another (usually Si) at the temperatures needed for preparation of certain components such as oxide films in integrated circuits. Thus a Ta–N film has been found to act as an efficient barrier between the metallizing alloy Al_3Ta and a silicon substrate, although SASAKI et al. [142] have shown that the barrier properties depend on the stoichiometry of the Ta–N film. These authors prepared structures of 30 nm of Ta–N on silicon, followed by 200 nm of Al_3Ta on top of the Ta–N, with the Ta–N intermediate having various stoichiometries. The composite structures were then heated to 500 and 600 °C, and analyzed by AES with depth profiling. The results for a Ta–N intermediate barrier formed by sputtering in an Ar + 5% N_2 mixture are shown in Figure 42. The profiles before heat treatment (Fig. 42 A) and after 550 °C treatment (Fig. 42 B) indicate sharp interfaces between the various layers, with little interdiffusion. After 600 °C heat treatment (Fig. 42 C), however, very substantial intermixing has taken place, and the interfaces have virtually disappeared. Clearly the Ta–N film has not acted as an effective diffusion barrier above 550 °C. On increasing the N_2 content in the sputtering gas to Ar + 10% N_2, and performing the same profiles, no intermixing was evident even at 600 °C, demonstrating that Ta–N can certainly be used as a diffusion barrier, but that care must be taken to achieve a minimum N : Ta stoichiometry. Ti–N is also used as a diffusion barrier between aluminum and silicon, and the analysis of Ti–N stoichiometry in contact holes in advanced integrated circuits has been studied by PAMLER and KOHLHASE [143]. Since the holes are themselves of 2–3-μm diameter, analysis of their interior surface requires special techniques.

The nature of *metallic contacts to compound semiconductors* has also been an interesting problem. Ohmic contacts must be made to semiconducting surfaces, otherwise circuits could not be fabricated. However, during fabrication, excessive interdiffusion must not occur, since otherwise the electrical characteristics of the semiconductor cannot be maintained. Among the many studies of the reactions of metallizing alloys with compound semiconductors are those of PROCOP et al. [144] on Au–Ge contacts to GaAs, of REIF et al. [145] on Ti–Pt–Au contacts to GaAs, and of STEIN et al. [146] on Au–Pt–Ti contacts to GaAs and InP.

AES has also been applied to a study of the oxidation of copper in printed circuit boards [147], of sulfur passivation of GaAs [148], of cleaning of germanium and GaAs with hydrogen dissociated by a noble-gas discharge [149], and of segregation and corrosion of silver layers electroplated onto light emitting diode (LED) lead frames as part of the final packaging [150].

27.3.6.3. Thin Films and Interfaces

The nature of the interface formed between very thin metallic films and substrates of various types has been studied extensively by AES, just as it has by XPS (see Section 27.2.6.5). The degree and extent of interaction and interdiffusion can be

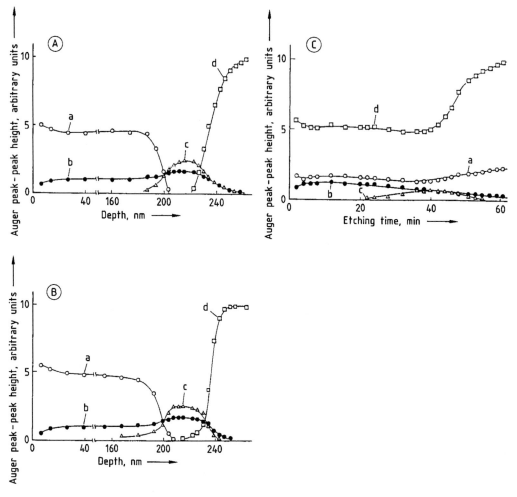

Figure 42. AES depth profiles through a structure consisting of the metallizing alloy Al$_3$Ta on top of a Ta–N diffusion barrier on a silicon substrate [142]
A) Before heat treatment; B) After heating to 550 °C; C) After heating to 600 °C
a) Aluminum; b) Tantalum; c) Nitrogen; d) Silicon

established by AES with depth profiling, although chemical information is normally absent, but the development of the interface is usually studied by continuous recording of Auger spectra during film deposition. Changes in the M$_{4,5}$VV Auger spectrum from a palladium substrate as europium was deposited on it are shown in Figure 43 [151]. Up to a coverage of about two monolayers, no change occurred in the position or shape of the palladium spectrum, but for greater thicknesses the spectrum became narrower, indicating a decrease in the width of the palladium valence band arising from filling of the band. Such filling is due to an inter- diffusion process between europium and palladium occurring even at room temperature.

In some cases, interaction of film with substrate does not occur at room temperature but occurs only after annealing at high temperature. A good example is that of a 10-nm-thick titanium film on an SiO$_2$ substrate. In Figure 44 [152], the Ti LMM Auger spectra are shown in the $N(E)$ (i.e., undifferentiated) mode, rather than the more usual differentiated $dN(E)/d(E)$ mode seen in Figures 41 and 43. Before annealing, the titanium film shows the typically sharp structure in the LMM region, but after annealing to 900 °C the

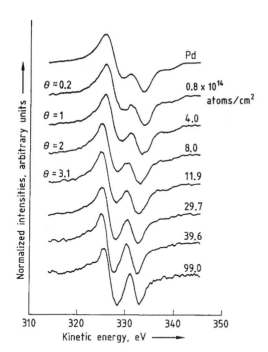

Figure 43. Changes in the Pd $M_{4,5}VV$ Auger spectrum during the deposition of increasing amounts of europium [151]
θ = coverage (monolayers)

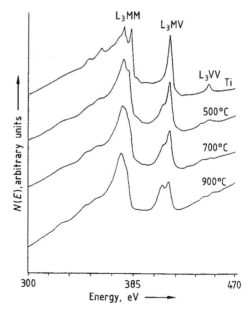

Figure 44. Changes in the Ti LMM Auger spectra, in the $N(E)$ mode, from a 10-nm titanium film on SiO_2 after annealing to increasingly higher temperatures [152]

spectra change character completely. The small L_3VV peak near 450 eV has disappeared; the L_3MV peak has reduced in intensity, with the appearance of an additional peak at lower kinetic energy; and the LMM peak has lost its fine structure and broadened. All these features, like the measured O:Ti ratio, are consistent with the appearance of TiO_2, indicating complete oxidation of titanium as a result of reaction with SiO_2.

Other interesting thin-film studies using AES have included the growth and interaction of silicon on nickel [153], and of palladium on silicon [154], and the preparation and characterization of radiofrequency (rf) sputtered TiN_x [155], WC [156], and Al_3Ta [157] films.

27.3.6.4. Superconductors

Although AES has not been applied quite as extensively as XPS to the new high-T_c superconducting oxides, the number of applications since their discovery has also been large. The majority of applications have consisted of routine depth profiling measurements aimed at checking the extent of reaction of various materials with the superconductor. Recently such measurements, routine though they may be, have become even more important in analyzing uniformity and concentration during attempts to prepare superconductors in the form of thin films. This is because the thin films must of course be deposited on a substrate and must also be annealed to achieve their correct superconducting properties. At the annealing temperature, reaction with the substrate may occur, which must be avoided if possible. A favored substrate, because of its alleged inertness, is $SrTiO_3$. Typical of the depth profiles recorded to check possible interaction are those shown in Figure 45 [158]. The $YBa_2Cu_3O_{7-x}$ film was deposited by ion beam codeposition from sources of Cu_2O, Y_2O_3, and $BaCO_3$, together with oxygen ion bombardment, onto an $SrTiO_3$ substrate held at 350 °C, to a thickness of 100 nm. Figure 45A shows the peak-to-peak Auger intensities as a function of sputter time, and Figure 45B the same after conversion of yttrium, barium, copper, and oxygen intensities to atomic concentrations. Notice in the upper profile the sharpness of the interface, with relatively little diffusion of strontium and titanium into the superconducting film and none of yttrium, barium, and copper into the $SrTiO_3$. In the lower profile, note the almost correct stoi-

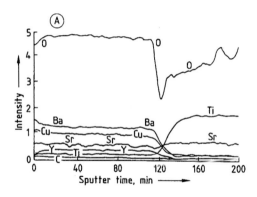

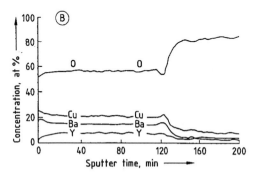

Figure 45. AES depth profiles through a superconducting $YBa_2Cu_3O_{7-x}$ film 100 nm thick on an $SrTiO_3$ substrate held at 350 °C [158]
A) Raw peak-to-peak (p-p) intensity data; B) Conversion of raw data to atomic concentrations

chiometry of the film and the uniformity through the film.

The effectiveness of so-called *buffer layers* has also been examined by AES; these are inert layers interposed between the superconducting film and a reactive substrate such as silicon. ZrO_2 is a much-used buffer material, and its role in preventing interdiffusion between $YBa_2Cu_3O_{7-x}$ films and silicon has been studied by AARNINK et al. [159] using not only AES but RBS as well. Another possibility is RuO_2, whose effectiveness was established by JIA and ANDERSON [160]. Other types of study of superconductors using AES have included the surfaces produced by fracture of sintered $YBa_2Cu_3O_{7-x}$ [161], a comparison of the surfaces of pure $YBa_2Cu_3O_{7-x}$ and its silver-added composite [162], and a quantitative comparison of superconducting bulk crystals and thin films [163].

27.3.6.5. Surface Segregation

The surface composition of alloys, compounds, and intermetallics may change profoundly during heat treatment as a result of segregation of either or both constituent and impurity elements. Surface enhancement may also occur through chemical forces during oxidation or other reaction. The mechanism of segregation is similar to that of grain boundary segregation already discussed, but of course at the free surface material is much more likely to be lost through reaction, evaporation, etc., so that excessive continued segregation can change the bulk properties as well as those of the surface. Nearly all such studies have been carried out on metals and alloys, and AES has been applied to the problem almost from its inception.

Sometimes the segregation of metallic and nonmetallic elements together to surfaces can result in the formation of surface compounds, as demonstrated by UEBING [164]. He heated a series of Fe–15 % Cr alloys containing either 30 ppm nitrogen, 20 ppm carbon, or 20 ppm sulfur to 630°–800 °C and obtained the typical Auger spectra shown in Figure 46. In each case an excess of both chromium and the nonmetal occurs in the spectrum, over that expected if the bulk concentration were maintained. Furthermore, some chemical information available from AES can be used here, because the fine structure associated with the nitrogen and carbon KLL spectra, if not that of sulfur, is characteristic of those elements as the nitride and carbide, respectively. Thus the cosegregation of chromium with nitrogen and with carbon has led to the formation of surface nitrides and carbides, with the same conclusion likely but not so certain, for sulfur. The surface nitride was identified as CrN and the carbide as CrC, the latter being hypothetical but probably stable as a surface compound.

Other recent surface segregation studies involving AES, often with depth profiling, have been carried out on the alloys Ag–Pd [165]; $Au_{0.7}Cu_{0.3}$ [166]; Fe–40Cr–3Ru, Fe–40Cr–3Pt, and Fe–40Cr, with a platinum overlayer [167]; Cr–5Pt and Cr–15Pt [168]; and phosphorus segregation in pure iron using XPS and LEED as well [169].

AES has also been used to study the implantation of nickel in silicon [170], the biological corrosion of condenser tubes [171], the surfaces

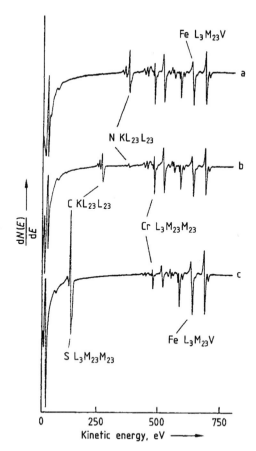

Figure 46. Formation of surface compounds on Fe – 15 % Cr alloys by cosegregation of chromium and a nonmetallic element [164]
a) Nitride; b) Carbide; c) Sulfide

of amorphous metals [172], [173], and superhard a-C:H coatings [174].

27.4. Static Secondary Ion Mass Spectrometry (SSIMS)

SSIMS originated with BENNINGHOVEN [175] in 1969 (i.e., at roughly the same time as AES). The prefix "static" was added by him to distinguish the technique from "dynamic" SIMS — the difference between the two lying in the incident current densities used, of the order of 1 – 10 nA/cm^2 for SSIMS but much higher for dynamic SIMS. In addition, dynamic SIMS usually operates in an imaging or scanning mode, whereas scanning SSIMS has only recently appeared.

SSIMS ranks with XPS and AES as one of the principal surface analytical techniques; its use, although expanding, is probably still rather less than that of the other two. Problems in interpretation continue, but its sensitivity for some elements is much greater than that of the other two techniques; in addition, a large amount of chemical information is available from it in principle.

27.4.1. Principles

A beam of positive ions irradiates a surface, leading to interactions that cause the emission of a variety of types of secondary particle, including secondary electrons, Auger electrons, photons, neutrals, excited neutrals, positive secondary ions, and negative secondary ions. SSIMS is concerned with the last two of these, *positive* and *negative secondary ions*. The emitted ions are analyzed in a mass spectrometer, giving rise to a positive or negative mass spectrum consisting of parent and fragment peaks characteristic of the surface. The peaks may be identifiable as arising from the substrate material itself, from contamination and impurities on the surface, or from deliberately introduced species adsorbed on the surface.

When a heavy energetic particle such as an argon ion at several kiloelectronvolts encounters a surface, it cannot be stopped immediately by the first layer of atoms but continues into the surface until it comes to a halt as a result of losing energy by atomic and electronic scattering. Along the way to its stopping place, the ion displaces some atoms from their normal positions in the solid structure, and as they recoil, these atoms in turn displace others, which also displace additional atoms, and so on, resulting in a complex sequence of collisions. Depending on the energy absorbed in an individual collision, some atoms are displaced permanently from their normal positions, whereas others return elastically after temporary displacement. The whole sequence is called a collision cascade and is illustrated schematically, along with some other processes, in Figure 47 [176]. As the cascade spreads out from the path of the primary ion, its effect may eventually reach atoms in the surface layer, and if by that stage enough kinetic energy is left in the collisional interaction, bonds may be broken and material leave the surface as atoms or clusters. The material is then said

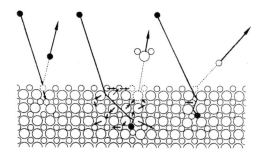

Figure 47. Schematic of various possible heavy-particle emission processes at a solid surface on ion bombardment [176]

to be sputtered; the above model of the sputtering process is from SIGMUND [177].

Although SIGMUND's collision cascade model is able to explain much about the parameters involved in the removal of material from a surface by sputtering, it cannot predict the degree of positive or negative ionization of the material. In truth, a unified theory of secondary ion formation in SSIMS does not yet exist, although many models have been proposed for the process. Since the secondary-ion yields (i.e., the probabilities of ion formation) can vary by several orders of magnitude for the same element in different matrices, or for different elements across the periodic table, lack of a means of predicting ion yields in given situations is a hindrance to both interpretation and quantification. Some theoretical and semitheoretical models have had success in predictions over very restricted ranges of experimental data, but they fail when extended further. This is not an appropriate place to describe the theoretical models; for further information see, e.g., [1].

27.4.2. Instrumentation

27.4.2.1. Ion Sources

Two types of source are in use according to the type of instrument employed for SSIMS. In the first, *positive ions from a noble gas* (usually argon) are produced either by electron impact or in a plasma created by a discharge. The ions are then extracted from the source region, accelerated to the chosen energy, and focused in an electrostatic ion optical column. An ion gun of this type using electron impact ionization is shown in Figure 39.

In the second type, *positive ions from a liquid metal* (almost always gallium) are produced in the

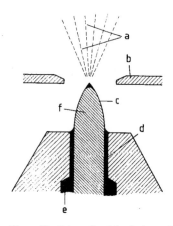

Figure 48. Schematic of the design and operation of a liquid-metal ion source (LMIS) [178]
a) Metal ions; b) Extractor; c) Liquid-metal film; d) Capillary tube; e) Liquid metal; f) Needle

manner shown schematically in Figure 48 [178]. A fine needle (f) (tip radius ≈ 5 μm) of refractory metal passes through a capillary tube (d) into a reservoir of liquid metal (e). Provided the liquid metal wets the needle, which can be ensured by fabrication procedures, the liquid is drawn through the tube and over the tip by capillary action. Application of 5–10 kV between the needle and an extractor electrode (b) a short distance away draws the liquid up into a cusp because of the combined forces of surface tension and electrostatic stress. In the region just above the cusp, ions are formed and are accelerated through an aperture in the extractor, from where they may be focused in an ion optical column onto the sample surface. Gallium is invariably used as the liquid metal because its melting point is 35 °C and it remains liquid at room temperature; other metals have been tried but require heating of the reservoir to melt them.

Argon-ion guns of the electron impact type operate in the energy range 0.5–5 keV at 1–10 μA, and provide optimum spot sizes on a surface of about 10 μm. The discharge-type argon-ion guns, often using the duoplasmatron arrangement, operate up to 10 keV, with currents variable up to 20 μA and, according to the particular focusing lens system used, can provide spot sizes down to about 2 μm. The *gallium liquid metal ion source* (*LMIS*) is what is known as a *high-brightness source*, which means that ion production occurs in a very small volume. As a result the ion beam can be focused to a fine spot, and at 8–10 keV a spot size of 0.2 μm can be achieved,

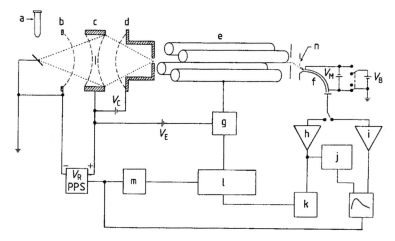

Figure 49. Experimental arrangement for SSIMS used by KRAUSS and GRUEN [179] employing a quadrupole mass spectrometer for mass analysis and a retarding-field analyzer for prior energy selection
a) Ion gun; b)–d) Lenses 1–3; e) Quadrupole mass spectrometer; f) Charge detecting electron multiplier; g) Quadrupole power supply; h) Pulse amplifier; i) Iso amplifier; j) Rate meter; k) Multichannel analyzer; l) Mass programmer; m) Sawtooth generator; n) Exit hole for neutrals

while at 30 keV it reduces to about 50 nm. All the ion gun optical columns provide deflection plates for rastering the ion beam over areas adjustable from many square millimeters to a few tens of square micrometers.

27.4.2.2. Mass Analyzers

As with ion sources, two types of ion mass analyzer are in use, and indeed each type of source goes with each type of analyzer. The argon-ion sources are used with quadrupole mass spectrometers and the gallium LMIS with the time-of-flight (TOF) mass spectrometer.

Quadrupole Mass Spectrometer. The quadrupole mass spectrometer (see also, →Mass Spectrometry) has been in use for many years as a residual gas analyzer (with an ionizing hot filament) and in desorption studies and SSIMS. It consists of four circular rods, or poles, arranged so that they are equally spaced in a rectangular array and exactly coaxial. Figure 49 indicates the arrangement, as depicted by KRAUSS and GRUEN [179]. Two voltages are applied to the rods, a d.c. voltage and an rf voltage. When an ion enters the space between the rods, it is accelerated by the electrostatic field, and for a particular combination of d.c. and rf voltages, the ion has a stable trajectory and passes to a detector. For other combinations of voltages, the trajectory diverges rapidly and the ion is lost either by hitting one of the poles or by passing between them to another part of the system. The mass resolution is governed by the dimensions of the mass spectrometer, the accuracy of construction, and the stability and reproducibility of the ramped voltage. The quadrupole mass spectrometer is compact, does not require magnets, and is entirely ultrahigh vacuum compatible—hence its popularity. It does have disadvantages, however, in that its transmission is low and decreases with increasing mass number.

Time-of-Flight Mass Spectrometer. Because the quadrupole mass spectrometer is unsuitable for the analysis of large molecules on surfaces or of heavy metals and alloys, the TOF mass spectrometer has been developed for use in SSIMS by BENNINGHOVEN and coworkers [180]. The TOF mass spectrometer works on the principle that the time taken for a charged particle such as an ion to travel in a constant electrostatic and magnetic field from its point of origin to a detector is a function of the mass-to-charge ratio of the ion. For the same charge, light ions travel the fixed distance more rapidly than heavy ions. For good mass resolution, the flight path must be sufficiently long (1–1.5 m), and very sophisticated high-frequency pulsing and counting systems must be employed to time the flight of the ion to within a few nanoseconds.

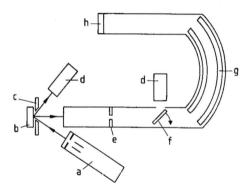

Figure 50. Schematic of the imaging time-of-flight SSIMS system designed by ECCLES and VICKERMAN [181]
a) Pulsed ion gun; b) Target; c) Extraction electrode; d) SED; e) Variable iris; f) Movable plate; g) Energy-compensated flight tube; h) Detector

The original design [180] used pulsed beams of argon ions, but with the commercial development of the LMIS (see Section 27.4.2.1) the capability of the TOF system has been extended significantly. The principle of LMIS operation allows the beam of ^{69}Ga$^+$ and ^{71}Ga$^+$ ions to be focused to a probe of 50-nm minimum diameter, while being pulsed at frequencies up to 20 kHz and rastered at the same time. Mass filtering selects either of the isotopes before the beam strikes the sample. When an LMIS is operated at beam currents of 10–100 pA, the possibility is opered up of performing scanning SSIMS and of producing secondary-ion images in selected masses. A system designed for such *TOF imaging SSIMS* is shown schematically in Figure 50, from ECCLES and VICKERMAN [181]. Pulsing of the ion source is achieved by rapid deflection of the ion beam across a small aperture, the pulse length being variable between 4 and 50 ns. The flight path length is 1.57 m, part of which includes an energy compensator to ensure that all ions of the same mass, but of different energies, arrive at the same time. Secondary electron detectors (SED) allow topographical images produced by ion-induced secondary electrons to be generated.

With such a TOF imaging SSIMS instrument, the useful mass range is extended beyond 10 000 amu; parallel detection of all masses occurs simultaneously; up to four secondary-ion images can be acquired simultaneously; and within each image, up to 20 mass windows (very narrow mass ranges) can be selected. The amount of data generated in a short time is enormous, and very sophisticated and capacious systems are required to handle and process the data.

27.4.3. Spectral Information

A SSIMS spectrum, like any other mass spectrum, consists of a series of peaks of varying intensity (i.e., ion current) appearing at certain mass numbers. The masses can be allocated on the basis either of atomic mass and charge, in the simple case of ions of the type $M^{n\pm}{}_i$, where both n and i can each equal 1, 2, 3,..., or of molecular mass and charge, in the more complex case of cluster ions of type $M_iX^{n\pm}{}_j$, where again n, i, and j can each equal 1, 2, 3,... In SSIMS, many of the more prominent secondary ions happen to have $n=i=1$, which makes allocation of mass numbers to fragments slightly easier. The masses can be identified as arising either from the substrate material itself, from deliberately introduced molecular or other species on the surface, or from contamination and impurities on the surface. Complications in allocation often arise from isotopic effects. Although some elements have only one principal isotope, in many others the natural isotopic abundance can cause difficulties in identification.

An example showing both the allocation of peaks and the potential isotopic problems appears in Figure 51, in which the positive and negative SSIMS spectra from a nickel specimen have been recorded by BENNINGHOVEN et al. [182]. The specimen surface had been cleaned by standard procedures, but SSIMS reveals many impurities on the surface, particularly alkali and alkaline-earth metals, for which it is especially sensitive. For reasons discussed in Section 27.4.4, the peak heights cannot be taken to be directly proportional to the concentrations on the surface; thus, for example, the similar heights of the Na$^+$ and ^{58}Ni$^+$ peaks should not lead to the conclusion that there is as much sodium as nickel on the surface. Note in Figure 51 the appearance of isotopes of nickel, potassium, and chlorine in their correct natural abundances and also, in the positive ion spectrum, peaks due to M_2^+ ions, also with isotopic effects. Since the intensity scale is logarithmic, the minor peaks are exaggerated, but identification problems do exist.

The relationship between what is recorded in a SSIMS spectrum and the chemical state of the surface is not as straightforward as in XPS and AES. Because of the large number of cluster ions that appear in any SSIMS spectrum from a mul-

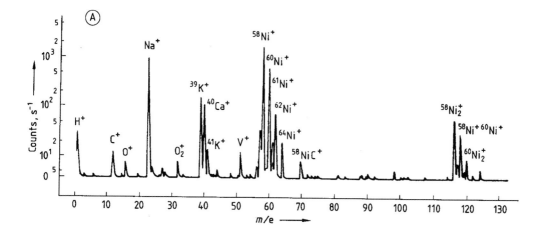
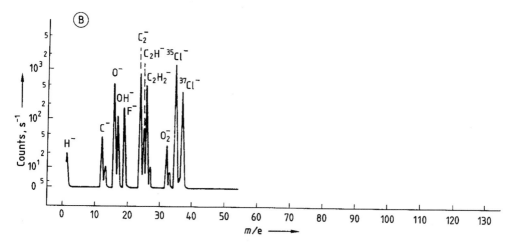

Figure 51. SSIMS spectra from a "cleaned" nickel target, under conditions of 3 keV primary argon ion energy and 5×10^{-9} A primary current [182]
A) Positive SSIMS spectrum; B) Negative SSIMS spectrum

ticomponent surface (e.g., during the study of a surface reaction), a great deal of chemical information is obviously available in SSIMS, potentially more than in XPS. The problem in using the information from a cluster ion lies in the uncertainty of knowing whether or not the cluster is representative of a group of the same atoms situated next to each other in the surface. For some materials, such as polymers, the clusters observed are definitely characteristic of the material, as seen in Figure 52 from BLETSOS et al. [183], where the SSIMS spectrum from polystyrene measured in a TOF mass spectrometer reveals peaks spaced at regular 104 mass unit intervals, corresponding to the polymeric repeat unit. On the other hand, for example during oxidation studies, many metal–oxygen clusters are normally found in both positive and negative secondary-ion spectra, whose relationship to the actual chemical composition is problematic. Figure 53 from McBREEN et al. [184] shows the SSIMS spectra recorded after interaction of oxygen with half a monolayer of potassium adsorbed on a silver substrate. In both positive and negative spectra K_iO_j clusters are found that the authors were unable to relate to the surface condition, although quite clearly much chemical information is available in the spectra. The use of pure and well-characterized standards can sometimes help in the interpretation of complex spectra.

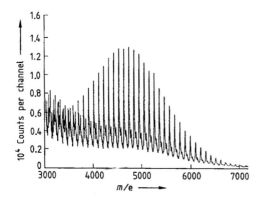

Figure 52. Positive SSIMS spectrum of polystyrene ($M_n = 5100$, $M_{n,\text{calc.}} = 4500$, $M_{n,\text{corr.}} = 5600$) from time-of-flight mass spectrometer [183]

27.4.4. Quantification

The secondary-ion current at a particular mass number from element A in a matrix B can be written as

$$I^{\pm}_{A(B)} = K \cdot S \cdot C_A \cdot \alpha^{\pm}_{A(B)} \tag{18}$$

where S is the total sputtering yield of B; C_A the concentration of A in B; and $\alpha^{\pm}_{A(B)}$ the probability of positive or negative ionization of element A in matrix B.

As for XPS and AES (Eq. 8 and Eq. 13, respectively) K is a constant containing the various instrumental and other factors that are maintained approximately fixed during an analysis.

The *sputtering yield* S can be measured from elemental and multielemental standards under various operational conditions and can therefore be known to reasonable accuracy for the material being analyzed by judicious interpolation between standards. However, the *ionization probability* α cannot be measured easily, and attempting to measure it for a set of standards and then trying to interpolate would be of no use, because it is very matrix and concentration dependent and can vary very rapidly. In fact, it represents the principal obstacle in achieving proper quantification in SSIMS.

Because of problems in the measurement of α, quantification is normally carried out either by relative sensitivity factor methods or by semi-empirical approximations to α based on one of the ionization theories.

Relative Sensitivity Factors. In SSIMS, unlike XPS and AES, elemental sensitivities are taken as

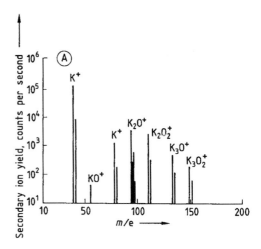

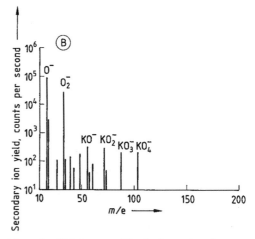

Figure 53. SSIMS spectra following interaction of oxygen with half a monolayer of potassium adsorbed on silver, at 300 K [184]
A) Positive SSIMS spectrum; B) Negative SSIMS spectrum

relative to the intensity of the signal from an internal reference element, rather than to that of a single chosen element such as fluorine or carbon. The internal reference would normally be the major constituent of the material. If the signal from the reference is $I^{\pm}_{R(B)}$, it can be written

$$I^{\pm}_{A(B)} / I^{\pm}_{R(B)} = \frac{K \cdot S \cdot C_A \cdot \alpha^{\pm}_{A(B)}}{K \cdot S \cdot C_R \cdot \alpha^{\pm}_{R(B)}} \tag{19}$$

or

$$I^{\pm}_{A(B)} / I^{\pm}_{R(B)} = \varrho_{AR} C_A / C_R \tag{20}$$

where ϱ_{AR} is the relative sensitivity factor of A with respect to R. Of course ϱ_{AR} contains matrix-dependent quantities, but their variations are damped to some extent by virtue of taking ratios, and in practice ϱ_{AR} is assumed constant for low concentrations of A (e.g., <1 at%). It can be evaluated from measurements on a well-characterized set of standards containing A in known dilute concentrations. The accuracy of the method is not as good as XPS and AES when relative sensitivity factors are used.

Semiempirical Approximation. The method that adopts a semiempirical approximation to α uses the *Andersen–Hinthorne local thermodynamic equilibrium (LTE)* model [185] for estimating the degree of ionization. The model does not take account of any details of individual ionization processes but assumes that the region at and near the surface involved in sputtering can be approximated by a dense plasma in local thermodynamic equilibrium. The plasma has an associated temperature T. The ratio of the concentrations of two elements X and Y in a matrix B, if sufficiently dilute, can then be written as

$$C_X/C_Y \sim \left(I^{\pm}_{X(B)}/I^{\pm}_{Y(B)}\right) \exp\left\{\left(I_{P(X)} - I_{P(Y)}\right)/kT\right\} \quad (21)$$

where $I^{\pm}_{X(B)}$, $I^{\pm}_{Y(B)}$ are the positive secondary-ion currents of X and Y, respectively, in B; and $I_{P(X)}$, $I_{P(Y)}$ the first ionization potentials of neutral atoms X and Y.

T is the only adjustable parameter in Equation (21). Measurement of the secondaryion currents of all components then provides a set of coupled equations for $C_X/C_Y, \ldots, C_N/C_Y$, which, together with the identity $C_X + C_Y + \ldots C_N = 100\%$, enables the concentrations to be found. The model does not have a sound physical basis, but the method has been found to be surprisingly accurate for some materials (e.g., steels). However, the values of T, the effective plasma temperature, that are necessary for agreement are sometimes ridiculously high.

27.4.5. Applications

In general, SSIMS has not been used for quantification of surface composition because of all the uncertainties described above. Its application has been more qualitative in nature, with emphasis on its advantages of high surface specificity, very high sensitivity for certain elements, and multiplicity of chemical information. Increasingly, it is being used in the high spatial resolution TOF imaging mode and also in parallel with a complementary technique, such as XPS, on the same system.

27.4.5.1. Oxide Films

The higher surface specificity of SSIMS compared to XPS and AES (i.e., 1–2 monolayers versus 2–8 monolayers) can be useful in establishing more precisely the chemistry of the outer surface than would be possible with the latter techniques. Although from details of the O 1s spectrum XPS could give the information that OH was present on a surface as well as oxide, and from the C 1s spectrum that hydrocarbons as well as carbides were present, only SSIMS could say which hydroxide and which hydrocarbons. In the growth of oxide films for various purposes (e.g., passivation or anodization), such information is valuable since it provides a guide to the quality of the film and the nature of the growth process.

One important area in which the properties of very thin films play a significant role is adhesion of aluminum to aluminum in the aircraft industry. For bonds of the required strength to be formed, the aluminum surfaces must be pretreated by anodizing them, which should create not only the right chemical conditions for adhesion but the right structure in the resultant oxide film as well. Films formed by anodizing aluminum in phosphoric acid under a.c. conditions have been studied by TREVERTON et al. [186] using SSIMS and XPS together. Knowledge of both the chemistry of the outer monolayers in terms of aluminum oxides and hydroxides, and of the nature and degree of any contamination introduced by the anodizing process, was important. Figure 54 from their paper compares the positive and negative SSIMS spectra after anodizing for 3 and 5 s at 20 V and 50 Hz. Both sets of spectra show that the longer anodizing treatment caused significant reduction in hydrocarbon contamination (peaks at 39, 41, 43, 45 amu in the positive spectrum, and 25, 41, 45 amu in the negative), and marked increases in oxide- and hydroxide-associated cluster ions at 44, 61, 70, and 87 amu in the positive spectrum and 43 and 59 amu in the negative. At higher masses, peaks allocated to $Al_2O_4H^-$, $Al_2O_3^-$, and $Al_3O_5^-$ were also identified. Residual PO_2^- and PO_3^- from the anodizing bath were reduced at the longer anodizing time. The low level of contaminant is important since the adhesive coating can then wet the surface

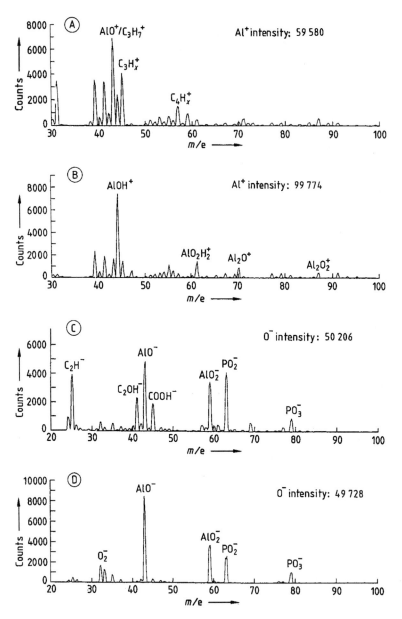

Figure 54. SSIMS spectra from oxide films formed on aluminum after anodization treatment in phosphoric acid for 3 or 5 s [186]
A) Positive SSIMS spectrum, 3 s; B) Positive SSIMS spectrum, 5 s; C) Negative SSIMS spectrum, 3 s; D) Negative SSIMS spectrum, 5 s

more effectively, leading to improved adhesion. In addition, the change in aluminum chemistry from AlO^+ (mass 43) to $AlOH^+$ (mass 44) on going from 3- to 5-s anodizing was significant, and not an effect that could have been established by XPS.

27.4.5.2. Interfaces

The protection of steel surfaces by paint depends significantly on the chemistry of the paint–metal interface. The system has many variables since the metal surface is normally pretreated in a variety of ways, including galvanizing and phos-

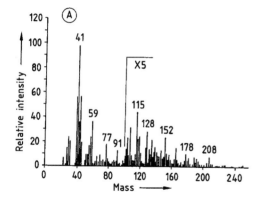

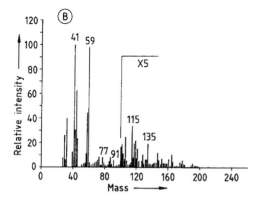

Figure 55. Positive SSIMS spectra from each side of the metal–paint interface on a hot-dip galvanized steel surface after accelerated cyclic corrosion testing [187]
A) Metal; B) Paint

phating. Indeed, several interfaces of importance probably exist, and corrosion protection may be a function of conditions at all of them. To study the effects of the variables without having to wait for natural weathering processes to cause corrosion, normal procedure is to apply an accelerated cyclic corrosion test (CCT) that involves scribing through the paint to the bare metal followed by salt spray, drying, and exposure to high humidity.

SSIMS has been applied by VAN OOIJ et al. [187] to study of the interfaces in a paint–metal system; both quadrupole and TOF SSIMS were employed, as well as XPS. Figure 55 from their paper shows the positive SSIMS spectra from the metal (Fig. 55 A) and paint (Fig. 55 B) sides of an interface exposed by soaking samples in N-methylpyrrolidone, which removes the paint by swelling but does not attack it chemically. The metal surface had been phosphate treated before paint application. Both spectra are identified in terms of organic materials; in other words the failure was in an organic layer close to, but not at, the metal surface. This identification is supported by the observations that the mass peak at 52 amu, characteristic of chromate, was negligible and that in the negative spectra the intensity of PO_x^- ions was also low. Note that epoxy-related peaks at 59 and 135 amu were observed on both sides, as well as aromatic fragments at 77, 91, 115, 128, 152, and 178 amu, the latter being more intense on the metal side. When the spectra were compared with those from an unphosphated and painted metal surface that had suffered delamination, the characteristic epoxy peaks were not seen on the metal side, indicating that in the absence of the phosphate layer the adhesion of the epoxyurethane to the substrate was lost.

27.4.5.3. Polymers

Treatment of polymer surfaces to improve their properties with respect to wetting or water repulsion and to adhesion, is by now a standard procedure. The treatment is designed to change the chemistry of the outermost groups in the polymer chain, without affecting bulk polymer properties. Any study of the effects of treatment therefore requires a technique that is specific mostly to the outer atom layers, which is why SSIMS has been used extensively in this area.

One of the favored forms of surface treatment is that of plasma etching. Depending on the conditions of the plasma discharge (i.e., nature of discharge gas, gas pressure, rate of gas flow, discharge power and frequency, and substrate polarity), plasma etching is able to alter the chemical characteristics of the surface over a wide range. By its nature a plasma discharge is also both highly controllable and reproducible. The effects of plasma treatment on the surface of polytetrafluoroethylene (PTFE) have been studied by MORRA et al. [188] using combined SSIMS and XPS. They found, as evidenced in Figure 56, that not only did the plasma change the surface chemistry, but the surface emission fragments also changed the plasma discharge conditions temporarily. Thus the untreated (Fig. 56 A) and 15-min oxygen plasma treated (Fig. 56 C) positive-ion spectra are virtually identical, whereas the 0.5-min oxygen plasma treated spectrum (Fig. 56 B) is quite different. The initial breakup of the PTFE surface produced hydrocarbon fragments which reacted with the oxygen to produce water vapor, thus changing the local nature of the discharge

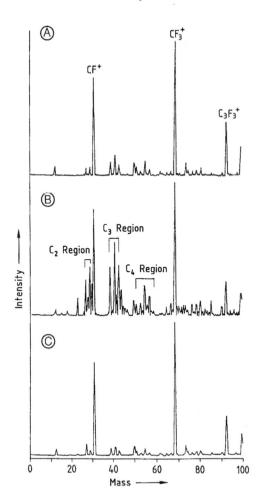

Figure 56. Positive SSIMS spectra from PTFE [188]
A) Untreated; B) 0.5-min oxygen plasma treatment;
C) 15-min oxygen plasma treatment

of the reactions of surfaces with gases and vapors must be viewed with caution, but in combination with other surface techniques it can provide valuable additional information. The parallel techniques are most often XPS, TDS, and LEED, and the complementary information required from SSIMS normally refers to the nature of molecules on surfaces and with which, if any, other atoms they are combined.

A typical SSIMS spectrum of an organic molecule adsorbed on a surface is that of thiophene on ruthenium at 95 K, shown in Figure 57, from the study of Cocco and Tatarchuk [193]. The exposure was only 0.5 Langmuir (i.e., 5×10^{-7} torr · sec $\triangleq$ 37 Pa · s) and the principal positive ion peaks are those from ruthenium, consisting of a series of seven isotopic peaks around 102 amu. However, ruthenium–thiophene complex fragments are found at ca. 186 and 160 amu, each showing the same complicated isotopic pattern, indicating that even at 95 K interaction between the metal and the thiophene had occurred. In addition, thiophene and protonated thiophene peaks are observed at 84 and 85 amu, respectively, with the implication that no dissociation of the thiophene had taken place. The smaller masses are those of hydrocarbon fragments of different chain lengths.

SSIMS has also been used in the study of the adsorption of propene on ruthenium [194], the decomposition of ammonia on silicon [195], and the decomposition of methanethiol on nickel[196].

Recently SSIMS has been applied in the TOF–SSIMS imaging mode to the study of very thin layers of organic materials [197], polymeric insulating materials [198], and carbon fiber and composite fracture surfaces [199]. In these studies a spatial resolution of ca. 200 nm in mass-resolved images was achieved.

gas. The important deduction to be made is that one of the vital parameters in plasma treatment is the length of time in contact with the plasma; in this case, too long a contact time resulted in no surface changes.

Other SSIMS studies of polymer surfaces have included perfluorinated polyether [189], low-density polyethylene [190], poly(ethylene terephthalate) [191], and the oxidation of polyetheretherketone [192].

27.4.5.4. Surface Reactions

Because of the inherently destructive nature of ion bombardment, the use of SSIMS alone in study

27.5. Ion Scattering Spectroscopies (ISS and RBS)

Analysis of surfaces by ion scattering has a long history since several techniques are included according to the energy of the primary ion. They have been classified by Armour [200] as *low-energy ion scattering* (LEIS) for the range 100 eV to 10 keV, *medium-energy ion scattering* (MEIS) for the range 100 to 200 keV, and *high-energy ion scattering* (HEIS) for the range 1 to 2 MeV. However, LEIS is more often called *ion scattering*

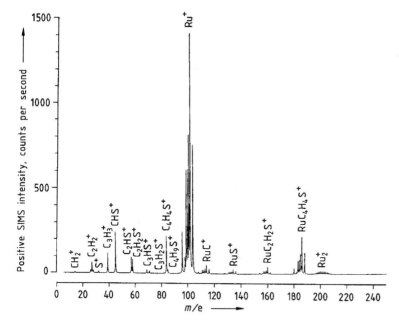

Figure 57. Positive SSIMS spectrum from Ru (0001) after exposure to 0.5 L thiophene at 95 K [193]

spectroscopy (ISS), whereas HEIS is always called *Rutherford backscattering spectroscopy* (RBS). ISS is invariably surface specific, while RBS is normally surface sensitive.

27.5.1. Ion Scattering Spectroscopy (ISS)

Although treated here as one of the major surface analytical techniques, ISS, like SSIMS, is used much less extensively than XPS or AES. Again, like SSIMS, it is often used in conjunction with other techniques and is valuable as a complement to the others for its two advantages of high surface specificity and ease of quantification.

27.5.1.1. Principles

A beam of inert gas ions, usually He^+ or Ne^+, of typical energy 0.5–4 keV and current 10^{-8}–10^{-9} A, is directed at a surface, and the energies of the ions of the same mass and charge that are scattered from the surface are analyzed. Because inert gas ions have a very strong electron affinity, the probability is high that on collision with a surface the majority will be neutralized. After two collisions the number of original ions left will be negligible. Thus analysis of scattered ion energies will refer mostly to those ions that have suffered only one collision (i.e., from the outermost layer of atoms). Hence ISS is very surface specific. In addition, with primary ions of low mass and energy, the ion–atom interaction time is much shorter than the characteristic time of vibration of an atom in a solid, so that the collision can be approximated as that between the ion and a free atom. This means that binary collision (or "billiard-ball") theory can be applied, in which case the relationship between the energy E_1 after collision and the energy E_0 before collision, of the ion of mass M_1 with a surface atom of mass M_2, is

$$E_1/E_0 = \{1 + (M_2/M_1)\}^{-2} \left[\cos\theta \pm \left\{(M_2/M_1)^2 - \sin^2\theta\right\}^{1/2}\right]^2 \quad (22)$$

where θ is the scattering angle (i.e., angle between incident and scattered beams).

If θ is chosen to be 90°, then Equation (22) reduces to the simple relationship

$$E_1/E_0 = (M_2 - M_1)/(M_2 + M_1) \quad (23)$$

Equation (22) shows that if the energy and mass of the incident ion, and the scattering angle, are all fixed, E_1 depends only on M_2, the mass of

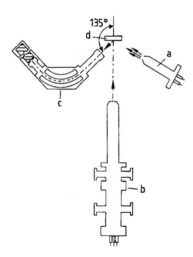

Figure 58. Experimental arrangement of ISS system used by MARTIN and NETTERFIELD [201]
a) Electron flood gun; b) Ion gun; c) ISS analyzer; d) Sample

the surface atom. Thus the energy spectrum measured in ISS can be converted directly into a mass spectrum, providing an elemental compositional analysis of the surface. Because only the mass of the surface atom is involved, no chemical state information is available.

Note from Equations (22) and (23) that as M_2 increases, the ratio E_1/E_0 tends to unity, which means that for a light ion such as He^+, heavy elements are difficult to resolve in the energy spectrum. ISS is therefore most suitable for the analysis of light elements on surfaces. The mass resolution can be improved by increasing the mass of the ion (e.g., by going to Ar^+ or Xe^+), but then problems of sputtering intrude. Another way of improving the resolution is by using scattering angles >90°, and in most ISS geometries the angle is $\approx 135°$.

27.5.1.2. Instrumentation

As with other surface-specific techniques, the vacuum requirements for ISS are stringent, vacua better than 10^{-8} Pa being needed. With the single monolayer specificity mentioned above, good UHV conditions are vital.

The basic instrumental requirements are an ion source, a sample target, and an energy analyzer. The source is normally a commercial ion gun, for example that shown in Figure 39, operated at a few kiloelectronvolts and with beam currents of He^+ or Ne^+ of 1–100 nA. For scattered ion energy analysis, a standard CHA or CMA with reversed polarity can be used, or alternatively a special analyzer may be constructed. The basic arrangement is shown in Figure 58 from MARTIN and NETTERFIELD [201]; also shown is an auxiliary electron source used for charge neutralization by flooding the surface with low-energy electrons. In particularly sophisticated experiments the incident ion beam is mass-filtered before striking the target, and both target and analyzer are rotatable around the point of impact to allow angular resolution.

Similar to developments in SSIMS, significant improvements in ISS have been obtained by using *time-of-flight techniques*. The additional sensitivity available with a TOF system allows much lower incident ion current densities to be employed, thereby reducing the effects of beam damage to the surface. Such damage is of course negligible when using He^+, but if heavier ions such as Ar^+ are required for better mass resolution, the current density must be reduced. Another major advantage of a TOF energy analyzer is that it enables energy analysis not only of ions but of neutrals ejected from the surface, since the flight time is independent of charge state. This has led to the introduction of a structural technique called *noble-gas impact collision ion scattering spectroscopy* with neutral detection (NICISS), by NIEHUS [202].

27.5.1.3. Spectral and Structural Information

Spectra in ISS are recorded in the form of scattered intensity versus the ratio E_1/E_0, so that the range of the latter will be from 0 to 1. Some typical examples are shown in Figure 59, from BEUKEN and BERTRAND [203], for a set of Al_2O_3, MoS_2, and Co_9S_8 standards used by them in their study of cobalt–molybdenum sulfide catalysts. The conditions — He^+ at 2 keV and 15 nA, and a scattering angle of 138° — are also typical. Acquisition time per spectrum was about 5 min, and an electron flood gun was used for charge neutralization.

As seen in Figure 59, the E_1/E_0 ratio can be converted directly to mass and therefore to elemental identification. Notice also the disadvantage mentioned above, that as $E_1/E_0 \rightarrow 1$, the mass separation becomes poor; i.e., if cobalt ($M=59$) and molybdenum ($M=96$) had appeared in the same spectrum they would have only just been resolved

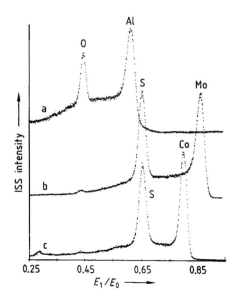

Figure 59. Examples of ISS spectra for a set of standard materials used in the study of cobalt–molybdenum sulfide catalysts [203]
a) Al_2O_3; b) MoS_2; c) Co_9S_8

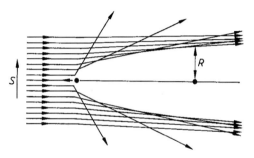

Figure 60. Ion trajectories for a parallel beam of light ions hitting an atom [204]

under the conditions. However, the resolution for light elements is good.

In addition to elemental compositional information, ISS can also be used to provide information about surface structure, and in this application it is complementary to LEED. The reason for the structural possibility is indicated schematically in Figure 60, which shows the ion trajectories calculated by VAN DER VEEN [204] for a parallel beam of ions hitting an atom at the surface. Most ions are scattered in the forward direction, with only a few making an interaction strong enough to be scattered back toward the detector. Thus a "shadow cone" arises behind the atom, R being the radius of the cone at the position of the next atom in the row. Since R is of the same order as interatomic spacing, atoms in a row lying immediately behind the scattering atom are "invisible" to the primary beam. S is the so-called impact parameter, which depends on ion energy and distance of nearest approach between ion and atom.

If the experimental geometry is varied, by varying either the angle of primary ion incidence or the target azimuthal angle, while the scattering angle is kept constant, the shadow cone is effectively swept through the rows of atoms behind a surface atom. The scattered intensity will then vary periodically according to whether the scattering is mostly from surface atoms or whether at certain angles contributions to the scattering come from atoms further along a row. If the near-surface crystallinity is good, then the variations in angular intensity can be used to derive or confirm a surface structure. The technique is particularly valuable when using scattering angles close to 180°, since the problem of multiple scattering (i.e., successive scattering off two or more atoms) is minimized, and the dependence of the scattered intensity on the angle of incidence is a direct representation of the shape of the shadow cone. This variation of ISS is called *impact collision ion scattering spectroscopy* (ICISS). Figure 61 from NIEHUS and COMSA [205] compares the ICISS measurements from a Cu (110) crystal surface before (Fig. 61 A) and after adsorption of oxygen (Fig. 61 B). Various models had been proposed for the structure of the oxygen-reacted surface, but only the "missing row" model shown on the lower right can explain the observed angular dependence.

27.5.1.4. Quantification and Depth Information

If the scattering process is considered as a simple binary collision, the intensity I_A^+ of primary ions scattered from atoms of type A in the surface is given by

$$I_A^+ = I_0^+ N_A \sigma_A (1 - P) S \cdot \Delta\Omega \cdot T \tag{24}$$

where I_0^+ is the intensity of the primary beam; N_A the density of scattering atoms A; σ_A the scattering cross section of atoms A; P the neutralization probability of a primary ion; S the shadowing factor; $\Delta\Omega$ the acceptance angle of the analyzer; and T the transmission function of the analyzer.

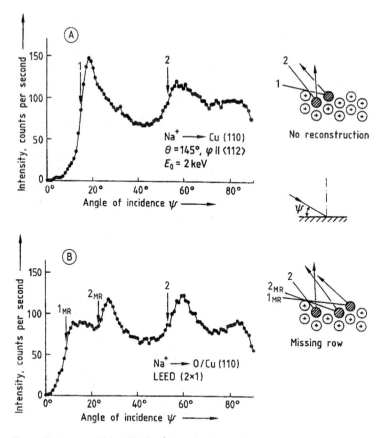

Figure 61. Impact collision ISS of Na$^+$ scattering from a Cu (110) surface before (A) and after (B) surface reconstruction due to oxygen adsorption, as a function of angle of incidence [205]

In an analysis, the primary ion current, the experimental geometry, and the transmission function are kept constant, so that to relate I_A^+ to N_A requires knowledge of σ_A and P.

Scattering Cross Section. To evaluate σ_A, information is needed about the physical nature of the interaction between the ion and the atom. In the ion energy range used in ISS the interaction can be considered in terms of purely repulsive potential (i.e., a Coulomb repulsion with inclusion of some screening, since some electronic screening occurs during the interaction). The general form of the interaction potential is thus

$$V(r) = \frac{Z_1 Z_2 e^2}{r} \varphi(r/a) \qquad (25)$$

where Z_1, Z_2 are the atomic numbers of ion and scattering atom; r is the ion–atom distance; and a the characteristic screening length.

The most widely used screening function is that due to MOLIÈRE [206] using Thomas–Fermi statistics, which is approximated by

$$\varphi(r/a) = \varphi(x) = 0.35 \exp(-0.3x) \\ + 0.55 \exp(-1.2x) + 0.1 \exp(-6x) \qquad (26)$$

with

$$a = 0.4685 \left[Z_1^{2/3} + Z_2^{2/3} \right]^{-1/2} \qquad (27)$$

Thus $V(r)$ can be computed and hence the socalled scattering integral

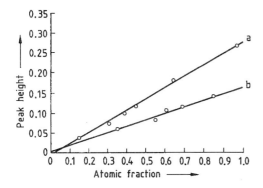

Figure 62. Calibration of the ISS peak heights of gold and silver against the known bulk compositions of a series of Au–Ag alloy standards [210]
a) Au; b) Ag

$$\psi = \pi - 2p \int_{r_{min}}^{\infty} dr/r^2 \{1 - (p^2/r^2) - V(r)/E\}^{1/2} \quad (28)$$

where $\psi = \theta + \sin^{-1}\{(\sin\theta)/A\}$, θ being the scattering angle, and $A = M_2/M_1$; p is the impact parameter; and $E = \{M_2/(M_1 + M_2)\}E_0$.

From Equation (28) can be calculated the differential scattering cross section for scattering into unit solid angle, which is

$$d\sigma(\theta)/d\Omega = (p/\sin\theta)(dp/d\theta) \quad (29)$$

and hence the cross section.

Neutralization Probability. The neutralization probability P in Equation (24) is not quite so straightforward to compute. As stated earlier, neutralization is a highly probable event, because the approaching ion represents a deep potential well compared to the work function of a solid. Based on the theory of neutralization by Auger electron ejection developed by HAGSTRUM [207], the probability of an ion escaping neutralization has the form

$$(1 - P) = \exp\left\{\left(\frac{1}{V_{IT}} + \frac{1}{V_{ST}}\right)V_c\right\} \quad (30)$$

where V_{IT}, V_{ST} are the components of velocities of initial and scattered ions, respectively, normal to the surface; and V_c is the characteristic velocity.

V_c is dependent on the electron transition rate, which in turn depends, among other things, on the chemical state of the surface. It is approximated by

$$V_c = \int_{r_0}^{\infty} R(r)\,dr \quad (31)$$

where $R(r)$ is the transition rate and r_0 the distance of closest approach of ion to surface.

V_c is thus a complex quantity, and difficult to estimate, which makes $(1-P)$ difficult to calculate. However, most estimates indicate that $(1-P) \approx 10^{-3}$.

Quantification. As with the other surface analytical techniques, in principle one can therefore quantify by calculation, and indeed in a few cases, such as the quantitative study of potassium segregation to the V_6O_{13} (001) surface carried out by DE GRYSE et al. [208], such an approach has been successful. In general, however, the uncertainties in some of the quantities in Equation (24) make such a calculation too difficult, and quantification is based on standards. Thus, BRONGERSMA et al. [209], in their study of surface enrichment of copper in a Cu–Ni alloy, calibrated the ISS intensities against pure copper and nickel targets, whereas NELSON [210] used gold and silver elemental standards in ISS analyses of a range of Au–Ag alloys. He found a linear relationship between the surface concentrations of the alloys after prolonged sputtering to equilibrium composition, and the bulk concentrations measured by an electron microprobe, as shown in Figure 62. BEUKEN and BERTRAND [203] reduced each of the spectra from the standard materials in Figure 59 to components consisting of Gaussian peaks and inelastic tails, and then added the components in different proportions to synthesize the observed spectra from their sulfided Co–Mo catalysts. The result of such synthesis when applied to the ISS spectrum from the catalyst after heating to 800 °C is given in Figure 63. From such data processing the relative concentrations of the metallic elements on the catalyst surface after various heat treatments could be derived very accurately.

Depth Profiling. Unlike SSIMS, ISS is not inherently destructive under normal conditions of operation (i.e., He^+ at a few kiloelectronvolts and 1–10 nA) since the sputtering rate is negligible. If information in depth is required, then surface erosion must be introduced, as for XPS and AES. One obvious way of performing depth profiling is to use the same primary ion for both ISS and sputtering, but at quite different power densities for the two operations. Thus ROSSI et al.

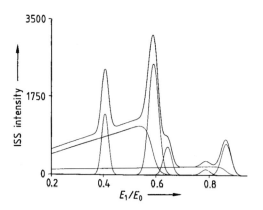

Figure 63. Curve synthesis method applied to the ISS spectra from the catalyst CoMoS–γ-Al$_2$O$_3$ after heat treatment at 800 °C [203]

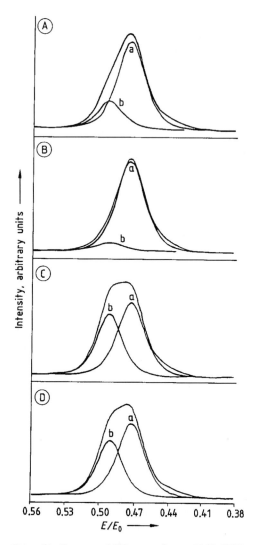

Figure 64. Sequence of ISS spectra from an Fe53–Ni47 alloy passivated in 1 mol/L NaOH ($t_p = 5$ min, $E_p = 0.44$ V), during depth profiling [211]
The peaks have been resolved into iron and nickel contributions after subtraction of an inelastic background
A) 1 min; B) 5 min; C) 15 min; D) 30–120 min; a) Iron; b) Nickel

[211] used Ne$^+$ at 45 nA and 3 keV (i.e., 1.35×10^{-4} W) for ISS, and at 300 nA and 5 keV (i.e., 1.5×10^{-3} W) for sputtering, in their study of passive film formation on Fe53–Ni. Figure 64 shows the Fe–Ni regions of ISS spectra obtained by them after successively longer sputtering times. Inelastic backgrounds have been subtracted from the spectra, and the peaks have been resolved into iron and nickel contributions. After 5 min sputtering, the iron contribution is greatly enhanced, indicating an Fe-rich layer within the passive film, about 1 nm below the surface. On prolonged sputtering the bulk composition is reached.

Another method of depth profiling with ISS, used by KANG et al. [212], is to bombard the surface simultaneously with He$^+$ and Ar$^+$ ions. ISS is performed using He$^+$, while the surface is eroded with Ar$^+$. Because the gases were mixed together in the same ion source, a duoplasmatron, the energies of measurement and sputtering were the same, in the range 1.5–2.0 keV. According to the mixing ratio used, the relative currents of He$^+$ and Ar$^+$ could be varied.

The much smaller sampling depth of ISS, of the order of one atomic layer, compared to those of XPS and AES, implies that the depth resolution in ISS during profiling will be better. This improved depth resolution has been demonstrated by HAEUSSLER [213] in comparisons of ISS, XPS, and AES profiles of the same materials.

27.5.1.5. Applications

Because of its high surface specificity relative to XPS and AES, most published applications of ISS have been concerned with basic experiments on clean surfaces, usually of single crystals. The negligible sputtering rate when using low-energy He$^+$ also means that gas–solid interactions can be studied without worry about alteration of the surface during measurement.

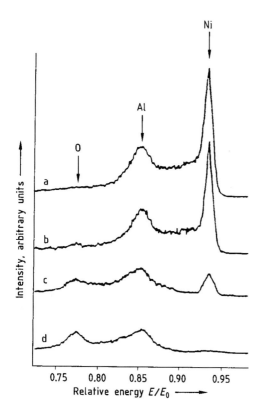

Figure 65. ISS spectra from Ni₃Al (001) after exposure to oxygen at 2×10⁻⁵ Pa and 700 °C [214]
a) Clean; b) 1-s Exposure; c) 15-s Exposure; d) 100-s Exposure

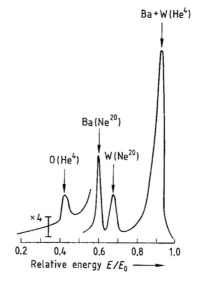

Figure 66. ISS spectra from a tungsten surface containing oxygen and barium, using a mixture of He⁺ and Ne⁺ to analyze for oxygen as well as separate barium and tungsten [215]
×4 = amplification factor

Typical of recent basic studies has been that of SHEN et al. [214] of oxygen adsorption on Ni₃Al (001). The intermetallic compound Ni₃Al is known to have excellent oxidation resistance in air at high temperature, forming an adherent aluminum oxide film, so information about the initial stages of oxidation is of interest. The Ni₃Al (001) surface was cleaned by Ne⁺ ion bombardment, and its stability as a function of temperature checked by ISS, which showed that up to 900 °C the Ni–Al ratio remained unchanged. Figure 65 shows a selection of ISS spectra from the alloy surface as a function of oxygen exposure at 700 °C, using 2-keV He⁺ at current densities between 2×10⁻⁸ and 8×10⁻⁸ A/cm⁻². The oxygen peak increases, as expected, and both the aluminum and the nickel peaks decrease. However, the nickel peak intensity decreases at a much faster rate than that of aluminum, and the peak eventually disappears. Plots of normalized nickel and oxygen peak intensities indicated that the nickel peak disappeared when one monolayer of Al₂O₃ was formed, confirming that the protective mechanism operates in the earliest stages of interaction. The authors also used ISS structurally to establish that the growing oxide had no order and was not in coherence with the underlying crystal.

A more technological application to which ISS is particularly suited is that of the surface composition of dispenser cathodes. These cathodes are believed to achieve optimum operation when an approximately monolayer coverage of barium and oxygen exists at the surface. When using XPS and AES, establishing that any barium signal comes only from the outer atomic layer of the surface is not easy, but the single-layer sensitivity of ISS should be able to provide that information. A problem in applying ISS to the cathode system is that the active barium–oxygen layer is supported on a tungsten substrate, and that both barium and tungsten are heavy elements whose resolution using He⁺ would be difficult. On the other hand, oxygen must also be analyzed. The problem was solved by MARRIAN et al. [215] by performing ISS with a mixture of He⁺ and Ne⁺, so that oxygen would be analyzable by the He⁺, while the use of Ne⁺ would help separate barium and tungsten. A spectrum from their paper is given in Figure 66, showing that the analyses using the mixture did in fact

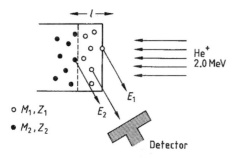

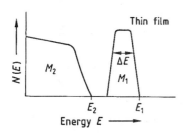

Figure 67. Illustration [219] of the principle of Rutherford backscattering spectroscopy (RBS) from a solid consisting of a thin film of thickness l composed of atoms of mass M_1 and atomic number Z_1 situated on a substrate of atoms of mass M_2 and atomic number Z_2

provide the analytical information needed. The authors were able to establish the existence of fractional monolayer coverage of barium and oxygen on the cathode surface.

Other recent applications of ISS have included the study of thin ZrO_2 films [201], a comparison of various techniques for cleaning InP substrates before epitaxial growth [216], analysis of conversion coatings on aluminum alloys [217], and study of alkali-induced sintering of an iron oxide monolayer on platinum [218].

27.5.2. Rutherford Backscattering Spectroscopy (RBS)

RBS has been in use since the 1960s for bulk materials analysis, and increasingly since the 1970s for thin-film analysis, particularly in the semiconductor field. The overall number and range of applications are enormous, but those dealing specifically with surface problems are less numerous. Nevertheless RBS ranks along with SSIMS and ISS as one of the major surface analytical techniques.

27.5.2.1. Principles

As in ISS, elemental compositional analysis is achieved by measuring the energy losses suffered by light primary ions in single binary collisions. Although in ISS the primary species is usually He^+, in RBS either H^+ or He^+ is used exclusively, and at very much higher energies, typically 1–2 MeV. Consequently, due to the penetration of the primary ion, both outer atomic layers and subsurface layers of a material are accessible; the surface sensitivity comes when the material in the outer few monolayers consists of heavy elements.

The energy of the backscattered ion, and hence elemental analysis, is governed by the same relationship (Eq. 22) already described for ISS. However, in RBS the ion penetrates the target and may be backscattered by the target atoms at any point along its path. Because of its very high energy, the ion can lose energy by collisional interactions with electrons and plasmons along either or both incoming and outgoing paths, which are straight lines. Thus, instead of the single peak observed by scattering from a particular element in the first atom layer in an ISS spectrum, the additional losses in energy cause a much broadened peak to be observed in an RBS spectrum. In RBS the energy spectrum is in fact a superposition of mass (i.e., elemental) analysis and depth analysis.

The principles have been illustrated elegantly by VAN OOSTROM [219], as shown in Figure 67. There a thin film consisting of atoms of mass M_1, atomic number Z_1, and thickness l is situated on a substrate of atoms of mass M_2 and atomic number Z_2. The scattered ion energies from M_1 and M_2 are E_1 and E_2, respectively. If $M_1 > M_2$, then the energy $E_1 > E_2$, in which case the peak corresponding to M_1 is well separated from M_2, and the peak width ΔE is related directly to l. Because the atoms M_1 are in the surface, the threshold E_1 will be given by Equation (22) as in ISS, but since the atoms M_2 are below the thin film the threshold E_2 may be lower than that given by Equation (22) if energy losses occur in the film. The energy spectrum produced by scattering from atoms M_2 will start at the threshold E_2 and then increase into a very broad band since M_2 forms the substrate of effectively infinite depth. If $M_1 < M_2$, the peak due to M_1 merges with the broad band from M_2, and may or may not appear as a much smaller feature on top of the band.

The above makes clear that for surface analysis by RBS, conditions must be such that $M_1 \gg M_2$ for

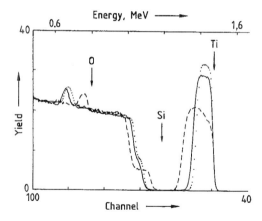

Figure 68. RBS spectra obtained from a sandwich structure of 110-nm titanium on top of 90 nm SiO$_2$ on a silicon substrate [221]
... = Before heating; —— = After heating to 575 °C;
- - - - = After heating to 800 °C

the peak due to M_1 to be completely resolved. Under these circumstances, the sensitivity of RBS is comparable to AES and XPS (i.e., of the order of 10^{-5} atomic layers). As in ISS, no chemical information is available.

27.5.2.2. Instrumentation

He$^+$ or H$^+$ ions with energies in the megaelectronvolt region are produced in particle accelerators. These may take the form of a tandem electrostatic Van de Graaff accelerator; a cyclotron, which uses electromagnetic confinement to a circular orbit; or a linear accelerator, in which the ion is accelerated linearly across gaps between drift tubes by electromagnetic fields. The reader is referred to [220] for greater detail about accelerator technology and about accelerator-based methods of analysis. Accelerators normally produce ions in several charge states, so that the required ion is chosen by magnetic selection and often by velocity selection as well. The purified ion beam passes immediately into a UHV environment via differentially pumped apertures, since the vacuum environment in most accelerators is not that of UHV. For general analysis, in which channeling of the beam along particular crystal directions is not present, ions strike the specimen at normal incidence. When structural information is sought, however, in the same manner as in ISS, the specimen may be rotated about the point of ion impact to vary the angle of incidence about channeling directions.

Detection of backscattered ions is normally performed by a silicon surface barrier detector at high scattering angles for increased surface sensitivity, as indicated in Figure 67. The detector position is often variable about the specimen position to allow various scattering angles to be chosen.

The ion beams used in RBS can be focused to small spots if required, although conventional analysis would normally be performed over an area of ≈ 1 mm^2. Spot sizes of 5–10 µm can be obtained by suitable lens systems, but scanning at such high energies is too difficult. Selected area analyses using small spot beams is carried out by specimen manipulation.

27.5.2.3. Spectral and Structural Information

An RBS spectrum, like that of an ISS spectrum, consists of the energy distribution of backscattered primary ions. Unlike ISS, however, the energies of the backscattered ions are governed not only by the mass of the scattering atom but also by the depth of that atom.

An example of the nature of the compositional information obtainable from an RBS spectrum is based on Figure 68, from the work of KUIPER et al. [221]. They grew a 90-nm-thick film of SiO$_2$ on a silicon substrate, then deposited a 110-nm-thick film of titanium on top of the SiO$_2$, and heated the sandwich to various temperatures. In Figure 68 the dotted line is the spectrum recorded before heating; the solid line, after heating to 575 °C for 1 h; and the dashed line, after heating to 800 °C for 1 h. The vertical arrows are positioned at the backscattered energies expected for atoms of the indicated elements if they were in the surface layer.

For the titanium film, some titanium atoms always exist in the outer layer, but on heating to successively higher temperatures, titanium is found at progressively greater depths (i.e., it is redistributing). The width of the peak before heating corresponds exactly to the thickness of the deposited film. In the case of silicon, at no stage do any silicon atoms appear in the outer layer, although after heating to 800 °C the silicon edge is near the surface. The double edge of the silicon signal is due to the fact that, after heating to 800 °C, silicon is in two chemical states, one as a silicide in combination with titanium, and the other in the underlying SiO$_2$, which has reduced in thickness to 14 nm. After 800 °C heating, the ox-

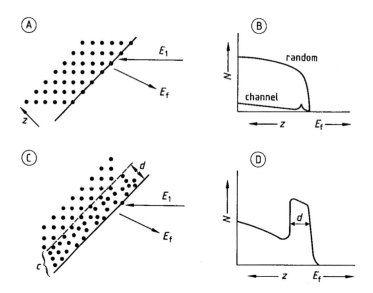

Figure 69. Illustration of RBS expected according to direction of incident beam [222]
A) Beam along a channeling (open) direction; B) Spectra expected from either nonchanneling (random) or channeling (open) directions; C) Beam incident on an amorphous layer of thickness d on surface; D) Spectrum expected from amorphous layer

ygen atoms have mostly moved nearer to the surface, with little change in their distribution after 575 °C treatment.

For relatively simple situations such as that of Figure 68, a spectrum simulation program can be used, in which the adjustable parameters are the stoichiometries of the likely compounds and their distribution with depth. Thus after 575 °C heating, the simulation gave the result that the surface consisted of titanium on top of a layer of $TiO_{0.23}$ overlaying 72 nm of SiO_2, with negligible silicide. After 800 °C, however, the fit was to a layer of Ti_2O_3 on top of the silicide Ti_5Si_3, covering the remaining 14 nm of SiO_2. The results agreed with depth profiling by AES performed at the same time.

RBS can be used to detect and measure the thickness of an amorphous layer on an otherwise crystalline substrate. The method is illustrated schematically in Figure 69, from WERNER [222]. If the incident ion beam is directed onto the crystal in an "open" or channeling direction (Fig. 69 A), it will penetrate more deeply into the crystal than if it is aligned along a nonchanneling or "random" direction. Thus the overall backscattered intensity in the channeling direction will be much lower than in the random direction, and in addition the small surface peak arising from collisions at the surface will be visible, as indicated in schematic spectra in Figure 69 B. Should the surface be covered with an amorphous layer (Fig. 69 C), no channeling is possible in the layer, so that more collisions will take place with atoms near the surface, and the intensity of the backscattered signal from the surface will increase (for the same incident beam direction), as shown schematically in Figure 69 D.

An application of this method of obtaining surface structural information has been demonstrated by BHATTACHARYA and PRONKO [223]. Implantation in compound semiconductors for production of the required electrical characteristics is usually accompanied by crystalline damage, and in particular a surface amorphous layer may be formed. Figure 70 from their paper shows the results of the RBS channeling measurements on GaAs implanted with silicon before and after annealing at 950 °C. The upper spectrum (a) is taken with the ion beam in a nonchanneling or "random" direction, and has the expected structureless shape. When the beam was aligned with a channeling direction, the spectrum before annealing (b) was also of high intensity and revealed a broad peak near the edge, while after annealing the intensity dropped substantially and a much narrower near-edge peak remained (c). The broad peak in the unannealed state showed that the surface had become amorphous to a thickness of

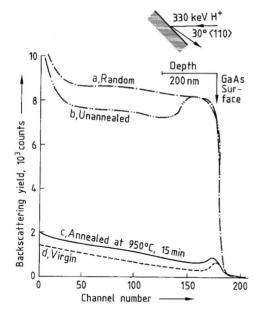

Figure 70. RBS spectra from GaAs implanted with silicon before and after annealing at 950 °C (120 keV Si$^+$–GaAs, dose: 5×10^{15} cm^{-2}) [223]
Uppermost spectrum is taken in a nonchanneling direction; the others are in channeling directions

≈ 140 nm, while the spectrum after annealing resembled closely that of the crystalline unimplanted crystal (d). Further structural information was obtained by recording the backscattered intensity as a function of angle around the GaAs [100] and [110] crystal directions, from which it was possible to calculate that about 70 % of the implanted silicon occupied substitutional lattice sites.

27.5.2.4. Quantification

The intensity of the current of high-energy ions backscattered from surface atoms of type A is very similar to Equation (24) for ISS, except that of course the neutralization probability is not relevant. Equation (24) then becomes for RBS

$$I_A^+ = I_o^+ N_A \sigma_A S \cdot \Delta\Omega \cdot T \tag{32}$$

where the symbols have the same meanings as before. Here σ can be derived from the Rutherford differential scattering cross section dσ given by

$$d\sigma = \frac{Z_1 Z_2 e^2}{4E^2} \cdot \frac{\cos\theta - \{1 - [(M_1/M_2)\sin\theta]^2\}^{1/2}}{\sin^4\theta \{1 - [(M_1/M_2)\sin\theta]^2\}^{1/2}} \tag{33}$$

θ being the scattering angle. For a particular primary ion and fixed experimental conditions, the differential cross section is thus a function only of the mass and atomic number of the scattering atom A, and can be calculated. Thus in principle, all the terms in Equation (32), except the required N_A, are known. In practice, the effective solid angle $\Delta\Omega$ of the analyzer is not well known, and using standards against which to calibrate the peak in the RBS spectrum arising from surface atoms (the so-called surface peak) is more common than attempting to derive calculated compositions. Such a peak can be seen clearly in the lower spectra of Figure 70.

Calibration is achieved by either implanting into or depositing on a light element such as silicon, a known amount of a much heavier element. Then if N_S is the known density of the standard element, one can write

$$N_A = (\sigma_S/\sigma_A) \cdot (I_A^+/I_S^+) N_S \tag{34}$$

given the same experimental conditions of measurement on the standard and the unknown. Then, in terms of the usual relative sensitivity factors, Equation (34) can be written as

$$N_A = R(A{:}S) I_A^+ \tag{35}$$

The accuracy depends on being able to measure the areas of the surface peaks in both standard and unknown with sufficient accuracy, which is often easy. Accuracies in RBS under favorable conditions can be better than 1 %.

27.5.2.5. Applications

Since RBS has operated as a major compositional and structural technique for bulk analysis for about 30 years, the number of published applications in that area is now enormous. Its applications in the field of specifically surface studies have not been so numerous, but they are increasing in number with the advent of improved vacuum technology. Many have been of the type already discussed in Section 27.5.2.3, in which the reaction or interdiffusion of one or more heavy metals on a semiconductor surface has been studied.

Typical of such studies is that of NAVA et al. [224] on the interaction of an Ni–Pt alloy with crystalline silicon. The system is a particularly favorable one since both nickel and platinum are much heavier than silicon, and platinum is in turn much heavier than nickel. Alloy films about

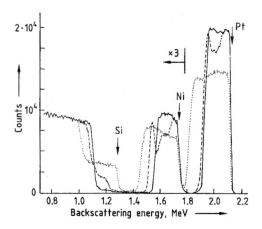

Figure 71. RBS spectra of a 145-nm-thick Ni–Pt alloy film on silicon before and after annealing [224]
— = Before annealing; --- = Annealed at 400 °C; ... = Annealed at 700 °C

145 nm thick were deposited on silicon by codeposition of nickel and platinum in approximately equal amounts. Annealing of the film and substrate was then carried out at progressively increasing temperature. Figure 71 shows the RBS spectra before annealing (solid line), after 30 min at 400 °C (dashed line), and after 30 min at 700 °C (dotted line). As before, the vertical arrows indicate the energy positions of the backscattered edges to be expected if the indicated elements are in the surface layer.

As shown, the as-deposited film contained both platinum and nickel atoms in the surface layer, and covered the silicon completely, since no silicon atoms appeared at the surface layer energy position. The sharpness of the trailing edges of the platinum and nickel peaks, as well as that of the leading silicon edge, showed that the interface between film and substrate was itself sharp. The film composition, based on relative peak area under the platinum and nickel peaks, was $Ni_{55}Pt_{45}$. Annealing at 400 °C causes little movement of the platinum, but segregation of the nickel into the interface, coupled with movement of silicon in the opposite direction (i.e., also into the interface). The compound formed there is identified as NiSi from the silicon step height. At 700 °C platinum has also shifted considerably, along with further movement of nickel, until they are in fact completely intermixed. The silicon edge has now moved up to the surface position, indicating that it is mixed with both platinum and nickel. Careful quantification establishes that both PtSi and NiSi have been formed in approximately equal quantities. Confirmation of the formation of these compounds came from X-ray diffraction. Thus the alloy–silicon reaction occurred first by the formation of NiSi, with a Pt-rich layer left above it, followed at slightly higher temperature by the reaction of platinum with diffusing silicon, and finally at the highest temperatures by complete mixing of PtSi and NiSi.

Similar metal–Si reaction studies have been performed by DE NIJS and VAN SILFHOUT [225] on Ti–Si, by FASTOW et al. [226] on pulsed ion beam irradiation of Ni–Cr on silicon, by ELLWANGER et al. [227] of $Pt_{0.4}Ni_{0.6}$ on silicon, and by OBATA et al. [228] of TiC on carbon and molybdenum.

RBS has also been used to characterize palladium and tin catalysts on polyetherimide surfaces [229], titanium nitride thin films [230], silicon oxynitride films [231], and silicon nitride films [232], and to study the laser mixing of Cu–Au–Cu and Cu–W–Cu thin alloy films on Si_3N_4 substrates [233], and the annealing behavior of GaAs after implantation with selenium [234].

27.6. Scanning Tunneling Methods (STM, STS, AFM)

Scanning tunneling microscopy (STM) was invented by BINNIG and ROHRER in 1983 [235], for which they received the Nobel Prize. The ability to produce images related directly to topography on an atomic scale was a major step forward for surface science, which until then had had to deduce surface atomic structure from techniques such as LEED and ISS. Since the appearance of STM, many sister techniques using similar principles have been developed for the study either of special materials, of surfaces in nonvacuum ambients, or of the surface distribution of effects other than purely topographical.

27.6.1. Principles

In *scanning tunneling microscopy* a metal tip is brought so close to a conducting surface (of the order of 0.5–2.0 nm) that an electron tunneling current becomes measurable at applied voltages in the range 2 mV to 2 V. The tip is then scanned over the surface. The tunneling current is monitored while scanning and used in either of two *mode*). In the simplest, one-dimensional view, the dependence of tunneling current I_T on applied voltage V_T and tip–surface separation s is given by the approximation

$$I_T \approx V_T \exp\left(-A\varphi^{1/2}s\right) \tag{36}$$

where A is a constant and φ is the effective local work function. Equation (36) shows that for a typical work function value of about 4.5 eV, the current changes by about an order of magnitude for each change in the separation of 0.1 nm, when close to the surface. Therefore in the constant current mode of operation the tip follows the surface contours by rising or falling, whereas in the constant height mode the current rises or falls. In either case an image related to surface topography is obtained by constructing a map from line scans displaced progressively in one direction by equal amounts. The principle is illustrated schematically in Figure 72, taken from HANSMA and TERSOFF [236] and HANSMA et al. [237].

For constant local work function φ, Equation (36) predicts a linear dependence of I_T on V_T, but this is obeyed only at very low values of V_T, since the equation is approximated from a more complex and rigorous expression. At higher V_T, the one-dimensional tunneling model on which Equation (36) is based breaks down, and the three-dimensional electronic structure of the surface region (i.e., the local band structure) must be considered. The tunneling current must then be calculated by first-order perturbation theory; if the assumptions are made that the tip can be replaced by a point and that the tip wave functions are localized (i.e., a constant density of surface states), then the current is given by

$$I_T \sim \sum_s \|\psi_s(r)\|^2 \delta(E_s - E_F) \tag{37}$$

where ψ_s is the wave function of surface state s; E_s the energy associated with ψ_s; and E_F the Fermi energy. The quantity on the right of Equation (37) is simply the surface local density of states (LDOS). Thus the tunneling current is proportional to the surface LDOS at the position of the point. For a fixed applied voltage V_T, the image in STM is therefore not just a function of topographical variations, but is essentially an image of constant surface LDOS. If the tip is now positioned over a particular point on the surface, and the current recorded as a function of V_T, the I–V dependence will mirror the LDOS, that is, peaks

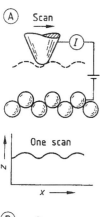
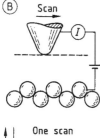

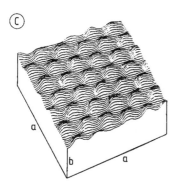

Figure 72. Illustration of modes used in STM [236], [237]
A) Constant current mode; B) Constant height mode; C) Surface topographical-spectroscopic image produced by scanning along progressive incremental distances in the y direction
a) 1.3 nm; b) 0.5 nm

ways: in one, a feedback circuit changes the height of the tip above the surface in order to maintain the current constant (*constant current mode*); in the other, the height of the tip is kept constant while variations in current are recorded (*constant height*

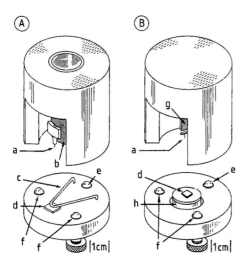

Figure 73. Schematics of the designs for an STM and an AFM used by HANSMA et al. [237]
A) STM; B) AFM
a) Tip; b) Single-tube *xyz* piezo; c) Sample clip; d) Sample; e) Fine advance screw; f) Coarse advance screws; g) Spring deflector sensor; h) *xyz* Piezo

will occur near saddle points in the surface band structure, along with features due to localized states (e.g., so-called dangling bonds). The $I-V$ dependence is therefore providing a tunneling spectroscopy, and according to whether $V_T < 0$ or $V_T > 0$, occupied or unoccupied states, respectively, are explored. Once a spectral feature has been found at a particular value of V_T, if the applied voltage is set to that value and the tip is scanned as before, a two-dimensional distribution of that feature is obtained. This is known as *scanning tunneling spectroscopy* (STS).

Because STM and STS are basically the same, in that topography and electronic distribution at the surface both contribute in varying proportions to the image that is actually seen, other methods of recording pure topography have been sought. In addition, STM (STS) requires that the surface be conducting, whereas the topography of many nonconducting surfaces would be of interest. In 1986, BINNIG et al. [238] calculated that a cantilever could be made with a spring constant lower than that of the equivalent spring between atoms. Thus the vibrational frequencies of atoms in a molecule or solid combined with the atomic mass lead to interatomic spring constants of about 10 N/m, compared with spring constants of 0.5 – 1 N/m in narrow strips of soft metal such as aluminum. A cantilever with a very fine tip virtually in contact with the surface, with actual applied forces in the range of $10^{-13} - 10^{-6}$ N, and a means of sensing and monitoring the tip deflection, will therefore measure tip – surface interactions arising from van der Waals, electrostatic, magnetic, and other weak forces. This is known as *atomic force microscopy* (AFM). It does not depend on the conducting nature of the surface, and the information is purely topographical; a vacuum environment is not necessary, and indeed its application in a variety of media has led to a number of variants of the technique.

27.6.2. Instrumentation

All *scanning tunneling microscopes* have been constructed as variations on the schematic shown in Figure 72. Although several commercial instruments are available, the basic construction is relatively simple, so that many individual designs have appeared in the literature. An example is shown in Figure 73 A, from HANSMA et. al. [237]. The sample (d) is fixed to a table mounted on three advance screws (e, f). The tip (a) is attached to the end of a piezoelectric translator (b), the tripod arrangement of Figure 72 being replaced by a segmented tube since the latter is more rigid and therefore less sensitive to vibration. The sample is brought near the tip first by manual operation of the coarse advance screws while the tip is viewed through a microscope, until the separation is about 10 μm. Then a stepping motor drives the fine advance screw to bring the sample more slowly toward the tip, the final approach being monitored via the tunneling current itself. Since the ends of the coarse advance screws form the pivotal axis for the fine screw, the motion of the latter is reduced by a factor of 10, allowing very fine approach adjustment.

The tip is scanned linearly across the surface by ramped voltage applied to the *x* piezosegment, with progressive shifts in the *y* direction to create a rastered pattern of movement. Distance to the sample surface is controlled by a voltage applied to the *z* segment, the voltage being determined in a feedback circuit by the tunneling current. Images are accumulated by multiple scans, each scan consisting of the voltage applied to the *z* segment as a function of that applied to the *x*, and displaced successively by *y* shifts.

Fabrication of tips does not require any difficult technology. The reason for this is that any fine point fabricated by mechanical or electrochemical

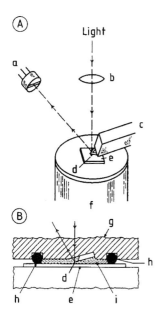

Figure 74. Cantilever deflection sensing used in AFM [239]
A) Principal scheme; B) Example of use in a liquid cell
a) Photodiode; b) Lens; c) Cantilever; d) Tip; e) Sample;
f) *xyz* Translator; g) Plexiglas; h) O-ring; i) Liquid

means will not be smooth on an atomic scale, but will have many "microtips" of near-atomic dimensions. When the tip is brought close enough to the surface, the tunneling current automatically selects the microtip that is nearest. Occasionally tips compete, or undesirable whiskers may be present, in which case an increase in voltage beyond the tunneling range will usually rectify the situation by removal of one or the other.

The other constructional requirement is rigidity. Low-frequency vibrations such as those found in buildings must not reach the tip–sample region; so, as indicated in Figure 73, components must be robust and rigid in order that such vibrations are damped.

In an *atomic force microscope* (an example is given in Fig. 73 B), the deflection of the cantilever spring is normally sensed by reflection of a laser beam from a mirror on the back of the cantilever into a two-segment photodiode, as shown in Figure 74 from WEISENHORN et al. [239]. The sample is positioned on a piezoelectric translator, and the signal from the photodiode is sent to a feedback circuit controlling the voltage applied to the z piezoelement, which moves the sample up or down in response to small changes in cantilever deflection. In this way an almost constant force can be maintained on the tip, which is important when imaging vulnerable biological material. For the latter the surface is scanned at the smallest force compatible with tip contact. Rastering to produce images is performed by the x and y piezo-elements of the translator.

Since the AFM does not use tunneling currents to sense the deflection of the cantilever and does not require a vacuum environment for operation, it has been used very successfully under water and other liquids. Figure 74 B shows the type of liquid cell employed by HANSMA and coworkers [239]–[243]. The tip with its mirror is attached to the underside of a strip of poly-(methyl methacrylate) that also carries a circular groove for a small O-ring. Once the sample has made contact with the O-ring, the cell is sealed, but the O-ring material is sufficiently flexible to allow movement of the sample under the tip in the usual way. A recent modification [244] of the design allows the fluid in the cell to be circulated in order to simulate flow conditions, while inserted electrodes allow cyclic voltammetry to be performed.

27.6.3. Spatial and Spectroscopic Information

Since the inception of STM, and latterly of AFM, a large number of elegant and often beautiful images have appeared in the literature, increasingly in color to emphasize the contrast. Although atomic resolution seems to have been achieved frequently with *STM*, care must be taken in the interpretation of images because of the overlap of topographical and electronic distribution variations. Thus in Figure 75, from EBERT et al. [245], imaging of the InP (110) surface reveals the positions either of the phosphorus atoms (A) or the indium atoms (B), depending on the applied voltage. In (A), at $V_T = -2.6$ V, the occupied state density is mapped, and in (B), at $V_T = +2.2$ V, the unoccupied state density. In both cases the resolution is atomic, but the spatial localization of the states is quite different.

With *AFM*, the images are unambiguous, provided the tip is very sharp so that multiple microtip effects are absent. Several materials have been used as tests of resolution, a favorite being freshly cleaved, highly oriented pyrolytic graphite, in which the average distance between carbon atoms is 0.14 nm. Calculations by GOULD et al. [246] show that every surface carbon atom should be imaged with equal intensity, and Figure 76 from

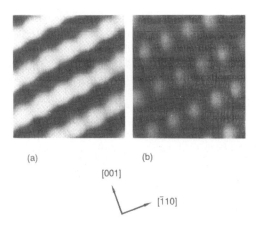

(a)   (b)

[001] [110̄]

Figure 75. STM images of the InP (110) surface over an area of 2×2 nm² [245]
A) Tunneling voltage –2.6 V; B) Tunneling voltage +2.2 V. Images at the two voltages reveal the occupied and unoccupied state densities, respectively

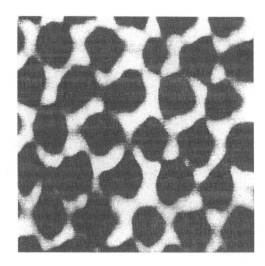

Figure 76. AFM image of graphite in air [247] Image size 1.17 nm. Carbon atoms can be seen as darker regions around central holes

GOULD et al. [247] confirms this prediction. The field of view in the figure is 1.17 nm across, and individual carbon atoms can be seen as alternate light and dark regions around each hole.

According to the applied voltage range, as indicated above for InP, STM, or rather, STS, can give information on both occupied and unoccupied surface states. The ST spectra over an extended range of V_T are therefore likely to be comparable with those from other techniques, although the comparison may not be exact because of the very localized nature of the tunneling that takes place in STS. Only if the bulk band structure is similar to the surface density of states (DOS) is agreement likely, and this has been found by KAISER and JAKLEVIC [248] for the clean Pd (111) surface. In Figure 77 from their paper, the $I-V$ spectrum from –2.4 to +2.4 V (Fig. 77 A) is compared with the experimentally determined band structure (Fig. 77 B) and the UPS spectrum (Fig. 77 C). The STS peaks at –1.4 and +1.0 V correspond to the positions of the d-band edge at about –1.2 eV and of the saddle point at +0.94 eV in B, while in the UPS spectrum (Fig. 77 C) surface-state peaks are found at about –0.5 and –2.1 eV, which match the peaks at –0.6 and –2.1 V. The minor feature in A near +0.5 V may also be a surface state. STS has also been compared with IPES [249] over the unoccupied range of surface states of graphite.

27.6.4. Applications

The ability of *AFM* to provide topographical images of all types of surface at spatial resolutions varying from the atomic to the micrometer scale, in air, vacuum, or under liquids, has led to a great variety of applications. They can be divided roughly into inorganic and organic (biological) applications.

An example of the way in which AFM has been applied to *inorganic material* surfaces is provided by the study of GRATZ et al. [250] on the dissolution of quartz in potassium hydroxide solution. Such a study forms part of a general attempt to understand the phenomena of the dissolution and growth of crystal surfaces, either of natural origin or grown synthetically. The authors etched optical-quality natural quartz for 4 h at 148 °C in KOH solution and then examined the surface with AFM, in this case in ethanol, which was found to keep the surface clean. Quartz is an insulator and could not therefore be studied by STM.

The etching created prismatic etch pits with smooth floors. Figure 78 shows some of the results. Image A, of area 1490×1490 nm², reveals the contours in the pit; the height separation from top to bottom of the pit is about 20 nm. The *c* and *a* axes of the quartz are indicated by the long and short arms of the cross, respectively. In image B, one quadrant of the pit is shown at higher magnification and in three dimensions, the field of

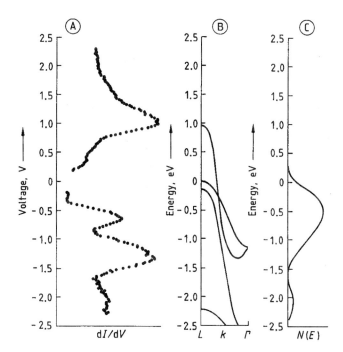

Figure 77. Comparison of
A) STS spectrum; B) Experimental band structure; and C) UPS spectrum, of palladium (111) [248]L, k, Γ = directions of symmetry in three-dimensional space

view being 650×650 nm². On the vertical scale the range is only 14 nm. The ledges visible in B are shown in C to be of only 0.7-nm height, to be atomically flat, and to be about 30 nm wide. Thus the etch pit can be described as an elongated amphitheatre, with a series of flat terraces of individual height only a few tenths of a nanometer. The results are said to be in agreement with the classical model of crystal dissolution.

Other recent inorganic material applications have included synthetic ultrafiltration membranes in air and in water [251], clay mineral surfaces [252]–[253], binding of molecules to zeolite surfaces [239], and electrochemical deposition of copper on gold in sulfate and perchlorate solutions [244].

Both AFM and STM have been able to image *biological materials* without apparent damage, which might be expected for AFM but is surprising for STM. In fact the imaging conditions used in STM must be kept as weak as possible to avoid damage. Thus, to obtain the image shown in Figure 79, from ARSCOTT and BLOOMFIELD [254], of a series of parallel fibers of Z-DNA, tunneling current and voltage levels were 1–2 nA and –100 to +100 mV, respectively. The characteristic left-handed twist of the Z-DNA helix is obvious in the image, and the measured helical periodicity of 4.19 nm is only 8% less than the crystallographic value, indicating minimal contraction due to dehydration. The measurements were performed in air.

STM has also successfully imaged phosphorylase from muscle [255] and biomembranes [256]–[257], while AFM has produced images of proteins in a buffer solution [258], immunoglobulin [259]–[260], DNA [261], [252], and bacteria [263].

Although many surface analytical techniques, including some of those already described, can provide valuable information about the nature of surface reactions, the information is invariably averaged over a larger number of reaction sites. With STS, reactions and reactivity can be studied at selected individual, even atomic, surface sites, which is an exciting capability. The observations of WOLKOW and AVOURIS [264] on the reaction of NH_3 with the Si(111) surface illustrate the possibilities elegantly. Figure 80 from their paper shows a series of images of the Si (111)–(7×7) surface. In (A), the clean surface is imaged at +0.8 V, revealing the characteristic hexagonal sur-

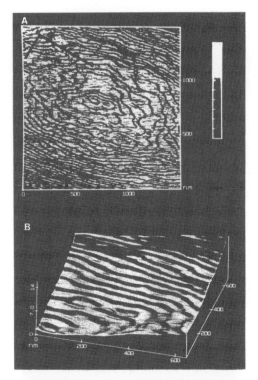

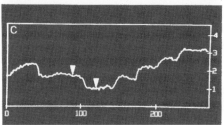

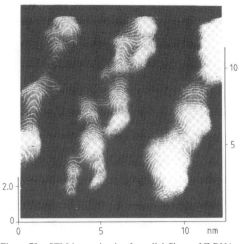

Figure 78. AFM images of an etch pit formed on quartz by etching in KOH at 148 °C [250]
The images were recorded under ethanol
A) Vertical view showing contours; B) Three-dimensional view of one pit quadrant; C) Profile across pit floor; height between cursors is 0.7 nm

Figure 79. STM image in air of parallel fibers of Z-DNA [254]
Tunneling current and voltage were kept very low to minimize damage

face structure, while in (B), the same tunneling voltage of +0.8 V is used, but after exposure to 1 Langmuir (i.e., 10^{-6} Torr · s or 1.33×10^{-4} Pa · s) of NH_3. In (B), about half the surface atoms seem to be missing. If the tunneling voltage is increased to +3.0 V, as in (C), the missing atoms reappear, but with different contrast from (A). The apparent disappearance is not real, but is an electronic effect due to the reaction of some of the surface atoms with NH_3; the basic structure is preserved, changes being those in the local DOS at certain atoms.

The accepted model for the (7×7) reconstruction of the Si (111) surface, that of TAKAYANAGI et al. [265], has silicon atoms in three special positions in the unit cell, those (in position A) underneath holes in a regular array in the outer layer, and others (B and C) present as adatoms (i.e., situated on top of the outer layer). Atoms B are adjacent to the holes at A; atoms C are in the center of the side of the unit cell. With STS the tip can be stationed over each of these atomic sites in turn, and the differences in local DOS recorded, both before and after reaction with NH_3. Figure 81 A shows the spectra from the clean surface, Figure 81 B, after the reaction. The quantity (dI/dV)/(I/V) is proportional to the local DOS.

Spectra from the clean surface show a very strong band at ≈ -0.8 eV from atoms A, suggesting a fully occupied dangling bond surface state. Since atoms B and C have much weaker occupied bands, the implication is that charge transfer occurs from B and C to A, with more being contributed by C than B. Both B and C have stronger unoccupied bands, confirming the transfer. After reaction with NH_3, the strong occupied dangling bond state at A has vanished, and at B (the dashed line) the small occupied band has also been lost. Elimination of these states accounts for the apparent disappearance of reacted atoms from the image in Figure 80 B. The other spectra in Figure 81 B, the solid lines for B and C, are for atoms of those types not themselves reacted but adjacent to reacted A atoms. In both cases, but particularly for

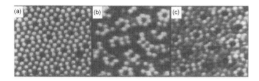

Figure 80. Series of STM images of the Si(111)–(7×7) surface [264]
A) Clean surface at 0.8 V; B) After exposure to 1 L of NH$_3$, also at 0.8 V; C) Same as B), but at 3.0 V

C, a transfer of intensity has occurred from the unoccupied to the occupied states, suggesting that during reaction a charge transfer takes place from A to C, and to a lesser extent B, allowing A to react with NH$_3$.

The above experiment demonstrates clearly how STS can follow changes in electronic distribution, and differences in reactivity, on an atomic scale — an unique capability. Although the Si (111) surface is probably one of the best candidates for such a demonstration, being atomically flat and possessing strong dangling bond states, the possibilities that are opened up by this technique in general are very clear.

STS has also been applied to a study of the unoccupied surface states of graphite [249], hydrogen-like image states on clean and oxygen-covered nickel and on gold epitaxed on silicon (111) [266], [267], the superconducting energy gap in Nb$_3$Sn [268], the electronic structure of the InP (110) surface [245], and copper phthalocyanine adsorbed on Cu (100) [269].

27.7. Other Surface Analytical Methods

27.7.1. Ultraviolet Photoelectron Spectroscopy (UPS)

In its principle of operation, UPS is similar to XPS, in that a surface is irradiated with photons and the energies of the ejected photoelectrons are analyzed. The physical relationship involved is the same as that of Equation (1) for XPS, but in UPS the energy $h\nu$ of the exciting photons is much lower since the photons are derived from a gaseous discharge that produces hard UV radiation. The gas normally used is helium, which, depending on pressure conditions in the discharge, will provide line sources of energy 21.21 eV (He I) or 40.82 eV (He II) with very narrow linewidths

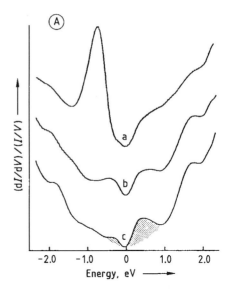

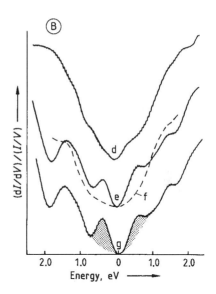

Figure 81. STS spectra recorded from the Si(111)–(7×7) surface [264]
A) In the clean condition; B) After reaction with 1 L of NH$_3$
a) Atom A; b) Atom B; c) Atom C; d) Atom A; e) Atom B not itself reacted but situated next to reacted atoms; f) Reacted atom B; g) Atom C not itself reacted but situated next to reacted atoms

(≈ 20 meV). Because of these low exciting energies, the binding energies E_B that appear in Equation (1) refer not to core levels, as in XPS, but to shallow valence band levels and to other shallow levels, such as those of adsorbates, near the valence band. For energy analysis, the CHA is again

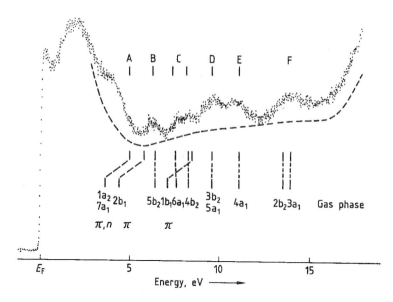

Figure 82. UPS He II spectrum from 5 Langmuirs pyridine adsorbed on Pd (111), with probable molecular orbital assignments ($\alpha = 60°$, $\theta = 50°$) [270]
The dashed line is part of the spectrum from clean Pd (111)
A – F = designation of spectral position of peaks

used, but the pass energies employed for UPS are much lower than in XPS since better energy resolution is required.

UPS is not an analytical technique in the sense that it can provide quantified elemental concentrations on a surface as XPS and AES can, but it is a powerful tool in basic research and is another of the group of most frequently used techniques. It is used in two ways. In its *angle-integrated* form it has been employed very extensively in studying the interactions of gaseous molecules with surfaces, usually in combination with other techniques such as XPS, AES, LEED, and TDS. For complex molecules, interpretation of the UPS spectra during interaction is often possible in terms of the molecular orbital structure of the free molecule, as demonstrated in Figure 82 from NETZER and MACK [270]. The other way in which it has been used is in *angle-resolved UPS (ARUPS)*. In this form the energy analyzer, a CHA of small radius, is mounted in such a way mechanically that it can traverse the space around the sample in both polar and azimuthal directions, while the angle of photon incidence is kept constant. The angle of incidence is then changed by rotation of the sample. If the sample is a single crystal with a clean surface, the dispersion of spectral features (i.e., their movement along the energy axis and their increase or decrease in magnitude) can be used theoretically to plot the band structure in the surface region. The changes in surface-associated band structure during reaction with gaseous molecules can then be followed, leading to additional information about the nature of the reaction.

27.7.2. Light-Spectroscopic Methods (RAIRS, SERS, Ellipsometry, SHG, IBSCA, GDOS)

The light-spectroscopic techniques have one thing in common: information is provided by them in the form of light, visible or IR, that has either had its spectroscopic properties changed by reflection from a surface or is emitted from atoms in excited states ejected from the surface. Four of the techniques (RAIRS, SERS, ellipsometry, SHG) also use light as the incident probe, the other two (IBSCA, GDOS) use incident ions.

Reflection – Absorption Infrared Spectroscopy. In RAIRS, IR radiation is reflected from a plane surface through an adsorbed layer, the reflected light losing intensity at those frequencies at which the light frequency coincides with a vibrational mode at the surface. The vibrations may be those within the adsorbed species itself, or they may arise from interaction with surface atoms.

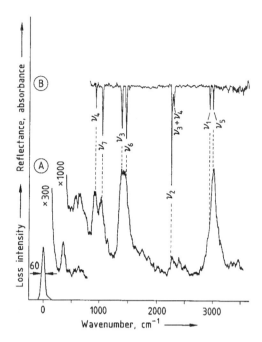

Figure 83. Comparison of HREELS and RAIRS vibrational spectra of CH_3CN adsorbed on Cu (111) at 100 K [271]
A) HREELS; B) RAIRS

The substrate surface must be a good reflector. The incident IR beam is plane polarized and is arranged to strike the surface near grazing incidence, typically 80–82°; both plane polarization and grazing incidence are required theoretically. For adequate signal-to-noise ratio in the reflected signal, the absorption cross section of the adsorbed layer at the characteristic frequencies must be high enough to be able to use good energy resolution.

Most earlier research using RAIRS was carried out with dispersive spectrometers to analyze the reflected light, but Fourier transform methods have largely taken over now, leading to the acronym FT-RAIRS. The advantages of using FT methods are, firstly, that either much improved signal-to-noise ratios can be obtained for the same acquisition time or data can be acquired faster for the same signal-to-noise ratio; and, secondly, that all spectral features are recorded simultaneously.

FT-RAIRS is closely related to high-resolution electron energy loss spectroscopy (HREELS) (see Section 27.7.3), and they are often used together in the same system. The two techniques are complementary, since FT-RAIRS has high energy resolution (≈ 2 cm^{-1}) and speed of acquisition, but its effective spectral lower limit is ≈ 800 cm^{-1} due to the transmission characteristics of the IR transparent optical components, while HREELS can cover the entire spectral range down to 0 cm^{-1} but has inferior energy resolution ($\approx 20-60$ cm^{-1}). A comparison of spectra obtained by the two techniques is shown in Figure 83, from ERLEY [271], for the adsorption of CH_3CN on Cu (111) at 100 K. The bands labeled v_3, v_6, v_1, and v_5 can be resolved by FT-RAIRS but not by HREELS, but HREELS can record vibrational bands at wavenumbers below the FT-RAIRS cutoff.

Surface-Enhanced Raman Scattering. Conventional Raman spectroscopy is beginning to be used more often in the study of very thin surface films, with the availability of improved detectors, but it still suffers from low intensity in the scattered signal. The discovery by FLEISCHMANN et al. [272] of the *surface enhancement effect* has enabled the Raman scattering effect to be applied more widely, even though restrictions exist. When the surfaces of certain metals, particularly silver, copper, and gold, are roughened on an atomic or near-atomic scale, the intensity of the scattered light is enhanced by factors of $10^5 - 10^6$. Roughening is usually performed electrochemically and is subject to some variation. Various theories have been put forward for the effect, without general agreement as yet. Despite the restrictions on the nature of the substrate, SERS has found very many applications, some of which have been carried out in UHV (e.g., the reactions of N_2 and CO_2 on silver, copper, and potassium surfaces at 40 K) [273]. For a thorough review of the physics of the SERS phenomenon and a discussion of applications, see [274].

Ellipsometry is a method of measuring the thicknesses, and changes in thickness, of very thin films on surfaces, to very high sensitivity, of the order of 5×10^{-3} nm. Polarized light, of wavelength within the UV to IR range of the spectrum, is specularly reflected from a surface with one or more layers of refractive index different from that of the substrate. The latter must be flat and highly reflecting. After the reflection the light is no longer plane polarized but elliptically polarized. The parameters measured are $\tan \psi$, the amplitude ratio of the resolved components of the electric vector of the reflected light parallel to and perpendicular to the plane of incidence, and $\cos \Delta$, the phase difference of the two components. *Fixed-*

photon-energy ellipsometry can provide only a measure of the thickness, but if the elliptical polarization parameters are measured as a function of the light energy and the angle of incidence, the technique is known as *spectroscopic ellipsometry*. The observed dispersion with energy can be related to the dielectric loss function; since the energy dependence of loss functions varies from one material to another, the contributions of different surface layers can in principle be distinguished. Ellipsometry in either form is therefore not a compositional analytical technique, but it can provide valuable additional physical information when used with one of the compositional techniques such as AES or XPS.

Optical second harmonic generation (SHG) is another light reflection technique of fairly recent origin that is being applied increasingly in UHV, although it can be used in almost any medium. Like ellipsometry, to which it is related, SHG is very sensitive to the interfacial regions between thin films and substrates, and can be used to study adsorption, molecular orientation, and surface structure. If a linearly polarized wave at a frequency ω is incident on a surface whose dielectric constant is different from that of the ambient medium, then a nonlinear source polarization is induced at the surface, which radiates at frequency 2ω in both reflected and transmitted directions. Inversion symmetry dictates that the second-order nonlinear susceptibility vanishes in the bulk of the solid medium, but at the surface the inversion symmetry is broken, leading to electric dipole contributions to the second-order nonlinearity. By using a laser pulsed at high frequency, time-resolved reactions at surfaces with time scales in the range $10^{-14} - 10^{-6}$ s can be studied. Enhancement of the SHG response can be achieved by tuning the laser to an electronic transition in the substrate or an optical transition of an adsorbate. The technique has recently been reviewed at length [275].

Ion Beam Spectrochemical Analysis. As remarked earlier, ion irradiation of a surface produces visible light photons as well as many other types of particle. The light emission arises from the deexcitation of ions and excited neutrals, and is characterized by a spectrum consisting of a series of sharp lines at wavelengths associated with the type of ionized or excited atom. Spectral analysis thus provides elemental identification by reference to standard emission wavelength tables.

The technique is called ion beam spectrochemical analysis (IBSCA) but has several other names as well. Typical operating conditions include bombardment by Ar^+ ions at energies in the range $2 - 20$ keV, with current densities of $10 - 100$ $\mu A/cm^2$; the light emitted just above the target surface is then focused onto the entrance slit of a scanning optical monochromator, which analyzes the light spectrum. Spatial resolution depends on the type of ion source being used, and can be of the order of $2 - 5$ μm with a suitable type such as a duoplasmatron. Because emission takes place from particles ejected from the surface layer, the surface specificity is similar to SSIMS (i.e., one atom layer). Quantification can in principle be carried out theoretically (e.g., based on the Andersen–Hinthorne LTE model [185] mentioned in Chapter 27.4) but is normally performed by relative intensity measurements, that is, by measuring the intensity of emission from a pure elemental standard under the same instrumental conditions of analysis as used for an unknown. Applications have included studies of the oxidation of chromium [276], the effect of segregation of carbon and sulfur to the surfaces of Fe–Cr–Mo crystals [277], the reaction between a thin PbO film and fused silica [278], and the protonation of lithium silicate glasses [279].

Glow Discharge Optical Spectroscopy. Unlike many of the other techniques described in this section, GDOS is widely used industrially and instruments are available commercially. In the great majority of applications, however, GDOS is used not as a surface-specific analytical technique, but as a rapid and quantifiable technique for determination of elemental concentration profiles through films often tens of micrometers thick. GDOS has certain similarities to IBSCA. The surface to be analyzed is made the cathode in a diode-type discharge cell, whose design originated with GRIMM [280], and argon is flowed through the cell at pressures of $10^3 - 10^4$ Pa. Application of a d.c. voltage of ≈ 1 kV between an anode and the sample cathode establishes a glow discharge. Positive ions produced in the discharge sputter the sample surface and erode it continuously, most of the sputtered material leaving the surface as neutral atoms. The emitted atoms pass into the discharge, in which they are excited by multiple collisions, and the excited states then decay by the emission of visible light, whose spectrum is analyzed in a multichannel spectrophotometer. According to the chosen operating conditions, the rate of erosion of

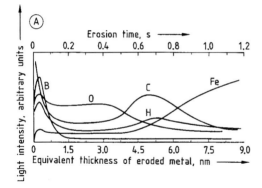

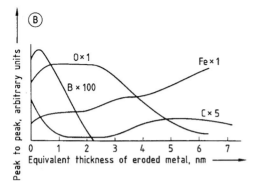

Figure 84. Comparison of depth profiles through a passive anodic film on iron [282]
A) GDOS; B) AES

the sample can be varied between 30 nm/min and 1 μm/min. The inherent sensitivity of the technique is high so that, if required, thin surface layers could be analyzed.

In arriving at a quantification of GDOS intensities, one must realize that as a result of the rapid uniform sputtering of the sample surface, both the surface composition (due to preferential sputtering) and the surface geometry (due to crater formation) are changing during the analysis. The latter change alters the glow discharge operating conditions since during sputtering the cathode surface is receding from the discharge region. BENGTSON [281] has shown how to obtain reasonable quantification on the basis of empirical expressions for the excitation function of the discharge and the sputtering rate in the discharge. If the intensity I_λ of an emission at wavelength λ from an atom of type A is measured, then the intensity is related to the concentration c_A of A by

$$I_\lambda = k_{\lambda A}\, c_A\, C\, i^2\, (V - V_0)^{x(\lambda, A)} \qquad (38)$$

where $k_{\lambda A}$ is a constant containing instrumental and atomic factors; C a constant dependent on cell geometry; i the bombardment current; V the applied cell voltage; V_0 the discharge threshold voltage, also dependent on geometry; and $x(\lambda, A)$ a constant characteristic of the emission line.

Equation (38) can be used to measure the concentration of an element in an unknown material if the emission parameters from a standard are known, since by taking the ratio of currents from unknown and standard, some of the parameters in the expression disappear. For elements in low concentration in alloys or compounds the accuracy is good, typically ±5 %.

Although GDOS is normally applied as a very fast profiling technique, leading to the publication of many profile diagrams of films up to ≈20 μm thick, its operating principles indicate that it is inherently surface-specific, and it can indeed be operated at a much lower erosion rate to study the near-surface region. GDOS also has an advantage over most of the other principal surface analytical techniques in that it can analyze for hydrogen. This ability and also its operation in a surface-specific manner are demonstrated in Figure 84, from BERNERON and CHARBONNIER [282], where the profiles by GDOS (A) and by AES (B) of an anodic passive film on iron are compared. The film was only ≈5 nm thick. GDOS shows a pile-up of hydrogen at the film–metal interface and is able to detect boron (from the boric acid buffer) beyond the point at which AES cannot. The profiles are qualitatively similar. Note that GDOS gives purely elemental information, with no direct chemical information available.

Other near-surface applications have included the implantation profiles of phosphorus in iron and FeO [283]; TiN and copper coatings on steel [284]; chromate layers on aluminum [285]; and implantation profiles of phosphorus and boron in iron and nickel [286].

27.7.3. Electron Energy Loss Methods (ELS, CEELS, HREELS, IETS)

Energy Loss Spectroscopy and Core-Electron Energy Loss Spectroscopy. The techniques of ELS and CEELS can be discussed together because they use the same experimental arrangement, basically that of conventional AES. A beam of primary electrons interacts with electrons in surface

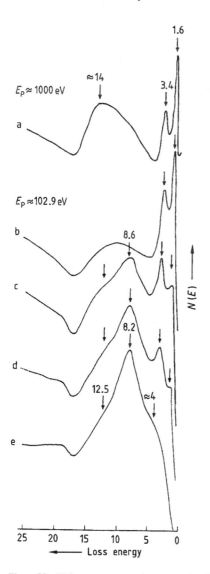

Figure 85. ELS spectra at two primary energies (E_p) recorded during interaction of hydrogen with the Ce (001) surface [287]
a) +900 L H_2; b) +900 L H_2; c) +600 L H_2; d) +300 L H_2; e) Clean

atoms, and the energy losses suffered by the primary electrons are measured by recording the appropriate part of the secondary electron spectrum in an energy analyzer. In *ELS* the losses are relatively small being due to the energies required to stimulate electronic transitions within the valence band of the solid, and collective oscillations of the conduction electrons (plasmon oscillations). For each solid the plasmons have energies character-

istic of the bulk structure and of the surface structure, the surface plasmon having theoretically an energy lower than that of the bulk by a factor of $2^{1/2}$. In *CEELS* the primary electron interacts with a core level, usually but not always leading to ionization, and the losses are in general much greater than in ELS. Thus in ELS the observed losses are normally within 30–50 eV of the primary energy, whereas in CEELS they may be several hundred electronvolts away.

Since all loss features are by definition associated with the primary energy, features characteristic of the surface can be compared to those of the bulk by variation of the primary energy (i.e., by effectively varying the electron escape depth). Such variation is standard practice in ELS, in which the range of primary energies used is typically 100–1000 eV. In CEELS, on the other hand, although the relation between primary energy and escape depth still applies, the primary energy must be high enough to ionize the core level with adequate efficiency, as discussed in Section 27.3.2.2 for AES. Thus for CEELS the primary energy range would be the same as for conventional AES (i.e., 2–5 keV).

Applications. ELS cannot be used to extract compositional information, but it has been used extensively in conjunction with other techniques, particularly AES and UPS, in basic studies of reactions occurring at clean or otherwise well-characterized surfaces. Changes in both plasmon and valence band losses occurring during reaction, particularly as a function of primary energy, provide useful information complementary to that offered by the other techniques. A typical example is shown in Figure 85, from the work of ROSINA et al. [287], for the interaction of H_2 with the clean Ce (001) surface. A primary energy of 102.9 eV was used for most of the measurements, to ensure maximum surface sensitivity, with a switch to 1000 eV occasionally to emphasize more bulk-related features. In the clean surface spectrum (curve e) the surface plasmon at 8.2 eV is dominant, with a shoulder at ≈ 12.5 eV due to the bulk plasmon and another shoulder at ≈ 4 eV probably arising from an intrinsic surface state. On hydrogenation, two additional features at 1.6 and 3.4 eV appear simultaneously with suppression of the 4-eV shoulder. The new features increase with H_2 exposure, and become dominant at 900 L and a primary energy of 102.9 eV. When the primary energy is changed to 1000 eV, the bulk plasmon is seen to have shifted to ≈ 14 eV from ≈ 12.5 eV. Both the features at 1.6 and 3.4 eV and the new position of the bulk plasmon are characteristic of the formation of CeH_2. The increase in the 1.6-eV loss relative to that at 3.4 eV in going to the higher primary energy is indicative of a hydride phase increasing in hydrogen concentration with depth. This particular example is also important because it illustrates that ELS can give information about surface interactions with H_2 that neither XPS nor AES can give.

The many other recent applications of ELS include the interaction of oxygen with titanium [288], [289]; of oxygen [290] and of hydrogen [291] with Co–Ti; of oxygen with tin [292]; of pyrazine with silver [293]; and of ytterbium films with aluminum [294].

CEELS has been applied more to the study of core-electron excitation to localized final atomic states than to excitation to the continuum (i.e., complete ionization). Core-level ionization thresholds can in principle be used to obtain an elemental composition, but the spectral features are weak and the analysis is performed far better by XPS. On the other hand, where shallow, highly localized final states exist, CEELS has been able to provide useful information. Such localized states are plentiful throughout the lanthanide rare-earth series, in some cases being very close to the valence band. Figure 86, from STRASSER et al. [295], shows some examples of the fine structure to be found in CEELS above the main $4d$ excitation threshold, in the lighter lanthanides. The structure arises from strong coupling between the $4d$ holes left after excitation and the $4f$ electrons or holes, the $4f$ states being highly localized. Where the $4f$ states are close to the valence band, their occupancy is dependent on the atomic valence, and valence changes can be monitored exactly by recording the $4d$ CEELS spectra during treatments such as oxidation that usually alter the valence. This analytical variation has been explored exhaustively by NETZER and coworkers [296]–[298].

High-Resolution Electron Energy Loss Spectroscopy. HREELS has already been mentioned in the description of RAIRS as being a vibrational spectroscopy. In this technique, electrons of very low primary energy, typically 3–5 eV, are directed at a surface carrying adsorbed molecules. As the electrons approach closely enough, they interact with the molecular field and are reflected (i.e., scattered) having lost discrete amounts of energy corresponding to characteristic vibrational frequencies, either within the molecules or between the molecules and the surface. Analysis of the scattered electron energy distribution then reveals the energy losses, and hence the vibrational frequencies. If the molecules are attached to the surface only loosely (*physisorption*), many of the characteristic vibrations within a molecule are similar to those observable in the gas phase. These consist of various stretching and deformation modes of vibration. In addition, other modes appear as a result of attachment to the surface. In the gas phase a molecule can translate and rotate freely, but if one end is fixed to a surface it can no longer do so, and therefore certain translational and rotational modes of vibration become observable that cannot be observed in the free molecule. Should the molecule be attached strongly to the surface (*chemisorption*), perhaps with the loss of reactive groups, and certainly with a transfer of electronic charge, then the characteristic free-molecule stretching vibrations can become so distorted that they no longer resemble those seen in the gas phase. Thus the HREELS spectrum from a complex polyatomic molecule adsorbed on a surface can itself be very complex and contain a great deal of information concerning the way in which the molecule has reacted with the surface and also with neighboring molecules. The orientation of the molecule can often be deduced as well, particularly when HREELS is used with other techniques such as LEED, UPS, and TDS.

Because the energy widths of the characteristic vibrational energies are small, < 0.1 meV, for optimum energy resolution one must achieve as narrow an energy spread in the incident electron beam as possible. In all HREELS equipment, therefore, extreme measures are taken to reduce the spread in a variety of ways. Most importantly, the incident electrons are energy selected in a monochromator that is basically an electron energy analyzer used in reverse (see, for example, [299]). Monochromators properly constructed can reduce the energy spread in the incident beam to 2–3 meV, but to maintain that width in the beam at the sample surface requires the complete elimination of stray magnetic and electrostatic fields around the sample. Thus very careful screening of the entire HREELS analytical arrangement is necessary. Many experiments are also carried out at low temperature to reduce the thermal broadening of characteristic vibrations.

The above description indicates that the applications of HREELS have been entirely basic and concerned with the interactions of molecules with clean or well-characterized surfaces. A very typi-

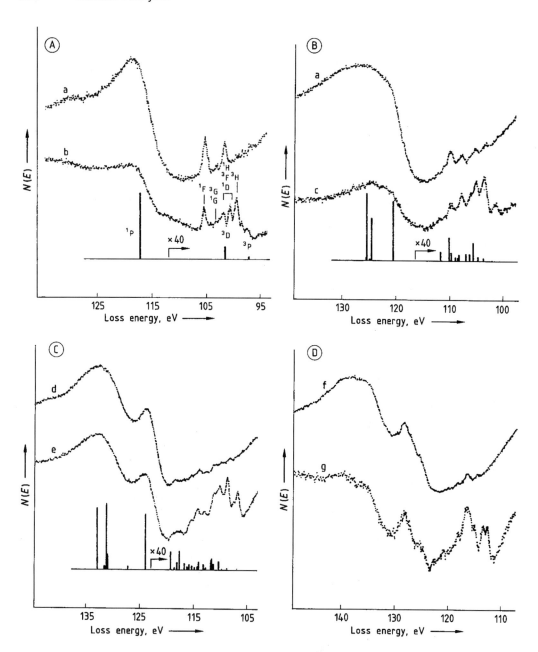

Figure 86. The $4d \to 4f$ CEELS spectra of the light lanthanides, at two primary energies [295]
A) Lanthanum; B) Cerium; C) Praeseodymium; D) Neodymium
a) $E_p = 1595$ eV; b) 315 eV; c) 350 eV; d) 1750 eV; e) 325 eV; f) 1725 eV; g) 290 eV
^1P, ^1F, etc; are final-state spectroscopic terms labeling the loss peaks

cal example of the HREELS spectra observed during increasing coverage of an adsorbed molecule is shown in Figure 87, from BARTKE et al. [300], for the adsorption of NO on Nb (100) at 20 K. The energy loss scales are expressed both in millielectronvolts and reciprocal centimeters for convenience, and the energy resolution, as governed by the width of the primary beam, is ≤ 4 meV.

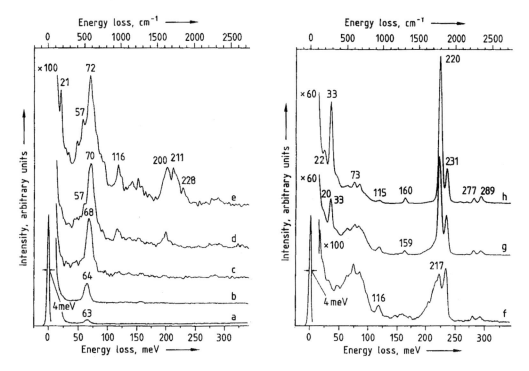

Figure 87. HREELS spectra recorded during the adsorption of NO on Nb (110) at 20 K (primary energy 2.5 eV, $I_0 = I_s = 52°$) [300]
Intensities are normalized to the elastic peak, and spectra were taken in the specular reflection mode
a) 0 L; b) 0.25 L; c) 0.5 L; d) 1 L; e) 2 L; f) 4 L; g) 8 L; h) 16 L; L = Langmuir (10^{-6} Torr · s)

The loss spectrum from the "clean" surface (curve a) is featureless apart from a tiny peak at 63 meV due to atomic oxygen, undetectable by AES. Initial adsorption of NO causes the peak to grow and others to appear; with increasing coverage the oxygen peak (arising from lattice Nb–O vibrations) shifts to 72 meV. The new peaks at 57 and 116 meV are ascribed to atomic nitrogen bonded in a "bridged" position (i.e., across two Nb atoms) and to atomic oxygen bonded on top of a niobium atom, respectively. Further new peaks at 200, 211, and 228 meV, after 2 Langmuirs adsorption, are associated with stretching vibrations within the NO molecule adsorbed on top of a niobium atom. The slight differences in loss energy are explained by different angles of adsorption of the NO (i.e., tilted or bent). With further NO adsorption (curves f–h) the NO stretching vibration shifts to 220 meV and becomes very strong, with a simultaneous decrease in the atomic vibration, indicating a preponderance of molecular adsorption. Additional peaks at 231 meV and at 22 and 33 meV are attributed to vibrations within an ON–NO dimer on the surface. Thus, with correct interpretation, a few spectra such as those in Figure 87 can provide a large amount of information on the course of a reaction and the nature of the entities actually on the surface.

HREELS has also been used recently to study the adsorption systems 2,3-dimethyl-2-butene and 2-butyne on Ni (111) [301]; straight-chain alcohols on Ag (100) [302]; CO and potassium on Cu (111) [303]; O_2 on cleaved and sputtered graphite [304]; CO on copper-covered Ni (111) [305]; O_2 on Fe (100) [306]; and CO on Ru ($10\bar{1}0$) [307]. Many of these systems are models for catalytic reactions of technological interest.

Inelastic Electron Tunneling Spectroscopy. IETS is unique in that it is entirely surface specific, but does not require a vacuum environment and has almost never been performed in UHV. Its principle is relatively simple. If two metals are separated by a thin (≈ 3 nm) insulating layer, and a voltage is applied across them, electrons can tunnel from one to the other through the insulator, and if no energy is lost by the electrons, the process is known as elastic tunneling. However, should discrete impurity states occur in the interface between either metal and the insulator, or molecules exist in such an interface that possess characteristic vibrational energies, then the tunneling electron can give up some of this energy either to the state or to the vibrational mode, before reaching the other metal. This is called inelastic tunneling. Obviously, the applied voltage must be greater than the state or vibrational energies. If the current across the metal–insulator–metal sandwich is recorded as a function of applied voltage, the current increases as the threshold for each state or vibrational mode is crossed. The increases in current are in fact very small, and for improved detectability, the current is double differentiated with respect to voltage, thereby providing in effect a vibrational spectrum that can be compared directly with free-molecule IR and Raman spectra.

The metal–insulator–metal sandwich is known as a tunnel junction, and its preparation is all-important. The standard junction consists of an aluminum strip ca. 60–80 nm thick and 0.5–1.0 mm wide deposited in a very good vacuum onto scrupulously clean glass or ceramic, the surface of the strip then being oxidized either thermally or by glow discharge. The resultant oxide layer is extremely uniform and about 3 nm thick. Introduction of adsorbed molecules onto the oxide layer, or "doping" as it is called, is then effected either by immersion in a solution followed by spinning to remove excess fluid, or by coating from the gas phase. The final stage in preparation is deposition of the second metal (invariably lead) of the sandwich; this deposition is carried out in a second vacuum system (i.e., not the one used for aluminum deposition), the final thickness of lead being about 300 nm, and the width of the lead strip being the same as that of the aluminum strip. The reason for using lead is that all IETS measurements are carried out at liquid-helium temperature, 4.2 K, to optimize the energy resolution by reducing the contribution of thermal broadening to the linewidth, and of course lead is superconducting at that temperature.

Normally not one, but several, junctions are fabricated simultaneously, since the likelihood of junction failure always exists. After fabrication the electrical connections are made to the metal electrodes, the junctions are dropped into liquid helium, and measurements are commenced at once.

Although insulators other than aluminum oxide have been tried, aluminum is still used almost universally because it is easy to evaporate and forms a limiting oxide layer of high uniformity. To be restricted therefore to adsorption of molecules on aluminum oxide may seem like a disadvantage of the technique, but aluminum oxide is very important in many technical fields. Many catalysts are supported on alumina in various forms, as are sensors, and in addition the properties of the oxide film on aluminum metal are of the greatest interest in adhesion and protection.

Penetration of moisture into an adhesion bond to aluminum is one of the reasons for ultimate failure of such a bond, and the problem is minimized by treatment of the aluminum oxide surface with a "hydration inhibitor" before adhesion. Some of the most successful inhibitors have been aminophosphonates, the suggestion being that P–O–Al bonds are formed that prevent the ingress of moisture. The interaction of various phosphorus acids with alumina has been studied by RAMSIER et al. [308]. Figure 88 compares the IETS spectra (spectra b) with the IR absorption spectra (spectra a) for phosphonic acid (A) and phosphinic acid (B). In both cases the strong IR peaks at 979 and 1185 cm^{-1} for phosphinic acid and at 946 and 1179 cm^{-1} for phosphonic acid, characteristic of P–OH and P=O stretching vibrations, respectively, are completely absent in the IETS spectra. These absences indicate that the acids are adsorbed on the alumina in resonance-stabilized forms reached via a condensation mechanism in which a water molecule is formed, which then disperses into solution.

Other applications of IETS have included complexation reactions of $CoBr_2$ and $CoCl_2$ on silane-modified alumina [309], silylation of plasma-grown aluminum oxide [310], the effect of hydrogen on the growth of MgO [311], and adsorption of silane adhesion promoters on aluminum oxide [312].

27.7.4. Appearance Potential Methods (SXAPS, AEAPS, DAPS)

Appearance potential methods all depend on detecting the threshold for ionization of a shallow core level and the fine structure near the threshold, differing only in the way in which detection is performed. In all of them the primary electron energy is ramped upward from near zero to whatever is appropriate to the sample material, while

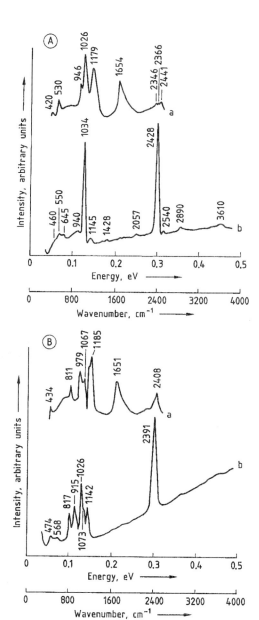

Figure 88. Comparison of IR absorption spectra in water with IETS spectra when adsorbed on aluminum oxide [308], of
A) Phosphonic acid; B) Phosphinic acid
a) IR spectrum; b) IETS spectrum

the primary current to the sample is kept constant. As the incident energy is increased, it passes through successive thresholds for the ionization of core levels of atoms in the surface. An ionized core level, as discussed earlier, can recombine by the emission either of a characteristic X-ray photon or of an Auger electron.

Soft X-Ray Appearance Potential Spectroscopy. In SXAPS the X-ray photons emitted by the sample are detected, normally by letting them strike a photosensitive surface from which photoelectrons are collected, but also— with the advent of X-ray detectors of increased sensitivity— by direct detection. Above the X-ray emission threshold from a particular core level the excitation probability is a function of the densities of unoccupied electronic states. Since two electrons are involved, the incident and the excited, the shape of the spectral structure is proportional to the self-convolution of the unoccupied state densities.

Auger Electron Appearance Potential Spectroscopy. Due to the emission of an Auger electron as an alternative to soft X-ray photons, the total secondary electron yield will show an increase as an ionization threshold is crossed. It is the total secondary yield that is monitored in AEAPS, in a rather simple experimental arrangement. The secondary current arises not just from the Auger electrons themselves, but also from inelastic scattering of Auger electrons produced at greater depths below the surface. Since the yield change is basically a measure of the probability of excitation of a core-level electron to an empty state above the Fermi level, the fine structure above the threshold will be similar to that seen in SXAPS. One of the problems in AEAPS is that of a poorly behaved background, which means that SXAPS is preferred.

Disappearance Potential Spectroscopy. Crossing an ionization threshold means that electrons are lost from the primary beam as a result of ionization of a core hole. Thus if the reflected current of electrons at the primary energy, more usually termed the *elastically reflected current*, is monitored as a function of energy, it should show a sharp decrease as a threshold is crossed. This is the principle of operation of DAPS. It is in a sense the inverse of AEAPS, and indeed if spectra from the two techniques from the same surface are compared, they can be seen to be mirror images of each other. Background problems occur also in DAPS.

The principal advantages of AEAPS and DAPS over SXAPS is that they can be operated

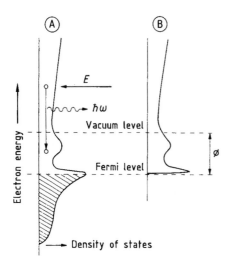

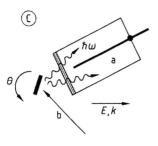

Figure 89. Schematic of the principles of IPES (A) and BIS (B) [313]; C) Experimental arrangement
a) Geiger–Müller counter; b) Electron beam

at much lower primary electron currents, thus causing less disturbance to any adsorbed species.

27.7.5. Inverse Photoemission (IPES, BIS)

Irradiation of a surface with electrons leads to the emission not only of X-ray photons of energies characteristic of the material, but also of a continuous background of photons called *bremsstrahlung radiation*. If a detector is set to detect only those background photons of a particular energy, and the primary electron energy is ramped upward from zero, then the variations in photon flux should mirror the variations in the densities of unoccupied electron states. The process is called inverse photoemission because it is clearly the opposite of ordinary photoemission as in XPS and UPS, and is illustrated in Figure 89 from WOODRUFF et al. [313]. Although the name inverse photoemission spectroscopy (IPES) should apply to all forms of the inverse photoemission technique, in fact it is confined by usage to the form in which the energy at which the photons are detected is in the UV region, typically with $hv = 9.7$ eV. If on the other hand a crystal monochromator normally used to provide monochromatized X-rays for XPS is used in reverse, with a detector placed where the electron source is in Figure 3, then detection is at the X-ray energy $hv = 1486.6$ eV, and the technique is called bremsstrahlung isochromat spectroscopy (BIS).

Since IPES maps the densities of unoccupied states, it is related to other techniques that do the same (i.e., STS and SXAPS). When used in conjunction with a technique that maps the densities of occupied surface states, such as UPS or ELS, a continuous spectrum of state density from occupied to unoccupied can be obtained. Just as in UPS, in which angular resolution allows elucidation of the three-dimensional occupied band structure, so in IPES angular resolution allows mapping of the three-dimensional unoccupied band structure. This version is called *KRIPES* (i.e., *K-resolved IPES*).

The various versions of inverse photoemission have been applied to purely basic problems, just as with UPS and ELS. Recent typical examples include the bonding of H_2 on Ni (110) [314]; the oxidation of a titanium thin film [315]; the effects on H_2 chemisorption of a monolayer of nickel on Cu (111) [316]; unoccupied electronic structure in La_2CuO_4 [317] and in $Bi_2Sr_2CaCu_2O_8$ [318]; and KRIPES studies of the Ge (113) (2×1) [319] and $TiN_{0.83}$ (100) [320] surfaces.

27.7.6. Ion Excitation Methods (IAES, INS, MQS, SNMS, GDMS)

Ion (Excited) Auger Electron Spectroscopy. Auger emission following creation of a core hole by electron or photon irradiation has been described under AES and XPS. Incident ions can also create core holes, so that Auger emission as a result of ion irradiation can also occur, giving rise to the technique of IAES. The ions used are normally noble-gas ions, but protons and α-particles have occasionally been used as well. In addition to the normal Auger features found in AES and XPS spectra, peaks are found in IAES spectra arising from Auger transitions apparently taking place in atoms or clusters sputtered from the surface. These peaks do not always coincide with those found in gas-phase excitation, and others are found that are not present in gas-phase measurements. Because of the complexity of the spectra, IAES cannot be used directly as an analytical technique, but it is very useful in basic physics experiments that study Auger processes occurring in excited atoms. Good examples of ion-excited Auger spectra and their interpretation can be found in [321]–[327].

Ion Neutralization Spectroscopy. The technique of INS is probably the least used of all those described here, because of experimental difficulties, but it is also one of the physically most interesting. Ions of He^+ of a chosen low energy in the range 5–10 eV approach a metal surface and within an interaction distance of a fraction of a nanometer form ion–atom pairs with the nearest surface atoms. The excited quasi molecule so formed can deexcite by Auger neutralization. If unfilled levels in the ion fall outside the range of filled levels of the solid, as is the case for He^+, then an Auger process can take place in which an electron from the valence band of the solid fills the core hole in the ion and the excess energy is given up to another valence electron, which is then ejected. Since either of the valence electrons can come from anywhere within the valence band, the observed energy spectrum reflects the local density of states at the solid surface, but is a self-convolution of the LDOS and of transition probabilities across the valence band. The technique of INS was originated by HAGSTRUM [328], who has given a very detailed exposition of the complex mathematical treatment necessary for deconvolution of the spectra [329], [330]. Because the technique of STS is also capable of extracting the LDOS, the results from INS and STS should be comparable, but such a comparison does not seem to have been attempted yet. HAGSTRUM has used INS extensively to study nickel, both the differences in the various low-index faces and the effects of adsorption of the chalcogen elements oxygen, sulfur, and selenium [331]–[333].

Metastable Quenching Spectroscopy. MQS is in a sense an extension of INS. Instead of He^+ ions, helium atoms in metastable states are used in the incident beam, at the same low energies. The excited singlet state $He^*(2^1S_1)$ has an energy of 20.62 eV and a lifetime of 2×10^{-2} s, and so is suitable. It is produced by expanding helium gas at high pressure through a nozzle into a cold cathode discharge sustained by combined electrostatic and magnetic fields. The high fields prevent ions and fast neutrals from leaving the discharge, and the beam is then nearly all $He^*(2^1S_1)$. As the metastable ion approaches the surface, either of two mechanisms can lead to deexcitation, both resulting in Auger emission similar to INS. If the excited level in the atom *can resonate with empty states at the Fermi level* of the surface, electron transfer from the atom to the surface can occur, leading to resonance ionization of the helium atom. Auger neutralization then takes place as in INS, and the resultant spectrum is again a self-convolution of the LDOS. If, however, the *excited level cannot resonate with empty surface states*, then direct Auger deexcitation can occur, in which the hole in the inner shell of the metastable helium atom is filled from a surface state of the sample, followed by ejection of the excited electron from the helium atom. The process is also called Penning ionization. In the latter process, only one electron is ejected, and the resultant spectrum is thus an unconvoluted reflection of the LDOS.

MQS was first demonstrated by CONRAD et al. [334], and the mechanisms involved in deexcitation have been discussed by CONRAD et al. [335] and by BOZSO et al. [336]. Even the brief description given above makes clear that during reaction at a surface, leading to changes in the surface electronic structure, the situation can easily change from resonance to nonresonance of the excited atomic level with surface states, in which case the deexcitation then switches from resonance ionization + Auger neutralization to Penning ionization. The resultant spectrum then changes in character from one similar to INS to one similar to UPS. Such a switch has been observed by BOZSO et al. [337]. Like INS, MQS is highly surface specific. Its applications have been entirely basic, for example, the adsorption of NH_3 on Ni (111) [338] and Ni (110) [339]; of CO, PF_3, NH_3, C_2H_2, and C_6H_6 on palladium and copper surfaces [340]; of CO on copper and on copper-covered ruthenium [341]; and of potassium on hydrogen-covered ruthenium [342].

Secondary Neutral Mass Spectrometry and Glow Discharge Mass Spectrometry. SNMS and GDMS are grouped together because of the close similarities in principle between them. In both techniques the surface to be analyzed is sputtered and the contribution to the sputtered flux of secondary neutrals is measured. They differ in the methods used to measure the secondary neutral flux. In *GDMS*, as in GDOS, atoms are sputtered

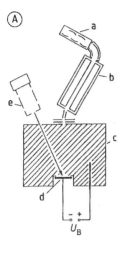

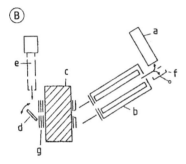

Figure 90. Schematics of designs for SNMS [344]
A) Direct bombardment mode via noble-gas ions in the plasma, or separate bombardment mode via rastered ion beam traversing the plasma; B) External bombardment mode via external ion gun
a) Multiplier; b) Quadrupole mass spectrometer; c) High-frequency plasma; d) Sample; e) Ion gun; f) Faraday cup; g) Electrical diaphragm

via a glow discharge in which the surface is the cathode, and on entering the discharge region, many of the atoms are ionized. The ions are then extracted from the discharge and passed via differential pumping stages into a region at much lower pressure where they can be mass analyzed. Either magnetic or quadrupole mass spectrometers are used for mass analysis. In *SNMS*, three different approaches to postionization of neutrals have been adopted. The first uses *electron impact ionization* (i.e., direct ionization of neutrals by an electron beam from a hot filament as they leave the surface). Secondary ions produced as a result of sputtering are deflected out of the secondary particle flux before they reach the postionization region. If the ion deflection and hot filament are both switched off, the instrument can be used as a normal SSIMS analyzer. The second postionization approach is that of *plasma ionization* in a high-frequency, low-pressure gas discharge (i.e., under conditions very different from the higher pressure discharge used for GDMS and GDOS). Sputtering can be performed either by accelerating positive ions from the discharge to the sample surface or by an external ion gun. In either case, neutrals produced by sputtering pass into the plasma where some are ionized and are thus extracted directly into a quadrupole mass spectrometer. Again, with an external ion gun and without the plasma, SSIMS can be performed. This variation of SNMS has been pioneered by OECHSNER and coworkers [343], and Figure 90 shows schematically some of the experimental arrangements developed by OECHSNER [344]. Thirdly, postionization can be achieved by *high-intensity laser beams*, involving either resonant or nonresonant multiphoton ionization. In the resonant form, a tunable dye laser is tuned to exactly the correct wavelength to excite a chosen atom to an excited state; while in that state, interaction with a second photon leads to ejection of an electron (i.e., ionization). In the nonresonant form, excimer lasers of such high power are used that even if the laser frequency is a long way off resonance for many atoms, adequate ionization efficiency is achieved. The latter variations of SNMS have been given their own acronyms, surface analysis by resonance ionization of sputtered atoms (*SARISA*) and surface analysis by laser ionization (*SALI*), respectively.

Like SSIMS and GDOS, both SNMS and GDMS have been developed commercially, and GDMS has also had wide industrial application. Most of the applications have, like GDOS, consisted of rapid profiling of surface films many micrometers thick, with special emphasis on the depth distribution of trace elements, and would not be regarded as surface specific. Nevertheless, again like GDOS, GDMS is in principle a surface-specific technique and has occasionally been used [345] in an ultrahigh-vacuum system and at very low erosion rates.

One of the attractions of SNMS and GDMS is the relative ease and accuracy of quantification using standard materials and relative sensitivity factors. For most elements, particularly the metallic and semiconducting elements, the spread of sensitivity factors is less than an order of magnitude, comparable to the situation in XPS. In addition, the actual sensitivities are significantly

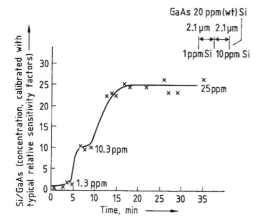

Figure 91. Profile by GDMS through two layers of silicon-doped GaAs grown epitaxially on bulk GaAs [346]

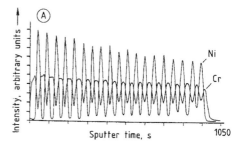

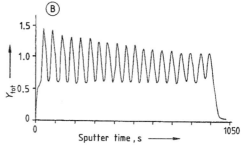

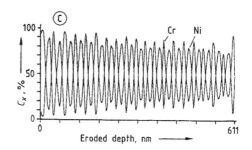

Figure 92. Profile by SNMS of multilayer Ni–Cr sandwich system, with layer thickness of 15 nm [347]
A) Raw intensity data; B) Time-dependent total sputter yield; C) Plot of concentration versus depth evaluated from A) and B)

higher than in XPS or AES, with analysis at the part-per-million level being routine and at the part-per-billion level being quite possible. A typical profile using GDMS to detect low concentrations is shown in Figure 91 from SANDERSON et al. [346], through two layers of silicon-doped GaAs grown epitaxially on bulk GaAs, the silicon levels being different in each layer and in the bulk. Each epilayer can be distinguished easily in terms of the silicon concentration at the part-per-million level, and the semiquantitatively derived silicon concentrations agree well with the amounts known to be incorporated during growth.

A demonstration of both the high depth resolution obtainable in SNMS and the quantification of the technique is given by Figure 92 from WUCHER et al. [347]. It shows a profile through a sandwich structure consisting of 39 alternate layers of chromium and nickel, each 15 nm thick, recorded using a commercial instrument with both bombardment and neutral postionization provided by a high-frequency plasma with an applied voltage of 515 V. Figure 92 A shows the raw intensity data; Figure 92 B, the measured total sputter yield, in which the chromium yield is markedly lower than that of nickel; and Figure 92 C, the intensities corrected for differing sputter yields and then quantified. The sharpnesses of the interlayer interfaces allowed the thicknesses of the individual layers to be determined with a standard deviation of only 10%.

GDMS has been applied extensively in the steel industry for trace-element analysis and for profiling surface films and coatings. Some examples are given in [348], e.g., detection of uranium and thorium in concentrations of a few parts per billion in aluminum; of some 16 trace elements in concentrations of fractions of a part per billion in indium; of more than 25 impurity elements in a high-temperature alloy in concentrations ranging from 0.2 to 2000 ppm; and of 5 nonmetallic impurity contents in low-alloy steels. Good accuracy was achieved throughout. Applications of SNMS have included determination of hydrogen profiles in a-Si:H [349]; analysis of insulators such as SiO_2, Si_3N_4, PTFE, and glasses [350]; the study of aminoacid layers on metals [351]; depth profiles of silicides [352]; analysis of bulk polymers [353];

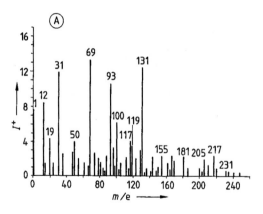
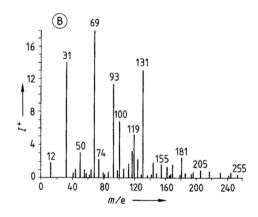

Figure 93. Comparison of SSIMS (A) and FABMS (B) positive secondary-ion spectra from PTFE [357]

thin-film systems [354]; and Cu–Al alloy [355] by SALI, and various metallic and nonmetallic materials by SARISA [356].

27.7.7. Neutral Excitation (FABMS)

Fast-atom bombardment mass spectrometry is very similar to SSIMS in practice, the only difference being that instead of using positively charged ions as the primary probe, a beam of energetic neutral atoms is used. Secondary ions are emitted as in SSIMS and analyzed in a mass spectrometer, usually of the quadrupole type. The beam of fast atoms is produced by passing a beam of ions through a charge-transfer cell, which consists of a small volume filled with argon to a pressure of about 10^2 Pa. Charge transfer occurs through resonance between fast argon ions and argon atoms with thermal energy, with efficiencies of 15–20% if geometric and pressure conditions are optimized. Residual ions are removed by electrostatic deflection.

FABMS has two advantages over SSIMS, both arising from the use of neutral rather than charged particles. Firstly, little or no surface charging of insulating materials occurs, so that organics such as polymers can be analyzed without the need to employ auxiliary electron irradiation to neutralize surface charge. Secondly, the extent of beam damage to a surface, for the same particle flux, is much lower using FABMS than SSIMS, thus allowing materials such as inorganic compounds, glasses, and polymers to be analyzed with less worry about damage introducing ambiguity into the analysis. An example illustrating these advantages is given in Figure 93 from MICHAEL and STULIK [357].

There the SSIMS (Fig. 93 A) and the FABMS (Fig. 93 B) secondary positive ion spectra are compared for PTFE; the surface of the PTFE has to be coated with a gold film for SSIMS measurement but not for FABMS. More fragment ions occur in the SSIMS spectrum than in FABMS, and in the FABMS spectrum the size of the peak at $m/e = 12$ (C^+) relative to the principal peak at $m/e = 69$ (CF_3^+) is much smaller than in SSIMS, indicating much less damage of the polymer surface during FABMS analysis.

FABMS was developed as an alternative to SSIMS by VICKERMAN and coworkers; for a detailed description, see [358]–[362]. Applications have included analyses of tin and lead oxides [363], of poly(methyl methacrylate)–PTFE sandwich structures [364], of multicomponent glasses [358], of silica [358], and of phosphate- and chromate-treated aluminum [360].

27.7.8. Atom Probe Field-Ion Microscopy (APFIM, POSAP, PLAP)

The atom probe field-ion microscope (APFIM) and its subsequent developments, the *position-sensitive atom probe* (POSAP) and the *pulsed laser atom probe* (PLAP), have the ultimate sensitivity in compositional analysis (i.e., single atoms). The APFIM was developed by MÜLLER et al. [365], having grown out of *field-ion microscopy* (*FIM*) which also originated earlier with MÜLLER [366].

FIM is purely an imaging technique in which the specimen in the form of a needle with a very fine point (radius 10–100 nm) is at low temperature (liquid nitrogen or helium) and surrounded by a noble gas (He, Ne, or Ar) at 10^{-2}–10^{-3} Pa. A

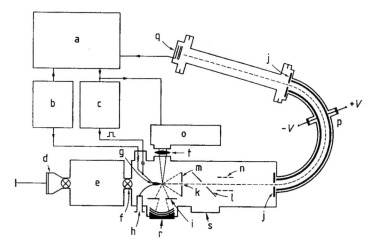

Figure 94. Energy-compensated time-of-flight mass spectrometer for APFIM measurements [367]
a) Computer-controlled fast digital CAMAC system; b) d.c. High voltage; c) High-voltage pulse; d) Airlock; e) Preparation chamber; f) Isolation value; g) Specimen; h) Cryostat; i) Variable iris; j) Herzog electrode; k) Probe aperture; l) Mirror; m) Channel plate; n) Einzel lens; o) Pulsed laser; p) Energy-compensating lens; q) Single-atom detector; r) Atom probe imaging viewpoint; s) FIM viewpoint; t) Focusing lens

fluorescent screen or a microchannel plate is situated a few centimeters from the needle. A high positive voltage is applied to the needle, which causes noble-gas atoms approaching the needle to be ionized over points of local field enhancement (i.e., over prominent atoms at the surface); the ions are then repelled from these points and travel in straight lines to the screen or plate, where an atomic image of the needle tip is formed. This is called a field-ion image, and the voltage at which the image is optimized is called the *best imaging voltage*. The latter is in the range 5–20 kV.

Atom Probe Field-Ion Microscopy. If the voltage is raised further above the imaging voltage, the cohesive energy that binds atoms to the surface can be exceeded, and atoms can be removed by what is termed *field evaporation*. The evaporation field required is a function of the sample material and the crystallographic orientation of the needle. The removal of atoms is the basis of APFIM and its daughter techniques. If the evaporation field is pulsed with very short (≤ 10 ns) voltage pulses of extremely sharp rise times (≈ 1 ns), and the time taken for a field-evaporated atom to travel from the tip to a detector is measured, then the mass-to-charge ratio of the ionized atom can be established; i.e., elemental identification of the individual atom is possible. The region on the tip from which atoms are removed is selected by rotating or tilting the tip until the desired region, as viewed in the FIM image, falls over an aperture of about 2-mm diameter in the fluorescent screen. Most field-evaporated atoms will strike the screen, but those from the selected area will pass through the aperture into the time-of-flight mass spectrometer. For an aperture of 2 mm, the area analyzed on the tip is ca. 2-nm in diameter.

A modern instrument using an energy-compensated TOF mass spectrometer is shown in Figure 94 from MILLER [367]. Atoms may also be removed from a tip by focusing a pulsed laser onto it, as an alternative to field evaporation, leading to the variation of APFIM called the *pulsed laser atom probe* (*PLAP*).

Position-Sensitive Atom Probe. Pulsed operation of the APFIM leads to analysis of all atoms within a volume consisting of a cylinder of ca. 2-nm diameter along the axis of the tip and aperture, with single atomic layer depth resolution but no indication of just where within that volume any particular atom originated. The advent of position-sensitive detectors has allowed APFIM to be extended so that three-dimensional compositional variations within the analyzed volume can be determined. The development has been pioneered by SMITH and coworkers [368], [369], and is called position-sensitive atom probe (POSAP). The aperture and single-ion detector of APFIM are replaced by a wide-angle double channel plate, with a position-sensitive anode just behind the plate. Field- or laser-evaporated ions strike the channel plate, releasing an electron cascade, which is accelerated toward the anode. The impact position is located by the division of electric charge between three wedgeand-strip electrodes. From this position the point of origin of the ion on the tip surface can be determined, since the ion trajectories are radial. Thus after many evaporation pulses, leading to removal of a volume of material from the tip, both the identities and the positions of all atoms within that volume can be mapped in three dimensions. Because an evaporated volume may contain many thousands of atoms, the data collection and handling capabilities must be particularly sophisticated.

POSAP is the only technique available for identifying and locating precipitates, second phases, particles, and interfaces on an atomic scale, and has therefore had considerable application in metallurgical and semiconductor problems. A typical example of a POSAP three-dimensional analysis is given by CEREZO [370] for the ferrite phase of an aged duplex stainless steel CF3 containing 21% chromium and 9% nickel as well as manganese, silicon, molybdenum, and carbon. The Fe–Cr system has a spinodal region in which it is thermodynamically unstable with respect to small compositional fluctuations, leading to chromium-rich and iron-rich phases forming a complex interconnected morphology. Figure 95 from CEREZO shows a series of sections by POSAP, each image being about 15 nm in diameter, with a vertical separation between sections of about 3 nm. The images are in chromium, on a gray scale, with white representing ≤ 5 at% Cr, and black ≥ 50 at% Cr. The sequence shows clearly that the chromium-rich phase is interconnected in depth. From such a sequence, with the help of powerful data handling, a three-dimensional reconstruction can be derived, as shown in Figure 96, where the chromium isosurface for concentrations about halfway between the maximum and minimum values is illustrated. The view is the same as in Figure 95 (i.e., along the direction of analysis). The isosurface represents the approximate interface between the chromium-rich and chromium-depleted regions, and shows clearly the nature of the interconnected structure with an average width of only 2–3 nm.

POSAP has also been applied to the phase chemistry of AlNiCo 2 magnet alloys; the precipitation of copper in iron-based alloys; surface segregation in ceramic oxide superconductors; hetero- structures in compound semiconductors [369]; and Pt–Rh catalysts and multiquantum well structures [371]. Applications of *APFIM* have included finely dispersed carbides in a 14Co–10Ni–2Cr–1Mo high-strength steel [372]; the distribution of vanadium atoms in the ordered alloy Fe–Co–2V [373]; the location of hafnium atoms in the ordered alloy Ni_3Al [367]; the formation of precipitates and segregation of elements in a nickel-based superalloy [374]; and metal–semiconductor and Si–SiO_2 interfaces [375].

27.7.9. Desorption Methods (ESD, ESDIAD, TDS)

Electron-Stimulated Desorption and Electron-Stimulated Desorption Ion Angular Distribution. Electron irradiation of a surface, particularly one covered with one or more adsorbed species, can give rise to many types of secondary particles, including positive and negative ions. In ESD and ESDIAD, the surface is irradiated with electrons of energies in the range of 100–1000 eV, and the ejected positive-ion currents of selected species are measured in a mass spectrometer. If the angular distribution of the secondary ions is also measured, either by display on a screen or by using position-sensitive detection, then electron-stimulated desorption (ESD) becomes electron-stimulated desorption ion angular distribution (ESDIAD).

The electron desorption techniques are not used, and probably cannot be used, for compositional analysis, but they provide valuable information on the nature of electronic interactions leading to the breaking of bonds and, in the angle-resolved form, on the geometry of surface molecules and the orientation of broken bonds. The primary electrons do not, at the energies employed, succeed in breaking molecular or surface-to-molecule bonds or in knocking ions out of the surface directly, but the process is one of initial electronic excitation. An electron is absorbed by a surface–adsorbate complex or an adsorbed molecule itself, leading to excitation to an excited state by a Franck–Condon process. If the excited state is antibonding and the molecule or radical is already far enough from the surface, desorption can occur. Since return to the ground state after excitation is a much more probable process, the cross sections for ion desorption are low, $10^{-20} - 10^{-23}$ cm^2. If core-level ionization is involved in the initial interaction with the incident

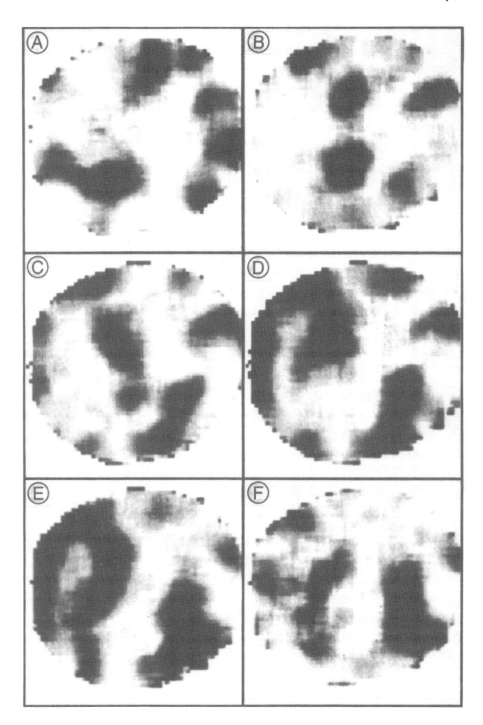

Figure 95. Sequence of images A – F obtained in POSAP by CEREZO [370] from the ferrite phase of an aged duplex stainless steel CF3
Each image is 15 nm in diameter, and the amount removed between each successive image corresponds to a vertical separation of ca. 4 nm
White areas represent ≤5 at % Cr, dark areas ≥50 at % Cr

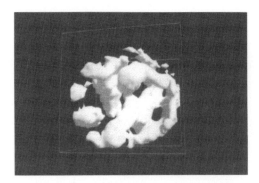

Figure 96. Three-dimensional reconstruction of data of Figure 95 [370]
The isosurface has been drawn through all points halfway between maximum and minimum chromium concentrations. The interconnected structure is Cr-rich

electron, rather than valence levels as in the Franck–Condon-type excitation, then a desorption mechanism based on Auger decay has been proposed by KNOTEK and FEIBELMAN [376]. The core level left behind after interatomic Auger decay creates a positive ion, which is then expelled by the repulsive Madelung potential.

The most common ions observed as a result of electron-stimulated desorption are atomic (e.g., H^+, O^+, F^+), but molecular ions such as OH^+, CO^+, H_2O^+, and CO_2^+ can also be found in significant quantities following adsorption of H_2O, CO, CO_2, etc. Substrate metallic ions have never been observed, which means that ESD is not applicable for surface compositional analysis. The most important application of ESD in the angularly resolved form ESDIAD is in determining the *structure and mode of adsorption* of adsorbed species. This derives from the fact that the ejection of positive ions in ESD is not isotropic. Instead the ions desorb along certain directions only, characterized by the orientation of the molecular bonds that are broken by electron excitation. This orientational dependence is exemplified in Figure 97 from MADEY [377], illustrating the H^+ ESDIAD patterns and suggested models for the adsorption of H_2O on Ag (110). Series A is for adsorption of H_2O on clean silver, B for adsorption on silver with a preadsorbed bromine layer, and C for adsorption on silver with a preadsorbed oxygen layer. The top row consists of three-dimensional perspective plots of the ion intensities; the second row shows two-dimensional intensity contour maps; and the bottom row shows schematic models. On clean silver, the random emission centered about the surface normal indicates that the adsorbed H_2O is disordered locally, with no preferred orientation of the broken bonds. When bromine is present, the H_2O molecules are believed to form hydration shells around each bromine atom, leading to pronounced ordering and hence preferred bond orientation, which is reflected in the very distinct four-beam ESDIAD pattern seen in B. With preadsorbed oxygen, another type of ordering is found, caused by the dissociation of H_2O and reaction with the adsorbed oxygen to form adsorbed OH species. These OH radicals are thought to be tilted along (001) azimuths, giving rise to the intense two-beam pattern, with much weaker beams along the ($1\bar{1}0$) azimuths, seen in C. The weaker beams are due to OH-stabilized H_2O in low concentration.

ESDIAD is obviously not a diffraction technique such as LEED, but it gives direct information about surface structure in real space. The sensitivity is to local bonding geometry, and long-range order is not necessary as in LEED. ESDIAD is especially sensitive to the orientation of hydrogen atoms in surface complexes, which is difficult to observe by any other technique.

Thermal desorption spectroscopy (TDS), sometimes called temperature-programmed desorption (TPD), is simple in principle. A gas or mixture of gases is allowed to adsorb on a clean metal foil for a chosen length of time; then, after the gas is pumped away, the foil is heated at a strictly linear rate to a high temperature, during which the current of a particular ion or group of ions is monitored as a function of temperature. The ion masses are selected in a quadrupole mass spectrometer. As the binding energy thresholds of the adsorbed species on the surface are crossed, peaks in the desorbed ion current appear at characteristic temperatures. From the characteristic temperatures and the shape of the desorption peak above the threshold, the activation energies for desorption can be obtained, along with information about the nature of the desorption process. The mass spectrum from the mass spectrometer, of course, provides information about the species that actually exist on the surface after adsorption.

Although simple in principle, experimental artifacts that are possible in TDS must be avoided. Thus, ions accepted by the mass spectrometer must originate only from the surface of the foil, and in addition, the temperature distribution across the foil should be uniform to avoid the overlapping of desorption processes occurring at different tem-

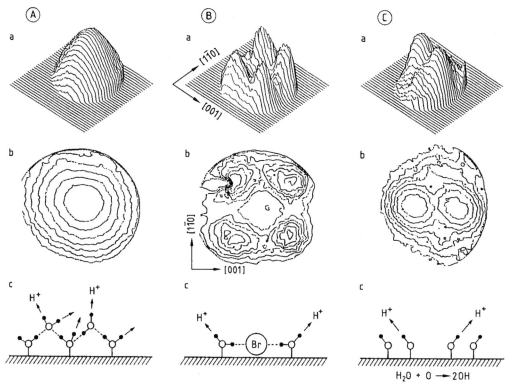

Figure 97. ESDIAD pattern in H⁺ from Ag (100) [377]
A) Water-adsorbed; B) Water on preadsorbed Br; C) Water on preadsorbed O
a) Three-dimensional perspective plots; b) Two-dimensional contour maps; c) Schematic models

peratures. To ensure that these experimental requirements are met, the angle of acceptance into the spectrometer is restricted by placing a drift tube with an aperture between the foil and the spectrometer, and the foil itself is usually in the form of a long thin ribbon, from whose center section only are ions accepted. In addition, the heating rate must be sufficiently fast that the desorbed species accepted into the mass spectrometer is characteristic of the desorption process, but not so fast that the pressure rise around the foil is too great.

With correct experimental procedure, TDS is straightforward to use and has been applied extensively in basic experiments concerned with the nature of reactions between pure gases and clean solid surfaces. The majority of these applications have been catalysis related (i.e., performed on surfaces acting as models for catalysts), and in every case TDS has been used with other techniques such as UPS, ELS, AES, and LEED. To a certain extent it is quantifiable, in that the area under a desorption peak is proportional to the number of ions of that species desorbed in that temperature range, but measurement of the area is not always easy if several processes overlap.

Figure 98 from KRÜGER and BENNDORF [378] shows some typical desorption spectra, following the adsorption of ethylene oxide on Ag (110) at 110 K at exposures increasing from 0.22 to 13.3 Langmuirs. The oxidation of ethylene to ethylene oxide is an industrial process for which silver has important catalytic properties. At very low coverage, only a single mass 29 (the major fragment from the cracking of C_2H_4O in the spectrometer ionization source) desorption peak is found at 170 K, which increases to saturation at about 3 Langmuir, and shifts down to 160 K. Beyond 3 Langmuir exposure, peaks b and c appear at lower temperatures, and c in particular continues to increase with exposure without reaching saturation. For the latter reason, peak c is obviously due to multilayer condensed ethylene oxide. Peaks a and b originate from the adsorption of ethylene oxide in two different sites on the Ag (110) surface, site a being filled first, followed

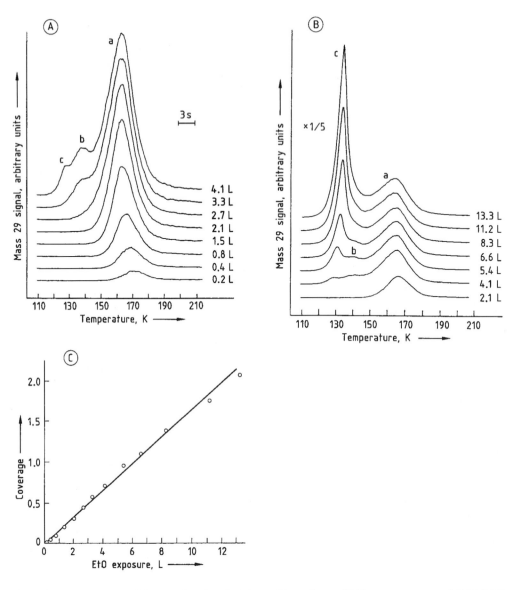

Figure 98. Desorption spectra following adsorption of ethylene oxide on silver at 110 K up to an exposure of 13.3 L [378]
A) 0.2–4.1 L; B) 2.1–13.3 L; C) Relation between surface coverage and exposure of ethylene oxide

by b when all a sites are occupied. Figure 98 C shows the total integrated area of all peaks as a function of exposure, a strictly linear dependence indicating that the occupancy of sites a and b and condensation into a multilayer all have equal probabilities. Data such as these, when taken with measurements from complementary techniques, are invaluable in establishing reaction mechanisms.

27.7.10. Diffraction Methods (LEED, RHEED) (see also →Structural Analysis by Diffraction)

Both LEED and RHEED are probes of surface structure that depend on the existence of long-range order of periodic structure in the surface. They do not provide compositional information and are always used along with other techniques such as AES that can analyze for composition.

LEED is probably the oldest surface-specific technique, having been discovered by DAVISSON and GERMER in 1927 [379].

Low-Energy Electron Diffraction. In LEED a collimated and monoenergetic beam of electrons of variable energies between about 20 and 1000 eV strikes the surface of a single crystal at normal incidence. An energy selector, almost universally consisting of a set of spherical grids centered on the specimen, repels all secondary electrons except those elastically scattered (i.e., those at the primary energy). The elastically scattered electrons are then accelerated to a spherical fluorescent screen, on which the diffraction pattern can be seen. Measurement of diffraction spot intensity can be made either with a photometer or with a movable Faraday cup.

Diffraction occurs due to the wave nature of electrons. The waves represented by the incident electron beam interfere as a result of reflection by a periodic crystal structure in such a way that the reflected beams are anisotropic in direction and are in fact concentrated in certain directions related directly to the crystal structure. Where the reflected beams are intercepted by the fluorescent screen, a spot pattern results. The purely geometric relationship between the observed pattern and the surface structure is governed by the well-known Bragg equation, which connects the directions of the interference maxima with the surface lattice spacing and the primary energy. The observed diffraction pattern is a direct representation of the reciprocal lattice of the surface. Since the electron wavelength is of the same order of magnitude as the surface lattice spacing, only a few diffraction spots appear at low primary energies (<100 eV), but as the energy is increased, further spots appear and move with energy toward the normally deflected (i.e., undiffracted) beam. At the same time, the existing spots vary in intensity, appearing and disappearing on the fluorescent screen as they move with energy inward along radial directions.

The simple geometric, or kinematic, representation of the LEED pattern is valid only for scattering from those atoms actually in the surface (i.e., exposed to the incident electron beam). However, the incident beam has a finite depth of penetration, which of course increases with increasing energy; therefore the periodicity of the crystal normal to the surface must also be taken into account. This can be explored by recording the intensity of a particular diffraction spot as a function of primary electron energy (i.e., wavelength). The normally reflected beam will have intensity maxima due to the vertical periodicity, which are called primary Bragg peaks if the maxima are predicted by the Bragg equation, and secondary Bragg peaks if they are not. The latter are caused by multiple scattering effects, which can be accounted for only in a dynamic theory of scattering. The latter is complicated and can be applied only by using computer simulation, in which a feasible structure for the surface is guessed, from which diffraction intensities are calculated and compared with experiment. In this way a surface structure can be solved completely, but ambiguities cannot be eliminated in most cases.

Reflection High-Energy Electron Diffraction. The same diffraction mechanism is used in RHEED, but the experimental arrangement is quite different, and the observed pattern differs as well. Much higher primary electron energies are used than in LEED, in the range of 15–30 keV, and the primary beam is directed onto the crystal surface at near grazing incidence, at angles $<5°$. As a result, the diffraction pattern is a mixture of spots and streaks. For such low angles of incidence the surface must be very smooth and flat, which is a limitation of the technique, but the requirement also means that any growth feature appearing above the original surface is easily identified since it produces its own characteristic diffraction pattern in transmission. In a variation of RHEED, the angle of the primary beam is varied backward and forward through a few degrees, which is called *rocking*, by analogy with the rocking patterns recorded in X-ray diffraction. Analysis of surface structure, with the assistance of sophisticated data processing, is less ambiguous when the rocking variation is used.

Applications. In virtually all basic work on interactions with, and reactions on, single-crystal surfaces over the last 50 years *LEED* has been employed to provide essential structural information. The structures proposed would in general be ambiguous, in the sense of the existence of several possibilities, if LEED were used on its own, but when the LEED data are taken in conjunction with those from other techniques applied at the same time, they have proved invaluable. The number of publications that have included LEED is probably greater than for any of the other techniques described here.

RHEED has also been used in basic studies, although to a far lesser extent because of its inherent restrictions, but in recent years it has found technological application as well. Semiconductor surfaces are usually flat, and epitaxial films grown on them are often flat as well. In those circumstances RHEED has proved to be a useful tool in the quality control of the structural condition of a semiconductor substrate and then of the perfection or otherwise of any films or layers grown or deposited on it. Thus each of the large UHV systems built for molecular beam epitaxy (MBE) contains a RHEED diagnostic system as one of the standard inspection techniques of the epilayer during and after growth.

27.8. Summary and Comparison of Techniques

Important surface analytical techniques are summarized and compared in Table 5.

27.9. Surface Analytical Equipment Suppliers

By now a large number of suppliers of surface analytical equipment can be found; a few are able to provide any combination of several techniques; the majority specialize in just one or two.

Supplier	Techniques			
Kratos Analytical Ltd., Barton Dock Road, Urmston, Manchester M31 2LD, United Kingdom	XPS SSIMS	AES UPS	SAM ISS	
Ulvac Corporation, 10-3, 1-Chome Kyobashi, Chuo-Ku, Tokyo 104, Japan	AES	SSIMS	LEED	
Riber, 133–137, Boulevard National, F-92503 Rueil Malmaison, B.P. 231, France	XPS SAM	AES	SSIMS	
V. G. Microtech, Bellbrook Business Park, Bell Lane, Uckfield, East Sussex TN22 1BR, United Kingdom	XPS HREELS SNMS UPS AFM	AES SSIMS ISS BIS APFIM	SAM GDMS STM	
Perkin Elmer, Physical Electronics Division, 6509 Flying Cloud Drive, Eden Prairie MN 55344, United States	XPS SSIMS	AES UPS	SAM	
SPECS GmbH, Voltastrasse 5, 13355 Berlin, Germany	SNMS	SSIMS		
Scienta Instruments AB, Seminariegatan 33H, S-752 28 Uppsala, Sweden	XPS			
Omicron Vakuumphysik GmbH, Idsteinerstrasse 78, 65232 Taunusstein, Germany	STM	AFM	LEED	
Oxford Applied Resarch Ltd., Crawley Mill, Witney, Oxon. OX8 5TJ, United Kingdom	RHEED			

Supplier	Techniques		
Staib Instrumente, Obere Hauptstrasse 45, 85354 Freising, Germany	RHEED		
Princeton Research Instruments, Inc., P.O. Box 1174, Princeton NJ 08542, United States	LEED		
National Electrostatics Corp., Graber Road, Box 310, Middleton WI 53562, United States	RBS		
J. A. Woollam Co., 650 J Street, Suite 39, Lincoln NE 68508, United States	ELL		
Rudolph Research, One Rudolph Road, Box 1000, Flanders NJ 07836, United States	ELL		
Sub Monolayer Science, 486 Ellis Street, Mountain View CA 94043, United States	SSIMS		
McAllister Technical Services, West 280 Prairie Avenue, Coeur d'Alene ID 83814, United States	STM		
Park Scientific Instruments, 1171 Borregas Avenue, Sunnyvale CA 94089, United States	STM	AFM	
Topometric, 1505 Wyatt Drive, Santa Clara CA 95054, United States	STM	AFM	
Digital Instruments, 6780 Cortona Drive, Santa Barbara CA 93117, United States	STM	AFM	
Burleigh Instruments Inc., Burleigh Park, Fishers NY 14453, United States	STM		
LK Technologies Inc., 3910 Roll Avenue, Bloomington IN 47403, United States	STM		
W. A. Technology Ltd., Chesterton Mills, French's Road, Cambridge CB4 3NP, United Kingdom	STM	AFM	LEED

Table 5. Summary and comparison of the more important techniques of surface analysis

Technique	Primary probe	Secondary particle(s)	Quantity measured	Elemental range	Type of information	Depth of information	Lateral resolution	Sensitivity, at %	Ease of quantification	Insulator analysis	Destructive	UHV environment
XPS	X-ray photons (Al or Mg K_α)	e^- (photo.)	E	all except H, He	elemental, chemical	4–8 monolayers	1 mm (standard), 5 µm (imaging)	10^{-4}	good	yes	hardly at all	yes
XAES	X-ray photons (Al or Mg K_α)	e^- (Auger)	E	all except H, He	Auger parameter	4–8 monolayers	1 mm	10^{-4}	N/A*	yes	hardly at all	yes
AES	e^- (5–10 keV)	e^- (Auger)	E	all except H, He	elemental	2–6 monolayers	0.5 µm	10^{-4}	moderate	difficult	often	yes
SAM	e^- (20–50 keV)	e^- (Auger)	E	all except H, He	elemental	2–6 monolayers	30 nm	10^{-4}	moderate	no	often	yes
SSIMS	n^+ (Ar$^+$, Ga$^+$)	n^+, n^-, clusters	M	all	chemical	1–2 monolayers	1 mm (standard), 50 nm (LMIS)	10^{-2}–10^{-6}	poor	difficult	yes	yes
ISS	n^+ (He$^+$, Ne$^+$)	n^+ (He$^+$, Ne$^+$)	E	all except H, He	elemental, structural	1–2 monolayers	≈ 1 mm	10^{-4}	good	yes	not with He$^+$, possibly with Ne$^+$	yes
RBS	n^+ (H$^+$, He$^+$)	n^+ (H$^+$, He$^+$)	E	high Z on low Z	elemental, depth distribution	1 monolayer–2 µm	5 µm–1 mm	10^{-6}	good	yes	no	no
UPS	UV photons (He I, He II)	e^- (photo.)	E	(adsorbed molecules)	chemical (valence band)	2–10 monolayers	≈ 2 mm	10^{-3}	N/A*	yes	no	yes
RAIRS	IR photons	IR photons (reflected)	wavenumber	(adsorbed molecules)	chemical (vibrational)	1–2 monolayers	≈ 1 mm	10^{-4}	N/A*	no	no	no
SERS	laser photons	photons (scattered)	wavenumber	restricted	chemical (vibrational)	10–100 nm	≈ 0.5 mm	10^{-6}	N/A*	no	no	yes
GDOS	n^+ (Ar$^+$)	$n^* \to h\nu$	wavelength	all	elemental (profile)	10 nm–5 µm (profile)	none	10^{-6}	good	yes	yes	no
ELS	e^- (50–2000 eV)	e^- (loss)	E	(adsorbed molecules)	chemical	1–6 monolayers	≈ 0.5 mm	10^{-3}	N/A*	yes	no	yes
HREELS	e^- (2–5 eV)	e^- (loss)	E	(adsorbed molecules)	chemical (vibrational)	1 monolayer	≈ 1 mm	10^{-3}	N/A*	yes	no	yes

Table 5. (continued)

Technique	Primary probe	Secondary particle(s)	Quantity measured	Elemental range	Type of information	Depth of information	Lateral resolution	Sensitivity, at %	Ease of quantification	Insulator analysis	Destructive	UHV environment
IETS	(high field)	e^- (loss)	I (tunneling)	(adsorbed molecules)	chemical (vibrational)	1 monolayer	N/A*	10^{-3}	moderate	no	no	no
IPES	e^- (ramped)	photons (UV or X-ray)	photon flux	all, except H, He	chemical (unoccupied states)	1–6 monolayers	0.5–5 μm	10^{-3}	N/A*	yes	possibly	yes
SNMS	n^+ (Ar^+)	n^0	M	all	elemental	1–2 monolayers	≈1 mm	10^{-2}–10^{-6}	moderate	yes	yes	yes
GDMS	n^+ (Ar^+)	$n^* \to n^+$	M	all	elemental	10 nm–5 μm (profile)	none	10^{-6}	good	yes	yes	no
APFIM	(high field) (laser)	n^+	M	all	elemental	1 monolayer	2 nm	(1 atom)	very good	no	yes	yes
POSAP	(high field)	n^+	M	all	elemental	1–100 monolayers (profile)	2–20 nm	(1 atom)	very good	no	yes	yes
STM, STS	(high field)	e^-	I (tunneling)	all	topographic, chemical	1 monolayer	1 atom	(1 atom)	N/A*	no	possibly	yes
AFM	mechanical force		mechanical deflection	all	topographic	1 monolayer	1 atom	(1 atom)	N/A*	yes	no	no
LEED	e^- (200–1000 eV)	e^- (elastic)	I (diffracted)	all	structural	1–4 monolayers	≈0.5 mm	10^{-3}	N/A*	no	no	yes
RHEED	e^- (15–30 keV)	e^- (elastic)	I (diffracted)	all	structural	1–30 monolayers	30 nm	10^{-3}	N/A*	no	possibly	yes
TDS	T (ramped)	n^+, n^-	I (desorbed)	(adsorbed molecules)	chemical	1–2 monolayers	≈2–3 mm	10^{-4}	moderate	no	yes	yes

* N/A = not applicable.

27.10. References

[1] J. C. Rivière: *Surface Analytical Techniques*, Oxford University Press, Oxford 1990.
[2] M. P. Seah, D. Briggs in D. Briggs, M. P. Seah (eds.): *Practical Surface Analysis*, 2nd ed., **vol. 1**, J. Wiley & Sons, Chichester 1990, p. 13.
[3] D. M. Hercules, S. H. Hercules, *J. Chem. Educ.* **61** (1984) 402–409.
[4] M. J. Higatsberger, *Adv. Electron. Electron. Phys.* **56** (1981) 291–358.
[5] H. W. Werner, *Mikrochim. Acta Suppl.* **8** (1979) 25–50.
[6] A. E. Morgan, H. W. Werner, *Phys. Scr.* **18** (1978) 451–463.
[7] J. Tousset, *Le Vide* **33** (1978) 201–211.
[8] G. B. Larrabee, *Scanning Electron Micros.* part I 1977, 639–650.
[9] A. Benninghoven, *Appl. Phys.* **1** (1973) 3–16.
[10] K. Siegbahn et al.: *ESCA: Atomic, Molecular, and Solid State Structure Studied by Means of Electron Spectroscopy*, Almqvist and Wiksells, Uppsala 1967.
[11] P. Auger, *J. Phys. Radium* **6** (1925) 205–209.
[12] M. P. Seah, W. A. Dench, *SIA Surf. Interface Anal.* **1** (1979) 2–11.
[13] M. P. Seah in J. M. Walls (ed.): *Methods of Surface Analysis*, Cambridge University Press, Cambridge 1989.
[14] P. W. Palmberg, *J. Electron Spectrosc. Relat. Phenom.* **5** (1974) 691–703; *J. Vac. Sci. Technol.* **12** (1975) 379–384.
[15] P. Coxon, J. Krizek, M. Humpherson, I. R. M. Wardell, *J. Electron Spectrosc. Relat. Phenom.* **51/52** (1980) 821–836.
[16] A. Barrie, *Chem. Phys. Lett.* **19** (1973) 109–113.
[17] M. Scrocco, *Chem. Phys. Lett.* **63** (1979) 52–56.
[18] M. Campagna, G. K. Wertheim, Y. Baer in L. Ley, M. Cardona (eds.): *Photoemission in Solids II*, Springer Verlag, Berlin 1979, chap. 4.
[19] G. K. Wertheim, D. N. E. Buchanan, N. V. Smith, M. M. Traum, *Phys. Lett. A* **49A** (1974) 191–192.
[20] J. H. Scofield, *J. Electron. Spectrosc. Relat. Phenom.* **8** (1976) 129–137.
[21] R. F. Reilman, A. Msezane, S. T. Manson, *J. Electron Spectrosc. Relat. Phenom.* **8** (1976) 389–394.
[22] M. P. Seah in D. Briggs, M. P. Seah (eds.): *Practical Surface Analysis*, 2nd ed., **vol. I**, J. Wiley & Sons, Chichester 1990, p. 222.
[23] D. Briggs, M. P. Seah (eds.): *Practical Surface Analysis*, 2nd ed., **vol. I**, J. Wiley & Sons, Chichester 1990, p. 635.
[24] P. Brüesch, K. Müller, A. Atrens, H. Neff, *Appl. Phys. A* **A38** (1985) 1–18.
[25] S. Hoffmann in D. Briggs, M. P. Seah (eds.): *Practical Surface Analysis*, 2nd ed., **vol. I**, J. Wiley & Sons, Chichester 1990, p. 143.
[26] C. D. Wagner, *Faraday Discuss. Chem. Soc.* **60** (1975) 291–300.
[27] C. D. Wagner, L. H. Gale, R. H. Raymond, *Anal. Chem.* **51** (1979) 466–482.
[28] C. D. Wagner in D. Briggs, M. P. Seah (eds.): *Practical Surface Analysis*, 2nd ed., **vol. I**, J. Wiley & Sons, Chichester 1990, p. 595.
[29] J. M. Walls (ed.): *Methods of Surface Analysis*, Cambridge University Press, Cambridge 1989.
[30] D. P. Woodruff, T. A. Delchar: *Modern Techniques of Surface Analysis*, Cambridge University Press, Cambridge 1988.
[31] J. F. Watts: *An Introduction to Surface Analysis by Electron Spectroscopy*, Oxford University Press, Oxford 1990.
[32] G. Ertl, J. Küppers: *Low Energy Electrons and Surface Chemistry*, 2nd ed., VCH, Weinheim 1985.
[33] D. Briggs (ed.): *Handbook of X-Ray and Ultraviolet Photoelectron Spectroscopy*, Heyden, London 1977.
[34] J. M. López Nieto, V. Cortés Corberán, J. L. G. Fierro, *SIA Surf. Interface Anal.* **17** (1991) 940–946.
[35] T. A. Patterson, J. C. Carver, D. E. Leyden, D. M. Hercules, *J. Phys. Chem.* **80** (1976) 1700–1708.
[36] G. Morea et al., *J. Chem. Soc. Faraday Trans.* **85** (1989) 3861–3870.
[37] A. Cimino, D. Gazzoli, M. Inversi, M. Valigi, *SIA Surf. Interface Anal.* **10** (1987) 194–201.
[38] A. Katrib et al., *Surf. Sci.* **189/190** (1987) 886–893.
[39] K. Balakrishnan, J. Schwank, *J. Catal.* **127** (1991) 287–306.
[40] B. L. Gustafson, P. S. Wehner, *Appl. Surf. Sci.* **52** (1991) 261–270.
[41] Yong-Xi Li, J. M. Stencel, B. H. Davis, *Appl. Catal.* **64** (1990) 71–81.
[42] G. Moretti, *SIA Surf. Interface Anal.* **17** (1991) 745–750.
[43] M. A. Baltanas, J. H. Onuferko, S. T. McMillan, J. R. Katzer, *J. Phys. Chem.* **91** (1987) 3772–3774.
[44] V. Di Castro, C. Furlani, M. Gargano, M. Rossi, *Appl. Surf. Sci.* **28** (1987) 270–278.
[45] V. Lindner, H. Papp, *Appl. Surf. Sci.* **32** (1988) 75–92.
[46] M. C. Burrell, G. A. Smith, J. J. Chera, *SIA Surf. Interface Anal.* **11** (1988) 160–164.
[47] D. L. Cocke, M. S. Owens, R. B. Wright, *Langmuir* **4** (1988) 1311–1318.
[48] J. L. G. Fierro, J. M. Palacios, F. Tomas, *SIA Surf. Interface Anal.* **13** (1988) 25–32.
[49] C. S. Kuivila, J. B. Butt, D. C. Stair, *Appl. Surf. Sci.* **32** (1988) 99–121.
[50] F. Lange, H. Schmelz, H. Knözinger, *J. Electron Spectrosc. Relat. Phenom.* **57** (1991) 307–315.
[51] J. Z. Shyu, K. Otto, *Appl. Surf. Sci.* **32** (1988) 246–252.
[52] K. Asami, K. Hashimoto, *Langmuir* **3** (1987) 897–904.
[53] P. Marcus, I. Olefjord, *Corros. Sci.* **28** (1988) 589–602.
[54] Shin-Puu Jeng, P. H. Holloway, C. D. Batich, *Surf. Sci.* **227** (1990) 278–290.
[55] P. J. Bremer et al., *SIA Surf. Interface Anal.* **17** (1991) 762–772.
[56] M.-P. Lahalle, R. Guillaumont, G. C. Allen, *J. Chem. Soc. Faraday Trans.* **86** (1990) 2641–2644.
[57] B.-P. Zhang et al., *Corros. Sci.* **33** (1992) 103–112.
[58] H. Yoshioka et al., *Corros. Sci.* **32** (1991) 313–325.
[59] R. B. Diegle et al., *J. Electrochem. Soc.* **135** (1988) 1085–1092.
[60] C. G. Pantano, R. K. Brow, *J. Am. Ceram. Soc.* **71** (1988) 577–581.
[61] Chin-An Chang, Yong-Kil Kim, A. G. Schrott, *J. Vac. Sci. Technol. A* **8** (1990) 3304–3309.
[62] F. J. Boerio, D. J. Ondrus, *J. Adhes.* **22** (1987) 1–12.
[63] J. N. Andersen et al., *J. Phys. Condens. Matter* **1** (1989) 7309–7313.
[64] A. A. Galuska, *Appl. Surf. Sci.* **40** (1989) 19–31, 33–40, 41–51.
[65] W. J. van Ooij, A. Sabata, A. D. Appelhans, *SIA Surf. Interface Anal.* **17** (1991) 403–420.

[66] Y.-H. Kim, Y. S. Chaug, N. J. Chou, J. Kim, *J. Vac. Sci. Technol. A* **5** (1987) 2890–2893.
[67] P. Steiner et al., *Z. Phys. B* **67** (1987) 497–502.
[68] C. Ziegler et al., *Fresenius J. Anal. Chem.* **341** (1991) 308–313.
[69] Y. Hwu et al., *Appl. Phys. Lett.* **59** (1991) 979–981.
[70] Y. Hwu et al., *Appl. Phys. Lett.* **57** (1990) 2139–2141.
[71] Q. Y. Ma et al., *J. Vac. Sci. Technol. A* **9** (1991) 390–393.
[72] R. P. Vasquez, M. C. Foote, B. D. Hunt, L. Bajuk, *J. Vac. Sci. Technol. A* **9** (1991) 570–573.
[73] H. Behner, G. Gieres, B. Sipos, *Fresenius J. Anal. Chem.* **341** (1991) 301–307.
[74] W. A. M. Aarnink, J. Gao, H. Rogalla, A. van Silfhout, *J. Less Common Met.* **164/165** (1990) 321–328.
[75] C. C. Chang et al., *J. Appl. Phys.* **67** (1990) 7483–7487.
[76] G. Frank, C. Ziegler, W. Göpel, *Phys. Rev. B. Condens. Matter* **43** (1991) 2828–2834.
[77] S. Tanaka et al., *Jpn. J. Appl. Phys.* **30** (1991) L 1458–L 1461.
[78] S. Kohiki et al., *Appl. Phys. A* **50** (1990) 509–514.
[79] S. Kohiki et al., *J. Appl. Phys.* **68** (1990) 1229–1232.
[80] M. A. Sobolewski, S. Semancik, E. S. Hellman, E. H. Hartford, *J. Vac. Sci. Technol. A* **9** (1991) 2716–2720.
[81] G. Subramayam, F. Radpour, V. J. Kapoor, G. H. Lemon, *J. Appl. Phys.* **68** (1990) 1157–1163.
[82] K. Tanaka, H. Takaki, K. Koyama, S. Noguchi, *Jpn. J. Appl. Phys.* **29** (1990) 1658–1663.
[83] H. Yamanaka et al., *Jpn. J. Appl. Phys.* **30** (1991) 645–649.
[84] T. Suzuki et al., *Phys. Rev. B Condens. Matter* **42** (1990) 4263–4271.
[85] A. Fujimori et al., *Phys. Rev. B Condens. Matter* **42** (1990) 325–328.
[86] M. Klauda et al., *Physica C* **173** (1991) 109–116.
[87] K. Okada, Y. Seino, A. Kotani, *J. Phys. Soc. Jpn.* **60** (1991) 1040–1050.
[88] P. Adler, H. Buchkremer-Hermanns, A. Simon, *Z. Phys. B Condens. Matter* **81** (1990) 355–363.
[89] K. H. Young, E. J. Smith, M. M. Eddy, T. W. James, *Appl. Surf. Sci.* **52** (1991) 85–89.
[90] S. Gokhale et al., *Surf. Sci.* **257** (1991) 157–166.
[91] N. Aoto, E. Ikawa, N. Endo, Y. Kurogi, *Surf. Sci.* **234** (1990) 121–126.
[92] M. Niwano et al., *J. Vac. Sci. Technol. A* **9** (1991) 195–200.
[93] J. Onsgaard, M. Christiansen, F. Ørskov, P. J. Godowski, *Surf. Sci.* **247** (1991) 208–214.
[94] M. Kawasaki, G. J. Vandentop, M. Salmeron, G. A. Somorjai, *Surf. Sci.* **227** (1990) 261–267.
[95] M. Alonso et al., *J. Vac. Sci. Technol. B* **8** (1990) 955–963.
[96] M. Tang et al., *J. Vac. Sci. Technol. B* **8** (1990) 705–709.
[97] R. Klauser, M. Oshima, H. Sugahara, Y. Murata, *Surf. Sci.* **242** (1991) 319–323.
[98] U. K. Kirner, K. D. Schierbaum, W. Göpel, *Fresenius J. Anal. Chem.* **341** (1991) 416–420.
[99] F. S. Ohuchi, M. Kohyama, *J. Am. Ceram. Soc.* **74** (1991) 1163–1187.
[100] N. D. Shinn, C. H. F. Peden, K. L. Tsang, P. J. Berlowitz, *Phys. Scr.* **41** (1990) 607–611.
[101] S. Harel, J.-M. Mariot, E. Beauprez, C. F. Hague, *Surf. Coat. Technol.* **45** (1991) 309–315.
[102] G. Beamson, A. Bunn, D. Briggs, *SIA Surf. Interface Anal.* **17** (1991) 105–115.
[103] W. F. Stickle, J. F. Moulder, *J. Vac. Sci. Technol A* **9** (1991) 1441–1446.
[104] A. Naves de Brìto et al., *SIA Surf. Interface Anal.* **17** (1991) 94–104.
[105] A. Chikoti, B. D. Ratner, D. Briggs, *Chem. Mater.* **3** (1991) 51–61.
[106] W. Pamler, F. Bell, L. Mühlhoff, H. J. Barth, *Mater. Sci. Eng. A* **139** (1991) 364–371.
[107] L. L. Cao, Y. M. Sun, L. Q. Zheng, *Wear* **140** (1990) 345–357.
[108] S. Noël, L. Boyer, C. Bodin, *J. Vac. Sci. Technol. A* **9** (1991) 32–38.
[109] A. Osaka, Y.-H. Wang, Y. Miura, T. Tsugaru, *J. Mater. Sci.* **26** (1991) 2778–2782.
[110] S. M. Mukhopadhyay, S. H. Garofalini, *J. Non Cryst. Solids* **126** (1990) 202–208.
[111] E.-T. Kang, D. E. Day, *J. Non Cryst. Solids* **126** (1990) 141–150.
[112] Z. Hussain, E. E. Khawaja, *Phys. Scr.* **41** (1990) 939–943.
[113] R. K. Brow, Y. B. Peng, D. E. Day, *J. Non Cryst. Solids* **126** (1990) 231–238.
[114] B. A. van Hassel, A. J. Burggraaf, *Appl. Phys. A* **52** (1991) 410–417.
[115] P. Prieto, L. Galán, J. M. Sanz, *Surf. Sci.* **251/252** (1991) 701–705.
[116] K. Oyoshi, T. Tagami, S. Tanaka, *J. Appl. Phys.* **68** (1990) 3653–3660.
[117] Z.-H. Lu, A. Yelon, *Phys. Rev. B Condens. Matter* **41** (1990) 3284–3286.
[118] A. Carnera et al., *J. Non Cryst. Solids* **125** (1990) 293–301.
[119] R. W. Bernstein, J. K. Grepstad, *J. Appl. Phys.* **68** (1990) 4811–4815.
[120] A. Ermolieff et al., *Semicond. Sci. Tech.* **6** (1991) 98–102.
[121] S. V. Hattangady et al., *J. Appl. Phys.* **68** (1990) 1233–1236.
[122] Maria Faur et al., *SIA Surf. Interface Anal.* **15** (1990) 641–650.
[123] H. Hasegawa et al., *J. Vac. Sci. Technol. B* **8** (1990) 867–873.
[124] L. A. Harris, *J. Appl. Phys.* **39** (1968) 1419–1427, 1428–1431.
[125] R. E. Weber, W. T. Peria, *J. Appl. Phys.* **38** (1967) 4355–4358.
[126] P. W. Palmberg, G. K. Bohm, J. C. Tracy, *Appl. Phys. Lett.* **15** (1969) 254–255.
[127] E. Casnati, A. Tartari, C. Baraldi, *J. Phys. B* **15** (1982) 155–167.
[128] D. Landolt, H. J. Mathieu, *Oberfläche Surf.* **21** (1980) 8–15.
[129] L. E. Davis et al.: *Handbook of Auger Electron Spectroscopy*, Physical Electronic Ind., Minnesota 1976.
[130] (a) E. Kny, *J. Vac. Sci. Technol.* **17** (1980) 658–660.(b) J. Kleefeld, L. L. Levinson, *Thin Solid Films* **64** (1979) 389–393.(c) M. A. Smith, L. L. Levinson, *Phys. Rev. B Solid State* **16** (1977) 1365–1369.
[131] M. Gryzinski, *Phys. Rev.* **138** (1965) A 305–321.
[132] S. Ichimura, R. Shimizu, *Surf. Sci.* **112** (1981) 386–408.
[133] R. Payling, *J. Electron Spectros. Relat. Phenom.* **36** (1985) 99–104.
[134] S. Mroczkowski, D. Lichtman, *J. Vac. Sci. Technol. A* **3** (1985) 1860–1865.
[135] T. Sato et al., *SIA Surf. Interface Anal.* **14** (1989) 787–793.

[136] M. P. Seah, *SIA Surf. Interface Anal.* **9** (1986) 85–98.
[137] R. Hamminger, G. Grathwohl, F. Thümmler, *J. Mater. Sci.* **18** (1983) 353–364.
[138] H. Goretzki, *Fresenius Z. Anal. Chem.* **329** (1987) 180–189.
[139] T. H. Chuang, *Mat. Sci. Eng.* **A 141** (1991) 169–178.
[140] D. J. Nettleship, R. K. Wild, *SIA Surf. Interface Anal.* **16** (1990) 552–558.
[141] C. Setti, *Mikrochim. Acta* **1** (1987) 437–444.
[142] K. Sasaki, A. Noya, T. Umezawa, *Jpn. J. Appl. Phys.* **29** (1990) 1043–1047.
[143] W. Pamler, A. Kohlhase, *SIA Surf. Interface Anal.* **14** (1989) 289–294.
[144] M. Procop, H. Raidt, B. Sandow, *Phys. Status Solidi A* **99** (1987) 573–580.
[145] A. Reif, P. Streubel, A. Meisel, D. Zeissig, *Phys. Status Solidi A* **122** (1990) 331–340.
[146] S. Stein et al., *Fresenius J. Anal. Chem.* **341** (1991) 66–69.
[147] A. Manara, V. Sistori, *SIA Surf. Interface Anal.* **15** (1990) 457–462.
[148] J. Shin, K. N. Geib, C. W. Wilmsen, Z. Lilliental-Weber, *J. Vac. Sci. Technol. A* **8** (1990) 1894–1898.
[149] S. V. Hattangady et al., *Mater. Res. Soc. Symp. Proc.* **165** (1990) 221-226.
[150] A. Hupfer, J. Albrecht, D. Dietrich, *Fresenius J. Anal. Chem.* **341** (1991) 439–444.
[151] F. Bertran et al., *Surf. Sci. Lett.* **245** (1991) L163–L169.
[152] X. Wallart et al., *J. Appl. Phys.* **69** (1991) 8168–8176.
[153] V. M. Bermudez, *Surf. Sci. Lett.* **230** (1990) L155–L161.
[154] H. Roux, N. Boutaoui, M. Tholomier, *Thin Solid Films* **172** (1989) 141–148.
[155] B. J. Burrow, A. E. Morgan, R. C. Ellwanger, *J. Vac. Sci. Technol. A* **4** (1986) 2463–2469.
[156] K. Machida, M. Enyo, I. Toyoshima, *Thin Solid Films* **161** (1988) L91–L95.
[157] K. Sasaki, A. Noya, T. Umezawa, *Jpn. J. Appl. Phys.* **27** (1988) 1190–1192.
[158] A. Gauzzi, H. J. Mathieu, J. H. James, B. Kellett, *Vacuum* **41** (1990) 870–874.
[159] W. A. M. Aarnink et al., *Appl. Surf. Sci.* **47** (1991) 195–203.
[160] Q. X. Jia, W. A. Anderson, *Appl. Phys. Lett.* **57** (1990) 304–306.
[161] J. Colino, J. L. Sacedon, L. Del Olmo, J. L. Vincent, *J. Vac. Sci. Technol. A* **8** (1990) 4021–4025.
[162] Udayan De, S. Natarajan, E. W. Seibt, *Physica C* **183** (1991) 83–89.
[163] E. W. Seibt, A. Zalar, *Mater. Lett.* **11** (1991) 1–5.
[164] C. Uebing, *Surf. Sci.* **225** (1990) 97–106.
[165] F. Reniers, M. Jardinier-Offergeld, F. Bouillon, *SIA Surf. Interface Anal.* **17** (1991) 343–351.
[166] S. Nakanishi et al., *Surf. Sci. Lett.* **247** (1991) L215–L220.
[167] M. J. van Staden, J. P. Roux, *Appl. Surf. Sci.* **44** (1990) 271–277.
[168] R. F. Visser, J. P. Roux, *Appl. Surf. Sci.* **51** (1991) 115–124.
[169] W. Arabczyk, H.-J. Müssig, F. Storbeck, *Surf. Sci.* **251/252** (1991) 804–808.
[170] A. Schönborn, H. Bubert, E. H. de Kaat, *Fresenius J. Anal. Chem.* **341** (1991) 241–244.
[171] J.-R. Chen et al., *Appl. Surf. Sci.* **33/34** (1988) 212–219.
[172] S. K. Sharma, S. Hofmann, *Appl. Surf. Sci.* **51** (1991) 139–155.
[173] S. Badrinarayanan, A. B. Mandale, S. Sinha, *Surf. Coat. Technol.* **34** (1988) 133–139.
[174] R. Hauert, J. Patscheider, R. Zehringer, M. Tobler, *Thin Solid Films* **206** (1991) 330–334.
[175] A. Benninghoven, *Z. Phys.* **220** (1969) 159–180.
[176] J.-C. Pivin, *J. Mater. Sci.* **18** (1983) 1267–1290.
[177] P. Sigmund in N. H. Tolk, J. C. Tully, W. Heiland, C. W. White (eds.): *Inelastic Ion-Surface Collisions*, Academic Press, New York 1977.
[178] P. D. Prewett, D. K. Jefferies, *J. Phys. D* **13** (1980) 1747–1755.
[179] A. R. Krauss, D. M. Gruen, *Appl. Phys.* **14** (1977) 89–97.
[180] P. Steffens et al., *J. Vac. Sci. Technol. A* **3** (1985) 1322–1325.
[181] A. J. Eccles, J. C. Vickerman, *J. Vac. Sci. Technol. A* **7** (1989) 234–244.
[182] A. Benninghoven, K.-H. Müller, M. Schemmer, P. Beckman, *Proc. Int. Conf. Solid Surf.* **3rd** (1977) 1063–1066.
[183] I. V. Bletsos, D. M. Hercules, A. Benninghoven, D. Greifendorf in A. Benninghoven, R. J. Colton, D. S. Simons, H. W. Werner (eds.): *SIMS V*, Springer Verlag, Berlin 1986.
[184] P. H. McBreen, S. Moore, A. Adnot, D. Roy, *Surf. Sci.* **194** (1988) L112–L118.
[185] C. A. Andersen, J. R. Hinthorne, *Anal. Chem.* **45** (1973) 1421–1438.
[186] J. A. Treverton et al., *SIA Surf. Interface Anal.* **15** (1990) 369–376.
[187] W. J. van Ooij, A. Sabata, A. D. Appelhans, *SIA Surf. Interface Anal.* **17** (1991) 403–420.
[188] M. Morra, E. Occhiello, F. Garbassi, *SIA Surf. Interface Anal.* **16** (1990) 412–417.
[189] D. E. Fowler, R. D. Johnson, D. van Leyen, A. Benninghoven, *SIA Surf. Interface Anal.* **17** (1991) 125–136.
[190] G. J. Leggett, D. Briggs, J. C. Vickerman, *SIA Surf. Interface Anal.* **17** (1991) 737–744.
[191] W. D. Ramsden, *SIA Surf. Interface Anal.* **17** (1991) 793–802.
[192] D. J. Pawson et al., *SIA Surf. Interface Anal.* **18** (1992) 13–22.
[193] R. A. Cocco, B. J. Tatarchuk, *Surf. Sci.* **218** (1989) 127–146.
[194] B. H. Sakakini et al., *Surf. Sci.* **271** (1992) 227–236.
[195] X.-L. Zhou, C. R. Flores, J. M. White, *Surf. Sci. Lett.* **268** (1992) L267–L273.
[196] M. E. Castro, J. M. White, *Surf. Sci.* **257** (1991) 22–32.
[197] D. Briggs, M. J. Hearn, *SIA Surf. Interface Anal.* **13** (1988) 181–185.
[198] D. Briggs et al., *SIA Surf. Interface Anal.* **15** (1990) 62–65.
[199] M. J. Hearn, D. Briggs, *SIA Surf. Interface Anal.* **17** (1991) 421–429.
[200] D. G. Armour, *Vacuum* **31** (1981) 417–428.
[201] P. J. Martin, R. P. Netterfield, *SIA Surf. Interface Anal.* **10** (1987) 13–16.
[202] H. Niehus, *Appl. Phys. A* **53** (1991) 388–402.
[203] J.-M. Beuken, P. Bertrand, *Surf. Sci.* **162** (1985) 329–336.
[204] J. F. van der Veen, *Surf. Sci. Rep.* **5** (1985) 199–287.
[205] H. Niehus, G. Comsa, *Surf. Sci.* **140** (1984) 18–30.
[206] G. Molière, *Z. Naturforsch.* **2 A** (1947) 133–143.

[207] H. D. Hagstrum, *Phys. Rev.* **123** (1961) 758–765.
[208] R. De Gryse, J. Landuyt, L. Vandenbroucke, J. Vennik, *SIA Surf. Interface Anal.* **4** (1982) 168–173.
[209] H. H. Brongersma, M. J. Sparnaay, T. M. Buck, *Surf. Sci.* **71** (1978) 657–666.
[210] G. C. Nelson, *J. Vac. Sci. Technol.* **13** (1976) 974–975.
[211] A. Rossi, C. Calinski, H.-W. Hoppe, H.-H. Strehblow, *SIA Surf. Interface Anal.* **18** (1992) 269–276.
[212] H. J. Kang, R. Shimizu, T. Okutani, *Surf. Sci.* **116** (1982) L 173–L 178.
[213] E. N. Haeussler, *SIA Surf. Interface Anal.* **2** (1980) 134–139.
[214] Y. G. Shen, D. J. O'Connor, R. J. MacDonald, *SIA Surf. Interface Anal.* **17** (1991) 903–910.
[215] C. R. K. Marrian, A. Shih, G. A. Haas, *Appl. Surf. Sci.* **24** (1985) 372–390.
[216] S. J. Hoekje, G. B. Hoflund, *Thin Solid Films* **197** (1991) 367–380.
[217] H. Puderbach, *Fresenius Z. Anal. Chem.* **314** (1983) 260–264.
[218] G. H. Vurens, D. R. Strongin, M. Salmeron, G. A. Somorjai, *Surf. Sci.* **199** (1988) L 387–L 393.
[219] A. van Oostrom, *Vacuum* **34** (1984) 881–892.
[220] T. W. Conlon, *Contemp. Phys.* **26** (1985) 521–558.
[221] A. E. T. Kuiper, M. F. C. Willemsen, J. C. Barbour, *Appl. Surf. Sci.* **35** (1988–1989) 186–198.
[222] H. W. Werner in H. Oechsner (ed.): *Thin Film and Depth Profile Analysis,* Springer Verlag, Berlin 1984.
[223] R. S. Bhattacharya, P. P. Pronko, *Appl. Phys. Lett.* **40** (1982) 890–892.
[224] F. Nava et al., *Thin Solid Films* **89** (1982) 381–385.
[225] J. M. M. de Nijs, A. van Silfhout, *Appl. Surf. Sci.* **40** (1990) 349–358.
[226] R. Fastow et al., *J. Vac. Sci. Technol. A* **5** (1987) 164–168.
[227] R. C. Ellwanger, A. E. Morgan, W. T. Stacy, Y. Tamminga, *J. Vac. Sci. Technol. B* **1** (1983) 533–539.
[228] T. Obata et al., *Thin Solid Films* **87** (1982) 207–214.
[229] M. C. Burrell, G. A. Smith, J. J. Chera, *SIA Surf. Interface Anal.* **11** (1988) 160–164.
[230] H. Z. Wu et al., *Thin Solid Films* **191** (1990) 55–67.
[231] A. E. T. Kuiper, S. W. Koo, F. H. P. M. Habraken, Y. Tamminga, *J. Vac. Sci. Technol. B* **1** (1983) 62–66.
[232] F. H. P. M. Habraken et al., *J. Appl. Phys.* **53** (1982) 404–415.
[233] Z. L. Wang, J. F. M. Westendorp, F. W. Saris: "Ion Beam Modification of Materials," in B. Biasse, G. Destefanis, J. P. Gailliard (eds.): *Proc. 3rd Int. Conf.*, Grenoble 1982.
[234] R. S. Bhattacharya et al., *J. Phys. Chem. Solids* **44** (1983) 61–69.
[235] G. Binnig, H. Rohrer, *Surf. Sci.* **126** (1983) 236–244.
[236] P. K. Hansma, J. Tersoff, *J. Appl. Phys.* **61** (1987) R 1–R 23.
[237] P. K. Hansma, V. B. Elings, O. Marti, C. E. Bracker, *Science (Washington D.C.)* **242** (1988) 209–216.
[238] G. Binnig, C. F. Quate, C. Gerber, *Phys. Rev. Lett.* **56** (1986) 930–933.
[239] A. L. Weisenhorn et al., *Science (Washington D.C.)* **247** (1990) 1330–1333.
[240] B. Drake et al., *Science (Washington D.C.)* **243** (1989) 1586–1589.
[241] H. G. Hansma et al., *Clin. Chem. (Winston Salem N.C.)* **37** (1991) 1497–1501.
[242] S. Manne, H.-J. Butt, S. A. C. Gould, P. K. Hansma, *Appl. Phys. Lett.* **56** (1990) 1758–1759.
[243] H.-J. Butt et al., *J. Struct. Biol.* **105** (1990) 54–61.
[244] S. Manne et al., *Science (Washington D.C.)* **251** (1991) 183–186.
[245] Ph. Ebert, G. Cox, U. Poppe, K. Urban, *Surf. Sci.* **271** (1992) 587–595.
[246] S. A. C. Gould, K. Burke, P. K. Hansma, *Phys. Rev. B Condens. Matter* **40** (1989) 5363–5366.
[247] S. A. Gould et al., *Ultramicroscopy* **33** (1990) 93–98.
[248] W. J. Kaiser, R. C. Jaklevic, *IBM J. Res. Dev.* **30** (1986) 411–416.
[249] B. Reihl, J. K. Gimzewski, J. M. Nicholls, E. Tosatti, *Phys. Rev. B Condens. Matter* **33** (1986) 5770-5773.
[250] A. J. Gratz, S. Manne, P. K. Hansma, *Science (Washington D.C.)* **251** (1991) 1343–1346.
[251] P. Dietz et al., *Ultramicroscopy* **35** (1991) 155–159.
[252] H. Hartman et al., *Clays Clay Miner.* **38** (1990) 337–342.
[253] H. Lindgreen et al., *Am. Mineral.* **76** (1991) 1218–1222.
[254] P. G. Arscott, V. A. Bloomfield, *Ultramicroscopy* **33** (1990) 127–131.
[255] R. D. Edstrom et al., *Ultramicroscopy* **33** (1990) 99–106.
[256] K. A. Fisher et al., *Ultramicroscopy* **33** (1990) 117–126.
[257] J. A. N. Zasadzinski et al., *Science (Washington D.C.)* **239** (1988) 1013–1015.
[258] A. L. Weisenhorn et al., *Biophys. J.* **58** (1990) 1251–1258.
[259] H. G. Hansma et al., *Clin. Chem. (Winston Salem N.C.)* **37** (1991) 1497–1501.
[260] J. N. Lin et al., *Langmuir* **6** (1990) 509–511.
[261] H. G. Hansma et al., *J. Vac. Sci. Technol. B* **9** (1991) 1282–1284.
[262] A. L. Weisenhorn et al., *Langmuir* **7** (1991) 8–12.
[263] H.-J. Butt et al., *J. Struct. Biol.* **105** (1990) 54–61.
[264] R. Wolkow, Ph. Avouris, *Phys. Rev. Lett.* **60** (1988) 1049–1052.
[265] K. Takayanagi, Y. Tanishiro, M. Takahashi, S. Takahashi, *J. Vac. Sci. Technol. A* **3** (1985) 1502–1506.
[266] G. Binnig et al., *Phys. Rev. Lett.* **55** (1985) 991–994.
[267] A. Baratoff et al., *Surf. Sci.* **168** (1986) 734–743.
[268] S. A. Elrod, A. L. de Lozanne, C. R. Quate, *Appl. Phys. Lett.* **45** (1986) 1240–1242.
[269] P. Sautet, C. Joachim, *Surf. Sci.* **271** (1992) 387–394.
[270] F. P. Netzer, J.-U. Mack, *Chem. Phys. Lett.* **95** (1983) 492–496.
[271] W. Erley, *J. Electron Spectros. Relat. Phenom.* **44** (1987) 65–78.
[272] M. Fleischmann, P. J. Hendra, A. J. McQuillan, *Chem. Phys. Lett.* **26** (1976) 163–174.
[273] W. Akemann, A. Otto, *Surf. Sci.* **272** (1992) 211–219.
[274] A. Otto, I. Mrozek, H. Grabhorn, W. Akemann, *J. Phys. Condens. Matter* **4** (1992) 1143–1212.
[275] G. L. Richmond, J. M. Robinson, V. L. Shannon, *Prog. Surf. Sci.* **28** (1988) 1–70.
[276] R. J. MacDonald, P. J. Martin, *Surf. Sci.* **67** (1977) 237–250.
[277] J. C. Hamilton, R. J. Anderson, *Surf. Sci.* **149** (1985) 81–92.
[278] H. Bach, F. G. K. Baucke, *J. Am. Ceram. Soc.* **65** (1982) 527–533.
[279] F. G. K. Baucke, H. Bach, *J. Am. Ceram. Soc.* **65** (1982) 534–539.
[280] W. Grimm, *Naturwissenschaften* **54** (1967) 586; *Spectrochim. Acta Part B* **23 B** (1968) 443–448.

[281] A. Bengtson, *Spectrochim. Acta Part B* **40B** (1985) 631–639.
[282] R. Berneron, J. C. Charbonnier, *SIA Surf. Interface Anal.* **3** (1981) 134–141.
[283] J. Pons-Corbeau, *SIA Surf. Interface Anal.* **7** (1985) 169–176.
[284] A. Bengtson, L. Danielson, *Thin Solid Films* **124** (1985) 231–236.
[285] A. Quentmeier et al., *Mikrochimica Acta Suppl.* **11** (1985) 89–102.
[286] J. Takadoum et al., *SIA Surf. Interface Anal.* **6** (1984) 174–183.
[287] G. Rosina, E. Bertel, F. P. Netzer, *Phys. Rev. B Condens. Matter* **34** (1986) 5746–5753.
[288] B. M. Biwer, S. L. Bernasek, *Surf. Sci.* **167** (1986) 207–230.
[289] E. Bertel, R. Stockbauer, T. E. Madey, *Surf. Sci.* **141** (1984) 355–387.
[290] J. C. Rivière, F. P. Netzer, G. Rosina, *SIA Surf. Interface Anal.* **18** (1992) 333–344.
[291] F. P. Netzer, J. C. Rivière, G. Rosina, *Phys. Rev. B* **38** (1988) 7453–7460.
[292] C. M. Stander, *Appl. Surf. Sci.* **16** (1983) 463–468.
[293] A. Otto, B. Reihl, *Surf. Sci.* **178** (1986) 635–645.
[294] J. Onsgaard, I. Chorkendorff, O. Ellegaard, O. Sørensen, *Surf. Sci.* **138** (1984) 148–158.
[295] G. Strasser, G. Rosina, J. A. D. Matthew, F. P. Netzer, *J. Phys. F Met. Phys.* **15** (1985) 739–751.
[296] F. P. Netzer, G. Strasser, J. A. D. Matthew, *Solid State Commun.* **45** (1983) 171–174.
[297] G. Strasser, F. P. Netzer, *J. Vac. Sci. Technol. A* **2** (1984) 826–830.
[298] G. Strasser, E. Bertel, J. A. D. Matthew, F. P. Netzer in P. Wachter, H. Boppart (eds.): *Valence Instabilities*, North Holland Publ., Amsterdam 1982.
[299] M. Nishijima, S. Masuda, H. Kobayashi, M. Onchi, *Rev. Sci. Instrum.* **53** (1982) 790–796.
[300] T. U. Bartke, R. Franchy, H. Ibach, *Surf. Sci.* **272** (1992) 299–305.
[301] A. Fricke, H. Graupner, L. Hammer, K. Müller, *Surf. Sci.* **272** (1992) 182–188.
[302] Q. Dai, A. J. Gellman, *Surf. Sci.* **257** (1991) 103–112.
[303] S. Bao et al., *Surf. Sci.* **271** (1992) 513–518.
[304] M. J. Nowakowski, J. M. Vohs, D. A. Bonnell, *Surf. Sci. Lett.* **271** (1992) L351–L356.
[305] X. H. Feng et al., *J. Chem. Phys.* **90** (1989) 7516–7522.
[306] J.-P. Lu, M. R. Albert, S. L. Bernasek, D. J. Dwyer, *Surf. Sci.* **215** (1989) 348–362.
[307] G. Lauth, T. Solomun, W. Hirschwald, K. Christmann, *Surf. Sci.* **210** (1989) 201–224.
[308] R. D. Ramsier, P. N. Henriksen, A. N. Gent, *Surf. Sci.* **203** (1988) 72–88.
[309] K. W. Hipps, U. Mazur, *Surf. Sci.* **207** (1989) 385–400.
[310] P. N. T. van Velzen, M. C. Raas, *Surf. Sci.* **161** (1985) L605–L613.
[311] M. C. Gallagher, Y. B. Ning, J. G. Adler, *Phys. Rev. B Condens. Matter* **36** (1987) 6651–6656.
[312] D. M. Brewis et al., *SIA Surf. Interface Anal.* **6** (1984) 40–45.
[313] D. P. Woodruff, P. D. Johnson, N. V. Smith, *J. Vac. Sci. Technol. A* **1** (1983) 1104–1110.
[314] G. Rangelov, N. Memmel, E. Bertel, V. Dose, *Surf. Sci.* **236** (1990) 250–258.
[315] R. Konishi, S. Ikeda, T. Osaki, H. Sasakura, *Jpn. J. Appl. Phys.* **29** (1990) 1805–1806.

[316] K. H. Frank, R. Dudde, H. J. Sagner, W. Eberhardt, *Phys. Rev. B Condens. Matter* **39** (1989) 940–948.
[317] N. B. Brookes et al., *Phys. Rev. B Condens. Matter* **39** (1989) 2736–2739.
[318] T. Watanabe et al., *Physica C* **176** (1991) 274–278.
[319] J. M. Nicholls, B. Reihl, *Surf. Sci.* **218** (1989) 237–245.
[320] P. L. Wincott, K. L. Håkansson, L. I. Johansson, *Surf. Sci.* **211/212** (1989) 404–413.
[321] E. W. Thomas, *Vacuum* **34** (1984) 1031–1044.
[322] R. Whaley, E. W. Thomas, *J. Appl. Phys.* **56** (1984) 1505–1513.
[323] R. A. Baragiola, L. Nair, T. E. Madey, *Nucl. Inst. Methods Phys. Res.* **B58** (1991) 322–327.
[324] R. H. Milne, E. A. Maydell, D. J. Fabian, *Appl. Phys. A* **52** (1991) 197–202.
[325] H. Brenten, H. Müller, D. Kruse, V. Kempter, *Nucl. Inst. Methods Phys. Res.* **B58** (1991) 328–332.
[326] C. Benazeth, N. Benazeth, M. Hou, *Surf. Sci.* **151** (1985) L137–L143.
[327] M. Nègre, J. Mischler, N. Benazeth, *Surf. Sci.* **157** (1985) 436–450.
[328] H. D. Hagstrum, *Phys. Rev.* **150** (1966) 495–515.
[329] H. D. Hagstrum, G. E. Becker, *Phys. Rev. B Solid State* **4** (1971) 4187–4202.
[330] H. D. Hagstrum, G. E. Becker, *Phys. Rev. B Solid State* **8** (1973) 1592–1603.
[331] H. D. Hagstrum, G. E. Becker, *J. Chem. Phys.* **54** (1971) 1015–1032.
[332] G. E. Becker, H. D. Hagstrum, *Surf. Sci.* **30** (1972) 505–524.
[333] H. D. Hagstrum, G. E. Becker, *J. Vac. Sci. Technol.* **14** (1977) 369–371.
[334] H. Conrad et al., *Phys. Rev. Lett.* **42** (1979) 1082–1086.
[335] H. Conrad et al., *Surf. Sci.* **100** (1980) L461–L466.
[336] F. Bozso et al., *J. Chem. Phys.* **78** (1983) 4256–4269.
[337] F. Bozso et al., *Surf. Sci.* **136** (1984) 257–266.
[338] F. Bozso et al., *Surf. Sci.* **138** (1984) 488–504.
[339] L. Lee et al., *Surf. Sci.* **165** (1986) L95–L105.
[340] W. Sesselmann et al., *Surf. Sci.* **146** (1984) 17–42.
[341] G. Rocker, H. Tochihara, R. M. Martin, H. Metiu, *Surf. Sci.* **181** (1987) 509–529.
[342] G. H. Rocker et al., *Surf. Sci.* **208** (1989) 205–220.
[343] H. Oechsner, E. Stumpe, *Appl. Phys.* **14** (1977) 43–47.
[344] H. Oechsner, *Nucl. Inst. Methods Phys. Res.* **B33** (1988) 918–925.
[345] J. W. Coburn, E. Taglauer, E. Kay, *Appl. Phys.* **45** (1974) 1779–1786.
[346] N. E. Sanderson et al., *Microchim. Acta* **1** (1987) 275–290.
[347] A. Wucher, H. Oechsner, F. Novak, *Thin Solid Films* **174** (1989) 133–137.
[348] P. M. Charalambous, *Steel Res.* **5** (1987) 197–203.
[349] J. Sopka, H. Oechsner, *J. Non Cryst. Solids* **114** (1989) 208–210.
[350] H. Oechsner, *Scanning Microscopy* **3** (1989) 411–418.
[351] A. Benninghoven, M. Kempken, P. Klüsener, *Surf. Sci.* **206** (1988) L927–L933.
[352] P. Beckmann, M. Kopnarski, H. Oechsner, *Mikrochim. Acta Suppl.* **11** (1985) 79–88.
[353] U. Schühle, J. B. Pallix, C. H. Becker, *J. Vac. Sci. Technol. A* **6** (1988) 936–940.
[354] J. B. Pallix, C. H. Becker, N. Newman, *J. Vac. Sci. Technol. A* **6** (1988) 1049–1052.

[355] E. Kawatoh, R. Shimizu, *Jpn. J. Appl. Phys.* **30** (1991) 832–836.
[356] D. M. Gruen et al., *Nucl. Inst. Methods Phys. Res.* **B 58** (1991) 505–511.
[357] R. Michael, D. Stulik, *App. Surf. Sci.* **28** (1987) 367–381.
[358] D. J. Surman, J. A. van den Berg, J. C. Vickerman, *SIA Surf. Interface Anal.* **4** (1982) 160–167.
[359] A. Brown, J. C. Vickerman, *Analyst (London)* **109** (1984) 851–857.
[360] A. Brown, J. C. Vickerman, *SIA Surf. Interface Anal.* **6** (1984) 1–14.
[361] A. Brown, J. A. van den Berg, J. C. Vickerman, *Spectrochim. Acta Part B* **40 B** (1985) 871–877.
[362] A. Brown, *Eur. Spectros. News* **81** (1988) 13–21.
[363] W. Unger, W. Pritzkow, *Int. J. Mass. Spec. Ion Process.* **75** (1987) 15–26.
[364] R. Michael, D. Stulik, *Appl. Surf. Sci.* **28** (1987) 53–62.
[365] E. W. Müller, J. A. Panitz, S. B. McLane, *Rev. Sci. Instrum.* **39** (1968) 83–86.
[366] E. W. Müller, *Z. Phys.* **131** (1951) 136–142.
[367] M. K. Miller, *Int. Mat. Rev.* **32** (1987) 221–240.
[368] A. Cerezo, T. J. Godfrey, G. D. W. Smith, *Rev. Sci. Instrum.* **59** (1988) 862–866.
[369] A. Cerezo et al., *J. Micros. (Oxford)* **154** (1989) 215–225.
[370] A. Cerezo, *Vacuum* **42** (1991) 605–611.
[371] A. Cerezo et al., *EMSA Bull.* **20** (1990) 77–83.
[372] Li Chang, G. D. W. Smith, G. B. Olson, *J. Phys. Colloq.* **C 2, Suppl. 3, 47** (1986) 265–275.
[373] S. S. Brenner, M. K. Miller, *J. Met.* **35** (1983) 54–63.
[374] G. D. W. Smith, *Metals Handbook,* **10** (1986) 9th ed., 583–602.
[375] A. Cerezo, G. D. W. Smith, *Materials Forum* **10** (1987) 104–116.
[376] M. L. Knotek, P. J. Feibelman, *Phys. Rev. Lett.* **40** (1978) 964–967.
[377] T. E. Madey, *Science (Washington D.C.)* **234** (1986) 316–322.
[378] B. Krüger, C. Bendorff, *Surf. Sci.* **178** (1986) 704–715.
[379] C. J. Davisson, L. H. Germer, *Phys. Rev.* **30** (1927) 705–740.

28. Chemical and Biochemical Sensors

KARL CAMMANN, Institut für Chemo- und Biosensorik, Münster, Germany (Chap. 28, 28.4, 28.5, Sections 28.2.2, 28.2.3.1.3, 28.2.3.1.4, 28.2.3.4, 28.2.4, 28.2.5)

BERND ROSS, Institut für Chemo- und Biosensorik, Münster, Germany (Sections 28.2.3, 28.2.3.1, 28.2.3.1.2)

ANDREAS KATERKAMP, Institut für Chemo- und Biosensorik, Münster, Germany (Sections 28.2.3.2, 28.3.3.2)

JÖRG REINBOLD, Institut für Chemo- und Biosensorik, Münster, Germany (Sections 28.2.2.2, 28.2.3.3)

BERND GRÜNDIG, SensLab GmbH - Bioelektrochemische Sensoren, Leipzig, Germany (Sections 28.3.1 – 28.3.2.3, 28.3.5)

REINHARD RENNEBERG, Hong Kong University of Science & Technology - Department of Chemistry, Hong Kong, PR China (Sections 28.3.1, 28.3.2.4 – 28.3.4)

28.	Chemical and Biochemical Sensors 951	
28.1.	Introduction to the Field of Sensors and Actuators 952	
28.2.	Chemical Sensors 953	
28.2.1.	Introduction 953	
28.2.2.	Molecular Recognition Processes and Corresponding Selectivities . . 959	
28.2.2.1.	Catalytic Processes in Calorimetric Devices 959	
28.2.2.2.	Reactions at Semiconductor Surfaces and Interfaces Influencing Surface or Bulk Conductivities . . . 960	
28.2.2.3.	Selective Ion Conductivities in Solid-State Materials 966	
28.2.2.4.	Selective Adsorption – Distribution and Supramolecular Chemistry at Interfaces 967	
28.2.2.5.	Selective Charge-Transfer Processes at Ion-Selective Electrodes (Potentiometry) 968	
28.2.2.6.	Selective Electrochemical Reactions at Working Electrodes (Voltammetry and Amperometry) . 968	
28.2.2.7.	Molecular Recognition Processes Based on Molecular Biological Principles 969	
28.2.3.	Transducers for Molecular Recognition: Processes and Sensitivities 969	
28.2.3.1.	Electrochemical Sensors 969
28.2.3.2.	Optical Sensors 997
28.2.3.3.	Mass-Sensitive Devices 1003
28.2.3.4.	Calorimetric Devices 1026
28.2.4.	Problems Associated with Chemical Sensors .1029
28.2.5.	Multisensor Arrays, Electronic Noses, and Tongues 1030
28.3.	**Biochemical Sensors (Biosensors)** .1032
28.3.1.	Definitions, General Construction, and Classification 1032
28.3.2.	Biocatalytic (Metabolic) Sensors. .1033
28.3.2.1.	Monoenzyme Sensors 1034
28.3.2.2.	Multienzyme Sensors 1036
28.3.2.3.	Enzyme Sensors for Inhibitors – Toxic Effect Sensors 1039
28.3.2.4.	Biosensors Utilizing Intact Biological Receptors 1039
28.3.3.	Affinity Sensors – Immuno-Probes.1040
28.3.3.1.	Direct-Sensing Immuno-Probes without Marker Molecules 1041
28.3.3.2.	Indirect-Sensing Immuno-Probes using Marker Molecules 1043
28.3.4.	Whole-Cell Biosensors 1048
28.3.5.	Problems and Future Prospects . . .1049
28.4.	**Actuators and Instrumentation** . . .1051
28.5.	**Future Trends and Outlook** 1052
28.6.	**References** 1053

28.1. Introduction to the Field of Sensors and Actuators

When a person suffers the loss of one of the five senses it is recognized as a serious obstacle to carrying out everyday activities. In the case of blindness and deafness the expression "disabled" is even used as an indication of the severity of the loss, just as with those whose limbs (legs or arms) or other crucial body components are incapable of functioning properly. In the animate world generally, sensing and then acting or reacting is closely linked with the very essence of life, where input from some type of sensory system always precedes intelligent action. From the earliest days of mankind it was the availability of a powerful "coping mechanism" that constituted the driving force leading first to intelligent action and later to the unbiased accumulation of knowledge.

Any intellectual information structure capable of producing theories is dependent upon a reliable input of information. Imperfect observations, or perceptual processes subject to frequent errors or some bias, will never provide a satisfactory basis for working theories subject to verification by others. In this respect analytical chemistry can be regarded as one of the cognitive means available to the chemist interested in ascertaining the composition of the material world and the distribution and arrangement of atoms and molecules within a particular sample. Since both the nature and the three-dimensional arrangement of atoms in space determine the characteristics of matter, reliable information with respect to both is crucial for understanding the whole of the material world. This observation is equally valid for the inorganic, the organic, and also the living world. A high level of structured information is essential to the work of most natural scientists. The ability to describe in detail a complex chemical process (e.g., the combustion of fuel in an engine or waste in a municipal incinerator, the synthesis of minerals deep within the earth, or the course of biochemical reactions taking place in living systems) is crucially dependent on the quality of the information available. High technology is impossible without detailed information on the state of the system in question, as has been amply demonstrated in semiconductor technology, where knowledge of the purity of the materials and the state of cleanliness in the production plant is absolutely essential.

Complex and especially dynamic systems involving chemical processes are often described in terms of simplified models. The more chemical information is available, the better will be the model — and consequently the more complete the understanding and control over the system. Chemical analysis constitutes the primary source of information for characterizing the various states of such a system.

As a result of a tremendous increase in the demand for analytical chemical information, the number of analyses conducted for acquiring information about particular samples is increasing dramatically. It is thought that worldwide more than 10^9 analyses are now performed every day — with a double-digit growth rate. The latter because quality control has become a worldwide topic, agreed upon in many international trade contracts. Many complex systems are governed by time-dependent parameters, which makes it essential that as much information as possible be acquired within a particular period of time so that the various reactions taking place within the system can be followed, and to provide insight into the dynamics of the mechanisms involved. Devices designed to permit continuous measurement of a quantity are called *sensors*. Various types of sensors have been developed for monitoring many important *physical* parameters, including temperature, pressure, speed, flow, etc. The number of potential interferences is rather limited in these cases, so it is not surprising that the corresponding sensors offer a level of reliability considerably higher than that for sensors that measure *chemical* quantities (e.g., the concentration of a certain species, the activity of an ion, or the partial pressure of a gaseous material).

There are various major fields of application for chemical sensors: environmental studies, quality control of chemically produced compounds and processed food and biomedical analysis, especially for medical diagnostics. In the field of environmental analysis there is a great demand for the type of continuous monitoring that only a sensor can provide, since the relevant parameter for toxicological risk assessment is always the *dose* (i.e., a concentration multiplied by an exposure time). The study of environmental chemistry also depends upon a continuous data output that provides not only baseline levels but also a reliable record of concentration outbursts. *Data networks* can be used to accumulate multiple determinations of a particular analyte regardless of whether these are acquired simultaneously or at different times within a three-dimensional sampling space (which may have dimensions on the order of miles), thereby providing information about concentration gradients and other factors essential for pinpointing the location of an emission source. The inevitable exponential increase in the number of chemical analyses associated with this need for maximum information threatens to create serious financial and also ecological problems, since most classical analytical chemical techniques themselves produce a certain amount of chemical waste.

Our society is characterized by a population with an average age that is continuously increasing. This has stimulated the widespread biomedical application of chemical sensors for monitoring the health of the elderly without confining them to hospitals. Sensors for determining the concentrations of such key body substances as glucose, creatinine, or cholesterol could prove vitally important in controlling and maintaining personal health, at the same time enhancing the quality of the patient's life. The same principle underlies sensor applications in intensive care wards. Sensors also offer the potential for significant cost reductions relative to conventional laboratory tests, and they can even improve the quality of the surveillance because continuous monitoring can lower the risk of complications. For example, the monitoring of blood potassium levels can give early warning of the steady increase that often precedes an embolism, providing sufficient time for clinical countermeasures.

It should be noted at the outset, however, that the development costs associated with a chemical sensor are typically many orders of magnitude higher than those for a sensor measuring a physical quantity. A satisfactory return on investment therefore presupposes either mass production in the case of low-cost sensors or special applications that would justify an expensive sensor.

Another important field of application for chemical sensors is *process control*. Here the sensor is expected to deliver some crucial signal to the actuators (valves, pumps, etc.) that control the actual process. Fully automated process control is feasible with certain types of feedback circuits. Since key chemical parameters in many chemical or biochemical processes are not subject to sufficiently accurate direct determination by the human senses, or even via the detour of a physical parameter, sensor technology becomes the key to automatic process and quality control. Often the performance of an entire industrial process depends on the quality and reliability of the sensors employed. Successful adaptation of a batch process to *flow-through reactor production technology* is nearly impossible without chemical sensors for monitoring the input, the product quality, and also — an increasingly important consideration — the waste. *Municipal incinerators* would certainly be more acceptable if they offered sensor-based control of SO_2, NO_x, and, even more important, the most hazardous organic compounds, such as dioxines. Compared with the high construction costs for a plant complete with effective exhaust management, an extra investment in sensor research can certainly be justified.

These few examples illustrate the increasing importance of chemical sensor development as a key technology in various closed-looped sensor/actuator-based process control applications. Rapid, comprehensive, and reliable information regarding the chemical state of a system is indispensable in many high technology fields; for this reason there is also an increasing need for intelligent or "smart" sensors capable of controlling their own performance automatically.

Finally, there is an urgent and growing demand for analytical chemists and engineers skilled in producing reliable chemical sensors delivering results that are not only reproducible but also accurate (free of systematic errors attributable to interferences from the various sample matrices).

28.2. Chemical Sensors [1]–[9]

28.2.1. Introduction

Until the introduction in the 1980s of the lambda-(λ) probe [10]–[12] for oxygen measurement in the exhaust systems of cars employing catalytic convertors (see Section 28.2.2.3), the prototype for the most successfully applied chemical sensor was certainly the electrochemical pH electrode, introduced over 50 years ago and discussed in detail in Section 28.2.3.1.1. This device can be used to illustrate the identification and/or further optimization of all the critical and essential features characterizing sensor performance, including reversibility, selectivity, sensitivity, linearity, dynamic working range, response time, and fouling conditions.

Electrochemical electrodes for chemical analysis have actually been known for over 110 years (the Nernst equation was published as early as 1889), and they can therefore be considered as the forerunners of chemical sensors. In general, a chemical sensor is a small device placed directly

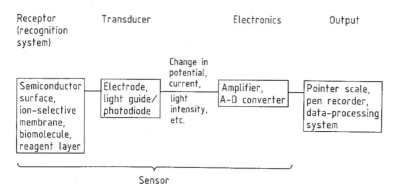

Figure 1. Schematic arrangement of a typical chemical or biochemical sensor [13]

on or in a sample and designed to produce an electrical signal that can be correlated in a specific way with the concentration of the *analyte*, the substance to be determined. If the analyte concentration varies in the sample under investigation, the sensor should faithfully follow the variation in both directions; i.e., it should function in a reversible way, and it should deliver analytical data at a rate greater than the rate of change of the system (quasi-reversible case). From the standpoint of practical applications, complete and fast reversibility is relatively unimportant so long as measurements (even readings of such as from a dosimeter) are acquired with sufficient rapidity compared with changes in the analyte concentration that are to be followed. Another restriction arises
from the demand for a reagent-free measurement: an extremely selective or specific cooperation between the analyte atom, ion, or molecule and the surface of a receptor-like sensor element must then be the sole basis for gaining information about the analyte concentration.

The general construction principle of a chemical sensor is illustrated in Figure 1. The most important part is the *sensing element* (receptor) at which the molecular or ionic recognition process takes place, since this defines the overall selectivity of the entire sensor. The sensing-element surface is sometimes covered by an additional surface layer acting as a protective device to improve the lifetime and/or dynamic range of the sensor, or to prevent interfering substances from reaching the sensing-element surface. The analyte recognition process takes place either at the surface of the sensing element or in the bulk of the material, leading to a concentration-dependent change in some physical property that can be transformed into an electrical signal by the appropriate transducer. The transducer cannot itself improve the selectivity (exceptions are transducer like electrochemical electrodes which intrinsically deliver an analyte proportional signal), but it is responsible for the sensitivity of the sensor, and it functions together with the sensing element to establish a dynamic concentration range. The electrical signal of the transducer is usually amplified by a device positioned close to the transducer or even integrated into it. If extensive electrical wiring is required between the sensor and the read-out–control unit, the electrical signal might also be digitized as one way of minimizing noise pick-up during signal transmission. Several chemical signals produced by multiple sensing elements, each connected to its own specific transducer, can be transmitted via modern multiplexing techniques, which require only a two-wire electrical connection to the control unit also delivering the electrical power ("Bus"-technique in German).

If a biochemical mechanism (mostly enzymatic catalysis, immuno-chemical reaction or complementary DNA hybridization) is used in the molecular recognition step, the sensor is called a *biosensor*. Systems of this type are dealt with in Chapter 28.3.

Key features of chemical sensors subject to specification are, in order of decreasing importance:

1) Selectivity
2) Limit of detection (LOD) for the analyte
3) Accuracy (trueness — a new English term — more important than reproducibility)
4) Sensitivity (slope of the calibration curve)
5) Dynamic response range
6) Stability

7) Response time
8) Reliability (maintenance-free working time)
9) Lifetime

Selectivity. The *selectivity* of a chemical sensor toward the analyte is often expressed in terms of a dimension that compares the concentration of the corresponding interfering substance with an analyte concentration that produces the same sensor signal. This factor is obtained by dividing the sensitivity of the sensing device toward the corresponding interfering substance (= slope of the calibration curve; see below) by the sensitivity for the analyte. Typical selectivities range from 10^{-3} to $<10^{-12}$ the last extremely small value stands for the sodium error of classical pH-glass-electrodes. In order to estimate the error attributable to limited selectivity the concentration of the interferent must be multiplied by the selectivity coefficient. If, for example, a sodium-selective glass electrode shows a typical selectivity coefficient toward potassium ions on the order of 10^{-3}, a thousand-fold excess of potassium- over sodium-ion concentration would produce an equivalent sensor signal, therefore doubling the reading and leading to an error of 100%. It should be noted, however, that such selectivities are themselves concentration- and matrix-dependent and often depends on the way they are determined. This means that near the limit of detection of a particular sensor an interfering species will disrupt the measurement of an analyte to a greater extent than it will at higher concentrations. Therefore, response factors can rarely be used for correction purposes.

The selectivity is the most important parameter associated with a chemical sensor since it largely determines the trueness of the analytical method. Trueness (or *accuracy*) is more than simply precision, repeatability, or reproducibility: it represents agreement with the true content of the sample. This is the concentration found independent of the method, the time, the analyst, the laboratory, the country, etc. and is demanded in international trade agreements as comparability. All the other parameters mentioned are subject to systematic error, which can only be recognized if an analyte is determined by more than one method. Since selectivity is always limited, all chemical sensors are prone to report higher concentrations than a sample actually contains. In the case of environmental analysis this positive error can be regarded as a safety margin if relevant interferents are also present. In such a situation the sensor acts as a probe for establishing when a sample should be taken to be analyzed in the laboratory. The reproducibility of chemical sensor measurements is often well below 5% relative.

However, it should also be noted that sensing elements are subject to poisoning, resulting in a diminished response for the analyte. Sensor poisons are substances that either cover the sensing surfaces, thereby influencing the analyte recognition process or, in the case of catalytic sensors, decrease the activity of the catalyst. Since these fouling effects are not linear, predictable, or mathematically additive, great care is necessary whenever data from a sensor array are treated by the methods of mathematical data management (e.g., pattern recognition, neural networks, etc.). The influence of sensor poisoning can be completely different and unpredictable for every single sensing element. There is no simple theory that permits the transfer of poisoning effects with respect to one sensing element to other parts of a sensing array as a way of avoiding recalibration. Thus, any attempt to overcome insufficient selectivity by the use of a sensor array with multiple sensor elements displaying slightly differing selectivities for the analyte and the interferents, followed by the application of chemometrics to correct for selectivity errors, is very dangerous in the context of unknown and variable sample matrices. Unfortunately, this is usually the situation in environmental analysis. In the learning and/or calibration phase of sensor-array implementation it is essential that the sample matrix be understood. In the case of a known matrix and a closed system, and if the number of possible interferents is >5, the number of calibration steps (involving different mixtures of the interfering compounds, selected to cover all possible concentration and mixture variations) required to provide reliable information on the 5-dimensional response space of the sensor array for a $5+N$ array is extremely large (usually several hundred). With real samples the time necessary for calibrating the sensors might exceed the period of stability. Since with open environmental systems the type and number of interferents is not known, any attempt to correct for inadequate sensor selectivity by use of a sensor array increases the risk of systematic errors. This assertion is valid for every type of mathematical data treatment (pattern recognition, neural networks, etc.) for samples of unknown and/or variable matrices. However, for trace analysis in the sub-ppm range all analytical methods (with the exception of isotope dilution) are subject to systematic errors, as has been demonstrated with interlaboratory comparison tests using even the most sophisticated techniques [14]. Therefore one should not expect an accuracy of better than 1–5 % relative from sensor measurements.

Limit of Detection (LOD) for the Analyte. In analytical chemistry the *limit of detection* (LOD) is exceeded when the signal of the analyte reaches at least three times the general noise level for the reading. No quantitative measurement is possible at this low concentration; only the presence of the analyte can be assumed with a high probability (>99 %) in this concentration range. The range of quantification ends at ten times the LOD. Therefore, the LOD of an analytical instrument must always be ten times smaller than the lowest concentration to be quantified. It is worth mentioning in this context that LOD values reported by most instrument manufacturers disregard the influence of interfering compounds. Furthermore, the actual LOD is often worse than published values with simple mono-analyte solutions when an analyte must be determined within its characteristic matrix (e.g., a pollution site, urine, or blood).

In some cases with extremely sensitive sensors a zero-point calibration might be difficult to perform, because traces of the analyte would always be present or might easily be carried into the calibration process by solvents or reagents. The lowest *measurable* level is then called the *blank value*. The LOD is then defined as three times the standard deviation of the blank value, expressed in concentration units (not in the units of the signal, as is sometimes improperly suggested).

Normally the LOD becomes worse as a sensor ages.

Sensitivity. As mentioned above, the *sensitivity* of a sensor is defined by the signal it generates, expressed in the concentration units of the substance measured. This corresponds to the slope of the corresponding calibration curve when the substance is the analyte, or to the so-called response curve for interferents. With some sensors the sensitivity rises to a maximum during the device's lifetime. A check of the sensitivity is therefore a valuable quality-assessment step. Intelligent sensors are expected to carry out such checks automatically in the course of routine performance tests. Since in most cases the sensitivity depends on such other parameters as the sample matrix, temperature, pressure, and humidity, certain precautionary measurements are necessary to ensure that all these parameters remain constant both during calibration and in the analysis of real samples.

Dynamic Response Range. The *dynamic response range* is the concentration range over which a calibration curve can be described by a single mathematical equation. A potentiometric sensing device follows a logarithmic relationship, while amperometric and most other electrochemical sensors display linear relationships. Both types of signal-to-concentration relationship are possible with optical sensors; in absorption measurements it is the absorbance with its logarithmic base that is the determining factor, whereas a fluorescence measurement can be described by a linearized function. The broader the concentration range subject to measurement with a given sensing system, the less important are dilution or enrichment steps during sample preparation. The measurement range is limited by the LOD at the low concentration end, and by saturation effects at the highest levels. Modern computer facilities make it possible also to use the nonlinear portion of a calibration curve as a way of saving sample preparation time, but analytical chemists are very reluctant to follow this course because experience has shown that the initial and final parts of a calibration curve are generally most subject to influence by disturbances (electronic and chemical). A good chemical sensor should function over at least one or two concentration decades. Sensors with excellent performance characteristics include the lambda oxygen probe and the glass pH electrode, both of which cover analyte concentration ranges exceeding twelve decades!

Stability. Several types of signal variation are associated with sensors. If the signal is found to vary slowly in two directions it is unlikely to be regarded as having acceptable stability and reproducibility, especially since it would probably not be subject to electronic correction.

Output variation in a single direction is called *drift*. A steadily drifting signal can be caused by a drifting zero point (if no analyte is present) and/or by changes in sensor sensitivity (i.e., changes in the slope of the calibration curve). Drift in the sensor zero-concentration signal (*zero-point drift*) can be corrected by comparison with a signal produced by a sensor from the same production batch that has been immersed in an analyte-free solution.

When drift observed in a sensor signal is attributable exclusively to zero-point drift, and there has been no change in the sensitivity, a one-point calibration is permissible: that is, either the instrumental zero point must itself be adjusted prior to the measurement, or else the sensor signal for a particular calibration mixture must be recorded. Assuming there is no change in the slope of the calibration curve, this procedure is equivalent to effecting a parallel shift in the calibration curve. If the sensor is routinely stored in a calibrating environment between measurements then such a one-point calibration becomes particularly easy to perform.

Special attention must be directed toward the physical and chemical mechanisms that cause a sensor signal to drift. Differential techniques are satisfactory only if the effect to be compensated has the same absolute magnitude and the same sign with respect to both devices. A consideration often neglected in potentiometry is proper functioning of the reference electrode. Several authors [15]–[17] have suggested that in ion-selective potentiometry a "blank membrane" electrode (one with the same ion-selective membrane composition as the measuring electrode but without the selectivity-inducing electroactive compound) would create the same phase-boundary potential for all interfering compounds as the measuring electrode itself. This is definitely not the case, however. At the measuring electrode, analyte ions establish a relatively high exchange-current density, which fixes the boundary potential. The effect of a particular interfering ion will depend on its unique current–voltage characteristics at such an interface. Therefore, interfering ions may establish different phase-boundary potentials depending on the particular mixed potential situations present at the membranes in question, which means that a differential measuring technique will almost certainly fail to provide proper compensation.

If both the zero point of a sensor and its sensitivity change with time, a two-point calibration is necessary. The combination of a changing slope for the calibration curve and a change in the intercept is generally unfavorable, but this is often the situation with catalytic gas sensors. If the slope of the calibration curve also depends on the sample matrix, only standard addition techniques will help.

The stability of a chemical sensor is usually subject to a significant aging process. In the course of aging most sensors lose some of their selectivity, sensitivity, and stability. Some sensors, like the glass pH electrode, can be rejuvenated, while others must be replaced if certain specifications are no longer fulfilled.

Response Time. The *response time* is not defined in an exact way. Some manufacturers prefer to specify the time interval over which a signal reaches 90 % of its final value after a ten-fold concentration increase, while authors sometimes prefer the 95 % or even 99 % level. The particular percentage value chosen represents a pragmatic decision, since most signal–time curves follow an exponential increase of the form:

$$\text{signal} \times (1 - e^{-kt})$$

where the true final value is unknown and/or will never be reached in a mathematical sense. Therefore, specification in terms of the time constant k is clearer. This corresponds to the time required for a signal to reach about 67 % of its final value, which can be ascertained without waiting for a final reading: as the slope of the curve ln (signal) vs. time. If the response is constant and independent of the sample matrix, this equation can be used to calculate a final reading immediately after the sensor has been introduced into the sample.

Typical response times for chemical sensors are in the range of seconds, but some biosensors require several minutes to reach a final reading. At the other extreme, certain thin-film strontium titanate oxygen sensors show response times in the millisecond range.

The response time for a sensor is generally greater for a decreasing analyte concentration than for an increasing concentration. This effect is more pronounced in liquids than in gaseous samples. Both the surface roughness of the sensor and/or the dead volume of the measuring cell have some influence on the response time. Small cracks in the walls of a measuring cell can function as analyte reservoirs and diminish the rate of analyte dilution.

With certain sensors the response time can also depend on the sample matrix. In the presence of strongly interfering substances the response time for a chemical sensor might increase as a result of an increase in the time required to reach final equilibrium.

Reliability. There are various kinds of reliabilities. One involves the degree of trust that can be placed in an analytical result delivered by a particular chemical sensor. Another is a function of the real time during which the sensor actually performs satisfactorily without a breakdown and/or need for repair.

There is no way to judge fairly the analytical reliability of a chemical sensor, since this depends strongly upon the expert ability of the analyst to choose a suitable sensor and a suitable sample preparation routine in order to circumvent predictable problems.

With respect to the first generation of sensors, the ion-selective electrodes, the experience of many users has led to the following "reliability hit list", with the most reliable devices cited first: glass pH electrodes, followed by fluoride, sodium glass, valinomycin-based potassium, sulfide, and iodide electrodes, and culminating in the electrodes for divalent ions. In a general way, sometimes very personal judgements regarding reliability have always been linked to a particular sensor's specificity toward the analyte in question.

Reliability can be improved considerably if analytical conclusions are based on measurements obtained by different methods. For example, it is possible to determine sodium either with an appropriate glass-membrane electrode or with a neutral carrier-based membrane electrode. If both give the same analytical results, the probability of a systematic error is very low. Likewise, potassium might be measured using selective membrane electrodes with different carrier molecules. This is of course different from analysis based on a sensor array, in which the selectivities of the individual sensors differ only slightly, and different also in the sense that no correction procedure is invoked in case two results are found to disagree. Use of the Nernst–Nikolsky equation (see Section 28.2.3.1.1), together with the introduction of additional electrodes for determining the main interferents, has been shown to be an effective way of correcting errors in the laboratory with synthetic samples. Since selectivity numbers (i.e., parameters like the selectivity coefficient) are influenced by the sample matrix through its ionic strength, content of surface-active or lipophilic compounds, or interfering ions, it is dubious whether data from real environmental samples should be corrected by this method.

The length of time over which a chemical sensor can be expected to function reliably can be remarkably great (in the range of years), as in the case of glass-membrane or solid-state membrane electrodes, the lambda probe (Section 28.2.2.3), and the Taguchi gas sensor (see Section 28.2.2.2). On the other hand, a biosensor that depends upon a cascade of enzymes to produce an analyte signal will usually have a short span of proper functioning (a few days only). Potentiometric ion-selective membrane electrodes and optical chemosensors have lifetimes of several months. In the case of biosensors it should be noted that anything that changes the quaternary space-orientation of the recognition biomolecules will destroy the proper functioning of the sensor. Enzymes can be influenced by such factors as pH, certain heavy-metal ions, certain inhibitors, and high temperature (resulting in denaturation).

Lifetime. As already mentioned above, some rugged solid-state sensors have *lifetimes* of several years. Certain sensors can also be regenerated when their function begins to deteriorate. The shortest lifetimes are exhibited by biosensors. Ion-selective electrodes and optical sensors based on membrane-bound recognition molecules often lose their ability to function by a leaching-out effect. In optical sensors the photobleaching effect may also reduce the lifetime to less than a year. On the other hand, amperometric cells work well for many years, albeit with restricted selectivities.

Problems associated with inadequate lifetimes are best overcome with mass-produced miniaturized replacement sensors based on inexpensive materials. Minimizing replacement costs may well represent the future of biosensors. Installation of a new sensor to replace one that is worn-out can also circumvent surface fouling, interfering layers of proteins, and certain drift and poisoning problems. The integrated optical system IOS developed at the ICB Muenster and described in Section 28.3.3.2 has shown for the first time how to work with pre-calibrated low-cost immuno-chips with long storage capability.

Comparison of Sensor Data with that Obtained by Traditional Analytical Methods. It is extremely difficult to compare sensor data with traditional data; indeed, generalizations of this type are rarely possible anywhere in the field of analytical chemistry. Sensor developers are often confronted with the customer's tendency to consider use of a sensor only if all else has failed. This means that the most adverse conditions imaginable are sometimes proposed for the application of a chemical sensor. It is also not fair to compare a device costing a few dollars with the most expensive and sophisticated instrumentation available, nor is it appropriate to compare the performance of a chemical sensor with techniques involving time-consuming separations. In this case only the corresponding detector should be compared with the sensor. Sometimes the very simple combination of a selective chemical sensor with an appropriate separation technique is the most effective way to obtain the redundant data that offer the highest reliability.

Chemical sensors are generally superior to simple photometric devices because they are more selective or faster (as in the case of optical sensors based on the photometric method), more flexible, more economical, and better adapted to continuous sensing. The latter advantage can also be achieved by traditional means via flow-through measuring cells, but this leads to a waste of material and sample, and also to the production of chemical waste. Reagent-free chemical sensors show their greatest advantages in continuous-monitoring applications, in some of which they are called upon to fulfill a control function without necessarily reporting an analytical result, as in the case of the lambda probe described below (Section 28.2.2.3).

28.2.2. Molecular Recognition Processes and Corresponding Selectivities

In any sensing element the functions to be fulfilled include sampling, sample preparation, separation, identification, and detection. Therefore, successful performance in these tasks largely determines the quality of the chemical sensor as a whole. Selective recognition of an analyte ion or molecule is not an easy task considering that today there are more than five million known compounds and a real sample may contain hundreds of potentially interfering compounds. Recognition can be accomplished only on the basis of unique characteristics of the analyte in question. Several different recognition processes are relevant to the field of chemical sensors, ranging from energy differences (in spectroscopy) to thermodynamically determined variables (in electrochemistry), including kinetic parameters (in catalytic processes). The most specific interactions are those in which the form and the spatial arrangement of the various atoms in a molecule play an important role. This is especially true with biosensors based on the complementary (lock-and-key) principle. Here the analyte molecule and its counterpart have exactly complementary geometrical shapes and come so close together that they interact on the basis electrostatic interaction or with weak van der Waals interaction forces. Apart from biomolecules, supramolecular chemistry has been given increased attention [18], [19].

28.2.2.1. Catalytic Processes in Calorimetric Devices

Pellistors are chemical sensors for detecting gaseous compounds that can be oxidized by oxygen. A catalyst is required in this case because the activation energy for splitting a doubly-bonded oxygen molecule into more reactive atoms is too high for the instantaneous "burning" of oxidizable molecules. In most cases platinum is the catalyst of first choice because of its inertness. The principle underlying oxidizable-gas sensors involves catalytic burning of the gaseous analyte, which leads to the production of heat that can in turn be sensed by various temperature-sensitive transducers. Often what is actually measured is the increase in electrical resistance of a metal wire heated to an elevated temperature (ca. $300-400\,°C$) by the current flowing through it. However, it is also possible to use more sensitive semiconductor devices or even thermopiles in order to register temperature changes of $< 10^{-4}\,°C$.

In order to understand the selectivity displayed by a pellistor toward various flammable compounds it is necessary to consider the elementary steps in the corresponding catalytic oxidation of analyte at the catalyst surface. Since this is in fact a surface reaction, various adsorption processes play a dominant role. First, oxygen must be adsorbed and chemisorbed, permitting the oxygen double bond to be weakened by the catalyst. Then the species to be oxidized must also be adsorbed onto the same catalyst surface where it can subsequently react with the activated oxygen atoms. The process of adsorption may follow one of two known types of adsorption isotherms, as reflected in the calibration curves for these devices. An equilibrium consisting of adsorption, catalyzed reaction, and desorption of the oxidized product leads to a constant signal at a constant analyte concentration. The sensor response function is influenced by any change in the type or number of active surface sites, since this in turn affects both the adsorption processes and the catalytic efficiency. Likewise, compounds that have an influence on any of the relevant equilibria will also alter the calibration parameters. Especially problematic are strongly adsorbed oxidation products, which lower the turnover rate of analyte molecules and thus the sensitivity of the gas sensor.

Given the sequence of events that must occur when an organic molecule is oxidized by atmospheric oxygen, thereby delivering the heat that is actually to be measured, one can readily understand the importance of changes in the adsorption and desorption equilibria. The selectivity observed with such a calorimetric device arises not because some gas reactions are associated with larger enthalpy changes than others, but rather because those gas molecules that exhibit the most rapid adsorption and desorption kinetics are associated with the highest turnover rates. The latter of course depend on molecular-specific heats of ad- and desorption, which have a major influence on the overall reaction kinetics. A change in these specific heats always results in a corresponding change in sensor selectivity. Consequently, any change in the catalyst material, its physical form, or its distribution within the mostly ceramic pellet will in turn alter the gas selectivity and sensitivity. The same is true for variations in the working temperature. On the other hand, nonreacting compounds can also influence a sensor's response and thereby the calibration function if they in some way affect one of the relevant equilibria and/or the catalytic power of the catalyst. Catalyst poisons such as hydrogen sulfide or organic silicon compounds show a strongly detrimental effect on sensor response.

Summarizing the selectivity characteristics of these calorimetric gas sensors, any gas will be subject to detection if it can be catalytically oxidized with a high turnover rate by atmospheric oxygen at elevated temperature. From a kinetic point of view, smaller molecules like carbon monoxide or methane are favored. In contrast to biosensors there is here no precise molecular "tight-fit" recognition of the geometrical form of the analyte. Thus, the selectivity of this type of gas sensor is rather limited. Differences in the often reaction-rate controlled adsorption and desorption processes for different oxidizable gas molecules are not sufficiently large to allow selective detection of only a single compound. However, this is not necessarily a disadvantage in certain sensor applications, such as detecting the absence of explosive gases (especially important in the mining industry) or carbon monoxide in an automobile garage. In the latter case any positive error resulting from gasoline interference could in fact be regarded as providing a safety margin, since unburned gasoline should be absent from such locations as well.

28.2.2.2. Reactions at Semiconductor Surfaces and Interfaces Influencing Surface or Bulk Conductivities

Introduction. Since the early 1960s it has been known that the electrical conductance of certain semiconductor materials such as binary and ternary metal oxides (e.g., SnO_2 [20], ZnO [21], Fe_2O_3 [22]—all of which are n-semiconductor materials—and CuO or NiO [23]—p-semiconductors) depends on the adsorption of gases on their solid surfaces. The underlying principle here involves a transfer of electrons between the semiconductor surface and adsorbed gas molecules, together with charge transduction in the interior of the material. Typical gases detected by semiconducting devices include oxidizable substances such as hydrogen, hydrogen sulfide, carbon monoxide, and alkanes (SnO_2, ZnO, etc.), as well as reducible gases like chlorine, oxygen, and ozone (NiO, CuO).

In 1967 both SHAVER [24] and LOH [25] described effects achievable with oxide semiconductors modified by the addition of noble metals (e.g., Pt, Pd, Ir, Rh), and since that time the sensitivity and selectivity of semiconductor sensing devices has been significantly enhanced. Intense efforts in this direction, coupled with the further addition of metal oxides [26]–[28], resulted in widespread application of semiconductor gas sensors beginning in the 1970s.

One of the earliest SnO_2 sensors, designed by N. TAGUCHI, is referred to as the "Figaro sensor" (see below). Sensors of this type make it possible to detect as little as 0.2 ppm of an oxidizable compound such as carbon monoxide or methane [29]. Nevertheless, certain details of the associated sensing mechanism are still not fully understood theoretically. An important aim of current research is to overcome limitations of the present generation of sensors, especially instability, irreproducibility, and nonselectivity [30].

Construction and Characterization of Semiconductor Sensors. Semiconductor gas sensors are characterized by their simple construction. A schematic overview of the construction principle of a homogeneous semiconducting gas sensor is provided in Figure 2A [31]. Sensor operation is based on a change in the surface resistance (or conductance) of an oxidic microcrystalline semiconductor in the presence of interacting gases. A time-dependent record reflecting transient expo-

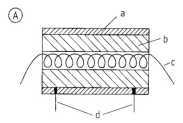

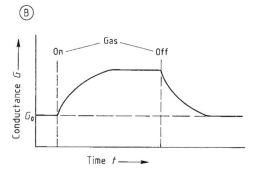

Figure 2. A) Schematic drawing of a homogeneous semiconducting gas sensor; B) Result of transient exposure of the sensor to a gas that increases the conductance [31]
a) Semiconductor; b) Ceramic; c) Heater; d) Contacts
(with permission from Elsevier, Amsterdam)

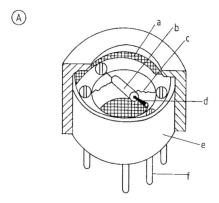

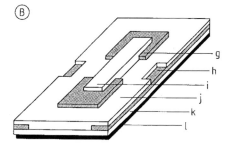

Figure 3. Design characteristics of gas sensors (with permission from VDE-Verlag, Berlin)
A) Figaro Type TGS 813 [32]; B) SnO_2 thick-film type [13]
a) Stainless steel screen; b) Sensor element; c) Leads; d) Heater; e) Epoxy resin; f) Nickel contact; g) Gold; h) Heat contact; i) SnO_2; j) Insulating layer; k) Heater; l) Substrate

sure of a sensor to a gas leading to an increase in conductance is illustrated in Figure 2 B. The sensitivity of a semiconductor sensor is strongly affected by its operating temperature, which is normally in the range 200–400 °C. Chemical regeneration of the oxide surface is possible by a reheating process.

A commercial sensor of the "Figaro" type (TGS 813) is shown in Figure 3 A; 400 000 such sensors had already been sold in 1988 [32].

Miniaturization leads to a more modern version of the SnO_2 sensor, normally prepared by thick-film techniques (Fig. 3 B) in which thin SnO_2 films, insulator layers (SiO_2), and integrated heating films are sputtered onto silicon substrates. This approach is compatible with high rates of heating and low-cost production [13].

Working Principles and Theory. The mechanisms responsible for semiconductor gas-sensor operation can be divided into two classes. The first class involves changes in bulk conductance (transducer function), while the second relies on changes in surface conductance (receptor func-

tion). The physical phenomena associated with these two mechanisms are shown schematically in Figure 4 A [33]; Figure 4 B addresses the same problem at the microstructural level [34].

The description of functional principles that follows relates directly to n-semiconductors, but its application to p-semiconductors is straightforward. In a first step, oxygen molecules from the air form a layer of more or less strongly adsorbed (chemisorbed/ionisorbed) oxygen molecules at the surface, resulting in a local excess of electrons. In other words, oxygen acts as a *surface acceptor*, binding electrons from the surface space–charge layer. With respect to the principal energy states (levels) of the electrons in the surface space–charge layer, ionisorption results in a decrease in the electron concentration and an increase in the electronic energy (Fig. 5) [32].

Subsequent reaction with reducing gases (e.g., CH_4 or CO) leads to an increase in charge density (and therefore an increase in conductivity), asso-

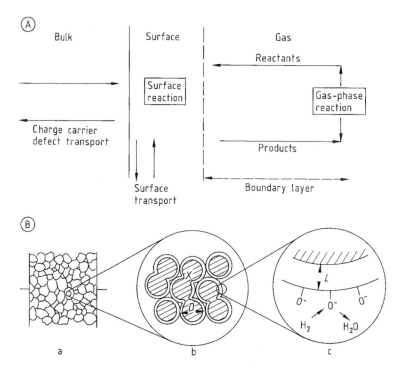

Figure 4. A) Illustration of the physical and chemical phenomena involved in the transduction of a change in the gaseous atmosphere into a change in conductance of a ceramic oxide (with permission from Adam Hilger, Bristol); B) Surface reactions (receptor function) may occur between adsorbed species and defects in the solid (transducer function), between different adsorbed species, or between gas molecules and adsorbed species (or any or all of these together), and the surface reaction may involve a catalyst [33], [34] (with permission from Elsevier, Amsterdam)
a) Element (output resistance change); b) Microstructure (transducer function); c) Surface (receptor function)
D = Particle size; X = Neck size; L = Thickness of space–charge layer

ciated with three possible mechanisms [23], [31], [35]:

1) Adsorption of the reducing molecules as donors, causing electrons to be shifted into the conductance band of the oxide
2) Reaction of the reducing molecules with ionisorbed oxygen under conditions leading to the production of bound electrons
3) Reduction of oxidic oxygen by the reducing molecules, resulting in oxygen vacancies which act as donors, thereby increasing the conductance

Cases 1) and 2) alter the amount of charge stored in the *surface states*, and therefore the amount of charge of opposite sign in deeper parts of the region. For a theoretical derivation of the relationship between the conductance of a semiconducting oxide layer and the composition of the gaseous surroundings, the following facts must be considered [30], [36]:

Oxygen must be present; that is, these sensors respond only to nonequilibrium gas mixtures containing both combustible gases (CO, hydrocarbons, H_2, etc.) and oxygen.

There exists a temperature of maximum response; that is, the relative change in conductivity upon introduction of a combustible gas increases with increasing temperature, but falls to zero at sufficiently high temperatures.

With respect to the relationship between conductivity (σ) and gas partial pressure, Equation (1) has been found to apply:

$$\sigma \sim p^\beta \tag{1}$$

where p is the partial pressure of the combustible gas and β is generally in the range 0.5–1.0 depending on the mechanism.

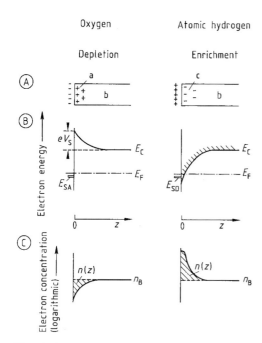

Figure 5. Space–charge layer at the surface of ZnO after exposure to oxygen and hydrogen (with permission from VDE-Verlag, Berlin)
A) Vacuum; B) Band structure in the vicinity of the conducting edge; C) Concentration n (z) of electrons in the conductance band [32]
a) Ionized donors; b) Crystal; c) Electrons
E_{SA} = Acceptor level; E_{SD} = Donor level; n_B = Bulk electron density; z = Distance from surface; eV_S = Surface energy barrier height; E_C = Conductance band edge; E_F = Fermi level

This microscopic model is limited to idealized thin films (10–100 nm of, for example, SnO_2). Here the gas–solid interactions can be described in terms of the electronic surface states, which are related to energy levels in the band gap model (Fig. 5). Other surface phenomena involved are lattice defect points, trace impurities, and material segregations near the surface. Crystal dislocations and adsorbed atoms or molecules also play an important role in states at the gas–solid interface. As far as n-semiconductors are concerned, both adsorbed hydrogen and oxygen vacancies can function as surface donors, whereas ionisorbed oxygen is a surface acceptor. A surface potential difference develops as a consequence of negatively charged adsorbed oxygen (O_2^- and O^{2-}) and positively charged oxygen vacancies within the space–charge region below the surface layer. Increasing the amount of negatively charged oxygen at the surface causes the surface potential to increase up to the Fermi level (the highest level occupied by electrons). This defines the surface potential and represents the surface state, but the precise magnitude of the potential is a function of the oxygen partial pressure. The distribution of the various adsorbed oxygen species depends on the temperature, and is influenced by the presence of hydrogen and other gaseous compounds. With respect to the two possible charged oxygen species, O_2^- and O^-, it can be assumed that only O^- is reactive, and that the rate of interconversion of the species is low compared to the rate of the surface combustion reaction, consistent with the following kinetic scheme [33], [37]:

$$O_2(g) \rightleftarrows O_2(s) \rightleftarrows O_2^- \rightleftarrows O^- \rightleftarrows O^{2-}$$

with $2OH^- \uparrow\downarrow H_2O$ above O^- and $\downarrow R$ giving RO.

At low temperature the adsorbed species is mainly O_2^-, which is converted into O^- when the temperature is increased above 450 K.

If a combustible gas R reacts with the adsorbed oxygen species, a steady state occupancy Θ of the surface state is established, which is less than the equilibrium occupancy in air. The following mechanism can be assumed:

$$n' + 1/2\, O_2(g) \underset{k_{-1}}{\overset{k_1}{\rightleftarrows}} O^-(s)$$

$$R(g) + O^-(s) \xrightarrow{k_2} RO(g) + n'$$

where n' denotes a conductance electron.

Necessary conditions for sensitivity with respect to the partial pressure p of the combustible gas are then:

$$k_{-1} \ll k_2 p(R) \quad \text{and} \quad k_1 p(O_2)^{1/2} \ll k_2 p(R) \qquad (2)$$

in which case

$$\Theta = \frac{k_1 p(O_2)^{1/2}}{k_{-1} + k_1 (O_2)^{1/2} + k_2 p(R)} \qquad (3)$$

The third case described above is somewhat different: here the observed effect is related to a change in bulk conductance as a function of stoichiometric changes in oxygen activity in the interior of the crystal lattice. The change in conductivity can be described by the relationship [33]:

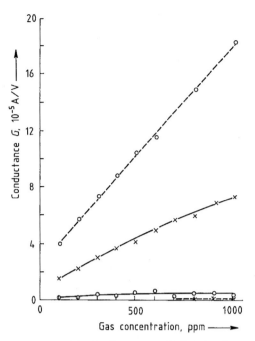

Figure 6. Sensitivity data from commercial SnO_2-based sensors, with sintered layers containing stabilizing additives like SiO_2 (conductance as a function of CO and CH_4 concentration in air); the Type 711 sensor probably operates at a lower temperature and incorporates different catalytically active additives than the Type 813 sensor [31], [38] (with permission from Elsevier, Amsterdam)
----- Sensor Type 711; ——— Sensor Type 813;
×= CH_4; o = CO

$$\sigma = \sigma_0 \exp(E_A/kT) p(O_2)^{\frac{1}{n}} \qquad (4)$$

where σ_0 denotes the electrical conductivity in air, k denotes the Boltzmann constant, T is the temperature in Kelvin, E_A is an activation energy, and the sign and value of n depend on the nature of the point defects arising when oxygen is removed from the lattice. At low temperature ($< 500\,°C$) the rate-determining step is interfacial combustion, whereas oxygen vacancies within the lattice dominate at higher temperatures below ca. 1000 °C.

Thus, the observed overall conductance represents a combination of surface effects (both electronic and ionic) together with grain-boundary and volume-lattice effects. The principles elaborated above have been applied widely in the control of oxygen in combustion processes at high temperatures, as in automobile engines.

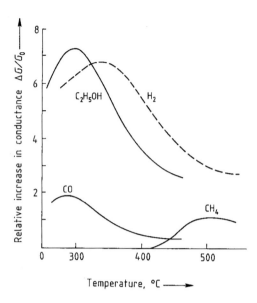

Figure 7. Sensor response to reducing gases (1000 ppm in air) as a function of temperature (sintered layer of SnO_2, thickness 0.05 mm, 0.05 wt% Sb; based on [39], [31]) (with permission from Elsevier, Amsterdam)

Parameters. Under a constant set of conditions (temperature, air, pressure, humidity, etc.) the ratio of the conductance in sample air to that in pure air (G/G_0) is proportional to the analyte concentration (Fig. 6). Apart from the concentration of adsorbed gases, other parameters, both physical and chemical, may also have an influence on the conductance and the optimum working conditions. Considering both components of the conductance change described above, the most important factors influencing conductance are:

1) The microstructure of the semiconducting particles and the surface composition, both of which are characterized by the crystallite size D, the grain-size distribution, and the coagulation structure, all subject to some control through the inclusion of additives.
2) The working temperature of the material, which has a unique optimum (maximum) value for a particular material composition, and must therefore be determined separately for every analyte gas and every substrate composition (Fig. 7).

Tin dioxide has become the favorite material for sensor applications of this type because of its simple preparation and high sensitivity at low operating temperature. Commercially available sensors are usually sintered $\geq 700\,°C$ to ensure sufficient

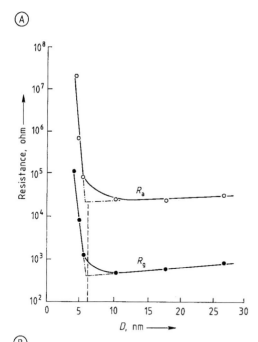

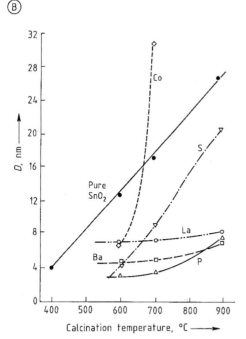

Figure 8. A) Influence of crystallite size (D) on electrical resistance for an SnO_2-based sensor in dry air (R_a) and in air containing 800 ppm H_2 (R_g) at 300 °C (elements sintered at 400 °C); B) Changes in SnO_2 crystallite size as a function of calcination temperature for pure SnO_2 samples and SnO_2 impregnated (5 %) with oxides of various foreign metals [34] (with permission from Elsevier, Amsterdam)

mechanical strength; the cross-sectional diameter of the SnO_2 particles is typically ≥ 20 nm. Unfortunately there is a sharp decrease in gas sensitivity (corresponding to a decrease in the resistance of the SnO_2) when the particle size D increases beyond about 6 nm (Fig. 8 A) [34]. Additives such as barium, phosphorus, or lanthanum oxides can be used to restrict thermal growth of the SnO_2 particles to a limit of ca. 10 nm (Fig. 8 B) [34].

Various approaches have been applied to interpreting the parameters influencing gas sensitivity. Models that relate gas sensitivity to grain size (D) and the geometry of contacts between the SnO_2 particles include the intergrain model and the neck model. An extended approach taking into account both models has been developed by YAMAZOE and MIURA [34], in which gas sensitivity is related to electron charge transfer within the microstructure of the polycrystalline elements (cf. Fig. 4 B).

Selectivity. The selectivity of all semiconductor gas sensors is very limited, but it can be optimized within narrow limits by choosing the best operating temperature and the most appropriate dopant. As in the case of pellistors, overall selectivity is controlled by the ultimate rate-determining step; i.e., the compound that shows the highest oxidation rate determines the selectivity. As is generally the case, the slowest step controls the overall reaction rate. In order to construct a sensor that selectively transforms only the analyte into another product, side reactions must be prevented between the compound that is actually detected (here oxygen) and compounds other than the analyte. This is nearly impossible to achieve in the case of a considerable excess of interfering compounds. Since the physicochemical phenomenon underlying heterogeneous catalysis is based on adsorption–desorption steps and subject to the kinetics of the heterogeneous chemical reaction itself, no known mechanism displays exceptional selectivity. It is not generally possible for a solid-state surface to recognize a single class of molecules because the forces involved in the ad- and desorption steps are rather strong, and are influenced only by overall molecular design and not by details of molecular structure, in contrast to the situation with biocatalysts. It must also be kept in mind that every compound can be considered as a potential interferant, since it might influence the heterogeneous rate constant for catalysis even if it does not itself react with oxygen. This also means that every catalyst poison, such as silicones or even dust blocking the active surface,

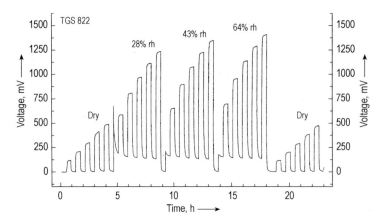

Figure 9. Effect of the relative air humidity on the response of an SnO_2 semiconductor sensor (Taguchi Type TGS 822) on the calibration cycle with ethanol as analyte. The ethanol concentrations were varied from 13, 26, 65 ppm and the relative humidity between dry, 23 %, 43 %, and 64 % rh.

Table 1. Humidity interference of SnO_2-based gas sensors. Analyte, ethanol; relative humidity rh; S = sensitivity /slope of calibration curve at 12.5 ppm ethanol; norm = factor of sensitivity increase in corresponding humid air (values in parenthesis)

Ethanol, ppm	rh, %	S, mV/ppm		
		TGS 800	TGS 813	TGS 822
13	0.00 ± 0.00	14.368 (1.00)	3.568 (1.00)	9.571 (1.00)
26	22.23 ± 0.42	57.752 (4.02)	13.850 (3.88)	47.423 (4.95)
39	42.13 ± 1.18	61.776 (4.30)	15.488 (4.34)	52.851 (5.52)
52	64.43 ± 0.93	63.932 (4.45)	16.801 (4.71)	54.275 (5.67)
65	0.00 ± 0.00	13.978 (0.97)	3.330 (0.93)	9.296 (0.97)

is likely to interfere in a measurement, which in turn limits the possibility of using chemometric methods to correct for errors caused by lack of selectivity.

Humidity effect. Humidity often influences both the zero point of a Taguchi sensor and the sensitivity, which makes corrections difficult as shown in Figure 9.

Table 1 demonstrates that this is the general behavior of semiconductor gas sensors. The slope of the calibration curve can vary by up to a factor of 5. In this table, three different Taguchi sensors (TGS 800, TGS 813, and TGS 822) were carefully compared [257] with respect to their humidity interference using ethanol as the analyte.

This behavior has to be taken into account if these devices are used in electronic noses!

28.2.2.3. Selective Ion Conductivities in Solid-State Materials

The lambda probe, based on selective oxygen ion diffusion through solid yttrium-doped ZrO_2 above 400 °C, is an extremely selective sensor for gaseous oxygen. Frequently used for measuring the residual oxygen content in exhaust gases from internal combustion engines, the lambda probe is primarily a potentiometric device that functions as a concentration cell in which gaseous oxygen is in equilibrium with lattice oxygen in the solid electrolyte (Fig. 10). A higher concentration of oxygen on one side of the membrane leads to a potential difference which, according to the Nernst equation, is proportional to the ratio of the two oxygen concentrations. Likewise the fluoride ion-selective electrode based on selective fluoride-ion diffusion through solid-state (crystalline) LaF_3 is an extremely selective sensor for fluoride ions in solution.

Selectivity in both of these cases is controlled by the size and the charge of the diffusing ion. The

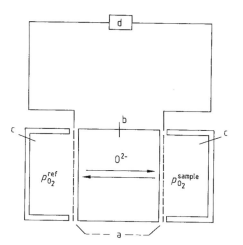

Figure 10. Schematic representation of a lambda probe operated as a potentiometric sensor
a) Porous electrode; b) ZrO_2, doped; c) Oxygen; d) Voltage measurement system

diffusion process takes place via discrete ion jumps to nonoccupied lattice positions and/or from and to interstitial positions. Ion selection here can be regarded as a filtering process, with the additional requirement for an appropriate ionic charge to prevent overall electroneutrality. Because the analyte molecules and ions involved are rather small, they can be separated quite readily from larger molecules or ions. Thus, the solid-state material functions like an ideal permselective membrane. Any solid state material characterized by exceptionally high ionic conductivity, such as Ag_2S (which has extremely high Ag^+-ion conductivity), will be a very good sensor for that ion. In the case of the ion-selective membrane electrodes used as Nernstian potentiometric sensors in solution, however, more detailed consideration is necessary (see Section 28.2.3.1.1).

Since the analyte recognition process is here rather straightforward, and because the filtering effect is accompanied by additional restrictions (appropriate charge and multiple straining), the selectivities obtained through this approach to measuring are extremely high. Many of the compounds that interfere do so not through direct participation in the solid-state diffusion process but rather by affecting the analyte concentration on the sensor surface by other means (e.g., reactions with the analyte, etc.). Thus, with the help of appropriate catalysts and high temperatures the λ-probe can also be used for the detection of other analytes that are oxidized at the surface and thereby reduce the amount of adsorbed oxygen.

28.2.2.4. Selective Adsorption–Distribution and Supramolecular Chemistry at Interfaces

Supramolecular chemistry has been a source of fascinating new molecular recognition processes. At the active center of a biocatalyst (enzyme) and/or in the binding region of a large antibody macromolecule, the recognition process is attributable mainly to a perfect bimolecular approach of complementary structures. Association or binding occurs only if the relatively weak intermolecular forces involved (dipole–dipole and van der Waals forces) are further supported by a uniquely favorable bimolecular approach that encourages interatomic interactions. A selective reaction of this type requires that the host molecule be carefully tailored so that it matches the guest molecule perfectly. Molecular modeling with attention to the activation energies for closest approach is an especially valuable tool in the design of synthetic host molecules for particular analytes. The analyte can be either a charged or an uncharged species. In the case of an uncharged molecule, however, selective binding of the analyte to the host molecule must be detected with the aid of a mass-sensitive transducer or by following the changes of certain optical parameters.

Further research in the field of supramolecular chemistry for sensors seems warranted given the prospect for improved selectivity based on custom-tailored synthetic guest molecules for the selective binding of particular host molecules (analytes). Even differentiation between enantiomers may be achievable. The latter could be demonstrated by employing known stationary phases used in gas chromatography for the separation of enantiomers also on the surface of mass sensitive transducers. During recent years it has been claimed that the technique of molecular imprinting polymers (MIP) is able to perform a similiar high selectivity. While this may be achievable in chromatographic separations due to the high repetition of multiple interface equilibria, it has not yet been convincingly shown for a chemical sensor with a single "hit and recognize" event. The rigidity of the polymer backbone of the imprint makes it highly unlikely that the same recognition power as flexible biomolecules with their "induced-fit" approach can be achieved.

28.2.2.5. Selective Charge-Transfer Processes at Ion-Selective Electrodes (Potentiometry)

An important aim of quantitative analytical chemistry is the selective detection of specific types of ions in aqueous solution. An electrode capable of selectively detecting one type of ion is called an *ion-selective electrode* (ISE). The operating principle of a particular ISE depends upon the nature of the interaction between ions in solution and the surface of that electrode. Consistent with thermodynamic distribution, the ion in question crosses the phase boundary and interacts with the electrode phase after first discarding its hydrate shell (new studies have shown that this may not always be the case in a quantitative sense, since water can be detected in ion-selective PVC membranes), which results in a net flow of ions across the boundary. Ideally, the counterion would not take part in this flow, so a charge separation would develop, causing the counterion to remain in the neighboring border area. This is almost the situation when there will be no net energy gain by this transfer due to some supramolecular binding, and a distribution because of the lipophilicity of counterions should be less pronounced. The resulting electrode potential (difference) between the interior of the electrode and the bulk of the sample solution can be described thermodynamically and kinetically—again, in the ideal case—by the Nernst equation.

The selectivity of such a potentiometric device is that of an ideal Nernstian sensor, the potential of which remains stable irrespective of the fact that some current may be flowing through the sensor interface. An ideal Nernstian sensor is characterized by a current–voltage curve that is very steep and nearly parallel to the current axis.

The selectivity of a potentiometric ion-selective interface is controlled largely by interfacial charge-transfer kinetics and not by the stability constant for reaction of the analyte ion with the electroactive compound (e.g., valinomycin in the case of potassium ion). The exchange-current density for the analyte ion relative to interfering ions determines the extent of potentiometric selectivity. Any factor that increases the former (conditioning of the electrode with the measuring ion, ion-pair formation by lipophilic counterions, etc.) leads to a better and more selective Nernstian sensor. The ideal electroactive compound apparently behaves like a selective charge-transfer catalyst for the analyte ion. Advantages that might perhaps be gained from tailored host molecules for accelerating the transfer of analyte ions into the surface of the electrode have not yet been fully investigated. Recognition of an analyte ion depends upon size and charge as well as the speed with which the ion loses its solvent sphere prior to entering the electrode phase. Therefore, the electroactive compounds most suitable for embedding in the membrane of an ion-selective electrode (usually consisting of PVC containing a plasticizer) are those that play an active role in successive replacement of the often firmly held hydrate shells of the analyte ions. Recent results in the author's institute clearly demonstrate that the thermodynamic complex stability constants determined in homogeneous solution (e.g. via NMR) do not mirror the potentiometric selectivity! A newly synthesized compound for sodium ions, as an example, turned out to yield a very selective PVC membrane for $Pb(OH)^+$ ions [40]. While in theory supramolecular recognition worked perfectly well in homogeneous solutions, at the interface a different mechanism obviously plays a dominant role.

28.2.2.6. Selective Electrochemical Reactions at Working Electrodes (Voltammetry and Amperometry)

Depending on the potential range at the working electrode, as many as about ten different electrode processes can be separated voltammetrically based on the typical resolution of a polarographic curve, which requires a difference voltage of about 200 mV for the half-wave potential and a concentration-dependent limiting current. Selectivity in the process is therefore rather limited, and depends on such thermodynamic parameters as the reaction enthalpy for reduction or oxidation at the electrode surface. The potential of the working electrode determines whether a particular compound (analyte) can be oxidized or reduced, and compounds with similar redox characteristics might interfere.

The problem of interference is of course greatest when extremely positive or negative working electrode potentials are applied at the working electrode, since at high potential the small differences between electrochemical reaction rates become negligible. The "resolution" of a typical current–voltage curve is diminished if the electrochemical reaction is kinetically hindered, resulting in so-called irreversible current steps that are less steep. Irreversible electroactive species require a greater overpotential before the diffusion-controlled limiting current plateau is reached, therefore increasing the chance that an interfering compound will become electrochemically active. The limited resolution power of voltammetry can be improved by the use of chemically modified electrode surfaces. If a surface layer or a thin membrane is introduced between the phase boundary this additional layer will increase the selectivity if only the analyte can pass through this layer.

In certain biosensors an irreversible heterogeneous reaction is transformed into a homogeneous redox reaction via a very reversible redox system. The latter can be transformed back at the working electrode by a much lower overpotential, thereby reducing the chance of co-oxidation or co-reduction of interfering compounds. Since the limiting current is strictly proportional to the analyte concentration and also to the concentrations of interfering compounds, chemometric data treatment and/or differential measurements may be used to correct for errors.

28.2.2.7. Molecular Recognition Processes Based on Molecular Biological Principles

Metabolic biosensors are based on special enzyme-catalyzed reactions of the analyte. The recognition process, also known as the lock-and-key principle, is extremely selective, permitting differentiation even between chiral isomers of a molecule. Recognition in this case demands a perfect geometrical fit as well as the appropriate dipole and/or charge distribution to permit binding of the analyte molecule inside the generally much larger biomolecule (host). Since fitting is associated with all three dimensions, a very large biomolecule with its stabilized structure is more effective than the smaller host molecules commonly encountered in supramolecular chemistry. The selectivity of a specific biocatalytic reaction can be further improved by ensuring detection of only the transformed analyte or a stoichiometric partner molecule through a chemical transducer located behind the recognition layer. In a certain sense, this type of biosensor represents the highest possible level of selectivity, and comparable results would be difficult to achieve by other means.

Immunosensors based on mono- or polyclonal antibodies constitute the second best choice with respect to high selectivity. Depending on the amount of effort expended in screening and choosing the desired antibodies from among the millions of different antibody molecules available, the selectivity can be remarkably high. On the other hand, it is also possible to choose antibody molecules that bind only to a certain region (epitope) of the analyte molecule, leading to biomolecules capable of recognizing all members of some class of compounds sharing similar molecular structures. This type of sensor, though it cannot be calibrated, is very valuable for screening purposes used to detect the presence of an only certain class of substance . The application of antibody molecules has been considerably simplified by the large-scale preparation of monoclonal versions with precisely identical features. It has even become possible to construct immunosensor arrays based on monoclonal antibody molecules with exactly matched selectivities. Results obtained from such arrays must be evaluated by modern methods of pattern recognition.

In order to overcome the limited stability of large protein molecules, research is currently in progress to isolate only the binding region of the F_{ab} part (antibody-binding fragment) of an antibody molecule. Smaller molecules are of course likely to display somewhat more limited recognition ability, however, with selectivities approaching those of supramolecular host molecules synthesized in the traditional way. Recent research has also focussed on recombinant antibodies produced from large protein libraries without the need for any animal immunization.

28.2.3. Transducers for Molecular Recognition: Processes and Sensitivities

28.2.3.1. Electrochemical Sensors

Electrochemical sensors constitute the largest and oldest group of chemical sensors. Although many such devices have reached commercial maturity, others remain in various stages of development. Electrochemical sensors will be discussed here within the broadest possible framework, with electrochemistry interpreted simply as any interaction involving both electricity and chemistry. Sensors as diverse as enzyme electrodes, high-temperature oxide sensors, fuel cells, and surface-conductivity sensors will be included. Such sensors can be subdivided based on their mode of operation into three categories: *potentiometric* (measurement of voltage), *amperometric* (measurement of current), and *conductometric* (measurement of conductivity) [13], [41].

Electrochemistry implies the transfer of charge from an electrode to some other phase, which may be a solid or a liquid. During this process chemical changes take place at the electrodes, and charge is conducted through the bulk of the sample phase. Both the electrode reactions themselves and/or the charge transport phenomenon can be modulated chemically to serve as the basis for a sensing process. Certain basic rules apply to all electrochemical sensors, the cardinal one being the requirement of a closed electrical circuit. This means that an electrochemical cell must consist of at least two electrodes, which can be regarded from a purely electrical point of view as a sensor electrode and a signal return.

Another important general characteristic of electrochemical sensors is that charge transport within the transducer portion of the sensor and/or inside the supporting instrumentation (which constitutes part of the overall circuit) is always electronic. On the other hand, charge transport in the sample under investigation may be electronic, ionic, or mixed. In the latter two cases electron transfer takes place at the electrode–sample interface, perhaps accompanied by electrolysis, and the corresponding mechanism becomes one of the most critical aspects of sensor performance.

The overall current – voltage relationship is complex, and it can vary as the conditions change. The relationship is affected primarily by the nature and concentration of the electroactive species, the electrode material, and the mode of mass transport. A total observed current can be analyzed in terms of its cathodic and anodic components. If the two currents are equal in magnitude but opposite in sign, there must be an *exchange current* passing through the electrode-surface/sample interface.

Both the charge-transfer resistance and the exchange-current density are critically important parameters in the operation of most electrochemical sensors, since they reflect the kinetics of an electrode reaction. These parameters are directly proportional to the electrode area, so the smaller the electrode the higher will be the resistance, all other parameters remaining unchanged. Therefore, it is the electrochemical process taking place at the smaller electrode that determines the absolute value of the current flowing through the entire circuit. The auxiliary electrode will begin to interfere only if its charge-transfer resistance becomes comparable to that of the working electrode. Microelectrodes represent one approach to avoiding such interferences (see Section 28.2.3.1.2).

On the other hand, in zero-current potentiometry the relative sizes of the two electrodes is immaterial. Acquiring useful information in this case requires only that the potential of the working electrode be measured against a well-defined and stable potential from a reference electrode. Any foreign potential inadvertently present within the measuring circuit can contaminate the information, so it is mandatory that the reference electrode be placed as close to the working electrode as is practically possible. Thus, in an amperometric (or conductometric) measurement the source of information can be localized by choosing a small working electrode, whereas it cannot be localized with zero-current potentiometric measurements.

28.2.3.1.1. Self-Indicating Potentiometric Electrodes

Fundamental Considerations. Within the context of this chapter "potentiometry" is understood to mean the measurement of potential differences across an indicator electrode and a reference electrode under conditions of zero net electrical current. Such a measurement can be used either for determining an analyte ion directly (direct potentiometry) or for monitoring a titration (see below).

In recent years potentiometry has proven to be well suited to the routine analysis of a great number and variety of analytes [5], [40], [42]–[60]. Ion-selective electrodes (ISEs) are commercially available for many anions and cations, as indicated in Tables 4–7. Other analytes can be determined using ISEs in an indirect way. Chapter 28.3 deals with various types of potentiometric biosensors. A special advantage of ion-selective potentiometry is the possibility of carrying out measurements even in microliter volumes without any loss of analyte.

At present, most ISEs unfortunately do not provide absolute selectivity for a single ion. This fact requires that one have access to very detailed information regarding the nature of samples to be analyzed so that interferences can be either eliminated or otherwise taken into account.

The theory of (ion-selective) potentiometry has already been introduced in Section 28.2.2.5. Here it is necessary only to remind the reader of the Nernst equation and the Nernst–Nikolsky equation, both of which are used in the evaluation of potentiometric measurements.

The *Nernst equation* is valid only under ideal conditions (with no interfering ions, etc.):

$$E = E^0 + (RT/zF) \ln a_M \tag{5a}$$

$$= E^0 + (0.059/z) \log a_M \quad (5b)$$

The *Nernst–Nikolsky equation* on the other hand takes into consideration the influence of interfering ions on the potential of an ISE:

$$E = E^0 + (RT/z_M F)\ln\left[a_M + \sum K_{MI}(a_I)^{z_M/z_I}\right] \quad (6)$$

where

E	= potential of the ISE, measured with zero net current
E^0	= standard potential of the ISE
R	= gas constant (8.314 J K^{-1} mol^{-1})
T	= absolute temperature (K)
F	= faraday constant (96 485 C/mol)
a_M	= activity of the ion to be measured
a_I	= activity of an interfering ion
z_M, z_I	= electrical charges of the measured and interfering ions
K_{MI}	= selectivity coefficient

The value of K_{MI} depends both on the activity a_M and on the particular combination of analyte ion and interfering ion. Recently BAKKER [48] suggested another equation which described the EMF versus concentration behavior of an ISE with a higher accuracy. The selectivity coefficients are determined here without conditioning the electrode with the measured ion, which yields constant selectivity coefficients. The new equation describes the situation when a monovalent analyte ion I$^+$ is interfered by a divalent ion J^{2+} especially well, and *vice versa*:

$$E = E_I^0 + \frac{RT}{F}\ln\left(\frac{a_I(IJ)}{2} + \frac{1}{2}\sqrt{a_I(IJ)^2 + 4a_J(IJ)\left(K_{IJ}^{pot}\right)^2}\right) \quad (6a)$$

for $z_I = 2$ and $z_J = 1$

$$E = E_I^0 + \frac{RT}{F}\ln\left(\sqrt{a_I(IJ) + \frac{1}{4}K_{IJ}^{pot}a_I(IJ)^2} + \sqrt{\frac{1}{4}K_{IJ}^{pot}a_I(IJ)^2}\right) \quad (6b)$$

HORVAI [49] developed another more general equation for ion-selective ion-exchange membranes, which also considers the concentration dependency of the selectivity coefficient:

$$E \cong \epsilon_I^{\geq} + \frac{\varrho}{\zeta_1 \Phi}\ln\alpha_1(IJ) + \frac{\varrho}{\zeta_1 \Phi}\varkappa_{IJ}\alpha_I(IJ)^{\zeta_J/\zeta_I}\alpha_J(IJ) \quad (6c)$$

One can see here, separated from each other, the response of the ISE towards the pure analyte ion I and (as the last term) the EMF deviation caused by the interfering Ion J. Note that there is no logarithm in the last term and also that the last term depends both on the interfering and primary ion concentrations.

Types of Ion-Selective Electrodes. Simple metal-ion electrodes such as those based on silver, gold, platinum, etc., will not be dealt with in this discussion since they are used mainly for indicating purposes in potentiometric titrations. Apart from these, the following types of ISEs can be distinguished:

1) Glass-membrane electrodes
2) Solid-membrane electrodes
3) Liquid-membrane electrodes (including PVC-membrane electrodes)
4) ISEs based on semiconductors (ion-selective field-effect transistors, etc.)

These four types will be the subject of more detailed consideration.

Glass-Membrane Electrodes [50]. The best-known glass-membrane electrode is the pH electrode. Its most important component is a thin glass membrane made from a sodium-rich type of glass. Depending on the intended application the membrane may take one of several geometric forms: spherical, conical, or flat. For applications in process streams, special pH electrodes have been devised that are resistant to high pressures, and sterilizable pH electrodes are available for biotechnological applications.

Examples of the different types of pH electrodes are illustrated in Figure 11. The *spherical* type is most frequently used for the direct measurement of pH or for acid–base monitoring. It is robust and appropriate for most routine applications. *Conical* electrodes can be used as "stick-in electrodes" for pH measurements in meat, bread, cheese, etc. A conical membrane can easily be cleaned, which is important if measurements are to be made in highly viscous or turbid media. A *flat* membrane facilitates pH measurements on surfaces, such as on human skin.

The pH electrode must be combined with a reference electrode [51] that provides a constant potential for completing the electrical circuit. A suitable reference system is frequently integrated into the body of the pH electrode itself. Combined electrodes of this type are particularly easy to

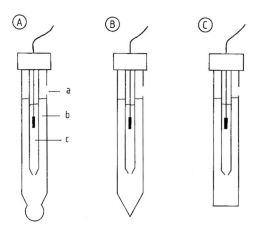

Figure 11. Various types of glass-membrane pH electrodes (with permission from Springer-Verlag, Heidelberg)
A) Spherical membrane; B) Conical membrane; C) Flat membrane
a) Fill hole; b) Inner filling solution (buffer); c) Reference system

handle. pH Measurements at > pH 10 by means of glass electrodes are subject to interference by alkalimetal ions, especially Na^+. The selectivity coefficient K_{MI} for interference by the sodium ion is 10^{-13} in special lithium containing glasses with a reduced alkali error. This means that a 10^{13}-fold concentration of sodium is required to produce the same potential as a given concentration of protons.

It is important that the surface of a glass membrane be swollen before measurements are taken. Prior to first use the electrode should therefore be allowed to swell overnight in a 3 mol/L KCl solution. For the same reason the electrodes should also be stored in a solution of this type between measurements and after measurements have been conducted in nonaqueous solvents. Here, the choice of a double junction reference electrode with the same solvent as in the measuring solution and added organic salt for sufficient conductivity is important. In the case of endpoint detection a platinum/ferrocene redox element without a liquid junction could also be used.

Apart from the pH-selective glass-membrane electrodes, other commercially available glass-membrane electrodes are responsive to sodium ions, although the selectivity of a pNa electrode is much lower than that of a pH electrode. Approximate selectivity coefficients K_{MI} in this case are 10^3 for interference by H^+ and 10^{-3} for interference by Li^+, K^+, and Cs^+. However, an interference by H^+ ions is seldom serious since by appropriate pH adjustment one can easily shift the H^+ ion concentration to a range where it does not influence the electrode potential.

Solid-Membrane Electrodes. The sensing element of a solid-membrane electrode consists of a material showing ionic conductivity for the particular ions that are to be determined. Such a device may be manufactured from a single crystal, as in the well-known fluoride-selective electrode, where the sensing element is a single crystal of lanthanum fluoride doped with europium(II) fluoride to lower the electrical resistance. Alternatively, a membrane can be produced by grinding the crystalline sensor material together with an appropriate additive (PTFE powder, silicone rubber), followed by solidification either by application of high pressure or by addition of a cross-linking agent. It is thus essential that the crystal particles stay in contact which each other. In addition to these so-called heterogeneous membranes, the homogeneous membranes based on silver sulfide are better known. They consist of Ag_2S which is freshly precipitated, washed, dried, and pressed into a pellet with a traditional KBr-IR press. Since Ag_2S is rarely stoichiometric it shows a high Ag^+ ion conductivity leading to a perfect Nernstian-sensor for that ion. Because the standard exchange current density at its suface is much higher than on a silver metal surface (overpotential due to crystallization polarization) it senses lower Ag^+ ion concentrations much better. By incorporating about 30 weight % of a certain silver halide into this pellet one obtains the corresponding halide sensitive membrane. Note, the same result would be obtained if the appropriate silver halide were to be added as a powder to the measuring solution. That means that it is still the Ag^+ ion which determines the potential via the corresponding solubility product. If certain metal sulfides such as CuS, PbS, or CdS are added the membrane senses those metal ions via two solubility products and an intermediate S^{2-} ion-concentration. Thus, Ag^+ ions must be totally absent as well as compounds oxidize S^{2-}.

The membrane is then fixed to the electrode body. Electrical contact is established either by a conducting adhesive or by an internal reference system. Figure 12 illustrates schematically the construction of two types of solid-membrane electrodes. A detailed compilation of commercially available solid-state electrodes is provided in Tables 4 – 7 at the end of this section.

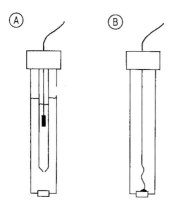

Figure 12. Two types of solid-membrane electrodes (with permission from Springer-Verlag, Heidelberg)
A) With internal filling solution and reference system;
B) With solid contact

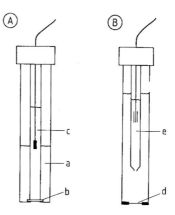

Figure 13. Two types of liquid-membrane electrodes (with permission from Springer-Verlag, Heidelberg)
A) Porous-membrane electrode; B) PVC-membrane electrode
a) Reservoir for liquid ion carrier; b) Porous membrane; c) Reference system; d) PVC membrane; e) Reference system

Generally speaking, interference must be anticipated from any ion that tends to react with the membrane material to form compounds less soluble than the membrane itself. Thus, the chloride electrode, which consists of a $AgCl/Ag_2S$ membrane, is poisoned by mercury(II) ions. Due to the solid state of the membrane the reaction rate is rather low, so a poisoned membrane can be repaired by carefully polishing it with an appropriate polishing powder (as described in instruction manuals furnished with the electrodes).

Liquid-Membrane Electrodes. The electroactive phase in this type of electrode consists of a liquid that has been stabilized mechanically. For this purpose the electroactive components may be dissolved in a highly viscous and apolar organic solvent such as one of the higher alcohols or paraffin. A porous material (e.g., commercial membrane filters such as Sartorius cellulose nitrate with pore diameters of about 0.45 μm), are initially made hydrophobic with, e.g., hexamethyl disilazane and then soaked with this solution and introduced in place of the solid membrane in an electrode of the type described above. More frequently, however, the liquid is not fixed in a porous material, but rather embedded into a high molecular mass polymer like PVC. In this case the plasticizer (normally present at a concentration of about 66 % by weight) acts as a solvent for the electroactive component (the concentration, which seldom exceeds 1 %!). Simplified construction schemes for these two types of electrodes are illustrated in Figure 13.

For applications like those discussed in Sections 28.2.2.4 and 28.2.2.5 the electroactive component of a liquid-membrane electrode would consist of a (charged) ion exchanger or an (uncharged) ion carrier. In both cases the electroactive component must have a rather high affinity for the ion to be determined, but at the same time display the greatest possible selectivity with respect to interfering ions. Table 2 lists ion exchangers and ion carriers frequently used in liquid-membrane ISEs. For an excellent and almost exhaustive compilation of ISEs based on this principle see the reviews of BAKKER, BÜHLMANN, and PRETSCH [59], [60].

The first part of these publications deals with the most recent theoretical treatment while the second part describes in detail nearly all known membrane compositions for nearly every ion of particular importance. It should be noted that PVC-based ISEs are especially easy to construct in the laboratory just before they are to be used. The Fluka company offers a large collection of electroactive compounds (including the most famous ones with their characteristic ETH number), the right PVC and plasticizer and also construction information. Recently reasons for the typical detection limits in the range of μM solutions have been found [61]. If the internal filling solution contains too high a concentration of the measured ion it will be distributed through the entire membrane and can be exchanged with interfering ions at the measuring surface and thereby produce a certain concentration in the sample solution. Thus, in case of trace determinations the inner concentration of the measured ion should be $< $ μM.

Table 2. Example of ion exchangers and ion carriers frequently encountered in liquid-membrane electrodes

Ion	Electroactive compound
Cations	
K^+	valinomycine
NH_4^+	nonactine – monactine
Li^+	lithium carrier, e.g., ETH 1644 (Fluka)
Ca^{2+}	calcium salt of didecylphosphonic acid
Anions	
NO_3^-	$Ni(o\text{-phen})_3^{2+}$
ClO_4^-	$Fe(o\text{-phen})_3^{2+}$
BF_4^-	$Fe(o\text{-phen})_3^{2+}$

However, care is needed not to loose the internal measured ion by adsorption processes. Thus, an appropriate ion-buffer leading to a constant low concentration is recommended.

For anion determinations, most PVC-membrane based ISEs follow the so-called Hofmeister series, which is: the more lipophilic anion is always more favored by the organic membrane and therefore sensed better. However, there are exceptions to this general rule which are also most interesting from a theoretical point of view, especially when a supramolecular carrier is used and more than the typical one lipophilic salt addition is performed. Because of this, owing to the incorporation of a lipophilic cation and anion, a carbonate selective ISE could be constructed with decreased interference from salicylate by five orders of magnitude [62]. This addition surprisingly increases the charge transfer resistance of the ISE so that it can be measured besides the high bulk resistance in series. However, the ratio of the transfer resistances for the analyte and the interfering ions changes too, in favor of the measured ion!

ISEs Based On Semiconductors (ISFETs). Devices based on ion-selective field-effect transistors (ISFETs) are sufficiently distinctive to warrant separate treatment in Section 28.2.3.1.4.

Instrumentation. The discussion here of instrumentation required for potentiometric measurements will be limited to a brief overview of suitable reference electrodes [51]. The conventional reference electrode consists of an electrode of the second kind, like the Ag/AgCl/KCl system. Much effort has been devoted to the development of miniaturized reference systems suited to direct integration into ISFETs, but most systems of this type fail to fulfill all the requirements for a "true" reference electrode with a well-defined and stable potential difference relative to the standard hydrogen electrode. The greatest problem is the short distances needed for the liquid junction, leading to an influx of unwanted compounds within too short a time period. Reference electrodes with gel-stabilized internal electrolytes are produced for applications in process analysis.

Table 3 provides an overview of the most frequently used reference electrodes.

Evaluation of Potentiometric Results. The aim of all methods for the evaluation of potentiometric data is the computation of substrate activities from voltage changes observed between an ISE and a reference electrode. Generally speaking, three different approaches are possible: (1) the direct potentiometric measurement of activities, (2) monitoring during the course of a titration, and (3) flow-injection analysis (FIA).

A determination by *direct potentiometric measurement* is accomplished either by calibrating the electrode with solutions of known concentration, or by using the techniques of standard addition or standard subtraction. Since calibration with standard solutions usually involves solutions of pure salts, thorough knowledge of the samples to be analyzed is a prerequisite to avoiding erroneous results due to interfering ions present in the sample matrix. Influences attributable to other substances present in the sample that do not interfere directly with the determination but produce changes in ionic strength can be minimized by adding a total ionic strength adjusting buffer ("TISAB solution") to both standard solutions and samples. Standard-addition or -subtraction methods should be invoked whenever it is not possible to make appropriate standard solutions (owing to a very complex matrix, for example). Figure 14 illustrates the evaluation of a standard-addition analysis.

Monitoring a titration by means of an ISE has the advantage over direct potentiometric measurement that the accuracy is determined largely by the titration reaction and not simply by the calibration function of the electrode. This is especially important for ISEs with changing or unknown slopes. For this reason the determination of a surfactant should always be carried out as a titration and not as a direct measurement.

Flow-Injection Analyses with ISEs are especially easy to carry out. If the ISE has a flat surface it can be connected to the liquid junction of a reference electrode via a small strip of filter paper. This paper strip is then held by a clamp and the carrier stream of the FIA system is applied just

Table 3. Common reference electrodes

Electrode	Potential vs. NHE	Useful temperature range, °C	Interferences
Hg/Hg$_2$Cl$_2$/KCl (0.1 mol/L)	+0.334 V		
Hg/Hg$_2$Cl$_2$/KCl (1.0 mol/L)	+0.280 V	15–70	complexing agents (CN$^-$), S^{2-}, strong oxidizing or reducing agents
Hg/Hg$_2$Cl$_2$/KCl (satd.)	+0.241 V		complexing agents (CN$^-$), S^{2-}, strong oxidizing or reducing agents
Ag/AgCl/KCl (0.1 mol/L)	+0.290 V		complexing agents (CN$^-$), S^{2-}, strong oxidizing or reducing agents
Ag/AgCl/KCl (1.0 mol/L)	+0.222 V	15–110	complexing agents (CN$^-$), S^{2-}, strong oxidizing or reducing agents
Ag/AgCl/KCl (satd.)	+0.197 V		complexing agents (CN$^-$), S^{2-}, strong oxidizing or reducing agents
Hg/Hg$_2$SO$_4$/K$_2$SO$_4$ (0.5 mol/L)	+0.682 V	15–70	complexing agents (CN$^-$), S^{2-}, strong oxidizing or reducing agents
Hg/Hg$_2$SO$_4$/K$_2$SO$_4$ (satd.)	+0.650 V		
Tl(Hg)/TlCl/KCl (3.5 mol/L)	−0.575 V	15–120	S^{2-}, strong oxidizing agents; no interference from complexing agents

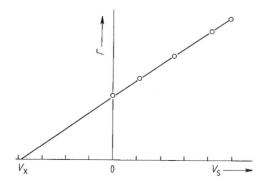

Figure 14. Plot for the evaluation of a standard-addition analysis
V_s = Volume of standard solution; V_x = Volume of analyte
$\Gamma = pE \cdot V_{total}$, with $pE = 10^{\Delta E \cdot (z/S)}$ (ΔE = Potential difference; z = Charge of the ion, including sign; S = Slope of the electrode, including sign)
The unknown concentration of the analyte c_x is
$c_x = (-c_s \cdot V_x)/V_{initial}$, where c_s = Concentration of the standard solution and $V_{initial}$ = Initial volume of the analyte

before the measuring ISE, passing over its surface and later over the reference electrode. New levels of ultra trace analysis can be reached — as already mentioned — by never conditioning the ISE membrane with higher concentrations of the measured ion either from the outside or from the inside! This automatically leads to a differential FIA operating mode in which the injected sample segment is first passed over one of the ISE surfaces and then with a matched retarding coil to the other side of the ISE membrane. Figure 15 shows results and the experimental set-up with two external reference electrodes, the ISE membrane and the redarding coil which should direct the sample segment to the second site of the ISE-membrane exactly at the moment it leaves the first side [63]. In ultra trace analysis the carrier solution should contain only traces of the measured ion in order to get a stable voltage reading. Note that the injection volume in Figure 15 is in the μL range. Thus, such a simple and low cost approach allows the determination of pg amounts as it is demonstrated in Figure 16.

Commercially Available ISEs. Tables 4–7 provide an overview of ISEs commercially available at the time. ISEs described only in the literature, and which must therefore be selfconstructed, have not been included in the tables. Instructions for the preparation of PVC-membrane electrodes can be found in the literature [4], [5], [15].

28.2.3.1.2. Voltammetric and Amperometric Cells

Information is obtained with this type of electrochemical sensor from either the combined current/potential – concentration relationship (voltammetry) or from the current – concentration relationship alone (amperometry).

A voltammetric measurement is accomplished by scanning the potential difference across an electrochemical cell containing a working electrode, a reference electrode and usually an auxiliary electrode separated from the sample solution with a diaphragma from one preset value to another and recording the cell current as a function of the applied potential. A curve so generated is known as a *voltammogram*. Amperometric meas-

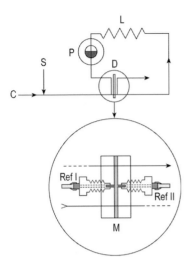

Figure 15. Differential flow-injection analysis with one single ISE membrane and two reference electrodes in an electrolytic closed loop.
An injector block used in liquid chromatography injects about 500 µL of sample into the carrier stream. The volume of the carrier stream loop from one side of the ISE membrane to the opposite side including the volume of both membrane sides is also approximately 500 µL. Thus, when the sample segment is fully washed out of the first membrane side the first front of it reaches the backside. Through this a positive and a negative analyte peak are obtained, which look like a second derivative of a peak.
C = carrier solution; S = sample (introduction with injector loop); P = pump; L = loop adapted to different sample volumes; D = detector; M = ISE membrane with two spacers with a flow channel on each side; Ref I and Ref II = commercial reference electrodes with threads

urements are made by recording the current flow in a cell at a single applied potential.

The essential operational function of a transducer for this purpose is the transfer of one or more electrons to or from the transducer surface (= working electrode). The flow of these electrons is what constitutes the output signal. Voltammetric and amperometric devices are also capable of conferring a degree of selectivity on the overall sensing process by the proper choice of the working electrode potential against a reference electrode.

Furthermore they are relatively simple: in the most elementary case, the transducer (electrochemical cell) consists of nothing more than two electrodes immersed in a suitable electrolyte. A more complex arrangement might involve the use of a three-electrode cell mentioned above, with the advantages that the reference half-cell is not disturbed (polarized) by the current flowing and that the ixR drop can be compensated by such a set-up.

In spite of the inherent diagnostic advantages of voltammetry, a transducer based on this technique represents a rather cumbersome approach to sensing, mainly because of the electronic circuitry required to scan the applied potential, the time needed for a single scan, and the evaluation algorithm identifying the correct current peak to be measured. Accordingly, most sensing applications involve cells operating in an amperometric (fixed potential) mode.

Voltammetry. When a slowly changing potential is applied to an electrode immersed in an electrolyte solution containing a redox species, a current will be observed to flow as soon as the applied potential reaches a certain value. This current arises from a heterogeneous electron transfer between the electrode and the redox couple, resulting in either oxidation or reduction of the electroactive species. At a sufficiently oxidizing or reducing overpotential the magnitude of the current may become a function of mass transfer of the redox species to the electrode. In a well-stirred solution so-called quasi-stationary current–voltage curves with typical current steps and plateaus are obtained. The half-step (wave) potential is characteristic of the type of species being electrochemically oxidized or reduced, the height of the current steps represents the concentration of it. Without stirring, current peaks are obtained, since the diffusion towards the working electrode surface is no longer sufficient to maintain a constant current. Redox couples that give rise to symmetrical current peaks separated by $58/z$ mV (at 25 °C) with cyclic voltage ramps in unstirred solutions are frequently referred to as *reversible couples*. If the rate of electron transfer between the redox couple and the electrode is high compared with the rate of mass transfer, then the electrode reaction is reversible. Under these circumstances the concentration ratio for oxidized and reduced forms of the couple at the electrode surface is described by the Nernst equation [Equations (5a) and (5b)].

If electron transfer between the redox species and electrode is very slow (kinetical controlled) relative to the mass transport of solution species to the electrode, then the observed current will not be a function of mass transport. In this case, the low rate of electron transfer results in a concentration ratio of the two forms of the redox couple at the electrode surface that no longer conforms to the Nernst equation. Current–voltage curves of these

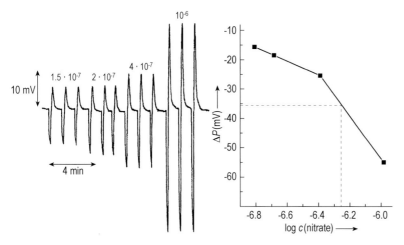

Figure 16. The calibration plot in the lower ppb range shown was constructed with peak-to-peak values resulting in a double Nernstian response. Because of the FIA detection the ISE membrane is never equilibrated with higher concentrations of the analyte ion.
Details: nitrate-selesctive ISE membrane: 1% Tetradodecylammoniumbromide; 50 mol% of the first is potassium-tetrakis-(4-chlorophenyl)-borat; 49% 2-.NPOE; 50% PVC
Carrier solution: 10^{-3} M $Li_2SO_4 + 10^{-7}$ (!) M KNO_3

Table 4. Commercially available ISEs with selectivity for cations

ISE type	MT[a]	Concentration range, mol/L[b]	Useful pH range[b]	Temp. range, °C[b]	Membrane resistance, MΩ[b,c]	Suppliers[d]	Interfering ions[b,e] in order of decreasing interference; miscellaneous remarks
pNa^+	G	$2 \cdot 10^{-6}-1$	pH = pNa + 4	0–60	100–200	CIMORWΩ	H^+, K^+, NH_4^+, Li^+; Ag^+ must be absent
pK^+	P	$10^{-6}-1$	2–12	0–40	10–20	CIMORWΩ	Rb^+, Cs^+, NH_4^+, Tl^+, H^+, Ag^+, $Tris^+$, Li^+, Na^+; cationic surfactants must be absent
pCa^{2+}	P	$5 \cdot 10^{-7}-1$	2.5–11	0–40	1–4	CIMORWΩ	Pb^{2+}, Hg^{2+}, H^+, Sr^{2+}, Fe^{2+}, Cu^{2+}, Ni^{2+}, NH_4^+, Na^+, $Tris^+$, Li^+, K^+, Ba^{2+}, Zn^{2+}, Mg^{2+}; cationic surfactants must be absent
pCd^{2+}	S	$10^{-7}-10^{-1}$	2–12	0–100	≤1	MOWΩ	Cu^{2+}, Hg^{2+}, Ag^+ must be absent; Fe^{2+}/Fe^{3+} interfere if present in concentrations $>c\,[Cd^{2+}]$
pPb^{2+}	S	$10^{-6}-10^{-1}$	4–7	0–100	≤1	MORWΩ	Cu^{2+}, Hg^{2+}, Ag^+ must be absent; Fe^{2+}/Fe^{3+} and Cd^{2+} interfere if present in concentrations $>c\,[Pb^{2+}]$
pCu^{2+}	S	$10^{-8}-10^{-1}$	2–12	0–80	≤1	CMORW	Hg^{2+}, Ag^+, S^{2-} must be absent; Cl^-, Br^-, Fe^{3+}, Cd^{2+} interfere if $c > c\,[Cu^{2+}]$; Anions forming complexes with Cu^{2+} (Hal^-, OAc^-) interfere; this ISE is sensitive to light, so all measurements should be carried out in the dark
pAg^+	S	$10^{-7}-10^{-1}$	2–12	0–80	≤1	CIMORWΩ	Hg_2^{2+}, Hg^{2+}, proteins
pNH_4^+	P	$<10^{-6}-1$	4–7	0–50	≤10	IR	cationic surfactants must be absent; interferences from K^+, Rb^+, H^+, Cs^+, Na^+/Li^+, Sr^{2+}, Ba^{2+}, Mg^{2+}

[a–h] See Table 7.

Table 5. Commercially available ISEs with selectivity for anions

ISE type	MT[a]	Concentration range, mol/L[b]	Useful pH range[b]	Temp. range, °C[b]	Membrane resistance MΩ[b,c]	Suppliers[d]	Interfering ions[b,e] in order of decreasing interference; miscellaneous remarks
pF$^-$	S	10^{-6} – sat. solu.	5–11	0–100	≈ 0.2/≤1	CIMORWΩ	pH < 5, pH > 11
pCl$^-$	S	$5 \cdot 10^{-5}$ – 1	2–12	0–50	≤ 0.1	CIMORWΩ	CN$^-$, I$^-$, Br$^-$, S$_2$O$_3^{2-}$, NH$_3$, OH$^-$, metal ions forming complexes with Cl$^-$; S^{2-} must be absent
pBr$^-$	S	$5 \cdot 10^{-6}$ – 1	2–12	0–50	≤ 0.1	CIMORWΩ	CN$^-$, I$^-$, S$_2$O$_3^{2-}$, NH$_3$, Cl$^-$, metal ions forming complexes with Br$^-$; S^{2-} must be absent
pI$^-$	S	10^{-7} –>1	2–12	0–80	≤ 0.1	CIMORWΩ	S^{2-}, CN$^-$, Br$^-$, Cl$^-$, S$_2$O$_3^{2-}$, metal ions forming complexes with I$^-$
pCN$^-$	S	< 10^{-6} – 10^{-2}	11–13	0–80	≤ 30/≤0.5	CIMORWΩ	S^{2-} must be absent; interferences from I$^-$, CrO$_4^{2-}$, S$_2$O$_3^{2-}$, Br$^-$, Cl$^-$, metal ions forming complexes with CN$^-$
pSCN$^-$	S	$5 \cdot 10^{-6}$ – 1	2–12	0–50	≤ 0.1	MOW	S^{2-} ≈ I$^-$, Br$^-$, CN$^-$, S$_2$O$_3^{2-}$, Cl$^-$, OH$^-$
pClO$_4^-$	P	$7 \cdot 10^{-6}$ – 1		0–40		O	not sorted: I$^-$, NO$_3^-$, Br$^-$, ClO$_3^-$, CN$^-$, NO$_2^-$, HCO$_3^-$, CO$_3^{2-}$, Cl$^-$, H$_x$PO$_4^{x+3-}$, OAc$^-$, F$^-$, SO$_4^{2-}$
pNO$_3^-$	P	10^{-5} – 1	3–12	0–50	1	CIMORWΩ	anionic surfactants must be absent; interference from SCN$^-$ ≈ MnO$_4^-$, NO$_2^-$ ≈ ClO$_4^-$, I$^-$, Br$^-$, HCO$_3^-$, F$^-$ ≈ SO$_4^{2-}$
pBF$_4^-$	P	$7 \cdot 10^{-6}$ – 1	2.5–11	0–40		MOW	ClO$_4^-$, I$^-$, ClO$_3^-$, CN$^-$, Br$^-$ ≈ NO$_2^-$, NO$_3^-$, Cl$^-$, PO$_4^{3-}$, OAc$^-$, F$^-$, SO$_4^{2-}$
pS^{2-}	S	10^{-6} – 10^{-1}	12–14	0–80		CIMORWΩ	Hg$_2^{2+}$, Hg^{2+}, proteins

[a–h] See Table 7.

Table 6. Commercially available ISEs with selectivity for gases

ISE type	MT[a]	Concentration range, mol/L[b]	Useful pH range[b]	Temp. range, °C[b]	Membrane resistance, MΩ[b,c]	Suppliers[d]	Interfering ions[b,e] in order of decreasing interference; miscellaneous remarks
pNH$_3$	M	< 10^{-6} – $5 \cdot 10^{-2}$	alkaline soln.	0–50		IMOWΩ	all types of detergents and wetting agents must be absent; interferences from volatile amines
pCO$_2$	M	$7 \cdot 10^{-6}$ – $2 \cdot 10^{-2}$	acidic soln.	0–50		IOΩ	all types of detergents and wetting agents must be absent; interferences from SO$_2$, NO$_x$, H$_2$S
pNO$_x$	M	10^{-6} – $5 \cdot 10^{-3}$	acidic soln.	0–50		IO	all types of detergents and wetting agents must be absent; interferences from SO$_2$, CO$_2$, volatile carbonic acids
pCl$_2$	M	10^{-7} – $3 \cdot 10^{-4}$		0–50		O	oxidizing agents

[a–h] See Table 7.

Table 7. Commercially available ISEs with selectivity for miscellaneous species

ISE type	M-T[a]	Concentration range, mol/L[b]	Useful pH range[b]	Temp. range, °C[b]	Membrane resistance, MΩ[b,c]	Suppliers[d]	Interfering ions[b,e] in order of decreasing interference; miscellaneous remarks
Hard[f]	P	$7 \cdot 10^{-6} - 1$		0–50		O	not sorted: Na^+, Cu^{2+}, Zn^{2+}, Fe^{2+}, Ni^{2+}, Sr^{2+}, Ba^{2+}, K^+
Surf[g]	P	$\approx 10^{-6:h}$	1–13	0–40		MO	oppositely charged bulky ions interfere with the titration reaction; high concentrations of nonionic surfactants may interfere

[a] MT = Membrane type; G = Glass-membrane electrode; P = PVC-membrane electrode; S = Solid-state membrane electrode; M = gas-permeable membrane electrode. Glass and solid-state membranes are more resistant to certain organic solvents (acetone, methanol, benzene, dioxane, etc.). PVC-membrane electrodes must not be allowed to come into contact with any organic solvent, because this dramatically reduces the lifetime of the electrode.
[b] Mean values as provided by the suppliers.
[c] Only if different suppliers provide very different values for membrane electrical resistance are two values listed in the table.
[d] This list was compiled in Spring 1993 and should not be regarded as complete; C = Ciba Corning; I = Ingold Messtechnik; M = Metrohm; O = Orion; R = Radiometer; W = WTW; Ω = Omega.
[e] A very detailed compilation of selectivity coefficients is available in [23].
[f] Hard: water hardness; i.e., divalent cations.
[g] Surf = ionic surfactants.
[h] This type of ISE should be used only for the monitoring of titrations; the value cited refers to the mimimum titratable concentration of an anionic surfactant (pure solution).

so-called irreversible processes show the corresponding current peaks more to be separated and of different heights and are of limited analytical utility.

An intermediate situation arises when the electron-transfer and mass-transport rates are comparable. Such quasi-reversible electrode reactions are quite common, and their analytical utility depends to a large extent on careful control of the mass-transfer rate in the electroanalytical method used for their study.

Instruments suitable for voltammetry and amperometry consist of three basic components: a wave-form generator, some form of potential control, and an electrochemical cell. Modern electroanalytical systems employ a three-electrode arrangement for the electrochemical cell. A device called a *potentiostat* is used to maintain a programmed or fixed potential difference between the two current-carrying electrodes (the working electrode and the auxiliary electrode) relative to a third electrode (reference electrode), the function of which is to provide a fixed potential reference in the cell [64], [65].

Both potentiostats and waveform generators have benefited substantially in recent years from the introduction of operational amplifiers, as well as from the availability of desktop computers. Electrochemical cell design is now wellestablished both for static systems and for detectors situated in flowing reagent streams.

Despite the fact that solid electrodes have been in use for electroanalytical purposes for some time and are well documented, selection of an appropriate electrode material and its preparation before use remains an area of considerable interest and great controversy among electroanalytical chemists [66]. Solid electrodes for analytical use should always be prepared in a rigorous and reproducible manner. This generally involves polishing the electrode physically with successively finer grades of carborundum or diamond paste. The polished electrode should then be rinsed thoroughly to remove as completely as possible all traces of the polishing materials.

Many different applied potential waveforms have been employed for analytical purposes, but only a few are of mainstream significance [67], [68]. These are presented in Table 8.* → = Point at which the current is measured; τ = Portion of the staircase during which pulses of width $\tau/2$ are superimposed; E_{sw} = Square-wave amplitude.

In *linear-sweep* voltammetry the current is recorded while the potential of the working electrode is swept from one selected limit to another at a rate between 1 mV/s and 1000 mV/s. In *cyclic* voltammetry, two linear sweeps are recorded, as shown in Table 8. The distinguishing feature of cyclic voltammetry is that electrogenerated species formed in the forward sweep are subject to the reverse electrochemical reaction in the return sweep [69]. Figure 17 shows a typical cyclic vol-

Table 8. Voltammetric techniques (adapted from [26])

Technique	Applied potential*	Measured signal	Comments
Linear-sweep voltammetry	(ramp of E vs t)	(peak of i vs E)	used mainly for diagnostic purposes; current measured throughout the scan
Cyclic voltammetry	(triangular E vs t)	(duck-shaped i vs E)	reversal of the potential at the end of the forward scan makes this a powerful diagnostic tool; current is measured throughout the scan
Differential-pulse voltammetry	(pulsed ramp E vs t)	(peak of i vs E)	the displayed signal is the difference between the current just prior to and at the end of the applied pulse; low scan rates, but good limits of detection
Square-wave voltammetry	(E_{sw}, τ, ΔE square wave vs t)	(peak of i vs E)	the displayed signal is the difference between currents measured on the forward and reverse pulses; rapid scans and good detection limits

tammetric curve together with the important parameters. In both methods, as the potential is swept to the electroactive region of the redox couple, the current response rises to a peak value before decaying. This decay is caused by depletion of the electroactive species in the zone close to the electrode surface, which means that the diffusion zone spreads out further into the bulk solution.

Pulse and square-wave voltammetry [70] are much better candidates for incorporation into a chemical sensor than linear-sweep or cyclic voltammetry, primarily because of the possibility of discriminating between faradaic and capacitive currents. When a potential pulse is applied to an electrode, the capacitive current that flows is proportional to the magnitude of the pulse, and it decays exponentially with time. The faradaic current, on the other hand, decays according to the square root of time. Figure 18 illustrates the expected decreases in faradaic and capacitive currents, showing that the capacitive current decreases faster than the faradaic one. This gives the opportunity of sampling the current only after the capacitive one can be neglected.

Proper selection of the measuring time permits a signal-to-noise ratio to be improved dramatically. This characteristic is exploited in normal-pulse, differential-pulse, and square-wave voltammetry. In the first of these techniques it represents the only mechanism for decreasing the effect of capacitive current. Further elimination of the capacitive current can be achieved in differential-pulse voltammetry by limiting the duration of the applied pulse and by subtracting the current observed immediately prior to the imposition of the pulse. Increased rejection of the charging cur-

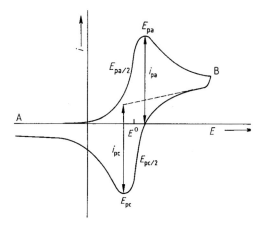

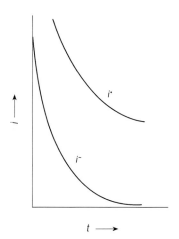

Figure 17. Typical potential vs. current (E vs. i) curve from a cyclic voltammetric experiment with a reversible redox couple
A) Starting potential; B) Reverse potential
E^0 = Standard potential; E_{pa} = Anodic peak potential; E_{pc} = Cathodic peak potential; $E_{pa/2}$ = Anodic half-wave potential; $E_{pc/2}$ = Cathodic half-wave potential; i_{pc} = Cathodic peak current; i_{pa} = Anodic peak current

Figure 18. Decrease in faradaic and capacitive current with elapsed time from the imposition of a potential pulse
$i^{\bullet}$ = Faradaic current; $i^{\sim}$ = Capacitive current

rent leads to improved detection limits; indeed, this variant of voltammetry permits concentrations as low as $10^{-8} - 10^{-7}$ mol/L to be measured quite readily. In order to avoid problems associated with the electroactive species, the delay between application of successive pulses must be approximately one-half second. This in turn imposes a limit on the scan rate in differential-pulse voltammetry, thereby decreasing its usefulness in sensor applications.

The technique of square-wave voltammetry offers greater promise as a voltammetric method for probing selective chemistry because of the high rate at which the corresponding scan can be executed. The analytical signal in this technique constitutes a difference between the current for the forward pulse and the current for the reverse pulse. Because of the large amplitude of the square wave in a reversible reduction, a reduced electroactive species formed at the electrode during the forward pulse is reoxidized by the reverse pulse. Consequently, the sensitivity of the method is enhanced relative to differential-pulse voltammetry.

This element of speed is crucial to square-wave voltammetry, because, like all voltammetric techniques based on pulse waveforms, the measured current is proportional to $t^{-1/2}$. However, in contrast to the other pulsed voltammetric techniques, square-wave voltammetry causes very little of the depletion that gives rise to distortion of the current–voltage waveform. Accordingly, square-wave voltammetry is uniquely capable of benefiting from high scan rates. A typical compromise frequency for the waveform is 200 Hz. Coupled with a dE value of 10 mV, this gives rise to a scan rate of 2 V/s. The scan rate is limited by a concomitant increase in the capacitive current. At a solid electrode, the square-wave method is no better than pulsed-voltammetric techniques with respect to rejection of the capacitive current.

Another significant advantage of square-wave technology is the possibility of rejecting a wide range of background currents. With respect to capacitive currents this is achieved by the subtraction of two currents in a manner analogous to differential-pulse voltammetry. Slowly varying capacitive currents that arise when surface groups reorganize on certain types of solid electrodes are also eliminated provided the rate of variance is sufficiently low.

Amperometry. Amperometry has traditionally been concerned with maintaining a fixed potential between two electrodes, but pulsed techniques have recently attracted considerable attention as well. The applied potential at which current measurements are made is usually selected to correspond to the mass-transport-limited portion of the corresponding voltammetric scan. Table 9 summarizes the various amperometric methods.

Theoretically, the current obtained in a quiescent solution at a conventional set of electrodes gradually decays to zero according to the Cottrell equation [71]:

Table 9. Amperometric methods (adapted from [26])

Method	Applied potential	Measured signal	Comments
Chronoamperometry			the electrode potential is pulsed to a region in which the analyte is electroactive; a decaying current reflects the growth of the diffusion layer
Pulsed amperometry			the electrode potential is pulsed briefly to a region in which the analyte is electroactive; between pulses the diffusion layer may be eliminated by forced or natural convection
Pulsed amperometric detection			electrode conditioning, analyte sorption, and catalytic electrooxidation are all promoted by the use of this waveform; the current is measured only during the last part of the cycle (region A)

$$i_j = nFAC(D/\pi t)^{1/2} \qquad (7)$$

where t is the elapsed time from application of the potential pulse.

The observed decrease in current is due to a slow spread of the diffusion layer out into the bulk solution, with a concomitant decrease in the concentration gradient. In practice, this process continues for ca. 100 s, after which random convection processes in the solution take over, putting an end to further movement of the diffusion layer. Waiting nearly two minutes to obtain a steady-state current is not a particularly attractive alternative. Accordingly, in amperometric measurements for sensor applications the spread of the diffusion layer is controlled by invoking one or more of the following mechanisms:

1) *Convective Diffusion.* From a practical point of view, convective diffusion can be achieved in two ways: by moving the solution relative to the electrode, or moving the electrode relative to the solution. Of the systems developed to move the electrode, only one merits consideration here, the *rotating disc electrode* (RDE), but there are few possibilities for realizing this electrode in a sensor.

2) *Imposition of a Physical Barrier in the Form of a Membrane.* There are four primary advantages to covering the electrodes of an amperometric device with a membrane permeable only to the analyte: (a) poisoning of the electrodes by electroactive or surface active species is limited; (b) the resolution of the system is enhanced, because extraneous electroactive species that might otherwise undergo electron transfer at the electrode are excluded; (c) the composition of the electrolyte occupying the space between the membrane and the electrodes remains constant; and (d) the membrane forms a physical barrier to prevent the diffusion layer from spreading into the bulk solution.

The limiting current obtained at a membrane-covered amperometric device is a function of time, reaching in due course a steady-state value. The current–time transient cannot be described by a single equation, because different transport mechanisms operate during this time interval. For a typical membrane-covered amperometric oxygen detector with an electro-

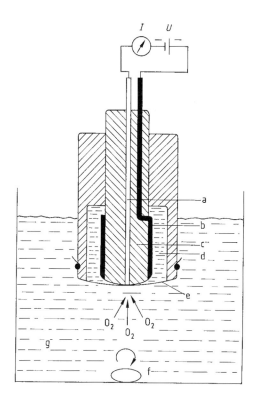

Figure 19. Schematic diagram of a Clark-type oxygen sensor
a) Cathode; b) Anode; c) Insulator; d) Electrolyte; e) Membrane; f) Stirrer; g) Sample for analysis

lyte thickness of 10 µm, a membrane thickness of 20 µm, an electrode radius >2 mm, and a diffusion coefficient for the analyte two orders of magnitude lower in the membrane than in the electrolyte solution, it can be shown that the current–time transient can be described by three equations. For times >20 s after the imposition of the potential pulse to the electroactive region, the limiting current is given by

$$i_j = nFAC(P_m/d) \tag{8}$$

where P_m is the permeability coefficient of the membrane and d is its thickness.

An important consequence of this equation is that the limiting current is a function of the permeability of the membrane. Unfortunately, membrane properties such as permeability are themselves a function of time and ambient conditions, particularly temperature. Accordingly, there are advantages to operating membrane-covered detectors away from this steady-state condition. For a typical detector, the limiting current < 100 ms after imposition of the measurement potential pulse is given by

$$i_j = nFAC(K_b/K_0)(D_e/\pi t)^{1/2} \tag{9}$$

where t is elapsed time since imposition of the measuring potential, D_e is the diffusion coefficient for the analyte in the electrolyte, and K_b is the distribution coefficient for the analyte at the internal electrolyte–membrane interface. The ratio K_b/K_0 expresses the salting-out effect of the electrolyte solution.

In effect this equation describes the diffusion-limited current arising within the electrolyte layer. In other words, the diffusion layer has not had sufficient time to spread to the membrane, and so the limiting current is completely independent of the membrane. This has the advantage that alterations in membrane characteristics, as well as alterations in ambient conditions that might affect membrane performance, have no effect whatsoever on the limiting current. One of the factors to which this applies is stirring. Under steady-state conditions, stirring of some type is necessary to prevent changes that would otherwise be caused by the diffusion layer moving out into the bulk solution, although the rate of such stirring is not critical. When transient measurements are made, no stirring whatsoever is required. The true situation regarding transient measurements with amperometric devices is actually not quite so simple as depicted here; for example, there is also a need to take into account contributions from the capacitive current.

The classic amperometric membrane-covered device is the Clark oxygen sensor [72], which consists of a platinum working electrode, an Ag/AgCl reference electrode, a KCl electrolyte solution, and an oxygen-permeable membrane made of teflon or silicone rubber. Figure 19 is a schematic diagram of such a sensor. Dissolved oxygen passes through the permeable membrane, after which it is reduced at the working electrode. The current flow is proportional to the amount of oxygen present in the solution. The following reactions take place at the surface of the working electrode:

$$O_2 + 2H_2O + 2e \longrightarrow [H_2O_2] + 2OH^-$$

$$[H_2O_2] + 2e \longrightarrow 2OH^-$$

A separate polarographic wave corresponding to the reduction of hydrogen peroxide is observed only with a mercury cathode, which produces two waves for the reduction of oxygen. At other electrodes (Pt, Au, C) only one fourelectron wave is obtained, corresponding to complete reduction to four hydroxide ions [73]. The potential E of the working electrode (at the diffusion-limiting current plateau) usually lies between -0.6 and -0.9 V vs. Ag/AgCl.

Progress in semiconductor technology during recent years has made it possible to miniaturize the classical oxygen sensor [74]–[77] for use as a transducer in several chemical and biochemical sensors.

3) *Pulsing the Electrode from a Region of Electroinactivity to a Region of Transport-Controlled Electroactivity, after which the Potential is Restored to the Electroinactive Region.* For electrochemical sensors without membranes, applying a potential pulse from a region of electroinactivity to a region of electroactivity has several advantages. For example, the duty cycle of the pulse can be arranged in such a way that the electrode spends the majority of its time at the electroinactive potential. Under these circumstances electrode reactions give rise to products that poison the electrode, but these products are of minimal significance because they are produced at a very low rate. Moreover, natural processes of convection are able to redistribute the electroactive species during the time that the electrode is at the rest potential, thus limiting the spread of the diffusion layer into the bulk solution. The drawback to this technique is the appearance of a capacitive current. Limiting the size of the pulse decreases the contribution of this current, but a small pulse from a rest potential to the electroactive potential is achievable only for reversible electrode reactions. The alternative is to allow sufficient time for the capacitive current to decay away between imposition of the potential pulse and measurement of the current.

4) *Pulsed Amperometric Detection* [78]. This is a relatively new technique that has made possible the direct amperometric determination of many compounds formerly considered unsuitable for analysis in this way. The corresponding waveform is illustrated in Table 9. This rather complex waveform serves several purposes. On platinum, for example, and at the extreme anodic potential, surface-confined oxidation products that tend to poison the electrode are desorbed. This process occurs concurrently with the formation of PtO_2 on the surface. At the cathodic potential the surface oxide is subsequently reduced, and a clean platinum surface is regenerated. In some cases, sorption of the analyte occurs at this cathodic potential, so the analytical signal is recorded at an intermediate potential. Several different mechanisms can give rise to this signal. Electroactive species that are not adsorbed at the electrode may undergo conventional oxidative electrode reactions. Alternatively, electroactive species that are adsorbed at the electrode during the cathodic swing of the cleaning cycle may undergo either conventional oxidative desorption or oxidative desorption catalyzed by an oxide species on the surface of the electrode. For many compounds the latter reaction is much more favorable from a thermodynamic point of view than oxidation in the absence of a catalyst. This type of reaction occurs most readily when both the analyte and the oxide are confined to the surface. Finally, adsorbed electroinactive analytes may also be detected by virtue of their ability to block the electrode sites at which the PtOH formation occurs. Consequently, any faradaic current arising from the oxidation of platinum is diminished in the presence of adsorbed electroinactive analyte.

5) *Microelectrodes.* Electroanalytical applications using microelectrodes have become more common in recent years [79]–[83] as silicon technology has opened new possibilities for constructing electrodes with micrometer and submicrometer dimensions [84], [85]. The rate of electrolysis at such electrodes is approximately equivalent to the rate of diffusion. Linear-sweep voltammetry in this case gives rise to the same S-shaped current–voltage curves observed with rotating disc or dropping mercury electrodes. The time-dependent current arising at a microelectrode is given by the equation for the chronoamperometric current at a conventional electrode under spherical diffusion conditions:

$$i_t = nFACD\left[\left\{1/(tD\pi)^{1/2}\right\} + \{1/r\}\right] \qquad (10)$$

where r is the electrode radius. Indeed, a sphere or hemisphere represents a good model for the diffusion zone that surrounds such an electrode, and because of enhanced mass transport to the electrode, a steady-state current is achieved very rapidly once a potential pulse has been applied to the electrode.

Although the diffusion zone is large in this case relative to the electrode surface, it is small compared to the zone surrounding a conventional electrode. This small size means that the faradaic current associated with a microelectrode is relatively immune to the effects of convection in the bulk solution, and the current in a flowing stream is independent of the rate of flow. Decreased effects due to capacitance and resistance at microelectrodes, coupled with high mass-transport rates, make it possible for electrochemical measurements to be obtained in this way in cells containing highly resistive solutions. In fact, provided sufficient free carrier is available to charge the double layer, microelectrodes can be used for direct electrochemical measurements in the gas phase [86]. One drawback of employing working electrodes with diameters less than about 5 µm are the small currents which will flow through such small interfaces. Current measurements in the pA range are prone to electromagnetic interferences (e.g., AC pick-up) and thus, the electric noise level can be a limiting factor. This problem can be solved by employing an array of microelectrodes all at the same potential and separated from each other by about ten times their diameter. Such arrays have been constructed and have shown all the benefits of single microelectrodes [87], [88].

28.2.3.1.3. Conductance Devices

Conductance is used as a sensor signal in two different areas of sensor technology. On one hand it is possible to determine the electrolytic conductance of electrolytes, which permits conclusions to be reached on a wide range of electrolyte properties. Moreover, partial pressures of gaseous compounds can be determined in the gas phase by examining their influence on the electronic conductance of such materials as semiconducting oxides [89]–[92].

Conductometric Sensors for Monitoring Electrolytic Conductance. *Introduction.* Electrolytic conductance serves as a useful signal of the electrical conductivity of mainly aqueous electrolyte solutions, and *conductometry* has been developed into an electrochemical analytical method based on measuring the conductance in electrolyte solutions. All ions in such a solution act as current carriers as a function of their mobility, charge, and concentration. For this reason conductivity measurements are inherently nonselective with respect to any system to which a changing voltage difference has been applied and the diffusion current is measured. Despite this disadvantage, conductivity measurements are indispensable sources of information, especially for monitoring the chemical purity of water samples. In pollution control, conductometric data provide a reliable measure of the total content of ionic pollutants, which is sometimes all that is required. Advantages of conductivity detectors include their universal applicability, simplicity, low price, and wide effective concentration range. Conductivity measurement is one of the most commonly used electrochemical techniques in the control of industrial chemical processes. Conductivity detectors are also employed in ion chromatography and HPLC as a way of detecting separated ions in the eluate from a column [93]. Furthermore, conductometric analyzers are used in the monitoring of atmospheric pollutants such as acidic or alkaline gases (e.g., HCl, CO_2, SO_2, SO_3, NH_3) after these have been transferred as ions into aqueous solution [94],[95]. Conductivity measurements are nondestructive with respect to the sample, and they are always advantageous when electrolytes must be determined in a medium of low self-conductivity.

Definitions. The electrolytic conductivity σ of a solution is defined as the conductance G of the electrolyte between two electrodes, each with an area A, separated by a distance d. The conductivity σ is then given by

$$\sigma = G \cdot d/A \qquad (11)$$

The traditional unit of electrolytic conductivity, Siemens per centimeter (S/cm), has recently been replaced by S/m [96], but the new unit has not yet been accepted widely by the instrument-producing industry, nor by the users of conductivity measurement devices. The English-language literature has historically made use of the "mho" (reciprocal ohm), which is another way of expressing the conductance G. The quotient d/A (cm^{-1}) is defined as the cell constant k, and depends on the geometry of the cell. The electric field is never homogeneous over the cross section of a real cell, so it is not in fact possible to calculate a cell constant k directly from geometric dimensions; it

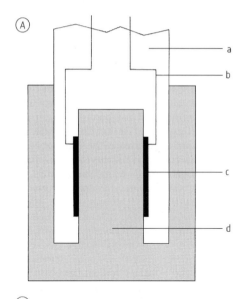

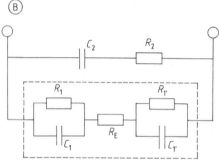

Figure 20. Depiction of a two-electrode cell (A) together with the corresponding equivalent circuit (B)
The dashed rectangle includes all network components involving the electrolyte in contact with the electrodes.
a) Cell body; b) Leads; c) Electrode; d) Sample
R_E = Electrolyte resistance; R_1, $R_{1'}$ = Charge-transfer resistances at the interface electrode–electrolyte; C_1, $C_{1'}$ = Capacitance of the electrochemical double layer; R_2, C_2 = Shunt circuit in the cell

must instead be determined with calibration solutions of known conductivity [97].

The temperature dependence of σ can be deduced from ionic mobilities. This dependence is expressed in terms of the temperature coefficient $\alpha = (d\sigma/dT)(1/\sigma)$, which generally has a positive value between 1 and 3 % per degree Celsius. Knowledge of α in turn makes it possible to calculate conductivities at temperatures other than the temperature of measurement. Whenever the need arises to determine the conductivity of a solution with a very low electrolyte concentration it is essential also to take into account the self-dissociation of pure water, which has a conductivity $\sigma_{25} = 0.05483$ µS/cm.

Measurement Principles. The equipment used for measuring conductance usually consists of a conductivity sensor, a measuring attachment, and a temperature-compensation unit. Conductance values are obtained by establishing either the current flow between a set of electrodes subjected to a constant AC voltage, or the current induced in the secondary coil of a transformer connected to the primary coil by the electrolyte of interest. Capacitive coupled cells have been developed as well, but because of their expensive construction they are no longer of interest [98]–[101].

Two-Electrode Cells. Two-electrode cells are equipped with a pair of electrodes across which both the voltage and the current can be measured. Use of an AC voltage source is a convenient way of avoiding polarization effects at the electrode/electrolyte interface and electrolyte decomposition. Figure 20 illustrates a typical construction together with the corresponding equivalent circuit. The advantage of this design is the simplicity of both the cell and the measuring attachment.

Disadvantages are associated with the need for measuring both the current and the voltage at a single electrode, which may lead to electrode polarization and a measured conductivity lower than the true value. The resistance contributed by a cell in contact with a sample must also be taken into consideration in conjunction with other resistance and capacitance values. At high conductivities ($\sigma > 1$ mS/cm) the resistance of the connecting wires is another significant factor, and electrode reactions can cause polarization of the electrode–electrolyte interface. These problems can be overcome most effectively with an AC voltage oscillating at 1–5 kHz and electrode materials with minimum polarizability due to a large surface area, such as platinum covered with a layer of platinum black. At low conductivities ($\sigma < 10$ µS/cm) measured conductivity values may be too high because of capacitive–ohmic shunts caused by the cell body or the connecting wires ("Parker effect"). Contamination by adsorbed ions is also a problem with porous electrodes. The measurement frequency should therefore be kept as low as possible, and the electrodes should be smooth. Resistive precipitates at the electrode surface are sometimes a cause of artificially low conductivity.

For the reasons mentioned above, the "cell constant" is in fact constant only over a limited conductivity range. Suitable values for the cell constant depend on the anticipated level of conductivity:

Expected conductivity	Cell constant
< 0.1 µS/cm – 1 mS/cm	0.1 cm^{-1}
1 µS/cm – 100 mS/cm	1.0 cm^{-1}
10 µS/cm – 1 S/cm	10 cm^{-1}

Four-Electrode Cells. The four-electrode cell was developed as a way of overcoming the problems inherent in two-electrode cells. Two current electrodes in this case generate an electric field, and two separate voltage electrodes are provided for measuring the voltage drop with the aid of a high-impedance amplifier. This design is insensitive to polarization effects, and it also avoids interference from insulating films or precipitates because these are formed outside the zone of measurement. The result is a single system that can be used to measure any conductivity > 1 µS/cm.

Inductively Coupled Cells. The principle underlying inductively coupled conductivity cells is very simple. Two coils of a transformer are shielded from each other magnetically such that the only coupling loop is formed by an insulating tube filled with the sample. An AC voltage in the primary coil induces an AC current in the sample, the magnitude of which depends on the dimensions of the loop and the conductivity of the sample. The voltage induced in the second coil depends only on the input voltage of the primary coil and the conductance of the sample. Typical working frequencies are 80 – 200 Hz. An advantage of these cells is that they permit contactless measurement. Very high conductivities (>1 S/cm) can be measured without polarization errors because of the contactless decoupling of the signal. Even thin resistive films within the tube fail to falsify the measured conductivity. Another advantage associated with cells of this type is the possibility of encapsulating the entire device in a polymer offering high chemical resistance. Inductively coupled conductivity cells are thus of special interest for measurements in highly conducting and aggressive media [102].

Ion-Selective Conductometric Microsensors. The greatest problem in miniaturizing potentiometry was the lack of a longer working reference electrode. The development of ion-selective conductometric microsensors mainly for cations such as H^+, Li^+, Na^+, K^+, NH^{4+}, Ca^{2+}, Cd^{2+}, Pb^{2+}, Hg^{2+}, and Ag^+ thus far in the ICB Muenster overcomes this problem [103]. In principle, they are partly based on the ion-selective optodes introduced by the late Prof. Simon who showed that the measured ion can also be selectively extracted into an ionophore-based membrane. In order to maintain electroneutrality in the bulk of the membrane a co-immobilized indicator dye looses a proton which passes into the sample solution and thereby changes the color. The influx of the measured cation, which will be complexed by the neutral ionophore and leads to an even larger cation complex in the PVC membrane, is electrically compensated by an outflux of protons that dissociate from a neutral lipophilic acid. This exchange of protons from the membrane against the measured cation will cause the membrane conductivity to change. This can easily be followed by a planar interdigital electrode array onto which a corresponding membrane is drop-coated. In order to keep the lipophilic acid in the membrane in its neutral state and in order to control the interfacial ion-exchange processes in a reproducible way, all measurements must be performed in a buffer of an appropriately chosen and constant pH. Figure 21 shows the measuring set-up and Figure 22 some results of a valinomycin based potassium-selective membrane. The length A of the interdigital electrode fingers is only about 1 mm and α the width of a finger and spacing lies in the region of about 10 µm. The membrane thickness is δ in the range 0.5 – 5 µm. The membrane cocktail — similar to the one in case of ISE membranes — can be brought onto the interdigital electrode array by dip- or drop-coating. U_{in} an alternating voltage in the range 10 – 100 mV and is applied via the corresponding output of a lock-in amplifier. U_{out} at the load resistor is fed to the input of the lock-in amplifier. The in-phase component of the complex impedance is measured and evaluated. In general, the characteristics of those bulk conductive microsensors resemble those of the corresponding ISE membranes with slight variations in the selectivity pattern. The advantage is that the accuracy of their measurements does not drop with increasing charge of the measured cation. With a membrane thickness in the region of 1 µm the response time is in the second range. A difference compared with ISE membranes lies in the increased concentration of ionophore that is added up to about 5% by weight. These ion-selective conductometric microsensors do not need any reference electrode and can be scaled down to the µm range. The lifetime depends on the lipophilicity of the ionophore and plasticizer. Normally they work for up to several weeks and show a relatively small drift. The further advantages are that they are easily

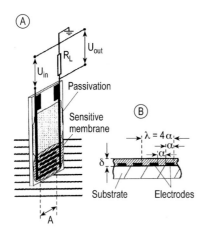

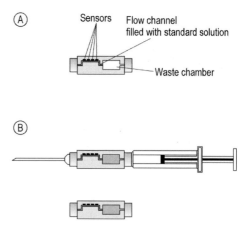

Figure 21. Schematic drawing of the construction and measuring circuit of ion-selective conductometric microsensors not requiring any reference electrodea) General view; b) Cross section

Figure 23. Measuring principle of a cartridge system using microsensors for the case of medical "point of care" blood electrolyte measurements
A) Storage of the cartridge with a standard solution; recall of data; measuring of the standard solution for plausibility and quality controls.
B) Drawing of the blood sample; displacement of standard; disconnection of the cartridge; insertion of the cartridge into the analyzer; measuring of the undiluted blood sample.

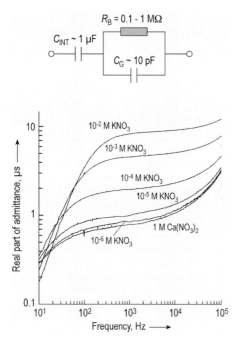

Figure 22. Simplified electronic equivalent circuit (measured is R_B = bulk membrane resistance) and impedance spectra of the real part (admittance) in a 1 M $Ca(NO_3)_2$ (!) supporting electrolyte and increasing K^+ concentrations with a valinomycin-based potassium conductometric microsensor. Note that even μM amounts of K^+ ions in a well conducting electrolyte solution can be sensed

produced by CMOS technology and open also the way to the detection of higher charged analyte ions for which the Nernst equation only allows a small mV change per decade concentration change. Despite the fact that their selectivity pattern is more similar to the one of the corresponding optodes they lack the drawback of the latter with respect of chromo- or fluorophore bleaching. Furthermore, in case of light absorption measurements a compromise between sensitivity (depending on the thickness) and response time must be made.

Because of the excellent sensitivity, ease of fabrication, and storage capabilities these microsensors are ideally suited for medical bedside monitoring or "point of care" devices. Figure 23 shows a novel cartridge developed at the ICB Muenster for blood electrolyte measurements. In comparison with existing devices no open blood transfer with a certain risk of infection is needed. In another special version of the cartridge, a valve allows the continuous uptake of blood in the syringe needing only the first 100 μL for the electrolyte. The content of the syringe can then be analyzed as usual. The evaluation is performed by putting the cartridge in a hand-held readout unit. Note, prior to blood sampling the same cartridge has to be checked by the same readout unit for quality control reasons and re-calibration.

Conductometric Sensors for Monitoring Gaseous Compounds. *Introduction.* Gaseous compounds in the gas phase are detected with chemically sensitive devices that respond electrically to changes in conductance. BRATTAIN and BARDEEN (using germanium, in 1952) [104] and HEILAND (zinc oxide, 1954) [105] were the first to observe modulation of the electrical conductance of a solid-state device by gases from the surrounding atmosphere. The most commonly used material for conductometric solid-state gas sensors is now tin dioxide [106]. As noted previously (Section 28.2.2.2) commercial sensors based on polycrystalline SnO_2 ("Taguchi sensors") [107] are produced by companies such as Figaro Inc., for a wide range of industrial and domestic applications, including gas monitors, leak detectors, and alarm systems for toxic or inflammable gases. Apart from SnO_2, an n-semiconductor, many other inorganic semiconducting oxides can serve as sensitive materials in gas sensors, including TiO_2, ZnO, and Fe_2O_3 with various additives and dopants [13], [108]. Certain organic semiconductors such as phthalocyanines with appropriate central metal atoms, or polymers like polypyrrole or polysiloxane, also show sensitivity to gaseous compounds. These sensors are often referred to as homogeneous or heterogeneous semiconductors, or as dielectric sensors. Detailed lists of materials sensitive to detectable gaseous compounds together with the corresponding operating temperatures are provided in [89]–[92], [108]. The design principle upon which such sensors are based is quite simple: a sensitive semiconducting sample is placed between two metallic contacts, and the combination is equipped with a resistance heater capable of maintaining the appropriate working temperature, usually between 100 °C and 700 °C. Semiconductor sensors have dimensions in the range of millimeters, and are easily fabricated by either thick- or thin-film technology [109].

Heterogeneous Semiconductor Sensors. The response of a heterogeneous semiconductor sensor is based on changes in surface conductance as mentioned in the introduction, mainly due to changes in the free electron concentration within a thin surface layer of the semiconductor as a consequence of charge exchange with adsorbed species accompanying chemisorption or heterogeneous reactions [33] (see Section 28.2.2.2). The most commonly used semiconducting materials are the two very stable oxides SnO_2 [110], [111] and ZnO [112]. Because the conductance change is a surface phenomenon, the ratio of surface to bulk material should be made as large as possible (using thin layers) to maximize the effect. For sintered polycrystalline materials the grain size should also be as small as possible.

At a given temperature the number of gas molecules adsorbed at a semiconductor surface — hence the conductance — correlates exponentially with the partial pressure of the gas. Oxidizing gaseous compounds such as oxygen, nitrogen dioxide, and chlorine cause a resistance increase by extracting electrons from the conduction band of an n-type semiconductor, while reducing gases like hydrogen, carbon monoxide, methane, hydrocarbons, or alcohols lead to a resistance decrease since they are themselves oxidized by the adsorbed oxygen. Humidity decreases the resistance of an n-type semiconductor, causing considerable interference.

Response times for sensors of this type are usually in the range of seconds or minutes. For some gases it has been found that catalytic effects at the semiconductor surface can have a significant impact on the response. Thus, sensor properties such as response and selectivity can be improved by adding catalytically active metals like palladium, rhodium, silver, or platinum. This doping of the semiconductor surface is one important way of modifying the selectivity of a sensor [113]–[115], but another possibility for adjusting sensor properties for a particular application is careful selection of the working temperature [116].

The influence of temperature on the selectivity of a sensor depends on the fact that each gas has its own *characteristic temperature:* a temperature at which the reversible relationship between gas concentration and the concentration of electrons at the semiconductor surface (and therefore the conductance) reaches a maximum [110]. Working temperatures for SnO_2-based sensors usually lie between 200 °C and 450 °C depending on the application.

Various techniques are used for the fabrication of semiconductor sensors. Conductance sensors from structurized sintered polycrystalline ceramics can be produced by thick- or thin-film technology. Chemically sensitive materials in the form of single crystals or whiskers can be attached to electrodes by thin- or thick-film techniques as well. Mass production of sensors requires that the resulting devices be characterized by a defined level of conductance. For example, the conductance of polycrystalline SnO_2 can be adjusted by subsequent thermal treatment >800 °C under a controlled partial pressure of oxygen. Another approach to defined conductance involves doping the semiconductor with antimony or fluorine. The reproducibility and stability of a sensor signal is generally diminished by irreversible changes in the defect structure of the semiconductor surface.

Homogeneous Semiconductor Sensors. The effectiveness of this type of semiconductor sensor depends on changes in bulk conductance. The stoichiometry and concentration of bulk point defects, and thus the conductance of such materials, is a function of the partial pressure of a gaseous analyte. Conductance changes arise through reversible changes in small deviations from ideal stoichiometry, which in turn influence the concentrations of bulk point defects and free electrons [33]. The deviation from ideal stoichiometric composition is usually very small; for example, in $Cu_{2-x}O$ the deficit x in the Cu ratio is on the order of 0.001. Since conductance and deviation from ideal stoichiometry are proportional, a measured conductance can be used to determine a corresponding partial pressure. Regarding the kinetics of these sensors, an important factor is diffusion of a compound like oxygen or the corresponding lattice defects that compensate for nonideal stoichiometry within the bulk of the semiconductor. Any gaseous reaction involving point defects at the surface of the semiconductor must also be appropriately fast. Both requirements can be met through the combination of high temperature and a catalytic coating (e.g., Pt).

Most bulk-conductance sensors consist of semiconducting oxides the conductance of which changes with the partial pressure of oxygen. They are used primarily for monitoring oxygen partial pressures in the air or in combustion gases. Titanium dioxide (TiO_2) is one of the most commonly used homogeneous sensor materials for the determination of oxygen partial pressures [117]. At 900 °C and an oxygen partial pressure of 0.1 MPa, TiO_2 has precisely the stoichiometric composition. At lower oxygen partial pressures TiO_2 has an oxygen deficit x expressed by the formula TiO_{2-x}. This oxygen deficit specifies the number of unoccupied oxygen sites and the corresponding excess of trivalent titanium in the lattice. Both defects act as electron donors, thereby contributing to the concentration of free electrons in the conductance band of the semiconductor. The oxygen deficit, and thus the conductance, varies inversely with the partial pressure of oxygen in a characteristic way. Thus, the conductance σ_e of an n-type semiconducting oxide like TiO_{2-x}, CeO_{2-x}, or Nb_2O_{5-x} is given by the equation:

$$\sigma_e = \mathrm{const}\,(T) \cdot p(O_2)^{-1/m}, \qquad (12)$$

where $\mathrm{const}\,(T)$ is a temperature-dependent constant and m usually varies between 4 and 8 depending on the nature of the disorder in the oxide structure.

On the other hand, p-type semiconducting oxides like $Cu_{2-x}O$, $Ni_{1-x}O$, and $Co_{1-x}O$ have a deficit x of metal relative to ideal stoichiometry. This deficit increases with increasing oxygen partial pressure. Ionic lattice defects occurring in oxides of this type act as electron acceptors. In this case the concentration of the defects, and therefore the conductance, increases with increasing oxygen partial pressure according to the expression:

$$\sigma_e = \mathrm{const}\,(T) \cdot p(O_2)^{1/m} \qquad (13)$$

Here again, m usually varies between 4 and 8.

The strong exponential dependence of $\mathrm{const}\,(T)$ on temperature is a great disadvantage with homogeneous semiconductor sensors, one that can lead to large measurement errors. Nevertheless, appropriate sensor design makes it possible to compensate for the temperature dependence. In the case of a TiO_2 combustion-gas sensor, for example, the conductance of an oxygen-sensitive TiO_2 sample can be compared with that of a second sample displaying the same response to temperature but coated in such a way as to eliminate its sensitivity to gas composition. The conductance difference between the two samples is then a function only of the gas composition, not of fluctuations in the temperature.

An important requirement for any metal oxide under consideration for use as a homogeneous semiconducting sensor is that it display a sufficiently high bulk diffusion coefficient $\tilde{D}$ for the conductance-determining species. Response times t for oxide layers with a thickness y can be estimated from the Einstein equation:

$$y^2 = 2\tilde{D} \cdot t \qquad (14)$$

For example, a Cu_2O layer with a thickness of 1 µm reaches a response time of ca. 1 ms, corresponding to a diffusion coefficient of ca. 10^{-5} cm^2 s^{-1}, only upon heating to 700–800 °C. The diffusion coefficient can be altered by doping, but it also depends on the extent of deviation from ideal stoichiometry.

A number of oxides (e.g., TiO_2, CeO_2, Nb_2O_5, CoO) or titanates with a perovskite structure (Ba/Sr/CaTiO$_3$) have been successfully tested for application as extremely fast responding oxygen sensors [118]–[120] allowing individual cylinder control in automobile engines. Most of these investigations involved determination of the oxygen partial pressure in exhaust gases from combustion processes at partial pressures between 10 and 10^{-18} kPa. At low oxygen partial pressures and high temperature, certain oxides (e.g., Cu_2O, CoO, NiO) are unsuitable because of their reduction to pure metal. Sensors based on TiO_2 ceramics are by far the most highly developed devices. These consist of porous sintered ceramic materials with a platinum surface doping, and are used to determine air–fuel ratios in the exhaust gases of combustion engines.

Apart from TiO_2, a number of other semiconducting oxides might be useful as sensor materials for special applications over limited ranges of oxygen partial pressure, as in monitoring the level of oxygen in inert gases. There is also a good possibility of optimizing various ternary and higher oxide combinations for sensor application by suitable choice of composition with respect to stability, diffusion, and surface properties [121]–[123]. Thus, barium and strontium ferrates show conductances with considerably less temperature dependence relative to TiO_2 or CoO.

Organic Conductometric Sensors. For several years, certain organic semiconducting polymers have attracted attention in the search for new chemical sensors [124]. The possibility of utilizing conductivity changes induced by gases adsorbed on organic materials is interesting for a number of reasons. For example, it might permit the detection of very low levels of atmospheric pollutants. Moreover, organic materials are much more easily modified than inorganic materials with respect to such characteristics as sensitivity, working temperature, and selectivity. The greatest disadvantage of organic materials in gas detection is that they are usually very poor conductors, and conductivity measurements would be correspondingly difficult. Also, organic materials are thermally unstable, so it is often impossible to use them at temperatures at which gas–solid interactions proceed rapidly and reversibly.

Phthalocyanines constitute one group of organic materials with thermal stability >400 °C, and they are usually considered to have semiconducting rather than insulating properties. Thin layers of phthalocyanine complexes based on metals like lead [125] and copper [126] have been particularly carefully studied as potential gas sensors [127]. The conductance of these films is a very sensitive function of gases with electron-acceptor properties, including nitrogen dioxide, chlorine, and iodine. Films of phthalocyanines with various central metal atoms can be deposited on substrates with a planar interdigital electrode array by sublimation or screen printing. Changing the nature of the central metal atom makes it possible to adjust the sensitivity and selectivity of a phthalocyanine film in accordance with the proposed application [128]–[133].

Another organic semiconductor with gas-sensitive properties is polypyrrole. Chemiresistors with ambient-temperature gas sensitivity to both electron-donating and electron-accepting gases have been achieved using thin polypyrrole films prepared by electrochemical polymerization on electrodes. Sensitivities have been investigated toward such gases as nitrogen dioxide [134], ammonia [135], and methanol [136], [137]. A wide variety of polymers of this type is available, including substituted polypyrroles, polythiophenes, polyindoles, and polyanilines. This is leading the way to combining more of these sensors in sensor arrays and using chemometrics to evaluate unknown smells (electronic noses).

Humidity Sensors. Humidity sensors are devices that respond to changes in water-vapor pressure by a change in electrical resistance, capacitance, or a combination of the two as reflected in the impedance. The sensitive material might be an electrolyte, an organic polymer, or a metal oxide subject to a change in ionic–electronic conductivity or capacitance with respect to humidity as a consequence of physical adsorption or absorption of water molecules. In general these devices offer a very cost-effective continuous or spot-check approach to monitoring humidity in air and other gases [138], [139].

Electrolyte humidity sensors. Electrolyte humidity sensors are based on an electrolyte solution (e.g., the hygroscopic LiCl) the ion conductivity of which changes as a result of evaporation or condensation of water in response to changes in the relative humidity of the surrounding atmosphere. The fabrication of these sensors is quite straightforward. A porous substrate or an organic binder is impregnated with the electrolyte in such a way as to prevent electrolyte from flowing out of the sensor even at very high humidity. Contact is maintained with platinum electrodes on both sides of the substrate. Even thin films of potassium metaphosphate or barium fluorite are suitable for humidity measurements. Despite their slow response and limited linear range these sensors have an outstanding reputation due to good reproducibility, long-term stability, and low cost.

Ceramic Humidity Sensors. The first widely-used ceramic humidity sensor was the aluminum oxide sensor, in which aluminum is oxidized anodically in an acidic solution to form a porous oxide film, sometimes less than 1 μm thick, that displays the necessary hygroscopicity. Adsorption of water molecules causes a decrease in resistance and an increase in capacitance, which is reversed by subsequent desorption. Other materials are also used in humidity sensors, including ZnO with added Li_2O and V_2O_5, colloidal Fe_3O_4, α-Fe_2O_3, mixed oxides of TiO_2 and SnO_2, $MgCr_2O_4$–TiO_2, TiO_2–V_2O_5, $ZnCr_2O_4$–$LiZnVO_4$, $MgAl_2O_3$, $MgFe_2O_4$, or H_3PO_4–$ZrSiO_2$, as well as $PbCrO_4$ with added alkali oxide [140]. Sometimes the signal of such a sensor is subject to drift as a result of changes in the surface structure or contamination from other adsorbed gases. Nevertheless, reproducibility can be restored by heat cleaning.

The most important considerations with respect to sensor characteristics are surface properties (hydrophilic–hydrophobic), pore-size distribution, and electrical resistance. To ensure adequate sensitivity and response a sensor of this type should consist of a very thin film of porous ceramic with a porosity >30 %. The contacting electrodes may be interdigitated or porous sandwich-type structures made from noble metals (e.g., Pd, Pt, Au) and so constructed that they do not obstruct the pores of the oxide film. Humidity affects not only the resistance of a porous ceramic but also its capacitance by extending the surface area in contact with the electrodes. The high dielectric constant of adsorbed water molecules also plays an important role.

Polymeric Humidity Sensors. Certain polymers are capable of taking up water as a function of the relative humidity of the surrounding atmosphere, leading to changes in such mechanical and electrical properties as volume, resistance, and dielectric characteristics. A number of polymers have been used as sensitive materials in this regard, including ion-exchange resins, polymer electrolytes, hydrophilic polymers, and hygroscopic resins [141]. Hydrophilic polymers [e.g., poly(ethylene oxide)–sorbitol] swell by absorption of water. If such a polymer is filled (mixed) with carbon or metal powder of very small particle size, the resistance observed depends on the distance and the number of particles that are in contact with each other. Uptake of water leads to swelling of the polymer and disturbance of the ohmic contacts between conducting particles, which in turn increases the resistance. For this reason the resistance correlates with humidity. Useful materials include cellulose esters and polyalcohols.

The resistance of polymers with dissociable groups (e.g., sulfonates, amines, amides, or even hydroxides) also depends on humidity. The resistance of such polymers in fact decreases with absorption of water because of ionic dissociation and an increase in ionic conduction.

The absorption of water by polymers is especially suited to the development of a capacitive humidity sensor based on the high dielectric constant of water, since capacitance depends on the area, the thickness, and the dielectric constant of a dielectric. Uptake of water causes an increase in the dielectric constant followed by an increase in capacitance. Polyimide is a suitable sensitive material, but so are certain inorganic ceramics such as Al_2O_3 [142] and low-density Ta_2O_5, which change their capacitance with the formation of water dipoles. Low-density Ta_2O_5 layers can be formed by anodic oxidation of sputtered tantalum films, whereas polyimide films are deposited by spin coating and subsequent polymerization.

Conducting Polymer-Based Gas Sensors. The above mentioned ion-selective conductometric microsensors can easily be transformed into gas sensors by omitting the ionophore and adding an organic salt for a certain ground conductivity instead. These PVC/plasticizer membranes are sensitive to reaction with certain gases that can enter the membrane phase and alter the conductivity of the organic ions either by increasing their dissociation from an ion-pair situation or by increasing their diffusional property due to membrane swelling. Figure 24 gives an example of when the salt and the plasticizer are present simultaneously. The selectivity for a certain gas can be modulated by changing the type and concentration of the latter. Therefore, much more variations are possible compared with other polymer-based gas sensors based only on their known gas chromatographic distribution behavior. A typical membrane composition is: about 50 % PVC, 30–35 % plasticizer, the remainder being the organic salt such as tetraalkylammonium halides. The compounds are as usual dissolved in tetrahydrofurane and then drop-coated on the interdigital electrode array with similar dimensions as in the above mentioned case of the conductometric ion-selective microsensors.

Gases that can be detected in the lower ppm range with response time of a few seconds are so far: tetrachloroethene, ethanol, benzene, toluol,

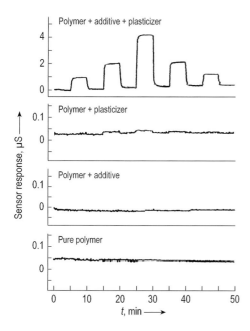

Figure 24. Variations of the sensor membrane composition. Sensor response against different concentrations of ethanol (500 ppm, 1000 ppm, 2000 ppm).
Polymer: polyisobutylene; plasticizer: ortho-nitrophenyloctylether; organic salt: tetraoctylammoniumbromide

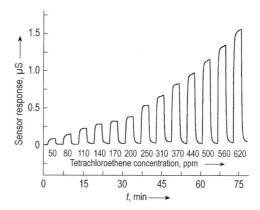

Figure 25. Polymer membrane sensor response against tetrachloroethene. The membrane contains polyethylenoxide, 2-fluorophenyl-2-nitrophenylether, tetraoctylammoniumbromide

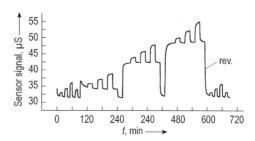

Figure 26. Humidity effect on a conductometric polymer gas sensor for ethanol.
Response toward: 350, 500, and 700 ppm ethanol at 10 %, 40 %, and 60 % relative humidity; membrane composition, 54 % polyepichlorhydrin, 36 % o-nitrophenyl-octyl-ether, 10 % tetradodecylammoniumbromide

and xylene. The measurement technique is exactly the same as with the conductometric ion sensors for application in liquids (Fig. 21). Compared with other gas-sensing devices the measurements are performed at room temperature. Thus, the energy consumption is very low, which will open the way to hand-held instruments. Unfortunately the optimization of the membrane composition for one gaseous analyte has yet to be on a purely empirical base. First studies at the ICB Muenster showed that the sensitivity and selectivity could be changed by many orders of magnitude [143], [144]. One example of a membrane with a certain selectivity towards tetrachloroethene to be deployed in the dry cleaning business is shown in Figure 25. Other examples are shown together with electronic nose developments.

Here the relevant limits of 25 ppm room concentration and 270 ppm in the cleaner before door is opened can easily be detected. Compared with the semiconductor gas sensors, humidity only changes the zero point for the measurements not the slope of the calibration curve. Figure 26 shows the humidity effect on a conductometric ethanol microsensor. Thus any placement of the sensor in a location with similar humidity but zero analyte concentration will correct for that.

In order to solve the humidity problem of most gas sensors the ICB Muenster developed and instrument called "Air-Check" in which a short but efficient pre-sampling on Tenax material takes place. With this the sensitivity could be increased as needed, water no longer interferes and the selectivity could also be increased by analyzing the time course of the rapid thermal desorption step. The whole adsorption–desorption cycle can be performed automatically within less than 1 min!

28.2.3.1.4. Ion-Selective Field-Effect Transistors (ISFETs)

As described in Section 28.2.3.1.1, with conventional ion-selective electrodes (ISEs) the ion-

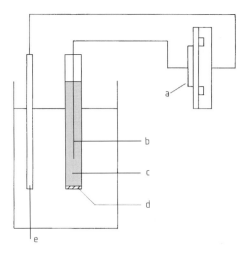

Figure 27. Schematic diagram of a membrane-electrode measuring circuit and cell assembly
a) MOSFET; b) Internal reference half cell; c) Internal filling solution; d) Ion-selective membrane; e) Reference electrode

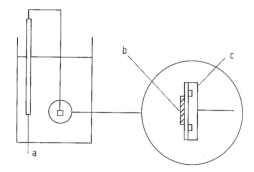

Figure 28. Schematic diagram of an ISFET measuring circuit and cell assembly
a) Reference electrode; b) Ion-selective membrane; c) ISFET

selective membrane is placed between the analyte and the internal reference solution. An internal reference electrode makes electrical connection to the internal reference solution, and the measurement system is completed by a second reference electrode in contact with the analyte solution. In any modern pH meter the internal reference electrode is connected to the gate metal of a metal-oxide semiconductor field-effect transistor (MOSFET), which acts as an impedance-converting device (Fig. 27). Normally, a conventional ISE functions very well. However, in certain applications problems can arise due to large size, high cost, and sensitive construction.

These problems might be resolved by the ion-selective field-effect transistor (ISFET). In an ISFET the ion-selective membrane is placed directly on the gate insulator of the field-effect transistor; alternatively, the gate insulator itself, acting as a pH-selective membrane, may be exposed to the analyte solution (Fig. 28). If one compares an ISFET with the conventional ISE measurement system, the gate metal, connecting leads, and internal reference system have all been eliminated. An ISFET is a small, physically robust, fast potentiometric sensor, and it can be produced by microelectronic methods, with the future prospect of low-cost bulk production. More than one sensor can be placed within an area of a few square millimeters. The first ISFETs were described independently by BERGVELD [142] and MATSUO, ESASHI, and INUMA [145]. The Institute of Microtechnology of the University of Neuchatel is very active in this field in recent days [146].

ISFET Operation. The ISFET can be regarded as a special type of MOSFET. To understand the operation of an ISFET it is therefore necessary first to understand the operation of a MOSFET.

An n-channel MOSFET (Fig. 29) consists of a p-type silicon substrate with two n-type diffusions, the source (c) and the drain (d). The structure is covered with an insulating layer (b), on top of which a gate electrode (a) is deposited over the area between the source and the drain. A voltage is applied to the gate that is positive with respect to both the source and the bulk material. This voltage produces a positive charge at the gate electrode and a negative charge at the surface of the silicon under the gate. The negative charge is formed by electrons, which are the minority carriers in the substrate, and this in turn creates a conducting channel between the source and the drain. The density of electrons in the channel, and thus the conductivity of the channel, is modulated by the applied gate voltage (V_G). If a constant voltage (V_D) is applied between the source and the drain a change in the conductivity of the channel leads to a change in current between source and drain. The drain current I_D in the unsaturated region ($V_D < V_G - V_T$) of a MOSFET is equal to:

$$I_D = \mu C_{ox} \frac{W}{L}\left[(V_G - V_T)V_D - \frac{1}{2}V_D^2\right] \quad (15)$$

where

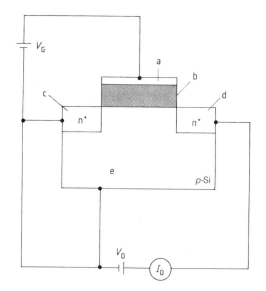

Figure 29. Schematic diagram of an n-channel MOSFET
a) Gate metal; b) Insulator; c) Source; d) Drain; e) Bulk

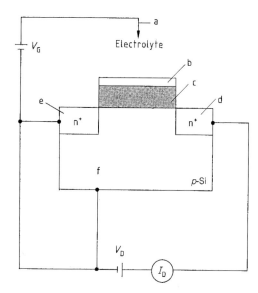

Figure 30. Schematic diagram of an ISFET
a) Reference electrode; b) Ion-selective membrane; c) Insulator; d) Drain; e) Source; f) Bulk

μ = Electron mobility in the channel
C_{ox} = Oxide capacitance per unit area
W/L = Ratio of channel width to length
V_T = Threshold voltage

A detailed description of the MOSFET is provided in [147].

In an ISFET (Fig. 30) the gate metal is replaced by a selective membrane and an electrolyte solution, or simply by the electrolyte solution, which is in contact with a reference electrode. The potential of the gate with respect to the substrate is the sum of the potentials of the various electronic and electrochemical processes that occur in the external path between the silicon substrate and the gate; i.e., the reference electrode half-cell potential E_{ref}, the solution–membrane or solution–insulator interfacial potential φ_M, and the metal–silicon contact potential. These voltages are added to the externally applied voltages V_g, and the resulting voltage has the same significance and function as that defined in the theory of MOSFET operation. V_{T*}, the threshold voltage of the ISFET, can be described as:

$$V_{T*} = V_T - E_{ref} - \varphi_M \tag{16}$$

From Equation (15) we obtain the following relationship:

$$I_D = \mu C_{ox} \frac{W}{L}\left[(V_G - V_T + E_{ref} - \varphi_M)V_D - \frac{1}{2}V_D^2\right]. \tag{17}$$

φ_M, the solution–membrane or solution–insulator interfacial potential, depends on the activity of the ions in the analyte solution according to the Nernst–Nikolsky equation [Eq. (6), Section 28.2.3.1.1] for ISFETs with an additional membrane, or according to a similar equation (see below) for ISFETs with a solution–insulator interface.

If all parameters in Equation (17) apart from φ_M are kept constant, changes in I_D are caused exclusively by changes in the activity of ions in the analyte solution. Thus, in an ISFET, the drain current is modulated by the ionic activity in the analyte solution. A detailed description of ISFET operation is presented in [148].

pH-Sensitive ISFETs. Gate insulators such as Si_3N_4, Al_2O_3, and Ta_2O_3 have been found to be pH sensitive, so FETs in which bare gate insulators of this type are exposed to a solution respond to changes in pH. The pH response can be explained by the "site-binding theory." This model assumes that there are ionizable binding sites present at the surface of an insulator, and that these can exhibit amphoteric behavior with respect to the potential-determining ion (i.e., H^+):

$$A-OH \rightleftharpoons A-O^- + H_S^+$$

$$A-OH + H_S^+ \rightleftharpoons A-OH_2^+$$

According to this model the pH dependence of φ_M is given by the following equation:

$$\varphi_M = 2.3 \frac{kT}{q} \left[\frac{\beta}{\beta+1} \right] (pH_{pzc} - pH) \qquad (18)$$

under the condition that $\beta \gg q\varphi_M/kT$. pH_{pzc} is the pH value for which $\varphi_M = 0$; β is a parameter that reflects the chemical sensitivity of the outer gate insulator, with a value dependent on the density of hydroxyl groups and the surface reactivity. Surfaces with a high β value show Nernstian behavior. For a detailed description of the site-binding model, see [148].

ISFETs Sensitive to Other Ions. Deposition of an ion-selective membrane on top of the gate insulator opens the way to measurement of ions other than H^+. Most of the sensitive materials for ISE applications described in Section 28.2.3.1.1 have been used in conjunction with ISFETs, providing sensors covering a wide variety of species. Thus, ISFETs with solid membranes have been described with AgCl–AgBr membranes sensitive to Ag^+, Cl^-, and Br^- [149], for example, or with LaF_3 membranes sensitive to fluoride [150]. Ion-sensitive polymer-matrix membranes (liquid membranes with a polymer matrix) have also been used as sensitive membranes for ISFETs. The first experiments involved mainly PVC membranes [151], [152], but these membranes show poor adhesion and poor mechanical strength. To improve membrane adhesion, modified PVC was utilized as a matrix material [153]. The use of silicones as matrix materials has made it possible to prepare very durable ISFETs with polymer-matrix membranes [154]. Other reports describe the use of photopolymerized polymers as matrix materials [155], [156].

Problems. The commercial development of ISFETs has been slow despite intensive research. From the standpoint of broad commercialization the ISFET is in competition with conventional ISEs with respect to price, performance, and reliability, and it is essential that some means be devised for producing ISFETs by commercially acceptable methods. Reaching this stage requires that several problems be solved, however.

ISFETs have so far been used primarily in conjunction with conventional macro reference electrodes. This fact greatly limits the potential benefits to be gained from miniaturization of the sensor. An optimal reference electrode for ISFETs should be miniaturizable and display long-term stability, and it should be subject to fabrication with microelectronic techniques. Attempts have been made to miniaturize macro reference electrodes [157], or to use modified surfaces with extremely low surface-site densities as reference FETs [158], but to date the above-mentioned requirements have not been met.

A further problem relates to encapsulation. The ion-sensitive gate area is exposed to the analyte solution, but all other parts of the sensor must be insulated from the solution. In most cases encapsulation is carried out by hand using epoxy resins, but it is impossible to produce ISFETs at low cost by this technique. Both photolithographic methods [159], [160] and an electrochemical method [161] have been proposed for the solution of this problem. One very promising development is the backside-contacted ISFET [162]. The electrical contacts in this case are protected from the analyte solution by means of O-rings, thereby circumventing the need for resins.

As noted previously, ISFETs selective to ions other than H^+ are prepared by deposition of an additional membrane on top of the gate insulator. If such ISFETs are to be commercialized this membrane must be deposited by a mass-production technique, which will be difficult to achieve with polymer membranes. One possible solution is the development of photolithographically patternable ion-sensitive membranes [163], but such membranes have so far been developed only for a few ions, and problems in the photolithographic deposition of more than one type of membrane on a single chip have not been resolved. The interface between an insulator and a sensitive membrane is also ill-defined in a thermodynamic sense, but this problem was solved by introducing a hydrogel between the sensing membrane and the insulator. Futhermore, ISFETs suffer from drift. The influence of this effect would be eliminated if the sensor were to be used in flow-injection analysis (FIA), so successful applications can be anticipated for ISFETs in the field of dynamic measurement [164].

Only if all the problems cited are successfully resolved will the ISFET live up to the promises

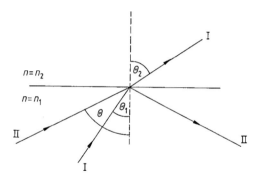

Figure 31. Refraction and total internal reflection
n = Refractive index

that have been held out for it for more than two decades: low cost, multifunctionality, and a robust and reliable construction.

28.2.3.2. Optical Sensors

28.2.3.2.1. Fiber-Optical Sensors

In recent years the use of optical fibers has become increasingly important in the field of chemical and biochemical sensors. This is clearly due to the fact that fiber-optical sensors combine the strengths of well-known bulk optical analytical methods with the unique advantages of optical fibers. More specifically:

1) An enormous technology base already exists for familiar optical methods of chemical analysis. There is probably no class of chemical analyte that has not at some time been the subject of optical determination through absorption or fluorescence spectroscopy. Methods also exist for labeling a target analyte with an appropriate chromophore or a fluorophore, a common practice in immunoassays.
2) Fiber-optic techniques offer increased sensitivity relative to conventional bulk optic approaches. For example, evanescent wave (EW) spectroscopy is significantly more sensitive than bulk attenuated total reflection spectroscopy. Fiber-optical EW spectroscopy is the only technique suitable for use with highly absorbing or scattering media.
3) The interrogating light in a fiber-optics system remains guided, so no coupling optics are required in the sensor region, and an all-fiber approach is feasible. This represents a sharp contrast to bulk optics, where proper optical adjustment is extremely critical. Fiber optics also provides a good solution to the problem of wide separation between the sensing area and the detector, as in the in vivo monitoring of medical parameters.
4) Recent developments in optical communications and integrated optics have led to optical materials that permit significant miniaturization of the sensing and detecting devices.
5) Since the measurement signal is optical, electrical interference is avoided, and the absence of electrical connectors ensures safe operation in hazardous environments, such as those containing explosive vapors.

The light inside an optical fiber is guided by the principle of total internal reflection, as described by Snell's law and depicted in Figure 31. A ray of light striking the interface between two media with differing refractive indices n_1 and n_2 is refracted (ray I) according to the relationship $\sin(\Theta_1) \times n_1 = \sin(\Theta_2) \times n_2$. If the condition $n_1 > n_2$ is fulfilled, total internal reflection will be observed for several angles of incidence. Thus, at an angle close to the horizontal the incident beam will no longer be refracted into the second medium, but will instead be totally reflected internally at the interface (ray II in Fig. 31). The *critical angle of incidence* $\Theta_c = \arcsin(n_2/n_1)$, which defines the onset of total internal reflection, is determined by the refractive indices of the two media, and whenever $\Theta > \Theta_c$ light is guided inside the medium characterized by n_1.

The general structure of an optical fiber is illustrated in Figure 32. Light entering from the left is confined within the core, which is surrounded by a cladding of lower refractive index (n_{clad}). The whole arrangement is surrounded by a nonoptical jacket, usually made from plastic and designed to provide both stabilization and protection. The fibers themselves can be constructed from a wide variety of transparent materials with differing refractive indices, including various glasses, fused silica, plastics, and sapphire. Most commercially available fibers have glass cores with n_{core} in the vicinity of 1.48 and diameters ranging from 5–200 µm; typical claddings have $n_{\text{clad}} < 1.48$ and thicknesses of 50–1000 µm. An optical fiber is usually characterized by its numerical aperture $NA = (n_{\text{core}}^2 - n_{\text{clad}}^2)^{1/2}$. The value of NA determines the cone angle Θ_c within which incident light will become entrapped by the fiber, where $\Theta_c = \arccos(NA)$; $\Theta_{\max} = \arcsin(NA/n_{\text{env}})$

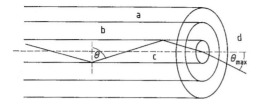

Figure 32. Structure of an optical fiber
a) Jacket; b) Cladding, $n = n_{clad}$; c) Core, $n = n_{core}$; d) Environment, $n = n_{env}$

for the cone angle Θ_{max} with which light leaves the fiber at its end (see Fig. 32).

The simple ray-optical description of light-guiding presented above fails to account for several phenomena important with respect to fiber-optical sensors. The most important of these is the *evanescent wave*. The intensity of light that is totally reflected internally does not in fact fall abruptly to zero at the core–cladding interface. Instead, the electromagnetic field intensity decays exponentially as a function of distance from the interface, thereby extending into the medium of lower refractive index. This field is called the *evanescent field*, and its penetration depth d is defined by the distance from the interface within which it decays to a fraction 1/e of its value at the interface. The angle of incidence Θ strongly influences the penetration depth. Values for d vary from 50–200 nm, increasing as Θ decreases; d reaches its maximum when Θ is very close to the critical angle.

Fiber-optical sensors can be constructed in either of two fundamentally different optical configurations with respect to the sensing area. The first is a wave-guide-binding configuration, in which recognition occurs at the surface of the fiber core via the evanescent field. This is achieved by removing the jacket and cladding from the fiber and causing the analyte molecules to bind to the core surface, as in the example described below. Such an arrangement is described as an *evanescent wave sensor* or *intrinsic fiber-optical sensor*. The resulting sensitivity depends greatly on the value of NA [165].

Alternatively, the fiber-end configuration might be considered, in which analyte is released for recognition purposes into the illuminated space at the end of the fiber. In this case the fiber acts only as a light pipe, resulting in what is known as an *extrinsic fiber-optical sensor*.

Both configurations entail similar optical systems: a spectral light source for excitation (e.g., a

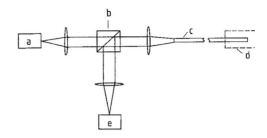

Figure 33. General configuration of an optical detection system for a fiber-optical sensor
a) Source; b) Beam splitter; c) Fiber; d) Sensing area; e) Detector
The excitation system consists of a xenon or tungsten lamp together with a monochromator or optical filter, an LED, or a laser, whereas the light detection system might be a monochromator or filter with a photomultiplier or photodiode.

xenon or tungsten lamp equipped with a monochromator or optical filter, an LED, or a laser) and a spectral light-detection device consisting of a monochromator or filter together with a photomultiplier or photodiode. A schematic diagram applicable to the optical systems for these two sensor configurations is presented in Figure 33.

More detailed information regarding the physical principles of fiber-optical chemical and biochemical sensors and the corresponding detection systems is available in [166] and [167].

Biochemical Sensors. Antibodies are widely used as recognition elements in fiber-optical biosensor applications. This is due to the high specificity of the binding reaction between an antibody and an analyte (antigen). To determine this reaction optically, labeling agents such as fluoresceine, rhodamine, or phycoerythrine are used, all of which show absorption and emission bands in the visible region of the spectrum. Linear amplification can be achieved by binding more than one label. Depending on the circumstances, either the antigen or the antibody might be tagged (see below).

The evanescent wave (intrinsic) configuration is most important for biosensors because it permits the antibody or antigen to be immobilized directly on the core surface, and it extends the path for possible interaction between the light and the recognition elements. Accordingly, only tagged molecules inside the evanescent field are subject to excitation. The amount of light energy absorbed, or the light emitted from these molecules and collected by the fiber, gives rise to the measurement signal. In contrast to the extrinsic sensor

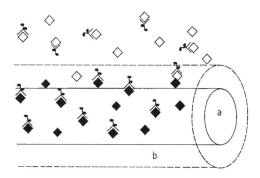

Figure 34. A fiber-optical immunosensor, in which labeled antibody molecules are gradually replaced by free analytes, causing the fluorescence signal to decrease
a) Core; b) Region of evanescent wave
◆ Fixed analyte; ◇ Free analyte;

Tagged antibody

approach, the intrinsic arrangement therefore reduces noise that would otherwise result from free labeled molecules or intrinsic fluorescence within the environment [167]. Despite these advantages, however, extrinsic biosensors have also been constructed, with binding devices placed in front of the fiber as a way of fixing the recognition elements. These devices can be either glass plates or transparent membranes. This offers the advantage of easy removal and replacement of the recognition elements [168].

Figure 34 provides an example of a biosensor of the intrinsic type. Here the antigen has been immobilized at the fiber core surface where it has been allowed to bind with a labeled antibody. Energy from the evanescent wave is absorbed by the labels, generating a detectable fluorescence signal. Addition of unlabeled analyte to the surrounding solution leads to competition for the binding sites. This permits some of the labeled antibody molecules to diffuse away from the surface, causing the fluorescence signal to decrease. The same general approach is also applicable to immobilized antibodies and tagged antigens [169].

Competition between a labeled antigen and an unlabeled analyte can also be effected in an external column, where the labeled antigen has previously been bound to an antibody. Once the competition reaches equilibrium any labeled antigen that has been released can be detected with antibody immobilized on a fiber. In this case the fluorescence signal would be expected to increase with an increased concentration of analyte.

Enzymes are in some cases also useful for recognition and labeling purposes. For example, glucose oxidase has been used to determine glucose on the basis of free oxygen that is consumed, where the disappearance of oxygen is measured with a fiber-optical oxygen subsensor [170]. Labeling of antibodies or antigens with urease or alkaline phosphatase and subsequent addition of a fluoro- or chromogenic substrate leads to detectable production of fluorophores or chromophores. The signal in this case is subject to multiple enhancement since it is proportional to the concentrations of both labeled antigen and antibody as well as to incubation time.

Chemical Sensors. Recognition with a *chemical sensor* is achieved via an analyte-specific reaction involving chemical compounds located inside the sensing area. The reaction must be one that is accompanied by changes in absorbance or fluorescence characteristics. Several examples are described below. A common physical arrangement utilizes a fiber-end approach and two single fibers or two fiber bundles, one for guiding light into the sensing area and the other to carry it away. The chemical reagents are confined within a transparent membrane or a small segment of the fiber itself, near its end and shielded by a membrane.

A particularly wide variety of fiber-optical sensors has been developed for measuring pH values, and many of these sensors are also used as subsensors for other analytes. Sensors for pH determination can be divided into two general types. The first relies on colorimetric acid–base indicators to produce changes in absorbance. For example, the dye phenol red can be copolymerized with acrylamide and bisacrylamide to yield dyed polyacrylamide microspheres. These are in turn packed into a length of cellulose dialysis tubing placed at the end of a pair of fibers that can serve as a vehicle for monitoring changes in absorbance [171]. Bromothymol blue immobilized on styrene–divinylbenzene copolymer has been similarly employed [172]. Alternatively, fluorometric acid–base indicators can be used to detect changes in fluorescence intensity that accompany protonation. Thus, one pH sensor was based on fluorescence quenching by H^+ of the excited state of immobilized fluoresceinamine [173]. 8-Hydroxypyrene-1,3,6-trisulfonate (HPTS) fixed on ethylene vinyl acetate has also been invoked as a fluorescence pH sensor [174].

Chemical sensors for other ions have been reported as well, including the halides and

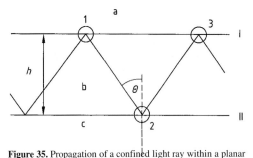

Figure 35. Propagation of a confined light ray within a planar waveguide mounted on a substrate
Most substrate materials are glasses or crystals with a refractive index of 1.4–2.4, and the cover material is air; the thickness of the film ranges from 0.1 to several micrometers.
a) Cover, $n = n_c$; b) Film, $n = n_f$; c) Substrate, $n = n_s$

28.2.3.2.2. Integrated Optical Chemical and Biochemical Sensors

Generally speaking, integrated optical devices can be constructed using techniques similar to those for manufacturing semiconductor devices. This offers great potential for cost reduction and facilitates the miniaturization often required in sensor applications. Furthermore, such sensors retain all the advantages of the well-known fiber-optical sensors.

The propagation of a light ray confined within a planar dielectric film (or *waveguide*) is depicted in Figure 35. Guidance of the light is possible only if total internal reflection occurs at the interfaces film–cover (I) and film–substrate (II). This condition is met if the refractive index of the film n_f is greater than that of both the cover n_c and the substrate n_s. Moreover, the angle of incidence Θ must be greater than the critical angles for total internal reflection at the interfaces I (Θ_{cI}) and II (Θ_{cII}), given by the expressions $\sin(\Theta_{cI}) = n_c/n_f$ and $\sin(\Theta_{cII}) = n_s/n_f$. To fulfill the self-consistency condition within the waveguide, the phase shift of the light wave as it travels from point 1 to point 3 must be an integral multiple m of 2π, where m is called the *mode number*.

$$\left(\frac{2\pi}{\lambda}\right) \cdot L(n_f, h, \Theta) + \Phi_I(n_f, n_c, \Theta) + \Phi_{II}(n_f, n_s, \Theta) = 2\pi m \qquad (19)$$

Φ_I and Φ_{II} are the phase shifts caused by reflection at the interfaces I and II, and $(2\pi/\lambda)L$ is the shift accompanying travel along the optical path $L = 2h\cos(\Theta)$. The values of n_f, n_c, n_s, and m are usually fixed by the fabrication process for the waveguide, and λ is established by the light source. The value of Θ can therefore be calculated from Equation (19). It is useful to define an *effective refractive index* n_{eff} for the guided mode, defined as $n_{eff} = n_f \sin(\Theta) = n_{eff}(n_f, n_c, n_s, h, m, \lambda)$, which, apart from λ, depends only on the material and geometric parameters of the waveguide. The parameter n_{eff} can be used to treat all types of light-propagation effects within the waveguide.

Even though light is totally reflected internally at the interfaces, its intensity does not fall to zero outside the waveguide. Just as with optical fibers, the small part of the guided light located outside the waveguide is referred to as the evanescent field, with a penetration depth ranging from 100–300 nm depending on the wave-guide performance. A more detailed mathematical treat-

pseudohalides [175]. A sensor of this type is based on the dynamic quenching of fluorescence emission from glass-immobilized, heterocyclic, acridinium and quinolinium indicators. The fluorometric process involved is known to obey the Stern–Volmer equation [176]. Al(III) ions have been determined with morin (3,5,7,2′,4′-pentahydroxyflavone) [177]. Thus, aluminum displaces the hydrogen ion when it binds to morin, leading to a change in fluorescence properties. Be(II) has also been determined using morin [178]. Furthermore, reports have appeared on sensors for Mg(II), Zn(II), and Cd(II) based on quinoline-8-ol sulfonate immobilized electrostatically on an anion-exchange resin [179].

Gas sensors for carbon dioxide [180] and ammonia [181] in aqueous solution have been constructed on the basis of a gas-permeable membrane and an optical pH subsensor. Gases diffusing through the membrane produce a change in the pH in the sensing region of a fiber-optical pH sensor. Sulfur dioxide [182] and oxygen [183] sensors take advantage of the quenching of the energy transfer (Förster transfer) between excited pyrene and perylene, both of which can be immobilized on a silicone matrix. Other O_2 sensors rely on the effect of fluorescence quenching as described by the Stern–Volmer equation. These generally involve metal complexes like tris(2,2′-bipyridine)ruthenium(II) [184]. In most cases fluorescence quenching also leads to a shortening of the fluorescence lifetime, which can be determined, for example, by phasesensitive detection [185]. The advantage of this method is that fluorescence lifetime is independent of the fluorophore concentration, and there is no signal drift due to bleaching.

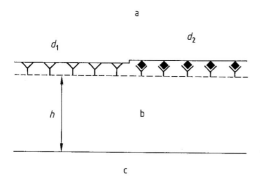

Figure 36. Specific antibody–antigen binding at the surface of a planar waveguide, causing the thickness of the adlayer to change
a) Cover, $n = n_c$; b) Film, $n = n_f$; c) Substrate, $n = n_s d_1$, d_2 = Layer thicknesses before and after binding

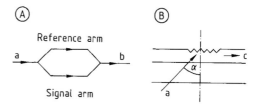

Figure 37. General arrangement of a Mach–Zehnder interferometer (A) and an input grating coupler (B)
In A only the signal arm is exposed to an active chemical layer while the reference arm is covered, whereas in B the grating is provided with an active layer and exposed to the sample.
a) Input beam; b) Output beam; c) Incoupled beam

ment of planar and strip waveguides is presented in [186].

Waveguides intended as chemical or biochemical transducers can be operated in either of two modes. A comprehensive physical description of these operating modes and the corresponding efficiencies is provided in [187].

First, it is possible to use a waveguide for acquiring refractive index data. This can be accomplished by coating the waveguide with a reactive but transparent chemical layer of a thickness greater than the penetration depth of the evanescent field. In this case the layer can be treated as if it were a cover material. Any chemical reaction with the layer will result in a change in the refractive index n_c, which in turn leads to a change in the effective refractive index n_{eff}. This approach has been used successfully in the production of sensors for CO_2 and SO_2 [188].

The adsorption and desorption of molecules is depicted in Figure 36 as an example of the second operational mode. The waveguide in this case is coated with a thin chemical layer consisting, for example, of an antibody. When the antibody–antigen binding reaction occurs, the effective thickness [187] of the entire waveguide increases, which again causes a change in the effective refractive index of the guide. Direct immunosensors based on this method have often been reported in the literature [189], [190].

Detecting a change in n_{eff} is a problem that can be solved in various ways. An interferometer approach analogous to the Mach–Zehnder interferometer [191], [192], the Fabry–Perot interferometer [188], or the difference interferometer [193] has often been used. The general arrangement of a Mach–Zehnder interferometer is illustrated in Figure 37 A. The diameter of the strip waveguide is usually on the order of 5 μm, with a distance between the two arms of < 1 mm, and a total length for the device of < 3 cm. Only the signal arm with its reactive chemical layer is exposed; the reference arm is covered by a dielectric film with a refractive index lower than that of the waveguide. The output power P_{out} of the device depends on the extent of the light phase shift $\Delta\Phi = (2\pi/\lambda) L\, n_{eff}$ between the signal arm and the reference arm; e.g., $P_{out} = P_{in} \cos^2(\Delta\Phi/2)$.

Grating couplers can also be used to determine the value of n_{eff}, as depicted schematically in Figure 37 B. For efficient light *incoupling* the angle α must be consistent with the expression $n_{eff} = n_0 \sin(\alpha) + l(\lambda/\Lambda)$, where l is the diffraction order and Λ the grating period. According to this relationship, a change in n_{eff} leads to a change in the angle α that will reinforce light guided within the waveguide. A grating coupler can also be used in an *outcoupling* arrangement. In this case it is the angle for the outcoupled light that would conform to the grating equation. A detailed description of grating-coupler methods is presented elsewhere [194].

28.2.3.2.3. Surface Plasmon Resonance

Transducers for optical sensing systems in chemical and biochemical sensors have also been developed based on the sensitivity of surface plasmon resonance (SPR) devices. Since first described in 1983 [195], SPR transducers have been used in many different applications, including immunoassays [195], [196] and sensors for gases [195], [197] and liquids [198].

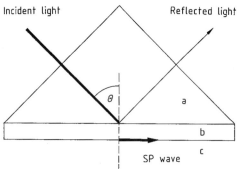

Figure 38. Attenuated total reflection coupler used to excite surface plasmons at the metal–dielectric interface; excitation is observed as a minimum in the intensity of the reflected light by scanning the angle of incidence
a) Prism; b) Metal; c) Dielectric

Electron charges located at a metal–dielectric interface can be caused to undergo coherent fluctuations known as *surface plasma oscillations* or *surface plasmons* (SP). The frequency ω of such longitudinal oscillations is related to the wave vector k_{sp} by the dispersion relationship.

$$k_{sp} = \left(\frac{\omega}{c}\right) \cdot \sqrt{\frac{\epsilon_M \cdot \epsilon_D}{\epsilon_M + \epsilon_D}} \qquad (20)$$

where $\epsilon_M(\omega)$ is the permittivity of the metal and $\epsilon_D(\omega)$ that of the surrounding dielectric (cf. Fig. 38); c is the velocity of light in a vacuum. The permittivity of the metal ($\epsilon_M = \epsilon_{M1} + i \cdot \epsilon_{M2}$) is complex and the value of $\epsilon_D = n_D^2$ is real, where n_D is the refractive index of the dielectric. A surface plasmon can be excited only if the conditions $\epsilon_{M1} < 0$ and $|\epsilon_{M1}| > \epsilon_{M2}$ are fulfilled. For example, gold, silver, and palladium are potentially applicable metals with negative values of ϵ_{M1} for frequencies corresponding to visible light [199].

SP excitation can be accomplished in several ways. A very simple and often used approach is illustrated in Figure 38. RATHER [199] has described this as the *attenuated total reflection* (ATR) coupler. Here the base of a prism is coated with a thin metal film approximately 50 nm thick. Incident monochromatic light is reflected at this interface in such a way that a projection of the light vector k_0 on the interface becomes $k_x = k_0 n_p \sin(\Theta)$. If the resonance (Re) condition $k_x = Re\{k_{sp}\}$ is fulfilled, the evanescent field of the incident light excites an SP at the metal–dielectric interface, although this is only possible with light polarized parallel to the plane of incidence (p-polarized). The excitation is observed as a sharp minimum in the intensity of the reflected light, and it can be detected by scanning the incident angle.

Using such an ATR device, small changes in $\epsilon_D = n_D^2$ and the build-up of a thin layer at the metal–dielectric interface can easily be detected by measuring the shift in the *resonance angle* Θ.

For an immunosensing application the metal layer would be coated with appropriate antibodies, either immobilized covalently or embedded into a matrix layer. Upon exposure to a specific antigen the resonance angle should shift relative to that observed in an analyte-free environment due to the increase in layer thickness that accompanies binding of the analyte. For example, a thickness change of 0.1 nm results in a 0.01° change in the resonance angle, a shift that can be measured quite accurately (see also Section 28.3.3.1).

If an ATR device is to be used for chemical sensing, the metal surface must be coated with a thick dielectric layer of a reactive chemical. Chemical reaction within the layer then produces changes in n_D that can be detected by the corresponding shift in Θ.

28.2.3.2.4. Reflectometric Interference Spectroscopy

Gauglitz introduced reflectometric interference spectroscopy (RIFS) into the field of chemical and biochemical sensing [200], [201]. Figure 39 shows the underlying optical principle. The interference of the schematically shown partial light rays 2 and 3 leads to an intensity modulation of the overall reflected light spectra. This must be happening within the so-called coherence length of light after leaving the substrate. The coherence length of light using, e.g., a tungsten light bulb in the wavelength range of 400–1000 nm is about 20 μm. Therefore, the thickness of the interference causing layer has to be less than 20 μm and has to exceed the wavelength of the used light to give a measurable interference pattern. A layer thickness between of 0.5 and 10 μm with $n_2 \approx 1.4$ fits well this demand.

Light rays which undergo multiple reflections at the different interfaces contribute little because of their lower intensity. The scattered light should be as low as possible as with light from other sources. The equation for the interference pattern is:

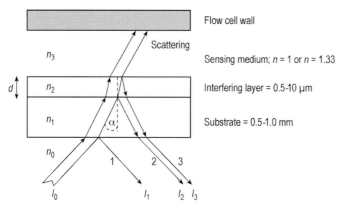

Figure 39. Schematic sketch of the principles of RIFS in a flow through set-up. Schematically shown light rays I_2 and I_3 can be re-collected with a fiber optic and lead to a monochromator with a CCD detector

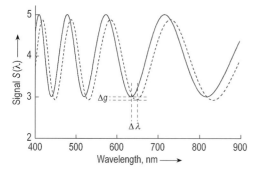

Figure 40. Observed interference pattern of RIFS normally recorded via a CCD array; comprehensive software [202] is needed to gain the needed information from this spectrum

$$I_{\text{all}}(\lambda) = I_1 + I_2(n_2) + I_3(n_2) + 2 \\ \cdot (I_2(n_2) \cdot I_3(n_2))^{1/2} \cdot \cos\{(4\pi/\lambda) \cdot (n_2 \cdot d)\} \tag{20a}$$

Figure 40 shows a typical interference pattern after an alteration of the refractive index and the thickness of the interference layer. The shift $\Delta\lambda$ is caused by Δd and Δn_2 whereas the chance in the modulation depth Δg is caused by Δn_2 only.

Thus, the basic response of an RIFS based polymer or immunosensor is the alteration of the optical pathway. For the evaluation of the interference pattern several algorithms have been developed. After the determination of the minima and maxima the interference order k is determined using a series of calculations and curve fitting techniques using newly developed software. Before a measurement is carried out a reference spectrum is obtained from an uncoated sensor including all spectral information except for the interference pattern. All further spectra obtained are divided by the reference spectrum to obtain the natural interference pattern.

28.2.3.3. Mass-Sensitive Devices

28.2.3.3.1. Introduction

Mass changes are usually detected with the aid of a gravimetric device, such as a balance. Ideally, balance sensitivity can be extended down into the microgram range, but the desire for further improvements in sensitivity and miniaturization of the experimental setup has led to the development of other mass-sensitive devices: the so-called quartz microbalance (QMB), which is the best-known system, and other, more recent devices such as the surface acoustic wave oscillator (SAW), the Lamb wave oscillator (LW), and the acoustic plate-mode oscillator (APM), which represent the present state of the science (Section 28.2.3.3.2). These devices are all based on the *piezoelectricity* of solids, (see Section 28.2.3.3.3), a phenomenon that permits the generation of vibrations or the propagation of waves. A piezoelectric substrate, usually a quartz plate, is stimulated to vibration by an oscillating electric circuit. This corresponds to an interconversion of electrical and mechanical vibrations (deformations of the crystal), resulting in a device that can be regarded as either an oscillator or a resonator.

Bulk waves, surface waves, and waves of other types (see Section 28.2.3.3.3) are associated with characteristic resonance frequencies corresponding to the most stable frequencies of vibration. Such a resonance frequency is itself highly dependent on the mass of the oscillating plate. A slight change in the mass of the plate will result in a considerable change in the resonance frequency, which can be measured very accurately. An oscillator of this type can therefore be used as a mass-sensitive transducer. Other factors that may influence the observed frequency shift include changes in the viscosity and density of surface layers as well as such physical parameters as forces, temperature, and pressure.

To construct a mass-sensitive sensor the surface of a piezoelectric material must be coated with an appropriate chemically or biochemically reactive layer capable of interacting with the proposed analyte and — ideally — recognizing it as well (see Section 28.2.3.3.4). Such an interaction causes the mass of the crystal to increase, resulting in a measurable shift in the resonance frequency. A relationship thus exists between the amount of analyte adsorbed or absorbed and the frequency shift of the oscillating system. Resonance frequency changes can be established very accurately, so very small changes in mass can be detected. A theoretical approach applicable to simple cases is described in [203].

To date, mass-sensitive devices have been used mainly for sensing analytes in the gas phase; they are also applicable in principle to liquids, but this presupposes a more sophisticated physical arrangement, a much more complicated signal control, and very careful interpretation. Many published papers have dealt with the potential use of QMB, SAW, and, more recently, APM and LW devices as mass-sensitive detectors in chemical and biochemical sensors, but no such device has yet established itself commercially — with the exception of QMBs employed for thin-layer thickness measuring, which cannot be regarded as chemosensors in the strictest sense of the term [204].

Historical Development. The first report concerning the use of piezoelectric resonance from a quartz crystal in a gravimetric microbalance is by J. STRUTT (Lord RAYLEIGH), who in 1885 described a shift in resonance frequency accompanying an infinitesimal mass change in a mechanical oscillator [205]. Fundamental studies into the mass sensitivity of quartz resonators and their potential application for measuring the thickness of thin layers were reported in 1959 by SAUERBREY [203], [206]. KING was the first to use a quartz microbalance "QMB" (in 1964) as a gas-phase detector in an analytical application [207], [208]. Numerous more recent papers discuss the fundamentals [209]–[216] and applicability [217]–[220] of bulk acoustic wave (BAW) piezoelectric resonators as chemical and biochemical sensors in the gas phase as well as in liquids, but physical devices for the measurement of pressure, temperature, or distance have attracted greater commercial interest [221], [222]. Nevertheless, quartz microbalances based on bulk acoustic waves do find frequent use in analytical research and development.

RAYLEIGH in an early paper [205] also noted the existence of waves that are propagated only in relatively thin surface layers, but it was not until 1979 that WOHLTJEN and DESSY adapted these so-called surface acoustic waves (SAWs) for incorporation into a chemical gas sensor [223]–[225]. The chief advantages of SAWs are increased sensitivity relative to BAWs of lower resonant frequencies and a greater potential for miniaturization. On the other hand, SAWs are associated with a variety of circuitry problems resulting from the relatively high operating frequencies required for adequate sensor sensitivity. A modification of this type of wave was introduced by CHANG and WHITE [226], [227], who adapted a SAW sensor by creating an exceptionally thin membrane-like piezoelectric region in the substrate directly beneath the acoustic path. This in turn led to the production of *Lamb waves*, which proved to be particularly suitable for chemical sensor applications, opening the way to the so-called plate-mode oscillator, a device representing the present state of the art.

28.2.3.3.2. Fundamental Principles and Basic Types of Transducers

Solids and other vibrationally active mechanical structures are capable of being stimulated by electrical, thermal, or optical means either to vibration or to the further propagation of waves. Piezoelectric solids, preferably quartz (SiO_2), are especially susceptible to emission and reception of mechanical vibrations and waves. The anisotropic nature of quartz supports various types of waves, each with its characteristic frequency, amplitude, and propagation properties.

Survey of the Different Types of Piezoelectric Substrates. A summary of the various types of transducers involving piezoelectric solids, includ-

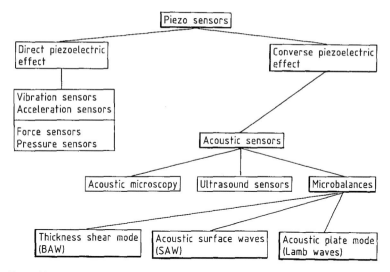

Figure 41. Selected applications of piezoelectric transducers with respect to both the direct piezoelectric effect (active transducer) and the converse piezoelectric effect (passive transducer) [222] (with permission from Springer-Verlag, Heidelberg)

ing their chemical and physical applications, is presented in Figure 41. Acoustic gravimetric sensors can be categorized generally on the basis of their design, their wave mode (BAW, SAW, APM, LW), and their frequency range. Figure 42 offers a schematic overview of the principles and design characteristics of various piezoelectric transducers. Criteria for establishing the usefulness of a particular vibrational system for the determination of mass include [209]:

1) The system should be readily excitable, preferably by electrical means
2) The system must lend itself to coupling with a frequency- (or period-) measuring device without significant disturbance
3) The system must be associated with sharply defined resonant frequencies so that the frequency (or period) of vibration can be determined precisely within a reasonably short period of time
4) In order to provide the required mass sensitivity, the mass-induced change in resonant frequency must be greater than the instability level of the resonant frequency and such uncertainties in the frequency (or period) as may accompany the measurement technique employed
5) Resonant frequency changes due to environmental disturbances (e.g., fluctuations in temperature or pressure, the presence of electrical or magnetic fields, or external mechanical stress) must be small relative to those caused by mass changes
6) An equation must be available in analytical form relating mass changes to the corresponding shifts in resonant frequency

Bulk Acoustic Wave Transducers. Waves of the bulk acoustic type (BAWs) can be invoked for microgravimetric applications through the use of either AT- or BT-cut plates derived from a single crystal of α-quartz, where "AT" and "BT" refer to cuts with specific orientations relative to the main axis of the crystal (see below and [229]). Figure 43 A represents a schematic diagram of such a quartz crystal with its associated electrodes and sensing layers. This diagram also indicates how a species to be detected might be absorbed into the sensing layer. Both sides of the quartz are in contact with electrodes. The corresponding electrodes are usually 300–1000 nm thick and 3–8 mm in diameter, and are made of gold, nickel, silver, or aluminum [217]. Introduction of oscillating electrical energy through an appropriate electrical circuit causes the mechanical structure to begin vibrating. In the case of AT-cut quartz the crystal vibrates in the thickness shear mode, as shown in Figure 43 B (see also Fig. 58).

Theory. Microgravimetric application of acoustic devices presupposes a quantitative relationship between an observed relative shift in the resonance frequency and an added mass. In the case of a quartz plate vibrating in a liquid an

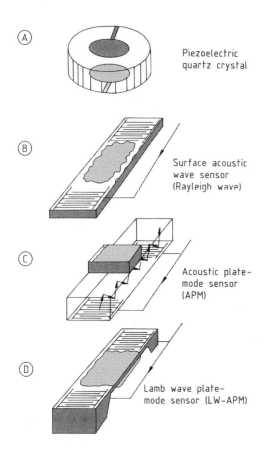

Figure 42. Schematic overview of acoustic gravimetric devices, illustrating the principles of wave propagation and design [228] (with permission from VDE-Verlag, Berlin) A) BAW; B) SAW; C) APM; D) LW substrates

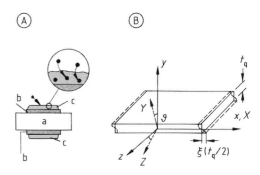

Figure 43. Schematic diagram of a quartz crystal augmented with electrodes and sensing layers, illustrating the way in which the species to be detected is absorbed into the sensing layer (A) and the ideal thickness shear mode of a quartz plate (B) [230], [203]
a) Quartz crystal; b) Electrodes; c) Sensing layers x, y, z = Coordinates of the quartz plate; X, Y, Z = (IEEE-standard) coordinates for a single crystal of α-quartz, where X is the polar axis and Z is the optical axis; t_q = Thickness of the quartz plate, where the wavelength $\lambda = 2\,t_q$; ξ = Shear amplitude; ϑ = Angle of rotation

$$\Delta f_q = -f_q \frac{\Delta m}{m} \qquad (21)$$

which finally provides the following expression for an AT-cut crystal:

$$\Delta f_q = -2.3 \times 10^6 f_q^2 \frac{\Delta m}{A} \qquad (22)$$

where f_q is expressed in megahertz, Δm in grams, and A in square centimeters. Since f_q for a quartz crystal is typically 10 MHz, and a frequency change of 0.1 Hz can be readily detected by modern electronics, it is possible in this way to detect mass changes of as little as about 10^{-10} g/cm² in the gas phase (!) [230].

Lu has developed a model involving propagation velocities for sound waves in the substrate (for quartz, v_q) and in the sensing layer (v_f) [216]. Reflection and refraction occur at the interface between the crystal (q) and the film (f) (Fig. 44) analogous to the optical reflection and refraction observed at a boundary between two materials with different optical densities. The shear velocities in the quartz crystal and the film coating, respectively, are:

$$\text{quartz} \quad v_q = \sqrt{\frac{\mu_q}{\varrho_q}} \qquad (23)$$

$$\text{film} \quad v_f = \sqrt{\frac{\mu_f}{\varrho_f}}$$

extended approach is required with respect to changes in the viscosity and density. The first useful relationship of this type was developed by SAUERBREY, who treated the added mass as an "added thickness" of the oscillator. The relationship itself was based on a quartz crystal vibrating in its thickness shear mode in the gas phase (Fig. 44). The observed oscillation frequency (f_q) is inversely proportional to the thickness t_q of the crystal; $f_q = N/t_q$, where N is a frequency constant ($N = 0.168$ MHz cm for AT-cut quartz at room temperature). The mass of the crystal is $m = \varrho A t_q$, where ϱ is the crystal density and A its cross-sectional area. In this derivation the mass increment due to a foreign mass Δm is treated as an equivalent change in the mass of the crystal. Invoking the approximation $\varrho_q = \varrho_f$ leads to the equation:

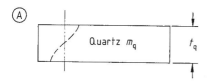
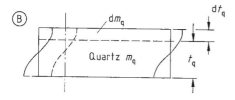
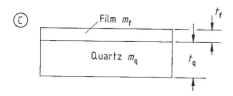

Figure 44. Shear wave (with permission from Plenum Publishing, New York)
A) In a quartz crystal; B) After an incremental increase in the crystal thickness; C) With the addition of a film m_f, t_q = Mass and thickness of the quartz; m_f, t_f = Mass and thickness of the film

Given a knowledge of the shear moduli μ_f and μ_q, Equation (23) can be used to calculate the shear velocity of the wave. Ideal wave-propagation behavior is observed with a crystal upon which such a film has been deposited provided there is a negligible amount of frictional loss. In this case the resonance condition is valid, which leads to Equation (24):

$$\tan\left(\frac{\pi \cdot f_c}{f_q}\right) = \left(\frac{\varrho_f \cdot v_f}{\varrho_q \cdot v_q}\right) \tan\left(\frac{\pi \cdot f_c}{f_f}\right) \qquad (24)$$

where $f_c = \omega/2\pi$ is the resonance frequency of the crystal upon which material has been deposited. The terms $\varrho_f \cdot v_f$ and $\varrho_q \cdot v_q$ are the acoustic impedances Z_f and Z_q of the film and crystal, respectively. Their ratio, $Z = Z_f/Z_q$, is an important parameter with respect to acoustic matching of the materials. For optimum resonance conditions Z should be as close as possible to unity, because its value affects the resilience of the whole assembly, and maximum stability corresponds to $Z \sim 1$ [231].

For example, Z has a value of 1.08 for a piezoelectric crystal covered with aluminum electrodes (shear-mode impedance of SiO_2: $Z_{SiO_2} = 8.27 \times 10^6$ kg s m^{-2}; that of Al: $Z_{Al} = 8.22 \times 10^6$ kg s m^{-2}), indicative of the good match associated with this material combination. Deposition of gold electrodes on an SiO_2 surface leads to a Z-value of 0.381, which is barely acceptable (shear-mode impedance of gold: $Z_{Au} = 23.2 \times 10^6$ kg s m^{-2}) [231].

From a practical point of view it is important to develop a relationship involving more experimentally common parameters influencing the frequency response. The frequency response of a piezoelectric sensor in the presence of a gas at a particular pressure can be related generally to three effects: the hydrostatic effect (p), the impedance effect (x), and the sorption effect (m):

$$-\frac{\Delta f}{f_0} = \left(\frac{\Delta f}{f}\right)_p + \left(\frac{\Delta f}{f}\right)_x + \left(\frac{\Delta f}{f}\right)_m \qquad (25)$$

For gravimetric measurements and sensor applications, the latter effect (that is, to added mass) is the most important, and the other two effects can be viewed as nonspecific interferences. Because both of these terms are three orders of magnitude smaller than the mass term, the mass relationship alone is sufficient for characterizing measurements in the gas phase. However, situations do exist in which the other two effects cannot be ignored. For instance, if the pressure is changed from that of the surrounding air to a vacuum the first term is no longer negligible, and in the case of liquids any change in viscosity or density alters the impedance, so this term must be taken into consideration.

Other theoretical approaches relevant to applications in the gas phase and dealing with thickness-shear mode acoustic wave quartz sensors are presented in [213], [232]–[237].

Bulk Acoustic Waves in Liquids. When a quartz crystal is placed in a liquid, there is a significant change in both the density and the viscosity of the surrounding medium. Both become much larger than in the gas phase, and the consequences must be taken into account with respect to the second term in Equation (25). Thus, the Sauerbrey equation is no longer applicable if the viscoelastic properties of the contacting medium (liquid and/or film) change during the course of an experiment. The liquid now represents an additional mass load, which produces coupling between the substrate elastic shear wave and the

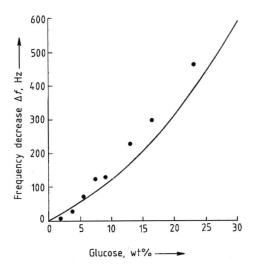

Figure 45. The effect of kinematic viscosity on resonance frequency as a function of glucose concentration (in water) for a 5-MHz quartz crystal
—— = Calculated; ● = Measured (with permission from Am. Chem. Soc., Washington)

liquid. KANAZAWA derived the first theoretical relationship linking the properties of a liquid to the frequency response of a piezoelectric crystal. In a liquid medium, the shear motion of the crystal causes motion to occur in the adjacent layer of solvent molecules (usually water). Up to a layer thickness of approximately 1 μm the solvent molecules move with a certain degree of "slip," which corresponds to a phase shift in the third and subsequent layers. Therefore, vibrational energy from the oscillator is continuously dissipated into the liquid. If only one side of the crystal is coupled to the liquid, the result is a standing wave perpendicular to the substrate surface. For liquids with Newtonian behavior this wave can be described theoretically as a damped wave. As is usual for this type of wave, a decay constant k is used to describe the magnitude of the damping. In this case the constant k can be related to $(\omega/2\,v)^{1/2}$, where v is the kinematic viscosity. The damping distance (limit of wave propagation) is on the order of micrometers, while the frequency shift can be expressed [238] as:

$$\Delta f = f_0^{3/2} \left(\frac{v}{\pi \mu_q \varrho_q} \right)^{1/2} \qquad (26)$$

Figure 45 shows the effect of microscopic kinematic viscosity at the crystal–liquid interface,

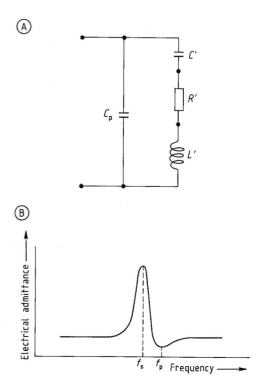

Figure 46. Equivalent electrical circuit (A) and resonance frequency of a mounted piezoelectric crystal oscillator (B) [231], [239]
C_p = Capacitance of the mounted crystal; C', L' = Rigidity and mass of the added material; R' = Equivalent to the mechanical loss, which also represents the acoustic (mechanical) load; f_s, f_p = Series and parallel resonance frequencies
(with permission from Plenum Publishing, New York)

where the magnitude of the observed frequency shift has been plotted against the corresponding increase in density for a glucose solution.

The frequency shift for quartz is highly dependent on changes in the viscosity and density of a surrounding liquid. This situation can be discussed in a straightforward way on the basis of the *quartz equivalent circuit* [231], [239], shown in Figure 46, which consists of the following elements:

1) The $L'-C'$ combination, which determines the resonance frequency f_m of the motional arm
2) A parallel capacitance C_p, caused by the electrodes on the quartz and any stray capacitance
3) A resistive component R'

The resistance R' is a measure of mechanical losses from the vibrating quartz, and it therefore reflects the viscoelastic properties of the contact-

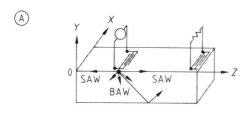

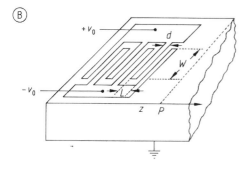

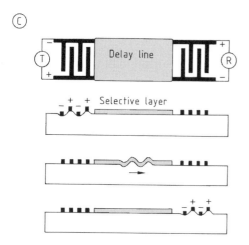

Figure 47. Interdigital metal pattern of a uniform transducer; the IDT behaves like a sequence of ultrasonic sources or receivers [247]
A) An SAW device with input and output transducers; B) A uniform transducer defined by half-spatial period L, electrode width d, and aperture width W; the observation line P used for derivation of the impulse response is at a distance $z+L$ from the first source; C) Schematic diagram of an SAW sensor with a transmitter T, a receiver R, and a chemically selective layer deposited in the form of a delay line [231] (with permission from Plenum Publishing, New York)

ing liquid [240], [241] or a deposited film [242]. Surface roughness also influences the value of R' [243]. This is often not considered in certain applications. It is highly unlikely that in the case of a directly sensed immuno-surface-reaction the binding events are uniformly smeared out. Likewise, immobilized peptide molecules may alter the tertiary shape by changing the ionic strength, resulting in a different surface roughness. Changes in the vibrating mass correspond to changes in the self-inductance L' [244], [245]. Thus, the resonance frequency of the motional arm (f_m; see Fig. 46) changes in proportion to the mass. In deriving Equation (26) KANAZAWA considered only those L' values for the motional arm that contribute to a frequency shift, but in a real experiment any change in $\Delta R'$ is normally accompanied by a change in L' (typical values for R' in air and in water are 7 Ω and 350 Ω, respectively, for a 6 MHz quartz crystal). Thus, when a quartz crystal is immersed in liquids of different viscosities, both L' and R' can be expected to vary. Use of an impedance analyzer makes it possible to determine the increase in R' associated with operation of a quartz crystal in air versus a liquid. The frequency shift of the motional arm can be calculated from the values of C' and L' using the formula $f_m = 1/[2\pi(L' C')^{1/2}]$ [239].

SAW Transducers. Surface acoustic wave (SAW) sensors have been designed with sensitivity to many different physical quantities, including force, acceleration, hydrostatic pressure, electric field strength, dew point, and gas concentration [246]. The operating principle of an SAW for use with gases is conceptually quite simple. A surface acoustic wave — a periodic deformation perpendicular to the material surface — is transmitted across the surface of some appropriate solid that has been exposed to the atmosphere subject to analysis. With a homogeneous substrate, usually a piezoelectric solid like quartz or $LiNbO_3$, the required SAW (sometimes also called a Rayleigh wave) is generated by means of an interdigital transducer (IDT).

In contrast to a quartz microbalance, the requisite IDT components are deposited only on one side of the crystal. These take the form of planar, interleaved, metal electrode structures, (with permission from Plenum Publishing, New York) where adjacent electrodes are supplied with equal but opposite potentials (Fig. 47). Application of a time-varying r. f. potential causes the crystal to undergo physical deformations, and if these are confined to the surface region of the crystal, the result is a surface acoustic wave (Fig. 47 B) [248]. Such vibrations will interfere constructively only if the distance $L/2$ between two adjacent "fingers" (see Fig. 37 A) is equal to one-half the elastic

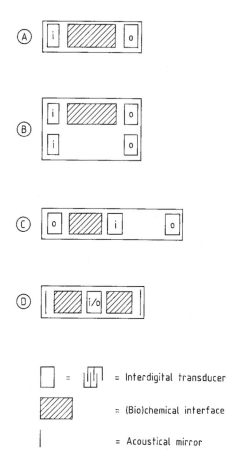

Figure 48. Various possible sensor configurations for SAW chemical sensors [230]
A) Single delay line; B) Dual delay line; C) Three-transducer device; D) Resonator
i = Input transducer; o = Output transducer

wavelength. The frequency $f = v_R/L$ that corresponds to such a cumulative effect is called the *synchronous frequency* or *resonance frequency* (f) [230].

The waves that are generated have their mechanical and electrical components in a single plane, the *sagittal* plane, which is perpendicular to the substrate surface. The phase velocity (v_R) and the amplitude (A) of an SAW are determined by elastic, piezoelectric, dielectric, and conductive properties of the substrate as well as by its mass. If one of these parameters can be modulated in an appropriate way by the quantity of interest, a sensing device can be constructed. Such a modulation could be induced in the transducer itself and/or in the transmission region (waveguide or *delay line*). With layered substrates the physical properties of each layer as well as the thicknesses of the layers determine the phase velocity and the amplitude of the resulting SAW [230]. Various energy-loss mechanisms also operate on the propagating SAW: scattering loss due to finite grain size, thermoelastic loss due to nonadiabatic behavior of the acoustic conductor, viscous loss caused by the dissipation of energy in a direction perpendicular to the transducer–environment interface, and hysteresis absorption due to irreversible coupling between SAW energy and an adsorbate [231]. Any addition of mass during the sensing step may change the magnitude of one or more of these phenomena, and may also lead to changes in v_R and A, especially a frequency shift (Δf) or a relative frequency shift ($\Delta f/f$). An exact description of the physical processes involved in the generation, propagation, and detection of SAWs is complex, and is available from such sources as [248], [249].

The considerations presented above provide the basis for using SAWs in gas-monitoring devices, because observed frequency shifts are proportional to added mass. However, the characteristic frequency of an SAW is also sensitive to changes in temperature or pressure. The ideal substrate would be one with a zero temperature delay coefficient but a high piezoelectric coupling effect. Despite extensive materials research efforts, it is currently necessary to choose between these parameters; e.g., ST-cut quartz is the preferred substrate if a zero temperature coefficient is most important, and cuts of lithium niobate are selected for high piezoelectric coupling.

Sensor Configurations. A single SAW sensor may sometimes suffice for a gas sensing application (Fig. 48 A), but often the sensors display undesirable sensitivity to such effects as temperature, pressure, and ambient humidity. Overcoming these problems requires the establishment of a relative signal output. For example, VETELINO et al. have described a dual-line SAW gas sensor [230], [250], [251]. One delay line in this case is used for measuring, while the other acts as a reference (see Fig. 48 B). The advantage of a dual delay-line configuration is that it transforms the output signal into a relative change in oscillator frequency, which can be attributed only to the effects of the analyte. D'AMICO et al. have described a system based on a three-transducer delay line (see Fig. 48 C). This device also has two dual-delay lines, but the paths are established in this case by one input IDT and two output IDTs. Again, one path of the delay line is coated and

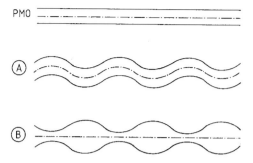

Figure 49. Lamb waves propagating in a thin plate-mode oscillator (PMO) [226], [231] (with permission from Plenum Publishing, New York)
A) Symmetric waves; B) Antisymmetric waves

the other remains uncoated as a way of minimizing the effects of both temperature and pressure [252].

The first use of an *SAW resonator* sensor was reported by MARTIN et al. [253]. The corresponding resonator incorporated a single IDT located between two acoustically reflecting mirrors. The distance between the mirrors was so adjusted as to ensure constructive interference between successive reflections, resulting in a maximum for one particular frequency (Fig. 48 D).

Theory. The SAW velocity v_R can be affected by many factors, each of which is associated with a potential sensor response:

$$\frac{\Delta v}{v_R} = \frac{1}{v_R}\left(\frac{\partial v}{\partial m}\Delta m + \frac{\partial v}{\partial c}\Delta c + \frac{\partial v}{\partial \sigma}\Delta \sigma + \frac{\partial v}{\partial \epsilon}\Delta \epsilon + \frac{\partial v}{\partial T}\Delta T + \frac{\partial v}{\partial P}\Delta P + \ldots \right) \quad (27)$$

where v_R is the phase velocity unperturbed by such external factors as m, c, σ, and ε (the mass, elastic constant, electrical conductivity, and dielectric constant of the solid medium, respectively), as well as T and p, the environmental temperature and pressure [246]. Three types of films have been considered for use as reactive layers:

1) Nonconductive isotropic overlay film
2) Electrically conducting overlay film
3) Metal-oxide semiconducting film

A theoretical treatment can be developed for each of these film types with respect to its properties, involving, in the nonconductive case, the thickness t_f, the mass density ϱ, λ (the Lamé constant), and μ (the shear modulus).

Regarding the relationship $\Delta v/v = -\Delta f/f$ between SAW velocity and the variation in the SAW dual-delay line oscillator frequency, perturbation analysis leads to the following result for the relative frequency shift [249]:

$$\Delta f = (k_1 + k_2)f_0^2 t_f \varrho - k_2 f_0^2 t_f \left[\frac{4\mu(\lambda + \mu)}{v_r^2(\lambda + 2\mu)}\right] \quad (28)$$

where k_1 and k_2 are material constants for the SAW substrate. Additional information with respect to these equations is available from [224], [246], [249].

Thus, the film mass per unit area is $t_f \varrho$. If chemical interaction does not alter the mechanical properties of the film, the second term in Equation (28) can be neglected, in which case the frequency shift Δf can be attributed exclusively to the added mass:

$$\Delta f = (k_1 + k_2)f_0^2 t_f \varrho \quad (29)$$

Coating a polymer film 1 µm thick onto a quartz SAW sensor operating at 31 MHz should therefore cause a frequency shift $\Delta f = -130$ kHz [231].

Plate-Mode Oscillators (PMOs). *Lamb Wave Oscillators.* In all the SAW devices discussed above, the acoustic wave propagates in a slab of material whose thickness is infinitely larger than the wavelength λ of the propagating wave. When the thickness of the plate is reduced to such an extent that it becomes comparable to λ, the entire plate becomes involved in the periodic motion, producing a symmetric and antisymmetric *Lamb wave* (LW) (Fig. 49). This behavior is observed in plate-mode oscillators with thicknesses of a few micrometers. A cross-sectional view through a typical PMO is shown in Figure 50.

Certain performance factors make the LW-PMO potentially attractive as a chemical sensor. Both surfaces contribute to the signal, so the observed sensitivity is greater than for a corresponding SAW device. However, the most important advantage follows from the fact that the velocity of the lowest-order wave in the antisymmetric mode is much lower than in the case of the corresponding SAW oscillator. This may be important for applications involving mass-sensing in liquids, which are problematic with high-frequency SAW devices. As the frequency of the Lamb wave decreases below the velocity of the compressional wave in the liquid, energy loss in the perpendicular direction decreases as well. A

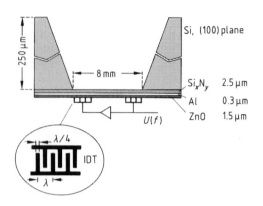

Figure 50. Cross section through a Lamb wave PMO, including the corresponding interdigital transducers [228]

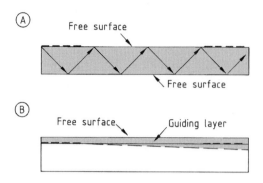

Figure 51. Cross-sectional view of (A) APM and (B) Love plate devices [254]

Lamb wave also has a greater frequency dispersion than the corresponding surface acoustic wave, resulting in increased resolution and sensitivity.

However, there are certain practical problems. The fragility of the thin plate and its consequent sensitivity to external pressure is particularly serious. Performance analyses as well as details regarding fabrication of the device can be found in the original paper [227].

Acoustic Plate-Mode Oscillators and Love Plate Devices [254]. These two types of acoustic PMOs, like BAW sensors, are characterized by surface-guided, shear-horizontally (SH) polarized waves, which have a horizontally polarized component of displacement (see also Fig. 58). In principle the performance of such APMs is similar to that of the Lamb wave sensors described above, but the thickness of the substrate is greater, so waves propagate from the surface into the bulk of the material. Energy is confined mainly in the bulk of the plate as the wave propagates through multiple reflections, generating displacements in both the upper and lower surfaces. The Love plate sensor includes a waveguide structure, whereby the SH wave is confined within an elastic layer deposited on a SAW substrate capable of supporting SH waves. Figure 51 provides cross-sectional views of APM and Love plate devices.

Acoustic sensors based on SH waves have become rather widespread in recent years due to their ability to function in liquid media.

Relative Sensitivities of BAW, SAW, LW, APM, and Love Plate Devices. Because of excitation and propagation differences among the various transducers described, the mass-deposition sensitivity of the several devices would be expected to vary considerably, with some geometries proving more attractive than others for particular sensing applications. Table 10 offers a performance comparison for these transducers. Table 10 also summarizes the calculated frequency-to-mass sensitivities for the various devices. Sensitivity variations can be explained on the basis of energy confinement on the sensing surface and a particular device's operating frequency. A decrease in the thickness of the substrate leads to a sensitivity increase, which can be realized with either an acoustic plate-mode or a Lamb wave device. These are therefore associated with the highest sensitivities, followed by the Love plate, which also has a very thin layer. Less sensitivity is available with BAW devices, for which the sensitivity depends on the thickness of the crystal.

The above statements have to be corrected because of very recent developments. With the commercial availability of a 155 MHz BAW quartz at reasonable prices the QMB should be more favored. According to the Sauerbrey equation the new BAW device shows a much higher sensitivity than the traditional 10 MHz QMB. Table 11 summarized recent data of the different noise levels published in the literature. By this it becomes evident that the new 155 MHz device will be the transducer of choice when an extremely sensitive gas sensor transducer is needed. This sensor also shows almost no drift. Thus drift compensation methods often used with SAW devices are no longer needed. The price of only a few Euro for this mass-produced unit will make it highly competitive in sensor applications.

Table 10. Comparison of sensitivities and other characteristics of acoustic gravimetric sensors [228], [255], [256]; international symbols (short form) have been used to denote specific cuts of the various materials

Acoustic wave type	Sensitivity formula, cm^2/µg	Calculated sensitivity, cm^2/µg	Frequency, MHz	Thickness, µm	Material
Bulk wave resonator (BAW)	$S_m = -\dfrac{2}{\varrho\lambda} = -\dfrac{1}{\varrho t_f}$	−23	10	165	AT-cut quartz
Surface (Rayleigh, SAW)	$S_m = -\dfrac{K(\sigma)}{\varrho\lambda}$ $0.8 < K < 2.2$ $0 < \sigma < 0.5$	−129 −516	100 400	NA* NA*	ST-cut quartz, LiNbO$_3$ $K = 1.16$; $\sigma = 0.35$
Lamb plate (LW), A_0 mode	$S_m = \dfrac{1}{\varrho t_f}$	−450 −951	4.7 2.6	3 (not given)	ZnO on SiN
Acoustic plate mode (APM, SH)		−3000 −300 −30	NA* NA* NA*	1 10 100	ST-cut quartz
Love plate (transverse surface)		−37 −182	100 500	NA* NA*	LiNbO$_3$

* not available.

Table 11. Transducer noise of the new polymer-coated 155 MHz BAW in air in comparison with mass-sensitive an SAW-transducer of different operating frequencies from literature values. The integration times of the frequency counters were in the range between 1.0 s and 2.0 s; resonance frequency f_0

Transducer	f_0, MHz	Research group/[257]	Noise, Hz
AT-quartz			
QMW; self-build	155	Cammann, Reinbold [257]	0.2–1.0
SAW LiNbO$_3$			
Plessey Semiconductors	67	Wohltjen et. al.	9–10
SAW$_{quartz}$ covered with different Me(II)-	39		11
phthalocyanines, leading to	52	Nieuwenhuizen et al.	7
different noise	78		17–29
Xensor Integration, Delft	80	Göpel et al.	1–5
Microsensor Systems, Inc.	158	Zellers, Patrash et al.	11–15
Microsensor Systems, Inc.	158	Grate, Mc Klusty et al.	5–7
Sawtek, Inc. Orlando, FL	200	Grate, Mc Klusty et al.	3–10
Microsensor Systems, Inc.	300	Grate, Mc Klusty et al.	26–55
Microsensor Systems, Inc.	400	Grate, Mc Klusty et al.	15–40
R2632 Siemens; self-build	433	Göpel et al.	1–5

28.2.3.3.3. Theoretical Background

Piezoelectricity. *The Piezoelectric Effect.* Because all acoustic gravimetric sensors are based on the phenomenon of piezoelectricity, it seems appropriate to discuss briefly the effect itself. Piezoelectricity was first observed by the CURIE brothers (JAQUES and PIERRE) in 1880 [258]. It is a reversible phenomenon, consisting of linear electromechanical interactions between mechanical and electrical properties in certain crystals (Fig. 52). The effect is generated, as already mentioned, by application of an AC potential to the piezoelectric material via contact electrodes. One can distinguish between bulk- and surface-generated AW (BAW and SAW, respectively) when the electric field is applied across the substrate or only at its surface, respectively.

A piezoelectric effect occurs only with those ionic crystalline solids whose crystals contain a polar axis along which the physical properties are not constant. The *direct* piezoelectric effect is observed when an applied force (F) produces an electrical polarization (P), and a *converse* (re-

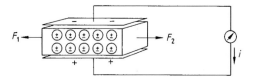

Figure 52. Schematic illustration of the way mechanical and electrical energy can be interconverted with the aid of an appropriate crystal [259]
F_1, F_2 = Mechanical forces; i = Electrical current

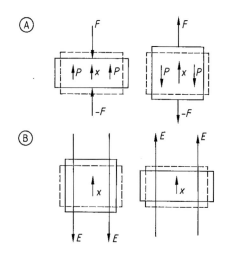

Figure 53. Schematic representation of the direct piezoelectric effect (A) and the converse (reciprocal) piezoelectric effect (B) for an X-quartz plate (right-handed quartz) [222]
F = Applied force; P = Polarization; E = Applied electrical field
(with permission from Springer-Verlag, Heidelberg)

ciprocal) piezoelectric effect results when an applied electrical field (E) induces a strain (S, deformation of the crystal) (Fig. 53).

A simplified view of piezoelectricity at the microscopic level is provided by the example of α-quartz (Fig. 54). Here a net dipole moment will arise if the hexagonal structure is stretched or compressed along a direction parallel (Fig. 54 B) or perpendicular (Fig. 54 C) to one of the three in-plane symmetry axes. When symmetry is achieved, the net dipole moment of the molecule vanishes.

Structural Aspects of Piezoelectric Solids. In order for a crystal to be piezoelectric it must be noncentrosymmetric; i.e., it cannot contain a center of symmetry. Solids are usually grouped for structural characterization purposes into seven crystal systems (from lowest to highest symmetry: triclinic, monoclinic, orthorhombic, tetragonal, trigonal, hexagonal, and cubic), which can in turn be divided into 32 point groups, depending upon point symmetry. Of these, 11 classes are centrosymmetric and 20 are piezoelectric. An exceptional case is class 432 from the cubic crystal system, because it is neither centrosymmetric nor piezoelectric. In addition to the 21 noncentrosymmetric classes, 11 classes can be distinguished on the basis that they possess no plane of symmetry. This means that they are associated with both right- and left-handed forms, which cannot be interconverted by simple rotation, a phenomenon known as enantiomorphism [261].

Table 12 summarizes on the basis of international point-group symbols the structural distribution with respect to piezoelectric classes, centrosymmetric classes, and enantiomorphism. For example, $LiNbO_3$, which belongs to the class $3m$, is piezoelectric, but not enantiomorphic, whereas α-quartz, in class 32, is both piezoelectric and enantiomorphic. The latter characteristic is illustrated in Figure 55.

Not every solid material associated with a piezoelectric class is suitable for practical applications, because in some cases the piezoeffect is too weak. Large piezoelectric coefficients are a prerequisite for an electromechanical transducer, like those found, for example, with α-quartz or $LiNbO_3$. Therefore, these are the solids most often used as mass-sensitive plates or substrates.

Theory and Physics of Piezoelectricity. The discussion that follows constitutes a very brief introduction to the theoretical formulation of the physical properties of crystals. If a solid is piezoelectric (and therefore also anisotropic), acoustic displacement and strain will result in electrical polarization of the solid material along certain of its dimensions. The nature and extent of the changes are related to the relationships between the electric field (E) and electric polarization (P), which are treated as vectors, and such elastic factors as stress T and strain (S), which are treated as tensors. In piezoelectric crystals an applied stress produces an electric polarization. Assuming the dependence is linear, the direct piezoelectric effect can be described by the equation:

$$P_i = d_{ijk} T_{jk} \quad \text{or} \quad P_i = e_{ijk} S_{jk} \tag{30}$$

The quantities d_{ijk} and e_{ijk} are known as the *piezoelectric strain* and *piezoelectric stress* coefficients, respectively.

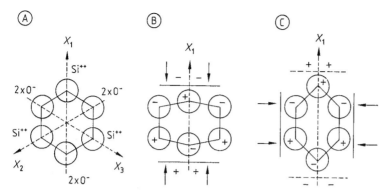

Figure 54. Physical interpretation of the piezoelectric effect [260]
A) Simplified microscopic model of a quartz crystal lattice; B) Longitudinal effect; C) Transverse effect
"Si^{++}" and "2 O$^-$" refer to centers of gravity (circles) for charges associated with the two types of atoms, where the tetrahedral "SiO$_4$" structure has been projected onto a plane (as a hexagon)

Table 12. Distribution of crystal structures (point groups) with respect to centrosymmetric, piezoelectric, and enantiomorphic classes [261]

Crystal system	Centrosymmetric classes	Piezoelectric classes	Classes with enantiomorphism
Triclinic	$\bar{1}$	1	1
Monoclinic	2/m	2, m	2
Orthorhombic	mmm	222, mm 2	222
Tetragonal	4/m, 4/mmm	4, $\bar{4}$, 422, 4 mm, 42 m	4, 422
Trigonal	$\bar{3}$, $\bar{3}$ m	3, 32, 3 m	3, 32
Hexagonal	6/m, 6/mmm	6, $\bar{6}$, 622, 6 mm, 6 m 2	6, 622
Cubic	m 3, m 3 m	23, $\bar{4}3$ m	23, 432

Conversely, if such a crystal is placed in an electrical field it will become deformed, a phenomenon known as the converse or reciprocal piezoelectric effect (Fig. 53 B). The contributions of stress and strain in this case can be expressed:

$$S_{jk} = d_{ijk} E_i \quad \text{or} \quad T_{ij} = e_{ijk} E_i \tag{31}$$

Since the stress and strain tensors are symmetric, the piezoelectric coefficients can be converted from tensor to matrix notation. Table 13 provides the piezoelectric matrices for α-quartz together with several values d_{ijk} and e_{ijk}, including the corresponding temperature coefficients [259], [261].

For sensor applications, the magnitudes of d_{ijk} and e_{ijk} should be as great as possible, and the temperature dependence should be as small as possible. Although lithium niobate displays larger stress and strain constants than α-quartz, the temperature influence in this case is also great. Therefore, α-quartz is the material most often selected for piezoelectric sensor applications. Because of the anisotropic behavior of a piezoelectric material, properties like resonance frequency and temperature dependence can be optimized by cutting a plate from a single crystal in a particular way. Figure 56 shows two possible cuts with respect to a natural α-quartz crystal. For use as acoustic gravimetric sensors, only AT- and BT-cut quartz plates are useful. These y-rotated cuts provide two different high-frequency plates that vibrate in a shear mode along an axis parallel to the major surface, as indicated previously (see also Fig. 43 B). Minimizing temperature effects requires that the plate be cut at a very precise orientation. The temperature dependence of the relative frequency for various cuts of quartz crystal is shown as a function of the angle of cut in Figure 57 [262].

Table 13. Piezoelectric matrices for α-quartz and LiNbO₃, together with the corresponding piezoelectric stress and strain coefficients [259], [261]

Classification	α-Quartz: Trigonal system, class 32		LiNbO₃: Trigonal system, class 3 m			
Matrix	$\begin{pmatrix} d_{11} & -d_{11} & 0 & d_{14} & 0 & 0 \\ 0 & 0 & 0 & 0 & -d_{14} & -2d_{11} \\ 0 & 0 & 0 & 0 & 0 & 0 \end{pmatrix}$		$\begin{pmatrix} 0 & 0 & 0 & 0 & d_{15} & -2d_{22} \\ -d_{22} & d_{22} & 0 & d_{15} & 0 & 0 \\ d_{31} & d_{31} & d_{33} & 0 & 0 & 0 \end{pmatrix}$			
Piezoelectric strain constants	d_{11}	d_{14}	d_{22}	d_{31}	d_{33}	d_{15}
	2.3	−0.67	21	−1	6	68
	2.31	−0.670	22.4	−1.2	18.8	78.0
	2.31	−0.727	20.8	−0.85	6.0	69.2
Temperature coefficients of piezoelectric strain constants (×10⁻⁴/°C)	$T_{d_{11}}$	$T_{d_{14}}$	$T_{d_{22}}$	$T_{d_{31}}$	$T_{d_{33}}$	$T_{d_{15}}$
	−2.0	17.7	2.34	19.1	11.3	3.45
	−2.15	12.9				
Piezoelectric stress constants	e_{11}	e_{14}	e_{22}	e_{31}	e_{33}	e_{15}
	0.173	0.04	2.5	0.2	1.3	3.7
	0.171	0.0403	2.43	0.23	1.33	3.76
Temperature coefficients of piezoelectric stress constants (×10⁻⁴/°C)	$T_{e_{11}}$	$T_{e_{14}}$	$T_{e_{22}}$	$T_{e_{31}}$	$T_{e_{33}}$	$T_{e_{15}}$
	−1.6	−14.4	0.79	2.21	8.87	1.47

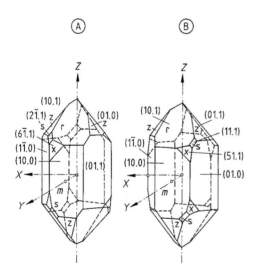

Figure 55. Major crystal surfaces of α-quartz together with their Bravais–Miller indices (hk.l) (with permission from The Institute of Electrical and Electronics Engineers, New York)
A) Left-handed form; B) Right-handed form; Cartesian coordinates are specified in accordance with the IEEE standard of 1987 [229]

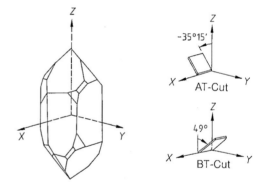

Figure 56. Oriented quartz crystal cuts in relation to the natural crystal [209] (with permission from Elsevier, Am-

presented in [266]. Interest has also developed in recent years in piezoelectric polymers, especially as a result of their strongly piezoelectric properties [222].

More detailed theoretical treatments of piezoelectric solids and their properties are provided by [222], [263]–[265]. Another more sophisticated approach based on dynamic quantum mechanics is

Vibrations and Waves. The theory of acoustic waves in solids is well understood, and many comprehensive descriptions exist [249], [267]. The description presented here should suffice to explain sensor function despite the fact that it is limited and qualitative.

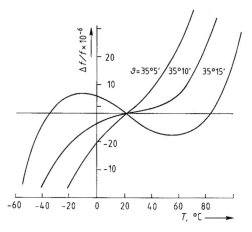

Figure 57. Temperature dependence of the relative vibrational frequency for different AT-cuts of a quartz crystal [262] (with permission from Elsevier, Amsterdam)

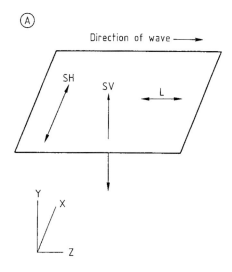

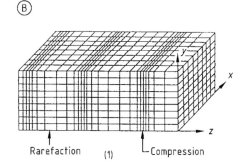

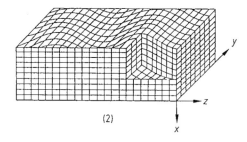

Figure 58. A) Schematic representation of longitudinal and shear displacement [256]; (with permission from Elsevier, Amsterdam) B) Grid diagrams for plane uniform waves propagating along the z axis in a material of infinite extent [248] (with permission from J. Wiley & Sons, New York): (1) Tensile motion (one dimension); (2) Shear motion (two-dimensional; y-polarized in this case)
L = Longitudinal (tensile) motion; SH = x-Polarized shear (horizontal); SV = y-Polarized shear (vertical)

Acoustic waves in solids consist of oscillatory motions of the material comprising the solid. The stress–strain relationship treated in the preceding section leads ultimately to two types of acoustic displacements: *longitudinal* or *tensile* motions that cause the dimensions of the solid body to change along a single direction without accompanying changes along the orthogonal directions, and *shear* motion, which produces changes in the shape of the solid body. Figure 58 illustrates these two types of motion [256].

In piezoelectric solids, an applied force generally creates both quasi-longitudinal and quasi-shear waves, where the acoustic displacement in a wave will be largely either longitudinal or shear. Pure longitudinal (L) and shear waves (SH = horizontally polarized shear waves; SV = vertically polarized shear waves) are generated if wave propagation occurs along certain crystallographic axes or with specific orientations. The generation of mixed wave types is also possible, but normally not desirable for sensor applications.

As far as microgravimetric sensors are concerned, two types of wave-propagation devices can be distinguished: those characterized by wave propagation perpendicular to the plate surface (BAW) and those in which the waves propagate along the surface (SAW) or in the bulk of the substrate (LW, APM). Transducers of the former type are normally resonators. A given device might be designed to utilize either longitudinal or shear waves propagated between the faces, as in microbalances based on AT- or BT-cut quartz plates operating in their SH high-frequency fundamental mode. Only acoustic waves satisfying boundary conditions for the resonator surfaces will propagate efficiently across such a device. The waves can be visualized as standing sinusoidal waves with displacement nodes at the resonator surfaces (see 28.2.3.3.2).

Transducers in the second category include wave delay-line oscillators (see Figs. 42 and 47), which are based primarily on two types of waves. One is a coupled linear combination of two wave-equation solutions, an L-wave and an SV-wave. In this wave, also called a Rayleigh wave, particle displacement takes the form of an elliptical motion in the plane perpendicular to the surface (sagittal plane) and containing the direction of propagation. Some displacement also occurs outside the sagittal plane, but the deviations are minor (Fig. 59). Two of the most important cuts providing these wave types, and which are used in chemical sensors, have been illustrated in Figure 56. Waveguides that rely on Rayleigh wave devices for their operation are often referred to as surface acoustic wave (SAW) devices, and they have been extensively employed as sensors.

The second type of wave in this context is a horizontal shear (SH) wave, whose acoustic displacement is in the plane of the waveguide. An SH wave generally travels at a different velocity than a Rayleigh wave, and for piezoelectric waveguides the SH wave becomes a true surface wave provided the acoustic impedance at the surface differs from that in the substrate bulk. Such a difference can be caused by piezoelectric stiffening of the SH mode, deposition of a layer of dissimilar material on the surface, the presence of a surface grating, or liquid in contact with the surface. Depending on their origin, SH surface waves are referred to in the literature as Bleustein–Gulayev (BG) waves, Stoneley waves, SH-SAW, or surface transverse waves. Pure surface transverse waves (shear waves) appear to show promise for applications in liquids, especially biosensor applications, because the dissipation of energy across the phase boundary into the liquid is approximately zero, which greatly reduces the generation of longitudinal sound waves in the liquid itself.

Plate-mode waves are analogous in form to Rayleigh and SH waves in a semi-infinite waveguide. Two groups of plate modes can be distinguished: the first consists of Lamb waves or flexural waves, where the acoustic displacement is a combination of longitudinal and vertical shear motion; a second family involves SH acoustic displacement [256]. Because of their frequency range (100 kHz to the GHz region) these sound waves fall in the ultrasound category, with a propagation velocity of 1–10 km/s (cf. electrical signals, which travel at 100 000–300 000 km/s).

Interdigitated Transducer (IDT). The IDT was invented by White and Voltmer in 1965. The IDT is a planar, interleaved metal structure at the surface of a highly polished piezoelectric substrate (see Figure 60). The adjacent electrodes are given equal potentials of the opposite sign. The resulting spatially periodic electric field produces a corresponding periodic mechanical strain pattern employing the piezoelectric effect. This gives rise to generation of surface acoustic waves (SAW), provided that the surface is stress free. In general both SAW and BAW may be generated by IDT.

The IDT behaves as a sequence of ultrasonic sources. For an applied sinusoidal voltage, all vibrations interfere constructively only if the distance $a/2$ between two adjacent fingers is equal to half the elastic wavelength. The frequency $F_0 = V_{AW}/\lambda = V_{AW}/a$ that corresponds to this cumulative effect is called the synchronous frequency or the resonance frequency. The bandwidth of an IDT is narrower, when there are more fingers. When IDT is used the AW velocity is determined by the plate material and orientation, while wavelength depends only on the ITD periodicity.

SAW are emitted in both opposite directions, which result in an inherent minimum of 3 dB transducer conversion loss at F_0. A minimum insertion loss of 6 dB is found for a delay line. The acoustic aperture A defines the effective region of transduction between two adjacent electrodes. The IDT is uniform if a constant aperture is obtained.

SAW component design is based on the application of an equivalent circuit model [29] using the values of the piezoelectric coupling coefficient of the material, F_0, and the static capacitance [30].

The frequency at which the AW device operates depends on:

– the acoustic wave velocity in the substrate material
– the IDT finger spacing (for SAW)
– the substrate (plate) thickness

Additionally, the type of AW generated by IDT depends on:

– the crystal cut
– the orientation of the IDT relative to the crystal cut

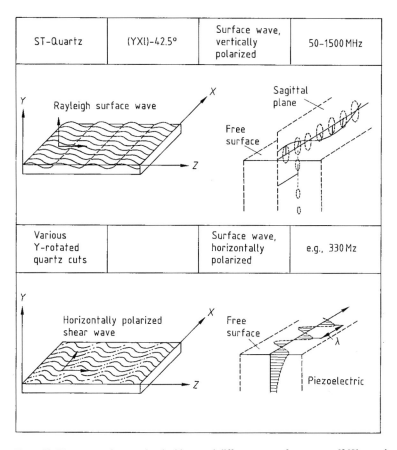

Figure 59. Wave properties associated with several different types of quartz cuts [260]; see also Figure 48

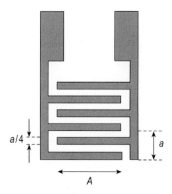

Figure 60. Layout of an Interdigitated Transducer (IDT) with period a and aperture A. The period determines the wavelength of the generated SAW $\lambda = a$. The aperture determines the width of the radiated beam and is typically 10–100 wavelengths in magnitude. The IDT length is $L = N\lambda$, when IDT contains N electrode pairs

– the substrate (plate) thickness
– the wave guiding mechanism (BAW, SAW, APM)

Acoustic wave devices (AWD). Today the family of AWDs is large. The overview shown in Figure 61 shows the basic types of the acoustic wave devices developed to date, indicating the polarization of the generated acoustic waves, and whether the devices can be operated in liquid.

28.2.3.3.4. Technical Considerations

The (Bio)Chemical Interface. A piezoelectric transducer is transformed into a mass-sensitive sensor by coating it with an appropriate (bio)chemical film or layer, which may be inorganic, organic, or even metallic in nature. Ideally the film should be strongly adherant and deposited in such a way as to obtain a plane surface. Concerning immobilized large biomolecules, certain doubts exist over the demand for a plane surface and the request for a strong adherence. It seems highly unlikely that an immobilized macromolecule will follow any surface movement of the transducer as a whole entity given its flexible structure. The thickness of such a layer is usually on the order of 1–100 μm depending upon the nature of the transducer. The chemically sensitive layer serves as an interface between the environment and a data-processing system, and it is responsible for both the sensitivity and the selectivity of the sensor. This layer should display the following properties:

1) Reversibility, sensitivity, and selectivity
2) Inertness with respect to chemical cross-influences, such as humidity or carbon dioxide
3) Ease of coating, strong adhesion, and long life

Most chemically sensitive layers are insulators, such as organic substances with special functionalities, but conducting materials such as metal films or conducting polymers are also useful. Examples include:

1) Donor–acceptor functionalities; e.g., carbocations [268], M(II)porphyrins [269], or betaines [270]
2) Host–guest binding systems (supramolecular compounds); e.g., molecular cavities [271], clathrate systems [272]
3) Pure metallic layers; e.g., palladium for the detection of hydrogen [273]

Interaction of the Chemical Layer with an Analyte. Interaction with an analyte can take many forms ranging from adsorption–absorption to chemisorption, including the compromise of coordination chemistry. The following types of chemical interactions can be distinguished [274]:

1) *Absorption*
   Here the analyte is distributed between the chemical interface (a liquid film, amorphous solid, or polymer, analogous to the stationary phase in gas chromatography) and the surrounding medium. In the case of a gas interacting with the chemical interfacial layer, once equilibrium has been achieved the amount of analyte present in the layer is a function of the partition coefficient between the gas and the interface (as specified by Raoult's and Henry's laws) as well as the thickness and area of the interface. Because distribution tends to be determined mainly by the polarity of the chemical interface and the analyte, little selectivity is expected.

2) *Adsorption*
   The analyte in this case interacts only at the surface of the interfacial layer, and no chemical bonds are formed or broken. Attraction energies are usually in the range 0–10 kJ/mol (i.e., attributable to van der Waals forces), extending in the case of hydrogen bonds to as much as 40 kJ/mol. Adsorption may result from physical attraction between a nonreactive gaseous analyte and a metal surface, for example, or formation of a donor–acceptor association (H^+ or e^-). Adsorption is a universal

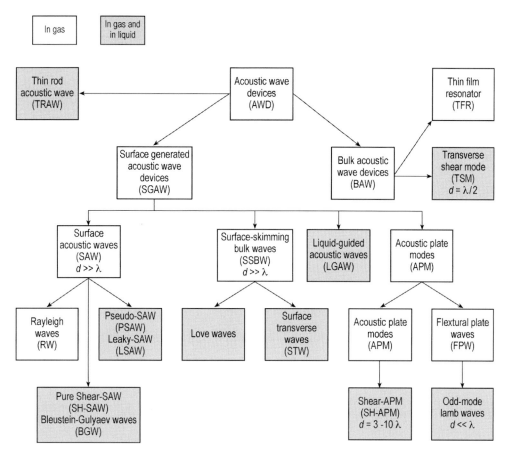

Figure 61. Flow diagram illustrating the basic types of acoustic wave devices. Here, d is the substrate (vibrating plate) thickness and l is the acoustic wave wavelength

phenomenon, and it is quite nonselective. The basic principles of gas adsorption have been described by LANGMUIR [275] and DE BOER [276].

3) *Chemisorption [277]*
   Chemisorption involves very strong interaction — up to ca. 300 kJ/mol — developed at the surface of the chemical interface, including the formation and breaking of chemical bonds. High selectivities can therefore be anticipated, but there is a simultaneous loss of reversibility. Because of the partially irreversible nature of bond formation, such a sensor would actually behave more like a dosimeter, although this might be advantageous in some applications.

4) *Coordination chemistry [274]*
   A compromise offering both selectivity and reversibility of analyte binding is presented by the area of coordination chemistry. A typical coordination compound consists of a central metal ion M surrounded by a neutral or charged (often organic) ligand. The extent of selectivity can often be influenced by the choice of the metal ion as well as by the choice of the ligand — taking into account both electronic and steric factors. Complexation with an analyte produces changes in the properties of the coordination compound, and these changes can be subject to detection. The selectivity is a function of structural, topological, and polarity parameters.

Other types of chemical interfaces have also been used for microgravimetric gas detection, including polymeric phthalocyanines [278], porphyrins [279], ferrocenes [280], metal

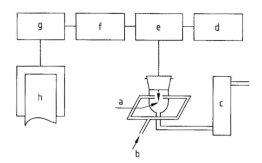

Figure 62. Typical experimental apparatus based on a piezoelectric quartz-crystal detector [217] (with permission from Elsevier, Amsterdam)
a) Piezoelectric quartz crystal; b) Sample; c) Flow meter; d) Power supply; e) Oscillator; f) Frequency counter; g) Digital-to-analog converter; h) Recorder

clusters [281], crown ethers [282], and cyclodextrins [283].

In many cases it has proven to be very difficult to find a selective chemical interface for a particular analyte, so the use of sensor arrays has been proposed. Numerous papers have dealt with this approach, invoking both BAW and SAW sensors. Pattern-recognition techniques employing increasingly powerful microprocessors are required to generate the appropriate signals [284].

Coating Methods. The stability of a (bio)chemical interface depends heavily on the attachment of the material to the surface of the sensor device. One or more of the following coating methods is employed according to the nature of the sensitive layer:

Smearing [224]
Spin coating [285]
Drop coating [103], [257]
Solvent evaporation [286]
Spraying [287]
Langmuir–Blodgett film formation [288]
Physical vapor deposition [289]
Sputtering [290]
Chemical immobilization [291]

The first five methods listed involve simple coating procedures in which the interfacial material (usually nonpolar) is dissolved in an organic solvent and then deposited directly onto the substrate surface. Evaporation of the solvent leaves a film that adheres to the surface of the substrate by physical bonds. The Langmuir–Blodgett technique makes it possible to achieve very thin films. This is a dipping procedure in which molecules are transferred with great accuracy from the surface of a liquid to the surface of the substrate. Even monolayers can be prepared in this way. More complicated methods include physical vapor deposition (PVD) and chemical vapor deposition (CVD). Very planar layers can be achieved in this way, but the reaction conditions are usually quite drastic, so the technique is essentially restricted to materials that are relatively inert, such as inorganic compounds or metals.

28.2.3.3.5. Specific Applications

Having reviewed the fundamental aspects of microgravimetric transducers and piezoelectric solids with respect to the generation and propagation of vibrations and waves, it is now appropriate to turn to the practical considerations of instrumentation and application, addressing first the BAW devices.

Bulk Acoustic Wave Sensors. *Experimental Arrangement.* A typical experimental setup involving a piezoelectric mass-sensitive quartz crystal detector is depicted schematically in Figure 62 [217]. The piezoelectric quartz crystal (a) is shown here inserted directly in a gas stream (b) with a flow velocity of ca. 10–100 mL/min, but a measuring cell containing a stationary atmosphere of the analyte gas is also useful. The oscillator (e) is usually powered by a regulated power supply (d; e.g., 5–15 V) that drives the quartz crystal. The frequency output from the oscillator is monitored with a frequency counter (f), which should be modified by a digital-to-analog converter (g) to permit the frequency data to be recorded.

Parameters that adversely affect the performance of a piezoelectric sensor include the mass of the chemical layer, built-in stress produced by the chemical layer, stress from the electrodes (clamping), and changes in temperature. A more sophisticated device such as a microprocessor-controlled dual-crystal instrument would be required to eliminate these undesirable effects [292].

The device described is designed for measurements in the gas phase; applications in a liquid medium demand an extended version, including a liquid-tight box, an oscilloscope or network analyzer, and provisions for more detailed signal analysis (impedance analysis). Improvements in the oscillating circuit are particularly important for obtaining satisfactory results with liquids [293]–[295]. A proper interpretation of any ob-

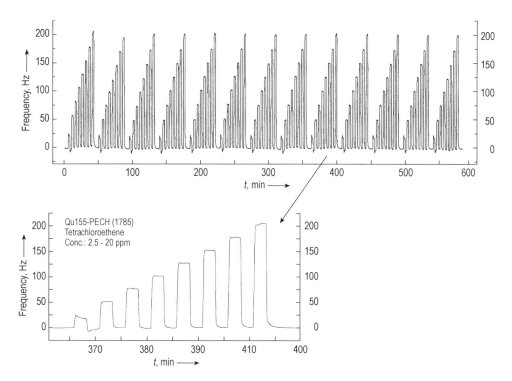

Figure 63. QMB-based gas sensing with polyepichlorohydrin (PECH) as polymer. The upper original recorder traces show 13 complete calibration cycles and demonstrates the very high reproducibility of this gas sensor in this low concentration range. The lower part shows the details of one cycle between 2.5 and 20 ppm tetrachloroethene. With a sensitivity of $\cong$ 10 Hz/ppm tetrachloroethene and an LOD of about 100 ppb this device is more sensitive than any Taguchi gas sensor or the FID considering its low noise, in the range of only about 0.5 Hz! Compared with a 25 MHz QMW the sensitivity could be increased by one order of magnitude [257]

served frequency change in liquid media still remains. Can the densitiy and the viscosity of the sample liquids be controlled so accurately (sometimes within the fifth digit after the decimal point) and does the surface roughness not change during the measurements increasing the viscoelastic effects?

Applications. This section will be limited to a brief consideration of the use of BAW sensors in gas-sensing devices and as sensors for liquid phases. Piezoelectric microbalances can also be used for monitoring such heterogeneous samples as aerosols and suspensions, as discussed in [262].

Gas-Phase Sensing. In principle, both inorganic and organic analytes are subject to detection. Figure 63 provides an example of results obtained from the novel 155 MHz piezoelectric quartz microbalance (QMB) resonator used to detect gaseous tetrachloroethylene (C_2Cl_4). The used absorption layer was polyepichlorhydrin (PECH). Observed response curves are illustrated as a function of time and for various analyte concentrations. [257].

A wide variety of solvent vapors can be detected with this new QMB with extremely high sensitivity (starting in the ppb range) and showing a dynamic working range of up to 4 decades with response times below 10 s, indicating the potential of this approach to gas-phase monitors and sensors for applications in environmental analysis or process control.

Table 14 provides an overview of the detection of selected gaseous analytes. Information regarding interfering gases can be found in the original literature. The effect of changes in the relative humidity of the environment tends to be a general problem.

Applications in Liquids. Apart from the classical sensors designed for detecting ions, piezoelectric crystal resonators are the preferred devices for conducting biochemical measurements in liquids. Figure 64 illustrates results for the detection

Table 14. Survey of selected reports on BAW sensors for inorganic and organic analytes [230]

Analyte	Range, ppm	Chemical interface	References
Inorganics			
CO	1–50	HgO/Hg→Au	[297]
CO_2	≤ 100 000	7,10-dioxa-3,4-diaza-1,5,12,16-hexadecatrol	[298]
$COCl_2$	8–200	methyltrioctylphosphonium dimethylphosphonate	[299]
HCN	13–93	bis(pentane 2,4 dionato)nickel(II)	[300]
Organics			
Nitroaromatics	0.001–100	carbowax 100	[301], [302]
Dimethylhydrazine	NR*	polybutadienes	[303]
Acetoin	0.008–0.120	tetrabutylphosphonium chloride	[304]
Formaldehyde	0.010–100	formaldehyde dehydrogenase	[305]
Vinylchloride	≤ 80	amine 220	[306]

* not reported.

of human immunodeficiency virus (HIV) antibodies by means of a synthetic HIV peptide (p 24) immobilized on a piezoelectric quartz sensor. The data refer to the frequency shift observed upon addition of authentic HIV antibody. The quartz device was operated in this case in its fun-damental thickness shear mode at 20 MHz, with an oscillator circuit stability of ca. 0.5 Hz. The quartz plate was coated with HIV peptide (p 24) dissolved in phosphate buffer at a dilution of 1 : 1000. Figure 65 shows the corresponding response as a function of different concentrations of monoclonal HIV antibodies. This device has not yet been brought onto the market despite complete automation and further improvements. Thus, some remarks concerning mass-sensitive transducers employed in liquids must be made, since it is still doubtful if the frequency change observed by passing a sample solution over such a sensor surface really reflects a mass change. Even if the user will manages to control the density and viscosity of the sample solutions within a narrow range determined by the required sensitivity (up to 5 digits after the point) a change in the surface roughness by non-homogeneously attached macromolecules during the immuno reaction may affect the viscoelastic effect. In addition, the complex structure of proteins/antibodies may contain a variable amount of occluded water and ions, which depends on the ionic strength and other factors that have to be carefully controlled to get the true analytical result.

Surface Acoustic Wave Sensors. Many reports have appeared regarding the use of SAWs as gas detectors. These reports also discuss problems associated with the choice of a coating, achieving adequate specificity, selecting a coating thickness, and depositing reproducible levels of polymer coating on the surface of the sensing element. The variety of coatings used is as diverse as the nature of the analytes, covering the range from simple polymers through metallocyanines to metal films [308]. The main drawback after the introduction of the 155 MHz QMB was the lack of availability and the price. Concerning the LOD, their inherent higher noise excludes, in general, the ppb range. The application of SAW sensors in liquid systems is still an active area of research, so this topic will not be treated. Information regarding liquid applications for mass-sensitive devices, especially SAW, APM, and LW devices, is provided in [309], [310].

Experimental Arrangement. A schematic representation of a SAW sensing device is provided in Figure 66. This particular resonator was used by D'AMICO et al. [311] for the detection of hydrogen. It consists of a dual delay-line structure in which one of the propagation paths (l_S) is coated with a thin palladium film to act as the sensing channel, while the other (l_R) is uncoated and serves as a reference.

Applications. The type of results to be expected with a variety of analytes in the gas phase is illustrated by the example of an NO_2-SAW sen-

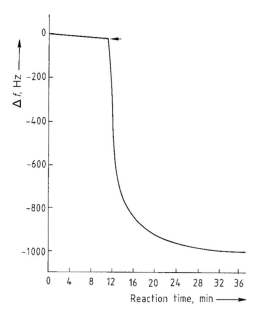

Figure 64. Response for the reaction of anti-HIV antibodies (dilution 1 : 1000) with adsorbed HIV peptide; the saturation value for the frequency shift is reached ca. 20 min after sample introduction (indicated by the arrow) [307] (with permission from Elsevier, Amsterdam)

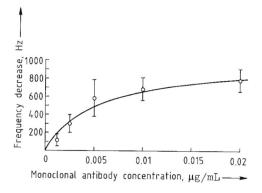

Figure 65. Frequency response for the reaction of HIV antibodies at several concentrations with adsorbed HIV peptide [307] (with permission from Elsevier, Amsterdam)

sor. The device in question was developed by NIEUWENHUIZEN and VENEMA [312], and had an operating frequency of 79 MHz with a metal-free phthalocyanine coating. A frequency change of ca. 670 Hz was observed upon exposure of the sensor to 88 ppm of nitrogen dioxide in air at 150 °C. Figure 67 A demonstrates the sensitivity of the device as a function of nitrogen dioxide concentration, while Figure 67 B shows the re-

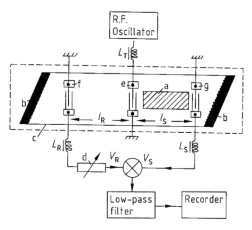

Figure 66. Schematic representation of a SAW hydrogen sensor [311] (with permission from Elsevier, Amsterdam)
a) Palladium film; b) Acoustic absorbers; c) YZ-LiNbO$_3$; d) "Stretched line," an adjustable resistance for matching the frequency of V_R to that of V_S; e) Input transducer T; f) Output transducer T_R; g) Output transducer T_S; l_R, l_S = Propagation path (l_R is the distance between T_R and T; l_S between T_S and T); L_T, L_R, L_S = Series inductors (establishing a fixed e.g., 50 Ω electric input–output impedance); V_R, V_S = Output voltage of the dual delay line

sponse of a typical sensor as a function of time with alternating exposure to nitrogen dioxide and air. Possible interferences were also investigated, as described in a similar report published earlier by the same authors [313].

A summary of selected papers describing the use of SAWs for gas sensing is provided in Table 15, subdivided according to the type of analyte.

28.2.3.3.6. Conclusions and Outlook

Mass-sensitive devices offer great promise as chemical sensors, particularly since they rely on the most basic physical effect that accompanies the interaction of one chemical substance with another: the mass effect.

Both BAWs and SAWs have already been successfully applied to the monitoring of many gases, vapors, and more recently, liquids. One of the most critical aspects of chemical sensor research is the search for a suitable chemical interface. The selectivity and sensitivity of mass-sensitive devices can be improved by the use of more carefully designed polymers and other special coating materials. Correction algorithms might also be developed to compensate for variations due to changes in humidity, temperature, pressure, etc. Humidity

Table 15. Survey of selected reports on SAW gas sensors [308] (with permission from Adam Hilger, Bristol)

Analyte	Sensor material	Frequency, MHz	Temperature, °C	Chemical interface	References
Inorganics					
H_2	$LiNbO_3$	75		Pd	[311], [314]
H_2	quartz	23		Pd	[315]
Halogens	quartz	50–80	30–150	Cu–PC[a]	[316]
H_2O	$LiNbO_3$	75	0–100	polymer	[317], [318]
NO_2	$LiNbO_3$	150	70	Cu–PC[a]	[319]
Organics					
DMMP[c]	quartz	31–300	20–80	polymers	[320]
Styrene	quartz	30		Pt complex	[286]
Cyclopentadiene	quartz	31	25	PEM[b]	[321]
Various	ZnO	109	24	ZnO	[322]–[324]
Various	quartz	158	35	polymers	[325]

[a] PC = Phthalocyanine.
[b] PEM = Poly(ethylene maleate).
[c] DMMP = Dimethyl methylphosphonate.

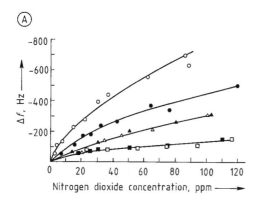

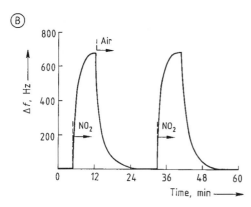

Figure 67. A) Sensitivity of an NO_2-SAW resonator at 120 °C (●, □, △) and 150 °C (○, □, △) with three different metal-free pthalocyanine layers: PC 42 (●, ○), PC 46 (□, □) and PC 47 (△, △) [312]; B) Sensor output as a function of time at 150 °C (PC 42, 88 ppm of NO_2) (with permission from Elsevier, Amsterdam)

poses one of the greatest problems with respect to most sensing devices. Better design of the electronic oscillator circuits might also increase the stability of the signal in response to a gas or vapor, thereby improving the signal-to-noise ratio and raising the level of sensitivity.

Research on chemical sensors tends to be unusually multidisciplinary. Such new techniques as the use of Love plates and LW-APM make it possible to combine the advantages of microcompatibility, high sensitivity, and suitability for work in both the gas and liquid phases. In particular, the new 155 MHz BAW could open up new horizons in gas sensor developments and electronic noses. The higher sensitivity of this transducer now allows dynamic ranges larger than 4 orders of magnitude with response ranges in the seconds range. Taking the first derivative of the absorption or desorption signal this device can be used as a more or less selective GC detector [257]. A 5 ppm peak of tetrachlorethene can then be sensed with a signal-to-noise ratio of about 3000.

Nevertheless, anyone proposing to conduct investigations in this area should be aware of the complexity of interpreting data accumulated from multicomponent gas mixtures over a prolonged period of time, and of parameter identification and interpretation in the case of liquids [230], [308].

28.2.3.4. Calorimetric Devices

Introduction. Most chemical as well as enzyme-catalyzed reactions are accompanied by changes in enthalpy. For this reason, calorimetric

Table 16. Molar reaction enthalpies for selected processes [328], [329]

Reaction	Examples	Enthalpy $-\Delta H_R$, kJ/mol
Oxidation	methane; O_2	800
Neutralization	NaOH, HCl	55
Protonation	tris(hydroxymethyl) aminomethane; H^+	47
Enzyme catalysis	enzyme: catalase; substrate: H_2O_2	100
	enzyme: glucose oxidase; substrate: glucose	80

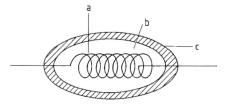

Figure 68. Pellistor device
a) Metal resistor; b) Inactive material; c) Catalyst

transducers represent a universally applicable approach to chemical or biochemical sensors.

Thermal devices for detecting an enthalpimetric effect (either exo- or endothermic) are insensitive to changes in optical properties of the sample, so colored solutions and turbidity do not constitute interferences. There is also no need for a detectable reaction product, whereas amperometric sensors, for example, require the formation of some product subject to transformation at an electrode. Amplification can be achieved by immobilization of a catalyzing reagent that enhances the rate of enthalpy change. Thermal transducers are especially suitable for continuous measurements in flow systems. An important disadvantage, however, is the requirement for thermostatic control of the device. For this reason calorimetric systems tend to be rather large. In addition, calorimetric sensors are relatively complicated to use [326], [327]. Table 16 provides molar enthalpies for several important types of chemical reactions.

The measurement signal in this case, ΔT, is related to the change in enthalpy ΔH, but also to the heat capacity of the system c_s, according to the following equation:

$$\Delta T = n\Delta H/c_s \quad (37)$$

n = Moles of product

This means that the heat capacity of the sensor itself should be minimized to ensure sensitive measurements. A temperature change can easily be transduced into an electrical signal. The usual measuring device is based on a reference thermal transducer incorporated into a Wheatstone bridge.

Thermal Transducers Commonly Utilized in Chemical and Biochemical Sensors. *Metal-Resistance Thermometers.* The electrical resistance of many metals rises sharply with increasing temperature. Resistors made of platinum (called "Pt 100" because they provide a resistance of 100 Ω at 0 °C) are often used for temperature measurement in the range from −220 °C to +750 °C. Other metals like nickel or copper are also applicable for temperatures ≤ 150 °C. Self-heating, caused by a current flow, interferes with the measurement, and the response time of a metal resistor is relatively long (ca. 5 s). The advantages of such resistors are high sensitivity and long-term stability. Platinum resistance devices are especially common in catalytic chemical sensors. A reducing gas may be oxidized in the presence of a heated catalyst (e.g., Rh, Pd, or Pt) in order to increase the reaction rate. In 1962 BAKER [330] described a device called a *pellistor* that effectively separates the catalyst from the platinum wire (see Fig. 68). The detection limit for methane with this device was found to be 20 ppm.

Thermistors. These are semiconductor resistances with temperature coefficients sufficiently high to make them suitable for use in temperature measurement. Many semiconductors have negative temperature coefficients (NTC), which means that their resistance decreases with increasing temperature. Thermistors with reproducible temperature coefficients are difficult to produce, and self-heating within the sensor is always a problem. On the other hand, thermistors are usually both inexpensive and sensitive, and they can be used over a wide temperature range (0 – 1000 °C).

The thermistor most commonly used in chemical or biochemical sensors, known as an *enzyme thermistor* (ET), was designed in 1974 by MOSBACH et al. [331]; it consists of an enzyme reactor

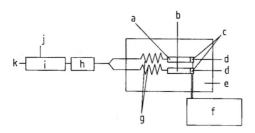

Figure 69. Enzyme thermistor system (with permission from Academic Press, Orlando)
a) Reference reactor; b) Enzyme reactor; c) Thermistors; d) Waste; e) Insulation; f) Wheatstone bridge; g) Heat exchangers; h) Pump; i) Injector; j) Sample; k) Buffer

specially adapted for use in a flow system. A split-flow arrangement was suggested in 1976 [332], whereby one portion of the analyte solution serves as a reference and flows through a blank column, while another passes through the enzyme reactor itself. The reference signal is subtracted from the signal produced by the enzyme reactor (Fig. 69) [333]. Samples are analyzed at a rate of ca. 20/h. Since the enzyme reaction and the detecting system are separated in space, an enzyme sensor of this type cannot be strictly interpreted as a "biosensor" according to the definition provided in Section 28.3.

Many other devices featuring lower detection limits have been developed in recent years, including a four-channel enzyme-based thermistor that permits the detection of four substrates simultaneously [334]. Another approach involves fabrication of a thermal microbiosensor on a silicon chip. In this case the sensor consists of a micromachined enzyme reactor, inlet and outlet flow channels, and a microthermistor [335].

Thermocouples/Thermopiles. Thermocouples operate on the basis of the Seebeck effect. If wires fabricated from two different metals or semiconductors are soldered together to form a circuit, any temperature difference that exists between the joined points leads to a measurable potential difference, the magnitude of which depends on the extent of the temperature difference and the materials involved. The use of several thermocouples connected in series (*thermopiles*) increases the sensitivity of the sensor, but if a single thermocouple is damaged the entire sensor is affected. It is more difficult to miniaturize thermoelectric sensors than resistance devices, and the long-term stability of thermocouples is often not good. A typical response time is less than one second, and sensors of this type can be used over a temperature range from $-200\,°C$ to $+1600\,°C$.

Thermal devices based on thermocouples have been extensively investigated in recent years. GUILBEAU et al. described in 1987 a thermoelectric sensor for the measurement of glucose [336]. This device incorporated an antimony–bismuth thin-film thermopile with enzyme immobilized on the side containing the active junctions but not on that of the reference junctions. Exposing the sensor to a solution containing the substrate causes the enzyme-catalyzed reaction to take place, resulting in heat exchange that is detected by the thermoelectric transducer. At the ICB Muenster, a thermopile produced with microsystem technology in the "Institut für physikalische Hochtechnologie" IPHT Jena (Germany) with the extreme sensitivity of about 10^{-4} K has been used for constructing a GOD-based glucose sensor which does not need any calibration. Figure 70 shows the construction of the thermopile with the microsystem technology.

Figure 71 shows the experimental set-up for absolute glucose sensing via the total heat of this specific GOD-based enzymatic reaction, which is strictly proportional to the number of moles reacting [337]. The heat measured in the diagram ΔH versus the glucose concentration follows the theoretical predictions. Thus, no calibration has to be performed, and no standard is needed!

Summary. Most chemical calorimetric sensors in use today are based on the pellistor device, and are designed for monitoring gases; sensor systems of this type have been subjected to intensive investigation [338]. In general, however, sensor determination of the heat of a reaction is rare, primarily because of the availability of other, more sensitive transducers. On the other hand, thermometric biosensors are applicable to a wide range of analytes, including enzyme substrates, enzymes, vitamins, and antigens. Many applications have been reported in clinical analysis, process and fermentation control, and environmental analysis [339]. Thermal biosensors are seldom used in industry because they are difficult to control and relatively expensive. Universal applicability and new techniques for fabricating microthermosensors suggest that these may find wider use in the future.

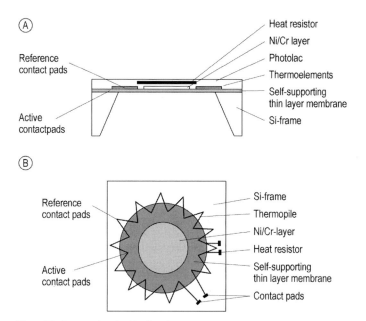

Figure 70. Construction details of a microsystem technology approach towards thermopiles. The extremely sensitive thermopile was produced in the "Institut für physikalische Hochtechnologie" IPHT - Jena

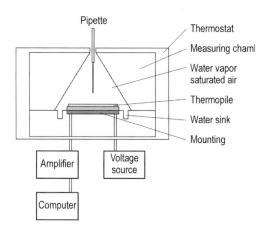

Figure 71. Measuring cell for absolute calorimetric GOD-based glucose determinations. The opening must be minimized because otherwise the water evaporates too fast and interferes with the measurement by its heat of evaporation. In order to accelerate the enzymatic reaction peroxidase was added simultaneously. As soon as the drop of blood was added the temperature–time course was recorded and integrated. The glucose concentration can then be directly calculated via the published value for ΔH for this reaction as Figure 72 demonstrates

28.2.4. Problems Associated with Chemical Sensors

All nonbiochemical sensors suffer from a lack of that selectivity which would permit their wide application without detailed knowledge regarding the matrix composition of the samples under investigation. Application is therefore restricted to those areas in which the sample matrix is thoroughly understood and not subject to change. Compared to the selectivity problem, all other problems (e.g., stability, sensitivity, lifetime) are relatively unimportant, since they can be resolved by incorporating the sensor into some type of intelligent instrument capable of carrying out all necessary corrections, recalibrations, and performance tests. In some cases insufficient selectivity of a chemical sensor can be circumvented by integrating a selective chemical reaction into the instrumental design. In this way many interfering compounds are eliminated and/or the analyte of interest is selectively transformed into a species capable of being detected unambiguously. A simple approach to this end takes advantage of the flow-injection analysis (FIA) technique.

Any chemometric method of error correction with respect to lack of selectivity must be checked very carefully, especially if it is to be applied to an open system with varying and/or unknown interfering compounds. In some cases a prior chemical separation and/or selection step is required.

Despite the superb selectivity or even specificity of biochemical sensors for recognizing particular analyte molecules, interfering compounds must be dealt with even here in real sample applications. The compounds in this case do not disturb the actual measurement of the analyte; instead they influence the recognition process. For this reason it is important to ensure the use of a stable and well-defined host molecule with an analyte-fitting geometry that is not subject to change. Anything that might alter the host would influence analyte recognition. Thus, any compound or measuring condition capable of changing the tertiary structure of the sensitive protein skeleton of some biomolecule used for recognition purposes would disturb the measurement: first by modifying the selectivity, but also by altering the sensitivity. It is well established that most selective biomolecules are very stable only in a physiological environment similar to the one in which they were produced; even changes in the ionic strength can influence the tertiary structure of a delicate protein macromolecule. Furthermore, certain metal ions (e.g., Hg^{2+}, Pb^{2+}, Ag^+, Cu^{2+}) are known to be bound to sulfur containing molecules (or destroy a stabilizing S–S bridge in a protein) and thus degrade recognition and thus strongly interfere. A similar effect is produced by surface-active molecules (detergents). Bearing these complications in mind, the superb selectivity of biochemical sensors must be evaluated in a broader perspective.

28.2.5. Multisensor Arrays, Electronic Noses, and Tongues

Theoretically, and under ideal conditions, a sensor array containing several sensors with slightly different selectivities and sensitivities towards the analyte and its main interferences should make it possible to compensate for errors caused by the limited selectivity of each individual sensor. The task resembles the solving of mathematical equations with several unkowns. There must be as many independent equations as unknowns. In sensorics this means: one additional but slightly different sensor is required for each interfering compound. The task then becomes one of solving n equations for n unknowns, which can be accomplished with an algorithm for matrix calculations and an electronic calculator. However, there are several restrictions that limit the use of this compensating technique.

First, the interfering components must be known in advance so that one can prepare a set of calibration mixtures for obtaining the individual analytical functions for each sensor under variable interference (matrix) concentrations. The approach is therefore limited to a closed sample location, and it presupposes complete information regarding potential interferents within the volume in question. This may prove applicable in certain production control situations, but certainly not in cases where the sample matrix is not sufficiently known a priori, as in the case of most environmental and some clinical analyses. Such precautions cannot be circumvented even with the use of the most sophisticated modern approaches to pattern-recognition analysis and/or so-called adaptive neural-network treatment. All potentially interfering compounds must be known in advance and introduced into the array under study during the so-called learning phase through a permutation of all possible types of mixtures. If for example one analyte is to be determined in a concentration range of 1 to 100 ppm and only five interfering compounds in a similar concentration range are present, and the concentration range should be checked for linearity by five concentrations between 0 and 100 ppm one has a permutation power of about a faculty of 10. Much time may be consumed in the preparation of the necessary mixtures and in performing all the measurements. In the end, the drift associated with most sensors and the limited lifetimes of biochemical sensors could render the whole procedure meaningless!

Second, the linearity and true independence of all effects cannot be guaranteed. In some cases of only one sensor poisoning (e.g. catalyst poisoning) all analytical functions (selectivities and sensitivities) obtained during prior calibration and/or learning are rendered invalid for every individual sensor. How is a sensor array to distinguish between irreversible poisoning effects and the normally reversible disturbances attributable to ordinary interferents? How should it differentiate between the effects of a positive error (where a portion of the signal is due to interferents) and general matrix effects resulting in diminished sensitivity? Especially in environmental trace analysis (<10 ppm) each sample is very likely to contain more than ten interfering compounds. Intimate knowledge of each, as well as the availability of appropriate standards to permit extensive permutation at various concentrations presupposes the

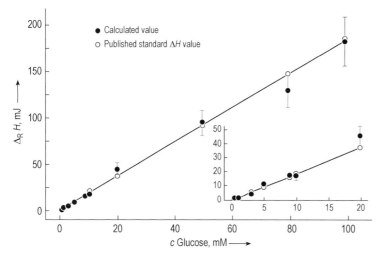

Figure 72. Several measurements with known glucose concentration are plotted on a diagram calculated from the "theoretical" ΔH values for this aliquot. Note the thermopile and the set-up was not optimized further. The only drawback is the fact that both enzymes are lost after each determination as they are part of the receptor buffer solution. This is also the case in traditional enzymatic analysis with photometric evaluation. Here, only a so-called spectral-line photometer can work without further calibration

preparation of countless mixtures to be utilized in the training process for the sensor array. It must also be recognized that even trace amounts of certain compounds may totally disrupt the analyte recognition process.

Taking all this into consideration, as well as the fact that most chemists lack fundamental experience with sensor arrays in real open-environmental trace analysis, considerable caution is warranted with respect to this approach to increasing the selectivity of sensors and correcting for interferants. In general, application of a simple and rapid preliminary separation step will be more productive with respect to the goal of achieving a reliable analytical result. Literature examples of successful sensor-array applications related to the identification of beverages or the characterization of odors [340]–[342] should not be extrapolated to the quantitative analysis of a single analyte. Many of the sensor-array applications (just identification) described in the literature [343]–[351] involve problems that could have been solved by much simpler means, such as exact density measurement, possible now with extreme accuracy. Differentiation among various samples with similar compositions in the context of a closed system (i.e., without the variability associated with an open matrix) can usually be achieved more economically on the basis of simple physical measurements. Nevertheless, electronic noses and tongues on the market have found some application despite the humidity interference of the first category in certain food processing processes, in which they sense any deterioration in the process (e.g., a rotten tomato in ketchup processing or a bad coffee bean in the roasting process. For those applications in the gas phase the ICB Muenster recently introduced a novel electronic nose based on the new sensor class of conductometric gas microsensors (polymer + plasticizer + organic salt). It has already been mentioned that the permutation of polymer, plasticizer, and organic salt together with their concentration allows the construction of an indefinite number of different membranes with different selectivities. This is shown in Figure 73.

The identifying power of an array of only five of such sensors is shown in Figure 74 [144], [352]. The intended application area was the detection of BETX (benzene, ethylbenzene, toluene and the different xylenes) near gasoline stations for reasons of industrial hygiene. The advantages of this electronic nose using only one type of sensor principle is the low energy consumption (conductivity measurements at room temperature) making handheld devices feasable and the fast exchange of the sensor array in the form of a cheap sensor card not larger than a business card [353].

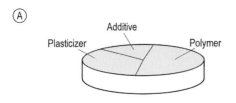

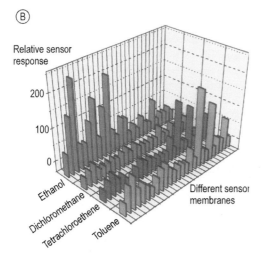

Figure 73. General composition of gas selective conductometric polymer membranes (A) and selectivity pattern against several organic solvents (B).

The above mentioned electronic nose was further improved by incorporating an extremely fast enrichment and water elimination step (in the seconds range) and increasing the resolving power following the time course of a fast thermal desorption step. Since the air sampling time, the desorption time and the number of conductometric gas sensors can be individually set, a very versatile and extremely sensitive electronic nose (called: Air-Check) with sensitivities in the ppb range and almost no interference by humidity was developed in the ICB Muenster. It will be positioned in clean rooms for the purpose of surveillance.

28.3. Biochemical Sensors (Biosensors)

28.3.1. Definitions, General Construction, and Classification

Biosensors are in general small devices based on a direct spatial coupling between a biologically active compound and a signal transducer equipped with an electronic amplifier (Fig. 75). This definition of an I.U.P.A.C. working group is unfortunately very specific. The crucial thing here is the demand for a "direct spatial coupling" of biomolecules and transducers. This might have influenced the development of further commercial available biosensors since much time has been spent on biomolecule immobilization and stabilization oriented towards biosensors that could be transported and stored without a need for a refrigerator. The intimate contact of the less stable biological recognition element with a transducer surface requires either one shot devices and throwing both away after one deteriorates or a reproducible surface rejuvenation with the biological receptor by the user. The latter is the problem since the user is not an expert in controlled coverage of surfaces. Concerning the life and working time and ease of exchanging the more labile biological recognition element, devices which separate the recognition space from the detection space show certain advantages though they cannot be named a biosensor. It is, e.g., quite feasible to employ the labile biomolecules with or without the required co-factors or reagents in a freeze-dried state in a cartridge to be activated only shortly before its use. This cartridge is then to be used upstream of an optical, electrochemical or calorimetric flow through cell. Keeping this cell for longer use allows more sophisticated constructions. With modern microsystem technology the whole device could be rather small and difficult to visualize at a first glance in that the demanded intimate contact for being a biosensor is missing.

Biological systems at various levels of integration are used to recognize a particular substance that is to be determined in a specific way. Different types of biologically sensitive materials can be used for selective recognition, including enzymes, multienzyme systems, organelles, photosensitve membranes from plants and bacteria, membranes generally, protoplasts, whole intact cells, tissue slices, antibodies, lectins, DNA, and transport protein or receptor systems isolated from cell membranes. The macromolecular biological compound is ordinarily immobilized in close proximity to the transducer surface, thereby facilitating direct or mediated signal transfer to the transducer. Besides fixing the receptor on the transducer surface this immobilization also serves the important function of stabilizing the biological material.

The first indicating step is specific complex formation between the immobilized biologically

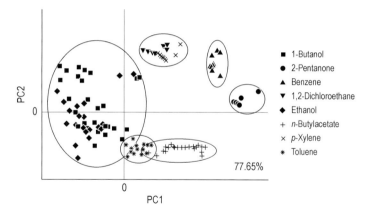

Figure 74. Identifying important BETX compound in the ppm range with only five conductometric gas sensors (ICB-Muenster patent) via principal component analysis (PCA), preliminary results [352]

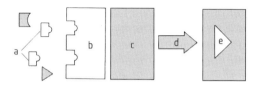

Figure 75. The basic principles of a biosensor
a) Analyte; b) Biological receptor; c) Transducer; d) Electrical signal; e) Signal processing

active compound (in a broader sense, the receptor R) and the analyte (A). This selective and usually reversible interaction produces secondary chemical or physical signals that are recognizable by a suitable electrochemical, optical, enthalpimetric, conductometric, or piezoelectric transducer. The transducer of choice depends on the nature of the molecular-level binding affinity with respect to the analyte, which may cause changes in such parameters as the redox state of the receptor compounds, the concentrations of electrochemically or optically active substances that are consumed or produced, or heat, mass, or electromagnetic radiation (see Table 17) [354]–[357].

Depending upon the mechanism of biochemical interaction between the receptor and the analyte, two basic types of biosensors can be distinguished: *biocatalytic* (metabolic) sensors, and *bioaffinity* sensors (Table 18).

Among the analytes detectable with biosensors are carbohydrates, organic acids, alcohols, phenols, amines, heterocyclic compounds, proteins, enzymes, cofactors, inorganic ions, hormones, vitamins, peptides, drugs, xenobiotics, microorganisms, gases, and pesticides. Analyte concentrations measurable with various types of biosensors range from 10^{-7} mol L^{-1} (biocatalytic sensors) to as little as 10^{-15} mol L^{-1} (affinity sensors). Based on their level of integration, biosensors can be subdivided into three generations:

1) *First Generation.* The receptor is entrapped between or bound to membranes, and the combination is fixed on the surface of an appropriate transducer or known chemical sensor.

2) *Second Generation.* The receptor is bound covalently to the transducer's surface, thereby eliminating the need for a semipermeable membrane.

3) *Third Generation.* The receptor is bound to an electronic device that transduces and amplifies the signal (e.g., the gate of a field-effect transistor, an array of amperometric ultramicroelectrodes, or interdigitated microelectrodes), utilizing the techniques of microelectronics and micromechanics.

It should be mentioned that this classification is not also meant to be an order of increasing performance. In general rugged systems have more advantages than complicated and expensive constructions.

28.3.2. Biocatalytic (Metabolic) Sensors

Metabolic biosensors are based on the specific recognition of some enzyme substrate and its subsequent chemical conversion to the corresponding product(s). The biocatalytic reaction itself accomplishes specific binding between the active site of

Table 17. Transducer systems used in the design of biosensors

Transducer	Indicating a change in:	Caused by:
Amperometric electrodes	electron transport	enzyme-catalyzed oxidation or reduction of an analyte
Potentiometric electrodes, ISFETs, MOSFETs, conductometric devices	ion concentration	enzyme-catalyzed oxidation, hydrolysis, or cleavage of C–C, C–O, C–N, or other bonds
Calorimetric devices (thermistors, thermopiles)	enthalpy	enzyme-catalyzed reactions
Piezoelectric crystals	mass	bioaffinity reaction (between Ab–Ag, Ab–hapten, an analyte and receptor protein, or analyte and lectin
Fluorometric fiber optics	fluorescing concentration of enzyme substrates/products or immunolabels	enzyme and bioaffinity reactions
Waveguides	optical path (adlayer)	bioaffinity reaction
Devices based on surface plasmon resonance	resonance (adlayer)	bioaffinity reaction

Table 18. Biosensor specification according to the nature of the primary sensing reaction [356]

Biocatalytic (metabolic) sensors		Bioaffinity sensors	
Indicating reaction:*			
$E \rightarrow A$ RA $E \leftarrow P$		$R \rightarrow A$ RA	
Analytes	Receptors	Analytes	Receptors
Substrate	enzyme	protein	dye
Cosubstrate	enzyme system	glycoprotein	lectin
Cofactor	organelle	cells	antigen
Prosthetic group	microorganism	antigen/hapten	antibody
	tissue slice	antibody	RNA/DNA
		neurotransmitter	receptor protein
		pesticide	
		enzyme (system)	

*E = Enzyme; A = Analyte; P = Product; R = Receptor; RA = Receptor–analyte complex.

the enzyme and the substrate (recognition process), conversion of the substrate to product, and release of the product from the active center, which in turn leads to regeneration of the binding affinity and catalytic state of the enzyme.

The linear measuring range for metabolic sensors depends on the "apparent" Michaelis–Menten constant (K_{Mapp}) of the immobilized enzyme (not free in solution and approachable from all directions, thus, "apparent"), which describes a particular substrate concentration associated with half the maximum rate of the enzyme reaction. The reaction rate of an immobilized receptor enzyme is controlled by the rate of substrate and product diffusion through both the semipermeable membrane covering the enzyme and the layer of immobilized enzyme itself, resulting in a higher K_M value than that anticipated under conditions of kinetic control. Enzymes, coupled enzyme sequences, organelles, microorganisms, or tissue slices can all be used as receptor materials for metabolic sensors. Under certain conditions, cosubstrates, effectors (inhibitors, activators), and enzyme activities can also be determined by metabolic sensors.

28.3.2.1. Monoenzyme Sensors

Enzyme sensors rely on an immobilized enzyme as the biologically active component. Many biocatalytic reactions, especially those of oxidoreductases, hydrolases, and lyases, are associated with the consumption or formation of electroactive substrates or products in concentrations

Table 19. Common configurations for monoenzyme sensors

Enzyme class	Detectable compound involved in the enzyme reaction	Typical transducer
Oxidoreductases		
Oxidases	O_2, H_2O_2	pO_2 electrode, noble-metal–carbon redox electrode
Oxygenases	O_2, quinones, catechol	pO_2 electrode, (amperometric) carbon electrode
NAD(P)-dependent dehydrogenases	NAD(P)H/NAD(P)	(amperometric) mediator-modified redox electrode, fluorometer, photometer
PQQ-dependent dehydrogenases	PQQ/PQQH$_2$	
Reductases	NAD(P)H/NAD(P) Heme-, Cu-, Mo-, FeS-containing centers	
Hydrogenases	H_2, NAD(P)H, NAD(P), FADH$_2$, FMNH$_2$, FeS-, Ni-containing centers	
Hydrolases	NH$_3$/(NH$_4$)	(potentiometric) ammonia-sensitive electrode
	CO_2	(potentiometric) pCO_2 electrode
	H^+	pH electrode, ISFETs, coulometer, conductometric and enthalpimetric devices
	redox-active products (phenolic, thio-group-containing compounds)	(amperometric) redox electrodes

proportional to the analyte concentration. For this reason most commercially available biosensors are enzyme sensors based on electrochemical transducers.

Typical electrochemically detectable (co-) substrates and products include oxygen, hydrogen peroxide, hydrogen ion, ammonia, carbon dioxide, reduced or oxidized cofactors, and redox-active (oxidized or reduced) prosthetic groups in oxidoreductases, all of which can easily be converted into electrical signals by suitable transducers (e.g., amperometric or potentiometric electrodes or conductometric sensors, Table 19).

Immobilization of an enzyme in or at an artificial matrix or a membrane usually stabilizes the biocompound, thereby increasing the half-life for activity. Techniques of enzyme immobilization include physical or chemical adsorption, ionic and covalent bonding (perhaps to functionalized transducer surfaces), cross-linking by bifunctional agents, and entrapment in such polymer matrices as natural or artificial gels and conducting polymers. The first biosensor was described by CLARK and LYONS in 1962, who introduced the term "enzyme electrode" [358]. In this first enzyme sensor, an oxidase enzyme in a sandwich membrane was placed next to a platinum redox electrode. A platinum anode polarized at +0.6 V (vs. SCE) responded to peroxide produced by the resulting enzyme reaction with the substrate. Application of a polarization voltage of -0.6 V permitted the amperometric detection of oxygen consumption. The primary target substrate was β-D-glucose (Fig. 76). Stoichiometric turnover of the glucose oxidase-catalyzed reaction yields a sensor response proportional to the glucose concentration up to the apparent K_M value of the immobilized enzyme. This in turn led to the first glucose analyzer for measuring glucose in dilute blood on the basis of an amperometric enzyme sensor. A commercial analog, the Yellow Springs Instrument Model 23 YSI, appeared on the market in 1974. The same indicating technique has more recently been applied to many other oxygen-mediated oxidase systems, which constitute the most widely exploited enzyme classes for the construction of biosensors.

Amperometric Biosensors Using Artificial Electron Mediators. Amperometric biosensors for monitoring oxidase-catalyzed oxygen consumption or the production of hydrogen peroxide are, under certain circumstances, restricted in their use because of a limited availability of dissolved oxygen. This may necessitate careful pretreatment of the sample with rigorous exclusion of oxygen-consuming compounds, as well as application of a high polarization voltage, which may cause interfering currents in complex matrices.

It should be possible to overcome such problems by direct electron transfer between the redox-active prosthetic group of an enzyme and the elec-

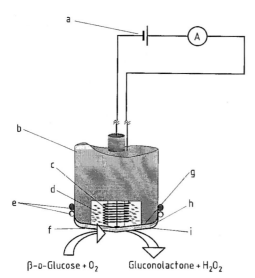

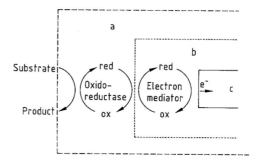

Figure 76. Configuration of an amperometric enzyme electrode for glucose determination
a) Polarization voltage (−600 mV or +600 mV vs. Ag/AgCl); b) Electrode body; c) Ag/AgCl pseudo-reference electrode; d) Electrolyte solution; e) O-Rings; f) Platinum working electrode; g) O_2/H_2O_2-permeable membrane; h) Glucose oxidase (immobilized layer); i) Semipermeable membrane

Figure 77. Operating principle of a mediator-modified amperometric biosensor, involving an oxidase or a dehydrogenase (in the case of a reductase, electron transfer would proceed from the electrode to the analyte substrate)
a) Amperometric enzyme sensor; b) Mediator-modified electrode; c) Redox electrode

trode surface. In a few cases, especially ones involving small redox proteins such as cytochrome c, efficient electron transfer has been achieved in this way with promoters that direct the active redox centers to the electrode surface. Macromolecular redox proteins exhibit rather low rates of heterogeneous electron transfer because according to the Marcus theory tunneling is not possible over the long distance from the active center to the electrode surface. Thus, the transfer is often accompanied by a large overpotential, but the use of redox mediators may increase the electron-transfer rate dramatically. These low-molecular-mass (in)organic redox species undergo facile reversible redox reactions with both the prosthetic groups of the enzyme and the electrode surface according to the general scheme depicted in Figure 77. The resulting electron shuttle reduces the overvoltage of the biological redox system to the formal potential of the mediator. Suitable redox mediators are redox systems which show a high exchange current density and include quinoid redox dyes, quinones, organometallic compounds, and fulvalenes (see Table 20) [359]–[363].

In constructing such a biosensor the redox mediators can be immobilized at the electrode surfaces in various ways, including chemisorption or covalent bonding, chemical or physical deposition of a polymer film containing the mediator molecules, or physical incorporation of the redox mediators (which are only slightly water soluble) directly into the electrode material. One ingenious approach involves covalent binding of the mediator to the electrode via an appropriate *spacer*, or electrical "wiring" of the enzyme through an electron-relaying redox polymer network. In both cases electron transfer from the redox centers to the electrode surface is thought to occur by electron tunneling [364]–[366]. The use of a redox mediator increases the dynamic range of a sensor, minimizes interferences, and provides an indicator reaction that is usually independent of the oxygen content of the matrix. Furthermore, such mediators undergo reversible redox reactions with redox-active coenzymes, dehydrogenases, reductases, and hydrogenases, as well as the redox-active products of enzyme-catalyzed reactions, which opens many new possibilities for incorporating these classes of oxidoreductases into new types of biosensors.

28.3.2.2. Multienzyme Sensors

Coupled enzyme reactions mimic certain metabolic situations in organelles and cells that have themselves already been exploited in biosensors [367], [368]. The use of enzyme sequences facilitates:

1) Broadening the scope of detectable analytes
2) Improving selectivity
3) Increasing both sensitivity and dynamic range
4) Eliminating interferences

Table 20. Redox mediators leading to efficient electron transfer with redox proteins

Mediator type	Examples	Structures
Quinones	benzoquinone	(benzoquinone structure)
Quinodimethanes	tetracyanoquinodimethane (TCNQ)	(TCNQ structure)
Tetrathiafulvalene (TTF) and its derivatives		(TTF structure)
Metallocenes and their derivatives	ferrocene	(ferrocene structure)
Phthalocyanines (Co, Fe, Mn, Cu)	cobalt phthalocyanine (Co-PC)	(Co-PC structure)
Quinoid redox dyes		
Phenazines	N-methylphenazonium (NMP^+) salts	(NMP structure)
Phenoxazines	meldola blue	(meldola blue structure)
Phenothiazines	methylene blue	(methylene blue structure)
Indamines	toluylene blue	(toluylene blue structure)
Viologens	methyl viologen	(methyl viologen structure)
Complex compounds		
Ruthenium complexes		$[Ru(NH_3)_6]^{3+/2+}$, $[Ru(en)_3]^{2+/3+}$
Osmium complexes		$[Os(bpy)_2Cl]^{3+/2+}$
Cobalt complexes		$[Co(bpy)_3]^{3+/2+}$, $[Co(en)_3]^{2+/3+}$
Iron complexes		$[Fe(CN)_6]^{3+/4+}$, $[Fe(phen)_3]^{2+/3+}$

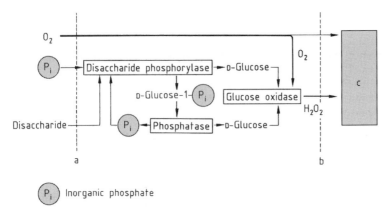

Figure 78. Enzyme sequence for the detection of inorganic phosphate, based on the principle of chemical signal amplification [369]
a) Dialysis membrane; b) H_2O_2/O_2 permeable membrane; c) Working electrode

Depending on the target analyte, sample matrix, required measuring range, etc., enzymes can be coupled to produce reaction sequences operating either in series or in parallel. An alternative approach to realizing multienzyme sensors is the use of cell complexes, whole cells, or tissue slices comprising physiologically integrated enzyme systems (see below).

Enzyme Sequences Operating in Series. The number of substances that can be determined directly by monoenzyme sensors is limited, because many enzyme-catalyzed reactions involve no physicochemically detectable substances. To form readily detectable compounds it is often necessary, therefore, to couple together several different enzyme reactions in a sequential way. The analyte of interest is first converted by some specifically catalyzing enzyme into a product that is itself convertable in the course of one or more subsequent enzyme reactions involving detectable secondary (co)substrates or products. Biosensors based on such enzyme sequences have been developed for the determination of many substrates, including disaccharides (sucrose, maltose, cellulose, lactose), cholesterol, creatinine, and nucleic acids [356].

With certain enzyme sequences it is possible to achieve chemical signal amplification through the accumulation of a particular product by cycling the substrates or cofactors. The sensitivity of some biosensors can be enhanced in this way by as much as three orders of magnitude. Signal-amplifying enzyme systems have been based on oxidase–dehydrogenase, dehydrogenase–transferase, and phosphorylase–phosphatase–oxidase sequences, permitting the determination, for example, of lactate, ATP, and phosphate at the picomole per liter level (Fig. 78) [357]–[369].

Enzyme Sequences Operating in Parallel. Multienzyme sequences operating in parallel are used for the elimination of interfering agents, expansion of the dynamic measuring range, and (in principle) the indirect detection of an electrochemically inactive reaction catalyzed by one of the enzymes in a substrate competition. *Anti-interference* enzyme systems are the most important. Substances interfering with an indicating reaction can be eliminated as one way of improving the analytical quality of the measurement by covering the immobilized enzyme with an anti-interference enzyme layer, in which a second enzyme converts the disturbing substance into an inert product. This strategy has made it possible to remove such interfering agents as ascorbic acid and endogenous glucose [367].

Biosensors Based on Integrated Biological Enzyme Systems. Cellular organelles and membranes, as well as whole cells or tissue slices from animal or plant sources comprising physiologically coupled multienzyme systems are often employed as receptor materials in biosensors based on electrochemical detection. Native physiologically integrated enzyme systems display enhanced stability due to their in vivo environment. Biological material of this type offers the potential for low cost and simple preparation. Furthermore, this opens the way to analytes that are converted to electrochemically detectable products not by simple enzyme reactions, but rather by complex multienzyme reactions within an intact biological structure. Analytes subject to such receptor systems include chloroaromatics, caprolactam, naphthalene, tryptophan, several hormones, and acetic acid [354], [370].

28.3.2.3. Enzyme Sensors for Inhibitors – Toxic Effect Sensors

Sensors for inhibitors are based on competitive inhibition between a substrate and an inhibitor for an enzyme binding site, inhibition of an enzyme reaction by product accumulation, irreversible binding of an inhibitor to an active binding site, or redox-active heme or SH groups of enzymes that lead to the blocking of enzyme activity.

When thylakoid membranes, protoplasts, or whole cells are used as receptor materials, various inhibitors cause metabolic "short circuits," or inhibition of the (photosynthetic) electron-transport system, which results in a decrease in such metabolic parameters as oxygen consumption or production, photocurrent development, dehydrogenase activity, or fluorescence. In contrast to metabolic sensors, the first sensing step here is a result of binding of the inhibitor to an active binding site or redox center on the enzyme. Sensors of this type usually operate at low enzyme-loading factors and with substrate saturation in order to ensure a highly sensitive signal of limitation in the kinetically controlled reaction due to the presence of an inhibitor. Examples of such inhibitors (analytes) include toxic gases, pesticides, chloroorganic compounds, phenols, and heavy-metal ions [371], [372]. To detect toxic effects on vital enzymes in a reproducible manner, to include also synergistic or antagonistic effects, is only possible with biosensors or bio-tests. This advantage as a fast warning device is still not fully used. Too often biosensors are developed as if they should compete with instrumental analysis. Given the pace of development in this area, especially concerning miniaturization, speed, and versatility, the biosensor as a mono-analytical device will have problems in gaining a larger market share. However, as a toxic warning device triggering some screening and sampling it is without competition.

28.3.2.4. Biosensors Utilizing Intact Biological Receptors

Molecular receptors are cellular proteins (often membrane-bound) that bind specific chemicals in such a way as to produce a conformational change in the protein structure. This conformational change in turn triggers a cellular response: opening of an ion channel, for example, or secretion of an enzyme. In general, molecular receptors can be distinguished from other types of receptors (including larger multicomponent systems) on the basis of their composition, which is a single protein, though this protein may contain more than one subunit. Among the important receptors known to date are receptors for hormones (control agents for a wide range of cellular and body functions), amino acids, insulin, and neurochemical transmitters, in addition to receptors capable of binding synthetic bioactive chemicals, such as drugs [373], [374].

Molecular receptors have two important properties relevant to their possible incorporation into biosensor devices: intrinsic signal amplification and high specificity. The specificity is a consequence of a highly evolved binding region within the receptor that is fully optimized for binding one particular ligand by a variety of electrostatic, hydrogen bonding, van der Waals, and hydrophobic forces.

Limitations affecting the study and use of molecular receptors include:

1) Relatively difficult experimental procedures for the isolation of the receptor proteins (even prior to purification), which are both time-consuming and expensive, and often require animal sources
2) An absence of techniques for obtaining more than very small amounts of pure receptors
3) The rapid loss of biological function that usually follows isolation

Molecular biology has brought considerable progress in the isolation and purification of receptors, while advances in artificial membrane technology and reconstitution have recently reduced the problem of instability.

Comparatively few biosensors of this type have been reported [373], [374], primarily because of problems associated with the successful isolation of molecular receptors (in sufficient quantity and purity) and their subsequent stabilization in artificial environments in such a way that appreciable biological activity is retained.

A clear exception, however, is a fiber-optic sensor based on evanescent fluorescence developed by ELDEFRAWI et al. [374], [375] that contains nicotinic acetylcholine receptor (nAChR). This was the first neurotransmitter receptor to be purified and characterized in vitro, and it is involved in chemical signalling. The sensor in question consisted of nAChR immobilized on the surface of a quartz fiber. Determination of the concentration of acetylcholine and other natural neurotransmitters (e.g., nicotine) was accomplished by measuring the extent to which they inhibited the binding of a standard solution of a fluorescence-labeled toxin (fluorescein isothiocyanate-α-bungarotoxin, FITC-α-BGT). Binding determinations were based on measurements of fluorescence intensity in the region of the evanescent wave (that is, at the fiber–solution interface, see Section 28.2.3.2.1), where the nAChR was immobilized. The sensor had an acceptable signal-to-noise ratio of 10^2 and an analysis time < 10 min (based on the initial rate of signal increase rather than a steady-state value), and it was capable of operating under conditions of high ionic strength. However, it suffered from the problem of non-regeneration of the sensing element because of the irreversibility of FITC-α-BGT binding to the nAChR.

A similar receptor-based biosensor using electrochemical detection was reported for the detection and measurement of riboflavin (vitamin B_2) [373], [374]. This device is based on the competition for aporiboflavin-binding protein (apoRBP) between riboflavin present in the sample subject to analysis and a riboflavin analogue bound in a membrane. In the absence of riboflavin all the apoRBP is bound to the analogue on the membrane. When riboflavin solution is added, the fivefold greater affinity of apoRBP for its natural cofactor (riboflavin) results in displacement of apoRBP from the membrane into the solution, causing a change in the electrical potential of the membrane. A riboflavin measurement range of $0.1 - 2 \times 10^{-6}$ mol/L was reported.

Further progress in molecular receptor-based sensors will be a function of advances in molecular biology with respect to enhancing the quantity, purity, and stability of receptors that can be used in vitro.

28.3.3. Affinity Sensors – Immuno-Probes

Affinity sensors take advantage of physicochemical changes that accompany complex formation (e.g., changes in layer thickness, refractive index, light absorption, or electrical charge). Once a measurement has been made the initial state must be regenerated by splitting off the complex. Because of this they cannot be used for continuous monitoring nor to follow declining concentrations of the analyte. Therefore, they cannot be named biosensors into the strict sense of the I.U.P.A.C. definition to be "reversible". Thus, they are named immuno-probes. Since the splitting of the complex can be very fast (e.g., with pH < 2.0 or chaotropic reagents) and with the process being automated (e.g., in FIA-systems) the response time of an immuno system can reach less than 1 min and, thus, becomes comparable to enzymatic biosensors working under diffusion control.

Examples include immunosensors utilizing antibodies or antigens for recognition purposes, lectins, and "true" biological receptors in matrix-bound form (see Section 28.3.2.4).

Immunosensors Based on Antibody Molecules. Biosensors based on the antigen–antibody interaction are called *immuno-probes*. The high specificity and high sensitivity that typifies an antigen–antibody (Ag–Ab) reaction has been used in a vast range of laboratory-based tests incorporating antibody components (immunoassays). The analytes detected and measured have included many medical diagnostic molecules, such as hormones (e.g., pregnancy-related steroids), clinical disease markers, drugs (therapeutic and abused), bacteria, and such environmental pollutants as pesticides. A crucial distinction must be made between an immunoassay and an immuno-probe, however, with respect to the technology relevant to the analysis. The ultimate aim is a low-cost, disposable immuno-probe, with the entire device being used only once. For example, it has now proven possible in practice to develop an instrument containing a disposable "sensor chip" that can be used for Ab or Ag tests.

Two different types of immuno-probes have been described:

1) *Direct immunosensors* register immunochemical complex formation at the transducer surface via electrochemical, mass or optical changes. The advantage of a direct immunosensor is that measurement of the antigen–an-

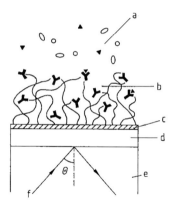

Figure 79. Operating principle of an SPR immunosensor
a) Bulk solution; b) Carboxymethylated dextran matrix;
c) Linker layer, HO–(CH$_2$)$_n$–SH; d) Gold film; e) Glass substrate; f) Light
(with permission from GEC Journal of Research, Essex)

tibody interaction can be accomplished immediately without any need for additional antibodies or markers (enzymes, fluorescent labels, etc.).

2) *Indirect immunosensors* are based (like homogeneous and heterogeneous immunoassays) on labeling of one of the partners in an immune interaction. One class of indirect immunosensors derived mainly from enzyme immunoassays (EIA) encompasses the enzyme immunosensors. These combine the high selectivity of an immunoreaction with the high sensitivity characteristic of the enzymatic amplification effect. However, the enzyme can affect the molecular recognition step because of its large size and needs in general special attention (cooling) during storage. In the case of the widely used assay technique of Enzyme Linked Immuno Sorbent Assay (ELISA) the use of enzymes generates two further manipulations: a) the addition of a suitable sustrate and b) the addition of a reagent which stops the enzymatic reation. This is not necessary when using modern fluorescence markers.

28.3.3.1. Direct-Sensing Immuno-Probes without Marker Molecules

Among direct immunosensors the most favored are optical sensors, because they incorporate the advantages of fiber optics or optoelectronic transducers. The nonelectrical properties of optical systems offer important safety advantages for medical, industrial, and military applications. Remote sensing is clearly advantageous, for example, for explosives and toxins. Optical-fiber probes are also mechanically flexible, small, inexpensive, and disposable.

The detection of an evanescent field in an optical waveguide (see Section 28.2.3.2.1) is a highly sensitive technique especially suitable for monitoring the antibody – antigen interaction [373], [374]. One partner of the immuno pair (e.g., the antibody) is immobilized at the surface of the waveguide, and its reaction with the other partner in the test sample is monitored. An evanescent wave generally penetrates a distance of several hundred nanometers extending the full length of an antibody – antigen complex, so in contrast to conventional techniques there is no need to carry out a prior separation of the non-specific biological components. There is no need for washing steps.

The technique of surface plasmon resonance (SPR) represents a very sensitive way to follow an immuno-reaction on an interface, and its feasibility has been adequately demonstrated [376]. As noted previously (Section 28.2.3.2.3), SPR is a phenomenon produced by a beam of light directed onto a glass – metal interface, usually a glass prism in contact with a layer of gold or silver (Fig. 79). At some specific angle (the resonance angle) a component of the electromagnetic light wave propagates in the metal along the plane of the interface, taking the form of surface plasmons. The resonance angle is quite sensitive to changes in refractive index and dielectric constant at the interface up to a distance of several hundred nanometers from the actual metal surface, with the sensitivity decreasing exponentially as a function of distance from the surface. Immobilization of an antibody on the surface causes a measurable shift in the resonance angle, and binding of a high-molecular antigen to the immobilized antibody leads to a further change. For typical biological systems this binding-induced shift in resonance angle (expressed in resonance units) is approximately linearly proportional to the concentration of bound antigen (or antibody if antigen has been pre-immobilized).

The SPR-based biosensor systems BIAcore and BIAlite, developed by Pharmacia Biosensor AB and now available commercially, represent a breakthrough in immunosensor technology, particularly in the use of SPR techniques. Apart from high cost, the only limitation associated with this technique is that the sensitivity depends on the molecular mass of the adsorbed layer, which in turn controls the optical thickness, so low concentrations of small molecules (molecular mass < 250, including most haptens) are unlikely to be measurable directly without additional labeling. In addition to SPR, Mach–Zehnder interferometers [377], RIFS [201], [202], and so-called grating couplers [378] are also used for optical immunosensing. An interesting step towards miniaturization was introduced by the development of practicable fiber optical SPR systems by JORGENSON AND YEE [379] and KATERKAMP et al. in the ICB Muenster [380], [381]. The latter device was so small and low-cost that application as a hand-held refractometer with a resolution in the fifth digit after der decimal point was feasible. The fiber optical SPR tips were rapidly exchangeable and bending of the fiber had no negative influence on the result.

All the above mentioned transducers measure changes in the refractive index and layer thickness together! With respect to quantitative assays one should bear in mind that those techniques only sense a mean layer thickness. Thickness extensions that occur because of large macromolecules are leveled out. Futhermore, the change of the refractive index inside the evalescent field (within the surface layer) may also depend on ions or water molecules entrapped in the proteins. The latter is dependent on the ionic strength. Thus, changes in the ionic strength of the sample solution may also produce a signal. In addition, if antibody molecules are immobilized on the transducer surface, how can anyone differentiate between an immuno-reaction with the analyte or a change in the tertiary and quarternary structure of the antibody molecule caused by other effects giving rise to a change in the thickness of the layer?

Alternative transducer technologies that have been applied in immunosensor research include piezoelectric and electrochemical systems. *Piezoelectric* immunosensors [383], [384] (see Section 28.2.3.3) tend to suffer from significant levels of nonspecific binding to the piezoelectric substrate, which makes accurate analyte quantification difficult. However, this type of immunosensor may have an important role to play in the search for traces of volatile drugs in security monitoring, and it offers the advantages of relatively small size and low manufacturing cost. The main disadvantage from the standpoint of immunosensing is that reproducible measurements have so far been achieved only in the gas phase, whereas the Ab–Ag interaction is restricted to aqueous solution. As has already been mentioned above, working with mass-sensitive transducers in liquids gives rise to doubts about the correct interpretation of frequency shifts. Here, the same as in case of the optical transducers, they need to be differentiated from real mass changes. Any structural changes of large protein molecules which induce associated changes of entrapped water and ions will result in a measurable signal. Despite the fact that these frequency changes have nothing to do with specifically bound analyte molecules these signals can be very reproducible. Without a thorough and convincing validation with real samples, the analytical accuracies are questionable. This is not to say that qualitative binding studies or determination of association constants could not be measured. All transducers needing no label compound are ideal for screening purposes. They are mostly used for the detection of specific receptor functions.

Potentiometric immunosensors [356] (see Section 28.2.3.1.1) are unfortunately associated with a low signal-to-noise ratio as a result of the low charge density on most biomolecules relative to such background interferences as ions. These sensors also show a marked dependence of signal response on sample conditions, including pH and ionic strength. Similarly, *immunoFET* devices (see Section 28.2.3.1.4), which measure very small changes in an electric field when an Ab–Ag binding reaction occurs, tend to suffer from practical problems associated with membrane performance and various artifacts. The theoretical basis of the sometimes observed voltage change remains unclear. The only plausible explanation for the fact is that clearly the binding of a neutral antigen molecule to a large antibody molecule with multiple charges (depending on the pH and isoelectric point) may change the charge distribution at the interface or may lead to a different mixed potential. If the whole immuno-reaction producing an analyte proportional signal is not watched carefully and controlled by convincing validation studies, a certain bias can easily be overlooked by all so called direct sensing techniques.

Main Drawback of all Direct Sensing Techniques. It has become a well known fact that the sensitivity (or LOD) of all direct sensing techniques is not any signal-to-noise ratio nor any theoretical mass-loading sensitivity! It is the lack of control over non-specific binding processes, which are not caused by the very specific molecular interaction under study. All direct sensing immuno-techniques cannot differentiate between competing surface processes leading to adsorbed particles with various heats of adsorption. Even if working with very hydrophilic surfaces to prevent this, the ultimate LOD of such an immuno-probe depends, according to the laws of Analytical Chemistry, on the standard deviation of such non-specific binding (LDO = 3 × σ_{blank})! This renders many extreme sensitive transducers useless for practical applications in real world samples.

Amperometric or fluorescence-based immuno-technology (immuno-electrodes or capillary fill devices with a fluorescence label, for example) has shown more promising results at the research prototype stage than other types of transducers. One preferred technique continues to be the indirect amperometric immunosensor approach described below, which uses enzymes as markers. The second emerging technique is based on fluorescence dyes which are now commercially available together with a simple covalently binding protocol (one to two steps only). Fluorescence can be as sensitive as enzyme-based amperometry especially if modern fluorophores are used with excitation wavelengths in the near IR and the emission shifted about 100 nm to higher wavelengths. Around 600–800 nm is an observation window for biological systems since no natural fluorescence (e.g., tryptophane) occurs in that region. Furthermore, cheap laser diodes of around 680 nm with high energy are available.

28.3.3.2. Indirect-Sensing Immuno-Probes using Marker Molecules

The most satisfactory marker molecules for use in this context have been fluorophores and enzymes. As regards the sensors themselves, several fundamental types exist.

Reactor immuno-systems are based on a cartridge or microcolumn packed with appropriate immobilized antibodies (which serve as so-called catching antibodies). As in the case of enzyme immunoassays, the competition, sandwich, and replacement principles are all subject to exploitation. The amount of bound (or replaced) antigen is determined via the extent of enzyme activity remaining after substrate addition, which can be established with an electrochemical or optical sensor. For many applications (on-line measurements, coupling with flow-injection analysis), repeated and automated measurements are required. Because of the extremely high loading of catching antibodies, reactor immunosensors have the advantages of high sensitivity, regenerability, and reproducibility. On average they can be reused 50–100 times without loss of activity. The US Naval Research Laboratories were very successful with respect to sensitivity, selectivity, reliability, and simplicity with several prototypes of reactor-based systems they developed under the guidance of Frances Ligler [382]. The developed immuno-systems allow field determinations and have been extensively tested and validated in the field of environmental analysis. In this respect, it is worthwhile to mention that the US EPA has adopted immuno-kits for certain important analytes in the environment, e.g. gasoline spills. If the fluorophore is carefully selected sensitivities in the ppb range with a FIA-like set-up and response times under 1 min are possible!

Membrane immuno-probes employ antibodies immobilized directly at the surface of a sensor, or at a membrane covering the sensor, as shown in Figure 80. An example of the use of this type of immunosensor is illustrated schematically in Figure 81. The analyte to be detected is an early myocardial infarction marker, the fatty acid binding protein (FABP). A modified Clark electrode is covered by a membrane bearing immobilized monoclonal catching antibodies directed against FABP. After binding of the analyte and washing, a secondary polyclonal antibody is added to the solution. This secondary antibody is conjugated with glucose oxidase (GOD). The next step is incubation followed by further washing, after which glucose is added to the measuring solution. Glucose is oxidized by GOD in an amount directly proportional to the amount of FABP, yielding H_2O_2, the concentration of which is monitored with an amperometric sensor, as shown in Figure 82. This system was developed at the ICB Muenster with the aim of developing a rapid test for a myocardial infarct to be performed actually in the ambulance with only one drop of blood. It works fully automatically. One drawback was, however, the use of a labile enzyme. Despite the fact that GOD is the most stable enzyme the experts know today, some precautions had to be taken and also the need for substrate addition and stopping of the reaction needed computer-controlled timing.

One example of a nice commercial immunosensor is produced by Serono Diagnostics. This fluorescence-based evanescent-wave immunosensor [374], [375], [376] incorporates a novel capillary-fill design, and is shown in Figure 83.

The system consists of two glass plates separated by a narrow capillary gap of 100 mm. The

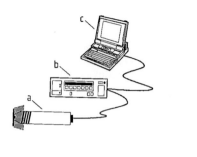

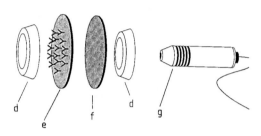

Figure 80. Design principle of an immunosensor system
a) Immunosensor; b) Electronic unit; c) Computing unit;
d) O-Rings; e) Catching antibody membrane; f) Dialysis membrane; g) Pt vs. Ag/AgCl electrode

lower plate acts as an optical waveguide and contains on its surface an immobilized layer of antibodies. The system benefits from its capillary-fill system, by means of which a fixed volume of sample is drawn into the space between the plates regardless of the volume of the bulk sample (a blood droplet for example). Delivery of a highly reproducible volume of sample eliminates a significant source of error in analyte measurement. The Serono immunosensing format constitutes a competitive assay, though a sandwich-type assay could also be used. The upper plate contains on its surface a layer of fluorescence-labeled analyte trapped in a water-soluble matrix. On addition of sample the labeled analyte is released into solution, where it competes for antibody sites on the lower plate with any unlabeled analyte present in the sample. After a fixed incubation period the fluorescence signal of the evanescent waves is measured and related to analyte concentration.

This technology does have shortcomings, however, some of which are related to the practical problems associated with carrying out measurements on whole blood. These include poor capillary flow because of the "stickiness" of blood compared with water, increased incubation time resulting from slow dissolution of the labeled an-

alyte matrix, and the presence of fluorescent and highly colored molecules in blood, such as hemoglobin. Drawbacks include the necessity for an incubation period of several minutes (due to Ag–Ab interaction) and increased manufacturing costs (each analyte to be tested requires a dedicated sensor, which involves labeling and immobilization prior to physical construction).

While immunological analysis with one of the different ELISA methods works very well and reliably in the laboratory, the fabrication of pre-calibrated small stick-like probes has usually failed until recently. The main problem was the lack of information one has about the surface density of active immunopartners and a way of reproducibly manufacturing those single-use devices with antibodies or analyte molecules bound to the surface. Thus, one or more calibration solutions to be run simultaneously on all surfaces with immobilized biomolecules were always necessary. Only if full control of the surface density and state (active or not, freely accessible or not, etc.) were be given pre-calibrated immuno-probes for such fields other than clinical diagnostic and environmental field analysis with quantitative results would be feasible.

The Integrated Optical System — IOS. The solution to overcoming the main drawbacks of all hitherto developed biosensors, with the exception of the glucose sensor (poor storage capability), is a single-use sensor chip on which the labile biomolecules could be kept active forever by freeze-drying them. The solution to overcoming the pre-calibration problems is to employ a kinetic data evaluation method. The solution to overcoming the mostly diffusion limited immuno-reaction at an interface is to reduce the Nernstian diffusion layer thickness significantly so that differences in the association kinetics become apparent. To overcome poor sensitivity the solution is to use stable fluorophores with high quantum yield and emission in the near IR. In order to overcome washing steps, the evanescent field method for excitation and back-coupling has to be employed used. The IOS-chip (developed at the ICB Muenster) consists of a plastic baseplate with a refracting index of about 1.50. The mostly aqueous samples lie in the range of about 1.33. Thus, if the interface plastic base sample solution is illuminated, total reflection takes place and the light is reflected back into the plastic chip material. However, the reflected light does not fall abruptly to zero in the sample phase but fades exponentially (evanescent field). The evanescent field reaches out into the

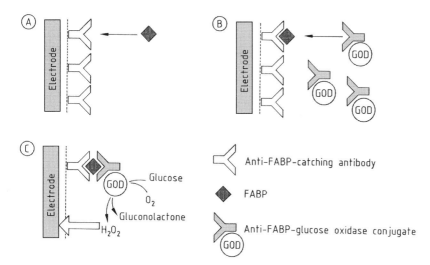

Figure 81. Schematic representation of the operation of an immunosensor for detecting the fatty acid binding protein (FABP)
A) Analyte binding step; B) Introduction of a secondary polyclonal antibody conjugated with glucose oxidase (GOD);
C) Determination of GOD activity after addition of glucose

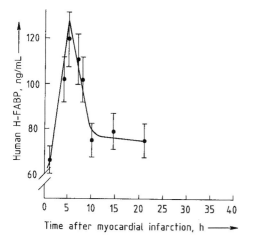

Figure 82. Analytical results obtained with the human heart FABP sensor described in Figure 81 as a function of time after onset of a myocardial infarction

sample medium only about 100 nm. The first 20 nm contain about 20 % of the optical power radiated onto the interface. The dimensions of most antibodies lie in a range between 5 and 7 nm. This means that only those fluorophore-labeled antibodies that are located directly on the surface are excited to fluorescence. Labeled but not specifically bound antibodies in the sample volume above the interface are not excited. Together with further improvements the IOS technology was pat-ented wordwide by the ICB Muenster (prior art: DE 19628002 A2/DE 19711281 A2 inventors: Katerkamp, Meusel, Trau).

The plastic base chip, onto which a diode laser beam (λ near the absorption wavelength of the fluorophore) is located, is directed from underneath. The corresponding antibodies or antigens are immobilized on the chip (about 30×10 mm) surface in a simple and straightforward batch process of several thousands or more. Then, in order to reduce the Nernstian diffusion layer thickness a very thin double adhesive spacer (about 50 µm or lower) and a cut-out sample flowing channel is put onto the side with the immobilized molecules. A plastic cover with two openings for the sample in- and outlet is put on forming the necessary narrow detection channel for the hydrodynamics. The sample is brought into the measuring channel by suction forces applied at the outlet within the apparatus. However, on the sample inlet side of the chip, an easy to change container (dwell) is located. On its wall all necessary reagents, buffer, etc. are freeze-dried. The type of recognition molecule is determined by the method. In general all known ELISA methods are feasible with higher speed and sensitivity. Of particular interest are generic chip designs in which, e.g., a labelled antibody (e.g., with a Cy5 kit) reacts with the antigen of the sample in the small dwell at the inlet and is then filtered into the latter via a membrane filter with immobilized antigen. Thus, the

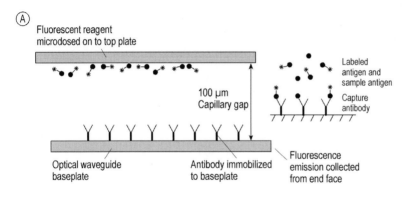

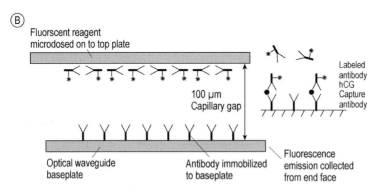

Figure 83. Principle of the patented Capillary-Fill Device. The upper part shows schematically a competitive immuno-assay; the lower part a sandwich assay. The optical waveguide is always the lower baseplate. Here a diode laser beam is used to produced an evanescent field in the sample inside the capillary

excess antibodies are not passing through since they are captured on the filter surface. However, the antigen–antibody complex reaches the optical channel after application of a slight vacuum on the outlet side. Here a catching layer for general antibody recognition makes the whole device generic! The dynamic measurement of the slope of the increasing fluorescence signal falling onto a simple optical detector below the chip is important. In the described operational mode this slope strictly depends on the concentration of the antigen. This slope can be measured in a few seconds! In cases of extremely low levels, the measuring time can be increased. By this kinetic evaluation and viewing of the whole active area the inactivation of receptor molecules on the surface is more or less evened-out. The slope depends much more on the concentration of the partners and the constant hydrodynamics.

This IOS technology has recently been licensed to several companies in the field of diagnostics and will change the future of biosensing because of its inherent advantages:

- no washing and separation steps are needed for performing solid phase immunoassays in this arrangement
- reduction in measuring time from several hours to a few seconds
- dramatic increase in sensitivity ($<$ ppb)
- automatic computer control of the whole process:
  1) insert the analyte specific sensor chip in the read-out unit
  2) fill the sample container (sample volume needed for slope determination $<$ 200 µL)
  3) start the measurement process (chip moves into read-out unit and the outlet is punctured (closed previously to prevent uncontrolled intake by capillary forces) and the constant metering pump forces the sample through the measuring channel.

The low-cost chip manufactured in mass-produced plastic injection molding can also contain several functional layers for optimum sample treatment:

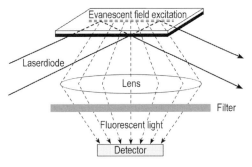

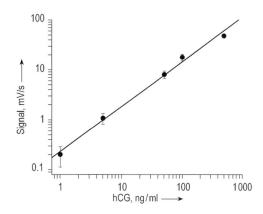

Figure 84. Optical set-up and construction of the IOS-chip. The laser light from a low-cost laser diode is directed towards the lower plastic baseplate of the immuno-probe chip. The evanescent field is produced within the flow channel produced by the spacer and the cover with sample entrance and outlet holes. The fluorescence of labeled molecules bound to the surface of the baseplate is collected, after passing an interference filter corresponding to the fluorescence wavelength, by a low-cost detector

Figure 86. Calibration curve for the pregnancy hormone hCG. Each measurement was made with a new IOS sensor. The change of the fluorescence signal is plotted vs. the concentration

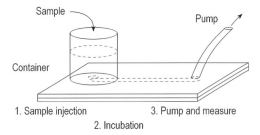

Figure 85. Simplicity of the whole measuring process. In the sample container (small dwell) all necessary reagents are present. After sample injection the reagents (with appropriate antibodies) dissolve in the sample and after a variable incubation for the immunological reaction the IOS-chip is automatically transported into the readout instrument where the outlet is punctured and a pump starts to suck the sample through the narrow measuring channel

- adjustment of pH and/or ionic strength
- filtration of sample (with or without analyte coating)
- separation of cellular blood
- components
- release of further immuno reagents or hydrodynamic puffers.

The Figure 84 shows the IOS chip design and Figure 85 the simple optical set-up. In Figure 86 a typical calibration plot for the quantification of the important pregnancy hormone hCG is presented. Note: This calibration curve was obtained by using single-use sensors at the corresponding concentration. The error brackets show the standard deviation obtained in each case with five different IOS sensors from one production batch!

What is evident from that calibration curve is its extremely high dynamic range, not typical in the ELISA techniques! The reproducibility can be brought well below 5% relative, which is outstanding in the trace analytical range. The matrix effect between an aqueous buffer solution and urine can be about 10% (less) and between buffer and serum about 20% (less) relative. Both can be compensated for with sufficient accuracy.

Outlook for the IOS technology-DNA-Chip. The IOS technology developed at the ICB Muenster is a base technology with high market orientation, with only as much sophisticated high-tech as needed, optimized for ease of production and quality control, increased production units by generic approach (one base chip for antibody capturing can be used for a wide number of different analytes depending only on the sample container at the inlet). All these advantages can even be topped by the fact that with another optical system delivering a focussed image of sensing spots onto a CCD camera, multi-analyte detection is feasible! Even further, fluorescence-based hybridization events are also sensed with the multi-analyte IOS technology. The upper limit is only given by patent violation reasons and lies at 400 detection spots. This application can further trigger mass-production and has the great advantage compared with other DNA-arrays that it quantifies automatically in real time!

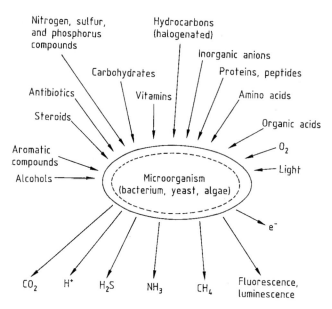

Figure 87. Microorganisms as receptor systems in biosensors useful for detecting specific groups of compounds

28.3.4. Whole-Cell Biosensors

Biological cells can be regarded as small membrane-bound structures containing a high concentration of chemicals, including enzymes, nucleic acids, ions, many types of protein, small organic molecules, and a host of others. A major challenge in exploiting whole (intact) cells in biosensor research is the problem of achieving selective transduction of a specific biochemical event or process within the context of this complex cellular chemistry [357], [370], [373], [374], [386] (Fig. 87).

The sensitivity and specificity of a particular cell with respect to a particular chemical varies enormously depending on the type, source, and environment of the cell and the precise nature of the chemical itself. It is not possible to provide any universal guidelines beyond such generalizations as the observation that most mammalian cells respond to a range of highly toxic chemicals, including cyanides. Even in this case it is known that hydrogen cyanide can interact with a number of different cellular components (such as electron transport proteins and enzymes) because of its small molecular size (which facilitates diffusion through biological membranes) and the ability of the cyanide anion to bind strongly to a wide variety of biomolecules.

The stability of biological cells is of critical importance in biosensor applications. In general, cells derived from higher organisms are rather difficult to isolate and maintain in a viable form for the extended periods of time required in most sensor formulations. The most biologically robust cells are derived from less complex organisms, particularly yeast, bacteria, and algae. Cells from these organisms can generally be stabilized in an in vitro environment more easily than complex cells; indeed, many of the organisms themselves are unicellular (yeast, bacteria, algae), leading to biosensors incorporating truly "living" biological components.

Natural cell differentiation and the growth characteristics of such whole-cell biosensors sometimes cause problems associated with signal drift over time. Nevertheless, there are potential advantages to be gained by exploiting whole cells in a sensor format, including:

1) Versatility (cells respond to a wide range of chemicals)
2) The presence in situ of biological structures supporting multicomponent or multistep biological reaction sequences useful for detection purposes and ensuring optimum enzyme activity (many enzymes lose activity when isolated in vitro)

3) Low preparation cost; i.e., growth in culture

For the reasons outlined above, nearly all whole-cell biosensors reported to date have employed relatively simple bacterial or algal cells [372], [387]–[390]. Existing whole-cell biosensor technology has also been based exclusively on electrochemical transduction of cellular responses.

The increasing interest in whole-cell biosensors since the first reports by DIVIES [387] and by RECHNNITZ [388] has led to the development of a large number of analyte-specific amperometric and potentiometric sensors. In these devices an immoblized layer of whole cells (entrapped in a gel, for example, or membrane-supported) is placed at the surface of an electrode. The cells then function as an enzyme source, where the enzyme of interest is present at a high relative concentration and is effectively able to function in its natural optimized environment. Bacterial cells in particular may contain very large amounts of particular enzymes relative to all others, especially if they are grown on a single source of carbon or nitrogen.

KARUBE [389] has pioneered in the development of many bacterial electrodes based on this principle using various bacteria depending on the target analyte. Ion-selective electrodes for NH_3, O_2, CO_2, H_2S and H^+ have all been used in conjunction with immobilized whole cells. RIEDEL et al. [390] have recently demonstrated that preincubation of certain bacterial electrodes with the desired analyte (substrate) can enhance the sensitivity of a sensor toward that chemical by a factor of as much as 25. This induction approach may prove to be widely applicable. The shelf-life of a whole-cell electrochemical sensor can extend up to several weeks with fully optimized storage conditions (low temperature, for example). Microbe thermistors (sensors that respond to the heat evolved during bacterial metabolism of a substrate) have also been developed, but these present problems once again with respect to analyte specificity.

A significant advance in whole-cell biosensor technology accompanied the development by RAWSON et al. [372] of a mediated electrochemical sensor employing a layer of blue-green algae (cyanobacteria). These authors demonstrated that a low molecular mass redox-active mediator, such as $Fe(CN)_6^{3-}$ or benzoquinone, could accept electrons from the photosynthetic energy chain of a cyanobacterium and relay them, via diffusion, to an electrode. The sensor proved capable of detecting aqueous herbicides on the basis of their inhibition of electron transport in the cells. This resulted in a dramatic fall in current, since fewer electrons reached the mediator and hence the electrode. The advantage of this approach to the use of whole cells is that it permits reasonably specific detection of a group of chemicals (herbicides, in this case) rather than limited-range detection of a single substance.

The principal application at the present time for cell sensors is in the measurement of biochemical oxygen demand (BOD) in wastewater. For this purpose yeast and/or bacterial cells (mainly the yeasts *Trichosporon* sp. or *Issatchenkia* and the bacterium *Rhodococcus erythropolis* [370], [386]) are immobilized on top of a Clark oxygen electrode and the oxygen consumption is measured. The great advantage of BOD sensors is a short measurement time (20 s to several minutes) compared with the classical BOD test in wastewater treatment, which requires five days. The main disadvantage is that a BOD sensor reflects only low molecular mass and easily assimilable substrates. High molecular mass substrates such as proteins and starch cannot penetrate the protective dialysis membrane of the sensor. Pretreatment with hydrolytic enzymes could solve this problem. In general, however, BOD sensors, represent a promising approach to the control of wastewater facilities.

28.3.5. Problems and Future Prospects

Compared with traditional sensors and classical methods of analysis, biosensors offer an attractive alternative for the rapid, selective, and highly sensitive assay of a wide variety of analytes. Further advantages of this cheap and economical tool emerge with the possibility of direct measurements within complex matrices, demanding only a rudimentary sample pretreatment, and the prospect of an uncomplicated measuring procedure that avoids the use of toxic reagents and organic solvents. Potential fields of application for biosensors are suggested in Table 21.

Current biosensor research is concerned with a variety of unsolved problems posing barriers to further expansion within the analytical market. Until recently, biosensors were available for only a few analytes (roughly two dozen). This is mainly a consequence of insufficient reliability associated with poor stability of the biomaterial, including a dependence on physical and chemical parameters in the environment and interferences within the transducers. In the case of implantable biosensors

Table 21. Potential fields of application for biosensors

Field of application	Biosensor opportunities
Medicine	patient self-control, home health care (glucose, lactate, creatinine, phenylalanine, histamine)
	in vivo analysis
	long-term in vivo control of metabolites and drugs
	control element (biotic sensor) for artifical prosthesis and organs
	rapid analysis at intensive-care units
	surface imaging of organs during implantation
	bedside monitoring
Clinical chemistry	diagnostics for metabolites, drugs, enzymes, vitamins, hormones, allergies, infectious diseases, cancer markers, pregnancy, etc.
	laboratory safety
Environmental protection	pollution control
	monitoring/screening of toxic compounds in water supplies, solid and liquid wastes, soil and air (e.g., pesticides, inorganic ions, explosives, oils, PAHs, PCBs, microorganisms, volatile vapors and gases)
	self control of industrial companies and farms
	alarm systems for signaling hazardous conditions
	determination of organic load (BOD)
Chemical, pharmaceutical, food industry	monitoring and control of fermentation processes and cell cultivation (substrates, metabolites, products)
	food quality control (screening/detection of microbial contaminations; estimation of freshness, shelf life; olfactory qualities and flavor; rancidity; analysis of fats, proteins, carbohydrates in food)
	study of efficiency of drugs
	detection of leakage and hazardous concentrations of liquids and gases in buildings and mine shafts
	indoor air quality checks
	location of oil deposits
Agriculture	quality control of soils
	estimation of degradation/rottage (e.g., of biodegradable waste, or in wood or plant storage)
	rapid determination of quality parameters of milk
Military	detection of chemical and biological warfare agents (e.g., nerve gases, pathogenic bacteria, viruses)

for in vivo monitoring of metabolites or feedback control of artificial organs (e.g., the pancreas), a lack of biocompatibility with the body environment, organic interferences, and inadequate stability of the biological receptor have so far prevented long-term implantation. Nevertheless, with respect to rapid, sensitive, and selective low-cost determination of a great variety of relevant compounds no suitable alternative to the principle of biosensors is apparent at this time.

Constructing biosensors with greater reliability requires a better understanding of the basic principles of enzyme catalysis and immunochemical reactions, protein structures, pathways for receptor-based signal amplification, and the interfacial behavior of biocompounds at the artificial transducer surface. Biosensor research must therefore be directed toward the following points:

1) Protein engineering, with the aim of enhancing the stability of enzymes, catalytic antibodies, and bifunctional enzymes
2) Screening stable enzymes from microorganisms (extremophiles)
3) Utilization of antibody fragments (Fab, Fab'), bispecific antibodies, and single-chain and single-domain antibodies (tailored through genetic engineering and its expression in bacteria)
4) Recombinant antibodies
5) Modification of the efficiency, specificity, and modulation of antibodies by site-directed mutagenesis
6) Use of DNA libraries for the production of monoclonal antibodies to meet specific demands of the desired transducer

7) Selection and specification of biological receptor systems applicable to the quantification of olfaction and taste
8) Manipulating enzymes and cofactors by electrically wiring redox-active centers to transducer surfaces, utilizing the principles of both molecular electronics and biological electron transfer
9) Investigation of interfacial effects and interactions between biocompounds and transducer surfaces at the molecular and atomic levels by exploiting new methods of surface analysis (e.g., scanning microscopy techniques, fast-atomic-bombardment mass spectrometry, laser-assisted mass spectrometry, time-of-flight secondary ion mass spectrometry, Fourier-transform infrared spectrometry, ellipsometry, X-ray photoelectron spectroscopy, and electron microscopy)
10) Materials research, including biocompatible membranes (based on compounds found within the body) and redox-active electrically conducting (metal)organics and conjugated polymers
11) Development of (multifunctional) silicon-based ultramicroelectrodes covalently linked to biomaterials
12) Preventing the fouling of the biological compounds through microbiological degradation and attacks from proteases

Concerning methodological applications of biosensors, most of the recent developments relate to enzyme electrodes integrated into laboratory analyzers for the determination of low molecular mass compounds in the medical and biotechnological fields. A further important line of development is directed toward disposable biosensors for screening and semiquantitative analysis in environmental control, patient home diagnostics, and food-quality control. Direct testing would avoid the time-consuming procedures of sample storage and transfer as well as expensive laboratory analyses. A medium-term application is expected to be the use of biosensors in (quasi)continuously operating random-process analyzers based on micromachined fluidics elements, together with low-maintenance systems for the control and monitoring of water supplies and clinical bedside monitoring. Finally, a long-term target is the development of highly reliable implantable biosensors for long-term implantation. Biosensor applications in the field of instrumentation should also contribute to the solution of many important problems in conventional measuring techniques, opening major new areas to modern analysis.

28.4. Actuators and Instrumentation

The instrumentation associated with chemical sensors is rather simple and inexpensive, especially in the case of electrochemical transducers and the new IOS technology developed at the ICB Muenster. All that is normally required is a high-ohmic voltmeter (pH meter), a conductometer, and/or polarographic equipment. Fiber-optical devices, on the other hand, can be quite expensive depending on quality requirements with respect to the monochromatic light source. Spectrophotometers with fiber-optical inputs and outputs are already commercially available. A laser source is required in certain cases of sensitive fluorescence measurements with small-volume samples. Exploiting the modern techniques of acoustic wave modulation also requires more than a simple frequency counter. For example, a network analyzer can be of great help in following the shift in a resonance curve caused by interaction of an analyte with the transducer surface.

The highest costs are usually related to development of a selective analyte recognition layer, perhaps the most important part of a complete chemical sensor. Transducer technology is much more highly developed; indeed, there seems to be little need here for further improvement without equivalent improvements in the selectivity of sensor elements.

One of the most important areas of current sensor research is microsystem technology (mst). A microsystem is defined as one too small to be fabricated by traditional miniaturized mechanical techniques, especially one that is dependent on modern mass-production techniques involving silicon or LIGA technology. A *system* must, by definition, consist of at least two interactive processes. A typical example is the antiblocking brake system (ABS) in an automobile, in which acceleration sensors are capable of detecting a blocked wheel and selectively controlling the brake power to that wheel via actuators.

Examples of analogous complete systems in chemistry are still rare. The best example is perhaps the "artificial pancreas," in which a glucose sensor measures the actual blood-glucose concentrations in order to control an appropriate actuator, the insulin pump. Bedside devices for this purpose have operated successfully for many years, but all attempts at implantation have so far failed, mainly because of the limited lifetime of the glucose biosensor and biofouling of the biosensor membrane by macrophages and other body constituents.

A complete system providing both a sensor and an actuator would be ideal in the field of process control, but because of a lack of truly reliable chemical sensors on the market the concept has not been widely implemented. One exception relates to the analytical method of coulometry, a technique that offers great potential for delivering chemical compounds to a controlled reaction. Especially attractive in this context is the method of constant-current coulometry, which can be carried out with an end-point sensor and a coulometric actuator for maintaining a generator current until the end-point has been reached. In this case both of the required devices can be miniaturized and constructed with the same technology.

28.5. Future Trends and Outlook

Surveys that have been conducted on the subject of chemical sensors all predict a growing market during the late 1990s and the early years of the 21st century [391]–[393]. Anticipated developments in automation, environmental protection, health care, and high technology ensure a strong demand for sensors. However, devices of this type must become increasingly reliable, must survive a thorough validation, and become even simpler to use. Japanese researchers see the first area for mass application of sensors to be the health care industry, especially with respect to the elderly. The demographic prospect of an increasing number of retirees and a lack of sufficient care centers may trigger the use of many sensors for monitoring the health of people living for as long as possible in their own homes. It is now clear that the application area of health care will dominate the application of biosensors. The "point of care" idea will replace large central laboratories to which the labile biological sample has to be transported. The reliabilty, built-in validation checks, and ease of use will allow their use by the normal physician who will have then the result within a few minutes. While the numbers of DNA arrays used in the genom research area will become saturated shortly, the use of simple arrays for looking for specific protein expression patterns will increase. Combining bio-informatics with the latter in finding super-marker molecules for extremely early tumor diagnostics will be possible. This leads to the corresponding mass-production of these diagnostic chips.

With respect to environmental chemistry, controls over the emission of organic compounds will undoubtedly be strengthened in the future. For example, municipal incinerators without provisions for the effective measurement of all hazardous organic compounds will soon be strictly forbidden everywhere. It seems likely that a large fraction of the moneys previously spent for defense purposes will be redirected toward improvements in continuous measurement technology. Sensors will be required to compete in this respect with such fast separation methods as capillary electrophoresis or flash chromatography, which permit the separation of most compounds within only a few minutes.

One critical remark is in order regarding papers published in the field of chemical sensors. Many such reports reflect a lack of input from qualified analytical chemists during the development and testing stages. Insufficient understanding is often apparent with respect to the analytical process itself and the difficulties involved in obtaining a truly reliable result; indeed, many sensor researchers seem to be unaware of such basic problems as sample–matrix effects. Anyone who claims to have developed a new sensor should be required to demonstrate its effectiveness with a real sample under real conditions. The latter is required and called validation by international quality standards (ISO 25). Publications that fail to reveal this most important feature of a sensor are also essentially worthless. Chemical sensors are analytical devices, so they should be subject to the same standards of characterization as other instrumental tools. Quality-control considerations mandate critical evaluation of any new molecular sensor presented to the scientific community. Every report concerning a new sensor development should clearly delineate the crucial features, with none omitted.

The most important parameters for characterizing or validating a chemical sensor or a biosensor include:

1) *Selectivity.*

The selectivity with respect to important known interferents for the particular application in question should always be expressed numerically; expressions like "relative selectivity" toward the analyte, or "not affected by typical concentrations of interfering compounds" are inappropriate.

2) *Sensitivity and Limit of Detection (LOD).*
   The slope of a calibration curve expressed in the appropriate units (e.g. microamps, millivolts, or hertz) per millimole or micromole should always be provided both for a matrix-free solution and for a sample in its matrix. The corresponding LOD should also be reported in both cases. In the absence of a blank the signal 3 × larger than the noise determines the LOD. The limit of quantification is typical 10 × LOD.

3) *Stability of the Analytical Function (Drift).*
   A zero-point drift value should be provided for a sensor in a matrix-free solution and also with respect to real samples involving a matrix. The same applies to the stability of the slope (varying sensor sensitivity).

4) *Response Time.*
   Since response time is also a function of the sample matrix it should be measured both in pure calibration solutions and with real samples. Generally two different response times are observed, one for a concentration rise and one for decreasing analyte concentration, with the latter usually slightly greater. Since IUPAC has not yet recommended a specific procedure for measuring this parameter, one should express the response time in the form of a time constant (=1/e of the final value), since this is the approach taken with electronic circuits. Only in the case of a sensor that never reaches its final value should other approaches be considered for characterizing the rate of response to a changing analyte concentration.

5) *Sensor Lifetime.*
   Especially in the field of biosensors the lifetime of a sensor in contact with a calibration solution or a real sample, or during storage in a refrigerator with an optimal conditioning solution, is of considerable interest, and it should always be mentioned. Conditions leading to rapid sensor fouling should also be identified in order to permit appropriate precautions to be taken. Examples include the fouling of certain enzymes by inhibitors or heavy metals, or microbiological attack. Any use of conditioning solutions containing metal complexation reagents, inhibitor scavengers, sodium azide, etc., should be clearly described.

Other important information might include total power consumption, auxiliary equipment costs, total elapsed time between failures, temperature compensation measures, or maintenance requirements.

Finally, sensor developers must remain aware of progress within instrumental analytical chemistry generally, especially in the area of fast high-resolution chromatography. Indeed, progress has already been so great in the latter field that more than 20 anions (including phosphate) can now be separated completely in less time than would be required for a response from a multienzyme phosphate biosensor! The serious challenge this poses to sensor advocates will be obvious.

28.6. References

[1] W. Göpel, J. Hesse, J. N. Zemel (eds.): *Sensors – A Comprehensive Survey*, **vols. 1 – 8 + updates**, VCH Verlagsgesellschaft, Weinheim 1989.
[2] T. Seiyama, S. Yamauchi (eds.): *Chemical Sensor Technology*, **vols. 1 + 4**, Elsevier Science Publisher B, Amsterdam 1988.
[3] R. W. Murray et al.: "Chemical Sensors and Microinstrumentation," *ACS Symp. Ser.* **403** (1989).
[4] J. Janata: *Principles of Chemical Sensors*, Plenum Press, New York 1989.
[5] K. Cammann, H. Galster: *Das Arbeiten mit ionenselektiven Elektroden*, 3rd ed., Springer Verlag, Berlin 1996.
[6] J. Janata, R. Huber: *Solid State Chemical Sensors*, Academic Press, New York 1985.
[7] K. Cammann et al.: *Analytiker Taschenbuch*, **vol. 15**, Springer-Verlag, Berlin – Heidelberg 1997, pp. 3 – 40.
[8] J. Reinbold, K. Cammann: *Analytiker Taschenbuch*, **vol. 16**, Springer-Verlag, Berlin – Heidelberg 1997, pp. 3 – 42.
[9] K. Cammann (ed.): *Instrumentelle Analytische Chemie*, Spektrum-Verlag, Heidelberg 2000.
[10] T. Takeuchi, *Sens. Actuators* **14** (1988) 109.
[11] G. Hötzel, H. M. Wiedemann, *Sens. Rep.* **4** (1989) 32.
[12] J. P. Pohl, *GIT Fachz. Lab.* **5** (1987) 379.
[13] K. Cammann et al., *Angew. Chem.* **103** (1991) 519; *Angew. Chem. Int. Ed. Engl.* **30** (1991) 516 – 539.
[14] W. Masing: *Handbuch der Qualitätssicherung*, Hanser Verlag, München 1988.
[15] K. Cammann: *Working with Ion-Selective Electrodes*, Springer Verlag, Berlin 1979.
[16] A. K. Covington: *Ion Selective Electrode Methodology*, vols. I and II, CRC Press, Boca Raton, Fla., 1979.
[17] H. Freiser (ed.): *Ion Selective Electrodes in Analytical Chemistry*, Plenum Press, New York 1979.
[18] K. Schierbaum, *Sens. Actuators B* **18 – 19** (1994) 71 – 76.
[19] J. Reinbold et al., *Sens. Actuators B* **18 – 19** (1994) 77 – 81.
[20] N. Taguchi, JP 4 5 38 200, 1962.
[21] T. Seiyama, A. Kato, K. Fujiishi, M. Nagatani, *Anal. Chem.* **34** (1962) 1502.
[22] Y. Nakatani, M. Sakai, M. Matsuoka in T. Seyama et al. (eds.): *Proc. Int. Meet. Chem. Sensors (Fukuoka, Japan)*, Elsevier, Amsterdam, Sept. 1983, pp. 147 – 152.

[23] S. R. Morrison: "Semiconductor Gas Sensors," *Sens. Actuators* **2** (1982) 329.
[24] P. J. Shaver, *Appl. Phys. Lett.* **11** (1967) 255.
[25] J. C. Loh, JP 4 3 28 560; FR 1 545 292, 1967.
[26] N. Yamazoe, Y. Krakowa, T. Seiyama: "Effects of Additives on Semiconductor Gas Sensors," *Sens. Actuators* **4** (1983) 283.
[27] S. Matsushima, Y. Teraoka, N. Miura, N. Yamazoe, *Jpn. J. Appl. Phys.* **27** (1988) 1798.
[28] M. Nitta, S. Kanefusa, S. Ohtani, M. Haradome, *J. Electr. Mater.* **13** (1984) 15.
[29] A. Chiba: "Development of the TGS Gas Sensor," in S. Yamauchi (ed.): *Chemical Sensor Technology*, vol. 4, Elsevier, Amsterdam 1992, pp. 1–18.
[30] T. A. Jones: "Characterisation of Semiconductor Gas Sensors," in P. T. Moseley, B. C. Tofield (eds.): *Solid State Gas Sensors*, Adam Hilger, Bristol 1987.
[31] G. Heiland, D. Kohl, *Sens. Actuators* **8** (1985) 227.
[32] W. Weppner: "Halbleiter-Sensoren" in R. Grabowski (ed.): *Sensoren und Aktoren*, VDE-Verlag, Berlin 1991, pp. 167–183.
[33] D. E. Williams: "Characterization of Semiconductor Gas Sensors," in P. T. Moseley, B. C. Tofield (eds.): *Solid State Gas Sensors*, Adam Hilger, Bristol 1987.
[34] N. Yamazoe, N. Miura: "Some Basic Aspects of Semiconductor Gas Sensors," in S. Yamauchi (ed.): *Chemical Sensor Technology*, vol. 4, Elsevier, Amsterdam 1992, pp. 19–41.
[35] G. Heiland: "Homogeneous Semiconducting Gas Sensors," *Sens. Actuators* **2** (1982) 343.
[36] H. Windischmann, P. Mark, *J. Electrochem. Soc.* **126** (1979) 627.
[37] S. Chang, *J. Vac. Sci. Technol.* **17** (1980) 366.
[38] J. Watson: "The Tin Dioxide Gas Sensor and Its Applications," *Sens. Actuators* **5** (1984) 29.
[39] N. Komori, S. Sakai, K. Konatsu in T. Seyama et al. (eds.): *Proc. Int. Meet. Chem. Sensors (Fukuoka, Japan)*, Elsevier, Amsterdam, Sept. 1983, p. 57.
[40] B. Ahlers: *Dissertation*, University of Muenster 1996.
[41] M. L. Hitchmann, H. A. O. Hill, *Chem. Br.* **22** (1986) 1177.
[42] D. Ammann: *Ion Selective Microelectrodes – Principles, Design and Application*, Springer Verlag, Berlin 1986.
[43] P. L. Bailey: *Analysis with Ion-Selective Electrodes*, Heyden & Sons, London 1976.
[44] R. Bock: "Nachweis- und Bestimmungsmethoden," in *Methoden der Analytischen Chemie*, vol. 2, part 2, Verlag Chemie, Weinheim 1984, pp. 33–83.
[45] K. Cammann: "Fehlerquellen bei Messungen mit ionenselektiven Elektroden," in *Analytiker-Taschenbuch*, vol. 1, Springer Verlag, Berlin 1980.
[46] R. A. Durst: "Ion-Selective Electrodes," *NBS Special Publication No. 314*, 1969.
[47] S. Ebel, W. Parzefall: *Experimentelle Einführung in die Potentiometrie*, Verlag Chemie, Weinheim 1975.
[48] E. Bakker, *Trends Anal. Chem.* **16** (1997) 252–260.
[49] G. Horvai, *Trends Anal. Chem.* **15** (1997) 260–266.
[50] G. Eisenman: *Glass Electrodes for Hydrogen and Other Cations, Principles and Practice*, Marcel Dekker, New York 1967.
[51] D. J. G. Ives, J. Janz: *Reference Electrodes*, Academic Press, New York 1961.
[52] R. Koryta: *Ion Selective Electrodes*, Cambridge University Press, Cambridge 1975.
[53] T. S. Ma, S. S. M. Hassan: *Organic Analysis Using Ion-Selective Electrodes*, 2 vols., Academic Press, New York 1982.
[54] G. J. Moody, J. D. R. Thomas: *Selective Ion Electrodes*, Merrow Publishing, Watford 1977.
[55] F. Oehme: *Ionenselektive Elektroden: CHEMFETs – ISFETs – pH-FETs, Grundlagen, Bauformen und Anwendungen*, Hüthig Verlag, Heidelberg 1991.
[56] E. Pungor: *Ion-Selective Electrodes*, Akadémiai Kiadó, Budapest 1973.
[57] F. Honold, B. Honold: *Ionenselective Elektroden*, Birkhäuser Verlag, Basel 1991.
[58] K. Umezawa, Y. Umezawa: *Selectivity Coefficients for Ion Selective Electrodes*, University of Tokyo Press, Tokyo 1983.
[59] E. Bakker, P. Bühlmann, E. Pretsch: "Carrier-Based Ion-Selective Electrodes and Bulk Optodes. 1. General Characteristics," *Chem. Rev.* **97** (1997) 3083–3132.
[60] P. Bühlmann, E. Pretsch, E. Bakker: "Carrier-Based Ion-Selective Electrodes and Bulk Optodes. 2. Ionophores for Potentiometric and Optical Sensors," *Chem. Rev.* **98** (1998) 1593–1687.
[61] Y. Mi, S. Mathison, R. Goines, A. Logue, A. Bakker, *Anal. Chim. Acta* **397** (1999) 103–111.
[62] A. Schwake: *Dissertation*, University of Muenster 1999.
[63] F. Zuther: *Dissertation*, University of Muenster 1997.
[64] A. J. Bard, L. R. Faulkner: *Electrochemical Methods*, J. Wiley, New York 1980.
[65] A. M. Bond, H. B. Greenhill, I. D. Heritage, J. B. Reust, *Anal. Chim. Acta* **165** (1984) 209.
[66] A. J. Bard: *Electroanalytical Chemistry*, vol. 9, Marcel Dekker, New York 1976.
[67] G. Henze, R. Neeb: *Elektrochemische Analytik*, Springer Verlag, Berlin 1985.
[68] K. Cammann: "Elektrochemische Verfahren," in H. Naumer, W. Heller (eds.): *Untersuchungsmethoden in der Chemie*, Thieme Verlag, Stuttgart 1986.
[69] P. T. Kissinger, W. R. Heinemann, *J. Chem. Educ.* **60** (1983) 702.
[70] J. Osteryoung, J. J. O'Dea in A. J. Bard (ed.): *Electroanalytical Chemistry*, **vol. 14**, Marcel Dekker, New York 1986, pp. 209–308.
[71] P. Vadgama, G. Davis, *Med. Lab. Sci.* **42** (1985) 333.
[72] L. C. Clark, US 2 913 386, 1958.
[73] H. Suzuki et al., *Sens. Actuators B* **2** (1990) 297.
[74] S. C. Cha, M. J. Shao, C. C. Liu, *Sens. Actuators B* **2** (1990) 239.
[75] W. Sansen et al., *Sens. Actuators B* **1** (1990) 298.
[76] H. Suzuki et al., *Sens. Actuators B* **1** (1990) 528.
[77] H. Suzuki et al., *Sens. Actuators B* **1** (1990) 275.
[78] D. S. Austin et al. *J. Electroanal. Chem. Interfacial Electrochem.* **168** (1984) 227.
[79] R. M. Wightman, *Anal. Chem.* **53** (1981) 1125.
[80] T. Hepel, J. Osteryoung, *J. Electrochem. Soc.* **133** (1986) 752.
[81] L. Zhaohui et al., *J. Electroanal. Chem. Interfacial Electrochem.* **259** (1989) 39.
[82] C. L. Colyer, J. C. Myland, K. B. Oldham, *J. Electroanal. Chem. Interfacial Electrochem.* **263** (1989) 1.
[83] J. Osteryoung, S. T. Singleton, *Anal. Chem.* **61** (1989) 1211.
[84] R. E. Howard, E. Hu: *Science and Technology of Microfabrication*, Materials Research Society, Pittsburgh 1987.
[85] C. D. Baer, N. J. Stone, D. A. Sweigart, *Anal. Chem.* **58** (1988) 78.

[86]   J. Ghoroghchiana et al., *Anal. Chem.* **58** (1988) 2278.
[87]   H. Meyer et al., *Anal. Chem.* **67** (1995) 1164–1170.
[88]   M. Wittkamp, G. Chemnitius, K. Cammann, M. Rospert, M. Mokwa, *Sens. Actuators B* **43** (1997) 87–93.
[89]   W. Göpel, J. Hesse, J. N. Zemel: *Sensors,* vol. 2, VCH Verlagsgesellschaft, Weinheim 1991.
[90]   W. Göpel, J. Hesse, J. N. Zemel: *Sensors,* vol. 3, VCH Verlagsgesellschaft, Weinheim 1991.
[91]   P. T. Moseley, B. C. Tofield: *Solid State Gas Sensors,* Adam Hilger, Bristol 1987.
[92]   P. T. Moseley, J. O. W. Norris, D. E. Williams: *Techniques and Mechanisms in Gas Sensing,* Adam Hilger, Bristol 1991.
[93]   H. Emons, G. Jokuszies, *Z. Chem.* **28** (1988) 197.
[94]   R. Kalvoda, *Electroanalysis* **2** (1990) 341.
[95]   F. Opekar, A. Trojanek, *Anal. Chim. Acta* **203** (1987) 1.
[96]   ISO/TC 147 (draft):*Electrolytic Conductivity,* Beuth Verlag, Berlin.
[97]   DIN 38 404:*Deutsche Verfahren zur Wasser-, Abwasser- und Schlammuntersuchung,* VCH Verlagsgesellschaft, Weinheim 1993.
[98]   F. Oehme in W. Göpel, J. Hesse, J. N. Zemel (eds.): *Sensors,* vol. 2, VCH Verlagsgesellschaft, Weinheim 1991.
[99]   F. Oehme, R. Bänninger: *ABC der Konduktometrie,* Polymetron AG, Mönchaltorf (Switzerland).
[100] K. Rommel, *VGB Kraftwerkstech.* **65** (1985) 417.
[101] K. Rommel: *Kleine Leitfähigkeitsfibel,* Wissenschaftlich-Technische Werke, Weilheim 1988.
[102] F. Oehme, M. Jola, *Betriebsmeßtechnik,* Hüthig Verlag, Heidelberg 1982.
[103] M. Niggemann: *Dissertation,* University of Muenster 1999.
[104] W. H. Brattain, J. Bardeen, *Bell Syst. Tech. J.* **32** (1952) 1.
[105] G. Heiland, *Z. Phys.* **138** (1954) 459.
[106] J. Watson, *Sens. Actuators B* **8** (1992) 173.
[107] N. Taguchi, US 3 644 795, 1972.
[108] D.-D. Lee, D.-H. Choi, *Sens. Actuators B* **1** (1990) 231.
[109] B. C. Tofield in P. T. Moseley, B. C. Tofield (eds.): *Solid State Gas Sensors,* Adam Hilger, Bristol 1987.
[110] R. Lalauze, C. Pijolat, S. Vincent, L. Bruno, *Sens. Actuators B* **8** (1992) 237.
[111] J. Watson, *Sens. Actuators* **5** (1984) 29.
[112] R. M. Geatches, A. V. Chadwick, J. D. Wright, *Sens. Actuators B* **4** (1991) 467.
[113] S. R. Morrison, *Sens. Actuators* **12** (1987) 425.
[114] S. Matsushima et al., *Sens. Actuators B* **9** (1992) 71.
[115] T. Maekawa et al., *Sens. Actuators B* **9** (1992) 63.
[116] S. Nakata, H. Nakamura, K. Yoshikawa, *Sens. Actuators B* **8** (1992) 187.
[117] U. Kirner et al., *Sens. Actuators B* **1** (1990) 103.
[118] N. Li, T.-C. Tan, *Sens. Actuators B* **9** (1992) 91.
[119] U. Lampe, M. Fleischer, H. Meixner, *Sens. Actuators B* **17** (1994) 187–196.
[120] N. Koshizaki, K. Yasumoto, K. Suga, *Sens. Actuators B* **9** (1992) 17.
[121] M. Egashira, Y. Shimizu, Y. Takao, *Sens. Actuators B* **1** (1990) 108.
[122] P. T. Moseley, D. E. Williams, *Sens. Actuators B* **1** (1990) 113.
[123] E. Ye. Gutman, I. A. Myasnikov, *Sens. Actuators B* **1** (1990) 210.
[124] R. S. Tieman, W. R. Heineman, J. Johnson, R. Seguin, *Sens. Actuators B* **8** (1992) 199.
[125] Y. Sadaoka, T. A. Jones, W. Göpel, *Sens. Actuators B* **1** (1990) 148.
[126] S. Dogo, J. P. Germain, C. Maleysson, A. Pauly, *Sens. Actuators B* **8** (1992) 257.
[127] J. W. Gardner, M. Z. Iskandarani, B. Bott, *Sens. Actuators B* **9** (1992) 133.
[128] M. Trometer et al., *Sens. Actuators B* **8** (1992) 129.
[129] J. D. Wright, *Prog. Surf. Sci.* **31** (1989) 1.
[130] T. A. Temofonte, K. F. Schoch, *J. Appl. Phys.* **65** (1989) 1350.
[131] S. Kanefusa, M. Nitta, *Sens. Actuators B* **9** (1992) 85.
[132] W. Göpel, K. D. Schierbaum, D. Schmeisser, D. Wiemhöfer, *Sens. Actuators* **17** (1989) 377.
[133] B. Bott, S. C. Thorpe in P. T. Moseley, J. O. W. Norris, D. E. Williams (eds.): *Techniques and Mechanisms in Gas Sensing,* Adam Hilger, Bristol 1991.
[134] T. Hanawa, S. Kuwabata, H. Yoneyama, *J. Chem. Soc. Faraday Trans. 1* **84** (1988) 1587.
[135] C. Nylander, M. Armgarth, I. Lundström, *Anal. Chem. Symp. Ser.* **17** (1983) 203.
[136] P. N. Bartlett, P. B. M. Archer, S. K. Ling-Chung, *Sens. Actuators* **19** (1989) 125.
[137] P. N. Bartlett, P. B. M. Archer, S. K. Ling-Chung, *Sens. Actuators* **19** (1989) 141.
[138] H. Arai, T. Seiyama: "Humidity Control," in W. Göpel, J. Hesse, J. N. Zemel (eds.): *Sensors,* vol. 2, VCH Verlagsgesellschaft, Weinheim 1991.
[139] A. K. Michell in P. T. Moseley, J. O. W. Norris, D. E. Williams (eds.): *Techniques and Mechanisms in Gas Sensing,* Adam Hilger, Bristol 1991.
[140] T. Nenov, S. Yordanov, *Sens. Actuators B* **8** (1992) 117.
[141] Y. Sakai, Y. Sadaoka, H. Fukumoto, *Sens. Actuators* **13** (1988) 243.
[142] P. Bergveld, *IEEE Trans. Biomed. Eng.* **BME-17** (1970) 70.
[143] K. Buhlmann: *Dissertation,* University of Muenster 1997.
[144] K. Buhlmann, B. Schlatt, K. Cammann, A. Shulga, *Sens. Actuators B* **49** (1998) 156–165.
[145] T. Matsuo, M. Esashi, K. Inuma, *Proc. Dig. Joint Meeting,* Tohoku Sect. IEEEJ, Oct. 1971.
[146] J. E. Moneyron, A. De Roy, J. P. Besse, *Solid State Ionics* **46** (1991) 175.
[147] S. M. Sze: *Physics of Semiconductor Devices,* Wiley, New York 1981.
[148] P. Bergveld, A. Sibbald: *Analytical and Biomedical Applications of Ion Sensitive Field Effect Transistors,* Elsevier, Amsterdam 1988.
[149] Y. G. Vlasov, D. E. Hackleman, R. P. Buck, *Anal. Chem.* **51** (1979) 1579.
[150] W. Moritz, I. Meierhöffer, L. Müller, *Sens. Actuators* **15** (1988) 211.
[151] S. D. Moss, J. Janata, C. C. Johnson, *Anal. Chem.* **47** (1975) 2238.
[152] A. Sibbald, P. D. Whalley, A. K. Covington, *Anal. Chim. Acta* **159** (1984) 47.
[153] D. J. Harrison et al., *J. Electrochem. Soc.* **135** (1988) 2473.
[154] P. D. van der Wal et al., *Anal. Chim. Acta* **231** (1990) 41.
[155] A. van den Berg, A. Griesel, E. Verney-Norberg, *Sens. Actuators B* **4** (1991) 235.
[156] R. W. Catrall, P. I. Iles, J. C. Hamilton, *Anal. Chim. Acta* **169** (1985) 403.
[157] R. Smith, D. C. Scott, *IEEE Trans. Biomed. Eng.* **BME 33** (1986) 83.

[158] A. van den Berg, P. Bergveld, D. N. Reinhoudt, E. J. R. Sudhölter, *Sens. Actuators* **8** (1985) 129.

[159] N. J. Ho, J. Kratochvil, G. F. Blackburn, J. Janata, *Sens. Actuators* **4** (1983) 413.

[160] C. Dumschat et al., *Sens. Actuators B* **2** (1990) 271.

[161] K. Domansky, J. Janata, M. Josowicz, D. Petelenz, *Analyst (London)* **118** (1993) 335.

[162] H. H. van den Vlekkert et al., *Sens. Actuators* **14** (1988) 165.

[163] C. Dumschat et al., *Anal. Chim. Acta* **243** (1991) 179.

[164] P. Bergveld, *Sens. Actuators B* **4** (1991) 125.

[165] W. F. Love, L. J. Button, R. E. Slovacek: "Optical Characteristics of Fiberoptic Evanescent Wave Sensors," in D. L. Wise, L. B. Wingard (eds.): *Biosensors with Fiberoptics,* Humana Press, Clifton, N.J., 1991.

[166] S. J. Lackie, T. R. Glass, M. J. Block: "Instrumentation for Cylindrical Waveguide Evanescent Fluorosensors," in D. L. Wise, L. B. Wingard (eds.): *Biosensors with Fiberoptics,* Humana Press, Clifton, N.J., 1991.

[167] I. M. Walczak, W. F. Love, T. A. Cook, R. E. Slovacek: "The Application of Evanescent Wave Sensing to a High-sensitivity Fluoroimmunoassay," *Biosensors Bioelectron.* **7** (1992) 39–47.

[168] B. J. Tromberg, M. J. Sepaniak, T. Vo-Dinh, G. D. Griffin: "Fiberoptic Probe for Competitive Binding Fluoroimmunoassay," *Anal. Chem.* **59** (1987) 1226–1230.

[169] R. B. Thompson, F. S. Ligler: "Chemistry and Technology of Evanescent Wave Biosensors," in D. L. Wise, L. B. Wingard (eds.): *Biosensors with Fiberoptics,* Humana Press, Clifton, N.J., 1991.

[170] W. Trettnak, P. J. M. Leiner, O. S. Wolfbeis: "Fiberoptic Glucose Biosensor with an Oxygen Optrode as the Transducer," *Analyst (London)* **113** (1988) 1519–1523.

[171] I. J. Peterson, S. R. Goldstein, R. V. Fitzgerald, D. K. Buckhold: "Fiberoptic pH Probe for Physiological Use," *Anal. Chem.* **52** (1980) 864–869.

[172] G. F. Kirkbright, R. Narayanaswamy, N. A. Welti: "Studies with Immobilized Chemical Reagents Using a Flow-cell for the Development of Chemically Sensitive Fiberoptic Devices," *Analyst (London)* **109** (1984) 12–19.

[173] L. A. Saari, W. R. Seitz: "pH Sensor Based on Immobilized Fluoresceinamine," *Anal. Chem.* **54** (1982) 821–823.

[174] Z. Zhujun, W. R. Seitz: "A Fluorescence Sensor for Quantifying pH in the Range from 6.5 to 8.5," *Anal. Chim. Acta* **160** (1984) 47–55.

[175] E. Urbano, M. Offenbacher, O. S. Wolfbeis: "Optical Sensor for Continuous Determination of Halides," *Anal. Chem.* **56** (1984) 427–430.

[176] O. S. Wolfbeis, E. Urbano: "Fluorescence Quenching Method for Determination of Two or Three Components in Solution," *Anal. Chem.* **55** (1983) 1904–1907.

[177] L. A. Saari, W. R. Seitz: "Immobilized Morin as Fluorescence Sensor for Determination of Aluminum(III)," *Anal. Chem.* **55** (1983) 667–670.

[178] L. A. Saari, W. R. Seitz: "Optical Sensor for Beryllium Based on Immobilized Morin Fluorescence," *Analyst* (London) **109** (1984) 655–657.

[179] Z. Zhujun, W. R. Seitz: "A Fluorescence Sensor for Aluminum(III), Magnesium(II), Zinc(II) and Cadmium(II) Based on Electrostatically Immobilized Quinolin-8-ol-sulfonate," *Anal. Chim. Acta* **171** (1985) 251–258.

[180] Z. Zhujun, W. R. Seitz: "A Carbon Dioxide Sensor Based on Fluorescence," *Anal. Chim. Acta* **170** (1984) 209–216.

[181] M. A. Arnold, T. J. Ostler: "Fiberoptic Ammonia Gas Sensing Probe," *Anal. Chem.* **58** (1986) 1137–1140.

[182] A. Sharma, O. S. Wolfbeis: "Fiberoptic Fluorosensor for Sulfur Dioxide Based on Energy Transfer and Exciplex Quenching," *Proc. SPIE Int. Soc. Opt. Eng.* **990** (1989) 116–120.

[183] A. Sharma, O. S. Wolfbeis: "Fiberoptic Oxygen Sensor Based on Fluorescence Quenching and Energy Transfer," *Appl. Spectrosc.* **42** (1988) 1009–1011.

[184] O. S. Wolfbeis, M. J. P. Leiner, H. E. Posch: "A New Sensing Material for Optical Oxygen Measurement, with Indicator Embedded in an Aqueous Phase," *Mikrochim. Acta* 1986, no. 3, 359–366.

[185] K. W. Berndt, J. R. Lakowicz: "Electroluminescent Lamp-Based Phase Fluorometer and Oxygen Sensor," *Anal. Biochem.* **201** (1992) 319–325.

[186] "Integrated Optics" in T. Tamir (ed.): *Topics in Applied Physics,* Springer Verlag, Berlin 1975.

[187] K. Tiefenthaler, W. Lukosz: "Sensitivity of Grating Couplers as Integrated-optical Chemical Sensors," *J. Opt. Soc. Am. B Opt. Phys.* **B 6** (1975) 209–220.

[188] A. Brandenburg, R. Edelhauser, F. Hutter: "Integrated Optical Gas Sensors Using Organically Modified Silicates as Sensitive Films," *Sens. Actuators B* **11** (1993) 361–374.

[189] D. Schlatter et al.: "The Difference Interferometer: Application as a Direct Immunosensor," in: *The Second World Congress on Biosensors,* Oxford, Elsevier Advanced Technology, 1992, pp. 347–355.

[190] Ph. Nellen, K. Tiefenthaler, W. Lukosz: "Integrated Optical Input Grating Couplers As Biochemical Sensors," *Sens. Actuators B* **15** (1988) 285–295.

[191] R. G. Heidemann, R. P. H. Kooyman, J. Greve: "Performance of a Highly Sensitive Optical Waveguide Mach–Zehnder Interferometer Immunosensor," *Sens. Actuators B* **10** (1993) 209–217.

[192] N. Fabricius, G. Gaulitz, J. Ingenhoff: "A Gas Sensor Based on an Integrated Optical Mach–Zehnder Interferometer," *Sens. Actuators B* **7** (1992) 672–676.

[193] Ch. Fattinger, H. Koller, P. Wehrli, W. Lukosz: "The Difference Interferometer: A Highly Sensitive Optical Probe for Molecular Surface-Coverage Detection," in: *The Second World Congress on Biosensors,* Oxford, Elsevier Advanced Technology, 1992, pp. 339–346.

[194] K. Tiefenthaler: "Integrated Optical Couplers As Chemical Waveguide Sensors," *Adv. Biosensors* **2** (1992) 261–289.

[195] B. Liedberg, C. Nylander, I. Lundström: "Surface Plasmon Resonance for Gas Detection and Biosensing," *Sens. Actuators* **4** (1983) 299–304.

[196] P. B. Daniels, J. K. Deacon, M. J. Eddowes, D. Pedley: "Surface Plasmon Resonance Applied to Immunosensing," *Sens. Actuators* **14** (1988) 11–17.

[197] J. Gent et al.: "Optimization of a Chemooptical Surface Plasmon Resonance Based Sensor," *Appl. Opt.* **29** (1990) 2343–2849.

[198] K. Matsubaru, S. Kawata, S. Minami: "Optical Chemical Sensor Based on Surface Plasmon Measurement," *Appl. Opt.* **27** (1988) 1160–1163.

[199] H. Rather: *Surface Plasmons,* Springer Verlag, Berlin 1988.

[200] G. Gauglitz et al.: DE 4 200 088 A1, 1993.

[201] A. Brecht, J. Ingenhoff, G. Gauglitz, *Sens. Actuators B* **6** (1992) 96–100.
[202] G. Kraus, G. Gauglitz, *Fresenius' J. Anal. Chem.* **349** (1994) no. 5, 399–402.
[203] G. Sauerbrey, *Z. Phys.* **155** (1959) 206–222.
[204] H. K. Pulker, J. P. Decostered in C. Lu, A. W. Czanderna (eds.): "Applications of Piezoelectric Quartz Crystal Microbalances," *Methods and Phenomena,* vol. 7, Elsevier, Amsterdam 1984, Chap. 3.
[205] Lord Rayleigh, *Proc. London Math. Soc.* **17** (1885) 4–11.
[206] G. Sauerbrey, *Z. Phys.* **178** (1964) 457.
[207] W. H. King, Jr., *Anal. Chem.* **36** (1964) 1735–1739.
[208] W. H. King, Jr., US 3 164 004, 1965.
[209] A. W. Czanderna, C. Lu: "Applications of Piezoelectric Quartz Crystal Microbalances," in C. Lu, A. W. Czanderna (eds.): *Methods and Phenomena,* vol. 7, Elsevier, Amsterdam 1984. chap. I.
[210] M. Thompson et al. *Analyst (London)* **116** (1991) 881–890.
[211] L. Wimmer, S. Hertl, J. Hemetsberger, E. Benes, *Rev. Sci. Instrum.* **55** (1984) 605–609.
[212] C.-S. Lu, O. Lewis, *J. Appl. Phys.* **43** (1972) 4385–4390.
[213] J. G. Miller, D. I. Bolef, *J. Appl. Phys.* **39** (1968) 5815–5816.
[214] E. Benes, *J. Appl. Phys.* **56** (1984) 608–626.
[215] D. R. Denison, *J. Vac. Sci. Technol.* **10** (1973) 126–129.
[216] C.-S. Lu, *J. Vac. Sci. Technol.* **12** (1975) 578–583.
[217] G. G. Guilbault: "Applications of Piezoelectric Quartz Crystal Microbalances," in C. Lu, A. W. Czanderna (eds.): *Methods and Phenomena,* vol. 7, Elsevier, Amsterdam 1984, chap. 8.
[218] H. Beitnes, K. Schrøder, *Anal. Chim. Acta* **158** (1984) 57–65.
[219] A. Kindlund, H. Sundgren, I. Lundström, *Sens. Actuators* **6** (1984) 1–17.
[220] G. G. Guilbault: *Ion-Selective Electrode Review,* vol. 2, Pergamon Press, Oxford 1980, pp. 3–16.
[221] R. M. Langdon: "Resonator Sensors—a Review," *J. Phys. E* **18** (1985) 103–115.
[222] J. Tichy, G. Gautschi: *Piezoelektrische Meßtechnik,* Springer Verlag, Berlin 1980.
[223] H. Wohltjen, R. Dessy, *Anal. Chem.* **51** (1979) 1458–1464.
[224] H. Wohltjen, R. Dessy, *Anal. Chem.* **51** (1979) 1465–1470.
[225] H. Wohltjen, R. Dessy, *Anal. Chem.* **51** (1979) 1470–1478.
[226] C. T. Chang, R. M. White: "Excitation and Propagation of Plate Mode in an Acoustically Thin Membrane," *Proc. IEEE Ultrasonics Symp.* 1982, 295–298.
[227] R. M. White, P. W. Wicher, S. W. Wenzel, E. T. Zellers: "Plate-Mode Ultrasonic Oscillator Sensors," *IEEE Trans. Ultrason. Dev. Ferroelectr. Freq. Contr.* **UFFC-34** (1987) 162–171.
[228] T. G. Giesler, J.-U. Meyer: *Fachbeilage Mikroperipherik,* vol. 6 (1992) XLVI.
[229] IEEE Standard on Piezoelectricity (ANSI/IEEE Std. 176-1987), The Institute of Electrical and Electronics Engineers, New York 1987.
[230] M. S. Nieuwenhuizen, A. Venema: *Sensors – A Comprehensive Survey,* vol. II, VCH Verlagsgesellschaft, Weinheim 1991, p. 652.
[231] J. Janata: *Principles of Chemical Sensors,* Plenum Press, New York 1989, pp. 55–80.
[232] J. G. Miller, D. I. Bolef, *J. Appl. Phys.* **39** (1968) 45.
[233] K. H. Behrndt, *J. Vac. Sci. Technol.* **8** (1971) 622.
[234] A. P. M. Glassford in A. M. Smith (ed.): *Progress in Astronautics and Aeronautics,* **vol. 56,** American Institute of Aeronautics and Astronautics, New York 1977, p. 175.
[235] A. P. M. Glassford, *J. Vac. Sci. Technol.* **15** (1978) 1836.
[236] V. Mecea, R. V. Bucur, *Thin Solid Films* **60** (1979) 73.
[237] R. A. Crane, G. Fischer, *J. Phys. D* **12** (1979) 2019.
[238] K. K. Kanazawa, J. G. Gordon, *Anal. Chem.* **57** (1985) 1770.
[239] C. Fruböse, K. Doblhofer, C. Soares, *Ber. Bunsen-Ges. Phys. Chem.* **97** (1993) 475–478.
[240] R. Beck, U. Pittermann, K. G. Weil, *Ber. Bunsen-Ges. Phys. Chem.* **92** (1988) 1363–1368.
[241] F. Eggers, Th. Funck, *J. Phys. E* **20** (1987) 523–530.
[242] A. Glidle, R. Hillman, S. Bruckenstein, *J. Electroanal. Chem. Interfacial Electrochem.* **318** (1991) 411–420.
[243] R. Beck, U. Pittermann, K. G. Weil, *J. Electrochem. Soc.* **139** (1992) no. 2, 453–461.
[244] R. Beck, U. Pittermann, K. G. Weil, *Ber. Bunsen-Ges. Phys. Chem.* **92** (1988) 1363–1368.
[245] D. Salt: *Handbook of Quartz Crystal Devices,* van Nostrand Reinhold, New York 1987.
[246] Z. P. Khlebarov, A. I. Stoyanova, D. I. Topalova, *Sens. Actuators B* **8** (1992) 33–40.
[247] V. M. Ristic: *Principles of Acoustic Devices,* J. Wiley and Sons, New York 1983.
[248] P. Hauptmann, *Sens. Actuators A* **25** (1991) 371–377.
[249] B. A. Auld: *Acoustic Fields and Waves in Solids,* vol. II, J. Wiley and Sons, New York 1973.
[250] J. F. Vetelino, D. L. Lee, WO 83 1511, 1983.
[251] A. Bryant, D. L. Lee, J. F. Vetelino, *Ultrason. Symp. Proc.* 1981, 171–174.
[252] A. D'Amico, *Sens. Actuators* **3** (1982/83) 31–39.
[253] S. J. Martin, S. S. Schwartz, R. L. Gunshor, R. F. Pierret, *J. Appl. Phys.* **54** (1983) 561–569.
[254] E. Gizeli, A. C. Stevenson, N. J. Goddard, C. R. Lowe: *Chemical Sensors,* Waseda University Int. Conference Center, Tokyo, Sept. 13–17, 1992.
[255] P. Hauptmann: *Sensoren, Prinzipien und Anwendungen,* Hanser Verlag, München 1990.
[256] G. J. Bastiaans in S. Yamauchi (ed.): *Chemical Sensor Technology,* Elsevier, Amsterdam 1992, pp. 181–204.
[257] J. Reinbold: *Dissertation,* University of Muenster 2000.
[258] J. Curie, P. Curie, *Bull. Soc. Min. Paris* **3** (1880) 90.
[259] R. A. Heising: *Quartz Crystal for Electrical Circuits – Their Design and Manufacture,* van Nostrand, New York 1947, p. 16.
[260] S. Büttgenbach: *Mikromechanik,* Teubner, Stuttgart 1991, p. 39.
[261] L. E. Halliburton, J. J. Martin in E. A. Gerber, A. Ballato (eds.): *Precision Frequency Control,* vol. 1, Acoustic Resonators and Filters, Academic Press, New York 1985.
[262] M. H. Ho: "Applications of Quartz Crystal Microbalances in Aerosol Measurements," in C. Lu, A. W. Czanderna (eds.): "Applications of Piezoelectric Quartz Crystal Microbalances," *Methods and Phenomena,* vol. 7, Elsevier, Amsterdam 1984.
[263] W. G. Cady: *Piezoelectricity,* McGraw Hill, New York 1946.

[264] J. F. Nye: *Physical Properties of Crystals,* Oxford University Press, London 1957.
[265] J. C. Brice: "Crystals for Quartz Resonators," *Rev. Mod. Phys.* **57** (1985) 105.
[266] M. Born, K. Huang: *Dynamical Theory of Crystal Lattices,* Clarendon Press, Oxford 1954.
[267] E. K. Sittig: "Acoustic Wave Devices," in *Encyclopedia of Physical Science and Technology,* vol. 1, Academic Press, New York 1987, pp. 83–109.
[268] F. L. Dickert, F. Vonend, H. Kimmel, G. Mages, *Fresenius Z. Anal. Chem.* **333** (1989) 615–618.
[269] A. W. Snow, J. R. Griffith, N. P. Marullo, *Macromolecules* **17** (1984) 1614.
[270] F. L. Dickert, *Chem. Unserer Zeit* **26** (1992) 138–143.
[271] B. A. Cavic-Vlasak, L. J. Rajakovic, *Fresenius J. Anal. Chem.* **343** (1992) 339–347.
[272] E. Weber, A. Ehlen, C. Wimmer, J. Bargon, *Angew. Chem.* **105** (1993) 116–117.
[273] I. Lundström, M. S. Shivaraman, C. M. Stevenson, *J. Appl. Phys.* **46** (1975) 3876–3881.
[274] M. S. Nieuwenhuizen, W. Barendsz, *Sens. Actuators* **11** (1987) 45–62.
[275] I. Langmuir: "The Adsorption of Gases on Plane Surfaces of Glass, Mica and Pt," *J. Am. Chem. Soc.* **40** (1918) 1361.
[276] J. H. de Boer: *The Dynamical Aspects of Adsorption,* Academic Press, Oxford 1953.
[277] A. Clark: *The Chemisorptive Bond,* Academic Press, New York 1974.
[278] F. M. Moser, A. L. Thomas: *Phthalocyanine Compounds,* Reinhold Publ. Co., New York 1963, chap. 8.
[279] L. F. Barringer, D. P. Rillema, J. H. Ham IV, *J. Inorg. Biochem.* **21** (1984) 195.
[280] J. R. Lenhard, R. W. Murvay: "Chemically Modified Electrodes," *J. Am. Chem. Soc.* **100** (1978) 7880.
[281] J. S. Miller (ed.), *ACS Symp. Ser.* **182** (1981) part 15.
[282] M. T. Reed et al.: "Thermodynamic and Kinetic Data for Cation Macrocyclic Interactions," *Chem. Rev.* **85** (1985) 271.
[283] D. W. Armstrong, *J. Liq. Chromatogr.* **7, Suppl. 2** (1984) 353–376.
[284] M. S. Nieuwenhuizen, A. Venema: "Mass-Sensitive Devices," in W. Göpel et al. (eds.): *Sensors,* vol. II, chap. 13, VCH Verlagsgesellschaft, Weinheim 1991, pp. 657–658.
[285] J. G. Brace, T. S. Sanfelippo, S. Joshi, *Sens. Actuators* **14** (1988) 47–68.
[286] E. T. Zellers, R. M. White, S. M. Rappaport, S. W. Wenzel, *Proc. Int. Conf. Sens. Actuators 4th* 1987, Tokyo: IEE Japan, pp. 459–461.
[287] M. S. Nieuwenhuizen, A. J. Nederlof, *Sens. Actuators B* **2** (1990) 97–101.
[288] A. W. Snow et al., *Langmuir* **2** (1986) 513–519.
[289] A. J. Ricco, S. J. Martin, T. E. Zipperian, *Sens. Actuators* **8** (1985) 105–111.
[290] A. Byrant et al., *Sens. Actuators* **4** (1983) 105–111.
[291] M. Janghorbani, H. Freund, *Anal. Chem.* **45** (1973) 325–332.
[292] A. Kindlund, H. Sundgren, I. Lundström, *Sens. Actuators* **6** (1984) 1.
[293] D. Soares, *J. Phys. E,* in press.
[294] C. Barnes, *Sens. Actuators A* **29** (1991) 59–69.
[295] S. Bruckenstein, M. Shay, *Electrochim. Acta* **30** (1985) 1295–1300.
[296] F. L. Dickert, A. Haunschild, V. Maune, *Sens. Actuators B* **12** (1993) 169–173.
[297] M. Ho, G. G. Guilbault, E. P. Scheide, *Anal. Chem.* **52** (1982) 1998–2002.
[298] J. M. Jordan, Dissertation, University of New Orleans 1985.
[299] A. Suleiman, G. G. Guilbault, *Anal. Chim. Acta* **162** (1984) 97–102.
[300] J. F. Alder, A. E. Bentley, P. K. P. Drew, *Anal. Chim. Acta* **182** (1986) 123–131.
[301] Y. Tomita, M. H. Ho, G. G. Guilbault, *Anal. Chem.* **51** (1979) 1475–1478.
[302] J. A. O. Sanchez-Pedreno, P. K. P. Drew, J. F. Alder, *Anal. Chim. Acta* **182** (1986) 285–291.
[303] G. M. Varga, Jr., US NTIS AD Report 780171/5 GA, 1974.
[304] E. C. Hahn, A. Suleiman, G. G. Guilbault, J. R. Cananaugh, *Anal. Chim. Acta* **197** (1987) 195–202.
[305] G. G. Guilbault, *Anal. Chem.* **55** (1983) 1682–1684.
[306] M. J. van Sant, Dissertation, University of New Orleans 1985.
[307] C. Kößlinger et al., *Biosens. Bioelectron.* **7** (1992) 397–404.
[308] C. G. Fox, J. F. Alder: "Surface Acoustic Wave Sensors for Atmospheric Gas Monitoring," in P. T. Moseley, J. Norris, D. E. Williams (eds.): *Techniques and Mechanism in Gas Sensing,* Adam Hilger, Bristol 1991, chap. 13.
[309] J. Auge, P. Hauptmann, F. Eichelbaum, S. Rösler, *Sens. Actuators,* in press.
[310] J. W. Grate, S. J. Martin, R. M. White, *Anal. Chem.* **65** (1993) 987–996.
[311] A. D'Amico, A. Palma, E. Verona, *Sens. Actuators* **3** (1982/83) 31–39.
[312] A. Venema et al., *Sens. Actuators* **10** (1986) 47.
[313] A. Venema et al., *Electron. Lett.* **22** (1986) 184.
[314] A. D'Amico, A. Palma, E. Verona, *Appl. Phys. Lett.* **41** (1982) 300–301.
[315] A. D'Amico, M. Gentilli, P. Veradi, E. Verona, *Proc. Int. Meet. Chem. Sens. 2nd* 1986, 743–746.
[316] A. W. Snow et al., *Langmuir* **2** (1986) 513–519.
[317] J. G. Brace, T. S. Sanfelippo, S. Joshi, *Sens. Actuators* **14** (1988) 47–68.
[318] J. G. Brace, T. S. Sanfelippo, *Proc. 4th Int. Conf. Sens. Actuators* 1987, 467–470.
[319] M. S. Nieuwenhuizen, A. J. Nederlof, M. J. Vellekoop, A. Venema, *Sens. Actuators* **19** (1989) 385–392.
[320] H. Wohltjen, *Sens. Actuators* **5** (1984) 307–325.
[321] A. W. Snow, H. Wohltjen, *Anal. Chem.* **56** (1984) 1411–1416.
[322] S. J. Martin, S. S. Schwartz, R. L. Gunshor, R. F. Pierret, *J. Appl. Phys.* **54** (1983) 561–569.
[323] S. J. Martin, K. S. Schweizer, S. S. Schwartz, R. L. Gunshor, *Ultrason. Symp. Proc.* 1984, 207–212.
[324] A. J. Ricco, S. J. Martin, T. E. Zipperian, *Sens. Actuators* **8** (1985) 319–333.
[325] J. W. Grate et al., *Anal. Chem.* **60** (1988) 869–875.
[326] K. Mosbach, *Biosens. Bioelectron.* **6** (1991) 179.
[327] P. T. Walsh, T. A. Jones in W. Göpel et al. (eds.): *Sensors: A Comprehensive Survey,* **vol. 2,** VCH Verlagsgesellschaft, Weinheim 1991, p. 529.
[328] A. F. Hollemann, E. Wiberg: *Lehrbuch der anorganischen Chemie,* De Gruyter, Berlin 1985.
[329] B. Danielsson, F. Winquist in A. E. Cass (ed.): *Biosensors: A Practical Approach,* Oxford University Press, Oxford 1989, p. 191.
[330] A. R. Baker, GB 892 530, 1962.

[331] K. Mosbach et al., *Biochim. Biophys. Acta* **364** (1974) 140.
[332] B. Mattiasson et al., *Anal. Lett.* **9** (1976) 217.
[333] B. Danielsson et al., *Appl. Biochem. Bioeng.* **3** (1981) 103.
[334] H. G. Hundeck et al., *GBF Monogr. Ser.* **17** (1992) 321.
[335] B. Xie et al., *Sens. Actuators B* **6** (1992) 127.
[336] E. J. Guilbeau et al., *Trans. Am. Soc. Artif. Intern. Organs* **33** (1987) 329.
[337] G. Steinhage: *Dissertation*, University of Muenster 1996.
[338] M. G. Jones, T. G. Nevell, *Sens. Actuators* **16** (1989) 215.
[339] B. Danielsson et al., *Methods Enzymol.* **137** (1988) 181.
[340] S. Wold et al.: *Food Research and Data Analysis*, Applied Sciences, Barking 1983, pp. 147–188.
[341] T. Nakamoto, K. Fukunishi, T. Moriizumi, *Sens. Actuators B* **1** (1989) 473.
[342] K. Ema et al., *Sens. Actuators B* **1** (1989) 291.
[343] J. R. Stetter, P. C. Jurs, S. L. Rose, *Anal. Chem.* **58** (1986) 860.
[344] U. Weimar et al., *Sens. Actuators B* **1** (1990) 93.
[345] W. P. Carey et al., *Sens. Actuators* **9** (1986) 223.
[346] D. Wienke, K. Danzer, *Anal. Chim. Acta* **184** (1986) 107.
[347] B. R. Kowalski, *Anal. Chem.* **47** (1975) 1152.
[348] B. R. Kowalski, *Anal. Chem.* **52** (1980) 112R.
[349] E. Frank, B. R. Kowalski, *Anal. Chem.* **54** (1982) 232R.
[350] K. R. Beebe, B. R. Kowalski, *Anal. Chem.* **59** (1987) 1007R.
[351] D. W. Osten, B. R. Kowalski, *Anal. Chem.* **57** (1985) 908.
[352] B. Schlatt: *Diplomarbeit*, University of Muenster 1996.
[353] A. Schulga, K. Cammann in G. Henze, M. Köhler, J. P. Lay (eds.): *Umweltdiagnostik mit Mikrosystemen*, Wiley-VCH, Weinheim 1999, pp. 14–32.
[354] A. F. P. Turner, I. Karube, G. S. Wilson (eds.): *Biosensors – Fundamentals and Applications*, Oxford University Press, Oxford 1987.
[355] H.-L. Schmidt, R. Kittsteiner-Eberle, *Naturwissenschaften* **73** (1986) 314–321.
[356] F. Scheller, F. Schubert: *Biosensors, Techniques and Instrumentation in Analytical Chemistry*, vol. 11, Elsevier, Amsterdam 1992.
[357] E. A. H. Hall: *Biosensors*, Open University Press, Buckingham 1990.
[358] L. C. Clark, Jr., C. Lyons: *Ann. N. Y. Acad. Sci.* **102** (1962) 29–45.
[359] J. J. Kulys, A. S. Samalius, G. J. S. Sviermikas, *FEBS Lett.* **114** (1980) 7–10.
[360] M. J. Eddows, H. A. O. Hill, *Biosci. Rep.* **1** (1981) 521–532.
[361] M. L. Fultz, R. A. Durst, *Anal. Chim. Acta* **140** (1982) 1–18.
[362] L. Gorton, *J. Chem. Soc. Faraday Trans. 1* **82** (1986) 1245–1258.
[363] B. Gründig et al., in F. Scheller, R. D. Schmid (eds.): "Biosensors: Fundamentals, Technologies and Applications," *GBF Monogr.* **17** (1992) 275–285.
[364] Y. Degani, A. Heller, *J. Phys. Chem.* **91** (1987) no. 6, 1285–1289.
[365] Y. Degani, A. Heller, *J. Am. Chem. Soc.* **111** (1989) 2357–2358.
[366] W. Schuhmann, T. J. Ohara, H.-L. Schmidt, A. Heller, *J. Am. Chem. Soc.* **113** (1991) no. 4, 1394–1397.
[367] F. W. Scheller, R. Renneberg, F. Schubert, in S. Colowick, N. O. Kaplan, K. Mosbach (eds.): *Methods in Enzymology*, **vol. 137**, Academic Press, San Diego 1988, pp. 29–44.
[368] F. Schubert, D. Kirstein, K. L. Schröder, F. W. Scheller, *Anal. Chim. Acta* **169** (1985) 391–406.
[369] A. Warsinke, B. Gründig, DE 4 227 569, 1993.
[370] K. Riedel, B. Neumann, F. Scheller, *Chem. Ing. Tech.* **64** (1992) no. 6, 518–528.
[371] M. H. Smit, A. E. G. Cass, *Anal. Chem.* **62** (1990) 2429–2436.
[372] D. M. Rawson, A. J. Willmer, *Biosensors* **4** (1989) 299–311.
[373] R. S. Sethi: *Biosensors & Bioelectronics* **9** (1994) 243–264.
[374] M. P. Byfield, R. A. Abuknesha: "Biochemical Aspects of Biosensors," *GEC J. Res.* **9** (1991) 97–117.
[375] M. E. Eldefrawi et al., *Anal. Lett.* **21** (1988) 1665–1680.
[376] S. Lofas, B. Johnsson, *J. Chem. Soc. Chem. Commun.* 1990, 1526–1528.
[377] F. Brosinger et al., *Sens. Actuators B* **44** (1997) 350–355.
[378] D. Kröger, A. Katerkamp, R. Renneberg, K. Cammann, *Biosensors & Bioelektronics* **13** (1998) 1141–1147.
[379] WO 9416312 A1, 1994 (R. C. Jorgenson, S. S. Yee).
[380] A. Katerkamp et al., *Mikrochim. Acta* **119** (1995) (1–2) 63–72.
[381] M. Niggemann et al., *Proc. SPIE-Int. Soc. Opt. Eng.* (1995) 2508. Chemical, Biochemical, and Environmental Fiber Sensors VII, 303–311.
[382] U. Narang, P. R. Gauger, F. S. Ligler, *Anal. Chem.* **69** (1997) no. 14, 2779–2785.
[383] J. E. Roederer, G. J. Bastiaans, *Anal. Chem.* **55** (1983) 2333–2336.
[384] G. G. Guilbault, J. H. Luong, *J. Biotechnol.* **9** (1988) 1–10.
[385] M. P. Byfield, R. A. Abuknesha, *GEC J. of Res.* **9** (1991) 108.
[386] M. Hikuma et al. *Biotechnol. Bioeng.* **21** (1979) 1845–1850.
[387] C. Diviès, *Ann. Microbiol. (Paris)* **126A** (1975) 175–186.
[388] G. A. Rechnitz, *Science* **214** (1981) 287–291.
[389] I. Karube, S. Suzuki, *Ion Sel. Electrode Rev.* **6** (1984) 15–58.
[390] K. Riedel, R. Renneberg, F. Scheller, *Anal. Lett.* **53** (1990) 757–770.
[391] Frost & Sullivan: *The European Market for Industrial Gas Sensors* and The European Market for Portable and Transportable Analytical Instruments, Frost & Sullivan Eigenverlag, Frankfurt 1992.
[392] INFRATEST: *Chemische und biochemische Sensoren – Marktübersicht*, Infratest Industria, 1992.
[393] MIRC: *The European Market for Process Control Equipment and Instrumentation*, Frost & Sullivan Eigenverlag, Frankfurt 1992.

29. Microscopy

ANDRES KRIETE, Klinikum der Justus-Liebig-Universität Gießen, Institut für Anatomie und Zellbiologie, Gießen, Federal Republic of Germany (Chap. 29)

HEINZ GUNDLACH, Carl Zeiss, Jena, Federal Republic of Germany (Chap. 29)

SEVERIN AMELINCKX, Universiteit Antwerpen (RUCA), Antwerpen, Belgium (Sections 29.2.1, 29.2.2)

LUDWIG REIMER, Physikalisches Insitut Universität Münster, Münster, Federal Republic of Germany (Section 29.2.3)

29.	Microscopy	.1061
29.1.	**Modern Optical Microscopy**	.1061
29.1.1.	Introduction	.1061
29.1.2.	Basic Principles of Light Microscopy	.1062
29.1.2.1.	Optical Ray Path	.1062
29.1.2.2.	Imaging Performance and Resolution	.1063
29.1.2.3.	Characteristics and Classification of Lenses	.1063
29.1.2.4.	Eyepieces and Condensers	.1065
29.1.3.	Illumination and Contrast Generation	.1065
29.1.3.1.	Optical Contrast Generation	.1065
29.1.3.2.	Fluorescence Microscopy	.1067
29.1.4.	Inverted Microscopy	.1069
29.1.5.	Optoelectronic Imaging	.1069
29.1.6.	Confocal Laser Scanning Microscopy	.1070
29.1.6.1.	Basic Principles	.1070
29.1.6.2.	Imaging Performance	.1071
29.1.6.3.	Instrumentation	.1072
29.1.6.4.	Imaging Modalities and Biomedical Applications	.1074
29.1.7.	Computer Applications in Digital Microscopy	.1075
29.1.7.1.	Image Analysis	.1076
29.1.7.2.	Visualization	.1077
29.2.	**Electron Microscopy**	.1077
29.2.1.	Introductory Considerations	.1077
29.2.2.	Conventional Transmission Electron Microscopy (CTEM)	.1078
29.2.2.1.	Introduction	.1078
29.2.2.2.	Scattering by Atoms: Atomic Scattering Factor	.1078
29.2.2.3.	Kinematic Diffraction by Crystals	.1078
29.2.2.4.	Dynamic Diffraction by Crystals	.1081
29.2.2.5.	Operating Modes of the Electron Microscope	.1084
29.2.2.6.	Selected-Area Electron Diffraction (SAED)	.1085
29.2.2.7.	Diffraction Contrast Images	.1086
29.2.2.8.	Convergent Beam Diffraction	.1088
29.2.2.9.	High-Resolution Electron Microscopy	.1091
29.2.2.10.	Scanning Transmission Microscopy	.1095
29.2.2.11.	Z-Contrast Images	.1096
29.2.2.12.	Analytical Methods	.1096
29.2.2.13.	Specimen Preparation	.1100
29.2.2.14.	Applications to Specific Materials and Problems	.1100
29.2.3.	Scanning Electron Microscopy	.1115
29.2.3.1.	Introduction	.1115
29.2.3.2.	Instrumentation	.1116
29.2.3.3.	Electron–Specimen Interactions	.1119
29.2.3.4.	Image Formation and Analysis	.1121
29.2.3.5.	Elemental Analysis	.1124
29.3.	**References**	.1125

29.1. Modern Optical Microscopy

29.1.1. Introduction

Optical microscopy goes back as far as the 16th century, when magnifiying glasses and optical lenses became available. ANTONI VAN LEEUWENHOEK (1623–1723) observed structures lying beyond the resolution limit of the eye, such as bacteria, with a 270 fold magnifying, single lens. The exact design of the microscope was established from a theory of image formation developed by ABBE in 1873 [1]. With the introduction of apochromatic lenses and oil immersion, a theoret-

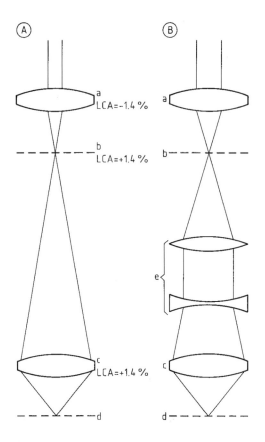

Figure 1. Schematic optical path for objectives of finite length; compensating eyepieces correct for chromatic aberrations
A) Conventional microscope with finite optics; B) Microscope with finite optics and telan optics for insertion of optical components
a) Eyepiece; b) Primary intermediate image; c) Objective; d) Object plane; e) Telan system (incident light fluorescence, Optovar, Bertrand lens)

in particular it can remove out-of-focus blur. Thus, an optical, noninvasive sectioning capability is achieved in thick specimens. The final breakthrough of confocal microscopy in biomedicine and material science boosted the development of a variety of commercial confocal laser scanning microscopes. Highly developed optoelectronic components such as laser light sources and sensitive detectors, as well as the application of specific fluorescent dyes, enabled wide acceptance of this new imaging technique together with effective use of powerful computers to digitally store, visualize, and analyze three-dimensional data.

29.1.2. Basic Principles of Light Microscopy

29.1.2.1. Optical Ray Path

The rigid *stand* of the microscope is the "carrying" element, with mechanical and optical parts attached. The *objective* is the central optical unit, generating an intermediate image that is magnified by the *eyepiece* for visual observation [2]–[4]. *Condenser* and *collector* optimize specimen illumination. Lens designs corrected for a *finite length* form a mirrored, magnified, aerial (intermediate) image, which is projected infinity by the eyepiece (Fig. 1). The distance between the back focal plane of the objective and the primary image plane (which is the front focal plane of the eyepiece) is defined as the *optical tube length*. The *mechanical tube length*, measured from the mounting plane of the objective to the mounting plane of the eyepiece, is 160 mm for currently used microscopes. Any aberrations present in the intermediate image, such as the lateral chromatic aberration (LCA), are compensated by the eyepiece. Optical–mechanical components such as the Bertrand lens, or epi-illuminating reflectors mounted within the optical path require a so-called telan system. Objectives of *infinity-corrected designs* generate an infinite image projected into a real image by the tube lens (Fig. 2). The infinity color-corrected system (ICS) optic is an example of how the tube lens together with the objective can correct aberrations, particularly chromatic ones [5]. The intermediate image is monochromatic; consequently, photographic and video recording do not require compensating lenses. Moreover, the path between lens and tube lens may be used by all optical–mechanical components for various illuminating and contrast generating methods without telan optics (Fig. 2).

ical resolution limit of 0.2 µm became available at this early stage. Light microscopy advanced further by new lens designs, optical contrast enhancement methods, and application of fluorescent and immunofluorescent dyes, also in combination with inverted microscope designs. Optoelectronic methods, such as video techniques, have further broadened the spectrum of light microscopic applications. Finally, laser scanning techniques and confocal imaging allowed the investigation of microscopic structures three-dimensionally. Compared with a standard microscope, the confocal microscope has enhanced lateral and axial resolution, as well as it has improved contrast, and

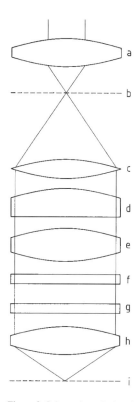

Figure 2. Schematic optical path for objectives corrected for infinity with color-corrected intermediate image
a) Eyepiece; b) Plane of intermediate image; c) Tube lens; d) Bertrand lens slider; e) Optovar 1 X, 1.25 X, 1.6 X, reflector for incident light; f) λ/4 plate, plate, compensators; g) Differential interference contrast (DIC) slider; h) Objective; i) Specimen plane

29.1.2.2. Imaging Performance and Resolution

The image generation in microscopy cannot be explained sufficiently by means of geometric optics alone; wave optics must also be considered [1], [6]. Imaging a point source generates a diffraction pattern in the focal plane. The central parts of this Airy disk contain most of the intensity; therefore, the diameter of this kernel can be defined as the smallest resolvable image element (Fig. 3 A–D). The diameter of the kernel d_k is given by

$$d_k = 1.22 \lambda / n \cdot \sin\alpha \tag{1.1}$$

with λ being the wavelength, n the refractive index of the material, α the angle between the optical axis and the marginal beam of the light cone that enters the objective, and $n \cdot \sin\alpha$ the numerical aperture (NA).

The resolution limit for two points is reached when the maximum of the first diffraction spot coincides with the first minimum of the second one. ABBE developed this theory and proved it experimentally in his historical papers [1]. The resolution is calculated from the available numerical aperture and the wavelength λ by

$$d_r = 1.22 \lambda / NA_{obj} + NA_{cond} \tag{1.2}$$

As an example, a lens and condenser of $NA = 1.4$ and $\lambda = 530$ nm give a smallest separation of $d = 200$ nm. Resolution can be improved either by a higher NA, which is difficult to obtain in lens design, or by using shorter wavelength, as in ultraviolet microscopy [7].

Both the size of the diffraction spot and the resolution of the eye define the utilizable magnification. Since the resolution of the eye is ca. 2–4 arcminutes, total magnification should not exceed 500–1000 times the numerical aperture. Magnification above this limit does not deliver any additional information, but may be used for automatic measuring procedures or counting.

29.1.2.3. Characteristics and Classification of Lenses

The imperfections of optical lenses give rise to various distortions, either monochromatic or chromatic, that are visible in the diffraction pattern. Monochromatic distortions include aberration, curvature, astigmatism, and coma. Chromatic aberrations are caused by differences in the refractive index of glasses. These types of errors are minimized by combining several pieces of glass with different form and refractive index, or by using diaphragms. The three main categories of lenses are

1) Achromatic objectives
2) Fluorite or semiapochromatic objectives
3) Apochromatic objectives

Fluorite and apochromatic lenses are mostly designed as plane (i.e., corrected for a flat field of view). Computers allow calculation of the diffraction pattern under various imaging conditions to optimize imaging performance for a wide range of applications. The present advantage in modern lens design and production is indicated by the image quality of the ICS Plan Neofluar class. At a field-of-view number of 25, color photography, as well as all optical contrast enhancement methods, are supported with achromatic correction

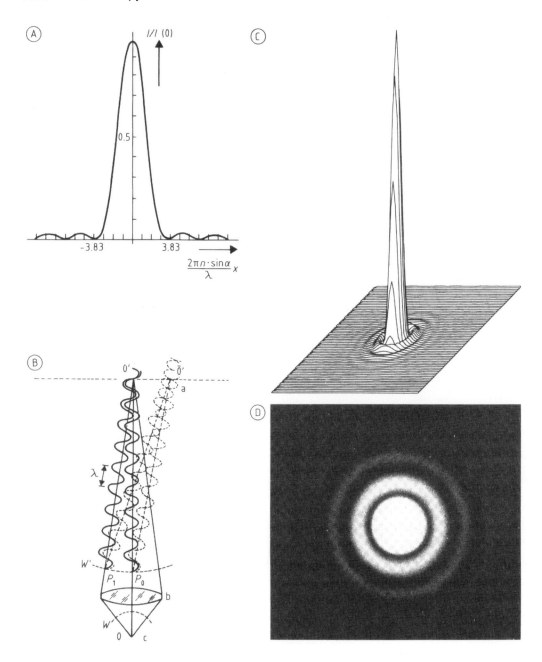

Figure 3. A) and B) Intensity distributions in the image of a luminous point; a) Image plane; b) Lens; c) Objects plane; C) Computed intensity distribution; D) Airy disk

comparable to apochromatic lenses with a good transmission down to UV at 340 nm. These objectives are generally applicable for all fluorescent methods (see Sections 29.1.3.2, 29.1.6.4). Plan apochromates feature a still higher aperture: chromatic rendition is improved, as well as correction at the field-of-view boundary; therefore these objectives are very suitable for color microphotography. In addition, a complete correction not only at the focus level but also for a limited range above

29.1.2.4. Eyepieces and Condensers

The main characteristics of *eyepieces* are magnification and type of correction, such as color compensation and correction of viewing angle. The correction of the *condensers* must match the properties of the lenses. Special or combined condensers are required for certain methods of contrast generation or for longer working distance with inverted microscopes.

29.1.3. Illumination and Contrast Generation

Illumination and contrast are central points in the discussion of imaging performance. Without sufficient contrast, neither the required magnification nor a maximal resolution can be obtained. A. KÖHLER developed optical illumination for microscopy in 1893 [8]. The main advantage of Köhler illumination is that in spite of the small filament of the light bulb, the entire area of the lamp field stop, as well as the illuminated part of the specimen, have a uniform luminance (Fig. 4). The twofold ray path of diaphragms and lenses suggests that adjustment of the condenser iris, being the illumination aperture, controls resolution, contrast, and resolution depth. This principle is also important for correct setup of the confocal microscope as well as electron microscopy.

29.1.3.1. Optical Contrast Generation

For the following considerations the principles of wave optics are used, which describe image formation as a result of diffraction and interference.

Dark-Field Illumination. In dark-field illumination the illuminating aperture is always set higher than the lens aperture. Only light waves being diffracted by the microstructures are captured by the lens. Objects appear as bright diffraction patterns on a dark background.

Phase Contrast. Based on ABBE's theory the Dutch physicist FRITS ZERNIKE (Nobel prize in 1953) developed phase contrast imaging in 1935, to convert the invisible phase shift within the specimen into recognizable contrast [9]. A phase plate located in the lens modifies the phase and amplitude of the direct (iluminating) wave in such a way that a phase shift of $\lambda/2$ is obtained (Fig. 5). The

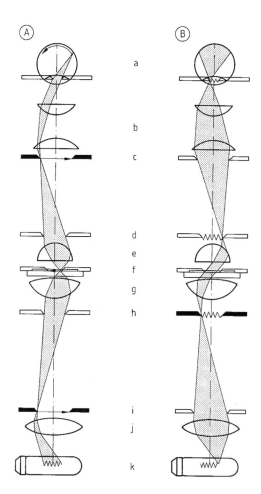

Figure 4. Schematic of Köhler illumination
A) Image-forming ray path; B) Illuminating ray path
a) Eye; b) Eyepiece; c) Eyepiece field stop; d) Exit pupil of objective; e) Objective; f) Specimen; g) Condenser; h) Aperture diaphragm; i) Lamp field stop; j) Light collector; k) Light source

and below the focal plane is achieved. For specific applications, the following lenses are also available:

1) Multi-immersion objectives of type Plan Neofluar, corrected for use with oil, glycerol, or water, with or without a cover glass
2) Objectives for specimen without cover slide
3) Objectives with a long working distance
4) Objectives with iris diaphragm
5) Objectives for phase contrast investigation
6) Objectives, strain free for polarized light microscopy
7) Ultrafluars, corrected between 230 and 700 nm

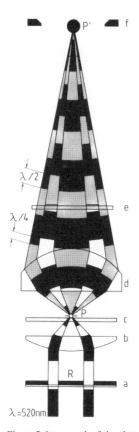

Figure 5. Image path of the phase contrast microscope, converting phase shifts into a noticeable contrast
a) Illuminating diaphragm ring; b) Condenser; c) Specimen generating a phase shift; d) Objective plate; e) Phase plate; f) Intermediate image

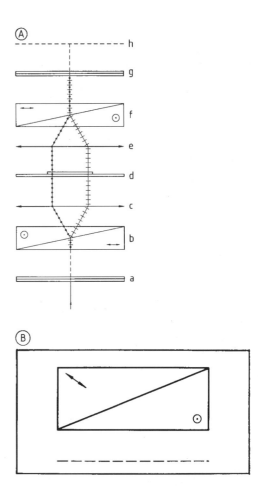

Figure 6. A) Image path of a two-beam interference microscope; B) Nomarski prism with direction of optical axes and position of interference plane indicated
a) Polarizer (45°); b) First Wollaston prism; c) Condenser; d) Specimen; e) Objective; f) Second Wollaston prism; g) Analyzer (135°); h) Intermediate image
Contrast is generated by a phase shift of the polarized and prism-split illuminating beams, which are combined by the second prism

amplitudes of the direct diffracted waves are equalized, and by destructive interference, dark-phase objects appear on a bright background. The application of this technique ranges from routine diagnosis to cell biology research [6], [10]. Today this method is also used advantageously in combination with epifluorescence. The specimen is adjusted under phase contrast before switching to fluorescence light in order to avoid bleaching, or both imaging modes are combined to match the fluorescent signal with the morphological image.

Nomarski Differential Interference Contrast (DIC). In 1955 the physicist GEORGE NOMARSKI simplified the two-beam interference microscope in a way that it became available for routine microscopy [11] (Fig. 6). DIC uses modified wollaston prisms (Fig. 6B), lying outside the focal areas of condenser and objective [11]–[13]. Normal lenses are used; the contrast can be adjusted by shifting the lens-sided prisms. As in phase contrast, differences in the optical path length are visualized. Objects appear as a relief, and thick specimens lack the halo-typical pattern of phase contrast imaging. Full aperture of both the objective and the condenser allow optical sectioning (Fig. 7). Combined with inverted microscopes (see Section 29.1.4) and contrast-enhanced video microscopy (see Section 29.1.5), ultrastructures in cells and tissues can be imaged. Certain applica-

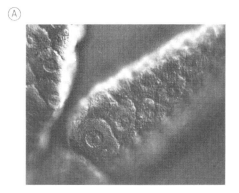

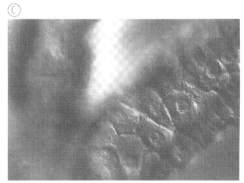

Figure 7. A)–C) Optical sectioning of salivary gland of drosophila by DIC Nomarski (Specimen courtesy of L. Rensing, Universität Bremen, Germany)

tions combine DIC with phase contrast. For routine applications, phase contrast lenses may give better results at lower magnification. The DIC image is well suited to be combined with a fluorescent image as well.

Polarization and Reflection Contrast Microscopy. Polarization microscopy is used preferentially for certain applications in cell research,

to enhance double diffraction (anisotropic) structures such as muscle fibers or the mitotic apparatus. Quantitative studies require strain-free polarization lenses for the polarized light used. With the help of reflection contrast microscopy [14], [15], visible interference is generated at epi-illumination, for example, to study adhesive properties of living cells on glass surfaces [16]. Moreover, this method may be combined with phase contrast or Nomarski DIC contrast. For the evaluation of immunogoldmarked specimens, immunogold staining (IGS), reflection as well as transmission contrast is feasible.

29.1.3.2. Fluorescence Microscopy

Basic Principles. The development of immunofluorescence greatly influenced medical practice and basic biomedical research [17]–[20]. Many biomedical specimens emit *autofluorescence* if illuminated properly. Today a wide range of fluorescent microstructures can be excited by application of fluorochromes (*secondary fluorescence*). Immunofluorescence is based on antigen–antibody connectivity [21]. This couples the high specificity of the immune reaction with the extreme sensitivity of fluorescence microscopy. Various markers for antibodies exist and derivates of fluorescein, mainly fluorescein–isothiocyanate (FITC), are the preferred fluorochromes. For this kind of microscopy, special lamps and filters are required. Light sources used include high-pressure bulbs as well as lasers. Fluorescence microscopy is used either in transmission or in epi-illumination; the latter combines phase contrast and Nomarski DIC [22], [23]. Fluorescence microscopic objectives with high aperture and a good transmission between 340 and 700 nm are preferred. Multi-immersion lenses of plan-type Neofluar have been designed specifically for such applications, as well as low-magnification lenses featuring high apertures (primary magnification 40, *NA* 1.3 oil). Plan apochromates are more suitable for illumination in visible light. The brightness b of fluorescent images increases proportionally to the square of the numerical aperture of lens L and condenser C, and decreases inversely with the square of the total magnification M

$$b \sim \frac{NA_L^2 + NA_C^2}{M^2} \qquad (1.3)$$

This makes ocular lenses with low primary magnification an ideal choice. At epi-illumination the

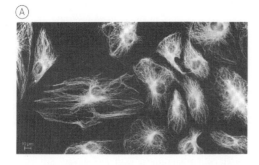

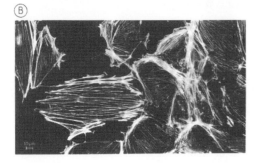

Figure 8. Double fluorescence micrograph of epithelial kidney cells
A) Microtubules, fluorescent dye FITC; B) Actin filament, fluorescent dye tetramethylrhodamine-B-isothiocyanate (TRITC)

brightness depends on the power of four of the aperture, since the lens acts as the condenser as well:

$$b \sim \frac{NA_L^4}{M^2} \tag{1.4}$$

In the 1980s, fluorescence methods became a major tool for sensitive and highly specific detection in biomedical research and medical diagnosis.

Markers. The development of reagents based on fluorescence includes the use of biological molecules: antibodies, RNA, DNA, proteins, and other macromolecules and physiologic indicators [17]–[19]. These methods have enabled new findings in cell biology, molecular biology, developmental biology, and cancer research. The cytoskeleton (Fig. 8) is responsible for the cell shape, cell division, capacity of the cell to move, and the intracellular transport of cell organelles [24], [25]. Multiparameter fluorescence microscopy uses a broad selection of fluorescent dye molecules to examine several properties or components of a single cell simultaneously (Fig. 9).

A series of new probes suitable for this purpose has been developed by extending the usable spectrum to the near infrared (Fig. 10). *Fluorescence-based in situ hybridization* (FISH) is rapidly being recognized as a powerful tool for detecting and quantifying genetic sequences in cells and isolated chromosomes [26]. FISH is based on the highly specific binding of pieces of DNA, labeled with a fluorescent probe, to unique sequences of DNA or RNA in interphase nucleus or metaphase chromosomes. Recent developments enable the simultaneous visualization of three or more DNA regions (chromosome painting) [27].

Dyes and Applications. The green fluorescence protein (GFP) is a strongly fluorescent reporter molecule that stems from the jellyfish *Aquorea victoria* [28]. It has a fluorescein-like characteristic, it requires no cofactors or substrates, and is species-independent. GFP engineering has produced different color mutants which have been used for gene expression, tracing of cell lineage, and as fusion tags to monitor protein localization [29].

Time-Resolved Methods. Special methods and dyes have been developed to observe functional phenomena in living cells. Fluorescence lifetime imaging microscopy (FLIM) monitors fluorecence lifetime as a function of environmental parameters of cellular components. These include Ca^{2+}/pH-sensitive indicators, also called biosensors [19]. Typically a fluorescent dye is chemically linked to a macromolecule, creating an analogue of the natural molecule. The fluorescent analogue is specifically designed to measure chemical properties of the macromolecule, instead of tracking its location within the cell. Time-gated multichannel plates (MCPs) are used for time- and spatially resolved measurements.

Fluorescence recovery after photobleaching (FRAP) and the measurement of the fluorescence resonance energy (FRET), which allow the diffusional mobility of cellular components and their interaction at the molecular level to be monitored are other varieties of time-resolved methods [30]. A novel method is fluorescence correlation spectroscopy (FCS), which measures the statistical fluctuations of fluorescence intensity within a confocally illuminated volume. Correlation analysis allows the concentration of particles and their diffusion to be determined [31], [32]. FCS has proved

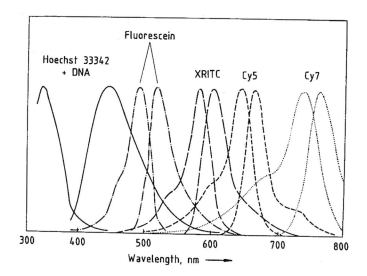

Figure 9. Excitation and emission curves of various fluorochromes
Filter curves depict dyes Hoechst 33342 (suitable for DNA), FITC (for actin), rhodamine (endosomes), and cyanine 5 (mitochondria)
(Courtesy of L. TAYLOR, University of Pittsburgh, United States)

be useful in the study of enzyme kinetics, DNA/protein interactions, receptor/ligand interactions, and cellular measurements [33]. Molecular interactions can be studied without interfering with the natural binding conditions, since only the reference substances in competition experiments need to be fluorescently labeled.

29.1.4. Inverted Microscopy

Inverted microscopes were originally designed for the evaluation and counting of plankton and sediments [34]. The first inverted research microscope which included photomicrography, microphotometry, cinemicrography, and video imaging was the Axiomat [4], [35]–[37]. Inverted microscopes in U-design (Fig. 11) with a rigid stage are of great advantage for investigations in cell and developmental biology, as well as in cell physiology experiments that are performed in optical microscopes [37]–[40]. Identification and selective micropreparation on live nuclear components in oocytes of *Xenopus* in the DIC Nomarski technique facilitate highresolution optical microscopy in different planes of the specimen (optical sectioning), which can be isolated by selective micropreparation for subsequent analysis in the electron microscope [41].

In the life cycle of the green alga *Acetabularia*, high-resolution phase and DIC microscopy have made part of the endoplasmic reticulum and lampbrush chromosomes visible [42], [43]. The distinct advantage of inverted compared to upright microscopes is the unhindered application of cell and tissue culture chambers in conjunction with micromanipulators, micropipettes, microelectrodes (patch clamp method), and injection systems for experimental cell research. Computer-controlled microinjection can be performed automatically [44]. Manipulation of cells, organelles, and genomes is also possible by means of laser microbeam and optical trapping [25].

29.1.5. Optoelectronic Imaging

Electronic imaging methods such as video microscopy and laser scanning microscopy can extend resolution slightly, but the ability to extend the detectability of subresolved structures and subresolution movements is more important. A video camera can generally be mounted to the same port employed for normal photomicrography. In *video-enhanced contrast microscopy* (VEC) (Fig. 12), high-resolution video cameras based on a vidicon tube are used. This principle of adjusting gain, and offset and digital image processing is also used in

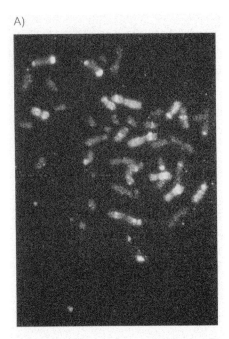

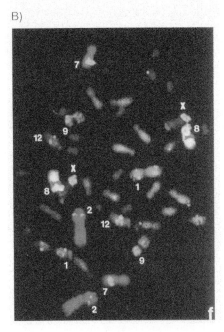

Figure 10. A) Micrograph by triple exposure of color slide film using conventional fluorescence microscopy; B) Pseudocolored digital image recorded with the CCD camera for chromosome 1, 2, 7, 8, 9, 12 and the human X chromosome Green = FITC; Red = TRITC; Blue = DAPI
(Courtesy of T. Cremer, Heidelberg, Germany and Oxford University Press)

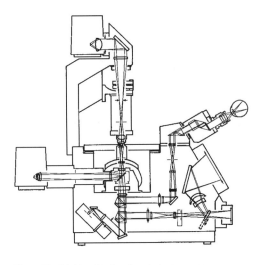

Figure 11. Light path of an inverted microscope

laser scanning microscopes (see Section 29.1.6). Digital imaging and image processing make it possible to see structures that are smaller than the resolution limit [12], [45]–[48]. By means of this method, single microtubules can be observed [49], [50]. The size of each microtubule is only one-tenth of the visible-light resolution limit. In the 1980s, solid-state cameras were introduced into microscopy [51], [52]. Most of them are based on a *charge coupled device* (CCD), which produces digital data directly for computer processing and image analysis. For low light level (LLL) or video intensification microscopy (VIM), i.e., weak fluorescence or small FISH signals (see Fig. 10 B), cooled, sensitive CCD cameras for black and white or color are available now and have replaced tube cameras [53]. Digital imaging techniques have gained increasing attention as supplements to conventional epifluorescence and photomicrography [54]. *Microphotometry* permits spectrophotometric analysis of substances, as well as statistical and kinetic measurement and compilation of two-dimensional intensity profiles. [40].

29.1.6. Confocal Laser Scanning Microscopy

29.1.6.1. Basic Principles

The fundamentals of confocal microscopy go back to 1959, when MINSKY described a new kind of microscopic apparatus [55]. The historical de-

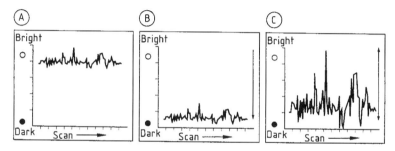

Figure 12. Principle of video-enhanced contrast microscopy
A) Original image; B) Offset (brightness); C) Contrast gain

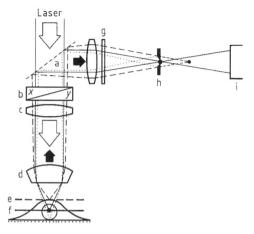

Figure 13. Ray path in the confocal microscope
a) Dichroic beam splitter; b) Scanner; c) Tube lens; d) Objective; e) Specimen; f) Focal plane; g) Filter; h) Pinhole; i) Detection system (photomultiplier)

velopment of confocal microscopy is reviewed elsewhere [56], [57]. Two different kinds of confocal instruments are available: the tandem scanning microscope and the laser scanning microscope. The *tandem scanning microscope* uses normal light chopped by a rotating Nipkow disk [58]. The light beams reflected by the specimen cross the rotating disk on its opposite side (thus tandem), and due to the symmetric pattern of holes, out-of-focus blur is removed. This system allows real-time and real-color imaging, but is limited in sensitivity because of the fixed size of the pinholes.

The more comonly applied *confocal laser scanning microscope* (CLSM) features a variable pinhole and allows adjustment of magnification by modifying scan angles. A schematic example of the principle of confocal laser scanning is given in

Figure 13. The laser light is focused via the scanner (b) through the tube lens (c) and the objective (d), and illuminates a small spot in the specimen (e). Emitted light emanating from the focal plane and the planes above and below (dotted and dashed lines) is directed via the scanner to the dichroic beam splitter (a) where it is decoupled and directed onto a photomultiplier (i). A pinhole (h) in front of the photomultiplier is positioned at the crossover of the light beams emerging from the focal point. This plane corresponds to the intermediate image of the Köhler illumination described in Section 29.1.3. Light emanating from above and below the focal point has its crossover behind and before the pinhole plane so that the pinhole acts as a spatial filter. Numerous papers elucidate the basic aspects of confocal image formation [59]–[64].

Since only a small specimen point is recorded, the entire image of a specimen is generated by either moving the specimen (*stage scanning*) or moving the beam of light across the stationary specimen (*beam scanning*). The latter is utilized in all commercial instruments (Fig. 14). Theoretical considerations have shown that both scanning forms give identical resolution. The optical sectioning capability at lateral scanning opens a way to image a three-dimensional structure with a slight axial stepping of the stage (see Fig. 15).

29.1.6.2. Imaging Performance

The imaging characteristic in confocal microscopy is given by the properties of the specimen weighted by the applicable spatial confocal response function. An explanation based on wave optics suggests that the point spread functions of the illuminating path and the detection path (resembling the photon detection possibility) are con-

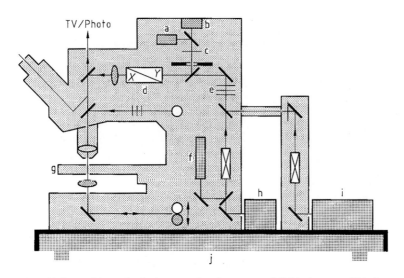

Figure 14. Setup of the confocal microscope, based on a research light microscope (Zeiss)
a), b) Photomultiplier tubes; c) Confocal variable pinhole; d) Scanners; e) Barrier filter; f) He–Ne laser (543 nm, 633 nm); g) Stage; h) Ar laser (488 nm, 514 nm); i) UV laser (350–360 nm); j) Antivibration table
Light of internal (He–Ne) and external (Ar ion or UV) lasers passes the scanner unit; switchable mirrors and barrier filters permit multiwavelength operation. Additional barrier filters are located between the confocal variable pinhole and the beam splitter adjacent to the photomultipliers.

volved, consequently resulting in a narrow point resolution [65].

Describing image formation of an reflecting object and the corresponding coherent imaging properties must be distinguished from a fluorescent object viewed in incoherent imaging because in the latter the illuminating wavelength and the detected wavelength are different. In terms of resolution, the Rayleigh criterion determines the minimum resolvable distance between two points of equal brightness (see Eq. 2). For *incoherent imaging* using a lens with a certain numerical aperture NA at a wavelength λ, the *lateral resolution* d_r obtainable in confocal imaging is [62]:

$$d_r \text{ (conf.)} = 0.46\lambda/NA \qquad (1.5)$$

With a wavelength of 514 nm (green argon laser line) and a numerical aperture of 1.4, this equation gives a resolution of $d_r = 157$ nm. Compared to conventional imaging (see Eq. 1.2), the resolution is ca. 32 % higher [66].

An approximation of the *axial resolution d_z* is given by

$$d_z \text{ (conf.)} = 1.4\lambda/NA^2 \qquad (1.6)$$

The values (514 nm and NA = 1.4) give a d_z of 367 nm.

One way to measure the axial response of a confocal microscope is to record light intensity from a thin layer with infinite lateral extension. In a conventional microscope, no difference — and therefore no depth discrimination — are obtained. However, in confocal microscopy, a strong falloff occurs if the object is out of focus. The free width at half maximum (FWHM) of such a function is proportional to the optical section thickness [64], [67]. The optical section thickness also depends on the diameter of the pinhole. The optimum pinhole size D with a maximum deterioration of 10 % in reflection is given by [63]

$$D = 0.95\lambda M/NA \qquad (1.7)$$

with magnification M, λ, and D given in micrometers. In fluorescence microscopy the size has to be reduced 10–30 %, depending on the ratio of excitation to emission wavelengths. Another way to describe the imaging performance is by the modulation transfer function (MTF); such measured and theoretically derived three-dimensional functions have been published [68], [69].

29.1.6.3. Instrumentation

Confocal microscopic equipment is available as an add-on to a standard microscope or in the

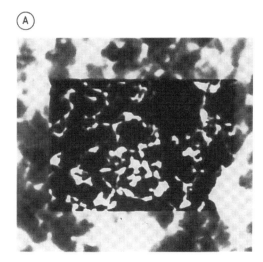

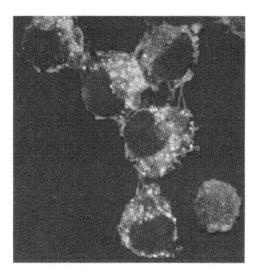

Figure 16. Multichannel confocal image of macrophages (J774, mouse cell line)
Extrathyroidal release of thyroid hormones from thyroglobulin is marked by fluorescence dye (first channel, green); actin cytoskeleton is labeled with rhodamine-phallodin (second channel, red). Morphological information (third channel), providing a dark background in the cells, is detected in transmission.
(Permission of K. Brix, V. Herzog, Institut für Zellbiologie, Bonn, Germany)

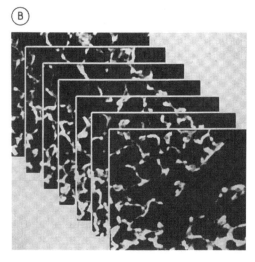

Figure 15. A) Conventional versus confocal imaging (inserted region of interest) of a lung structure in fluorescence; B) Multislice image sequence obtained by optical sectioning

form of more fully integrated and stable devices, including computer control equipment. Both upright and inverted types of microscopes are available. The inverted confocal microscope allows bigger specimens to be studied and offers easy handling of micromanipulating or injecting devices (see also Section 29.1.4). Beam scanning is obtained by rotating galvanometric mirrors. Axial control is realized by a stepping motor drive or a piezoelectric translator. Most instruments are equipped with several detectors to record multi-fluorescence events simultaneously. Some also offer transmission detectors. Signals from different channels can then be mixed electronically (see Fig. 16). Recently, confocal transmission microscopy has also been studied. Improvement of contrast and resolution depending on the pinhole size of a thin specimen has been documented in a confocal transmission arrangement using differential interference contrast [70].

Variations of the confocal design have been developed, such as bilateral scanning in real time with a scanning slit [71]. Because of the fast scanning, TV detectors such as a sensitive CCD camera are best suited. Slit-scan confocal microscopes offer a better signal-to-noise ratio and real-time imaging capabilities but a slightly reduced resolution compared to a conventional CLSM.

Various types of lasers can be linked to the microscope [72], but up to now only continuous-wave gas lasers have been used routinely. This includes argon-ion lasers having excitation lines at 488 and 514 nm and helium–neon lasers with a major line at 633 nm and a weaker line at 543 nm. The use of UV lasers requires achromatic lenses and specifically designed scanning mirrors. Lasers

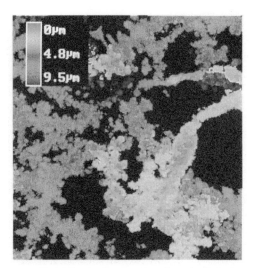

Figure 17. Golgi-stained neuron: three-dimensional display of a confocal data set acquired in reflection
Color depth coding visualizes the three-dimensional neuron; various spines of different size give a brushlike impression. (Specimen courtesy of B. FORTMANN, Institut für Ernährungswissenschaften, Gießen, Germany)

used in the UV range are argon-ion UV or argon-ion tunable lasers, as well as krypton-ion and helium–cadmium lasers. Depending on the emission lines of the fluorescent dyes, various filters are used in front of the photomultiplier, characterized by quantum efficiency and spectral response. The total sensitivity of a system is given by the transfer efficiency of the various optical components [73] and, in addition, the quantum efficiency of the detector and its associated electronics [74].

Simplified types of laser scanning microscopes, such as the *fiber scanning optical microscope* have also been proposed [68]. Here the fiber can act both as a light delivering source and as a pinhole. A totally different physical approach to reduce out-of-focus blur is based on a physical phenomenon called *double-photon (2 p) excitation*, whereby fluorescence is generated by simultaneous absorption of two photons of long wavelength at a single molecule. The necessary energy emitted by pulse lasers is present only at the focal plane; thus a "confocal" effect is produced. At turbid tissue media, the signal level under 2 p-excitation drops much faster than that under single-photon excitation, although image resolution is higher in the former case [75], [76].

The diffraction pattern of the illuminating beam is elongated axially compared to the lateral direction (see Eq. 1.5 versus 1.6). This anisotropy is caused basically by the limited numerical aperture of the microscopic lens. An optimal diffraction pattern would be achieved by an unlimited, 360° aperture. This concept is known as the *4-pi microscope*. Prototypes use two facing lenses or two lenses oriented at an angle (theta microscopy). The object in focus is illuminated from two sides, and fluorescence emanating from the object is detected through both lenses [77].

29.1.6.4. Imaging Modalities and Biomedical Applications

Imaging Modalities. Confocal contrast generation include reflectance, rescattering, and fluorescence. The first results in confocal imaging were obtained on unstained tissue in *reflectance* [78]. Reflectance is also the preferred imaging modality used in industrial inspection of semiconductors, often combined with methods such as optical beam induced conductivity (OBIC).

In biomedical science, the study of unstained tissue is limited due to reflectance. The reflectance must be intensified by metal impregnation (e.g., Golgi stain); one such example is given in Figure 17. Other examples include peroxidase–labeling with nickel intensification in neurobiology [79] or silver-stained nucleolar organizer region-associated proteins (AgNORs). Gold immunolabeling has also been used [80]. In *rescattering*, variations in the refractive index can give rise to reasonable contrast. One particular example is in ophthalmology [81] where structures in the transparent cornea such as cell nuclei, keratocytes, neurons, or the fibers of the lens can be visualized (see Fig. 18).

Fluorescence is the most commonly used imaging mode in confocal microscopy, such as, applications in cell biology [82], [83], which include the use of autofluorescence, specific dyes in combination with antibodies, and in situ hybridization. Comprehensive reviews of available fluorochromes are published elsewhere [80], [84]–[87]. UV lasers broaden the spectrum of available fluorescence dyes; even the autofluorescence present in reduced pyridine nucleotides [NAD(P)H] can be monitored [88].

Biomedical Applications. Optical sectioning devices, being noninvasive and having low radiation damage, are ideal for studying living specimen, organs, and tissues. With confocal microscopes, structures can be observed in vivo at full three-dimensional microscopic resolution [87].

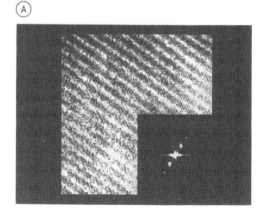

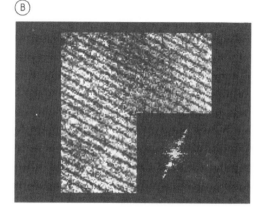

Figure 18. Confocal images of the eye's lens fibers, resulting from rescattering at microstructures with different indices of refraction
A) Imaging through the 400 μm transparent cornea;
B) Imaging with cornea removedThe inserted Fourier power spectrum documents the loss in spatial frequency content.
(Specimen courtesy of B. R. MASTERS, USUHS, Bethesda, United States)

Small organisms or organisms in an early stage of development can be observed, as in embryogenesis [89], [90]. Concerning individual structures, morphological changes, such as the movement of cell compartments (see Fig. 19), growth of neurons [91], or synaptic plasticity in the retina, have also been studied [92]. Moreover, confocal arrangements allow optical trapping (i.e., user-controlled movement of small spheres filled with biochemical solution into and within cells). A new application of confocal microscopy is in clinical imaging, which includes the observation of

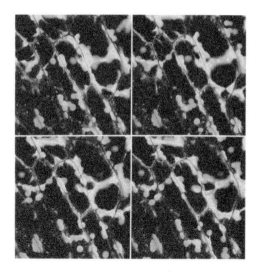

Figure 19. Application of confocal microscopy to plant cells (*Egeria densa Planchon*), in vivo
Cell boundaries and the movement of chloroplasts are visualized by autofluorescence (green) and FITC labeling (red). (Permission of I. DAHSE, Institut für Biophysik, Jena, Germany)

wound healing, flow processes in veins, and opthalmology [81], [86] (see Fig. 18).

29.1.7. Computer Applications in Digital Microscopy

Computer applications in digital microscopy include image storage in databases, compression, enhancement, analysis, and visualization. Digital analysis applied to microscopical data sets has the aim of automatically extracting structures of interest out of digitized images and then quantifying them. This includes field-specific or stereological, geometrical or object-specific, and intensity-related parameters. In the case of 3D images, typically generated by a confocal imaging technique, a computer graphical 3D visualization of the data stack is performed, which allows inspection of the original data and subsequent image-processing operations.

Different digital image formats are used in microscopy, for example, the TIFF format. In 1999, a microscopic imaging standard was established by the ARC-NEMA Institute as part of the DICOM standard (Digital Imaging and Communication in Medicine). This format, DICOM-VL (Visible Light), regulates storage of instrumental

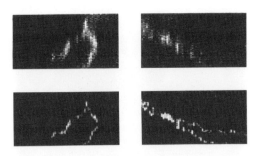

Figure 20. Projectional side views of rat hippocampal dendrites stained with Lucifer yellow
Top: Original confocal data set; Bottom: Restored data following 100 iterations of the maximum likelihood blind deconvolution algorithm
(Permission of T. HOLMES, Rensselaer Insitute, Troy, United States)

and specimen parameters besides the image information. This fosters the use of microscopic images in telemicroscopic applications and in medical networks.

29.1.7.1. Image Analysis

Image quality assurance is an important control mechanism for image data to be analyzed quantitatively. Appropiate techniques include sampling [93], focus setting techniques [94], image quality measurement [95], and calibration routines [96]. Some of the undesired phenomena which influence or prevent successful application of image analysis require preprocessing. This includes attenuation of the laser beam in thick specimens, photobleaching or saturation of the fluorescence, autofluorescence, and background noise, and the distortions caused by the embedding media [80], [86], [97]. The loss of resolution in thick specimens is shown in Figure 18. The correction of such phenomena is often limited by the underlying theoretical models selected, since the effects are highly object dependent.

Undesired effects in fluorescence confocal microscopy can be corrected by ratio imaging. Ratioing (i.e., the compensation of two channels by digital division) can substantially reduce cross-talk effects in double detectors [63]. Environmental markers indicating the pH value or the level of free calcium ions (Ca^{2+}) also require ratio imaging to compensate for structural densities or section thickness [19].

Due to the anisotropic resolution of the data sets, various ways of improving axial resolution are available; *deconvolving methods* use the point spread function of the optics, which has to be measured first [98]. The algorithms can be differentiated further by their ability to incorporate noise. The classical filter is the Wiener filter; new developments include the iterative *maximum likelihood estimation* [99]. This method was extended with a blind deconvolution algorithm, using certain constraints to model the point spread function, which is a promising approach for biomedical users (see Fig. 20). However, calculation times are still in the range of 1 h for some hundred interations. A different approach is the *nearest-neighbor deconvolution* developed for bright-field imaging [100], by taking into account the blur present in the planes above and below the focal plane. More recently, this method has been contrasted with a no-neighbor deconvolution method [101]. Besides directly improving data, isotropy can also be obtained by interpolating the lateral dimension [80]. Transformation of data from a cubic format into other lattices has also been proposed, based on mathematical morphology [102].

Sets in digital image processing to enhance, filter, segment, and identify structures are frequently applied to three-dimensional microscopic data sets. This can be done either section by section or by using the corresponding three-dimensional extension of digital filters, such as the Gauss, Laplace, Sobel, and Median filter or noise removal algorithms [103]–[106]. Segmentation often requires interaction of the user, since automatic segmentation algorithms generally rely on the intensities of the voxels only. Once the binary images have been obtained, binary morphological operations using structuring elements for erosion and dilatation can be performed [107].

Digital measurements address the size and volume of structures, quantities, or forms and relation between volumetric subtleties. To measure volumes, all voxels that identify a certain structure are summed up. It is particularly difficult to measure the surface of structures, and most methods offer an estimation based only on stereological methods. Object counting in volumes usually requires the definition of guard boxes, which are moved throughout the volume. Objects falling into such boxes are evaluated by following certain stereological rules to avoid any bias. Stereological considerations are also used to estimate spatial statistics, such as numerical densities or nearest-neighbor distances; such methods have been used extensively in three-dimensional cytometry [80].

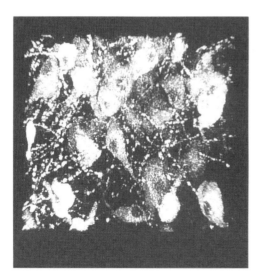

Figure 21. Volume rendering of a confocal sequence
The neural network of immunohistochemically stained cells, together with connecting fibers in the neural optical tract of hamsters is visualized by this projection
(Specimen courtesy of H. KORF, Zentrum für Morphologie, Frankfurt, Germany)

29.1.7.2. Visualization

Visualization of serial sections is the most frequently used computer technique in confocal microscopy, independent of specific applications [83], [108], [109]. Most commercial instruments have three-dimensional software implementations ready for the user. Three-dimensional objects are displayed on the computer screen by staking up the individual sections. Such techniques have been reviewed in several articles [110]–[114]. Besides surface rendering based on contours, volumetric representations that rely on the intensities of the voxels are the preferred techniques in confocal microscopy. A comparative assessment of both techniques is given in [115]. Reconstructions are presented three-dimensionally on a computer graphics screen by using either visual cues, animation, of stereoscopic displays.

One method used in *volume rendering* is to define imaginary rays through the volume (*raycasting techniques*) [116], [117]. The way of casting is accomplished by the definition of certain projection geometry. Given a central projection, the angles of the rays are defined by the matrix of the screeen and the center of projection. A common implementation traverses the data in a front-to-back order, which has the advantage that the algorithm can stop when a predefined property of the accumulation process has been reached. This class of methods is also referred to as *image order rendering*, since the matrix of the screen is used as the starting point for the accumulation process. Because of the projection geometry, such a ray-firing technique does not necessarily pierce the center of a voxel, and an interpolation has to be taken into account. To improve the output of raycasting renderings, preprocessing of the voxel scenes may be performed as, for example, in the renditions of complex chromatin arrangements [106]. Modalities available in image order rendering include maximum projection (Fig. 21), integration, and surface mode (Fig. 18).

A much simpler geometry is realized in a parallel projection, such as in *object order rendering*. Here the volume is rotated according to the current viewing angle and then traversed in a back-to-front order [118]. During traversal, each voxel is projected to the approriate screen pixel where the contributions are summed or blended, a situation that is referred to as compositing. During the accumulation process other properties of voxels besides the intensities may be considered, such as opacity, color, gradient values [119], [120], or the simulation of fluorescence processes (SFP) [121]. However, since volume rendering relies entirely on voxel attributes, it has some disadvantages if analytical representations are required. Therefore, elements of computer graphics are combined with volumetric renditions. Examples can be found in vector descriptions of volumetric data sets [122] or in geometric attributes expressing local properties and topologies that can be embedded in the three-dimensional reconstruction [113], [120], [123]. The vast amount of data available in four-dimensional studies initiated new concepts in computing and visualization that utilize special data compression techniques [124].

29.2. Electron Microscopy

29.2.1. Introductory Considerations [138]

There are two main classes of electron microscopy (EM) techniques. In the first class, the electron probe is a stationary beam incident along a fixed direction. This incident beam can be parallel [conventional transmission electron microscopy (CTEM), high-resolution transmission electron microscopy (HRTEM), high-voltage transmission electron microscopy (HVTEM), selected-area electron diffraction (SAED)] or convergent [convergent-beam electron diffraction (CBED), convergent-beam electron microscopy (CBEM)]. The resolution is determined by the quality of the imaging optics behind the specimen. Instruments implementing these techniques are conceptually related to classical light microscopes.

In the second class of methods, a fine electron probe is scanned across the specimen, and transmitted electron (TE) [scanning transmission electron microscopy (STEM)] or the desired excited signal such as secondary electrons (SE) or backscattered electrons (BSE) [scanning electron microscopy (SEM)], and/or AE [Auger Electron spectroscopy (AES)/scanning Auger microscopy (SAM)] is selected, detected, and the signal amplified and used to modulate the intensity of another electron beam which is scanned synchronously eith the first over the screen of a TV monitor. The stationary beam methods are based on image formation processes, whereas the scanning methods are essentially "mapping" techniques. Their resolution is mainly determined by the probe size, i.e., by its electron probe formation optics. The magnification is geometric; it is determined by the ratio of the areas scanned by the electron beam on the screen and synchronously by the electron probe on the specimen.

Analytical microscopes (AEM) have been developed which make it possible to implement a number of imaging modes, diffraction modes, and analytical modes in a single instrument and often also to apply them to a single specimen. The synergism of these tools has the potential to provide detailed structural and chemical information on the nanoscale. There is a trend of extending stationary beam probe analytical tools such as X-ray microanalysis, Auger electron spectroscopy (AES), and electron energy loss spectroscopy (EELS) into mapping methods which make it possible to image spatial distributions of chemical elements.

29.2.2. Conventional Transmission Electron Microscopy (CTEM)

29.2.2.1. Introduction

The images produced in transmission electron microscopy are essentially due to local diffraction phenomena; absorption contrast plays only a minor role. Not only is electron diffraction responsible for the image formation; it also allows establishment of the orientation relationship between direct space, as observed in the image, and reciprocal space as imaged in the diffraction pattern (see → Structural Analysis by Diffraction). Images and the corresponding diffraction patterns are equally important for a detailed interpretation; both should always be produced from the same selected area, with the specimen orientation unchanged. For this reason, most commercial microscopes have the ability to switch easily from the diffraction mode to the imaging mode and vice versa (see Section 29.2.2.5).

29.2.2.2. Scattering by Atoms: Atomic Scattering Factor

Electrons are scattered by atoms as a result of Coulomb interaction with the nucleus and the electron cloud. The atomic scattering factor $f_e(\theta)$ thus contains two terms of opposite sign

$$f_e(\theta) = \frac{me^2 \lambda}{2h^2}[z - f_x(\theta)]/\sin^2\theta \qquad (2.1)$$

where z is the atomic number; $f_x(\theta)$ the scattering factor for X rays; m the electron mass; e the electron charge; λ the wavelength of the electrons; and h Planck's constant [139].

The first term clearly relates to the nucleus, whereas the second term is due to the electron cloud. The interaction with matter is stronger ($\times 10^4$) for electrons than for X rays or neutrons, which interact only with the electron cloud or with the nucleus, respectively. As a result multiple scattering will not be negligible in electron diffraction experiments. Moreover, electron scattering is oriented mainly in the forward direction. Tables of $f_e(\theta)$ for different atoms are given in [140].

29.2.2.3. Kinematic Diffraction by Crystals

29.2.2.3.1. Lattice, Reciprocal Lattice

The amplitude diffracted by an assembly of atoms results from interference of the waves scattered by the atoms (i.e., the scattered amplitude in any given direction is obtained by summing the amplitudes scattered in that direction by the individual atoms and taking into account the phase differences due to path differences resulting from the geometry of the assembly). For a physical and mathematical description of the important terms

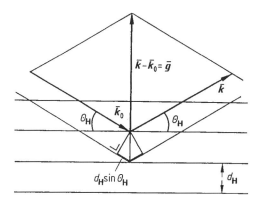

Figure 22. The derivation of Bragg's diffraction law in direct space

lattice and reciprocal lattice, see [141] and → Structural Analysis by Diffraction.

29.2.2.3.2. Geometry of Diffraction

Bragg's Law [142]. Like other diffraction phenomena in periodic structures, electron diffraction can be described in either direct or reciprocal space. Bragg's diffraction condition describes diffraction in direct space. A diffracted beam will be formed whenever the incident beam encloses with the set of lattice planes an angle θ_H, such that the path difference $2 d_H \sin \theta_H$ between waves diffracted by successive lattice planes is an integer n of wavelengths λ of the radiation used:

$$2 d_H \sin \theta_H = n\lambda \quad (2.2)$$

where **H** denotes the Miller indices of the plane. In Figure 22 this path difference is indicated by a thicker line. For 100-kV electrons, $\lambda = 4$ pm, and as a result, the Bragg angles θ_H are very small — a few degrees at most.

Ewald Construction [143]. Diffraction conditions can also be formulated in terms of reciprocal space as

$$\bar{k} = \bar{k}_0 + \bar{g} \quad (2.3)$$

where $\bar{k}_0$ is the wave vector of the incident electron beam and $\bar{k}$ the wave vector of the diffracted beam, where $|\bar{k}| = |\bar{k}_0| = 1/\lambda$; $\bar{g}$ is a reciprocal lattice vector. Equations (2.2) and (2.3) have the same physical content. They can be obtained by expressing the conservation of energy and of momentum of the incident electrons on scattering. More details on Ewald's construction are given in [144] and in → Structural Analysis by Diffraction.

Diffraction by a Thin Foil. Since electrons are strongly "absorbed" in solids the specimen must be a thin foil (< 300 nm); otherwise no diffracted beams will be transmitted. The number of unit cells along the normal to the foil is thus finite, whereas along directions parallel to the foil plane, the number of unit cells can be considered infinite. In such specimens the diffraction conditions are relaxed, and diffraction also occurs for angles of incidence deviating somewhat from the Bragg angle. In reciprocal space this relaxation results in a transformation of the sharp reciprocal lattice nodes into thin rods (so-called relrods) perpendicular to the foil plane. The Ewald sphere intersects such a rod to produce a diffracted beam. The direction of this beam is obtained by joining the center of Ewald's sphere with this intersection point. The resulting diffraction pattern can to a good approximation be considered to be a planar section of the reciprocal lattice. Indexing of the diffraction spots is simple and unambiguous.

The distance $\bar{s}_g$ by which the lattice node G is missed by Ewald's sphere is called the *excitation error*. It is a vector along the direction of the foil normal joining the reciprocal lattice node G to the intersection point with Ewald's sphere. It is positive when pointing in the sense of the incident beam and negative in the opposite case.

Column Approximation [145]. The small magnitude of the Bragg angles causes the electrons to propagate along narrow columns parallel to the beam direction. The intensity observed in a point at the exit face of the specimen is thus determined by the amplitude scattered by the material present in the column located at that point. In a perfect foil the intensity does not depend on the considered column, but when defects are present, it does.

Kinematic Rocking Curve. The amplitude of the beam scattered by a foil of thickness z_0 is obtained by summing the amplitudes scattered by the volume elements along a column perpendicular to the foil surface and taking into account the phase differences due to their depth position in the column. For a perfect foil the intensity of the scattered beam, I_s for reflection **H** is given by

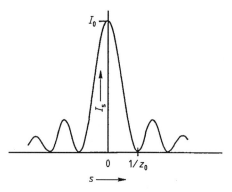

Figure 23. Plot of scattered intensity I_s versus excitation error, i.e., the "rocking curve" according to the kinematic approximation

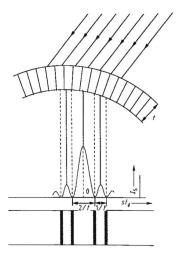

Figure 24. Formation of bent extinction contours by a cylindrically bent foil

$$I_s(s, z_0) = F_H^2 \sin^2 \pi s z_0 / (\pi s)^2 \qquad (2.4)$$

where F_H is the structure factor [127] with indices **H**.

The dependence of I_s on s (i.e., on the direction of the incident beam and on the foil thickness z_0) is called the *rocking curve* (see Fig. 23). The separation of the zeros is $1/z_0$, and the full width of the central peak is $2/z_0$.

Bent Contours [146]. For a curved foil, s varies along the specimen; the loci of constant s, imaged as bright and dark fringes, are called bent extinction contours. They form as shown schematically in Figure 24 for a cylindrically bent foil.

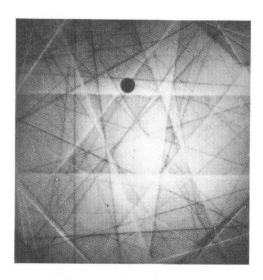

Figure 25. Kikuchi line pattern produced by a silicon foil Note the parallel pairs of bright (excess) and dark (deficiency) lines. The deficiency lines are close to the center. The excess lines are separated in angle by $2\theta_H$ from the corresponding deficiency lines

Thickness Contours. The thickness dependence of I_s is periodic with period $1/s_{\bar{g}}$: the period thus diverges for $s_{\bar{g}} = 0$ (i.e., for the exact Bragg condition). The loci of equal thickness are also imaged as lines of equal intensity; they are called thickness extinction contours or wedge fringes.

Kikuchi Lines [147]. In rather thick specimens a second type of diffraction pattern becomes prominent, which consists mainly of lines rather than spots. These lines occur in parallel pairs; the line closest to the center of the diffraction pattern is darker, whereas the one further from the center is brighter than the background (Fig. 25). These lines are the intersections of Ewald's sphere with rather flat double cones having semiapex angles equal to $90° - \theta_H$, where θ_H is the Bragg angle associated with the pair of lines ($\theta_H \approx 1-2°$). The angular separation of the bright-dark line pair is $2\theta_H$.

If the Bragg condition for a set of lattice planes is satisfied exactly, the bright Kikuchi line associated with that set of planes will pass through the corresponding Bragg spot, whereas the dark one will pass through the origin. The separation of a Bragg spot from its corresponding Kikuchi line is a measure of the excitation error s of that reflection. As a result, the Kikuchi line pattern provides a means to measure s [148].

29.2.2.4. Dynamic Diffraction by Crystals

29.2.2.4.1. General Considerations

According to kinematic theory the intensity of diffraction spots is proportional to the square of the structure factor (Eq. 4). This simple proportionality is lost in reality because of multiple diffraction effects, which are taken into account adequately in dynamic theory [149].

Although kinematic theory does not predict the intensities of diffraction spots correctly, it is useful because the geometric features of diffraction are well described. For large s or very small foil thicknesses the relative spot intensities are qualitatively reproduced. For $s=0$, kinematic theory breaks down since a diffracted beam with diffraction vector $\bar{g}_1$ may then acquire an amplitude comparable to that of the incident beam, and thus act as an incident beam and give rise to diffraction in the reverse sense (i.e., with diffraction vector $-\bar{g}_1$).

For most orientations of the incident beam, many diffracted beams are excited simultaneously and a section of reciprocal space is imaged. However by carefully orienting a sufficiently thick specimen the intensity of a single diffracted beam can be maximized: this is called a *two-beam case*. Such specimen orientations should be used, whenever possible, in the quantitative study of defects.

29.2.2.4.2. Basic Equations

The two-beam dynamic theory describes the interplay between transmitted and diffracted beams by means of a set of coupled differential equations of the form [128], [150].

$$dT/dz = (\pi i/t_{-g}) S \exp 2\pi i s z \quad (2.5a)$$

$$dS/dz = (\pi i/t_g) T \exp -2\pi i s z \quad (2.5b)$$

where T and S are the complex amplitudes of the transmitted and scattered beams, respectively; z measures the depth in the foil. The exponential factors take into account the growing phase shift due to the excitation error, with increasing distance z behind the entrance face. The factor i takes into account the phase jump on scattering; t_g and t_{-g} are called the extinction distances. The extinction distance, which has the dimension of length, is a measure of the strength of the reflection $\bar{g}$. For low-order reflections of elemental metals it is of the order of a few tens of nanometers, but for weak superstructure reflections it may be several hundred nanometers. Whether a foil is "thick" or "thin" depends on the number of extinction distances in the foil thickness. Kinematic theory is valid only for foils with a thickness that is a fraction of an extinction distance.

Absorption can phenomenologically be accounted for by assuming the extinction distances to be complex [151]. Formally $1/t_g$ is replaced by $1/t_{\bar{g}} + i/\tau_g$, where τ_g is called the absorption length. Empirically, it is found that τ_g ranges from $5 t_{\bar{g}}$ to $15 t_{\bar{g}}$. The substitution can be made, either in the Equation set (2.5) or directly in the solution of this set.

29.2.2.4.3. Dynamic Rocking Curve

By ignoring absorption and taking into account the fact that $T=1$ and $S=0$ at $z=0$ (entrance face), integration of Equation set (2.5) gives

$$T(s,z) = [\cos\pi\sigma z - i(s/\sigma)\sin\pi\sigma z]\exp\pi i s z \quad (2.6a)$$

$$S(s,z) = (i/\sigma t_g)\sin\pi\sigma z \exp\pi i s z \quad (2.6b)$$

and thus

$$I_s = SS^* = (1/\sigma t_g)^2 \sin^2 \pi\sigma z \quad (2.7)$$

$$I_T = 1 - I_s \quad (2.8)$$

where

$$1/\sigma = t_g/\left(1 + s^2 t_g^2\right)^{1/2} \quad (2.9)$$

For $s=0$ this reduces to $\sigma = 1/t_g$ (i.e., the depth period is now $t\bar{g}$ and for large s: $\sigma = s$). The divergence of kinematic theory for $s=0$ is now removed. The depth variation of the transmitted and scattered beams, leading to thickness extinction contours, is represented schematically in Figure 26 for $s=0$ as well as for $s \neq 0$.

29.2.2.4.4. Anomalous Absorption, Bormann Effect [151], [152]

Taking absorption into account gives more complex rocking curves of the type shown in Figure 27 for a foil of thickness $t = 3 t\bar{g}$ and $\tau_g = 10 t\bar{g}$. The curve for the diffracted beam (b) is symmetric, whereas the curve representing the transmitted beam (a) is asymmetric in s. For the same absolute value of s, the transmitted intensity is larger for $s>0$ than for $s<0$. This asymmetry is known as

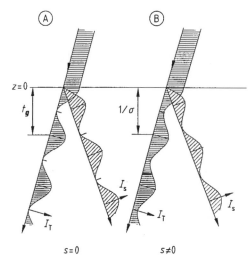

Figure 26. Interplay between incident and diffracted beam according to the dynamic theory
A) Bragg's condition is satisfied exactly, the depth period is t_g; I_S and I_T pass periodically through zero; B) $s \neq 0$ the depth period is now $1/\sigma_g$; I_T varies periodically but does not pass through zero; I_S and I_T are complementary since absorption was neglected

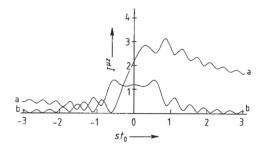

Figure 27. Rocking curves for transmitted (a) and scattered (b) beams according to the two-beam dynamic theory, with anomalous absorption taken into account

the Bormann effect; it was demonstrated initially for X-ray diffraction under dynamic conditions [152]. The steep change of I_T with s in the vicinity of $s=0$ causes a large sensitivity of I_T to small changes in orientation.

29.2.2.4.5. Lattice Fringes [153]

The Electron Wave Function at the Exit Face. The amplitude of the transmitted wave for a unit incident wave can be represented by $T \exp 2\pi i \bar{K} \cdot \bar{r}$, where $\bar{K}$ is the wave vector of the incident wave inside the crystal (i.e., corrected for refraction at the vacuum–crystal interface). Similarly the diffracted wave can be represented by $S \exp 2\pi i (\bar{K} + \bar{g}) \cdot \bar{r}$, where $\bar{g}$ is the diffraction vector (i.e., the reciprocal lattice vector corresponding to the excited reflection). The wave function of the electrons emerging at the exit face of the foil is then

$$\psi = \left(T + S \exp 2\pi i \bar{g} \cdot \bar{r} \right) \exp 2\pi i \bar{K} \cdot \bar{r} \qquad (2.10)$$

The observed intensity

$$I_\psi \equiv \psi \psi^* = I_T$$
$$+ I_S + 2\sqrt{I_T I_S} \sin(2\pi g x + \varphi) \qquad (2.11)$$

with

$$\tan \varphi = (s/\sigma) \tan \pi \sigma t; \qquad (2.12)$$

x is measured along the exit surface in the direction of $\bar{g}$: further, the following abbreviations are used: $I_T \equiv T T^*$; $I_S \equiv S S^*$. This expression represents sinusoidal fringes with a period $1/|\bar{g}|$; they can be considered as forming an image of the lattice planes $\bar{g}$.

29.2.2.4.6. Faulted Crystals

Planar Interfaces. Planar interfaces are imaged as fringes parallel to the foil surface; their characteristics depend on their geometric features. *Translation interfaces* (stacking faults, out-of-phase boundaries, etc.) separate two crystal parts related by a parallel translation $\bar{R}$ (Fig. 28 A), and *domain boundaries* separate domains that differ slightly in orientation (i.e., for which the excitation errors are different $\Delta s = s_1 - s_2$; (Fig. 28 B) [154].

Diffraction Equations for Faulted Crystals. The displacement over R_0 ($R_0 =$ constant) of the exit part of a column with respect to the entrance part (Fig. 28 A), the two parts being separated by a fault plane, can be accounted for by noting that a supplementary phase change $\alpha = 2\pi \bar{g} \cdot \bar{R}$ occurs on diffraction but not on transmission (see Fig. 29). The path difference Δ between diffracted waves 1 and 2 is $\Delta = 2 R_0 \sin \theta_H$, where θ_H is the Bragg angle and hence $2 d_H \sin \theta_H = \lambda$ with $d_H = 1/|\bar{g}|$. This gives for the phase change $\alpha = (2\pi/\lambda) \Delta = 2\pi g R_0$, where R_0 is in general the component of $\bar{R}$ along $\bar{g}$ (i.e., $g R_0 = \bar{g} \cdot \bar{R}$).

The equations describing diffraction by the displaced exit part are obtained by the substitution

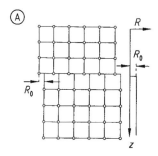

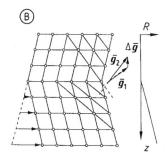

Figure 28. Schematic of two different types of interfaces
A) Translation interfaces with constant displacement vector $\bar{R}$; B) Domain boundary between two regions for which the excitation error is slightly different; the displacement R increases with increasing distance from the interface (the homologous diffraction vectors $\bar{g}_1$ and $\bar{g}_2$ are slightly different: $\Delta \bar{g} = \bar{g}_2 - \bar{g}_1$)

$S \rightarrow S \exp -i\alpha$ in Equations (2.5). The amplitudes T and S emerging from a column intersecting such an interface are then obtained by integrating the Equation (2.5) set along a column down to the level of the interface and subsequently further down to the exit face, using the adapted form of Equation (2.5) (Fig. 29).

The effect on T and S of the presence of a defect described by a displacement field $\bar{R}(\bar{r})$ (for instance due to a dislocation) (Fig. 29) can be taken into account in Equations (2.5) by replacing the constant s by an effective local value s_{eff} which now depends on z, i.e., on the depth along the columns, and which is given by $s_{\text{eff}} = s + \bar{g}(d\bar{R}/dz)$. This can be shown rigorously but it can also be understood intuitively by the following geometrical considerations (Fig. 29 B).

Let the diffraction vector of the considered set of lattice planes be $\bar{g}$ at the level z below the entrance face. At level $z+dz$ the displacement $\bar{R}$ has increased by $d\bar{R}$ and as a result the local vector $\bar{g}$ is rotated over a small angle $\delta\theta \approx \tan\theta = dR/dz$. The change in s corresponding to this rotation of $\bar{g}$

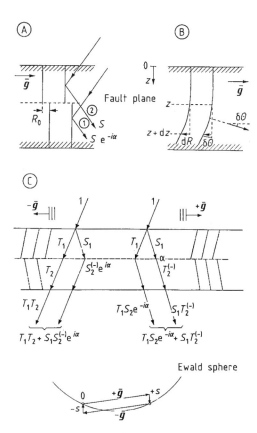

Figure 29. Diffraction by a faulted crystal
A) Effect of a translation interface with vector $\bar{R}_0$; B) Effect of a displacement field $\bar{R}(\bar{r})$; C) Schematic of the expressions for transmitted and scattered beam amplitude for a foil containing a general interface parallel to the foil surfaces

over $\delta\theta$ is given by $\delta s = g\delta\theta$, i.e., by $g(dR/dz)$ or somewhat more general by $d(\bar{g} \cdot \bar{R})/dz$, expressing that only displacements which change the orientation of the vector $\bar{g}$ are operative.

In the particularly simple case of a domain boundary as modeled in Figure 28 B, $R = kz$ (k = constant) in the exit part. Hence, at the level of the domain boundary, s changes abruptly into $s + k$ along the integration column.

Diffraction by a Crystal Containing a Planar Interface [155]. The amplitudes of the transmitted and scattered beams for a foil containing a planar interface, with displacement vector $\bar{R}$ (Fig. 29 B), can be formulated as

$$T(s_1,s_2,z_1,z_2,\alpha) = T_1(s_1,z_1)T_2(s_2,z_2)$$
$$+ S_1(s_1,z_1)S_2(-s_2,z_2)\exp i\alpha \qquad (2.13)$$

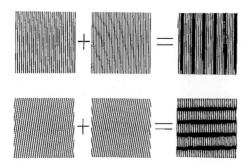

Figure 30. Optical analogue for the formation of moiré fringes in a sandwich crystal
(Upper) The two crystals are parallel, but the lattice parameters are slightly different; (Lower) The two crystal parts exhibit a slight orientation difference

$$S(s_1,s_2,z_1,z_2,\alpha) = T_1(s_1,z_1)S_2(s_2,z_2)$$
$$\cdot \exp(-i\alpha) + S_1(s_1,z_1)T_2(-s_2,z_2) \quad (2.14)$$

with $\alpha = 2\pi\bar{g} \cdot \bar{R}$. These relations can also be derived from the adapted set of Equation (2.5).

The first equation states that the amplitude of the transmitted beam T results from the interference between the doubly transmitted beam (i.e., the beam transmitted by the first part T_1 and subsequently transmitted by the second part: $T_1 \times T_2$) and the doubly scattered beam $S_1 \times S_2^{(-)} \exp i\alpha$. The factor $\exp i\alpha$ takes into account the phase shift on the amplitude S_2 scattered by the second part, resulting from the translation of the second part of the foil.

A similar interpretation can be given to the second expression. The amplitudes $T_1(s_1,z_1)$, $T_2^{(-)}$, $S_1^{(-)}$, $S_2^{(-)}$ are given by Equations (2.6 a) and (2.6 b). The indices 1 and 2 refer to the entrance and exit parts, respectively: the minus superscript means that in the corresponding Equations (2.6 a) and (2.6 b), s must be replaced by $-s$, because diffraction occurs from the negative side of the set of lattice planes (i.e., the diffraction vector is $-\bar{g}$).

If $s_1 = s_2$ but $\alpha \neq 0$, Equations (2.13) and (2.14) refer to a pure translation interface. If $s_1 \neq s_2$ and $\alpha = 0$, the expressions are applicable to a domain boundary. For mixed interfaces, $s_1 \neq s_2$ as well as $\alpha \neq 0$. If the interfaces are parallel to the foil surfaces, transmitted and scattered intensities are constant over the specimen area, but they vary in a pseudoperiodic fashion with the thicknesses z_1 and z_2 of entrance and exit parts, with $z_1 + z_2 = z_0$ (total thickness). The expressions describing this vari-

ation explicitly are obtained by substituting Equations (2.6 a) and (2.6 b) into (2.13) and (2.14).

If the interfaces are inclined with respect to the foil surfaces this depth variation is displayed as a pattern of fringes parallel to the foil surfaces in the area where the two wedge-shaped crystal parts overlap.

The fringe profiles of α-fringes ($s_1 = s_2$; $\alpha \neq 0$) and δ-fringes ($\Delta s = s_1 - s_2 \neq 0$; $\alpha = 0$) in bright field (BF) and dark field (DF), exhibit different properties. The symmetry properties allow the fringe patterns due to the two kinds of interfaces to be distinguished. The properties of α-fringes with $\alpha = \pi$ are singular [156].

29.2.2.4.7. Moiré Patterns [157]–[159].

The superposition of two identical crystal foils having a small orientation difference θ (rotation moiré) or of two parallel crystal foils with a small difference in lattice parameters (parallel moiré) gives rise to a fringe pattern, which can to a good approximation be considered as the coincidence pattern of the lattice fringes corresponding to the diffraction vectors $\bar{g}_1$ and $\bar{g}_2$ active in the two crystal foils.

The fringes are parallel to the average direction $\bar{g}$ of the two diffraction vectors $\bar{g}_1$ and $\bar{g}_2$ in the case of rotation moiré patterns and perpendicular to the common direction of the two diffraction vectors in the case of a parallel moiré. From the moiré spacing Δ, small differences can be deduced in the lattice parameters d_1 and d_2 of the two components of the sandwich since $\Delta = d_1 d_2/(d_1 - d_2)$, or the small orientation difference $\theta = |\Delta\bar{g}|/|\bar{g}|$ in the case of a rotation moiré. Figure 30 shows an optical analogue for the two types of moiré fringes produced by the superposition of two line patterns.

29.2.2.5. Operating Modes of the Electron Microscope [128], [130], [131]

29.2.2.5.1. Microscope Optics

The ray paths in an electron microscope are shown in Figure 31, according to the geometrical optics approximation, for the two main operating modes: high-magnification, high-resolution imaging, and selected area diffraction.

The microscope is essentially a three-lens system: an objective lens (c), an intermediate lens (f), and a projector lens (g). Each lens may be a composite lens. A movable selector aperture (b) is present in the image plane of the objective lens; a second aperture (d) is placed at the objective lens, close to the back focal plane. With the first

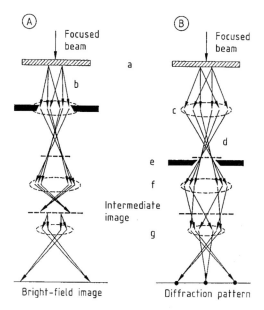

Figure 31. Schematic of the beam path in an electron microscope according to the geometrical optics approximation
A) High-magnification, high-resolution mode; B) Selected area diffraction mode
a) Specimen; b) Objective aperture; c) Objective lens; d) Field-limiting aperture; e) Gaussian image plane; f) Intermediate lens; g) Projector lens

aperture a small area (≤ 1 µm) of the image (i.e., of the specimen) is selected, with the second aperture a single beam is selected or a number of the image-forming diffracted beams.

The characteristics of the objective lens are crucial because they determine to a large extent the image resolution and the contrast. The other two lenses mainly provide the desired magnification.

Although magnetic lenses normally rotate the image about the optical axis, in microscopes designed at the end of the 1980s, these rotations are compensated by a suitable device, and the image and the diffraction pattern have the same orientation.

29.2.2.5.2. High-Resolution, High-Magnification Mode

In the first (Fig. 31) imaging ray path the beam produced by a source (tungsten filament, LaB_6, field emission gun; see also Section 29.2.3.2.1) and collimated by a condenser lens system, is scattered by the object, and an image is formed in the image plane of the objective lens. By means of the selector aperture, an area of the specimen is selected and magnified by the intermediate lens, which is focused on the image plane of the objective lens and provides a magnified image in the image plane of the intermediate lens. This image serves as the object for the projector lens, which forms the final image either on a fluorescent screen, on a photographic plate, or on an image intensifier followed by a TV camera. In the latter case — which is virtually a requirement for successful high-resolution work — the image can be viewed on a TV monitor and the final adjustments performed before the image is recorded photographically. Electronic recording on tape is of course possible directly via the TV signal; this mode is used mainly for so-called in situ dynamic studies.

29.2.2.5.3. Diffraction Mode (Figure 31 B)

When operating in the diffraction mode the intermediate lens (f) is weakened (i.e., its focal length is enlarged) so as to cause the back focal plane of the objective lens (c) to coincide with the object plane of the projector lens (g). This produces a magnified image of the diffraction pattern. In doing so, the selected area is not changed since only the intermediate lens current is changed; the diffraction pattern is thus representative of the selected area. The area selected in the diffraction mode is much larger than the field of view under high-resolution conditions.

As the electron beam passes through the specimen, various interactions occur next to diffraction: characteristic X rays are produced, the electrons suffer energy losses, etc. These effects can be used for analytical applications (see Section 29.2.2.12).

29.2.2.6. Selected-Area Electron Diffraction (SAED)

ED methods (SAED, CBED) are a useful complement to X-ray diffraction, especially when the material is available only as small microcrystals. Then only powder diffractometry can be applied when using X rays. However, even fine powders usually contain single crystals of a sufficient size to produce a single-crystal ED pattern that allows the approximate (but unambiguous) determination of the lattice parameters. These approximate lattice parameters in turn allow unambiguous indexing of the powder diffraction pattern and sub-

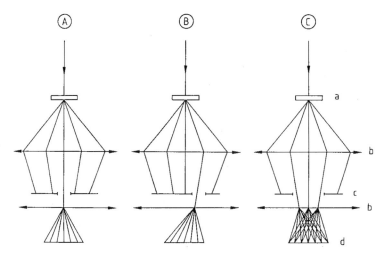

Figure 32. Schematic representation of different imaging modes
A) Bright-field diffraction contrast mode; B) Dark-field diffraction contrast mode; C) High-resolution imaging mode
a) Specimen; b) Lens; c) Aperture; d) Screen

sequently make possible a precise lattice parameter measurement. In addition, ED patterns often exhibit weak reflections due to superstructures, which are not visible in X-ray powder diffraction.

The spatial geometry of diffuse scattering can be reconstructed more easily from ED patterns than from X-ray diffraction patterns, since ED patterns are planar sections of reciprocal space, which is not the case for X rays.

29.2.2.7. Diffraction Contrast Images [128], [148]

29.2.2.7.1. Imaging Modes

Diffraction contrast images are usually obtained under two-beam conditions. By means of the aperture placed close to the back focal plane of the objective lens, either the transmitted or the diffracted beam is selected and the corresponding diffraction spot is highly magnified (Fig. 32). The image obtained in this way is a map of the intensity distribution in the selected diffraction spot. The image made in the transmitted beam is a *bright-field image*, whereas the image made in the diffracted beam is a *dark-field image*. Since the electrons are confined to narrow columns, the latter can be considered the picture elements of this image. The amplitude of the transmitted (or scattered beam) associated with a point at the back surface of the specimen is obtained by summing the contributions of the volume elements along a column parallel to the incident beam centered on that point. The amplitude depends on how the excitation error s changes along the column, which in turn varies with changes in the orientation of the diffracting planes along the column. Strain fields cause such orientation changes and can thus be imaged in diffraction contrast. Diffraction contrast images do not reveal the crystal structure but are very sensitive to the presence of strain fields caused by defects such as dislocations.

29.2.2.7.2. Dislocation Contrast [125], [128].

Consider a foil (Fig. 33) containing an edge dislocation in E, with an orientation such that in the perfect parts of the foil the amplitude of the incident beam is approximately equally divided between the transmitted and the diffracted beams. To the left of the dislocation the considered lattice planes are locally inclined in such a way as to better satisfy Bragg's condition (i.e., s is smaller) than in the perfect parts; more electrons are diffracted than in the perfect part. A lack of intensity will then be noted left of the dislocation in the transmitted beam, leading to a dark line in the bright-field image. At the right of the dislocation, Bragg's law is less well satisfied than in the perfect part. This leads to a lack of intensity in the scattered beam and thus to a dark line in the dark-field image. The image is on one side of the dislocation: the side on which the dark line occurs in the bright-field image is called the *image side*. Ac-

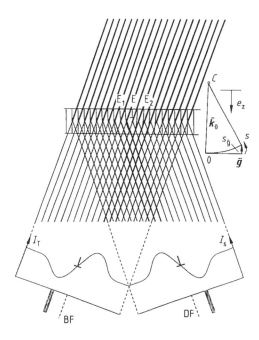

Figure 33. Schematic representation of the image formation at an edge dislocation in a thin foil
The line thickness is a measure of the beam intensity. The image profiles for bright-field (BF) and dark-field (DF) images are represented schematically along with diffraction conditions in the perfect part of the foil

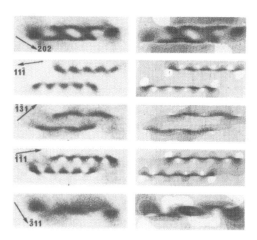

Figure 34. Examples of computed images (right) and the corresponding observed images (left) of the same dislocation by using different diffraction vectors [162]

29.2.2.7.4. Domain Textures

Phase transitions in crystals are usually accompanied by changes in symmetry, the space group of the low-temperature phase being a subgroup of the space group of the high-temperature phase. On cooling, the crystal breaks up into *orientation variants* (or twins) related by the lost symmetry elements of the high-temperature phase and *translation variants* (out-of-phase boundaries) related by the lattice translations lost on forming the superlattice [164]. The composite diffraction pattern of a domain texture is the superposition of the diffraction patterns due to individual domains. The diffraction pattern taken across a *reflection twin interface* contains one row of unsplit spots, which is perpendicular to the mirror plane. All other diffraction spots are split along a direction parallel to the unsplit row of spots [127].

In diffraction contrast images made in a cluster of *split diffraction spots*, orientation domains exhibit differences in brightness (i.e., domain contrast), especially for conditions close to $s=0$ where the intensity variation with s is steep (Fig. 35). The presence of translation variants is not reflected in the diffraction pattern, except by some diffuse scattering.

Images made in *unsplit reflections* do not reveal brightness differences due to orientation differences of the lattice, but they may exhibit domain contrast due to differences in the structure factor. This is the case for Dauphiné twins in quartz, which are related by a 180° rotation about

cording to this model the image side is where the lattice planes are tilted into the Bragg orientation.

The conditions for kinematic diffraction [160] are best approximated in the *weak-beam* method, which consists of making a dark-field image in a weakly excited diffraction spot. The dislocation image then consists of a narrow bright line on a darker background.

29.2.2.7.3. Extinction Conditions for Defects

Diffraction by a family of lattice planes that remain undeformed by the presence of the dislocation will not produce an image of the dislocation [128], [161].

Quantitative images based on dynamic theory can be computer simulated by using the methods developed in [162]. A comparison of simulated and observed images allows determination of all the relevant parameters of various dislocation types (Fig. 34). More details are given in [163].

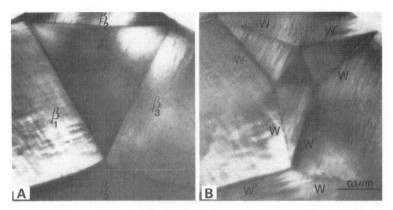

Figure 35. Domain texture in β-lead orthovanadate resulting from the $\gamma \to \beta$ transformation
A) The central triangular area still consists of the high-temperature γ-phase, whereas the other domains are in the low temperature β-phase; B) The same area as A) after additional cooling. The unstable triangular γ-area has been reduced further (Courtesy of C. Manolikas)

the threefold axis (Fig. 36). The two domains produce the same diffraction pattern since their lattices are common, but certain reflections have structure factors that differ in magnitude [165]. A dark-field image made in such a reflection will exhibit a different brightness in the two domains, which is called the structure factor contrast [128].

29.2.2.7.5. Interface Contrast and Domain Contrast

Interface contrast (i.e., an α-fringe pattern) is produced in the projected area of the interface even if the structure factors in the two domains have the same magnitude, but differ in phase by α; no domain contrast is produced under these conditions. Interface contrast is also produced at *orientation domain boundaries*; δ-fringes are produced in the projected area, since the s-values on both sides of the boundary are different in general [166]. Extinction of the δ-fringe pattern occurs if the s-values are the same for the two domains.

The presence or absence of domain contrast across a fringe pattern helps to distinguish the two types of interface. If the two domains on either side of the interface exhibit the same brightness for all reflections, the fringes must be due to a translation interface; they must exhibit α-character. Domains exhibiting a difference in brightness are separated by a domain boundary and the fringes are of the δ-type.

29.2.2.7.6. Strain Field Contrast

The strain field associated with precipitate particles in a matrix whose lattice parameters differ slightly from those of the precipitate can also be revealed as regions of different brightness [167]. For example, a spherical particle in an elastically isotropic matrix produces a spherically symmetric strain field. In certain parts of this strain field the vector $\bar{R}$ is locally perpendicular to $\bar{g}$; such areas will show up with the background brightness. In other areas, $\bar{g}$ is parallel to $\bar{R}$; such areas will show up with a brightness that is different from the background, either lighter or darker.

29.2.2.8. Convergent Beam Diffraction [132], [168]

29.2.2.8.1. Geometry of Convergent Beam Patterns

In the early 1980s, a class of applications based on the use of a convergent beam of electrons (convergent beam diffraction, CBD) was developed. Whereas parallel beams allow high spatial resolution, *high angular resolution* is more easily achievable with convergent beams. For these applications the electrons are incident along directions within a cone of revolution or along the surface of a cone (hollow cone method) having its apex in (or close to) the sample plane. Under these conditions, electrons are incident on the specimen under all possible directions within a

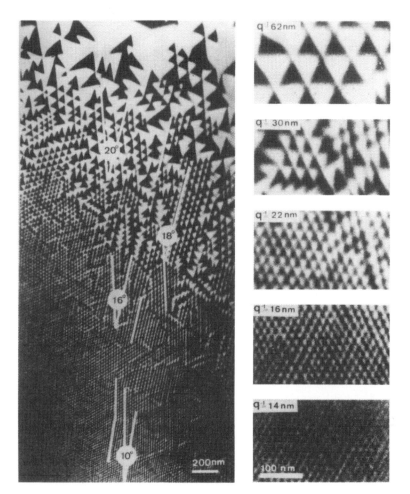

Figure 36. Dauphiné twin lattice in α-quartz close to the phase transition into the high-temperature β-phase viewed along the threefold axis
A temperature gradient is established across the specimen area. The α-phase is broken up in regular arrays of columnar domains whose size decreases with increasing temperature [165]

certain angular range, which should be of the order of the Bragg angles of the lowest-order reflections of the material under study. In practice, the semi-apex angle α varies from 2 to 10 mrad. Also, the corresponding diffracted beams for each reciprocal lattice node form cones. The intersections of these cones with the photographic plate are circular disks whose radii are proportional to the semiapex angle of the cone of incident directions (Fig. 37). If the specimen is perfectly flat and defect free, these disks are images of the angular intensity distribution in each particular reflection.

29.2.2.8.2. Point Group Determination [169], [170]

When the incident beam is parallel to a zone axis, the symmetry of the intensity distribution in the complete disk pattern, including the fine lines in the 000 disk, reflects the point symmetry of the crystal along this zone axis (Fig. 38).

Several techniques can be used to determine the point group [169]. The method developed by BUXTON et al. [170] is based on the use of zone axis patterns and of dark-field diffraction patterns obtained under exact two-beam conditions. By

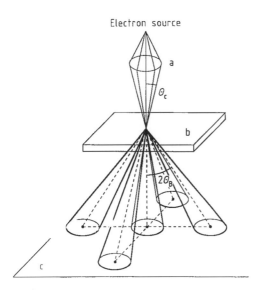

Figure 37. Production of a convergent beam electron diffraction pattern
a) Probe forming optics; b) Specimen; c) Film
The incident beam is convergent. Diffraction disks are centered on each diffraction spot.

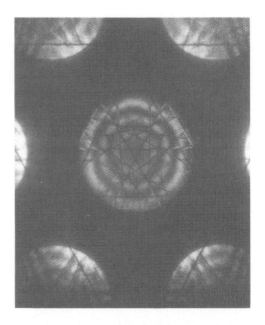

Figure 39. Deficiency HOLZ (high-order Laue zone) lines in the central disk of a convergent beam diffraction pattern of silicon [168]

A disk in the CBED pattern usually consists of broad fringes due to the dynamic interaction between beams belonging to the zero-order Laue zone (ZOLZ). Finer lines are formed by the interaction of high-order Laue zone reflections (HOLZ) with the zero-order beams (Fig. 39). The geometry of the HOLZ lines is determined by the accelerating potential and the lattice parameters; comparison of computer-simulated and observed patterns allows determination of one parameter if the other is known [171], [172].

29.2.2.8.3. Space Group Determination

Space group determination is based on the observation of the Gjønnes–Moodie lines [173] in certain "forbidden" diffraction disks. Reflections that are kinematically forbidden may appear as a result of double diffraction under multibeam dynamic diffraction conditions. If such forbidden spots are produced along pairs of different symmetry-related diffraction paths that are equally excited, the interfering beams may be exactly in antiphase for certain angles of incidence if the structure factors have opposite signs. Since the convergent beam disks are formed by beams with con-

Figure 38. Zone axis convergent beam electron diffraction pattern revealing the presence of a sixfold rotation axis [168]

choosing the appropriate optical conditions, these disks do not overlap and the exact Bragg spots remain in the centers of the corresponding disks. The optics required for the application of these methods are described in [171]. With increasing cone angle the disks start to overlap and the nature of the pattern changes, but the symmetry elements are conserved.

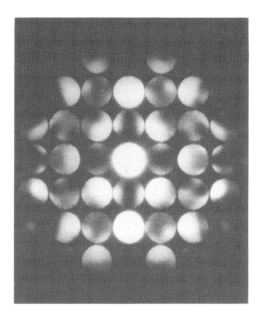

Figure 40. Convergent beam electron diffraction pattern exhibiting Gjønnes–Moodie lines due to the presence of glide mirror planes and twofold screw axis [168]

tinuously varying directions of incidence, this condition will always be satisfied somewhere in the disk. The locus of points for which the phase difference is exactly π is a dark fringe along a diameter of the disk for which the Bragg condition is satisfied exactly (Gjønnes–Moodie lines, Fig. 40).

29.2.2.8.4. Foil Thickness Determination

From the dark-field two-beam rocking curve (Section 29.2.2.8.1) observed in the disk corresponding to the diffracted beam, the specimen thickness can be deduced with high accuracy provided the accelerating voltage and the interplanar spacing of the excited reflection are known [174].

29.2.2.9. High-Resolution Electron Microscopy

In recent years, high-resolution electron microscopy has become an important tool in the study of complicated crystal structures and their defects. Whereas detailed considerations of electron optics are of only marginal importance in the case of diffraction contrast images, they become essential for a discussion of the atomic resolution structure images produced by crystals.

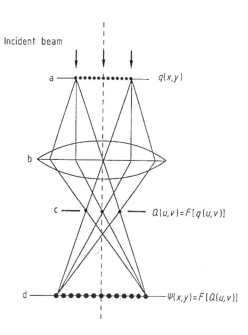

Figure 41. Image formation in an ideal microscope
a) Specimen; b) Objective lens; c) Back-focal plane, objective aperture; d) Image plane
The diffraction pattern $Q(u, v)$ is the Fourier transform of the object $q(x, y)$; the image is the inverse Fourier transform $\psi(x, y)$ of the diffraction pattern

29.2.2.9.1. Image Formation in an Ideal Microscope (Fig. 41)

Let the crystalline object be a thin foil characterized by a two-dimensional transmission function $q(x, y)$ that describes at each point of the exit surface of the specimen the amplitude and phase of the electron beams emerging from the column situated at (x, y) after dynamic diffraction in the foil. The diffraction pattern can be described to a good approximation as the Fourier transform $Q(u, v)$ of the object function $q(x, y)$. This diffraction pattern acts in turn as a source of Huyghens wavelets, which interfere to form the image, after linear magnification by the optical lens systems; the image is, in turn, the Fourier transform $\psi(x, y)$ of the diffraction pattern.

29.2.2.9.2. Image Formation in a Real Microscope

In real microscopes the situation is complicated by the presence of finite apertures and lens aberrations. The apertures truncate the Fourier

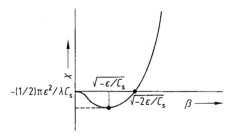

Figure 42. Dependence on β of the phase shift $\chi(\beta)$ of a beam enclosing an angle β with the optical axis of the microscope

transform by admitting only a finite number of beams to the image formation process. Lens aberrations depend on the angle β that a diffracted beam encloses with the optical axis of the microscope, hence they introduce angle-dependent phase shifts between diffracted beams. Moreover, they cause "blurring" of the image, a point of the object plane being represented as a disk in the image plane and vice versa.

Spherical Aberration. Spherical aberration causes electron beams enclosing an angle β with the optical axis to suffer a phase shift χ_S with respect to the central beam:

$$\chi_S = \frac{1}{2} \pi C_S \beta^4 / \lambda \qquad (2.15)$$

where C_S is the spherical aberration constant.

Defocus. Under exact Gaussian focusing conditions the contrast produced by an object that changes only the phase of the incident electron wave (i.e., a so-called phase grating) is minimal. This is the case for thin crystalline foils. Visual contrast improves when working in underfocused conditions. The phenomenon is in a sense similar to phase contrast in optical microscopy of phase objects. Defocusing causes a relative phase shift of the diffracted beams enclosing different angles with the central beam. The phase shift caused by a defocus distance ε with respect to the Gaussian focus is given by

$$\chi_D = \pi \epsilon \beta^2 / \lambda \qquad (2.16)$$

The total phase shift caused by spherical aberration and defocusing is thus:

$$\chi(\beta) = \left(\frac{1}{2} \pi C_S \beta^4 + \pi \epsilon \beta^2 \right) / \lambda \qquad (2.17)$$

$\chi(\beta)$ depends on β in the way shown in Figure 42, i.e., there is a flat minimum at $\beta_{\min} = (-\varepsilon/C_S)^{1/2}$ and a zero at $\beta = (-2\varepsilon/C_S)^{1/2}$.

Phase Grating Approximation [175]. The effect of this phase shift on the image parameter can be understood most easily by discussing the case of a pure phase grating, which is a reasonable model for a very thin crystalline foil. Such a specimen causes a phase shift χ of the electron wave on passing through the foil because the wavelength of the electron is different in vacuum and in the crystal. This phase shift in the point (x, y) can be written as $\sigma \varphi (x, y)$ where φ is the projected potential integrated over the sample thickness along the propagation direction of the electrons, and $\sigma = \pi/\lambda E$ (λ = wavelength of electron in vacuum; E = accelerating voltage). Ignoring absorption, the object function for a thin foil can be written as

$$q(x, y) = \exp i \sigma \varphi(x, y) \simeq 1 + i \sigma \varphi(x, y) \qquad (2.18)$$

Taking also into account the phase shift (Eq. 2.17) caused by the lens system on the diffracted beams, and neglecting absorption it can be shown by Fourier transformation [176] that the image intensity is finally given by

$$I(x, y) \simeq 1 - 2\sigma \varphi(x, y) \qquad (2.19)$$

whereby it was assumed that $\sin \chi = -1$ ($\cos \chi = 0$). The image contrast, defined as $(I - I_0)/I_0 = -2\sigma \varphi (x, y)$, where I_0 is the incident intensity, is thus directly proportional to the projected potential $\varphi(x, y)$ provided $\sin \chi = -1$.

Provided $\sin \chi = -1$, columns of large projected potential are imaged as dark dots. If $\sin \chi = +1$, the same columns would be imaged as bright dots.

Optimum Imaging Conditions. The most stringent requirement is $\sin \chi = -1$; this condition can only be met approximately and only in a limited β interval. The condition $\sin \chi = +1$ can be satisfied only for a limited number of discrete β-values.

The situation is somewhat comparable to positive and negative phase contrast in optical microscopy. The lens imperfections have been used to introduce a phase shift of $\pi/2$ in the same way as the quarter wavelength ring in the optical microscope. Only beams passing through the "window" in which the condition $\sin \chi = -1$ is met, interfere with the required phase relationship that causes the image to represent maxima in projected potential as dark areas.

From Equation (2.17) it is clear that $\chi(\beta)$ depends on the parameters C_S, ε, and λ. The condition $\sin \chi = -1$ can thus be met by adjusting one or several of these parameters. For a given instrument C_S and λ are fixed and the observer can meet this condition by optimizing the defocus ε. From Equation (2.17) it follows that χ adopts the essentially negative value $\chi = -\pi \varepsilon^2/(2 \lambda C_S)$ for $\beta = (-2\varepsilon/C_S)^{1/2}$. $\sin \chi$ will be -1 if $\pi \varepsilon^2/(2 \lambda C_S) = \pi/2$, i.e., for

$$\epsilon_S = -(\lambda C_S)^{1/2}. \qquad (2.20)$$

For the defocus $\varepsilon = \varepsilon_S$ the curve $\sin \chi$ versus β will exhibit a rather flat part in the region around

$$\beta = (-\epsilon/C_S)^{1/2} \qquad (2.21)$$

where $\sin \chi = -1$. The underfocus value ε_S corresponding to this optimum is called Scherzer defocus [177]. Most high resolution images are made under this defocus condition and in the thinnest part of the sample.

29.2.2.9.3. Resolution Limiting Factors

Aperture. The presence of an aperture admitting only beams that enclose an angle with the optical axis not exceeding β_A imposes a resolution limit, called the Abbe limit [178]. A point in object space is imaged as a circle with a radius

$$\varrho_A = 0.61\lambda/\beta_A$$

Chromatic Aberration. Instabilities ΔE in the high voltage E of the microscope cause a spread in the wavelength of the incident electron, which blurs the image. Variation ΔI in the lens current I similarly causes a spread in the focal length of the lenses, which also contributes to such blurring. Furthermore, the inelastic scattering of electrons in the specimen causes slight changes in the wavelength of electrons emerging from the specimen. The net result of these different phenomena can be described as an effective spread in the focal length of the objective lens:

$$\Delta f = C_C \left[(\Delta E/E)^2 + 4(\Delta I/I)^2 \right]^{1/2} \quad (2.22)$$

where C_C is the chromatic aberration constant; the corresponding disk of least confusion in object space is $\varrho_C = \beta \Delta f$. In concrete cases, $\Delta E/E \leq 10^{-6}$; $\Delta F \approx 10$ mm, $C_C \approx 10$ mm.

Astigmatism. Ideal lenses should have cylindrical symmetry. However, in reality, slight deviations may occur, leading to a dependence of the focal length on the azimuth of the ray path considered. Astigmatism can occur for various reasons, including inhomogeneities in the polepiece material, asymmetry in the windings, or a dirty aperture. As a result of astigmatism a point source is imaged as two different line foci at right angles to each other. At exact focus the image deformation disappears but the image is still blurred. Astigmatism can be corrected by a stigmator — a device (usually an octopole) that makes the lens appear perfectly cylindrical — by applying weak additional magnetic fields. Careful correction of astigmatism is essential for the production of high-quality, high-resolution images.

Beam Divergence. The intense illumination required for high-resolution imaging is obtained by a condenser lens system that produces a slightly conical beam with a typical apex angle of the order of 10^{-3} rad. This also leads to some image blurring since the final image is the superposition of images corresponding to the different directions of incidence.

Mechanical Instability. The obtainable resolution depends not only on the lens characteristics but also to a large extent on the mechanical stability of the microscope, which should be installed on a vibration-free heavy concrete block. In many cases the mechanical stability is actually the resolution-determining factor. The specimen holder should moreover be creep free, which is a problem for hot and cold stages. High-resolution work is therefore performed almost exclusively at room temperature.

Ultimate Resolution. The final disk of confusion due to the different nonmechanical blurring effects is given approximately by

$$\varrho = \left(\varrho_A^2 + \varrho_S^2 + \varrho_C^2 \right)^{1/2} \quad (2.23)$$

which depends on β. At low angles the aperture effect is usually dominant, whereas at high angles the spherical aberration is the limiting factor (for 100-kV microscopes, $C_S \approx 8.2$ mm and $C_C \approx 3.9$ mm). The optimum radius of least confusion is achieved for a certain β-value and the radius of confusion corresponding to that optimum is

$$\varrho_0 = 0.9\lambda^{3/4} C_S^{1/2} \quad (2.24)$$

Therefore, a gain in resolution can be achieved by reducing C_S and by reducing λ (i.e., using a higher voltage). At present, many microscopes used in material science and in solid-state chemistry operate at 200–400 kV.

At higher voltages, radiation damage to the specimen severely limits observation time. Advanced designs aim at reducing C_S to allow a decrease of the high voltage for a given resolution. Practical resolution limits are at present ca. 0.16 nm, but new developments tend to peak this value to below 0.10 nm.

Directly interpretable high-resolution images (i.e., having a direct relationship to either the projected lattice potential or the electron density) are formed by the interference between beams passing through the "window" in the image transfer function (ITF). As a result, only low-order reflections (i.e., low-order Fourier components of the lattice potential) generally contribute to the image, which limits the detail that can be resolved (irrespective of the point resolution) since fine details are carried by the high-order Fourier components. However, information transmitted beyond the window (i.e., by the rapidly oscillating part of the image transfer function) can also be used. This information can be extracted by computational techniques, for instance, by using as input a set of images made at a series of closely spaced defocus values [179].

29.2.2.9.4. Image Formation Models [180]

Fourier Model. High-resolution image formation is considered most conveniently as occurring in two steps. In the first step, electrons propagate through the crystal and produce a two-dimensional periodic electron distribution at the exit face, which images the projected lattice potential or the projected electron density. In a second step, this exit face acts as a planar distribution of point sources of spherical electron waves, which interfere behind the foil and form the diffracted beams that move in the lens system of the

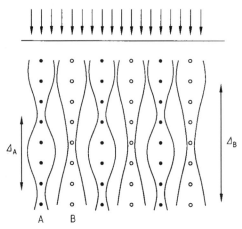

Figure 43. Channeling of electrons along columns of atoms parallel to the incident beam
Alternating focusing and defocusing occurs. The focusing distances Δ_A and Δ_B are different for A and B columns

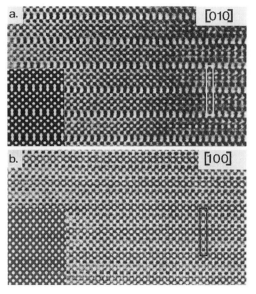

Figure 44. High-resolution image of the high-temperature superconductor $Y_2Ba_4Cu_7O_{15-\delta}$. The bright dots reveal the heavy atom columns as projected along two different zones. Simulated images are reproduced as insets. The projected unit cell is outlined (Courtesy G. Van Tendeloo). The images were made at optimum defocus. The two images refer to different specimens, with comparable thickness.

microscope. Interference of these beams produces the final image, which will be directly interpretable only if the beams interfere with the correct phases.

The resolution of the image depends on the spatial frequencies of the highest-order beams admitted by the selector aperture that still contribute to the image with the correct phase. Only beams passing through the window in the ITF contribute to a directly interpretable image.

The image will exhibit more detail as the order of included reflections increases, but it will give a faithful representation of the structure only if the Fourier components have correct relative phases; this puts a practical limit on the number of useful reflections.

Complete software packages for the simulation of high-resolution images are available commercially (see, e.g., [181]).

Channeling Model. An alternative model that emphasizes the particle nature of electrons provides a simple intuitive picture [182]. An atomic column parallel to the incident beam is a cylindrically symmetrical potential well, which acts on the moving incident electrons as a succession of alternating convergent and divergent lenses. At the entrance face of the foil the incident electron distribution is uniform, but in the first part of the foil the electrons are attracted toward the core of the atom columns and focusing occurs, transforming part of the potential energy of the electrons into kinetic energy. Subsequently the electrons repel one another and defocusing occurs. The focusing–defocusing motion occurs periodically as a function of the depth in the foil, the depth period being a function of the atomic number of the atoms in the column. The heavier the atoms are, the smaller is the depth period. As a result, for a foil of given thickness, one type of atom column will give rise to a relative maximum in the electron distribution, whereas another type may give rise to a relative minimum. These extremes are not necessarily imaged as extremes of the same nature. A maximum may be imaged either as a bright dot or as a dark dot, depending on the defocus and on the foil thickness (see Fig. 43). This model explains why different atom columns produce dots of differ-

ent brightness. The focusing and defocusing behavior was confirmed by computer simulations, which also showed that the depth periods are in the 4–10-nm range depending on the atomic number.

29.2.2.9.5. Image Interpretation [180]

Trial and Error. Digital simulation programs for HRTEM are usually based on the Fourier approach. They allow one to compute the image of a given structure projected along a given zone axis as a function of the foil thickness and of the imaging conditions (defocus) for a microscope with known characteristics (spherical aberration coefficient, accelerating voltage, beam convergence) (Fig. 44).

Identification of a structure or of a defect in a structure proceeds by "trial and error". A model is proposed, and a matrix of images that varies the two main independent variables (foil thickness and defocus) is computed.

These theoretical images are then compared with the observed images at the different thicknesses along a wedge-shaped part of the foil. A solution is considered acceptable when the corre-

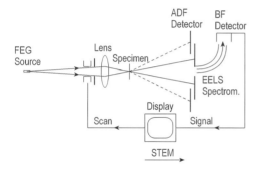

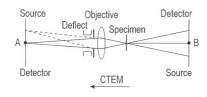

Figure 45. Scanning transmission electron microscopy
Top: Schematic representation of the essential components of a scanning transmission electron microscope (ADF: annular dark field; BF: bright field; FEG: field image gun)
Bottom: Illustrating the reciprocity relationship between the beampaths in STEM and CTEM microscopes; the electrons move in opposite senses (courtesy of J. Cowley)

spondence between observed and computed images is judged visually to be "good enough" for all available thicknesses at a constant defocus. This comparison allows one at the same time to estimate the foil thickness and the defocus; these quantities are usually not known a priori. Numerical criteria to quantify the goodness of fit have been proposed but have seldom been applied as yet. In recent developments, high-resolution images have been used to determine the chemical composition along individual columns in rather special circumstances, such as along the interfaces in synthetic layer structures of semiconductors grown by molecular beam epitaxy. The composition profile across an interface can be obtained at an atomic scale. The application of these methods requires a certain amount of a priori knowledge concerning the structure.

Direct Retrieval. Recently, "direct retrieval" methods have been developed which use as the input a series of images taken at closely spaced defocus values (focus variation method). If the microscope parameters describing the transfer function and its inverse are known, the projected wavefunction at the exit plane of the foil can be reconstructed. From this, the projected structure (the object) can be retrieved by using an analytical formulation of the channeling model. Knowing the projected structure along more than one zone axis allows one to reconstruct the three-dimensional structure. Less a priori knowledge is required than for the methods based on "trial an error".

29.2.2.10. Scanning Transmission Microscopy [183]

Transmission electron microscopy can also be performed with a scanning incident beam (STEM). In a STEM instrument a fine convergent electron probe, formed by demagnifying a small, but brilliant, electron source, is scanned over the specimen area of interest. The incident beam is focused on the specimen plane and a convergent electron diffraction pattern is formed in a plane behind the specimen. Parts of this diffraction pattern (CBED pattern) can be selected by an appropriate aperture, and the signal detected hereby gives rise to an electronic signal which is displayed on a TV monitor, the scan of which is operated synchronously with the probe scan. A bright field (BF) image is obtained when the directly transmitted beam is selected. If one or several beams outside the central beam are selected, a dark field (DF) image is produced.

High resolution is achieved by making the effective probe size as small as possible. For this purpose use is made of a field emission gun with an effective source diameter on the order of 5 nm; demagnification then allows a probe size on the order of 1 nm in diameter to be obtained. The beam current is of the order of 0.5 nA.

The essential electron optics of STEM and CTEM instruments are in principle not very different, even though the electrons travel in opposite senses in the two instruments. A comparison of the principles of STEM and CTEM is represented schematically in Figure 45. In CTEM the image signal is produced in parallel (i.e., simultaneously) in all parts of the image plane (e.g., a photographic plate), whereas in STEM the signal is generated in serial form, as a time-dependent electric signal, which makes it easy to apply on-line image processing. Like in CTEM various signals, excited by the passage of the electron beam through the specimen, can be displayed by using the appropriate detectors.

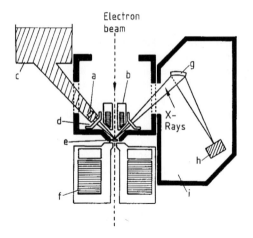

Figure 46. Specimen–detector configuration used in X-ray microanalysis
Left: Energy-dispersive X-ray system (EDX) with Si(Li) detector; Right: Wavelength-dispersive spectrometer (WDS) with bent crystal
a) Si(Li) detector; b) Mini lens; c) Liquid-nitrogen reservoir; d) X-ray window; e) Specimen; f) Objective lens; g) Crystal; h) Detector; i) Crystal spectrometer

29.2.2.11. Z-Contrast Images [184]

The commonly used imaging modes (CTEM, HRTEM) are based on interference and diffraction and rely strongly on coherently scattered electrons. However, simultaneously with the coherent scattering, incoherently scattered electrons also are produced by thermal diffuse scattering and in particular by Rutherford scattering. In the STEM mode these incoherently scattered electrons can be used to image atom columns, provided that the proper electron optics are available. A fine electron probe is obtained by focusing the incident convergent electron beam in the sample and scanning over the foil. Thereby, a significant fraction of the incoherently scattered electrons, emerging from the sample at relatively large scattering angles, is captured in an annular aperture and detected.

In the case of incoherent imaging conditions, the contrast transfer function is a monotonously decreasing function of the spatial frequency, whereas under coherent imaging conditions it is a rapidly oscillating function. This has important consequences. In the coherent case the brightness of a dot imaging a given atom column may change from bright to dark and vice versa as a function of defocus. In contrast, in the incoherent case, the relative brightness of a dot remains consistently the same, independent of defocus, i.e., there is no contrast reversal. Moreover, the dot brightness increases monotonously with increasing average Z value of the atoms in the column as a consequence of the contribution of the Rutherford scattering to the incoherently scattered electrons. Incoherent images can therefore be interpreted on an intuitive basis even for relatively large thickness. It can be shown that structure retrieval is in principle possible and simpler than in the case of coherent imaging. The method is therefore well suited to the study of geometric defects in crystals.

29.2.2.12. Analytical Methods

29.2.2.12.1. X-Ray Microanalysis [134]

Principle. An important feature of electron microscopy is the possibility of performing in the same instrument both a qualitative and a quantitative elemental analysis of the same small crystal fragment that produced the electron diffraction pattern and the image. The most commonly used method consists in principle of detecting the characteristic X rays excited by the high-energy electrons passing through the specimen.

For this purpose, an X-ray spectrometer, mounted sideways on the microscope column at the level of the specimen holder, captures the X rays emitted in all directions by the specimen within a certain solid angle centered around the chosen takeoff direction and transmitted through an X-ray transparent window (Fig. 46). The spectrum consists of the characteristic X-ray lines superposed on a continuous background of bremsstrahlung (see also → Surface Analysis, → Surface Analysis).

The physical basis of X-ray microanalysis is Moseley's law [185], which establishes a direct and systematic relation between the atomic number Z and the characteristic X-ray wavelength λ in a family of X-ray lines (e.g., associated with similar transitions in different elements):

$$\lambda = A/(Z - C)^2 \qquad (2.25)$$

A and C depend on the considered family; K_α, $K_\beta, \ldots$, $L_\alpha, \ldots$, X-ray lines are produced when an inner shell (e.g., K-shell) electron has been ejected from the atom by absorbing an amount of energy from an incident electron, exceeding the critical ionization energy associated with that shell. On deexcitation, this vacant level is filled by an electron originating from a higher-lying shell or subshell of the same atom. The transition is accompanied either by the emission of X rays, which is the phenomenon of interest here, or by the ejection of an Auger electron (→ Surface Analysis). Equation (2.25) shows that the atomic number can in principle be determined by measuring the characteristic X-ray wavelength.

Wavelength-Dispersive Spectrometry (WDS). Two types of X-ray spectrometers are in use: wavelength-dispersive spectrometers and energy-dispersive spectrometers.

In wavelength-dispersive spectrometers, dispersion is achieved by an analyzing crystal. Since the Bragg angle for a given set of lattice planes depends on the wavelength of the incident radiation, X rays of different wavelengths can be separated spatially by the crystal, and a spectrum can thus be analyzed by changing the orientation of the analyzing crystal by θ and detecting the amount of radiation passing through a slit whose angular position is rotated simultaneously over 2θ. Often the analyzing crystal, which acts as a grating, is cylindrically curved so that focusing along a line occurs. The detector, usually a gas proportional counter, is placed in the focus and connected to a single channel analyzer, followed by standard counting electronics.

Depending on the wavelength range to be detected, different analyzing crystals are necessary; provision is usually made on the equipment to allow switching in several crystals from a built-in carousel. Extension of this method to long-wavelength X rays is hampered by the absence of suitable crystals with large interplanar spacings. Synthetic layered structures, prepared by molecular beam epitaxy, and Langmuir–Blodgett layers have been used for this purpose since the end of the 1980s.

Energy-Dispersive Spectrometer (EDS). In energy-dispersive spectrometers, spectral analysis is based on measuring the X-ray energy rather than the wavelength. As a detector, a solid-state ionization chamber is used (e.g., a $p-i-n$ junction in silicon, an intrinsic germanium crystal, or a mercuric iodide crystal). The energy resolution is of the order of 150 eV or less for X-ray energies in the range 1 – 12 keV (Fig. 47). The silicon $p-i-n$ junction is produced by diffusing lithium (at elevated temperature) into a p-type silicon slab under a reverse bias, creating an intrinsic region via the exact compensation of fixed acceptors by mobile lithium donors, as a result of electromigration of the lithium ions. Under reverse bias an electric field is created in the broad intrinsic region, which acts as the sensitive volume. Such detectors must be kept permanently at liquid-nitrogen temperature since lithium ions migrate at room temperature. When an X-ray photon passes through the intrinsic region, its energy is dissipated almost entirely by the formation of electron–hole pairs, which are swept away by the reverse bias field and produce a voltage pulse across the detector crystal that is proportional to the energy of the incoming X-ray photon. The X-ray spectrum is then established by analysis of the pulse heights with a multichannel analyzer. For qualitative analysis, only the line positions are important, but for quantitative analysis the line shape and the peak heights are also used. Fundamental for the interpretation of EDS spectra is the assumption that all the energy of the X-ray photon is deposited in the detector (silicon escape peaks are ignored). The number of charges n created by a photon with energy E is then given by $n = E/E_i$, where $E_i = 3.6$ eV for silicon. Since the charge pulses to be detected are small, reducing the noise of the detector is essential; this is a second reason for cooling the detector. EDS has many advantages over WDS, the main ones being the much shorter acquisition time and the smaller minimum useful probe size. On the other hand, complications also occur that do not arise in WDS instruments; the major ones are escape peaks, pulse pileups, electron beam scattering, peak overlap, and window absorption effects. In practice, the latter prevents detection of light elements with $Z \leq 10$ (beryllium window) unless a windowless detector is used.

Quantitative Analysis. Quantitative analysis is based on the assumption that the intensity of the Lorentzian X-ray peak is proportional to the number of atoms of the species responsible for the peak. Since the theoretical relations are complicated and involve a large number of parameters, some of which are difficult to evaluate, well-characterized homogeneous standards are generally used for calibration, the composition of which has been determined by classical chemical macroscopic methods. These standards, which should also have similar shapes to the sample, are then measured in EDS equipment under the same well-defined and reproducible conditions as the unknown sample, and the spectra are compared. This calibration leads to the so-called k_i factors (coef-

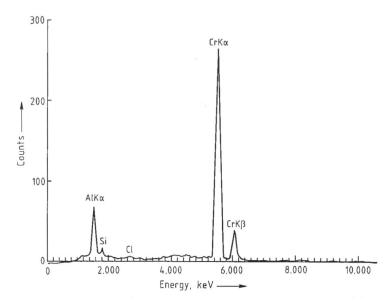

Figure 47. Example of X-ray spectrum obtained by means of an energy dispersive spectrometer
The material is a chromium aluminum alloy.

ficients of proportionality) for the different elements [186]:

$$C_{i\,\text{unknown}}/C_{i\,\text{standard}} = I_{i\,\text{unknown}}/I_{i\,\text{standard}} = k_i \quad (2.26)$$

where C_i are weight fractions.

Since k_i may depend on the particular matrices, Equation (2.26) must be corrected as follows [134], [187]:

$$C_{i\,\text{unknown}}/C_{i\,\text{standard}} = Z_i \times A_i \times F_i \quad (2.27)$$

where the matrix correction factors Z_i, A_i, and F_i still depend on the element under consideration; Z_i is the factor due to atomic number; A_i is due to X-ray absorption; and F_i to X-ray fluorescence. Other correction factors are due to the specimen geometry because X-ray production is depth dependent since it is proportional to the local electron energy. In the thin foils used in TEM the ZAF correction factors can often be neglected. X-ray microanalysis allows quantities of material of the order of 10^{-11} g to be analyzed; the concentration can be as low as 0.01 %, and the precision is of the order of 1 %.

For a more detailed discussion of the method and its technicalities, see [186]–[188].

Special Effects [189]. For foil orientations close to an exact Bragg orientation, the transmitted and scattered wave fields both propagate along the lattice planes. In a model crystal with a primitive structure, one wavefield has its maximum amplitude along planes coinciding with the atomic planes; the other has its maximum along the planes midway between atomic planes. For $s > 0$, the latter wavefield is excited, and therefore easy transmission occurs since the electrons in this wavefield avoid the atom cores (Bormann effect). For $s < 0$, the former wavefield is excited, but strongly absorbed, since the electrons pass close to the atom cores and can excite X-ray emission. X-ray emission is thus expected to be enhanced on the $s < 0$ side of a bent contour ($s = 0$).

In a crystal whose superstructure consists of alternating A and B atomic layers parallel to the considered lattice planes, which are close to the exact Bragg position, one type of wavefield is peaked along the planes of A atoms, whereas the other is peaked along the B-atom planes. Depending on the sign of s, one of the wavefields is enhanced relative to the other; hence the A and B atomic layers will be excited unequally. Under such conditions, X-ray microanalysis may produce erroneous results for the A/B ratio. Therefore, foil orientations should be avoided that cause an extinction contour to pass through the selected area due to anomalous transmission. On the other hand, this effect has been exploited to locate the site occupied by a third atomic species, in an AB superstructure, using a technique called ALCHEMI (atom location by channeling-enhanced microanalysis) [189].

29.2.2.12.2. Electron Energy Loss Spectrometry (EELS) (→ Surface Analysis) [190]

Principles [134], [191]. Apart from elastic scattering, which is responsible for electron diffraction, inelastic scattering events also occur as electrons pass through the foil. Inelastic processes can be caused by (1) single electron excitations,

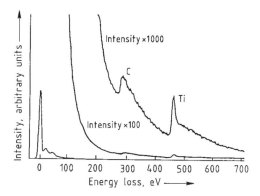

Figure 48. Example of EELS (electron energy loss spectrometer) spectrum of TiC
On the low-energy side the plasmon peak is visible. The fine structure on the high-energy end is caused by interference of waves reflected by the neighboring atoms

such as X-ray and Auger electron production; and (2) collective excitations, such as volume and surface plasma oscillations (plasmons) and phonons. Collective excitation can be revealed indirectly as characteristic energy losses of the incident electrons. The energy of elementary excitations (plasmons and phonons) is low compared to the incident electron energy (100–140 kV).

Plasmons have energies in the range of 10–20 eV, whereas phonon energies are of the order of 10 meV. Individual quantized plasmons are thus much easier to detect than phonons. Since the positions of the plasmon loss peaks are characteristic of the materials, they can be used as analytical tools for aluminum and magnesium, for example.

However, for chemical analysis, the absorption edges in EELS curves are more important (Fig. 48); they reflect the absorption phenomena leading to X-ray production and exhibit a fine structure on the high-energy side. This fine structure is referred to as EXELFS (extended energy loss fine structure) [192], which is the analogue of EXAFS (extended X-ray absorption fine structure). The steep rise on the low-energy side of the absorption edge is due to the excitation of inner-shell electrons and characterizes the element.

The fine structure is produced by the electron wave originating from an inner shell that is partially back-reflected by the surrounding atoms, which leads to a modulation of the excitation probability of inner-shell electrons. This fine structure can therefore provide information not only on the chemical nature of the absorbing atom, but also on the number of nearest neighbors and their distance.

As an analytical tool, EELS is in a sense complementary to X-ray microanalysis and makes use of the same type of phenomena. EELS works well for light elements, whereas X-ray microanalysis is much more difficult to apply to elements with $Z \leq 11$.

Spectrometers. The simplest system uses a magnetic prism to produce spatial separation of electrons with different energies. The spectrum is scanned by means of a slit, and a phosphor coupled to a photomultiplier is used as the detection system.

One of the most advanced types of equipment is the ω electron spectrometer [193]. It consists of four magnetic prisms that bend the incident beam back to its original direction; it can be mounted on an electron microscope column behind the specimen where inelastic scattering is produced. Slits allow a certain energy range to be selected from the spectrum. The spectrometer can thus also act as a filter and can produce, e.g., zero-loss images in which the blurring due to inelastic scattering is eliminated.

Since the 1980s *X-ray microanalysis* based on energy-dispersive spectrometers has become a routine technique. Equipment as well as data processing hardware and software are available commercially as optional accessories for most microscopes. In the late 1990s the range of applications of EELS spectroscopy was considerably expanded and extended to nanoscale samples. The commercial availability of user-friendly equipment compatible with most transmission electron microscopes has greatly stimulated its use. Also the improvements in the interpretation methods and the development of the corresponding computer software greatly contributed to the expansion of its use. It is now possible to map the spatial distribution of selected features of the information contained in the EELS spectrum. In this way spatial distribution of different elements (also light elements) can be mapped with nanometer resolution, but for instance also the spatial distribution of different ionization states of the same element can be visualized. These advances were made possible by the use of arrays of diodes that allow parallel detection instead of the initially used serial detection.

29.2.2.13. Specimen Preparation

The objective of all specimen preparation methods is to obtain a crystal fragment that is sufficiently thin to transmit electrons with energies of 100 keV or more, and has a sufficiently large lateral size. The numerous methods available depend strongly on the type of material and observation method.

29.2.2.13.1. Diffraction Contrast Specimens

Specimens should have a thickness of a few hundred nanometers. If too thick, the foil is not transparent; if too thin, achieving a two-beam orientation is difficult; moreover, the sample is not representative of bulk material for a number of applications.

Electrically conducting bulk samples (metals and alloys) can generally be thinned by electropolishing; *bulk ceramic samples* are often chemically thinned and polished. The resulting sample is usually perforated in the center; the edges of the hole are then wedge shaped and have transparent edges.

Insulating thin foils may become charged and deflect the electron beam; this can be avoided by coating the specimen with a thin layer of amorphous carbon, which is electrically conducting.

Thin films of uniform thickness can be obtained by *vapor deposition* under vacuum or by *chemical deposition* techniques; such samples usually have microstructures that are different from those of samples prepared from bulk material.

Layered crystals exhibiting a pronounced cleavage plane, such as graphite or MoS_2, can be thinned by repeated cleavage to produce rather uniformly thin specimens.

A truly universal thinning method does not yet exist. However, *ion beam bombardment under grazing incidence* is applicable to many materials, in particular to brittle ones such as most semiconductors. In the first step, these materials are mechanically polished to produce slabs with dimples; subsequent ion beam thinning under grazing incidence then produces a hole whose edges are wedge shaped. Thinning can be achieved either from one side only or from both sides of the slabs; this is important, for instance, for quantum wells grown by molecular beam epitaxy. Surface damage left after ion bombardment can often be removed by a final chemical polish. Figure 49 schematically represents thinning methods used for semiconductor specimens.

29.2.2.13.2. High-Resolution Specimens

The various methods described above can also of course be used to prepare specimens for use in high-resolution imaging. However much thinner specimens (1 – 10 nm) are required than for diffraction contrast. Nevertheless, more rudimentary methods can often be used successfully, since much smaller lateral dimensions of the specimen can be tolerated. Specimens of brittle materials (e.g., oxide superconductors) are in many cases obtained simply by crushing the material in a quartz mortar under a protecting organic liquid. In such a way the crystal fragments are dispersed in a suitable liquid, and a drop of the suspension is deposited on a grid covered with a holey carbon film and allowed to dry. The edges of fragments protruding over the holes are potentially suitable specimens, but actually only a small fraction of the crystal fragments are sufficiently thin to produce high-quality images. The thickness of the specimen is often the resolution-limiting parameter. Materials that deform plastically when crushed at room temperature often behave as brittle when crushed while immersed in liquid nitrogen.

29.2.2.14. Applications to Specific Materials and Problems

The number and diversity of applications of different forms of electron microscopy have become so large that an exhaustive survey, even when restricted to chemical aspects, is almost impossible. In particular, its application to structural problems in solid-state chemistry has evolved explosively in recent years. The examples given in what follows attempt to show as large a diversity as possible.

29.2.2.14.1. Crystal Structures

Only in simple cases can an "ab initio" structural determination be made by electron microscopy and diffraction, essentially because in electron diffraction the intensity of the reflections is not simply proportional to the square of the structure amplitude, but also depends in a complicated manner on the orientation and the thickness of the sample.

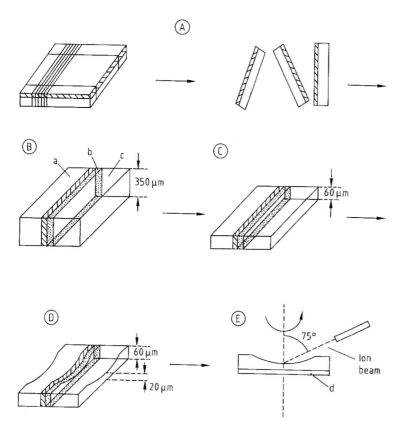

Figure 49. Preparation of a cross section of integrated circuits
A) Specimen is cut into thin slabs; B) Slabs are glued on a glass slide; C) Composite specimen is thinned mechanically; D) Thin specimen is dimpled; E) Final thinning is done by ion bombardment from one side while the specimen is rotated
a) Integrated circuit; b) Glue line; c) and d) Glass slide
(Courtesy of IMEC)

However, whereas X-ray and neutron diffraction provide only average structures (→ Structural Analysis by Diffraction), electron diffraction and microscopy can provide detailed information on local structures, especially when the average structure is known from other techniques. Also planar defects (i.e., local deviations from the average structure) can be studied to suggest possible new structures in which such defects occur periodically.

Since electron microscopy requires minute samples, the crystallography of materials that are available only as small particles (e.g., catalysts) can be studied in direct and reciprocal space. The early crystallographic studies on C_{60} and C_{70} crystals were for instance, performed by electron diffraction [194]. Even fine powders or fine-grained ceramics contain single crystal particles large enough to produce a single crystal electron diffraction pattern; therefore, the lattice parameters and the unit cell can be determined easily, albeit with moderate precision. Subsequently, X-ray powder diffraction patterns can be indexed unambiguously and the lattice parameters determined with high precision. Electron diffraction and especially high-resolution electron microscopy are complementary techniques to X-ray and neutron diffraction.

Imaging Modes for Various Materials. *Polytypes, Mixed-Layer Polytypes.* The structures of polytypes are usually based on the various stacking modes of closely packed layers of spheres. The stacking sequence can best be revealed by imaging the structure along the zone parallel to the direction of the close-packed rows of atoms. Figure 50 shows a foil of SiC – 15 R imaged along the close-packed rows by using three different modes; it

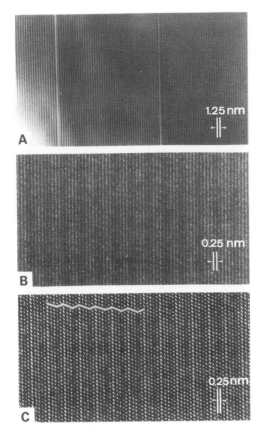

Figure 50. Imaging of the 15 R polytype of SiC along the closely packed rows of atoms
A) Two successive spots of the 0001 row contribute to the image; only the five-layer lamellae are imaged as lines; singular line spacings represent stacking faults; B) Cluster of five successive spots of the 0001 row contributes to the image; single SiC layers are revealed; each lamellae of A) is seen to contain five layers; C) Spots from parallel rows are allowed to contribute to the image; the stacking sequence ABCAC can be deduced from the two-dimensional image since close-packed atom columns are imaged as dots

illustrates clearly the effect on the image detail of increasing the number of beams participating in the image-forming process. In Figure 46 A, only two successive spots out of the densely populated row of 0001 spots are included; only the superspacing consisting of five-layered SiC lamellae is revealed as a line image. The isolated stacking faults are imaged as anomalously wide line spacings.

Including in the aperture, a cluster of at least five successive spots of the central row between the 0000 and 00015 reflections, but excluding parallel rows, reveals the individual SiC layers and the superperiod due to the five-layered lamellae as lines; it shows in particular that the superperiod contains five layers (Fig. 46).

Finally, when spots from neighboring dense rows are also collected, a two-dimensional dot pattern is obtained and the close-packed rows are imaged as bright dots; the stacking sequence ABCAC... is directly revealed (Fig. 46).

The same imaging mode is used for the study of mixed-layer compounds based on the epitaxy of close-packed layers [195].

Ordered Alloys. The diffraction patterns of ordered binary alloys, derived from the face-centered cubic (fcc) structure, consist of strong basic spots that would also be produced by the disordered fcc lattice, and of weaker superlattice spots characteristic of the superstructure. The first-order superlattice spots correspond to larger interplanar spacings and are thus closer to the origin than the basic spots. Therefore images can be produced by selecting superstructure spots next to the direct beam only and excluding basic spots. The image so obtained (*bright-field superlattice image*) reveals only the configuration of the minority atoms as bright dots, since these atoms determine the superlattice. This information is sufficient to determine completely the projection of the superstructure, provided the basic lattice is known.

An image can also be formed exclusively by means of the superstructure spots situated within a reciprocal unit cell of the basic lattice (*dark-field superlattice image*) [196]. This last mode produces the best-contrasted images of the minority atom sublattice (Fig. 51).

If basic and superlattice spots are included in the image-forming process the basic lattice is also revealed; minority and majority atoms are imaged as dots with a different brightness. Such images contain information that is redundant if only the superstructure is sought. Examples of superlattice images of ordered alloys obtained by using the bright-field superlattice mode are reproduced in Figure 52.

Long-Period Interface Modulated Structures. Many long-period structures are derived from simpler structures by the periodic introduction of planar translation interfaces. Typical examples are the long-period antiphase boundary structures in many ordering alloy systems such as Au–Mn, Cu–Au, and the shear structures derived from a simple transition-metal oxide structure such as WO_{3-x}, TiO_{2-x}. Such compounds produce charac-

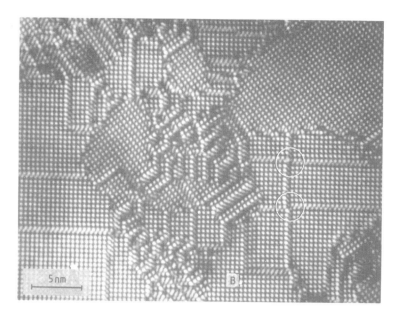

Figure 51. Dark-field superlattice image of Au$_4$Mn
The manganese atoms are represented as bright dots. Note the presence of two orientation variants and several translation variants [196], [198]

teristic diffraction patterns. In combination with their high-resolution images, their structures can usually be elucidated completely, including local deviations from the average structure, which often occur in such phases.

In a model [197], the superstructure is assumed to consist of identical slabs of basic structure with thickness D limited by planes normal to $\bar{e}_n$. Usually the slab thickness is equal to an integer number n of unit cell parameters a of the basic structure i.e., $D = na$, but this need not be the case. Successive slabs are separated by planar interfaces (stacking faults, anti-phase boundaries, discommensuration walls, etc.) with a displacement vector $\bar{R}$ and a unit normal $\bar{e}_n$.

The diffraction pattern of such a structure consists of clusters of equally spaced superstructure spots (spacing: $1/D$), called satellites, located around the positions of the basic spots. The intensity of the satellites decreases rapidly with separation of the satellite from the basic spot to which it belongs, which also determines its position. The positions $\bar{H}$ of satellites are given by the relation [197]

$$\bar{H} = \bar{h} + (m + \bar{h} \cdot \bar{R})\bar{q} \qquad (2.28)$$

(m = integer; order of the satellite; $\bar{q} = (1/D)\bar{e}_n$. The diffraction pattern thus exhibits main (or basic) spots at $\bar{h}$. With each basic spot $\bar{h}$ a linear sequence of equidistant satellites $m\bar{q}$ is associated. These sequences are perpendicular to the interfaces and they are shifted with respect to the positions of the basic spot over a fraction $\bar{h} \cdot \bar{R}$ of the intersatellite spacing; $\bar{q}$ is the wave vector of the modulation. Provided the basic structure is known, the long-period structure can be determined from the geometry of the diffraction pattern.

If the displacement vector $\bar{R}$ is parallel to the interfaces, the latter are termed conservative since the long-period structure has the same chemical composition as the basic structure.

If $\bar{R}$ is not parallel to the interfaces the superstructure may have a composition different from that of the basic structure; this is the case for shear structures, in which the interfaces perpendicular to $\bar{e}_n$ are generated by removing slabs of material with thickness $\bar{R} \cdot \bar{e}_n$, followed by closing the gaps so created by the displacement $\bar{R}$. If the removed slab has a composition different from that of the basic structure, the overall composition changes in the process.

The alloy, based on a face-centered cubic lattice, with ideal composition Au$_{22}$Mn$_6$, produces the diffraction pattern shown in Figure 53 A and

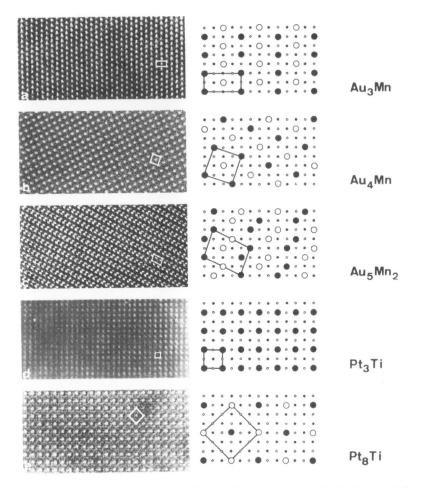

Figure 52. Various superstructures in ordered binary alloys imaged using the bright-field superlattice mode and compared to structural models

represented schematically in Figure 53 B. It consists of sequences of superstructure spots, marked as full dots in Figure 53 B, associated with the basic spots marked as small open disks in Figure 53 B, which are the spots that would be produced by the Au_4Mn structure. Both types of spots are present simultaneously in Figure 53 A because the selected sample contained a small area of Au_4Mn as well as an area exhibiting the $Au_{22}Mn_6$ structure. From the direction of the rows of superstructure spots, their spacing, and the fractional shifts, the model of Figure 53 D can be derived [198]. In this model, only the manganese columns are represented. The high-resolution image (Fig. 53 C) made in the bright-field superlattice mode can be compared directly with the model of Figure 53 D; only the minority atom columns are revealed as bright dots.

Molecular Crystals; Fullerites. Organic crystals are usually prone to ionization damage and decompose very rapidly under electron irradiation; they can thus be studied for only a short time (a few seconds) and only with a very low electron beam intensity. Transmission electron microscopy has, therefore, seldom been applied to organic crystals. However, the all-carbon molecules C_{60}, C_{70}, etc., (fullerenes) discovered at the end of the 1980s resist electron radiation fairly well. Early structural studies on the crystalline phases of fullerenes (fullerites) were performed mainly by electron microscopy because only small quantities of sufficiently pure material were available. At room

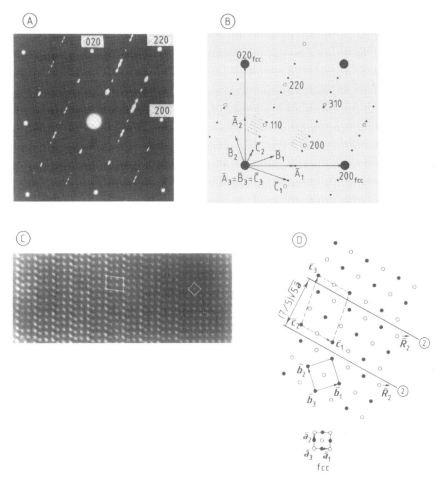

Figure 53. One-dimensional nonconservative long-period superstructure of the Au$_4$Mn structure, with theoretical composition Au$_{22}$Mn$_6$ [198]
A) Composite diffraction pattern along the [001] zone of Au$_4$Mn; B) Schematic representation of composite diffraction pattern (open dots indicated locations of basic spots, full dots are superstructure spots); C) High-resolution image of the superstructure (on the right, a small area of basic Au$_4$Mn structure is also visible); D) Model of the one-dimensional superstructure; only manganese columns are represented

temperature, C$_{60}$ crystals were found to have the fcc structure, often containing intrinsic stacking faults, twins, and other defects characteristic of low stacking fault energy fcc alloys [199]. Figure 54 shows, for instance, various faults in a crystal of C$_{60}$: an intrinsic fault containing a dipole of stair rod dislocations in S, and a Frank partial dislocation and its associated stacking fault in A. The diffraction effects and the microstructure caused by the orientational phase transition in C$_{60}$ at 255 K, from the room temperature fcc phase to the simple cubic low-temperature phase, were studied by means of electron microscopy [200] (Fig. 55).

Vapor-grown C$_{70}$ crystals were found to belong to two different structural types. About 1% of the vapor-grown crystals are hexagonally close packed; the rest are face-centered cubic. Many crystal fragments exhibit both crystal structures. The hexagonally close-packed crystals undergo several orientational phase transitions on cooling, leading to a monoclinic orientationally ordered superstructure at low temperature [201]. The fcc

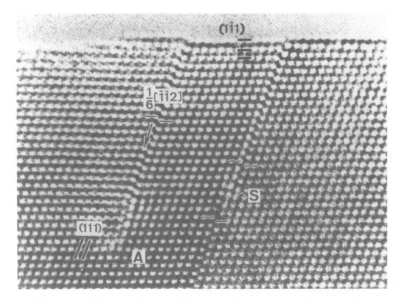

Figure 54. High-resolution image of a fragment of a C_{60} molecular crystal: bright dots represent rows of C_{60} molecules. Note the presence of a stepped intrinsic stacking fault (in S). Also a Frank-type partial dislocation and its associated stacking fault are present (in A) [199]

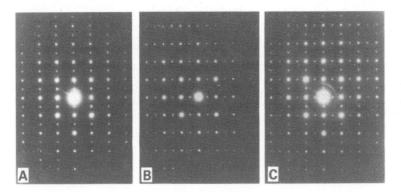

Figure 55. Diffraction patterns of the simple cubic phase of C_{60} along a cube zone [200]
A) One variant; B) Perpendicular variant to A); C) Both variants present

crystals undergo a single sluggish phase transition into a rhombohedral phase.

Quasi Crystals. Until 1984, one of the most firmly established fundamentals of crystallography was the postulate that crystals have three-dimensional translation symmetry, and that only symmetry rotation axes with multiplicities 1, 2, 3, 4, and 6 can occur because they are the only ones that are consistent with homogeneous space filling. However electron microscopy and electron diffraction have shown that 5-, 8-, 10-, and 12-fold symmetry axes occur in the structure of certain, usually rapidly cooled, alloy phases with complicated compositions [202], [203]. Careful tilting experiments have demonstrated the presence of icosahedral and dodecahedral symmetry groups in many of these phases. Their electron diffraction patterns consist of sharp spots, whose geometry reflects the above-mentioned symmetry elements, but they do not exhibit translation symmetry. The same conclusions follow from high-resolution images [204].

At present, numerous alloy systems are known to exhibit noncrystallographic symmetry elements and to lack three-dimensional periodicity; they are

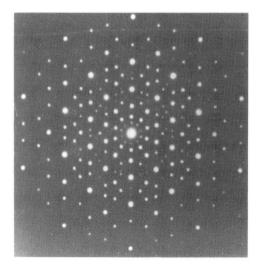

Figure 56. Diffraction pattern of the icosahedral phase of quenched Al–Mn alloy exhibiting fivefold symmetry (Courtesy of G. Van Tendeloo)

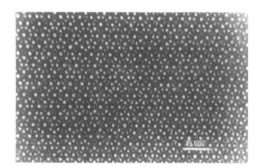

Figure 57. High-resolution electron micrograph of rapidly cooled Al–Mn alloy showing the absence of a periodic structure; pentagonal arrangements of dots can be observed

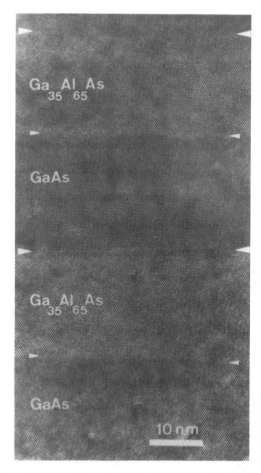

Figure 58. High-resolution image of quantum well consisting of alternation of pure GaAs and (GaAl)As

called quasi crystals. Figure 56 shows the diffraction pattern of a quasi crystal with composition Al–Mn exhibiting a fivefold axis, whereas Figure 57 shows the corresponding high-resolution image; pentagonal arrangements of bright dots are visible in many places.

Artificial Layer Structures and Semiconductor Devices. In recent years the growth of artificial layer structures, which can be used as quantum wells, has become an important research activity. The properties of such layered structures are determined to a large extent by the perfection of the interfaces, as well as the regularity of their spacing. Both parameters can be studied by means of high-resolution, cross-sectional samples. The steps in the preparation of cross-sectional samples are discussed in Section 29.2.2.13. The method is clearly destructive and cannot be used for fabrication control. However, it is the most direct way to calibrate other nondestructive methods.

Figure 58 shows a high-resolution image of a quantum well consisting of the alternation of GaAs and (GaAl)As. The positions of the interfaces are indicated by arrows. The interfaces are flat to within two or three atomic layers; no interfacial dislocations are observed. The composition profiles across such interfaces can in principle be studied by measuring the brightness of the dots representing the atom columns [205].

For the study of semiconductor devices, diffraction contrast images made in a high-voltage electron microscope (≈ 1000 kV) are often useful since thicker specimens can be tolerated than in a

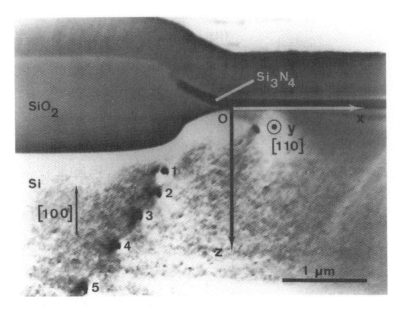

Figure 59. Cross section of field effect device
The procession of dislocation 1...5 results from stresses set up by the oxidation process (Courtesy of J. Van Hellemont)

conventional microscope (≈ 200 kV). An example is shown in Figure 59, which is the cross section of a field effect device. The different layers have been indicated along with the orientation of the silicon substrate. The black dots numbered 1–5 are images of dislocations seen end on, which were nucleated by the stresses set up as a result of the oxidation process. From the spatial distribution of the dislocations the magnitude of the stress can be estimated.

During crystal growth, small oxide particles are often formed in the interior of silicon single crystals. Such particles cause a compressive stress on the surrounding matrix. This may lead to prismatic punching, whereby disks of self-interstitials limited by loops of perfect dislocations are emitted, thereby relieving the stresses. These loops can glide along cylindrical surfaces whose cross section is determined by the size and shape of the particle, and whose generators are parallel to the different glide vectors of silicon, which are of type 1/2 [110]. The geometry of the process can be established with diffraction contrast images (see Fig. 60).

Carbon Nanotubes, Onions. The preparation of fullerene-containing soot by the vaporization of graphite electrodes in an electric arc leads simultaneously to the formation of very fine *hollow carbon needles* or "tubules" with diameters as low as 1 nm. Electron microscopy and electron diffraction have shown that these needles consist of concentric seamless graphene tubes (2–15 tubes) [206]. The most remarkable feature is the helical character of some of the tubes within a tubule, as deduced from fiber diffraction patterns (Fig. 61). The helical structure is a consequence of the stepwise increase in circumference of successive concentric tubes by πc (c = lattice parameter of 2H graphite). Since πc is not commensurate with the a-parameter of graphite the $c/2$ spacing between successive tubes and their seamless character can be reconciled only with the changing diameter if some of the tubes at least become helical. However, a detailed analysis of the fiber pattern has shown that in most tubules the majority of tubes are nonhelical [207].

Intense "in situ" electron irradiation of particles of carbon soot for tenths of minutes in the electron microscope produces spherical particles consisting of *onion-like* concentric spherical shells of graphene-like layers [208]. The microscope allows the concentric shell structure to be imaged during its formation.

High-T_c Superconductors. High-resolution electron microscopy and electron diffraction have recently been applied extensively in structural studies of high-T_c oxide superconductors. In particular, $YBa_2Cu_3O_{7-\delta}$ and the related 1–2–3

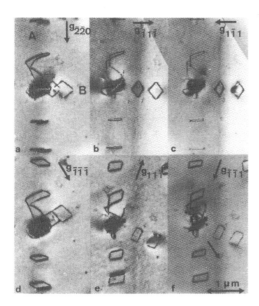

Figure 60. Oxide particle in the interior of a silicon single crystal
Stresses have been relieved by prismatic punching. The dislocation loops are imaged using different reflections (Courtesy of H. Bender)

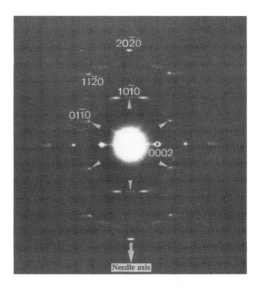

Figure 61. Electron diffraction (fiber) pattern produced by a graphite nanotube showing that certain graphene tubes are helically wound, whereas others remain normal
Note the outward streaking of the diffraction spots
(Courtesy of X. F. Zhang)

family of compounds have been studied in great detail. Numerous studies of superconducting compounds can be found in *Physica C* from 1989 onward. The generic 1-2-3 compound (YBa$_2$Cu$_3$O$_{7-\delta}$) has a structure derived from the triperovskite structure in which the succession of layers along the c-direction is CuO–BaO–CuO$_2$–Y–CuO$_2$–BaO, etc. The CuO layer consists of Cu–O–Cu–O– chains parallel to the b_0-direction; this feature breaks the tetragonal symmetry and reduces the symmetry to orthorhombic. The average structure was determined by X-ray and neutron diffraction, but several important structural features were discovered by electron microscopy. The oxygen deficiency was found to be accommodated by the absence of oxygen in a fraction of the chains in the CuO layer [209]. The structure in which one of two chains is alternately free of oxygen, the so-called $2a_0$ (or ortho II) superstructure, has an ideal composition with $\delta = 0.5$. It occurs in two orientation variants, corresponding to the (110) reflection twin texture of the 1-2-3 compound, which was also discovered by electron microscopy. The presence of the $2a_0$ structure was shown to be responsible for the 60 K plateau in the T_c versus composition curve [209]. Locally, structures with a period of $3a_0$ occur [210].

Much attention has been devoted to various substitutions in CuO layers, replacing copper ions by other metallic ions such as Fe, Co, etc., but also substituting by complex anions such as CO$_3^{2-}$, SO$_4^{2-}$, PO$_4^{3-}$, or NO$_3^-$. High-resolution electron microscopy has demonstrated that the substitution occurs in the CuO layers, and has allowed the resulting superstructures to be visualized. In the case of SO$_4^{2-}$ substitution, for instance, the superstructure has been shown not to be commensurate with the basic 1-2-3 lattice [211]. Figure 62 shows a view along the b_0-direction of the 1-2-3 matrix. The prominent, bright dot squares reveal the SO$_4^{2-}$-containing rows that replace CuO chains. Their arrangement is not periodic and can best be described as resulting from a concentration wave with a wave vector that is not commensurate with the 1-2-3 lattice along the a-direction.

Replacing copper by cobalt or gallium in the CuO layer results in the formation of corner-sharing chains of CoO$_4$ (GaO$_4$), tetrahedra along the [110] and [1$\bar{1}$0] directions of the 1-2-3 matrix lattice [211]. The resulting structure is still orthorhombic, but with a diagonal unit cell ($a_0 \approx b_0 \approx a_p \sqrt{2}$), where a_p is the lattice parameter of the basic perovskite. Electron microscopy [212] showed that a superstructure is formed, which is localized in the Co–O (Ga–O) layers, and in which alternating parallel CoO$_4$ (GaO$_4$) chains

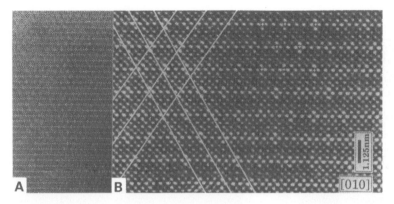

Figure 62. High-resolution image along the $[010]_O$ direction of the SO_4^{2-}-substituted 1–2–3 ($YBa_2[Cu_{3-x}(SO_4)_x]O_{7-\delta}$) compound The SO_4^{2-}-containing rows parallel to [010] are marked as small crosses of prominent bright dots. (Courtesy of T. Krekels) [211]

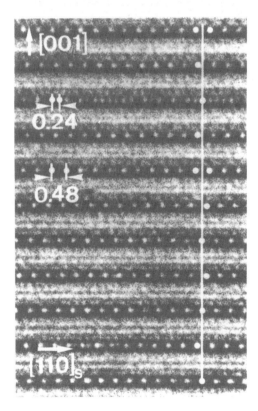

Figure 63. High-resolution image along the $[120]_p$ zone of the cobalt-substituted 1–2–3 compound $YBa_2Cu_2CoO_7$ Dark dots reveal indirectly the arrangement of chains of corner-sharing CoO_4 tetrahedra that replace the CuO layers. (Courtesy of T. Krekels) [212]

have mirror-related configurations, leading to doubling of the lattice parameter perpendicular to the chain direction. Figure 63 is a view of the cobalt-containing material along a $[120]_p$ direction, which allows the period doubling along a_0, as well as the possibility of polytypism along c_0, to be observed.

Superionic Conductors. Superionic conductors often consist of a stable rigid framework forming large channels along which ions can move easily. The channels in such structures can be made visible in high-resolution images made along a zone parallel to the channel direction. Figure 64 shows such a high-resolution image of the natural mineral hollandite with idealized composition $Ba_xMn_8O_{16}$ [213]. Octagonal channels, formed by interconnected strings of edge and corner-sharing MnO_6 octahedra, are imaged as octagons of black dots. These channels are occupied by barium ions, which are also visible as black dots in the centers of the octagons.

Polymers. Although most organic materials usually deteriorate rapidly from ionization damage in the electron beam, meaningful observations can be made on certain polymers. In Figure 65, fiber patterns of a poly(-p-phenylene) stretched six to seven times are reproduced (PPV, an alternating copolymer of p-phenylene and acetylene) at two different photographic exposures to reveal the details of the pattern close to the origin [214]. The direction of stretching is indicated by arrows. The patterns of the left column (A) and (C) refer to the pristine material. On doping with $FeCl_3$ this material becomes a good electrical conductor. The effect of $FeCl_3$ doping on the diffraction pattern is

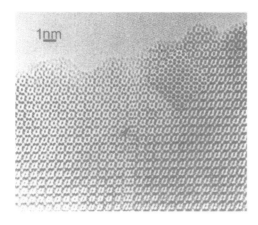

Figure 64. Structure of natural monoclinic hollandite $Ba_xMn_8O_{16}$ as viewed along the [010] zone
The image is compared with a simulated image in the inset [213].

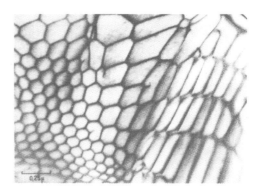

Figure 66. Network of undissociated dislocations in the (0001) plane of zinc

Figure 67. Dissociated dislocations in silicon viewed along the [110] zone in the high-resolution mode (Courtesy of H. Bender)

29.2.2.14.2. Defects

Dislocations. Dislocation configurations have been studied mainly by diffraction contrast brightfield imaging since this method is very sensitive to lattice strain fields. Under these conditions, dislocations appear as dark lines when two-beam diffraction conditions close to $s \approx 0$ are used. A network of dislocations in the basal plane (0001) of zinc is visible in Figure 66.

Images of dissociated dislocations (i.e., dislocation ribbons) have been used extensively as a means to deduce the stacking fault energy [125].

The separation between partial dislocations is usually measured on a weak beam image (see Section 29.2.2.7.2). The image, which is now a very narrow bright line on a darker background, is essentially kinematic. In the high-resolution image of Figure 67, dissociated dislocations in silicon are viewed along the [110] zone. Between the two partial dislocations an intrinsic stacking fault

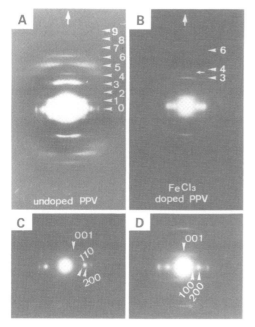

Figure 65. Fiber texture pattern of a stretched PPV polymer
A) and C) Pristine polymer; B) and D) After $FeCl_3$ doping (Courtesy of X. F. Zhang)

shown in the right column (B and D). Whereas in the undoped material the 001 diffraction vector encloses an obtuse angle with the row of h00 reflections, this angle becomes 90° in the doped material, which suggests that the symmetry changes from monoclinic to orthorhombic on doping.

Figure 68. High-resolution image of a stacking fault in 2H-wurtzite (Courtesy of H. Bender)

is present. Such configurations have been used to deduce stacking fault energies [125], [176].

Planar Defects. The diffraction contrast images of planar interfaces that intersect the foil surfaces consist of a set of fringes parallel to the closest surface with a depth period given by $1/\sigma_g$ *(which is equal to t_g for $s = 0$; see Section 29.2.2.4.6)*. If the planar interface is parallel to the foil surfaces, which is often the case in cleaved foils, the faulted area exhibits only a brightness difference.

High-resolution images can reveal stacking faults directly, especially in close-packed structures. In such structures the image made along a zone parallel to the close-packed rows reveals the stacking directly. A stacking fault in a wurtzite crystal is reproduced in Figure 68.

Domain boundaries and twins in planes that intersect the foil surfaces are imaged as δ-fringes (Section 29.2.2.4.6) [155]. The extinction condition is now $\Delta s = 0$ (or $\Delta \bar{g} = 0$!) which is satisfied for the family of planes common to the two domains. When the interfaces are nearly perpendicular to the foil plane the main effect is domain contrast (i.e., a brightness difference in the domains on either side of the interface) as shown in Figure 69, where the same twin boundaries in the high-temperature superconductor $YBa_2Cu_3O_{7-\delta}$ have been imaged in three different modes.

29.2.2.14.3. Small Particles

Like most other electron microscope techniques, transmission electron microscopy allows

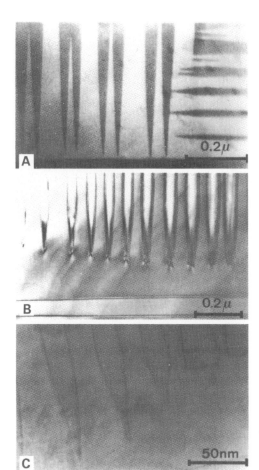

Figure 69. Twin boundaries in $YBa_2Cu_3O_{7-\gamma}$
(A) Domain contrast; (B) Interface contrast; (C) High-resolution imaging (Courtesy of H. W. Zandbergen)

the particle-size distributions and particle shapes of very fine powders to be studied. Such studies are important in several areas of research: including catalysis, magnetic recording, and photography. Transmission electron microscopy provides, in addition, the possibility of obtaining electron diffraction patterns of single particles as well as high-resolution images. In Figure 70 single particles of silver vapor deposited on an amorphous carbon substrate is shown. They are clearly not single crystals but aggregates of multiply twinned fcc crystallites having an overall icosahedral shape.

The topotactic dehydration reaction [215] transforming goethite into hematite was studied "in situ" at room temperature and shown to

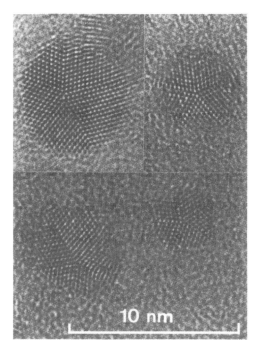

Figure 70. Small silver particles consisting of an icosahedral aggregate of multiply twinned face-centered cubic crystals (Courtesy of C. Goessens)

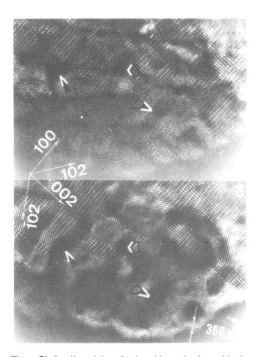

Figure 71. Small particles of twinned hematite formed in the electron microscope as a decomposition product of goethite [215]

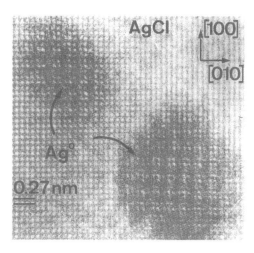

Figure 72. Silver chloride crystal covered by silver specks viewed along the cube zone

produce small twinned hematite crystals (Fig. 71). In the initial stages of the dehydration process, satellite spots appear in the diffraction pattern around the hematite spots; they were found to be due to a texture consisting of a periodic alternation of voids and hematite crystals, rather than to a long-period superstructure.

The crystal habit of the silver halide particles used in photographic emulsions is important since it is one of the parameters determining the photosensitivity of the film. The hexagonal and triangular (111) tabular crystals of AgCl were shown to be in fact multiply twinned on planes parallel to the habit plane. Depending on the parity of the number of twin lamellae the particle grows into either hexagonal or a triangular habit.

When silver halide crystals with a cube habit are exposed to electrons, photolytic metallic silver specks are formed "in situ" in an epitaxial relation to the silver halide substrate. This gives rise to moiré fringes parallel to the cube edges due to the difference in lattice spacing between AgCl and Ag (Fig. 72).

29.2.2.14.4. Surface Studies [216]

Crystal surfaces can be studied in a transmission electron microscope by different modes. It is possible to use an electron beam, in reflection under *grazing incidence*, magnifying the Bragg reflected beam due to planes roughly parallel to the surface. The surface of the sample must be almost parallel to the beam. The resulting image

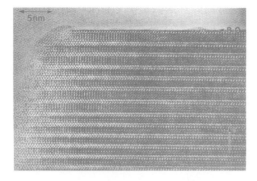

Figure 73. High-resolution image of the free cleavage surface of a superconductor ($Bi_2Sr_2Ca_1Cu_2O_x$)
Cleavage occurs between the two BiO layers [217]

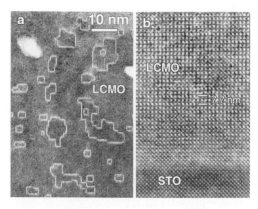

Figure 74. Thin film of $La_{1-x}Ca_xMnO_3$ on a $SrTiO_3$ (STO) single-crystal substrate
a) Plane view, low-resolution image revealing antiphase boundaries
b) High-resolution image of a cross section specimen (courtesy O. Lebedev)

is highly distorted in the sense that the magnification in a direction perpendicular to the beam is much lower than the magnification along the projection of the electron beam. This Bragg reflection mode can be used to image surface steps and to detect surface superstructures due to reconstruction or to adsorbed layers by diffraction. A very clean vacuum is required.

A second mode consists of observing the surface profile in transmission with the *electron beam parallel to the surface*. In this way, surface relaxations and surface reconstruction can be observed together with atomic resolution of the substrate. An example of the cleavage surface of a high-T_c superconductor is reproduced in Figure 73. It allows the layer within the structure to be located along which cleavage takes place [217].

29.2.2.14.5. Thin Epitaxial Layers

The development of modern miniaturized electronic devices relies heavily on the use of thin epitaxial films of which the relevant physical properties sensitively depend on their thickness and on their microstructure. In recent years high-resolution electron microscopy has extensively been applied for the primary characterization of such films. In particular various modes of accommodating the misfit between a single-crystal substrate and the epitaxial films have been discovered by this method.

Routine characterization, which necessarily has to be based on nondestructive methods can be "calibrated" by means of HRTEM, which is a destructive, but highly informative method. Examples are shown in Figure 74.

29.2.2.14.6. "In situ" Studies [218]

By using different types of stages the specimen chamber of the microscope can be transformed into a small laboratory. Heating and cooling stages make it possible to study "in situ" the changes in crystal structure and in microtexture accompanying phase transitions, by observing (1) the appearance or disappearance of superstructure spots in the diffraction pattern, and (2) the fragmentation with translation and/or orientation domains in the direct space image.

Environmental cells can be used to study in situ solid–gas reactions, such as oxidation processes or decomposition reactions.

Plastic deformation can be studied and the dislocation propagation observed in real time in a straining stage.

The creation of radiation damage by ionization or by electron–atom collisions can be studied in medium- and high-voltage microscopes. The formation of agglomerates of point defects, such as dislocation loops and stacking fault tetrahedra, can be observed during irradiation with the image-forming electrons. (Light atoms, such as oxygen, are readily displaced by electrons of 400 kV.)

Radiation ordering and disordering in alloy systems have successfully been studied in high-voltage microscopes at different temperatures.

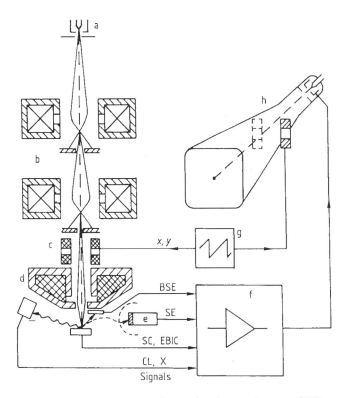

Figure 75. Schematic cross section of a scanning electron microscope (SEM)
a) Electron gun; b) Condenser lenses; c) Scan coils; d) Objective; e) Photomultiplier; f) Amplifier; g) Scan generator; h) Cathode-ray tube
BSE = Backscattered electrons; SE = Secondary electrons; SC = Specimen current; EBIC = Electron-beam-induced current; CL = Cathodoluminescence; X = X rays

29.2.3. Scanning Electron Microscopy

29.2.3.1. Introduction

In a scanning electron microscope (SEM) [219]–[221], a small electron probe 1–10 nm in diameter scans in a raster across the surface of the specimen (Fig. 75). The incident electrons are elastically and inelastically scattered by the specimen. *Elastic scattering* results in large scattering angles and zigzag electron trajectories. Therefore, a fraction of electrons can leave the specimen as backscattered electrons (BSE) (Fig. 76). The slowing down of electrons by *inelastic scattering* results in an electron range R. Electrons from the specimen atoms that are excited by inelastic scattering can leave the specimen as *secondary electrons* (SE) from a thin surface layer Λ_{SE} of ca. 1–10 nm. By convention, electrons in the energy spectrum (Fig. 77) with $E \leq 50$ eV are called SE. The secondary electrons consist of (1) SE1 excited by the primary electrons; (2) SE2 excited by BSE on their path through the surface; (3) SE3 are excited when BSE strike the lower polepiece; or flow (4) as SE4 through the polepiece bore (see Fig. 76). The ionization of inner atomic shells results in the emission either of characteristic X-ray quanta (X) or of Auger electrons (AE).

The *image* is formed by the signal of emitted secondary electrons, backscattered electrons, Auger electrons, absorbed speciment current (SC), or X-ray quanta, which modulate the intensity of a cathode-ray tube rastered in synchronism (Fig. 75). Conventional SEMs work with electron acceleration voltages of 5–30 kV, whereas a *low-voltage scanning electron microscope* (LVSEM) [222], [223] uses 0.5–5 kV.

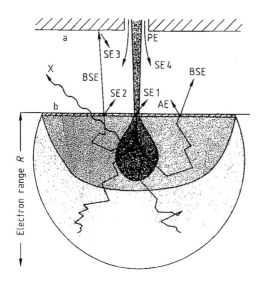

Figure 76. Excitation of backscattered electrons (BSE), Auger electrons (AE), and different groups of secondary electrons (SE) by primary electrons (PE)
Volumes of electron diffusion, BSE trajectories, and emitted X rays are shown with increasing density of gray levels
a) Polepiece; b) Specimen

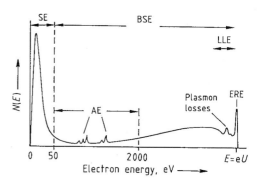

Figure 77. Schematic energy spectrum of emitted electrons with secondary (SE), backscattered (BSE), elastically reflected (ERE), low-loss (LLE), and Auger electrons (AE)

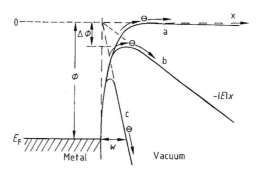

Figure 78. Potential at the cathode–vacuum interface for W and LaB$_6$ thermionic emission, ZrO–W Schottky emission, and field emission guns
a) Thermionic emission; b) Schottky emission; c) Field emission
Φ = Work function; E_F = Fermi energy

29.2.3.2. Instrumentation

29.2.3.2.1. Electron Guns

The following types of electron emitters are used in SEM electron guns.

Thermionic cathodes consist of a directly heated tungsten hairpin cathode at $T_c = 2500–3000$ K, or an indirectly heated pointed rod of lanthanum or cerium hexaboride (LaB$_6$, CeB$_6$) at 1400–2000 K. The electrons must overcome the work function Φ of 4.5 eV (W) or 2.7 eV (LaB$_6$) by thermal activation (Fig. 78, curve a). Between the cathode at the potential $-U$ and the grounded anode, a negatively biased Wehnelt electrode forms a crossover of diameter 20–50 µm (W) or 10–20 µm (LaB$_6$) as an effective electron source. The emitted electrons show an energy spread $\Delta E = 1–2$ eV (W) or 0.5–1 eV (LaB$_6$). A measure of the quality of an electron gun is the axial gun brightness β:

$$\beta = j/\pi\alpha^2 \approx j_c E/\pi k T_c \qquad (2.29)$$

where j is the current density, α the aperture angle, and E the electron energy. The axial gun brightness is constant for all points on the axis regardless of lenses and aperture diaphragms (j_c and T_c are current density and temperature, respectively, at the cathode). It increases proportionally to the electron energy E, with $\beta \approx 10^5$ A cm^{-2} sr^{-1} (where sr denotes steradian) for tungsten at $E = 20$ keV and about ten times higher values for LaB$_6$.

Schottky emission cathodes consist of zirconium-doped tungsten tips, with a radius of about 0.5–1 µm, coated with a ZrO layer. This layer decreases the work function from 4.5 to 2.7 eV. A Schottky emission cathode works with a higher electric field strength at the tip that decreases the work function by $\Delta\Phi$ (Schottky effect, Fig. 78, curve b) and concentrates the emission at the tip with a virtual electron source having a diameter of 15–20 nm. However, the electrons still have to

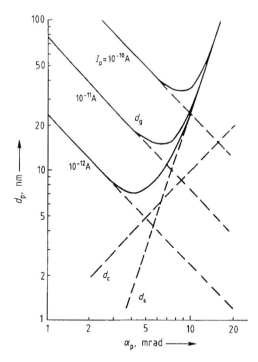

Figure 79. Double-logarithmic superposition of geometric probe size d_g for different probe currents I_p and aberration disks d_c and d_s versus the electron probe aperture α_p for a thermionic tungsten cathode
$E = 20$ keV; $\beta = 7 \times 10^4$ A cm^{-2} sr^{-1}; $C_s = 50$ mm; $C_c = 20$ mm; $\Delta E = 1$ eV

overcome the work function by thermal activation. The energy spread is 0.3–1 eV.

Field-emission guns (FEG) consist of tungsten tips with a radius of ca. 0.01–0.1 μm. An intermediate anode at ca. 1–2 kV extracts electrons by tunneling through the barrier of the work function, which is $w = 1-10$ nm in width (Fig. 78, curve c). Two types of FEG exist: (1) *cold FEGs* that work with the tip at room temperature, and the tip is flashed once a day, (2) *heated FEGs* work at 1800 K. The energy spread is only 0.2–0.4 eV for cold and 0.5–0.7 eV for heated FEGs. The gun brightness of FEGs $\beta \approx 10^7-10^8$ A cm^{-2} sr^{-1}.

29.2.3.2.2. Electron Probe Formation

The crossover of a thermionic gun or the virtual point sources of Schottky or field-emission guns are demagnified by two to three electromagnetic lenses (Fig. 75), so that a geometric probe size d_g is formed at the specimen with a diameter of 1–10 nm. A diaphragm of ca. 50–200 μm diameter in front of the last lens limits the electron-probe aperture α_p. The electron–probe current I_p, d_g, and α_p cannot be changed independently because of the conservation of gun brightness β in Equation (2.29):

$$I_p \approx \frac{\pi}{4} d_g^2 j_p = \frac{\pi^2}{4} \beta d_g^2 \alpha_p^2 \qquad (2.30)$$

The geometric probe size d_g is superposed by aberration disks of spherical (d_s) and chromatic (d_c) aberration and diffraction (d_d):

$$d_s = 0.5 C_s \alpha_p^3; \quad d_c = C_c (\Delta E/E) \alpha_p; \quad d_d = 0.6 \lambda / \alpha_p \qquad (2.31)$$

with the aberration constants C_s and C_c, respectively; ΔE is the energy spread of the electron gun, and λ is the deBroglie wavelength

$$\lambda = h/mv = 1.23/\sqrt{E} \qquad (2.32)$$

with λ in nanometers and E in kiloelectronvolts. An astigmatism of the last probe-forming lens can be compensated using a stigmator, which consists of magnetic quadrupoles.

The (quadratic) superposition of d_g, d_d, d_s, and d_c results in an effective probe diameter d_p (Fig. 79), where d_g and d_s dominate for high electron energies and thermionic guns, and d_d and d_c dominate for low-voltage SEM and field-emission guns. This superposition results in a minimum probe size between 1 and 10 nm at optimum apertures of tens of milliradians. An anticipated increase in resolution (decrease of electron probe diameter) requires a compensation of the chromatic aberration by using a combination of electrostatic and magnetic multipoles.

The low optimum probe aperture α_p of the SEM results in a large depth of field D_f:

$$D_f = \delta/\alpha_p = \Delta/\alpha_p M \qquad (2.33)$$

where $\Delta = \delta M \approx 0.1$ mm is the resolution and M the magnification on the cathode-ray tube (CRT) screen. As a consequence, an SEM has a two-order-of-magnitude greater depth of field than a light microscope even at low magnification.

To decrease the aberration constants C_s and C_c, the focal length of the last probe-forming lens can be reduced by a stronger excitation of the magnetic lens, which means a short working distance between specimen and polepiece for high

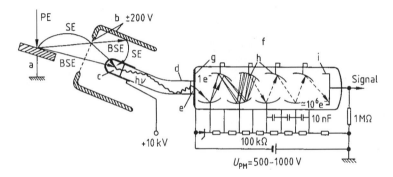

Figure 80. Everhart–Thornley detector for SE
a) Specimen; b) Collector grid and screen; c) Scintillator; d) Light pipe; e) Optical contact; f) Photomultiplier; g) Photocathode; h) Dynodes; i) Anode

resolution. The aberration constants can be decreased further if the specimen is positioned in the lens field; this makes through-lens detection of secondary electrons necessary.

29.2.3.2.3. Detectors

Secondary electrons are most frequently used for image formation and can be detected by a scintillator–photomultiplier combination (*Everhart–Thornley detector*, Fig. 80). A large fraction of slow SE is attracted by a positively biased grid (b) and accelerated to a yttrium–aluminum–garnet (YAG) scintillator or a P47 powder scintillator (c) biased at +10 kV. The photons generated are guided in a light pipe (d) to a photomultiplier (f). This SE detector shows a high signal-to-noise ratio and a large bandwidth up to 10 MHz. The collection field of an Everhart–Thornley detector mounted on one side of the specimen can disturb the electron probe in LVSEM. In this case, beam deflection and distortion can be avoided either by a combination of electrostatic and magnetic quadrupoles forming a Wien filter, by a retarding electrostatic lens, or by through-lens detection of SE.

Backscattered electrons have enough energy for direct production of a greater number of photons in a scintillator coupled via light pipe to a photomultiplier. However, the Everhart–Thornley detector cannot be used effectively for faster BSE because only a small fraction hits the scintillator area. Therefore, *annular or semiannular top detectors* collecting BSE with a high takeoff angle or *ring detectors* with low takeoff, are applied to make the best use of the angular characteristics of BSE. Alternatively, *semiconductor detectors* can be used. The BSE produce a large number $n = E/E_i$ of electron–hole pairs, where $E_i = 3.6$ eV is the mean energy per excitation in silicon. These charge carriers can be separated in a $p-n$ junction and form an electron beam-induced current (EBIC). BSE can also be detected by the *conversion of BSE to SE3* at the polepiece and other parts of the specimen chamber, when a negatively biased electrode around the specimen retards SE1 and SE2 from the specimen.

Detectors for LVSEM. Both scintillator and semiconductor detectors show a signal decreasing linearly with decreasing electron energy, and a threshold energy of ca. 1–5 keV. Therefore, their application in LVSEM requires postacceleration. As a recent alternative, *microchannel plates* (MCPS) can be used for the detection of BSE. These consist of a slice from a boule of tightly packed, fused tubes of lead-doped glass with a 10–20 μm inner diameter and a resistance of $10^8 - 10^9$ Ω over their length. Annular disks 3–4 mm thick can be mounted below the polepiece. The incident electrons produce SE at the inner tube wall, which are accelerated by a continuous voltage drop along the tube with a bias of 1 keV and are multiplied inside the MCP tubes by a multiplier-like action. The anode plate and a preamplifier at a potential of 1 kV are electrically insulated by an optical decoupler or by transmitting a modulated 30-MHz signal.

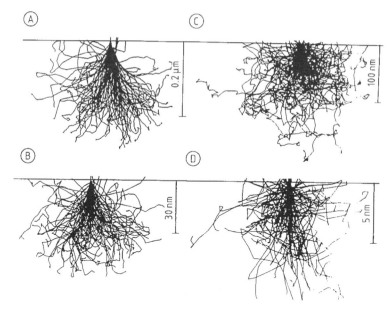

Figure 81. Monte Carlo simulations of 10- and 1-keV electron trajectories in carbon and gold
A) C, 10 keV, $\varrho = 1$ g/cm^3; B) C, 1 keV, $\varrho = 1$ g/cm^3; C) Au, 10 keV; D) Au, 1 keV

29.2.3.3. Electron–Specimen Interactions

29.2.3.3.1. Elastic and Inelastic Scattering

Elastic scattering of incident electrons results from the attractive Coulomb force of the nucleus screened by the atomic electron cloud with a differential Mott cross section of

$$\frac{d\sigma_M}{d\Omega} = r(\theta) \frac{eZ}{4(4\pi\epsilon_0)^2 m^2 v^4} \frac{1}{\sin^4(\theta/2)} \quad (2.34)$$

where $r(\theta)$ is the ratio between the Mott and Rutherford cross sections, Ω is the solid angle, Z the atomic number of the nucleus, ε_0 the vacuum permittivity, and m the mass of the electron [219]. The strong differences between Mott and Rutherford cross sections result from taking account of the spin–orbit coupling of electrons and solving the relativistic Dirac equation, whereas the Rutherford cross section is only a solution of the Schrödinger equation not containing the spin.

Inelastic scattering results in an excitation of electrons of the solid and a corresponding energy loss ΔE of the incident electron. Information about the differential inelastic cross section $d^2\sigma/d\Omega\,d(\Delta E)$ can be obtained from dielectric theory or experimental electron energy loss spectra (EELS; see Section 29.2.2.12.2) of high-energy electrons [224], [225].

A series of inelastic scattering processes with statistical energy losses results in a slowing down of electrons, which can be described by a mean energy loss per unit path length (Bethe stopping power S)

$$S = \left|\frac{dE_m}{ds}\right| = \frac{2\pi e^4 N_A \rho Z}{(4\pi\epsilon_0)^2 AE} \ln(1.166 E/J) \quad (2.35)$$

with the mean ionization potential $J \approx 12.5\,Z$. As energy decreases, fewer subshells are ionized, which can be described by a parabolic decrease of $1/S$ below $E/J = 6.3$.

29.2.3.3.2. Electron Diffusion

The decrease of mean electron energy E_m with increasing path length s of electron trajectories due to Equation (2.35) results in a mean total path (Bethe range R_B) that increases with increasing Z of the material. In low-Z material with less frequent large-angle scattering, R_B and the practical range R are equal, whereas in high-Z material with more frequent scattering, the trajectories are strongly curled and $R < R_B$, as demonstrated by Monte Carlo simulations in Figure 81. The range

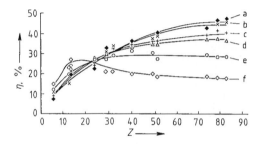

Figure 82. Backscattering coefficient η as a function of atomic number Z for different electron energies E
a) $E=5$ keV; b) $E=4$ keV; c) $E=3$ keV; d) $E=2$ keV; e) $E=1$ keV; f) $E=0.5$ keV

$$R \approx 6.6 E^{5/3} \quad (2.36)$$

is to a large extent independent of material, when R is measured in units of mass thickness (µg/cm²) and E in keV.

29.2.3.3.3. Emission of Secondary and Backscattered Electrons

The *backscattering coefficient* η describes the fraction of *primary electrons* leaving the specimen with an energy reduced by energy losses. Emitted electrons with energies of ≤ 50 eV are called secondary electrons (see Section 29.2.3.1) and are described by the *secondary electron yield* δ. Both quantities depend on electron energy E, atomic number Z of the specimen, and surface tilt angle ϕ ($\phi = 0$: normal incidence), and show characteristic energy and angular distributions that are important to the discussion of image formation in using BSE and SE signals.

Figure 82 shows the dependence of the *backscattering coefficient* η on atomic number Z for different electron energies E. For $E > 5$ keV, η increases monotonically with increasing atomic number. For multicomponent targets

$$\eta = \Sigma c_i \eta_i \quad (2.37)$$

shows a best fit to experiments, where c_i represents the mass fractions. This dependence of η on Z is responsible for the atomic number (compositional) contrast of the BSE signal for $E = 5 - 30$ keV (see Section 29.2.3.4.1). The backscattering coefficient η is approximately independent of E in the range 10-100 keV. Below 5 keV, η decreases for $Z > 30$ and increases for $Z < 30$

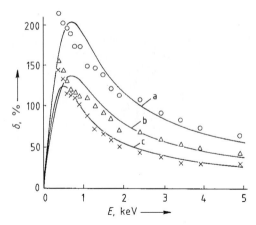

Figure 83. Decrease of SE yield δ with increasing electron energy E at normal incidence for
a) Au; b) Cu; c) C

with decreasing E. The reason for this is the Mott elastic cross section (Eq. 2.34). An increasing tilt angle ϕ of the specimen ($\phi = 0$: normal incidence) results in an increase of η.

The *secondary electron yield* δ shows a maximum for primary energies E of a few hundred electron volts and decreases $\sim E^{-0.8}$ for higher energies (Fig. 83). For $E > 5$ keV, the SE yield contains the contributions of SE1 excited by the primary electrons (PE) (Fig. 75) and of SE2 excited by the BSE. For $E < 5$ keV SE1 and SE2 can hardly be distinguished because the exit depths of SE and BSE have the same order of magnitude. The dependence of δ on the surface tilt angle ϕ can be approximated by $\delta \sim \sec^n \phi$, where the exponent n decreases from 1.3 (Be) to 1.1 (Al) and 0.65 (Au).

29.2.3.3.4. Specimen Charging and Damage

Insulating specimens show a *negative charging* by adsorbed electrons, when the total electron yield $\sigma = \eta + \delta$ is less than unity beyond a critical energy E_2 where σ changes from values larger than to values smaller than unity. Therefore, a conductive coating is necessary for working at high electron energies. With decreasing energy, σ can become greater than unity below E_2. In this case, more electrons leave the specimen as SE or BSE than are absorbed. The specimens becomes *positively charged* by a few volts only because SE of low energies will be retarded. Therefore, insulating specimens can be observed at low energies

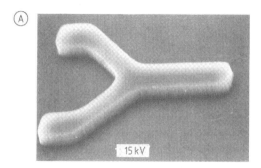

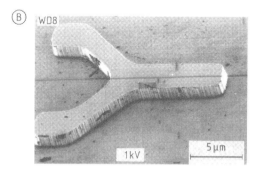

Figure 84. SE image of an etched structure (Y) on silicon with
A) 10-keV; B) 1-keV electrons

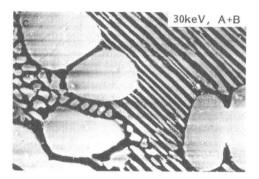

Figure 85. Example of material contrast of an Al–Ag eutectic alloy recorded with 30-keV electrons

without coating, although rough specimens can still show negative charging when, for example, SE cannot escape from holes.

Temperature-sensitive specimens can be damaged by *heat generation*. The increase of surface temperature at the electron probe decreases with decreasing energy, and such specimens can be observed better by LVSEM. Biological specimens are damaged by ionization, which results in a loss of mass and finally a polymerized carbon-enriched conglomerate within a layer of the order of the electron range R.

29.2.3.4. Image Formation and Analysis

29.2.3.4.1. Topographic and Material Contrast

Surface topography can be imaged with the SE signal, where the contrast is generated by the dependence of δ on the tilt angle ϕ of a surface element, and by shadowing effects caused by reduced SE collection from surfaces with normals opposite to the direction of SE collection or inside holes and trenches. This results in a contrast nearly equivalent to a light illumination from the detector, which, however, is disturbed by the diffusion contrast caused by SE2 generated by BSE at a greater distance from the electron impact. This typically results in bright zones near edges with a width of about the range R. Therefore, a low-voltage SEM at 1–5 keV decreases the diffusion effect and shows a better topography, as demonstrated in Figure 84. With in-lens operation and through-lens collection of SE by an Everhart–Thornley detector (ETD) inside the last lens, the SE signal becomes independent of azimuth, which is an advantage for the metrology of integrated circuits, but a disadvantage for recognizing the topography and distinguishing elevations and indentations, for example.

The dependence of δ and especially of η on atomic number Z (Fig. 82) leads to an atomic number or compositional contrast that can be used for the discrimination of phases with different mean atomic numbers, as demonstrated by a eutectic alloy shown in Figure 85.

29.2.3.4.2. Electron Channeling Effects

Electron waves entering a single crystal propagate as Bloch waves, which results in channeling effects and an orientation dependence of all electron–specimen interactions that are concentrated at the nuclei, such as large-angle scattering for backscattering or inner-shell ionization for X-ray and Auger electron emission.

When rocking an incident electron beam, the backscattering coefficient is modulated by a few percent. This results in an electron channeling

Figure 86. Electron channeling pattern of a (111) silicon surface recorded with 20-keV electrons

Figure 87. Superposed crystal orientation and magnetic contrast type I on polycrystalline cobalt

pattern (ECP) (Fig. 86) containg excess (bright) or defect (dark) Kikuchi lines (see Section 29.2.2.3.2) that are the intersections of the Kossel cones of apex $90° - \theta_B$ and an axis normal to the lattice planes of distance d. Therefore, such a pattern contains the crystal symmetry, and the distances of opposite Kikuchi lines include an angle $2\theta_B$, which fulfills the Bragg condition

$$2d \sin\theta_B = \lambda \qquad (2.8)$$

Scanning a polycrystalline specimen results in changes of brightness forming the crystal orientation or channeling contrast (Fig. 87).

Another type of channeling pattern is the electron backscattering pattern (EBSP) in which the intensity modulation can be observed on a fluorescent screen as the dependence of backscattering on takeoff direction. An EBSP has the advantage of covering a large angular range of about ±30°, whereas the rocking for ECP can be realized only with maximum ±2–3°.

29.2.3.4.3. Imaging and Measurement of Surface Potentials

A positively biased or charged part of the specimen retards low-energy SE and appears darker than the surrounding parts at ground potential. Negatively biased or charged parts appear brighter because more SE are repelled and can reach the collection field of the ETD. This results in the voltage contrast. However, no unique relation exists between brightness and bias because the SE signal depends on the surrounding potentials (near-field effect) and the collection efficiency of an ETD (far-field effect).

For quantitative measurement of surface potentials U_s on integrated circuits, a high extraction field strength (400–1000 V/cm) is needed at the surface, which decreases the influence of near fields by neighboring potentials. The potential is measured by the shift in the SE spectrum after passage through a spectrometer of the retarding or deflection type [226].

29.2.3.4.4. Imaging of Magnetic Fields

Magnetic fields of the specimen can act on primary, secondary, and backscattered electrons by the Lorentz force $F = e\, v \times B$. The *magnetic contrast type 1* is caused by the deflection of SE by external magnetic fields, which can be observed for magnetic recording media and uniaxial ferromagnetic materials. The best contrast is observed when only about half of the SE are collected. Then the signal intensity can change by 1–10% for opposite directions of the stray field. Figure 87 shows the superposition of magnetic contrast type 1 and the channeling contrast on a polycrystalline cobalt specimen.

Magnetic contrast type 2 is caused by the deflection of BSE in internal magnetic fields. However, sufficient contrast can be observed only for tilt angles ϕ of 40–50°, and the contrast increases with increasing energy, but is only a few per thousand for opposite directions of B. Another possibility for measuring external stray fields is the deflection of primary electrons, which pass at a short distance parallel to the surface.

29.2.3.4.5. Electron-Beam Induced Current

The mean number E/E_i of electron–hole pairs generated in semiconductors—with a mean formation energy of $E_i = 3.6$ eV in silicon, for example—normally recombine. The electric field inside depletion layers separates the charge carriers, and minority carriers can diffuse to the depletion layer and contribute to the charge collection I_{cc} or electron-beam induced current. Depletion layers can be formed by $p–n$ junctions parallel or perpendicular to the surface or by Schottky barriers formed by a nonohmic evaporated metal contact. Therefore, a scanning electron probe becomes a useful tool for qualitative and quantitative analysis of junctions and semiconductor parameters [227], which is demonstrated by the following examples:

1) Imaging the position and depth of depletion layers below conductive pads and passivation layers, by utilizing the increasing penetration depth (range) with increasing electron energy
2) Measuring the width of depletion layers and their increase with increasing reverse bias
3) Imaging sites of avalanche breakdown in depletion layers
4) Imaging lattice defects (dislocations, stacking faults) and dopant striations that actively influence the diffusion length, for example, by a Cottrell atmosphere of dopant atoms
5) Measuring the diffusion length L from a semilogarithmic plot of the EBIC signal [$\sim \exp(-x/L)$] versus the distance x of the electron probe from a perpendicular $p–n$ junction, and measuring the lifetime τ with a signal [$\sim \exp(-t/\tau)$] when chopping the electron beam
6) Measuring the surface recombination rate S from the variation of EBIC with increasing electron energy and depth of carrier formation

29.2.3.4.6. Cathodoluminescence

The excitation of light from materials stroked by electrons, known from fluorescent screens of TV tubes for example, is called cathodoluminescence (CL). In semiconductors with a direct band gap, electrons that are excited from the valence to the conduction band can recombine with the holes by emission of radiation, whereas semiconductors with an indirect band gap have a reduced probability of radiative recombination. In semiconductors and most inorganic materials, CL depends strongly on the concentration of dopants, which can either enhance CL by forming luminescence centers for radiative transitions or quench CL by forming centers of nonradiative transitions.

Organic specimens such as anthracene and plastic scintillator material, for example, can also show CL. Fluorescent dyes are used in biology for selective staining. However, all organic material is damaged by electron irradiation, and CL is quenched at incident charge densities a few orders of magnitude lower than those necessary for radiation damage of the crystal structure and loss of mass by ionization processes.

The CL signal can be detected by a photomultiplier or dispersively by using a spectrometer between specimen and detector. As in color TV, the signals from three detectors with color filters can be used for a real-color image [228].

29.2.3.4.7. Special Imaging Methods

The *specimen current mode* uses the current

$$I_s = I_p[1 - (\eta + \delta)] \quad (2.39)$$

to earth, which is complementary to the SE and BSE emission and can be used for imaging the compositional contrast of BSE or the magnetic contrast type 2.

In *environmental* SEM, the partial pressure of water or other gases can be increased near the specimen by differentially pumped diaphragms. This offers the possibility of observing wet specimens without drying, for example [229]. The production of ions in the gas also reduces the negative charging of the specimen by absorbed electrons in nonconductive material.

In the *thermal-wave acoustical mode*, the electron beam is chopped at frequencies in the 100-kHz to 5-MHz range and produces periodic specimen heating. The periodic changes in thermal expansion excite an acoustic wave that can be

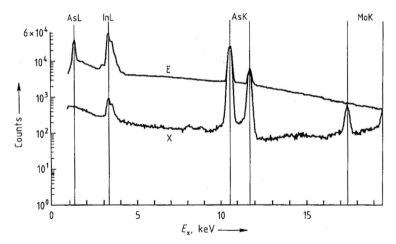

Figure 88. Energy-dispersive X-ray spectrum of InAs (logarithmic scale of counts per channel) recorded by excitation with 30-keV electrons (E) and X-ray fluorescence (X) using the molybdenum K radiation excited in a molybdenum foil target in front of the specimen

29.2.3.5. Elemental Analysis

29.2.3.5.1. X-Ray and Auger Electron Emission

Either ionization of an inner shell and the subsequent filling of the vacancy by an electron from higher energy states result in the emission of an X-ray quantum, or the energy is transferred to another electron in the higher shell, which leaves the atom as an Auger electron (see also → Surface Analysis). The sum of the X-ray fluorescence yield ω and the Auger electron yield is unity, and ω decreases strongly from near unity to low values with decreasing Z and increasing X-ray series $n = $ K, L, M. The X-ray quantum energies of a series increases approximately $\sim \sqrt{Z}$ (Moseley's law), which is basic for elemental analysis by X rays.

29.2.3.5.2. X-Ray Spectrometers

Two different types of X-ray spectrometer exist: wavelength- and energy-dispersive spectrometers (WDS and EDS, see Section 29.2.2.12.1), which can be used in either an X-ray microanalyzer (XRMA) or an SEM.

A *wavelength-dispersive spectrometer* can analyze the characteristic lines of all elements down to beryllium when special analyzing crystals with a large lattice-plane spacing are used for Bragg reflection. XRMA with electron energies between 10 and 60 keV can work with WDS for recording three different wavelengths simultaneously, whereas only one WDS is used in an SEM.

The electron–hole pairs produced in the *energy-dispersive spectrometer* are separated by a voltage drop of ca. 1 kV at the crystal and result in a charge pulse proportional to the X-ray quantum energy. This pulse is amplified by a charge-sensitive preamplifier and runs through a unit for pulse shaping so that it can be recorded in a multichannel analyzer (MCA). The number of counts in a channel that is proportional to the quantum energy is increased by one unit. This allows all quantum energies to be recorded simultaneously, and the growth of a spectrum can be followed on the screen of the MCA (Fig. 88). Therefore, EDS are often also used in XRMA to survey the emitted X-ray spectrum. The vacuum inside an EDS can be separated from the vacuum of the microscope by a 8–10-μm-thick beryllium foil, which absorbs the characteristic radiation of elements $Z \leq 11$ below $E_x = 1$ keV. An organic foil or a windowless detector can also be used to record the lines of carbon ($E_x = 250$ eV), oxygen, and nitrogen, for example.

WDS and EDS show the following characteristic differences. *WDS* requires very accurate orientation of the electron impact on the Rowland circle within an area a few micrometers in diameter. The specimen must be planar and is shifted mechanically to record a line scan or elemental map. The electron probe can, by contrast, be scanned over an area of a few millimeters without any loss of efficiency when an EDS is used. WDS analyze only one line with a resolution of about 10 eV. The proportional counter allows a high counting rate of about 10 000 counts per second (cps). Therefore, high probe currents and a large number of counts per unit time can be recorded. The energy resolution of an *EDS* is only about 100–200 eV, and a ten times greater fraction of the continuum below the characteristic X-ray peaks is counted compared to WDS. This decreases the signal-to-background ratio in EDS. The poorer resolution of an EDS can also result in an overlap of neighboring X-ray lines, whereas all lines can normally be separated by a WDS. An EDS must work with lower probe currents because all quanta are counted and only a count rate of ca. 10 000 cps guarantees no overlap of sequential pulses. Pulse-rejection electronics are used when the time between two pulses is shorter than the pulse-shaping time necessary for the MCA.

29.2.3.5.3. X-Ray Microanalysis

For a quantitative XRMA [230], the number n_a of counted characteristic X-ray quanta of an element a of concentration c_a in an alloy or compound is compared with the number n_s from a pure elemental standard ($c_s = 1$) or a compound of known composition. Only an approximate value for c_a can be obtained from the k ratio

$$k = n_a/n_s \approx c_a/c_s \qquad (2.40)$$

The exact measurement of concentration needs a *ZAF correction method* (see p. 1097). This considers the differences in the stopping power and the backscattering of electrons between specimen and standard (*Z correction*), in the depth distribution of X-ray generation and *absorption* (A) of X rays recorded under a takeoff angle of 40–60°, and in the excitation of X-ray quanta of the element of interest by characteristic X rays and continuum of the matrix (*fluorescence F*).

Special correction programs must be used, for example, for tilted specimens, thin film coatings, small particles, and biological specimens.

29.2.3.5.4. Special X-Ray Techniques

When the specimen is scanned in a raster or along a line, the pulses of selected characteristic X-ray lines can produce an elemental distribution profile or map, respectively.

In a *single crystal*, the excited X rays are Bragg diffracted at the lattice plane, and their isotropic angular characteristics show defect and excess Kossel lines of apex angle $90° - \theta_B$, which can be used for accurate measurement of lattice parameters and strains when the Kossel pattern is recorded on a photographic emulsion.

X-ray fluorescence analysis can be applied in an SEM when generating X rays in a thin foil, which stops the electrons, but the transmitted X rays can excite X-ray fluorescence in the specimen, which is recorded by an energy-dispersive spectrometer. The background of the spectrum is much lower than that from direct electron excitation (Fig. 88, curve X versus curve E), which results in a better signal-to-background ratio for trace elements. However, only larger areas 0.1–1 mm in diameter can be analyzed.

The small source of X rays generated by an electron probe on a bulk target can also be used for X-ray projection microscopy.

29.3. References

[1] E. Abbe: "Beiträge zur Theorie des Mikroskops und der mikroskopischen Wahrnehmung," *Arch. Mikrosk. Anat.* **9** (1873) 413.
[2] D. Gerlach: *Das Lichtmikroskop,* 2nd ed., Thieme Verlag, Stuttgart 1985.
[3] H. Gundlach: *Mikroskope. Handbuch Biomedizinische Technik,* **vol. 4,** Springer Verlag, TÜV Rheinland, (1991) pp. 1–20.
[4] K. Michel: "Axiomat von Zeiss, ein Mikroskop mit neurem Konzept," *Zeiss. Inf.* **20** (1973).
[5] F. Muchel: ""ICS"-A New Principle in Optics," *Zeiss Inf.* **30** (1988) 20–27.
[6] K. Michel: *Die Grundzüge der Theorie des Mikroskops,* 3rd ed., Wissenschaftliche Verlagsgesellschaft, Stuttgart 1981.
[7] A. Köhler: "Mikrophotographische Untersuchungen mit ultraviolettem Licht," *Z. Wiss. Mikrosk. Mikrosk. Tech.* **21** (1904) 129–165, 273–404.
[8] A. Köhler: "Ein neues Beleuchtungsverfahren für mikrophotographische Zwecke," *Z. Wiss. Mikrosk. Mikrosk. Tech.* **10** (1893) 433–440.
[9] F. Zernike: "Das Phasenkontrastverfahren bei der mikroskopischen Beobachtung," *Phys. Z.* **36** (1935) 848.
[10] H. Gundlach: "Neuere Entwicklungen und Anwendungen in der Lichtmikroskopie," *Gynäkologie* **23** (1990) 328–335.
[11] G. Nomarski: "Microinterferomètre Differentielle a Ondes Polarisées," *J. Phys. Radium* **16** (1955) 9.
[12] R. D. Allen et al.: "The Zeiss Nomarski Differential Interference Equipment for Transmitted-Light Microscopy," *Z. Wiss. Mikrosk. Mikrosk. Tech.* **69** (1969) 193–221.

[13] W. Lang: "Nomarski Differential Interference Contrast Microscopy," *Zeiss Inf.* **16** (1968) no. 70, 114–120.
[14] M. Francon: *Einführung in die neuen Methoden der Lichtmikroskopie*, Verlag Braun, Karlsruhe 1967.
[15] W. I. Patzelt: "Reflexionskontrast. Eine neue lichtmikroskopische Technik," *Mikrokosmos* **3** (1977) 78–81.
[16] S. Pentz et al.: "Darstellung des Adhäsionsverhaltens kultivierter Leberzellen an Glasoberflächen während der Mitose durch Reflexionskontrast-Mikroskopie," *Zeiss Inf.* **25** (1980) 91, 41–43.
[17] H. Gundlach et al.: "Immunfluorescance Microscopy with the New Zeiss Photomicroscope Axiophot," *Zeiss Inf.* **29** (1997) 36–39.
[18] A. J. Lacey: *Light Microscopy in Biology*, IRL Press, Oxford University, 1989.
[19] D. L. Taylor, M. Nederlof, F. Lanni, A. S. Waggoner: "The New Vision of Light Microscopy," *Am. Sci.* **80** (1992) 322–335.
[20] D. L. Taylor et al.: *Fluorescence Microscopy of Living Cells in Culture*, Academic Press, San Diego 1989.
[21] A. H. Coons et al.: "Immunological Properties of an Antibody Containing a Fluorescent Group," *Proc. Soc. Exp. Biol. Med.* **47** (1941) 200–202.
[22] J. S. Ploem: "The Use of a Vertical Illuminator with Interchangeable Dichroic Mirrors for Fluorescence Microscopy with Incident Light," *Z. Wiss. Mikrosk. Mikrosk. Tech.* **68** (1967) 129–143.
[23] L. Trapp: "Über Lichtquellen und Filter für die Fluoreszenzmikroskopie und über die Auflichtfluoreszenzmethode bei Durchlichtpräparaten," *Acta Histochemica* **7** (1965), pp. 327–338.
[24] W. W. Franke: "Different Intermediate-Sized Filaments Distinguished by Immunofluorescence Microscopy," *Proc. Nat. Acad. Sci. USA* **75** (1978) no. 10, 5034–5038.
[25] O. Greulich et al.: "The Light Microscope on its Way from an Analytical to a Preparative Tool," *J. Microsc. (Oxford)* **167** (1992) 127–151.
[26] P. Lichter et al.: "Analysis of Genes and Chromosomes by Nonisotopic in Situ Hybridization," *Gata* **8** (1991) no. 1, 24–35.
[27] Chr. Lengauer et al., "Chromosomal Barcodes produced by Multifluorescence in situ Hybridization with Multiple YAG Clones and Whole Chromosomes painting probes", in *Human Molecular Genetics*, 5th ed., vol. 2, Oxford University Press, pp. 505–512.
[28] M. Chalfie, Y. Tu, D. C. Prasher, "Green Fluorescent Protein as a Marker for Gene Expression", *Science* **263** (1994) 802–805.
[29] R. Heim, R. Y. Tsien, "Engineering Green Fluorescent Protein for Improved Brightness, Longer Wavelength and Fluorescence Resonance Energy Transfer", *Curr. Biol.* **6** (1996) 178–182.
[30] B. Herman, *Fluorescence Microscopy*, BIOS Scientific Publisher, Oxford, 1998.
[31] E. L. Elson, D. Magde, "Fluorescence Correlation Spectroscopy (I). Conceptual Basis and Theory", *Biopolymers* **13** (1974) 1–27.
[32] R. Rigler, J. Widengreen, Ü. Mets, "Interactions and Kinetics of Single Molecules as Observed by Fluorescence Correlation Spectroscopy", *Fluorescence Spectroscopy—New Methods and Applications*, Springer, Heidelberg 1992, 13–24.
[33] M. Eigen, R. Rigler, "Sorting Single Molecules: Application to Diagnostics and Evolutionary Biotechnology", *Proc. Natl. Acad. Sci. USA* **91** (1994) 5740–5747.

[34] H. Utermöhl: "Neue Wege in der quantitativen Erfassung des Planktons," *Verh. intern. Verein Limnol.* **5** (1931) 567–596.
[35] H. Gundlach et al.: "Mikrokinematographie mit dem Mikroskop "Axiomat"," *Res. Film* **9** (1976) no. 1, 30–36.
[36] H. Gundlach: *Die Anwendung von inversen Mikroskopen in der Zell- und Entwicklungsbiologie*, Suppl. GIT, Verlag E. Giebler, Darmstadt 1981, pp. 9–15.
[37] H. A. Tritthart et al.: "The Zeiss Axiomat Applied to the Examination of the Morphology and Physiology of Myocardial Cells in Culture," *Zeiss Inf.* **25** 1980 no. 90, 10–13.
[38] M. Horster et al.: "Application of Differential Interference Contrast with Inverted Microscopes to the In-vitro Perfused Nephron," *J. Microsc. (Oxford)* **117** (1979) 375–379.
[39] F. K. Möllring: "Inverse Mikroskopie IM 35 und ICN 405 für Biologie, Medizin und Metallographie," *Zeiss Inf.* **23** (1977) 18–19.
[40] H. Piller: *Microscope Photometry*, Springer, Berlin 1977.
[41] H. Gundlach et al.: "Identification and Selective Micropreparation of Live Nuclear Components with the Zeiss IM 35 Inverted Microscope," *Zeiss Inf.* **25** (1980), no. 91, 36–40.
[42] H. Spring et al.: "DNA Contents and Numbers of Nucleoli and Pre-rRNA-Genes in Nuclei of Gemetes and Vegetative Cells of Acetabularia Mediterranea," *Exp. Cell. Res.* **114** (1978) 203–215.
[43] H. Spring et al.: "Transcriptionally Active Chromatin in Loops of Lumpbrush Chromosomes at Physiological Salt Concentration as Revealed Electron Microscopy of Sections," *Eur. J. Cell Biol.* **24** (1981) 298–308.
[44] W. Ansorge et al.: "Performance of an Automated System for Capillary Microinjection into Living Cells," *J. Biochem. Biophys. Methods* **16** (1988) 283–292.
[45] R. D. Allen et al.: "Video-Enhanced-Contrast Polarization (AVEC-Pol) Microscopy: A New Method Applied to the Detection of Birefringence in the Motile Reticulopodial Network of Allogromia Laticollaris," *Cell Motil.* **1** (1981) 275–289.
[46] R. D. Allen et al.: "Video-Enhanced-Contrast, Differential Interference Contrast (AVEC-DIC) Microscopy: A New Method Capable of Analyzing Microtubule-Related Motility in the Reticulopodial Network of Allogromia Laticollaris," *Cell Motil.* **1** (1981) 291–302.
[47] S. Inoue: "Foundations of Confocal Scanning Imaging in Confocal Microscopy" in [60], pp. 1–14.
[48] S. Inoue et al.: "The Acrosonal Reaction of Thyone Sperm Head Visualized by High Resolution Video Microscopy," *J. Cell Biol.* **93** (1982) 812–819.
[49] J. H. Hayden et al.: "Detection of Single Microtubules in Living Cells: Particle Transport Can Occur in Both Directions Along the Same Microtubule," *J. Cell Biol.* **99** (1984) 1785–1793.
[50] T. Salmon et al.: "Video-Enhanced Differential Interference Contrast Light Microscopy," *Bio Techniques* **7** (1989) 624–633.
[51] D. J. Arndt-Jovin et al.: "Fluorescence Digital Imaging Microscopy in Cell Biology," *Science (Washington D.C.)* **230** (1985) 247–256.
[52] Y. Hiraoka et al.: "The Use of a Charge-Coupled Device for Quantitative Optical Microscopy of Biological Structures," *Science (Washington D.C.)* **238** (1987) 36–41.

[53] H. H. Sedlacek et al.: "The Use of Television Cameras Equipped with an Image Intensifier in the Immunofluorescence Microscopy," *Behring Inst. Mitt.* **59** (1976) 64–70.

[54] H. Gundlach et al.: *Electronic Photography-Technology, Systems and Applications in Microbiology. Proc. of Int. Symosium on Electronic Photography,* Cologne The Society for Imaging Sciene and Technology, Springfield, Virginia USA, 1992, pp. 60–65.

[55] M. Minsky: US 3013467, 1961.

[56] G. Cox: "Photons Under the Microscope. Marvin Minsky – the Forgotten Pioneer?," *Australian EM Newsletter* **38 no. 4,** (1993) 4–10.

[57] C. J. R. Sheppard: "15 Years of Scanning Optical Microscopy at Oxford," *Proc. Roy. Mic. Soc.* **25** (1990) 319–321.

[58] M. Petran, M. Hadravsky, M. D. Egger, R. Galambos: "Tandem-Scanning Reflected Light Microscope," *J. Opt. Soc. Am.* **58** (1968) 661–664.

[59] G. J. Brakenhoff, P. Blom, P. Barends: "Confocal Scanning Light Microscopy with High Aperture Immersion Lens," *J. Micros. (Oxford)* **117** (1979) 219–232.

[60] J. Pawley: *Handbook of Biological Confocal Microscopy,* 2nd ed. Plenum Press, New York 1995.

[61] C. J. R. Sheppard, A. Choundhury: "Image Formation in the Scanning Microscope," *Opt. Acta* **24** (1977) 1051–1073.

[62] C. J. R. Sheppard, T. Wilson: "Depth of Field in the Scanning Microscope," *Opt. Lett.* **3** (1978) 115–117.

[63] P. Wallen, K. Carlsson, K. Mossberg: "CLSM as a Tool for Studying the 3-D Morphology of Nerve Cells," in [109], pp. 110–143.

[64] T. Wilson: "Optical Sectioning in Confocal Fluorescence Microscopy," *J. Microsc. (Oxford)* **154** (1989) 143–156.

[65] S. Hell, E. Lehtonen, E. H. K Stelzer: "Confocal Fluorescence Microscopy: Wave Optics and Applications to Cell Biology", in [28], pp. 145–160.

[66] I. J. Cox, C. J. R. Sheppard, T. Wilson: "Superresolution in Confocal Fluorescent Microscopy," *Optik (Stuttgart)* **60** (1982) 391–396.

[67] T. Wilson (ed.): *Confocal Microscopy,* Academic Press, London 1990.

[68] C. J. R. Sheppard, M. Gu: "3-D Transfer Functions in Confocal Scanning Microscopy," in [109], pp. 251–280.

[69] M. Gu, *Principles of 3D Imaging in Confocal Microscopes,* World Scientific, Singapore, 1996.

[70] C. J. Cogswell, J. W. O'Bryan: "A High Resolution Confocal Transmission Microscope," *SPIE Proceedings* **1660** (1992) 503–511.

[71] G. J. Brakenhoff, K. Visscher: "Bilateral Scanning and Array Detectors," *J. Micros. (Oxford)* **165** (1990) 139–146.

[72] E. Gratten, M. J. van der Veen: "Laser Sources for Confocal Microscopy", in [60], pp. 69–98.

[73] K. S. Wells, D. R. Sandison, J. Strickler, W. W. Webb: "Quantitative Fluorescence Confocal Laser Scanning Microscopy," in [60], pp. 39–53.

[74] J. Art: "Photon Detectors for Confocal Microscopy," in [60], pp. 127–139.

[75] W. Denk, J. H. Strickler, W. W. Webb: "Two-photon laser scanning fluorescence microscopy", *Science (Washington D.C.)* **248** (1990) 73–76.

[76] M. Gu, X. Gan, A. Kisteman, M. G. Xu, *Appl. Phys. Lett.* **77** (2000) no. 10, 1551–1553.

[77] S. Lindek, E. H. K. Stelzer, S. W. Hell: "Two New High-Resolution Confocal Fluorescence Microscopies (4pi, theta) with One- and Two-Photon Excitation", in [60], pp. 445–458.

[78] M. D. Egger, M. Petran: "New Reflected Light Microscope for Viewing Unstained Brain and Ganglian Cells," *Science (Washington D.C.)* **157** (1987) 305–307.

[79] J. S. Deitch, K. L. Smith, J. W. Swann, J. N. Turner: "Parameters Affecting Imaging of HRP Reaction Product in the Confocal Scanning Laser Microscope," *J. Microsc. (Oxford)* **160** (1990) 265–278.

[80] J. P. Rigaut, S. Carvajal-Gonzalez, J. Vassy: "3-D Image Cytometry," in [109], pp. 205–237.

[81] B. R. Masters, A. Kriete, J. Kukulies: "Ultraviolet Confocal Fluorescence Microscopy of the In-Vitro Cornea: Redox Metabolic Imaging," *Applied Optics* **32** (1993) no. 4, 592–596.

[82] D. M. Shotton: "Confocal Scanning Optical Microscopy and its Applications for Biological Specimens," *J. Cell. Sci.* **94** (1989) 175–206.

[83] J. N. Turner: "Confocal Light Microscopy. Biological Applications (Special Issue)," *Electron Microsc. Tech.* **18** (1991) 1.

[84] R. P. Haugland: *Molecular Probes: Handbook of Fluorescent Probes and Research Chemicals,* Eugene Molecular Probes Inc., 1989.

[85] R. Y. Tsien, A. Waggoner: "Fluorophores for Confocal Microscopy: Photophysics and Photochemistry," in [60], pp. 169–178.

[86] A. Villringer, U. Dirnagl, K. Einhäupl: "Microscopical Visualization of the Brain in Vivo," in [109], pp. 161–181.

[87] R. Yuste, F. Lanni, A. Konnerth (eds.): *Imaging Neurons: A Laboratory Manual,* CSHL Press, Cold Spring Harbor, 2000.

[88] B. R. Masters: "Confocal Ocular Microscopy – a New Paradigm for Ocular Visualization," in [59], pp. 183–203.

[89] W. A. Mohler, J. G. White, "Stereo-4D Reconstruction and Animation from Living Fluorescent Specimens", *Biotechniques* **24** (1998) 1006–1012.

[90] R. G. Summer, P. C. Cheng: "Analysis of Embryonic Cell Division Patterns Using Laser Scanning Confocal Microscopy," in G. W. Bailey (ed.): *Proc. Annual meeting Electron Microsc. Soc. Am.* **47** (1989) 140–141.

[91] N. O'Rourke, S. E. Fraser: "Dynamic Changes in Optic Fiber Terminal Arbors Lead to Retinotopic Map Formation: An In-Vivo Confocal Microscopic Study," *Neuron* **5** (1990) 159–171.

[92] A. Kriete, H.-J. Wagner, "Computerized Spatio-temporal (4D) Representation in Confocal Microscopy: Application to Neuroanatomical Plasticity," *J. Micros. (Oxford)* **169** (1993) 27–31.

[93] I. T. Young, "Characterizing the Imaging Transfer Function", in D. L. Taylor, Y. L. Wang (eds.): *Methods in Cell Biology,* Academic Press, San Diego, 1989, pp. 1–45.

[94] R. A. Jarvis, "Focus Optimization Criteria for Computer Image Processing", *The Microscope* **24** (1976) 163–180.

[95] A. Kriete, "Image Quality Considerations in Computerized 2D and 3D Microscopies", in P. C. Cheng, T. H. Lin, W. H. Wu, J. L. Wu (eds.): *Multidimensional Microscopy,* Springer, New York, 1990, pp. 141–150.

[96] I. T. Young, Quantitative Microscopy, IEEE Eng. in Medicine and Biology, Jan./Feb. 1996, pp. 59–66.

[97] T. Visser, J. L. Oud, G. T. Brakenhoff: "Refractive Index and Axial Distance Measurements in 3-D Microscopy," *Optik (Stuttgart)* **90** (1991) 17–19.

[98] J.-A. Conchello, E. W. Hanssen: "Enhanced 3-D Reconstruction from Confocal Scanning Microscope Images. Deterministic and Maximum Likelihood Reconstrauctions," *App. Opt.* **29** (1990) no. 26, 3795–3804.

[99] T. J. Holmes, Y.-H. Liu: "Image Restoration for 2-D and 3-D Fluorescence Microscopy," in [109], pp. 283–323.

[100] D. A. Agard, J. W. Sedat: "Three-Dimensional Architecture of Polytene Nucleus," *Nature (London)* **302** (1983) 676–681.

[101] G. Wang, W. S. Liou, T. H. Lin, P. C. Cheng: "Image Restoration in Light Microscopy," in T. H. Lin, W. L. Wu, J. L. Wu (eds.): *Multidimensional Microscopy,* Springer Verlag, New York 1993, 191–208.

[102] F. Meyer: "Mathematical Morphology: 2-D to 3-D," *J. Microsc.* **165** (1992) 5–28.

[103] P. C. Cheng et al.: "3-D Image Analysis and Visualization in Light Microscopy and X-Ray Micro-Tomography," in [109], pp. 361–398.

[104] K. Mossberg, U. Arvidsson, B. Ulfhake: "Computerized Quantification of Immunoflourescence Labelled Axon Terminals and Analysis of Co-Localization of Neurochemicals in Axon Terminals with A Confocal Scanning Laser Microscope," *J. Histochem Cyctochem.* **38** (1990) 179–190.

[105] B. Roysam et al.: "Unsupervised Noise Removal Algorithms for 3-D Confocal Fluorescence Microscopy," *Micron and Micros. Acta* **23** (1992) in press.

[106] M. Montag et al.: "Methodical Aspects of 3-D Reconstruction of Chromatin Architecture in Mouse Trophoblast Giant Nuclei," *J. Microsc. (Oxford)* **158** (1990) 225–233.

[107] V. Conan et al.: "Geostatistical and Morphological Methods Applied to 3-D Microscopy," *J. Microsc. (Oxford)* **166** (1992) 169–184.

[108] N. S.White: "Visualization Systems for Multidimensional CLSM Systems", in [60] pp. 211–254.

[109] A. Kriete: *Visualization in Biomedical Microscopies 3-D Imaging and Computer Applications,* VCH-Verlagsgesellschaft, Weinheim 1992.

[110] H. Chen, J. W. Sedat, J. A. Adard: "Manipulation, Display and Analysis of Three-Dimensional Biological Images," in J. Pawley (ed.): *Handbook of Biological Confocal Microscopy,* Plenum Press, New York 1990, pp. 141–150.

[111] J. S. Hersh: "A Survey of Modeling Representations and their Application to Biomedical Visualization and Simulation," Conf. VBC, IEEE Comp. Soc. Press, Los Alamitos 1990, pp. 432–441.

[112] D. P. Huijsmanns, W. H. Lamers, J. A. Los, J. Strackee: "Toward Computerized Morphometric Facilities: a Review of 58 Software Packages for Computer-Aided 3-D Reconstruction, Quantification, and Picture Generation from Parallel Serial Sections," *Anat. Rec.* **216** (1986) 449–470.

[113] A. Kriete, P. C. Cheng (eds.): "3-D Microscopy (Special Issue)," *Computerized Medical Imaging and Graphics,* vol. 17, Plenum Press, N.Y. 1993, p. 8.

[114] R. Ware, V. LoPresti: "Three-Dimensional Reconstruction from Serial Sections," *Comp. Graph.* **2** (1975) 325–440.

[115] J. K. Udupa, H. M. Hung: "Surface Versus Volume Rendering: a Comperative Assessment," *VBC '90,* Proceedings IEEE, Atlanta 1990, pp. 83–91.

[116] S. D. Roth: "Ray-Casting for Solid Modeling," *Comput. Graphics Image Processing* **18** (1982) 109–144.

[117] H. K. Tuy, L. T. Tuy: "Direct 2-D Display of 3-D Objects," *IEEE CG & A* **4** (1984) 29–33.

[118] G. Frieder, D. Gordon, R. A. Reynolds: "Back-to-Front Display of Voxel-Based Objects," *IEEE CG & A* **5** (1985) no. 1, 52–60.

[119] R. A. Debrin, L. Carpenter, P. Hanrahan: "Volume Rendering," *Computer Graphics* **22** (1988) no. 4, 65–74.

[120] M. Levoy: "A Hybrid Ray-Trace for Rendering Polygon and Volume Data," *IEEE CG & A* **3** (1990) 33–40.

[121] H. T. M. van der Voort, G. J. Brakenhoff, M. W. Baarslag: "Three-Dimensional Visualization Methods," *J. Microsc. (Oxford)* **153** (1989) no. 2, 123–132.

[122] P.-O. Forsgren: "Visualization and Coding in Three-Dimensional Image Processing," *J. Microsc. (Oxford)* **159** (1990) no. 2, 195–202.

[123] M. Levoy et al.: "Volume Rendering in Radiation Treatment Planning," *VBC '90,* Proceedings IEEE, Atlanta 1990, pp. 4–10.

[124] A. Kriete, N. Klein, L. C. Berger, "Data Compression in Microscopy: a Comparative Study", *Proc. SPIE.* **3605** (1999) 158–168.

General References

[125] S. Amelinckx: "The Direct Observation of Dislocations," Suppl. 6 in F. Seitz, D. Turnbull (eds.): *Solid State Physics,* Academic Press, London 1964.

[126] S. Amelinckx, R. Gevers, J. Van Landuyt (eds): *Diffraction and Imaging Techniques in Material Science,* North-Holland Publishing Company, Amsterdam 1970, 1978.

[127] F. R. N. Nabarro (ed.): *Dislocation in Solids,* North-Holland, Amsterdam 1979.

[128] P. B. Hirsch, R. B. Nicholson, A. Howie, D. W. Pashley, M. J. Whelan: *Electron Microscopy of Thin Crystals,* Butterworths, London 1965.

[129] H. Bethge, H. Heydenreich (eds.): *Electronenmikroskopie in der Festkörperphysik,* Springer Verlag, Berlin 1982.

[130] J. C. H. Spence: "Experimental High Resolution Electron Microscopy," *Monographs on the Physics and Chemistry of Materials,* Oxford Science Publications, Clarendon Press, Oxford 1981.

[131] G. Thomas: *Transmission Electron Microscopy of Metals,* John Wiley and Sons, New York 1962.

[132] J. C. H. Spence, J. M. Zuo: *Electron Microdiffraction,* Plenum Press, New York 1992.

[133] R. W. Cahn, P. Haasen, E. J. Kramer (eds.): *Materials Science and Technology,* vol. 2 A, VCH Verlagsgesellschaft, Weinheim 1992.

[134] D. C. Joy, A. D. Romig, Jr., J. I. Goldstein (eds.): *Principles of Analytical Electron Microscopy,* Plenum Press, New York 1986.

[135] L. Reimer: "Transmission Electron Microscopy," *Springer Series in Optical Sciences,* 4th ed., Springer Verlag, Berlin 1997.

[136] J. M. Cowley (ed.): "Electron Diffraction Techniques," vols. 1 and 2, *Monographs on Crystallography 3,* International Union of Crystallography, Oxford University Press, Oxford 1992.

[137] P. G. Merli, M. Vittori Antisari (eds.): *Electron Microscopy in Materials Science,* World Scientific, Singapore 1992.

Specific References

[138] S. Amelinckx, D. Van Dyck, J. Van Landuyt, G. Van Tendeloo (eds.): *Handbook of Microscopy,* VCH, Weinheim 1997.
[139] N. F. Mott, H. S. W. Massey: *The Theory of Atomic Collisions,* Clarendon Press, Oxford 1949.
[140] J. A. Ibers, B. K. Vainshtein: "Scattering Amplitudes for Electrons," in K. Londsdale (ed.): *International Tables for X-Ray Crystallography,* vol. 3, Kynoch, Birmingham 1962.
[141] in [133], p. 250.
[142] W. L. Bragg, *Nature (London)* **124** (1929) 125.
[143] P. P. Ewald, *Ann. Phys. (Leipzig)* **54** (1917) 519.
[144] in [133], p. 251.
[145] S. Takagi, *Acta Crystallogr.* **15** (1962) 1311.
[146] R. D. Heidenreich, *J. Appl. Phys.* **20** (1949) 993.
[147] S. Kikuchi, *Jpn. J. Phys.* **5** (1928) 23.
[148] in [125], p. 125.
[149] H. A. Bethe, *Ann. Phys. (Leipzig)* **87** (1928) 55.C. H. MacGillavry, *Physica (Amsterdam)* **7** (1940) 329.
[150] A. Howie, M. J. Whelan, *Proc. Roy. Soc. London A,* **263** (1961) 217.A. Howie, M. J. Whelan, *Proc. Roy. Soc. London A,* **267** (1962) 206.
[151] H. Yoshioka, *J. Phys. Soc. Jpn.* **12** (1957) 628.H. Hashimoto, A. Howie, M. J. Whelan, *Proc. Roy. Soc. London A* **269** (1962) 80.
[152] G. Bormann, *Z. Phys.* **42** (1941) 157.G. Bormann, *Z. Phys.* **127** (1950) 297.
[153] H. Hashimoto, M. Mannami, T. Naiki, *Philos. Trans. R. Soc. London A* **253** (1961) 459.J. W. Menter, *Proc. Roy. Soc. London A* **236** (1956) 119.
[154] G. Van Tendeloo, S. Amelinckx, *Acta Crystallogr. Sect. A: Cryst. Phys. Diffr. Theor. Gen. Crystallogr.* **A 30** (1974) 431.
[155] S. Amelinckx, J. Van Landuyt, in [126], p. 107.H. Hashimoto, M. J. Whelan, *J. Phys. Soc. Jpn.* **18** (1963) 1706.P. B. Hirsch, A. Howie, M. J. Whelan, *Philos. Trans. R. Soc. London A* **252** (1960) 499.]P. B. Hirsch, A. Howie, M. J. Whelan, *Philos. Mag.* **7** (1962) 2095.
[156] C. M. Drum, M. J. Whelan, *Philos. Mag.* **11** (1965) 205.]J. Van Landuyt, R. Gevers, S. Amelinckx, *Phys. Status Solidi* **7** (1964) 519.
[157] G. A. Bassett, J. W. Menter, D. W. Pashley, *Proc. Roy. Soc. London A* **246** (1958) 345.]D. W. Pashley, J. W. Menter, G. A. Bassett, *Nature (London)* **179** (1957) 752.
[158] H. Hashimoto, R. Uyeda, *Acta Crystallogr.* **10** (1957) 143.
[159] R. Gevers, *Philos. Mag.* **7** (1963) 769.
[160] D. J. H. Cockayne, I. L. E. Ray, M. J. Whelan, *Philos. Mag.* **20** (1969) 1265.]D. J. H. Cockayne, M. J. Jenkins, I. L. E. Ray, *Philos. Mag.* **24** (1971) 1383.]R. de Ridder, S. Amelinckx, *Phys. Status Solidi B* **43** (1971) 541.
[161] S. Amelinckx, P. Delavignette, *J. Appl. Phys.* **33** (1962) 1458.
[162] P. Humble, in [126], p. 315.P. Humble, *Aust. J. Phys.* **21** (1968) 325.A. K. Head, *Aust. J. Phys.* **20** (1967) 557.A. K. Head, P. Humble, L. M. Clarebrough, A. T. Morton, G. T. Forwood: "Computed Electron Micrographs and Defect Identification," in: S. Amelinckx, P. Gevers, J. Nihoul (eds.): *Defects in Crystalline Solids,* vol. 7, North-Holland, Amsterdam 1973.
[163] in [133], p. 58.
[164] C. Boulesteix, J. Van Landuyt, S. Amelinckx, *Phys. Status Solidi A* **33** (1976) 595.
[165] G. Van Tendeloo, J. Van Landuyt, S. Amelinckx, *Phys. Status Solidi A* **33** (1976) 723.M. Snijkers, R. Serneels, P. Delavignette, R. Gevers, S. Amelinckx, *Chryst. Lattice Defects* **3** (1972) 99.
[166] R. Gevers, J. Van Landuyt, S. Amelinckx, *Phys. Status Solidi* **11** (1965) 689.M. J. Goringe, U. Valdré, *Proc. R. Soc. London A* **295** (1966) 192.
[167] M. F. Ashby, L. M. Brown, *Philos. Mag.* **8** (1963) 1083, 1649.
[168] J. W. Steeds, E. Carlino, in [137], p. 279.
[169] P. Goodman, *Acta Crystallogr. Sect. A: Cryst. Phys. Diffr. Theor. Gen. Crystallogr.* **31** (1975) 793;**31** (1975) 804.M. Tanaka, *J. Electron Microsc. Tech.* **13** (1989) 27.J. W. Steeds, R. Vincent, *J. Appl. Crystallogr.* **16** (1983) 317.
[170] B. F. Buxton, J. A. Eades, J. W. Steeds, G. M. Rackham, *Philos. Trans. R. Soc. London* **A 281,** 171.J. Eades, M. Shannon, B. Buxton in O. Johari (ed.): *Scanning Electron Microscopy,* I.I.I.R. Institute, Chicago 1983, p. 83.
[171] M. Tanaka, M. Terauchi, T. Kaneyama: *Convergent Beam Electron Diffraction I and II,* Jeol Ltd., Tokyo.
[172] N. S. Blom, E. W. Schapink, *J. Appl. Crystallogr.* **18** (1985) 126.
[173] J. Gjønnes, A. F. Moodie, *Acta Crystallogr.* **19** (1965) 6567.P. Goodman, G. Lempfuhl, *Z. Naturforsch. A* **19** (1964) 818.
[174] W. Kossel, G. Möllestedt, *Ann. Phys. Leipzig* **36** (1939) 113.C. H. MacGillavray, *Physica (Amsterdam)* **7** (1940) 329.
[175] G. R. Grinton, J. M. Cowley, *Optik (Stuttgart)* **34** (1971) 221.J. M. Cowley, S. Iijima: "The Direct Observation of Crystal Structures," in H. R. Wenk (ed.): Electron Microscopy in Mineralogy, Springer Verlag, Berlin 1976, p. 123.
[176] S. Amelinckx in [133], p. 5.
[177] O. Scherzer, *J. Appl. Phys.* **20** (1949) 20.
[178] E. Abbe, *Arch. Mikrosk. Anat.* **9** (1873) 413.
[179] D. van Dyck, M. Op de Beeck, *Proc. Int. Conf. Electron Microsc.* **12th** (1990) vol. 1, 64. D. van Dyck, W. Coene, *Ultramicroscopy* **15** (1989) 29. D. van Dyck, M. Op de Beeck, *Proc. Int. Conf. Electron Microsc.* **12th** (1990) 26.
[180] D. Van Dyck: "High Resolution Electron Microscopy" in [138] vol. I, Chap. 1.1.2, p. 353.
[181] J. M. Cowley, A. F. Moodie, *Acta Crystallogr.* **10** (1957) 609.
[182] D. van Dyck et al. in W. Krakow, M. O'Keefe (eds.): *Computer Simulation of Electron Microscope Diffraction and Images,* The Minerals, Metals and Materials Society, 1989, p. 107.
[183] J. M. Cowley: "Scanning Transmission Electron Microscopy" in [138] vol. II, Chap. 2.2, p. 563.
[184] S. J. Pennycook, D. E. Tesson, P. D. Nellist, M. F. Chisholm, N. D. Browning: "Scanning Transmission Electron Microscopy: Z-Contrast" in [138] vol. II, Chap. 2–3, p. 595.
[185] H. G. J. Moseley, *Philos. Mag.* **26** (1913) 1024;**27** (1914) 703.
[186] E. L. Hall, in [133], p. 147.
[187] M. H. Jacobs, J. Baborovska: *Electron Microscopy,* The Institute of Physics, London 1972, p. 136.

K. F. J. Heinrich: *Electron Beam X-Ray Microanalysis,* Van Rostrand, New York 1981.

[188] S. J. B. Reed: *Electron Microprobe Analysis,* Cambridge University Press, London 1975.

[189] J. C. Spence, H. Tafto, *J. Microsc. (Oxford)* **130** (1983) 147.

[190] C. Colliex: "Electron Energy Loss Spectrometry Imaging" in [138] vol. I, Chap. 1.3, p. 425.

[191] B. Jouffrey, in [137], p. 363.

[192] R. D. Leapman, L. A. Grunes, P. L. Fejes, J. Silcox, in B. K. Teo, D. C. Joy (eds.): *EXALFS Spectroscopy,* Plenum Press, New York 1981, p. 217.

[193] G. Zanchi, J. P. Perez, J. Sevely, *Optik* **43** (1945) 495. W. Pejas, H. Rose, *Electron Microscopy,* Microscopic Society of Canada, Toronto 1978, p. 44.
H. T. Pearce-Percy, D. Krahl, J. Jeager in D. G. Brandon (ed.): *Electron Microscopy,* **vol. 1,** Tal International, Jerusalem 1976, p. 348.

[194] G. Van Tendeloo, M. Op de Beeck, S. Amelinckx, J. Bohr, W. Krätschmer, *Europhys. Lett.* **15** (1991) no. 3, 295.

[195] S. Kuypers et al., *J. Solid State Chem.* **73** (1988) 192.

[196] S. Amelinckx, *Chim. Scr.* **14** (1978/1979) 197.

[197] J. van Landuyt, R. de Ridder, R. Gevers, S. Amelinckx, *Mater. Res. Bull.* **5** (1970) 353.

[198] G. van Tendeloo, S. Amelinckx, *Phys. Status Solidi A* **43** (1977) 553; **49** (1978) 337; **65** (1981) 431.

[199] S. Muto, G. Van Tendeloo, S. Amelinckx, *Philos. Mag.,* in press.

[200] G. van Tendeloo et al., *J. Phys. Chem.* **96** (1992) 7424.

[201] G. Van Tendeloo et al., *Europhys. Lett.* **21** (1993) 329. M. A. Verheijen et al., *Chem. Phys.* **166** (1992) 287.

[202] D. Schechtman, I. Blech, D. Gratias, J. W. Cahn, *Phys. Rev. Lett.* **53** (1984) 1951.

[203] C. Janot: "Quasicrystals, A Primer," *Monographs on the Physics and Chemistry of Materials,* Oxford Science Publishers, Oxford 1992.

[204] R. Penrose, *Bull. Inst. Math. Appl.* **10** (1974) 266.

[205] A. Ourmazd, D. W. Taylor, J. Cunningham, C. W. Tu, *Phys. Rev. Lett.* **62** (1989) no. 8, 933. A. Ourmazd, R. H. Baumann, M. Bode, Y. Kim, *Ultramicroscopy* **34** (1990) 237.

[206] S. Iijima, *Nature (London)* **354** (1991) 56. S. Iijima, T. Ichihashi, Y. Ando, *Nature (London)* **356** (1992) 776.

[207] X. F. Zhang et al., *J. Cryst. Growth* **130,** (1993) p. 36 ff. (1993).

[208] D. Ugarte, *Nature (London)* **359** (1992) 707.

[209] G. Van Tendeloo, H. W. Zandbergen, S. Amelinckx, *Solid State Commun.* **63** (1987) no. 5, 389–393; **63** (1987) no. 7, 603–606. H. W. Zandbergen, G. Van Tendeloo, T. Okabe, S. Amelinckx, *Phys. Status Solidi A* **103** (1987) 45–72.

[210] T. Krekels et al., *Appl. Phys. Lett.* **59** (1991) 3048. T. Krekels et al., *Solid State Commun.* **79** (1991) 607–614.

[211] T. Krekels et al., *Physica C* **210** (1993), 439–446.

[212] T. Krekels et al., *J. Solid State Chem.* (1993), in press.

[213] L. C. Nistor, G. Van Tendeloo, S. Amelinckx, *J. Solid State Chem.* **105** (1993), 313–335.

[214] X. F. Zhang, personal communication.

[215] F. Watari, J. van Landuyt, P. Delavignette, S. Amelinckx, *J. Solid State Chem.* **29** (1979) 137–150. F. Watari, P. Delavignette, S. Amelinckx, *J. Solid State Chem.* **29** (1979) 417–427.

[216] K. Yagi, in [136], vol. 2, p. 261. K. Yagi, *J. Appl. Crystallogr.* **20** (1987) 147. T. Hasegawa et al., *The Structure of Surfaces II,* Springer Verlag, Berlin 1987, p. 43.

[217] H. W. Zandbergen et al., *Physica C* **158** (1989) 155. H. W. Zandbergen, W. A. Groen, F. C. Mijlhoff, G. Van Tendeloo, S. Amelinckx, *Physica C* **151** (1988) 325.

[218] H. Fujita (ed.): *In-situ Experiments with High Voltage Microscope,* Research Center for High Voltage Electron Microscopy, Osaka 1985.

[219] L. Reimer: "Scanning Electron Microscopy. Physics of Image Formation and Microanalysis," *Springer Ser. in Opt. Sciences,* vol. 45, 2nd ed., Springer Verlag, Berlin 1998.

[220] O. C. Wells: *Scanning Electron Microscopy,* McGraw-Hill, New York 1974.

[221] D. B. Holt, M. D. Muir, P. R. Grant, I. M. Boswarva: *Quantitative Scanning Electron Microscopy,* Academic Press, London 1974.

[222] J. B. Pawley: "LVSEM for High Resolution Topographic and Density Contrast Imaging," *Adv. Electron. Electron. Phys.* **83** (1992) 203–274.

[223] L. Reimer: *Image Formation in Low-Voltage Scanning Electron Microscopy,* SPIE Press, Bellingham 1993.

[224] H. Raether: "Excitation of Plasmons and Interband Transitions by Electrons," *Springer Tracts Mod. Phys.* **88,** Springer, Berlin 1980.

[225] R. F. Egerton: *Electron Energy-Loss Spectroscopy in the Electron Microscope,* Plenum Publishing, New York 1986.

[226] E. Menzel, E. Kubalek, *Scanning* **5** (1983) 151–171.

[227] D. B. Holt, D. C. Joy: *SEM Microcharacterization of Semiconductors,* Academic Press, London 1989.

[228] G. V. Saparin: "Cathodoluminescence" in P. W. Hawkes, U. Valdrè (eds.): *Biophysical Electron Microscopy,* Academic Press, London 1990, pp. 451–478.

[229] G. D. Danilatos: "Foundations of Environmental Scanning Electron Microscopy," *Adv. Electron. Electron. Phys.* **78** (1988) 1–102.

[230] K. F. J. Heinrich: *Electron-Beam X-Ray Microanalysis,* Van Nostrand-Reinhold, New York 1981.

30. Techniques for DNA Analysis

WILSON J. WALL, Kidderminster, UK

30.	Techniques for DNA Analysis . . . 1131
30.1.	Introduction. 1131
30.1.1.	DNA Structure 1131
30.1.2.	Structure of DNA in Life. 1132
30.2.	Primary Molecular Tools for DNA Analysis. 1133
30.2.1.	Exonucleases 1133
30.2.2.	Endonucleases 1134
30.2.3.	Polymerases. 1134
30.2.4.	Ligases 1135
30.2.5.	Methylases. 1135
30.3.	Methods of DNA Detection 1135
30.3.1.	Bioluminescence. 1137
30.3.2.	Colorimetry 1137
30.3.3.	Electrochemiluminescence 1137
30.3.4.	Fluorescence 1137
30.3.5.	Denaturing Gradient Gel Electrophoresis (DGGE) 1139
30.3.6.	Single-Strand Conformation Polymorhism (SSCP). 1140
30.3.7.	Random Amplified Polymorphic DNA (RAPD). 1141
30.3.8.	Short Tandem Repeat (STR) Analysis 1141
30.3.9.	Single Nucleotide Polymorphism (SNP) Detection 1142
30.3.10.	Mitochondrial DNA Analysis. . . . 1142
30.3.11.	DNA Analysis and Bioinformatics 1143
30.4.	Applications of DNA Analysis . . . 1144
30.4.1.	General Principles of ARMS Analysis of DNA 1145
30.4.2.	Analysis of Dynamic Mutations . . 1146
30.4.3.	Using DNA Analysis to Determine Sex 1146
30.4.4.	Methods of Personal Identification 1147
30.4.5.	Bacterial Contamination of Water Supplies. 1148
30.4.6.	Adulteration of Food Stuffs 1149
30.5.	References 1150

30.1. Introduction

DNA can be viewed as a biological polymer made from very simple building blocks, but resulting in large-scale complexity. It is the complexity which lends this remarkable molecule its power to control cells, tissues, and ultimately organisms. It is also this complexity which is looked at when DNA is analyzed, for whatever reason and in whatever way. The length of the molecule gives a sense of scale. For example, each human chromosome contains DNA varying between 1.4 and 7.3 cm in length, depending on the chromosome. It should always be remembered, however, that this is a biological molecule and manipulation of it can have profound results, both physically and ethically.

30.1.1. DNA Structure

DNA is made up of a chain of nucleotides, with each nucleotide being made up of a deoxyribose sugar, a phosphate group, and a base. Variety within DNA stems from the bases, these can be either adenine or guanine (purines), or, cytosine or thymidine (pyrimidine). The phosphates and sugars provide the external backbone, from which the bases project. The backbone is constructed of sugars attached to each other by a phosphate ester being formed between the 3'-hydroxyl group of one sugar with the 5'-phosphate group on the next sugar, the DNA molecule is therefore polarized 5' to 3'.

The double helix structure is formed by two strands of DNA pairing in opposite polarities. The phosphate-sugar backbone is on the outside with the bases projecting into the middle. While the components of the individual strands are held together by covalent bonds, the two strands are held

Techniques for DNA Analysis

Figure 1. Normal base pairing found in DNA between adenine and thymine and between cytosine and guanine

Table 1. DNA content in pictograms of a range of species showing the wide variation that is found

Common name	Species	Nuclear DNA content, pg
Saccharomyces cerevisiae	Yeast	0.026
Drosophila	Fruit fly	0.1
Mus musculus	Mouse	2.5
Homo sapiens	Man	3.7
Avena sativa	Bread wheat	18.1
Protopterus	Lungfish	50
Fritillaria davisii	Lily	98.4

together by hydrogen bonds formed between the bases.

There are four bases associated with DNA, which are generally referred to by their initial letters. These are adenine (A), cytosine (C), thymidine (T), and guanine (G). Throughout this chapter the initial letters will be used, rather than the complete name. These bases key together from opposite strands in a very particular way. A will only pair with T and G will only pair with C. The association between these bases is controlled by the formation of hydrogen bonds, as shown in Figure 1.

It is the delicacy of the hydrogen bonds compared with covalent bonds which lends both functional performance and elegance to DNA. The conformation of the double stranded helix can be visualized as what you would see when viewing a right handed spiral staircase, with 10 steps (bases) per complete revolution and a diameter of approximately 2 nm. As a consequence of this structure it is not possible to separate the two strands without unwinding the helix.

It is a useful convention when dealing with DNA to ignore the sugars and phosphates and merely refer to the bases of a single strand. This is because the exact bonding of one strand with another allows us to determine, precisely, the complementary sequence of any given single strand.

Consequentially, when a DNA sequence is written out, it is conventional only to list the base sequence of a single strand, the other one being explicitly determined. It should always be remembered that genetic complexity is not directly related to DNA content of a cell. This can be broadly seen in Table 1.

The huge differences in DNA content are generally associated with variations in the amount of repeated sequences present. The complexity of the genome, that it the proportion of repeats compared with a unique sequence in any given organism is determined using Cot curves, Cot being the concentration (Co) of DNA in moles of nucleotide per liter × renaturation time (t) in seconds. The calculations for working out Cot values are complex, but are always presented in a uniform way. The values are plotted on a log scale resulting in a curved plot whose slope gives an indication of the level of repetition present in the DNA sample. Cot curves are based on the observation that DNA can be denatured by heat and will reanneal when the temperature is dropped. This observation also forms an important part of many other methods of DNA analysis.

By cutting DNA into short sections of about 400–600 bp (base pairs) and then denaturing these fragments the highly repetitious sequences will reanneal first, simply because the repeat sequences are more likely to come into contact with complementary sequence areas. Therefore the sequences are complementary. The more complex the genome is the longer it will take low copy, or single-copy sequences to find the right partner.

30.1.2. Structure of DNA in Life

DNA does not sit in cells in an uncontrolled manner. It is regulated in both its position and expression. These are carried out by proteins, genetically silent, but of huge importance. The proteins involved in structural control of DNA are some of the most highly conserved found in the living world. While it is not the remit of this chapter to detail the manner in which DNA is controlled within the cell, it is worth noting the first-order structure of chromatin, this being the term used to describe the DNA–protein complex.

Using osmotic shock it is possible to unwind chromatin to show the fundamental unit as a beaded string of about 10 nm in diameter. The beads are histone proteins with the DNA double helix wrapped around them. Histone proteins are almost exclusively associated with nuclear DNA. There are essentially five different histones associated with DNA, designated H1, H2A, H2B, H3, and H4. These have been highly conserved throughout evolutionary time, indicating their fundamental importance to the integrity and control of DNA. Changes to histone proteins by methylation, phosphorylation, etc., have effects on the charge of the protein, which can alter the interaction between histone and DNA. For example, acetylation of histone H4 results in the nucleosome core unfolding and is associated with transcriptionally active regions of DNA.

Beyond this level of organization the DNA–histone complex is repeatedly coiled such that the whole strand becomes shorter and shorter and more tightly controlled. This process is much like repeatedly winding an elastic band, but under much more controlled conditions. The ultimate step results in a chromosome, which during certain stages of the cell cycle is sufficiently well condensed to be visible with a light microscope, and using specific staining techniques, individual chromosomes can be identified and compared. Visualizing chromosomes is relatively easy in eukaryotes, that is higher organisms with a clearly defined nucleus. For prokaryotes, however, such as bacteria, the system is slightly different in that nucleic acid is not generally associated with histone proteins and the single chromosome is often circular. This reflects the simplicity of the genome. For example, humans, who do not have the largest genome in the animal kingdom contain approximately 3000 million base pairs, while a gut bacteria *Escherichia coli* contains 4 720 000 base pairs and a virus which infects *E. coli* with the designation ΦX174 contains 5386 bases of single-stranded DNA.

30.2. Primary Molecular Tools for DNA Analysis

With DNA being such a large molecule it is important to be able to manipulate it effectively. This can take any form of manipulation from the entire chromosome, or to just very small pieces of DNA of particular interest. To make these manipulations various enzymes are used which are a part of a cell's own ability to alter the activity of DNA and in some cases protect the cell from invasion by pathogens.

Enzymes used in DNA analysis come under a range of headings, depending on their type of activity. Some are highly specific in their action, some less so. It is therefore important to choose the right tool for the investigation being carried out.

The most important enzyme types used in molecular biology are:

– Exonucleases
– Endonucleases
– Polymerases
– Ligases
– Methylases

These will be dealt with separately and their function and activity explained. It is important to remember that all these enzymes occur naturally and serve very precise functions in the cell. Later on specific, practical, examples of their use will be described. First it would be worth looking at the conventions of nomenclature covering some of these enzymes. Exonucleases and endonucleases are collectively referred to as DNases, or more fully deoxyribonucleases. Enzymes which can degrade both DNA and RNA (RNA is a variety of nucleic acid which is chemically similar to DNA but serving a different biological function) are called simply nucleases. If a DNase cuts only one strand of a double stranded molecule it is called a nicking enzyme, and the resultant single stranded break is a nick. One such enzyme is Dnase I. These enzymes seem to be important in releasing tension in the double helix as it is unwound during replication.

30.2.1. Exonucleases

These enzymes degrade DNA from the ends and cannot themselves generally introduce breaks within a strand. For example, mung bean nuclease

Table 2. A range of enzymes and their cutting sites

Enzyme name	Restriction site
Nae I	5'...GCC↓GGC...3'
	3'...CGG↑CCG...5'
Nco I	5'...C↓CATGG...3'
	3'...GGTAC↑C...5'
Hae III	5'...GG↓CC...3'
	3'...CC↑GG...5'
Mbo I	5'...↓GATC...3'
	3'......CTAG↑...5'
Msp I	5'...C↓CGG...3'
	3'...GGC↑C...5'

a nuclease, as the name implies, extracted from Mung beans will progressively remove bases from both the 5' and 3' end of DNA where there is an overhang, resulting in two blunt ends, that is, strands of equal length of complementary sequences. In contrast, exonuclease III preferentially degrades recessed 3' ends, that is, where the complementary strand extends beyond the 3' end. If the 3' end overhangs more than 4 bases the enzyme cannot attack it. This property can be exploited by first nicking the DNA at specific sites with an enzyme and then using exonuclease III to create a single strand gap propagating in a specific direction.

30.2.2. Endonucleases

These are enzymes which cut DNA internally, rather than at the ends and are more correctly called restriction endonucleases. The activity of endonucleases can vary widely from enzyme to enzyme. Several hundred are now known, with very different operating conditions and sites at which they restrict, or cut, the DNA. Endonucleases recognize short sequences along the DNA, which may occur once every few hundred, or once every few thousand bases along the molecule. The frequency with which cutting sites can be found depends largely upon either how specific the sequence has to be, or how long the sequence has to be before it is recognized by the enzyme.

Most endonucleases recognize sites between 4 and 6 bases long and will either cut the two strands symmetrically, or the cut can be staggered. A range of such restriction endonucleases and their restriction sites are shown in Table 2.

The names are generally derived from the initials of the organism in which the particular enzyme was found, so for example all the restriction enzymes beginning *Eco* are derived from *Escherichia coli*. It is not possible to derive the species originator simply from the enzyme name.

Using any given endonuclease on a specific piece of DNA will always render the same number and size of fragments. With large genomes there can be variation between individuals, but with very small genomes, as found in viruses where there is no sexual reproduction, the variation in restriction sites between individuals is effectively zero. So using a restriction endonuclease on such material would result in fragments of an exactly known size. These could then be used to compare with unknown fragments to give information about the DNA under analysis. An example of this is the digestion of a small plasmid of 4361 base pairs, designated pBR322, which is usually found in *E. coli*. Digestion of this with the endonuclease *Msp* I always results in 26 fragments ranging in size from 622 base pairs (4.04×10^5 Da) to only 9 base pairs (0.06×10^5 Da).

30.2.3. Polymerases

These are enzymes which can take individual nucleotides and attach them to a progressively lengthening chain of DNA. In living cells they perform the essential function of DNA replication. This is also exploited in vitro to produce multiple copies of specific regions of DNA. In cells, control of polymerase is undertaken by a battery of associated enzymes. All known polymerases replicate in the $5' \Rightarrow 3'$ direction. When carried out in the laboratory a different approach has to be taken to make full use of the technique.

There are many different types of DNA polymerase, most of which are, like most enzymes, unstable at high temperatures. These enzymes are of a specific value in some circumstances. For example, T7 DNA polymerase comes from bacteriophage T7, which attacks *E. coli* by taking over cellular control and replicating its own DNA. This enzyme can be very useful for replicating long stretches of DNA quickly. Sometimes a short section of DNA needs to be replicated, but more than once, and for this thermostable enzymes are needed. The first of these was extracted from a deep sea species *Thermus aquaticus* and called *Taq*. Later on other forms of heat stable DNA polymerases were found from other species living around deep sea thermal vents. These later additions to the list of useable enzymes tend to be more accurate in their replication, making fewer errors. Examples of these include a DNA polymerase

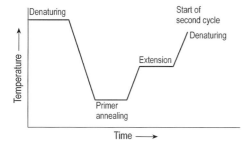

Figure 2. The Polymerase Chain Reaction temperature cycle. The exact temperatures depend upon the type of primers, AT rich primers denature more easily than GC rich primers

from *Thermococcus litoralis*, which lives in a marine environment at 98 °C and one extracted from a species of *Pyrococcus* recovered from a deep sea vent at 2010 m and capable of growth at 104 °C.

The value of these thermostable enzymes is that they can easily be used in a Polymerase Chain Reaction (PCR). This is a technique in which short sections of DNA are replicated exponentially. By knowing two short sequences of DNA on either side of the target DNA, which itself can be of a completely unknown sequence, it is possible to produce single stranded primers, complementary to the two sequences. At this stage the target DNA is double stranded, held together by hydrogen bonds. Exposure to heat will denature the DNA and result in single-stranded DNA. By cooling the sample at this stage, to a temperature at which the DNA will reanneal, some complementary strands will rejoin, but many of the primers will attach to the single stranded DNA. Increasing the temperature at this point to below the denaturing temperature allows the DNA polymerase to extend the strand by adding on bases that are also present in the solution. This cycle is repeated until enough target DNA is present for further analysis [1].

PCR is one of the most powerful methods of producing material for analysis from very small samples. If PCR was 100% efficient then after 20 cycles there would be 2^{20} times the number of starting molecules, each one with a primer at one end. It is these primers which can be tagged if required for later detection. Every primer detected, which is greater than the original primer mass, implies the presence of an amplified product. After approximately 25–30 cycles the enzyme limits the reaction and so, in conjunction with other factors, this results in a realistic 10^6 fold increase in target DNA. One such limiting factor is that as PCR proceeds and more complementary DNA is produced, so there is an increasing tendency for complementary strands to anneal, to the exclusion of the primers. Figure 2 illustrates the temperature cycles and the increase in amplified product.

30.2.4. Ligases

These are enzymes which are used to reattach strands of DNA together. Ligases come in many forms, some only able to work on single-stranded DNA or RNA, while some are able to work on duplex strands. Another major difference between ligases is the ability to reseal a single-strand nick created by a nicking enzyme. Although some ligases will only reattach blunt ends, single base overhangs or cohesive ends, that is two single-strand overhangs with exactly complementary sequences, all ligases operate in the same way. The enzyme catalyzes the formation of a phosphodiester bond between juxtaposed 5'-phosphate and 3'-hydroxyl termini of two adjacent oligonucleotides.

30.2.5. Methylases

These are enzymes which chemically modify DNA residues. In higher eukaryotes this usually takes the form of attachment of a methyl group to cytosine to form 5-methylcytosine. Usually the methylated residues are in C–G pairs. The level of methylation of eukaryote DNA shows a strong correlation with transcriptional inactivity. Lower eukaryotes and prokaryotes are also able to methylate adenine residues. There are no known methylases that work on guanine.

Methylases have recognition sites in which they work, so it is generally possible to tailor the methylation to a specific area of DNA. The value of this stems from methylated residues being resistant to attack from restriction endonucleases. This allows the bacteria carrying endonucleases for protection against invading viruses to protect their own DNA by methylation. It is this aspect of methylases which is important in DNA analysis.

30.3. Methods of DNA Detection

Once a DNA sequence of interest has been produced by PCR, or extracted from a cell, there are several further stages which can be used to determine much more about the DNA. There are broadly two aspects here which can be investigated, size and sequence. Often, for practical purposes, sizing alone is sufficient to detect, say, a mutation of medical importance [2]. In some cases, as we will see later, sizing of some specific fragments may be enough to allow determination of a sequence.

Sequencing is generally undertaken only when absolutely necessary because of cost and time constraints. The most well known of the sequencing efforts is the Human Genome Mapping Project (HGMP). This is intended to sequence the entire human genome, but it should not be imagined that the story ends there. The reading frame of much of the DNA is still unknown, it is also unknown what most of the sequences are for, if anything. Consequently, knowing a sequence is only the start.

There are several different ways in which DNA can be detected and sized. The majority of which involve separation of different sized DNA fragments under the influence of an electric current — electrophoresis. The medium for gel electrophoresis varies from simple agarose to polyacrilamide.

The earliest method introduced for detecting DNA fragments after electrophoresis, was the Southern blot. It involves transfer of the various discrete fragments to a solid support, where they are immobilized. This transfer is accomplished by capillary action in SSC (sodium chloride – sodium citrate) buffer. The capillary action transfers the DNA to a nylon or other artificial membrane, after which it is fixed by heating to approximately 80 °C. The transfer takes place on an horizontal bed, so the relative positions of the DNA does not alter. Reacting the DNA on the blot with probes, which are complimentary sequences to those which are under investigation, yields a series of bands that can be taken as representing the positions of those known, probe, sequences of DNA. These probes are not irreversibly bound to the membrane, unlike the DNA under investigation, so they can be removed and a different probe, or probes, hybridized.

Broadly, there are two different forms of probe, multilocus probes (MLP) and single locus probes (SLP). The difference between these two is that multilocus probes hybridize at several different sites on the genome, while single locus probes only hybridize to a single site. For this reason an MLP will give a multi-banded result, but an SLP will give a maximum of only two bands/probe/individual.

The reason for there being a maximum of two bands is that all higher organisms carry two copies of every gene, one from each parent. When using SLPs these may be both the same, homozygous, resulting in only a single band. However, when the two alleles are different, heterozygous, an SLP will produce two bands. Whichever of these systems, SLP or MLP, is used, indeed, whenever a gel separation system is used, an internal size standard is required. This takes the form of either an allelic ladder or DNA digested by a specific enzyme yielding fragments of known sizes. Whatever the sizing fragment type which is used, the same system applies to the comparison of fragment sizes. This is generally referred to as the Local Southern method. It takes into account the tendency of DNA fragments that are similar in mass, to migrate in a generally linear fashion relative to each other. It therefore requires at least three DNA mass markers which are close in size to the fragment of interest so that an accurate comparison can be made. It is important that the DNA mass markers are closely associated with the fragments under investigation because as they depart in mass the calculation of mass by comparison becomes increasingly inaccurate.

This technique of comparison of size of DNA fragments against a standard is still the best and most accurate method of determining the size of a piece of DNA when it is either fixed on a membrane, or travelling through a gel. This extends to other techniques which do not require the use of a membrane blot to transfer DNA to a solid substrate.

While the original method of DNA analysis required the transfer to a membrane of DNA that had been separated by gel electrophoresis before further analysis could be undertaken, more recent developments have rendered this unnecessary except in specific areas of research. However, automated systems still require a method of sizing which is both reliable and repeatable, and so DNA fragments of known mass are still the method of choice.

Techniques for detection of DNA on gels has also changed. The original method required the use of ^{32}P, which has a half-life of approximately 14 days. It would be substituted for a non-radioactive phosphoros atom in the DNA probe. When hybridized to the probe and exposed to X-ray film a darkening would take place on the film at the radioactive sites. Besides the short half-life there are two other major considerations which have spawned a considerable field of research into the non-isotopic detection of nucleic acids. These are the potential long-term health hazards and the problem of safe disposal of residual radioactive material from the assay. Most of the development of chemical detection techniques was based around immunoassay tests and then developed further.

Non-isotopic labeling to detect nucleic acid – probe hybrids can be carried out using either direct or indirect labeling methods [3]. Direct methods involve attaching a detectable label directly to the target, usually the probe, by a covalent bond. This is the technique now employed in machine based systems, but on gels it is still common to use an indirect method. This involves a hapten, a low molecular weight molecule usually of less than 1000, attached to the probe which can then be detected using a specific binding protein, these may be antibody based. Although the binding strategy must attempt to avoid involving the hydrogen bonds, which are essential for binding of probes to the target DNA, this is not always practicable and may partially compromise the efficacy of the system.

The sensitivity of hybridization assays is a function of the detection limit of the label, some of which are not only as sensitive as radiolabelled assays, but are much faster in exposure times to X-ray film. There are four broad categories of detection for non-isotopic labels. These are bioluminescence, colorimetry, electrochemiluminescence, and fluorescence.

30.3.1. Bioluminescence

This is essentially the same as chemiluminescence found in nature, for example in fireflies and certain types of fungi and bacteria. They involve the use of a luciferin substrate and an enzyme, luciferase. Because of the versatility and sensitivity of these systems, they can be used with either film or charge coupled device (CCD) cameras.

30.3.2. Colorimetry

Systems using colorimetry tend to be less sensitive than luminescent systems, but they do have some advantages. By producing a precipitate associated with the DNA – probe hybrid they are ideal to produce a permanent record of the distribution on a membrane.

30.3.3. Electrochemiluminescence

This is a very specialized technique in which an electrochemical reaction produces an excited state, after which decay to a ground state produces light. It can be readily appreciated that the need for both specialist electrochemical generation and detection facilities renders this sort of system both expensive and specialist.

30.3.4. Fluorescence

This is the system of choice now generally employed in automated equipment. Straightforward fluorescence detection of DNA on gels, by either film or CCD devices, is difficult because of intrinsic problems of noise. However, specific fluorochromes have enabled systems to be developed which utilize different methods of detection, rather than simply overlaying a gel with a photoreactive paper.

Fluorescence detection of specific dyes at specific wavelengths has revolutionized not only the detection of target alleles, but also personal identification, paternity details, and DNA sequencing. This technology has been implemented in many different systems using a range of equipment. A common method for detection of specific fragments is to attach a dye to either an oligonucleotide primer, for use in a PCR reaction, or a probe. By using different fluorochromes for different oligonucleotide primers it is possible to run several PCR reactions simultaneously in the same reaction vessel. Detection in these systems is carried out by inducing fluorescence with a laser, the data being captured and processed automatically as the DNA fragments pass in front of the laser. In this manner an image can be created from the data which broadly corresponds to the older autoradiograph image created from Southern blots and probes. However, this is not actually the most convenient method of data display. The graphic display in Figure 3 would be the more usual method of showing results of analysis.

By using four different dyes, red, blue, green and yellow, large numbers of reactions can be carried out in a single tube. It is also possible to put in size standards, which makes accurate sizing of fragments very much more precise. Figure 4 shows the emission spectra of the various dyes in commercial use. Even if two different fragments are the same molecular weight, if they fluoresce at different wavelengths they can be sep-

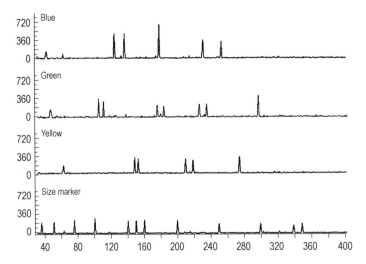

Figure 3. Graphic display of output from an automated DNA analyzer. By using loci of widely differing sizes in the same lane, they can use the same laser activated dye without risk of confusion. This is a composite image of data generated on the basis of when each fragment passes the laser window. It is therefore not a static image but based on a dynamic time scale. The larger molecular weight fragments are towards the right-hand side of the display. It should be noted that there is still a need for the size maker, at the bottom, which fluoresces red. Since all the alleles are separable, they can be run in a single column, the image being worked out by the associated computer. Image courtesy of PE Applied Biosystems

arated using filters. It should be noted that many of these dyes are proprietary products so the exact chemical constitution has not been released.

So far these methods have broadly been applicable to slab gels, but more recently capillary electrophoresis for DNA has been introduced. The main advantage of this system is speed and reproducibility. Arrays of capillaries are used instead of a single gel, but in general terms the technology is very similar and the output identical.

As new technologies are introduced there is a move away from gel based systems towards more exotic methods of DNA analysis. One such is the DNA chip. This is a technology still in its infancy, but potentially of great power and speed. DNA chips utilize the negative charge on DNA to move it about and concentrate it in target areas. These target areas have probes chemically bound to the chip. After several cycles of DNA binding to the chip, each with a different probe, the same process is repeated, but this time with the DNA to be analyzed. By concentrating the DNA around the probes, hybridization rapidly takes place. Detection of the bound DNA is carried out using a laser and arrays to detect the resulting fluorescence. As can be readily appreciated, although potentially very useful, the difficulty is in making the chips flexible enough to be applicable to any but the most specialist of areas of research and diagnos-

tics, where mass screening would make the system commercially viable.

One of the most useful techniques to help in understanding DNA and genes is DNA sequencing. This is quite different to the measurement of sizes of fragments, sequencing determines the precise arrangement of bases making up the DNA. This has received considerable public exposure with the advent of the Human Genome Mapping Project. The HGMP is an attempt to sequence the entire human genome by a combination of public and private collaboration. Although the implementation of new fluorescence based technology has speeded up and automated the process of DNA sequencing, the method used is still broadly based on the Sanger–Coulson method of chain terminating nucleotides. It is, therefore, really the detecting methods that have changed, rather than the underlying science. The alternative method of sequencing, chain degradation, has not been so widely employed in modern systems.

The chain terminating method of DNA sequencing requires single-stranded DNA, because this technique involves the production of a second complementary strand. Consequently, the DNA to be sequenced is normally cloned into a single-stranded vector such as a plasmid. This is done by cutting the gene of interest out, using nucleases and then ligating it back into the plasmid. By

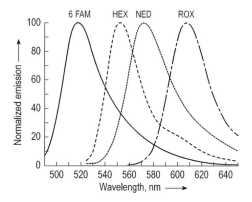

Figure 4. The emission spectra of four commonly used fluorescent dyes in automated DNA analyzers. Image courtesy of PE Applied Biosystems

having single-stranded DNA, both to be sequenced and in the vector, it is possible to produce a short oligonucleotide primer which will anneal to the plasmid and allow extension of the DNA using the gene under investigation as the template for the growing PCR product. The use of a short primer is necessary because most of the DNA polymerases which are used in sequencing reactions need a double stranded region from which to start synthesis.

When undertaking a chain terminating sequencing reaction, there has to be each of the four deoxynucleotides which can be incorporated into the growing strand. These are dATP, dTTP, dGTP, and dCTP. At this point the original method would require four reactions to be carried out separately. One with the addition of dideoxyATP (ddATP), one with ddTTP, one with ddGTP, and one with ddCTP. These are the chain terminators. Dideoxynucleotides can be incorporated into a growing chain as easily as a deoxynucleotide, but will block further strand synthesis. This blockage happens because the dideoxynucleotides lack the hydroxyl group at the 3'-position of the sugar, where the next nucleotide would normally be attached.

If ddCTP is added to the mixture, strand growth will cease when it is incorporated opposite a G on the template strand. However, by adjusting the concentrations of the reagents in the reaction mixture, termination will not always take place at the first template G, because an ordinary dCTP may be added by the enzyme instead. So a number of different fragments of different lengths are produced from this one reaction, each terminated by a ddCTP. When it was normal practice to radiolabel the dideoxynucleotides, they had to be run in separate reactions and run in adjacent columns on a sequencing gel, which was then blotted, autoradiographed, and the bands read horizontally across the gel lanes, with only one band in any position down the gel representing the base at that position on the strand. Sequencing gels are very thin polyacrilamide containing urea to denature the strands.

The modern trend is towards labeling the dideoxynucleotides with a fluorescent dye, each one fluorescing at a different wavelength [3]. Besides allowing the sequence reading part of the experiment to be carried out automatically, it also means that the four different reactions can be carried out in one tube, and with the sequence reading carried out in one lane of a gel [4]. This automation of processes helps to minimize errors which would otherwise appear. As can be appreciated, since the reaction is competitive between deoxynucleotides and dideoxynucleotides, there is a tendency for more low molecular weight products to be formed than high molecular weight ones. There is also a limit to the length of strand which can be accurately sequenced because as they get longer the percentage molecular weight difference between strands differing in length by a single base becomes progressively smaller. Figure 5 shows diagramatically how cycle sequencing is carried out using a circular plasmid and a thermostable DNA polymerase, so that the reaction can be set up and left to run without intervention until the results are transferred to an electrophoresis system.

In Figure 6 a traditional sequencing gel is represented using four lanes of a sequencing gel. It can be seen that as the molecular weight of the fragments gets larger, towards the top of the gel, they move more slowly and become increasingly difficult to separate. Using fluorescent dyes allows all the products to be run in the same gel lane. Although the DNA fragments still tend to "bunch" as their molecular weight increases, it is easier to separate the different products because of their different fluorescent wavelengths.

30.3.5. Denaturing Gradient Gel Electrophoresis (DGGE)

It is possible to determine the presence or absence of a DNA polymorphism without actually knowing anything about the nature of the poly-

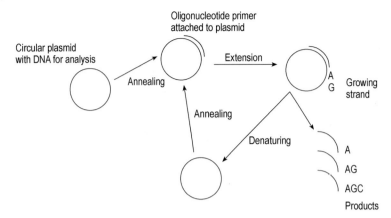

Figure 5. The circular plasmid, containing the DNA to be sequenced, short oligonucleotide primers complementary to a known section of DNA, and labeled nucleotides are reacted together. The primer attaches to the plasmid and then polymerase enzymes extend the primer until incorporation of a dideoxynucleotide terminates chain extension. Denaturing the products, followed by annealing of more primer results in accumulation of artificially produced single strands of DNA complementary to the original template. Overall there will be DNA strands varying throughout the total area of interest in steps of only one base, allowing for step by step sequencing

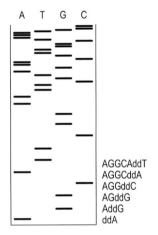

Figure 6. A sequencing gel produced from four separate reactions in a chain termination experiment. Note how these are read sequentially across the gel and as the molecular weights get larger band separation becomes more difficult. With different colored fluorescent dyes these reactions can be carried out in the same tube and sequenced on the same track of a gel, or simultaneously on a single capillary electrophoresis device

single-stranded DNA. Since single-stranded DNA travels at a slower rate than double-stranded DNA when electrophoresed through polyacrilamide gels, a denaturing gradient will enhance separation.

By having a gradient of a DNA denaturing agent through the gel, as the similar, but not identical, double-stranded DNA passes down the gel it will melt at different points, at which point the passage through the gel slows down and the original heteroduplexes can be seen to be different. Of course, it is not possible to say in what way the heteroduplexes are different, only that they are. The denaturing gradient can be introduced in a number of ways, such as formamide, urea, or a temperature gradient.

30.3.6. Single-Strand Conformation Polymorhism (SSCP)

Similar to DGGE, single-strand conformation polymorphism (SSCP) is a method which can be used to detect mutations and polymorphisms without characterizing them. SSCP analysis utilizes the principle that electrophoretic mobility is dependent on size, and shape of the molecule as well as the electric charge. Ideally DNA to be analyzed using SSCP is less than about 400 base pairs. By running a PCR across the area to be tested, enough DNA can be produced to analyze. The analysis is carried out by denaturing the PCR products and

morphism itself. This is done using DNA which has been through a PCR to produce enough material to be analyzed. The detection relies on the fact that DNA heteroduplexes, which differ from each other by only a single base pair, not of content, but of type, have measurably differing characteristics affecting the transition from double-stranded to then rapidly cooling them. This stops most of the single-stranded DNA reassociating with complementary sequences, but does allow the single-strands to fold into a conformation determined by the sequence itself. A difference of only a single base will alter the conformation and therefore the mobility within the gel, which is of a native, rather than a denaturing, type. Because the conformation can be affected by ion concentration and temperature it is important that the system is highly uniform throughout the procedure. This will mean controlling the temperature of the gel during the experiment as well as the solvent and ion concentrations.

30.3.7. Random Amplified Polymorphic DNA (RAPD)

Random amplified polymorphic DNA is in some ways a misleading name. What it does allow is detection of polymorphisms without any knowledge of the polymorphism, but also without any knowledge of the DNA sequence of the target organism. Indeed, the results can be of limited value because of this, but there are specific applications to which this technique can be put with great success. RAPD analysis requires the use of a single short oligonucleotide, usually about 10 bases long, the sequence of which is chosen at random. The primer will often anneal to genomic DNA during PCR at several different places, giving rise to several bands on subsequent electrophoresis, but if conditions are maintained between experiments the number of bands should remain the same for any individual.

The amplified fragments are highly dependent on the primer sequence and the origin of genomic DNA being looked at. So the results may be of interest, but without further analysis it is not possible to say what form the polymorphisms take or where in the genome they reside. One area of particular interest for RAPD use is in agriculture where it might be necessary to determine the sex of a plant in dioecious species, that is species where the male and female flowers occur on different plants. This is useful to stop the planting of more pollinating plants than is necessary, but, as is the case with date palms, until the trees mature and flower it is not possible to determine which trees are male and which female. So by using RAPD analysis differences may come to light between the sexes which can be exploited to determine the sex of the plant while still a seedling.

30.3.8. Short Tandem Repeat (STR) Analysis

Short tandem repeats (STR) are widely distributed throughout the genomes of virtually every higher organism that has been investigated. STRs are arrays of variable numbers of short repeated sequences. They are sometimes referred to as microsatellites being very small compared with other satellite DNA. Satellite DNA can be defined as DNA from a eukaryote that separates as a distinct fraction during gradient centrifugation in caesium chloride. The satellite fraction is made up of material of a distinct composition, having a preponderance of A+T or G+C. Some satellite DNA, such as the Alu sequence found in humans is made up of 300 bp repeats, but in STRs the repeat may be only 1, 2, 3, 4, 5, or 6 bases long. The repeat can be made up of any combination of bases, repeated tandemly.

STRs can be found scattered through the genome in most, if not all, genes. They are to be found in intron DNA, that is the interspersed noncoding sections which lie between the exons, or coding sequences. While it is true that introns are not transcribed, it would be incorrect to call them junk DNA. Although not as yet clear why, it seems that introns are necessary to maintain the integrity of the gene. They are also important in evolution, allowing changes to take place within the genome without compromising the metabolism of the cell.

The importance of STR analysis stems from the observation that while some STRs are constant in their repeat number across entire populations, others are highly variable in repeat number. This variability has been shown in all species looked at. It has also been shown that not all the different alleles occur at the same frequency in any given species. So for a specified STR some alleles will be commoner than others, the distribution frequently approximating to normal, with very small and very large repeat numbers being rarest. Because this is not always the case we can surmise that some other mechanism than simple random processes are affecting repeat numbers of STRs.

It is the inherent variability of STR repeats which makes them of value [5]. By using a PCR reaction across the STR, given that we know the amount of flanking DNA, the fragment can be accurately sized in terms of the number of repeats present. This means that an STR result is expressed not in terms of base pairs, but as a single figure denoting the number of repeats. STR analysis is one of the simplest of PCR processes as analysis of results can be carried out using small agarose gel electrophoresis systems. This would include a sizing ladder in a second lane. The entire gel can then be stained with ethidium bromide and fluoresced using ultraviolet radiation. This approach requires that unless widely separated by size only a single STR is analyzed per experiment. The alternative is to make use of the automated systems which can cope with several different STRs at a time using several different fluorescent markers.

STRs have several important uses for the biologist, among them are personal identification, which we will look at later and population studies of animal and plant species. This latter function of STR analysis is especially useful in endangered species where there is small residual genetic variation. Since STRs are inherited in a strictly Mendelian fashion, with mutation rates generally less than 1 in 10 000 meiotic events, it is possible to determine the level of inbreeding among a group. STRs can also be used to help breeding programs of endangered species by formulating family trees and aiding captive breeding to optimize outbreeding and reduce inbreeding suppression of fitness. Mutation of STRs most often result in a change in complete repeat number, but this is not always so, with occasional single bases being added in the middle of an STR. It has been suggested that mutation events only serve to increase the size of the STRs, but simply because reductions have not been seen it does not mean they do not occur. It is quite obvious that a continuos increase in STR size would eventually result in physical restraints on the genome as well as a reduced ability of the cell to function metabolically.

30.3.9. Single Nucleotide Polymorphism (SNP) Detection

Single nucleotide polymorphisms (SNPs) are represented in the genome as a single base change in a gene, or other DNA being looked at, although they may be considered less informative than other mutations that have several different variants at a single loci. This is because an SNP will normally be biallelic, that is it is found in either one of two forms only. However, with SNPs being found at a rate across the human genome varying from one every 500 base pairs to one every 5000 base pairs, depending upon which area of the genome is being looked at, there are potentially anything up to 3 million SNPs in the human genome. This makes them potentially very useful in a range of areas, especially population structure, pharmacogenetics, and complex genetic traits. These last conditions come under the general heading of polygenic and multifactorial inheritance. Examples of this include a surprising range of conditions that have not generally been thought of in the past to be genetically controlled because the number of genes involved and the interaction of these with the environment made it impossible using classical genetic techniques to analyze them. Such conditions as some forms of heart disease and a persons height can be included in complex genetic traits.

SNP analysis is another PCR based technique. It can be done in a variety of different ways but the most straightforward, although not the method with the greatest throughput, involves a modification of the sequencing reaction. By creating an initial PCR reaction which runs for approximately 100 bp across the known SNP, sufficient material can be generated for the determination of the nature of the SNP. This first PCR has to be carried out to make sure that enough target DNA is available. The next stage uses a primer immediately adjacent to the SNP, and dideoxynucleotides attached to a fluorescent marker. As soon as the next nucleotide is incorporated, which is in the SNP position, the sequence is terminated. When these products are run out on a gel there will be either one fluorescent band of a single color, if the individual is homozygous for that SNP, or two bands of different colors if the individual is heterozygous. Although the two products are essentially the same length in base pairs the difference in mass with the attached dyes is sufficient for them to be separated. If for any reason they are not, then electrophoresis mobility modifiers can be used. It can be appreciated that the first PCR reaction is necessary to produce enough material for the second round fluorescence to be detectable.

30.3.10. Mitochondrial DNA Analysis

Mitochondria are small self replicating subcellular organelles found in the cytoplasm of all eukaryotes. However, the situation is rather more complicated than that. Mitochondria are essential as they are the site of oxidative phosphorylation, the energy producing process in eukaryotes, and although they have their own DNA (mtDNA), only a minority of mitochondrial proteins are actually coded on the mitochondrial genome, most are produced by nuclear DNA. So although self-replicating, they are not autonomous. Generally, mitochondria also lack intron sequences within their genes.

The mitochondrial genome varies from about 15–20 kb of double stranded DNA with approximately 50 genes sequences. At this point it is worth considering mitochodria in general because it can impinge on interpretation of analysis of mtDNA results. Mitochondria are generally thought to be inherited only from the maternal line. This has been shown to be incorrect, but in a way which appears contradictory. In humans most of the mitochondrial genome originates from the maternal side: the ovum. However, it has been demonstrated that the mitochondria of the sperm can, and does, enter the ovum on fertilization. The question then arises as to how much of the mtDNA in the zygote comes from a maternal origin and how much is paternal. Generally speaking most testable mtDNA comes from the maternal side, although the sperm does in fact contribute approximately 100 mtDNA genomes to the ovum on fertilisation in mammals. Recent studies have shown that there is a selective destruction of the invading sperm mitochondria soon after fertilization.

Studies of three dimensional reconstructions of cells using serial sections and electron microscopy suggest that mitochondria either freely associate with each other or are, in any given cell, single entities. Certainly in yeasts it has been shown that mitochondria fuse and bud off from each other quite frequently. Either way, every cell contain several copies of the mtDNA genome which can be tested [6].

Being generally accepted as maternal in origin, mtDNA has the potential to enable us to track through generations via the maternal pathway of inheritance, rather than paternal. This differentiation has great significance in some applications [7].

Mitochodrial DNA comes in the form of a distorted circle, forming a three dimensional structure as it forms hydrogen bonds with itself. In humans mtDNA is generally about 16 570 base pairs long, but it is not the entire sequence which is interrogated when mtDNA is analyzed. It is a short section of highly variable sequence which is looked at, again by PCR.

The reason mtDNA analysis can be so valuable is that unlike nuclear DNA, mtDNA can be found in robust tissues such as bone shafts and hair shafts, which contain no nuclear DNA at all. These tissues, which are tough and resistant to decay, can harbor useful DNA signatures long after the soft tissues have decayed along with the nuclear DNA and become of no use as to the determination of origin of a sample.

It is the robust nature of mtDNA and relative ease with which it can be copied using a PCR reaction which makes it ideal for investigating ancient samples, such as mummified corpses and bones from ancient burial sites. It has also found favor for determining phylogeographic variability in ancient societies. It is not generally necessary to sequence the entire mtDNA because some areas are intrinsically highly variable. One such is the mitochondrial DNA control region with a high variation rate in the 360 nucleotides at the 5'-end of the sequence and to a lesser degree 200 nucleotides at the 3'-end of the control region. Other methods focus on a section of 780 bases of a variable region, which shows a variation rate of approximately 1 %. By concentrating on 12 sites within this region it becomes easier to produce a result and yet maintains the discriminatory power of the test. Mitochondrial DNA analysis has wide applications from familial identification to questioning species origin in food materials.

30.3.11. DNA Analysis and Bioinformatics

It has become ever more apparent that to analyze DNA from any organism increasingly requires fast and reliable computers. The sheer scope of trying to analyze something of the size of, say, the CF gene which spans over 250 kb can be enormous. It would also be extremely difficult without computers to have any chance of finding out whether a sequence has already been defined amongst all the available sequence data. It is often no longer possible to publish database information as tables or lists, the data are just too much for such a manual system to cope with [8].

When looking for something more precise than just sequence data, perhaps data on mutation rates or sequence variability, it is normal to employ a search engine. Although search engines designed for scientists are slowly becoming available, commercial search engines do not well serve the need of scientists. This is essentially because the software uses crawler programs, which index a page, and then jumps to a linked page and goes through the process again. Commercial web pages often have embedded words in HTML code for their

Table 3. Useful starting points for finding genetic data on the World Wide Web

Adress	Type Of Data Contained
http://www.gdb.org/hugo/	Home page of Human Genome Organization
http://www.expasy.ch/sprots/sprot-top.html	Home page of the Swiss-Prot database containing data relating to protein function and variation as well as domain structures
http://www.ncbi.nim.nih.gov/omin	Home page of the database Online Mendelian Inheritance in Man. This is an annotated listing of traits, both diseases and bening polymorphisms which are inherited in a Mendelian fashion. It also contains data regarding mitochondrial dissorders
http://www.incyte.com./products/pathoseq/pathoseq.html	This is a commercial database which contains data relating to sequence data of pathogenic organisms, primarily bacteria and fungi
http://www.ncbi.nim.nih.gov/genbank/genbankoverview.html	A primary depository for human sequence data, currently containing over 4.5 million sequences covering approximately 3 billion base pairs
http://www.ncbi.nim.nih.gov/snp/	A primary depository for SNPdata, currently containing in excess of 2.4 million SNPs

page, or use metatags, words invisible but present on the home page.

The unaudited nature of many web pages, both commercial and scientific is also a potential source of problems. If a laboratory sets up a web page that allows publication of any results they have, two things may occur, neither of which are desirable. The first is that erroneous data may be assumed correct when they are not and the second is that even if correct and valuable they may still be overlooked because of the construction of the web page. This second possibility is all the more distressing since the web has a primary aim of disseminating information as widely as possible. A primary feature of DNA analysis in all its forms is the need for different groups to be able to compare their results with other people working in the same area. Without unified databases this is not easily possible and can almost negate the potential value of having data in an electronic form.

Some large databases containing scientific data have the ability to check new data for consistency and contradiction with what is already available. This is a very valuable facility and underlines the need for well edited and definitive databases. Although there are good starting points for searching a particular area of data or data types, there are also some areas where fractioned data are so widely distributed that it becomes extremely difficult to navigate around the subject. There are for example, more than 150 different sites devoted to human mutations. If it were a simple case of the data being widely distributed, that would make the system complicated enough to deal with, but for many years nomenclature also varied so that it was not always entirely obvious if the same mutation was being referred to. Such was the potential for confusion that the Human Genome Organization (HUGO) has created the Mutation Database Initiative which has produced a consensus on nomenclature and computing conventions so that human mutation data can be cross-referenced with other types of data dealing with both other types of human genetic data and also data from other species. Table 3 lists some useful starting points for data mining.

30.4. Applications of DNA Analysis

One of the earliest applications of DNA analysis came about indirectly, through the agency of gene mapping [9]. Ordering genes on a chromosome or section of DNA can be carried out with no knowledge of the DNA, gene or gene product, such maps have been constructed since the beginning of the twentieth century by breeding experiments. Creation of such maps became ever more sophisticated to the point where a classical map gave way to data regarding the fundamental chemical nature of DNA.

It is not only by sequencing strands of DNA that we can gain information and knowledge from the genetic code. Indeed, it is well recognized that even when the entire human genome has been sequenced, this will only be the very first stage in a long story of research. Knowing a DNA sequence alone gives no indication of the reading frame, the length of the gene and perhaps more importantly, the nature of the protein product. This last point is particularly significant because even if an amino acid sequence is known from a gene, it may not be obvious what the final tertiary structure might be, and consequently what the protein does. A good example of this is the characterization of the cystic fibrosis gene, where the gene was known before the gene product or its action was known.

Also, it may not be necessary to know the entire sequence of a gene to produce a valid analysis of its significance and activity. Taking the case of cystic fibrosis (CF), this is the commonest severe autosomal recessive condition in Caucasian populations. Approximately 5 % of people are carriers, which results in close to 1 in 2000 live births being affected. For a long time it had been possible to predict the probability of carrier status from a simple family history, but it was not possible to be more precise than that for many years, until it became possible to use linkage analysis to associate specific DNA polymorphisms within a family. This is a process where a marker, that is any determinable DNA sequence, is usually transmitted with the disease gene of interest by simply being in very close proximity to the disease gene. The logic is therefore that if an individual carries the measurable DNA then they will also carry a defective gene.

Using this idea the CF gene was broadly localized to a specific area on human chromosome 7. Later and progressively detailed analysis of the area produced a candidate gene. This was the first time a gene had been sought and found with no knowledge of the protein product for which it coded. It was only found later, after the protein sequence had been determined, that in 68 % of all CF mutations a single 3 bp deletion of the gene, resulting in the loss of a phenylanaline residue at position 508 was responsible for the problem. This mutation was designated ΔF508, Δ for delete, F for phenyanaline, and 508 for the amino acid position. Since then more than 170 different variant mutations of the gene have been described.

To analyze DNA mutations directly in the CF gene a technique can be used which is widely applicable in many areas of DNA analysis and mutation detection. This is the amplification refractory mutation system (ARMS).

30.4.1. General Principles of ARMS Analysis of DNA

A basic ARMS test requires the use of three primers. These cover the area where a known mutation occurs There is one common primer and two other primers, one which we can refer to as the wild type, that is the normal, fully functional one and one which includes the mutation. The length of the PCR products is not in itself important, what is important is the sequence at the end of the mutation detection primers. It is for this reason that the sequence of the gene being studied is known.

If a primer has a mismatch at the 3'-end it is most unlikely to produce a PCR product. So by having one wild type primer and one mutation primer two PCR reactions will produce either one product, the homozygous wild type or normal, alternatively the homozygous affected, or two products indicating carrier status. This mismatch specificity is maintained by using a PCR enzyme which does not have a 3' to 5' proof reading ability. If an enzyme is used with this capacity to proof read, no sense can be made of the results. Because the PCR reaction is indicating a specific finding, it is usual to run another, unrelated, PCR reaction in the same tube simply to be certain that the PCR has worked; a form of quality control.

Increased specificity of the reaction can be created by introducing deliberate mismatches close to the 3'-end of the primer. It should always be remembered that the nearer to the 3'-end the introduced mismatch is the greater the destabilization, so although specificity may be increased, product yield is likely to go down.

While it was necessary to use separate PCR reactions, when there was no way in which PCR products could be differentiated individual reactions had to be carried out. With the advent of fluorescent tagging of products this situation has changed. It is now possible, using only a limited number of different fluorescent wavelengths, to carry out ARMS testing in a single reaction for more than 10 different known mutations. This is done by using a combination of different common primers which results in a series of differently sized products. So the fluorescence might appear as red, blue, yellow, red, green, blue ... and so on, essentially the same fluorophore is used several times, but has to come within a sandwich of other fluorophores to be accurately sized.

Another technique, now widely used to construct tests for conditions like CF where there are a large number of possible mutations, all resulting in essentially the same phenotype, is the PCR–oligonucleotide ligation assay (PCR/OLA). In this technique multiplex PCR reactions are followed by a multiplex oligonucleotide ligation reaction which is made up of three probes for each biallelic site to be looked at. One probe is common and is labeled with one of three fluorescent dye markers, the other two correspond to either the normal or mutant variant at that site. These interrogative probes are attached to various different numbers of electrophoresis mobility markers. These are generally pentaethyleneoxide units, which can be stacked to give products of the same number of nucleotides widely different mobility rates. So when the common probe is ligated to the specific probe it results in a unique combination of fluorescence and mobility. Using only four different fluorescent dyes, three for the probes and one as a lane size marker, it is possible to size a large number of different products in the same lane and determine the presence or absence of mutant or normal alleles at each site within the gene.

30.4.2. Analysis of Dynamic Mutations

There are a group of human conditions which have caused considerable consternation to those trying to define them clearly. These are the group of disorders now known to be characterized by dynamic mutations resulting in familial transmission, but in some conditions not necessarily in a clearly Mendelian fashion. In those that are clearly inherited in a dominant/recessive way, the age of onset may be extremely variable. This latter state could sometimes be mistakenly associated with closely linked polygenic inheritance or environmental factors. Although there have been relatively few of these conditions so far described, they are very important, either because of their frequency or severity. These include Fragile X syndrome, Huntington's Chorea and Myotonic dystrophy, where normal dominant inheritance of the condition is modified by a dynamic mutation in the 3'-untranslated region of the gene for myotonin protein kinase.

One of the worlds commonest genetic diseases and certainly the commonest form of familial mental retardation is Fragile X syndrome, affecting approximately 1 in 2500 children. The condition gets its name from the fragile site found at the end of the X chromosome, this being one of the two sex determining chromosomes, the other being the Y. Two X chromosomes results in a female phenotype but an X and a Y results in a male phenotype. For many years the only way of diagnosing this condition was cytogenetically, which was inaccurate, time consuming, and very expensive. Cytogenetic techniques gradually gave way to rather cumbersome pre-screening using molecular methods. Now, however, it is possible to gain a much clearer picture using PCR in a technique which is equally applicable to other dynamic mutations. Using Fragile X as our model we shall look at how dynamic mutations can be investigated.

Dynamic mutations are those which are represented not by a change in base composition, but a progressive lengthening of a sequence. This takes the form of progressive expansion of a trinucleotide repeat. In the case of Fragile X it is an increase in the number of polymorphic CCG repeats in the 5'-untranslated region of the gene. The number of repeats varies from about 6 to 55 in normal X chromosomes, but towards the upper limit of 55 the copy number becomes unstable between generations. Once over 55 repeats the chromosome is said to have a premutation, the rate of instability increases dramatically, so a dynamic mutation is one where once changed the DNA has an altered risk of further change. Once the number of copies goes beyond 230 it is a full mutation. It is therefore useful to know how many copies are present in premutation conditions. This is done by PCR reactions across the area of the trinucleotide repeat. With very large numbers of repeats in these conditions it may sometimes be necessary to produce products which are more than 2.5 kb long. This can be extremely difficult as the possibility of damage to target DNA is in proportion to its length. For this reason it is important to make a mild extraction of the DNA and not risk shearing it when using pipettes. It is also worth simultaneously using an internal control to check the voracity of the PCR reaction. Figure 7 shows the results of sizing the PCR products from three normal individuals, homozygous, heterozygous, and what is commonly referred to as the N+1 heterozygote where there is only one repeat difference between the two alleles. The other peaks represent the ApoE check on PCR and the sizing ladder used to measure the number of repeats.

30.4.3. Using DNA Analysis to Determine Sex

We have seen how RAPD systems can be used to determine the sex of dioecious plants, but there are other times when knowing the sex of an organism is just as important, but for other reasons. We shall take two examples, one human and one animal.

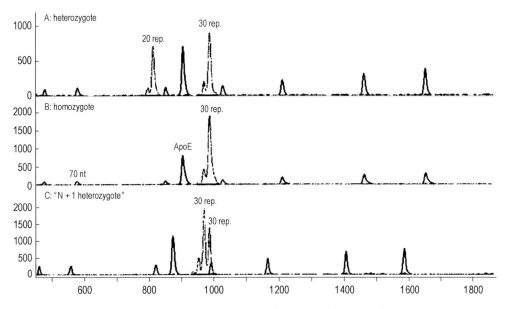

Figure 7. The automated output from measurement of repeat numbers in Fragile X analysis. The three traces represent heterozygote, homozygote, and N+1 heterozygote. The peak marked ApoE is a check that the PCR has run adequately and the other peaks are the sizing ladder. This technique is readily extended to other conditions involving this sort of dynamic repeat expansion. Image courtesy of PE Applied Biosystems

Humans differ chromosomally between the sexes in that females are the homogametic sex having two identical X chromosomes and males are the heterogametic sex having one X and one Y chromosome. There is a gene which codes for a protein associated with development of tooth enamel called amelogenin. By running a PCR across part of the gene two fragments are generated, one of 212 bases from the X chromosome and the other of 218 bases from the inactive gene on the Y chromosome. These are created by the same primers, so by running the products on a gel there will be two bands from a male and only one band from a female. Although the gene on the Y chromosome is inert, it does not seem to be susceptible to alteration, so this test is now routinely used on forensic samples to determine the sex of the originator of the sample.

Birds are quite different. Here it is the males which are homogametic and females heterogametic. Because of this difference the sex chromosomes are designated Z and W, with males being ZZ and females ZW. There is a highly conserved sequence on the sex chromosomes of birds which differ in only a minor way. By running a PCR reaction across the highly conserved region, enough product is produced for the second stage.

The difference between the two is a restriction site which can be cut by the restriction endonuclease *DdeI*. This operates by cutting at the sequence 5'CTNAG3'. The W chromosome produces a 73 bp fragment and the Z chromosome a 104 bp fragment, so there will be either one or two bands on a gel depending on the sex of the bird being tested. Because of the highly conserved nature of the tested sequence, it has been possible to use it in virtually any bird group other than the Ratites (emus, ostriches, etc). With so many bird species being identical between the sexes it has become important to be able to determine the sex of the birds in captivity so that breeding programs can make the best use of stocks [10].

30.4.4. Methods of Personal Identification

There have been many different methods of personal identification developed in the last 25 years based on DNA analysis. The more recently developed ones require more automated equipment but less human intervention, making interpretation of the results far less subjective. The first systems required relatively large quantities of material, usually blood, but as technology progressed so the amount of starting material has decreased so that it is now theoretically possible to determine with a good degree of certainty the generator of a sample. This is especially important in two areas of law, criminal cases and paternity disputes. Although there are specific methods used in such cases, this is more for continuity of service and agreement of acceptable methods, rather than one being better than another.

All methods which can discriminate between two sequences can in principle be used to discriminate between two individuals, but the level at which they can be relied upon does vary enormously. This reliability can stem just from factors such as inadequate discriminatory power of the test or subjective interpretation of the results. This latter situation being more important in earlier systems that were implemented rather than later ones.

The first methods of personal identification involved the use of mulitlocus (MLP) and single locus (SLP) probes. A sample of native DNA would be digested using an enzyme and then run out on a gel. This could then be probed to give several bands per individual in the case of MLPs, or one or two bands in the case of an SLP analysis. The major problem with these techniques is the size of the fragments which are probed. As the measured DNA can run into several kilobases, separation inevitably involves interpretation and agreement on methods of being able to say that a large band does in fact only comprise DNA of a particular type The possibility being that a band in what appears to be the position representing 14 kb, may in fact be 14.1 kb and therefore different, but not separable at this scale. It is these sorts of questions which helped to move techniques of personal identification along.

Although the ultimate personal identification method would be to sequence an individuals genome this is at the present time not a practical possibility. There are, however, two very good methods currently available which are routinely used. When used in this application they both utilize the same methodology. These methods are STR analysis and SNP. Both of these suffer from an intrinsically low power of discrimination. For example, while any given STR may have 10 or more possible alleles, they will not all be equally common, so a single STR may have a frequency within the population of 1 in 10, that is, you would expect 1 in 10 of the population to have that allele. Some may be much commoner than that. The way this is got round is by using a panel of STRs, either singly or in one PCR reaction mixture. By multiplying the individual frequencies it is possible to generate a composite value for the probability of finding the total combination of STRs by chance, therefore allowing us to generate a likelihood ratio for that profile coming from a specific individual. With SNP analysis this procedure is taken slightly further, because for any given SNP the number of possible variants is both smaller and fixed, but the method is essentially the same, production of an accumulate probability by analysis of several different SNPs.

As can be appreciated, when comparing, say, material from a body with a reference sample, mutation rates are unimportant. But when looking at questions of paternity it is vital to remember that mutation from one generation to another is a natural consequence of genetic variation. So although these systems may be relatively stable, mutations do occur. This is one reason that the number of sites looked at is usually much bigger than in situations where a sample is compared directly with an individual [11]. Another reason for this increase is that it gives a greater degree of certainty in the production of a family tree because only half the loci in the genome of a child comes from the father, the other half originates from the mother.

A very specialized, and very limited, method which can be employed in personal identification is mtDNA analysis. This procedure is limited in that although it is possible to link a child to a mother, it is not possible to link a child to a father using this technique, and neither is it possible to separate siblings using mtDNA. It has proved very useful, however, where remains have been found and only mtDNA has survived, typically this might be after burial or fire. Under these circumstances mtDNA analysis can be used to help authorities re-unite the right remains with the right family.

30.4.5. Bacterial Contamination of Water Supplies

A practical and very useful application of DNA analysis is in detecting coliform bacteria in water supplies. The same techniques can be extended to check for pollution of surface water and beaches by sewage. The importance of this is can easily be understood by looking at the range of pathogenic microorganisms which can be found in fecal material and can from time to time appear in areas where it is not wanted: *Salmonella* (causing typhoid fever), *Shigella* (giving rise to dysentery), and *Vibrio cholerae* (the causal agent of cholera). Although not associated in any way with fecal pollution another important contaminant is *Legionella pneumophila*, often present in air conditioning units and the cause of legionnaires disease. Indirect qualitative analysis of DNA from these organisms can be said to occur when an infection takes place. The toxins produced are proteins which are themselves transcribed from genes, but analytical techniques are now more sophisticated.

Although it has long been possible to detect these organisms, the traditional methods of microscopy and viable plate count are not necessary quantitative methods. Sensitivity is also inadequate with these systems, being approximately 10^4 cells/ml, which is far too low for sensitive issues of environmental contamination. A better and more sensitive way of assessing both the presence and quantity of these organisms is by the use of a PCR reaction coupled with a suitable probe. Sensitivity is now orders of magnitude greater than previously achievable, it being possible to detect far beyond the limits required by issues of public health. Thus it is possible to detect as little as 1–10 cells/ml, which equates to between 1 and 10 fg of DNA depending on the target organism.

For these environmentally important species a panel of very precise probes are used after DNA extraction and PCR of target DNA. The very high specificity of this technique is essential as there are frequently several different species in a sample, most of which will be benign. Because the detection is based on a presence/absence assay, the PCR products are not run out on a gel, but dot blotted onto a suitable membrane, usually nylon or one of the many proprietary brands available. Once immobilized onto the membrane a probe is used which is labeled at the 5'-end. These dot blots are then probed using the complementary sequence to the target. Various probes are used for this, depending upon the target. For *Legionella pneumophila* the probe is based on the macrophage infecting protein (*mip*), and for total coliform bacteria it is based upon a section of the *lacZ* gene of *Escherichia coli*.

30.4.6. Adulteration of Food Stuffs

Perhaps surprisingly, food labeling is not always as accurate as might be supposed, sometimes accuracy has to be checked. Any question which can be asked as to whether, for example, if a tin of tuna is tuna, if meat products are from the species that is stated on the product, whether for human consumption or pet food. It is even possible to determine whether pasta is correctly made from Durum wheat or from bread wheat. Durum wheat is a hard wheat, while bread wheat is soft, and this considerably alters the stability of pasta when it is cooked.

Given the importance of checking foods for adulteration this is one of the most rapidly developing areas of applied DNA analysis. The constraints are similar to environmental testing, such that it is essential to have an unequivocal standard against which to measure the result. This may seem self-evident, but if you are trying to decide whether a food is what it says it is, there is little point in comparing your results against a standard for which you do not have the provenance, or chain of custody. Mistakes on these lines have happened.

Generally, the methods of checking for authenticity of species origin are based on DNA targets within the cytochrome B gene. This is located on the mitochondrial DNA and has a high species specificity. Although autoclaving material is usually regarded as an adequate decontaminating procedure for material being used in DNA analysis involving PCR, this may not necessarily be so. For example, it is possible to determine the species of animal from which a pet food is produced, even though it has been rigorously cooked at high temperature and pressure. Similarly the checking of canned tuna for authenticity is important. This is because "tuna" covers a range of fish in several different genera, so to be sure that a can labelled "skipjack", really is of the genus *Katsuwonus* rather than *Thunnus* (tunny) or *Germo* (the albacore) it becomes necessary to use DNA analysis. In these cases, as with any cooked product, it is generally possible to produce a repeatable result, unless the food stuff has been cooked at high temperature in an acidic liquor, after which results are unreliable and frequently non-existent.

Whenever food adulteration is suspected, it should be remembered that PCR is a technique of extraordinary sensitivity. Any result must be quantified, the possibility of accidental contamination must always be considered.

30.5. References

[1] B. Budowle, R. Chakraborty, A. M. Giusti, A. J. Eisenberg, R. C. Allen, *Am. J. Hum. Genetics* **48** (1991) 137–144.
[2] C. M. Hearne, S. Ghsoh, J. A. Todd, *TIG* **8** (1992) 288–294.
[3] L. Kricka (ed.): *Nonisotopic DNA Probe Techniques,* Academic Press, London 1992.
[4] K. M. Sullivan, S. Pope, P. Gill, J. M. Robertson, *PCR Methods and Applications* **2** (1992) 34–40.
[5] C. P. Kimpton, N. J. Oldroyd, S. K. Watson, R. Frazier, P. E. Johnson, et al., *Electrophoresis* **17** (1996) 1283–1293.
[6] S. Anderson, A. T. Bankier, B. G. Barrell, M. H. L. deBruijn, A. R. Coulson, et al., *Nature (London)* **290** (1981) 457–465.
[7] E. Hagelberg, J. B. Clegg, *Proc. R. Soc. London* **B 252** (1993) 163–170.
[8] S. M. Maurer, R. B. Firestone, C. R. Scriver, *Nature (London)* **405** (2000) 117–120.
[9] D. Chadwick, G. Cardew (eds.): *Variation in the Human Genome,* Ciba Foundation Symposium 197, John Wiley and Sons, Chichester 1996.
[10] R. Griffiths, B. Tiwari, *Nature (London)* **375** (1995) 454.
[11] R. L. Alford, H. A. Hammond, I. Coto, C. T. Caskey, *Am. J. Hum. Genet.* **55** (1994) 190–195.

Index

α emitters 132
Abbe theory 1063
Abel inversion 639
aberration, optical lenses 1063, 1092
ablation, atomic spectroscopy 668 ff
absolute configuration, diffraction 400
absolute mass units 17
absorber, Mössbauer spectroscopy 564, 567 f
absorption
– acoustic mass sensors 1020
– radionuclides 134
– thermogravimetry 828, 831
– UV spectroscopy 422 ff, 430, 443
– X-ray fluorescence spectrometry 754
absorption bands, IR/Raman spectroscopy 475
absorption length 1081
absorption methods 151
absorption spectrometry 729, 741 f
abundance 19
accelerated cyclic corrosion test (CCT) 897
accelerating rate calorimeters 849
accelerator mass spectrometry (AMS) 602
acceptance quality level (AQL) 74 f
acceptors
– acoustic mass sensors 1020
– semiconductor sensors 961
accuracy 14
– atomic spectroscopy 698, 720
– chemical sensors 954 f
– chromatography 190
– mass spectrometry 581, 706
– UV spectroscopy 440
– voltammetry 809
– weighing 65
– X-ray fluorescence spectrometry 760
  see also: errors
acetic acid 80
acetone
– liquid chromatography 284
– organic trace analysis 99
– thin layer chromatography 336
acetylenes 478
achromatic objectives 1063
acid solutions 813
acidification, samples 80
acids
– nuclear magnetic resonance 526
– trace analysis 80 ff
acoustic plate mode oscillator (APM) 1003, 1012
acrylamide 349
acryloylmorpholine (ACM) 350 f
activated charcoal 102
activation analysis 6, 17, 767–784
– affinity chromatography 318
– radionuclides 128
activators determination 155
active hydrogen 144
active sampling, trace analysis 102
activities
– enzymes 148, 153
– radionuclides 128, 132
actuators 952, 1051

acylation 250
acylcholines 155
additive errors see: errors
adenine 1131
adhesion, surface analysis 852, 869
adiabatic calorimeters 840 ff
adsorbable organic halogen compounds (AOX) 119
adsorbents
– chromatography 176
– gas chromatography 201
adsorption
– acoustic mass sensors 1020
– semiconductor sensors 962, 967
– thermogravimetry 828
– trace analysis 81
adsorption agents 100
adsorption chromatography 6, 288 f
adsorption current 788
adsorption effects, trace losses 79
aerosols
– mass spectrometry 613
– Mössbauer spectroscopy 576
affinity biosensors 1033 f, 1040 ff
affinity chromatography 177, 316
affinity ligands 164
aflatoxins 99
agarose
– affinity chromatography 317
– electrophoresis 346 f
– immobilized enzymes 156
agglomerates, electron microscopy 1114
agglomerative methods, chemometrics 56
air–acetylene flame 676
air buoyancy 67
air combustion 87
air control 613
air samples 102
Alazin Red S 802
alcohol oxidase/dehydrogenase 154
alcohols 479
aldehydes 483
aliphatic amines 280
aliphatic ring compounds 477
aliphatic systems 525
aliquoting 81, 96
alkaloids 820
alkane groups
– IR/Raman spectroscopy 476
– nuclear magnetic resonance 526
– organic trace analysis 97
alkine solutions 813
alkyl groups 529 ff
alkylation 250
allocation of test objects (ALLOC) method 58
alloys, electron microscopy 1102
alternating current polarography (ACP) 794
alumina
– chromatography 176
– gas chromatography 208
– liquid chromatography 286
– thin layer chromatography 328
aluminium–manganese alloys 1107

^{26}aluminum   602
aluminum supports, thin layer chromatography   327
amalgams   800
amelogenin   1147
amides
– IR/Raman spectroscopy   479, 486
– nuclear magnetic resonance   529
– voltammetry   821
amino acids   305, 318
amino groups
– antibodies   159
– extraction   100
– liquid chromatography   288
– organic trace analysis   103
aminopropyl groups   329
amperometric cells   975
amperometric detectors, liquid chromatography   270, 276
amperometric metabolic sensors   1035
amperometric modes, molecular recognition   969
amperometry   968, 981
amplification refractory mutation systems (ARMS)   1145
amplifiers, UV spectroscopy   439
amyl groups, thin layer chromatography   335
analog-to-digital conversion
– mass spectrometry   610
– nuclear magnetic resonance   517
analysis of variances (ANOVA)   117
analyte concentration, organic trace analysis   102
analytical electron microscopy (AEM)   6, 1078
analytical quality control (AQC)   33 f
analyzers
– Auger electron spectroscopy   876
– mass spectrometry   597 ff
– static secondary ion mass spectrometry   891
– UV spectroscopy   435 f
– X-ray photoelectron spectroscopy   859
Anderson–Hinthorne local thermodynamic equilibrum (LTE) model   895
andosteron   582
androstanol derivatives   556
angle-resolved ultraviolet photoelectron spectroscopy (ARUPS)   918
anhydrides   484
anilines   280
anions
– extraction   100
– ion-selective electrodes   978
– mobile trace analysis   118 f
– voltammetry   815
anode materials, X-ray generation   383
anodic stripping voltammetry (ASV)   799
anthropogenic radioactivity   130
antiblocking brake system (ABS)   1051
antibodies   147, 158 ff
– affinity biosensors   1040
– affinity chromatography   321
– biochemical sensors   998
– fluorescence microscopy   1068
– immunosensors   969
anticoagulants   80
antigens
– affinity biosensors   1040
– affinity chromatography   321
– biochemical sensors   999
^{121}antimon   567
antioxidants   280

aperture, ideal microscopes   1093
apochromatic objectives   1063
apoenzyme reconstitution immunoassay system (ARIS)   161
appearance potential methods, surface analysis   6, 926 ff, 942
arcs, atomic spectroscopy   640 f, 668, 691
aromatic amines   280
aromatic compounds   163
aromatic hydroxyls   280
aromatic rings   477
aromatics test, liquid chromatography   289
arsenic   800, 818
artificial intelligence   8
artificial neural networks   472
aryl groups   529 ff
ashing, trace analysis   82, 87
Aspergillus niger   154
astigmatism   1063, 1093
ASTM standard   73
atmospheric aerosols   576
atmospheric pressure, wet digestion   85
atmospheric pressure chemical ionization (APCI)   589
atmospheric pressure combustion   88
atom probe field ion microscopy (APFIM)   6, 932
atom scattering factor, electron microscopy   1078
atomic absorption, diode lasers   741
atomic absorption spectrometry (AAS)   6, 17, 629, 673
– trace analysis   89, 95
atomic cells   639 ff
atomic emission   13
atomic emission detector (AED)   232, 239
atomic emission spectroscopy (AES)   113, 688
atomic fluorescence spectrometry (AFS)   113, 713 ff
atomic force microscopy (AFM)   6, 910, 943
atomic form factors   380
atomic number, X-ray fluorescence spectrometry   755
atomic resolution   373
atomic spectroscopy   627–726
– trace analysis   89
atomizers
– atomic spectroscopy   679, 715
– laser spectroscopy   733, 737, 740
atoms   11
atrazin   163, 169
attenuated total reflection (ATR)   492, 1002
Auger analyzer   855, 860, 867
Auger effect   756
Auger electron appearance potential spectroscopy (AEAPS)   927
Auger electron emission, scanning electron microscopy   1124
Auger electron production, EELS   1099
Auger electron spectroscopy (AES)   6, 17, 874 ff, 942, 1078
Auger maps   882
autofluorescence   1067
autoionization state, laser spectroscopy   739
automatic calibration, balances   65
automatic cold needle, gas chromatography   219
automatic sample inlets
– liquid chromatography   268
– thin layer chromatography   332
autooxidation   335
autoradiolysis   133
auxiliary electrode   786, 808, 808

Avena sativa 1132
avidin 164
Avogadro constant 671
Azaroff–Buerger method 379
azo compounds
– IR/Raman spectroscopy 479 f
– voltammetry 821

β-backscattering 6
β-carotin isomers 290
β emitters 132
Babington nebulizer 661
backflush techniques, column switching 306
background correction, atomic absorption spectroscopy 682
backscattered electrons (BSE) 1078, 1115, 1118
backscattering, radionuclides 134
bacterial growth 80
bacterial water contaminations 1148
bacteriochlorophyll 550
balances 63 ff
Balmer formula 629
band profile
– atomic spectroscopy 638
– chromatography 176, 181
– factor analysis 55
– UV spectroscopy 424, 427
bandbroadening, chromatography 186, 221
barbituric acids 821
Barker pulse 791
Bartlett test 44
base lines, Mössbauer spectroscopy 565
basic analytical research 8
basic principles
– activation analysis 768
– atomic absorption spectroscopy 629, 673
– atomic emission spectroscopy 688
– atomic fluorescence spectrometry 714
– Auger electron spectroscopy 874
– chromatography 173–198
– confocal laser scanning 1070
– electron energy loss spectrometry 1098
– fluorescence microscopy 1067
– ion scattering spectroscopy 899
– light microscopy 1062 ff
– magnetic force compensation 63
– mass spectrometry 580
– Mössbauer spectroscopy 561 ff
– multivariate methods 51
– nuclear magnetic resonance 511
– plasma mass spectrometry 704
– Rutherford backscattering spectroscopy 906
– scanning tunneling microscopy 910
– static secondary ion mass spectrometry 889
– X-ray microanalysis 1096
– X-ray photoelectron spectroscopy 854
basic sampling statistics 73 ff
batch inlets, mass spectrometry 586
Bayesian classification 58
beam deflection refractive index detector 271
beam divergence, ideal microscopes 1093
Beer's law 55, 59
bent contours, diffraction 1079
benzene, ethylbenzene, toluene, xylene (BETX) 1031

benzenes
– liquid chromatography 284
– nuclear magnetic resonance 524
– organic trace analysis 97
– thin layer chromatography 335
– UV spectroscopy 442
benzidines 280
benzyl 103
Bertrand lens 1062
^{10}beryllium 602
Bessel functions 398
Bethe stopping power 1119
beverages, trace analysis 80
bialkali photocathode 651
bifurcated fibers 448
binders, thin layer chromatography 327 ff
biocatalysts 967
biochemical interfaces 1020
biochemical sensors 10, 951–1060
biochemistry 545
bioinformatics, DNA analysis 1143
biological materials
– activation analysis 782
– scanning tunneling microscopy 915
biological methods 1
biological receptors 1039
bioluminescence
– DNA analysis 1137
– enzymes 151
biomedical samples, mass spectrometry 615
biomedical sciences, radionuclides 131
biopolymers 617
biosensors 1032 ff
– trace analysis 118
biosynthesis 131
Biot's law 429
biotic matrices 14
biotin 164
bisacrylpoperazine (BAP) 350 f
Blackman–Harris apodization function 468
blank values
– chemical sensors 956
– trace analysis 112 f
blending 72
Bloch waves 1121
blood samples 14
– activator analysis 156
– chemical sensors 1029
– storage 80
– wet digestion 85
blotting 362
Bohr–Einstein equation 421
Bohr theory 629
boiling point, mobile phases 284
Boltzmann constant 384, 964
Boltzmann law
– atomic spectroscopy 631, 639
– nuclear magnetic resonance 512
– UV spectroscopy 424
bonds, IR/Raman spectroscopy 478
Bormann effect 1081
Born–Oppenheimer approximation 427
boron
– Auger spectrum 877
– nuclear magnetic resonance 515
Bouman notation 635

bovine serum albumin (BSA)   159 ff
Boyle–Mariotte law   182
Bragg–Brentano diffractometer   390
Bragg diffraction   858
Bragg law   377 f, 383 f, 405, 412, 1079, 1086
– atomic spectroscopy   647
– electron microscopy   1113
– X-ray fluorescence spectrometry   754
Bragg reflection   412
Bravais–Miller indices   1026
Bravais lattice   379 ff, 398 f, 406 f
bremsstrahlung
– X-ray fluorescence spectrometry   754
– X-ray photoelectron spectroscopy   858
bremsstrahlung isochromat spectroscopy (BIS)   928
bright field image   1085 f, 1091
brittle fracture   852
broadbands
– infrared spectroscopy   467
– spin decoupling   531
– UV spectroscopy   430
broadening, atomic spectral lines   633
bromothyl blue   999
Büchi high-pressure adapter   90
Buckwheat anthrazite   830
Buerger camera   388
buffer layers, Auger electron spectroscopy   888
buffer solutions, voltammetry   812
buffers
– DNA analysis   1136
– enzymes   153
– thin layer chromatography   336
bulk acoustic wave (BAW) transducers   1005, 1022
bulk analysis   13
bulk conductivities, chemical sensors   960 ff
bulk loss, X-ray photoelectron spectroscopy   863
bulk metal analysis, atomic emission spectroscopy   702
bulk samples, electron microscopy   1100
buoyancy   67
burners, atomic absorption spectroscopy   676
butanol   284
butyl groups
– extraction   100
– liquid chromatography   288
– thin layer chromatography   335

calcium sulfate   329
calibration   7, 10, 13
– atomic emission spectroscopy   688
– atomic spectroscopy   642, 658
– balances   65
– chemical sensors   956
– chemometrics   45 ff
– chromatography   193
– instrument quality   32
– liquid chromatography   298
– mass spectrometry   610
– thin layer chromatography   342
– trace analysis   112 ff
– UV spectroscopy   443
– voltammetry   809, 977
calorimetric bomb   846
calorimetric devices   959, 1026
calorimetry   827, 836 ff
cameras   384

camphor   538 ff
candoluminescence   6
cantilever deflection, scanning tunneling microscopy   913
capacitively coupled microwave plasmas   699
capacity ratio, liquid chromatography   264
capillary electrophoresis (CE)   112
capillary gas chromatography (CGC)   6, 201 ff, 210, 252
capillary inlets
– gas chromatography   217
– immunoprobes   1045
capillary optics X-ray fluorescence spectrometry   764
capillary supercritical fluid chromatography (SFC)   252, 313 f
capillary zone electrophoresis (CZE)   345, 350, 363 ff
capillary zone electrophoresis interface, mass spectrometry   590
carbamates
– extraction   99
– postcolumn derivatization   305
carbon, Auger spectrum   880
carbon content, wet digestion   85
carbon dioxide   1000
carbon disulfides   102
carbon groups, thin layer chromatography   335
carbon nanotubes   1108
carbon paste electrode (CPE)   786, 806
[13]carbon nuclear magnetic resonance   6, 515, 520 f, 525, 541, 546
[14]carbon, mass spectrometry   602
carbonate salts   486
carbonyl compounds
– IR/Raman spectroscopy   483
– nuclear magnetic resonance   526
carboxyl groups
– affinity chromatography   318
– antibodies   159
– organic trace analysis   103
carboxylate salts   483
carboxylic acids
– immobilized enzymes   156
– IR/Raman spectroscopy   483
β-carotin isomers   290
carrier gas, chromatography   201
carrier selection, affinity chromatography   317
CARS   6
catabolites, mass spectrometry   613
catalist poison, semiconductor sensors   965
catalysis
– Mössbauer spectroscopy   576
– surface analysis   852, 867
catalytic currents   788
catalytic processes, calorimetric devices   959
catechol   802
catecholamines   280
cathode ray tube (CRT)   1117
cathodes, scanning electron microscopy   1116
cathodic sputtering   670
cathodic stripping voltammetry (CSV)   799, 818
cathodoluminescence   1122
cation adsorption, trace analysis   81
cation exchangers   100
cations
– ion exchange chromatography   293
– ion-selective electrodes   977
– mobile trace analysis   118 f
cell constants, diffraction   405

cellulose
- affinity chromatography   317
- polarimetry   457
- thin layer chromatography   328
cellulose acetate   346
cement concrete   832
centrosymmetric structures, crystal systems   1015
ceramic humidity sensors   992
ceramic samples, electron microscopy   1100
certified reference materials (CRMs)   34
^{252}Cf plasma desorption   595
chamber saturation, thin layer chromatography   337
channeling, scanning electron microscopy   1121
channeling model, ideal microscopes   1094
channeltron multiplier, mass spectrometry   608
characteristic radiation, X-ray fluorescence spectrometry   755
characteristic temperature, semiconductor sensors   989
charge coupled devices (CCD)
- atomic spectroscopy   651
- DNA analysis   1137
- microscopy   1070
- Raman spectroscopy   468
- UV spectroscopy   437
charge injection devices (CID)   651
charge transfer
- ion-selective electrodes   968
- voltammetry   786
charge transfer devices (CTD)   651
charged particle activation analysis (CPAA)   6, 780
charged specimen, scanning electron microscopy   1120
charring, thin layer chromatography   340
charts, quality assurance   33 f
chelate HPLC   18
chemical information, X-ray photoelectron spectroscopy   855
chemical ionization, mass spectrometry   589 ff, 617
chemical oxygen demand (COD) mobile trace analysis   119
chemical properties, chromatography   190
chemical reaction detector (CRD), liquid chromatography   302
chemical sensors   10, 951–1060
chemical shifts
- nuclear magnetic resonance   518, 523 ff
- X-ray photoelectron spectroscopy   862
chemical structure, nuclear magnetic resonance   541
chemical vapor deposition (CVD)   1022
chemically bonded phases, liquid chromatography   286, 289
chemically bonded silica, chromatography   176
chemically induced electron polarization (CIDEP)   551
chemically modified electrode (CME)   786, 807
chemiluminescence (CL)   151
chemisorption   1021
chemometrics   10, 19, 37–62
chevron multiplier   608
Chirasil-Val   212
chlorides   812
chlorinated hydrocarbons (CHC)   110
chlorine   743
chloroform   335
chlorophenols
- diode laser spectrometry   742
- extraction   99

chlorophyll   748
cholestane derivatives   556
cholesterol   953
cholines   155
chromatic distortions, light microscopy   1063
chromatograms, instrument quality   29
chromatographic media   156
chromatography
- basic principles   3 f, 173–198
- HPLC   18
- IR/Raman   497
- trace analysis   112
- UV spectroscopy   446
chromium   816
chromophores
- optical sensors   997
- UV spectroscopy   427
chronoamperometry   982
chronopotentiometry   798
circle diffractometers   390 f
circular dichroism (CD)
- IR/Raman spectroscopy   498
- UV spectroscopy   427 ff
citric acids   103
Clark oxygen sensor   983
classification
- biosensors   1032
- chemometrics   56 ff
- electrophoresis   346
- laser spectroscopy   728
- lenses   1063
- sample preparation   82
- trace analysis   110
cleaning
- LC samples   301
- UV optical cell   442
clean-room conditions, trace analysis   79
closed vessel combustion   88
cluster analysis
- chemometrics   56 f
- multivariate methods   53
coal samples, thermogravimetry   830
coating, acoustic mass sensors   1022 ff
cobalt–molybdenum–alumina system   868
cofactors, enzymes   148, 154 f
coherence, laser spectroscopy   728
coherent scattering
- atomic absorption spectroscopy   686
- X-ray fluorescence spectrometry   754
coiled open tubes, liquid chromatography   303
cold plasma ashing   82, 89 ff
cold vapor techniques, atomic absorption spectroscopy   681
collector, light microscopy   1062
collimators, atomic spectroscopy   645
color measurements   448
colorigenic/-less substrates   151
colorimetric acid-base indicators   999
colorimetric test strips   118
colorimetry   421, 1137
column approximation, diffraction   1079
column overloading, chromatography   196
column packing
- ion exchange chromatography   294
- liquid chromatography   285 ff, 292
- size exclusion chromatography   296

column switching, liquid chromatography 305
columns
– chromatography 175, 182
– gas chromatography 201 ff
– liquid chromatography 268, 288
coma, optical lenses 1063
combustion, trace analysis 87
combustion calorimeters 840, 846
competing substances, radionuclides 142
completeness, trace analysis 92
complex compounds, metabolic sensors 1037
complexation, voltammetry 812
complexing agents
– potentiometry 975
– stripping voltammetry 802 f
compounds
– diffraction 374
– organic 15
– trace analysis 111
– voltammetry 819
compressibility, gas phase 182
Compton ratio 776
Compton scattering 754
computer aided microscopy 1075
computer aided spectral analysis 489
computer-based analytical chemisty (COBAC) 8
computer connected UV spectroscopy 442
computer simulation, ESR spectra 554
computer system identification 27 ff
computer systems, mass spectrometry 610
computers 7 f
computing functions, weighing 66
concentration dependent distribution method (CDD), radionuclides 142
concentration determination, UV spectroscopy 443, 451
concentration sensitive detectors
– gas chromatography 231
– liquid chromatography 270
concentration sensitivity, trace analysis 111
concentrations 17
– chemometrics 38
– low 763
– trace analysis 80, 93, 102
concentric hemispherical analyzer (CHA) 859, 876
concentric nebulizers 660
conductance devices, chemical sensors 985
conducting polymer-based gas sensor 992
conducting samples, electron microscopy 1100
conductivity, chemical sensors 960 ff
conductometric detectors, liquid chromatography 274
conductometric modes, molecular recognition 969
conductometry 6
cone pattern, diffraction 388
confidence intervals 20
– chemometrics 41, 45 f
– sampling 73
confocal laser scanning microscopy (CLSM) 1070 ff
conformity declaration, EN 45014 24
conjugates, organic trace analysis 97
connecting tubes
– chromatography 189
– liquid chromatography 266
constant current/height mode, scanning tunneling microscopy 911
constituent elements 1, 15
consumer's risk 74

contamination
– analytical signals 50
– instrument quality 30
– organic trace analysis 97, 117
– radionuclides 135
– sampling 79
content ranges, trace analysis 110
continuous dynode electron multiplier 657
continuous flow fast atom bombardment 588
continuous flow isoelectric focusing 366
continuous radiation sources, UV spectroscopy 433
continuous sample inlet, liquid chromatography 263
continuous waves
– electron spin resonance 548
– laser spectroscopy 729
– nuclear magnetic resonance 513, 516 f
contrast
– light microscopy 1065
– scanning electron microscopy 1121
contrast images 1086
control, mass spectrometry 610
control charts, quality assurance 33 f
control samples 33
controlled rate thermal analysis (CRTA) 828
convective diffusion 982
conventional Auger electron spectroscopy 875
conventional transmission electron microscopy (CTEM) 1077 f
convergent beam diffraction 1088
cool on-column injection, gas chromatography 224
cool plasma conditions, mass spectrometry 707
Cooley–Tukey algorithm 467
cooling rates, thermogravimetry 828
core electron energy loss spectroscopy (CEELS) 878, 921 ff
core matrix 60
Corona equation, atomic spectroscopy 635
correlated spectroscopy (COSY) see: homonuclear chemical shift
correlation, multivariate methods 52
correlation via long-range couplings (COLOC) 539
corrosion 852, 867
corrosion products 575
cosmogenic radionuclides 602
Cot curves 1132
Cotrell equation 789, 981
Cotton effect 429
coulometric detectors 282
coulometry 6 f
coumarines 730
counter electrode 786, 808
counter tube method 389
coupled inharmonic oscillator, IR/Raman spectroscopy 474
coupled spectrometry 191 ff
couplings
– mass spectrometry 587
– nuclear magnetic resonance 519 ff
– optochemical sensors 1001
coupling chromatographic methods, trace analysis 112
coupling techniques, liquid chromatography 305
covalent linkage, immobilized enzymes 156
coverage factor 15
Craig model 178
creatinine 953

critical signal value, trace analysis 114
cross-flow nebulizers 660
cross section, ion scattering spectroscopy 902
crossed immunoelectrophoresis (CIE) 361
crosslinkers, electrophoresis 352
cryomagnets 515
crystal growth 381
crystal lattices see: lattices
crystal structures 1100 ff
crystal systems, acoustic mass sensors 1015
crystallite size, diffraction 410
crystallization, enzymes 153
CSEARCH 541
cubic spline algorithm 453
cumulated double bonds 478
cupferon 802
currents, voltammetry 788 ff
curvature, light microscopy 1063
cutting, DNA analysis 1132
cyano groups 288
cyanogen bromide (CMBr) 156
cyanopropyl 210 f
cyclic activation analysis 770
cyclic voltammetry 795, 980
cyclodextrin 328
cyclohexane 730
cyclohexyl
– extraction 100
– thin layer chromatography 331
cyclopentane 335
cylindrical mirror analyzer (CMA) 859, 876
cytochrome c 600
cytosine 1131
Czerny–Turner mounting 648, 674, 690

Daly detector 607
damage, scanning electron microscopy 1120
Darcy law 182
dark field illumination 1065
dark field image
– electron microscopes 1086
– scanning transmission microscopy 1095
data acquisition process 10
data banks
– crystallographic 409
– electrophoresis 358
– mass spectrometry 611
– nuclear magnetic resonance 541
data processing
– atomic spectroscopy 658 f
– chemometrics 50
– diffraction 392
– liquid chromatography 300
– mass spectrometry 610
– multivariate methods 51 ff
– UV spectroscopy 444
data system verification report 31
daughter ions, mass spectrometry 603
Dauphiné twin lattice 1089
de Broglie wavelength 1117
deactivation, UV spectroscopy 426
dead time, activation analysis 774
Debye–Hückel theory 346
Debye–Scherrer diffraction 385 f, 390, 413
Debye–Waller factor 573, 749

Debye path length 636
decay products, radionuclides 134
decays, Mössbauer spectroscopy 567
decision limit, trace analysis 113
decomposition
– thermogravimetry 828, 831
– trace analysis 90
deconvolving methods, digital microscopy 1076
deep-freezing 80
defective instruments 34
defects
– diffraction 410 ff
– electron microscopy 1087, 1111
definitions
– analytical procedures 13 ff
– biosensors 1032
– mass spectrometry 580
– mobile trace analysis 118 f
– radionuclides 127 f
– thermal analysis 827
defocus 1092
deformations, electron microscopy 1114
degrees of freedom, chemometrics 42, 55
dehydration
– electron microscopy 1112
– thermogravimetry 828
delay line, mass sensors 1010
delayed fluorescence 426
denaturating gradient gel electrophoresis (DGGE) 1139
density functions, chemometrics 38, 58
deoxyribose sugar 1131
depolarization spectroscopy 456
DEPT, nuclear magnetic resonance 537
depth profiling
– atomic spectroscopy 702, 712
– Auger electron spectroscopy 879 f
– ion scattering spectroscopy 901 ff
– mass spectrometry 619
– X-ray photoelectron spectroscopy 865 ff
derivative isotope dilution analysis 137
derivative spectroscopy 452
derivatization
– LC samples 301
– organic trace analysis 102
design qualification (DQ) 25
desorption 6
– mass spectrometry 594, 616 f
– surface analysis 934 ff
– thermogravimetry 828, 831
– trace analysis 81, 102
destructive visualization, thin layer chromatography 340
detail scans, X-ray photoelectron spectroscopy 863
detection limits
– activation analysis 778
– atomic absorption spectroscopy 677
– atomic fluorescence spectrometry 715
– atomic spectroscopy 719
– chemical sensors 954 f, 1053
– chemometrics 49
– immunosensors 1042
– laser spectroscopy 734, 740 ff
– mass spectrometry 621
– plasma mass spectrometry 706
– radionuclides 132
detection power 13

detectors
- atomic spectroscopy 649
- chromatography 189 ff
- gas chromatography 231 ff
- infrared spectroscopy 467
- liquid chromatography 269, 300
- mass spectrometry 607 ff
- scanning electron microscopy 1118
- supercritical fluid chromatography 312
- UV spectroscopy 436
- X-ray fluorescence spectrometry 758
deterministic methods 10
deuterated triglycerine sulfate (TDGS) 467
deuterium lamp technique
- atomic absorption spectroscopy 683
- UV spectroscopy 433
dextran 317
diallyltartardiamide (DATD) 349 ff
dialysis 101
dichloromethane 99
dichromates 144
dideoxynucleotide labeling 1139
Diels–Alder rearrangements 583
diethylaminoethyl-cellulose 157
differantial pulse voltammetry 980
differential interference contrast (DIC) 1066
differential pulse polarography (DPP) 793
differential scanning calorimetry (DSC) 828 ff, 841
differential temperature calorimeters 841 f
differential thermal analysis (DTA) 828 ff
differentiation, signal sharpening 52
diffracometers 384
diffraction
- electron microscopy 1078
- structure analysis 373–418
- X-ray fluorescence spectrometry 754
- X-ray photoelectron spectroscopy 858
diffraction contrast, electron microscopy 1086, 1100
diffraction gratings, atomic spectroscopy 647
diffraction methods, surface analysis 938 f
diffraction pattern
- electron microscopes 1085
- light microscopy 1063
diffuse reflectance infrared Fourier transform spectroscopy (DRIFT) 494
diffuse reflection 438, 448
diffusion
- amperometry 982
- scanning electron microscopy 1119
diffusion barriers, Auger electron spectroscopy 885
digestion 80 ff
digital microscopy, computer aided 1075
dihydroxyethylenebisacrylamide (DHEBA) 349 ff
dilute analysis, isotopes 709
dilution, liquid chromatography 264
dimethylacryamide (DMA) 351
diode array detector (DAD) 270 ff
diode arrays
- atomic spectroscopy 652 f
- UV spectroscopy 432
diode laser atomic absorption spectrometry (DLAAS) 742
diode lasers 741
diol groups
- extraction 100
- liquid chromatography 288
- thin layer chromatography 331

dioxanes
- laser spectroscopy 730
- thin layer chromatography 336
dioxins 110
diphenylpicylhydrazyl (DPPH) 552
dipole–dipole relaxation 522
dipole interactions 570
dipoles, UV spectroscopy 422
Dirac equation 1119
direct analysis of daughter ions (DADI) 603
direct chemical ionization 617
direct coupling, mass spectrometry 587
direct current polarography (DPC) 788
direct injection, gas chromatography 222
direct isotope dilution analysis 136
direct liquid introduction, mass spectrometry 588
direct methods, diffraction 375 f, 396 f
direct probes, mass spectrometry 585
direct retrieval, ideal microscopes 1095
direct solid sampling, atomic spectroscopy 667, 682, 710
direct trace analysis 94 ff
disappearance potential spectroscopy (DAPS) 927
discharges, atomic spectroscopy 629, 670
  see also: glow discharge
discontinuous electrophoresis 350 ff
discontinuous sample inlet 263
discriminant matrices 52
dislocations, electron microscopy 1086, 1111
dispersion
- surface plasmon resonance 1002
- UV spectroscopy 422
dispersive methods 12
dispersive Raman techniques 468
dispersive spectrometers 646
dispersive systems, X-ray fluorescence spectrometry 759 f
displacement chromatography 177, 195
- liquid 263
displacement substoichiometry 139
dissociation
- atomic spectroscopy 637
- enzymes 149
dissolution, trace analysis 80
distance matrices 52
distillation 81, 101
distortions, light microscopy 1063
distributed Bragg reflector (DBR) 731
distribution coefficients, organic trace analysis 98
distributions 6, 19
disulfides 488
DMD test 289
DNA analysis 20, 1131–1150
documentation, instrumentation 26 f
domain contrast, electron microscopes 1088
donors
- acoustic mass sensors 1020
- semiconductor sensors 962
Doppler broadening 634, 639
Doppler effect 564 f
double beam equipment, UV spectroscopy 431
double diffraction 1067
double focusing, magnetic/electrostatic sectors 656
double focusing mass spectrometry 598
double isotope dilution analysis 137
double photon excitation, confocal laser scanning 1074
double resonance, nuclear magnetic 531 f
drift correction, atomic spectroscopy 658

drifts, chemical sensors   956, 1053
drinking water, trace analysis   80
drop calorimeters   840
dropping mercury electrode (DME)
– radionuclides   144
– voltammetry   786, 805
Drosophilia   1132
Druyvenstein distribution   631
dry ashing
– trace analysis   82, 87
– voltammetry   810
dry digestion, trace analysis   87
dry fill packing, liquid chromatography   287
dry solutions
– atomic emission spectroscopy   692
– plasma mass spectrometry   712
drying
– atomic spectroscopy   666
– samples   80, 96
– thin layer chromatography   334
dual channel instruments, atomic absorption spectroscopy   675
dual electrode cell, liquid chromatography   282
dual wavelength spectroscopy   454
dump sites, mass spectrometry   614
duoplasmatron   920
duplex stainles steel   935
dwell time, gas chromatography   219
dye lasers   730
dye phenol red   999
dyes, fluorescence microscopy   1068
dynamic crystal diffraction   1081
dynamic headspace techniques   101
dynamic mechanical analysis (DMA)   828, 835
dynamic mutations, DNA analysis   1146
dynamic range, mass spectrometry   611
dynamic SIMS   6
dynamic split mode, supercritical fluid chromatography   312
dynode multipliers
– atomic electron spectroscopy   657
– mass spectrometry   609

Ebert mounting   648, 674, 690
eccentric load test   66
echelle gratings   648, 691
ecological behavior   5
efficiency, activation analysis   770, 774
eigenvalue problems   53
Einstein equation   990
Einstein transitions   632
elastic scattering
– electron–specimen interactions   1119
– scanning electron microscopy   1115
elastic stress   411
elctrochemical evaporation (ETE)   698, 710
electron stimulated desorption (ESD)   934
electron stimulated desorption ion angular distribution (ESDIAD)   934
electric field gradient (EFG)   571
electric hyperfine interaction, Mössbauer spectroscopy   561
electroactive compounds, ion-selective electrodes   974
electrochemical methods   18
– enzymes   152

– liquid chromatography   274
– trace analysis   89
electrochemical reactions, working electrodes   968
electrode probe formation   1117
electrodeless discharge lamps (EDL), laser spectroscopy   729
electrodes, voltammetry   785 ff
electrogravimetry   6
electroluminescence   1137
electrolyte humidity sensors   991
electrolytes, voltammetry   785
electrolytic conductivity detector (ELCD), gas chromatography   236
electrolytic deposition, inorganic trace materials   93
electromagnetic sector fields   597
electromagnetic spectra, vibrational   471
electrometry   828, 836
electron–hole recombination   730
electron–specimen interactions   1119
electron beam induced current (EBIG)   1118, 1123
electron beams, diffraction   382 f
electron bombardment   857
electron capture detector (ECD)
– chromatography   191
– gas chromatography   228, 232 f
– mass spectrometry   591
– organic trace analysis   102
– supercritical fluid chromatography   312
electron diffraction   6, 413
electron energy analyzers 859, 876
electron energy-loss spectrometry (EELS)   6, 16, 921, 942, 1098
electron excitation, surface analysis   853
electron guns   1116
electron impact, ion generation   590 f
electron impact ionization (EII)   930
electron impact mass spectrometer (EIMS)   582
electron ionization, gas chromatography   237
electron microprobe analysis (EMPA)   16
electron microscopy   1077 ff
electron multipliers   657
electron nuclear double resonance (ENDOR)   548
electron sources, Auger electron spectroscopy   875
electron spin echo envelope modulation (ESEEM)   551
electron spin exchange   554
electron spin resonance (ESR)   6, 12, 509, 512, 548 ff
electron temperature   639
electron tunneling spectroscopy   6
electronegativity   523
electronic data processing   7 f
electronic energy, UV spectroscopy   421
electronic noses/tongues   1030
electronic pressure control (EPC)   214, 255
electronic states, UV spectroscopy   421
electronic systems, UV spectroscopy   442
electrophoresis   6, 345–372
– DNA analysis   1139
electrospray interface, mass spectrometry   6, 587 ff, 596 f
electrothermal atomic absorption spectroscopy   678
electrothermal evaporation, atomic spectroscopy   664
element species, trace analysis   95
elemental analysis   15
– activation analysis   773 ff
– diode lasers   741
– factoring   60
– mass spectrometry   618

– scanning electron microscopy   1124
– X-ray photoelectron spectroscopy   854
elemental analysis voltammetry   814
elemental ion compositions, mass spectrometry   583
elementary analysis   15
ellipsometry   918 f
– reflectometry   449
– UV spectroscopy   439
elution
– affinity chromatography   317, 322
– chromatography   177
– gas chromatography   243
– liquid chromatography   263
– thin layer chromatography   332
elutropic order   283
emission, atomic spectroscopy   631
emission spectra, liquid chromatography   273
emission spectroscopy, IR/Raman   495
emulsions, atomic spectroscopy   649
EN 45001, quality assurance   23
EN 45014, conformity declaration   24
enantiomeric structures   1015
enantiomeric purity   544
enantioselective separation, lactones   215
encapsulation
– antibodies   166
– ion-selective field effect transistors   996
end-cut techniques, column switching   306
endonuclease   1133 f, 1147
energy dispersive spectrometry (EDS)   1097
energy dispersive X-ray fluorescence analysis (EDXRF)   122, 760
energy level diagram, UV spectroscopy   425
Engelhardt–Jungheim test   289
enkephalin   600
enrichment, LC samples   301
enthalpy   1027
entrance collimators   645
environmental chemistry   613
environmental materials, activation analysis   781
environmental pollutants   165
environmental research   3
environmental technology verification program (ETV)   118
environmental trace analysis   117 f
envolved gas analysis (EGA)   828, 833 f
enzme catalysis enthalpies   1027
enzymatic digestion   85
enzyme catalyzed reactions   969
enzyme commission numbers   154
enzyme immunoassay (EIA)   1041
enzyme linked immunosorbent assay (ELISA)   160 ff, 1041
enzyme multiplied immunoassay technique (EMIT)   161
enzyme sequences, metabolic sensors   1038
enzyme thermistors   1027
enzymes   147–172
– DNA analysis   1133
– semiconductor sensors   967
epi-illumination reflectors   1062, 1067
epithermal neutron activation analysis (ENAA)   769 f
epoxy resins   871
equiabsorption method, UV spectroscopy   455
equilibrum–dispersive model, chromatography   179, 195

equipment
– activation analysis   774
– electron energy loss spectrometry   1099
– gas chromatography   201
– installation/operation   25 ff
– liquid chromatography   266
– supercritical fluid chromatography   311
– surface analysis   940
– trace analysis   81, 97, 119
– UV spectroscopy   430, 438
errors
– activation analysis   772
– additive   47
– atomic spectroscopy   643, 658
– chemometrics   40, 43, 47 ff
– chromatography   194
– diffraction   399
– liquid chromatography   299
– radionuclides   133
– sampling   73, 78 f
– systematic   17
– thin layer chromatography   342
– trace analysis   111 ff
– UV spectroscopy   431
– weighing   66
– X-ray fluorescence spectrometry   760 f
ESCACOPE, X-ray photoelectron spectroscopy   861
esterification
– organic acids   5
– organic trace analysis   103
esters
– IR/Raman spectroscopy   485
– nuclear magnetic resonance   526
estimates
– chemometrics   40
– sampling   73
estrogens   280
ethanol
– laser spectroscopy   730
– liquid chromatography   284
ethanol determination, enzymes   154
etherification, organic trace analysis   103
ethers   479
ethyl groups   335
ethylbenzene   1031
ethylene glycol   730
ethylenediacrylate (EDA)   349 ff
ethylenediaminetetraacetic acid (EDTA)
– blood samples   80
– voltammetry   813, 820
Euclidean distances   57 f
Euclidean space   52 ff
Euler–Moivre theorem   394
Euler cradle   392
EURACHEM   14
evanescent field   1041
evanescent waves   997
evaporation
– atomic spectroscopy   664 ff, 698
– thin layer chromatography   332
evaporative light scattering detector (ELSD)   270, 273
Everhart–Thornley detector   1118
Ewald sphere   377, 1079
excess noise, atomic spectroscopy   643
exchange current, molecular recognition   970
exchange processes, electron spin resonance   554

excitation, surface analysis   853
excitation error, diffraction   1079
excitation spectra
– fluorimetry   447
– liquid chromatography   273
excitation spectroscopy   456
excitation temperature, atomic spectroscopy   639
exonucleases   1133
experimental data sets, multivariate methods   51
expert systems   8, 541
external cavity diode laser (ECDL)   731
external quality assurance, trace analysis   117
external reflection spectroscopy   491
extinction distances   1081, 1087
extraction–digestion techniques, trace analysis   85
extraction
– TLC samples   330
– trace analysis   93, 98
– voltammetry   814
extreme trace analysis   110
eyepiece, light microscopy   1062

Fabry–Perot diodes   731
Fabry–Perot interferometer   1001
factorial methods, chemometrics   53 ff, 60
Faraday cage   607
Faraday cup detector   656
Faraday current   786 ff
Faraday effect   686
Faraday law   7
Faraday modulator   438
fast atom bombardment mass spectroscopy (FABMS)   6, 595, 617, 932
fast Fourier transform (FFT) infrared spectroscopy   467
fast high-resolution capillary gas chromatography   254
fast neutron activation analysis (FNNA)   771
fast scanning detector (FSD), liquid chromatography   270 ff
Fastie–Ebert mounting   648
fatty acid ethyl esters   316
fatty acid methyl esters   207, 227 f
fatty acid phenyl esters   302
faulted crystals   1082
faults, electron microscopy   1112
^{57}Fe Mössbauer spectroscopy   561 f
Fermi contact field   571
Fermi level   963
ferrocene   1037
fiber optical sensors   118, 997
fiber optics   646
fiber scanning optical microscopy   1074
Fick's law   786 ff
field analytic technology encyclopedia (FATE)   118
field desorption mass spectrometry (FDMS)   595, 615
field emission guns (FEG)   1117
field ionization   592, 617
field ionization laser spectroscopy (FILS)   739
Figaro sensor   960
   see also: Taguchi sensor
figures of merit, atomic spectroscopy   642, 677, 680
film methods, diffraction   385
film thickness, UV spectroscopy   448, 451
filters
– atomic spectroscopy   652
– digital microscopy   1076

– fluorimetry   447
– Fourier transform   50
– liquid chromatography   266
fine structure
– electron energy loss spectrometry   1099
– UV spectroscopy   427
fingerprint bands   475
firefly luciferase   156
fission neutrons   769
fitting procedures, activation analysis   776
flame ionization detector (FID)   10
– gas chromatography   232
– mobile trace analysis   119
– supercritical fluid chromatography   309 ff
flame photometric detector (FPD)
– chromatography   191
– gas chromatography   232, 235
– organic trace analysis   103
flame techniques, trace analysis   90
flames
– atomic absorption spectroscopy   676
– atomic emission spectroscopy   691
– atomic spectroscopy   628, 641
– laser enhanced ionization spectrometry   717
– laser spectroscopy   733
flavin adenine dinucleotide   154
flavones   280
flexures, weighing   64
Flicker noise   643
Florisil
– liquid chromatography   291
– organic trace analysis   102
flow, gas chromatography   214
flow calorimeters   840
flow cell contamination   30
flow injection analysis (FIA)
– chemical sensors   1029
– potentiometry   974
– voltammetry   808
flow rate, chromatography   182
fluorenyl methoxycarbonyl chloride (FMOC)   302
fluorescein–isothiocyanate (FITC)   1067
fluorescence
– Auger electron spectroscopy   874
– confocal laser scanning   1074
– DNA analysis   1137
– Mössbauer spectroscopy   561 f
– UV spectroscopy   426, 438, 443
fluorescence analysis, solid solutions   748
fluorescence-based hybridization (FISH)   1068
fluorescence correlation spectroscopy (FCS)   1068
fluorescence lifetime imaging (LIM)   1068
fluorescence line narrowing (FLN)   749
fluorescence microscopy   1067
fluorescence polarization immunoassay (FPIA), homogeneous   162
fluorescence recovery after photobleaching (FRAP)   1068
fluorescence spectrometry
– enzymes   151
– laser-excited   732 ff
– X-ray   753–766
fluorescence-phosphorescence detector   270, 273
fluorescine   998
fluoride ion-selective electrode   966
fluorimetry   446 f

fluorine nuclear magnetic resonance 515, 527
fluorite objectives 1063
fluoroalkanes 335
fluorochlorohydrocarbons (FCHC) 110
fluorometric acid-base indicators 999
fluorophores
– immunoassays 148, 161
– optical sensors 997
fluoropolymers 862
flux gradients, activation analysis 772
focusion, gas chromatography 221
fodder plants 60
Forgy's method, 56
formazone 802, 818
formic acids 103
four-electrode cell 987
Fourier model
– ideal microscopes 1091 ff
– thermal analysis 837
Fourier transform 10
– atomic spectroscopy 652
– chemometrics 50 f
– diffraction 373 ff
– nuclear magnetic resonance 513, 517 f
Fourier transform infrared (FTIR) techniques 466 f
– chromatography 192
– gas chromatography 232, 238
– liquid chromatography 270, 274
– trace analysis 118
Fourier transform mass spectroscopy (FTMS) 581, 599, 606
Fourier transform Raman techniques 468
Fourier transform spectroscopy 6, 432
fraction ranges, trace analysis 110
fragmentation, organic mass spectrometry 583
Franck–Condon principle
– desorption 936
– UV spectroscopy 427
free fatty acid phases, gas chromatography 212, 227 ff
free induction decay (FID), nuclear magnetic resonance 517
freedom degrees, chemometrics 42, 55
freeze drying
– enzymes 153
– immunoprobes 1045
– samples 80
freezing 80
frequencies
– mass sensors 1010
– microwave ashing 84
– nuclear magnetic resonance 513
frequency histograms, chemometrics 38
frequency modulation, laser spectroscopy 741
friction 852
Friedel law 379, 400
fringes 1082, 1088
Fritilaria davisii 1132
fritted disk nebulizer 661
front techniques, column switching 306
frontal analysis, chromatography 177
fructose 152
full width at half maximum (FWHM)
– activation analysis 774
– confocal laser scanning 1072
fullerites 1104
full-scan mode, gas chromatography 237

functional groups 11
– affinity chromatography 317
– immobilized enzymes 156 f
– IR/Raman spectroscopy 475
– liquid chromatography 288
functional organization, analytical chemistry 4 f, 9
fundamental parameter algorithms, mobile trace analysis 123
furans 110
furnaces
– atomic absorption spectroscopy 665, 680
– atomic spectroscopy 641
– emission spectrometry 701
fused rocket crossed immunoelectrophoresis 361
fused silica open tubular (FSOT) columns 204
fusion, trace analysis 89
fuzzy clustering 57

g-factor, electron spin resonance 551
γ emitters 132
γ radiation 562 ff
γ-spectroscopy 6
galactosidase 162
gallium arsenide 1107
gallium liquid metal ion source 890
gas–solid adsorption equilibra 185
gas–solid/liquid chromatography 175
gas chromatography 6, 199–260
– organic acids 5
– trace analysis 112
gas chromatography/mass spectrometry interfaces 6, 586
gas detectors 774
gas impurities, lead-salt diode lasers 744
gas–liquid chromatography (GLC) 201
gas mixtures, flame atomic absorption spectroscopy 676
gas phase, electron diffraction 413
gas segmentation, open tubes 304
gas selectivity 978
gas sensors 1000
gas–solid chromatography (GSC) 201
gas temperature, atomic spectroscopy 639
gas volumetry 6
gaseous compounds, conductometric sensors 988
gases viscosity 183
Gauss filter 1076
Gaussian band 55
Gaussian distribution
– chemometrics 39 f
– sampling 73
Gaussian planes 394
Gaussian profile, chromatography 180 f, 187
GC detectors 231 ff
Geiger–Müller counter 928
gel electrophoresis interface 590
gel filtration chromatography (GFC) 6, 295
gel permeation chromatography (GPC) 6, 295
gels, thin layer chromatography 328
geometry, diffraction 1079, 88
germanium 492
germanium detectors 775
Gibbs free energy, chromatography 190
Gidding model 179
Gjonnes–Moodie lines 1091
glass 156
glass chambers 337

glass membranes, ion-selective electrodes   971
glass supports, thin layer chromatography   327
glass transition   831
glass vessel   97
glassy carbon electrode (GCE)   786, 806
glassy carbon vessels   81
glassy multielemental solid   852
global position system (GPS), mobile trace analysis   119
glow discharge mass spectrometry (GDMS)
   6, 17, 592, 710, 928 f, 943
glow discharge optical spectroscopy (GDOS)   918 ff, 942
glow discharge source, trace analysis   113
glow discharges
– atomic emission spectroscopy   700 f
– atomic spectroscopy   641, 652
– laser spectroscopy   734
glucose   152 ff, 953, 1029
glucose oxidase (GOD)   1043
glucuronide   97
glutaraldehyde   157
glycerol   610
goethite
– electron microscopy   1112
– Mössbauer spectroscopy   575
Golay cells   10
Golay equation   180
gold
– ion scattering spectroscopy   903
– valence band spectra   865
^{197}gold, Mössbauer spectroscopy   567
gold electrodes   277
Golgi stained neuron   1074
good analytical practice (GAP)   92
good laboratory practice (GLP)   7, 23, 92
good manufacturing practice (GMP)   23
goodness of fit (GOF), trace analysis   116
governmental regulations, weighing   69
gradient elution, liquid chromatography   263, 297
grain boudaries   16
grain boudary segregation   883
Gram–Schmidt vector   497
granulated gel layers   365
graphene like layers   1108
graphite   914
graphite crystal, diffraction   383
graphite electrodes   806
graphite electrothermal atomizers (GETA)   733
graphite furnaces
– atomic spectroscopy   665, 716
– diode lasers   742
– trace analysis   95
graphitized carbon   176, 208
grated decoupling, nuclear magnetic resonance   532
grating couplers
– affinity biosensors   1042
– optochemical sensors   1001
– UV spectroscopy   450
grating monochromators   434
gratings, atomic spectroscopy   647
gravimetric sensors, acoustic   1013
gravimetry   6
gravity   67
grazing angle reflection–absorption spectroscopy   491
grazing incidence
– electron microscopy   1113
– ion bombardment   1100

grid multiplier   608
group frequencies   474 ff
groups, functional   11
Grubbs tests   43
guanine   1131
Guinier diffraction   386, 391
gyromagnetic ratio   511, 514

Hadamard spectroscopy   432 f, 439
Hadamard transform   652
haematite   575
half lifes
– Mössbauer spectroscopy   567
– radionuclides   131, 770
Hall electrolytic conductivity detector (ELCD)   236
haloform components, diode laser spectrometry   742
halogenated aromatic hydrocarbons   163
Hamiltonians, Mössbauer spectroscopy   568
handling, defective instruments   34
hanging mercury drop electrode (HMDE)   786
haphazard sampling   72
hapten   148, 158 ff
hardware tests   28
harmonic oscillator   422
headspace injection, gas chromatography   6, 228 ff
headspace techniques, trace analysis   101
health risks   3
heart techniques, column switching   306
heat capacities, specific   845
heat flux calorimeter   842
heating, atomic spectroscopy   666
heating rates, thermogravimetry   828
heavy atom method, diffraction   395
heavy metals
– ELISA tests   163
– enzyme sensors   1039
Heisenberg uncertainty
– Mössbauer spectroscopy   567
– UV spectroscopy   423
helium   201
helix structure, DNA analysis   1131
hematite   1112
Henry–Dalton law   101
Henry constant   185
heptane   284
Hertz dipole   422 f
Herzog–Mattauch geometry   598, 609
heteregenous samples   72
heterogeneous multiphase systems   12
heterogeneous semiconductor sensors   989
heterogeneous systems, nuclear magnetic resonance   546 ff
heteronuclear chemical shift correlation (HETCOR)   539
heteronuclear multiple quantum coherence (HMQC)   539
heteronuclear spin decoupling   532
hexacyanoferrate   152
hexamethylene ammonium-
   hexamethylenedithiocarbamidate (HMA-HMCD)   93
hexane
– gas chromatography   227
– liquid chromatography   284
– organic trace analysis   99
Heyrovsky polarography   788
hierachic clustering, chemometrics   56
hierarchial ordered spheres of environment (HOSE)   541
high-energy ion scattering (HEIS)   898

high-frequency titration 6
high-performance liquid chromatography (HPLC) 6
– GC 200, 251
– instrument quality 28
– MS 6
– trace analysis 112 f
high-pressure ashing (HPA) 86
high-pressure digestion 86 f
high-pressure nebulizers 662
high-pressure shut down, liquid chromatography 265
high-purity materials 781
high-purity plastics 81
high-resolution electron loss spectroscopy (HREELS) 919 ff, 942
high-resolution mode, electron microscopes 1085
high-resolution nuclear magnetic resonance 512, 546 f
high-resolution specimen, electron microscopy 1100
high-resolution transmission electron microscopy (HRTEM) 1077
high-temperature superconductors 1108
highest occupied molecular orbital (HOMO) 422
hollandite 1111
hollow cathode lamps (HCL) 729, 741
hollow cathodes, atomic emission spectroscopy 701, 716
holoenzymes 150
holographic filters 468
Homo sapiens, DNA analysis 1132
homogeneity, variances 44
homogeneous fluorescence polarization immunoassay (FPIA) 162
homogeneous liposome immunoassay 161
homogeneous semiconductor sensors 990
homogenization
– radionuclides 131
– samples 81, 96
homonuclear chemical shift correlation (COSY) 537
homonuclear spin decoupling 531 ff
Hooke's law 469
hormones 158
horse radish peroxidase inhibition 169
host–guest binding systems, acoustic mass sensors 1020
Hotteling principal components 53
hot-wire detector 233
Hughes equation 398
human genome mapping project (HGMP) 1136
human immunodeficieny 1024
humidity sensors 991, 966
Huyghens wavelength 1091
hybridiaztion, fluorescence microscopy 1068
hybridoma 160
hybrids, mass spectrometry 606
hydrazines 280
hydride generation 663, 698
hydride techniques 681
hydrocarbons
– extraction 99
– fluorescence analysis 748
– trace analysis 110, 118
– voltammetry 821
hydrochloric acids 81 ff
hydrodynamic diagram, polycyclic hydrocarbons 278
hydrofluoric acids 81 ff
hydrogen 144
– gas chromatography 201
– nuclear magnetic resonance 511, 515, 523, 547
– semiconductor sensors 963

hydrogen–air flame 676
hydrogen bonds, DNA analysis 1132
hydrogen fluoride 143
hydrogen lamps 433
hydrogen peroxide
– enzymes 151 ff
– trace analysis 84
hydrogen shifts, mass spectrometry 583
hydrogenase 1035
hydrogenic stretches 502
hydrolysis 97
hydrophilic gels 348
hydrophilic modification, precoated plates 329
hydroxyl groups
– affinity chromatography 318
– antibodies 159
– organic trace analysis 103
hyper/hypochromic absorption 427
hyperfine interaction, Mössbauer spectroscopy 561, 568 f
hyperfine splitting, electron spin resonance 552
hyperfine structures, atomic spectroscopy 634
hyperplanes, chemometrics 58
hyphenated techniques 18
– gas chromatography 236
– liquid chromatography 307
– mass spectrometry 586 ff, 613

ice calorimeters 840
ideal imperfect crystal 374
ideal microscope 1091
ideal model, nonlinear chromatography 194
illumination
– atomic spectroscopy 645
– light microscopy 1065 ff
image dissector tubes, atomic spectroscopy 651
imaging 12
– atomic spectroscopy 645 ff
– confocal laser scanning 1071 ff
– diffraction 373 ff
– digital microscopy 1076
– electron microscopes 1086
– electron spin resonance 551
– ideal microscopes 1091 ff
– light microscopy 1063
– nuclear magnetic resonance 547
– scanning electron microscopy 1121
– vibrational spectroscopy 501
immobilization 1022
immobilized enzymes 156 f
immobilized metal ion affinity chromatography (IMAC) 320
immobilized pH gradients 362
immunoaffinity techniques 165 ff, 321
immunoassays 118, 147–172
immunoelectrophoresis 360
immunogold staining (IGS) 1067
immunoprobes 1040 f
immunoradiometric assay (IRMA) 161
immunosensors 969
– fiber-optical 999
imperfect tracers, radionuclides 132
implementation, chromatography 177
impurities 17
– chemometrics 48
– diffraction 407

- laser spectroscopy   727
- nuclear magnetic resonance   516
- radionuclides   134
- thin layer chromatography   335

incidence angle, optical sensors   997
indamines   1037
indicators, chemical sensors   999
^{129}indium, Mössbauer spectroscopy   567
individual gauge for localized orbitals (IGLO)   541
indoles   280
inductively coupled cells, chemical sensors   987
inductively coupled plasma–atomic emission spectrometry (ICP-AES)   113
inductively coupled plasma atomic spectroscopy   634, 641, 695
inductively coupled plasma–gas chromatography (ICP-GC)   6
inductively coupled plasma–mass spectrometry (ICP-MS)   6, 89, 95, 586, 593 f, 621, 706
- blood   85
- trace analysis   113

inductively coupled plasma–optical emission spectrometry (ICP-OES)   6, 17, 89, 95
inelastic electron tunneling spectroscopy (IETS)   921, 925, 943
inelastic scattering, electron–specimen interactions   1119
inert gases   132
infinity color corrected system (ICS)   1062
influence correction method, X-ray fluorescence spectrometry   762
infrared microscopy   6
infrared reflection–absorption spectroscopy (IRRAS)   491
infrared sensors   119
infrared spectra, vibrational   471
infrared spectrometry   192
infrared spectroscopy   6, 10, 465–508
Ingamel equation   73
inherent fluorescence, thin layer chromatography   340
inhibition methods, enzymes   149, 155
inhibitors, enzyme sensors   1039
injection systems, chromatography   188
inlet related discrimination, gas chromatography   218
inlets, atomic spectroscopy   660
inorganic analysis
- digestion   80 ff
- bulk acoustic waves   1024

inorganic compounds, diffraction   374
inorganic traces   110, 814
instability, ideal microscopes   1093
instrumental analysis   9
instrumental data sets, multivariate methods   51
instrumental direct methods, trace analysis   111
instrumental line shape (ILS), infrared spectroscopy   468
instrumental neutron activation analysis (INAA)   777
instrumentation
- atomic absorption spectroscopy   674
- atomic emission spectroscopy   690, 696
- atomic fluorescence spectrometry   714
- atomic spectroscopy   642
- Auger electron spectroscopy   875
- biosensors   1051
- calorimetry   839
- chromatography   188
- confocal laser scanning   1072
- diffraction   393
- gas chromatography   201

- ion exchange chromatography   293
- ion scattering spectroscopy   900
- mass spectrometry   580 f, 604 ff, 610 ff
- Mössbauer spectroscopy   565
- plasma mass spectrometry   704, 711
- potentiometry   974
- quality assurance   7, 23–36
- radionuclides   134
- Rutherford backscattering spectroscopy   907
- scanning electron microscopy   1116, 1124
- scanning tunneling microscopy   912
- static secondary ion mass spectrometry   890
- thermogravimetry   828 ff
- thin layer chromatography   337
- UV spectroscopy   430
- voltammetry   803 ff
- X-ray fluorescence spectrometry   757 ff
- X-ray photoelectron spectroscopy   856
  see also: equipment

integrated enzyme systems, metabolic sensors   1038
integrated optical system (IOS)   1044
integrated opto/biochemical sensors   1000
intensity
- laser spectroscopy   728
- Mössbauer resonance lines   573

interdigitated transducers (IDT)   1009, 1018
interfaces
- Auger electron spectroscopy   885
- biochemical   1020
- diffraction   1082
- electron microscopes   1088
- LC-MS   308
- mass spectrometry   585 ff
- semiconductor sensors   967
- static secondary ion mass spectrometry   896
- X-ray photoelectron spectroscopy   873

interference noise, atomic spectroscopy   643
interference pattern, reflectometry   449
interferences
- activation analysis   774
- flame atomic absorption spectroscopy   678 ff
- immunoassays   168
- ionization   636
- plasma mass spectrometry   706
- working electrodes   968

interferometers
- affinity biosensors   1042
- infrared spectroscopy   467
- optochemical sensors   1001

interferones   323
intermediate gel-crossed immunoelectrophoresis   361
internal laboratory reference materials   117
internal reflection, optical sensors   997
internal reflection spectroscopy   492
internal standards, X-ray fluorescence spectrometry   761
inverse calibration, chemometrics   47
inverse detection, nuclear magnetic resonance   516
inverse method (Q-matrix), UV spectroscopy   445
inverse photoemission spectrometry (IPES)   928, 943
inverse voltammetry   799
inverted microscopy   1069
ion adsorption, trace analysis   81
ion Auger electron spectroscopy (IAES)   928
ion beam bombardment   1100
ion beam spectrochemical analysis (IBSCA)   918 ff
ion chromatography   6

ion compositions, mass spectrometry   583
ion conductivities, solid state materials   966
ion cyclotron resonance (ICR)   599
ion detection, atomic spectroscopy   656
ion exchange, thin layer chromatography   328
ion exchange chromatography (IEC)   177, 293
ion excitation
– surface analysis   853
– atomic spectroscopy   657
ion generation, mass spectrometry   590 ff
ion kinetic spectrometry, mass selected   603
ion mobility spectrometer (IMS)   120
ion neutralization spectroscopy (INS)   6, 928 f
ion scattering spectroscopy (ISS)   898, 942
ion-selective conductometric cells   987
ion-selective electrodes (ISE)   968 ff
ion-selective field effect transistors (ISFETs)   993 ff
ion sources, static secondary ion mass spectrometry   890
ion storage devices   606
ion traps   601, 606
ionization
– atomic spectroscopy   635 f
– laser-enhanced   735
– mass spectrometry   589 ff, 616 f
ionization spectrometry, laser-anhanced   716
ionization temperature, atomic spectroscopy   639
ions   11
ions analysis, voltammetry   814
iron–chromium alloys   871
^{57}iron, Mössbauer spectroscopy   561, 567
irradiation facilities, activation analysis   769
irreversible thermodynamics   838
ISO 9001
– instrumentation   24
– weighing   69
ISO guide   14
ISO/IEC 17025, quality assurance   23
isocratic mode, liquid chromatography   263
isoelectric focusing (IEF)   346, 351, 365
isomer shifts, Mössbauer spectroscopy   568
isomers, nuclear magnetic resonance   525
isoperibolic calorimeters   840
isotachophoresis (ITP)   346, 358 f
isothermal calorimeters   839
isotherms, chromatography   180, 184 f
isotope contents, nuclear magnetic resonance   544
isotope dilution analysis (IDA)   6
– plasma mass spectrometry   709
– radionuclides   128 f, 136 ff
isotope dilution mass spectroscopy (IDMS)   7, 585
isotope effects   131
isotope exchange method (IEM)   131, 142
isotope ratio analysis   708
isotope-selective detection   743
isotopes   11
– mass spectrometry   601
– Mössbauer spectroscopy   567
isotopic dilution radioimmunoassay   161
isotopic spectra, electron spin resonance   554
isotropic neutron sources   771
iterative calculation   445

j-j coupling, X-ray photoelectron spectroscopy   854
J-modulated spin-echo pulse sequences   535

J-resolved spectra, nuclear magnetic resonance   537
James–Martin compressibility   182
jet impact nebulizers   662

K-band   548
K-matrix method, UV spectroscopy   444
k-means, chemometrics   56
k_0-method, activation analysis   772
k-nearest neighbor (KNN) method   57
K-shell, Mössbauer spectroscopy   562
Kaiser–Ehrlich limits   115
Kaiser varimax criterion   55
Kalman filter   445
Karplus curves   525
katharometers   233
kernels
– chemometrics   58
– Fourier transform   50
– light microscopy   1063
keto groups   443
ketones
– IR/Raman spectroscopy   483
– nuclear magnetic resonance   526
– voltammetry   821
kieselguhr   328
Kikuchi lines   1080
kinematic crystal diffraction   1078
kinetic currents, voltammetry   788
kinetics
– chromatography   180
– enzymes   148 f
Kjeldahl analysis   91
KLL series, Auger electron spectroscopy   855, 875
Knapp digestion   90
knitted open tube, liquid chromatography   304
Köhler–Milstein method   160
Köhler illumination   1065
Kohlrausch function   359
Koroleff method   91
Kovats retention index system   243
Kramers–Kronig relation
– IR/Raman spectroscopy   491
– UV spectroscopy   430
kryptonates, radioactive   143
Kubelka–Munk equation
– DRIFT   494
– UV spectroscopy   439, 448
Kuderma–Danish concentrator   102

L-band   548
λ probe   953, 966 f
labeling
– antibodies   166
– DNA analysis   1138
– electron spin resonance   555
– immunoassays   158
– radionuclides   141
laboratory means, standard deviation   45
lack of fit (LOF), trace analysis   116
lactams   487
lactones
– enantioselective separation   215
– IR/Raman spectroscopy   485 f
Lamb wave, mass sensors   1003, 1011

Lambert–Beer law
- atomic spectroscopy 633, 674
- laser spectroscopy 741
- liquid chromatography 271, 300
- UV spectroscopy 424 ff, 441 ff, 451 ff
- vibrational spectroscopy 472
Langmuir–Blodgett films 1022
Langmuir isotherm 180 ff, 195
lanthanide shift reagents (LSR), nuclear magnetic resonance 544
laser ablation 16, 669, 750
laser analytical spectroscopy 727–752
laser atomic absorption spectrometry (LAAS) 16
laser desorption, mass spectrometry 596, 616 f
laser enhanced ionization (LEI) 16, 716 f, 735 ff
laser excited atomic fluorescence spectrometry (LEAF) 32 ff
laser inductively coupled plasma–mass spectrometry (LICP-MS) 6
laser induced breakdown spectroscopy (LIBS) 750
laser induced fluorescence (LIF) 16, 742
- liquid chromatography 273
laser induced MS 113
laser ionization
- mass spectrometry 596
- surface analysis 930
laser optoacoustic spectroscopy (LOAS) 745 ff
laser sources, atomic emission spectroscopy 703
laser spectroscopy 6, 14
laser vaporization 16
lasers
- atomic spectroscopy 641
- lead salt diode 744
- UV spectroscopy 434
latent variable regression 59
lateral chromatic aberration (LCA) 1062
latex particle agglutination immunoassay 161
lattice contributions, Mössbauer spectroscopy 572
lattice fringes 1082
lattices
- diffraction 374
- electron microscopy 1078
Laue diffraction 386 f, 412, 1090
Laurell rocket technique 360
layered crystals, electron microscopy 1100
layers
- acoustic mass sensors 1020
- chemical sensors 954 f
- electron microscopy 1107 f, 1114 f
lead salt diode lasers 744
learning object samples 57
least-sqares refinement 399
leave-one-out method, chemometrics 57
lectins 321
legally binding analytical results 20
lenses, optical 1063
leuco-dyes 151
library see: data bank
lifetime, chemical sensors 955, 958, 1053
lifetime measurements, UV spectroscopy 456
ligands, affinity chromatography 318 ff
ligases 1133 ff
light scattering methods 161
light sources, UV spectroscopy 433
light spectroscopic methods, surface analysis 918 f
light transmission, liquid chromatography 284

limit of decision, trace analysis 113
limit of detection (LOD) 13
- atomic spectroscopy 719
- chemical sensors 954 f, 1053
- immunosensors 1042
- laser spectroscopy 734 ff, 740 ff
- plasma mass spectrometry 706
- atomic absorption spectroscopy 677, 680
limitations, chemometrics 48
limiting diffusion current, voltammetry 787
limiting quality level (LQL) 74
limits of procedures, trace analysis 114
line broadening 633
line light sources, UV spectroscopy 433
line narrowing 749
line pairs, atomic spectroscopy 633
linear chromatography 177 ff
linear regression 46, 58
linear sweep voltammetry (LSV) 795, 979
linearity, LC detectors 269
Lineweaver–Burk double-reciprocal method 149
linewidths, X-ray photoelectron spectroscopy 857
linkage methods, immobilized enzymes 156
linked scans, mass spectrometry 603
liposome immunoassay, homogeneous 161
liquid–liquid extraction, trace analysis 93, 98
liquid–solid chromatography 175, 185
liquid bulk acoustic waves 1007
liquid chromatography 261–326
liquid chromatography/mass spectroscopy interfaces 5897
liquid column chromatography 263
liquid crystals 212
liquid–liquid distribution, substoichiometric separation 138
liquid membranes, ion-selective electrodes 971 ff
liquid metal ion source (LMIS) 890
liquid mobile phase, chromatography 174 ff
liquid phases, gas chromatography 209
liquid secondary ion mass spectrometry (LSIMS) 595
liquid states, electron spin resonance 553
liquids viscosity 183
LMM series, Auger electron spectroscopy 855, 875
local density of states (LDOS) 911, 914
local thermal equilibrum (LTE) 632
logbook, qualification assurance 27
logic measurements 38
lonography 6
Lorentz band 55
Lorentz broadening 634
Lorentz distribution 562, 567
Lorentz polarization 381
losses
- organic samples 96
- radionuclides 135
- trace analysis 79
- wet digestion 85
- X-ray photoelectron spectroscopy 863
lot tolerance percent defective (LTPD) level 74
love plate devices 1012
low-concentration analysis, X-ray fluorescence spectrometry 763
low-energy electron diffraction (LEED) 938, 943
low-energy ion scattering (LEIS) 898
low light level (LLL) 1070

low-pressure discharges   670
low-resolution NMR   547
low-voltage scanning electron microscopy (LVSEM)   1115 ff
lowest unoccupied molecular orbital (LUMO)   422
luciferin   156
luminescence   134, 836
luminescence spectroscopy   456
lumped kinetic models, chromatography   180
lyophilization   80

Mach–Zehnder interferometers   1001, 1042
macrobalances   63
macromolecular ligands   321
macromolecules   11
macroprocedures   16
maghemite   575
magnesia   291
magnetic circular dichromism (MCD)   430
magnetic fields
– nuclear magnetic resonance   511
– scanning electron microscopy   1122
magnetic force compensation, weighing   63
magnetic hyperfine interaction   561, 570
magnetic induced optical activity   458
magnetic particle immunoassay   165
magnetic properties, nuclei   515
magnetic splitting, Mössbauer spectroscopy   568
magnetite   574
magnetization, nuclear magnetic resonance   513
magnetogyric ration   511, 514
magnetometry   828, 836
magnetooptical effects   430
magnetooptical rotatory dispersion (MORD)   430
magnetron   84
Mahalanobis/Manhattan distance   53, 58
malic acids   103
mapping, electrophoresis   356
markers
– affinity biosensors   1041
– microscopy   1067 f
Martin–Synge modelchromatography   179
mass–weight distinction   68
mass analyzers, static secondary ion mass spectrometry   891
mass balance models, chromatography   179 f
mass flow detectors, gas chromatography   231
mass selected ion kinetic spectrometry (MIKES)   603
mass selective detectors, liquid chromatography   270
mass sensitive devices   1003
mass sensitivity, trace analysis   111
mass spectrometers   654 f
mass spectrometry (MS)   6, 10 f, 579–626
– chromatography   191
– gas chromatography   232, 237
– liquid chromatography   307
– trace analysis   103, 113
master data files, instrument quality   29
mathematical techniques   2, 8, 37 ff
matrices   12
– electrophoresis   346
– factor analysis   60
– multivariate methods   52
matrix-assisted laser desorption (MALDI)   616
matrix-assisted laser desorption ionization time of flight mass spectroscopy (MALDI-TOF MS)   159, 350
matrix destruction, atomic spectroscopy   666
matrix effects, X-ray fluorescence spectrometry   760
Mattauch–Herzog multipliers   598, 609
Matthieu equation   598
maximum operation temperature range (MAOT), gas chromatography   210
Maxwell distribution   631, 639
Maxwell theory   380
McLafferty rearrangement   583 f
means
– chemometrics   41, 45
– trace analysis   114
measuring cells, voltammetry   808
MECA spectroscopy   6
mechanical methods, thermogravimetry   834
Median filter   1076
mediators, metabolic sensors   1035
medium energy ion scattering (MEIS)   898
megabore colums, gas chromatography   204
meldola blue   1037
membranes
– amperometry   982
– chemical sensors   954 f
– immunoprobes   1043
– ion-selective electrodes   971
– isoelectric focusing   368
– trace analysis   101
memory effects, plasma mass spectrometry   706
Mentha arvensis   247
mercury cadmium telluride (MCT) detector   467
mercury electrodes
– dropping   144
– voltammetry   785 ff
mercury reacting groups, voltammetry   821
mercury vapor analyzer   118
metabolic biosensors   1033 f
metabolites
– mass spectrometry   613
– organic trace analysis   97
metal ions determination   148, 155
metal oxide semiconductor field effect transitors (MOSFET)   994
metal oxides, semiconductor sensors   960
metal resistance thermometers   1027
metallic contacts, Auger electron spectroscopy   885
metallic layers, acoustic mass sensors   1020
metallocenes   1037
metalloenzymes   150, 155
metals, radioactive   144
metastable ions   603 ff
metastable quenching spectroscopy (MQS)   928 f
methanol
– gas chromatography   227
– laser spectroscopy   730
– liquid chromatography   284
– organic trace analysis   99
– thin layer chromatography   336
methyl groups
– IR/Raman spectroscopy   475
– nuclear magnetic resonance   545
methyl isopropyl ketone-xylene   93
methyl phenyl   209 f
methyl silicone   209 f
methyl viologen   1037
methylases   1133 ff

methylene groups
– gas chromatography 213 ff
– IR/Raman spectroscopy 475
Mettler balances 64
Michaelis–Menten reaction 148
Michelson interferometer 467
microanalysis/procedures 13, 15 f
microbalances 63
microbore columns 268
microcapillaries 330
microchannel plates (MCP) 608, 1118
microcracks 412
microelectrodes
– chemical sensors 984
– voltammetry 807
microencapsulation 157
microfraction range, trace analysis 110
microorganism receptor systems 1048
microphotometry 1070
microscopy 1061–1130
microthermal analysis 836
microwave ashing 82 ff
microwave digestion
– trace analysis 82 ff, 90
– voltammetry 811
microwave induced plasma (MIP) 6, 17
microwave plasma, atomic spectroscopy 641, 699
microwave spectra, vibrational spectroscopy 471
microwave spectroscopy 6
mid IR/Raman spectroscopy 502
Mie scattering 423
migration
– chromatography 174 f
– electrophoresis 345
Miller indices
– diffraction 380, 1079
– quartz 1016
mineral acids 80
minimum detectable absorption (MDA), laser spectroscopy 741
minor fraction range, trace analysis 110
misclassification, chemometrics 57
mitochondrial DNA analysis 1142
mixed layer polytypes, electron microscopy 1101
mixture spectra 55
mixtures, phases systems 12
MNN series, Auger electron spectroscopy 855, 875
mobile phases
– chromatography 174 ff, 182 ff
– gas chromatography 200 f
– liquid chromatography 263, 283 ff
– thin layer chromatography 334 f
mobile trace analysis 118 ff
mode number, optochemical sensors 1000
moderated neutrons 769
modes
– electron microscopes 1084
– gas chromatography 201, 237
– liquid–liquid chromatography 292
modified haptens 159
modifiers, organic trace analysis 99
modular arrangements, UV spectroscopy 436
modules, gas chromatography 201
Moiré fringes 1084
molar reaction enthalpies 1027
molecular absorption, diode lasers 744

molecular analysis 744
molecular bands, atomic spectroscopy 638
molecular crystals 1104
molecular imprinting polymers (MIP) semiconductor sensors 967
molecular recognition 959 ff, 969 f
molecular sieves see: zeolites 00
molecular spectroscopic methods 3
molecular symmetries, vibrational spectroscopy 473
molecular tools, DNA analysis 1133 f
molecular tumbling, electron spin resonance 553
molecules 11
molybdenum 813
molybdenum–samarium–oxygen system 868
moment analysis 181
moment orientation, nuclear magnetic resonance 512
monitoring, contaminants 117
monitoring and measurement technology program (MMTP) 118
monochromatic distortions, light microscopy 1063
monochromators, UV spectroscopy 434
monoclonal antibodies 159, 165, 322
monodisperse aerosol generator interface for chromatography (MAGIC) 588
monoenzyme metabolic sensors 1034
monomers, electrophoresis 351
monopole interactions, Mössbauer spectroscopy 568, 571
monospecific ligands 321
Mordant blue 802
Moseley's law
– scanning electron microscopy 1124
– X-ray fluorescence spectrometry 756
– X-ray microanalysis 1096
Mössbauer spectroscopy 6, 561–578
Mott cross section 1119
mounting, atomic spectroscopy 648
moving belt, mass spectrometry 308, 587
moving boundary electrophoresis (MBE) 345
muffle furnace 87
multianalyte methods, trace analysis 112
multichannel analyzers (MCA) 774
multichannel instruments 675
multichannel plates (MCPs), fluorescence microscopy 1068
multicompartment electrolyzers 368
multicomponent analysis, UV spectroscopy 444
multidimensional arrays, chemometrics 59
multidimensional capillary coupled gas chromatography (MDGCGC) 228, 244
multidimensional nuclear magnetic resonance 536
multidimensional scaling, chemometrics 53
multienzyme metabolic sensors 1036
multiexponential regression 46
multifactor plan, chemometrics 48
multilocus probes (MLP), DNA analysis 1136, 1148
multimodal high-performance liquid chromatography/ capillary GC 246
multimodal SFE–capillary GC 249
multinuclear NMR 6
multiphase systems, heterogenous 12
multiphoton ionization, mass spectrometry 594
multiple decomposition techniques, trace analysis 90
multiple ion detection (MID) 585
multiple isotope dilution analysis 137
multiple linear regression (MLR) 58 f
multiple point analysis, Auger electron spectroscopy 882

multiple sampling plan 75
multiple sector instruments, mass spectrometry 605
multiple splitting, X-ray photoelectron spectroscopy 864
multiplex detector UV spectroscopy 432
multiplex spectrometers, atomic 652
multiplicative errors, chemometrics 47
multipliers, mass spectrometry 608 f
multipulse Fourier transform nuclear magnetic resonance 534
multiquantum well (MQW) 731
multisensor arrays 1030
multistep procedures 5, 18
multivariate data analysis 444
multivariate regression 51, 58 ff
Mus musculus 1132
mutarotation 429
mutations, DNA analysis 1145 f
myeloma cell, antibodies 160

NAD(P)/NADH 150
NAD(P)H 1074
NAMAS 23
naphthacyl 103
narrow bands, UV spectroscopy 430
narrow scans, X-ray photoelectron spectroscopy 863
narrowbore columns, gas chromatography 204
natrium 515
natural radioactivity 130
near infrared gas sensors 744
near IR/Raman spectroscopy 502
nebulizers
– atomic emission spectroscopy 697
– atomic spectroscopy 660
– flame atomic absorption spectroscopy 677
needle injection, gas chromatography 218
negative chemical ionization 591
negative temperature coefficients (NTC) 1027
nephelometry
– immunoassay 161
– UV spectroscopy 456
Nernst–Nikolsky equation 958 ff, 970, 995
Nernst diffusion
– voltammetry 786 ff
– semiconductor sensors 966 ff, 970
Nernst glower 467
Neumann–Moore tests 43
neutral excitations, fast atom bombardment mass spectroscopy 932
neutral loss scans, mass spectrometry 604
neutralization enthalpies, calorimetric devices 1027
neutralization probability, ion scattering spectroscopy 903
neutron activation analysis 768 ff
neutron beams 382 f
neutron diffraction 6, 376, 412
neutron shadowing 772
neutron spectroscopy 6
Newton law 837
Newtonian behavior, mass sensors 1008
nickel-based superalloys 833
nickel–platinum alloys, Rutherford backscattering spectroscopy 909
Nier–Johnson geometry 598, 603
Niggli matrix 377 ff, 406 f
nitrates 812
nitric acids 81 ff

nitriles
– IR/Raman spectroscopy 478
– nuclear magnetic resonance 526, 529
nitriloacetic acid (NTA) 820
nitro groups 820
– IR/Raman spectroscopy 479 ff
– liquid chromatography 281, 288
nitroge
– gas chromatography 201
– nuclear magnetic resonance 515, 528
nitrogen phosphorus detector (NPD) 232 ff
nitrophenol esters 151
nitroso compounds 555
nitroso groups 820
nitrous oxide 676
noble gas impact collision ion scattering spectroscopy (NICISS) 900
noble metal electrodes
– liquid chromatography 277
– voltammetry 806
noble metals, semiconductor sensors 960
noise
– atomic spectroscopy 643, 689
– LC detectors 269
noise equivalent power (NEP) 469
noise removal algorithms, digital microscopy 1076
Nomarski differential interference contrast (DIC) 1066
nomenclatura
– atomic spectroscopy 635
– laboratory balances 63
– sampling 71
– X-ray spectrometry 757, 855
nonaqueous reversed-phase (NARP) chromatography 289
noncrystallinity 409
nondestructive visualization 340
nonhierachic clustering 56
nonisiotopic immunoassays 161
nonlinear chromatography 194 ff
nonlinear regression 46
nonlinearity, weighing 66
nonmetals, decomposition 91
nonparametric tests, chemometrics 43
nonsaturation analysis, radionuclides 142
normal distribution
– chemometrics 39 f
– sampling 73
normal phase, liquid-liquid chromatography 292
normal phase chromatography 186, 289
normal pulse polarography (NPP) 794
noses, electronic 1030
notch filters, Raman spectroscopy 468
nuclear activation analysis 128
nuclear hyperfine interaction 552
nuclear magnetic resonance (NMR) 6, 12, 509–560
nuclear Overhauser enhancement (NOE) 521, 529, 533 ff, 540
nuclear resonance fluorescence 561 f
nucleotide chains, DNA analysis 1131
nucleotides, mass spectrometry 616
null hypothesis, chemometrics 40
numerical algorithms, UV spectroscopy 453
numerical aperture 1063, 1093
Nutmeg oil 256
nutridition 4

object order rendering, digital microscopy 1077
objectives, light microscopy 1062 f
oblique transformations 55
octadecyl 100
octyl groups
– extraction 99 f
– liquid chromatography 288
off-resonance spin decoupling 533
oils analysis 206 f, 226 f, 245, 255
oligonucleotides 330
oligosaccharides
– mass spectrometry 616
– postcolumn derivatization 305
oncolumn injection 223
one-sided tests 41
onion-like concentric spherical shells 1108
online procedures, trace analysis 94
online process control, UV spectroscopy 450
online sample clean-up, column switching 306
OPA (o-phthalaldehyde) 302
open coupling, mass spectrometry 587
open tubes, postcolumn derivatization 303 f
operating characteristic (OC), sampling 74
operating modes, electron microscopes 1084
operating principle, magnetic force compensation 63
operating software, qualification assurance 27
operational conditions, radionuclides 131
operational qualification (OQ) 25
optical activity
– magnetic field-induced 458
– saccharides 152
optical beam induced conductivity (OBIC) 1074
optical components, UV spectroscopy 430
optical emission spectroscopy (OES) 6, 17
– chemometrics 47
optical methods, ionization 594
optical microscopy 1061–1077
optical path
– light microscopy 1062
– UV spectroscopy 435 ff
optical radiation, atomic spectroscopy 628
optical resonator, laser spectroscopy 730
optical rotatory dispersion (ORD) 427 ff
optical sensors 997
– UV spectroscopy 451
optical spectrometers, atomic 644
optics, electron microscopes 1084
optoelectronic imaging 1069
orange oil 248
orbitals
– nuclear magnetic resonance 518 f
– UV spectroscopy 421
– X-ray fluorescence spectrometry 756
ordered alloys 1102
organic acids 5
organic analytes
– bulk acoustic waves 1024
– liquid chromatography 280
– sample preparation 96 ff
organic binders 329
organic compounds 15
– diffraction 374
– IR/Raman spectroscopy 474 ff
organic conductometric sensors 991
organic dyes 730
organic mass spectrometry 583

organic molecules, fluorescence analysis 748
organic solutions, atomic emission spectroscopy 698
organic trace analysis 97, 814
organic traces 110, 119
organochlorine pesticides 99
organochloropesticide endrin 220
organolead compounds 242
organophosphorus compounds 155
organophosphorus pesticides 99
orientation, quartz 1016
orientation variants, electron microscopes 1087
orthogonal transformations 55
osmosis, reverse 101
osmotic shock 1133
Osteryoung pulse 793
outliers, chemometrics 43
ovens, supercritical fluid chromatography 311
overspotting, TLC samples 333
oxazines 730
oxidable groups 821
oxidation enthalpies 1027
oxidative mode, liquid chromatography 280
oxide films
– depth profiles 867
– static secondary ion mass spectrometry 895
oxidizing agents 975
oxidoreductase 150
oximetry 529, 556
oxine, stripping voltammetry 802
oxygen
– liquid chromatography 283
– nuclear magnetic resonance 515, 530
– radionuclides 144
– semiconductor sensors 961
oxygen combustion 88
oxygenase 1035
ozone
– radionuclides 143
– trace analysis 94

Paar microwave assisted pressurized decomposition (PMD) 90
packed bed reactors 303
packed column
– gas chromatography 201 ff, 217
– supercritical fluid chromatography 311
packing, chromatography 182
palm oil 206
paper chromatography (PC) 6, 263
paraffin wax 771
parallel factor analysis (PARAFAC) 60
parent ion scans 604
Parr bomb 88
partial least squares (PLS)
– chemometrics 59
– UV spectroscopy 445
– vibrational spectroscopy 472
particle beam, LC-MS 308
particle beam interface 588
particle counters 161
particle excitation, surface analysis 853
particle induced X-ray spectrometry (PIXE) 6, 16
particle size, liquid chromatography 285
partition chromatography, liquid–liquid 291
partition coefficients, chromatography 184

partitionmaking algorithms   56
Paschen–Runge mounting   648, 690
passivation   852, 867
passive sampling, trace analysis   102
Patric detector   609
Patterson synthesis   375, 394
peak area
– activation analysis   776
– chromatography   192 ff
– mass spectrometry   611
– thermogravimetry   831
– thin layer chromatography   342
– voltammetry   809
pellistors   959, 1027
Peltier element   651
Penning effect   696
pentanes   284
peppermint oil   245
peptide bonds   159
peptide fragments   615
peptides   305
perchlorates   812
perchloric acids   81 ff
perfluoalcoxy resin (PFA) vessels   81
perfluorokerosene (PFK)   581, 610
performance, balances   66
performance qualification (PQ)   25, 30 ff
Perkin–Elmer calorimeter   844
permeability   182
personal identification, DNA analysis   1147
Perspex   830
pesticides
– ELISA tests   163
– enzyme sensors   1039
– enzymes   155
– extraction   99
– gas chromatography   220, 257
– liquid chromatography   280
– mobile trace analysis   118 f
– postcolumn derivatization   305
pH, liquid chromatography   284
pH changes, UV spectroscopy   451
pH effects, enzymes   153, 157
pH sensitive ion-selective field effect transistors (ISFETs)   995
pharmaceutical agents   158
phase analysis, diffraction   394, 399, 406
phase contrast, light microscopy   1065
phase grating, ideal microscopes   1092
phase systems, chromatographic   176, 183 ff
phases, heterogeneous systems   12
phenacyl   1
phenazine methasulfate (PMS)   151
phenol red   999
phenols
– enzyme sensors   1039
– IR/Raman spectroscopy   479
– liquid chromatography   280
– voltammetry   821
phenyl
– extraction   100
– thin layer chromatography   331
phonon wing, fluorescence analysis   749
phosphate groups   1131
phosphorescence   426
phosphorus   515, 528

phosphorus-nitrogen selective detectors (PND)   103
photoacoustic spectroscopy (PAS)   6, 458
– IR/Raman   495
photocells   10
photochemical reactions   426
photodiode arrays   651
photodiodes   437
photoelectric absorption   754
photoelectron, X-ray fluorescence spectrometry   755
photoelectron spectroscopy   6
photographic emulsions   649
photoionization, mass spectrometry   594
photoionization detector (PID)
– gas chromatography   232, 236
– mobile trace analysis   119 f
photoionization scheme, laser spectroscopy   739
photolysis   85
photometer
– ELISA   163
– thermal analysis   836
– UV   430
photomultipliers
– atomic spectroscopy   650
– UV spectroscopy   436
photon activation analysis (PAA)   779
photon correlation spectroscopy   456
photon excitation   853
photon noise   689
photons, UV spectroscopy   421
photophysics   425
o-phthalaldehyde (OPA)   302
phthalocyanines
– conductometric sensors   991
– metabolic sensors   1037
phycoerythrine   998
physical methods   1, 5
physical organization, analytical laboratory   10 f
physical properties, laser spectroscopy   728, 732, 735
physical signals, chemometrics   38
physical vapor depostion (PVD)   1022
physicochemical processes   81
physiochemical properties, immobilized enzymes   157 f
physiological behavior   5
piezoelectric effect   1013
piezoelectric immunosensors   1042
piezoelectric structures, crystal systems   1015
piezoelectric substrates   1004
pixels, atomic spectroscopy   652
Plackett–Burman multifactors   48
Plan Neofluar   1063 ff
planar defects, electron microscopy   1112
planar interfaces, diffraction   1082
planar structures, immunoassays   166
Planck constant   384, 511, 1078
Planck's law   629
plasma ionization   592, 617
plasma mass spectrometry   704
plasma sources   694
plasma spectroscopy   6
plasma treatment, static secondary ion mass spectrometry   897
plasmajet   640 f, 694
plasmas, atomic spectroscopy   631, 640 f
plasmon loss peaks, X-ray photoelectron spectroscopy   863
plasmons, electron energy loss spectrometry   1099
plastic deformation   1114

plastic supports, thin layer chromatography   327
plasticizers   81, 97 ff
plate development, thin layer chromatography   337
plate height equation   187
plate mode oscillator   1003, 1011
plate models, chromatography   178 f
plate number
– gas chromatography   207
– liquid chromatography   264
platinum electrodes
– liquid chromatography   277
– voltammetry   786, 806
platinum-drug-modified biopolymers, mass spectrometry   617
plausibility tests   78
pneumatic nebulizers   660
point defects   1114
point groups
– crystal structures   1015
– diffraction   1089
– IR/Raman spectroscopy   474
point measurement   384
poisons
– chemical sensors   955
– trace analysis   84
Poisson distribution
– activation analysis   775, 778
– chemometrics   39
– chromatography   190
Poisson ratio   411
polar adsorbents, liquid chromatography   289
polarimetry   438, 457
polarity, gas chromatography   208
polarization, UV spectroscopy   422, 427, 435
polarization contrast microscopy   1067
polarography   6, 785–826
pollutants
– immunoassays   165
– organic trace analysis   97
polyacrylamide gels (PAG)   346 ff
polyamides
– liquid chromatography   291
– thin layer chromatography   328
polyamino acids   820
polychlorinated biphenyl (PCB)   118
polychromatic radiation   753
polyclonal antibodies   159, 165
– immunosensors   969
polycrystalline solids   852
polycrystalline specimens   403 f
polycyclic aromatic hydrocarbons (PAH)   750
polycyclic hydrocarbons   278
polyetheretherketone (PEEK)   266
polyethylene groups   610
polymer beads, gas chromatography   209
polymer binders   329
polymerase chain reaction (PCR)   20
polymerases   1133 f
polymeric humidity sensors   992
polymeric molecules   12
polymers
– diode laser spectrometry   743
– electron microscopy   1110
– IR/Raman spectroscopy   502
– molecular imprinting   967
– static secondary ion mass spectrometry   897

– thermogravimetry   830
– vibrational spectroscopy   472
polynominal regression   46
polynominal smoothing
– chemometrics   50
– UV spectroscopy   453
polynuclear aromatic hydrocarbons (PAH)   118
polypropylene (PP) vessels   81
polypyrrole   991
polysaccharides   317
polystyrenes   610
polytetrafluoroethylene (PTFE)
– atomic spectroscopy   668
– liquid chromatography   266
– static secondary ion mass spectrometry   897
– trace analysis   81, 86
– voltammetry   811
polytypes, electron microscopy   1101
populations, sampling   71
pore diameter, liquid chromatography   285
porositiy   182
porosity gradient gels   355
porous layer open tubular (PLOT) columns   201, 204, 209
porphyrins   748
position sensitive atom probe (POSAP)   932, 943
postanalysis   6
postcolumn derivatization   301 ff
potential clustering   57
potential functions, thermodynamic   845
potentiometric detectors   275
potentiometric immunosensors   1042
potentiometric modes, molecular recognition   969
potentiometric stripping analysis   785, 802, 818
potentiometry, ion-selective electrodes   968
potentiostats   979
powder pattern   405, 409
powders, atomic emission spectroscopy   702
power compensated calorimeters   841 f
preanalysis   6
precipitation   93
precision
– chemometrics   40
– chromatography   190
  see also: accuracy
precision balances   63
precision camera   388
precoated plates   327 ff
precolumn derivatization   301
precursor ion scans   604
preformed ions, mass spectrometry   595
preparation techniques, radionuclides   131
preparative chromatography   196
preparative electrophoresis   364
pressure digestion, trace analysis   82, 86
pressure programmed gas chromatography   214
preventative maintenance, instrument quality   32
principal component regression (PCR)   59
– UV spectroscopy   445
– vibrational spectroscopy   472
principal components analysis (PCA)   53 f
prism monochromators   434
probability distributions   39 f
probability sampling   72 f
probeheads, nuclear magnetic resonance   518
procedure limits, trace analysis   114

procedures, analytical chemistry 1–22
process analysis, UV spectroscopy 450
process conditions, mobile trace analysis 119
process control, chemical sensors 953
processing, weighing 66
producer's risk 74
product scans, mass spectrometry 604
propane–air flame 676
protein complexes 1133
proteins 148
– affinity chromatography 321 ff
– diffraction 375 f
– mass spectrometry 616, 620
proteome analysis 356
protocols, sampling 72
proton–proton coupling 525
proton noise decoupling 531
protonation enthalpies 1027
protons, nuclear magnetic resonance 511
Protopterus 1132
pulling, crystal growth 381
pulsation, liquid chromatography 267
pulse electron spin resonance 550
pulse techniques, voltammetry 791
pulsed amperometry 982 ff
pulsed laser atom probe (PLAP) 932
pulsed neutron irradiation 770
pumps, liquid chromatography 266, 300
purge and trap samplers 228 ff
purging, trace analysis 101
purification, LC solvents 283
purines
– DNA analysis 1131
– liquid chromatography 280
purity
– radionuclides 131, 134
– trace analysis 82
PVC membranes 971
pyridines
– confocal laser scanning 1074
– nuclear magnetic resonance 529
– thin layer chromatography 336
pyrimidines
– DNA analysis 1131
– liquid chromatography 280
pyrolysis
– gas chromatography 230
– mass spectrometry 586
Pythagorean theorem 52 f

Q-matrix method 445
quadrupole plasma mass spectrometry 708, 711
quadrupole splitting 568, 571 f
quadrupoles, mass spectrometry 598, 606, 654, 891
qualitative analysis
– atomic emission spectroscopy 688
– atomic spectroscopy 641, 672, 715
– chromatography 189 ff
– diffraction 405, 410
– gas chromatography 242
– traces 78
qualitative factors, weighing 68
quality assurance
– instrumentation 7, 23–36
– trace analysis 117

quality control, sampling 71, 74, 79
quantitative analysis 1
– atomic emission spectroscopy 688
– Auger electron spectroscopy 879
– chemometrics 38
– chromatography 192
– diffraction 405 ff
– gas chromatography 244
– ion scattering spectroscopy 901 ff
– liquid chromatography 298 ff
– mass spectrometry 584
– nuclear magnetic resonance 543
– polarimetry 428
– radionuclides 132
– Rutherford backscattering spectroscopy 909
– static secondary ion mass spectrometry 894
– thin layer chromatography 341
– traces 78, 115
– UV spectroscopy 443
– vibrational spectroscopy 471
– X-ray fluorescence spectrometry 761
– X-ray microanalysis 1097
– X-ray photoelectron spectroscopy 865
quantum cascade laser (QCL) 732
quartz 1015, 1026
quartz crystals 383
quartz equivalent circuit, mass sensors 1008
quartz microbalance (QMB) 1003, 1023
quartz vessels 81 ff
quasi-equilibrum theory, mass spectrometry 583
quasi-phase matching structures (QPMS), laser
    spectroscopy 732
quasicrystals 1106
quasilinear regression 46
quenching 134
quick freezing 80
quinones
– metabolic sensors 1037
– voltammetry 821

R test 399
radial density function (RDF) 411
radiant quantities, UV spectroscopy 425
radiation, atomic spectroscopy 628
radiation–matter interactions 421
radiation damage 1114
radiation induced effects 133
radiationless deactivation 426
radioactive kryptonates 143
radioactive metals 144
radioactive tracers 79, 128
radiochemical activation analysis 774, 777
radiochemical analysis 6, 19
radioimmunoassays (RIA) 128, 143, 160 ff
radioisotopes 148
radiometric titration 144
radionuclides
– analytical chemistry 19, 127–146
– cosmogenic 602
radiopolarography 144
radioreagent methods (RRM) 140
radiorelease methods 143
radiotracers 128
Raman bands 447
Raman microspectroscopy 6, 499

Raman spectroscopy 6, 465–508
Randles–Sevcik equation 799
random amplified polymorphic DNA (RAPD) 1140
random errors
– liquid chromatography 299
– trace analysis 114
– X-ray fluorescence spectrometry 760
random noise, atomic spectroscopy 643
random sampling 72
random variables, chemometrics 38 f
random walk model 179
rapid immunofiltration test hybritech ICON 168
rare earth atomic emission 634
rate model, chromatography 181
rate processes, nuclear magnetic resonance 544
ray casting techniques, digital microscopy 1077
Rayleigh scattering
– infrared spectroscopy 467
– laser spectroscopy 733
– UV spectroscopy 423
– X-ray fluorescence spectrometry 754
reaction calorimetry 847
reaction interfaces, mass spectrometry 587
reactive visualization, thin layer chromatography 341
reactivity 19
reactor immunosystems 1043
reactor neutrons, activation analysis 769
real sample analysis, laser spectroscopy 734 ff, 704 ff
receptors
– chemical sensors 954 ff
– metabolic sensors 1034
reciprocal lattice 377, 1078
recognition
– chemical sensors 954, 959 ff
– selective 236
recoil-free nuclear resonance fluorescence 561 ff
recombinant antibodies 159
recording methods
– diffraction 384
– UV spectroscopy 431
recycling free-flow focusing (RF3) 368
recycling isoelectric focusing 366
redox chemiluminescence detector (RCD) 236
redox mediators 1037
redox substoichiometry 139
reducible groups 821
reducing agents 975
reduction modes 281
reference compounds, mass spectrometry 581
reference electrodes
– potentiometry 975
– voltammetry 786
reflectance 1074
reflection–absorption spectroscopy 491
reflection
– optical sensors 997
– UV spectroscopy 423, 438, 443, 448
reflection absorption infrared spectroscopy (RAIRS) 6, 918, 942
reflection contrast microcopy 1067
reflection high-energy electron diffraction (RHEED) 938, 943
reflectometric interference spectroscopy (RIFS) 1002
reflectometry 448
refractive index
– internal reflection 492
– light microscopy 1063
– mobile phases 284
– optochemical sensors 1000
– UV spectroscopy 439
refractive index detector (RID) 270
refractory metal furnaces 666
regression models
– atomic spectroscopy 659
– chemometrics 45, 58
– trace analysis 116
– X-ray fluorescence spectrometry 762
regulations, weighing 69
rejectable quality level (RQL) 74
rejection number, sampling 75
relaxation
– nuclear magnetic resonance 513 f, 521
– UV spectroscopy 425
reliability, chemical sensors 955 ff
reliability–measurement uncertainty 14 f
relrods 1079
repeatability
– chemometrics 40
– sampling 72
– thermogravimetry 829
– trace analysis 113
– weighing 66
represenstative sampling 71
representative labeling, radionuclides 133
reproducibility
– chemometrics 40
– chromatography 190
– liquid chromatography 299
– trace analysis 113
– UV spectroscopy 440
– voltammetry 810
rescattering, confocal laser scanning 1074
residence time *see:* retention
resolution
– activation analysis 774
– chromatography 187
– confocal laser scanning 1072
– diffraction 1088 ff
– ideal microscopes 1093
– infrared spectroscopy 467
– light microscopy 1063
– liquid chromatography 264
– mass spectrometry 580 f, 593
– X-ray photoelectron spectroscopy 861
resonance absorption
– laser spectroscopy 729
– Mössbauer spectroscopy 561 f
– nuclear magnetic 512
resonance frequencies 1010
resonance ion mass spectrometry (RIMS) 743
resonance ionization mass spectrometry (RIMS) 16
resonance ionization spectroscopy (RIS) 16, 738
resonance methods, UV spectroscopy 450
resonance neutrons 769
response data
– chemical sensors 954 f, 1053
– gas chromatography 232
restrictors 313
resubstitution 57
retention data
– chromatography 182 ff, 189 f
– gas chromatography 200, 207, 243

retention time
- liquid chromatography   264
- methylene groups   215
retention time locking (RTL)   255
reverse isotope dilution analysis   137
reverse osmosis   101
reversed geometry, mass spectrometry   598
reversed phase chromatography   176, 186
- liquid-liquid   286, 289, 292
reversed phase partition development   339
reversed phase sorbents   328
reversible couples, voltammetry   976
reversible redox processes   796
rhodamines
- biochemical sensors   998
- laser spectroscopy   730
riboflavin   1040
Rice–Ramsberger–Kassel–Marcus (RRKM) theory   583
Rietveld method   407, 412
ring current effect   523
robots   7, 94
rocket immunoelectrophoresis   360
rocking curve, diffraction   1079
rocking vibrations, IR/Raman spectroscopy   475
rotating crystal pattern   387
rotating disk electrode (RDE)   982
rotating platinum electrode (RPE)   786
rotational energy   421
rotational hyperfine structures, atomic spectroscopy   637
rotational temperature, atomic spectroscopy   638
rotofor   366
routine analysis   8, 11
routine maintenance   30 f
routine use, instruments   26
Rowland circles   647
Russel–Saunders (L–S) coupling   630
Rutherford backscattering (RBS)   6, 17, 898, 906 f, 942
Rutherford cross section   1119
Rutherford scattering   1096
Rydberg constant   629
Rydberg states   739

3σ rule   48
saccharides
- optical activity   152
- postcolumn derivatization   305
Saccharimyces cerevisiae   1132
safety   3
- diffraction   392
- radionuclides   131
- thermal analysis   849
- trace analysis   84
Saha theory   635 f, 639 ff
salt solutions, UV spectroscopy   441
sample inlets
- atomic spectroscopy   660
- gas chromatography   215 ff
- liquid chromatography   263, 267, 300
- mass spectrometry   585 ff
- supercritical fluid chromatography   311
- thin layer chromatography   332
sample preparation
- activation analysis   776
- electron microscopy   1100
- immunoassays   160

- liquid chromatography   301
- thin layer chromatography   330
- trace analysis   77–108
- UV spectroscopy   435 f
- voltammetry   810
samples
- chemometrics   43, 57
- factor analysis   60
- quality control   33
- size   75
sampling   9, 71–76
- chromatography   193
- radionuclides   135
Sand equation   798
sandwich chamber   337
sandwich immunoassay   164
satellite lines, X-ray photoelectron spectroscopy   857 ff
saturation analysis, radionuclides   142
saturation transfer, electron spin resonance   557
Sayre equation   396
scales, weighing   63 ff
scaling, multidimensional   53
scanners, thin layer chromatography   342
scanning, mass spectrometry   603
scanning Auger microscopy (SAM)
   6, 875, 882 ff, 942, 1078
scanning calorimetry   841 ff
scanning electron microscopy (SEM)   882, 1078, 1115 ff
scanning transmission electron microscopy (STEM)   1078
scanning transmission microscopy (STM)   1095
scanning tunneling microscopy (STM)   6, 910 f, 943
scanning tunneling spectroscopy (STS)   910, 943
scattering
- atomic absorption spectroscopy   686
- electron–specimen interactions   1119
- electron microscopy   1078
- fluorimetry   447
- Mössbauer spectroscopy   564
- UV spectroscopy   422 f, 455
- X-ray fluorescence spectrometry   754
Scherrer equation   410
Schöniger combustion   88
Schottky emission cathodes   1116
Schrödinger equation   421
scintillation semiconductor detectors   774
scintillators
- atomic spectroscopy   650
- Mössbauer spectroscopy   566
screening, contaminants   117
sea level weights   68
second harmonic generation (SHG)   920
secondary electron detectors (SED)   892
secondary electron multipliers   10, 608
secondary electrons   1078, 1118
secondary ion mass spectroscopy (SIMS)   6, 16, 594
- trace analysis   113
secondary ions, mass spectrometry   889
secondary neutron mass spectrometry (SNMS)
   6, 17, 928, 943
sector–quadrupoles, hybrid instruments   606
sector field mass spectrormeteres   655
Seemann–Bohlin circles   386
segmentation, open tubes   304
segregation, Auger electron spectroscopy   883, 888
Seidel function   650
selected area electron diffraction (SAED)   1078, 1085

selected ion monitoring (SIM)   238
selected ion recording (SIR)   585
selection
– atomic spectroscopy   630
– mobile phases   284, 334
– radionuclides   132
– sample preparation   78
– vibrational spectroscopy   473
– X-ray fluorescence spectrometry   756
selective detectors, chromatography   191
selective inlets, gas chromatography   228
selective recognition, gas chromatography   236
selective spin decoupling, nuclear magnetic resonance   532
selectivity
– acoustic mass sensors   1020
– chemical sensors   954 f, 959, 1052
– chemometrics   48
– chromatography   174
– gas chromatography   208, 211
– liquid chromatography   264
– semiconductor sensors   965
selenium   800, 818
self-absorption, atomic spectroscopy   634
self-indicating potentiometric electrodes   970
self-modeling, mixture spectra   55
self-shielding, activation analysis   772
semiautomatic calibration, balances   65
semiconductor detectors   774
semiconductor devices   1107
semiconductor diode lasers   730
semiconductor gas sensors   960
semiconductor sensors, chemical   960, 989
semiconductors
– Auger electron spectroscopy   883
– ion-selective electrodes   971
sensitivity   13
– acoustic mass sensors   1012, 1020
– activation analysis   778
– atomic absorption spectroscopy   680
– chemical sensors   954 f, 1053
– chemometrics   47
– GC detectors   231
– LC detectors   269
– static secondary ion mass spectrometry   894
– X-ray photoelectron spectroscopy   865
sensors   10
– bio/chemical   951–1060
  see also: transducers
separation   10
– activation analysis   774
– chromatography   174, 184
– gas chromatography   200 ff, 207
– liquid chromatography   264, 288, 292
– mass spectrometry   587
– radionuclides   136
– trace analysis   80, 93, 97
Sephadex layers   365
sequential methods   9
sequential sampling   75
sequential spectrometers, atomic emission   690
sex determination, DNA analysis   1146
Seya–Namioka mounting   648
shake-up satellite, X-ray photoelectron spectroscopy   864
shear moduli   1006
Shewart charts   34
shifting, UV spectroscopy   427

shim coils   515
short tandem repeat (STR)   1141
Shpolskii method   748 f
signal amplification, UV spectroscopy   455
signal intensity, nuclear magnetic resonance   520
signal processing
– chemometrics   49 ff
– mass spectrometry   607 ff
signal to noise ratio (S/N)
– atomic spectroscopy   644
– nuclear magnetic resonance   517, 531, 539
– voltammetry   980
signal values, trace analysis   114
silica
– affinity chromatography   317
– chromatography   176
– gas chromatography   208
– immobilized enzymes   156
– liquid chromatography   285
– thin layer chromatography   328
silicate glasses   166
silicon, nuclear magnetic resonance   515, 531
^{32}silicon, mass spectrometry   602
silicon carbide
– Auger map   884
– diffraction   408
silicon intensified target (SIT) vidicon   651
silicone rubber   830
silicones   965
silver   903, 937
silylation   250
simple radioreagent methods (SRRM)   141
simulation of fluorescence processes (SFP)   177
simultaneous techniques   9
– atomic emission   690
– thermogravimetry   833
single beam equipment, UV spectroscopy   430
single beam spectrum, infrared spectroscopy   467
single crystal diffractometers   391
single diamond polarizers   493
single locus probes (SLP), DNA analysis   1136, 1148
single nucleotide polymorphism (SNP) detection   1142
single photon counting, UV spectroscopy   456
single-strand conformation polymorphism (SSCP)   1140
SISCOM data   612
site selection spectroscopy (SSS)   749
site specific natural isotope fractionation (SNIF)   544
size, diffraction   410
size exclusion chromatography (SEC)   177, 295
skimmers,   654
slits, atomic spectroscopy   646
slow molecular tumbling   553
slurry technique
– atomic spectroscopy   666
– liquid chromatography   287
Smith–Hieftje technique   685
Snell's law   423, 997
Sobel filter   1076
sodium chloride–sodium citrate buffer   1136
sodium dodecyl sulfate (SDS) electrophoresis   355
sodium nitrate determination   456
sodium salicylate scintillators   650
soft independent modeling of class analogy (SIMCA)   58
soft X-ray appearance potential spectroscopy (SXAPS)   927
software, qualification assurance   27 ff

soil
– mass spectrometry   614
– mobile trace analysis   118
sol–gel method, antibody encapsulation   166
solenoid, nuclear magnetic resonance   515
solid electrodes, liquid chromatography   277
solid membranes, ion-selective electrodes   971
solid phase extraction (SPE)
– gas chromatography   228
– thin layer chromatography   331
– trace analysis   93 f, 99
solid phase microextraction (SPME)   100
solid phases, gas chromatography   208
solid samples
– activation analysis   767, 776
– atomic emission spectroscopy   691
solid sampling, direct   667
solid solutions
– diffraction   407
– fluorescence analysis   748
solid state detectors, atomic spectroscopy   651
solid structures, diffraction   374
solids, nuclear magnetic resonance   546 ff
solubilization, wet digestion   85
solute thermal stability   250
solvent focusing, gas chromatography   221
solvent strength, thin layer chromatography   335
solvents
– fluorimetry   447
– laser spectroscopy   730
– liquid chromatography   283 ff
– TLC samples   330
sorbents, thin layer chromatography   327 f
sources
– atomic absorption spectroscopy   675 f
– atomic spectroscopy   629, 639 f
– Auger electron spectroscopy   875
– enzymes   153
– laser ablation   669
– Mössbauer spectroscopy   567
– static secondary ion mass spectrometry   890
– X-ray fluorescence spectrometry   757
– X-ray photoelectron spectroscopy   857
Soxhlet extraction   98, 102
space explorer, diffraction   388
space groups, diffraction   377, 381 ff, 405 f, 1090
spark ion source mass spectroscopy (SSMS)   113
sparks, atomic spectroscopy   640 f, 668, 691
spatial coherence, laser spectroscopy   728
spatial information, scanning tunneling microscopy   913
spatial resolution, X-ray photoelectron spectroscopy   861
species analysis   19 f
– mass spectrometry   618
– voltammetry   817
species concentrations, chemometrics   38
SpecInfo, nuclear magnetic resonance   541
spectral analysis   3
spectral editing, nuclear magnetic resonance   535
spectral information
– Auger electron spectroscopy   876
– ion scattering spectroscopy   900
– Rutherford backscattering spectroscopy   907
– scanning tunneling microscopy   913
– static secondary ion mass spectrometry   892
– X-ray photoelectron spectroscopy   862
spectral lines, broadening   633

spectral parameters
– electron spin resonance   551
– nuclear magnetic resonance   518 f
spectrometric methods, enzymes   150 f
spectroscopic methods   3
spectroscopic notation, X-ray photoelectron spectroscopy   854
specular reflection, UV spectroscopy   439
spin coating, acoustic mass sensors   1022
spin concentrations, electron spin resonance   552
spin decoupling   531 ff
spin echo pulse sequences   535
spin labeling, electron spin resonance   555
spin–lattice relaxation   513, 521
spin–spin coupling   519 f, 524
spin trapping   554
spinels   574
split diffraction spots   1087
split injection, gas chromatography   218
splitting
– Mössbauer spectroscopy   568
– X-ray photoelectron spectroscopy   864
splittless injection, gas chromatography   220
spontaneous emissions, UV spectroscopy   426
spraying   1022
sputtered atoms ionization   930
sputtered neutral mass spectrometry (SNMS)   593, 620
sputtering
– acoustic mass sensors   1022
– atomic spectroscopy   670
– static secondary ion mass spectrometry   894
square-wave voltammetry (SWV)   793, 980
stability, chemical sensors   954 ff, 1053
stabilization, samples   80, 96
stacking faults   1112
staining techniques, electrophoresis   362
standard additions
– atomic spectroscopy   658
– potentiometry   975
– trace analysis   79, 117
– voltammetry   809
standard deviation
– chemometrics   38 ff, 45 ff
– activation analysis   778
– atomic spectroscopy   642, 659
– gas chromatography   222
– sampling   73
– trace analysis   114
– weighing   66
standard methods, liquid chromatography   299
standard reference materials (SRMs)
– laser spectroscopy   737
– trace analysis   116 ff
standard temperatures, atomic spectroscopy   636
standard uncertainty   15
standardization
– activation analysis   771
– mass spectrometry   585
– weighing   69
– X-ray fluorescence spectrometry   761
Stark broadening   634 ff
static calorimeters   839
static headspace techniques   101
static mercury drop electrode (SMDE)   805
static secondary ion mass spectrometry (SSIMS)   6, 889 ff, 942

stationary magnetic fields   511
stationary phases
– chromatography   174 ff
– gas chromatography   200, 206 ff
– liquid chromatography   263, 285
statistical errors
– atomic spectroscopy   658
– X-ray fluorescence spectrometry   760
statistics
– chemometrics   37 ff
– chromatography   179
– sampling   73 f
steady state mode, laser spectroscopy   729
steam distillation   101
stereochemical analysis   16, 457
Sternberg analysis   188
steroid esters   251 f
steroids   443
stilbene   452
Stokes law   346
Stokes lines
– IR/Raman spectroscopy   472, 475 ff
– laser spectroscopy   733
storage, samples   80, 96
strain field contrast, electron microscopes   1088
strategic organization, analytical processes   9
stratified random sampling   72
stray electrons, X-ray photoelectron spectroscopy   857 f
stray light, UV spectroscopy   441
stress coefficients   1014
stripping chronopotentiometry (SCP)   802
stripping voltammetry   799
structure analysis   16, 457
– atomic spectroscopy   629
– diffraction   373–418
– electron microscopy   1100 f
– ion scattering spectroscopy   900
– mass spectrometry   582
– nuclear magnetic resonance   543
– Rutherford backscattering spectroscopy   907
– UV spectroscopy   443
structures, crystal systems   1015
Student's t-distribution   115
styrene divinylbenzene copolymers   201, 999
subequivalence method   139
subboiling process, trace analysis   81 ff
sublimation   828, 831
subnanogram amount handling   133
substance identification, UV spectroscopy   443
substoichiometric isotope dilution analysis   138 f
substrate labeled fluorescence immunoassay (SLFIA)   162
substrates determination, enzymes   154
sugar analysis   457
sulfatase   98
sulfates   812
sulfides   488
sulfonamides   280
sulfones
– IR/Raman spectroscopy   488
– UV spectroscopy   441
– voltammetry   821
sulfonic acid   288
sulfur containing compounds
– emission spectrum   241
– IR/Raman spectroscopy   488 ff
sulfur dioxide   143

sulfuric acids
– thin layer chromatography   340
– trace analysis   81 ff
superalloys   833
superconductors
– electron microscopy   1108
– X-ray photoelectron spectroscopy   871, 887
supercritical fluid chromatography (SFC)   6, 174, 308
– gas chromatography   200
– trace analysis   112
supercritical fluid extraction (SFE)
– gas chromatography   228
– trace analysis   94 f, 98 f
supercritical fluid/mass spectroscopy interface   588
superequivalence method   139
superionic conductors   1110
superlattice image   1102
supervised classification, chemometrics   57
supporting electrolytes, voltammetry   785, 812
suppressed conductivity detection   294
supramolecular chemistry   967
surface acceptor, semiconductor sensors   961
surface acoustic wave (SAW) mass sensors
    1003, 1009, 1024
surface analysis   12, 16 f, 851–950
– electron microscopy   1113
– mass spectrometry   619
surface analysis by laser ionization (SALI)   930
surface analysis by resonance ionization of sputtered atoms
    (SARISA)   930
surface area, liquid chromatography   285
surface conductivity   960 ff
surface enhanced Raman scattering (SERS)   918, 942
surface loss, X-ray photoelectron spectroscopy   863
surface phases, organic trace analysis   99
surface plasmon resonance (SPR)
– affinity biosensors   1041
– chemical sensors   1001
– UV spectroscopy   450
surface potentials, scanning electron microscopy   1122
surface reactions   898
surface segregation   888
surfactants   5
– ELISA tests   163
– immunoassays   160
– voltammetry   820
SURFER software, mobile trace analysis   122
survey spectrum, X-ray photoelectron spectroscopy   862
sweep codistillation, trace analysis   101
synchronization, Mössbauer spectroscopy   566
synchroton frequency   1010
synchroton radiation   412, 858
synchroton radiation photoelectron spectroscopy (SRPS)
    865
synchrotons   764
synthesis, radionuclides   131
synthetic polymers, affinity chromatography   317
syringe injectors
– gas chromatography   218, 224
– liquid chromatography   267
– thin layer chromatography   332
systematic errors   17
– chemometrics   47
– liquid chromatography   299
– sampling   78 f
– trace analysis   111 ff

T1,2 experiments 534 f
tag-along effect 195
Taguchi sensor 958 ff, 966
tailing, linear chromatography 181
tandem crossed immunoelectrophoresis 361
tandem mass spectrometry 604 ff
tandem scanning microscopy 1071
target testing, factor analysis 55
targets 11 f
tartaric acids 103
t-distribution 115
telan system 1062
tellurium 800, 818
temperature changes, UV spectroscopy 451
temperature drift, weighing 67
temperature influence, semiconductor sensors 989
temperature modulated differential scanning colorimetry 833, 838 f, 844 f
temperature programming
– atomic spectroscopy 666
– gas chromatography 213 f, 226
temperatures
– air combustion 87
– atomic spectroscopy 636, 639 ff
– enzymes 153
– flame atomic absorption spectroscopy 676
temporal coherence, laser spectroscopy 728
Tenax 102
tensammetry 795 f
terminology, sampling 71
terphenyls 730
test strips, immunoassay 166
testosterone compounds 251
tetrachlorodioxins 614
tetrachloromethane (TCM) 120
tetramethylethylenediamine (TEMED) 348
tetramethylpiperidine (TEMPO) 552
tetramethylsilane (TMS) 519
tetrathiafulvalene 1037
tetrazolium salts, enzymes 151
texture, diffraction 409
thallium 143
thallium iodide–thallium bromide (KRS-5), internal reflection 492
thermal analysis 827–850
thermal conductivity cells 10
thermal conductivity detector (TCD)
– gas chromatography 232 f
– mobile trace analysis 119
thermal desorption 228 ff
thermal desorption spectroscopy (TDS) 934 f, 943
thermal energy analyzer (TEA) 236
thermal focusing, gas chromatography 221
thermal ionization mass spectrometry (TIMS) 594, 619
thermal lens spectroscopy (TLS) 747
thermal neutron activation analysis (TNAA) 769 f
thermal solid sampling, atomic spectroscopy 667
thermal stability, solutes 250
thermal transduders 1027
thermally convective digestion 85 f
thermionic cathodes 1116
thermionic detectors
– chromatography 191
– gas chromatography 234
– supercritical fluid chromatography 312
thermistors 1027

thermobalance 828
thermochemistry 679
thermocouples 10, 1028
thermodilatometry 828, 835
thermodynamics 183 ff
thermogravimetry 498, 828 ff
thermoptometry 828, 835
thermospray techniques
– liquid chromatography 307
– mass spectrometry 6, 587 ff
thick film columns 202
thin epitaxial layers 1113
thin films, Auger electron spectroscopy 885
thin foils, diffraction 1079
thin layer cells, liquid chromatography 277
thin layer chromatography (TLC) 175, 263, 277, 327–344
thin layer chromatography interface 590
thin mercury film electrode (TFME) 786, 806
thiobarbiturates 800 ff
thiols
– IR/Raman spectroscopy 488
– liquid chromatography 280
– voltammetry 800 ff, 821
thiophilic ligands 320
Thomas–Fermi statistics 902
Thurstone factor analysis 55 f
thymidine, DNA analysis 1131
thyroglobulin, antibodies 159
Tian–Calvet calorimeter 841
time factors, trace analysis 92
time of flight (TOF)
– atomic mass spectrometry 655
– ion scattering spectroscopy 900
– mass spectrometry 581, 599, 606, 891
time resolution, laser spectroscopy 6, 728
time resolved FT-IR/FT-Raman spectroscopy 501
time resolving, fluorescence microscopy 1068
time series, sampling 71
time split injection, supercritical fluid chromatography 312
titanium dioxides 291, 990
titration
– potentiometric 974
– radiometric 144
TMS-clenbuterol 238
Tölg technique 91
toluene
– chemical sensors 1031
– IR/Raman spectroscopy 479
– liquid chromatography 284
– thin layer chromatography 335
toluylene blue 1037
tongues, electronic 1030
topochemical analysis 16
topography, scanning electron microscopy 1121
total ion chromatogram (TIC) 122
total organic carbon (TOC) analyzer 92, 119
total petroleum hydrocarbons (TPH) 118
total reflection X-ray fluorescence analysis (TXRFA) 6, 17, 113, 763
toxic effect sensors 1039
toxic metals 118 f
trace analysis 9, 17, 109–126
– chemometrics 48
– sample preparation 77–108

– voltammetry 814
– X-ray fluorescence spectrometry 762
trace elements 1, 18
trace enrichment 10, 306
trace losses *see:* losses 00
tracer labeling 129
transducers
– chemical sensors 954 f
– mass sensors 1004 ff
– metabolic sensors 1034
– molecular recognition 969
– surface plasmon resonance 1001
 see also: sensors
transferred plasmas, atomic emission spectroscopy 694
transflectance 491
transition elements, Auger spectrum 878
transition probability, UV spectroscopy 427
transition time chronopotentiometry 798
translational diffusion, nuclear magnetic resonance 545
translational energy, UV spectroscopy 421
transmission electron diffraction pattern 414
transmission spectroscopy 489
transmittance, vibrational spectroscopy 472
transverse magnetization 513
trapping
– atomic emission spectroscopy 696
– electron spin resonance 554
– gas chromatography 228 f
– trace analysis 101
Trend tests, chemometrics 43
trial-and-error, ideal microscopes 1094
trial-and-error method, diffraction 399
triallylcitrictriamide (TACT) 349 ff
trinitrotoluene 160
triple bonds, IR/Raman spectroscopy 478
triple electron spin resonance 548
triple quadrupoles 605
tritium labeling 129
tropolone 802
trueness
– chemical sensors 954 f
– chemometrics 47
– sampling 78
t-statistics 73
tubes
– chromatography 189
– light microscopy 1062
Tucker3 model 60
tumbling, electron spin resonance 553
tunability, laser spectroscopy 728, 730 ff
tungsten–halogen lamps 433
turbidimetry
– immunoassay 161
– UV spectroscopy 424, 455
twins 1087, 1112
twisting vibrations 475
two-electrode cell 986

Ulbricht sphere 448
ultrafiltration 101
ultrahigh vacuum (UHV), X-ray photoelectron
 spectroscopy 856
ultramicrobalances 63
ultramicroelementary analysis 15
ultrapure water 97

ultrasonic nebulization
– atomic emission spectroscopy 698
– atomic spectroscopy 663
– plasma mass spectrometry 709
ultratrace analysis 818
ultratrace elements, activation analysis 767
ultraviolet–visible (UV-VIS) absorption detector 270 f
ultraviolet–visible (UV-VIS) spectroscopy 12, 419–464
ultraviolet absorbance, thin layer chromatography 340
ultraviolet detector, quality 30
ultraviolet digestion 85
ultraviolet photoelectron spectroscopy (UPS)
 6, 865, 917 ff, 942
ultraviolet photolysis 94, 811
uncertainty 14 f
– radionuclides 134
– sampling 74
– trace analysis 79 f, 92
– weighing 66
units 17, 111
universal inlets *see:* sample inlets
unsplit reflections 1087
uracil derivatives 821
urine samples 80, 97

vaccines 323
vacuum electrothermal atomizers (VETA) 740
vacuum requirements, X-ray photoelectron
 spectroscopy 856, 875
valence contributions, Mössbauer spectroscopy 572
validation
– immunoassays 158
– trace analysis 113
valve injectors, liquid chromatography 268
van Deemter equation
– chromatography 180
– liquid chromatography 265
– supercritical fluid chromatography 309
vanadate 144
vanadyl acetylacetonate 553
vapor, mobile trace analysis 118
vapor charring, thin layer chromatography 341
vapor deposition
– chemical *see:* chemical vapor deposition
– TEM samples 1100
– physical *see:* physical vapor deposition
vaporization 828, 831
variables
– chemometrics 38 f
– multivariate methods 51
variance
– chemometrics 44
– trace analysis 114
vector method, diffraction 394
venetian blind multiplier 608
ventilation, trace analysis 79
verification process, software 30
vertical cavity surface emitting lasers (VCSEL) 731
vessel materials 81
vibrational circular dichroism, IR/Raman
 spectroscopy 498
vibrational energy, UV spectroscopy 421
vibrational hyperfine structures 637
vibrational spectroscopy 470 f, 489 f, 501
vibrations, acoustic mass sensors 1016

Index

vibronic coupling, fluorescence analysis 749
video-enhanced contrast microscopy (VEC) 1069
video intensification microscopy (VIM) 1070
vidicons 652
vinyl/vinylidene groups 477
viologens 1037
viscosity
– chromatography 183
– mobile phases 284
viscous liquids 176
visualization
– digital microscopy 1077
– thin layer chromatography 339 ff
vitamines
– liquid chromatography 280
– postcolumn derivatization 305
Voigt profile 634
volatile analytes 79, 96
volatile hydrides 709
volatile organic compounds (VOCs) 96, 118
volatility 250
volatilization
– atomic spectroscopy 660, 664
– trace analysis 88
voltammetry 785–826
– cells 975
– working electrodes 968
volume loss, X-ray photoelectron spectroscopy 863
volume rendering, digital microscopy 1077
volumetric analysis 7

wagging vibrations 475
Wahrhaftig diagram 584
wall-coated open tubular (WCOT) columns 201, 204
wall-jet cells 277
waste gases 110
waste materials 614
waste water
– factor analysis 55
– trace analysis 80, 94
– voltammetry 812
water
– activation analysis 781
– mass spectrometry 614
– sample losses 80
– thin layer chromatography 335
– trace analysis 80, 94, 118
– voltammetry 810, 818
wave function, diffraction 1082
wavelength, X-ray fluorescence spectrometry 755
wavelength dispersive spectrometry (WDS) 1097
wavelength dispersive systems, X-ray fluorescence spectrometry 759
wavelength modulation, laser spectroscopy 741
wavelength selection, UV spectroscopy 434, 440
wavenumber 471
waves, acoustic mass sensors 1016, 1019
weighing 63–70

Weissenberg camera 387
wet digestion
– trace analysis 82 ff
– voltammetry 810
Wheatstone bridge
– calorimetric devices 1027
– liquid chromatography 275
white gas cells 490
whole-cell biosensors 1048
Wickbold combustion 88
wide spectrum, X-ray photoelectron spectroscopy 862
widebore colums 204 f, 227
Wollaston prism 438
working electrodes
– amperometry 983
– liquid chromatography 277
– voltammetry 786, 807
working principle, semiconductor sensors 961

X band, electron spin resonance 548
X-ray crystallinity 410
X-ray crystallography 376 ff
X-ray diffraction 6
X-ray emission, scanning electron microscopy 1124
X-ray emission spectroscopy 6
X-ray fluorescence (XRF) 13, 17
– trace analysis 113, 118
X-ray fluorescence spectrometry (XRFS) 753–766
X-ray microanalysis 1096
X-ray notation, AES 855, 875
X-ray photoelectron spectroscopy (XPS) 6, 17, 854 ff, 942
X-ray tubes 382
X-charts, quality assurance 34
xenon lamps 434
xylenes
– chemical sensors 1031
– IR/Raman spectroscopy 480

yttrium-doped zirconia 966

Z-contrast images 1096
Z correction, scanning electron microscopy 1125
Zeeman effect
– atomic spectroscopy 630, 684
– Mössbauer spectroscopy 570
zeolites 201
zero hypothesis, chemometrics 40
zero order Laue zone 1090
zero phonon line (ZPL) 749
zero point drift 956
zero tracking, balances 66
^{119}zinc, Mössbauer spectroscopy 567
zinc selenide 492
zirconia 317
zone electrophoresis 345